HEAR

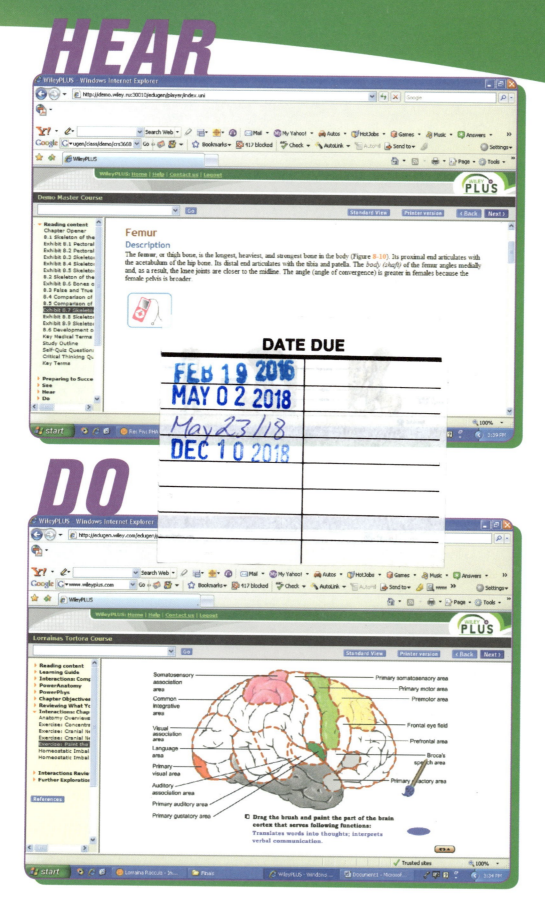

DO

- **Audio downloads keyed to significant illustrations within the text**
- **Mastering Vocabulary with audio glossary and flashcards**

- **Practice Quizzes**
- **Interactive and Animated Exercises**
- **Interactive Concept Maps**
- **Review Sheets linked to Animations**
- **Illustrated Anatomy Practice and Review**
- **Cadaver Practical Practice and Review**
- **Histology Practice and Review**
- **Negative Feedback Loop Exercises**

All of these resources work to create a course that motivates students every step of the way. And *WileyPLUS* also gives the instructor the tools to monitor student progress and engagement throughout the course. Interested? View details or take an interactive tour: www.wileyplus.com/experience

ANATOMY AND PHYSIOLOGY
FROM SCIENCE TO LIFE

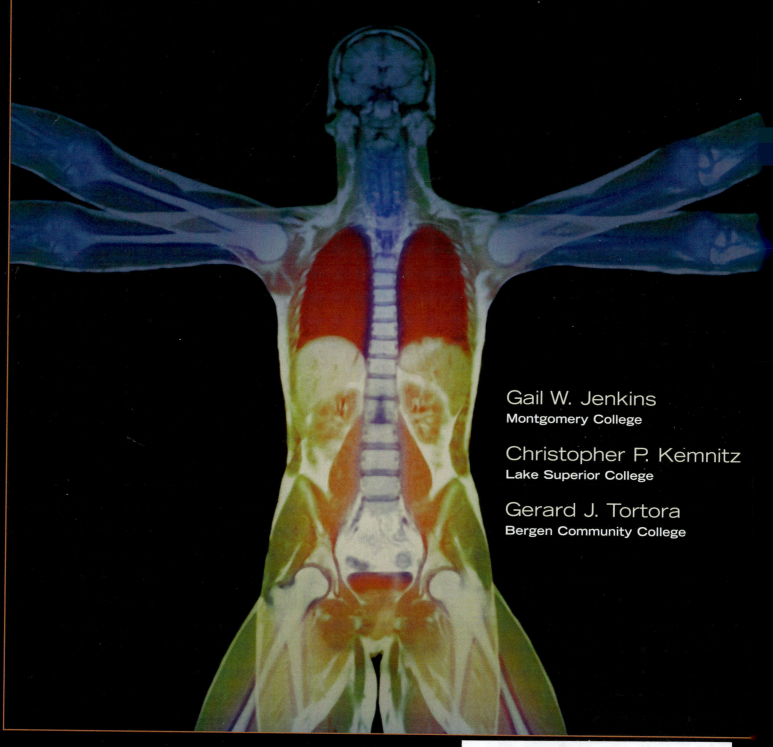

Gail W. Jenkins
Montgomery College

Christopher P. Kemnitz
Lake Superior College

Gerard J. Tortora
Bergen Community College

WILEY

John Wiley & Sons, Inc.

Executive Editor	Bonnie Roesch
Executive Marketing Manager	Clay Stone
Development Editors	Mary O'Sullivan and Karen Trost
Text Designer	Karin Gerdes Kincheloe
Senior Illustration Editor	Anna Melhorn
Senior Media Editor	Linda Muriello
Media Project Manager	Bonnie Roth
Photo Editor	Hilary Newman
Project Editor	Lorraina Raccuia
Production Manager	Lucille Buonocore
Senior Production Editor	Elizabeth Swain
Program Assistant	Lauren Morris
Outside Production Management	Suzanne Ingrao
Cover image	©Simon Fraser/Photo Researchers, Inc.
Cover design	Howard Grossman

The case stories that appear in this text are fictitious and not based on any specific real people or situations. The photographs used to illustrate the case stories do not depict the individuals described, nor do the individuals in the photographs have any of the conditions described.

Illustration and photo credits follow the Glossary.

This book was typeset by GGS Higher Education Resources, A Division of PreMedia Global, Inc. It was printed and bound by Quebecor Dubuque. The cover was printed by Quebecor Dubuque.

The paper in this book was manufactured by a mill whose forest management programs include sustained yield harvesting of its timberlands. Sustained yield harvesting principles ensure that the number of trees cut each year does not exceed the amount of new growth.

Library of Congress Cataloging-in-Publication Data

Jenkins, Gail W.
   Anatomy and physiology : from science to life / Gail W. Jenkins, Christopher P. Kemnitz, Gerard J. Tortora. -- 2nd ed.
      p. cm.
   Includes index.
   ISBN 978-0-470-22758-9 (cloth)
   1. Human anatomy. 2. Human physiology. I. Kemnitz, Christopher P. II. Tortora, Gerard J. III. Title.
   QP34.5.J475 2010
   612--dc22
                              2008036130

Printed in the United States of America
10 9 8 7 6 5 4 3 2 1

Gail W. Jenkins is a popular Professor of Biology at Montgomery College in Maryland, where she teaches human anatomy and physiology as well as general biology and microbiology. She received her bachelor's degree in botany from the University of California in Davis, with a minor in medical technology and completed a graduate biological sciences instructor credential program. Her master's degree in biological sciences was from California State University in Sacramento, where she focused on anatomy with research in neuroembryology conducted at the University of Kentucky College of Medicine in Lexington.

Gail is passionately devoted to assisting students in the learning process and in their preparation for health science vocations. She was the recipient of the Montgomery College Outstanding Faculty Award in 1999 and the 1997 National Institute for Staff and Organizational Development Excellence Award for Outstanding Contribution to Teaching and Learning from the University of Texas. At Montgomery College, Gail is Course Coordinator of the human anatomy and physiology curriculum, course curriculum liaison to college Health Science programs, and has served as the Chair of the Department of Biology, Physical Education and the Health Science; Chair of the Faculty Council (the faculty governance organization); Phi Theta Kappa Advisor, mentor for adjunct science faculty through the college Center for Teaching and Learning, and co-developed the Physical

(continues)

Christopher P. Kemnitz is a full-time nursing instructor at Lake Superior College in Duluth. Minnesota. Chris Kemnitz received his doctoral degree in biology from Marquette University in 1994, and completed his postdoctoral research while on active duty with the US Army from 1994 to 1997. He taught anatomy and physiology at the University of Wisconsin Superior for the past ten years. His research background includes interests in neuroscience and human performance. In 2005, he returned to school to complete a bachelors of science degree in nursing. He recently left his position as chair of the department of biology and earth sciences at UW Superior and took a position as a nursing instructor at Lake Superior College in Duluth, Minnesota where he teaches pathophysiology, nursing, and clinical instruction. He also works part-time as a neuroscience/trauma RN for St. Mary's Hospital in Duluth. Chris is a member of the Human Anatomy and Physiology Society, American Nurses Association, and Minnesota Nursing Association. Chris has always had an interest in interdisciplinary case-based and problem-based learning and has worked with colleagues across disciplines to develop and enhance learning. Chris, his wife Rebecca, and their three children Jenny, Diana, and Daniel live in Minnesota. Chris is an avid outdoorsman enjoying fishing, hunting, camping and water sports, not unusual for a Minnesotan!

Gerard J. Tortora is Professor of Biology and former Biology Coordinator at Bergen Community College in Paramus, New Jersey, where he teaches human anatomy and physiology as well as microbiology. He received his bachelor's degree in biology from Fairleigh Dickinson University and his master's degree in science education from Montclair State College. He is a member of many professional organizations, including the Human Anatomy and Physiology Society (HAPS), the American Society of Microbiology (ASM), American Association for the Advancement of Science (AAAS), National Education Association (NEA), and the Metropolitan Association of College and University Biologists (MACUB).

Above all, Jerry is devoted to his students and their aspirations. In recognition of this commitment, Jerry was the recipient of MACUB's 1992 President's Memorial Award. In 1996, he received a National Institute for Staff and Organizational Development (NISOD) excellent award from the University of Texas and was selected to represent Bergen Community College in a campaign to increase awareness of the contributions of community colleges to higher education.

Jerry is the author of several best-selling science textbooks and laboratory manuals, a calling that often requires an

(continues)

(Gail W. Jenkins continued)

Therapist Assistant Program and Associate of Science Degree in Science with concentrations in the Life Sciences, Physics, and Mathematics.

Gail is Treasurer and member of the Executive Committee of the Human Anatomy and Physiology Society (HAPS) and has been an active participant in annual and regional conference planning. Gail founded and directed a program at Stanford University's Department of Anatomy to integrate human cadaveric materials and medical imaging into pre-health science curricula, co-founded the Northern California Society of Anatomists, developed a hospital laboratory work/learn internship program for health science students, was a Federal Liaison Officer in Washington D.C. working with Congress for higher education funding, and served as an educational consultant to Stanford University's Advanced Media Research Group and to publishing and software companies.

To my students for their shared fascination and love of human body design and to my husband Don for his devotion, patience, and encouragement of my path in teaching. — G.W. J.

(Gerard J. Tortora continued)

additional 40 hours per week beyond his teaching responsibilities. Nevertheless, he still makes time for four or five weekly aerobic workouts that include biking and running. He also enjoys attending college basketball and professional hockey games and performances at the Metropolitan Opera House.

To Lynne, Gerard Jr. Kenneth, Anthony, and Drew – five wonderful children who never cease to amaze me and have made me the luckiest Dad on planet Earth. – G. J. T.

As active, "seasoned" teachers of anatomy and physiology, we find the mission of guiding our students toward an understanding of the complexities of the human body immensely rewarding. The study of anatomy and physiology is the foundation upon which a variety of gratifying health-related careers is built. Assisting students in recognizing the importance of this science as essential to their future professions and personal life is fulfilling. It is also challenging. Today's anatomy and physiology students bring to the classroom an increasing diversity in educational background, learning styles, abilities, and language skills. Many students entering an anatomy and physiology course have not yet developed a working comfort level of the critical thinking and problem-based learning skills that are essential for success in health science vocations. These include the ability to analyze and summarize presented information, extrapolate gained knowledge to newly encountered situations, and integrate isolated facts into retained conceptual understanding.

While we endeavor to help our students with the challenges that they face, we recognize that our colleagues face a number of obstacles as well. Those of us fortunate enough to teach anatomy and physiology strive to design sound curricula within the confines of limited course contact hours and instructional resources. We faculty are obligated to keep pace with an information overload driven by the ever-expanding needs of the health science programs into which our students will matriculate. Faced with these challenges, today's anatomy and physiology students, and their faculty, can become overwhelmed in the development of vital skills and by the rigors of the course textbook and curriculum.

NEW TO THIS EDITION

The second edition focuses on aiding critical thinking, conceptual understanding, and relevant application of knowledge when studying anatomy and physiology. All chapter case stories have been revised and streamlined to provide a clearer path for students in making the relevant links between factual knowledge and practical application. And, many illustrations and photos have been newly rendered in a more realistic, contemporary style to create an enriched visual environment for students.

Text Updates Every chapter in this edition of *From Science to Life* incorporates a host of improvements to both the text and the art developed by ourselves and suggested by reviewers, educators, or students. Chapter 12 has been completely rewritten in order to provide a clearer understanding of nervous tissue structure and function. This updated narrative is supported by eight new illustrations and several revised illustrations. Chapter 14 has been completely rewritten to facilitate student understanding of nerves, reflexes, and peripheral nervous system function. The updates include two new figures, many revised figures and a new table. Chapter 15 has been rewritten in order to clarify how the brain and spinal cord process sensory and motor information, and a discussion of integrative functions of the brain is now included. Chapter 20 has been rewritten to provide a clearer understanding of the structure of blood vessels, capillary networks, and the forces involved in circulating blood throughout the body. This updated narrative is supported by new orientation illustrations for all flow diagrams of arterial distribution and venous drainage in blood vessel figures, and three new cadaver photos of venous valves, the arch of the aorta, and the abdominal aorta. A new discussion and illustration have been added on pulse points too. Chapter 21 includes completely rewritten sections on adaptive immunity, cell-mediated immunity, and antibody-mediated immunity, along with updated illustrations.

Art and Design A textbook with beautiful illustrations or photographs on most pages requires a carefully crafted and functional design to assist students in making the most of the text's many features and outstanding art. Each page is carefully laid out to place related text, figures, and tables near one another, minimizing the need for page turning while reading a topic. Red print is used to indicate all callouts of figures and tables. Not only is the reader alerted to refer to the figure or table, but the red print also serves as a place locator for easy return to the narrative.

Studying human anatomy and physiology is both a visual and a descriptive enterprise. Because more and more students identify themselves as visual learners, it was a high priority during development to ensure that the illustrations in the text are as helpful to students as possible. For example, labeling facilitates quick orientation to represented structures without overcrowding the art. Each illustration found in *Anatomy and Physiology: From Science to Life* was carefully crafted and chosen to provide memorable understanding of the concepts being presented. The illustrations are rich with color. Colors are used in a consistent and meaningful manner throughout the text to emphasize structural and functional relations. For example, sensory structures, sensory neurons, and sensory regions of the brain are shades of blue, whereas motor structures are red. Sodium ions and their

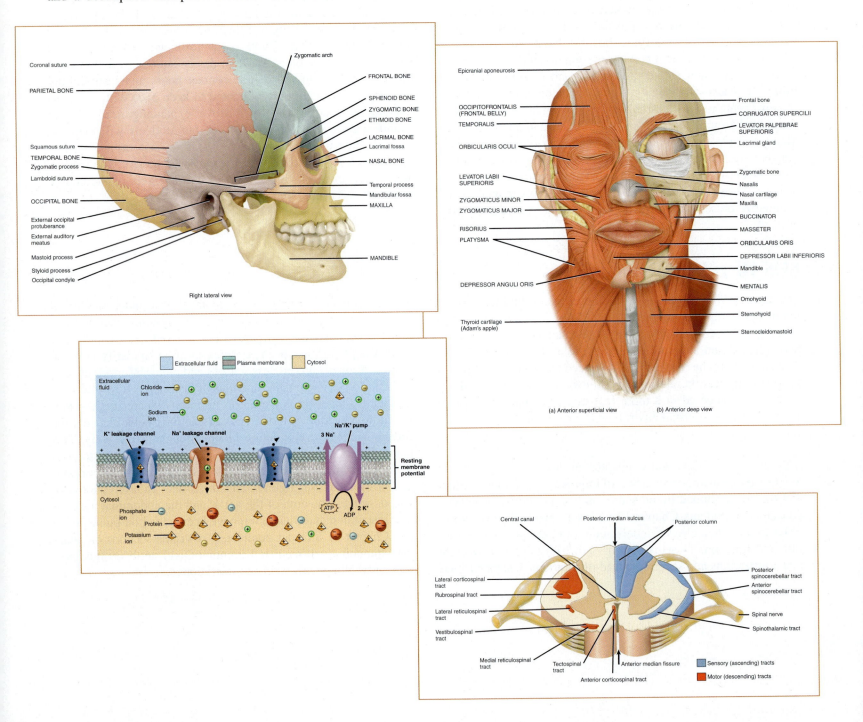

membrane channels are green, while potassium ions and membrane channels are yellow. Through careful deliberations with the illustrators and photo researchers, we have developed an outstanding art program that uniquely complements the text.

Continuing in this tradition, you will find exciting new three-dimensional illustrations gracing the pages of nearly every chapter in the text. Significantly, all of the illustrations in Chapters 7 and 8 on the skeleton are new, as well as all of the illustrations in Chapter 11 on muscles. These new illustrations are among the best that we have ever seen in any anatomy and physiology textbook and truly support the visual learner in meeting the challenge of learning so many anatomical structures. Equally important are

the numerous new illustrations depicting and clarifying physiological processes. See, for example, the eight new figures in Chapter 12 on membrane potentials.

Thoughtful revisions have been made to many of the figures depicting both anatomy and physiology throughout the text. These revisions include enhanced use of color for visual impact and to better engage students, and clarifying details for better understanding of processes. All figures showing transverse sections of the spinal cord have been recolored to better reflect gray and white matter (See Figures 13.18–13.19). Other examples include Figures 14.17–14.26 on cranial nerves and Figures 21.5 and 21.8 on immune processes.

Photographs The number of cadaver photos in this edition has been increased, and most previously existing photos have been replaced. These distinctive images were photographed by Mark Nielsen in his laboratory at the University of Utah. Many of the meticulous dissections are the work of his colleague (and

former student) Shawn Miller. Others are dissected by other students under Mark's guidance. The matching of these photographs with the line art brings your students that much closer to experiencing an actual cadaver lab.

(d) Anterior view

(b)

Audio Downloads An exciting new feature has been added to the illustration program for this edition. Audio downloads, linked to identified illustrations in each chapter, give the students the opportunity to hear while they study—as they would in lecture—about the importance and relevance of the structures or concepts that are depicted. These illustrations are identified in each chapter by a distinctive icon.

Complete Teaching and Learning Package

The second edition of *Anatomy and Physiology: From Science to Life* is accompanied by a host of dynamic resources designed to help you and your students maximize your time and energies. Please contact your Wiley representative for details about these and other resources or visit our website at www.wiley.com/college/jenkins

New! Clinical Connections
This manual helps students make the connections between the normal anatomy and physiology they are learning and the abnormal situations they might encounter in life or their future careers. Organized by body system, information on disease and disorders, medical tests and procedures, and medical terminology is included. *Clinical Connections* is included with the purchase of all new copies of this textbook.

REAL ANATOMY

New! Real Anatomy
Mark Nielsen and Shawn Miller of the University of Utah led a team of media and anatomical experts in the creation of this powerful new DVD, *Real Anatomy*. Their extensive experience in undergraduate anatomy classrooms and cadaver laboratories as well as their passion for the subject matter shine through in this new, user-friendly program with its intuitive interface. The 3-D imaging software allows students to dissect through numerous layers of a real three-dimensional human body to study and learn the anatomical structures of all body systems from multiple perspectives. Histology is viewed via a virtual microscope at varied levels of magnification. Professors can use the program to capture and customize images from a large database of stunning cadaver photographs and clear histology photomicrographs for presentations, quizzing, or testing.

Interactions: Exploring the Functions of the Human Body 3.0
by Lancraft and Frierson. Covering all body systems, this dynamic and highly acclaimed program includes anatomical overviews linking form and function, rich animations of complex physiological processes, a variety of creative interactive exercises, concept maps to help students make the connections, and animated clinical case studies. The 3.0 release boasts enhancements based on user feedback, including coverage of ATP, the building blocks of proteins, and dermatomes; a new overview on special senses; cardiac muscle;

and a revised animation on muscle contraction. *Interactions* is available in one DVD or in a web-based version, and is fully integrated into Wiley*PLUS*.

PowerPhys by Allen, Harper, Ivlev, and Lancraft. This exceptional resource includes ten self-contained lab modules for exploring physiological principles. Each module contains objectives with illustrated and animated review material, prelab quizzes, prelab reporting, data collection and analysis, and a full lab report with discussion and application questions. Experiments contain randomly generated data, allowing users to experiment multiple times but still arrive at the same conclusions. Available as a stand-alone product, *PowerPhys* is also bundled with every new copy of the Allen and Harper Human Anatomy and Physiology Laboratory Manual and is integrated into Wiley*PLUS*.

WILEY PLUS *WileyPLUS* is a powerful online tool that provides students and instructors with an integrated suite of teaching and learning resources in one easy-to-use website. With *WileyPLUS*, students will come to class better prepared for lectures, get immediate feedback and context-sensitive help on assignments and quizzes, and have access to a full range of interactive learning resources, including a complete online version of their text. A description of some of the resources available to students within *WileyPLUS* appears on the front endpapers of this text. Instructors benefit as well; *WileyPLUS* includes all the tools and resources you need to prepare and present dynamic lectures as well as assess student progress and learning. New within *WileyPLUS*, **Quickstart** is an organizing tool that makes it possible for you to spend less time preparing lectures and grading quizzes and more time teaching and interacting with students. Ask your sales representative to set you up with a test drive, or view a demo online.

Visual Library for Anatomy and Physiology 4.0
A cross-platform DVD includes all of the illustrations from the textbook in three formats: labeled, unlabeled, and unlabeled with leader lines. In addition, the *Visual Library* contains many additional illustrations and photographs that could easily enhance lecture or lab. Search for images by chapter or by using keywords.

Companion Websites
A dynamic website for students, rich with many activities for review and exploration, includes self-quizzes for each chapter, review exercises, and weblinks. An access code is bundled with each new text. A dedicated companion website for instructors provides many resources for preparing and presenting lectures. Additionally, this website provides a web version of the *Visual Library for Anatomy and Physiology*, additional critical thinking questions with answers, an editable test bank, a computerized test bank, transparencies on demand, and clicker questions. These websites can be accessed through www.wiley.com/college/jenkins.

A Brief Atlas of the Skeleton, Surface Anatomy, and Selected Medical Images

by Gerard J. Tortora. This atlas of stunning photographs provides a visual reference for both lecture and lab.

Human Anatomy and Physiology Laboratory Manual 3e

by Allen and Harper. This newly revised laboratory manual includes multiple activities to enhance the student laboratory experience. Illustrations and terminology closely match the text, making this manual the perfect companion. Each copy of the lab manual includes a CD with the **PowerPhys** simulation software for the laboratory. *WileyPLUS*, with a wealth of integrated resources including cat, fetal pig, and rat dissection videos, is also available for adoption with this laboratory. The *Cat Dissection Laboratory Guide* and a *Fetal Pig Laboratory Guide* are available to package at no additional cost with the main laboratory manual or as stand-alone dissection guides, depending on your course needs.

Photographic Atlas of the Human Body, Second Edition

by Gerard J. Tortora. Loaded with excellent cadaver photographs and micrographs, the high quality imagery can be used in the classroom, laboratory, or for study and review.

ACKNOWLEDGMENTS

We wish to especially thank several academic colleagues for their helpful contributions to this edition.

A talented group of educators has contributed to the high quality of the diverse supplementary materials that accompany this text. We wish to acknowledge each and thank them for their work. Thanks to Abdulai Barrie, Montgomery College; Heather Davis, University of Montana, Missoula; Barbara Mania-Farnell, Purdue University, Calumet; Janet Gaston, Troy University; Clare Hays, Metropolitan State College of Denver; Gary Heisermann, Salem State College; Jerri Lindsey, Tarrant County College; Scott Rahschulte, Ivy Tech Community College, Lawrenceburg; and James Rayburn, Jacksonville State University.

We would like to express our sincere thanks to our Executive Editor, Bonnie Roesch, who believed in the vision of the book and in us as its authors. Bonnie rode shotgun on the entire process and her expertise was invaluable. It is impossible to know how to thank Bonnie for all her guidance, creativity, and dedication in making our vision for this book become a reality. Bonnie has inspired and influenced every vital component of the book from front cover to end of book content. Bonnie, we can only say, "You're the best!"

We would also like to thank other vital members of our winning John Wiley & Sons, Inc. project team who often worked diligently days, evenings, and weekends to ensure the success of this book. Their commitment to and passion for this textbook is so obvious and appreciated. In particular, Mary O'Sullivan, our Developmental Editor, masterfully orchestrated this endeavor by helping us implement the revision plan, and managed the endless day-to-day details to ensure that the book's content, look, and feel reflect the needs of current anatomy and physiology curricula. Mary is uniquely able to balance warm compassion and considerate flexibility with efficient professionalism. Mary deserves personal thanks for the many pep talks and virtual pats on the back that kept us going. Karen Trost, our Freelance Developmental Editor, carefully edited and mapped the text and art, and effectively managed the numerous revisions and reviewer feedback. Karen misses nothing, and with accuracy and details,

reigns supreme! Thanks also to Suzanne Ingrao, Freelance Production Manager, whose determination, grit, and superb coordination organized all stages of the production of our book. We really value the talents of Harry Nolan, Director of Creative Services, and Karin Kincheloe, Freelance Designer, who laid out the pages to be visually attractive, pedagogically effective, and flow in a student-friendly style that makes the book come alive. Clay Stone, Executive Marketing Manager, inspired us by his tireless enthusiasm for this particular publication. Hilary Newman, our diligent photo researcher, brought faces to the case stories and transitioned our mental images into real world photographs. We're still amazed that Hilary always managed to quickly find exactly the right photo that our mind's eye envisioned. A special "thank-you" to Kelly Tavares for her help creating and coordinating the art logs. We must also express our thanks to Lorraina Raccuia, Project Editor, for her outstanding efforts in coordinating the numerous aspects of the development of supplements for the text. Special thanks also to Linda Muriello, our Senior Media Editor, who really gets all the high tech minutiae, and to Lauren Morris, our Editorial Assistant, who never failed to find the right answer every request. We are truly appreciative!

Special thanks from Gail Jenkins to those at Montgomery College: Dr. Brad Stewart, Vice President and Provost, for his ongoing support of this authoring project; Dr. Carole Wolin for the many collegial discussions about teaching methodologies; Donna Schena for her affection, understanding, and ability to find humor in all endeavors; Molly Cochran Clay, MS, CRNP, for writing the Chapter 5 case resources, and to all the faculty and students who served as a test bed that strongly influenced this edition.

Finally, we particularly want to express our gratitude to the following reviewers who took the time to read and evaluate the manuscript prior to production. Our reviewers generously provided their expertise to help us to continue to ensure the book's accuracy, clarity, and focus on the needs of today's anatomy and physiology instruction. Their insightful suggestions were appreciated and were well utilized in this edition of the book. Thank you for your invaluable contributions.

2nd Edition Reviewers

Barry Boatwright, *Gadsden State Community College*
Sonya Conway, *Northern Illinois University*
Kenneth A. Cutler, *North Carolina Central University*
Michelle Gibson, *Montcalm Community College*
Jason Hunt, *Brigham Young University—Idaho*
Daniel R. Jones, *Indiana Wesleyan University*
Elisabeth Martin, *College of Lake County*
Nichole McDaniel, *Bronx Community College CUNY*
Mark Meade, *Jacksonville State University*
Tony Morris, *Fairmont State University*
Betsy Peitz, *California State University at Los Angeles*
Brandon H. Poe, *Springfield Technical Community College*
Dawn Roberts, *Elms College*
Joseph Shostell, *Penn State University Fayette Campus*
Janice Webster, *Ivy Tech Community College*
Joe Wheeler, *North Lake College*

1st Edition Reviewers

Erin Amerman, *Santa Fe Community College*
Lynne Anderson, *Meridian Community College*
Ronald Beumer, *Armstrong Atlantic State University*
James A. Carson, *University of South Carolina*
James Tim Daniels, *Southern Arkansas University*
Abraham Flores, *Pontiff Catholic University of Puerto Rico*
Katrina Fogleman, *University of Michigan*
Purti Gadkari, *Wharton County Junior College*
Eric Genz-Mould, *Shoreline Community College*
D. Bruce Gray, *Simmons College*

Tim Grogan, *Valencia Community College*
Pramila Gurrala, *Wharton County Junior College*
Nora Herbert, *Red Rocks Community College*
Karen Hlinka, *Kentucky Community & Technical College*
Scott Jackson, *Asheville-Buncombe Technical College*
Joyce Ellen Lathrop-Davis, *The Community College of Baltimore County*
Jerri K. Lindsey, *Tarrant County College*
Michael C. Lyons, *Kirkwood Community College*
Angela Mason, *Tidewater Community College*
Susan Murphy-Walker, *Our Lady of the Lake University*
Linda Nichols, *Santa Fe Community College*
Tricia Paramore, *Hutchinson Community College*
Ratna Perikala, *Houston Community College Southwest*
Davonya J. Person, *Auburn University*
Danny Pincivero, *The University of Toledo*
Terrence J. Ravine, *University of South Alabama*
J. Orion Rogers, *Radford University*
Marilyn Shopper, *Johnson County Community College*
John Simmons, *Barton County Community College*
Susan A. Skambis, *Valencia Community College*
Yong Tang, *Front Range Community College*
Teresa Trendler, *Pasadena City College*
Richard Tsou, *Gordon College*
Leticia Vosotros, *Ozarks Technical Community College*
Kathleen Warren, *Daytona Beach Community College*
Brian W. Witz, *Nazareth College*
Fadi Zaher, *Gateway Technical College*
Samuel Zeakes, *Radford University*
Paul Zillgitt, *University of Wisconsin*

NOTE TO STUDENTS

Your book has a variety of special features that will make your time studying anatomy a more rewarding experience. These features have been developed based on feedback from students like you who have used previous editions of the text.

As you start to read each chapter, you will find that each one opens with a **Case Story** as an introduction. The case story returns at key intervals throughout the chapter to directly link the clinical scenario with the content of each concept module. The case stories develop along with the chapter content; together they offer you a constantly growing learning experience.

Richard's Story

The dream had been so pleasant, until that car began beeping nonstop. Richard had been on the beach with the warm sun beating down on him. As he began to regain awareness, he realized he'd been dreaming and recognized the car beeping for what it really was—the piercing squeal of a smoke detector. The attic bedroom of the old Minnesota farmhouse where he slept was unusually warm, and he smelled smoke.

"Mom, Dad, wake up! I smell smoke!" he screamed as he dashed down the narrow staircase to the second floor. He could feel the heat of the fire as he looked into his parents' bedroom. No one was there. "They must have gotten out," he said to himself. He dashed down the stairs to the first floor and saw his father battling in vain to put out the flames that crept along the old wooden floors and were consuming the curtains of the living room. "Dad, where's Mom and the girls?" Richard shouted.

"They're outside already! Get out of here! The place is burning like a pile of matches!"

"I'll get the fire extinguisher—it's in the kitchen!" called Richard.

As he dashed toward the kitchen door, his father yelled out, "No! Richard—stop!"

As Richard pushed on the thin wooden door to the kitchen, he felt the air suck past him into the kitchen; it was followed by a silence that seemed to last forever. Then the back draft hit him. He saw a wall of flame, and he felt searing heat and instant pain like nothing he had ever imagined.

In this chapter you will study the integumentary system. The skin, or integument, is an important organ system that we often take for granted until it is compromised. As you will see in Richard's story, the skin has many important functions that are crucial to maintaining the homeostasis of the human body.

At the end of the concept module, take time to try and answer the Checkpoint questions placed there. If you can answer them, then you are ready to move on. If you have trouble answering the questions, you may want to re-read the section before continuing.

ids. Because akage chan- tassium ions of the cell is e down their 1 more posi- embrane be- e membrane

.other factor iterior: Most e they are at- d large mol-

bring into the cell, they are *electrogenic*, which means they contribute to the relative negativity of the cell interior.

Checkpoint

17. How does the cell membrane of a neuron allow the cell to store electrical charges?
18. What are the three stimuli that result in ion channels opening or closing? What happens when these channels open?
19. How are leakage channels different from gated channels?
20. What two conditions allow maintenance of the resting membrane potential?

Studying the figures (illustrations that include artwork and photographs) in this book is as important as reading the text. To get the most out of the visual parts of this book, use the tools we have added to the figures to help you understand the concepts being presented. Start by reading the Legend, which explains what the figure is about. Next, study the Key Concept Statement, indicated by a "key" icon, which reveals a basic idea portrayed in the figure. Added to many figures you will also find an Orientation Diagram to help you understand the perspective from which you are viewing a particular piece of anatomical art. Selected figures include Functions boxes, brief summaries of the functions of the anatomical structure of the system shown.

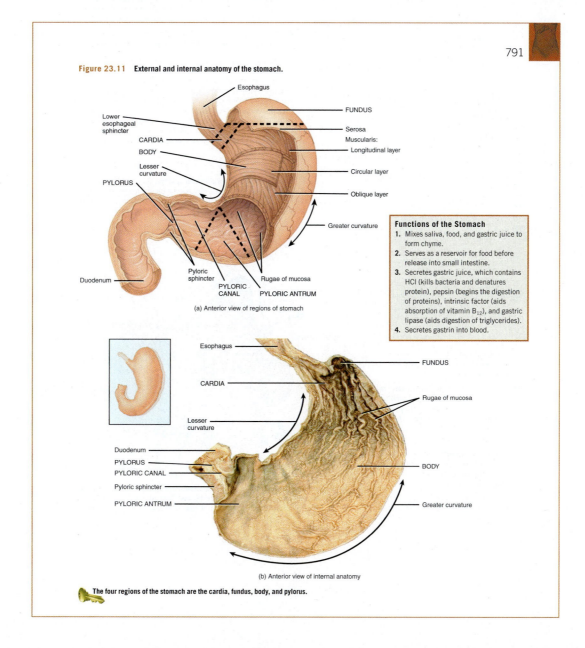

791

Figure 23.11 External and internal anatomy of the stomach.

Functions of the Stomach
1. Mixes saliva, food, and gastric juice to form chyme.
2. Serves as a reservoir for food before release into small intestine.
3. Secretes gastric juice, which contains HCl (kills bacteria and denatures protein), pepsin (begins the digestion of proteins), intrinsic factor (aids absorption of vitamin B_{12}), and gastric lipase (aids digestion of triglycerides).
4. Secretes gastrin into blood.

(a) Anterior view of regions of stomach

(b) Anterior view of internal anatomy

The four regions of the stomach are the cardia, fundus, body, and pylorus.

In each chapter you will find that several illustrations are marked with an icon that looks like an MP3 player. This is an indication that an audio download, which narrates and discusses the important elements of that particular illustration, is available for your study. You can access these downloads on the student companion website or through *WileyPLUS*.

Studying physiology requires an understanding of the sequence of processes. Correlation of sequential processes in text and art is achieved through the use of special numbered lists in the narrative that correspond to numbered segments in the accompanying figure. This approach is used extensively throughout the book to lend clarity to the flow of complex processes.

383

Chemical Synapses

Unlike the gap junctions in electrical synapses, the plasma membranes of presynaptic and postsynaptic neurons in a **chemical synapse** are close, but they are not connected directly (Figure 12.22). They are separated by the **synaptic cleft**, a space of 20–50 nm that is filled with interstitial fluid. Because action potentials cannot conduct across the synaptic cleft, an alternate, indirect form of communication occurs. In response to a nerve impulse, the presynaptic neuron releases a neurotransmitter that diffuses through the fluid in the synaptic cleft and binds to receptors in the plasma membrane of the postsynaptic neuron. The postsynaptic neuron receives the chemical signal and, in turn, produces a **postsynaptic potential**, a type of graded potential. Thus, the presynaptic neuron converts an electrical signal (nerve impulse) into a chemical signal (released neurotransmitter). The postsynaptic neuron receives the chemical signal and, in turn, generates an electrical signal (postsynaptic potential). The time required for these processes at a chemical synapse, a synaptic delay of about 0.5 msec, is the reason that chemical synapses relay signals more slowly than electrical synapses.

A typical chemical synapse transmits a signal as follows (Figure 12.22):

1 A nerve impulse arrives at the synaptic end bulb of a presynaptic axon.

2 The depolarizing phase of the nerve impulse opens **voltage-gated Ca^{2+} channels** in the membrane of synaptic end bulbs. Because calcium ions are more concentrated in the extracellular fluid, Ca^{2+} flows inward through the opened channels.

3 An increase in the concentration of Ca^{2+} inside the presynaptic neuron triggers exocytosis of some of the synaptic vesicles. As vesicle membranes merge with the plasma membrane, neurotransmitter molecules within the synaptic vesicles are released into the synaptic cleft. Each synaptic vesicle contains several thousand molecules of neurotransmitter.

4 The neurotransmitter molecules diffuse across the synaptic cleft and bind to **neurotransmitter receptors** in the postsynaptic neuron's plasma membrane. The receptor shown in Figure 12.22 is part of a ligand-gated channel (see Figure 12.10b).

5 Binding of neurotransmitter molecules to their receptors on ligand-gated channels opens the channels and allows particular ions to flow across the membrane.

6 As ions flow through the opened channels, the voltage across the membrane changes.

This change in membrane voltage is a **postsynaptic potential**. Depending on which ions the channels admit, the postsynaptic potential may be a depolarization or hyperpolarization. For example, opening of Na^+ channels allows inflow of Na^+, which causes depolarization. Opening of K^+ channels allows K^+ to move out, producing hyperpolarization.

7 When a depolarizing postsynaptic potential reaches threshold, it triggers an action potential in the postsynaptic neuron.

At most chemical synapses, only *one-way information transfer* can occur—from a presynaptic neuron to a postsynaptic neuron or to an effector, such as a muscle fiber or a gland cell. For example, synaptic transmission at a neuromuscular junction proceeds from a motor neuron to a skeletal muscle fiber (but not in the opposite direction). Only synaptic end bulbs of presynaptic neurons can release neurotransmitter, and only the postsynaptic neuron's membrane has the receptor proteins that can recognize and bind that neurotransmitter. As a result, action potentials move in one direction.

Figure 12.22   Signal transmission at a chemical synapse. Through exocytosis of synaptic vesicles, a presynaptic neuron releases neurotransmitter molecules. After diffusing across the synaptic cleft, the neurotransmitter binds to receptors in the plasma membrane of the postsynaptic neuron and produces a postsynaptic potential.

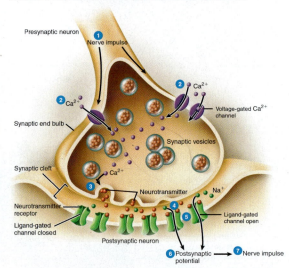

At a chemical synapse, a presynaptic neuron converts an electrical signal (nerve impulse) into a chemical signal (neurotransmitter release). The postsynaptic neuron then converts the chemical signal back into an electrical signal (postsynaptic potential).

At the end of each chapter is a **Concept and Resource Summary** that provides summaries of the key concepts from each section of the chapter, along with the appropriate online resources for further understanding of the concepts. **Understanding the Concepts** questions are more in-depth questions that ask you to apply the concepts you have studied in the chapter to specific situations. **Answers to the Checkpoint Questions** appear in an appendix at the end of the book so you can check your progress.

At times, you may require extra help to learn specific anatomical features of the various body systems. One way to do this is through the use of **mnemonics**, aids to help memory. Mnemonics are included throughout the text, some displayed in figures, tables, and some included within the text discussion. We encourage you to use not only the mnemonics provided, but also to create your own to help you learn the multitude of terms involved in your study of human anatomy.

632   Chapter 19 • The Cardiovascular System: The Heart

Concept and Resource Summary

Concept	Resources

Introduction

1. The heart functions to pump blood to all body cells.
2. The study of the heart and its diseases is **cardiology**.

Concept 19.1 The heart is located in the mediastinum and has a muscular wall covered by pericardium.

1. The heart is about the size of a closed fist and is located in the mediastinum of the thoracic cavity, where it rests on the diaphragm. It is situated obliquely, with the **apex** (pointed end) directed inferiorly to the left and the **base** positioned superiorly to the right.

See
• **Anatomy Overview – Cardiac Muscle**

valve. Contraction of the papillary muscles
...ushing up into the atrium. Damage to the AV
...into the atria when the ventricles contract.
...pulmonary trunk and aorta. These valves per-
...revent backflow of blood into the ventricles

...nation, then pumps oxygen-rich blood

...at: the **pulmonary circulation** and the sys-

...circulation. **Deoxygenated blood** returning
...h the two vena cavas and the coronary sinus.
...pumps blood into the pulmonary trunk to the
...aries, where it unloads CO_2 and picks up O_2.
...lation. Freshly **oxygenated blood** from pulmonary veins enters the left atrium, then the left ventricle. Ventricular contraction ejects blood into the aorta and its branches, which carry oxygenated blood to all parts of the body except for the air sacs of the lungs. In the systemic circulation, blood flows through arteries to smaller arterioles and then into *systemic capillaries* where nutrients, gases, and wastes are exchanged. Venules carry deoxygenated blood away into larger veins and back to the right atrium of the heart.

4. The heart itself must receive blood enriched with nutrients and gases, and have wastes removed. The wall of the heart has its own blood vessels called the **coronary (cardiac) circulation** that delivers blood to and from the myocardium.
5. The principal coronary vessels are the **left** and **right coronary arteries** that arise from the ascending aorta. The **left coronary artery** divides into the **anterior interventricular branch** serving both ventricles, and the **circumflex branch** serving the left atrium and ventricle. The **right coronary artery** branches to supply the right atrium. Branches of the right coronary artery form the **posterior interventricular branch** serving walls of both ventricles, and the **marginal branch** supplying the right ventricle.
6. Arteries of the coronary circulation have many **anastomoses** (connections) that allow detours for arterial blood if a main route becomes obstructed.
7. Deoxygenated blood from the myocardium drains into the posterior **coronary sinus** that empties into the right atrium. The **greater**, **middle**, **small**, and **anterior cardiac veins** carry blood into the coronary sinus.

See
• **Anatomy Overview – Heart Structures**
• **Anatomy Overview – Pulmonary Circulation**

Hear
• Figure 19.6 – Systemic and Pulmonary Circulations

Do
• Exercise – Drag and Drop Blood Flow

Concept 19.5 The cardiac conduction system coordinates heart contractions for effective pumping.

1. The heart has specialized cardiac muscle fibers, called **autorhythmic cells**, that are self-excitable (have inherent rhythmical electrical activity). The powerful contractions of **contractile fibers**, which comprise the majority of myocardial fibers, propel blood throughout the body.

...n system, which coordinates

...ted at the heart's pacemaker,
...agate throughout both atria;
...atrial septum; (3) the action
...g along the AV bundle, the
...rventricular septum; and (5)
...o the remainder of the ven-

See
• **Anatomy Overview – Heart Structures**
• **Animation – Cardiac Conduction**

Hear
• Figure 19.8 – The Conduction System of the Heart

Do
• Exercise – Sequence Cardiac Conduction

7. Chemicals that influence cardiac muscle and its contraction include: (1) the **hormones** epinephrine, norepinephrine, and thyroid hormones that increase heart rate and contractility; and (2) **cations** that modify heart rate and contractility. Elevated Na^+ or K^+ decreases heart rate and contractility, while elevated Ca^{2+} increases heart rate and contractility.
8. Other factors that influence heart rate include age, gender, physical fitness, and body temperature.

Other Chapter Resources

For additional information on clinical applications, medical disorders, and medical terminology, see Chapter 9 in the *Clinical Connections* manual packaged with your text.

Within *WileyPLUS*
• Chapter Quiz
• Visual Anatomy
• Mastering Vocabulary
• Audio Glossary
• Homeostatic Imbalance: The Case of the Man with a Pain in his Jaw
• Homeostatic Imbalance: The Case of the Man with Labored Breathing
• Web Links

• Understanding the Concepts

1. Which chamber of the heart has the thickest wall? Why?
2. Describe the sequence of heart chambers, heart valves, and blood vessels a drop of blood encounters from the time it flows into the right atrium until it reaches the aorta. When during its journey does the drop of blood contain the most oxygen? The least oxygen?
3. Why is blood flowing through the chambers within the heart unable to supply sufficient oxygen or remove enough carbon dioxide from the myocardium? What is the body's solution to this problem?
4. Describe the status of each of the four heart valves during ventricular filling. How would a defect in the atrioventricular valve affect ventricular filling?
5. Why must ventricular pressure be greater than arterial pressure during ventricular ejection?
6. Which events cause the AV valves to open and shut? Which events make the SL valves open and shut?
7. Will stroke volume increase or decrease in each of the following situations? Explain each of your answers.
   • Your blood pressure rises when you're angry.
   • An anesthetic decreases the strength of contraction of your ventricles.
   • When you are exercising, contraction of skeletal muscles returns more blood to the heart.
8. What mechanism does the cardiovascular center use to speed up or slow down heart rate?

Throughout the text we have included **pronunciations** and, sometimes, **word roots**, for many terms that may be new to you. These appear in parentheses immediately following the new words, and the pronunciations are repeated in the glossary at the back of the book. Look at the words carefully and say them out loud several times. Learning to pronounce a new word will help you remember it and make it a useful part of your medical vocabulary. Take a few minutes now to read the following pronunciation key, so it will be familiar as you encounter new words.

Pronounciation Key

1. The most strongly accented syllable appears in capital letters, for example, bilateral (bī-LAT-er-al) and diagnosis (dī-ag-NŌ-sis).

2. If there is a secondary accent, it is noted by a prime (′), for example, constitution (kon′-sti-TOO-shun) and physiology (fiz′-ē-OL-ō-jē). Any additional secondary accents are also noted by a prime, for example, decarboxylation (dē′-kar-bok′-si-LĀ-shun).

3. Vowels marked by a line above the letter are pronounced with the long sound, as in the following common words:

   ā as in *māke*   ō as in *pōle*
   ē as in *bē*   ū as in *cūte*
   ī as in *īvy*

4. Vowels not marked by a line above the letter are pronounced with the short sound, as in the following words:

   a as in *above* or *at*   o as in *not*
   e as in *bet*   u as in *bud*
   i as in *sip*

5. Other vowel sounds are indicated as follows:

   oy as in *oil*
   oo as in *root*

6. Consonant sounds are pronounced as in the following words:

   b as in *bat*   m as in *mother*
   ch as in *chair*   n as in *no*
   d as in *dog*   p as in *pick*
   f as in *father*   r as in *rib*
   g as in *get*   s as in *so*
   h as in *hat*   t as in *tea*
   j as in *jump*   v as in *very*
   k as in *can*   w as in *welcome*
   ks as in *tax*   z as in *zero*
   kw as in *quit*   zh as in *lesion*
   l as in *let*

Brief Contents

XV

Contents

3 THE CELLULAR LEVEL OF ORGANIZATION 56

4   THE TISSUE LEVEL OF ORGANIZATION 100

9   ARTICULATIONS   230

10   MUSCLE TISSUE   258

11 THE MUSCULAR SYSTEM   292

Joe's Story   292   **Introduction   293**

12 INTRODUCTION TO THE NERVOUS SYSTEM   356

Mary's Story   *356*   **Introduction**   **357**

15   SENSORY, MOTOR, AND INTEGRATIVE SYSTEMS   480

Sannu's Story   **480**   **Introduction**   **481**

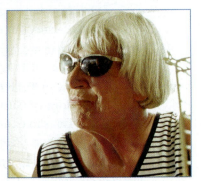

---

18   THE CARDIOVASCULAR SYSTEM: THE BLOOD   584

Ed's Story   584   **Introduction   585**

---

21 THE LYMPHATIC SYSTEM AND IMMUNITY 700

Marlene's Story 700   **Introduction 701**

---

24 THE URINARY SYSTEM   834

Andre's Story   834   **Introduction   835**

25 THE REPRODUCTIVE SYSTEMS AND DEVELOPMENT 874
Lou and Susan's Story 874 **Introduction 875**

ANATOMY AND PHYSIOLOGY
FROM SCIENCE TO LIFE

1

An Introduction to the Human Body

Jaime rifled anxiously through his closet, throwing clothes out onto the bed. "Amy, where's my blue scrub top?"

"It's in the closet," came her reply.

"I don't see it," he said.

Amy brushed past Jaime, reaching into the closet. "It's right here. Chill out, will you? I have to get ready for school," she said.

"Thanks, I'm just a little nervous. You know I need this job. I want to make a good impression," explained Jaime.

"It will be fine," said Amy. "Now get going."

Jaime put on the top, gave her a hug, and headed for the door. "Bye. See you in class tonight," he said, slamming the door on his way out.

By the time Jaime arrived at the radiology department, he was covered with a fine layer of perspiration. "Hi. I'm the new transporter," he explained to the woman sitting at the desk.

"Hi. I'm Ellen. Nice to meet you. Come on back," she said, opening the door. Ellen returned to her desk and finished organizing some files. "Most of what you'll need to know, you'll learn on the job. It's easy once you get the hang of it. You won't just be transporting patients. You'll get a chance to help the X-ray techs, develop X-ray films, and do scheduling, too. You can follow me around until you're comfortable with everything." Just then the phone rang. Ellen picked up the receiver, "Radiology, Ellen speaking. Okay. Patient's name? We have an opening at eleven o'clock. CT scan, C-spine. Gotcha. We'll see her then. Bye."

"That's a scan of the cervical spine, right?" asked Jaime.

"Yes. We schedule routine scans and X-rays here in this log. You write down the name, time, and procedure here," said Ellen. "Orders come through on the computer, too, but we always back everything up on paper."

A basic understanding of the human body is necessary not only for the health-care provider, as we will find following Jaime's story, but also for anyone who is concerned about their own health and well-being.

INTRODUCTION

In recent years the explosion of information in science and medicine has led to dramatic advances in the improvement of human health. To understand how your body maintains a healthy balance in all its systems, you must learn about its structures and how they function. Our exploration of the human body begins with an introduction to the disciplines of anatomy (structure) and physiology (function). We will then consider how living things are organized and explore the properties that we share with animals, plants, and even bacteria. The study of human anatomy and physiology will provide you with basic information about the human body and a common vocabulary that will help you speak about the body in a way that is understood by scientists and health-care professionals alike.

> *To understand how your body maintains a healthy balance in all its systems, you must learn about its structures and how they function.*

The sciences of anatomy and physiology are the foundations for understanding the structure and function of the human body. **Anatomy** (a-NAT-ō-mē; *ana-* = up; *-tomy* = process of cutting) is the science of body structures and the relationships among structures. **Physiology** (fiz´-ē-OL-ō-jē; *physio-* = nature; *-logy* = study of) is the science of body functions—how the body parts work. Because function can never be separated completely from structure, we can understand the human body best by studying anatomy and physiology together. In this book we will look at how each structure of the body is designed to carry out a particular function and how the structure of a part often determines the functions it can perform. For example, the bones of the skull join tightly to form a rigid case that protects the brain. The bones of the fingers, by contrast, are more loosely joined to allow a variety of movements.

1.1 The human body is composed of six levels of structural organization and contains eleven organ systems.

The structures of the human body are organized on several levels, similar to the way a language is organized into letters of the alphabet, words, sentences, paragraphs, and so on. Our exploration of the human body will extend from some of the smallest body structures and their functions to the largest structure—an entire person. From the smallest to the largest, six levels of organization will help you to understand anatomy and physiology: chemical, cellular, tissue, organ, system, and organismal (Figure 1.1).

Figure 1.1   Levels of structural organization in the human body.

The levels of structural organization are chemical, cellular, tissue, organ, system, and organismal.

1 Chemical level. This very basic level, which can be compared to the letters of the alphabet, includes includes **atoms**, the smallest components of a chemical element that retain the properties of the element, and **molecules**, two or more atoms joined together. Certain atoms, such as carbon (C), hydrogen (H), oxygen (O), nitrogen (N), phosphorus (P), and calcium (Ca), are essential for life. Two familiar examples of molecules found in the body are deoxyribonucleic acid (DNA), the genetic material passed from one generation to the next, and glucose ($C_6H_{12}O_6$), a sugar readily used by the body. Chapter 2 focuses on the chemical level of organization.

2 Cellular level. Molecules combine to form **cells**, the basic structural and functional units of an organism, and the smallest living units in the human body. Among the many types of cells in your body are muscle cells, nerve cells, and blood cells. The cellular level of organization is the focus of Chapter 3.

3 Tissue level. **Tissues** are groups of cells and the materials surrounding them that work together to perform a particular function. There are just four basic types of tissue in your body: *epithelial tissue*, *connective tissue*, *muscle tissue*, and *nervous tissue*. Note in Figure 1.1 that smooth muscle tissue consists of tightly packed smooth muscle cells. Chapter 4 describes the tissue level of organization.

4 Organ level. At this level different types of tissues join together to form body structures. **Organs** usually have a recognizable shape, are composed of two or more different types of tissues, and have specific functions. Examples of organs include the skin, bones, stomach, heart, liver, lungs, and brain. Figure 1.1 shows how several tissues make up the stomach. The stomach's outer covering, the *serous membrane*, is a layer of epithelial tissue and connective tissue that reduces friction when the stomach moves and rubs against other organs. Underneath are the *smooth muscle tissue layers*, which contract to churn and mix food and push it into the next digestive organ, the small intestine. The innermost lining of the stomach is an *epithelial tissue layer* that produces fluid and chemicals responsible for digestion in the stomach.

5 System level. A **system** consists of related organs that have a common function. One example is the digestive system, which breaks down and absorbs food. Its organs include the mouth, salivary glands, pharynx (throat), esophagus (food tube), stomach, small intestine, large intestine, liver, gallbladder, and pancreas. Sometimes an organ is part of more than one system. For example, the pancreas is part of both the digestive system and the hormone-producing endocrine system.

6 Organismal level. An **organism** is any living individual. All the parts of the human body functioning together constitute the total organism—a single living person.

In the chapters that follow, you will study the anatomy and physiology of the major body systems. Table 1.1 introduces the components and functions of these systems in the order they are discussed in the book. As you study the body systems, you will discover how they work together to maintain health, protect you from disease, and allow for reproduction of the species. You will also discover that all body systems influence one another.

As an example, consider how just two body systems—the integumentary and skeletal systems—cooperate to function at the organismal level. The integumentary system, which includes the skin, hair, and nails, protects all other body systems, including the skeletal system, which includes all the bones and joints of the body. The skin serves as a barrier between the outside environment and internal tissues and organs, such as those that make up the skeletal system. The skin also participates in the production of vitamin D, which the body needs to absorb the calcium in foods such as milk. Calcium is used to build bones and teeth. The skeletal system, in turn, provides support for the integumentary system, serves as a reservoir for calcium by storing it in times of plenty and releasing it in times of need, and produces white blood cells that help the skin resist invasion by disease-causing microbes such as bacteria and viruses.

Checkpoint

1. Two or more atoms join together to form what organizational structures?
2. Which levels of structural organization are you unable to see with your naked eye?
3. Which level of structural organization is composed of two or more different types of tissues that work together to perform a specific function?
4. How are organs and tissues different?

TABLE 1.1

The Eleven Systems of the Human Body

Integumentary System (Chapter 5)

Components: Skin, and structures derived from it, such as hair, nails, sweat glands, and oil glands.

Functions: Protects the body; helps regulate body temperature; eliminates some wastes; helps make vitamin D; and detects sensations such as touch, pain, warmth, and cold.

Hair

Skin and associated glands

Fingernails (and toenails)

Skeletal System (Chapters 6–9)

Components: Bones and joints of the body and their associated cartilages.

Functions: Supports and protects the body; provides a surface area for muscle attachments; aids body movements; houses cells that produce blood cells; stores minerals and lipids (fats).

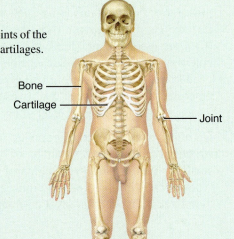

Bone

Cartilage

Joint

Muscular System (Chapters 10, 11)

Components: Muscles composed of skeletal muscle tissue, so named because it is usually attached to bones.

Functions: Produces body movements, such as walking; stabilizes body position (posture); generates heat.

Skeletal muscle

Tendon

Nervous System (Chapters 12–16)

Components: Brain, spinal cord, nerves, and special sense organs, such as the eye and ear.

Functions: Generates action potentials (nerve impulses) to regulate body activities; detects changes in the body's internal and external environment, interprets the changes, and responds by causing muscular contractions or glandular secretions.

Brain

Spinal cord

Nerve

Endocrine System (Chapter 17)

Components: Hormone-producing glands (pineal gland, hypothalamus, pituitary gland, thymus, thyroid gland, parathyroid glands, adrenal glands, pancreas, ovaries, and testes) and hormone-producing cells in several other organs.

Functions: Regulates body activities by releasing hormones, which are chemical messengers transported in blood from an endocrine gland to a target organ.

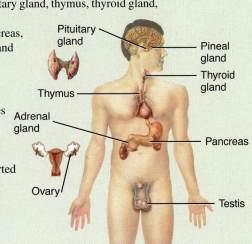

Pituitary gland

Pineal gland

Thyroid gland

Thymus

Adrenal gland

Pancreas

Ovary

Testis

Cardiovascular System (Chapters 18–20)

Components: Blood, heart, and blood vessels.

Functions: Heart pumps blood through blood vessels; blood carries oxygen and nutrients to cells and carbon dioxide and wastes away from cells and helps regulate acid–base balance, temperature, and water content of body fluids; blood components help defend against disease and repair damaged blood vessels.

Blood vessels:

Artery

Vein

Heart

Lymphatic System and Immunity (Chapter 21)

Components: Lymphatic fluid and vessels; also includes spleen, thymus, lymph nodes, and tonsils.

Functions: Returns proteins and fluid to blood; carries lipids from gastrointestinal tract to blood; includes structures where lymphocytes that protect against disease-causing organisms mature and proliferate.

Tonsil
Thymus
Thoracic duct
Spleen
Lymph node
Lymphatic vessel

Respiratory System (Chapter 22)

Components: Lungs and air passageways such as the pharynx (throat), larynx (voice box), trachea (windpipe), and bronchial tubes leading into and out of them.

Functions: Transfers oxygen from inhaled air to blood and carbon dioxide from blood to exhaled air; helps regulate acid–base balance of body fluids; air flowing out of lungs through vocal cords produces sounds.

Pharynx
Larynx (voice box)
Trachea (windpipe)
Bronchus
Lung

Digestive System (Chapter 23)

Components: Organs of gastrointestinal tract, a long tube that includes the mouth, pharynx (throat), esophagus, stomach, small and large intestines, and anus; also includes accessory organs that assist in digestive processes, such as the salivary glands, liver, gallbladder, and pancreas.

Functions: Achieves physical and chemical breakdown of food; absorbs nutrients; eliminates solid wastes.

Salivary gland
Pharynx
Mouth
Esophagus
Liver
Stomach
Gallbladder
Large intestine
Pancreas (posterior to stomach)
Small intestine
Anus

Urinary System (Chapter 24)

Components: Kidneys, ureters, urinary bladder, and urethra.

Functions: Produces, stores, and eliminates urine; eliminates wastes and regulates volume and chemical composition of blood; helps maintain the acid–base balance of body fluids; maintains body's mineral balance; helps regulate production of red blood cells.

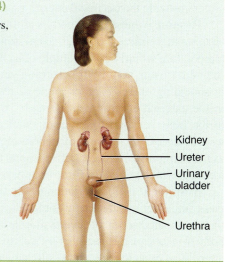

Kidney
Ureter
Urinary bladder
Urethra

Reproductive Systems (Chapter 25)

Components: Gonads (testes in males and ovaries in females) and associated organs (uterine tubes, uterus, and vagina in females and epididymis, ductus deferens, and penis in males).

Functions: Gonads produce gametes (sperm or oocytes) that unite to form a new organism; gonads also release hormones that regulate reproduction and other body processes; associated organs transport and store gametes.

Mammary gland
Uterine (fallopian) tube
Ovary
Uterus
Vagina
Penis
Testis
Ductus (vas) deferens
Seminal vesicle
Prostate

1.2   The human body carries on basic life processes that distinguish it from nonliving objects.

Basic Life Processes

All living organisms have certain characteristics that distinguish them from nonliving things. The following are the six most important life processes of the human:

Metabolism (me-TAB-ō-lizm) is the sum of all the chemical processes that occur in the body. One phase of metabolism is **catabolism** (ka-TAB-ō-lizm; *catabol-* = throwing down; *-ism* = a condition), the breaking down of complex chemical substances into simpler ones. The other phase of metabolism is **anabolism** (a-NAB-ō-lizm; *anabol-* = a raising up), the building up of complex chemical substances from smaller, simpler ones. For example, catabolism splits proteins in food into amino acids, which can be used during anabolism as the building blocks for new proteins that make up muscles and bones. In metabolic processes, oxygen taken in by the respiratory system and nutrients broken down in the digestive system provide the chemical energy to power cellular activities.

Responsiveness is the body's ability to detect and respond to changes in its internal or external environment. A decrease in body temperature would be an example of a change in the internal environment (inside your body). Turning your head toward the sound of squealing brakes is an example of responsiveness to a change in the external environment (outside your body). Different cells in the body respond to environmental changes in characteristic ways. Nerve cells respond by generating electrical signals known as nerve impulses. Muscle cells respond to nerve impulses by contracting, which generates force to move body parts.

Movement includes motion of the whole body, individual organs, single cells, and even tiny organelles inside cells. For example, the coordinated action of several muscles enables you to move your body from one place to another by walking or running. After you eat a meal that contains fats, your gallbladder (an organ) contracts and squirts bile into the gastrointestinal tract to help in the digestion of fats. When a body tissue is damaged or infected, certain white blood cells move from the blood into the affected tissue to help clean up and repair the area. Even inside individual cells, various cellular parts move from one position to another to carry out their functions.

Growth is an increase in body size. It may be due to an increase in (1) the size of existing cells, (2) the number of cells, or (3) the amount of material surrounding cells. In a growing bone, for example, mineral deposits accumulate between bone cells, causing the bone to grow in length and width.

Differentiation is the process in which unspecialized cells become specialized cells. Each type of cell in the body has a specialized structure and function. Specialized cells differ in structure and function from the unspecialized cells that gave rise to them. For example, red blood cells and several types of white blood cells differentiate from the same unspecialized cells in red bone marrow. Such unspecialized cells, which can divide and give rise to cells that undergo differentiation, are known as **stem cells**. Similarly, a single fertilized egg (ovum) undergoes tremendous differentiation to develop into a unique adult who is similar to, yet quite different from, either parent.

Reproduction refers to either (1) the formation of new cells for tissue growth, repair, or replacement or (2) the production of a new individual. Cell formation occurs continuously throughout life; life continues from one generation to the next through the production of a new individual through the fertilization of an ovum by a sperm cell.

Although not all of these processes occur all of the time in every cell of the body, when they cease to occur properly, cell death may occur. When cell death is extensive and leads to organ failure, the result is death of the organism. Clinically, loss of the heartbeat, absence of spontaneous breathing, and loss of brain functions indicate death in the human body.

Homeostasis

The French physiologist Claude Bernard (1813–1878) first proposed that the cells of many-celled organisms flourish because they live in the relative constancy of *le milieu interieur*—the internal environment—despite continual changes in the organisms' external environment. The American physiologist Walter B. Cannon (1871–1945) coined the term *homeostasis* to describe this dynamic constancy. **Homeostasis** (hō´-mē-ō-STĀ-sis; *homeo-* = sameness; *-stasis* = standing still) is the maintenance of relatively stable conditions in the body's internal environment. Through the continual interaction of all the body's regulatory processes, homeostasis ensures that the body's internal environment remains stable despite changes inside and outside the body. Each body structure, from the cellular level to the systemic level, contributes in some way to keeping the internal environment of the body within normal limits.

Homeostasis is a dynamic process. As conditions change, the body makes adjustments within a narrow range that is compatible with maintaining life. For example, the level of glucose in blood normally stays between 70 and 110 milligrams of glucose per 100 milliliters of blood.

Body Fluids

An important aspect of homeostasis is maintaining the volume and composition of **body fluids**, dilute, watery solutions containing dissolved chemicals that are found inside cells and surrounding them. The fluid within cells is **intracellular fluid** (*intra-* = inside). The fluid outside body cells is **extracellular fluid** (*extra-* = outside). Dissolved in the water of intracellular and extracellular fluids are oxygen, nutrients, proteins, and a variety of *ions* (electrically charged chemical particles). All these substances are needed to maintain life. The extracellular fluid that fills the narrow spaces between cells of tissues is known as **interstitial fluid** (in´-ter-STISH-al; *inter-* = between). As you progress with your studies, you will learn that the extracellular fluid within blood vessels is termed **blood plasma**, that within lymphatic vessels is called **lymph**, that in and around the brain and spinal cord is called **cerebrospinal fluid**, and that in joints is called **synovial fluid**.

As Bernard predicted, the proper functioning of body cells depends on precise regulation of the composition of their surround-

ing fluid. Because interstitial fluid surrounds all body cells, it is often called the body's *internal environment*. The composition of interstitial fluid changes as substances move back and forth between it and blood plasma. Such exchange of materials occurs across the thin walls of the smallest blood vessels in the body, the *blood capillaries*. This movement in both directions across capillary walls provides needed materials, such as glucose, oxygen, ions, and so on, to tissue cells. It also removes wastes, such as carbon dioxide, from interstitial fluid.

Checkpoint

5. What are some examples of movements that occur inside the human body?
6. Having a fever, high blood pressure, or low blood sugar are all examples of what process being out of balance?
7. Where would you find intracellular fluid, extracellular fluid, and interstitial fluid?
8. A spinal tap is a procedure in which fluid is drawn from the space between the membranes that surround the spinal cord. What type of body fluid is being extracted?

Return To Jaime's Story

Jaime felt like he had been running all day. He only had one hour between the end of his shift and his evening anatomy and physiology class. He thought about his first day as he hurried to class. Jaime had helped with several X-rays of different parts of the body, including the chest of a suspected tuberculosis case, the arm of a girl who had fallen, and the tummy of a child who had swallowed a coin. Ellen had shown him how to develop the X-ray film cassettes, and he tagged along with her as she transported patients. She explained most of the terms she was using to Jaime; anterior was the front of the body, posterior the back, lateral the side. It didn't seem too difficult, and he was enjoying learning on the job.

Amy was leaning against the wall talking with another student as Jaime arrived for class. "So. Save any lives today?" she asked as they settled into their seats in the back of the lecture hall.

"No, not exactly. But, I think I'm going to like working there. The X-ray techs are all nice, and Dr. Sebbo, the radiologist, bought everyone lunch today. I wasn't very hungry though. Right before lunch, I had to transport a patient with a urinary

catheter and while we were getting her into the wheelchair the line popped off the drainage bag. I got urine all over my shoes. It was really gross."

Amy made a face. "Ugh, that's nasty! Better get used to it, though."

"Yeah, I suppose. At least I didn't accidentally pull out her IV line." Jaime suddenly raised his hand to his face and sneezed. "I feel like I'm coming down with the flu. My neck feels swollen, and I have this sharp pain in my ribs. Do I feel warm to you?"

A. *Health-care workers like Jaime need to understand basic medical terminology, including anatomical and physiological terms. What are some everyday terms or phrases you use that come from the study of anatomy and physiology?*

B. *Jaime is helping move patients and develop X-rays. What levels of structural organization is he managing?*

C. *The patient Jaime transported had intravenous (IV) therapy; what body fluid type is blood plasma?*

D. *Would the urine from the patient's catheter be considered an intracellular or extracellular fluid?*

1.3 Homeostasis is controlled through feedback systems.

Homeostasis in the human body is continually being challenged. Disturbances come from the external environment in the form of physical insults such as the intense heat of a Texas summer or a lack of enough oxygen for that 2-mile run. Other disruptions originate in the internal environment, such as a blood glucose level that falls too low when you skip breakfast. Homeostatic imbalances may also occur due to psychological stresses in your social environment—the demands of work and school, for example. In most cases the disruption of homeostasis is mild and temporary, and the body responds quickly to restore balance in the internal environment. In some cases the disruption of homeostasis may be intense and prolonged, as in poisoning, overexposure to temperature extremes, severe infection, or major surgery. Under these circumstances, regulation of homeostasis may fail.

Fortunately, the body has many regulating systems that can usually bring the internal environment back into balance. Most often, the nervous system and the endocrine system, working together or independently, provide the needed corrective measures. The nervous system regulates homeostasis by sending messages in the form of *nerve impulses* to organs that can counteract changes from the balanced state. The endocrine system includes many glands that secrete messenger molecules called *hormones* into the blood. Nerve impulses typically cause rapid changes, and hormones usually work more slowly. However, both means of regulation work toward the same end, namely, the maintenance of homeostasis, usually through negative feedback systems.

Feedback Systems

The body can regulate its internal environment through many feedback systems. A **feedback system** is a cycle of events in which the status of a body condition is continually monitored, evaluated, changed, remonitored, reevaluated, and so on. Each monitored variable, such as body temperature, blood pressure, or blood glucose level, is termed a *controlled condition*. Any disruption that changes a controlled condition is called a *stimulus*. Three basic components make up a feedback system—a receptor, a control center, and an effector (Figure 1.2).

1. A **receptor** is a body structure that monitors changes in a controlled condition and sends input to a control center. Input is in the form of nerve impulses or chemical signals. Nerve endings in the skin that sense temperature are one of hundreds of different kinds of receptors in the body.

2. A **control center** in the body sets the range of values within which a controlled condition should be maintained, evaluates the input it receives from receptors, and generates output commands when they are needed. Output is in the form of nerve impulses or chemical signals. In our skin temperature example, the brain acts as the control center, receiving nerve impulses from the skin receptors and generating nerve impulses as output.

3. An **effector** is a body structure that receives output from the control center and produces a *response* or an effect that changes the controlled condition. Nearly every organ or tissue in the body can behave as an effector. When your body temperature drops sharply, thermoreceptors in the skin (receptors) send nerve impulses to your brain (control center) informing

your brain of the dropping body temperature; your brain then sends nerve impulses to your skeletal muscles (effectors) that cause you to shiver, which generates heat and raises your body temperature.

A group of receptors and effectors communicating with their control center forms a feedback system that can regulate a controlled condition in the body's internal environment. In a feedback system, the response of the system "feeds back" information to change the controlled condition in some way. If the response reverses the original stimulus, the system is operating by *negative feedback*. If the response enhances or intensifies the original stimulus, the system is operating by *positive feedback*.

Figure 1.2   Operation of a feedback system. The dashed return arrow symbolizes negative feedback.

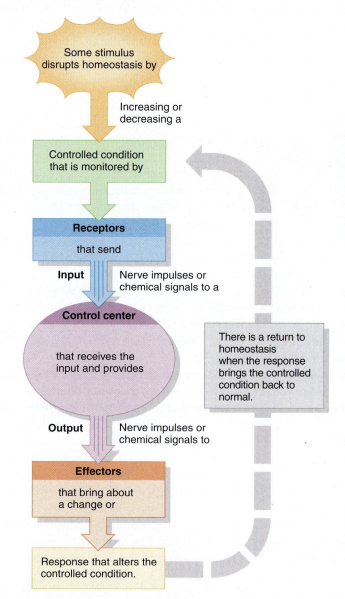

The three basic components of a feedback system are the receptors, a control center, and effectors.

Negative Feedback Systems

As noted, a **negative feedback system** or **loop** reverses a change in a controlled condition. First, a stimulus disrupts homeostasis by altering the controlled condition. The receptors that are part of the feedback system detect the change and send input to a control center. The control center evaluates the input and, if necessary, issues output commands to an effector. The effector produces a physiological response that is able to return the controlled condition to its normal state.

Consider one negative feedback system that helps regulate blood pressure. *Blood pressure* is the force exerted by blood as it presses against the walls of blood vessels. When the heart beats faster or harder, blood pressure increases. If a stimulus causes blood pressure (controlled condition) to rise, the following sequence of events occurs (Figure 1.3). *Baroreceptors* (the receptors), pressure-sensitive nerve cells located in the walls of certain blood vessels, detect the higher pressure. The baroreceptors send nerve impulses (input) to the brain (control center). The brain (control center) interprets the impulses and responds by sending nerve impulses (output) to the heart and blood vessels (the effectors). Heart rate decreases and blood vessels dilate (widen), which cause blood pressure to decrease (response).

This sequence of events quickly returns the controlled condition—blood pressure—to normal, and homeostasis is restored. Notice that the activity of the effector produces a result (a drop in blood pressure) that is opposite to the original stimulus (an increase in blood pressure). This is why it is called a negative feedback system.

Positive Feedback Systems

A **positive feedback system** operates similarly to a negative feedback system, except that the response strengthens or reinforces the change in one of the body's controlled conditions rather than reversing it. The control center provides commands to an effector, but the effector produces a response that adds to or *reinforces* the initial change in the controlled condition. The action of a positive feedback system continues until it is interrupted by some mechanism.

Normal childbirth provides a good example of a positive feedback system (Figure 1.4). The first contractions of labor (stimulus) push part of the baby into the cervix, the lowest part of the uterus, which opens into the vagina. Stretch-sensitive nerve cells (receptors) monitor the amount of stretching of the cervix (controlled condition). As stretching increases, they send more nerve impulses (input) to the brain (control center), which in turn releases the hormone oxytocin (output) into the blood. Oxytocin causes muscles in the wall of the uterus (effector) to contract even more forcefully. The contractions push the baby farther down the uterus, which

Figure 1.3 **Homeostatic regulation of blood pressure by a negative feedback system.** Note that the response is fed back into the system, and the system continues to lower blood pressure until there is a return to normal blood pressure (homeostasis).

If the response reverses the stimulus, a system is operating by negative feedback.

Figure 1.4   Positive feedback control of labor contractions during birth of a baby. The solid return arrow symbolizes positive feedback.

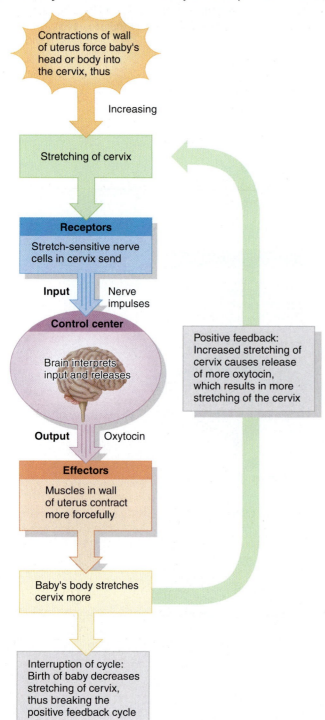

Contractions of wall of uterus force baby's head or body into the cervix, thus

Increasing

Stretching of cervix

Receptors
Stretch-sensitive nerve cells in cervix send

Input   Nerve impulses

Control center
Brain interprets input and releases

Positive feedback: Increased stretching of cervix causes release of more oxytocin, which results in more stretching of the cervix

Output   Oxytocin

Effectors
Muscles in wall of uterus contract more forcefully

Baby's body stretches cervix more

Interruption of cycle: Birth of baby decreases stretching of cervix, thus breaking the positive feedback cycle

🔑 **If the response enhances or intensifies the stimulus, a system is operating by positive feedback.**

stretches the cervix even more. The cycle of stretching, hormone release, and ever-stronger contractions is interrupted only by the birth of the baby. Then stretching of the cervix ceases and oxytocin is no longer released. Blood clotting is also an example of a positive feedback system.

These examples suggest some important differences between positive and negative feedback systems. Because a positive feedback system continually reinforces a change in a controlled condition, some event outside the system must shut it off. If the action of a positive feedback system is not stopped, it can "run away," and may even produce life-threatening conditions in the body. The action of a negative feedback system, by contrast, slows and then stops as the controlled condition returns to its normal state. Usually, positive feedback systems reinforce conditions that do not happen very often, and negative feedback systems regulate conditions in the body that remain fairly stable over long periods.

Homeostatic Imbalances

As long as all the body's controlled conditions remain within certain narrow limits, body cells function efficiently, negative feedback systems maintain homeostasis, and the body stays healthy. Should one or more components of the body lose their ability to contribute to homeostasis, however, the normal balance among all of the body's processes may be disturbed. If the homeostatic imbalance is moderate, a disorder or disease may occur; if it is severe, death may result.

A **disorder** is any abnormality of structure and/or function. **Disease** is a more specific term for an illness characterized by a recognizable set of signs and symptoms. **Symptoms** are subjective changes in body functions that are not apparent to an observer, for example, headache, nausea, and anxiety. **Signs** are objective changes that a clinician can observe and measure. Signs of disease can be either anatomical, such as swelling or a rash, or physiological, such as fever, high blood pressure, or paralysis. Specific diseases alter body structures and functions in characteristic ways, usually producing a recognizable cluster of signs and symptoms. A *local disease* affects one part or a limited region of the body; a *systemic disease* affects either the entire body or several parts of it.

Checkpoint

9. What two organ systems are largely responsible for maintaining homeostasis?
10. What is the name for the body structure that responds to the control center signal in a feedback system?
11. The response to a stimulus in either a negative or positive feedback system is initiated by what organ systems?
12. How are negative and positive feedback systems similar? How are they different? Which one bears primary responsibility for maintaining homeostasis?
13. Why do positive feedback systems that are part of a normal physiological response include a mechanism to stop them?

1.4 The human body is described using the anatomical position and specific terms.

Body Positions

Descriptions of any part of the human body assume that the body is in a specific position called the anatomical position. In the anatomical position, the subject stands erect facing the observer, with the head level and the eyes facing forward. The feet are flat on the floor and directed forward, and the upper limbs are at the sides

with the palms turned forward (Figure 1.5). With the body in the anatomical position, it is easier to visualize and understand how it is organized into various regions.

In the anatomical position, the body is upright. Two terms describe a reclining body: face down, it is in the **prone** position; face up, it is in the **supine** position.

Figure 1.5 The anatomical position. The common names and corresponding anatomical terms (in parentheses) are indicated for specific body regions. For example, the head is the cephalic region.

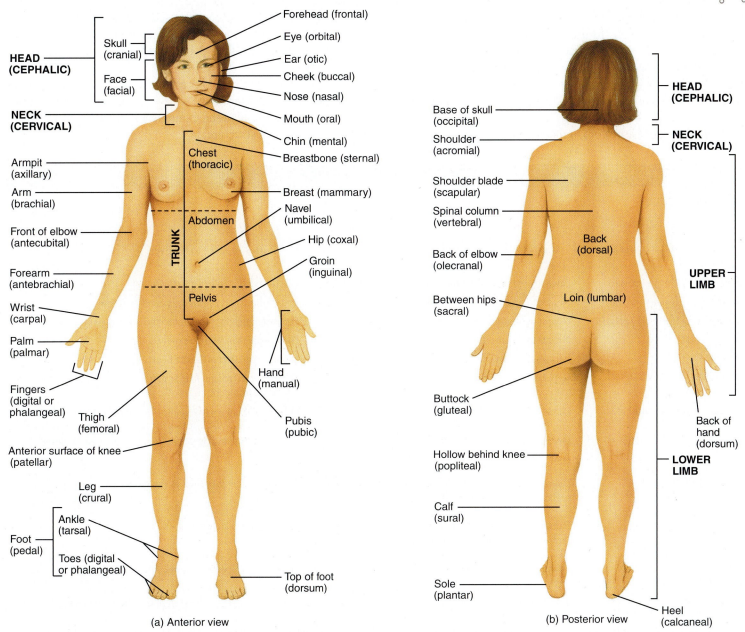

(a) Anterior view

(b) Posterior view

In the anatomical position, the subject stands erect facing the observer with the head level and the eyes facing forward. The feet are flat on the floor and directed forward, and the arms are at the sides with the palms facing forward.

Regional Names

The human body is divided into several major regions that can be identified externally. These are the head, neck, trunk, upper limbs, and lower limbs (Figure 1.5). The **head** consists of the skull and face. The **skull** encloses and protects the brain, whereas the **face** is the front portion of the head that includes the eyes, nose, mouth, forehead, cheeks, and chin. The **neck** supports the head and attaches it to the trunk. The **trunk** consists of the chest, abdomen, and pelvis. Each **upper limb** is attached to the trunk and consists of the shoulder, armpit, arm (portion of the limb from the shoulder to the elbow), forearm (portion of the limb from the elbow to the wrist), wrist, and hand. Each **lower limb** is also attached to the trunk and consists of the buttock, thigh (portion of the limb from the buttock to the knee), leg (portion of the limb from the knee to the ankle), ankle, and foot. The *groin* is the area on the front surface of the body marked by a crease on each side, where the trunk attaches to the thighs.

Figure 1.5 shows the common names of major parts of the body. The corresponding anatomical descriptive form (adjective) for each part appears in parentheses next to the common name. For example, if you receive a tetanus shot in your *buttock*, it is a *gluteal* injection. Because the descriptive form of a body part is usually based on a Greek or Latin word, it may look different from its common name. The Latin word for armpit, for example, is *axilla* (ak-SIL-a); thus, one of the nerves passing within the armpit is named the axillary nerve. You will learn more about the Greek and Latin word roots of anatomical and physiological terms as you read this book.

Directional Terms

To locate various body structures, anatomists use specific **directional terms**, words that describe the position of one body part relative to another. Most of the directional terms used to describe the human body can be grouped in pairs that have opposite meanings. For example, **superior** means toward the upper part of the body, and **inferior** means toward the lower part of the body. It is important to understand that directional terms have relative meanings; they only make sense when used to describe the position of one structure relative to another. For example, your knee is superior to your ankle, even though both are located in the inferior half of the body.

Table 1.2 and Figure 1.6 present the main directional terms. Study the directional terms in Table 1.2 and the example of how each is used. As you read the examples, look at Figure 1.6 to see the location of each structure.

TABLE 1.2		
Directional Terms		
DIRECTIONAL TERM	DEFINITION	EXAMPLE OF USE
Superior (soo´-PĒR-ē-or) **(cephalic or cranial)**	Toward the head, or the upper part of a structure.	The heart is superior to the liver.
Inferior (in´-FĒR-ē-or) **(caudal)**	Away from the head, or the lower part of a structure.	The stomach is inferior to the lungs.
Anterior (an-TĒR-ē-or) **(ventral)***	Nearer to or at the front of the body	The sternum (breastbone) is anterior to the heart.
Posterior (pos-TĒR-ē-or) **(dorsal)***	Nearer to or at the back of the body.	The esophagus is posterior to the trachea (windpipe).
Medial (MĒ-dē-al)	Nearer to the midline.†	The ulna is medial to the radius.
Lateral (LAT-er-al)	Farther from the midline.†	The lungs are lateral to the heart.
Intermediate (in´-ter-MĒ-dē-at)	Between two structures.	The transverse colon is intermediate between the ascending and descending colons.
Ipsilateral (ip-si-LAT-er-al)	On the same side of the body as another structure.	The gallbladder and ascending colon are ipsilateral.
Contralateral (CON-tra-lat-er-al)	On the opposite side of the body from another structure.	The ascending and descending colons are contralateral.
Proximal (PROK-si-mal)	Nearer to the attachment of a limb to the trunk; nearer to the origination of a structure.	The humerus is proximal to the radius.
Distal (DIS-tal)	Farther from the attachment of a limb to the trunk; farther from the origination of a structure.	The phalanges are distal to the carpals.
Superficial (soo´-per-FISH-al) **(external)**	Toward or on the surface of the body.	The ribs are superficial to the lungs.
Deep (internal)	Away from the surface of the body.	The ribs are deep to the skin of the chest and back.

* The terms *anterior* and *ventral* mean the same thing in humans. However, in four-legged animals *ventral* refers to the belly side and is therefore *inferior*. Similarly, the terms *posterior* and *dorsal* mean the same thing in humans, but in four-legged animals *dorsal* refers to the back side and is therefore *superior*.

†Recall that the **midline** is an imaginary vertical line that divides the body into equal right and left sides.

Figure 1.6 **Directional terms.**

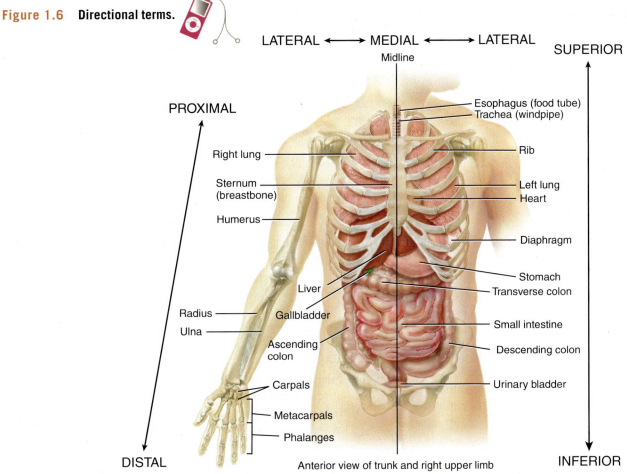

LATERAL ←→ MEDIAL ←→ LATERAL

Midline

SUPERIOR

PROXIMAL

- Right lung
- Sternum (breastbone)
- Humerus
- Liver
- Radius
- Ulna
- Gallbladder
- Ascending colon
- Carpals
- Metacarpals
- Phalanges

- Esophagus (food tube)
- Trachea (windpipe)
- Rib
- Left lung
- Heart
- Diaphragm
- Stomach
- Transverse colon
- Small intestine
- Descending colon
- Urinary bladder

DISTAL

Anterior view of trunk and right upper limb

INFERIOR

Directional terms precisely locate various parts of the body relative to one another.

Planes and Sections

You will also study parts of the body relative to **planes**—imaginary flat surfaces that pass through them (Figure 1.7). A **sagittal plane** (SAJ-i-tal; *sagitt-* = arrow) is a vertical plane that divides the body or organ into right and left sides. More specifically, when such a plane passes through the midline of the body or organ and divides it into *equal* right and left sides, it is called a **midsagittal (median) plane**. The *midline* is an imaginary vertical line that divides the body into equal left and right sides. If the sagittal plane does not pass through the midline but instead divides the body or organ into *unequal* right and left sides, it is called a **parasagittal plane** (*para-* = near). A **frontal (coronal) plane** divides the body or organ into front and back portions. A **transverse plane** divides the body or organ into upper and lower portions. A transverse plane may also be termed a *cross-sectional plane* or *horizontal plane*. Sagittal, frontal, and transverse planes are all at right angles to one another. An **oblique plane**, by contrast, passes through the body or organ at an angle between a transverse plane and a sagittal plane or between a transverse plane and a frontal plane.

Figure 1.7 **Planes through the human body.**

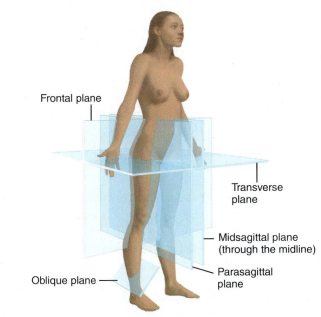

- Frontal plane
- Transverse plane
- Midsagittal plane (through the midline)
- Oblique plane
- Parasagittal plane

Right anterolateral view

Frontal, transverse, sagittal, and oblique planes divide the body in specific ways.

Figure 1.8   Planes and sections through different parts of the brain. The diagrams (top) show the planes, and the photographs (bottom) show the resulting sections. Note: The arrows in the diagrams indicate the direction from which each section is viewed. This aid is used throughout the book to indicate viewing perspectives.

(a) Transverse section   (b) Frontal section   (c) Midsagittal section

Planes divide the body in various ways to produce sections.

When you study a body region, you often view it in section. A **section** is a cut of the body or one of its organs made along one of the planes just described. It is important to know the plane of the section so you can understand the anatomical relationship of one part to another. Figure 1.8 indicates how three different sections—a *transverse section*, a *frontal section*, and a *midsagittal section*—provide different views of the brain.

Checkpoint

14. Describe the anatomical position in your own words.
15. Blood is typically drawn from the antecubital space. Where is this?
16. Is the esophagus anterior to the trachea? Is the urinary bladder medial to the ascending colon? Is the sternum lateral to the descending colon? Is the radius proximal to the humerus? Are the ribs superficial to the lungs?
17. Which directional terms can be used to specify the relationships between (1) the elbow and the shoulder, (2) the left and right shoulders, (3) the sternum and the humerus, and (4) the heart and the diaphragm?
18. What is the difference between a plane and a section?

Return To Jaime's Story

As Jaime sat in the crowded lecture hall, he felt a tickle in the back of his throat. He coughed, covering his mouth, then cleared his throat, but that didn't seem to help. He finally broke into an uncontrollable coughing fit. His face turned bright red and everyone was looking at him, but the instructor just kept lecturing, speaking louder to be heard over Jaime's coughing.

Jaime got up and bolted out the door of the lecture hall, coughing the whole way. He found a vending machine and bought an orange juice, which he gulped down. The juice was cold, and he shuddered involuntarily as he drank, even though he felt feverish.

Jaime made it through the rest of lecture without another coughing fit, but by the end of class, he felt worse. He went straight to bed once he got home, knowing he had to work the next day.

In the morning, he forced himself to get up, but still arrived late to find Terry, one of the nicer X-ray technicians, sitting at the desk. "Hey, partner. I think this is your job," said Terry.

"Sorry I'm late, Terry. I didn't sleep well last night."

"No big deal, kid. I'm going to need your help with this ten o'clock patient, though. Doc wants a bunch of new films on him. Motorcycle versus truck accident!" Jaime glanced at the X-ray films lying near Terry, "Those his films?" Jaime said.

"Yeah, check this out—he broke his neck, collar bone, humerus, and his sixth and seventh thoracic vertebrae. Look here! His dens is broken."

"What's a dens?" asked Jaime. "It's the part of the second cervical vertebra that the skull rotates on. You can see it in the anterior view," said Terry. "We'll probably have to get views of external and internal rotation on the arm, and an oblique view, too. You'll need to help position him; it's going to hurt... a lot. Why don't you see if this guy is ready now."

Between coughs, Jaime replied, "I'll call the nurses' station. Man, my chest hurts."

E.   The ability to sense our environment, such as the tickle Jaime felt in his throat or his feeling of warmth, is mediated by what structures?

F.   The instructor responded to Jaime's coughing by speaking louder. Is this a type of feedback mechanism? Explain your answer.

G.   Identify Jaime's specific signs and symptoms.

H.   If Terry and Jaime need to obtain a lateral view of the motorcycle rider's neck, from what direction would they take the X-ray picture?

I.   Where are the patient's cervical vertebrae located?

1.5 Body cavities are spaces within the body that help protect, separate, and support internal organs.

Bones, muscles, and ligaments separate the various body cavities from one another. Here we discuss several of the larger body cavities (Figure 1.9).

The **cranial cavity** is formed by the cranial bones and contains the brain. The **vertebral canal** is formed by the bones of the vertebral column (spine) and contains the spinal cord. Three layers of protective tissue, called **meninges** (me-NIN-jēz), line the cranial cavity and the vertebral canal.

The major body cavities of the trunk are the thoracic and abdominopelvic cavities. The **thoracic cavity** (thor-AS-ik; *thorac-* = chest), or chest cavity, is encircled by the ribs, the muscles of the chest, the sternum (breastbone), and the thoracic portion of the vertebral column (backbone) (Figure 1.10). Within the thoracic cavity are three smaller cavities: the **pericardial cavity** (per´-i-

KAR-dē-al; *peri-* = around; *-cardial* = heart), a fluid-filled space containing the heart, and two fluid-filled spaces called **pleural cavities** (PLOOR-al; *pleur-* = rib or side), each of which contains one lung. The central portion of the thoracic cavity is an anatomical region called the **mediastinum** (mē´-dē-as-TĪ-num; *media-* = middle; *-stinum* = partition). It is between the lungs and extends from the sternum to the vertebral column, and from the first rib to the diaphragm. The mediastinum contains all thoracic viscera except the lungs themselves. Among the structures in the mediastinum are the heart, esophagus, trachea, thymus, and several large blood vessels. The **diaphragm** (DĪ-a-fram = partition or wall) is a dome-shaped muscle that separates the thoracic cavity from the abdominopelvic cavity.

Figure 1.9   Body cavities. The dashed lines in (a) and (b) indicate the border between the abdominal and pelvic cavities.

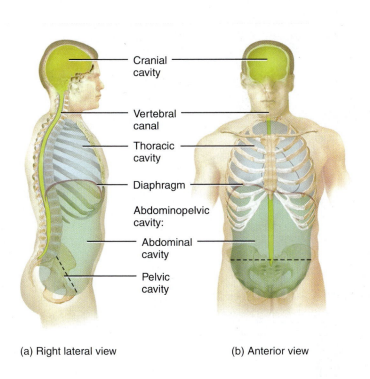

(a) Right lateral view   (b) Anterior view

CAVITY	COMMENTS
Cranial cavity	Formed by cranial bones and contains brain.
Vertebral canal	Formed by vertebral column and contains spinal cord and the beginnings of spinal nerves.
Thoracic cavity*	Chest cavity; contains pleural and pericardial cavities and mediastinum.
Pleural cavity	Each surrounds a lung; the serous membrane of the pleural cavities is the pleura.
Pericardial cavity	Surrounds the heart; the serous membrane of the pericardial cavity is the pericardium.
Mediastinum	Central portion of thoracic cavity between the medial walls of pleural cavities; extends from sternum to vertebral column and from first rib to diaphragm; contains heart, thymus, esophagus, trachea, and several large blood vessels.
Abdominopelvic cavity	Subdivided into abdominal and pelvic cavities.
Abdominal cavity	Contains stomach, spleen, liver, gallbladder, small intestine, and most of large intestine; the serous membrane of the abdominal cavity is the peritoneum.
Pelvic cavity	Contains urinary bladder, portions of large intestine, and internal organs of reproduction.

* See Figure 1.10 for details of the thoracic cavity.

The major cavities of the trunk are the thoracic and abdominopelvic cavities.

Figure 1.10   The thoracic cavity. The dashed lines indicate the borders of the mediastinum. Note: When transverse sections are viewed inferiorly (from below), the anterior aspect of the body appears on top and the left side of the body appears on the right side of the illustration.

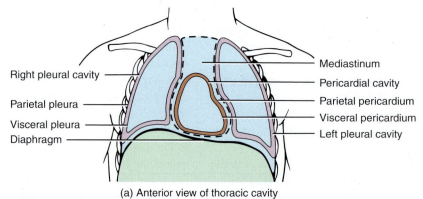

Right pleural cavity

Parietal pleura

Visceral pleura

Diaphragm

Mediastinum

Pericardial cavity

Parietal pericardium

Visceral pericardium

Left pleural cavity

(a) Anterior view of thoracic cavity

Transverse plane

View

ANTERIOR

Muscle

Heart

PERICARDIAL CAVITY

Right lung

Aorta

RIGHT PLEURAL CAVITY

Rib

Sternum (breastbone)

Thymus

Left lung

Esophagus (food tube)

Vertebral column (backbone)

LEFT PLEURAL CAVITY

POSTERIOR

(b) Inferior view of transverse section of thoracic cavity

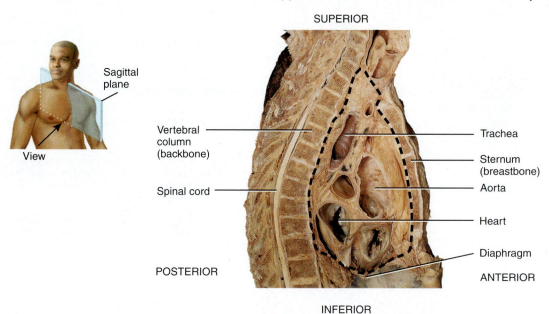

SUPERIOR

Sagittal plane

View

Vertebral column (backbone)

Spinal cord

Trachea

Sternum (breastbone)

Aorta

Heart

Diaphragm

POSTERIOR

ANTERIOR

INFERIOR

(c) Sagittal section of thoracic cavity

The thoracic cavity contains three smaller cavities and the mediastinum.

The **abdominopelvic cavity** (ab-dom´-i-nō-PEL-vik) extends from the diaphragm to the groin and is encircled by the abdominal wall and the bones and muscles of the pelvis (Figure 1.11). As the name suggests, the abdominopelvic cavity is divided into two portions, even though no wall separates them. The superior portion, the **abdominal cavity** (*abdomin-* = belly), contains the stomach, spleen, liver, gallbladder, small intestine, and most of the large intestine. The inferior portion, the **pelvic cavity** (*pelv-* = basin), contains the urinary bladder, portions of the large intestine, and internal organs of the reproductive system. Organs inside the thoracic and abdominopelvic cavities are called **viscera** (VIS-er-a).

Checkpoint

19. The diaphragm separates what two cavities?
20. In which cavities are the following organs located: urinary bladder, stomach, heart, small intestine, lungs, thymus, and liver? Use the following symbols for your response: T = thoracic cavity, A = abdominal cavity, or P = pelvic cavity.
21. Which of the following structures are contained in the mediastinum: right lung, heart, esophagus, spinal cord, trachea, ribs, thymus, left pleural cavity?
22. What is the name of the cavity that surrounds the heart? Which cavities surround the lungs?
23. What organ lies distinctly in both the abdominal and pelvic cavities?

Figure 1.11   **The abdominopelvic cavity.** The horizontal dashed line shows the approximate boundary between the abdominal and pelvic cavities.

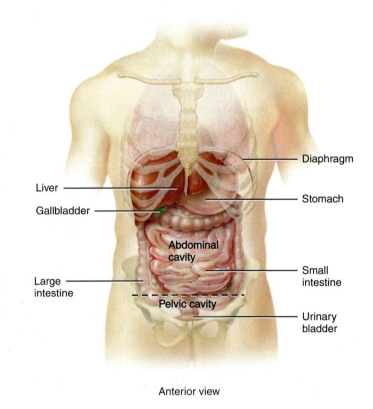

Anterior view

The abdominopelvic cavity extends from the diaphragm to the groin.

1.6 Serous membranes line the walls of body cavities and cover the organs within them.

Serous membranes are thin, slippery double-layered membranes that cover the viscera within the thoracic and abdominal cavities and line the walls of the thorax and abdomen. The parts of a serous membrane are (1) the *parietal layer* (pa-RĪ-e-tal), which lines the walls of the cavities, and (2) the *visceral layer*, which covers and adheres to the organs within the cavities. Serous membranes secrete a lubricating serous fluid into the space between the two layers. Serous fluid reduces friction, allowing the viscera to slide somewhat during movements—as when the lungs inflate and deflate during breathing.

The serous membrane of the pleural cavities is called the **pleura** (PLOO-ra) (see Figure 1.10a). The *visceral pleura* clings to the surface of the lungs, and the *parietal pleura* lines the chest wall. In between is the pleural cavity, filled with a small volume of serous fluid. The serous membrane of the pericardial cavity is the **pericardium** (see Figure 1.10a). The *visceral pericardium* covers the surface of the heart; the *parietal pericardium* lines the chest wall. Between them is the serous fluid-filled pericardial cavity. The **peritoneum** (per-i-tō-NĒ-um) is the serous membrane of the abdominal

cavity. The *visceral peritoneum* covers the abdominal viscera, and the *parietal peritoneum* lines the abdominal wall. Between them is the peritoneal cavity, which contains a small amount of lubricating fluid. Most abdominal organs are located in the peritoneal cavity. Some are located between the parietal peritoneum and the posterior abdominal wall. Such organs are said to be *retroperitoneal* (re´-trō-per-i-tō-NĒ-al; *retro* = behind). Examples include the kidneys, adrenal glands, pancreas, duodenum of the small intestine, ascending and descending colons of the large intestine, and portions of the abdominal aorta and inferior vena cava.

A summary of body cavities and their membranes is presented in the table included in Figure 1.9.

Checkpoint

24. What is the purpose of serous membranes?
25. Which part of the peritoneum covers the stomach? Which part lines the abdominal wall?
26. What do we mean when we say an organ is retroperitoneal?

1.7   The abdominopelvic cavity is divided into regions or quadrants.

To describe the location of the many abdominal and pelvic organs more easily, anatomists and clinicians use two methods of dividing the abdominopelvic cavity into smaller areas (Figure 1.12). In the first method, two horizontal and two vertical lines, aligned like a tick-tack-toe grid, partition this cavity into nine **abdominopelvic regions** (Figure 1.12a,b). The top horizontal line, the *subcostal line*, is drawn just inferior to the rib cage, across the inferior portion of the stomach; the bottom horizontal line, the *transtubercular line*,

Figure 1.12   **Regions and quadrants of the abdominopelvic cavity.**

(a) Anterior view showing abdominopelvic regions

(b) Anterior view showing location of abdominopelvic regions

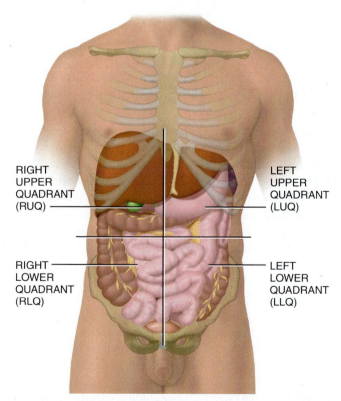

(c) Anterior view showing location of abdominopelvic quadrants

The nine-region designation is used for anatomical studies; the quadrant designation is used by clinicians to locate the site of pain, a tumor, or some other abnormality.

is drawn just inferior to the tops of the hip bones. Two vertical lines, the left and right *midclavicular lines*, are drawn through the midpoints of the clavicles (collar bones), just medial to the nipples. The four lines divide the abdominopelvic cavity into a larger middle section and smaller left and right sections. The names of the nine abdominopelvic regions are **right hypochondriac**, **epigastric**, **left hypochondriac**, **right lumbar**, **umbilical**, **left lumbar**, **right inguinal (iliac)** , **hypogastric (pubic)** , and **left inguinal (iliac)** .

The second method is simpler and divides the abdominopelvic cavity into **quadrants** (KWOD-rantz; *quad-* = one-fourth), as shown in Figure 1.12c. In this method, a transverse plane and a midsagittal plane are passed through the **umbilicus** (um-BI-lī-kus; *umbilic-* = navel) or *belly button*. The names of the abdominopelvic quadrants are **right upper quadrant (RUQ)**, **left upper quadrant (LUQ)**, **right lower quadrant (RLQ)**, and **left lower quadrant (LLQ)**. The nine-region division is more widely used for anatomical studies, and quadrants are more commonly used by clinicians for describing the site of abdominopelvic pain, a tumor, or other abnormality.

Checkpoint

27. In which abdominopelvic region is each of the following found: most of the liver, transverse colon, urinary bladder, spleen?
28. In which abdominopelvic quadrant would pain from appendicitis (inflammation of the appendix) be felt?
29. When anatomical terms refer to right and left, do we mean your right and left or those of the person you are facing? (*Hint: See Figure 1.12.*)

Jaime's Story: Epilogue and Discussion

▶▶▶ By the end of the day, Jaime still felt pretty bad. Terry took a good, long look at him and said, "You better get a chest X-ray yourself, kid. Sounds like you have pleurisy or walking pneumonia — or something worse!"

"Pleurisy—that's inflammation of the membrane around the lungs, right? I'll get it taken care of right away!" Jaime responded.

Jaime has found that his study of anatomy and physiology has direct application to his new job, and to his own health. Gaining knowledge about anatomy and physiology allows us to understand the relationship of body structures to their functions. As you read through this book keep in mind that, like Jaime's experience, what you are learning has meaning to you in both your professional life and your personal life. So far, you have learned how the body is organized, what characteristics determine life, and some basic anatomical terminology that names body parts and describes their locations. In Jaime's case, this basic information is helping him perform his job and allowing him to better understand his own health.

In each chapter we will explore a new story designed to illustrate the application of anatomical and physiological principles in your everyday life. Although the case stories themselves are fictional, the situations, illnesses, and characters are based on the experiences and stories of people just like you. ◀◀◀

J. Terry mentioned to Jaime the possibility of a condition called pleurisy. What cavity of the body is involved with this condition? What other terms are related to the prefix *pleur-*?

K. How did Jaime's understanding of anatomical terms help him in his job?

L. What organ system would be responding to attack Jaime's illness?

M. How do injury and disease differ? Does the body respond similarly to injury and disease?

N. Do you think the stress of Jaime's new job and school schedule has contributed to his illness? Why or why not?

Concept and Resource Summary

Concept	Resources

Introduction

1. **Anatomy** is the science of body structures and their relationships.
2. **Physiology** is the science of body functions.

Concept 1.1 The human body is composed of six levels of structural organization and contains eleven organ systems.

1. The six structural levels of body organization, from smallest to largest, are chemical, cellular, tissue, organ, system, and organismal.
2. The *chemical level* includes **atoms** and **molecules**.
3. The *cellular level* occurs when molecules build **cells**, the basic units of structure and function in living organisms.
4. Group of cells unite to form the *tissue level* composed of four basic types of tissues: epithelial, connective, muscle, and nervous.
5. Different tissues join together to form the *organ level*.
6. Different organs join together to form the *system level*.
7. All the systems form the largest level, the *organismal level*. An **organism** is a living individual.
8. There are eleven organ systems in the human body (see Table 1.1).

Concept

Concept 1.2 The human body carries on basic life processes that distinguish it from nonliving objects.

1. All living organisms undergo chemical processes referred to as **metabolism**, including **catabolism**, the breaking down of large chemical compounds to smaller ones, and **anabolism**, the building of large compounds from smaller ones.
2. Living organisms respond to changes in their environment (a characteristic called **responsiveness**), and they exhibit **movement** of their whole body or structures within their body.
3. **Growth**, an increase in body size, and **differentiation**, such as that of **stem cells** into specialized cells, are two additional basic life processes.
4. Living organisms undergo **reproduction** to form new cells for tissue growth or for production of a new individual.
5. **Homeostasis** is the maintenance of relatively stable internal body conditions despite changes that occur inside and outside the body.
6. Homeostasis maintains the volume and composition of **body fluids** inside and outside of cells.
7. The fluid inside cells is **intracellular fluid**, and the fluid outside of cells is **extracellular fluid. Interstitial fluid, blood plasma, lymph, cerebrospinal fluid**, and **synovial fluid** are extracellular fluids.

Concept 1.3 Homeostasis is controlled through feedback systems.

1. The body is continually confronted with internal and external stresses. The principal regulatory systems, the nervous system and endocrine system, work to provide corrective measures to achieve homeostasis and restore balance to the body's internal environment.
2. A **feedback system** is a cycle of events that monitor, evaluate, and change disruptions to a *controlled condition* in the body.
3. Three components comprise a feedback system: receptor, control center, and effector.
4. A **receptor** monitors changes and sends input to a **control center**, where input is evaluated and output commands are sent to **effectors** that produce a *response*.
5. Responses that reverse the original stimulus involve **negative feedback**.
6. Responses that intensify the original stimulus involve **positive feedback**.
7. The inability of one or more body components to contribute to homeostasis may result in disease or death.
8. A **disorder** is an abnormality of structure and/or function.
9. A **disease** refers to an illness accompanied by a set of clinically observable, measurable changes called **signs**, and subjective changes experienced by the patient referred to as **symptoms**.
10. Many diseases produce positive feedback loops that cause death.

See
- Animation – Communication, Regulation, and Homeostasis
- Animation – Negative Feedback Control of Blood Pressure
- Animation – Negative Feedback Control of Temperature
- Animation – Positive Feedback Control of Labor

Do
- Concepts and Connections – Regulation and Communication
- Concepts and Connections – Negative Feedback Loop
- Exercise – Waterways
- Concepts and Connections – Homeostasis

Concept 1.4 The human body is described using the anatomical position and specific terms.

1. In the anatomical position, the body is erect, facing forward, head level, eyes forward, feet flat and forward, and arms at the sides with palms forward.
2. The body is in the **prone** position when lying face down, and in the **supine** position when lying face up.
3. The human body is divided into several major regions identified externally. These are the **head, skull, face, neck, trunk, upper limbs,** and **lower limbs**.
4. In order to describe the position of one body part relative to another body part, specific **directional terms** must be used (see Table 1.2 and Figure 1.6).
5. The body and body parts can be cut along imaginary flat surfaces referred to as **planes**.
6. A **sagittal plane** divides the body or organ into right and left sections. When the plane passes directly through the midline resulting in equal right and left sides, it is a **midsagittal** or **median plane**. A **parasagittal plane** results in unequal right and left sides.
7. A **frontal** or **coronal plane** divides the body or organ into front and back sections.
8. A **transverse plane** results in upper and lower sections.
9. An **oblique plane** passes through the body or organ at an angle.
10. Body regions are often studied in sections. A **section** is one flat surface of the three-dimensional structure or cut along a plane for which is it named.

Hear
- Figure 1.5 – The Anatomical Position
- Figure 1.6 – Directional Terms

Concept 1.5 Body cavities are spaces within the body that help protect, separate, and support internal organs.

1. Body cavities are separated from one another by bones, muscles, and ligaments.
2. The **cranial cavity** houses the brain, and the **vertebral canal** contains the spinal cord.

Hear
- Figure 1.10 – The Thoracic Cavity

Concept	Resources

3. The **diaphragm** separates the superior thoracic cavity and the inferior abdominopelvic cavity.

4. The **thoracic cavity** houses three smaller cavities: the **pericardial cavity**, around the heart; and two **pleural cavities**, each surrounding a lung. Located between the lungs, extending from the neck to the diaphragm, is the **mediastinum**, which contains all thoracic viscera except the lungs.

5. The **abdominopelvic cavity** has a superior region, the **abdominal cavity**, containing the stomach, spleen, liver, gallbladder, small intestine, and most of the large intestine. The inferior region, the **pelvic cavity**, contains the urinary bladder, portions of the large intestine, and reproductive organs.

Concept 1.6 Serous membranes line the walls of body cavities and cover the organs within them.

1. The thoracic and abdominal cavity walls and viscera are covered with a thin, slippery **serous membrane**. The part of the serous membrane that lines cavity walls is the *parietal layer*, and the *visceral layer* covers the organs within the cavity.

2. The pleura, pericardium, and peritoneum are serous membranes.

3. The **pleura** is the serous membrane of the pleural cavities around the lungs; the **pericardium** is the serous membrane of the pericardial cavity around the heart; and the **peritoneum** is the serous membrane of the abdominal cavity around abdominal viscera.

4. Organs, such as the kidneys, that are located posterior to the parietal peritoneum are referred to as *retroperitoneal*.

See

• **Anatomy Overview – Serous Membrane**

Concept 1.7 The abdominopelvic cavity is divided into regions or quadrants.

1. Anatomists and clinicians use two methods of dividing the abdominopelvic cavity into smaller areas in order to describe the location of organs.

2. One method divides the cavity into nine **abdominopelvic regions: right hypochondriac, epigastric, left hypochondriac, right lumbar, umbilical, left lumbar, right inguinal (iliac), hypogastric (pubic),** and **left inguinal (iliac)**.

3. The other method divides the cavity into **quadrants: right upper quadrant (RUQ), left upper quadrant (LUQ) , right lower quadrant (RLQ) ,** and **left lower quadrant (LLQ)**.

Other Chapter Resources

For additional information on clinical applications, medical disorders, and medical terminology, see Chapter 1 in the *Clinical Connections* manual packaged with your text.

Within *WileyPLUS*

• **Chapter Quiz**
• **Visual Anatomy**
• **Mastering Vocabulary**
• **Audio Glossary**
• **Web Links**

• Understanding The Concepts

1. Referring to Table 1.1, which body systems help eliminate wastes? Which regulate body activities? Which protect the body? Which maintain acid–base balance in the body?

2. Which life process in the human body is utilized when any of the others occur?

3. Why would a weight-loss supplement that stimulates "metabolism" not necessarily help you lose weight? (Hint: Think about the definition of metabolism.)

4. Referring to Figure 1.3, what would happen to your heart rate if some stimulus caused your blood pressure to drop? Would this occur by positive or negative feedback?

5. Sometimes a synthetic form of the hormone oxytocin, called Pitocin, is given to pregnant women to induce labor. Referring to Figure 1.4, which part of the feedback system is bypassed when this drug is given? Explain your answer.

6. A television commercial advertising a new drug tells you to look for signs and symptoms of a certain condition. Is the advertiser using the terms *signs* and *symptoms* correctly? Explain your answer.

7. Would a sore throat be a symptom or a sign of a potential disease? Explain your answer.

8. On which part of your body would you find a plantar wart? The occipital bone of the skull? An inguinal hernia? Carpal tunnel syndrome?

9. Why are the pedals on a bicycle most likely called "pedals"?

10. Which plane divides the heart into anterior and posterior portions? The brain into unequal right and left portions? What plane would be used to visualize the appendix?

11. Why is meningitis, an inflammation of membranes in the cranial cavity and vertebral canal, particularly dangerous?

2

The Chemical Level of Organization

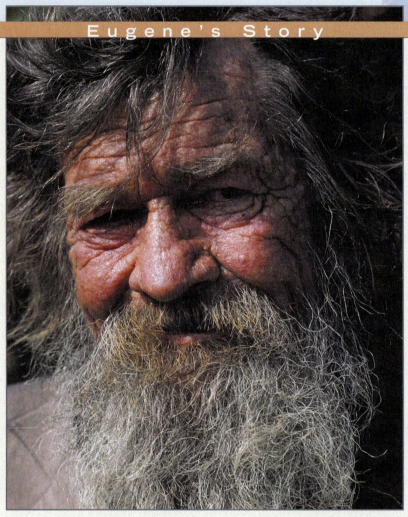

The doors to the emergency department swung open with a thud as the paramedics wheeled in an unconscious man on a gurney. "Is Dr. Kim working tonight?" asked one of the paramedics.

"Yes she is. Who do we have here?" responded the nurse.

"Well, Linda, this is Eugene. He was found passed out in the alley behind the Kozy Korner bar. Apparently he's had a bit too much to drink. We've brought him in before and Dr. Kim saw him. He'a a chronic alcoholic, diabetic too, I think," said the paramedic.

Linda touched the man's forehead, then gently lifted his eyelids. "Eugene, can you hear me? You're at the hospital. We need you to wake up!" Eugene mumbled and then coughed. "I don't think he's waking up anytime soon. Put him in bed two. I'll let Dr. Kim know he's here. You're in luck Eugene, Dr. Kim is on duty tonight." Just then Dr. Kim happened in.

She leaned over, looking at Eugene. "Hmm. Linda, get a CBC and blood chemistries started right away."

Just as the eyes are the windows to the soul, our blood is a window on the body's homeostasis. Blood chemistry analysis is used to assess various chemical products, both organic and inorganic, that are related to the function of the organs and organ systems.

Imbalances in blood chemistry can help in the diagnosis of specific disease states or other homeostatic imbalances; as you will learn in this chapter, they can also provide other important insights into the functions of the human body.

INTRODUCTION

The chemical level of organization, the lowest level of structural organization, consists of atoms and molecules. These small particles ultimately combine to form body organs and systems of astonishing size and complexity. In this chapter, we will consider how atoms bond together to form molecules and how atoms and molecules release or store energy in processes known as chemical reactions. We will also learn about the vital importance of water, which accounts for nearly two-thirds of body weight, in chemical reactions and the maintenance of homeostasis. Finally, we present several groups of molecules whose unique properties contribute to assembly of the body's structures or to powering the processes that enable you to live.

Because your body is composed of chemicals and all body activities are chemical in nature, it is important to become familiar with the language and basic ideas of chemistry to understand human anatomy and physiology.

Because your body is composed of chemicals and all body activities are chemical in nature, it is important to become familiar with the language and basic ideas of chemistry to understand human anatomy and physiology. **Chemistry** (KEM-is-trē) is the science of the structure and interactions of matter. All living and nonliving things consist of **matter**, which is anything that occupies space and has **mass**. Mass is the amount of matter in any object, which does not change. *Weight*, the force of gravity acting on matter, does change. When objects are farther from Earth, the pull of gravity is weaker; this is why the weight of an astronaut is close to zero in outer space.

2.1 Chemical elements are composed of small units called atoms.

Matter exists in three states: solid, liquid, and gas. *Solids*, such as bones and teeth, are compact and have a definite shape and volume. *Liquids*, such as blood plasma, have a definite volume and assume the shape of their container. *Gases*, like oxygen and carbon dioxide, have neither a definite shape nor volume. All forms of matter—both living and nonliving—are made up of a limited number of building blocks called **chemical elements**. Each element is a substance that cannot be split into a simpler substance by ordinary chemical means. Scientists now recognize 117 elements. Of these, 92 occur naturally on Earth. The rest have been produced from the natural elements using particle accelerators or nuclear reactors. Each element is designated by a **chemical symbol**, one or two letters of the element's name in English, Latin, or another language. Examples of chemical symbols are H for hydrogen, C for carbon, O for oxygen, N for nitrogen, Ca for calcium, and Na for sodium (*natrium* = sodium).*

Twenty-six different chemical elements normally are present in your body. Just four elements—oxygen, carbon, hydrogen, and nitrogen—make up about 96.2% of the body's mass. Eight others—calcium, phosphorus (P), potassium (K), sulfur (S), sodium (Na), chlorine (Cl), magnesium (Mg), and iron (Fe)—contribute 3.6% of the body's mass. An additional 14 elements, the *trace elements*, are present in tiny amounts—less than half a teaspoonful each—accounting for the remaining 0.2% of the body's mass. Several trace elements have important functions in the body. For example, iodine is needed to make thyroid hormones. The functions of some trace elements are unknown. Table 2.1 lists the main chemical elements that compose the human body.

Structure of Atoms

Each element is made up of **atoms**, the smallest units of matter that retain the properties and characteristics of an element. Atoms are extremely small. Two hundred thousand of the largest atoms would fit on the period at the end of this sentence. Hydrogen atoms, the smallest atoms, have a diameter less than 0.1 nanometer (0.1×10^{-9} m = 0. 0000000001 m); the largest atoms are only five times larger.

* The periodic table of elements, which lists all of the known chemical elements, can be found in Appendix B.

TABLE 2.1		
Main Chemical Elements in the Body		
CHEMICAL ELEMENT (SYMBOL)	% OF TOTAL BODY MASS	SIGNIFICANCE
MAJOR ELEMENTS		
Oxygen (O)	65.0	Part of water and many organic (carbon-containing) molecules; used to generate ATP, a molecule used by cells to temporarily store chemical energy.
Carbon (C)	18.5	Forms backbone chains and rings of all organic molecules: carbohydrates, lipids (fats), proteins, and nucleic acids (DNA and RNA).
Hydrogen (H)	9.5	Constituent of water and most organic molecules; ionized form (H^+) makes body fluids more acidic.
Nitrogen (N)	3.2	Component of all proteins and nucleic acids.
LESSER ELEMENTS		
Calcium (Ca)	1.5	Contributes to hardness of bones and teeth; ionized form (Ca^{2+}) needed for blood clotting, release of some hormones, contraction of muscle, and many other processes.
Phosphorus (P)	1.0	Component of nucleic acids and ATP; required for normal bone and tooth structure.
Potassium (K)	0.35	Ionized form (K^+) is the most plentiful cation (positively charged particle) in intracellular fluid; needed to generate action potentials.
Sulfur (S)	0.25	Component of some vitamins and many proteins.
Sodium (Na)	0.2	Ionized form (Na^+) is the most plentiful cation in extracellular fluid; essential for maintaining water balance; needed to generate action potentials.
Chlorine (Cl)	0.2	Ionized form (Cl^-) is the most plentiful anion (negatively charged particle) in extracellular fluid; essential for maintaining water balance.
Magnesium (Mg)	0.1	Ionized form (Mg^{2+}) needed for action of many enzymes, molecules that increase the rate of chemical reactions in organisms.
Iron (Fe)	0.005	Ionized forms (Fe^{2+} and Fe^{3+}) are part of hemoglobin (oxygen-carrying protein in red blood cells) and some enzymes (proteins that catalyze chemical reactions in living cells).
TRACE ELEMENTS	0.2	Aluminum (Al), boron (B), chromium (Cr), cobalt (Co), copper (Cu), fluorine (F), iodine (I), manganese (Mn), molybdenum (Mo), selenium (Se), silicon (Si), tin (Sn), vanadium (V), and zinc (Zn).

Dozens of different **subatomic particles** compose individual atoms. However, only three types of subatomic particles are important for understanding the chemical reactions in the human body: protons, neutrons, and electrons (Figure 2.1). The dense central core of an atom is its **nucleus**. Within the nucleus are positively charged **protons (p^+)** and uncharged (neutral) **neutrons (n^0)**. The tiny negatively charged **electrons (e^-)** move about in a large space surrounding the nucleus. They do not follow a fixed path or orbit but instead form a negatively charged "cloud" that envelops the nucleus (Figure 2.1a).

Even though their exact positions cannot be predicted, specific groups of electrons are most likely to move about within certain regions around the nucleus. These regions, called **electron shells**, are depicted as simple circles around the nucleus. Each electron shell can hold a specific number of electrons (Figure 2.1b). The first electron shell (nearest the nucleus) never holds more than 2 electrons. The second shell holds a maximum of 8 electrons, and the third can hold up to 18 electrons. The electron shells fill with electrons in a specific order, beginning with the first shell. For example, notice in Figure 2.2 that sodium (Na), which has 11 electrons total, contains 2 electrons in the first shell, 8 electrons in the second shell, and 1 electron in the third shell.

Figure 2.1   Two representations of the structure of an atom.
Electrons move about the nucleus, which contains neutrons and protons. (a) In the electron cloud model of an atom, the shading represents the chance of finding an electron in regions outside the nucleus. (b) In the electron shell model, filled circles represent individual electrons, which are grouped into concentric circles according to the shells they occupy.

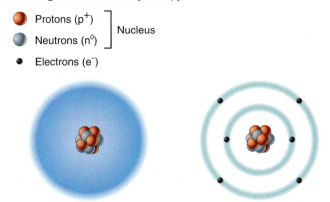

(a) Electron cloud model of carbon   (b) Electron shell model of carbon

🔑 **An atom is the smallest unit of matter that retains the properties and characteristics of its element.**

Figure 2.2   Atomic structures of several atoms that have important roles in the human body.

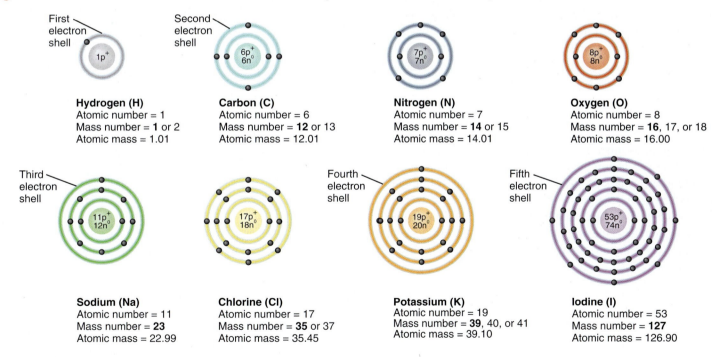

First electron shell
Second electron shell

Hydrogen (H)
Atomic number = 1
Mass number = **1** or 2
Atomic mass = 1.01

Carbon (C)
Atomic number = 6
Mass number = **12** or 13
Atomic mass = 12.01

Nitrogen (N)
Atomic number = 7
Mass number = **14** or 15
Atomic mass = 14.01

Oxygen (O)
Atomic number = 8
Mass number = **16**, 17, or 18
Atomic mass = 16.00

Third electron shell
Fourth electron shell
Fifth electron shell

Sodium (Na)
Atomic number = 11
Mass number = **23**
Atomic mass = 22.99

Chlorine (Cl)
Atomic number = 17
Mass number = **35** or 37
Atomic mass = 35.45

Potassium (K)
Atomic number = 19
Mass number = **39**, 40, or 41
Atomic mass = 39.10

Iodine (I)
Atomic number = 53
Mass number = **127**
Atomic mass = 126.90

Atomic number = number of protons in an atom
Mass number = number of protons and neutrons in an atom (boldface indicates most common isotope)
Atomic mass = average mass of all stable atoms of a given element in daltons

🔑 **The atoms of different elements have different atomic numbers because they have different numbers of protons.**

The most massive element present in the human body is iodine, which has a total of 53 electrons: 2 in the first shell, 8 in the second shell, 18 in the third shell, 18 in the fourth shell, and 7 in the fifth shell.

The number of electrons in an atom of an element always equals the number of protons. Because each electron and proton carries one charge, the negatively charged electrons and the positively charged protons balance each other. As a result, each atom is electrically neutral, meaning its total charge is zero.

Atomic Number and Mass Number

The *number of protons* in the nucleus of an atom is that atom's **atomic number**. Figure 2.2 shows that atoms of different elements have different atomic numbers because they have different numbers of protons. For example, oxygen has an atomic number of 8 because its nucleus has eight protons, and sodium has an atomic number of 11 because its nucleus has 11 protons.

The **mass number** of an atom is the sum of its protons and neutrons. Because sodium has 11 protons and 12 neutrons, its mass number is 23 (Figure 2.2). Although all atoms of one element have the same number of protons, they may have different numbers of neutrons. **Isotopes** are atoms of an element that have different numbers of neutrons and therefore different mass numbers. In a sample of oxygen, for example, most atoms have 8 neutrons, but a few have 9 or 10, even though all have 8 protons and 8 electrons. As you will discover shortly, the electrons of an atom determine the atom's chemical properties. Although the isotopes of an element have different numbers of neutrons, they have identical chemical properties because they have the same number of electrons.

Atomic Mass

The standard unit for measuring the mass of atoms and their subatomic particles is a **dalton**. A neutron has a mass of 1.008 daltons, and a proton has a mass of 1.007 daltons. The mass of an electron, at 0.0005 dalton, is almost 2000 times smaller than the mass of a neutron or proton. The **atomic mass** (also called the *atomic weight*) of an element is the average mass of all its naturally occurring isotopes. Typically, the atomic mass of an element is close to the mass number of its most abundant isotope.

Ions, Molecules, and Compounds

As we discussed, atoms of the same element have the same number of protons. The atoms of each element have a characteristic way of losing, gaining, or sharing their electrons when interacting with other atoms to achieve stability. The way that electrons behave enables atoms in the body to exist in electrically charged forms called ions, or to join with one another in complex combinations called molecules.

If an atom either *gives up* or *gains electrons*, it becomes an **ion** (Ī-on). An ion is an atom that has a positive or negative charge because it has unequal numbers of protons and electrons. An ion of an atom is symbolized by writing its chemical symbol followed by the number of its positive (+) or negative (−) charges. Thus, Ca^{2+} stands for a calcium ion that has two positive charges because it has lost two electrons.

When two or more atoms *share* electrons, the resulting combination is called a **molecule** (MOL-e-kūl). A *molecular formula* indicates the elements and the number of atoms of each element that make up a molecule. A molecule may consist of two atoms of the same kind, such as an oxygen molecule (Figure 2.3a). The molecular formula for a molecule of oxygen is O_2. The subscript 2 indicates that the molecule contains two atoms. Two or more different kinds of atoms may also form a molecule, as in a water molecule (H_2O). In H_2O one atom of oxygen shares electrons with two atoms of hydrogen.

A **compound** is a molecule that contains atoms of two or more different elements. Most of the atoms in the body are joined into compounds. Water (H_2O) and sodium chloride ($NaCl$), common table salt, are compounds. A molecule of oxygen (O_2) is not a compound because it consists of atoms of only one element. Thus, although all compounds are molecules, not all molecules are compounds.

Figure 2.3   Atomic structures of an oxygen molecule and a superoxide free radical.

(a) Oxygen molecule (O_2)   (b) Superoxide free radical (O_2^-)

Unpaired electron

🔑 **A free radical has an unpaired electron in its outermost electron shell.**

A **free radical** is an electrically charged atom or group of atoms with an unpaired electron in the outermost shell. A common example is superoxide, which is formed by the addition of an electron to an oxygen molecule (Figure 2.3b). Having an unpaired electron makes a free radical unstable, highly reactive, and destructive to nearby molecules. Free radicals become stable by either giving up an unpaired electron to, or taking an electron from, another molecule. In so doing, free radicals may break apart important body molecules.

Checkpoint

1. Which subatomic particles allow atoms to form ions or combine into molecules?
2. What is the charge of protons, neutrons, and electrons? Why is the charge of an electrically neutral atom zero?
3. What are the atomic number, mass number, and atomic mass of carbon? How are they related?
4. What are isotopes and free radicals?

2.2 Atoms are held together by chemical bonds.

The forces that hold together the atoms of molecules and compounds are **chemical bonds**. The likelihood that an atom will form a chemical bond with another atom depends on the number of electrons in its outermost shell, also called the **valence shell**. An atom with a valence shell holding eight electrons is *chemically stable*, which means it is unlikely to form chemical bonds with other atoms. Neon, for example, has eight electrons in its valence shell, and for this reason it does not bond easily with other atoms. The valence shell of hydrogen and helium is the first electron shell, which holds a maximum of two electrons. Because helium has two valence electrons, it too is stable and seldom bonds with other atoms. Hydrogen, on the other hand, has only one valence electron (see Figure 2.2), so it binds readily with other atoms.

The atoms of most biologically important elements do not have eight electrons in their valence shells. Under the right conditions, two or more atoms can interact in ways that produce a chemically stable arrangement of eight electrons in the valence shell of each atom. This chemical principle, called the *octet rule* (*octet = set of eight*), helps explain why atoms interact in predictable ways.

One atom is more likely to interact with another atom if doing so will leave both with eight valence electrons. For this to happen, an atom either empties its partially filled valence shell, fills it with donated electrons, or shares electrons with other atoms. The way that valence electrons are distributed determines which kind of chemical bond results. We will consider three kinds of chemical bonds: ionic bonds, covalent bonds, and hydrogen bonds.

Ionic Bonds

When atoms lose or gain one or more valence electrons, ions are formed. Positively and negatively charged ions are attracted to one another—opposites attract. The chemical bond formed by the attraction between ions of opposite electrical charge is called an **ionic bond**. Consider sodium and chlorine atoms, the components of common table salt. Sodium has one valence electron (Figure 2.4a). If sodium *loses* this electron, it is left with the eight electrons in its second shell, which becomes the valence shell.

Figure 2.4 **Ions and ionic bond formation.** (a) A sodium atom can attain the stability of eight electrons in its outermost shell by losing one valence electron; it then becomes a sodium ion, Na^+. (b) A chlorine atom can attain the stability of eight electrons in its outermost shell by accepting one electron; it then becomes a chloride ion, Cl^-. (c) An ionic bond may form between oppositely charged ions. (d) In a crystal of NaCl, each Na^+ is surrounded by six Cl^-. In (a), (b), and (c), the electron that is lost or accepted is colored red.

Electron donated

Atom · Ion

(a) Sodium: 1 valence electron

Electron accepted

Atom · Ion

(b) Chlorine: 7 valence electrons

(c) Ionic bond in sodium chloride (NaCl)

—Na^+
—Cl^-

(d) Packing of ions in a crystal of sodium chloride

🔑 An ionic bond is the force of attraction that holds together oppositely charged ions.

However, the total number of protons (11) now exceeds the number of electrons (10). As a result, the sodium atom becomes a **cation**, a positively charged ion. A sodium ion has a charge of +1 and is written Na^+. By contrast, chlorine has seven valence electrons (Figure 2.4b). If chlorine *gains* an electron from a neighboring atom, it will have eight electrons in its third electron shell. When this happens, the total number of electrons (18) exceeds the number of protons (17), and the chlorine atom becomes an **anion**, a negatively charged ion. The ionic form of chlorine is called a chloride ion. It has a charge of −1 and is written Cl^-. When an atom of sodium donates its sole valence electron to an atom of chlorine, the resulting positive and negative charges pull both ions tightly together into an ionic bond (Figure 2.4c). The resulting ionic compound is sodium chloride, or common table salt, written NaCl.

In general, ionic compounds exist as solids, with an orderly, repeating arrangement of the ions, as in a crystal of NaCl (Figure 2.4d). A crystal of NaCl may be large or small—the total number of ions can vary—but the ratio of Na^+ to Cl^- is always 1:1. In the body, ionic bonds are found mainly in teeth and bones, where they give great strength to these important structural tissues. An ionic compound that dissociates (breaks apart) into positive and negative ions in solution is called an **electrolyte** (e-LEK-trō-līt). Most ions in the body are dissolved in body fluids as electrolytes, so named because their solutions can conduct an electric current. As you will see in later chapters, electrolytes have many important functions. For example, they are critical for controlling water movement within the body, maintaining acid–base balance, and producing nerve and muscle impulses. Table 2.2 lists the names and symbols of the most common ions in the body.

TABLE 2.2

Common Ions and Ionic Compounds in the Body

CATIONS		ANIONS	
NAME	SYMBOL	NAME	SYMBOL
Hydrogen ion	H^+	Fluoride ion	F^-
Sodium ion	Na^+	Chloride ion	Cl^-
Potassium ion	K^+	Iodide ion	I^-
Ammonium ion	NH_4^+	Hydroxide ion	OH^-
Magnesium ion	Mg^{2+}	Bicarbonate ion	HCO_3^-
Calcium ion	Ca^{2+}	Oxide ion	O^{2-}
Iron (II) ion	Fe^{2+}	Sulfate ion	SO_4^{2-}
Iron (III) ion	Fe^{3+}	Phosphate ion	PO_4^{3-}

Covalent Bonds

When a **covalent bond** forms, neither of the combining atoms loses or gains electrons. Instead, the atoms form a molecule by *sharing* one, two, or three pairs of valence electrons. The larger the number of electron pairs shared between two atoms, the stronger is the covalent bond. Covalent bonds may form between atoms of the same element or between atoms of different elements. They are the most common chemical bonds in the body, and the compounds that result from them form most of the body's structures.

A **single covalent bond** results when two atoms share one electron pair. For example, a molecule of hydrogen forms when two hydrogen atoms share their single valence electrons (Figure 2.5a), which allows both atoms to have a full valence shell. A **double covalent bond** results when two atoms share two pairs of electrons, as happens in an oxygen molecule (Figure 2.5b). A **triple covalent bond** occurs when two atoms share three pairs of electrons, as in a molecule of nitrogen (Figure 2.5c). Notice in the *structural formulas* for covalently bonded molecules in Figure 2.5 that the number of lines between the chemical symbols for two atoms indicates whether the bond is a single (—), double (=), or triple (≡) covalent bond.

The same principles of covalent bonding that apply to atoms of the same element are also at work in covalent bonds between atoms of different elements. The gas methane (CH_4) contains covalent bonds formed between the atoms of two different elements, one carbon and four hydrogens (Figure 2.5d). The valence shell of the carbon atom can hold eight electrons but has only four of its own. The single electron shell of a hydrogen atom can hold two electrons, but each hydrogen atom has only one of its own. A methane molecule contains four separate single covalent bonds. Each hydrogen atom shares one pair of electrons with the carbon atom.

In some covalent bonds, two atoms share the electrons equally—one atom does not attract the shared electrons more strongly than the other atom. This type of bond is a **nonpolar covalent bond**. The bonds between two identical atoms always are nonpolar covalent bonds (Figure 2.5a–c). Another example of a nonpolar covalent bond is the single covalent bond that forms between carbon and each atom of hydrogen in a methane molecule (Figure 2.5d).

In a **polar covalent bond**, the sharing of electrons between two atoms is unequal—the nucleus of one atom attracts the shared electrons more strongly than the nucleus of the other atom. When polar covalent bonds form, the resulting molecule has a partial negative charge near the atom that attracts electrons more strongly. At least one other atom in the molecule then will have a partial positive charge. The partial charges are indicated by

Figure 2.5 Covalent bond formation. The red electrons are shared equally in (a)–(d) and unequally in (e). In writing the structural formula of a covalently bonded molecule, each straight line between the chemical symbols for two atoms denotes a pair of shared electrons. In molecular formulas, the number of atoms in each molecule is noted by subscripts.

DIAGRAMS OF ATOMIC AND MOLECULAR STRUCTURE	STRUCTURAL FORMULA	MOLECULAR FORMULA

(a) Hydrogen atoms → Hydrogen molecule H — H H_2

(b) Oxygen atoms → Oxygen molecule O = O O_2

(c) Nitrogen atoms → Nitrogen molecule N ≡ N N_2

(d) Carbon atom Hydrogen atoms → Methane molecule H—C—H (with H above and H below) CH_4

(e) Oxygen atom Hydrogen atoms → Water molecule O with H and H H_2O

In a covalent bond, two atoms share one, two, or three pairs of valence electrons.

a lowercase Greek delta with a minus or plus sign: δ^- or δ^+. An important example of a polar covalent bond in living systems is the bond between oxygen and hydrogen in a molecule of water (Figure 2.5e); in this molecule, the nucleus of the oxygen atom attracts the electrons more strongly than the nuclei of the hydrogen atoms, so the oxygen atom has a partial negative charge. Later in the chapter, we will see how polar covalent bonds allow water to dissolve many molecules that are important to life. Bonds between nitrogen and hydrogen and those between oxygen and carbon are also polar bonds.

Hydrogen Bonds

The polar covalent bonds that form between hydrogen atoms and other atoms can give rise to a third type of chemical bond, a hydrogen bond (Figure 2.6). A **hydrogen bond** forms when a hydrogen atom with a partial positive charge (δ^+) attracts the partial negative charge (δ^-) of neighboring electronegative atoms, most often larger oxygen or nitrogen atoms. Thus, hydrogen bonds result from attraction of oppositely charged parts of molecules, rather than from sharing of electrons as in covalent bonds or from the loss or gain of electrons as in ionic bonds. Hydrogen bonds are weak compared to ionic and covalent bonds. Thus, they cannot bind atoms into molecules. However, hydrogen bonds do establish important links between molecules or between different parts of a large molecule, such as a protein or nucleic acid (both discussed later in this chapter).

The hydrogen bonds that link neighboring water molecules give water considerable *cohesion*, the tendency of like particles to stay together. The cohesion of water molecules creates a very high **surface tension**, a measure of the difficulty of stretching or breaking the surface of a liquid. At the boundary between water and air, water's surface tension is very high because the water molecules are much more attracted to one another than they are attracted to molecules in the air. This is readily seen when a spider walks on water or a leaf floats on the surface of a pond or stream. The influence of water's surface tension on the body can be seen in the way it increases the work required for breathing. Because a thin film of watery fluid coats the air sacs of the lungs, each inhalation must have enough force to overcome surface tension as the air sacs stretch and enlarge when taking in air.

Even though single hydrogen bonds are weak, very large molecules may contain hundreds of these bonds. Acting collectively, hydrogen bonds provide considerable strength and stability and

Figure 2.6   Hydrogen bonding among water molecules. Each water molecule forms hydrogen bonds, indicated by dotted lines, with three to four neighboring water molecules.

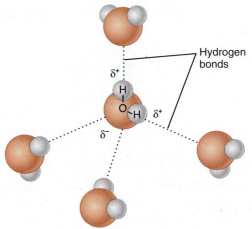

Hydrogen bonds

Hydrogen bonds occur because hydrogen atoms in one water molecule are attracted to the partial negative charge of the oxygen atom in another water molecule.

help determine the three-dimensional shape of large molecules. As you will see later in this chapter, a large molecule's shape determines how it functions.

Checkpoint

5.   Which electron shell is the valence shell of an atom, and what is its significance?

6.   What is the principal difference between an ionic bond and a covalent bond?

7.   What are hydrogen bonds?

2.3   Chemical reactions occur when atoms combine with or separate from other atoms.

A **chemical reaction** occurs when new bonds form or existing bonds break between atoms. Chemical reactions are the foundation of all life processes, and as we have seen, the interactions of valence electrons are the basis of all chemical reactions. Consider how hydrogen and oxygen molecules react to form water molecules (Figure 2.7). The starting substances—two H_2 and one O_2—are known as the **reactants**. The ending substances—two molecules of H_2O—are the **products**. The arrow in the figure indicates the direction in which the reaction proceeds. In a chemical reaction, the number of atoms of each element is the same before and after the reaction. However, because the atoms are rearranged, the reactants and products have different chemical properties. Through thousands of different chemical reactions, body structures are built and body functions are carried out. The term **metabolism** refers to all the chemical reactions occurring in the body.

Figure 2.7   The chemical reaction between two hydrogen molecules (H_2) and one oxygen molecule (O_2) to form two molecules of water (H_2O). Note that the reaction occurs by breaking old bonds and making new bonds.

The number of atoms of each element is the same before and after a chemical reaction.

Forms of Energy and Chemical Reactions

Each chemical reaction involves energy changes. **Energy** (*en-* = in; *-ergy* = work) is the capacity to do work. Two principal forms of energy are **potential energy**, energy stored by matter due to its position, and **kinetic energy**, the energy associated with matter in motion. For example, the energy stored in water behind a dam or in a person poised to jump down some steps is potential energy. When the gates of the dam are opened or the person jumps, potential energy is converted into kinetic energy. **Chemical energy** is a form of potential energy that is stored in the bonds of compounds and molecules. The total amount of energy present at the beginning and end of a chemical reaction is the same. Although energy can be neither created nor destroyed, it may be converted from one form to another. For example, some of the chemical energy in the foods we eat is eventually converted into various forms of kinetic energy, such as mechanical energy used to walk and talk. Conversion of energy from one form to another generally releases heat, some of which is used to maintain normal body temperature.

Figure 2.8 Activation energy.

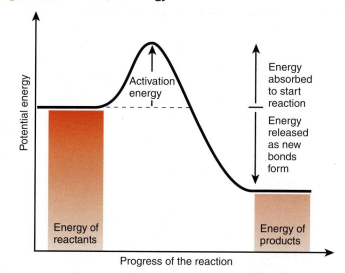

Activation energy is the energy needed to break chemical bonds in the reactant molecules so a reaction can begin.

Energy Transfer in Chemical Reactions

Chemical bonds represent stored chemical energy and chemical reactions occur when new bonds are formed or old bonds are broken between atoms. The *overall reaction* may either release energy or absorb energy. **Exergonic reactions** (*ex-* = out) release more energy than they absorb. By contrast, **endergonic reactions** (*end-* = within) absorb more energy than they release.

A key feature of the body's metabolism is the coupling of exergonic reactions and endergonic reactions. Energy released from an exergonic reaction is often used to drive an endergonic one. In general, exergonic reactions occur as nutrients, such as glucose, are broken down. Some of the energy released may be trapped in the covalent bonds of adenosine triphosphate (ATP), which we describe more fully later in this chapter. The energy transferred to the ATP molecules is then used to drive endergonic reactions needed to build body structures, such as muscles and bones. The energy in ATP is also used to do the mechanical work involved in the contraction of muscle or the movement of substances into or out of cells.

Activation Energy

Because particles of matter such as atoms, ions, and molecules have kinetic energy, they are continuously moving and colliding with one another. A sufficiently forceful collision can disrupt the movement of valence electrons, causing an existing chemical bond to break or a new one to form. The collision energy needed to break chemical bonds in the reactants is called the **activation energy** (Figure 2.8). This initial energy "investment" is needed to start a reaction. The reactants must absorb enough energy for their chemical bonds to become unstable and for their valence electrons to form new combinations. Then, as new bonds form, energy is released to the surroundings.

Both the concentration of particles and the temperature influence the chance that a collision will occur and cause a chemical reaction.

- *Concentration*. The more particles of matter present in a confined space, the greater is the chance that they will collide (think of people crowding into a subway car at rush hour). The concentration of particles increases when more are added to a given space or when the pressure on the space increases, which forces the particles closer together so that they collide more often.

- *Temperature*. As temperature rises, particles of matter move about more rapidly. Thus, the higher the temperature of matter, the more forcefully particles will collide, and the greater is the chance that a collision will produce a reaction.

Catalysts

As we have seen, chemical reactions occur when chemical bonds break or form after atoms, ions, or molecules collide with one another. However, body temperature and the concentrations of molecules in body fluids are far too low for most chemical reactions to occur rapidly enough to maintain life. Raising the temperature and

Figure 2.9   Comparison of energy needed for a chemical reaction to proceed with a catalyst (blue curve) and without a catalyst (red curve).

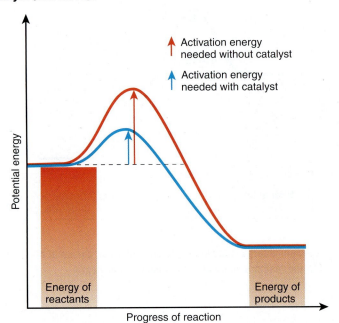

Catalysts speed up chemical reactions by lowering the activation energy required to initiate them.

the number of reacting particles of matter in the body could increase the frequency of collisions and thus increase the rate of chemical reactions, but doing so could also damage or kill the body's cells.

Substances called catalysts solve this problem. **Catalysts** are chemical compounds that speed up chemical reactions by lowering the activation energy needed for a reaction to occur (Figure 2.9). The most important catalysts in the body are enzymes, which we will discuss later in this chapter.

A catalyst does not alter the difference in potential energy between the reactants and the products. Rather, it lowers the amount of energy needed to start the reaction. For chemical reactions to occur, some particles of matter—especially large molecules—must not only collide with sufficient force, but they must "hit" one another at precise spots. A catalyst helps to properly orient the colliding particles. Thus, they interact at the spots that make the reaction happen. Although the action of a catalyst helps to speed up a chemical reaction, the catalyst itself is unchanged at the end of the reaction. A single catalyst molecule can repeatedly assist one chemical reaction after another.

Types of Chemical Reactions

As a chemical reaction takes place, the atoms of the reactants are rearranged to yield products with new chemical properties. In this section we will look at the types of chemical reactions that are most important to the operations of the human body discussed throughout the book.

Synthesis Reactions—Anabolism

The combination of two or more atoms, ions, or molecules to form new and larger molecules is called a **synthesis reaction**. The word *synthesis* means "to put together." A synthesis reaction can be expressed as follows:

$$\underset{\substack{\text{Atom, ion,}\\\text{or molecule A}}}{A} \quad + \quad \underset{\substack{\text{Atom, ion,}\\\text{or molecule B}}}{B} \quad \xrightarrow{\text{Combine to form}} \quad \underset{\text{New molecule AB}}{AB}$$

One example of a synthesis reaction is the reaction between two hydrogen molecules and one oxygen molecule to form two molecules of water (see Figure 2.7). Another example of a synthesis reaction is the formation of ammonia from nitrogen and hydrogen:

$$\underset{\substack{\text{One nitrogen}\\\text{molecule}}}{N_2} \quad + \quad \underset{\substack{\text{Three hydrogen}\\\text{molecules}}}{3H_2} \quad \xrightarrow{\text{Combine to form}} \quad \underset{\substack{\text{Two ammonia}\\\text{molecules}}}{2NH_3}$$

All the synthesis reactions that occur in your body are collectively referred to as **anabolism** (a-NAB-ō-lizm). Overall, anabolic reactions are usually endergonic because they absorb more energy than they release. Combining simple molecules like amino acids (discussed shortly) to form large molecules such as proteins is an example of anabolism.

Decomposition Reactions—Catabolism

Decomposition reactions split up large molecules into smaller atoms, ions, or molecules. A decomposition reaction is expressed as follows:

$$\underset{\text{Molecule AB}}{AB} \quad \xrightarrow{\text{Breaks down into}} \quad \underset{\substack{\text{Atom, ion,}\\\text{or molecule A}}}{A} \quad + \quad \underset{\substack{\text{Atom, ion,}\\\text{or molecule B}}}{B}$$

The decomposition reactions that occur in your body are collectively referred to as **catabolism** (ka-TAB-ō-lizm). Overall, catabolic reactions are usually exergonic because they release more energy than they absorb. The breakdown of large starch molecules into many small glucose molecules during digestion is an example of catabolism.

Exchange Reactions

Many reactions in the body are exchange reactions; they consist of both synthesis and decomposition reactions. One type of exchange reaction works like this:

$$AB + CD \longrightarrow AD + BC$$

The bonds between A and B and between C and D break (decomposition), and new bonds then form (synthesis) between A and D and between B and C. An example of an exchange reaction is

$$\underset{\substack{\text{Hydrochloric} \\ \text{acid}}}{HCl} + \underset{\substack{\text{Sodium} \\ \text{bicarbonate}}}{NaHCO_3} \longrightarrow \underset{\substack{\text{Carbonic} \\ \text{acid}}}{H_2CO_3} + \underset{\substack{\text{Sodium} \\ \text{chloride}}}{NaCl}$$

Notice that the ions in both compounds have "switched partners": The hydrogen ion (H^+) from HCl has combined with the bicarbonate ion (HCO_3^-) from $NaHCO_3$, and the sodium ion (Na^+) from $NaHCO_3$ has combined with the chloride ion (Cl^-) from HCl.

Reversible Reactions

Some chemical reactions proceed in only one direction, from reactants to products, as previously indicated by the single arrows pointing to the right. Other chemical reactions may be reversible. In a **reversible reaction**, the products can revert to the original reactants. A reversible reaction is indicated by two half-arrows pointing in opposite directions:

$$AB \underset{\text{Combines to form}}{\overset{\text{Breaks down into}}{\rightleftarrows}} A + B$$

Some reactions are reversible only under special conditions:

$$AB \underset{\text{Heat}}{\overset{\text{Water}}{\rightleftarrows}} A + B$$

In that case, whatever is written above or below the arrows indicates the condition needed for the reaction to occur. In these reactions, AB breaks down into A and B only when water is added, and A and B react to produce AB only when heat is applied. Often, different enzymes guide the reactions in opposite directions.

Checkpoint

8. What is the relationship between reactants and products in a chemical reaction? What do we call chemical reactions occurring in the body?
9. Why is the reaction illustrated in Figure 2.8 exergonic?
10. How do catalysts affect activation energy?
11. Define anabolism and catabolism. Which are usually endergonic reactions?
12. What are reversible reactions?

2.4 Inorganic compounds include water, salts, acids, and bases.

Most of the chemicals in your body exist in the form of compounds. Biologists and chemists divide these compounds into two principal classes: inorganic compounds and organic compounds. In general, **inorganic compounds** lack carbon, are structurally simple, and are held together by ionic or covalent bonds. They include water; many salts, acids, and bases; and two carbon-containing compounds, carbon dioxide (CO_2) and bicarbonate ion (HCO_3). Water makes up 55–60% of a lean adult's total body mass; all other inorganic compounds add an additional 1–2%. **Organic compounds**, by contrast, always contain carbon, usually contain hydrogen, and always have covalent bonds. Most are large molecules and many are made up of long chains of carbon atoms. Organic compounds make up the remaining 38–43% of the human body.

Water

Water is the most important and abundant inorganic compound in all living systems. Although you might be able to survive for weeks without food, without water you would die in a matter of days. Nearly all the body's chemical reactions occur in a watery medium. Water has many properties that make it an indispensable compound for life. The most important property of water is its polarity—the uneven sharing of valence electrons that confers a partial negative charge near the one oxygen atom and two partial positive charges near the two hydrogen atoms in a water molecule (see Figure 2.5e). This property alone makes water an excellent solvent for other ionic or polar substances, gives water molecules cohesion (the tendency to stick together), and allows water to resist temperature changes.

Water as a Solvent

In medieval times people searched in vain for a "universal solvent," a substance that would dissolve (change from a solid to a liquid state) all other materials. They found nothing that worked as well as water. Although it is the most versatile solvent known, water is not the universal solvent sought by medieval alchemists. If it were, no container could hold it because it would dissolve all potential containers! What exactly is a solvent? A **solvent** is a liquid or gas in which some other substance, called a **solute**, has been dissolved. The combination of solvent plus solute is called a **solution**. Usually there is more solvent than solute in a solution. For example, your sweat is a dilute solution of water (the solvent) plus small amounts of salts (the solutes).

The versatility of water as a solvent for ionized or polar substances is due to its polar covalent bonds and its bent shape, which allows each water molecule to interact with several neighboring ions or molecules. Solutes that are charged or contain polar covalent bonds are **hydrophilic** (*hydro-* = water; *-philic* = loving), which means they dissolve easily in water. Common examples of hydrophilic solutes are sugar and salt. Molecules that contain mainly nonpolar covalent bonds, by contrast, are **hydrophobic** (*-phobic* = fearing). They are not very water soluble. Examples of hydrophobic compounds include animal fats and vegetable oils.

To understand the dissolving power of water, consider what happens when a crystal of a salt such as sodium chloride (NaCl) is placed in water (Figure 2.10). The slightly negative oxygen atom in water molecules attracts the sodium ions (Na^+), and the slightly positive hydrogen atoms in water molecules attract the chloride ions (Cl^-). Soon, water molecules surround and separate some Na^+ and Cl^- from each other at the surface of the crystal, breaking the ionic bonds that held NaCl together. The water molecules surrounding the ions also lessen the chance that ions with opposite charges will come together and reform an ionic bond.

The ability of water to form solutions is essential to health and survival. Because water can dissolve or suspend so many different substances, it is an ideal medium for metabolic reactions. Water enables dissolved reactants to collide and form products. Water is the solvent that carries nutrients, oxygen, and wastes throughout

Figure 2.10   How water molecules dissolve salts and polar substances. When a crystal of sodium chloride is placed in water, the slightly negative oxygen end (red) of water molecules is attracted to the positive sodium ions (Na^+), and the slightly positive hydrogen portions (gray) of water molecules are attracted to the negative chloride ions (Cl^-).

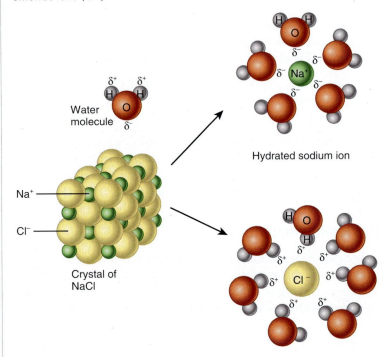

🔑 Water is a versatile solvent because its polar covalent bonds, in which electrons are shared unequally, create positive and negative regions.

the body. Water also dissolves waste products, which allows them to be flushed out of the body in the urine.

Water in Chemical Reactions

Water serves as the medium for most chemical reactions in the body, and it participates as a reactant or product in certain reactions. During digestion, for example, decomposition reactions break down large nutrient molecules into smaller molecules by the addition of water molecules. This type of reaction is called **hydrolysis** (hī-DROL-i-sis; *-lysis* = to loosen or break apart). Hydrolysis reactions enable dietary nutrients to be absorbed into the body. By contrast, when two smaller molecules join to form a larger molecule in a **dehydration synthesis reaction** (*de-* = from, down, or out; *-hydra-* = water), a water molecule is one of the products formed. For example, Figure 2.20 shows such reactions occurring during the synthesis of proteins.

Thermal Properties of Water

In comparison to most substances, water can absorb or release a relatively large amount of heat with only a slight change in its own temperature. The reason for this property is the large number of hydrogen bonds in water. As water absorbs heat energy, some of the energy is used to break hydrogen bonds. Less energy is then left over to increase the motion of water molecules, which would increase the water's temperature. The large amount of water in the body thus lessens the impact of changes in the environmental temperature, thereby helping maintain the homeostasis of body temperature.

Water also requires a large amount of heat to change from a liquid to a gas. When the water in sweat evaporates from the surface of the skin, it removes a large quantity of heat, providing an excellent cooling mechanism.

Water as a Lubricant

Water is a major component of saliva, mucus, and other lubricating body fluids. Lubrication is especially necessary in the thoracic and abdominal cavities, where internal organs touch and slide over one another. It is also needed at joints, where bones, ligaments, and tendons rub against one another. Inside the gastrointestinal tract, mucus and other watery secretions moisten foods, which aids their smooth passage through the digestive system.

Solutions, Colloids, and Suspensions

A **mixture** is a combination of elements or compounds that are physically blended together but not bound by chemical bonds. For example, the air you are breathing is a mixture of gases that includes nitrogen, oxygen, argon, and carbon dioxide. Three common liquid mixtures are solutions, colloids, and suspensions.

Once mixed together, solutes in a **solution** remain evenly dispersed among the solvent molecules. Because the solute particles in a solution are very small, a solution appears transparent.

A **colloid** differs from a solution mainly because of the size of its particles. The solute particles in a colloid are large enough to scatter light, just as water droplets in fog scatter light from a car's headlight beams. For this reason, colloids usually appear translucent or opaque. Milk is an example of a liquid that is both a colloid and a solution: The large milk proteins make it a colloid, whereas the calcium salts, milk sugar (lactose), ions, and other small particles are in solution.

The solutes in both solutions and colloids do not settle out and accumulate at the bottom of the container. In contrast, the suspended material in a **suspension** may mix with the liquid or suspending medium for some time, but eventually it will settle out. Blood is an example of a suspension. When freshly drawn from the body, blood has an even, reddish color due to the suspension of red blood cells. After blood sits for a while in a test tube, red blood cells settle out of the suspension and drift to the bottom of the tube (see Figure 18.1a). The upper layer, the liquid portion of blood, appears pale yellow and is called blood plasma. Blood plasma is another liquid that is both a solution of ions and other small solutes and a colloid due to the presence of larger plasma proteins.

Inorganic Acids, Bases, and Salts

Many inorganic compounds can be classified as acids, bases, or salts. When inorganic acids, bases, or salts dissolve in water, they **dissociate** (dis-SŌ-sē-āt), or separate, into ions surrounded by water molecules. An **acid** is a substance that dissociates into one or more **hydrogen ions (H^+)** and one or more anions (Figure 2.11a). A **base** dissociates into one or more **hydroxide ions (OH^-)** and one or more cations (Figure 2.11b).

A **salt**, when dissolved in water, dissociates into cations and anions, neither of which is H^+ or OH^- (Figure 2.11c). In the body, salts such as potassium chloride are important electrolytes that carry electrical currents (ions flowing from one place to another), especially in nerve and muscle tissues. The ions of salts also provide many essential chemical elements in body fluids such as blood, lymph, and interstitial fluid.

Figure 2.11 Dissociation of inorganic acids, bases, and salts.

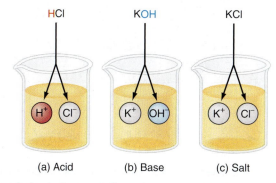

(a) Acid (b) Base (c) Salt

Dissociation is the separation of inorganic acids, bases, and salts into ions in a solution.

Acids and bases react with one another to form salts. For example, the reaction of hydrochloric acid (HCl) and potassium hydroxide (KOH), a base, produces the salt potassium chloride (KCl) and water (H_2O). This exchange reaction can be written as follows:

$$HCl + KOH \longrightarrow H^+ + Cl^- + K^+ + OH^- \longrightarrow KCl + H_2O$$
Acid   Base   Dissociated ions   Salt   Water

Acid–Base Balance: The Concept of pH

To ensure homeostasis, intracellular and extracellular fluids must contain balanced quantities of acids and bases. The more hydrogen ions (H^+) dissolved in a solution, the more acidic is the solution; the more hydroxide ions (OH^-) dissolved in a solution, the more basic (alkaline) is the solution. The chemical reactions that take place in the body are very sensitive to even small changes in the acidity or alkalinity of the body fluids in which they occur. Any departure from the narrow limits of normal H^+ and OH^- concentrations greatly disrupts body functions.

A solution's acidity or alkalinity is expressed on the **pH scale**, which extends from 0 to 14 (Figure 2.12). This scale is based on the concentration of hydrogen ions in a solution. The midpoint of

the pH scale is 7, where the numbers of H^+ and OH^- are equal. A solution with a pH of 7, such as pure water, is neutral—neither acidic nor alkaline. A solution that has more H^+ than OH^- is an **acidic solution** and has a pH below 7. A solution that has more OH^- than H^+ is a **basic (alkaline) solution** and has a pH above 7. A change of one whole number on the pH scale represents a *tenfold* change in the number of H^+. At a pH of 6, there are 10 times more H^+ than at a pH of 7. Put another way, a pH of 6 is 10 times more acidic than a pH of 7, and a pH of 9 is 100 times more alkaline than a pH of 7.

Maintaining pH: Buffer Systems

Although the pH of body fluids may differ, as we have discussed, the normal limits for each fluid are quite narrow. Table 2.3 shows the pH values for certain body fluids compared with those of common household substances. Homeostatic mechanisms maintain the pH of blood between 7.35 and 7.45, which is slightly more basic than pure water. If the pH of blood falls below 7.35, a condition called acidosis occurs, and if the pH rises above 7.45, it results in a condition called alkalosis; both conditions can seriously compromise homeostasis. Saliva is slightly acidic, and semen is slightly basic. Because the kidneys help remove excess acid from the body, urine can be quite acidic.

Even though strong acids and bases are continually taken into and formed by the body, the pH of fluids inside and outside cells remains almost constant. One important reason is the presence of **buffer systems**, which function to convert strong acids or bases into weak acids or bases. Strong acids (or bases) ionize easily and contribute many H^+ (or OH^-) to a solution. Therefore, they can change pH drastically, which can disrupt the body's metabolism. Weak acids (or bases) do not ionize as much and contribute fewer H^+ (or OH^-). Hence, they have less effect on the pH. **Buffers** are chemical compounds that convert strong acids or bases into weak acids or bases by removing or adding protons (H^+).

One important buffer system in the body is the **carbonic acid–bicarbonate buffer system**. Carbonic acid (H_2CO_3) can act as a weak acid, and the bicarbonate ion (HCO_3^-)

Figure 2.12   The pH scale. A pH below 7 indicates an acidic solution—more H^+ than OH^-. $[H^+]$ = hydrogen ion concentration; $[OH^-]$ = hydroxide ion concentration.

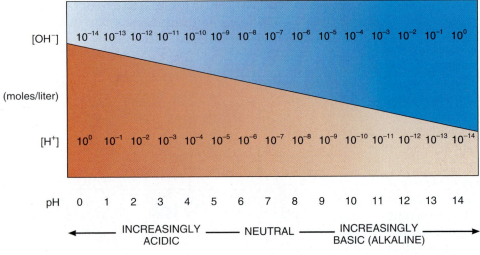

The lower the numerical value of the pH, the more acidic is the solution because the H^+ concentration becomes progressively greater. A pH above 7 indicates a basic (alkaline) solution; that is, there are more OH^- than H^+. The higher the pH, the more basic is the solution.

TABLE 2.3

pH Values of Selected Substances

SUBSTANCE	PH VALUE
• Gastric juice (found in the stomach)	1.2–3.0
Lemon juice	2.3
Vinegar	3.0
Carbonated soft drink	3.0–3.5
Orange juice	3.5
• Vaginal fluid	3.5–4.5
Tomato juice	4.2
Coffee	5.0
• Urine	4.6–8.0
• Saliva	6.35–6.85
Milk	6.8
Distilled (pure) water	7.0
• Blood	7.35–7.45
• Semen (fluid containing sperm)	7.20–7.60
• Cerebrospinal fluid (fluid associated with nervous system)	7.4
• Pancreatic juice (digestive juice of the pancreas)	7.1–8.2
• Bile (liver secretion that aids fat digestion)	7.6–8.6
Milk of magnesia	10.5
Lye (sodium hydroxide)	14.0

• Denotes substances in the human body.

can act as a weak base. Hence, this buffer system can compensate for either an excess or a shortage of H^+. For example, if there is an excess of H^+ (an acidic condition), HCO_3^- can function as a weak base and remove the excess H^+, as follows:

$$\underset{\text{Hydrogen ion}}{H^+} + \underset{\text{Bicarbonate ion (weak base)}}{HCO_3^-} \longrightarrow \underset{\text{Carbonic acid}}{H_2CO_3}$$

If there is a shortage of H^+ (an alkaline condition), by contrast, H_2CO_3 can function as a weak acid and provide needed H^+ as follows:

$$\underset{\substack{\text{Carbonic acid}\\\text{(weak acid)}}}{H_2CO_3} \longrightarrow \underset{\text{Hydrogen ion}}{H^+} + \underset{\text{Bicarbonate ion}}{HCO_3^-}$$

Checkpoint

13. How do inorganic compounds differ from organic compounds?
14. What functions does water perform in the body?
15. What is the difference between a solvent and a solute?
16. How do hydrophilic molecules differ from hydrophobic molecules?
17. What is hydrolysis?
18. Which pH is more acidic, 6.82 or 6.91?
19. Why are buffers important in maintaining homeostasis?

Return to Eugene's Story

As the paramedics followed Linda to bed two, Dr. Kim gave instructions. "Linda, get an IV started right away. I think I'd like to get some glucose and B-complex vitamins into Eugene right away. We'll need ABGs (arterial blood gases), too. With his history, Eugene's acid–base balance may be off, especially with the alcohol consumption. Let's get a head and neck CT scan. He may have fallen and hit his head."

Linda helped the paramedics transfer Eugene to the bed. She then began checking Eugene's vital signs: pulse rate 110/min, blood pressure 135/80, respirations 25/min, and blood oxygen saturation 97%.

It didn't take long for the lab results to return. Dr. Kim quickly scanned the results.

"His electrolytes are all off. Sodium, potassium, magnesium, and phosphorus levels are all below normal. Linda, keep an eye on his heart monitor and be alert for any sign of seizure activity. Electrolyte disturbance is common in chronic alcoholics, especially when there is damage to the liver. Eugene was in with a bout of pancreatitis a couple of months ago. Might be another flare up. His arterial blood gases look okay, but he is a bit acidotic."

"Linda, let's get some bicarbonate into Eugene; his blood pH is 7.28. It may be from the acute alcohol intoxication, but there may be something else going on.

A. What are electrolytes and why is the doctor concerned about them with respect to Eugene's heart and brain?

B. Why would administering bicarbonate to Eugene be advisable if his blood pH was too low? What is the bicarbonate going to do to adjust Eugene's blood pH?

C. Compile the information you currently know about Eugene. What are his vitals? What history do you have? What exam results do you have so far? What further information would help you determine if Eugene's loss of consciousness was from a chemical imbalance or something else?

2.5 Organic molecules are large carbon-based molecules that carry out complex functions in living systems.

Inorganic compounds are relatively simple. Their molecules have only a few atoms and cannot be used by cells to perform complicated biological functions. Many organic molecules, by contrast, are relatively large and have unique characteristics that allow them to carry out complex functions. Important categories of organic compounds include carbohydrates, lipids, proteins, nucleic acids, and adenosine triphosphate (ATP).

Organic compounds always contain carbon. Carbon has several properties that make it particularly useful to living organisms. It can form bonds with one to thousands of other carbon atoms to produce large molecules that can have many different shapes. Due to this property of carbon, the body can build many different organic compounds, each of which has a unique structure and function. Moreover, the large size of most carbon-containing molecules and the fact that some do not dissolve easily in water make them useful materials for building body structures.

Organic compounds are usually held together by covalent bonds. Carbon has four electrons in its outermost (valence) shell. It can bond covalently with a variety of atoms, including other carbon atoms, to form rings and straight or branched chains. Other elements that most often bond with carbon in organic compounds are hydrogen, oxygen, nitrogen, sulfur, and phosphorus.

Because organic molecules often are big, there are shorthand methods for representing their structural formulas. Figure 2.13 shows two ways to indicate the structure of the sugar glucose, a molecule with a ring-shaped chain of carbon atoms that has several hydroxyl groups attached.

Small organic molecules can combine into very large molecules called **macromolecules** (*macro-* = large). Macromolecules are usually **polymers** (*poly-* = many; *-mers* = parts). A polymer is a large molecule formed by the covalent bonding of many identical or similar small building-block molecules called **monomers** (*mono-* = one). Usually, the reaction that joins two monomers is a dehydration synthesis. In this type of reaction, a hydrogen atom is removed from one monomer and a hydroxyl group is removed from the other to form a molecule of water (see Figure 2.14). Macromolecules such as carbohydrates, lipids, proteins, and nucleic acids are assembled in cells via dehydration synthesis reactions.

Figure 2.13   Alternative ways to write the structural formula for glucose.

All atoms written out   Standard shorthand

🔑 **In standard shorthand, carbon atoms are understood to be at locations where two bond lines intersect, and single hydrogen atoms are not indicated.**

Checkpoint

20.   Why is carbon an essential component of all organic molecules?
21.   How do polymers and monomers differ?

2.6 Carbohydrates function as building blocks and sources of energy.

Carbohydrates include sugars, glycogen, starches, and cellulose. Even though they are a large and diverse group of organic compounds and have several functions, carbohydrates represent only 2–3% of your total body mass. In humans and animals, carbohydrates function mainly as a source of chemical energy for generating ATP needed to drive metabolic reactions. Only a few carbohydrates are used for building structural units. One example is deoxyribose, a type of sugar that is a building block of deoxyribonucleic acid (DNA), the molecule that carries inherited genetic information.

Carbon, hydrogen, and oxygen are elements found in carbohydrates. Although there are exceptions, carbohydrates generally contain one water molecule for each carbon atom. This is the reason they are called carbohydrates, which means "watered carbon." The three major groups of carbohydrates, based on their size, are monosaccharides, disaccharides, and polysaccharides (Table 2.4).

TABLE 2.4

Major Carbohydrate Groups

TYPE OF CARBOHYDRATE	EXAMPLES
Monosaccharides (Simple sugars that contain from 3 to 7 carbon atoms.)	Glucose (the main blood sugar). Fructose (found in fruits). Galactose (in milk sugar). Deoxyribose (in DNA). Ribose (in RNA).
Disaccharides (Simple sugars formed from the combination of two monosaccharides by dehydration synthesis.)	Sucrose (table sugar) = glucose + fructose. Lactose (milk sugar) = glucose + galactose. Maltose = glucose + glucose.
Polysaccharides (From tens to hundreds of monosaccharides joined by dehydration synthesis.)	Glycogen (the stored form of carbohydrates in animals). Starch (the stored form of carbohydrates in plants and the main carbohydrate in food). Cellulose (part of cell walls in plants that cannot be digested by humans but aids movement of food through the intestines).

Figure 2.14 **Structural and molecular formulas for the monosaccharides glucose and fructose, and the disaccharide sucrose.** In dehydration synthesis (read from left to right), two smaller molecules, glucose and fructose, are joined to form a larger molecule of sucrose. Note the loss of a water molecule. In hydrolysis (read from right to left), the addition of a water molecule to the larger sucrose molecule breaks the disaccharide into two smaller molecules, glucose and fructose.

Glucose ($C_6H_{12}O_6$)   Fructose ($C_6H_{12}O_6$)

Dehydration synthesis

Hydrolysis

Sucrose ($C_{12}H_{22}O_{11}$)   Water

🔑 **Monosaccharides are the monomers used to build carbohydrates.**

Monosaccharides and Disaccharides: The Simple Sugars

Monosaccharides and disaccharides are known as **simple sugars**. The monomers of carbohydrates, **monosaccharides** (mon´-ō-SAK-a-rīds; *sacchar-* = sugar), contain from three to seven carbon atoms. They are designated by names ending in *-ose*. In your body, the principal function of the monosaccharide glucose is to serve as a source of chemical energy for generating the ATP that fuels metabolic reactions. Ribose and deoxyribose are monosaccharides used to make ribonucleic acid (RNA) and deoxyribonucleic acid (DNA), which are described later in the chapter.

A **disaccharide** (dī-SAK-a-rīd; *di-* = two) is a molecule formed from the combination of two monosaccharides by dehydration synthesis. For example, molecules of the monosaccharides glucose and fructose combine to form a molecule of the disaccharide sucrose (table sugar), as shown in Figure 2.14. Notice that the formula for sucrose is $C_{12}H_{22}O_{11}$, not $C_{12}H_{24}O_{12}$, because a molecule of water is removed as the two monosaccharides are joined.

Disaccharides can also be split into smaller, simpler molecules by hydrolysis. A molecule of sucrose, for example, may be hydrolyzed into its components, glucose and fructose, by the addition of water. Figure 2.14 also illustrates this reaction.

Polysaccharides

The third major group of carbohydrates is the **polysaccharides** (pol´-ē-SAK-a-rīds). Each polysaccharide molecule contains tens or hundreds of monosaccharides joined through dehydration synthesis reactions. Unlike simple sugars, polysaccharides usually are insoluble in water and do not taste sweet. The main polysaccharide in the human body is **glycogen**, which is made entirely of glucose monomers linked to one another in branching chains (Figure 2.15). Glycogen is stored in cells of the liver and in skeletal muscles. **Starches** are polysaccharides formed from glucose by plants. They are found in foods such as pasta and potatoes and are the major carbohydrates in the diet. Like disaccharides, poly-

saccharides such as glycogen and starches can be broken down into monosaccharides through hydrolysis reactions. For example, when the blood glucose level falls, liver cells break down glycogen into glucose and release it into the blood, making it available to body cells, which break glucose down to synthesize ATP. **Cellulose** is a polysaccharide formed from glucose by plants that cannot be digested by humans; it provides bulk to help eliminate feces.

Figure 2.15 **Part of a glycogen molecule, the main polysaccharide in the human body.**

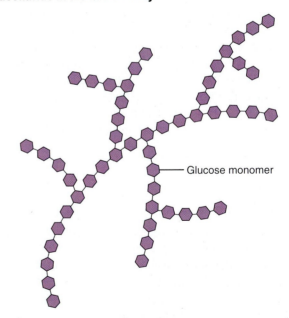

Glucose monomer

🔑 **Glycogen is made up of glucose monomers and is the stored form of carbohydrate in the human body.**

Checkpoint

22. What is the primary function of carbohydrates?
23. What are the structural differences among monosaccharides, disaccharides, and polysaccharides?
24. Which body cells store glycogen?

2.7   Lipids are important for cell membrane structure, energy storage, and hormone production.

A second important group of organic compounds is **lipids** (*lip-* = fat). Lipids make up 18–25% of body mass in lean adults. Like carbohydrates, lipids contain carbon, hydrogen, and oxygen. The proportion of electronegative oxygen atoms in lipids is usually smaller than in carbohydrates, so there are fewer polar covalent bonds. As a result, most lipids are insoluble in polar solvents such as water; they are *hydrophobic*.

The diverse lipid family includes fatty acids, triglycerides (fats and oils), phospholipids (lipids that contain phosphorus), steroids (lipids that contain rings of carbon atoms), and fat-soluble vitamins (vitamins A, D, E, and K). Table 2.5 introduces the various types of lipids and highlights their roles in the human body.

Fatty Acids

Among the simplest lipids are the **fatty acids**, which are used to synthesize triglycerides and phospholipids. Fatty acids can be either saturated or unsaturated. A **saturated fatty acid** contains only *single covalent bonds* between carbon atoms. Because they lack double bonds, each carbon atom is *saturated with hydrogen atoms* (see, for example, palmitic acid in Figure 2.16a). An **unsaturated fatty acid** contains one or more *double covalent bonds* between carbon atoms. Thus, the fatty acid is not completely saturated with hydrogen atoms (see, for example, oleic acid in Figure 2.16a). The unsaturated fatty acid has a kink (bend) at the site of the double bond. If the fatty acid has just one double bond in the hydrocarbon chain, it is *monounsaturated* and it has just one kink. If a fatty acid has more than one double bond in the hydrocarbon chain, it is *polyunsaturated* and it contains more than one kink.

Triglycerides

The most plentiful lipids in your body and in your diet are the **triglycerides** (trī-GLI-cer-īds; *tri-* = three). A triglyceride consists of two types of building blocks, a single glycerol molecule and three fatty acid molecules. The three-carbon **glycerol** molecule forms the "backbone" of a triglyceride (Figure 2.16b,c). Three fatty acids are attached by dehydration synthesis reactions, one to each carbon of the glycerol backbone. The reverse reaction, hydrolysis, breaks down a single molecule of a triglyceride into three fatty acids and glycerol.

TABLE 2.5	
Types of Lipids in the Body	
TYPE OF LIPID	FUNCTIONS
Fatty acids	Used to synthesize triglycerides and phospholipids or catabolized to generate adenosine triphosphate (ATP).
Triglycerides (fats and oils)	Protection, insulation, energy storage.
Phospholipids	Major lipid component of cell membranes.
Steroids	
Cholesterol	Minor component of all animal cell membranes; precursor of bile salts, vitamin D, and steroid hormones.
Bile salts	Needed for digestion and absorption of dietary lipids.
Vitamin D	Helps regulate calcium levels in the body; needed for bone growth and repair.
Adrenocortical hormones	Help regulate metabolism, resistance to stress, and salt and water balance.
Sex hormones	Stimulate reproductive functions and sexual characteristics.
Eicosanoids (prostaglandins and leukotrienes)	Have diverse effects on modifying responses to hormones, blood clotting, inflammation, immunity, stomach acid secretion, airway diameter, lipid breakdown, and smooth muscle contraction.
Other lipids	
Carotenes	Needed for synthesis of vitamin A, which is used to make visual pigments in the eyes. Also function as antioxidants.
Vitamin E	Promotes wound healing, prevents tissue scarring, contributes to the normal structure and function of the nervous system, and functions as an antioxidant.
Vitamin K	Required for synthesis of blood-clotting proteins.
Lipoproteins	Transport lipids in the blood, carry triglycerides and cholesterol to tissues, and remove excess cholesterol from the blood.

Triglycerides can be either solids or liquids at room temperature. A **fat** is a triglyceride that is a solid at room temperature. The fatty acids of a fat are mostly saturated. Because these saturated fatty acids lack double bonds in their hydrocarbon chains, they can closely pack together and solidify at room temperature. A fat that mainly consists of saturated fatty acids is called a **saturated fat**. Although saturated fats occur mostly in meats (especially red meats) and nonskim dairy products (whole milk, cheese, and butter), they are also found in a few plant products, such as cocoa butter, palm oil, and coconut oil. Diets that contain large amounts of saturated fats are associated with disorders such as heart disease and colorectal cancer.

An **oil** is a triglyceride that is a liquid at room temperature. The fatty acids of an oil are mostly unsaturated. Recall that unsaturated fatty acids contain one or more double bonds in their hydrocarbon chains. The kinks at the sites of the double bonds prevent the unsaturated fatty acids of an oil from closely packing together and solidifying. The fatty acids of an oil can be either monounsaturated or polyunsaturated. **Monounsaturated fats** contain triglycerides that mostly consist of monounsaturated fatty acids. Olive oil, peanut oil, canola oil, most nuts, and avocados are rich in triglycerides with monounsaturated fatty acids.

Polyunsaturated fats contain triglycerides that mostly consist of polyunsaturated fatty acids. Corn oil, safflower oil, sunflower oil, soybean oil, and fatty fish (salmon, tuna, and mackerel) contain a high percentage of polyunsaturated fatty acids. Both monounsaturated and polyunsaturated fats are believed to decrease the risk of heart disease.

Triglycerides are the body's most highly concentrated form of chemical energy. Triglycerides provide more than twice as much energy per gram as do carbohydrates and proteins. Our capacity to store triglycerides in adipose (fat) tissue is unlimited for all practical purposes. Excess dietary carbohydrates, proteins, fats, and oils all have the same fate: They are deposited in adipose tissue as triglycerides.

Figure 2.16 **Fatty acid structure and triglyceride synthesis.** Shown in (a) are the structures of a saturated fatty acid and an unsaturated fatty acid. Each time a glycerol and a fatty acid are joined in dehydration synthesis (b), a molecule of water is removed. The fatty acids vary in length and in the number and location of double bonds between carbon atoms (C=C). Shown here (c) is a triglyceride molecule that contains two saturated fatty acids and a monounsaturated fatty acid. The kink (bend) in the oleic acid occurs at the double bond.

(a) Structures of saturated and unsaturated fatty acids

Palmitic acid ($C_{15}H_{31}COOH$) (Saturated)

Oleic acid ($C_{17}H_{33}COOH$) (Monounsaturated)

(b) Dehydration synthesis involving glycerol and a fatty acid

Glycerol molecule

Palmitic acid ($C_{15}H_{31}COOH$)

Fatty acid molecule

H_2O

Ester linkage

(c) Triglyceride (fat) molecule

Palmitic acid ($C_{15}H_{31}COOH$) + H_2O (Saturated)

Stearic acid ($C_{17}H_{35}COOH$) + H_2O (Saturated)

Oleic acid ($C_{17}H_{33}COOH$) + H_2O (Monounsaturated)

One glycerol and three fatty acids are the monomers of triglycerides.

Phospholipids

Like triglycerides, **phospholipids** have a glycerol backbone and two fatty acid chains attached to the first two carbons (Figure 2.17a). Attached to the third carbon, however, a phosphate group (PO_4^{3-}) links a small charged group to the backbone of the molecule. This portion of the molecule (the "head") is polar and can form hydrogen bonds with water molecules (Figure 2.17b). The two fatty acids (the "tails"), by contrast, are nonpolar and can interact only with other lipids. Phospholipids line up tails to tails in a double row to make up much of the membrane that surrounds each cell (Figure 2.17c).

Figure 2.17   Phospholipids. (a) In the synthesis of phospholipids, two fatty acids attach to the first two carbons of the glycerol backbone. A phosphate group links a small charged group to the third carbon in glycerol. In (b), the circle represents the polar head region, and the two wavy lines represent the two nonpolar tails. Double bonds in the fatty acid hydrocarbon chain often form kinks in the tail.

(a) Chemical structure of a phospholipid

(b) Simplified way to draw a phospholipid

(c) Arrangement of phospholipids in a portion of a cell membrane

🔑 **Phospholipids are the main lipids in cell membranes.**

Steroids

The structure of **steroids** differs considerably from that of triglycerides. Steroids have four rings of carbon atoms (colored gold in Figure 2.18). Cholesterol is the steroid from which other steroids are synthesized by body cells (Figure 2.18a). In the body, the commonly encountered steroids are cholesterol, estrogens, testosterone, cortisol, bile salts, and vitamin D. Cholesterol is needed for cell membrane structure; estrogens and testosterone are required for regulating sexual functions; cortisol is necessary for maintaining normal blood sugar levels; bile salts are needed for lipid digestion and absorption; and vitamin D is related to bone growth.

Checkpoint

25. Why are most lipids insoluble in water?
26. How do saturated, monounsaturated, and polyunsaturated fats differ in structure? What are dietary sources of each type of triglyceride?
27. How does a phospholipid differ structurally from a triglyceride? Which portion of a phospholipid is hydrophilic, and which portion is hydrophobic?
28. What is the importance to the body of triglycerides, phospholipids, and steroids?

Figure 2.18   Steroids. All steroids have four rings of carbon atoms.

(a) Cholesterol

(b) Estradiol (an estrogen or female sex hormone)

(c) Testosterone (a male sex hormone)

(d) Cortisol

🔑 **Cholesterol is the starting material for synthesis of other steroids in the body.**

2.8 Proteins are amino acid complexes serving many diverse roles.

Proteins are large molecules that contain carbon, hydrogen, oxygen, and nitrogen. Some proteins also contain sulfur. A normal, lean adult body is 12–18% protein. Much more complex in structure than carbohydrates or lipids, proteins have many roles in the body and are largely responsible for the structure of body tissues. Enzymes are proteins that speed up most biochemical reactions. Other proteins work as "motors" to drive muscle contraction. Antibodies are proteins that defend against invading microbes. Some hormones that regulate homeostasis also are proteins. Table 2.6 describes several important functions of proteins.

Amino Acids and Polypeptides

The monomers of proteins are **amino acids** (a-MĒ-nō). Each of the 20 different amino acids has three important functional groups attached to a central carbon atom (Figure 2.19a): (1) an amino group ($—NH_2$), (2) an acidic carboxyl group (—COOH), and (3) a side chain (R group). At the normal pH of body fluids, both the amino group and the carboxyl group are ionized (Figure 2.19b). The different side chains give each amino acid its distinctive chemical identity (Figure 2.19c).

Figure 2.19 **Amino acids.** (a) In keeping with their name, amino acids have an amino group (shaded blue) and a carboxyl (acid) group (shaded red). The side chain (R group) is shaded gold and is different in each type of amino acid. (b) At pH close to 7, both the amino group and the carboxyl group are ionized. (c) Glycine is the simplest amino acid; the side chain is a single H atom. Cysteine is one of two amino acids that contain sulfur (S). The side chain in tyrosine contains a six-carbon ring. Lysine has a second amino group at the end of its side chain.

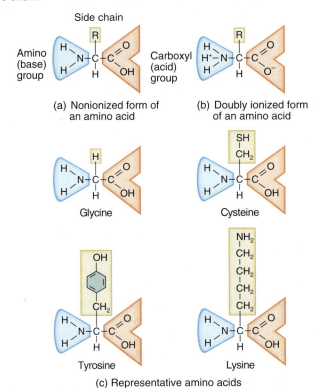

(a) Nonionized form of an amino acid

(b) Doubly ionized form of an amino acid

(c) Representative amino acids

🔑 **Body proteins contain 20 different amino acids, each of which has a unique side chain.**

TABLE 2.6

Functions of Proteins

TYPE OF PROTEIN	FUNCTIONS
Structural	Form structural framework of various parts of the body. *Examples:* Collagen in bone and other connective tissues, and keratin in skin, hair, and fingernails.
Regulatory	Function as hormones that regulate various physiological processes; control growth and development; as neurotransmitters, mediate responses of the nervous system. *Examples:* The hormone insulin, which regulates blood glucose level, and a neurotransmitter known as substance P, which mediates sensation of pain in the nervous system.
Contractile	Allow shortening of muscle cells, which produces movement. *Examples:* Myosin and actin.
Immunological	Aid responses that protect body against foreign substances and invading pathogens. *Examples:* Antibodies and interleukins.
Transport	Carry vital substances throughout body. *Example:* Hemoglobin, which transports most oxygen and some carbon dioxide in the blood.
Catalytic	Act as enzymes that regulate biochemical reactions. *Examples:* Salivary amylase, sucrase, and ATPase.

A protein is synthesized in stepwise fashion—one amino acid is joined to a second, a third is then added to the first two, and so on. The covalent bond joining each pair of amino acids is a **peptide bond**. It always forms between the carbon of the carboxyl group (—COOH) of one amino acid and the nitrogen of the amino group ($—NH_2$) of another. As the peptide bond is formed, a molecule of water is removed (Figure 2.20), making this a dehydration synthesis reaction. Breaking a peptide bond, as occurs during digestion of dietary proteins, is a hydrolysis reaction (Figure 2.20).

When two amino acids combine, a **dipeptide** results. Adding another amino acid to a dipeptide produces a **tripeptide**. Further additions of amino acids result in the formation of a chainlike **peptide** (4–9 amino acids) or **polypeptide** (10–2000 or more amino acids). Proteins are polypeptides that contain as few as 50 to thousands of amino acids.

Figure 2.20   Formation of a peptide bond between two amino acids during dehydration synthesis. When two amino acids are chemically united by dehydration synthesis (read from left to right), the resulting covalent bond between them is called a peptide bond. The peptide bond is formed at the point where water is lost. Here, the amino acids glycine and alanine are joined to form the dipeptide glycylalanine. Breaking a peptide bond occurs by hydrolysis (read from right to left).

🔑 **Amino acids are the monomers used to build proteins.**

Because each variation in the number or sequence of amino acids can produce a different protein, a great variety of proteins is possible. The situation is similar to using an alphabet of 20 letters to form words. Each different amino acid is like a letter, and their various combinations give rise to a seemingly endless diversity of words (peptides, polypeptides, and proteins).

Levels of Structural Organization in Proteins

Proteins exhibit four levels of structural organization. The **primary structure** is the unique sequence of amino acids that are linked by covalent peptide bonds to form a polypeptide chain (Figure 2.21a).

The **secondary structure** of a protein is the repeated twisting or folding of neighboring amino acids in the polypeptide chain (Figure 2.21b). The secondary structure of a protein is stabilized by hydrogen bonds, which form at regular intervals along the polypeptide backbone.

The **tertiary structure** (TUR-shē- er´-ē) refers to the three-dimensional shape of a polypeptide chain. Each protein has a unique tertiary structure that determines how it will function. The tertiary folding pattern may allow amino acids at opposite ends of the chain to be close neighbors (Figure 2.21c). Several types of bonds can contribute to a protein's tertiary structure, including S—S covalent bonds called *disulfide bridges*, hydrogen bonds, ionic bonds, and hydrophobic interactions. Some parts of a polypeptide are attracted to water (hydrophilic), and other parts are repelled by it (hydrophobic). Because most proteins in our body exist in watery surroundings, the folding process places most amino acids with hydrophobic side chains in the central core, away from the protein's surface.

In those proteins that contain more than one polypeptide chain (not all of them do), the arrangement of the individual polypeptide chains relative to one another is the **quaternary structure** (KWA-ter-ner´-ē; Figure 2.21d). The bonds that hold polypeptide chains together are similar to those that maintain the tertiary structure.

Different proteins have different architectures and different three-dimensional shapes. A protein's unique shape permits it to interact with other molecules to carry out a specific function. In practically every case, the function of a protein depends on its ability to recognize and bind to some other molecule. Thus, a hormone binds to a specific protein on a cell in order to alter its function, and an antibody protein binds to a foreign substance (antigen) that has invaded the body.

Homeostatic mechanisms maintain the temperature and chemical composition of body fluids, which allow body proteins to keep their proper three-dimensional shapes. If a protein encounters an altered environment, it may unravel and lose its characteristic shape (secondary, tertiary, and quaternary structure). This process is called **denaturation**. Denatured proteins are no longer functional. Although in some cases denaturation can be reversed, a frying egg is a common example of permanent denaturation. In a raw egg the soluble egg-white protein (albumin) is a clear, viscous fluid. When heat is applied to the egg, however, the protein denatures, becomes insoluble, and turns white. Just as we are not able to "unfry" an egg, proteins can no longer revert back to their original shapes after they have been permanently denatured.

Enzymes

In living cells, most catalysts are protein molecules called **enzymes** (EN-zīms). The names of enzymes usually end in the suffix *-ase*. All enzymes can be grouped according to the types of chemical reactions they catalyze. For example, *oxidases* add oxygen, *kinases* add phosphate, *dehydrogenases* remove hydrogen, *ATPases* split ATP, *anhydrases* remove water, *proteases* break down proteins, and *lipases* break down triglycerides.

Enzymes catalyze specific reactions. They do so with great efficiency and with many built-in controls. Three important properties of enzymes are as follows:

1. **Enzymes are highly specific**. Each particular enzyme binds only to specific **substrates**—the reactant molecules on which the enzyme acts. Of the more than 1000 known enzymes in your body, each has a characteristic three-dimensional shape with a specific surface configuration, which allows it to recog-

Figure 2.21   Levels of structural organization in proteins. (a) The primary structure is the sequence of amino acids in the polypeptide. (b) Common secondary structures include alpha helixes and beta pleated sheets. For simplicity, the amino acid side groups are not shown here. (c) The tertiary structure is the overall folding pattern that produces a distinctive, three-dimensional shape. (d) The quaternary structure in a protein is the arrangement of two or more polypeptide chains relative to one another.

Amino acids

Peptide bond

Polypeptide chain

(a) Primary structure
(amino acid sequence)

Hydrogen bond

Alpha helix

(b) Secondary structure
(twisting and folding of neighboring amino acids, stabilized by hydrogen bonds)

Beta pleated sheet

(c) Tertiary structure
(three-dimensional shape of polypeptide chain)

(d) Quaternary structure
(arrangement of two or more polypeptide chains)

🔑 **The unique shape of each protein permits it to carry out specific functions.**

nize and bind to certain substrates. In some cases, the part of the enzyme that catalyzes the reaction, called the **active site**, is thought to fit the substrate like a key fits in a lock. In other cases the active site changes its shape to fit snugly around the substrate once the substrate enters the active site.

Not only is an enzyme matched to a particular substrate, it also catalyzes a specific reaction. From among the large number of diverse molecules in a cell, an enzyme must recognize the correct substrate and then take it apart or merge it with another substrate to form one or more specific products.

2. Enzymes are efficient. Under optimal conditions, enzymes can catalyze reactions at rates that are from 100 million to 10 billion times more rapid than those of similar reactions occurring without enzymes. The number of substrate molecules that a single enzyme molecule can convert to product molecules in one second is generally between 1 and 10,000 and can be as high as 600,000.

3. Enzymes are subject to cellular controls. The rate of synthesis and the concentration of enzymes at any given time are under the control of a cell's genes. Substances within the cell may either enhance or inhibit the activity of a given enzyme. Many enzymes have both active and inactive forms in cells. The rate at which the inactive form becomes active or vice versa is determined by the chemical environment inside the cell.

Enzymes lower the activation energy of a chemical reaction by decreasing the "randomness" of the collisions between molecules. They also help bring the substrates together in the proper orientation so that the reaction can occur. Figure 2.22 illustrates how an enzyme works:

1. The substrates make contact with the active site on the surface of the enzyme molecule, forming a temporary compound called the **enzyme–substrate complex**. In this reaction, the two substrate molecules are the disaccharide sucrose and water.

2. The substrate molecules are transformed by the rearrangement of existing atoms, the breakdown of the substrate molecule, or the combination of several substrate molecules into the products of the reaction. Here the products are two monosaccharides: glucose and fructose.

3. After the reaction is completed and the reaction products move away from the enzyme, the unchanged enzyme is free to attach to other substrate molecules.

Figure 2.22   How an enzyme works.

1. Enzyme and substrate come together at active site of enzyme, forming an enzyme–substrate complex

2. Enzyme catalyzes reaction and transforms substrate into products

3. When reaction is complete, enzyme is unchanged and free to catalyze same reaction again on new substrates

🔑 **An enzyme speeds up a chemical reaction without being altered or consumed.**

Checkpoint

29. How are the components of proteins different from those in carbohydrates or lipids?
30. What is a peptide bond? How many peptide bonds would there be in a tripeptide?
31. What is the difference between a peptide, a polypeptide, and a protein?
32. List the important properties of enzymes.
33. Describe the mechanism by which enzymes are able to speed up the rate of chemical reactions.

Return to Eugene's Story

"Were you able to get the CT scan arranged?" asked Dr. Kim.

Linda responded, "Yes, they're sending up someone to transport him to radiology right now."

"Good," she replied. "Let's try to wake him up again. Give me a hand, Linda. Hey, Eugene, can you hear me? Wake up!" Dr. Kim gently nudged Eugene's arm. Linda shook his other arm.

"Huh? What do you want?" asked Eugene lethargically.

"Eugene, I'm Dr. Kim. Someone found you passed out. How do you feel? Are you in any pain?"

Eugene blinked slowly and replied, "I don't know, my stomach was hurting before, but I don't feel it much now."

Dr. Kim looked at Linda. "Might be his pancreas again. Call radiology and have them also get an ultrasound of the pancreas."

The radiology transport aide came in shortly after Eugene woke up and took him for his tests. Several hours passed before he was returned to the emergency department from radiology. Linda made sure Eugene was comfortable before she checked in with Dr. Kim. "So how did Eugene's tests come back, Dr. Kim?"

"Well, he has an enlarged fatty liver and an inflamed pancreas, but no head trauma. Let's draw some more blood and check his electrolytes again. Linda,

when you get a chance, please call someone from Social Services. We should try to get him into a rehabilitation program. Eugene's going to be in a lot of pain when the alcohol wears off. Plus he'll have the withdrawal to deal with. He's done some serious damage to his body."

Linda sighed as she glanced at Eugene's chart. "Wow, his liver enzymes are really elevated, lipase too, and his blood sugar was really low when he came in. Why is his cholesterol level so high, Dr. Kim? He seems so malnourished."

D. Linda mentioned Eugene's elevated cholesterol. Which kind of lipid is cholesterol? What other molecules related to cholesterol may be affected by an imbalance of cholesterol in Eugene's body? How might this be related to Eugene's low blood sugar?

E. Dr. Kim finds that Eugene's blood lipase levels are elevated; lipase is an enzyme released by the pancreas that assists in breaking down triglycerides. Which type of reaction would lipase normally be catalyzing?

F. Investigate the function of the liver and pancreas. What types of enzymes or organic molecules do these two organs produce? What are the functions of these molecules in the body?

2.9 Nucleic acids contain genetic material and function in protein synthesis.

Nucleic acids (noo-KLĒ-ic), so named because they were first discovered in the nuclei of cells, are huge organic molecules that contain carbon, hydrogen, oxygen, nitrogen, and phosphorus.

Nucleic acids are of two varieties. The first, **deoxyribonucleic acid (DNA)** (dē-ok´-sē-rī-bō-nū-KLĒ-īk), forms the inherited genetic material inside each human cell (Figure 2.23). In humans,

Figure 2.23   DNA molecule. (a) A nucleotide consists of a base, a five-carbon sugar, and a phosphate group. (b) The paired bases project toward the center of the double helix. The structure is stabilized by hydrogen bonds (dotted lines) between each base pair. There are two hydrogen bonds between adenine and thymine and three between cytosine and guanine.

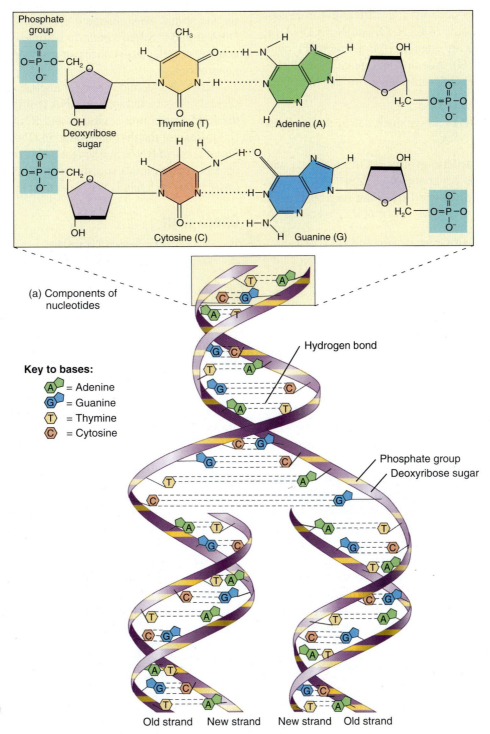

(a) Components of nucleotides

Key to bases:
- A = Adenine
- G = Guanine
- T = Thymine
- C = Cytosine

(b) Portion of a DNA molecule

Old strand New strand   New strand Old strand

Nucleotides are the monomers of nucleic acids.

each **gene** is a segment of a DNA molecule. Our genes determine the traits we inherit, and by controlling protein synthesis, they regulate most of the activities that take place in body cells throughout our lives. When a cell divides, its hereditary information passes on to the next generation of cells. **Ribonucleic acid (RNA)**, the second type of nucleic acid, relays instructions from the genes to guide each cell's synthesis of proteins from amino acids.

A nucleic acid is a chain of repeating monomers called **nucleotides**. Each nucleotide of DNA consists of three parts (Figure 2.23a):

1. *Nitrogenous base.* DNA contains four different nitrogenous bases, which contain atoms of C, H, O, and N. In DNA the four **nitrogenous bases** are adenine (A), thymine (T), cytosine (C), and guanine (G). Adenine and guanine are larger, double-ring bases, whereas thymine and cytosine are smaller, single-ring bases.

2. *Five-carbon sugar.* A five-carbon sugar called **deoxyribose** attaches to each base in DNA.

3. *Phosphate group.* Phosphate groups (PO_4^{3-}) alternate with deoxyribose to form the backbone of each of the two DNA strands; the bases project inward from the backbone chain (Figure 2.23b).

In 1953, F. H. C. Crick of Great Britain and J. D. Watson, a young American scientist, published a brief paper describing how these three components might be arranged in DNA. Their insights into data gathered by others led them to construct a model so elegant and simple that the scientific world immediately knew it was correct! In the Watson–Crick **double helix** model, DNA resembles a spiral ladder (Figure 2.23b). Two strands of alternating phosphate groups and deoxyribose sugars form the uprights of the ladder. Paired bases, held together by hydrogen bonds, form the rungs. Adenine always pairs with thymine, and cytosine always pairs with guanine. Each time DNA is copied, as when living cells divide to increase their number, the two strands unwind. Each strand serves as the template or mold on which to construct a new second strand. Any change that occurs in the base sequence of a DNA strand is called a *mutation*. Some mutations can result in the death of a cell, cause cancer, or produce genetic defects in future generations.

RNA, the second variety of nucleic acid, differs from DNA in several respects. In humans, RNA is single-stranded. The sugar in the RNA nucleotide is **ribose**, and RNA contains the base uracil (U) instead of the thymine found in DNA. Cells contain three different kinds of RNA: messenger RNA, ribosomal RNA, and transfer RNA. Each has a specific role to perform in carrying out the instructions coded in DNA (described in Chapter 3).

Checkpoint

34. What are the functions of DNA and RNA?
35. What are the components of a nucleotide?

2.10 Adenosine triphosphate (ATP) is the principal energy-transferring molecule in living systems.

Adenosine triphosphate (a-DEN-ō-sēn trī-FOS-fāt) or **ATP** is the "energy currency" of living systems. ATP transfers the energy liberated in exergonic catabolic reactions to power cellular activities that require energy (endergonic reactions). Among these cellular activities are muscular contractions, movement of chromosomes during cell division, movement of structures within cells, transport of substances across cell membranes, and synthesis of larger molecules from smaller ones. As its name implies, ATP consists of three phosphate groups attached to adenosine, a unit composed of adenine and the five-carbon sugar ribose (Figure 2.24).

When a water molecule is added to ATP, the third phosphate group (PO_4^{3-}), symbolized by Ⓟ in the following discussion, is removed, and the overall reaction liberates energy. The enzyme that catalyzes the hydrolysis of ATP is called *ATPase*. Removal of the third phosphate group produces a molecule called **adenosine diphosphate (ADP)** in the following reaction:

$$\underset{\substack{\text{Adenosine} \\ \text{triphosphate}}}{\text{ATP}} + \underset{\text{water}}{\text{H}_2\text{O}} \xrightarrow{\text{ATPase}} \underset{\substack{\text{Adenosine} \\ \text{diphosphate}}}{\text{ADP}} + \underset{\substack{\text{Phosphate} \\ \text{group}}}{Ⓟ} + \underset{\text{Energy}}{\text{E}}$$

The energy supplied by the breakdown of ATP into ADP is constantly being used by the cell. As the supply of ATP at any given time is limited, a mechanism exists to replenish it: The enzyme *ATP synthase* catalyzes the addition of a phosphate group to ADP in the following reaction:

$$\underset{\substack{\text{Adenosine} \\ \text{diphosphate}}}{\text{ADP}} + \underset{\substack{\text{Phosphate} \\ \text{group}}}{Ⓟ} + \underset{\text{Energy}}{\text{E}} \xrightarrow{\text{ATP synthase}} \underset{\substack{\text{Adenosine} \\ \text{triphosphate}}}{\text{ATP}} + \underset{\text{water}}{\text{H}_2\text{O}}$$

Figure 2.24 Structures of ATP and ADP. "Squiggles" (~) indicate the two phosphate bonds that can be used to transfer energy. Energy transfer typically involves hydrolysis of the last phosphate bond of ATP.

Adenosine diphosphate (ADP)

Adenosine triphosphate (ATP)

ATP transfers chemical energy to power cellular activities.

As you can see from this reaction, energy is required to produce ATP. The energy needed to attach a phosphate group to ADP is supplied mainly by the catabolism of glucose in a process called cellular respiration, which you will learn more about in Chapters 10 and 23.

Checkpoint

36. What are some cellular activities that depend on energy supplied by ATP?

37. What is the composition of ATP?

Eugene's Story: Epilogue and Discussion

▶▶▶Eugene, a man with a history of chronic alcohol abuse, was brought to the emergency room in an inebriated and unresponsive state. He was found unconscious. Vital signs indicated a pulse rate of 110/min, blood pressure of 135/80, respirations of 25/min, and blood oxygen saturation of 97%. He was placed immediately on IV saline with glucose supplementation. His blood chemistry results indicated that he has hyponatremia (low blood sodium), hypokalemia (low blood potassium), low blood magnesium (hypomagnesemia), and low blood phosphorus levels (hypophosphatemia). His blood pH was found to be acidotic, and bicarbonate was administered to bring his blood pH back to normal limits. No head trauma was found during Dr. Kim's examination. However, Eugene's liver was enlarged (hepatomegaly) and his pancreas was inflamed

(pancreatitis). Further investigation of Eugene's blood chemistries revealed low blood albumin and elevated blood lipase levels, indicating liver and pancreatic damage.

As this story illustrates, damage to the body and its organs and organ systems leads to imbalances in homeostasis at the chemical level of organization. This is an important concept to keep in mind as you progress through your studies and your career. An understanding of the chemical level of organization of the human body is needed to deal with and treat disease appropriately. ◀◀◀

G. *Understanding chemistry is important for truly understanding homeostasis. Provide examples from Eugene's story that illustrate the relationship between chemical homeostasis and homeostasis of organ systems and the human organism.*

Concept and Resource Summary

Concept	Resources

Introduction

  1. The lowest level of structural organization of the body is the chemical level, which consists of atoms that build molecules.
  2. All living and nonliving things consist of **matter**, anything that occupies space and has **mass**.
  3. **Chemistry** is the science of the structure and interactions of matter.

Concept	Resources	

Concept 2.1 Chemical elements are composed of small units called atoms.

1. Matter exists as solids, liquids, and gases.
2. All matter is made up of building blocks called **chemical elements**, which are substances that cannot be split into simpler substances by ordinary chemical means. Of the 117 elements, only four elements— oxygen, carbon, hydrogen, and nitrogen—constitute about 96% of body mass.
3. **Atoms** are the building blocks of elements.
4. Many **subatomic particles** make up an atom. The three most important types include positively charged **protons**, uncharged **neutrons**, and negatively charged **electrons**.
5. The protons and neutrons are located in the atomic core, or **nucleus**; the electrons form a moving cloud around the nucleus in regions referred to as **electron shells**.
6. The atoms of each element have a unique **atomic number**, the number of protons in an atom's nucleus. The **mass number** of an atom is the sum of its protons and neutrons. All atoms of a particular element have the same atomic number but can have different mass numbers due to differences in numbers of **neutrons**; such atoms are called **isotopes**.
7. The mass of an atom is measured in **daltons**. The **atomic mass** (atomic weight) of an element is close to the mass number of its most abundant isotope.
8. Atoms can lose, gain, or share electrons by interacting with other atoms. An **ion** is an atom that has a positive or negative charge due to the loss or gain of electrons.
9. Two or more atoms that share electrons combine to form a **molecule**. A **compound** is a molecule that contains two or more different elements.
10. A **free radical** is an electrically charged atom or group of atoms with an unpaired electron in its outermost shell.

See
- **Animation – Atomic Structure and the Basis of Bonds**

Concept 2.2 Atoms are held together by chemical bonds.

1. **Chemical bonds** are forces that hold together atoms of molecules and compounds. The number of electrons in the **valence** (outermost) **shell** influences formation of chemical bonds.
2. Atoms become charged ions when they gain electrons, resulting in negatively charged **anions**, or when they lose electrons, resulting in positively charged **cations**. Oppositely charged atoms can form an **ionic bond**.
3. An ionic compound that dissociates into positive and negative ions in solution is called an **electrolyte**.
4. A **covalent bond** forms between atoms that share one, two, or three pairs of valence electrons, resulting in a **single covalent bond**, a **double covalent bond**, or a **triple covalent bond**, respectively.
5. Atoms that share electrons equally form a **nonpolar covalent bond**; atoms that share electrons unequally form a **polar covalent bond**.
6. A **hydrogen bond** forms between hydrogen atoms with a partial positive charge and, typically, oxygen or nitrogen atoms with a partial negative charge. Hydrogen bonds between water molecules create the high **surface tension** of water.

See
- **Animation – Chemical Bonding**

Hear
- **Figure 2.6 –** **Hydrogen Bonding among Water Molecules**

Do
- **Exercise – Bond Boulevard**
- **Concepts and Connections – Chemical Bonds**

Concept 2.3 Chemical reactions occur when atoms combine with or separate from other atoms.

1. A **chemical reaction** involves formation of new bonds between atoms or breaking of existing bonds. In chemical reactions, the starting substances are the **reactants** and the ending substances are the **products**.
2. All of the chemical reactions of the body are collectively referred to as **metabolism**.
3. There are two principal forms of **energy**, the capacity to do work: **potential energy** is stored energy, and **kinetic energy** is the energy of motion. **Chemical energy** is the potential energy stored in chemical bonds.
4. A chemical reaction that releases energy is **exergonic**; one that absorbs energy is **endergonic**. These two types of reactions are often coupled. Released energy is transferred to the formation of ATP molecules, which the body can use to drive endergonic reactions and to perform mechanical work.
5. Atoms, ions, and molecules have kinetic energy and continuously move and collide with one another. **Activation energy** is the collision energy needed to break chemical bonds. Temperature and the concentration of the particles influence the chance that a collision will produce a reaction.
6. **Catalysts** are chemical compounds, such as enzymes, that speed up chemical reactions by lowering the activation energy needed for a reaction to occur.
7. Important chemical reactions in the body include **synthesis reactions** (**anabolism**), **decomposition reactions** (**catabolism**), exchange reactions (exchange of the atoms of different molecules among compounds), and **reversible reactions** (in which the products can revert to reactants under the right conditions).

See
- **Animation – Types of Reactions and Equilibrium**

Do
- **Exercise – Reaction Race**

Concept	Resources

Concept 2.4 Inorganic compounds include water, salts, acids, and bases.

1. Compounds can be divided into two principal classes. **Inorganic compounds** are structurally simple and lack carbon; **organic compounds** always contain carbon and are typically large molecules.
2. **Water** is an important and abundant inorganic compound in the body. Its polarity makes it an excellent solvent in body fluids. A **solvent** is a liquid or gas in which the solutes of a **solution** dissolve. Solutes that are charged or contain polar covalent bonds are **hydrophilic** and dissolve in water; molecules with nonpolar covalent bonds are **hydrophobic** and are not very water soluble.
3. **Hydrolysis** reactions in the body involve the breakdown of large nutrient molecules into smaller molecules by the addition of water; **dehydration synthesis reactions** involve the removal of a water molecule as two smaller molecules form a larger molecule.
4. Water can absorb and release a large amount of heat with only a slight change in its own temperature. It also allows evaporative cooling because it requires a large amount of heat to vaporize water from a liquid (in sweat) to a gas.
5. Water is a lubricant in saliva, mucus, and other body fluids.
6. A **mixture** is a combination of elements or compounds. The three common liquid mixtures are solutions, colloids, and suspensions. Solute particles in a **solution** are small and evenly dispersed and do not settle out. A **colloid** has larger solute particles that do not settle out. A **suspension** has large solute particles that do settle out.
7. In water, inorganic acids, bases, and salts separate into ions (**dissociate**). **Acids** dissociate into **hydrogen ions (H^+)** and anions. **Bases** dissociate into **hydroxide ions (OH^-)** and cations. **Salts** dissociate into cations and anions that are not H^+ or OH^-. Salts in the body are electrolytes that carry electrical currents.
8. The acidity or alkalinity of a solution is expressed on a **pH scale** from 0 to 14 that is based on the number of hydrogen ions in a solution. At pH 7, H^+ and OH^- are equal and the solution is neutral. **Acidic solutions** have more H^+ than OH^-, and **basic (alkaline) solutions** have more OH^- than H^+. A change of one whole number on the pH scale represents a 10-fold change in the concentration of H^+ ions.
9. Body fluids have normal pH ranges that must be maintained by homeostatic mechanisms. **Buffers** are chemical compounds that help stabilize the pH of a solution by adding or removing protons (H^+). **Buffer systems** in body fluids convert strong acids or bases into weak acids and bases. The **carbonic acid–bicarbonate buffer system** functions in the homeostasis of blood pH, which has a normal range of 7.35 to 7.45.

Concept 2.5 Organic molecules are large carbon-based molecules that carry out complex functions in living systems.

1. Organic compounds always contain carbon atoms, are often covalently bonded to form long carbon chains, and are bonded to other atoms, particularly hydrogen, oxygen, and nitrogen.
2. Small organic molecules can combine into large **macromolecules** such as **polymers**, which are formed by covalent bonding of identical or similar **monomers**.

Concept 2.6 Carbohydrates function as building blocks and sources of energy.

1. **Carbohydrates** include sugars, starches, glycogen, and cellulose. Carbon, oxygen, and hydrogen are the elements in carbohydrates; the ratio of hydrogen to oxygen atoms is usually 2:1. The three major groups of carbohydrates are monosaccharides, disaccharides, and polysaccharides.
2. **Simple sugars**, **monosaccharides** and **disaccharides**, contain from three to seven carbon atoms per monomer and are the building blocks of carbohydrates.
3. The monosaccharide glucose is a major source of chemical energy for generating ATP. Other monosaccharides include ribose and deoxyribose.
4. Disaccharides such as sucrose are formed by dehydration synthesis from two monosaccharide molecules.
5. **Polysaccharides** (complex carbohydrates) contain many monosaccharides joined by dehydration synthesis reactions. They typically differ from monosaccharides and disaccharides because they don't taste sweet and are insoluble in water. **Glycogen**, which is made of glucose monomers, is stored in liver and muscle cells. Other polysaccharides are **starches** and **cellulose** from plants in the diet.

See

- Anatomy Overview – Common Biomolecules: Water
- Animation – Polarity and Solubility of Molecules
- Animation – Water and Fluid Flow
- Animation – Acids and Bases
- Anatomy Overview – Common Biomolecules: Blood Gases
- Anatomy Overview – Common Biomolecules: Electrolytes

Hear

- Figure 2.20 – Formation of a Peptide Bond between Two Amino Acids during Dehydration Synthesis.

Do

- Exercise – Destination: Acid/Base Balance

Do

- Concepts and Connections – Organic Molecules

See

- Anatomy Overview – Carbohydrates

Concept

Concept 2.7 Lipids are important for cell membrane structure, energy storage, and hormone production.

1. **Lipids** have carbon, hydrogen, and some oxygen. They are insoluble in water. Major lipids include tri-glycerides, phospholipids, steroids, fatty acids, and fat-soluble vitamins (A, D, E, and K). (See Table 2.5.)

2. **Triglycerides** include **fats** and **oils**. They serve as a highly concentrated form of chemical energy in the body. **Glycerol** and three **fatty acids** are the building blocks of triglycerides.

3. **Saturated fats** are usually solid at room temperature and contain only single covalent bonds between fatty acid carbon atoms. **Monounsaturated fats** contain fatty acids with one double covalent bond be-tween two fatty acid carbon atoms. **Polyunsaturated fats** contain more than one double covalent bond between fatty acid carbon atoms.

4. **Phospholipids** are found in all plasma membranes arranged in a double row. They have glycerol, two fatty acid chains, and a polar phosphate group.

5. **Steroids** have four rings of carbon atoms. Body steroids include cholesterol, estrogens, testosterone, cortisol, bile salts, and vitamin D.

See
- **Anatomy Overview - Lipids**

Concept 2.8 Proteins are amino acid complexes serving many diverse roles.

1. **Proteins** are large, complex molecules that contain carbon, hydrogen, oxygen, and nitrogen, and many contain sulfur. Proteins have diverse roles in the body including hormones, catalysts, and antibodies.

2. **Amino acids** are monomers that serve as the building blocks of proteins; there are 20 different amino acids. Each amino acid has a central carbon atom to which the following are attached: (1) an amino group ($-NH_2$); (2) a carboxyl group ($-COOH$), and (3) a side chain (R).

3. Amino acids are bonded by **peptide bonds** during dehydration synthesis, resulting in **dipeptides**, **tri-peptides**, and **polypeptides**.

4. The **primary structure** of a protein is its unique sequence of amino acids forming a polypeptide chain. The **secondary structure** is the twisting or folding of amino acids in the polypeptide chain into alpha helixes and beta pleated sheets. The **tertiary structure** involves further folding that determines the func-tion of the protein. The **quaternary structure** occurs in proteins having two or more polypeptide chains and involves bonds joining the polypeptide chains.

5. When proteins are in certain unfavorable environments, they may unravel and lose their unique shapes and functions in a process called **denaturation**.

6. Most catalysts in body cells are protein molecules called **enzymes**.

7. Enzymes have **specificity**; they can only bind to specific **substrates**. The **active site** of an enzyme is the part that catalyzes the reaction. Another property of enzymes is **efficiency**; they can catalyze reactions at extremely rapid rates. Cellular **controls**, such as genes, affect the rate of synthesis and the concentration of enzymes.

8. Enzymes lower the activation energy of a chemical reaction and work in a three-step process: (1) the **enzyme–substrate complex** forms, (2) substrate molecules are transformed into products, and (3) un-changed enzymes release the reaction products and become free to catalyze the next reaction.

See
- **Anatomy Overview – Proteins**
- **Animation – Enzyme Functions and ATP**

Hear

- **Figure 2.22 – How an Enzyme Works**

Do
- **Exercise – Enzyme Anticipation**

Concept 2.9 Nucleic acids contain genetic material and function in protein synthesis.

1. **Nucleic acids** include **deoxyribonucleic acid (DNA)**, which forms the genetic material of cells, and **ribonucleic acid (RNA)**, which functions in protein synthesis.

2. A nucleic acid is a chain of monomers called **nucleotides**.

3. Each nucleotide in DNA consists of three parts: a **nitrogenous base** (adenine, thymine, cytosine, or gua-nine); the five-carbon sugar **deoxyribose**; and a **phosphate group**. DNA has two strands of nucleotides in a **double helix**.

4. RNA is single stranded. Each of its nucleotides consist of three parts: a **nitrogenous base** (adenine, cyto-sine, guanine, uracil); the five-carbon sugar **ribose**; and a **phosphate group**.

5. There are three types of RNA: messenger RNA, ribosomal RNA, and transfer RNA.

See
- **Anatomy Overview – Nucleic Acids**

Concept	Resources
	See

Concept 2.10 Adenosine triphosphate (ATP) is the principal energy-transferring molecule in living systems.

- Animation – Enzyme Functions and ATP

1. **Adenosine triphosphate (ATP)** is the "energy currency" of cells. ATP transfers the energy liberated in exergonic catabolic reactions to energy-requiring cellular activities, including endergonic reactions and mechanical work.
2. ATP is composed of adenine, ribose, and three phosphate groups.
3. ATPase catalyzes the hydrolysis of ATP to ADP and phosphate, whereas ATP synthase catalyzes the addition of a phosphate group to ADP to form ATP.

Other Chapter Resources

For additional information on clinical applications, medical disorders, and medical terminology, see Chapter 2 in the *Clinical Connections* manual packaged with your text.

Within *WileyPLUS*

- Chapter Quiz
- Visual Anatomy
- Mastering Vocabulary
- Audio Glossary
- Homeostatic Imbalance: The Case of Muscles and More
- Web Links

• Understanding the Concepts

1. Carbon dioxide (CO_2) and nitrogen (N_2) are gases involved in respiratory physiology. Would each of these gases best be described as a molecule or a compound? Explain your answer.

2. Which atom in a water molecule has a greater ability to attract electrons? Why?

3. Will the element potassium (K) be more likely to form an anion or a cation? Why? (Hint: See Figure 2.2 for the atomic structure of K.)

4. What form of energy is found in an apple you just ate? A vase sitting on a high shelf? A vase falling off a high shelf? How is energy conserved in each of these scenarios?

5. Are you creating a solution, colloid, or suspension in each of the following situations? Explain each of your answers.
   - As you stir a teaspoon of table sugar into a glass of iced tea, you can see sugar crystals spinning around undissolved.
   - After stirring a teaspoon of table sugar into a cup of hot tea, you notice that the liquid appears clear, indicating the table sugar crystals have dissolved in the tea.
   - While preparing a salad dressing, the vinegar and oil appear to stay mixed together as long as you keep vigorously stirring, but if you quit, the vinegar separates from the oil and settles to the bottom of the container.

6. The compound $CaCO_3$ (calcium carbonate) dissociates into a calcium ion (Ca^{2+}) and a carbonate ion (CO_3^{2-}). Is it an acid, a base, or a salt? What about H_2SO_4, which dissociates into two H^+ and one SO_4^{2-}? Explain your answers.

7. Do all proteins have a tertiary structure? Do they all have a quaternary structure?

8. Which levels of structural organization are lost when a protein is denatured?

9. Referring to Figure 2.22, why can't sucrase catalyze the formation of sucrose from glucose and fructose?

10. In the reaction catalyzed by ATP synthase, identify the substrates and products. Is this an exergonic or endergonic reaction? Explain your answer.

3

The Cellular Level of Organization

Joseph's Story

The bright summer sun shone down on Joseph and his son Marcus as they began setting up camp for a weekend father–son getaway. Joseph had been looking forward to it for weeks. He'd been busy working long hours and the stress of his job was wearing on his body. He had felt tightness in his chest, he couldn't sleep, his mind raced, and he had begun smoking again after three smoke-free years. He needed this time off. "Dad, let's take a break and play catch for a while," said Marcus.

"Okay, Marc, but take it easy on your old dad." said Joseph. Marcus was only 12 years old and wasn't much of a match for his 38-year-old dad, but liked to think he was. "Pitch that ball in here, slugger," said Joseph as he squatted down to catch the ball. After he threw the ball as hard as he could, Marcus thought his father was acting odd. Joseph grunted; as he reached for the ball, his face contorted and he collapsed on the ground in a heap.

"Dad, come on, knock it off. Quit fooling around and throw me the ball. You didn't even try to catch that one!" Marcus jogged toward Joseph. "Dad, get up." Marcus ran the last few steps to Joseph's side, and saw that his father's face was ashen gray. He wasn't breathing. Marcus panicked and screamed. "Dad! Someone help! Help me! Something's wrong with my dad!"

Most of us tend to think of humans at the organismal level. Who we are is more than simply a collection of chemicals, cells, and organ systems. However, the cellular level is essential to the homeostasis of the entire organism. In Joseph's case, you will see that each cell of his body is fighting a battle for survival.

INTRODUCTION

In Chapter 1, you learned that the human body has various levels of organization, and Chapter 2 helped you understand the chemistry of the human body. In this chapter you will see how atoms and molecules are organized to form the structures comprising cells and to perform cellular activities.

Cells are living structural and functional units enclosed by a membrane. About 200 different types of specialized cells carry out a multitude of unique biochemical or structural roles that support homeostasis and contribute to the many coordinated functional capabilities of the human body. **Cell biology** is the study of cellular structure and function. As you study the various parts of a cell and their relationships to each other, you will learn that cell structure and function are intimately related. In this chapter, you will also learn that cells carry out a dazzling array of chemical reactions to create and maintain life processes—in part, by isolating specific types of chemical reactions within specialized cellular structures.

About 200 different types of specialized cells carry out a multitude of unique biochemical or structural roles that support homeostasis and contribute to the many coordinated functional capabilities of the human body.

3.1   The principal parts of a cell are the plasma membrane, the cytoplasm, and the nucleus.

Figure 3.1 provides an overview of the typical structures found in body cells. Most cells have many of the structures shown in this diagram, but no one cell has all of them. For ease of study, we divide the cell into three main parts: plasma membrane, cytoplasm, and nucleus.

- The **plasma membrane** forms the cell's flexible outer surface, separating the cell's internal environment (inside the cell) from the external environment (outside the cell). This selective barrier, which regulates the flow of materials into and out of a cell, helps establish and maintain the appropriate environment for normal cellular activities. The plasma membrane also plays a key role in communication among cells and between cells and their external environment.

- The **cytoplasm** (SĪ-tō-plasm; -*plasm* = formed or molded) consists of all the cellular contents between the plasma membrane

and the nucleus. **Cytosol** (SĪ-tō-sol), the fluid portion of cytoplasm, contains water, dissolved solutes, and suspended particles. The cytosol is the site of chemical reactions needed to maintain cell structures and allow cell growth. Surrounded by cytosol are several different types of **organelles** (or-ga-NELZ = little organs). Each type of organelle has a characteristic shape and specific functions. Examples include the cytoskeleton, ribosomes, endoplasmic reticulum, Golgi complex, lysosomes, peroxisomes, and mitochondria.

- The **nucleus** (NOO-klē-us = nut kernel) is a large organelle that houses most of a cell's DNA. Within the nucleus, each **chromosome** (*chromo-* = colored), a single molecule of DNA associated with several proteins, contains thousands of hereditary units called **genes** that control most aspects of cellular structure and function.

Figure 3.1   Typical structures found in body cells.

Sectional view

🔑 **The cell is the basic, living structural and functional unit of the body.**

3.2 The plasma membrane contains the cytoplasm and regulates exchanges with the extracellular environment.

The **plasma membrane**, a flexible yet sturdy barrier that surrounds and contains the cytoplasm of a cell, is best described by using a structural model called the *fluid mosaic model*. According to this model, the molecular arrangement of the plasma membrane resembles an ever-moving sea of fluid lipids that contains a mosaic of many different proteins (Figure 3.2). Some proteins float freely like icebergs in the lipid sea; others are anchored at specific locations like boats at a dock. The membrane lipids allow passage of several types of lipid-soluble molecules but act as a barrier to the entry or exit of charged or polar substances. Some of the proteins in the plasma membrane allow movement of polar molecules and ions into and out of the cell. Other proteins act as receptors or link neighboring cells together.

The Lipid Bilayer

The basic structural framework of the plasma membrane is the **lipid bilayer**, two back-to-back layers made up of three types of lipid molecules—primarily phospholipids with smaller amounts of cholesterol and glycolipids (Figure 3.2). The bilayer arrangement occurs because the lipids have both polar and nonpolar parts. The head of each **phospholipid**, a lipid that contains phosphorus (see Figure 2.17), is polar and *hydrophilic* (*hydro-* = water; *-philic* = loving). The two fatty acid "tails" of the phospholipid are nonpolar and *hydrophobic* (*-phobic* = fearing). Because "like seeks like," the phospholipid molecules orient themselves into a bilayer, with their hydrophilic heads facing outward toward watery fluid—cytosol on the inside and extracellular fluid on the outside. The hydrophobic fatty acid tails point toward one another and away from water, forming a nonpolar, hydrophobic core in the membrane's interior.

Cholesterol molecules are interspersed among the other lipids in both layers of the membrane. Cholesterol molecules are steroids with an attached OH (hydroxyl) group. The tiny —OH$^-$ group (see Figure 2.18a), which is the only polar region of cholesterol, forms hydrogen bonds with the polar heads of phospholipids and glycolipids. The majority of cholesterol is nonpolar and fits among the fatty acid tails of the phospholipids and glycolipids.

Figure 3.2   The fluid mosaic arrangement of lipids and proteins in the plasma membrane.

Membranes are fluid structures because the lipids and many of the proteins are free to rotate and move sideways in the bilayer.

Glycolipids are lipids with attached carbohydrate groups that form a polar head; their fatty acid tails are nonpolar. Glycolipids appear only in the membrane layer that faces the extracellular fluid, which is one reason the two sides of the bilayer are asymmetric, or different.

Arrangement of Membrane Proteins

Membrane proteins are classified as integral or peripheral according to whether they are firmly embedded in the plasma membrane (Figure 3.2). **Integral proteins** extend into or through the lipid bilayer among the fatty acid tails and are firmly embedded in it. Most integral proteins are *transmembrane proteins*, which means that they span the entire lipid bilayer and protrude into both the cytosol and extracellular fluid. Like membrane lipids, integral membrane proteins have hydrophilic regions that protrude into either the watery extracellular fluid or the cytosol, and hydrophobic regions that extend among the fatty acid tails.

As their name implies, **peripheral proteins** are not as firmly embedded in the membrane. They associate more loosely with the polar heads of membrane lipids or with integral proteins at the inner or outer surface of the membrane.

Many membrane proteins are **glycoproteins**, proteins with carbohydrate groups attached to the ends that protrude into the extracellular fluid. The carbohydrate portions of glycolipids and glycoproteins form an extensive sugary coat called the **glycocalyx** (glī-kō-KAL-iks). Because the pattern of carbohydrates in the glycocalyx varies from one cell to another, the glycocalyx acts like a molecular "signature" that enables cells to recognize one another. For example, the ability of white blood cells to detect a "foreign" glycocalyx helps our bodies destroy invading organisms during the immune response. In addition, the glycocalyx enables cells to adhere to one another in some tissues and protects cells from being digested by enzymes in the extracellular fluid. The hydrophilic properties of the glycocalyx attract a film of fluid to the surface of many cells. This action makes red blood cells slippery as they flow through narrow blood vessels, and protects cells that line the airways and the gastrointestinal tract from drying out.

Functions of Membrane Proteins

Generally, the types of lipids in cellular membranes vary only slightly. In contrast, the membranes of different cells and various intracellular organelles have remarkably different assortments of proteins that determine many of the membrane's functions (Figure 3.3).

- Some integral membrane proteins form **ion channels**, *pores* or holes through which specific ions, such as potassium ions (K^+), can pass into or out of the cell. Most ion channels are *selective*; they allow only a single type of ion to pass through.

Figure 3.3   Functions of membrane proteins.

☐ Extracellular fluid   ▦ Plasma membrane   ☐ Cytosol

Ion channel (integral)
Allows specific ion (●) to move through water-filled pore. Most plasma membranes include specific channels for several common ions.

Pore

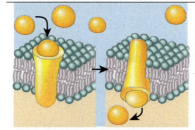

Carrier (integral)
Transports specific substances (●) across membrane by changing shape. For example, amino acids, needed to synthesize new proteins, enter body cells via carriers. Carrier proteins are also known as *transporters*.

Receptor (integral)
Recognizes specific ligand (▽) and alters cell's function in some way. For example, antidiuretic hormone binds to receptors in the kidneys and changes the water permeability of certain plasma membranes.

Ligand

Substrate
Products

Enzyme (integral and peripheral)
Catalyzes reaction inside or outside cell (depending on which direction the active site faces). For example, lactase protruding from epithelial cells lining your small intestine splits the disaccharide lactose in the milk you drink.

Linker (integral and peripheral)
Anchors filaments inside and outside the plasma membrane, providing structural stability and shape for the cell. May also participate in movement of the cell or link two cells together.

MHC protein

Cell-identity marker (glycoprotein)
Distinguishes your cells from anyone else's (unless you are an identical twin). An important class of such markers are the major histocompatibility (MHC) proteins.

🔑 **Membrane proteins largely reflect the functions a cell can perform.**

- Other integral proteins act as **carriers**, which change shape to selectively move a polar substance or ion from one side of the membrane to the other. Carriers are also known as *transporters*.

- Integral proteins called **receptors** serve as cellular recognition sites. Each type of receptor recognizes and binds to a specific type of molecule. For instance, insulin receptors bind to the hormone insulin. A specific molecule that binds to a receptor is called a *ligand* (LĪ-gand; *liga* = tied) of that receptor.

- Some integral proteins are **enzymes** that catalyze specific chemical reactions at the inside or outside surface of the cell.

- Integral proteins may also serve as **linkers**, which anchor proteins in the plasma membranes of neighboring cells to one another or to protein filaments inside and outside the cell. Peripheral proteins also serve as enzymes and linkers.

- Membrane glycoproteins and glycolipids often serve as **cell-identity markers**. They may enable a cell to recognize other cells of the same kind during tissue formation or to recognize and respond to potentially dangerous foreign cells. The ABO blood type markers are one example of cell-identity markers. When you receive a blood transfusion, the blood type must be compatible with your own.

In addition, peripheral proteins help support the plasma membrane, anchor integral proteins, and participate in mechanical activities such as moving materials and organelles within cells, changing cell shape in dividing and contracting cells, and attaching cells to one another.

Membrane Fluidity

Membranes are fluid structures; that is, most of the membrane lipids and many of the membrane proteins easily rotate and move sideways in their own half of the bilayer. Neighboring lipid molecules exchange places about 10 million times per second and may wander completely around a cell in only a few minutes! Membrane fluidity depends on the number of double bonds in the fatty acid tails of the lipids that make up the bilayer. Each double bond puts a "kink" in the fatty acid tail (see Figure 2.17), which increases membrane fluidity by preventing lipid molecules from packing tightly in the membrane.

Membrane fluidity is an excellent compromise for the cell; a rigid membrane would lack mobility, and a completely fluid membrane would lack the structural organization and mechanical support required by the cell. Membrane fluidity allows interactions to occur within the plasma membrane, such as the assembly of membrane proteins. It also enables the movement of the membrane components responsible for cellular processes such as cell movement, growth, division, and secretion, and the formation of cellular junctions. Fluidity allows the lipid bilayer to self-seal if torn or punctured. When a needle is pushed through a plasma membrane and pulled out, the puncture site seals spontaneously, and the cell does not burst. This property of the lipid bilayer allows fertilization of an oocyte (egg cell) by injecting a sperm cell through a tiny syringe to help infertile couples conceive a child. It also permits removal and replacement of a cell's nucleus in cloning experiments, such as the one that created Dolly, the famous cloned sheep.

Despite the great mobility of membrane lipids and proteins in the bilayer, they seldom flip-flop from one half of the bilayer to the other because it is difficult for hydrophilic parts of membrane molecules to pass through the hydrophobic core of the membrane. This difficulty contributes to the asymmetry of the membrane bilayer.

Membrane Permeability

When a structure is referred to as *permeable*, it means that substances are allowed to pass through it; substances are *not* allowed to pass through an *impermeable* structure. The permeability of the plasma membrane to different substances varies. Plasma membranes permit some substances to pass more readily than others, a property called **selective permeability**.

The lipid bilayer portion of the membrane is permeable to nonpolar, uncharged molecules such as oxygen, carbon dioxide, and steroids; impermeable to ions and large, uncharged polar molecules such as glucose; and *slightly* permeable to small, uncharged polar molecules such as water and urea, a waste product from the breakdown of amino acids. Water and urea are thought to pass through the lipid bilayer in the following way. As the fatty acid tails of membrane phospholipids and glycolipids randomly move about, small gaps briefly appear in the hydrophobic environment of the membrane's interior. Water and urea molecules are small enough to move from one gap to another until they succeed in crossing the membrane.

Transmembrane proteins that act as channels and carriers increase the plasma membrane's permeability to a variety of ions and uncharged polar molecules that, unlike water and urea molecules, cannot cross the lipid bilayer unassisted. Channels and carriers are very selective. Each one helps a specific molecule or ion cross the membrane. Macromolecules, such as proteins, are so large that they are unable to pass across the plasma membrane except by endocytosis and exocytosis (discussed later in this chapter).

Gradients across the Plasma Membrane

The selective permeability of the plasma membrane allows a living cell to maintain different concentrations of certain substances on either side of the plasma membrane. A **concentration gradient** is a difference in the concentration of a chemical from one place to another, such as from the cytosol side to the extracellular fluid side of the plasma membrane. For example, oxygen molecules and sodium ions are more concentrated in the extracellular fluid than in the cytosol; the opposite is true of carbon dioxide molecules and potassium ions.

The plasma membrane also creates a difference in the distribution of positively charged ions and negatively charged ions on the two sides of the plasma membrane. Typically, the inner surface of the plasma membrane is more negatively charged and the outer surface is more positively charged. A difference in electrical charges between two regions is called an **electrical gradient**.

As you will see shortly, the concentration gradient and electrical gradient are important because they help move substances across the plasma membrane. In many cases a substance will move across a plasma membrane *down its concentration gradient*. In other words, a substance will move "downhill" to reach equilibrium, from where it is more concentrated to where it is less concentrated. Similarly, a positively charged substance will tend to move toward a negatively charged area, and a negatively charged substance will tend to move toward a positively charged area. The combined influence of the concentration gradient and the electrical gradient on movement of a particular ion is referred to as its **electrochemical gradient**.

Checkpoint

1.   What are the three principal parts of a cell? What are their functions?
2.   How do hydrophobic and hydrophilic regions govern the arrangement of membrane lipids in a bilayer?
3.   What is the glycocalyx?
4.   Which substances can diffuse through the lipid bilayer? Which cannot?
5.   Explain in your own words the concept of selective permeability of the plasma membrane.
6.   What factors contribute to an electrochemical gradient?

3.3   Transport of a substance across the plasma membrane occurs by both passive and active processes.

Substances generally move across cellular membranes via transport processes that can be classified as passive or active, depending on whether they require cellular energy. In **passive processes**, a substance moves down its concentration or electrical gradient to cross the membrane using only its own kinetic energy (energy of motion). Kinetic energy is intrinsic to the particles that are moving; there is no input of energy from the cell. Simple diffusion is an example of a passive process. In **active processes**, cellular energy is used to drive the substance "uphill" against its concentration or electrical gradient; the cellular energy is usually in the form of ATP. Active transport is an example of an active process.

Passive Processes

The Principle of Diffusion

Learning why materials diffuse across membranes requires an understanding of how diffusion occurs in a solution. **Diffusion** (di-FŪ-zhun; *diffus-* = spreading) is a passive process in which the random mixing of particles in a solution occurs because of the particles' kinetic energy. Both the *solutes*, the dissolved substances, and the *solvent*, the liquid that does the dissolving, undergo diffusion. If a particular solute is present in high concentration in one area of a solution and in low concentration in another area, solute molecules will diffuse toward the area of lower concentration— they move *down their concentration gradient*. After some time, the particles become evenly distributed throughout the solution and the solution is said to be at equilibrium. The particles continue to move about randomly due to their kinetic energy, but their concentrations do not change.

For example, when you place a crystal of dye in a water-filled container (Figure 3.4), the color is most intense in the area closest to the dye because its concentration is higher there. At increasing distances, the color is lighter and lighter because the dye concentration is lower. Some time later, the solution of water and dye will have a uniform color, because the dye molecules and water molecules have diffused down their concentration gradients until they are evenly mixed in solution—they are at equilibrium.

Figure 3.4   The principle of diffusion. At the beginning of our experiment, a crystal of dye is placed in a cylinder of water and dissolves (a) and then diffuses from the region of higher dye concentration to regions of lower dye concentration (b). At equilibrium (c), the dye concentration is uniform throughout, although random movement continues.

Beginning (a)   Intermediate (b)   Equilibrium (c)

🔑 **In diffusion, a substance moves down its concentration gradient.**

In this simple example, no membrane was involved. Substances may also diffuse through a membrane if the membrane is permeable to them. Several factors influence the diffusion rate of substances across plasma membranes:

- **Steepness of the concentration gradient.** The greater the difference in concentration between the two sides of the membrane, the higher is the rate of diffusion. When charged particles are diffusing, the steepness of the electrochemical gradient determines the diffusion rate across the membrane.

- **Temperature.** The higher the temperature, the faster is the rate of diffusion. All of the body's diffusion processes occur more rapidly in a person with a fever.

- **Mass of the diffusing substance.** The larger the mass of the diffusing particle, the slower is its diffusion rate. Smaller molecules diffuse more rapidly than larger ones.

- **Surface area.** The larger the membrane surface area available for diffusion, the faster is the diffusion rate. For example, the air sacs of the lungs have a large surface area available for diffusion of oxygen from the air into the blood. Some lung diseases, such as emphysema, reduce the surface area. This slows the rate of oxygen diffusion and makes breathing more difficult.

- **Diffusion distance.** The greater the distance over which diffusion must occur, the longer it takes. Diffusion across a plasma membrane takes only a fraction of a second because the membrane is so thin. In pneumonia, fluid collects in the lungs; the additional fluid increases the diffusion distance because oxygen must move through both the built-up fluid and the membrane to reach the bloodstream.

Now that you have a basic understanding of the nature of diffusion, we will consider three types of diffusion: simple diffusion, facilitated diffusion, and osmosis.

Simple Diffusion

Simple diffusion is a passive process in which substances move freely through the lipid bilayer of the plasma membranes of cells without the help of membrane transport proteins (Figure 3.5). Nonpolar, hydrophobic molecules move across the lipid bilayer through the process of simple diffusion. Such molecules include oxygen, carbon dioxide, and nitrogen gases; fatty acids, steroids, and fat-soluble vitamins (A, D, E, and K); small alcohols; and ammonia. Small, uncharged polar molecules such as water, urea, and small alcohols also pass through the lipid bilayer by simple diffusion. Simple diffusion through the lipid bilayer is important in the movement of oxygen and carbon dioxide between blood and body cells, and between blood and air within the lungs during breathing. It also is the route for absorption of some nutrients and excretion of some wastes by body cells.

Facilitated Diffusion

Solutes that are too polar or highly charged to move through the lipid bilayer by simple diffusion can cross the plasma membrane by a passive process called **facilitated diffusion**. In this process, an integral membrane protein assists a specific substance across the membrane. The integral membrane protein can be either a membrane channel or a carrier.

Figure 3.5   Types of diffusion.

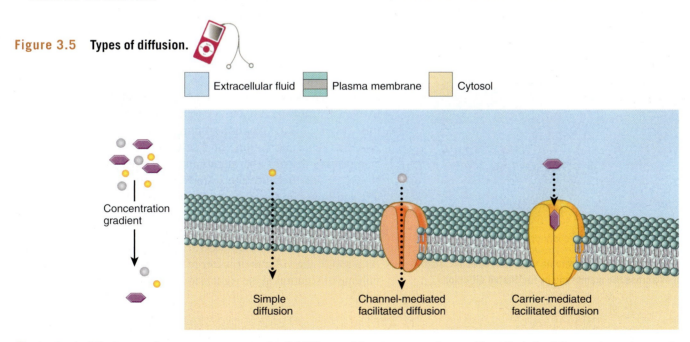

| Extracellular fluid | Plasma membrane | Cytosol |

Concentration gradient

Simple diffusion

Channel-mediated facilitated diffusion

Carrier-mediated facilitated diffusion

🔑 In simple diffusion, a substance moves across the lipid bilayer of the plasma membrane without the help of the membrane transport proteins. In facilitated diffusion, a substance moves across the lipid bilayer aided by a channel protein or a carrier protein.

Channel-Mediated Facilitated Diffusion

In **channel-mediated facilitated diffusion**, a solute moves down its concentration gradient across the lipid bilayer through a membrane channel (Figure 3.5). Most membrane channels are ion channels, integral transmembrane proteins that allow passage of small ions that are too hydrophilic to penetrate the nonpolar interior of the lipid bilayer. Each ion can diffuse across the membrane only at certain sites. In typical plasma membranes, the most numerous ion channels are selective for K^+ (potassium ions) or Cl^- (chloride ions); fewer channels are available for Na^+ (sodium ions) or Ca^{2+} (calcium ions). Diffusion of ions through channels is generally slower than free diffusion through the lipid bilayer because channels occupy a smaller fraction of the membrane's total surface area than lipids. Still, diffusion through channels is a very fast process: More than a million potassium ions can flow through a K^+ channel in one second!

A channel is said to be gated when part of the channel protein acts as a "gate," changing shape in one way to open the pore and in another way to close it (Figure 3.6). Some gated channels randomly alternate between the open and closed positions; others are regulated by chemical or electrical changes inside and outside the cell. When the gates of a channel are open, ions diffuse into or out of cells down their electrochemical gradients. The plasma membranes of different types of cells may have different numbers of ion channels and thus display different permeabilities to various ions.

Carrier-Mediated Facilitated Diffusion

In **carrier-mediated facilitated diffusion**, a carrier is used to move a solute down its concentration gradient across the plasma membrane (see Figure 3.5). The solute binds to a specific carrier on one side of the membrane and is released on the other side after the carrier undergoes a change in shape. The solute binds more often to the carrier on the side of the membrane with a higher concentration of solute. Once the concentration is the same on both sides of the membrane, solute molecules bind to the carrier on the cytosol side and move out to the extracellular fluid as rapidly as they bind to the carrier on the extracellular fluid side and move into the cytosol.

Osmosis

Osmosis (oz-MŌ-sis) is a type of diffusion in which there is net movement of a solvent through a selectively permeable membrane. Like the other types of diffusion, osmosis is a passive process. In living systems, the solvent is water, which moves by osmosis across plasma membranes from an area of *higher water concentration* to an area of *lower water concentration*. Another way to understand this idea is to consider the solute concentration: In osmosis, water moves through a selectively permeable membrane from an area of *lower solute concentration* to an area of *higher solute concentration*. During osmosis, water molecules pass through a plasma membrane in two ways: (1) by moving through the lipid bilayer via simple diffusion and (2) by moving through integral membrane proteins that function as water channels.

Osmosis occurs only when a membrane is permeable to water but is not permeable to certain solutes. A simple experiment can demonstrate osmosis. Consider a U-shaped tube in which a selectively permeable membrane separates the left and right arms of the tube (Figure 3.7). A volume of pure water is poured into the left arm, and the same volume of a solution containing a solute that cannot pass through the membrane is poured into the right arm (Figure 3.7a). Because the water concentration is higher on the left and lower on the right, net movement of water molecules—osmosis—occurs from left to right, so that the water is moving down its concentration gradient. At the same time, the membrane prevents diffusion of the solute from the right arm into the left arm. As a result, the volume of water in the left arm decreases, and the volume of solution in the right arm increases (Figure 3.7b).

Figure 3.6 Channel-mediated facilitated diffusion of potassium ions (K^+) through a gated K^+ channel. A gated channel is one in which a portion of the channel protein acts as a gate to open or close the channel's pore to the passage of ions.

Details of the K^+ channel

Channels are integral membrane proteins that allow specific, small ions to pass across the membrane by facilitated diffusion.

Figure 3.7 **Principle of osmosis.** Water molecules move through the selectively permeable membrane; the solute molecules in the right arm cannot pass through the membrane. (a) As the experiment starts, water molecules move from the left arm into the right arm, down the water concentration gradient. (b) After some time, the volume of water in the left arm has decreased and the volume of solution in the right arm has increased. At equilibrium, there is no net osmosis: Hydrostatic pressure forces just as many water molecules to move from right to left as osmosis forces water molecules to move from left to right. (c) If pressure is applied to the solution in the right arm, the starting conditions can be restored. This pressure, which stops osmosis, is equal to the osmotic pressure.

(a) Starting conditions   (b) Equilibrium   (c) Restoring starting conditions

🔑 **Osmosis is the movement of water molecules through a selectively permeable membrane.**

You might think that osmosis would continue until no water remained on the left side, but this is not what happens. In this experiment, the higher the column of solution in the right arm becomes, the more pressure it exerts on its side of the membrane. Pressure exerted in this way by a liquid, known as **hydrostatic pressure**, forces water molecules to move back into the left arm. Equilibrium is reached when just as many water molecules move from right to left due to the hydrostatic pressure as move from left to right due to osmosis (Figure 3.7b).

To further complicate matters, the solution with the impermeable solute also exerts a force, called the **osmotic pressure**. The osmotic pressure of a solution is proportional to the concentration of the solute particles that cannot cross the membrane—the higher the solute concentration, the higher is the solution's osmotic pressure. Consider what would happen if a piston were used to apply more pressure to the fluid in the right arm of the tube in Figure 3.7. With enough pressure, the volume of fluid in each arm could be restored to the starting volume, and the concentration of solute

in the right arm would be the same as it was at the beginning of the experiment (Figure 3.7c). The amount of pressure needed to restore the starting condition equals the osmotic pressure. So, in our experiment osmotic pressure is the pressure needed to stop the movement of water from the left tube into the right tube. Notice that the osmotic pressure of a solution does not produce the movement of water during osmosis. Rather, it is the pressure that would *prevent* such water movement.

Normally, the osmotic pressure of the cytosol is the same as the osmotic pressure of the interstitial fluid outside cells. Because the osmotic pressure on both sides of the plasma membrane (which is selectively permeable) is the same, cell volume remains relatively constant. When body cells are placed in a solution having a different osmotic pressure than cytosol, however, the shape and volume of the cells change. As water moves by osmosis into or out of cells, their volume increases or decreases. A solution's **tonicity** (*tonic* = tension) is a measure of the solution's ability to change the volume of cells by altering their water content.

Any solution in which a cell—for example, a red blood cell (RBC)—maintains its normal shape and volume is an **isotonic solution** (*iso-* = same) (Figure 3.8a). The concentrations of solutes that cannot cross the plasma membrane are the same on both sides of the membrane in this solution. For instance, a 0.9% NaCl solution, called a *normal saline solution*, is isotonic for RBCs. The RBC plasma membrane permits the water to move back and forth, but it behaves as though it is impermeable to Na⁺ and Cl⁻, the solutes. When RBCs are bathed in 0.9% NaCl, water molecules enter and exit at the same rate, allowing the RBCs to keep their normal shape and volume.

A different situation results if RBCs are placed in a **hypotonic solution** (*hypo-* = less than), a solution that has a *lower* concentration of solutes than the cytosol inside the RBCs (Figure 3.8b). In this case, water molecules enter the cells faster than they leave, causing the RBCs to swell and eventually to burst. The rupture of RBCs in this manner is called **hemolysis** (hē-MOL-i-sis; *hemo-* = blood; *-lysis* = to loosen or split apart); the rupture of other types of cells is referred to simply as **lysis**. Pure water is very hypotonic and causes rapid hemolysis.

A **hypertonic solution** (*hyper-* = greater than) has a *higher* concentration of solutes than does the cytosol inside RBCs (Figure 3.8c). One example of a hypertonic solution is a 2% NaCl solution. In such a solution, water molecules move out of the cells faster than they enter, causing the cells to shrink. Such shrinkage of cells is called **crenation** (kre-NĀ-shun)

Active Processes

Active Transport

Some polar or charged solutes that must enter or leave body cells cannot cross the plasma membrane through any form of passive transport because they would need to move "uphill," against their concentration gradients. Such solutes may be able to cross the membrane by a process called **active transport**. Active transport is considered an active process because energy is required for carrier proteins to move solutes across the membrane against a concentration gradient. Two sources of cellular energy can be used to drive active transport: (1) Energy obtained from hydrolysis of adenosine triphosphate (ATP) is the source in primary active transport; and (2) energy stored in an ionic concentration gradient is the source in secondary active transport. Solutes actively transported across the plasma membrane include several ions, such as Na⁺, K⁺, H⁺, Ca²⁺, I⁻ (iodide ions), and Cl⁻; amino acids; and monosaccharides. (Note that some of these substances also cross the membrane via facilitated diffusion when the proper channel proteins or carriers are present.)

Primary Active Transport

In **primary active transport**, energy derived from hydrolysis of ATP changes the shape of a carrier protein, which "pumps" a substance across a plasma membrane against its concentration gradient. Indeed, carrier proteins that carry out primary active transport are often called **pumps**. A typical body cell expends about 40% of the ATP it generates on primary active transport.

The most prevalent primary active transport mechanism expels sodium ions (Na⁺) from cells and brings potassium ions (K⁺) in. Because of the specific ions it moves, this carrier is called the **sodium–potassium pump**. Because a part of the sodium–potassium pump acts as an *ATPase*, an enzyme that hydrolyzes ATP, another name for this pump is **Na⁺/K⁺ ATPase**. All cells have thousands of sodium–potassium pumps in their plasma membranes. These sodium–potassium pumps maintain a low concentration of Na⁺ in the cytosol by pumping them into the extracellular fluid against the Na⁺ concentration gradient. At the same time, the pumps move K⁺ into cells against the K⁺ concentration

Figure 3.8   Tonicity and its effects on red blood cells (RBCs). The arrows indicate the direction and degree of water movement into and out of the cells. One example of an isotonic solution for RBCs is 0.9% NaCl.

Isotonic solution   Hypotonic solution   Hypertonic solution

(a) Illustrations showing direction of water movement

Normal RBC shape   RBC undergoes hemolysis   RBC undergoes crenation

SEM

(b) Scanning electron micrographs (all 15,000x)

🔑 **Cells placed in an isotonic solution maintain their shape because there is no net water movement into or out of the cell.**

Figure 3.9   The sodium–potassium pump (Na$^+$/K$^+$ ATPase) expels sodium ions (Na$^+$) and brings potassium ions (K$^+$) into the cell.

Sodium–potassium pumps maintain a low intracellular concentration of sodium ions.

gradient. Because K$^+$ and Na$^+$ slowly leak back across the plasma membrane down their electrochemical gradients, the sodium–potassium pumps must work nonstop to maintain a low concentration of Na$^+$ and a high concentration of K$^+$ in the cytosol.

Figure 3.9 depicts the operation of the sodium–potassium pump.

① Three Na$^+$ in the cytosol bind to the pump protein.

② Binding of Na$^+$ triggers the hydrolysis of ATP into ADP, a reaction that also attaches a phosphate group Ⓟ to the pump protein. This chemical reaction changes the shape of the pump protein, expelling the three Na$^+$ into the extracellular fluid. Now the shape of the pump protein favors binding of two K$^+$ in the extracellular fluid to the pump protein.

③ The binding of K$^+$ triggers release of the phosphate group from the pump protein. This reaction again causes the shape of the pump protein to change.

④ As the pump protein reverts to its original shape, it releases K$^+$ into the cytosol. At this point, the pump is again ready to bind three Na$^+$, and the cycle repeats.

The different concentrations of Na$^+$ and K$^+$ in cytosol and extracellular fluid are crucial for the maintenance of normal cell volume and for the ability of some cells to generate electrical signals such as action potentials. Recall that the tonicity of a solution is proportional to the concentration of its solute particles that cannot penetrate the membrane. Because sodium ions that diffuse into a cell or enter through secondary active transport are immediately pumped out, it is as if they never entered. In effect, sodium ions

behave as if they cannot penetrate the membrane. Thus, sodium ions are an important contributor to the tonicity of the extracellular fluid. A similar condition holds for K$^+$ in the cytosol. By helping to maintain normal tonicity on each side of the plasma membrane, the sodium–potassium pumps ensure that cells neither shrink nor swell due to the movement of water by osmosis out of or into cells.

Secondary Active Transport

In **secondary active transport**, the energy stored in a Na$^+$ or H$^+$ (hydrogen ions) concentration gradient is used to drive other substances across the membrane against their own concentration gradients. Because a Na$^+$ or H$^+$ gradient is established by primary active transport, secondary active transport indirectly uses energy obtained from the hydrolysis of ATP.

The sodium–potassium pump maintains a steep concentration gradient of Na$^+$ across the plasma membrane. As a result, the sodium ions have stored or potential energy, just like water behind a dam. Accordingly, if there is a route for Na$^+$ to leak back in, some of the stored energy can be converted to kinetic energy (energy of motion) and used to transport other substances *against their concentration gradients*. In essence, secondary active transport proteins harness the energy in the Na$^+$ concentration gradient by providing routes for Na$^+$ to leak into cells. In secondary active transport, a carrier protein simultaneously binds to Na$^+$ and another substance and then changes its shape so that both substances cross the membrane at the same time. If these carriers move two substances in the same direction they are called **symporters** (*sym-* = same); **antiporters**, in contrast, move two substances in opposite directions across the membrane (*anti-* = against).

Figure 3.10   Secondary active transport mechanisms. (a) Antiporters carry two substances across the membrane in opposite directions. (b) Symporters carry two substances across the membrane in the same direction.

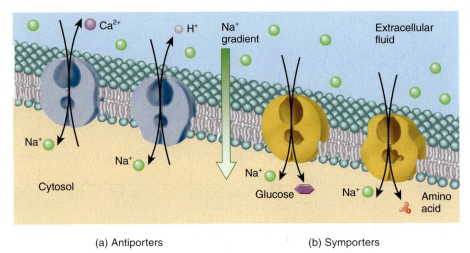

(a) Antiporters   (b) Symporters

🔑 **Secondary active transport mechanisms use the energy stored in an ionic concentration gradient (here, for Na⁺). Because primary active transport pumps that hydrolyze ATP maintain the gradient, secondary active transport mechanisms consume ATP indirectly.**

Plasma membranes contain several antiporters and symporters that are powered by the Na⁺ gradient. For example, the concentration of calcium ions (Ca^{2+}) is low in the cytosol because Na^+/Ca^{2+} antiporters eject calcium ions (Figure 3.10a). Likewise, Na^+/H^+ antiporters help regulate the cytosol's pH (H^+ concentration) by expelling excess H^+. By contrast, dietary glucose and amino acids are absorbed into cells that line the small intestine by Na^+/glucose and Na^+/amino acid symporters (Figure 3.10b). In each case, sodium ions are moving down their concentration gradient while the other solutes move "uphill," against their concentration gradients. Keep in mind that all these symporters and antiporters can do their jobs because the sodium–potassium pumps maintain a low concentration of Na^+ in the cytosol.

Transport in Vesicles

A **vesicle** is a small, spherical membrane sac. A variety of substances are transported in vesicles from one structure to another within cells. Vesicles also import materials into and out of the cell. During **endocytosis** (*endo-* = within), materials move *into* a cell in a vesicle formed from the plasma membrane. In **exocytosis** (*exo-* = out), materials move *out of* a cell by the fusion of vesicles with the plasma membrane. Both endocytosis and exocytosis require energy supplied by ATP. Thus, transport in vesicles is an active process.

Endocytosis

Receptor-mediated endocytosis is a highly selective type of endocytosis by which cells take up specific ligands. (Recall that ligands are molecules that bind to specific receptors.) A vesicle forms after a receptor protein in the plasma membrane recognizes

and binds to a particular substance in the extracellular fluid. Cells take up cholesterol-containing low-density lipoproteins (LDLs), transferrin (an iron-transporting protein in the blood), some vitamins, antibodies, and certain hormones by receptor-mediated endocytosis. Receptor-mediated endocytosis of LDLs (and other ligands) occurs as follows (Figure 3.11):

❶ **Binding.** On the extracellular side of the plasma membrane, an LDL particle that contains cholesterol binds to a specific receptor in the plasma membrane to form a receptor–LDL complex. The receptors are integral membrane proteins that are concentrated in regions of the plasma membrane called *clathrin-coated pits*. Here, a protein called *clathrin* attaches to the membrane on its cytoplasmic side. Many clathrin molecules come together, forming a basketlike structure around the receptor–LDL complexes that causes the membrane to invaginate (fold inward).

❷ **Vesicle formation.** The invaginated edges of the membrane around the clathrin-coated pit fuse and a small piece of the membrane pinches off. The resulting vesicle, known as a *clathrin-coated vesicle*, contains the receptor–LDL complexes.

❸ **Uncoating.** Almost immediately after it is formed, the clathrin-coated vesicle loses its clathrin coat to become an *uncoated vesicle*. Clathrin molecules either return to the inner surface of the plasma membrane or help form coats on other vesicles inside the cell.

❹ **Fusion with endosome.** The uncoated vesicle quickly fuses with a processing vesicle known as an endosome. Once in an endosome, the LDL particles separate from their receptors.

Figure 3.11 **Receptor-mediated endocytosis of a low-density lipoprotein (LDL) particle.**

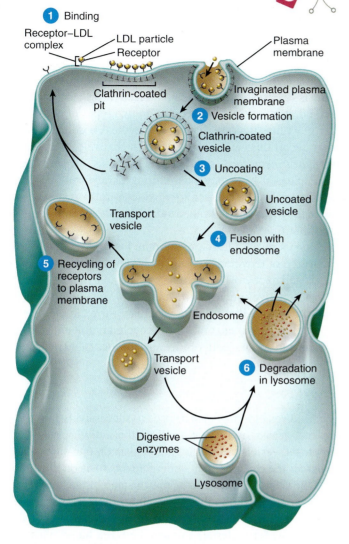

1 Binding
Receptor–LDL complex
LDL particle
Receptor
Plasma membrane
Clathrin-coated pit
Invaginated plasma membrane
2 Vesicle formation
Clathrin-coated vesicle
3 Uncoating
Uncoated vesicle
Transport vesicle
4 Fusion with endosome
5 Recycling of receptors to plasma membrane
Endosome
Transport vesicle
6 Degradation in lysosome
Digestive enzymes
Lysosome

🔑 **Receptor-mediated endocytosis imports materials that are needed by cells.**

5 *Recycling of receptors to plasma membrane.* Most of the receptors accumulate in elongated protrusions of the endosome. These pinch off, forming transport vesicles that return the receptors to the plasma membrane.

6 *Degradation in lysosomes.* Other transport vesicles, which contain the LDL particles, bud off the endosome and soon fuse with a lysosome. Lysosomes contain many digestive enzymes that break down the larger molecules into smaller molecules that can be utilized by the cell.

Phagocytosis (fag′-ō-sī-TŌ-sis; *phago-* = to eat) is a form of endocytosis in which the cell engulfs large solid particles, such as worn-out cells, whole bacteria, or viruses (Figure 3.12). Only a few body cells, termed **phagocytes**, are able to carry out phagocytosis. Two main types of phagocytes are *macrophages*, located

Figure 3.12 **Phagocytosis.** Pseudopods surround a particle, and the membranes fuse to form a phagosome.

Plasma membrane
Pseudopods
Microbe
Receptor
Lysosome
Digestive enzymes
Phagosome
Fusion of lysosome and phagosome
Digestion by lysosomal enzymes
Residual body

(a) Diagram of the process

Pseudopod
Microbe
Pseudopod
White blood cell

TEM about 3700x
(b) White blood cell engulfs microbe

TEM about 3700x
(c) White blood cell destroys microbe

🔑 **Phagocytosis is a vital defense mechanism that helps protect the body from disease.**

in many body tissues, and *neutrophils*, a type of white blood cell. Phagocytosis begins when the particle binds to a plasma membrane receptor on the phagocyte, causing it to extend **pseudopods** (SOO-dō-pods; *pseudo-* = false; *-pods* = feet), projections of its plasma membrane and cytoplasm. Pseudopods surround the particle outside the cell, and the membranes fuse to form a vesicle called a *phagosome*, which enters the cytoplasm. The phagosome fuses with one or more lysosomes, and lysosomal enzymes break down the ingested material. Undigested materials in the phagosome usually remain indefinitely in a vesicle called a residual body. The process of phagocytosis is a vital defense mechanism that helps protect the body from disease. Through phagocytosis, macrophages dispose of invading microbes and billions of aged, worn-out red blood cells every day; neutrophils also help rid the body of invading microbes. Pus is a mixture of dead neutrophils, macrophages, and tissue cells and fluid in an infected wound.

Most body cells carry out **bulk-phase endocytosis**, also called pinocytosis (pi-nō-sī-TŌ-sis; *pino-* = to drink), a form of endocytosis in which tiny droplets of extracellular fluid are taken into the cell (Figure 3.13). No receptor proteins are involved; all solutes dissolved in the extracellular fluid are brought into the cell. During bulk-phase endocytosis, the plasma membrane folds inward and forms a vesicle containing a droplet of extracellular fluid. The vesicle detaches or "pinches off" from the plasma membrane and enters the cytosol. Within the cell, the vesicle fuses with a lysosome, where enzymes degrade the engulfed solutes. The re-

sulting smaller molecules leave the lysosome to be used elsewhere in the cell. Bulk-phase endocytosis is common in absorptive cells in the intestines and kidneys.

Exocytosis

In contrast with endocytosis, which brings materials into a cell, **exocytosis** releases materials from a cell. All cells carry out exocytosis, but it is especially important in two types of cells: (1) secretory cells that liberate digestive enzymes, hormones, mucus, or other secretions; and (2) nerve cells that release substances called *neurotransmitters* (see Figure 12.22). In some cases, wastes are also released by exocytosis. During exocytosis, membrane-enclosed vesicles called secretory vesicles form inside the cell, fuse with the plasma membrane, and release their contents into the extracellular fluid.

Segments of the plasma membrane lost through endocytosis are recovered or recycled by exocytosis. The balance between endocytosis and exocytosis keeps the surface area of a cell's plasma membrane relatively constant. Membrane exchange is quite extensive in certain cells. In your pancreas, for example, the cells that secrete digestive enzymes can recycle an amount of plasma membrane equal to the cell's entire surface area in 90 minutes.

Transcytosis

Transport in vesicles may also be used to successively move a substance into, across, and out of a cell. In this active process, called **transcytosis**, vesicles undergo endocytosis on one side of a cell, move across the cell, and then undergo exocytosis on the opposite side. As the vesicles fuse with the plasma membrane, the vesicular contents are released into the extracellular fluid. Transcytosis occurs most often across the endothelial cells that line blood vessels and is a means for materials to move between blood plasma and interstitial fluid. For instance, when a woman is pregnant, some of her antibodies cross the placenta into the fetal circulation via transcytosis.

Table 3.1 summarizes the processes by which materials move into and out of cells.

Figure 3.13   Bulk-phase endocytosis. The plasma membrane folds inward, forming a vesicle.

Plasma membrane

Formation of vesicle

Lysosome

Fluid and dissolved solutes in vesicle

Fusion of lysosome and vesicle

Digestion by lysosmal enzymes

Digested solutes

🔑 **Most body cells carry out bulk-phase endocytosis, the nonselective uptake of tiny droplets of extracellular fluid.**

Checkpoint

7. What is the key difference between passive and active transport?
8. Which factors can increase the rate of diffusion?
9. What is osmotic pressure?
10. What is the key difference between primary and secondary active processes?
11. How do symporters and antiporters carry out their functions?
12. What is transcytosis?

TABLE 3.1

Transport of Materials Into and Out of Cells

TRANSPORT PROCESS	DESCRIPTION	SUBSTANCES TRANSPORTED
Passive Processes	Movement of substances down a concentration gradient until equilibrium is reached; do not require cellular energy in the form of ATP.	
Diffusion	The movement of molecules or ions down a concentration gradient due to their kinetic energy until they reach equilibrium.	
Simple diffusion	Passive movement of a substance down its concentration gradient through the lipid bilayer of the plasma membrane without the help of membrane transport proteins.	Nonpolar, hydrophobic solutes: oxygen, carbon dioxide, and nitrogen gases; fatty acids; steroids; and fat-soluble vitamins. Polar molecules such as water, urea, and small alcohols.
Facilitated diffusion	Passive movement of a substance down its concentration gradient through the lipid bilayer by transmembrane proteins that function as channels or carriers.	Polar or charged solutes: glucose, fructose, galactose, some vitamins, and ions such as K^+, Cl^-, Na^+, and Ca^{2+}.
Osmosis	Passive movement of water molecules across a selectively permeable membrane from an area of higher water concentration to an area of lower water concentration.	Solvent: water in living systems.
Active Processes	Movement of substances against a concentration gradient; requires cellular energy in the form of ATP.	
Active Transport	Active process in which a cell expends energy to move a substance across the membrane against its concentration gradient by transmembrane proteins that function as carriers.	Polar or charged solutes.
Primary active transport	Active process in which a substance is moved across the membrane against its concentration gradient by pumps (carriers) that use energy supplied by hydrolysis of ATP.	Na^+, K^+, Ca^{2+}, H^+, I^-, Cl^-, and other ions.
Secondary active transport	Coupled active transport of two substances across the membrane using energy supplied by a Na^+ or H^+ concentration gradient maintained by primary active transport pumps. Antiporters move Na^+ (or H^+) and another substance in opposite directions across the membrane; symporters move Na^+ (or H^+) and another substance in the same direction across the membrane.	Antiport: Ca^{2+}, H^+ out of cells. Symport: glucose, amino acids into cells.
Transport in Vesicles	Active process in which substances move into or out of cells in vesicles that bud from the plasma membrane; requires energy supplied by ATP.	
Endocytosis	Movement of substances into a cell in vesicles.	
Receptor-mediated endocytosis	Ligand–receptor complexes trigger infolding of a clathrin-coated pit that forms a vesicle containing ligands.	Ligands; transferrin, low-density lipoproteins (LDLs), some vitamins, certain hormones, and antibodies.
Phagocytosis	"Cell eating"; movement of a solid particle into a cell after pseudopods engulf it to form a phagosome.	Bacteria, viruses, and aged or dead cells.
Bulk-phase endocytosis	"Cell drinking"; movement of extracellular fluid into a cell by infolding of plasma membrane to form a vesicle.	Solutes in extracellular fluid.
Exocytosis	Movement of substances out of a cell in secretory vesicles that fuse with the plasma membrane and release their contents into the extracellular fluid.	Neurotransmitters, hormones, and digestive enzymes.
Transcytosis	Movement of a substance through a cell as a result of endocytosis on one side and exocytosis on the opposite side.	Substances, such as antibodies, across endothelial cells. This is a common route for substances to pass between blood plasma and interstitial fluid.

Return to Joseph's Story

As he lay unconscious in the park, inside his heart, Joseph's cardiac muscle cells struggled to survive the lack of blood flow caused by his atherosclerosed coronary arteries. Joseph's family had a history of vascular disease, but he hadn't thought about it much. His weight had climbed; he always found an excuse to justify eating the fatty foods he loved. Now, the cells in his heart struggled to work. As the flow of life-giving blood slowed, along with the oxygen, glucose, and essential ions it carried that were required for the heart to pump, each cell continued working, rapidly using up its ATP supply. Carbon dioxide levels inside the cells rose, and the pH began to drop; the mitochondria had no more oxygen and glucose to make essential ATP energy. Eventually, the cells began to die. Active transport pumps shut down the plasma membranes of Joseph's heart. The cells became leaky; sodium slowly began to leak into the cells, and potassium leaked out. As the chemical gradients diminished, Joseph's heart convulsed in an uncontrolled spasm of arrhythmic electrical activity.

Time was running out. Joseph's cells needed oxygen. "What's going on?!" shouted a tall man as he ran toward Marcus.

"My dad—he's not breathing. He just fell down!" Tears ran down Marcus' face as the man knelt by Joseph, then turned. "Jenny, call 911!" The man's friend pulled out her cell phone and called for help.

A.   List Joseph's risk factors and create a brief summary of the information you have so far. Identify how his risk factors would affect cellular function.

B.   Assuming Joseph's heart has stopped, what cellular processes and membrane functions are going to be affected by the loss of oxygen, blood glucose, and waste removal?

3.4 Cytoplasm consists of the cytosol and organelles.

Cytoplasm consists of all the cellular contents between the plasma membrane and the nucleus. It has two major components: (1) the cytosol and (2) organelles, tiny structures that perform different functions in the cell.

Cytosol

The **cytosol** (**intracellular fluid**) is the fluid portion of the cytoplasm that surrounds organelles (see Figure 3.1) and constitutes about 55% of total cell volume. Although it varies in composition and consistency from one part of a cell to another, cytosol is 75–90% water plus various dissolved and suspended components. Among these are different types of ions, glucose, amino acids, fatty acids, proteins, lipids, ATP, and waste products. Some cells also contain various organic molecules that aggregate into masses for storage. Examples include *lipid droplets* that contain triglycerides, and clusters of glycogen molecules called *glycogen granules* (see Figure 3.1).

The cytosol is the site of many chemical reactions required for a cell's existence. For example, enzymes in cytosol catalyze *glycolysis,* a series of chemical reactions that produce ATP from glucose. Other types of cytosolic reactions provide the building blocks for maintenance of cell structures and for cell growth.

Organelles

Organelles are specialized structures within the cell that have characteristic shapes and perform specific functions in cellular growth, maintenance, and reproduction. Despite the many chemical reactions going on in a cell at any given time, there is little interference among reactions because they are confined to different organelles. Each type of organelle has its own set of enzymes that carry out specific reactions, and serves as a functional compartment for specific biochemical processes. The numbers and types of organelles vary in different cells, depending on the cell's function. Although they have different functions, organelles often cooperate to maintain homeostasis. Even though the nucleus is a large organelle, it is discussed in a separate section because of its special importance in directing the life of a cell.

The Cytoskeleton

The **cytoskeleton** is a network of protein filaments that extends throughout the cytosol (see Figure 3.1). Three types of filamentous proteins contribute to the cytoskeleton's structure. In order of their increasing diameter, these structures are microfilaments, intermediate filaments, and microtubules.

Microfilaments, the thinnest elements of the cytoskeleton, are composed of the protein *actin* and are most prevalent at the periphery of a cell (Figure 3.14a). Microfilaments help generate cellular movements. Microfilaments are involved in muscle contraction, cell division, and cell locomotion, such as occurs in the migration of embryonic cells during development, the invasion of tissues by white blood cells to fight infection, or the migration of skin cells during wound healing.

Microfilaments also provide much of the mechanical support that is responsible for the basic strength and shapes of cells. They anchor the cytoskeleton to integral proteins in the plasma mem-

Figure 3.14 Cytoskeleton.

(a) Microfilament

Microvillus

MICROFILAMENTS

Nucleus

Microfilaments

Nucleus

IFM 1500x

INTERMEDIATE FILAMENTS

(b) Intermediate filament

Nucleus

Intermediate filaments (green)

IFM 800x

MICROTUBULES

Centrosome

(c) Microtubule

Microtubules (green)

Nucleus

IFM 500x

Functions
1. Serves as a scaffold that helps to determine a cell's shape and to organize the cellular contents.
2. Aids movement of organelles within the cell, of chromosomes during cell division, and of whole cells such as phagocytes.

🔑 The cytoskeleton is a network of three types of protein filaments that extend throughout the cytoplasm: microfilaments, intermediate filaments, and microtubules.

brane. Microfilaments also provide mechanical support for cell extensions called **microvilli** (*micro-* = small; *-villi* = tufts of hair; singular is *microvillus*), nonmotile, microscopic fingerlike projections of the plasma membrane. Within each microvillus is a core of parallel microfilaments that supports it. Because they greatly increase the surface area of the cell, microvilli are abundant on cells involved in absorption, such as the epithelial cells that line the small intestine.

As their name suggests, **intermediate filaments** are thicker than microfilaments but thinner than microtubules (Figure 3.14b). Several different proteins can compose intermediate filaments. They are exceptionally strong and are found in parts of cells sub-

ject to mechanical stress, help stabilize the position of organelles such as the nucleus, and help attach cells to one another.

The largest of the cytoskeletal components, **microtubules** are long, unbranched hollow tubes composed mainly of the protein *tubulin*. The assembly of microtubules begins in an organelle called the centrosome (discussed shortly). Microtubules grow outward from the centrosome toward the periphery of the cell (Figure 3.14c). Microtubules help determine cell shape. They also function in the movement of organelles such as secretory vesicles, of chromosomes during cell division, and of specialized cell projections, such as cilia and flagella.

Centrosome

The **centrosome**, located near the nucleus, consists of two components: a pair of centrioles and pericentriolar material (Figure 3.15a). The two **centrioles** are cylindrical structures, each composed of nine clusters of three microtubules (triplets) arranged in a circular pattern (Figure 3.15b). The long axis of one centriole is at a right angle to the long axis of the other (Figure 3.15c). Surrounding the centrioles is *pericentriolar material* (per´-e-sen´-trē-Ō-lar) composed of the protein *tubulin,* used as organizing centers for growth of the mitotic spindle and for microtubule formation in nondividing cells. As you will learn shortly, the mitotic spindle plays a critical role in cell division. During cell division, centrosomes replicate so that succeeding generations of cells continue to have the capacity to divide.

Cilia and Flagella

Microtubules are the dominant components of cilia and flagella, which are motile projections of the cell surface (Figure 3.16). **Cilia** (SIL-ē-a = eyelashes; singular is *cilium*) are numerous, short, hairlike projections that extend from the surface of the cell (see Figure 3.1 and Figure 3.16d). Each cilium contains a core of 20 microtubules surrounded by plasma membrane (Figure 3.16a). The microtubules are arranged such that one pair in the center is surrounded by nine clusters of two fused microtubules (doublets). Each cilium is anchored to a *basal body* just below the surface of the plasma membrane. A basal body is similar in structure to a centriole and functions in initiating the assembly of cilia and flagella.

A cilium displays an oarlike pattern of beating; it is relatively stiff during the power stroke (oar digging into the water), but more flexible during the recovery stroke (oar moving above the water preparing for a new stroke) (Figure 3.16b). The coordinated movement of many cilia on the surface of a cell causes the steady movement of fluid along the cell's surface. Many cells of the respiratory tract, for example, have hundreds of cilia that help sweep foreign particles trapped in mucus away from the lungs. The movement of cilia is paralyzed by nicotine in cigarette smoke. For this reason, smokers cough often to remove foreign particles from their airways. Cells that line the uterine (fallopian) tubes also have cilia that sweep oocytes toward the uterus, and females who smoke have an increased risk of ectopic (outside the uterus) pregnancy.

Flagella (fla-JEL-a = whip; singular is *flagellum*) are similar in structure to cilia but are typically much longer. Flagella usually move an entire cell. A flagellum generates forward motion along its axis by rapidly wiggling in a wavelike pattern (Figure 3.16c). The only example of a flagellum in the human body is a sperm cell's tail, which propels the sperm toward the oocyte in the uterine tube (Figure 3.16c).

Ribosomes

Ribosomes (RĪ-bō-sōms; -*somes* = bodies) are the sites of protein synthesis. The name of these tiny organelles reflects their high content of one type of ribonucleic acid, **ribosomal RNA (rRNA)**.

Figure 3.15   Centrosome.

> **Functions**
> 1. Builds microtubules in nondividing cells.
> 2. Forms the mitotic spindle during cell division.

(a) Details of a centrosome

Pericentriolar material

Centrioles

Microtubules (triplets)

(b) Arrangement of microtubules in centrosome

Pericentriolar material

SEM

Longitudinal section   Transverse section

(c) Centrioles

Located near the nucleus, the centrosome consists of a pair of centrioles and pericentriolar material.

Figure 3.16 Cilia and flagella.

Functions
1. Cilia move fluids along a cell's surface.
2. A flagellum moves an entire cell.

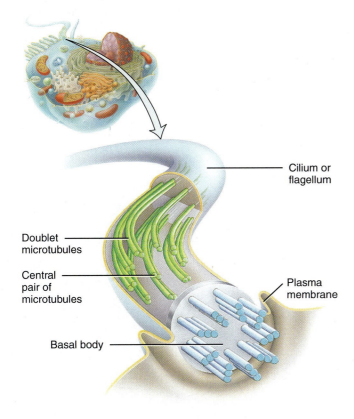

Cilium or flagellum

Doublet microtubules

Central pair of microtubules

Plasma membrane

Basal body

(a) Arrangement of microtubules in a cilium or flagellum

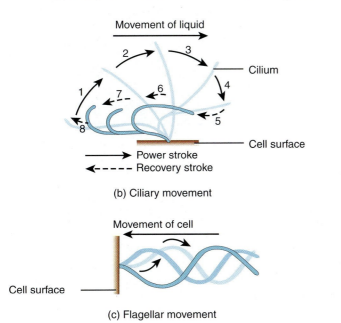

Movement of liquid

Cilium

Cell surface

⟶ Power stroke
⟵---- Recovery stroke

(b) Ciliary movement

Movement of cell

Cell surface

(c) Flagellar movement

Cilia

SEM 3000x

(d) Cilia lining the trachea

Flagellum

SEM 4000x

(e) Flagellum of a sperm cell

🔑 **A cilium contains a core of microtubules with one pair in the center surrounded by nine clusters of doublet microtubules.**

Figure 3.17   Ribosomes.

Large subunit   Small subunit   Complete functional ribosome

Details of ribosomal subunits

Functions
1. Attached ribosomes synthesize proteins destined for insertion in the plasma membrane or secretion from the cell.
2. Free ribosomes synthesize proteins used in the cytosol.

 Ribosomes are the sites of protein synthesis.

Structurally, a ribosome consists of two subunits, one about half the size of the other (Figure 3.17). The large and small subunits are made separately in the nucleolus, a spherical body inside the nucleus. Once produced, the large and small subunits exit the nucleus separately, then come together in the cytoplasm.

Some ribosomes are attached to the outer surface of the nuclear membrane and to an extensively folded membrane called the endoplasmic reticulum. These ribosomes synthesize proteins destined for specific organelles, for insertion in the plasma membrane, or for export from the cell. Other ribosomes are "free" or unattached to other cytoplasmic structures. Free ribosomes synthesize proteins used in the cytosol. Ribosomes are also located within mitochondria, where they synthesize mitochondrial proteins.

Endoplasmic Reticulum

The **endoplasmic reticulum** (en´-dō-PLAS-mik re-TIK-ū-lum; -*plasmic* = cytoplasm; *reticulum* = network) or **ER** is a network of membranes in the form of flattened sacs or tubules (Figure 3.18). The ER extends throughout the cytoplasm and is so extensive that it constitutes more than half of the membranous surfaces within the cytoplasm of most cells.

Cells contain two distinct forms of ER that differ in structure and function. **Rough ER** is continuous with the nuclear envelope (membrane around the nucleus) and usually is folded into a series of flattened sacs. The outer surface of rough ER is studded with ribosomes, the sites of protein synthesis. Proteins synthesized by ribosomes attached to rough ER enter spaces within the ER for processing and sorting. In some cases, enzymes attach the proteins

Figure 3.18   Endoplasmic reticulum.

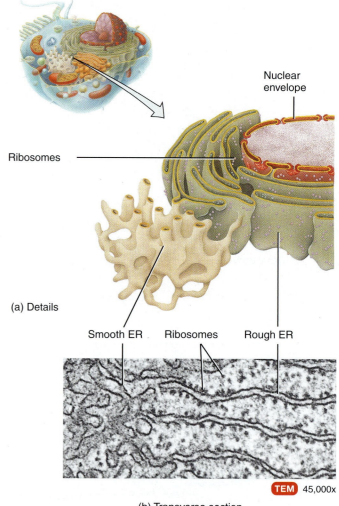

Nuclear envelope

Ribosomes

(a) Details

Smooth ER   Ribosomes   Rough ER

TEM 45,000x

(b) Transverse section

Functions
1. Rough ER synthesizes glycoproteins and phospholipids that are transferred into cellular organelles, inserted into the plasma membrane, or secreted during exocytosis.
2. Smooth ER synthesizes fatty acids and steroids, inactivates or detoxifies drugs and other potentially harmful substances, converts glucose 6-phosphate to glucose, and stores and releases calcium ions.

 The endoplasmic reticulum is a network of membrane-enclosed sacs or tubules that extend throughout the cytoplasm and connect to the nuclear envelope.

to carbohydrates to form glycoproteins. In other cases, enzymes attach the proteins to phospholipids, also synthesized by rough ER. These molecules may be incorporated into the membranes of organelles, inserted into the plasma membrane, or secreted via exocytosis. Thus rough ER produces secretory proteins, membrane proteins, and many organellar proteins.

Smooth ER extends from the rough ER to form a network of membrane tubules (Figure 3.18). Unlike rough ER, smooth ER appears "smooth" because it lacks ribosomes on the outer surfaces of its membrane. Because of the lack of ribosomes, smooth ER does not synthesize proteins, but it does synthesize fatty acids and steroids, such as estrogens and testosterone. In liver cells, the smooth ER helps release glucose into the bloodstream and inactivate or detoxify a variety of drugs or potentially harmful substances, such as alcohol, pesticides, and *carcinogens* (cancer-causing agents). In liver, kidney, and intestinal cells the smooth ER removes the phosphate group from glucose 6-phosphate, which allows the "free" glucose to enter the bloodstream. In muscle cells, the calcium ions (Ca^{2+}) that trigger contraction are released from the sarcoplasmic reticulum, a form of smooth ER.

Golgi Complex

Most of the proteins synthesized by ribosomes attached to rough ER are ultimately transported to other regions of the cell. The first step in the transport pathway is through an organelle called the **Golgi complex** (GOL-jē). It consists of 3 to 20 **cisternae** (sis-TER-nē = cavities; singular is *cisterna*), small, flattened membranous sacs with bulging edges that resemble a stack of pita bread (Figure 3.19). The cisternae are often curved, giving the Golgi complex a bowl-like shape. Most cells have several Golgi complexes; Golgi complexes are more extensive in cells that secrete proteins, a clue to the organelle's role in the cell.

The cisternae at the opposite ends of a Golgi complex differ from each other in size, shape, and enzymatic activity. The convex **entry face** is a cisterna that faces the rough ER. The concave **exit face** is a cisterna that faces the plasma membrane. Sacs between the entry and exit faces are called **medial cisternae**. Transport vesicles (described shortly) from the ER merge to form the entry face. From the entry face, the cisternae are thought to mature, in turn becoming medial and then exit cisternae.

Figure 3.19   Golgi complex.

Functions
1. Modifies, sorts, packages, and transports proteins received from the rough ER.
2. Generates secretory vesicles that discharge processed proteins via exocytosis into extracellular fluid.
3. Forms membrane vesicles that ferry new molecules to the plasma membrane.
4. Produces transport vesicles that carry molecules to other organelles, such as lysosmes.

Transport vesicle from rough ER

Entry face

Medial cisterna

Transfer vesicles

Exit face

Secretory vesicles

TEM 65,000x

(b) Transverse section

(a) Details

Most proteins synthesized by ribosomes attached to rough ER pass through the Golgi complex for processing.

Different enzymes in the cisternae of the Golgi complex modify, sort, and package proteins for transport to different destinations. The entry face receives and modifies proteins produced by the rough ER. The medial cisternae add carbohydrates to proteins to form glycoproteins, and lipids to proteins to form lipoproteins. The exit face modifies the molecules further and then sorts and packages them for transport to their destinations.

Proteins arriving at, passing through, and exiting the Golgi complex do so through maturation of the cisternae and exchanges that occur via transfer vesicles (Figure 3.20):

1 Proteins synthesized by ribosomes on the rough ER are surrounded by a piece of the ER membrane, which eventually buds from the membrane surface to form **transport vesicles**.

2 Transport vesicles move toward the entry face of the Golgi complex.

3 Fusion of several transport vesicles creates the entry face of the Golgi complex and releases proteins into its lumen (space).

4 The proteins move from the entry face into one or more medial cisternae. Enzymes in the medial cisternae modify the proteins to form glycoproteins, and lipoproteins. **Transfer vesicles** that bud from the edges of the cisternae move specific enzymes back toward the entry face and move partially modified proteins toward the exit face.

5 The products of the medial cisternae move into the lumen of the exit face.

6 Within the exit face cisterna, the products are further modified and are sorted and packaged.

7 Some of the processed proteins leave the exit face in **secretory vesicles**. These vesicles deliver the proteins to the plasma membrane, where they are discharged by exocytosis into the extracellular fluid.

8 Other processed proteins leave the exit face in **membrane vesicles** that deliver their contents to the plasma membrane for incorporation into the membrane. In doing so, the Golgi complex adds new segments of plasma membrane as existing segments are lost and modifies the number and distribution of membrane molecules.

Figure 3.20   Processing and packaging of proteins by the Golgi complex.

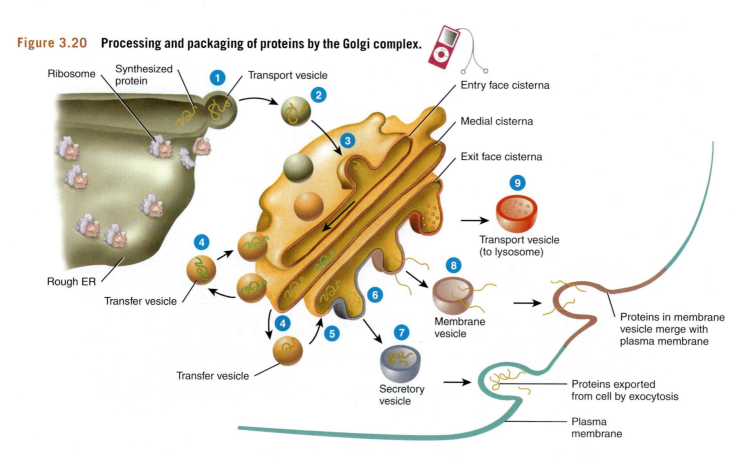

Ribosome · Synthesized protein · **1** Transport vesicle · **2** · **3** · Entry face cisterna · Medial cisterna · Exit face cisterna · **9** · Transport vesicle (to lysosome) · Rough ER · Transfer vesicle · **4** · **4** · **5** · **6** · **7** · **8** · Membrane vesicle · Transfer vesicle · Secretory vesicle · Proteins in membrane vesicle merge with plasma membrane · Proteins exported from cell by exocytosis · Plasma membrane

🔑 **All proteins exported from the cell are processed in the Golgi complex.**

9 Finally, some processed proteins leave the exit face in transport vesicles that will carry the proteins to another cellular destination. For instance, transport vesicles ferry digestive enzymes to lysosomes; the structure and functions of these important organelles are discussed next.

Lysosomes

Lysosomes (LĪ-sō-sōms; *lyso-* = dissolving; *-somes* = bodies) are membrane-enclosed vesicles that form from the Golgi complex (Figure 3.21). Inside, as many as 60 kinds of powerful enzymes can break down a wide variety of molecules once lysosomes fuse with

Figure 3.21   Lysosomes.

Functions
1. Digest substances that enter a cell via endocytosis and transport final products of digestion into cytosol.
2. Digest worn-out organelles (autophagy).
3. Destroy entire cell (autolysis).
4. Dissolve extracellular materials.

Digestive enzymes

(a) Lysosome

Lysosomes

TEM 11,700x

(b) Lysosomes in the cytoplasm

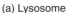 **Lysosomes contain several types of powerful digestive enzymes.**

vesicles formed during endocytosis. Because lysosomal enzymes work best at an acidic pH, the lysosomal membrane includes active transport pumps that import hydrogen ions (H^+). Thus, the lysosomal interior has a pH of 5, which is 100 times more acidic than the pH of the cytosol (pH 7). The lysosomal membrane also includes carriers that move the final products, such as glucose, fatty acids, and amino acids, into the cytosol.

Lysosomal enzymes also help recycle worn-out cell structures. A lysosome can engulf another organelle, digest it, and return the digested components to the cytosol for reuse. The process by which entire worn-out organelles are digested is called **autophagy** (aw-TOF-a-jē; *auto-* = self; *-phagy* = eating). In autophagy, the organelle to be digested is enclosed by a membrane derived from the ER to create a vesicle which then fuses with a lysosome. Lysosomal enzymes may also destroy the entire cell, a process known as **autolysis** (aw-TOL-i-sis). Autolysis occurs in some pathological conditions and is also responsible for the tissue deterioration that occurs immediately after death.

As we just discussed, most lysosomal enzymes act within a cell. However, some operate in extracellular digestion. One example occurs during fertilization. The head of a sperm cell releases lysosomal enzymes that aid its penetration of the oocyte by dissolving the protective coating around the oocyte in a process called the acrosomal reaction (see Concept 25.6).

Peroxisomes

Another group of organelles similar in structure to lysosomes, but smaller, are the **peroxisomes** (pe-ROKS-i-sōms; *peroxi-* = peroxide; *somes* = bodies; see Figure 3.1). Peroxisomes contain several *oxidases,* enzymes that can oxidize (remove hydrogen atoms from) various organic substances. For instance, amino acids and fatty acids are oxidized in peroxisomes as part of normal metabolism. In addition, enzymes in peroxisomes oxidize toxic substances. Thus, peroxisomes are very abundant in the liver, where detoxification of alcohol and other damaging substances occurs. A by-product of the oxidation reactions is hydrogen peroxide (H_2O_2), a potentially toxic compound. However, peroxisomes also contain the enzyme *catalase,* which decomposes H_2O_2. By both producing and degrading H_2O_2 within the same organelle, peroxisomes protect other parts of the cell from the toxic effects of H_2O_2.

Proteasomes

Although lysosomes degrade proteins delivered to them in vesicles, proteins within the cytosol also require disposal at certain times in the life of a cell. Continuous destruction of unneeded, damaged, or faulty proteins is the function of tiny structures called **proteasomes** (PRŌ-tē-a-sōms = protein bodies). For example, proteins that are part of metabolic pathways are degraded after they have accomplished their function. A typical body cell contains many thousands of proteasomes, in both the cytosol and the nucleus.

Mitochondria

Because they are the site of most ATP production, **mitochondria** (mī-tō-KON-drē-a; *mito-* = thread; *-chondria* = granules; singular is *mitochondrion*) are referred to as the "powerhouses" of the cell. A cell may have as few as a hundred or as many as several thousand mitochondria, depending on the activity of the cell. Active cells, such as those found in the muscles, liver, and kidneys, which use ATP at a high rate, have a large number of mitochondria.

A mitochondrion consists of two membranes, an **outer mitochondrial membrane** and an **inner mitochondrial membrane**, with a small fluid-filled space between them (Figure 3.22). Both membranes are similar in structure to the plasma membrane. The inner mitochondrial membrane is arranged in a series of folds called **cristae** (KRIS-tē = ridges). The central fluid-filled cavity of a mitochondrion, enclosed by the inner mitochondrial membrane, is the **matrix**. The elaborate folds of the cristae provide an enormous surface area for the chemical reactions that are part of the aerobic phase of *cellular respiration,* the reactions that produce most of a cell's ATP. The enzymes that catalyze these reactions are located on the cristae and in the matrix of the mitochondria.

Mitochondria self-replicate, a process that occurs during times of increased cellular energy demand or before cell division. Synthesis of some of the proteins needed for mitochondrial functions occurs on the ribosomes that are present in the mitochondrial matrix. Mitochondria even contain their own DNA with a small number of genes; these genes control synthesis of the proteins that build mitochondrial components.

Checkpoint

13. Which cytoskeletal components help form the structure of microvilli? Which help form centrioles, cilia, and flagella?
14. What is the functional difference between cilia and flagella?
15. Where are subunits of ribosomes synthesized? Where are they assembled?
16. What are the structural and functional differences between rough and smooth ER?
17. What are the three general destinations for proteins that leave the Golgi complex?
18. What are autophagy and autolysis?
19. What happens on the cristae and in the matrix of mitochondria?

Figure 3.22   Mitochondria.

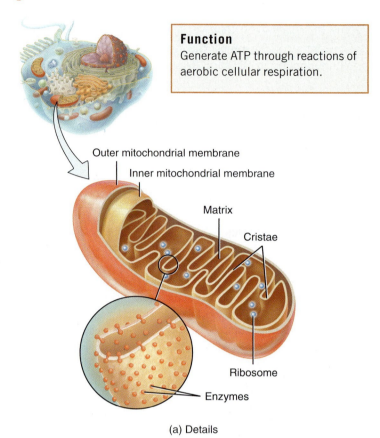

Function
Generate ATP through reactions of aerobic cellular respiration.

Outer mitochondrial membrane
Inner mitochondrial membrane
Matrix
Cristae
Ribosome
Enzymes

(a) Details

Outer mitochondrial membrane
Inner mitochondrial membrane
Matrix
Cristae

TEM   80,000x

(b) Transverse section

Within mitochondria, chemical reactions of aerobic cellular respiration generate ATP.

3.5 The nucleus contains nucleoli and genes.

The **nucleus** is a spherical or oval-shaped structure that usually is the most prominent feature of a cell (Figure 3.23). Most cells have a single nucleus, although some, such as mature red blood cells, have none. In contrast, skeletal muscle cells and a few other types of cells have multiple nuclei. A double membrane called

the **nuclear envelope** separates the nucleus from the cytoplasm. Both layers of the nuclear envelope are lipid bilayers similar to the plasma membrane. The outer membrane of the nuclear envelope is continuous with rough ER. Many openings called **nuclear pores** extend through the nuclear envelope. Each nuclear pore consists

Figure 3.23   Nucleus.

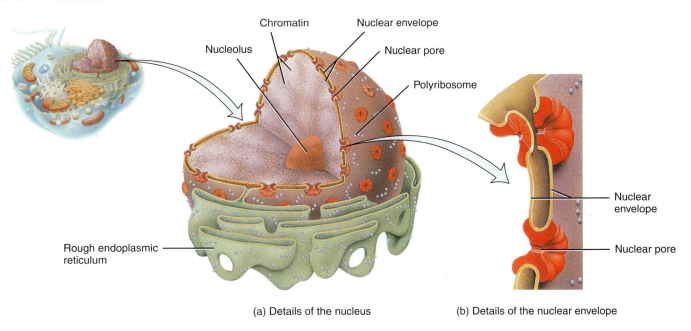

(a) Details of the nucleus

(b) Details of the nuclear envelope

about 10,000x **TEM**

(c) Transverse section of the nucleus

Functions
1. Controls cellular structure.
2. Directs cellular activities.
3. Produces ribosomal subunits in nucleoli.

🔑 The nucleus contains most of the cell's genes, which are located on chromosomes.

of a circular arrangement of proteins surrounding a large central opening that is about 10 times wider than the pore of a channel protein in the plasma membrane.

Nuclear pores control the movement of substances between the nucleus and the cytoplasm. Small molecules and ions move through the pores passively by diffusion. Most large molecules, such as RNAs and proteins, pass through the nuclear pores by an active transport process in which the molecules are recognized and selectively transported through the nuclear pore into or out of the nucleus. For example, proteins needed for nuclear functions move from the cytosol into the nucleus; newly formed RNA molecules move from the nucleus into the cytosol in this manner.

Inside the nucleus are one or more spherical bodies called **nucleoli** (noo-KLĒ-ō-lī; singular is *nucleolus*) that function in producing ribosomes. Each nucleolus is simply a cluster of protein, DNA, and RNA; it is not enclosed by a membrane. Nucleoli are the sites of synthesis of rRNA and assembly of ribosomal subunits. Nucleoli are quite prominent in cells that synthesize large amounts of protein, such as muscle and liver cells.

Within the nucleus are most of the cell's hereditary units, called **genes**, which control cellular structure and direct cellular activities. Genes are arranged along **chromosomes** (*chromo-* = colored). Human somatic (body) cells have 46 chromosomes, 23 inherited from each parent. Each chromosome is a long molecule of DNA that is coiled together with several proteins (Figure 3.24). This complex of DNA, proteins, and some RNA is called **chromatin**. The total genetic information carried in a cell or an organism is its **genome.**

In cells that are not dividing, the chromatin appears as a diffuse, granular mass (Figure 3.23c). Electron micrographs reveal that chromatin has a beads-on-a-string structure. Each bead is a **nucleosome** and consists of double-stranded DNA wrapped twice around a core of eight proteins called **histones**, which help organize the coiling and folding of DNA (Figure 3.24). The string between the beads is **linker DNA**, which holds adjacent nucleosomes together. In cells that are not dividing, another histone promotes coiling of nucleosomes into a larger diameter **chromatin fiber**, which then folds into large loops. Just before cell division takes place, however, the DNA replicates (duplicates) and the loops condense even more, forming a pair of **chromatids**. As you will see shortly, during cell division a pair of chromatids constitutes a chromosome.

The main parts of a cell, their descriptions, and their functions are summarized in Table 3.2.

Figure 3.24   Packing of DNA into a chromosome in a dividing cell. When packing is complete, two identical DNA molecules and their histones form a pair of chromatids, which are held together by a centromere.

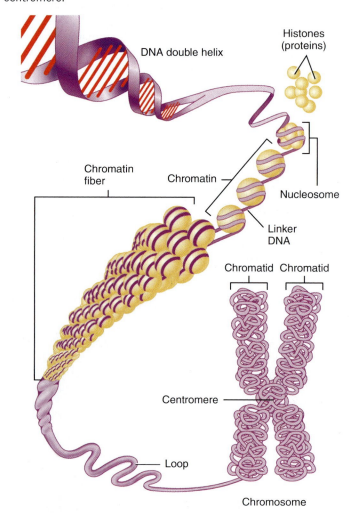

DNA double helix

Histones (proteins)

Chromatin fiber

Chromatin

Nucleosome

Linker DNA

Chromatid   Chromatid

Centromere

Loop

Chromosome

🔑 A chromosome is a highly coiled and folded DNA molecule that is combined with protein molecules.

Checkpoint

20.   How do large particles enter and exit the nucleus?
21.   How is DNA packed inside the nucleus?
22.   What are the components of a nucleosome?

TABLE 3.2

Cell Parts and Their Functions

PART	DESCRIPTION	FUNCTIONS
PLASMA MEMBRANE	Fluid-mosaic lipid bilayer (phospholipids, cholesterol, and glycolipids) studded with proteins; surrounds cytoplasm.	Protects cellular contents; makes contact with other cells; contains channels, carriers, receptors, enzymes, cell-identity markers, and linker proteins; mediates the entry and exit of substances.
CYTOPLASM	Cellular contents between the plasma membrane and nucleus—cytosol and organelles.	Site of all intracellular activities except those occurring in the nucleus.
Cytosol	Composed of water, solutes, suspended particles, lipid droplets, and glycogen granules.	Medium in which many of cell's metabolic reactions occur.
Organelles	Specialized structures with characteristic shapes.	Each organelle has specific functions.
Cytoskeleton	Network of three types of protein filaments: microfilaments, intermediate filaments, and microtubules.	Maintains shape and general organization of cellular contents; responsible for cellular movements.
Centrosome	A pair of centrioles plus pericentriolar material.	The pericentriolar material contains tubulins, which are used for growth of the mitotic spindle and microtubule formation.
Cilia and flagella	Motile cell surface projections that contain 20 microtubules and a basal body.	Cilia move fluids over a cell's surface; flagella move an entire cell.
Ribosome	Composed of two subunits containing ribosomal RNA and proteins; may be free in cytosol or attached to rough ER.	Protein synthesis.
Endoplasmic reticulum (ER)	Membranous network of flattened sacs or tubules. Rough ER is covered by ribosomes and is attached to the nuclear envelope; smooth ER lacks ribosomes.	Rough ER synthesizes glycoproteins and phospholipids that are transferred to cellular organelles, inserted into the plasma membrane, or secreted during exocytosis. Smooth ER synthesizes fatty acids and steroids; inactivates or detoxifies drugs; removes the phosphate group from glucose 6-phosphate; and stores and releases calcium ions in muscle cells.
Golgi complex	Consists of 3–20 flattened membranous sacs called cisternae; structurally and functionally divided into entry face, medial cisternae, and exit face.	Entry face accepts proteins from rough ER; medial cisternae form glycoproteins, glycolipids, and lipoproteins; exit face modifies the molecules further, then sorts and packages them for transport to their destinations.
Lysosome	Vesicle formed from Golgi complex; contains digestive enzymes.	Fuses with and digests contents of endosomes, pinocytic vesicles, and phagosomes and transports final products of digestion into cytosol; digests worn-out organelles (autophagy), entire cells (autolysis), and extracellular materials.
Peroxisome	Vesicle containing oxidases (oxidative enzymes) and catalase (decomposes hydrogen peroxide); new peroxisomes bud from preexisting ones.	Oxidizes amino acids and fatty acids; detoxifies harmful substances, such as alcohol; produces hydrogen peroxide.
Proteasome	Tiny barrel-shaped structure that contains proteases (proteolytic enzymes).	Degrades unneeded, damaged, or faulty proteins by cutting them into small peptides.
Mitochondrion	Consists of an outer and an inner mitochondrial membrane, cristae, and matrix; new mitochondria form from preexisting ones.	Site of aerobic cellular respiration reactions that produce most of a cell's ATP.
NUCLEUS	Consists of a nuclear envelope with pores, nucleoli, and chromosomes, which exist as a tangled mass of chromatin in interphase cells.	Nuclear pores control the movement of substances between the nucleus and cytoplasm, nucleoli produce ribosomes, and chromosomes consist of genes that control cellular structure and direct cellular functions.

Sectional view

3.6 Cells make proteins by transcribing and translating the genetic information contained in DNA.

Although cells synthesize many chemicals to maintain homeostasis, much of the cellular machinery is devoted to synthesizing large numbers of diverse proteins. The proteins in turn determine the physical and chemical characteristics of cells and the organisms formed from them. Some proteins help assemble cellular structures such as the plasma membrane, the cytoskeleton, and other organelles. Others serve as hormones, antibodies, and contractile elements in muscle tissue. Still others act as enzymes, regulating the rates of the numerous chemical reactions that occur in cells, or as transporters, carrying various materials in the blood. Just as genome means all of the genes in an organism, **proteome** (PRŌ-te-ōm) refers to all of an organism's proteins.

In **gene expression**, a gene's DNA is used as a template for synthesis of a specific protein (Figure 3.25). First, in a process aptly named transcription, the information encoded in a specific region of DNA is *transcribed* (copied) to produce a specific molecule of RNA (ribonucleic acid). In a second process, referred to as translation, the RNA attaches to a ribosome, where the information contained in RNA is *translated* into a corresponding sequence of amino acids to form a new protein molecule.

The genetic information carried in DNA and RNA is stored as sets of three-nucleotide sequences, each set representing a specific amino acid. The three-nucleotide sequence in DNA is called a **base triplet**. Each DNA base triplet is transcribed as a complementary sequence of three RNA nucleotides, called a **codon**. A given codon specifies a particular amino acid.

Transcription

Transcription occurs in the nucleus. During transcription, genetic information in a sequence of DNA base triplets is used as a template to copy a complementary sequence of codons. Three types of RNA are made from the DNA template:

1. **Messenger RNA (mRNA)**, the traveling copy of a gene, enters the cytoplasm to direct the synthesis of a specific protein.

2. **Ribosomal RNA (rRNA)** joins with ribosomal proteins to make ribosomes.

3. **Transfer RNA (tRNA)** binds to an amino acid and holds it in place on a ribosome until it is incorporated into a protein during translation. One end of the tRNA carries a specific amino acid, and the opposite end consists of a triplet of nucleotides called an **anticodon** (see Figure 3.28). The tRNA anticodon attaches to the mRNA codon by pairing of complementary bases. Specific tRNA binds to only one of the 20 different amino acids.

The enzyme **RNA polymerase** catalyzes transcription of DNA. However, the enzyme must be instructed where to start the transcription process and where to end it. A special sequence of nucleotides called a **promoter** is a segment of DNA near the beginning of a gene (Figure 3.26a).

During transcription, bases pair in a complementary manner: The bases cytosine (C), guanine (G), and thymine (T) in the DNA template pair with guanine, cytosine, and adenine (A), respectively, in the RNA strand (Figure 3.26b). However, adenine in the DNA template pairs with uracil (U), not thymine, in RNA:

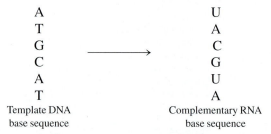

Template DNA base sequence		Complementary RNA base sequence
A		U
T		A
G	→	C
C		G
A		U
T		A

Transcription of the DNA strand ends at another special nucleotide sequence called a **terminator**, which specifies the end of the gene (Figure 3.26a). When RNA polymerase reaches the terminator, the enzyme detaches from the transcribed RNA molecule and the DNA strand. Once synthesized, mRNA passes through a pore in the nuclear envelope to reach the cytoplasm, where it participates in the next step in protein synthesis, translation.

Figure 3.25 Overview of gene expression. Synthesis of a specific protein requires transcription of a gene's DNA into RNA and translation of RNA into a corresponding sequence of amino acids.

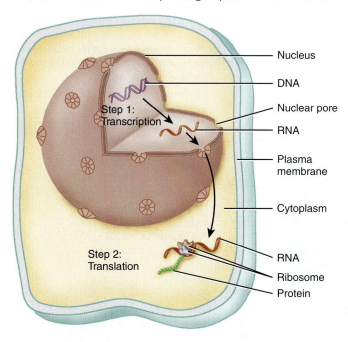

Transcription occurs in the nucleus; translation occurs in the cytoplasm.

Figure 3.26 **Transcription.** DNA transcription begins at a promoter and ends at a terminator.

(a) Overview

Gene

Promoter

DNA

Terminator

Newly synthesized RNA

(b) Details

RNA polymerase

Codon

RNA nucleotides

DNA strand being transcribed

Base triplet

Direction of transcription

Newly synthesized mRNA

mRNA

Nuclear pore

Nuclear envelope

Cytoplasm

Key:

A = Adenine

G = Guanine

T = Thymine

C = Cytosine

U = Uracil

 During transcription, the genetic information in DNA is copied to RNA.

Figure 3.27 **Translation.** During translation, an mRNA molecule binds to a ribosome. Then the mRNA nucleotide sequence specifies the amino acid sequence of a protein.

P site   A site

Large subunit

Protein

Small subunit

mRNA

mRNA binding site

(a) Components of a ribosome and their relationship to mRNA and protein during translation

(b) Interior view of tRNA binding sites

Ribosomes have a binding site for mRNA and a P site and an A site for attachment of tRNA.

Translation

In the process of **translation**, the nucleotide sequence in an mRNA molecule specifies the amino acid sequence of a particular protein. Ribosomes in the cytoplasm carry out translation. The small subunit of a ribosome has a *binding site* for mRNA; the large subunit has two binding sites for tRNA molecules, a *P site* and an *A site* (Figure 3.27). The first tRNA molecule bearing its specific amino acid attaches to mRNA at the P site. The A site holds the next tRNA molecule bearing its amino acid. Translation occurs in the following way (Figure 3.28):

❶ An mRNA molecule binds to the small ribosomal subunit at the mRNA binding site. A special tRNA carrying the first amino acid, called *initiator tRNA*, binds to the start codon (AUG) on mRNA, where translation begins. The tRNA anticodon (UAC) attaches to the mRNA codon (AUG) by pairing between the complementary bases.

❷ Next, the large ribosomal subunit attaches to the small ribosomal subunit–mRNA complex, creating a functional ribosome. The initiator tRNA, with its amino acid, fits into the P site of the ribosome.

❸ The anticodon of another tRNA with its attached amino acid pairs with the second mRNA codon at the A site of the ribosome.

❹ A peptide bond is formed between the amino acid carried by the initiator tRNA and the tRNA next to it.

Figure 3.28   Protein elongation and termination of protein synthesis during translation.

Large
subunit

P site

Initiator tRNA

A site

Small
subunit

2 Large and small ribosomal
subunits join to form a functional
ribosome and initiator tRNA
fits into P site.

Amino acid

tRNA

P site

Anticodon

A site

mRNA

Codons

3 Anticodon of incoming tRNA pairs
with next mRNA codon at A site.

Amino acid
(methionine)

Initiator tRNA

Anticodon

mRNA

mRNA
binding
site

Small
subunit

Start codon

1 Initiator tRNA attaches to a
start codon.

4 Amino acid on tRNA at P site
forms a peptide bond with
amino acid at A site.

New
peptide
bond

Stop codon

6 Protein synthesis stops when
the ribosome reaches stop
codon on mRNA.

5 tRNA at P site leaves ribosome,
ribosome shifts by one codon;
tRNA previously at A site is now
at the P site.

mRNA
movement

Key:

A = Adenine

G = Guanine

C = Cytosine

U = Uracil

mRNA

Growing
protein

Complete protein

tRNA

Summary of movement of ribosome along mRNA

🔑 **During protein synthesis the small and large ribosomal subunits join to form a functional ribosome.**

5 After peptide bond formation, the tRNA at the P site detaches from the ribosome, and the ribosome shifts the mRNA strand by one codon. The tRNA in the A site, now bearing the dipeptide, shifts into the P site, allowing another tRNA with its amino acid to bind to a newly exposed codon at the A site. Steps **3** through **5** occur repeatedly, and the protein lengthens progressively.

6 Protein synthesis ends when the ribosome reaches a stop codon at the A site, which causes the completed protein to detach from the final tRNA. When the tRNA vacates the A site, the ribosome splits into its large and small subunits.

Protein synthesis progresses at a rate of about 15 peptide bonds per second. As the ribosome moves along the mRNA and before it completes synthesis of the whole protein, another ribosome may attach behind it and begin translation of the same mRNA strand. The simultaneous movement of several ribosomes along the same mRNA molecule allows translation of one mRNA into several identical proteins, significantly speeding up protein synthesis.

Checkpoint

23. What is the difference between transcription and translation?
24. What are the functions of mRNA, rRNA, and tRNA?
25. Which roles do the P and A sites serve?

Return to Joseph's Story

Joseph lay motionless except for the chest compressions being applied by the stranger who had come to help. Minutes had passed, but with each cycle of CPR, some fresh oxygen diffused into Joseph's lungs, and some carbon dioxide was forced out. Even with these efforts, many cells in Joseph's body had begun to die. The lack of ATP had not only affected the plasma membrane pumps, it also meant that special calcium ATPases had stopped moving calcium from the cytosol into the endoplasmic reticulum of his cardiac muscle cells. As the intracellular calcium levels rose, they caused proteases to spill into the interior of the cell, attacking the cytoskeleton. Lysosomal enzymes normally bound safely inside vesicles began to digest the plasma membranes and the membranes of the organelles.

After what seemed an eternity to Marcus, the paramedics finally arrived. His dad's face had regained some color, but he was still unconscious. The paramedics quickly did their work, restarting Joseph's heart with an electrical jolt from their portable defibrillator. Stabilizing Joseph as best they could, they all headed off to the emergency department.

Joseph's wife Sally had been contacted by the Sheriff's Department and was waiting at the hospital when they brought him in. Marcus joined her in the waiting room, where they waited for over half an hour before the doctor finally came

to talk with them. "Your husband has suffered a massive heart attack. Right now he's stable and responsive, but he's very fatigued. From what we can tell, he has a blockage of one of the coronary arteries. We'll need to unblock the vessel to reestablish blood flow to the heart."

Sally put her arm around Marcus. "Is he going to be alright?" she asked.

"Well, the body has an amazing capacity to heal itself," said the doctor. "There are probably parts of the heart muscle that have been permanently damaged, but what we need to do right now is to provide oxygen and blood to the cells of the heart that are still alive."

C. Which intracellular organelles have membranes as part of their structure? How would the breakdown of the membranes of these structures affect the function of Joseph's heart cells?

D. Two important pieces of information—the instructions Joseph's body needs to repair itself and his predisposition for vascular disease—are both contained within the cell on which structures?

E. Joseph's heart attack has caused the function of his cells to change. What types of proteins in the cell membrane were involved in the homeostatic imbalances of his heart cells?

3.7 Cell division allows the replacement of cells and the production of new cells.

Most cells of the human body undergo **cell division**, the process by which cells reproduce themselves. The two types of cell division—somatic cell division and reproductive cell division—accomplish different goals for the organism.

A **somatic cell** (*soma* = body) is any cell of the body other than a gamete (sperm or oocyte) or any precursor cell destined to become a gamete. In **somatic cell division**, a cell undergoes a nuclear division called **mitosis** (mī-TŌ-sis; *mitos* = thread) and a cytoplasmic division called **cytokinesis** (sī´-tō-ki-NĒ-sis; *cyto* =

cell; *kinesis* = movement) to produce two identical cells, each with the same number and kind of chromosomes as the original cell. Somatic cell division replaces dead or injured cells and adds new cells during tissue growth.

Reproductive cell division is the mechanism that produces gametes (sperm and ova), the cells needed to form the next generation of sexually reproducing organisms. This process consists of a special two-step division called **meiosis**, in which the number of chromosomes in the nucleus is reduced by half.

Somatic Cell Division

The **cell cycle** (Figure 3.29) is an orderly sequence of events by which a somatic cell duplicates its contents and divides into two identical cells. Human cells, such as those in the brain, stomach, and kidneys, contain 23 pairs of chromosomes, for a total of 46. Because somatic cells contain two sets of chromosomes, they are called **diploid cells** (DIP-loyd; *dipl-* = double; *-oid* = form), symbolized **2n**. One member of each chromosome pair is inherited from each parent. The two chromosomes that make up each pair are called **homologous chromosomes** (hō-MOL-ō-gus; *homo-* = same); they contain similar genes arranged in the same (or almost the same) order. When examined under a light microscope, homologous chromosomes generally look very similar. The exception to this rule is one pair of chromosomes called the **sex chromosomes**, designated X and Y. In females the homologous pair of sex chromosomes consists of two large X chromosomes; in males the pair consists of an X chromosome and a smaller Y chromosome.

Figure 3.29 The cell cycle. Not illustrated is cytokinesis, division of the cytoplasm, which usually occurs during late anaphase of the mitotic phase.

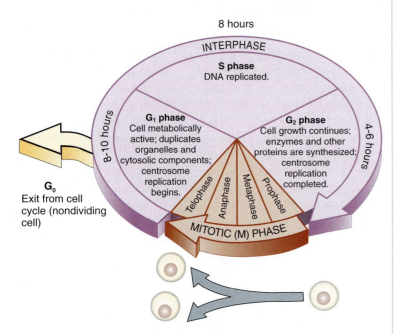

🔑 **In a complete cell cycle, a cell duplicates its contents and divides into two identical cells.**

When a cell reproduces, it must replicate (duplicate) all its chromosomes to pass its genes to the next generation of cells. The cell cycle consists of two major periods: interphase, when a cell is not dividing, and the mitotic (M) phase, when a cell is dividing (Figure 3.29).

Interphase

During **interphase** the cell replicates its DNA. It also produces additional organelles and cytosolic components in anticipation of cell division. Interphase is a state of high metabolic activity; it is during this time that the cell does most of its growing. Interphase consists of three phases: G_1, S, and G_2 (Figure 3.29). The S stands for *synthesis* of DNA. Because the G phases are periods when there is no activity related to DNA duplication, they are thought of as *gaps* or interruptions in DNA duplication.

The **G_1 phase** is the interval between the mitotic phase and the S phase. During G_1, the cell is metabolically active; it replicates most of its organelles and cytosolic components but not its DNA. Replication of centrosomes also begins in the G_1 phase. Virtually all the cellular activities described in this chapter happen during G_1. For a cell with a total cell cycle time of 24 hours, G_1 lasts 8 to 10 hours. However, the duration of this phase is quite variable. It is very short in many embryonic cells or cancer cells. Cells that remain in G_1 for a very long time, perhaps destined never to divide again, are said to be in the **G_0 phase**. Most nerve cells are in the G_0 phase. Once a cell enters the S phase, however, it is committed to go through cell division.

The **S phase**, the interval between G_1 and G_2, lasts about 8 hours in a 24-hour cell cycle. During the S phase, DNA replication occurs. As a result, the two identical cells formed during cell division have the same genetic material. When DNA replicates during the S phase, its helical structure partially uncoils, and the two strands separate at the points where hydrogen bonds connect base pairs (Figure 3.30). Each exposed base of the old DNA strand then pairs with the complementary base of a newly synthesized nucleotide. A new DNA strand takes shape as chemical bonds form between neighboring nucleotides. The uncoiling and complementary base pairing continues until each of the two original DNA strands is joined with a newly formed complementary DNA strand. The original DNA molecule has become two identical DNA molecules.

The **G_2 phase**, the interval between the S phase and the mitotic phase, lasts 4 to 6 hours of a 24-hour cell cycle. During G_2, cell growth continues, enzymes and other proteins are synthesized in preparation for cell division, and replication of centrosomes is completed.

Figure 3.30 **Replication of DNA.** The two strands of the double helix separate by breaking the hydrogen bonds (shown as dotted lines) between nucleotides. New, complementary nucleotides attach at the proper sites, and a new strand of DNA is synthesized alongside each of the original strands. Arrows indicate hydrogen bonds forming again between pairs of bases.

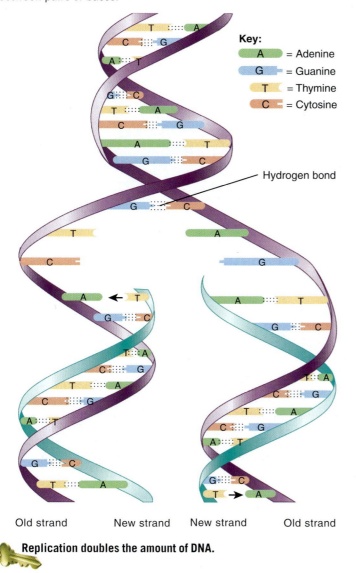

Key:
- A = Adenine
- G = Guanine
- T = Thymine
- C = Cytosine

Hydrogen bond

| Old strand | New strand | New strand | Old strand |

Replication doubles the amount of DNA.

A microscopic view of a cell during interphase shows a clearly defined nuclear envelope, a nucleolus, and a tangled mass of chromatin (Figure 3.31a). Once a cell completes its activities during the G_1, S, and G_2 phases of interphase, the mitotic phase begins.

Mitotic Phase

The **mitotic (M) phase** of the cell cycle consists of a nuclear division (mitosis) and a cytoplasmic division (cytokinesis) to form two identical cells. The events that occur during mitosis and cytokinesis are plainly visible under a microscope because chromatin condenses into discrete chromosomes.

Nuclear Division: Mitosis

Mitosis, as noted earlier, is the distribution of two sets of chromosomes into two separate nuclei. The process results in the exact partitioning of genetic information. For convenience, biologists divide the process into four stages: prophase, metaphase, anaphase, and telophase. However, mitosis is a continuous process; one stage merges directly into the next.

1 **Prophase.** During early prophase, the chromatin fibers condense and shorten into chromosomes that are visible under the light microscope (Figure 3.31b). The condensation process may prevent entangling of the long DNA strands as they move during mitosis. Because DNA replication took place during the S phase of interphase, each prophase chromosome consists of a pair of identical, double-stranded chromatids. A constricted region called a **centromere** holds the chromatid pair together. At the outside of each centromere is a protein complex known as the **kinetochore** (ki-NET-ō-kor). Later in prophase, tubulins in the pericentriolar material of the centrosomes start to form the **mitotic spindle**, a football-shaped assembly of microtubules that attach to each kinetochore (Figure 3.31b). The mitotic spindle is responsible for the separation of chromatids to opposite poles of the cell. As the microtubules lengthen, they push the centrosomes to the poles (ends) of the cell so that the spindle extends from pole to pole. Then the nucleolus disappears and the nuclear envelope breaks down.

2 **Metaphase.** During metaphase, the microtubules of the mitotic spindle align the centromeres of the chromatid pairs at the exact center of the mitotic spindle (Figure 3.31c). This midpoint region is called the *metaphase plate*.

3 **Anaphase.** During anaphase, the centromeres split, separating the two members of each chromatid pair, which move toward opposite poles of the cell (Figure 3.31d). Once separated, the chromatids are termed chromosomes. As the chromosomes are pulled by the microtubules of the mitotic spindle during anaphase, they appear V-shaped because the centromeres lead the way, dragging the trailing arms of the chromosomes toward the pole.

Figure 3.31   Cell division: mitosis and cytokinesis. Begin the sequence at **1** at the top of the figure and read clockwise to complete the process.

1

Centrosome:
  Centrioles
  Pericentriolar material
Nucleolus
Nuclear envelope
Chromatin
Plasma membrane
Cytosol

LM all at 700x

(a) INTERPHASE

6

2

(f) IDENTICAL CELLS IN INTERPHASE

5

Kinetochore

Centromere

Chromosome
(two chromatids
joined at
centromere)

Mitotic spindle
(microtubules)

Fragments of
nuclear envelope

Early   Late

(b) PROPHASE

Metaphase plate

3

Cleavage furrow

(e) TELOPHASE

(c) METAPHASE

4

Cleavage furrow

Chromosome

Late   Early

(d) ANAPHASE

🔑 **In somatic cell division, a single diploid cell divides to produce two identical diploid cells.**

4 **Telophase**. The final stage of mitosis, telophase, begins after chromosomal movement stops (Figure 3.31e). The identical sets of chromosomes, now at opposite poles of the cell, uncoil and revert to the threadlike chromatin form. A nuclear envelope forms around each chromatin mass, nucleoli reappear in the identical nuclei, and the mitotic spindle breaks up.

Cytoplasmic Division: Cytokinesis

As noted earlier, division of a cell's cytoplasm and organelles into two identical cells is called **cytokinesis**. This process begins in late anaphase with the formation of a **cleavage furrow**, a slight indentation of the plasma membrane, and is completed after telophase. The cleavage furrow usually appears midway between the centrosomes and extends around the periphery of the cell (**5** Figure 3.31e). Microfilaments that lie just inside the plasma membrane form a contractile ring that pulls the plasma membrane progressively inward. The ring constricts the center of the cell like a belt around the waist, and ultimately pinches it in two. Because the plane of the cleavage furrow is always perpendicular to the mitotic spindle, the two sets of chromosomes end up in separate cells. When cytokinesis is complete there are two new and separate cells, each with equal portions of cytoplasm and organelles and identical chromosomes, and interphase begins (**6** Figure 3.31f).

The sequence of events can be summarized as

$$G_1 \longrightarrow S \text{ phase} \longrightarrow G_2 \text{ phase} \longrightarrow \text{mitosis} \longrightarrow \text{cytokinesis}$$

Table 3.3 summarizes the events of the cell cycle in somatic cells.

Reproductive Cell Division

In the process called sexual reproduction, a new organism results from the union of two gametes, one produced by each parent. If gametes had the same number of chromosomes as somatic cells, the number of chromosomes would double at fertilization (the union of the nuclei of two gametes). **Meiosis** (mī-Ō-sis; *mei-* = lessening; *-osis* = condition of), the reproductive cell division that occurs in the gonads (ovaries and testes), produces gametes in which the number of chromosomes is reduced by half. As a result, gametes contain a single set of 23 chromosomes and thus are **haploid cells** (HAP-loyd; *hapl-* = single symbolized **n**). Fertilization restores the diploid number of chromosomes.

TABLE 3.3

Events of the Somatic Cell Cycle

PHASE	ACTIVITY
Interphase	Period between cell divisions; chromosomes not visible under light microscope.
G₁ phase	Metabolically active cell duplicates organelles and cytosolic components; replication of chromosomes begins. (Cells that remain in the G₁ phase for a very long time, and possibly never divide again, are said to be in the G₀ phase.)
S phase	Replication of DNA and centrosomes.
G₂ phase	Cell growth, enzyme and protein synthesis continues; replication of centrosomes complete.
Mitotic phase	Parent cell produces identical cells with identical chromosomes; chromosomes visible under light microscope.
Mitosis	Nuclear division; distribution of two sets of chromosomes into separate nuclei.
Prophase	Chromatin fibers condense into paired chromatids; nucleolus and nuclear envelope disappear; each centrosome moves to an opposite pole of the cell.
Metaphase	Centromeres of chromatid pairs line up at metaphase plate.
Anaphase	Centromeres split; identical sets of chromosomes move to opposite poles of cell.
Telophase	Nuclear envelopes and nucleoli reappear; chromosomes resume chromatin form; mitotic spindle disappears.
Cytokinesis	Cytoplasmic division; contractile ring forms cleavage furrow around center of cell, dividing cytoplasm into separate and equal portions.

Meiosis

Unlike mitosis, which is complete after a single round, meiosis occurs in two successive stages: **meiosis I** and **meiosis II**. During the interphase that precedes meiosis I, the chromosomes of the diploid cell replicate. As a result of replication, each chromosome consists of two sister (genetically identical) chromatids, which are attached at their centromeres. This replication of chromosomes is similar to the one that precedes mitosis in somatic cell division.

Meiosis I

Meiosis I, which begins once chromosomal replication is complete, consists of four phases: prophase I, metaphase I, anaphase I, and telophase I (Figure 3.32a). Prophase I is an extended phase in which the chromosomes shorten and thicken, the nuclear envelope and nucleoli disappear, and the mitotic spindle forms. Two events that are not seen in mitotic prophase occur during prophase I of meiosis. First, the two sister chromatids of each pair of homologous chromosomes pair off, an event called **synapsis** (Figure 3.32b). The resulting four chromatids are called a **tetrad**. Second, parts of the chromatids of two homologous chromosomes may be exchanged with one another. Such an exchange between parts of nonsister (genetically different) chromatids is termed **crossing-over** (Figure 3.32b). This process, among others, permits an exchange of genes between chromatids of homologous chromosomes. Due to crossing-over, the resulting cells are genetically unlike each other and genetically unlike the starting cell that produced them. Crossing-over results in *genetic recombination*—that is, the formation of new combinations of genes—and accounts for part of the great genetic variation among humans and other organisms that form gametes via meiosis.

In metaphase I, the tetrads formed by the homologous pairs of chromosomes line up along the metaphase plate of the cell, with homologous chromosomes side by side (Figure 3.32a). During anaphase I, the members of each homologous pair of chromosomes separate as they are pulled to opposite poles of the cell by the microtubules attached to the centromeres. The paired chromatids, held by a centromere, remain together. (Recall that during mitotic anaphase, the centromeres split and the sister chromatids separate.) Telophase I and cytokinesis of meiosis are similar to telophase and cytokinesis of mitosis. The net effect of meiosis I is that each resulting cell contains the haploid number of chromosomes because it contains only one member of each pair of the homologous chromosomes present in the starting cell.

Meiosis II

The second stage of meiosis, meiosis II, also consists of four phases: prophase II, metaphase II, anaphase II, and telophase II (Figure 3.32a). These phases are similar to those that occur during mitosis; the centromeres split, the sister chromatids separate and move toward opposite poles of the cell, and a nuclear envelope forms around each chromatin mass.

In summary, meiosis I begins with a diploid starting cell and ends with two cells, each with the haploid number of chromosomes. During meiosis II, each of the two haploid cells formed during meiosis I divides, resulting in four haploid gametes that are genetically different from the original diploid starting cell.

Figure 3.33 compares the events of meiosis and mitosis.

Checkpoint

26. What is the difference between somatic and reproductive cell division? What is the importance of each?
27. What are the major events of each phase of interphase?
28. What are the major events of each stage of the mitotic phase of the cell cycle?
29. When does cytokinesis begin during the cell cycle?
30. How does anaphase I of meiosis differ from anaphase of mitosis and anaphase II of meiosis?
31. What is the significance of meiosis I?

Figure 3.32 **Meiosis, reproductive cell division.** Details of events are discussed in the text.

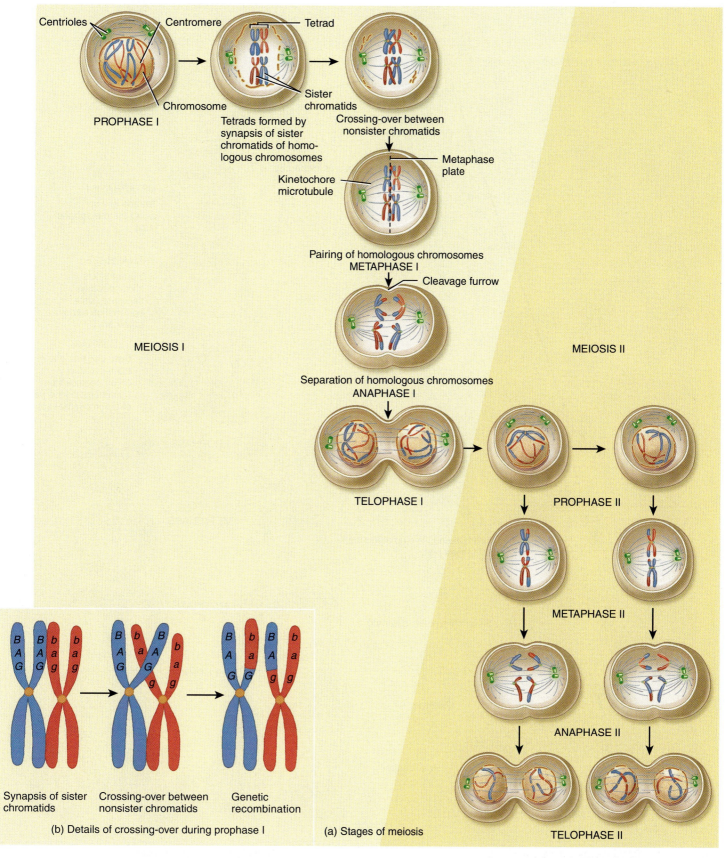

Centrioles

Centromere

Tetrad

Chromosome

Sister chromatids

PROPHASE I

Tetrads formed by synapsis of sister chromatids of homologous chromosomes

Crossing-over between nonsister chromatids

Kinetochore microtubule

Metaphase plate

Pairing of homologous chromosomes
METAPHASE I

Cleavage furrow

MEIOSIS I

MEIOSIS II

Separation of homologous chromosomes
ANAPHASE I

TELOPHASE I

PROPHASE II

METAPHASE II

ANAPHASE II

TELOPHASE II

B A G B A G b a g b a g

Synapsis of sister chromatids

Crossing-over between nonsister chromatids

Genetic recombination

(b) Details of crossing-over during prophase I

(a) Stages of meiosis

In reproductive cell division, a single diploid starting cell undergoes meiosis I and meiosis II to produce four haploid gametes that are genetically different from the starting cell that produced them.

Figure 3.33 Comparison between mitosis (left) and meiosis (right) in which the starting cell has two pairs of homologous chromosomes.

MITOSIS

MEIOSIS

Starting cell

Crossing-over

2n
Chromosomes already replicated

PROPHASE I

Tetrads formed by synapsis

METAPHASE I
Tetrads line up along the metaphase plate

ANAPHASE I
Homologous chromosomes separate (sister chromatids remain together)

TELOPHASE I
Each cell has one of the replicated chromosomes from each homologous pair of chromosomes (n)

PROPHASE II

Chromosomes line up along the metaphase plate

METAPHASE II

Sister chromatids separate

ANAPHASE II

Cytokinesis

TELOPHASE II

Resulting cells

2n   2n   n   n   n   n

Somatic cells with diploid number of chromosomes (not replicated)

Gametes with haploid number of chromosomes (not replicated)

The phases of meiosis II and mitosis are similar.

Joseph's Story: Epilogue and Discussion

▶▶▶ Joseph, a 38-year-old male smoker with a history of hypertension, has suffered a heart attack. He has several risk factors associated with cardiovascular disease, including a family history of heart disease, hypertension, and poor diet. As we have learned in this chapter, his heart attack directly impacted cellular processes. Oxygen is required by cells to produce ATP, the energy currency of the cell. Without ATP, oxygen, and nutrients, multiple cellular processes affecting homeostasis can shut down; water balance, ionic equilibrium, and pH balance are all key factors in the homeostasis of individual cells. Joseph was fortunate that oxygen flow to the cells of his body was maintained by CPR. The damage to the cellular processes discussed in this story can cause irreversible cell death in the brain within a matter of minutes if oxygen is not available. As you continue your studies of anatomy and physiology, remember that each tissue and organ system of the human body is composed of individual cells, each of which functions to keep you alive. Health and disease are ultimately determined at the cellular and molecular level. ◀◀◀

F. Why was reestablishing oxygen flow to Joseph's body so important? What processes would be affected by lack of oxygen?

G. Consider once again Joseph's health history and risk factors. What could you suggest to Joseph to reduce his risk for another heart attack?

H. Explain why Joseph's heart failed based on what you have learned so far about the function of cells in the human body.

Concept and Resource Summary

Concept	Resources

Introduction

1. **Cells** are the basic, living structural and functional units of the body.
2. **Cell biology** is the study of cellular structure and function.

Concept 3.1 The principal parts of a cell are the plasma membrane, the cytoplasm, and the nucleus.

1. The cell is separated from the external environment by its **plasma membrane**, a flexible surface that serves as a selective barrier, regulating exchanges of material with the extracellular environment and facilitating communication with other cells.
2. The **cytoplasm** is the region between the plasma membrane and nucleus; its two components are the cytosol and organelles. The **cytosol** is the fluid portion containing water, dissolved solutes, and suspended particles. The cytosol surrounds several types of **organelles**, which have specific shapes and carry out specific cellular functions.
3. The cell's DNA is housed in a large organelle called the **nucleus**. Each **chromosome** in the nucleus is a DNA molecule associated with several proteins. **Genes** within the DNA molecule are hereditary units that control cellular structure and function.

See
- **Anatomy Overview – Cell Structure and Function**

Hear
- **Figure 3.1 – Typical Structures Found in Body Cells**

Concept 3.2 The plasma membrane contains the cytoplasm and regulates exchanges with the extracellular environment.

1. The **plasma membrane** is a flexible barrier that surrounds the cell's cytoplasm. It has a *fluid mosaic model* structure consisting of a sea of fluid lipids containing a mosaic of different proteins.
2. The **lipid bilayer** contains **phospholipids, cholesterol**, and **glycolipids**. The phospholipids have a hydrophilic, polar phosphate "head," and a hydrophobic, nonpolar part containing two long fatty acid "tails." The hydrophilic heads face outward and the hydrophobic tails point toward one another in the membrane's interior. Cholesterol molecules are scattered throughout the lipid layers. Glycolipids are associated with the lipid layer facing the extracellular fluid.
3. There are two types of membrane proteins. **Integral proteins** extend into the membrane; many of these are transmembrane proteins, extending all the way through the membrane. **Peripheral proteins** are associated with the polar heads of membrane lipids or with integral proteins at the inner or outer membrane surfaces. **Glycoproteins** protrude into the extracellular fluid; the carbohydrate portions of the glycoproteins and glycolipids form the **glycocalyx**. The glycocalyx functions in cell-to-cell recognition, cell adherence, and attraction of a film of surface fluid around the cell.

See
- **Animation – Membrane Functions**
- **Anatomy Overview – Plasma Membrane**

Hear
- **Figure 3.2 – The Fluid Mosaic Arrangement of Lipids and Proteins in the Plasma Membrane**

Do
- **Exercise – Paint a Cell Membrane**
- **Concepts and Connections – Membrane Functions**

Concept

4. The specific types of membrane proteins vary considerably among cells. Proteins determine many membrane functions, including the following: forming **ion channels**, acting as **carriers**, serving as **receptors**, acting as **enzymes**, serving as **linkers**, and acting as **cell-identity markers**.

5. Most of the membrane lipids and many membrane proteins rotate and move laterally in the bilayer. Membrane fluidity is increased by the numbers of double bonds in the fatty acid tails of the lipids. Membrane fluidity allows interactions to occur within the membrane; enables cell movement, growth, division, secretion, and formation of cellular junctions; and helps the membrane to self-seal if punctured.

6. Plasma membranes have **selective permeability**; the lipid bilayer is permeable to nonpolar, uncharged molecules and slightly permeable to small, uncharged polar molecules, but it is impermeable to ions and large, uncharged polar molecules.

7. Transmembrane proteins acting as channels and carriers increase permeability to ions and uncharged polar molecules that cannot cross the lipid bilayer.

8. Selective permeability of the plasma membrane allows the living cell to maintain **concentration gradients**, different concentrations of certain ions and molecules on either side of the plasma membrane. Typically, the inner surface of the plasma membrane is more negatively charged and the outer surface is more positively charged due to a difference in distribution of charged ions; this creates an **electrical gradient**. The combined influence of the concentration gradient and the electrical gradient on movement of a particular ion is referred to as its **electrochemical gradient**.

Concept 3.3 Transport of a substance across the plasma membrane occurs by both passive and active processes.

1. **Diffusion** is a passive transport process in which the solutes and solvents are driven by their own kinetic energy to move from an area of higher concentration toward an area of lower concentration *down their concentration gradient* until evenly distributed. They continue to move about due to kinetic energy but their concentrations do not change. Factors that affect diffusion rates include steepness of the concentration gradient, temperature, mass of the diffusing substance, surface area of the membrane, and diffusion distance.

2. Nonpolar, hydrophobic molecules diffuse freely through the lipid bilayer without help of the membrane transport proteins. Water and urea are uncharged polar molecules that diffuse in small amounts through the lipid bilayer.

3. Integral transmembrane proteins form *ion channels* for passage of small, hydrophilic ions. Many ion channels are gated; they either open and close randomly or are regulated by chemical and electrical changes inside and outside the cells. Carriers can also ferry solutes across the plasma membrane by changing their shape.

4. The net movement of a solvent through a selectively permeable membrane is called **osmosis**. Water molecules move by osmosis down their concentration gradient, across plasma membranes. Osmosis only occurs when a membrane is permeable to water but not permeable to certain solutes.

5. **Tonicity** is the measure of a solution's ability to change the volume of cells by altering their water content. **Isotonic solutions** allow a cell to maintain normal shape and volume. In **hypotonic solutions**, which have a lower concentration of solutes than the cell's cytosol, water molecules enter the cell faster than they leave it, causing the cell to swell and undergo **lysis**. Because **hypertonic solutions** have a higher concentration of solutes than the cell's cytosol, water molecules leave the cell faster than they enter, causing the cell to shrink or undergo **crenation**.

6. **Facilitated diffusion** is a passive movement in which a polar or highly charged solute utilizes channels or carriers to cross the other side of the plasma membrane as it moves down its concentration gradent.

See

- Animation – Transport Across the Plasma Membrane

Hear

- **Figure 3.5 – Types of Diffusion**
- **Figure 3.8 – Tonicity and Its Effects on Red Blood Cells**
- **Figure 3.11 – Receptor-Mediated Endocytosis of a Low-Density Lipoprotein (LDL) Particle**

Do

- **Exercise – Animation Membrane Function Match Up**
- **Exercise – Create a Transport Condition**
- **Exercise – Membrane Transport Match Up**
- **Exercise – Osmosis – Move It!**
- **Concepts and Connections – Membrane Transport Processes**

Concept

7. **Active transport** occurs when polar or charged solutes move against their concentration gradients using carrier proteins and cellular energy. In **primary active transport**, hydrolysis of ATP supplies energy to change the shape of the carrier protein, which "pumps" the substance against its concentration gradient. One such carrier, called the **sodium–potassium pump**, hydrolyzes ATP to expel Na^+ and bring K^+ into the cell. In **secondary active transport**, the energy stored in a Na^+ or H^+ concentration gradient is used to drive other substances across the membrane against their own gradients.

8. **Endocytosis** is the movement of materials into a cell within a **vesicle** formed from the plasma membrane. The three types of endocytosis are receptor-mediated endocytosis, phagocytosis, and bulk-phase endocytosis (pinocytosis). **Receptor-mediated endocytosis** is a highly selective process in which ligands must attach to specific membrane receptors for vesicular transport into the cell. **Phagocytosis** occurs when the **phagocyte** (cell) engulfs large, solid particles or whole cells. In **bulk-phase endocytosis**, the plasma membrane folds inward to form a vesicle around tiny droplets of dissolved substances that then enter the cytoplasm.

9. **Exocytosis** is the movement of vesicle-packaged materials out of the cell by fusion of the vesicle with the plasma membrane; the contents are then released into the extracellular fluid.

10. **Transcytosis** involves endocytosis of vesicles on one side of a cell, movement of the vesicles across the cell, and exocytosis on the other side.

Concept 3.4 Cytoplasm consists of the cytosol and organelles.

1. The cytoplasm consists of all the cellular contents between the plasma membrane and the nucleus; it is made up of the cytosol and organelles. The **cytosol (intracellular fluid)**, the fluid portion of the cytoplasm that surrounds the organelles, consists of water with dissolved and suspended components. The cytosol is the site of many chemical reactions. **Organelles** are structures with specific shapes that perform specific cell functions.

2. The **cytoskeleton** acts as a structural framework, helps determine cell shape, helps organize cellular contents, and aids movement within the cell and of whole cells. The cytoskeleton consists of microfilaments, intermediate tubules, and microtubules. **Microfilaments** supply mechanical support for cell strength and shape, and generate movements. **Intermediate filaments** help give stability to the position of organelles. **Microtubules** help determine cell shape, move organelles, and are components of cilia and flagella.

3. Near the nucleus is the **centrosome**, composed of two **centrioles** and pericentriolar material, which plays a role in growth of the mitotic spindle in cell division.

4. **Cilia** and **flagella** are motile extensions of the cell surface that contain microtubules surrounded by plasma membrane. Cilia are numerous, short, hairlike projections with an oarlike pattern of beating that moves fluids across the cell surface. Flagella are single, long, whiplike structures that move an entire cell.

5. **Ribosomes**, composed of one small subunit and one larger subunit, are the sites of protein synthesis. Ribosomes attached to the surface of the endoplasmic reticulum synthesize proteins for insertion on the plasma membrane or for export from the cell, whereas free, unattached ribosomes synthesize proteins used in the cytosol.

6. The **endoplasmic reticulum (ER)** is a membranous network of tubules throughout the cytoplasm with connections to the nuclear envelope. **Rough ER** is studded with ribosomes proteins for cell secretion or use in the cell. **Smooth ER** synthesizes fatty acid and steroids, detoxifies drugs, and stores calcium in muscle cells.

7. The **Golgi complex** consists of small, flattened, stacked membranous sacs, or **cisternae**. Enzymes of the Golgi complex modify, sort, and package proteins in vesicles for transport out of the cell, to the plasma membrane, or to other locations within the cell.

8. The cytoplasm contains a number of different vesicular organelles. **Lysosomes** are membrane-enclosed vesicles, containing digestive enzymes used to break down worn-out organelles (**autophagy**) or cells (**autolysis**). **Peroxisomes** contain oxidases for oxidizing fatty acids, amino acids, and toxic substances.

9. **Mitochondria** have a smooth outer membrane and an inner membrane arranged in folds called **cristae**. Within mitochondria most of the cell's ATP is produced during the aerobic phase of cellular respiration. Mitochondria can self-replicate and have their own DNA.

See
- Anatomy Overview – Cell Structure and Function

Hear
- **Figure 3.20** – Processing and Packaging of Proteins by the Golgi Complex

Do
- Exercise – Concentrate on Cellular Functions
- Exercise – Target Practice
- Concepts and Connections – Human Cell
- Concepts and Connections – Membranous Organelles

Concept	Resources

Concept 3.5 The nucleus contains nucleoli and genes.

1. The most prominent organelle is the usually single **nucleus**. The **nuclear envelope** is a double-layered membrane, pierced with **nuclear pores**. One or more **nucleoli**, which function in producing ribosomes, are located in the nucleus. The nucleus is the cell's control center and houses the cell's hereditary units, called **genes**, arranged along **chromosomes**. All of the genetic information of a cell or organism is called its **genome**.
2. Chromosomes are long DNA molecules coiled with proteins. Each beadlike unit of DNA wrapped around a core of eight **histone** proteins is called a **nucleosome**. Between nucleosomes are strings of **linker DNA**. The entire complex of DNA and proteins is called **chromatin**. Prior to cell division, the DNA replicates and condenses into linear chromosomes in which there are two identical **chromatids** per chromosome.

See
- Anatomy Overview – The Nucleus

Concept 3.6 Cells make proteins by transcribing and translating the genetic information contained in DNA.

1. The abundant, functionally diverse population of proteins of an organism is called the **proteome**.
2. In **gene expression**, a gene's DNA is the template for synthesis of a specific protein.
3. Genetic information carried in DNA and RNA occurs in sets of three-nucleotide sequences called a **base triplet** in DNA and a **codon** in transcribed RNA. Each set of three nucleotides represents a specific amino acid in the protein undergoing synthesis.
4. Protein synthesis is facilitated by specific RNA molecules in two steps called transcription and translation. **Transcription** occurs in the nucleus, where RNA polymerase catalyzes transcription of DNA into **messenger RNA (mRNA)**. Bases are paired in a complementary manner. Newly synthesized mRNA passes through a nuclear pore to the cytoplasm, where it participates in translation.
5. **Translation** is carried out on ribosomes, where amino acids are joined to form a specific protein. The small ribosome subunit has a binding site for mRNA; the large subunit has two binding sites for **transfer RNA (tRNA)** that carry amino acids to the ribosomes.
6. Translation begins when mRNA binds to the small ribosomal subunits; the large ribosomal subunit attaches, forming a complete ribosome. As tRNAs shuttle in appropriate amino acids by following the codons on the mRNA, peptide bonds form a growing polypeptide. When the ribosome reaches a "stop" codon on the mRNA, protein synthesis stops and the protein detaches.

See
- Animation – Protein Synthesis

Hear
- Figure 3.28 – Protein Elongation and Termination of Protein Synthesis During Translation

Do
- Exercise – Protein-producing Processes
- Exercise – Synthesize and Transport a Protein
- Concepts and Connections – Protein Synthesis

Concept 3.7 Cell division allows the replacement of cells and the production of new cells.

1. Most body cells undergo **cell division** to reproduce themselves. In **somatic cell division**, somatic cells (body cells other than gametes) undergo a nuclear division called **mitosis** and a cytoplasmic division called **cytokinesis**, resulting in two identical cells with the same number and type of chromosomes as the original cell. Gametes are produced by **reproductive cell division** or **meiosis**, a two-step process that reduces the chromosome number by half.
2. The **cell cycle** is the sequence of events from the time a somatic cell is formed until it divides. It consists of two parts: interphase, when the cell is not dividing; and the mitotic (M) phase, when the cell is dividing.
3. Somatic cells are called **diploid cells** because they contain two sets of chromosomes. The two chromosomes of each pair are called **homologous chromosomes;** one member of each pair is inherited from each parent. Human cells have 23 pairs of chromosomes, including the one pair called the **sex chromosomes**. During cell reproduction, the cell must replicate all its chromosomes to pass them on to the newly formed cells.

See
- Animation – The Cell Cycle and Division Processes

Hear
- Figure 3.33 – Comparison Between Mitosis and Meiosis

Do
- Exercise – Mitosis Matchup
- Exercise – Meiosis Matchup
- Exercise – Cell Cycle Quiz

Concept   Resources

4. During **interphase**, the cell is metabolically active. During the **G₁ phase**, the cell replicates organelles and centrosomes. During the the **S phase**, DNA replication occurs. The DNA helix uncoils, and each exposed base of the old DNA strand pairs with a complementary base of a newly synthesized nucleotide. At the end of replication, each identical DNA molecule is composed of one old and one newly synthesized strand. The **G₂ phase** is the time when growth continues and final preparation for cell division occurs.

5. The **mitotic (M) phase** is a time of nuclear division and cytokinesis, and results in two identical cells. **Mitosis** is the distribution of two sets of chromosomes into two separate nuclei. It occurs in four stages: **prophase, metaphase, anaphase,** and **telophase**.

6. Late in mitosis the process of cytoplasmic division, called **cytokinesis**, takes place. After completion of cytokinesis there are two new and separate cells, with equal portions of cytoplasm and organelles and identical chromosomes.

7. Sexual reproduction produces an organism from the union of two gametes, one from each parent. Gametes must be produced with half the chromosome number; otherwise the chromosome number would double each time fertilization occurred. **Meiosis**, or reproductive cell division, consists of two consecutive nuclear divisions resulting in four daughter cells; each of the daughter cells is a **haploid cell** containing a single set of 23 chromosomes.

8. During the interphase preceding meiosis I, the chromosomes of the diploid cell replicate. **Meiosis I** consists of prophase I, metaphase I, anaphase I, and telophase I. During prophase I, the two sister chromatids of each pair of homologous chromosomes pair off during **synapsis**. **Crossing-over** of parts of the chromatids may occur; these new combinations of genes allow for genetic variability among the offspring. Homologus pairs of chromosomes line up in metaphase, then separate in anaphase I. Cytokinesis occurs in telophase I. Meiosis I results in two cells, each containing a haploid number of chromosomes.

9. **Meiosis II** consists of prophase II, metaphase II, anaphase II, and telophase II. These phases result in the separation of the paired chromatids of the chromosomes in each of the two cells produced in meiosis I, with a final product of four haploid gametes that are genetically different from the original diploid cell.

Other Chapter Resources

For additional information on clinical applications, medical disorders, and medical terminology, see Chapter 2 in the *Clinical Connections* manual packaged with your text.

Within *WileyPLUS*

- Chapter Quiz
- Visual Anatomy
- Mastering Vocabulary
- Audio Glossary
- Homeostatic Imbalance: The Case of the Girl with Fruity Breath
- Web Links

UNDERSTANDING THE CONCEPTS

1. "The proteins present in a plasma membrane determine the functions that a membrane can perform." Is this statement true or false? Explain your answer.

2. When stimulating a cell, the hormone insulin first binds to a protein in the plasma membrane. This action best represents which membrane protein function?

3. In Figure 3.7, will the fluid level in the right arm rise until the water concentrations are the same in both arms? Why or why not?

4. Will a 2% solution of NaCl cause hemolysis or crenation of RBCs? Explain your answer.

5. In which ways are endocytosis and exocytosis similar? How are they different?

6. What does cytoplasm have that cytosol lacks?

7. If you observed that a cell did not have a centrosome, what could you predict about its capacity for cell division?

8. If a DNA template has the base sequence AGCT, what will be the mRNA base sequence?

9. Why is it crucial that DNA replication occurs before cytokinesis in somatic cell division?

10. How does crossing-over affect the genetic content of the haploid gametes?

4

The Tissue Level of Organization

A light snow fell in the valley, dusting the ground. It was cold, with little wind. Tina and Eric climbed cautiously up the uneven mountain trail, stepping over large, jagged rocks. "Let's rest up on those big boulders. We'll get a good view of the valley from there," said Eric.

He bounded up the trail to where two large rocky outcrops emerged from the side of the hill, took off his pack, and began digging in the side pocket for his binoculars. "Wow, what a view!" exclaimed Tina, sitting down next to Eric as he scanned the horizon.

"We should see some mountain goats up here. They like this area. What in the world is that? Tina, take a look. Does that look like a plane to you?"

Tina reached for the glasses and adjusted the focus. "It is! I can see the registration numbers on the wings. But I don't

remember anything being reported about a plane going down around here."

The plane rested against a large tree. Its nose was smashed in; one wing was broken off, the other bent at an unnatural angle. "I think there's someone in there," Tina whispered.

"We need to call the police," said Eric. "Come on, Tina. Let's go!"

Two hikers discover a wrecked plane, a body, and a mystery. In this chapter, you will study the tissue level of organization. In this story, we will follow the investigation of a mysterious event, and learn how an understanding of anatomy and physiology at the tissue level is used by medical investigators to uncover clues that aid in understanding crime scenes and unexplained deaths.

INTRODUCTION

As you learned in Chapter 3, a cell is a complex collection of compartments, each of which carries out a host of biochemical reactions that make life possible. However, few cells function as isolated units in the human body. Instead, cells usually work together in groups called tissues. A **tissue** is a group of similar cells that function together to carry out specialized activities and usually have a common embryonic origin. The structure and properties of a specific tissue are influenced by factors such as the nature of the extracellular material that surrounds the tissue cells and the connections between the cells that compose the tissue.

Cells usually work together in groups called tissues.

Tissues may be hard, semisolid, or even liquid in their consistency, a range exemplified by bone, fat, and blood. In addition, tissues vary tremendously with respect to the kinds of cells present, how the cells are arranged, and the types of fibers present, if any.

Histology (his'-TOL-ō-jē; *hist-* = tissue; *-ology* = study of) is the science that deals with the study of tissues. A **pathologist** (pa-THOL-ō-gist; *patho-* = disease) is a physician who specializes in laboratory studies of cells and tissues to help other physicians make accurate diagnoses. One of the principal functions of a pathologist is to examine tissues for any changes that might indicate disease.

4.1 Human body tissues can be classified as epithelial, connective, muscle, or nervous.

Body tissues can be classified into four basic types according to their function and structure:

- **Epithelial tissue** covers body surfaces and lines hollow organs, body cavities, and ducts. It also forms glands.

- **Connective tissue** protects and supports the body and its organs. Various types of connective tissue bind organs together, store energy reserves as fat, and help provide immunity against disease-causing organisms.

- **Muscle tissue** generates the physical force needed to make body structures move and generates body heat.

- **Nervous tissue** detects changes in a variety of conditions inside and outside the body and responds by generating nerve impulses that activate muscular contractions and glandular secretions.

Epithelial tissue and most types of connective tissue are discussed in detail in this chapter. However, only the general features of bone and blood (connective tissues) are introduced here; detailed discussions will be presented in Chapters 6 and 18, respectively. Similarly, we acquaint you with the structure and function of muscle tissue and nervous tissue here but will cover them in more detail in Chapters 10 and 12, respectively.

Normally, most cells within a tissue remain anchored to other cells, to basement membranes (described shortly), and to connective tissues. A few cells, such as phagocytes, move freely through the body, searching for invaders. Most epithelial cells and some muscle and nerve cells are tightly joined into functional units by points of contact between their plasma membranes, called **cell junctions**. Some cell junctions form tight seals between cells, like a zip-lock closure at the top of a plastic storage bag, that prevent substances from passing between the cells. This fusion is important for tissues that line the stomach, intestines, and urinary bladder because it prevents the contents of these organs from leaking out. Other cell junctions anchor cells to one another so that they do not separate while performing their functions. Still other cell junctions provide channels for ions and molecules to pass from cell to cell within a tissue, allowing communication between cells and enabling nerve or muscle impulses to spread rapidly among cells.

Checkpoint

1. Define a tissue.
2. What are the four basic types of body tissues?
3. Which type of tissue connects the stomach to the underlying pancreas? Responds to a sudden sound? Moves an arm?
4. Why are cell junctions important?

4.2 Epithelial tissue covers body surfaces, lines organs and body cavities, or secretes substances.

An **epithelial tissue** (ep-i-THĒ-lē-al), or **epithelium** (plural is *epithelia*), consists of cells arranged in continuous sheets, in either single or multiple layers. The cells are closely packed with little intercellular space between adjacent plasma membranes and are held tightly together by many cell junctions. Therefore, epithelial tissue is an excellent protective barrier for body surfaces, such as the skin, or can provide a mechanism for separating and isolating substances within the body, such as urine in the tubules of the kidney.

The various surfaces of epithelial cells often differ in structure and have specialized functions. The **apical (free) surface** of an epithelial cell faces the body surface, a body cavity, or the lumen (interior space) of a hollow organ or duct (Figure 4.1). Apical surfaces may contain cilia or microvilli. The **lateral surfaces** of an epithelial cell face the adjacent cells on either side and may contain cell junctions. The **basal surface** of an epithelial cell is opposite the apical surface and adheres to extracellular materials. In subsequent discussions about epithelia with multiple layers, the term *apical layer* refers to the most superficial layer of cells; *basal layer* is used to refer to the deepest layer of cells.

Figure 4.1   Surfaces of epithelial cells and the structure and location of the basement membrane.

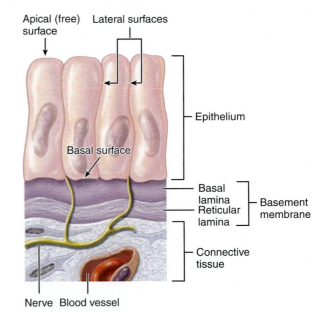

The basement membrane is found between epithelium and connective tissue.

The **basement membrane** is a thin extracellular layer that commonly consists of two layers, the basal lamina and reticular lamina. The *basal lamina* (*lamina* = thin layer) is closer to the epithelial cells and is secreted by them. It contains proteins such as collagen (described shortly) and laminin, as well as glycoproteins. The laminin molecules in the basal lamina attach epithelial cells to the basement membrane. The *reticular lamina* is deep to the basal lamina and contains fibrous proteins produced by underlying connective tissue cells called fibroblasts. The basement membrane functions as a point of attachment and support for the overlying epithelial tissue. It forms a surface along which epithelial cells migrate during growth or wound healing, restricts passage of larger molecules between epithelium and connective tissue, and participates in filtration of blood in the kidneys.

Epithelial tissue has its own nerve supply but is **avascular** (*a-* = without; *-vascular* = vessel)—it lacks its own blood supply. The blood vessels that bring in nutrients and remove wastes are located in the underlying connective tissue. Exchange of substances between epithelium and connective tissue occurs by diffusion.

Because epithelial tissue forms boundaries between the body's organs, or between the body and the external environment, it is repeatedly subjected to physical stress and injury. Fortunately, epithelial tissue has a high rate of cell division and is constantly renewing and repairing itself by sloughing off dead or injured cells and replacing them with new ones. Epithelial tissue plays many different roles in the body, the most important of which are protection, filtration, secretion, absorption, and excretion.

Epithelial tissue may be divided into two types. **Covering and lining epithelium** forms the outer covering of the skin and some internal organs. It also forms the inner lining of blood vessels, ducts, and body cavities, and the interior of the respiratory, digestive, urinary, and reproductive systems. **Glandular epithelium** constitutes the secreting portion of glands such as the thyroid gland, adrenal glands, and sweat glands.

Covering and Lining Epithelia

Covering and lining epithelial tissues are classified according to two characteristics: the arrangement of cells into layers and the shapes of the cells (Figure 4.2).

1. ***Arrangement of cells in layers.*** The cells are arranged in one or more layers depending on the functions the epithelium performs:

    a. *Simple epithelium* is a single layer of cells that functions in diffusion, filtration, secretion (the production and release of substances such as mucus, sweat, or enzymes), and absorption (the intake of fluids or other substances by cells, such as digested food from the intestinal tract).

    b. *Pseudostratified epithelium* appears to have multiple layers of cells because the cell nuclei lie at different levels and not all cells reach the apical surface. Cells that do extend to the apical surface may contain cilia; others (goblet cells) secrete mucus. Pseudostratified epithelium is actually a simple epithelium and all cells rest on the basement membrane.

    c. *Stratified epithelium* (*stratum* = layer) consists of two or more layers of cells that protect underlying tissues in locations where there is considerable wear and tear.

Figure 4.2   Cell shapes and arrangement of layers for covering and lining epithelium.

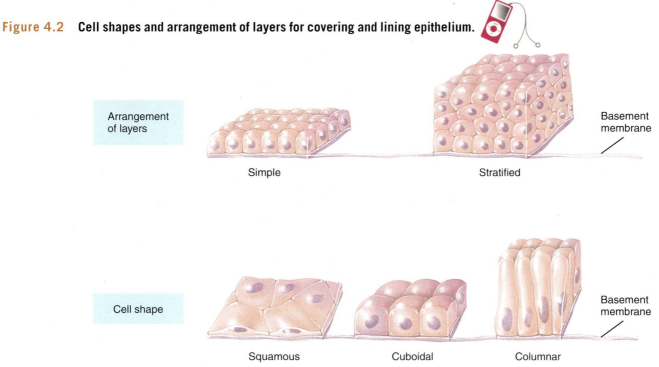

Cell shape and arrangement of layers are the bases for classifying covering and lining epithelium.

2. **Cell shapes.** The cells vary in shape depending on their function.

    a. *Squamous cells* (SKWĀ-mus = flat) are arranged like floor tiles and are thin, which allows for the rapid movement of substances through them.

    b. *Cuboidal cells* are as tall as they are wide and are shaped like cubes or hexagons. They may have microvilli at their apical surface and function in either secretion or absorption.

    c. *Columnar cells* are much taller than they are wide and protect underlying tissues. Their apical surface may have cilia or microvilli, and they often are specialized for secretion and absorption.

    d. *Transitional cells* change shape, from flat to cuboidal and back, as organs such as the urinary bladder stretch (distend) to a larger size and then collapse to a smaller size.

The common covering and lining epithelial tissues are described in the following sections and illustrated in Table 4.1. The tables throughout this chapter contain figures consisting of a photomicrograph, a corresponding diagram, and an inset that identifies a major location of the tissue in the body. Along with the illustrations are descriptions, locations, and functions of the tissues.

Simple Epithelia

Simple Squamous Epithelium

This tissue consists of a single layer of flat cells that resembles a tiled floor when viewed from their apical surface (Table 4.1A).

TABLE 4.1

Epithelial Tissues

Covering and Lining Epithelia

A. Simple Squamous Epithelium

Description: Single layer of flat cells; centrally located nucleus.

Locations: Lines heart, blood vessels, lymphatic vessels, air sacs of lungs, and glomerular capsules of kidneys; forms epithelial layer of serous membranes, such as the peritoneum.

Functions: Filtration and diffusion; secretion in serous membranes.

Peritoneum

LM   243x

Surface view of simple squamous epithelium
of mesothelial lining of peritoneum

Flat nucleus of simple
squamous cell

Connective tissue

Muscular tissue

LM   700x

Small
intestine

Sectional view of simple squamous epithelium of small intestine

Simple squamous
cell

Basement membrane

Connective tissue

Simple squamous epithelium

Covering and Lining Epithelia

B. Simple Cuboidal Epithelium

Description: Single layer of cube-shaped cells; centrally located nucleus.

Locations: Covers surface of ovary, lines kidney tubules and smaller ducts of many glands, and makes up secreting portion of some glands such as thyroid gland and ducts of some glands such as pancreas.

Functions: Secretion and absorption.

Sectional view of simple cuboidal epithelium of intralobular duct of pancreas

LM 330x

Simple cuboidal epithelium

C. Nonciliated Simple Columnar Epithelium

Description: Single layer of nonciliated rectangular cells with nuclei near base of cells; contains absorptive cells with microvilli; contains goblet cells in some locations.

Locations: Lines gastrointestinal tract from stomach to anus, ducts of many glands, and gallbladder.

Functions: Secretion and absorption.

Sectional view of nonciliated simple columnar epithelium of lining of jejunum of small intestine

LM 675x

Nonciliated simple columnar epithelium

(continues)

TABLE 4.1

Epithelial Tissues (continued)

Covering and Lining Epithelia

D. Ciliated Simple Columnar Epithelium

Description: Single layer of ciliated rectangular cells with nuclei near base of cells; contains goblet cells in some locations.

Locations: Lines a few portions of upper respiratory tract, uterine (fallopian) tubes, uterus, and central canal of spinal cord.

Function: Moves mucus and other substances by ciliary action.

Sectional view of ciliated simple columnar epithelium of uterine tube

Ciliated simple columnar epithelium

E. Pseudostratified Columnar Epithelium

Description: Not a true stratified tissue; nuclei of cells are at different levels; all cells are attached to basement membrane but not all reach apical surface. Cilia and goblet cells are present.

Locations: Lines the airways of most of upper respiratory tract, larger ducts of many glands, and portions of male reproductive tract.

Functions: Secretion and movement of mucus by ciliary action.

Sectional view of pseudostratified columnar epithelium of trachea

Pseudostratified columnar epithelium

Covering and Lining Epithelia

F. Stratified Squamous Epithelium

Description: Several layers of cells; cuboidal to columnar cells in deep layers; squamous cells form the apical layers; cells from the basal layer replace surface cells as they are lost.

Locations: Keratinized variety forms superficial layer of skin; nonkeratinized variety lines wet surfaces, such as lining of the mouth, esophagus, and vagina, and covers the tongue.

Function: Protection.

Vagina

Stratified squamous epithelium

Connective tissue

LM 200x

Flattened squamous cell at apical surface

Basement membrane

Connective tissue

Sectional view of stratified squamous epithelium of vagina

Stratified squamous epithelium

G. Transitional Epithelium

Description: Appearance is variable (transitional); shape of cells in apical layer ranges from squamous (when stretched) to cuboidal (when relaxed).

Locations: Lines urinary bladder and portions of ureters and urethra.

Function: Permits distention.

Urinary bladder

Lumen of urinary bladder

Nucleus of transitional cell

Transitional epithelium

Connective tissue

LM 400x

Apical surface

Basement membrane

Connective tissue

Relaxed transitional epithelium

Sectional view of transitional epithelium of urinary bladder in relaxed state

(continues)

TABLE 4.1

Epithelial Tissues (continued)

Glandular Epithelia

H. Endocrine Glands

Description: Secretory products (hormones) diffuse into blood after passing through interstitial fluid.

Locations: Examples include pituitary gland, thyroid gland, adrenal gland, pancreas, ovaries, testes.

Function: Produce hormones that regulate various body activities.

Thyroid gland

Blood vessel

Hormone-producing (epithelial) cell

Stored precursor of hormone

LM 500x

Sectional view of endocrine gland (thyroid gland)

Endocrine gland (thyroid gland)

I. Exocrine Glands

Description: Secretory products released into ducts.

Locations: Sweat, oil, and earwax glands of the skin; salivary glands, which secrete into mouth cavity; and pancreas, which secretes into the small intestine.

Function: Produce substances such as sweat, oil, earwax, saliva, or digestive enzymes.

Skin

Secretory portion of sweat gland

Lumen of duct of sweat gland

Nucleus of secretory cell of sweat gland

Basement membrane

LM 300x

Sectional view of the secretory portion of an exocrine gland (sweat gland)

Exocrine gland (sweat gland)

The nucleus of each cell is a flattened oval or sphere and is centrally located. Simple squamous epithelium is present at sites where filtration occurs, such as blood filtration in the kidneys, or where diffusion takes place, such as diffusion of oxygen into blood vessels of the lungs. It is not found in body areas that are subject to wear and tear.

The simple squamous epithelium that lines the heart, blood vessels, and lymphatic vessels is known as **endothelium** (*endo-* = within; *-thelium* = covering); the type that forms the epithelial layer of serous membranes (such as the peritoneum) is called **mesothelium** (*meso-* = middle).

Simple Cuboidal Epithelium

The cuboidal shape of the cells in this tissue (Table 4.1B) is obvious when the tissue is sectioned and viewed from the side. Cell nuclei are usually round and centrally located. Simple cuboidal epithelium is found in organs such as the thyroid gland and kidneys and performs the functions of secretion and absorption.

Simple Columnar Epithelium

When viewed from the side, the cells appear rectangular with oval nuclei in a row near the base of the cells. Simple columnar epithelium exists in two forms: nonciliated simple columnar epithelium and ciliated simple columnar epithelium.

Nonciliated simple columnar epithelium contains two types of cells: columnar epithelial cells with microvilli at their apical surface and goblet cells (Table 4.1C). **Microvilli** are microscopic fingerlike projections of the plasma membrane that increase cell surface area. Their presence increases the rate of absorption by the cell. **Goblet cells** are modified columnar epithelial cells that secrete mucus at their apical surfaces. Before it is released, mucus accumulates in the upper portion of the cell, causing that area to bulge out and giving the entire cell a goblet or wine glass appearance. Secreted mucus serves as a lubricant and protective covering for the linings of the digestive, respiratory, reproductive, and most of the urinary tracts.

Ciliated simple columnar epithelium contains columnar epithelial cells with cilia at their apical surface and goblet cells (Table 4.1D). The cilia help move oocytes expelled from the ovaries through the uterine (fallopian) tubes into the uterus and move cerebrospinal fluid through passageways in the central nervous system.

Pseudostratified Columnar Epithelium

As noted earlier, this tissue appears to have several layers because the nuclei of the cells are at various depths (Table 4.1E). Even though all the cells are attached to the basement membrane in a single layer, some cells do not extend to the surface. When viewed from the side, these features give the false impression of a multilayered tissue—thus the name *pseudostratified* epithelium (*pseudo-* = false). The cells of pseudostratified columnar epithelium that extend to the surface either secrete mucus (goblet cells) or bear cilia. The mucus forms a film over the airway surface that traps foreign particles. The cilia beat in unison, moving the mucus and any foreign particles superiorly toward the throat, where it can be coughed up and swallowed or spit out.

Stratified Epithelia

In contrast to simple epithelium, stratified epithelium has two or more layers of cells. Thus, it is more durable and can better protect underlying tissues. The name of the specific kind of stratified epithelium depends on the shape of the cells in the apical layer. The ducts of some glands and part of the urethra contain stratified cuboidal and stratified columnar epithelium, but the two common forms of stratified epithelia we investigate here are stratified squamous epithelium and transitional epithelium.

Stratified Squamous Epithelium

Cells in the apical layer of this type of epithelium are flat; in the deep layers, cells vary in shape from cuboidal to columnar (Table 4.1F). The basal (deepest) cells continually undergo cell division. As new cells grow, the cells of the basal layer are pushed upward toward the apical layer. As they move farther from the deeper layers and from their blood supply in the underlying connective tissue, the cells become dehydrated, shrunken, harder, and then die. At the apical layer, the cells lose their cell junctions and are sloughed off, but they are replaced as new cells continually emerge from the basal layer.

Stratified squamous epithelium exists in both keratinized and nonkeratinized forms. In *keratinized stratified squamous epithelium*, the apical layers are partially dehydrated and contain a layer of **keratin**, a tough, fibrous protein that helps protect the skin and underlying tissues from heat, microbes, and chemicals. Keratinized stratified squamous epithelium forms the superficial layer of the skin. *Nonkeratinized stratified squamous epithelium*, which is found, for example, lining the mouth and esophagus, does not contain keratin in the apical layers and remains moist. Both types of stratified squamous epithelia form the first line of defense against microbes.

Transitional Epithelium

This type of stratified epithelium is present only in the urinary system and has a variable appearance. In its relaxed (unstretched) state (Table 4.1G), cells in the apical layer of transitional epithelium tend to be large and rounded. As the tissue is stretched, cells in the apical layer become flatter and more squamous shaped. Because of its elasticity, transitional epithelium is ideal for lining hollow structures that are subjected to expansion from within, such as the urinary bladder. This tissue allows urinary tract organs to stretch to hold a variable amount of fluid without rupturing.

Glandular Epithelia

The function of glandular epithelia is secretion, which is accomplished by glandular cells that often lie in clusters deep to a covering and lining epithelium. A **gland** may consist of a single cell or a group of cells that secrete substances into ducts, onto a surface, or into the blood. All glands of the body are classified as either endocrine or exocrine.

The secretions of **endocrine glands** (*endo-* = within; *-crine* = secretion) (Table 4.1H) enter the interstitial fluid and then diffuse directly into the bloodstream without flowing through a duct. These secretions, called *hormones*, regulate many metabolic and physiological activities to maintain homeostasis. The pituitary, thyroid, and adrenal glands are examples of endocrine glands. Endocrine glands will be described in detail in Chapter 17.

Exocrine glands (*exo-* = outside) (Table 4.1I) secrete their products into ducts (tubes) that empty onto the surface of a covering and lining epithelium such as the skin surface or the lumen of a hollow organ. The secretions of exocrine glands include mucus, sweat, oil, earwax, saliva, and digestive enzymes. Exocrine glands are classified as unicellular or multicellular. As the name implies,

unicellular exocrine glands are single celled. Goblet cells are unicellular exocrine glands that secrete mucus directly onto the apical surface of a lining epithelium rather than into ducts. Most exocrine glands are **multicellular glands**, composed of many cells that form a distinctive microscopic structure or macroscopic organ. Examples of multicellular exocrine glands are sudoriferous glands, which produce sweat to help lower body temperature, and salivary glands, which secrete saliva.

As you will learn in later chapters, some glands of the body, such as the pancreas, ovaries, and testes, are mixed glands that contain both endocrine and exocrine tissue.

Checkpoint

5.   Which surface of an epithelial cell would face the lumen of an internal organ?
6.   What are the functions of the basement membrane? What structures produce each of the two basement membrane layers?
7.   What are the three ways in which cells in covering and lining epithelia may be layered? What are their three basic cell shapes?
8.   Where are endothelium and mesothelium located? Which type of epithelium composes each of them?
9.   What are the functions of goblet cells, microvilli, and cilia?
10.   Why is keratin needed in the epithelium of skin?

4.3 Connective tissue binds organs together, stores energy reserves as fat, and helps provide immunity.

Connective tissue is one of the most abundant and widely distributed tissues in the body. In its various forms, connective tissue has a variety of functions. It binds together, supports, and strengthens other body tissues; protects and insulates internal organs; compartmentalizes structures such as skeletal muscles; is the major transport system within the body (blood, a fluid connective tissue); is the major site of stored energy reserves (adipose, or fat, tissue); and is the main source of immune responses.

General Features of Connective Tissue

Connective tissue consists of two basic elements: extracellular matrix and cells. A connective tissue's **extracellular matrix** is the material located between its widely spaced cells. The extracellular matrix consists of protein fibers and ground substance, the material between the cells and the fibers. The extracellular matrix is usually secreted by the connective tissue cells and determines the tissue's qualities. For example, the extracellular matrix of cartilage is firm but pliable. By contrast, the extracellular matrix of bone is hard and not pliable.

In contrast to epithelia, connective tissues do not occur on body surfaces. Also unlike epithelia, connective tissues usually are highly vascular— they have a rich blood supply. Exceptions include cartilage, which is avascular, and tendons, which have a scanty blood supply. Like epithelia, all connective tissues except cartilage have a nerve supply.

Connective Tissue Cells

Each major type of connective tissue contains an immature class of cells with a name ending in *blast*, which means "to bud or sprout." These immature cells are called *fibroblasts* in loose and dense connective tissue, *chondroblasts* in cartilage, and *osteoblasts* in bone. Blast cells retain the capacity for cell division and secrete the extracellular matrix that is characteristic of the tissue. In carti-

lage and bone, once the extracellular matrix is produced, the blast cells differentiate into mature cells with names ending in *-cyte*, namely chondrocytes and osteocytes. Mature cells have reduced capacities for cell division and extracellular matrix formation and are mostly involved in maintaining the matrix.

The types of cells present in connective tissues vary according to the type of tissue and include the following (Figure 4.3):

- **Fibroblasts** (FĪ -brō-blasts; *fibro-* = fibers) are present in several connective tissues, where they usually are the most numerous connective tissue cells. Fibroblasts migrate through the connective tissue, secreting the fibers and ground substance of the extracellular matrix.

- **Macrophages** (MAK-rō-fā-jez; *macro-* = large; *-phages* = eaters) engulf bacteria and cellular debris by phagocytosis.

- **Plasma cells** secrete antibodies, proteins that attack or neutralize foreign substances in the body.

- **Mast cells** produce histamine, a chemical that dilates small blood vessels as part of the inflammatory response, the body's reaction to injury or infection.

- **Adipocytes**, also called fat cells, are connective tissue cells that store triglycerides (fats).

- **White blood cells** are not found in significant numbers in normal connective tissue. However, in response to certain conditions they migrate from blood into connective tissues to mediate immune system responses.

Connective Tissue Extracellular Matrix

Each type of connective tissue has unique properties, based on the specific extracellular matrix materials between the cells. The extracellular matrix consists of ground substance and fibers.

Figure 4.3   Representative cells and fibers present in connective tissue.

Macrophage

Collagen fiber

Elastic fiber

Ground substance

Blood vessel

Adipocyte

Reticular fiber

Fibroblast

White blood cell

Mast cell

White blood cell

Plasma cell

Fibroblasts are usually the most numerous connective tissue cells.

Ground Substance

As mentioned earlier, the **ground substance** is the component of a connective tissue between the cells and fibers. It may be fluid, semifluid, gelatinous, or calcified. The ground substance supports cells, binds them together, stores water, and provides a medium through which substances are exchanged between the blood and cells. It plays an active role in how tissues develop, migrate, proliferate, and change shape, and in how they carry out their metabolic functions.

Ground substance contains water and an assortment of large organic molecules, many of which are complex combinations of polysaccharides and proteins. The various polysaccharides are collectively referred to as **glycosaminoglycans** (glī-kos-a-mē´-nō-GLĪ-kans) or **GAGs**. One of the most significant properties of GAGs is that they trap water, making the ground substance more jellylike. An important example of GAGs is **hyaluronic acid** (hī´-a-loo-RON-ik), a viscous, slippery substance that binds cells together, lubricates joints, and helps maintain the shape of the eyeballs. White blood cells, sperm cells, and some bacteria produce *hyaluronidase*, an enzyme that breaks apart hyaluronic acid, thus causing the ground substance of connective tissue to become more liquid. The ability to produce hyaluronidase helps white blood cells move more easily through connective tissues to reach sites of infection and aids penetration of an oocyte by a sperm cell during fertilization. It also accounts for the rapid spread of bacteria through connective tissues.

Fibers

Fibers in the extracellular matrix strengthen and support connective tissues. Three types of **fibers** are embedded in the extracellular matrix between the cells: collagen fibers, elastic fibers, and reticular fibers.

Collagen fibers (*colla* = glue) are very strong and resist pulling forces, but they are not stiff, which allows tissue flexibility. Different types of collagen fibers in various tissues have slightly different properties. For example, the collagen fibers found in cartilage attract more water molecules than those in bone, which gives cartilage a more cushioning consistency. Collagen fibers often occur in bundles lying parallel to one another (Figure 4.3). The bundle arrangement gives the tissue great strength. Chemically, collagen fibers consist of the protein **collagen**. This is the most abundant protein in your body, representing about 25% of the total protein. Collagen fibers are found in most types of connective tissues, especially bone, cartilage, tendons (attach muscle to bone), and ligaments (attach bone to bone).

Elastic fibers, which are smaller in diameter than collagen fibers, branch and join together to form a network within a tissue. Elastic fibers contain the protein *elastin*, which allows them to stretch up to 150% of their relaxed length without breaking. Equally important, elastic fibers have the ability to return to their original shape after being stretched, a property called **elasticity**. Elastic fibers are plentiful in skin, blood vessel walls, and lung tissue.

Reticular fibers (*reticul-* = net) consist of collagen arranged as fine, branching, interwoven fibers that provide support in the walls of blood vessels and form a network around the cells in some tissues, such as areolar connective tissue, adipose tissue, and smooth muscle tissue. Reticular fibers are much thinner than collagen fibers and form branching networks. Like collagen fibers, reticular fibers provide support and strength. Reticular fibers are plentiful in reticular connective tissue, which forms the **stroma** (= bed or covering) or supporting framework of many soft organs, such as the spleen and lymph nodes.

Types of Connective Tissue

Classifying connective tissues is not always clear-cut due to the diversity of cells and extracellular matrix present, and the differences in their relative proportions. However, for purposes of our discussion we can classify connective tissues into two basic groups: (1) embryonic connective tissue and (2) mature connective tissue. **Embryonic connective tissue** is present primarily in the *embryo*, the developing human from fertilization through the first two months of pregnancy, and in the *fetus*, the developing human from the third month of pregnancy to birth. **Mature connective tissue** is present after birth. The five types of mature connective tissue are (1) loose connective tissue, (2) dense connective tissue, (3) cartilage, (4) osseous tissue, and (5) liquid connective tissue (blood and lymph).

Loose Connective Tissue

In **loose connective tissue** the fibers are loosely intertwined and many cells are present. The types of loose connective tissue in the body include areolar connective tissue, adipose tissue, and reticular connective tissue.

Areolar Connective Tissue

One of the most widely distributed connective tissues in the body is **areolar connective tissue** (a-RĒ-ō-lar; *areol-* = a small space). It contains several kinds of cells, including fibroblasts, macrophages, plasma cells, mast cells, and adipocytes (Table 4.2A). All three types of fibers—collagen, elastic, and reticular—are arranged randomly throughout the tissue. Combined with adipose tissue, areolar connective tissue forms the *subcutaneous layer*, the layer of tissue that attaches the skin to underlying tissues and organs.

Adipose Tissue

Adipose tissue is a loose connective tissue in which the cells, called **adipocytes** (*adipo-* = fat), are specialized for storage of triglycerides (fats) (Table 4.2B). Because the cell fills up with a single, large triglyceride droplet, the cytoplasm and nucleus are pushed into a thin ring around the periphery of the cell. Adipose tissue is a good insulator and, therefore, reduces heat loss through the skin. It is a major energy reserve and supports and protects various organs.

Reticular Connective Tissue

Reticular connective tissue consists of fine, interlacing reticular fibers and reticular cells (Table 4.2C). Reticular connective tissue forms the stroma (supporting framework) of the liver, spleen, and lymph nodes. Additionally, reticular fibers in the spleen filter blood and remove worn-out blood cells. Reticular fibers in lymph nodes filter lymph and remove bacteria.

Dense Connective Tissue

Dense connective tissue contains more fibers but fewer cells than loose connective tissue. There are three types: dense regular connective tissue, dense irregular connective tissue, and elastic connective tissue.

Dense Regular Connective Tissue

In this tissue, bundles of collagen fibers are *regularly* arranged in parallel patterns that provide the tissue with great strength to withstand pulling along the axis of the fibers (Table 4.2D). Fibroblasts appear in rows between the fiber bundles. In the body, the tissue is silvery white and tough, yet somewhat pliable. Examples include tendons and most ligaments.

TABLE 4.2

Mature Connective Tissues

Loose Connective Tissues

A. Areolar Connective Tissue

Description: Consists of fibers (collagen, elastic, and reticular) and several kinds of cells (fibroblasts, macrophages, plasma cells, adipocytes, and mast cells) embedded in a semifluid ground substance.

Locations: Subcutaneous layer deep to skin; superficial region of dermis of skin (deep to the epidermis); mucous membranes; and around blood vessels, nerves, and body organs.

Functions: Strength, elasticity, and support.

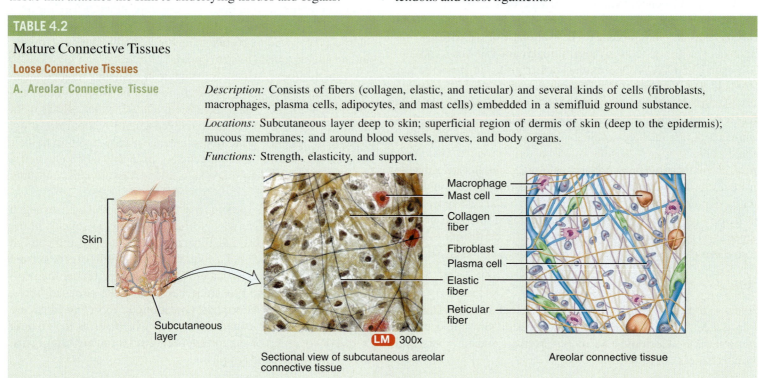

Skin

Subcutaneous layer

LM 300x

Sectional view of subcutaneous areolar connective tissue

Macrophage
Mast cell
Collagen fiber
Fibroblast
Plasma cell
Elastic fiber
Reticular fiber

Areolar connective tissue

Loose Connective Tissues

B. Adipose Tissue

Description: Consists of adipocytes, cells specialized to store triglycerides (fats) as a large centrally located droplet; nucleus and cytoplasm are peripherally located.

Locations: Subcutaneous layer deep to skin, around heart and kidneys, yellow bone marrow, and padding around joints and behind eyeball in eye socket.

Functions: Reduces heat loss through skin; serves as an energy reserve, supports, and protects.

Heart

Fat

Nucleus of adipocyte

Cytoplasm

Fat-storage area of adipocyte

Blood vessel

Plasma membrane of adipocyte

Adipose tissue

LM 300x

Sectional view of adipose tissue showing adipocytes

C. Reticular Connective Tissue

Description: Consists of network of interlacing reticular fibers and reticular cells.

Locations: Stroma (supporting framework) of liver, spleen, and lymph nodes.

Functions: Forms stroma of organs; filters and removes worn-out blood cells in spleen and microbes in lymph nodes.

Lymph node

Nucleus of reticular cell

Reticular fiber

Reticular connective tissue

LM 400x

Sectional view of reticular connective tissue of a lymph node

(continues)

Dense Irregular Connective Tissue

This tissue contains collagen fibers that are *irregularly* arranged (Table 4.2E). It is found in parts of the body where pulling forces are exerted in various directions such as in the dermis of the skin, which is deep to the epidermis, or the pericardium around the heart. Heart valves, the perichondrium (the membrane surrounding cartilage), and the periosteum (the membrane surrounding bone) are also dense irregular connective tissues.

Elastic Connective Tissue

Branching elastic fibers predominate in elastic connective tissue (Table 4.2F). Fibroblasts are present in the spaces between the fibers. Elastic connective tissue is quite strong and can recoil to its original shape after being stretched. Elasticity is important to the normal functioning of lung tissue, which recoils as you exhale, and elastic arteries, which recoil between heartbeats to help maintain blood flow.

TABLE 4.2

Mature Connective Tissues (continued)

Dense Connective Tissues

D. Dense Regular Connective Tissue

Description: Consists mainly of collagen fibers arranged in parallel bundles and fibroblasts present in rows between the fiber bundles.

Locations: Forms tendons (attach muscle to bone), most ligaments (attach bone to bone), and aponeuroses (sheetlike tendons that attach muscle to muscle or muscle to bone).

Function: Provides strong attachment between various structures.

LM 200x

Sectional view of dense regular connective tissue of a tendon

Dense regular connective tissue

E. Dense Irregular Connective Tissue

Description: Consists predominantly of collagen fibers, randomly arranged, and a few fibroblasts.

Locations: Fasciae (tissue around muscles and other organs), dermis of skin, periosteum of bone, perichondrium of cartilage, joint capsules, membrane capsules around various organs (kidneys, liver, testes, lymph nodes), pericardium of heart, and heart valves.

Function: Provides strength.

LM 275x

Sectional view of dense irregular connective tissue of reticular region of dermis

Dense irregular connective tissue

Cartilage

Cartilage consists of a dense network of collagen fibers and elastic fibers firmly embedded in *chondroitin sulfate*, a gel-like component of the ground substance. Cartilage can endure considerably more stress than loose and dense connective tissues. The strength of cartilage is due to its collagen fibers, but its *resilience* (ability to assume its original shape after deformation) is due to chondroitin sulfate.

The cells of mature cartilage, called **chondrocytes** (KON-drō-sīts; *chondro-* = cartilage), are found within spaces called **lacunae** (la-KOO-nē = little lakes; singular is *lacuna*) in the extracellular matrix. Chondrocytes may occupy lacunae singly or in groups. A covering of dense irregular connective tissue called the **perichondrium** (per´-i-KON-drē-um; *peri-* = around) covers the surface of most cartilage. Unlike other connective tissues, cartilage has no blood vessels or nerves, except in the perichondrium. Since carti-

Dense Connective Tissues

F. Elastic Connective Tissue

Description: Consists predominantly of freely branching elastic fibers and fibroblasts present in spaces between elastic fibers.

Locations: Lung tissue, trachea, bronchial tubes, walls of some arteries, and ligaments supporting penis and vertebrae.

Function: Allows stretching of various organs.

Aorta

Heart

Nucleus of fibroblast

Elastic fibers

LM 435x

Sectional view of elastic connective tissue of aorta

Elastic connective tissue

Cartilage

G. Hyaline Cartilage

Description: Consists of bluish-white, shiny ground substance with fine, pale collagen fibers and many chondrocytes; most abundant type of cartilage.

Locations: Ends of long bones; anterior ends of ribs; parts of nose, larynx, trachea, bronchi; and embryonic and fetal skeleton.

Functions: Provides smooth surfaces for movement at joints, as well as flexibility and support.

Trachea

Lacuna containing chondrocyte

Nucleus of chondrocyte

Ground substance

LM 205x

Sectional view of hyaline cartilage of trachea

Hyaline cartilage

(continues)

---

lage has no blood supply, it heals slowly following an injury. Three types of cartilage are recognized: hyaline cartilage, fibrocartilage, and elastic cartilage.

- *Hyaline Cartilage.* This type of cartilage contains prominent chondrocytes in lacunae, but collagen fibers are so fine that they are not visible with ordinary staining techniques (Table 4.2G). The ground substance is a resilient gel and appears in the body as a bluish-white, shiny substance. Hyaline cartilage is the most abundant cartilage in the body. It affords flexibil-

ity and support and, at joints, reduces friction and absorbs shock. Hyaline cartilage is the weakest of the three types of cartilage.

- *Fibrocartilage.* Chondrocytes are scattered among clearly visible bundles of collagen fibers within the extracellular matrix of this type of cartilage (Table 4.2H). Fibrocartilage combines strength and rigidity and is the strongest of the three types of cartilage. One location of fibrocartilage is the intervertebral discs, the disk-shaped structures between the vertebrae (backbones).

TABLE 4.2

Mature Connective Tissues (continued)

Cartilage

H. Fibrocartilage

Description: Consists of chondrocytes scattered among bundles of collagen fibers within the extracellular matrix.

Locations: Pubic symphysis (point where hip bones join anteriorly), intervertebral discs (discs between vertebrae), menisci (cartilage pads) of knee, and portions of tendons.

Functions: Support and fusion.

Tendon of quadriceps femoris muscle

Patella (knee cap)

Portion of right lower limb

Nucleus of chondrocyte

Collagen fibers in ground substance

Lacuna containing chondrocyte

LM 1100x

Sectional view of fibrocartilage of tendon

Fibrocartilage

I. Elastic Cartilage

Description: Consists of chondrocytes located in a network of threadlike elastic fibers within the extracellular matrix.

Locations: Lid on top of larynx (epiglottis), part of external ear (auricle), and auditory tubes.

Functions: Gives highly flexible support and maintains shape.

Auricle of ear

Perichondrium

Nucleus of chondrocyte

Elastic fiber in ground substance

Lacuna containing chondrocyte

LM 400x

Sectional view of elastic cartilage of auricle of ear

Elastic cartilage

• *Elastic Cartilage.* In this tissue, chondrocytes are located among a threadlike network of elastic fibers within the extracellular matrix (Table 4.2I). Because of the large number of elastic fibers, this tissue is highly flexible and resilient. Elastic cartilage provides elasticity and maintains the shape of certain structures, such as the external ear.

The locations of cartilage are summarized in Figure 4.4.

Osseous Tissue (Bone)

Cartilage, joints, and bones make up the skeletal system. The skeletal system supports soft tissues, protects delicate structures, and works with skeletal muscles to generate movement. Bones store calcium and phosphorus; house red bone marrow, which produces blood cells; and contain yellow bone marrow, a storage site for triglycerides. Bones are organs composed of

Osseous Tissue

J. Compact Bone

Description: Consists of osteocytes in lacunae and an extracellular matrix arranged in concentric rings or thin columns.

Location: Bones.

Functions: Support, protection, storage; houses blood-forming tissue; serves as levers that act together with muscle tissue to enable movement.

Femur

Osteocyte

Calcified extracellular matrix

Lacuna

Lacuna

LM 550x

Sectional view of osseous tissue of femur (thigh bone)

Details of an osteocyte

Liquid Connective Tissue

K. Blood

Description: Consists of blood plasma and formed elements: red blood cells, white blood cells, and platelets.

Locations: Within blood vessels and within the chambers of the heart.

Functions: Red blood cells transport respiratory gases; white blood cells carry on phagocytosis and contribute to immune system responses; platelets are essential for blood clotting.

Platelet

White blood cell

Red blood cell

Blood plasma

Red blood cells

White blood cells

Platelets

LM 1230x

Blood in blood vessels

Blood smear

several different connective tissues, including **bone** or **osseous tissue** (OS-ē-us). Osseous tissue is classified as either spongy or compact, depending on how its extracellular matrix and cells are organized. The extracellular matrix of **spongy bone** is arranged in lattice-like thin columns, whereas that of **compact bone** is arranged in concentric rings (Table 4.2J). Mature bone cells, called **osteocytes**, are located within lacunae in the extracellular matrix. The extracellular matrix of osseous tissue contains crystallized mineral salts (mostly calcium phosphates), which give bones their hardness, and collagen fibers, which give bones their strength. Chapter 6 presents osseous tissue histology in more detail.

Liquid Connective Tissue

In a liquid connective tissue, the cells are suspended in a liquid extracellular matrix. The two types of liquid connective tissues are blood and lymph.

Figure 4.4   Cartilage locations in the adult.

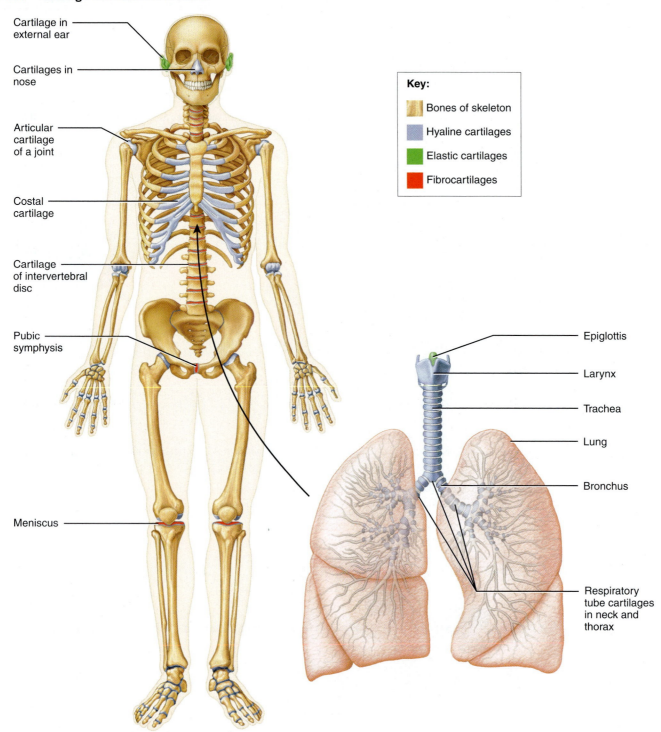

Cartilage in external ear

Cartilages in nose

Articular cartilage of a joint

Costal cartilage

Cartilage of intervertebral disc

Pubic symphysis

Meniscus

Key:

Bones of skeleton

Hyaline cartilages

Elastic cartilages

Fibrocartilages

Epiglottis

Larynx

Trachea

Lung

Bronchus

Respiratory tube cartilages in neck and thorax

🔑 **Cartilage provides flexible, resilient support.**

Blood

Blood is a connective tissue with a liquid extracellular matrix called **blood plasma** (Table 4.2K). Suspended in the blood plasma are formed elements—red blood cells, white blood cells, and platelets. **Red blood cells** transport oxygen to body cells and re- move carbon dioxide from them. **White blood cells** are involved in phagocytosis, immunity, and allergic reactions. **Platelets** par- ticipate in blood clotting. Blood plasma consists mostly of water with a wide variety of dissolved substances such as nutrients, wastes, plasma proteins, and hormones. The details of blood will be considered in Chapter 18.

Lymph

Lymph is a connective tissue that flows in lymphatic vessels. It consists of several types of cells in a clear liquid extracellular matrix that is similar to blood plasma, but with much less protein. Lymph circulates body fluids, transports lipids, and helps defend the body against pathogens. The composition of lymph varies from one part of the body to another. For example, lymph leaving lymph nodes includes many lymphocytes, a type of white blood cell, whereas lymph from the small intestine has a high content of newly absorbed dietary lipids. The details of lymph will be considered in Chapter 21.

Checkpoint

11. What are the general functions of connective tissue?
12. How do connective tissues differ from epithelia?
13. What do fibroblasts, chondroblasts, and osteoblasts have in common?
14. Which connective tissue cells store fats? Have a role in fighting infection (immunity)? Secrete fibers and ground substance?
15. Which fibers allow skin to snap back after being stretched? Keep tendons and ligaments from pulling apart? Form a netlike internal framework for filtering organs such as the spleen?
16. Which kind of cartilage is found in intervertebral discs? In the external ear? In the nose?

Return to John Doe's Story

The sheriff and several deputies had cordoned off the area of the crash. The medical examiner, FAA inspectors, and EMS personnel had all arrived, forming a small, busy crowd. The medical examiner, Dr. Greg Diego, didn't waste any time. He photographed the crash scene and body from various angles, performed a quick examination of the body, and collected samples from the body and the area around it.

"Well, Doc, any idea how long he's been here?" the sheriff asked, holding his nose with one hand.

"I'll have to perform an autopsy to determine exact time of death. Been dead a while, several weeks at least. He was probably alive when he crashed, though. Bruises don't form on dead people. Any idea who he is?"

The sheriff sighed. "Not yet. The FAA inspectors made some calls; they're checking the registration numbers."

Dr. Diego nodded his head toward the plane. "Can you give me a hand getting the body out?"

As Dr. Diego grasped the body by the wrists, he noticed that the skin slipped and pulled away from the underlying connective tissue. The mountain weather and decay of the corpse had taken their toll on the tissues.

Back at the morgue, Dr. Diego examined insect larvae he had taken from the body at the crash site. Their developmental stage told Diego that the John Doe had been dead for approximately 14 to 16 days.

Dr. Diego's assistant helped position the dead man on the autopsy table. The man's skin was discolored to a gray-bluish tinge, indicating early decay. Diego noted no external wounds and some bruising around the face and neck, indicating that the victim sustained injuries in the crash.

He made a few notes on his clipboard, then reached for what looked like a long, very sharp bread knife. "Let's see what we can find out on the inside."

A. What types of tissues form the surface of the dead man's skin? How does the structure of the skin normally protect the body?

B. As the body was being moved, skin slid away from the underlying connective tissue. What is the normal function of connective tissue? Would the slippage of the dead man's skin be due to decomposition of cell junctions or underlying connective tissues?

C. Would the bruises on the dead man have formed in the epithelia or in the underlying connective tissue? Explain your answer.

4.4 Membranes cover the surface of the body, line body cavities, and cover organs.

Membranes are flat sheets of pliable tissue that cover or line a part of the body. The combination of an epithelial layer and an underlying connective tissue layer constitutes an **epithelial membrane**. The principal epithelial membranes of the body are mucous membranes, serous membranes, and the cutaneous membrane. Another kind of membrane, a **synovial membrane**, lines joint cavities and contains connective tissue but no epithelial layer.

Epithelial Membranes

Mucous Membranes

A **mucous membrane** or **mucosa** (mū-KŌ-sa) lines a body cavity that opens directly to the exterior. Mucous membranes line the entire digestive, respiratory, and reproductive tracts, and much of the urinary tract. They consist of both a lining layer of epithelium and an underlying layer of connective tissue (Figure 4.5a).

The epithelial layer of a mucous membrane is an important feature of the body's defense mechanisms because it is a barrier that microbes and other pathogens have difficulty penetrating. Usually, cell junctions tightly connect the cells, so materials cannot leak in between them. Goblet cells and other cells of the epithelial layer of a mucous membrane secrete mucus, and this slippery fluid prevents the cavities from drying out. Mucus also traps particles in the respiratory passageways and lubricates food as it moves through the gastrointestinal tract. In addition, the epithelial layer secretes some of the enzymes needed for digestion and is the site of food and fluid absorption in the gastrointestinal tract. The epithelia of mucous membranes vary greatly in different parts of the body. For example, the mucous membrane of the small intestine is nonciliated simple columnar (see Table 4.1C), and the large airways to the lungs are pseudostratified columnar (see Table 4.1.E).

The connective tissue layer of a mucous membrane, is areolar connective tissue and called the **lamina propria** (LAM-i-na PRŌ-prē-a) (*propria* = one's own), so named because the layer (lamina) belongs to the mucous membrane. The lamina propria supports the epithelium, binds it to the underlying structures, and allows some flexibility of the membrane. It also holds blood vessels in place and protects underlying muscles from abrasion or puncture. Oxygen and nutrients diffuse from the lamina propria to the epithelium; carbon dioxide and wastes diffuse in the opposite direction.

Serous Membranes

A **serous membrane** (*serous* = watery) or **serosa** lines a body cavity that does not open directly to the exterior, and it covers the organs that lie within the cavity. Serous membranes consist of areolar connective tissue covered by mesothelium (simple squamous epithelium) (Figure 4.5b). Serous membranes have two layers:

The layer attached to the cavity wall is called the **parietal layer** (pa-RĪ-e-tal; *pariet-* = wall); the layer that covers and attaches to the organs inside the cavity is the **visceral layer** (*viscer-* = body organ). The mesothelium of a serous membrane secretes **serous fluid**, a watery lubricating fluid that allows organs to glide easily over one another or to slide against the walls of cavities.

The serous membrane lining the thoracic cavity and covering the lungs is the **pleura**; the parietal pleura lines the wall of the thoracic cavity, whereas the deep layer immediately covering the lungs is the visceral pleura. The serous membrane lining the heart cavity and covering the heart is the **pericardium**. The serous membrane lining the abdominal cavity and covering the abdominal organs is the **peritoneum**.

Cutaneous Membrane

The **cutaneous membrane**, or **skin**, covers the surface of the body and consists of a superficial portion called the epidermis and a deeper portion called the dermis (Figure 4.5c). The skin will be described in detail in Chapter 5.

Synovial Membranes

Synovial membranes (sin-Ō-vē-al; *syn-* = together, referring here to a place where bones come together) line the cavities of freely movable joints (Figure 4.5d). Like serous membranes, synovial membranes line structures that do not open to the exterior. Unlike mucous, serous, and cutaneous membranes, they lack epithelium and are therefore not epithelial membranes. Synovial membranes are composed of areolar connective tissue with a variable amount of adipocytes, elastic fibers, and collagen fibers. Synovial membranes secrete **synovial fluid** into the joint cavity (space between the bones). Synovial fluid lubricates and nourishes the cartilage covering the bones at movable joints and contains macrophages that remove microbes and debris from the joint cavity. Besides lining joint cavities, synovial membranes line cushioning sacs, called *bursae*, and *tendon sheaths*, for example, in the hands and feet, thus easing the movement of muscle tendons.

Checkpoint

17. What is an epithelial membrane?
18. Describe the location in the body and the function of each type of membrane.
19. What functions are served by mucus? Which membranes secrete mucus? Which cells produce the mucus that protects the nasal cavity linings?
20. Which part of the serosa covers the stomach and intestines? What is its function?
21. Where are synovial membranes found? What is the function of the synovial fluid they secrete?

Figure 4.5   Membranes.

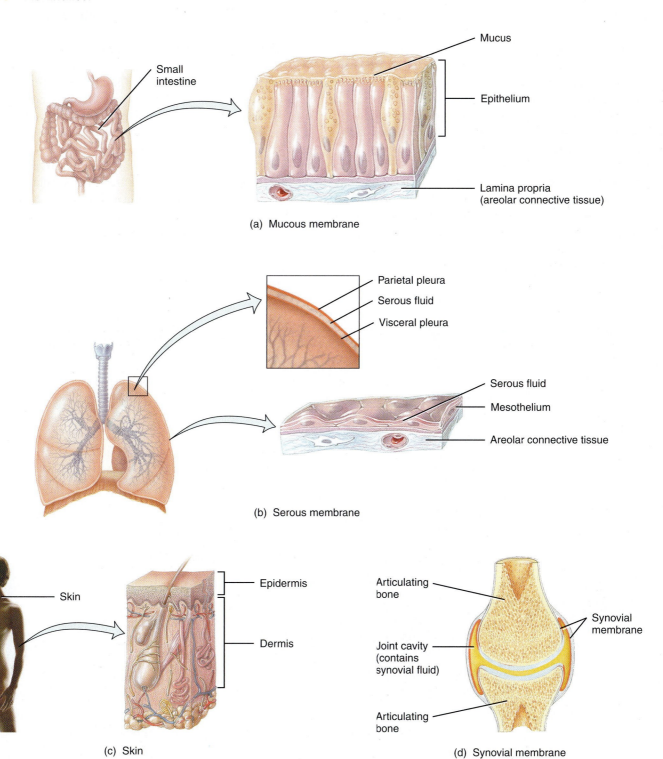

(a) Mucous membrane

(b) Serous membrane

(c) Skin

(d) Synovial membrane

A membrane is a flat sheet of pliable tissue that covers or lines a part of the body.

4.5   Muscle tissue generates the physical force needed to make body structures move.

Muscle tissue consists of elongated cells called *muscle fibers* that can use ATP to generate force. As a result, muscle tissue produces body movements, maintains posture, and generates heat. Based on its location and certain structural and functional features, muscle tissue is classified into three types: skeletal, cardiac, and smooth.

Skeletal muscle tissue is usually attached to bones and *striated*; it contains alternating light and dark bands called *striations* that are visible under a light microscope (Table 4.3A). A single skeletal muscle cell is very long [up to 30–40 cm (about 12–16 in.) in your longest muscles], roughly cylindrical in shape, and has many nuclei located at the periphery of the cell. Within a whole skeletal muscle, the individual muscle fibers are parallel to each other. Skeletal muscle tissue is *voluntary* because it can be made to contract or relax by conscious control.

Cardiac muscle tissue forms most of the wall of the heart (Table 4.3B). Like skeletal muscle, it is striated. Unlike skeletal muscle tissue, however, it is *involuntary*; its contractions are not consciously controlled. Cardiac muscle fibers are branched and usually have only one centrally located nucleus, although an occasional cell has two nuclei. The fibers attach end to end to one another by transverse thickenings of the plasma membrane called **intercalated discs** (*intercalat-* = to insert between). Intercalated discs contain cell junctions that strengthen cardiac muscle tissue and hold the fibers together during their vigorous contractions. The cell junctions also provide a route for quick conduction of electrical impulses from one cardiac muscle fiber to the next throughout the heart.

TABLE 4.3

Muscle Tissues

A. Skeletal Muscle Tissue

Description: Long, cylindrical, striated fibers with many peripherally located nuclei; voluntary control.

Location: Usually attached to bones by tendons.

Functions: Voluntary movement, posture, heat production.

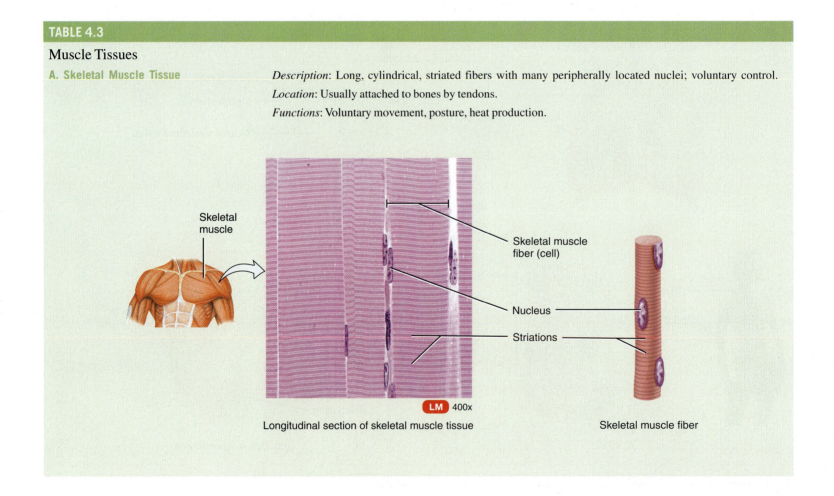

Skeletal muscle

Skeletal muscle fiber (cell)

Nucleus

Striations

LM 400x

Longitudinal section of skeletal muscle tissue

Skeletal muscle fiber

B. Cardiac Muscle Tissue

Description: Branched striated fibers with one or two centrally located nuclei; contains intercalated discs; involuntary control.

Location: Heart wall.

Function: Pumps blood to all parts of the body.

Heart

Nucleus

Striations

Cardiac muscle fiber (cell)

Intercalated disc

LM 600x

Longitudinal section of cardiac muscle tissue

Cardiac muscle fibers

C. Smooth Muscle Tissue

Description: Spindle-shaped (thickest in middle and tapering at both ends), nonstriated fibers with one centrally located nucleus; involuntary control.

Locations: Iris of the eyes and walls of hollow internal structures such as blood vessels, airways to the lungs, stomach, intestines, gallbladder, urinary bladder, and uterus.

Functions: Constriction of blood vessels and airways, propulsion of foods through gastrointestinal tract, contraction of urinary bladder and gallbladder.

Artery

Smooth muscle

Smooth muscle fiber (cell)

Nucleus of smooth muscle fiber

LM 350x

Longitudinal section of smooth muscle tissue

Smooth muscle fiber

Smooth muscle tissue is *nonstriated* (lacks striations), hence the term *smooth* (Table 4.3C). Smooth muscle fibers are small, thickest in the middle, and tapering at each end. They contain single, centrally located nuclei. Like cardiac muscle tissue, smooth muscle fibers are usually *involuntary*. Smooth muscle tissue is located in the walls of hollow internal structures such as blood vessels, airways to the lungs, the stomach, intestines, gallbladder, and urinary bladder. Its contractions help constrict or narrow the lumen of blood vessels, physically break down and move food along the gastrointestinal tract, move fluids through the body, and eliminate wastes. Such muscle tissues can produce powerful contractions as many smooth muscle fibers contract in unison. Chapter 10 will provide a detailed discussion of muscle tissue.

Checkpoint

22. List the functions of muscle tissue.
23. If you were presented with prepared slides, what characteristics could you use to identify skeletal muscle tissue? Cardiac muscle tissue? Smooth muscle tissue?
24. Which muscle tissue moves food through the intestines? Produces heart contractions? Moves bones?
25. Although movement of food through the gut is involuntary, initiation of swallowing is voluntary. Which type of muscle tissue lets you swallow a bite of pizza?
26. What are the functions of intercalated discs?

4.6 Nervous tissue consists of neurons and neuroglia.

Despite the awesome complexity of the nervous system, it consists of only two principal types of cells: neurons and neuroglia. **Neurons** (*neur-* = nerve, nervous tissue, nervous system), or nerve cells, are sensitive to various stimuli. They convert stimuli into electrical nerve impulses and conduct these impulses to other neurons, to muscle tissue, or to glands. Most neurons consist of three basic parts: a cell body and two types of cell processes—dendrites and axons (Table 4.4). The **cell body** contains the nucleus and

TABLE 4.4

Nervous Tissue

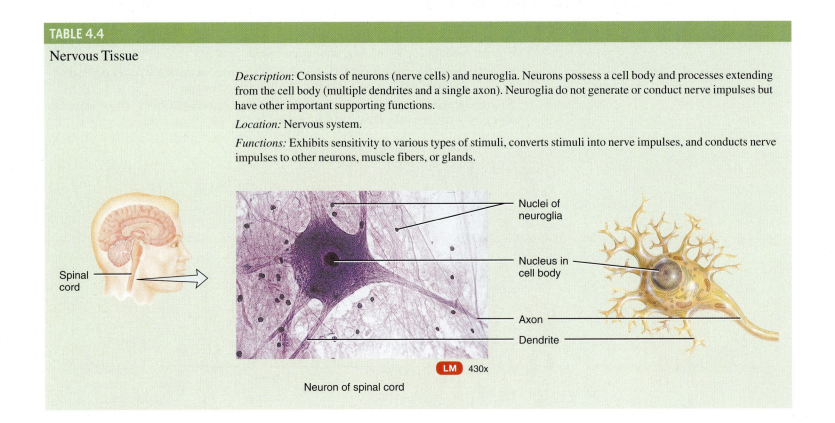

Description: Consists of neurons (nerve cells) and neuroglia. Neurons possess a cell body and processes extending from the cell body (multiple dendrites and a single axon). Neuroglia do not generate or conduct nerve impulses but have other important supporting functions.

Location: Nervous system.

Functions: Exhibits sensitivity to various types of stimuli, converts stimuli into nerve impulses, and conducts nerve impulses to other neurons, muscle fibers, or glands.

Spinal cord

Nuclei of neuroglia

Nucleus in cell body

Axon

Dendrite

LM 430x

Neuron of spinal cord

other organelles. **Dendrites** (*dendr-* = tree) are tapering, highly branched, and usually short cell processes. They are the major receiving or input portion of a neuron. The **axon** (*axo-* = axis) of a neuron is a single, thin, cylindrical process that may be very long. It is the output portion of a neuron, conducting nerve impulses toward another neuron or to some other tissue.

Neuroglia (noo-RŌG-lē-a; *-glia* = glue) are generally smaller than neurons. Even though neuroglia do not generate or conduct nerve impulses, these cells serve the important functions of supporting, nourishing, and protecting delicate neurons. The detailed structure and function of neurons and neuroglia will be considered in Chapter 12.

Checkpoint

27. What part of a neuron forwards nerve impulses to the neuron cell body? What part carries nerve impulses to other cells?
28. How do neurons differ from neuroglia?
29. Why would a glioma (a cancer of neuroglial cells) not necessarily disrupt the transmission of electrical impulses within the nervous system?

Return to John Doe's Story

Dr. Diego made a large Y-shaped incision along the dead man's trunk. The underlying connective tissue made a moist tearing sound as he peeled back the skin and underlying muscles of the chest and abdominal wall.

Using an electric saw, Dr. Diego made a cut from the bottom of the rib cage up each side of the body, severing the ribs, muscles, and connective tissue. Lifting off the chest plate, he exposed the membranes covering the heart and lungs.

Dr. Diego grasped the membrane surrounding the heart, tearing it to expose the heart. He cut the pulmonary trunk vessels where blood had once traveled to the lungs. "The thoracic cavity is full of blood. Tentatively, I'd say he died as a result of internal bleeding sustained from trauma in the crash." Dr. Diego's assistant leaned over the body to look. "Plane versus mountain, mountain wins. But what caused him to lose control of the plane and crash?"

Dr. Diego grasped the windpipe with one hand and cut above the larynx, severing the esophagus and trachea in one smooth motion. The hyaline cartilage rings of the trachea provided a relatively stable handle to grip as he pulled the organs down and out of the chest.

Dr. Diego cut the dome-shaped diaphragm next and pulled the abdominal organs out, handing them to his assistant. Clotted blood dripped off the organs as he carried the tissues to a dissecting table near the sink.

Dr. Diego's assistant began dissecting the mass of organs, while Diego began the process of removing John Doe's brain. With a quick swipe of his scalpel, Diego cut through the skin of the head and pulled the skin forward over the man's face. A reciprocating bone saw quickly sawed through the skull, exposing the membranes around the brain.

Dr. Diego carefully lifted out the brain, severing the spinal cord just below the brain stem, and suspended it in a large jar filled with preservative.

D. *Why do the cartilaginous structures in the airway make a good "handle" to grab onto in removing the lungs and heart?*

E. *As the medical examiner pulls out the thoracic and abdominal organs, the membranes within the body cavity provide little resistance. Why? What are these membranes called and what are they made of?*

F. *The dead man's body cavities contain clotted blood released when organs ruptured during the plane crash. What cells might you expect to find in blood? Which component of blood participates in blood clot formation?*

G. *Dr. Diego was careful to preserve the brain tissue, and didn't dissect and analyze it immediately during his preliminary autopsy. Why is nervous tissue soft and fragile?*

4.7 The ability of an injured tissue to repair itself depends on the extent of damage and the regenerative ability of the injured tissue.

Tissue repair is the process that replaces worn-out, damaged, or dead cells. In adults, each of the four basic tissue types (epithelial, connective, muscle, and nervous) has a different capacity for replenishing cells lost by damage, disease, or other processes.

Epithelial cells, which endure considerable wear and tear (and even injury) in some locations, have a continuous capacity for renewal. In some cases, immature, undifferentiated cells called **stem cells** divide as needed to replace lost or damaged cells. For example, stem cells reside in protected locations in the epithelia of the skin and gastrointestinal tract to replenish cells sloughed from the apical layer, and stem cells in red bone marrow continually provide new red and white blood cells and platelets. In other cases, mature, differentiated cells can undergo cell division; an example is endothelial cells in blood vessels.

Some connective tissues also have a continuous capacity for renewal. One example is osseous tissue, which has an ample blood supply. Other connective tissues, such as cartilage, can replenish cells less readily in part because of a poor blood supply.

Muscle tissue has a relatively poor capacity for renewal of lost cells. Skeletal muscle fibers do not divide rapidly enough to replace extensively damaged muscle fibers. Cardiac muscle fibers do not undergo mitosis to form new cells. However, stem cells do migrate into the heart from the blood. There, they can differentiate and replace a limited number of cardiac muscle fibers and endothelial cells lining heart blood vessels. Smooth muscle fibers can proliferate to some extent, but they do so much more slowly than the cells of epithelial or connective tissues.

Nervous tissue has the poorest capacity for renewal. Although experiments have revealed the presence of some stem cells in the brain, they normally do not undergo mitosis to replace damaged neurons. Discovering why is a major goal of researchers, who continually seek new ways to encourage repair of damaged or diseased nervous tissue.

During tissue repair, new cells originate by cell division from the **parenchyma**, cells that constitute the functioning part of a tissue or organ, or from the **stroma**, the supporting connective tissue. If parenchymal cells accomplish the repair, **tissue regeneration** is possible, and a near-perfect reconstruction of the injured tissue may occur. However, if fibroblasts of the stroma are active in the repair, the replacement tissue will be a new connective tissue. The fibroblasts synthesize collagen fibers and other extracellular matrix materials that aggregate to form scar tissue, a process known as **fibrosis**. Because scar tissue is not specialized to perform the functions of the parenchymal tissue, the original function of the tissue or organ is impaired.

When tissue damage is extensive, as in large, open wounds, cells from both the parenchyma and stroma are active in repair. During this process, fibroblasts manufacture new collagen fibers to provide structural strength. Blood capillaries sprout new buds to supply needed materials to the healing tissue. The actively growing connective tissue, called **granulation tissue**, forms across the wound or surgical incision to provide a supporting framework (stroma) through which epithelial cells migrate to fill the open area. The newly formed granulation tissue also secretes a fluid that kills bacteria.

Tissue repair is affected by nutrition and blood circulation. Nutrition is vital because the healing process places a great demand on the body's store of nutrients. Adequate protein in the diet is important because most of the structural components of a tissue are proteins. Several vitamins also play a direct role in wound healing and tissue repair. For example, vitamin C directly affects the normal production and maintenance of extracellular matrix materials, especially collagen fibers. Vitamin C also strengthens and promotes the formation of new blood vessels. In a person with vitamin C deficiency, even superficial wounds fail to heal, and the walls of the blood vessels become fragile and are easily ruptured. Proper blood circulation is essential to transport oxygen, nutrients, antibodies, and many defensive cells to the injured site during tissue repair. Blood also plays an important role in the removal of bacteria, foreign bodies, and debris, elements that would otherwise interfere with healing.

Checkpoint

30. Rank epithelial, connective, muscle, and nervous tissues in order from most to least able to fully repair.
31. Why does osseous tissue heal faster than cartilage?
32. What is the difference between the parenchyma and the stroma?
33. How are stromal and parenchymal repair of a tissue different?
34. What is the importance of granulation tissue?

John Doe's Story: Epilogue and Discussion

▶▶▶ Two weeks have passed since the autopsy, enough time for John Doe's brain to harden in the fixative. Dr. Diego can now finish his autopsy. He hadn't found any evidence of foul play, and no specific cause of death in the John Doe case. The sheriff had called a few days earlier and said the plane had not failed mechanically. So they were leaning toward pilot error as a cause of the crash.

Dr. Diego put on rubber gloves and lifted the jar containing John Doe's brain from a shelf. Opening the jar, he carried brain over to the metal dissecting table and sectioned the brain into 1-cm slices. As the pieces of brain tissue began to form stacks, he looked at one of them closely and said to himself, "Well! Now there's something you don't see every day." Circular chunks of dead fibrous tissue of various sizes were scattered throughout the man's brain like holes in a sponge. Some of the circular cysts were hard nodules. One nodule had been cut through, showing a small white mass inside a hollow shell.

Dr. Diego recognized the pathology. This was neurocysticerosis, the result of an intracranial infection of a parasitic tapeworm, *Taenia solium*, within the brain. It wasn't common in the United States; on the other hand, it wasn't unheard of. Certain forms of the parasite can be ingested in undercooked pork. Because humans are not the normal host for the parasite, we seem to confuse the beast. In unfamiliar territory, the tapeworm can escape through the intestinal wall and end up in the brain, forming cysts. The body responds to the invasion by sealing the tapeworm within a shell of scar tissue that eventually hardens into scarlike nodules.

John Doe most likely had lost consciousness and crashed. "Well, it looks like the mystery is solved," said Dr. Diego. ◀◀◀

H. *What types of cells would be found in the dead man's brain? Why didn't the man's brain regenerate new tissue to replace that damaged by the parasite?*

I. *How is an understanding of tissues essential to the job performance of a medical examiner (pathologist) like Dr. Diego?*

Concept and Resource Summary

Concept

Resources

Introduction

1. A **tissue** is a group of similar cells that have the same specialized function and embryonic origin.
2. The study of tissues is called **histology.**
3. A physician who studies cells and tissues is a **pathologist.**

Concept 4.1 Human body tissues can be classified as epithelial, connective, muscle, or nervous.

1. **Epithelial tissue** covers body surfaces, lines cavities, and forms glands.
2. **Connective tissue** binds organs together, stores energy, and participates in body defense.
3. **Muscle tissue** produces the physical force for body movements.
4. **Nervous tissue** detects and responds to changes in internal and external body conditions.
5. **Cell junctions** occur at points of contact between the plasma membranes of epithelial cells and can form tight seals, anchor cells together, or form channels for cell-to-cell passage of materials.

Concept 4.2 Epithelial tissue covers body surfaces, lines organs and body cavities, or secretes substances.

1. **Epithelial tissue**, or **epithelium**, is composed of a continuous sheet of cells that forms an excellent protective barrier on body surfaces and passageways. All epithelia have a superficial **apical (free) surface** facing a body cavity, lumen, or surface, and an opposing, deep **basal surface** that adheres to an extracellular **basement membrane.** The basement membrane attaches the epithelial tissue to the underlying connective tissue.
2. The functions of epithelium, which is **avascular** and relies on underlying connective tissue blood vessels for obtaining nutrients and removing wastes, include protection, filtration, secretion, absorption, and excretion. There are two types of epithelia: **covering and lining epithelium** and **glandular epithelium.**
3. Two characteristics are used to classify covering and lining epithelium: arrangement of cells in layers and cell shapes. The cells of covering and lining epithelium can be arranged as a single layer called *simple epithelium*; two or more layers called *stratified epithelium*; or a single layer that appears to have multiple layers called *pseudostratified epithelium*. The cell shapes of covering and lining epithelia include *squamous cells, cuboidal cells, columnar cells*, and *transitional cells* that change shape from cuboidal to flat in response to pressure.
4. **Simple squamous epithelium** is a single layer of broad, flat cells, located where filtration and diffusion occur. The simple squamous epithelium lining blood vessels, the heart, and lymphatic vessels is called **endothelium;** the simple squamous epithelium of serous membranes is called **mesothelium**.
5. **Simple cuboidal epithelium** is a single layer of cube-shaped cells, located where secretion and absorption occur.
6. **Simple columnar epithelium** is a single layer of tall, column-shaped cells that can be ciliated or nonciliated. *Nonciliated simple columnar epithelium* contains cells that have apical **microvilli** for increased absorption, and mucus-secreting **goblet cells** for lubrication. *Ciliated simple columnar epithelium* contains cells with apical cilia that sweep material across their surfaces and goblet cells.
7. **Pseudostratified columnar epithelium** is composed of a single layer of cells attached to the basement membrane but appears falsely stratified because the nuclei occur at various depths. Cells that reach up to the apical surface have cilia or are goblet cells.

See
- **Anatomy Overview – Epithelial Tissues**

Hear
- **Figure 4.2 – Cell Shapes and Arrangement of Layers for Covering and Lining Epithelium**

Concept   **Resources**

8. **Stratified squamous epithelium** has flat apical layer cells and basal layer cells that continually divide, get pushed upward, die, slough off, and get replaced by new cells. It can be divided into two types: The *keratinized type* contains **keratin**, a tough, fibrous protein; the *nonkeratinized type* remains moist.

9. **Transitional epithelium** stretches in response to pressure, causing the apical cells to become flat and then return to a larger, rounded appearance when pressure subsides.

10. **Glandular epithelia** form **glands** that produce secretions onto a surface, into a duct, or into the blood. Body glands are either endocrine glands or exocrine glands. **Endocrine glands** secrete *hormones* that enter interstitial fluid and then the blood. The secretions of **exocrine glands** enter ducts that empty secretions onto the skin surface or into the lumen of an organ.

Concept 4.3 Connective tissue binds organs together, stores energy reserves as fat, and helps provide immunity.

1. **Connective tissue** is widely distributed and abundant in the body, and is usually vascular. Its functions include binding, supporting, and strengthening other tissues, and protecting, insulating, and compartmentalizing structures. The two structural components of all connective tissues are *cells* and an *extracellular matrix*. The **extracellular matrix** is composed of protein fibers embedded in a ground substance.

2. **Fibroblasts** are the most numerous connective tissue cells; they secrete the fibers and ground substance of the matrix. Other connective tissue cells include phagocytic **macrophages**, antibody-secreting **plasma cells**, inflammatory **mast cells**, fat-storing **adipocytes**, and defensive **white blood cells.**

3. The **ground substance** supports and binds together cells embedded in it; allows for material exchange between blood and cells; and participates in tissue development and function. The composition of the ground substance includes water, proteins, and polysaccharides called **glycosaminoglycans.**

4. Three types of **fibers** can be embedded in the extracellular matrix: collagen fibers, elastic fibers, and reticular fibers. **Collagen fibers** are very strong, flexible, and composed of the protein **collagen.** They often occur in tightly packed parallel bundles. **Elastic fibers** are composed of the protein elastin, and can stretch and recoil back to shape. **Reticular fibers** consist of collagen and form delicate, branching networks that can form a supporting framework of soft organs.

5. There are five types of **mature connective tissue:** (1) loose connective tissue, (2) dense connective tissue, (3) cartilage, (4) bone tissue, and (5) liquid connective tissue (blood and lymph).

6. **Loose connective tissue** has loosely intertwined fibers and many cells. It includes areolar connective tissue, adipose tissue, and reticular connective tissue.

7. **Areolar connective tissue** is widely distributed in the body. It contains various cells including fibroblasts, macrophages, plasma cells, mast cells, and adipocytes. Collagen, elastic fibers, and reticular fibers are scattered throughout the tissue.

8. **Adipose tissue** is composed of **adipocytes** that store triglycerides (fat). It is an insulator and the body's major energy reserve.

9. **Reticular connective tissue** is made of fine interlacing reticular fibers and cells. It forms the filtering stroma of the liver, spleen, and lymph nodes.

See
* **Anatomy Overview – Connective Tissues**

Hear
* **Figure 4.3 –** **Representative Cells and Fibers Present in Connective Tissue**

Concept   **Resources**

10. **Dense connective tissue** has more fibers and fewer cells than loose connective tissue. It includes dense regular connective tissue, dense irregular connective tissue, and elastic connective tissue. **Dense regular connective tissue**, which contains *regularly* arranged parallel bundles of collagen fibers, forms the body's tendons and ligaments. **Dense irregular connective tissue** contains *irregularly* arranged collagen fibers and is found in the dermis, pericardium, heart valves, perichondrium, and periosteum.

11. **Elastic connective tissue** contains branching elastic fibers. It can stretch and recoil to its original shape and is important to the functioning of the lungs and elastic arteries.

12. **Cartilage** has a dense network of collagen and elastic fibers embedded in a gel-like ground substance. Mature cartilage contains **chondrocytes** found in **lacunae** in the extracellular matrix. Cartilage is avascular but is surrounded by a vascular **perichondrium.** There are three types of cartilage: hyaline cartilage, fibrocartilage, and elastic cartilage. **Hyaline cartilage** contains prominent chondrocytes and not visible collagen fibers, and provides flexible support and shock absorption. **Fibrocartilage**, the strongest type, is composed of chondrocytes scattered among visible bundles of collagen fibers. **Elastic cartilage**, which is elastic and resilient, contains chondrocytes scattered among elastic fibers.

13. **Osseous tissue** contains **osteocytes** within lacunae of an extracellular matrix containing crystallized mineral salts and collagen fibers. Osseous tissue is a primary component of bones. Bones store minerals, form the skeleton, and work with skeletal muscles to produce movement.

14. **Liquid connective tissue** includes blood and lymph, and has cells suspended in a liquid extracellular matrix. Blood has an extracellular matrix called **plasma** in which the formed elements (**red blood cells, white blood cells**, and **platelets**) are suspended. **Lymph** has a clear extracellular matrix similar to plasma, along with several types of cells including lymphocytes.

Concept 4.4 Membranes cover the surface of the body, line body cavities, and cover organs.

See

1. **Membranes** are flat sheets of flexible tissue that cover or line a body structure.

2. **Epithelial membranes** are composed of an epithelial layer and an underlying connective tissue layer. Mucous membranes, serous membranes, and the cutaneous membrane are epithelial membranes.

3. **Synovial membranes**, located in joints, contain connective tissue but lack an epithelial layer.

4. Body cavities that open to the exterior are lined with a **mucous membrane.** The epithelial layer of the mucosa functions as a protective barrier where goblet cells secrete mucus to prevent dehydration, trap particles, and allow lubrication. The underlying connective tissue layer of the mucosa is called the **lamina propria** and is composed of a supporting areolar connective tissue.

5. Body cavities not opening to the exterior are lined with a **serous membrane** that also covers cavity organs. Serous membranes have two layers: the **parietal layer** attached to the cavity wall and a **visceral layer** attached to cavity organs. **Serous fluid**, secreted by the mesothelium, is a lubricant that prevents organ friction. Serous membranes include the **pleura** lining the thoracic cavity and covering the lungs; the **pericardium** lining the heart cavity and covering the heart; and the **peritoneum** lining the abdominal cavity and covering abdominal organs.

6. The **skin**, or **cutaneous membrane**, is composed of an epithelial layer, the *epidermis*, and a connective tissue layer, the *dermis*.

7. **Synovial membranes** lack epithelium, are composed of areolar connective tissue, and secrete a lubricant called **synovial fluid.** Synovial membranes line cavities of freely movable joints, and also line *bursae* and *tendon sheaths*.

• Anatomy Overview – Epithelial Membranes

Concept

Concept 4.5 Muscle tissue generates the physical force needed to make body structures move.

1. **Muscle tissue** is composed of elongated cells called *muscle fibers*.
2. Muscle fibers generate the force to produce body movements, maintain posture, and generate heat.
3. There are three types of muscle tissue: skeletal, cardiac, and smooth.
4. **Skeletal muscle tissue** is attached to bones and is striated, multinucleated, and **voluntary**.
5. **Cardiac muscle tissue** forms most of the heart wall and is involuntary.
6. Cardiac muscle cells are striated, have one nucleus and have *intercalated discs*.
7. **Smooth muscle tissue** is *nonstriated*, with small tapered muscle fibers containing one central nucleus per cell; it is *involuntary* and is located in the walls of hollow structures.

See

- **Anatomy Overview – Muscle Tissue**

Concept 4.6 Nervous tissue consists of neurons and neuroglia.

1. The cells that make up nervous tissue are neurons and neuroglia.
2. **Neurons**, or nerve cells, respond to stimuli and conduct electrical nerve impulses to other neurons, muscle tissue, or glands.
3. The three parts of most neurons are the **cell body**; short cell processes called **dendrites**; and a single, elongated **axon**.
4. **Neuroglia** are smaller than neurons and function to support, nourish, and protect neurons.

See

- **Anatomy Overview – Nervous Tissue**

Concept 4.7 The ability of an injured tissue to repair itself depends on the extent of damage and the regenerative ability of the injured tissue.

1. Tissue repair is the process of replacing worn-out, damaged, or dead cells with healthy ones.
2. Epithelial cells have the capacity for continuous renewal.
3. **Stem cells**, found in some tissues, are immature, undifferentiated cells that divide as needed for tissue repair.
4. Connective tissues have different capacities for cell repair, from complete renewal to minimal repair. Muscle tissue has poor capacity for cell renewal. Nervous tissue has the poorest capacity for cell renewal.
5. When the functioning part of a tissue or organ, called the **parenchyma**, participates in cell repair, fully functioning **tissue regeneration** is possible.
6. When the supporting connective tissue, or **stroma**, of a tissue or organ is involved in repair, *fibrosis* occurs as fibroblasts produce scar tissue, impairing the function of the tissue or organ.
7. Extensive tissue damage involves repair by both the parenchyma and the stroma, resulting in new blood capillaries and a **granulation tissue** that spreads across the open wound.
8. Good nutrition and blood circulation are vital to tissue repair.

Other Chapter Resources

For additional information on clinical applications, medical disorders, and medical terminology, see Chapter 2 in the *Clinical Connections* manual packaged with your text.

Within *WileyPLUS*

WILEY PLUS
- **Chapter Quiz**
- **Visual Anatomy**
- **Mastering Vocabulary**
- **Audio Glossary**
- **Web Links**

Understanding the Concepts

1. If epithelial tissue is avascular, how do nutrients and oxygen get to the tissue and waste products leave?

2. Why is it important for epithelial tissue to repair itself quickly?

3. During an anatomy and physiology exam, you are asked to identify epithelial tissue from prepared slides. Based on the observations below, give the name and functions of each.

   - A single layer of cells that are about as high as they are wide.

   - A tissue that appears to contain multiple layers of cells because nuclei are at different levels. Cilia and goblet cells are visible.

   - A single layer of tall cells with nuclei near the basal surface of the cells. Goblet cells are occasionally visible.

   - Multiple layers of cells with flat cells in the apical layer.

   - A single layer of flat cells with centrally located oval or round nuclei.

   - Multiple layers of cells with bulging, rounded cells in the apical layer.

4. How do stratified and pseudostratified epithelia differ? (*Hint:* Pseudo *means "false."*)

5. How do the prefixes *exo-* and *endo-* give you a clue to the function of exocrine and endocrine glands?

6. A friend tells you that he has been diagnosed with *hyperhydrosis*, a disorder characterized by excessive sweating. He asks you if the glands secreting the excess sweat also secrete hormones. Explain your answer.

7. During an anatomy and physiology exam, you are asked to view and identify connective tissues from prepared slides. What is the name and function of each of the following connective tissues?

   - Cells appear empty in the center with nuclei visible in an outer ring of cytoplasm.

   - Cells inside lacunae; extracellular matrix contains bundles of thick fibers.

   - Cells inside lacunae; concentric rings of extracellular matrix.

   - Flattened cells located between densely packed, parallel bundles of thick fibers.

   - Random arrangement of several fiber and cell types.

8. What type of connective tissue connects skin to underlying structures? (*Hint: It is found in the subcutaneous layer.*)

9. What type of connective tissue can be found in the lungs, allowing them to expand during inhalation and snap back to a smaller size during exhalation?

10. How do the extracellular matrixes of cartilage, bone, and blood differ?

11. Why does mucosa stay attached to underlying structures?

12. How do synovial membranes and epithelial membranes differ in structure?

13. Why are stem cells of such intense interest to many scientists?

14. Why might replacement of damaged organ tissue with scar tissue compromise the function of an organ?

5

The Integumentary System

The dream had been so pleasant, until that car began beeping nonstop. Richard had been on the beach with the warm sun beating down on him. As he began to regain awareness, he realized he'd been dreaming and recognized the car beeping for what it really was—the piercing squeal of a smoke detector. The attic bedroom of the old Minnesota farmhouse where he slept was unusually warm, and he smelled smoke.

"Mom, Dad, wake up! I smell smoke!" he screamed as he dashed down the narrow staircase to the second floor. He could feel the heat of the fire as he looked into his parents' bedroom. No one was there. "They must have gotten out," he said to himself. He dashed down the stairs to the first floor and saw his father battling in vain to put out the flames that crept along the old wooden floors and were consuming the curtains of the living room. "Dad, where's Mom and the girls?" Richard shouted.

"They're outside already! Get out of here! The place is burning like a pile of matches!"

"I'll get the fire extinguisher—it's in the kitchen!" called Richard.

As he dashed toward the kitchen door, his father yelled out, "No! Richard—stop!"

As Richard pushed on the thin wooden door to the kitchen, he felt the air suck past him into the kitchen; it was followed by a silence that seemed to last forever. Then the back draft hit him. He saw a wall of flame, and he felt searing heat and instant pain like nothing he had ever imagined.

In this chapter you will study the integumentary system. The skin, or integument, is an important organ system that we often take for granted until it is compromised. As you will see in Richard's story, the skin has many important functions that are crucial to maintaining the homeostasis of the human body.

Skin and its accessory structures, hair, nails, various glands, and sensory receptors make up the **integumentary system** (in-teg-ū-MEN-tar-ē; *inte-* = whole; *-gument* = body covering). Of all the body's organs, none is more easily inspected or more exposed to infection, disease, and injury than the skin. Although the skin's location makes it vulnerable to damage from trauma, sunlight, microbes, and pollutants in the environment, it has protective features that ward off such damage. Because of its visibility, skin reflects our emotions and some aspects of normal physiology, as evi-

> Of all the body's organs, none is more easily inspected or more exposed to infection, disease, and injury than the skin.

denced by frowning, blushing, and sweating. Changes in skin color may indicate homeostatic imbalances in the body. For example, a bluish skin color indicates hypoxia (oxygen deficiency at the tissue level) and is one sign of heart failure as well as other disorders. Abnormal skin eruptions or rashes such as chickenpox, cold sores, or measles may reveal systemic infections or diseases of internal organs. Other conditions may be limited to the skin, such as warts, age spots, or pimples. So important is the skin to people's image that many spend a great deal of time and money to restore it to a more normal or youthful appearance.

5.1 Skin is composed of a superficial epidermis and a deeper dermis, and is anchored by the hypodermis.

The **skin**, or **cutaneous membrane**, covers the external surface of the body. It is the largest organ of the body in surface area and weight. In adults, the skin covers an area of about 2 square meters (22 square feet) and weighs 4.5–5 kg (10–11 lb), about 16% of total body weight. It ranges in thickness from 0.5 mm (0.02 in.) on the eyelids to 4.5 mm (0.18 in.) on the heels. However, over most of the body it is 1–2 mm (0.04–0.08 in.) thick.

The skin helps regulate body temperature, serves as a water-repellent and protective barrier between the external environment and internal tissues, contains sensory nerve endings that provide sensory information about the surrounding environment, excretes a small amount of salts and several organic compounds, and helps to synthesize the active form of vitamin D.

Structurally, the skin consists of two principal parts (Figure 5.1). The superficial, thinner portion, which is composed of *epithelial tissue*, is the **epidermis** (ep´-i-DERM-is; *epi-* = above). The deeper, thicker *connective tissue* portion is the **dermis**.

Figure 5.1   Components of the integumentary system.

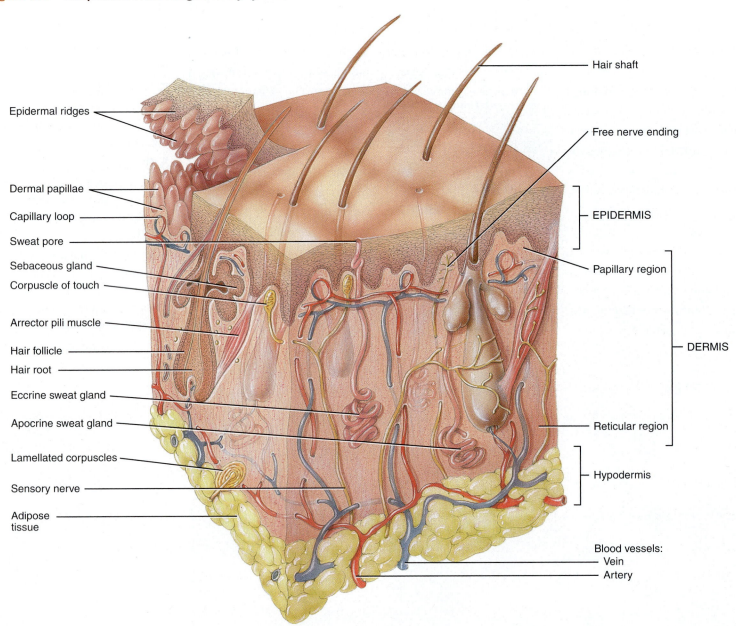

(a) Sectional view of skin and subcutaneous layer

Deep to the dermis, but not part of the skin, is the **hypodermis** (*hypo-* = below), or **subcutaneous layer**. This layer consists of areolar and adipose tissues. Fibers that extend from the dermis anchor the skin to the hypodermis, which in turn attaches to underlying fascia, the connective tissue around muscles and bones. The hypodermis serves as a storage depot for fat and contains large blood vessels that supply the skin. This region (and sometimes the dermis) also contains nerve endings called **lamellated (pacinian) corpuscles** (pa-SIN-ē-an) that are sensitive to pressure.

Although the skin over the entire body is similar in structure, there are quite a few local variations related to thickness of the epidermis, strength, flexibility, distribution and type of hair, density and types of glands, pigmentation, vascularity (blood supply), and innervation (nerve supply). Based on structural and functional properties, we recognize two major types of skin: thin (hairy) and thick (hairless).

Thin skin covers all parts of the body except for the palms, palmar surfaces of fingers, plantar surface of toes, and soles. Its epidermis is thin, just 0.10–0.15 mm (0.004–0.006 in.). Even though thin skin has hair and sebaceous (oil) glands, it has fewer sudoriferous (sweat) glands than thick skin. Thin skin also has a sparser distribution of sensory receptors than thick skin.

Functions
1. Regulates body temperature.
2. Stores blood.
3. Protects body from external environment.
4. Detects cutaneous sensations.
5. Excretes and absorbs substances.
6. Synthesizes vitamin D.

EPIDERMIS

Papillary region

DERMIS

Reticular region

Sebaceous gland

Hair root

Hair follicle

LM 60x

(b) Sectional view of skin

Sweat pores

Epidermal ridges

(c) Epidermal ridges and sweat pores

🔑 The integumentary system includes the skin, hair, nails, glands, and sensory receptors.

Thick skin covers the palms, palmar surfaces of fingers, plantar surfaces of toes, and soles. Its epidermis is relatively thick, 0.6–4.5 mm (0.024–0.18 in.). Thick skin lacks hair and sebaceous glands, and has more sudoriferous glands than thin skin. Sensory receptors are also more densely clustered in thick skin. Table 5.1 summarizes the features of thin and thick skin.

Checkpoint

1. List the structures of the integumentary system.
2. Which tissues compose the hypodermis?
3. How does skin on the bottom of the feet differ from that on the face? Why is that difference a survival advantage?

TABLE 5.1

Comparison of Thin and Thick Skin

FEATURE	THIN SKIN	THICK SKIN
Distribution	All parts of the body except palms, palmar surface of fingers, plantar surface of toes, and soles.	Palms, palmar surface of fingers, plantar surface of toes, and soles.
Epidermal thickness	0.10–0.15 mm (0.004–0.006 in.).	0.6–4.5 mm (0.024–0.18 in.).
Epidermal strata	Stratum lucidum essentially lacking; thinner strata spinosum and corneum.	Thick strata lucidum, spinosum, and corneum.
Hair follicles and arrector pili muscles	Present.	Absent.
Sebaceous glands	Present.	Absent.
Sudoriferous glands	Fewer.	More numerous.
Sensory receptors	Sparser.	Denser.

5.2 The layers of the epidermis include the stratum basale, stratum spinosum, stratum granulosum, stratum lucidum, and stratum corneum.

Cells of the Epidermis

The **epidermis** is composed of keratinized stratified squamous epithelium. It contains four principal types of cells: keratinocytes, melanocytes, Langerhans cells, and Merkel cells (Figure 5.2). **Keratinocytes** (ker-a-TIN-ō-sīts; *keratino-* = hornlike; *-cytes* = cells), the most numerous epidermal cells, are arranged in four or five layers and produce **keratin**, a tough, fibrous protein that helps protect the skin and underlying tissues from heat, microbes, and chemicals. Keratinocytes also produce lamellar granules, which release a waterproofing sealant.

Figure 5.2   Layers of the epidermis.

(a) Four principal cell types in epidermis

(b) Photomicrograph of a portion of the skin

Most of the epidermis consists of keratinocytes, which produce the protein keratin (protects underlying tissues) and lamellar granules (contain a waterproof sealant).

Melanocytes (MEL-a-nō-sīts; *melano-* = black) are found in the deepest layer of the epidermis and produce the pigment melanin. **Melanin** is a brown-black pigment that contributes to skin color and absorbs damaging ultraviolet (UV) light. Long, slender projections of melanocytes extend between the keratinocytes and transfer melanin granules to them. Once inside keratinocytes, the melanin granules cluster to form a protective veil over the nucleus on the side toward the skin surface. In this way they shield the nuclear DNA from UV light. Although keratinocytes gain some protection from melanin granules, melanocytes themselves are particularly susceptible to damage by UV light.

Langerhans cells (LANG-er-hans) constitute a small fraction of the epidermal cells. They participate in immune responses mounted against microbes that invade the skin and are easily damaged by UV light.

Merkel cells are the least numerous of the epidermal cells. They are located in the deepest layer of the epidermis, where they contact the flattened process of a sensory neuron (nerve cell), a dermal structure called a **Merkel disc**. Merkel cells and their associated Merkel discs function together in the sensation of touch.

Strata of the Epidermis

Several distinct layers of keratinocytes in various stages of development form the epidermis (Figure 5.2). In thin skin covering most regions of the body, the epidermis has four strata, or layers: stratum basale, stratum spinosum, stratum granulosum, and a thin stratum corneum (see Table 5.1). Thick skin covers the body where exposure to friction is greatest, as in the fingertips, palms, and soles. The epidermis of thick skin has five layers: stratum basale, stratum spinosum, stratum granulosum, stratum lucidum, and a thick stratum corneum.

The deepest layer of the epidermis, the **stratum basale** (ba-SA-lē; *basal-* = base), is composed of a single row of cuboidal or columnar keratinocytes, some of which are *stem cells* that undergo cell division to continually produce new keratinocytes. Melanocytes and Merkel cells (with their associated Merkel discs) are scattered among the keratinocytes of the basal layer.

Superficial to the stratum basale is the **stratum spinosum** (spi-NŌ-sum; *spinos-* = thornlike), arranged in 8 to 10 layers of many-sided keratinocytes fitting closely together. Cells in the more superficial portions of this layer become somewhat flattened. When cells of the stratum spinosum are prepared for microscopic examination, they shrink and pull apart so that they appear to be covered with thornlike spines (hence the name "spinosum"), although they are rounded and larger in living tissue. Langerhans cells and projections of melanocytes are present in this layer.

At about the middle of the epidermis, the **stratum granulosum** (gran-ū-LŌ-sum; *granulos-* = little grains) consists of three to five layers of flattened keratinocytes that are undergoing **apoptosis** (ap-ō-TŌ-sis = a falling off), an orderly, genetically programmed cell death in which the nucleus fragments before the cells die. As their nuclei break down during apoptosis, the keratinocytes of the stratum granulosum can no longer carry on vital metabolic reac-

tions, and they die. Thus, the stratum granulosum marks the transition between the deeper, metabolically active strata and the dead cells of the more superficial strata. A distinctive feature of this layer is the presence of membrane-enclosed **lamellar granules** within the keratinocytes. Lamellar granules release a lipid-rich secretion that fills the spaces between cells of the stratum granulosum, stratum lucidum, and stratum corneum. The lipid-rich secretion acts as a water-repellent sealant, retarding loss of water and entry of foreign materials.

The **stratum lucidum** (LOO-si-dum; *lucid-* = clear) is present only in thick skin. It consists of three to five layers of clear, flat, dead keratinocytes that contain large amounts of keratin.

The **stratum corneum** (COR-nē-um; *corne-* = hard or hooflike) consists of 25 to 30 layers of dead, flat keratinocytes. These cells are continuously shed and replaced by cells from the deeper strata. The interior of the cells contains mostly keratin. Between the cells are lipids from lamellar granules that help make this layer an effective water-repellent barrier. Its multiple layers of dead cells also help to protect deeper layers from injury and microbial invasion.

Keratinization and Growth of the Epidermis

Newly formed cells in the stratum basale are slowly pushed to the surface. As the cells move from one epidermal layer to the next, they accumulate more and more keratin in a process called **keratinization** (ker-a-tin-i-ZĀ-shun). Eventually the keratinized cells slough off and are replaced by underlying cells that in turn become keratinized. The whole process by which cells form in the stratum basale, rise to the surface, become keratinized, and slough off takes about four weeks in an average epidermis of 1.0 mm (0.04 in.) thickness.

Table 5.2 summarizes the distinctive features of the epidermal strata.

TABLE 5.2

Epidermal Strata

STRATUM	DESCRIPTION
Basale	Deepest layer, composed of a single row of cuboidal or columnar keratinocytes; stem cells undergo cell division to produce new keratinocytes; melanocytes, Langerhans cells, and Merkel cells are scattered among the keratinocytes.
Spinosum	Multiple rows of many-sided keratinocytes; includes projections of melanocytes and Langerhans cells.
Granulosum	Multiple rows of flattened keratinocytes, in which nuclei are beginning to degenerate; cells contain darkly staining keratin and lamellar granules, which release a lipid-rich, water-repellent secretion.
Lucidum	Present only in thick skin of fingertips, palms, and soles; consists of multiple rows of clear, flat, dead keratinocytes with large amounts of keratin.
Corneum	Multiple rows of dead, flat keratinocytes that contain mostly keratin.

Checkpoint

4. Which tissue type makes up the epidermis?
5. What is the function of lamellar granules? Of melanin?
6. Which epithelial cells can sense that someone is touching your skin? Help protect you from the damaging rays of the sun? Guard against surface bacteria? Help waterproof your skin?
7. Which epidermal layer includes stem cells that continually undergo cell division?
8. Which visual characteristics allow you to distinguish each epidermal layer?

5.3   The dermis contains blood vessels, nerves, sensory receptors, hair follicles, and glands.

The second, deeper part of the skin, the **dermis**, is composed mainly of connective tissue containing collagen and elastic fibers. Blood vessels, nerves, glands, and hair follicles are embedded in dermal tissue. Based on its tissue structure, the dermis can be divided into a superficial papillary region and a deeper reticular region (see Figure 5.1).

The **papillary region** makes up about one-fifth of the thickness of the total dermis (see Figure 5.1). It consists of areolar connective tissue containing fine elastic fibers. Its surface area is greatly increased by small, fingerlike projections called **dermal papillae** (pa-PIL-ē = nipples). These nipple-shaped structures indent the epidermis and some contain **capillary loops** (blood capillaries). Some dermal papillae also contain touch receptors called **corpuscles of touch**, or **Meissner corpuscles**, nerve endings that are sensitive to touch. Also present in the dermal papillae are **free nerve endings**, which initiate signals that produce sensations of warmth, coolness, pain, tickling, and itching.

The **reticular region** (*reticul-* = netlike) is attached to the hypodermis and consists of dense irregular connective tissue containing bundles of collagen and some coarse elastic fibers. Between the fibers are hair follicles, nerves, sebaceous (oil) glands, and sudoriferous (sweat) glands. The combination of collagen and elastic fibers in the reticular region provides the skin with strength, **extensibility** (ability to stretch), and **elasticity** (ability to return to its original shape after stretching). The extensibility of skin can be readily seen around joints and in pregnancy and obesity. However, extreme stretching may produce small tears in the dermis, causing *striae* (STRĪ-ē = streaks), or stretch marks, that are visible as red or silvery white streaks on the skin surface.

The surfaces of the palms, fingers, soles, and toes are marked by series of ridges and grooves. They appear either as straight lines or as a pattern of loops and whorls, as on the tips of the fingers. These **epidermal ridges** are downward projections of the epidermis into the dermis between the dermal papillae (see Figure 5.1). Epidermal ridges and dermal papillae join together like complementary teeth of a zipper forming an extremely strong connection.

This strengthens the skin against shearing forces (forces that laterally shift in relation to each other) that attempt to seperate the epidermis from the dermis. Epidermal ridges increase the surface area of the epidermis and thus increase the grip of the hand or foot by increasing friction. Because the ducts of sweat glands open on the distal tops of the epidermal ridges as sweat pores, the sweat and ridges form *fingerprints* (or *footprints*) upon touching a smooth object. The epidermal ridge pattern is genetically determined and is unique for each individual. Normally, the epidermal ridge pattern does not change during life, except to enlarge, and thus can serve as the basis for identification through fingerprints or footprints.

Table 5.3 summarizes the structural features of the regions of the dermis.

TABLE 5.3	
Dermal Regions	
REGION	DESCRIPTION
Papillary	The superficial portion of the dermis; consists of areolar connective tissue with elastic fibers and dermal papillae that house capillaries, corpuscles of touch, and free nerve endings.
Reticular	The deeper portion of the dermis; consists of dense irregular connective tissue with bundles of collagen and elastic fibers, adipocytes, hair follicles, nerves, sebaceous glands, and sudoriferous glands.

Checkpoint

9.   Which type of connective tissue forms the papillary and reticular regions of the dermis?
10. What are the functions of fibers in the reticular region of the dermis?
11. How could you explain that a slight paper cut to the skin does not bleed?

Richard was unconscious for almost 4 days after the fire. His mother, father, and two sisters came to visit him in the burn unit every day, and his mother stayed overnight. Richard's burns were serious, and the doctor was very cautious about Richard's prognosis. "Mrs. Anderson, your son has severe burns over one-third of his body. His hands and face have second-degree partial thickness burns. This type of injury damages the dermis of the skin but doesn't completely destroy it; it's very painful but should heal fairly well given enough time. I am more concerned about the burns on his chest and stomach. Those burns are third-degree full thickness, involving all the layers of the skin down to the underlying subcutaneous layers. In these types of injuries the tissue is damaged down into the underlying muscle. They often require skin grafting and a long recovery time."

Mrs. Anderson thought about the fleece top that Richard always wore to bed. "The attic gets cold, Mom," Richard had said. The flammable material had caught fire after the explosion and melted into Richard's skin, burning his chest severely.

The nurses cleaned Richard's burns and applied antibiotic ointments daily, and kept IV fluids and pain medication flowing into his body. His legs and arms were propped up on pillows to keep them higher than his torso. Blisters formed on the outer edges of the burns as the wounds began to form scabs.

A. *The doctor describes Richard as having two types of burns, "partial thickness" and "full thickness." Based on what you have learned about the skin, explain why a partial thickness burn is extremely painful and why it would heal faster than a full thickness burn.*

B. *Hospital staff members continually administer intravenous fluids, clean his wounds, administer antibiotics, and give him pain medication. Based on these observations, what functions of Richard's skin have been compromised? What other functions of skin would be of concern to the medical staff?*

5.4 Skin color is a result of the pigments melanin, carotene, and hemoglobin.

Melanin, carotene, and hemoglobin are three pigments that give skin a wide variety of colors. The amount of **melanin**, which is located mostly in the epidermis, causes the skin's color to vary from pale yellow to reddish-brown to black. Melanocytes, the melanin-producing cells, are most plentiful in the epidermis of the penis, nipples of the breasts, area just around the nipples (areolae), face, and limbs. They are also present in mucous membranes. Because the *number* of melanocytes is about the same in all people, differences in skin color are due mainly to the *amount of melanin* the melanocytes produce. In some people, melanin tends to accumulate in patches called *freckles*. As a person ages, *age (liver) spots* may develop. These flat blemishes look like freckles and range in color from light brown to black. Like freckles, age spots are accumulations of melanin. A round, flat, or raised area that represents a benign localized overgrowth of melanocytes and usually develops in childhood or adolescence is called a *nevus* (NĒ-vus), or a *mole*.

Exposure to ultraviolet (UV) light stimulates melanin production, which gives the skin a tanned appearance and protects the body against further UV radiation. Melanin absorbs UV radiation and prevents damage to DNA in epidermal cells. Thus, within limits, melanin serves a protective function. However, repeatedly exposing the skin to a large amount of UV light may cause skin cancer. A tan is lost when the melanin-containing keratinocytes are shed from the stratum corneum.

Carotene (KAR-ō-tēn; *carot-* = carrot) is a yellow-orange pigment that gives egg yolks and carrots their color. It is the precursor of vitamin A, which is used to synthesize pigments needed for vision. Carotene stored in fatty areas of the skin after eating large amounts of carotene-rich foods can turn skin orange.

Dark-skinned to black individuals have large amounts of melanin in the epidermis. Consequently, the epidermis has a dark pigmentation and skin color ranges from yellow to reddish-brown black. Light-skinned individuals have little melanin in the epidermis. Thus, the epidermis appears translucent and skin color ranges from pink to red depending on the oxygen content of the blood moving through capillaries in the dermis. The red color is due to **hemoglobin**, the oxygen-carrying pigment in red blood cells.

There are enormous variations in skin color among different populations. Individuals who live in the tropics (near the equator), where the sun's rays are intense, have evolved darkly pigmented skin; people who inhabit the subtropics and temperate regions have moderately pigmented skin and have the ability to tan; and individuals who live in regions near the poles are very light-skinned and burn easily.

Checkpoint

12. If the number of melanocytes is the same in peoples of all races, what causes their skin color to differ so significantly?

13. What do liver spots and freckles have in common?

14. What are the three pigments in the skin and how do they contribute to skin color?

5.5 The functions of hair, skin glands, and nails include protection and body temperature regulation.

The accessory structures of the skin include hair, glands, and nails. During embryological development, these structures develop from the epidermis. Together, the skin and accessory structures compose the integumentary system.

Hair

Hairs, or **pili** (PĪ-lī), are present on most skin surfaces except the palms, palmar surfaces of the fingers, soles, and plantar surfaces of the toes. In adults, hair is usually most heavily distributed across the scalp, in the eyebrows, in the axillae (armpits), and around the external genitalia. Genetic and hormonal influences largely determine the thickness and pattern of hair distribution.

Although the protection it offers is limited, hair on the head guards the scalp from injury and the sun's rays. It also decreases heat loss from the scalp. Eyebrows and eyelashes protect the eyes from foreign particles, as does hair in the nostrils and in the external ear canal. Touch receptors associated with hair follicles (hair root plexuses) are activated whenever a hair is even slightly moved. Thus, hairs also function in sensing light touch.

Anatomy of a Hair

Each hair is composed of columns of dead, keratinized epidermal cells bonded together by extracellular proteins. The **shaft** is the superficial portion of the hair, most of which projects from the surface of the skin (Figure 5.3a). The transverse-sectional shape of the shafts of hairs varies. Straight hair is round in transverse section; wavy hair is oval; and curly hair is kidney shaped. The **root** is the portion of the hair deep to the shaft that penetrates into the dermis, and sometimes into the hypodermis. The shaft and root both consist of three concentric layers (Figure 5.3c, d). The inner *medulla* is composed of cells containing pigment granules and air spaces. The middle *cortex* consists of cells that contain pigment granules in dark hair but mostly air in gray or white hair. The *cuticle* of the hair, the outermost layer, consists of a single layer of thin, flat cells that are the most heavily keratinized. Cuticle cells are arranged like shingles on the side of a house, with their free edges pointing toward the free end of the hair (Figure 5.3b).

Surrounding the root of the hair is the **hair follicle**, which is made up of an external root sheath and an internal root sheath (Figure 5.3c, d). The *external root sheath* is a downward continuation of the epidermis. The *internal root sheath* is produced by the hair matrix (described shortly) and forms a tubular sheath of epithelium between the external root sheath and the hair. The dense dermis surrounding the hair follicle is called the **dermal root sheath**.

The base of each hair follicle is enlarged into an onion-shaped structure, the **bulb** (Figure 5.3c). This structure houses a nipple-shaped indentation, the **papilla** of the hair, which contains many blood vessels that nourish the growing hair follicle. The bulb also contains a germinal layer of cells called the **hair matrix**, the site of hair cell division. Hence, hair matrix cells are responsible for the growth of existing hairs, and they produce new hairs when old hairs are shed. Hair matrix cells also give rise to the cells of the internal root sheath.

Sebaceous (oil) glands (discussed shortly) and a bundle of smooth muscle cells are also associated with hairs (Figure 5.3a). The smooth muscle is called **arrector pili** (a-REK-tor PĪ-lī; *arrect-* = to raise). It extends from the superficial dermis of the skin to the dermal root sheath around the hair follicle. In its normal position, hair emerges at an angle to the surface of the skin. Under physiological or emotional stress, such as cold or fright, autonomic nerve endings stimulate the arrector pili muscles to contract, which pulls the hair shafts perpendicular to the skin surface. This action causes "goose bumps" or "gooseflesh" because the skin around the shaft forms slight elevations.

Surrounding each hair follicle are nerve endings, called **hair root plexuses**, that are sensitive to touch (Figure 5.3a). The hair root plexuses generate nerve impulses if the hair shaft is moved, which happens, for example, when an insect bumps into a hair as it crawls across your arm.

Hair Growth

Each hair follicle goes through a growth cycle, which consists of a growth stage and a resting stage. During the **growth stage**, cells of the hair matrix divide. As new cells from the hair matrix are added to the base of the hair root, existing cells of the hair root are pushed upward and the hair grows longer. While the cells of the hair are being pushed upward, they become keratinized and die. In time, the growth of the hair stops and the **resting stage** begins. After the resting stage, a new growth cycle begins. The old hair root falls out or is pushed out of the hair follicle as a new hair begins to grow in its place. Scalp hair grows for 2 to 6 years and rests for about 3 months. At any time, about 85% of scalp hairs are in the growth stage. Visible hair is dead, but until the hair is pushed out of its follicle by a new hair, portions of its root within the scalp are alive.

Hair Color

The color of hair is due primarily to the amount and type of melanin in its cells. Melanin is synthesized by melanocytes scattered in the hair matrix of the bulb and passes into cells of the cortex and medulla (Figure 5.3c). Dark-colored hair contains mostly true melanin (brown to black). Blond and red hair contain variants of melanin (yellow to red) in which there is iron and more sulfur. Gray hair contains only a few melanin granules because of a progressive decline in the production of melanin. White hair results from the lack of melanin and the accumulation of air bubbles in the hair shaft.

Figure 5.3 Hair.

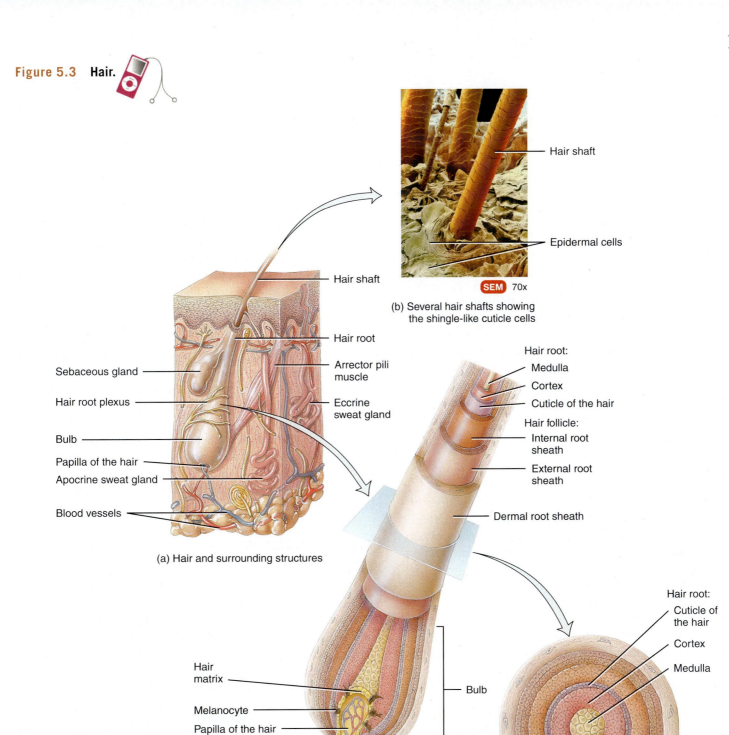

Hair shaft

Epidermal cells

SEM 70x

(b) Several hair shafts showing the shingle-like cuticle cells

Hair shaft

Hair root

Arrector pili muscle

Sebaceous gland

Eccrine sweat gland

Hair root plexus

Bulb

Papilla of the hair

Apocrine sweat gland

Blood vessels

(a) Hair and surrounding structures

Hair root:

Medulla

Cortex

Cuticle of the hair

Hair follicle:

Internal root sheath

External root sheath

Dermal root sheath

Hair matrix

Melanocyte

Papilla of the hair

Blood vessels

Bulb

(c) Frontal section of hair root

Hair follicle:

Internal root sheath

External root sheath

Hair root:

Cuticle of the hair

Cortex

Medulla

Dermal root sheath

(d) Transverse section of hair root

🔑 **Hairs are growths of epidermis composed of dead, keratinized cells.**

Glands of the Skin

Recall from Chapter 4 that glands are epithelial cells that secrete a substance. Several kinds of exocrine glands are associated with the skin: sebaceous (oil) glands, sudoriferous (sweat) glands, and ceruminous glands. Mammary glands, which are specialized sudoriferous glands that secrete milk, are discussed in Chapter 25 along with the female reproductive system.

Sebaceous Glands

Sebaceous glands (se-BĀ-shus; *sebace-* = greasy), or **oil glands**, with few exceptions, are connected to hair follicles (see Figures 5.1 and 5.3a). The secreting portions of the glands lie in the dermis and open into the hair follicles or directly onto a skin surface. Sebaceous glands are found in the skin over all regions of the body except the palms and soles. Sebaceous glands secrete an oily substance called **sebum** (SĒ-bum), a mixture of triglycerides, cholesterol, proteins, and inorganic salts. Sebum coats the surface of hairs and helps keep them from drying and becoming brittle. Sebum also prevents excessive evaporation of water from the skin, keeps the skin soft and pliable, and inhibits the growth of certain bacteria.

Sudoriferous Glands

There are 3 to 4 million **sudoriferous glands** (soo´-dor-IF-er-us; *sudori-* = sweat; *-ferous* = bearing), or **sweat glands**, that release sweat, or *perspiration*, onto the skin surface through pores or into hair follicles. Sudoriferous glands are divided into two main types: eccrine and apocrine.

Eccrine sweat glands (EK-rin; *eccrine* = secreting outwardly) are much more common than apocrine sweat glands. They are distributed throughout most of the skin and are most numerous in the skin of the forehead, palms, and soles. The secretory portion of eccrine sweat glands is located mostly in the deep dermis (sometimes in the upper hypodermis). The excretory duct projects through the dermis and epidermis and ends as a pore at the surface of the epidermis (see Figures 5.1 and 5.3a). The sweat produced by eccrine sweat glands (about 600 mL per day) consists of water, ions (mostly Na^+ and Cl^-), urea, uric acid, ammonia, amino acids, glucose, and lactic acid. The main function of eccrine gland sweat is to help regulate body temperature through evaporation. As sweat evaporates, large quantities of heat energy leave the body surface. Eccrine sweat also plays a small role in eliminating wastes such as urea, uric acid, and ammonia. However, the kidneys are the primary excretor of wastes from the body.

Apocrine sweat glands (AP-ō-krin; *apo* = separated from) are found mainly in the skin of the axilla (armpit), groin, areolae (pigmented areas around the nipples) of the breasts, and bearded regions of the face in adult males. The secretory portion of these sweat glands is located mostly in the hypodermis, and their excretory ducts open into hair follicles (see Figures 5.1 and 5.3a). Their secretion is slightly viscous compared to eccrine secretions and contains the same components as eccrine sweat plus lipids and proteins. Eccrine sweat glands start to function soon after birth, but apocrine sweat glands do not begin to function until puberty. Sweat secreted from apocrine sweat glands is odorless. However, skin bacteria soon metabolize its components, causing apocrine sweat to have a musky odor that is often referred to as body odor. Apocrine sweat glands secrete sweat during sexual activities. In addition, both apocrine and eccrine sweat glands are active during emotional sweating.

Table 5.4 compares the features of eccrine and apocrine sweat glands.

Ceruminous Glands

Modified sweat glands in the external ear, called **ceruminous glands** (se-RŪ-mi-nus; *cer-* = wax), produce a waxy secretion. Their excretory ducts open either directly onto the surface of the external auditory canal (ear canal) or into ducts of sebaceous glands. The combined secretion of the ceruminous and sebaceous glands is a yellowish material called **cerumen**, or earwax. Cerumen, together with hairs in the external auditory canal, provides a sticky barrier that impedes the entrance of foreign bodies.

TABLE 5.4

Comparison of Eccrine and Apocrine Sweat Glands

FEATURE	ECCRINE SWEAT GLANDS	APOCRINE SWEAT GLANDS
Distribution	Throughout skin; particularly numerous on forehead, palms, and soles.	Skin of axilla, groin, areolae, and bearded regions of face.
Location of secretory portion	Mostly in deep dermis.	Mostly in hypodermis.
Termination of excretory duct	Surface of epidermis.	Hair follicle.
Secretion	Less viscous; consists of water, ions (Na^+, Cl^-), urea, uric acid, ammonia, amino acids, glucose, and lactic acid.	More viscous; consists of the same components as eccrine sweat glands plus lipids and proteins.
Functions	Body temperature regulation, waste removal, stimulated by emotional stress.	Stimulated by emotional stress and sexual excitement.
Onset of function	Soon after birth.	Puberty.

Figure 5.4 **Nails.** Shown is a fingernail.

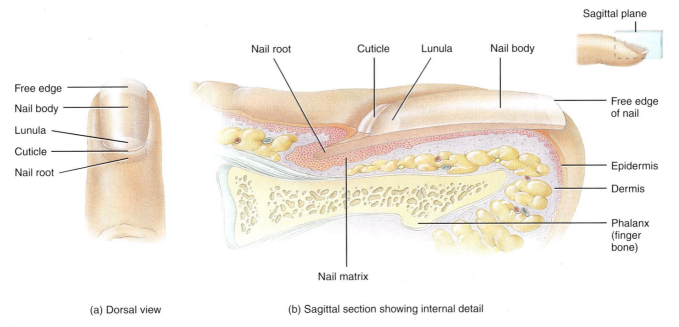

(a) Dorsal view (b) Sagittal section showing internal detail

🔑 **Nails grow by transformation of nail matrix cells into nail cells.**

Nails

Nails are plates of tightly packed, hard, keratinized epidermal cells that form a clear, solid covering over the dorsal surfaces of the distal portions of fingers and toes. Nails help us grasp and manipulate small objects, provide protection against trauma to the ends of the fingers and toes, and allow us to scratch various parts of the body.

Each nail consists of a nail body, a free edge, and a nail root (Figure 5.4). The **nail body** is the visible portion of the nail, the **free edge** is the distal of the body that may extend past the end of the finger or toe, and the **nail root** is the proximal portion that is buried in a fold of skin. Most of the nail body is pink because of blood in the underlying blood capillaries. The free edge is white because there are no underlying capillaries. The whitish semilunar area near the nail root is called the **lunula** (LOO-noo-la = little moon). It appears whitish because the thickened nail matrix (described shortly) obscures the vascular tissue underneath. The **cuticle**, a narrow band of epidermis that extends from the proximal border of the nail, consists of stratum corneum.

The epithelium deep to the nail root is known as the **nail matrix**. Nail growth occurs by the transformation of superficial matrix cells into nail cells. In the process, the harder outer layer is pushed distally toward the free edge. The average growth in the length of fingernails is about 1 mm (0.04 in.) per week. The growth rate of toenails is somewhat slower.

Checkpoint

15. What are the functions of hair?
16. List the layers of a hair from the innermost component to the outermost component.
17. What purpose does the papilla of the hair serve?
18. How are goose bumps produced?
19. What are the functions of sebum, eccrine sweat, and cerumen?
20. What are the functions of nails? Why are they so hard?

5.6 Skin damage sets in motion a sequence of events that repairs the skin to its normal (or near-normal) structure and function.

When skin is damaged, a sequence of events repairs the skin to its normal (or near-normal) structure and function. There are two types of wound-healing processes, depending on the depth of the injury. Epidermal wound healing occurs following wounds that affect only the epidermis; deep wound healing occurs following wounds that penetrate the dermis.

Epidermal Wound Healing

Even though the central portion of an epidermal wound may extend to the dermis, the edges of the wound usually involve only slight damage to superficial epidermal cells. Common types of epidermal wounds include abrasions, in which a portion of skin has been scraped away, and minor burns.

In response to an epidermal injury, stratum basale cells of the epidermis (also called basal cells) surrounding the wound break contact with the basement membrane. The cells then enlarge and migrate across the wound (Figure 5.5a). The cells appear to migrate as a sheet until advancing cells from opposite sides of the wound meet. When basal cells encounter one another, they stop migrating due to a cellular response called **contact inhibition**. Migration of the basal cells stops completely when each is finally in contact with other basal cells on all sides.

As the basal cells migrate, a hormone called *epidermal growth factor* stimulates basal stem cells to divide and replace the ones that have moved into the wound. The relocated basal cells divide to build new strata, thus thickening the new epidermis (Figure 5.5b).

Figure 5.5   Skin wound healing.

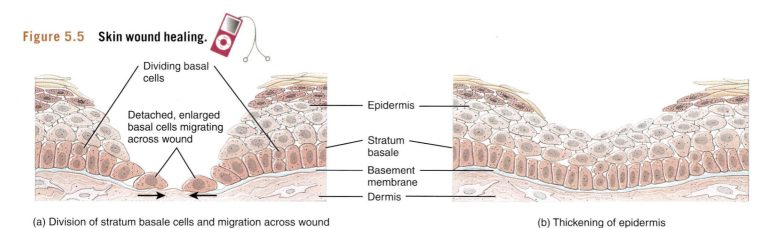

(a) Division of stratum basale cells and migration across wound

(b) Thickening of epidermis

Epidermal wound healing

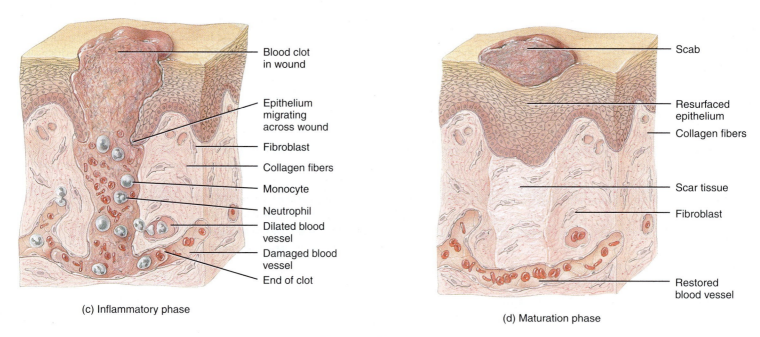

(c) Inflammatory phase

(d) Maturation phase

Deep wound healing

In an epidermal wound, the injury is restricted to the epidermis; in a deep wound, the injury extends deep into the dermis.

Deep Wound Healing

Deep wound healing occurs when an injury extends to the dermis and subcutaneous layer. Because multiple tissue layers must be repaired, the healing process is more complex than in epidermal wound healing. In addition, scar tissue is formed, and the healed tissue loses some of its normal function. Deep wound healing occurs in four phases: an inflammatory phase, a migratory phase, a proliferative phase, and a maturation phase.

During the **inflammatory phase**, bleeding produces a blood clot in the wound. The blood clot loosely unites the wound edges (Figure 5.5c). As its name implies, this phase of deep wound healing involves **inflammation**, a process that prepares the wound for repair by helping to eliminate microbes, foreign material, and dying tissue. Inflammation also increases the permeability and diameter of local blood vessels (vasodilation), enhancing delivery of helpful cells. These include white blood cells called neutrophils and monocytes (which develop into macrophages), which phagocytize microbes in the wound area, and mesenchymal cells, which develop into fibroblasts.

The three phases that follow do the work of repairing the wound. In the **migratory phase**, the clot dries into a scab, and epithelial cells migrate beneath the scab to bridge the wound. Deep to the epithelial cell bridge, fibroblasts migrate through the clot and begin synthesizing collagen fibers and glycoproteins in the dermis. Damaged blood vessels begin to regrow. During this phase, the fibrous connective tissue filling the wound, which is destined to become scar tissue, is called **granulation tissue**. The **proliferative phase** is characterized by extensive growth of epithelial cells beneath the scab, deposition by fibroblasts of collagen fibers in random patterns, and continued growth of blood vessels. Finally, during the **maturation phase**, the scab sloughs off once the epidermis has been restored to normal thickness. In the dermis, the granulation tissue is developing into scar tissue as collagen fibers become more organized, fibroblasts decrease in number, and blood vessels are restored to normal (Figure 5.5d).

The process of scar tissue formation is called **fibrosis**. Sometimes, so much scar tissue is formed during deep wound healing that a raised scar—one that is elevated above the normal epidermal surface—results. If such a scar remains within the boundaries of the original wound, it is a **hypertrophic scar**. If it extends beyond the boundaries into normal surrounding tissues, it is a **keloid scar**. Scar tissue differs from normal skin in that its collagen fibers are more densely arranged; it also has less elasticity and fewer blood vessels, and it may not contain the same number of hairs, skin glands, or sensory structures as undamaged skin. Because of the arrangement of collagen fibers and the scarcity of blood vessels, scars usually are lighter in color than normal skin.

Checkpoint

21. During wound healing, which epidermal strata can divide to replace lost strata?
22. How does inflammation assist in wound healing?
23. What is fibrosis? Why doesn't epidermal wound healing result in fibrosis?
24. What is a keloid scar?

Return to Richard's Story

The healing process had been painful and seemed very slow, at least to Richard. Every day became a monotonous, excruciating routine: the cleansing of the burns, the debridement (scraping and removal) of the dead tissue, and the administration of the IV antibiotics and nausea-inducing pain killers. Richard sometimes felt like dying in the fire would have been less of an ordeal.

His compression bandages tightly hugged the burns and gave him some comfort, so he couldn't really feel as much on his chest and stomach as he did on his face and arms. The pain eventually began to subside and his wounds began to heal.

"How bad is this going to look? You know, when it finally heals," he had asked the nurse. "

It's different with each person," she said. "Richard, they can do really great things with skin grafts these days. You have good granulation tissue at the edges of the wounds, and with the skin grafts, well, we'll have to wait and see."

It wasn't what Richard wanted to hear. "Well, maybe I'll have a really hairy chest, and that will cover up the scars." The nurse shook her head.

C. Why is it unlikely that Richard will be able to grow hair to cover the scars on his chest?

D. What takes place during the first phase of deep wound healing? Relate this process to what Richard has experienced during the initial stages of his burn healing.

E. The nurse notes that Richard has granulation tissue forming at the edges of his wounds. Will the formation of granulation tissue lead to normal appearing and normal functioning skin as Richard heals? Explain your answer.

5.7 Skin regulates body temperature, protects underlying tissues, provides cutaneous sensations, excretes body wastes, and synthesizes vitamin D.

Now that you have a basic understanding of the structure of the skin, you can better appreciate the numerous functions of the integumentary system.

Regulation of Body Temperature

The skin contributes to the homeostatic regulation of body temperature in two ways: by liberating sweat at its surface and by adjusting the flow of blood in the dermis. In response to high environmental temperature or heat produced by exercise, eccrine sweat production increases; the resulting evaporation of sweat from the skin surface helps lower an elevated body temperature to normal. In addition, blood flow through the skin increases, which increases the amount of heat radiated from the body. In response to low environmental temperature, body heat is conserved through a decrease in the production of eccrine sweat. Also, blood vessels in the dermis constrict (become narrow), which decreases blood flow through the skin and reduces heat loss from the body. The dermis houses an extensive network of blood vessels that carry 8 to 10% of the total blood flow in a resting adult. For this reason, the skin acts as a *blood reservoir*.

Protection

The skin provides **protection** to the body in various ways. Keratin protects underlying tissues from microbes, abrasion, heat, and chemicals, and the tightly interlocked keratinocytes resist invasion by microbes. Lipids released by lamellar granules inhibit evaporation of water from the skin surface, thus protecting the body from dehydration; they also retard entry of water across the skin surface when you bathe or swim. Sebum protects skin and hairs from drying out and contains bactericidal chemicals that kill surface bacteria. The acidic pH of perspiration retards the growth of some microbes. Melanin provides some protection against the damaging effects of ultraviolet light. Two types of cells carry out protective functions that are immunological in nature. Epidermal Langerhans cells (see Figure 5.2) alert the immune system to the presence of potentially harmful microbial invaders. Macrophages in the dermis phagocytize bacteria and viruses that manage to penetrate the skin surface. Hair and nails also have protective functions.

Cutaneous Sensations

Cutaneous sensations are those that arise in the skin. These include tactile sensations—touch, pressure, vibration, and tickling—as well as thermal sensations such as warmth and coolness. Another cutaneous sensation, pain, usually is an indication of impending or actual tissue damage. There are a wide variety of nerve endings and receptors distributed throughout the skin and hypodermis, including the Merkel discs, the corpuscles of touch, lamellated corpuscles, and hair root plexuses around each hair follicle (see Figures 5.1, 5.2, and 5.3). Chapter 15 provides more details on the topic of cutaneous sensations.

Excretion and Absorption

The skin has minor role in *excretion*, the elimination of substances from the body, and *absorption*, the passage of materials from the external environment into body cells. Despite the almost water-proof nature of the stratum corneum, about 400 mL of water evaporates through it daily. A sedentary person loses an additional 200 mL per day as sweat; a physically active person loses much more. Sweat is the vehicle for excretion of water and small amounts of salts, carbon dioxide, and two organic products of protein breakdown—ammonia and urea.

The absorption of water-soluble substances through the skin is negligible, but certain lipid-soluble materials do penetrate the skin. These include fat-soluble vitamins (A, D, E, and K), certain drugs, and oxygen and carbon dioxide gases. Toxic materials that can be absorbed through the skin include organic solvents such as acetone (in some nail polish removers) and carbon tetrachloride (dry-cleaning fluid); salts of heavy metals such as lead, mercury, and arsenic; and the toxins in poison ivy and poison oak. Since topical (applied to the skin) steroids, such as cortisone, are lipid soluble, they move easily into the papillary region of the dermis. Here, they exert their anti-inflammatory properties by inhibiting histamine production by mast cells (histamine contributes to inflammation).

Synthesis of Vitamin D

Synthesis of vitamin D requires activation of a precursor molecule in the skin by ultraviolet (UV) rays in sunlight. Enzymes in the liver and kidneys then modify the activated molecule, finally producing *calcitriol*, the most active form of vitamin D. Calcitriol is a hormone that aids in the absorption of calcium in foods from the gastrointestinal tract into the blood. Only a small amount of exposure to UV light (about 10 to 15 minutes at least twice a week) is required for vitamin D synthesis.

Checkpoint

25. How does the skin help regulate body temperature?
26. How does the skin serve as a protective barrier?
27. Which sensations arise from stimulation of neurons in the skin?
28. How is the skin involved in excretion? In nutrition?

▶▶▶ Richard's healing will be slow, and painful. Not just physically, but emotionally as well. As you have learned, the skin is an organ system with multiple functions. Burns, whether from fires, chemicals, radiation, or electricity, compromise those functions. More than 2 million people each year suffer burn injuries; of these, around 20,000 cases are as severe as Richard's, and over 10,000 people die each year as a result of severe burns.

The skin is not only the surface of the body we present to the world; it is also an important organ system with complex functions including protection, sen-sation, temperature regulation, water and salt balance, and even the synthesis of vitamin D. Being aware of the importance of your skin and protecting this important organ system is crucial to maintaining the overall homeostasis of the human body. ◀◀◀

F. Describe the possible long-term *consequences of Richard's injuries as they relate to each of the functions of the integu-ment described in Concept 5.7.*

Concept and Resource Summary

Concept	Resources

Introduction

1. The **integumentary system** is composed of the skin and its accessory structures, the hair, nails, glands, and sensory receptors.

Concept 5.1 Skin is composed of a superficial epidermis and a deeper dermis, and is anchored by the hypodermis.

1. The **skin**, the largest organ of the body, is a protective barrier that has sensory receptors, excretes excess salts, synthesizes vitamin D, and regulates body temperature.
2. The two principal parts of the skin are the upper, thin **epidermis** and the deeper, thicker **dermis**. Skin rests on the **hypodermis** that stores fat and has large blood vessels.
3. The two types of skin are thin skin and thick skin. **Thin skin** covers most of the body except for the palms, palmar surfaces of the fingers and toes, and the soles. **Thick skin** covers the palms, palmar sur-faces of the fingers and toes, plantar surface of toes, and the soles. Unlike thin skin, thick skin lacks hair and sebaceous glands, and it has more sudoriferous glands.

See
- **Anatomy Overview – The Skin**

Concept 5.2 The layers of the epidermis include the stratum basale, stratum spinosum, stratum granulosum, stratum lucidum, and stratum corneum.

1. The **epidermis** is composed of keratinized stratified squamous epithelium.
2. The most numerous epithelial cells are **keratinocytes** that produce the protective protein **keratin**. **Melano-cytes** are found only in the deepest epidermal layer and produce the brown-black skin pigment **melanin**. UV light stimulates production of melanin, which enters keratinocytes and protects their DNA from UV light damage. **Langerhans cells** defend against microbes, and **Merkel cells** function in the sensation of touch.
3. The **stratum basale** is the deepest epidermal layer. It contains keratinocytes, melanocytes, and Merkel cells. Some keratinocytes are stem cells that divide to make new keratinocytes.
4. The **stratum spinosum** is superficial to the stratum basale; the **stratum granulosum** is the middle layer, followed by the **stratum lucidum**, which is present only in thick skin. The most superficial layer is the **stratum corneum**. The protective, water-repellent stratum corneum has layers of continually shed dead keratinocytes that get replaced by cells pushing up from deeper strata.
5. **Keratinization** is a process in which new cells from the stratum basale gain more keratin as they move upward to the surface.

Hear
- **Figure 5.2 – Layers of the Epidermis**

Concept	Resources

Concept 5.3 The dermis contains blood vessels, nerves, sensory receptors, hair follicles, and glands.

1. The **dermis** contains blood vessels, nerves, sensory receptors, glands, and hair follicles.
2. The upper **papillary region** contains areolar connective tissue with fingerlike projections called **dermal papillae** that indent the epidermis. Epidermal ridges project down between dermal papillae, on skin of the palms, fingers, soles, and toes. These are the source of *fingerprints* and *footprints*.
3. Dermal papillae contain **capillary loops**, **corpuscles of touch**, and **free nerve endings** that sense temperature, pain, and itching.
4. The deeper **reticular region** rests on the hypodermis and contains dense irregular connective tissue, hair follicles, and glands.

See

• **Anatomy Overview – The Skin**

Concept 5.4 Skin color is a result of the pigments melanin, carotene, and hemoglobin.

1. **Melanin** gives the skin a pale yellow to reddish-brown to black color. The number of melanocytes is the same among different races, but the amount of melanin produced differs and leads to different skin colors.
2. **Carotene** is a yellow-orange pigment that can give skin an orange color during dietary excess.
3. **Hemoglobin**, an oxygen-carrying red pigment inside red blood cells in dermal blood vessels, gives skin a reddish hue.

Concept 5.5 The functions of hair, skin glands, and nails include protection and body temperature regulation.

1. **Hairs**, which are composed of dead, keratinized epidermal cells held together by extracellular proteins, are present on most skin surfaces except palms, palmar surfaces of fingers, soles, and plantar surfaces of toes. The thickness and pattern of hair distribution are influenced by hormones and genetics. Hairs offer a limited amount of protection from sunlight, heat loss, and foreign particles. Hairs also sense touch.
2. The **shaft** projects from the skin surface; the deeper **root** penetrates down into dermis.
3. The hair root is surrounded by the **hair follicle**. The expanded base of the hair follicle is the **bulb**, which houses a cluster of blood vessels called the **papilla**, and a germinal layer of dividing cells called the **hair matrix**. Sebaceous (oil) glands, **arrector pili**, and **hair root plexuses** are associated with hair follicles, which go through **growth stages** and **resting stages**. Melanocytes in the hair matrix produce the melanin for hair color.
4. **Sebaceous glands** occur all over the skin except the palms and soles and secrete an oily substance called **sebum** into hair follicles or onto the skin surface. Sebum lubricates the hair and skin, inhibits some bacteria, and prevents water loss.
5. The two types of **sudoriferous glands** are eccrine and apocrine sweat glands. **Eccrine sweat glands** are found throughout most of the skin; are most numerous on the forehead, palms, and soles; and help regulate body temperature through evaporative cooling. Eccrine sweat is secreted during emotional stress.
6. **Apocrine sweat glands**, which are found mainly in the axilla and groin, extend into the hypodermis and secrete through ducts into hair follicles during emotional stress and sexual excitement.
7. Ducts of **ceruminous glands** open into the external ear canal. The secretion of ceruminous glands combined with sebum is called **cerumen**, or earwax, which helps trap substances entering the ear.
8. **Nails**, hard, tightly packed keratinized epidermal cells that cover the dorsal, distal surfaces of fingers and toes, consist of a **nail body**, **free edge**, and **nail root**. A band of epidermis near the nail root is the **cuticle**, and a nearby whitish semilunar area is the **lunula**. Deep to the nail root is an area of actively dividing cells called the **nail matrix** from which the nail grows.

See

• **Anatomy Overview – Hair**
• **Anatomy Overview – Nails**

Hear

• **Figure 5.3 – Hair**

Concept 5.6 Skin damage sets in motion a sequence of events that repairs the skin to its normal (or near-normal) structure and function.

1. Two types of wound healing can occur when skin is damaged: epidermal wound healing and deep wound healing.
2. Healing of epidermal wounds begins as cells from the stratum basale migrate to cover the wound until stopped by **contact inhibition**. Epidermal growth factor stimulates basal cells to divide and replace migrating cells.

Hear

• **Figure 5.5 – Epidermal and Deep Wound Healing**

Concept   **Resources**

3. Deep wounds, extending into the dermis or hypodermis, result in scar tissue, and they involve a more complex healing process that occurs in four phases: the inflammatory phase, migratory phase, proliferative phase, and maturation phase.

4. Preparation for wound repair occurs during the **inflammatory phase**. A blood clot forms over the wound and **inflammation** rids the area of microbes and debris, followed by vasodilation, which facilitates an influx of phagocytic neutrophils, macrophages, and mesenchymal cells (which become fibroblasts).

5. During the **migratory phase**, epithelial cells migrate to form a repair bridge and fibroblasts migrate to fill in the wound with **granulation tissue**.

6. During the **proliferative phase**, epithelial cells, collagen fibers, and blood vessels grow.

7. The end of repair occurs during the **maturation phase**, when the scab falls off, the epidermis is restored, and **fibrosis** produces scar tissue in the dermis.

8. **Hypertrophic scars** remain within the confines of the wounds, whereas **keloid scars** extend into the surrounding tissue.

Concept 5.7 Skin regulates body temperature, protects underlying tissues, provides cutaneous sensations, excretes body wastes, and synthesizes vitamin D.

See

- **Anatomy Overview – The Skin and Disease Resistance**
- **Animation – Nonspecific Disease Resistance**

1. The skin participates in homeostatic regulation of body temperature through sweating with evaporative cooling, and adjusting blood flow in the dermis.

2. Skin provides **protection** in several ways: The keratinized epithelium is a barrier to microbes, chemicals, heat, and abrasions; lipid secretions prevent dehydration; sebum lubricates skin and hair, and is antibacterial; melanin protects against UV light damage; and Langerhans cells and macrophages defend against microbes.

3. The skin has many nerve endings and receptors for tactile, temperature, and pain sensations.

4. Sweat allows for *excretion* of water, salts, protein, urea, and carbon dioxide. *Absorption* of fat-soluble vitamins, drugs, oxygen, carbon dioxide, and toxic substances occurs through the skin.

5. UV light from sunlight activates a precursor to vitamin D needed by the liver and kidneys to produce *calcitriol*, an active form of vitamin D.

Other Chapter Resources

For additional information on clinical applications, medical disorders, and medical terminology, see Chapter 3 in the *Clinical Connections* manual packaged with your text.

Within *WileyPLUS*

- **Chapter Quiz**
- **Visual Anatomy**
- **Mastering Vocabulary**
- **Audio Glossary**
- **Web Links**

• Understanding the Concepts

1. Cells at the surface of skin are keratinized and dead. Why is this a survival advantage?

2. Which component of skin lets you know your hand is touching a glass of water? Helps you grip the wet glass? Informs you that the glass is cold?

3. Why does it hurt when you pluck a hair but not when you have a haircut? What makes your hair grow longer?

4. Distinguish between eccrine glands and apocrine glands. Which ones are associated with hair follicles?

5. Which part of a nail grows new nail? Which part shows underlying blood capillaries?

6. Would you expect an epidermal wound to bleed? Why or why not?

7. Which type of tissue initially fills a deep wound? What does this tissue eventually become?

6

Introduction to the Skeletal System

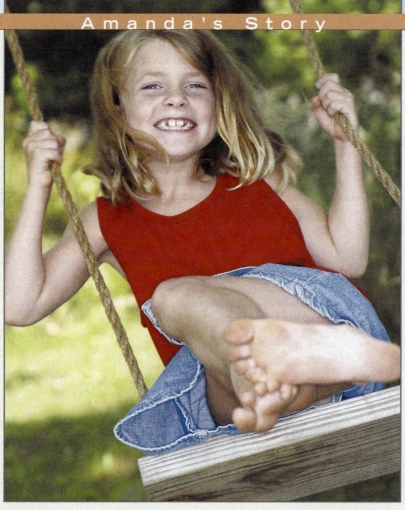

"Higher! Push me higher!" Amanda shrieked, giggling and leaning back on the swing. Her hair flung out behind her as she tipped her head back. The chain links holding the seat creaked noisily. Her friend Luke pushed as hard as he could, the arc of the swing growing with every push.

"Amanda! Get off that swing!" Amanda's mother appeared from out of nowhere, jogging toward the children. "Get off that swing now, young lady!"

"Aw, Mom. I'm not going to fall off." Amanda stopped pumping her legs, slowing the arc of the swing. Luke stood back, looking sheepish. He knew Amanda wasn't supposed to be on the swing, but he didn't really understand why.

Amanda dragged her feet, stopping the swing. She stood up slowly, shoulders drooping. Amanda's mom had good reason to be protective. Last summer Amanda had fallen while playing, and the results had been disastrous. Her femur was broken, badly. It wasn't even that bad a fall, but Amanda has osteogenesis imperfecta, also known as "brittle bone disorder."

Amanda is a healthy-looking 10-year-old girl. She wants to do all the things other girls do, but she carries a gene for a disease that is a constant threat and makes her parents wary of any excessive activity. Because of the numerous fractures associated with this skeletal disorder, physicians sometimes misdiagnose it as child abuse. Often, bones of an affected child can break just from the stress of learning to walk. In this chapter we will explore the skeletal system and the importance of bone and its homeostasis. As we will learn, our skeletal system not only provides us with support; it has numerous other important homeostatic functions.

INTRODUCTION

Despite their simple appearance, bones are complex and dynamic living tissues that are remodeled continuously—as you read these pages, new bone is being built in your body, and old bone is being broken down. A bone is composed of several different tissues working together: osseous tissue, cartilage, dense connective tissues, epithelium, adipose tissue, and

> *Despite their simple appearance, bones are complex and dynamic living tissues that are remodeled continuously.*

nervous tissue. For this reason, each individual bone is considered an organ. The entire framework of bones, their cartilages, along with ligaments and tendons, constitute the **skeletal system**. In this chapter we will explore the various components of bones, how bones perform their many functions, and how they develop and remodel.

6.1 Skeletal system functions include support, protection, movement, mineral homeostasis, blood cell production, and energy storage.

The skeletal system performs several basic functions:

- *Support.* The skeleton serves as the structural framework for the body by supporting soft tissues and providing attachment points for the tendons of most skeletal muscles.

- *Protection.* The skeleton protects many internal organs from injury. For example, cranial bones protect the brain, vertebrae protect the spinal cord, and the rib cage protects the heart and lungs.

- *Assistance in movement.* Most skeletal muscles attach to bones; when they contract, they pull on bones to produce movement. This function is discussed in detail in Chapter 10.

- *Mineral homeostasis.* Osseous tissue stores several minerals, especially calcium and phosphorus, which contribute to the strength of bone. Osseous tissue stores about 99% of the body's calcium. Bones can release minerals on demand into the bloodstream to maintain critical mineral homeostasis and to distribute minerals to other parts of the body.

- *Blood cell production.* A connective tissue called **red bone marrow** produces red blood cells, white blood cells, and platelets by a process called **hemopoiesis** (hēm-ō-poy-Ē-sis; *hemo-* = blood; *-poiesis* = making). Red bone marrow is present in certain bones, such as the pelvis, ribs, sternum (breastbone), vertebrae (backbones), skull, and ends of the arm bones and thigh bones.

- *Triglyceride storage.* **Yellow bone marrow** consists mainly of adipocytes, which store triglycerides. The stored triglycerides are a potential chemical energy reserve. In the newborn, all bone marrow is red and is involved in hemopoiesis. With increasing age, much of the bone marrow changes from red to yellow.

Checkpoint

1. Which minerals are stored in bones?
2. How do red and yellow bone marrow differ in function?

6.2 Bones are classified as long, short, flat, irregular, or sesamoid.

Almost all the bones of the body can be classified into five principal types: long, short, flat, irregular, and sesamoid (Figure 6.1).

Figure 6.1   Types of bones based on shape. The bones are not drawn to scale.

Long bone (humerus)

Flat bone (sternum)

Short bone (trapezoid, wrist bone)

Irregular bone (vertebra)

Sesamoid bone (patella)

🔑 The shapes of bones largely determine their functions.

Long bones have greater length than width and consist of a shaft and two extremities (ends). They are slightly curved for strength. A curved bone absorbs the stress of the body's weight at several different points so that the stress is evenly distributed. If such bones were straight, the weight of the body would be unevenly distributed and the bone would fracture easily. Long bones include those in the thigh (femur), leg (tibia and fibula), arm (humerus), forearm (ulna and radius), and fingers and toes (phalanges).

Short bones are somewhat cube-shaped and nearly equal in length and width. Examples of short bones are most wrist bones (carpals) and most ankle bones (tarsals).

Flat bones are generally thin, afford considerable protection, and provide extensive surfaces for muscle attachment. Flat bones include the cranial bones, which protect the brain; the breastbone (sternum) and ribs, which protect organs in the thorax; and the shoulder blades (scapulae).

Irregular bones, which have complex shapes and cannot be grouped into any of the above categories, include the backbones (vertebrae), hip bones, certain facial bones, and the heel bone (calcaneus).

Sesamoid bones (SES-a-moyd = shaped like a sesame seed) develop in certain tendons where there is considerable friction, tension, and physical stress, such as the palms and soles. They are not always completely ossified and measure only a few millimeters in diameter except for the two patellae (kneecaps), the largest of the sesamoid bones. Sesamoid bones vary in number from person to person except for the patellae, which are normally present in all individuals. Functionally, sesamoid bones protect tendons from excessive wear and tear, and they often change the direction of pull of a tendon, which improves the mechanical advantage at a joint.

Checkpoint

3. Which type of bone has a boxy shape? Which has an elongated shape?
4. Which type of bone provides a large, flat surface area for muscle attachment?
5. Which type of bone is small and located in tendons? What is its purpose?

6.3 Long bones have a diaphysis, a medullary cavity, epiphyses, metaphyses, and periosteum.

The structure of a bone may be analyzed by considering the components of a long bone such as the humerus (the arm bone) or the femur (the thigh bone) (Figure 6.2). A typical long bone consists of the following parts:

- The **diaphysis** (dī-AF-i-sis = growing between) is the bone's shaft, or body—the long, cylindrical, main portion of the bone.

- The **epiphyses** (e-PIF-i-sēz = growing own; singular is *epiphysis*) are the proximal and distal ends of the bone.

Figure 6.2   Parts of a long bone.

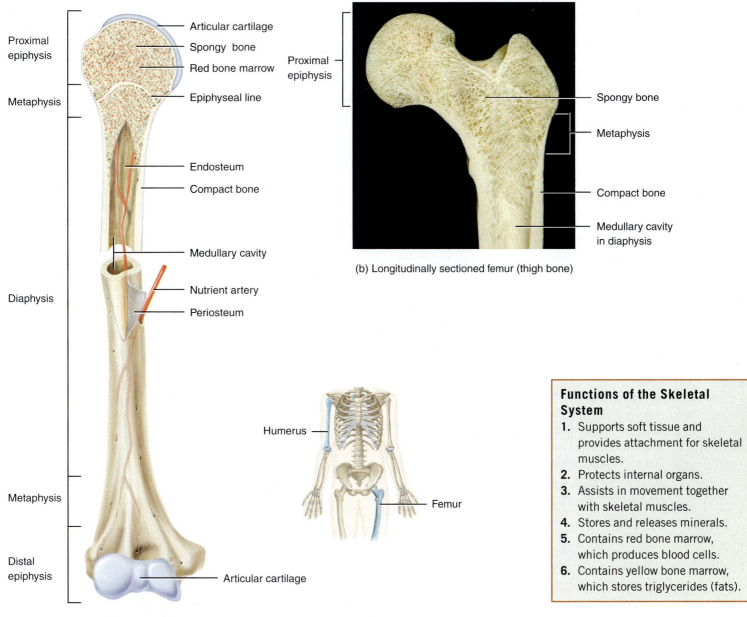

Proximal epiphysis

- Articular cartilage
- Spongy bone
- Red bone marrow

Metaphysis

- Epiphyseal line
- Endosteum
- Compact bone
- Medullary cavity

Diaphysis

- Nutrient artery
- Periosteum

Proximal epiphysis

- Spongy bone
- Metaphysis
- Compact bone
- Medullary cavity in diaphysis

(b) Longitudinally sectioned femur (thigh bone)

Humerus

Femur

Metaphysis

Distal epiphysis

- Articular cartilage

(a) Partially sectioned humerus (arm bone)

Functions of the Skeletal System

1. Supports soft tissue and provides attachment for skeletal muscles.
2. Protects internal organs.
3. Assists in movement together with skeletal muscles.
4. Stores and releases minerals.
5. Contains red bone marrow, which produces blood cells.
6. Contains yellow bone marrow, which stores triglycerides (fats).

A long bone is covered by articular cartilage at its proximal and distal epiphyses and by periosteum around the diaphysis.

- The **metaphyses** (me-TAF-i-sēz; *meta-* = between; singular is *metaphysis*) are the regions between the diaphysis and the epiphyses. In a growing bone, each metaphysis contains an **epiphyseal plate** (ep´-i-FIZ-ē-al), a layer of hyaline cartilage that allows the diaphysis of the bone to grow in length (described later in the chapter). When bone growth in length stops, the cartilage in the epiphyseal plate is replaced by osseous tissue and the resulting bony structure is known as the **epiphyseal line**.

- The **articular cartilage** is a thin layer of hyaline cartilage covering the part of the epiphysis where the bone forms an articulation (joint) with another bone. Articular cartilage reduces friction and absorbs shock at freely movable joints.

- The **periosteum** (per´-ē-OS-tē-um; *peri-* = around) is a tough sheath of dense irregular connective tissue that surrounds the bone surface wherever it is not covered by articular cartilage. The periosteum contains bone-forming cells that enable bone to grow in thickness. It also protects the bone, assists in fracture repair, helps nourish bone tissue, and serves as an attachment point for ligaments and tendons.

- The **medullary cavity** (MED-ū-lar´-ē; *medulla-* = marrow, pith), or **marrow cavity**, is the hollow, cylindrical space within the diaphysis that contains yellow bone marrow in adults.

- The **endosteum** (end-OS-tē-um; *endo-* = within), a thin connective tissue membrane that lines the medullary cavity, contains bone-forming cells.

Checkpoint

6.   Where are the diaphysis, epiphyses, and metaphyses each located on a long bone?

7.   Describe the location, composition, and function of the epiphyseal plate, articular cartilage, and periosteum.

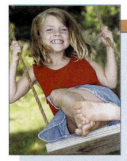

Return to Amanda's Story

Amanda sat at the kitchen table, pouting while her mother, Kaitlin, made lunch. Amanda's father, Karl, noticed the tension when he entered the room. He sat down at the table. "How are you, sweetheart?" He reached out and tousled Amanda's hair. She turned away, thrusting her lower lip forward.

"Karl, that daughter of yours was trying to break her neck again," Kaitlin said. "Will you please talk to her? She won't listen to me!"

Karl smiled. "I thought something was up with you two. Look, Amanda, your mom and I worry about you. You know you have to be careful."

Amanda glared at him and replied, "Why can't I have fun like normal kids?"

"We aren't saying you can't have fun. You just need to be more careful."

When Amanda was born with a broken arm and collarbone, the doctors diagnosed her with brittle bone disease. Her parents were devastated. By the time she was 8 years old, Amanda had broken over 20 bones, including her tibia, fibula, humerus, ulna, and femur. She had even fractured her skull when she hit her head on the monkey bars at school.

There is no cure for the disease, but most people with Type I osteogenesis imperfecta live fairly normal lives. The doctor advised that she undergo regular physical exams because Amanda's weakened bones might bow and not grow correctly. Every 6 months or so, the doctor checked Amanda's spine to see if it was growing straight and true. He listened to her heart and checked her hearing, because of the increased risk of heart problems and hearing loss associated with this bone disease. He also checked Amanda's joints and their range of movement, because the collagen imperfection that weakened her bones could also weaken the structure of her joints where tendons and ligaments attached.

The previous summer Amanda had broken her femur, requiring a metal device, called an intramedullary rod, to be placed inside the bone's shaft to stabilize and strengthen it. The painful, slow healing process had worried Kaitlin. What would happen as Amanda grew?

A.   Amanda's bones are weak and break easily due to an imperfection in the collagen structure of her connective tissues. Which functions of the skeletal system would be compromised in brittle bone syndrome?

B.   Using what you studied in the previous chapter and this chapter so far, imagine yourself as Amanda's surgeon. Visualize the layers of tissue you must cut through to fix Amanda's leg. Make a list of the tissue layers covering her femur, and describe the layers of bone tissue that you will encounter as you access the medullary cavity. Write a description of the surgical procedure using appropriate anatomical terms.

C.   Amanda's mother is concerned that the medullary rod in her femur will not grow with Amanda. What can you tell her about the location of the rod? Where does bone growth take place?

6.4   Osseous tissue can be arranged as compact bone tissue or spongy bone tissue.

Like other connective tissues, **osseous tissue** (OS-ē-us) contains an abundant extracellular matrix that surrounds widely separated cells. The extracellular matrix is about 25% water, 25% collagen fibers, and 50% crystallized mineral salts. The abundant mineral salts are mainly *hydroxyapatite* (primarily calcium phoshate and calcium hydroxide). As these mineral salts are deposited in the framework formed by the collagen fibers of the extracellular matrix, they crystallize and the tissue hardens. This process of

calcification (kal´-si-fi-KĀ-shun) is initiated by osteoblasts, the bone-building cells.

It was once thought that calcification simply occurred when enough mineral salts were present to form crystals. Now, however, we know that the process occurs only in the presence of collagen fibers. Mineral salts begin to crystallize between the collagen fibers, then mineral crystals accumulate around the collagen fibers.

Although a bone's *hardness* depends on the crystallized inorganic mineral salts, a bone's flexibility depends on its collagen fibers. Like reinforcing metal rods in concrete, collagen fibers and other organic molecules provide *tensile strength*, resistance to being stretched or torn apart.

Cells of Osseous Tissue

Four types of cells are present in osseous tissue: osteogenic cells, osteoblasts, osteocytes, and osteoclasts (Figure 6.3).

- **Osteogenic cells** (os´-tē-ō-JEN-ik; -*genic* = producing) are unspecialized stem cells (undifferentiated cells that can divide to produce new cells) derived from mesenchyme, the tissue from which almost all connective tissues are formed. They are the only bone cells to undergo cell division; the resulting cells develop into osteoblasts. Osteogenic cells are found in the inner portion of the periosteum, in the endosteum, and in the canals within bones that contain blood vessels.

- **Osteoblasts** (OS-tē-ō-blasts´; -*blasts* = buds or sprouts) synthesize and secrete collagen fibers and other organic components needed to build the extracellular matrix of osseous tissue, and they initiate calcification (described shortly). As osteoblasts surround themselves with extracellular matrix, they become trapped in their secretions and become osteocytes. (Note: Cells with the ending *blasts* in the name means that the cells secrete extracellular matrix.)

- **Osteocytes** (OS-tē-ō-sīts´; -*cytes* = cells), mature bone cells, are the most numerous cells in osseous tissue and maintain its daily metabolism, such as the exchange of nutrients and wastes with the blood. Like osteoblasts, osteocytes do not undergo cell division. (Note: Cells with the ending *cytes* in the name means that the cells maintain the tissue.)

- **Osteoclasts** (OS-tē-ō-clasts´; -*clast* = break) are huge cells derived from the fusion of as many as 50 monocytes (a type of white blood cell) and are concentrated in the endosteum. On the side of the cell that faces the bone surface, the osteoclast releases powerful lysosomal enzymes and acids that digest the protein and mineral components of the underlying extracellular matrix of osseous tissue. This breakdown of the extracellular matrix of osseous tissue, termed resorption (rē-SORP-shun), is part of the normal development, maintenance, and repair of bones. (Note: Cells with the ending *clast* in the name means that the cell breaks down extracellular matrix.) In response to certain hormones, osteoclasts help regulate blood calcium levels (see Concept 17.6).

Figure 6.3 **Types of cells in bone tissue.** Center and right photos: © Dr. Richard Kessel and Randy Kardon/Tissues & Organs/Visuals Unlimited.

Osteogenic cell (develops into an osteoblast)

Osteoblast (forms bone matrix)

Osteocyte (maintains bone tissue)

Osteoclast (functions in resorption, the breakdown of bone matrix)

SEM 1100x

SEM 9160x

SEM 5626x

Osteogenic cells develop into osteoblasts, which secrete the extracellular matrix of osseous tissue.

Types of Osseous Tissue

Osseous tissue is not completely solid but has many small spaces between its cells and extracellular matrix components. Some spaces are channels for blood vessels that supply bone cells with nutrients. Other spaces are storage areas for red bone marrow. Depending on the size and distribution of the spaces, osseous tissue may be categorized as compact or spongy (see Figure 6.2). Overall, about 80% of the skeleton is compact bone tissue, which is dense and has few spaces, and 20% is spongy bone tissue, which has larger spaces. Long bones consist mostly of compact bone tissue in their diaphyses but have considerable amounts of spongy bone tissue in their epiphyses. Short, flat, and irregular bones consist of spongy bone tissue except at the surface, where there is a thin protective plate of compact bone tissue.

Compact Bone Tissue

Compact bone tissue contains few spaces. It forms the external layer of all bones and makes up the bulk of the diaphyses of long bones (see Figure 6.2). Compact bone tissue provides protection and support and resists the stresses produced by weight and movement.

Blood vessels, lymphatic vessels, and nerves from the periosteum penetrate the compact bone tissue through transverse **perforating**, or **Volkmann's** (FŌLK-mans), **canals** (Figure 6.4a). The vessels and nerves of the perforating canals connect with those of the periosteum, **central** or **haversian canals**, and medullary cavity. The central canals run longitudinally through the bone. Around the central canals are **concentric lamellae** (la-MEL-ē)—rings of hard, calcified extracellular matrix much like the rings of a tree trunk. Between the lamellae are small spaces called **lacunae** (la-KOO-nē = little lakes; singular is *lacuna*), which contain osteocytes. Radiating in all directions from the lacunae are tiny **canaliculi** (kan´-a-LIK-ū-lī = small channels) filled with extracellular fluid. Inside the canaliculi are slender fingerlike processes of osteocytes (see inset at the right of Figure 6.4a). Neighboring osteocytes communicate via cell junctions. The canaliculi connect lacunae with one another and with the central canals. Thus, an intricate, miniature system of interconnected canals throughout the bone provides many routes for nutrients and oxygen to reach the osteocytes and for wastes to diffuse away. This is very important because diffusion through the lamellae is extremely slow.

Compact bone tissue is arranged in repeating structural units called **osteons** or **haversian systems** (ha-VER-shun) (Figure 6.4a). Each osteon consists of a central (haversian) canal with its concentrically arranged lamellae, lacunae, osteocytes, and canaliculi. Osteons in compact bone tissue are aligned in the same direction along lines of stress. In the shaft, for example, they are parallel to the long axis of the bone. As a result, the shaft of a long bone resists bending or fracturing even when considerable force is applied from either end. The osteons of a long bone can be compared to a stack of logs; each log is made up of rings of hard material, and together it requires considerable force to fracture them all. The lines of stress in a bone are not static. They change as a person learns to walk and in response to repeated strenuous physical activity, such as occurs when a person undertakes weight training. The lines of stress in a bone also can change in response to fractures or physical deformity. Thus, the organization of osteons changes over time in response to the physical demands placed on the skeleton.

The areas between osteons contain **interstitial lamellae**, which also have lacunae with osteocytes and canaliculi. Interstitial lamellae are fragments of older osteons that have been partially destroyed during bone rebuilding or growth. Lamellae that encircle the bone just beneath the periosteum or encircle the medullary cavity are called **circumferential lamellae**.

Spongy Bone Tissue

In contrast to compact bone tissue, **spongy bone tissue** does not contain osteons. Despite what the name seems to imply, the term "spongy" does not refer to the texture of the bone, only its appearance (Figure 6.4b,c). Spongy bone consists of lamellae that are arranged in an irregular lattice of thin columns of bone called **trabeculae** (tra-BEK-ū-lē = little beams; singular is *trabecula*). The spaces between the trabeculae help make bones lighter and are sometimes filled with red bone marrow, which contains numerous small blood vessels. Within each trabecula are osteocytes that lie in lacunae. Canaliculi radiate outward from the lacunae. Osteocytes obtain nutrients from the blood circulating through the blood vessels in the spaces between trabeculae.

At first glance, the trabeculae of spongy bone tissue may appear to be much less well organized than the osteons of compact bone tissue. However, the trabeculae of spongy bone tissue are precisely oriented along lines of stress, a characteristic that helps bones resist stresses and transfer force without breaking. Spongy bone tissue tends to be located where bones are not heavily stressed or where stresses are applied from many directions.

Figure 6.4 **Histology of compact and spongy bone tissue.** (a) Sections through the diaphysis of a long bone, from the surrounding periosteum on the right, to compact bone tissue in the middle, to spongy bone tissue and the medullary cavity on the left. The inset at upper right shows an osteocyte in a lacuna. (b and c) Details of spongy bone tissue.

(a) Osteons in compact bone and trabeculae in spongy bone

(b) Enlarged aspect of spongy bone trabeculae

(c) Details of a section of a trabecula

🔑 **Compact bone tissue is arranged in concentric circles around a central canal, whereas spongy bone tissue is arranged irregularly in trabeculae.**

Spongy bone tissue is light, which reduces the overall weight of a bone so that it moves more readily when pulled by a skeletal muscle. The trabeculae of spongy bone tissue support and protect the red bone marrow. Spongy bone tissue in the hip bones, ribs, sternum (breastbone), vertebrae (backbones), and the ends of long bones contains red bone marrow, the site of hemopoiesis (blood cell production) in adults.

Checkpoint

8. What is the composition of the extracellular matrix of osseous tissue?
9. Which cells turn into osteoblasts? Into osteocytes?
10. Which bone cell type is derived from white blood cells?
11. List the functions of osteoblasts, osteocytes, and osteoclasts.
12. Where in compact bone tissue are concentric lamellae, interstitial lamellae, and circumferential lamellae located?
13. Which kind of bone marrow is in spongy bone tissue?

6.5   Bones are richly supplied with blood vessels and nerves.

Bones are richly supplied with blood. Blood vessels, which are especially abundant in portions of bones containing red bone marrow, pass into bones from the periosteum. We consider the blood supply of a long bone such as the tibia (shin bone) shown in Figure 6.5.

Periosteal arteries (per´-ē-OS-tē-al), accompanied by nerves, enter the diaphysis through numerous **perforating canals** and supply blood to the periosteum and outer part of the compact bone tissue (see Figure 6.4a). Near the center of the diaphysis, a large **nutrient artery** enters the compact bone tissue through a hole called the **nutrient foramen** (see Figure 6.2). On entering the medullary cavity, the nutrient artery divides into proximal and distal branches that supply the inner part of compact bone tissue and the spongy bone tissue and red bone marrow as far as the epiphyseal plates (or lines). Some bones, like the tibia, have only one nutrient artery; others, like the femur (thigh bone), have several. **Metaphyseal arteries** and **epiphyseal arteries** enter the ends of a long bone and supply the red bone marrow and osseous tissue of the epiphyses.

Veins accompany their respective arteries to carry blood away from long bones. One or two **nutrient veins** exit in the diaphysis. Numerous **epiphyseal veins** exit long bones in the epiphyses, while **metaphyseal veins** exit in the metaphysis. Many small **periosteal veins** exit in the periosteum.

Nerves accompany the blood vessels that supply bones. The periosteum is rich in sensory nerves, some of which carry pain sensations. These nerves are especially sensitive to tearing or tension, which explains the severe pain resulting from a fracture or a bone tumor.

Figure 6.5   Blood supply of a mature long bone, the tibia (shin bone).

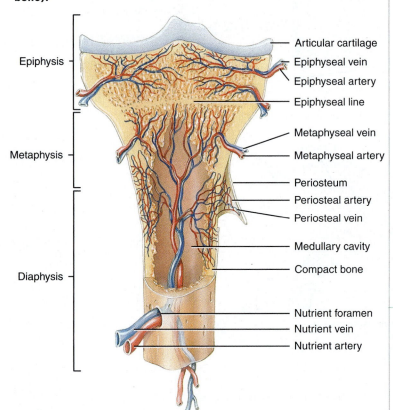

Epiphysis

Metaphysis

Diaphysis

- Articular cartilage
- Epiphyseal vein
- Epiphyseal artery
- Epiphyseal line
- Metaphyseal vein
- Metaphyseal artery
- Periosteum
- Periosteal artery
- Periosteal vein
- Medullary cavity
- Compact bone
- Nutrient foramen
- Nutrient vein
- Nutrient artery

🔑 Bones are richly supplied with blood vessels.

Checkpoint

14. Where do periosteal arteries enter a bone? Where does the nutrient artery enter a bone?
15. How does blood leave a bone?

Return to Amanda's Story

After lunch Amanda's dad suggested she go with him to the hardware store. Karl liked to include Amanda on errands and short trips, and Amanda enthusiastically tagged along.

Karl was distracted in the plumbing aisle when he noticed Amanda dodging in and out between the displays. As Amanda turned to peek at Karl from behind one of the sinks on display, her toe caught in a gap between the shelving boards. She toppled, face down, knocking into a porcelain sink and stand that crashed down on her. Karl ran to her, but he was too late. The sink hit the edge of her pelvis, painfully shattering the bone. "Daddy!" she screamed, letting loose a wail of fear and pain.

The commotion in the store was nothing compared to the bustling emergency room. The emergency room staff knew Karl and Amanda well from their many previous visits. The doctor pulled open the thin curtain that formed a temporary cubicle separating them from the other patients and waved for Karl to come over.

"Karl, Amanda's hip is broken in several places. The X-ray shows substantial bleeding from the rupture of blood vessels within the bone. We'll need to get her to surgery right away."

"Dad, what's going on?" asked Amanda, trying to overhear the adults.

"They're going to have to fix your hip like they did your leg last summer," said Karl as he walked back by her side.

"You mean more surgery? Can't I just get a cast or something?" Tears welled up in her eyes as she began sobbing.

D. *Bone healing is similar to bone growth and development, but rather than a membrane or cartilage model being the starting point, a blood clot first forms in the space of the break. After this first step, cells migrate into the site of the break and begin to produce osseous tissue. Describe in layman's terms how the bone will heal in a way that would be understood by Amanda's mother.*

E. *Amanda's father wants to know why these bone breaks are so painful. What can you tell him?*

6.6 The two types of bone formation are intramembranous ossification and endochondral ossification.

The process by which bones form is called **ossification** (os´-i-fi-KĀ-shun; *ossi-* = bone; *-fication* = making). Bone formation occurs in three principal situations: (1) the initial formation of bones in an embryo and fetus, (2) the growth of bones until their adult sizes are reached, (3) the remodeling and repair of bones.

We will first consider the initial formation of bones in an embryo and fetus. The embryonic "skeleton" is initially composed of mesenchyme shaped like bones. (Mesenchyme is the tissue from which other connective tissues arise.) These bone precursors are the site of ossification during the sixth week of embryonic development.

Bone formation follows one of two patterns. During **intramembranous ossification** (in´-tra-MEM-bra-nus; *intra-* = within; *-membran* = membrane), osseous tissue forms directly within mesenchyme arranged in sheetlike layers that resemble membranes. In the second method, **endochondral ossification** (en´-dō-KON-dral; *endo-* = within; *-chondral* = cartilage), osseous tissue forms within hyaline cartilage that develops from mesenchyme.

Intramembranous Ossification

Intramembranous ossification is the simpler of the two methods of bone formation. The flat bones of the skull and mandible (lower jawbone) are formed in this way. Also, the "soft spots" that help

Figure 6.6   Intramembranous ossification. Illustrations ❶ and ❷ show a smaller field of vision at higher magnification than illustrations ❸ and ❹.

❶ Development of ossification center

❷ Calcification

❸ Formation of trabeculae

❹ Development of the periosteum

🔑 **Intramembranous ossification involves the formation of a bone within mesenchyme arranged in sheetlike layers that resemble membranes.**

the fetal skull pass through the birth canal later harden as they undergo intramembranous ossification. Intramembranous ossification occurs as follows (Figure 6.6):

❶ *Development of the center of ossification.* Mesenchymal cells cluster together and differentiate, first into osteogenic cells and then into osteoblasts. The site of such a cluster is called an **ossification center**. Osteoblasts secrete the organic extracellular matrix of osseous tissue until they are surrounded by it.

❷ *Calcification.* After secretion of extracellular matrix stops, the cells, now called osteocytes, lie in lacunae and extend their narrow cytoplasmic processes into canaliculi that radiate in all directions. Within a few days, calcium and other mineral

salts are deposited and the extracellular matrix hardens as it calcifies. Mesenchymal cells continue to produce osteogenic cells, which form a peripheral layer of osteoblasts.

❸ *Formation of trabeculae.* The extracellular matrix develops into trabeculae that fuse with one another to form spongy bone tissue. Blood vessels grow into the spaces between the trabeculae and deposit connective tissue cells that differentiate into red bone marrow.

❹ *Development of the periosteum.* At the periphery, the mesenchyme condenses and develops into the periosteum. Within the periosteum, osteoblasts produce a thin layer of compact bone tissue, but spongy bone tissue remains in the center. Much of the newly formed bone is remodeled (destroyed and reformed) as the bone is transformed into its adult size and shape.

Endochondral Ossification

The replacement of cartilage by osseous tissue is called **endochondral ossification**. Although most bones of the body are formed in this way, the process is best observed in a long bone. It proceeds as follows (Figure 6.7):

① *Development of the cartilage model.* At the site where the bone is going to form, mesenchymal cells crowd together in the shape of the future bone, and then develop into chondroblasts. The chondroblasts secrete cartilage extracellular matrix, producing a **hyaline** cartilage model. A covering called the **perichondrium** (per´-i-KON-drē-um) develops around the hyaline cartilage model.

② *Growth of the cartilage model.* Once chondroblasts become deeply buried in the extracellular matrix of cartilage, they are called chondrocytes. The cartilage model grows by continual cell division of chondrocytes accompanied by further secretion of the cartilage extracellular matrix. As the cartilage model continues to grow, chondrocytes in its mid-region hypertrophy (increase in size). Some hypertrophied cells burst and release their contents, which calcify the surrounding extracellular matrix. Other chondrocytes within the calcifying cartilage die because nutrients can no longer diffuse quickly enough through the extracellular matrix. As these chondrocytes die, lacunae form and eventually merge into small cavities.

Figure 6.7 **Endochondral ossification.**

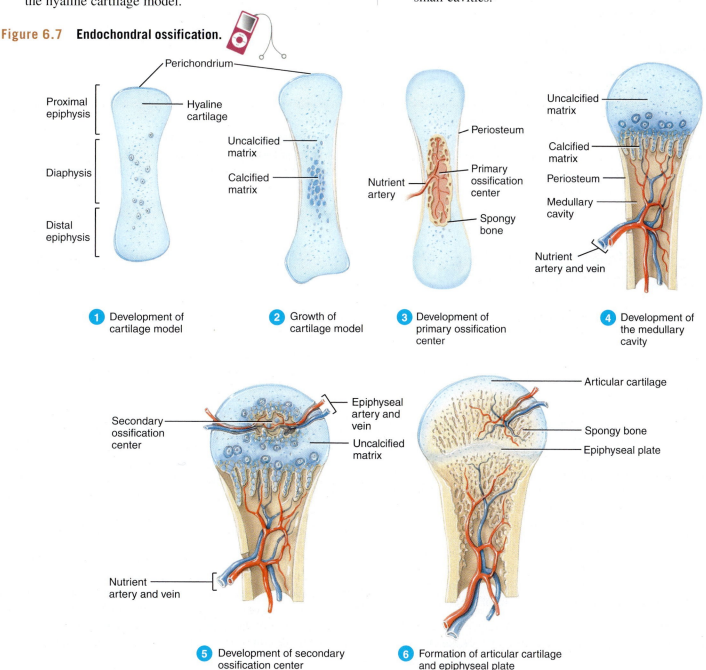

1 Development of cartilage model

2 Growth of cartilage model

3 Development of primary ossification center

4 Development of the medullary cavity

5 Development of secondary ossification center

6 Formation of articular cartilage and epiphyseal plate

During endochondral ossification, osseous tissue gradually replaces a hyaline cartilage model.

3 *Development of the primary ossification center.* A nutrient artery penetrates the perichondrium and the calcifying cartilage model through a nutrient foramen in the mid-region of the cartilage model, stimulating osteogenic cells within the perichondrium to differentiate into osteoblasts instead of chondroblasts. Deep to the perichondrium, the osteoblasts secrete a thin shell of compact bone tissue. Once the perichondrium starts to form osseous tissue, it is known as the **periosteum**. Near the middle of the model, periosteal capillaries grow into the disintegrating calcified cartilage. The capillaries induce growth of a **primary ossification center**, a region where osseous tissue will replace cartilage. Osteoblasts then begin to deposit osseous tissue extracellular matrix over the remnants of calcified cartilage, forming spongy bone trabeculae. Primary ossification spreads toward both ends of the cartilage model.

4 *Development of the medullary cavity.* As the primary ossification center grows toward the ends of the bone, osteoclasts break down some of the newly formed spongy bone trabeculae. This activity leaves a cavity, the medullary cavity, in the diaphysis. The medullary cavity is filled with red bone marrow by invading blood vessels. During growth of the skeleton, red bone marrow within the medullary cavity of long bones is progressively replaced by yellow bone marrow.

5 *Development of the secondary ossification centers.* When branches of the epiphyseal artery enter the epiphyses, **secondary ossification centers** develop, usually around the time of birth. Bone formation in the secondary ossification centers is similar to that in primary ossification centers. One difference, however, is that spongy bone tissue remains in the interior of the epiphyses (no medullary cavities are formed).

6 *Formation of articular cartilage and the epiphyseal plate.* The hyaline cartilage that covers the epiphyses becomes the articular cartilage. Prior to adulthood, hyaline cartilage remains between the diaphysis and epiphysis as the **epiphyseal plate**, which is responsible for the lengthwise growth of long bones.

Checkpoint

16. What are the major events of intramembranous ossification? Of endochondral ossification?
17. How do the primary and secondary ossification centers differ?
18. How do the medullary cavities in a long bone form?
19. Which ossification method would form a femur? Which would form most skull bones?

6.7 Bones grow longer at the epiphyseal plate and increase in diameter by the addition of new osseous tissue around the outer surface.

During childhood and adolescence, bones throughout the body grow in thickness, and long bones lengthen.

Growth in Length

The growth in length of long bones involves two major events: (1) growth of cartilage on the epiphyseal side of the epiphyseal plate and (2) replacement of cartilage on the diaphyseal side of the epiphyseal plate with osseous tissue. To understand how a bone grows in length, you need to know about the structure of the epiphyseal plate. The **epiphyseal plate,** a layer of hyaline cartilage in the metaphysis of a growing bone, consists of four zones (Figure 6.8).

• *Zone of resting cartilage.* This layer is nearest the epiphysis and consists of small, scattered chondrocytes. The cells do not function in bone growth (thus the term *resting*). Instead, they anchor the epiphyseal plate to the epiphysis of the bone.

• *Zone of proliferating cartilage.* Slightly larger chondrocytes in this zone are arranged like stacks of coins. These chondrocytes divide to replace those that die at the diaphyseal side of the epiphyseal plate.

• *Zone of hypertrophic cartilage* (hī-per-TRŌ-fik). This layer consists of large, maturing chondrocytes arranged in columns.

• *Zone of calcified cartilage.* The final zone of the epiphyseal plate is only a few cells thick and consists mostly of dead chondrocytes because the extracellular matrix around them has calcified. Osteoclasts dissolve the calcified cartilage, and osteoblasts and capillaries from the diaphysis invade the area. Osteoblasts lay down bone extracellular matrix, replacing the calcified cartilage by endochondral ossification. As a result, the zone of calcified cartilage becomes "new diaphysis" that is firmly cemented to the rest of the diaphysis of the bone.

The activity of the epiphyseal plate is the only way that the diaphysis can increase in length. As a bone grows, new chondrocytes are formed on the epiphyseal side of the plate, while old chondrocytes on the diaphyseal side of the plate are replaced by osseous tissue. In this way the thickness of the epiphyseal plate cartilage remains relatively constant, but the bone on the diaphyseal side increases in length.

(a) Radiograph showing the epiphyseal plate of the femur of a 3-year-old

Femur

EPIPHYSEAL PLATE

Tibia

Diaphyseal side

Zone of resting cartilage

Zone of proliferating cartilage

Zone of hypertrophic cartilage

Zone of calcified cartilage

Diaphysis

Epiphyseal side   **LM**   360x

(b) Histology of the epiphyseal plate

Figure 6.8   **The epiphyseal plate is a layer of hyaline cartilage in the metaphysis of a growing bone.** The epiphyseal plate appears as a dark band between whiter calcified areas in the radiograph (X-ray) shown in (a).

At about age 18 in females and 21 in males, the epiphyseal plates close. Rising hormone levels stimulate osteoblasts to produce osseous tissue faster than chondrocytes can produce epiphyseal plate cartilage. The epiphyseal plate becomes progressively narrower until it is replaced by a bony structure called the **epiphyseal line**. The appearance of the epiphyseal line signifies that the bone has stopped growing in length. The clavicle is the last bone to stop growing.

If a bone fracture damages the epiphyseal plate, the fractured bone may be shorter than normal once adult stature is reached. Because damaged cartilage is avascular, it repairs much more slowly than damaged osseous tissue, which is highly vascular. Epiphyseal plate damage accelerates ossification of epiphyseal plate cartilage, and bone growth in length stops.

Growth in Thickness

Bones grow in thickness (diameter) by the deposition of osseous tissue on the outer surface of the bone. At the bone surface, cells in the periosteum differentiate into osteoblasts. These osteoblasts secrete the extracellular matrix of osseous tissue, adding lamellae to the surface of the bone, and new osteons of compact bone tissue are formed. As the osteoblasts become surrounded by extracellular matrix, they develop into osteocytes. At the same time, the osseous tissue lining the medullary cavity is destroyed by osteoclasts in the endosteum. Thus, the medullary cavity enlarges as the bone increases in diameter.

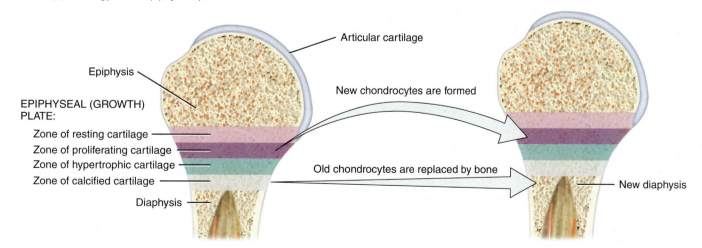

Articular cartilage

Epiphysis

EPIPHYSEAL (GROWTH) PLATE:

Zone of resting cartilage
Zone of proliferating cartilage
Zone of hypertrophic cartilage
Zone of calcified cartilage

Diaphysis

New chondrocytes are formed

Old chondrocytes are replaced by bone

New diaphysis

(c) Lengthwise growth of bone at epiphyseal plate

The epiphyseal plate allows the diaphysis of a bone to increase in length.

Factors Affecting Bone Growth

Normal bone growth in the young and maintenance of bones (bone remodeling, described shortly) depend on several factors. These include adequate dietary intake of minerals and vitamins. Large amounts of calcium and phosphorus are needed, as are smaller amounts of fluoride, magnesium, iron, and manganese. Vitamins A, C, B_{12}, D, and K are also needed. In addition, sufficient levels of several hormones and weight-bearing exercise (exercise that places stress on bones) are essential.

During childhood, the most important hormones that stimulate bone growth and remodeling are the insulinlike growth factors (IGFs), which are produced by osseous tissue and the liver. IGFs stimulate osteoblasts, promote cell division at the epiphyseal plate and in the periosteum, and enhance synthesis of the proteins needed to build new osseous tissue. Production of IGFs, in turn, is stimulated by human growth hormone from the pituitary gland. Thyroid hormones from the thyroid gland and insulin from the pancreas also promote bone growth.

At puberty, estrogens (sex hormones produced by the ovaries) and androgens (sex hormones such as testosterone produced by the testes) start to be released in larger quantities. These hormones increase osteoblast activity and synthesis of osseous tissue extracellular matrix and are responsible for the sudden "growth spurt" that occurs during the teenage years. Females produce much higher levels of estrogens than androgens, whereas males have higher levels of androgens than estrogens. Differences in sex hormone levels promote the typical femininization or masculinization of our skeletons. For example, estrogens account for the widening of the pelvis in females. Ultimately, sex hormones induce conversion of the epiphyseal plates into epiphyseal lines. At this point, elongation of bones ceases. Adult women are typically shorter than adult men because their higher levels of sex hormones stopped lengthwise growth of bones at an earlier age.

Checkpoint

20. Which activities of the epiphyseal plate account for the lengthwise growth of the diaphysis?
21. How does the epiphyseal line develop?
22. How do bones grow in diameter?
23. How does the medullary cavity enlarge during bone growth?
24. Which nutrients are important for bone growth and maintenance?
25. How does human growth hormone regulate bone growth?

6.8 Bone remodeling renews osseous tissue, redistributes bone extracellular matrix, and repairs bone injuries.

Even after bones have reached their adult shapes and sizes, they continue to be renewed. **Bone remodeling** is the ongoing replacement of old osseous tissue by new osseous tissue. During remodeling, osteoclasts destroy the extracellular matrix and then osteoblasts deposit new extracellular matrix. At any given time, about 5% of the total bone mass in the body is being remodeled. The renewal rate for compact bone tissue is about 4% per year and for spongy bone tissue is about 20% per year. Remodeling also takes place at different rates in various body regions. The distal portion of the femur is replaced about every four months. By contrast, bone in certain areas of the diaphysis of the femur will not be completely replaced during an individual's life.

Remodeling renews osseous tissue before deterioration sets in and heals an injured bone by replacing the damaged tissue with new osseous tissue. Remodeling has several other advantages. New osseous tissue is less brittle and stronger than old osseous tissue. Remodeling also redistributes bone extracellular matrix along lines of mechanical stress (muscle pull and gravity) so that the shape of a bone is altered to keep it strongest where there is the greatest need for support. Since a bone will remodel in response to the demands placed on it, subjecting a bone to heavy loads will produce a new bone that is thicker and therefore stronger than the old bone.

A delicate balance exists between the bone-destroying actions of osteoclasts and the bone-producing actions of osteoblasts. Should too much new osseous tissue be formed, the bones become abnormally thick and heavy. If too much mineral material is deposited in the bone, the surplus may form thick bumps, called *spurs*, on the bone that interfere with movement at joints. Alternatively, a loss of extracellular matrix weakens bones, and they may break, as occurs in osteoporosis, or they may become too flexible, as in rickets and osteomalacia. Abnormal acceleration of the remodeling process results in a condition called Paget's disease, in which the newly formed bones, especially the bones of the limbs, pelvis, lower vertebrae, and skull, become hard and brittle and fracture easily.

Checkpoint

26. What is bone remodeling? What roles do osteoblasts and osteoclasts play in the process?
27. Why is bone remodeling necessary when a person reaches adulthood?

Amanda's Story: Epilogue and Discussion

▶▶▶Amanda has a hereditary defect called osteogenesis imperfecta, a disease that causes a defect in the production of collagen. There are several types of collagen; the most abundant, type I collagen, is found in bones, the sclera (or whites) of the eye, ligaments, and teeth. Osteogenesis imperfecta interferes with the production of type I collagen. Collagen fibers give bones their flexibility and serve as a scaffold on which osseous tissue is built; it normally serves as a framework on which mineral salts crystallize. The structure of collagen fibers in individuals like Amanda alters the deposition of mineral salts, resulting in excessively brittle bones.

Typical problems seen in patients with osteogenesis imperfecta include fragile bones that fracture easily, skeletal deformities such as reduced adult height and scoliosis, dental defects due to inadequate enamel deposition, hearing loss due to deformity of middle ear bones, and hypermobility of joints due to loose ligaments. The disease can be categorized into four subtypes, from mild to severe.

The treatment for osteogenesis imperfecta varies based on an individual's age, symptoms, and the severity of the condition. In children under 2 years of age, the disease is usually treated as if the child had no specific disease. However, as the child grows and throughout life, the disease is dealt with symptomatically.

Amanda has a genetic condition with lifelong consequences. She will need to monitor her health and activities to avoid a lifetime of complications. Although the possibility is still years in the future, Amanda will also need to consider the likelihood that she might pass on the disease to her own children. ◀◀◀

F. What symptoms would you look for in a child that could indicate osteogenesis imperfecta? How could you explain osteogenesis imperfecta to Amanda in layman's terms?

G. What suggestions could you make to Amanda's parents regarding nutrition that might help Amanda?

H. What is the role of collagen fibers in Amanda's skeleton? Which of Amanda's cells produce the defective collagen fibers during bone remodeling?

Concept and Resource Summary

Concept	Resources

Introduction

1. Bones are dynamic living organs, composed of several different types of tissues. They undergo continuous remodeling.
2. All of the bones, their cartilages, along with ligaments and tendons, constitute the **skeletal system**.

Concept 6.1 Skeletal system functions include support, protection, movement, mineral homeostasis, blood cell production, and energy storage.

See
- **Anatomy Overview – Bone Structure and Tissues**

1. The skeleton supports soft tissues and provides attachment points for tendons of skeletal muscles.
2. The skeleton protects the internal organs.
3. Bones assist in movement by acting as levers for skeletal muscles.
4. Osseous tissue participates in storage and release of minerals, particularly calcium and phosphorus.
5. Certain bones contain **red bone marrow**, a connective tissue where blood cells are produced in a process called **hemopoiesis**.
6. **Yellow bone marrow** participates in the storage of triglycerides, a source of potential energy.

Concept 6.2 Bones are classified as long, short, flat, irregular, or sesamoid.

1. Bones are classified by shape into five principal types: long, short, flat, irregular, and sesamoid.

Concept 6.3 Long bones have a diaphysis, a medullary cavity, epiphyses, metaphyses, and periosteum.

Do
- **Exercise – Growing Long Bone**

1. The cylindrical shaft of a long bone is called the **diaphysis**. The space within the diaphysis that contains yellow marrow is called the **medullary cavity**.
2. The distal and proximal ends of a long bone are called the **epiphyses**. They are covered with **articular cartilage** composed of hyaline cartilage.
3. The regions where the diaphysis joins the epiphyses are called the **metaphyses**. In growing bone, each metaphysis has an **epiphyseal plate** essential for bone growth. Once growth stops, the epiphyseal plate becomes the **epiphyseal line**.
4. The **periosteum** is a dense irregular connective tissue sheath surrounding the bone, and the **endosteum** is a delicate connective tissue membrane lining the medullary cavity. The periosteum and endosteum contain bone-forming cells.

Concept

Concept 6.4   Osseous tissue can be arranged as compact bone tissue or spongy bone tissue.

1. **Osseous tissue** is a connective tissue containing an extracellular matrix of water, collagen fibers, and crystallized mineral salts called *hydroxyapatites*.
2. **Osteoblasts** initiate **calcification**, the process in which mineral salts are deposited around collagen fibers in the extracellular matrix, then crystallize into a hard tissue.
3. Osseous tissue contains four types of cells: osteogenic cells, osteoblasts, osteocytes, and osteoclasts.
4. **Osteogenic cells** are undifferentiated cells that can undergo cell division and give rise to osteoblasts.
5. **Osteoblasts** secrete the extracellular matrix of osseous tissue and then initiate calcification.
6. **Osteocytes** are mature bone cells that maintain the metabolism of osseous tissue.
7. **Osteoclasts** release enzymes and acids used in bone resorption, the breakdown of the extracellular matrix of osseous tissue for growth and repair of bones.
8. **Compact bone tissue** has few spaces, forms the external layer of all bones, and comprises most of the diaphysis of long bones. It is composed of **osteons**. Each osteon has a **central canal** where blood vessels, lymphatic vessels, and nerves are found. The central canal is surrounded by **concentric lamellae** with **lacunae** containing osteocytes that are connected to one another and to the central canal by radiating **canaliculi**.
9. **Spongy bone tissue** is composed of **trabeculae** consisting of irregular lamellae and osteocytes in lacunae. Spaces between trabeculae are filled with red bone marrow, a tissue that produces blood cells. Spongy bone predominates in short, flat, and irregularly shaped bones; most of the epiphyses of long bones; and a narrow rim around the medullary cavity of long bones.

See
- **Anatomy Overview – Compact Bone**
- **Anatomy Overview – Spongy Bone**

Hear
- **Figure 6.3 – Types of Cells in Bone Tissue**
- **Figure 6.4 – Histology of Compact and Spongy Bone Tissue**

Concept 6.5   Bones are richly supplied with blood vessels and nerves.

1. Bones are highly vascular, with blood vessels passing into them from the periosteum.
2. In a long bone, **periosteal arteries** and nerves enter the diaphysis through **perforating canals** to supply the periosteum and outer compact bone tissue. At least one large **nutrient artery** enters the diaphysis and branches in the medullary cavity. Bone also contains **metaphyseal** and **epiphyseal arteries**, **nutrient veins**, **periosteal veins**, and **epiphyseal veins**.

Concept 6.6   The two types of bone formation are intramembranous ossification and endochondral ossification.

1. The formation of bone, called **ossification**, occurs in the human embryo in areas where mesenchyme shaped like bone, provides the template for ossification.
2. The mandible and flat skull bones are formed by **intramembranous ossification**.
3. Intramembranous ossification begins when mesenchymal cells cluster to form an **ossification center**, where they differentiate into osteoblasts that secrete the organic extracellular matrix of osseous tissue.
4. **Calcification** occurs when calcium and mineral salts are deposited and harden. Osteocytes are then located in lacunae of the matrix.
5. Trabeculae form as spongy bone tissue, then blood vessels enter the area, and red bone marrow develops in spaces between trabeculae.
6. Development of the periosteum occurs at the periphery, then a thin layer of compact bone tissue develops at the surface covering the deeper spongy bone tissue.
7. Most of the bones of the body are formed when cartilage is replaced by osseous tissue during **endochondral ossification**.
8. Endochondral ossification begins when mesenchymal cells initiate the development of a hyaline cartilage model of a future bone. **Perichondrium** develops around the cartilage model, chondroblasts become chondrocytes, and growth of the cartilage model increases.
9. The **primary ossification center** begins in the mid-region of the cartilage bone model. Vascularization occurs and a **periosteum** forms. Osteoblasts begin to secrete extracellular matrix of spongy bone trabeculae.
10. The medullary cavity develops when osteoclasts break down some of the newly formed trabeculae.
11. Development of **secondary ossification centers** occurs in the epiphyses, but medullary cavities do not form. Hyaline cartilage remains as the articular cartilage and **epiphyseal plate**.

See
- **Animation – Bone Formation**

Hear
- **Figure 6.6 – Intramembranous Ossification**
- **Figure 6.7 – Endochondral Ossification**

Do
- **Exercise – Observe the Ossification**

Concept	Resources

Concept 6.7 Bones grow longer at the epiphyseal plate and increase in diameter by the addition of new osseous tissue around the outer surface.

1. Throughout childhood and adolescence, long bones grow in length when cartilage is replaced by osseous tissue on the diaphyseal side of the **epiphyseal plate**. The plate narrows when rising hormone levels, between the ages of 18 and 21, cause ossification that outpaces replenishment of cartilage. An **epiphyseal line** results as growth in length stops.
2. Bone grows in diameter as osteoblasts in the periosteum secrete extracellular matrix on the bone surface while the medullary cavity is enlarged by osteoclasts of the endosteum.
3. Factors that affect bone growth include adequate intake of calcium, phosphorus, magnesium, and vitamins A, B_{12}, C, D, and K; sufficient hormone levels; and adequate weight-bearing exercise.
4. During childhood, pituitary human growth hormone stimulates production of insulinlike growth factors that promote cell division at the periosteum and epiphyseal plate.
5. Thyroid hormones and pancreatic insulin also contribute to bone growth.
6. The growth spurt in adolescence is due to the sex hormones, estrogens and androgens, that stimulate ossification at the epiphyseal plate, leading to eventual conversion into an epiphyseal line.

See
- Animation – Bone Elongation and Bone Widening

Hear
- Figure 6.8 – The Epiphyseal Plate

Do
- Exercise – Bone Growth Sequencing
- Exercise – Concentrate on Bone Tissue
- Exercise – Regulation Sequences

Concept 6.8 Bone remodeling renews osseous tissue, redistributes bone extracellular matrix, and repairs bone injuries.

1. **Bone remodeling** continues throughout life as old osseous tissue gets replaced by less brittle, stronger, new osseous tissue.
2. Remodeling renews osseous tissue before deterioration occurs, heals injured bones, and redistributes bone extracellular matrix along lines of mechanical stress.

See
- Animation – Bone Remodeling
- Animation – Regulation of Blood Calcium
- Animation – Bone Dynamics and Tissue
- Animation – Regulation of Bone Growth

Other Chapter Resources

For additional information on clinical applications, medical disorders, and medical terminology, see Chapter 4 in the *Clinical Connections* manual packaged with your text.

Within *WileyPLUS*
- Chapter Quiz
- Visual Anatomy
- Mastering Vocabulary
- Audio Glossary
- Homeostatic Imbalance: The Case of the Brittle Woman and Back Pain
- Web Links

Understanding the Concepts

1. Why is osseous tissue considered a connective tissue?
2. What is the path a nutrient would travel through compact bone tissue from its diffusion out of a blood vessel in the periosteum to an osteocyte located within the second osteon in from the surface of the bone?
3. What effect would the gravity-free environment of space have on an astronaut's bones?
4. Compare and contrast the microscopic appearance, location, and function of spongy and compact bone.
5. Explain why the two ossification processes are called "intramembranous ossification" and "endochondral ossification."
6. As we approach adulthood, which hormone is primarily responsible for ending the elongation of bones? Explain your answer.
7. If radiographs of an 18-year-old basketball star show clear epiphyseal plates but no epiphyseal lines, is she likely to grow taller? Why or why not?
8. What would be the effect on your skeleton if your osteoclasts were more active than your osteoblasts? If your osteoblasts were more active than your osteoclasts?

7

The Axial Skeleton

The sun rose quickly. Hassan had been up early, excited by the finds of the previous day: several chunks of wood, evidence of an ancient campfire, animal bones, and the remnants of a leather sandal. To the untrained eye, it would have meant nothing, but Hassan and his team knew they were on the right track. No one had believed in him when he had decided to pursue this dig site, but Hassan had an uncanny knack for putting together clues.

Hassan stood over his makeshift table, closely scrutinizing a small bit of papyrus scroll. Just then the flap of his tent flew open.

"Hassan! We have found the tomb. You must see!" exclaimed his colleague Baru.

"What did you find?" Hassan asked excitedly.

"Come see for yourself," Baru replied.

The two men reached the excavation site quickly. Peering down into the hole, Hassan saw what looked like stacks of sticks. But as the dust settled, he could make out a jumble of bones. Human skeletons projected out of the ground; ribs, the rounded domes of skulls, and limbs were scattered in a jumbled heap. "They were only a few centimeters beneath the wood samples we found yesterday."

"Baru! You have found the guardians of Ishtamira. The tomb entrance will be near."

As he gazed into the pit, a profound sadness tempered Hassan's excitement. These builders, masons, and artisans had trusted their king and queen, and then were brutally sacrificed to be the sentinels of this tomb. But why? The only clues Hassan had to go on were a few scattered bits of papyrus scroll that told of a great plague and the desperate attempts of the ruler to appease his gods. Now that the tomb had been found, the ancient people would speak through their bones.

Baru, the team's medical anthropologist, looked at Hassan and smiled. "I can see that I will be very busy. Perhaps I should phone back to Cairo and have my graduate students come to assist."

"Yes, yes. There will be much to do. Come. Let's take a closer look."

As we have seen in Chapter 6, bones are living tissue that reflect the health and homeostasis of our bodies. Because skeletal remains may persist for thousands of years, it is possible to trace patterns of disease and nutrition, evaluate the effects of social and economic changes, and deduce patterns of reproduction and mortality. Skeletal remains also may reveal an individual's sex, age, height, and race. In this chapter and in Chapter 8 we will explore the anatomy of the skeleton and will learn how understanding its structure can serve as a tool for medical science and other disciplines.

Without bones, you could not survive. You would be unable to perform movements such as walking or grasping, and the slightest blow to your head or chest could damage your brain or heart. Because the skeletal system forms the framework of the body, a familiarity with the names, shapes, and positions of individual bones will help you locate and name many other anatomical features. For example, the radial artery, the site where the pulse is usually taken, is named for its closeness to the radius, the lateral bone of the forearm. The frontal lobe of

> *Because the skeletal system forms the framework of the body, a familiarity with the names, shapes, and positions of individual bones will help you locate and name many other anatomical features.*

the brain lies deep to the frontal (forehead) bone. Parts of certain bones also help locate structures within the skull and outline the lungs, heart, and abdominal and pelvic organs.

Movements such as throwing a ball, biking, and walking require interactions between bones and muscles. To understand how muscles produce different movements, you will learn where the muscles attach on individual bones as well as the types of joints acted on by the contracting muscles. The bones, muscles, and joints together form an integrated system called the **musculoskeletal** system.

7.1   Bones of the axial skeleton and appendicular skeleton have characteristic surface markings.

The adult human skeleton consists of 206 named bones. The skeletons of infants and children have more than 206 bones because some of their bones, such as the hip bones and some bones of the vertebral column (spine), fuse later in life.

There are two principal divisions in the adult skeleton: This chapter focuses on the **axial skeleton** and Chapter 8 will explore the **appendicular skeleton**. Refer to Figure 7.1 to see how the two divisions join to form the complete skeleton. The bones of the axial skeleton are shown in gray. The axial skeleton lies around the longitudinal *axis* of the human body, an imaginary vertical line that runs through the body's center of gravity from the head to the space between the feet. The axial skeleton consists of the skull bones, auditory ossicles, hyoid bone (see Figure 7.5), ribs, sternum (breastbone), and bones of the vertebral column. Functionally, the auditory ossicles, bones in the middle ear that vibrate in response to sound waves that strike the eardrum, are not part of either the axial or appendicular skeleton, but they are grouped with the axial skeleton for convenience. The appendicular skeleton contains the bones of the **upper** and **lower limbs**, plus the bones called **girdles** that connect the limbs to the axial skeleton. Table 7.1 presents the 80 bones of the axial skeleton and the 126 bones of the appendicular skeleton.

Bones have characteristic **surface markings**, structural features adapted for specific functions. Surface markings develop as the periosteum of a bone is pulled by tendons, ligaments, aponeuroses, and fasciae, stimulating new osseous tissue to be deposited as raised or roughened areas. Conversely, compression on a bone surface results in a depression. There are two major types of surface markings: (1) *depressions and openings*, which allow the passage of soft tissues (such as blood vessels and nerves) or form joints, and (2) *processes*, which are projections or outgrowths that either help form joints or serve as attachment points for connective tissue (such as ligaments and tendons). Table 7.2 describes the various surface markings and provides examples of each.

Checkpoint

1. Classify the following structures as part of the axial skeleton or the appendicular skeleton: skull, clavicle, vertebral column, shoulder girdle, humerus, pelvic girdle, and femur.
2. What is the function of depressions and openings? Which opening is shaped like a tube? A narrow slit? A shallow depression?
3. What is the function of processes? Which process has a flat surface? Projects from a narrowed part of a bone, or "neck"? Forms a pointed, narrow projection? Is large with a roughened surface?

TABLE 7.1

The Bones of the Adult Skeletal System

DIVISION OF THE SKELETON	STRUCTURE	NUMBER OF BONES	DIVISION OF THE SKELETON	STRUCTURE	NUMBER OF BONES
Axial Skeleton	**Skull**		**Appendicular Skeleton**	**Pectoral (shoulder) girdles**	
	Cranial bones	8		*Clavicle*	2
	Facial bones	14		*Scapula*	2
	Hyoid bone	1		**Upper limbs**	
	Auditory ossicles	6		*Humerus*	2
	Vertebral column	26		*Ulna*	2
	Thorax			*Radius*	2
	Sternum	1		*Carpals*	16
	Ribs	24		*Metacarpals*	10
	Subtotal =	**80**		*Phalanges*	28
				Pelvic (hip) girdle	
				Hip or coxal bone	2
				Lower limbs	
				Femur	2
				Patella	2
				Fibula	2
				Tibia	2
				Tarsals	14
				Metatarsals	10
				Phalanges	28
				Subtotal =	**126**
				TOTAL =	**206**

Figure 7.1 **Divisions of the skeletal system.** The axial skeleton is indicated in gray. Note the position of the hyoid bone in Figure 7.5.

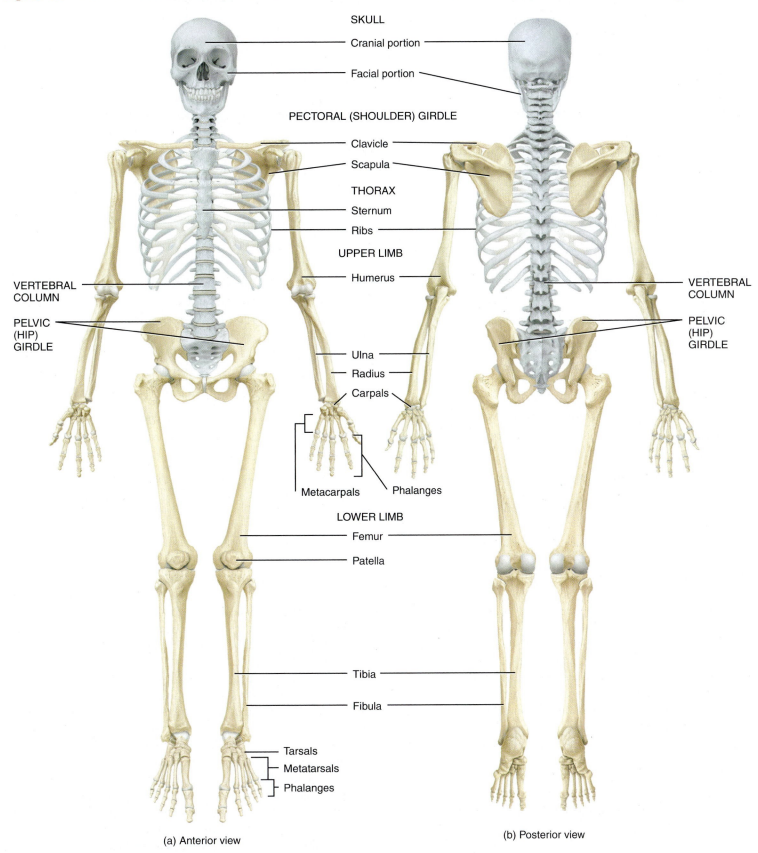

SKULL
— Cranial portion
— Facial portion

PECTORAL (SHOULDER) GIRDLE
— Clavicle
— Scapula

THORAX
— Sternum
— Ribs

UPPER LIMB
— Humerus

VERTEBRAL COLUMN

PELVIC (HIP) GIRDLE

— Ulna
— Radius
— Carpals
— Metacarpals
— Phalanges

VERTEBRAL COLUMN

PELVIC (HIP) GIRDLE

LOWER LIMB
— Femur
— Patella

— Tibia
— Fibula

— Tarsals
— Metatarsals
— Phalanges

(a) Anterior view

(b) Posterior view

The adult human skeleton consists of 206 bones grouped into axial and appendicular divisions.

TABLE 7.2

Bone Surface Markings

MARKING	DESCRIPTION	EXAMPLE
Depressions and Openings: Sites allowing the passage of soft tissue (nerves, blood vessels, ligaments, tendons) or formation of joints		
Fissure (FISH-ur)	Narrow slit between adjacent parts of bones through which blood vessels or nerves pass.	Superior orbital fissure of the sphenoid bone (Figure 7.11.)
Foramen (fō-RĀ-men; plural is **foramina**)	Opening (*foramen* = hole) through which blood vessels, nerves, or ligaments pass.	Optic foramen of the sphenoid bone (Figure 7.11).
Fossa (FOS-a)	Shallow depression (*fossa* = trench).	Coronoid fossa of the humerus (Figure 8.5a).
Sulcus	Furrow along a bone surface that accommodates a blood vessel, nerve, or tendon.	Intertubercular sulcus of the humerus (Figure 8.5a).
Meatus (mē-Ā-tus)	Tubelike opening (*meatus* = passageway).	External auditory meatus of the temporal bone (Figure 7.3).
Processes: Projections or outgrowths on bone that form joints or attachment points for connective tissue, such as ligaments and tendons.		
Processes that form joints:		
Condyle (KON-dil)	Large, round protuberance (*condylus* = knuckle) at the end of a bone.	Lateral condyle of the femur (Figure 8.13b).
Facet	Smooth flat articular surface.	Superior articular facet of a vertebra (Figure 7.17b).
Head	Rounded articular projection supported on the neck (constricted portion) of a bone.	Head of the femur (Figure 8.13c).
Processes that form attachment points for connective tissue:		
Crest	Prominent ridge or elongated projection.	Median sacral crest of the sacrum (Figure 7.20b).
Epicondyle	Projection above (*epi-* = above) a condyle.	Medial epicondyle of the femur (Figure 8.13a).
Line (linea)	Long, narrow ridge or border (less prominent than a crest).	Linea aspera of the femur (Figure 8.13b).
Spinous process	Sharp, slender projection.	Spinous process of a vertebra (Figure 7.16).
Trochanter (trō-KAN-ter)	Very large projection.	Greater trochanter of the femur (Figure 8.13b).
Tubercle (TOO-ber-kul)	Small, rounded projection (*tuber-* = knob).	Greater tubercle of the humerus (Figure 8.5a).
Tuberosity	Large, rounded, usually roughened projection.	Ischial tuberosity of the hip bone (Figure 8.10b).

7.2   The skull provides attachment sites for muscles and membranes, and protects and supports the brain and sense organs.

The **skull**, which contains 22 bones, rests on the superior end of the vertebral column. It includes two sets of bones: cranial bones and facial bones (Table 7.3). The **cranial bones** (*crani-* = brain case) form the **cranial cavity**, which encloses and protects the brain. The eight cranial bones are the frontal bone, two parietal bones, two temporal bones, the occipital bone, the sphenoid bone, and the ethmoid bone. Fourteen **facial bones** form the face: two nasal bones, two maxillae (or maxillas), two zygomatic bones, the mandible, two lacrimal bones, two palatine bones, two inferior nasal conchae, and the vomer. Figures 7.2 through 7.11 illustrate the bones of the skull from different viewing directions.

Besides forming the large cranial cavity, the skull also forms several smaller cavities that open to the exterior, including the nasal cavity and orbits (eye sockets). Certain skull bones also contain cavities called paranasal sinuses that are lined with mucous membranes and open into the nasal cavity. Other small cavities within the skull house the structures involved in hearing and equilibrium.

The cranial bones have other functions in addition to protecting the brain. Their inner surfaces attach to membranes (meninges) that stabilize the positions of the brain, blood vessels, and nerves. The outer surfaces of cranial bones provide large areas of attachment for muscles that move various parts of the head. The bones also provide attachment for some muscles that are involved in producing facial expressions such as the puckered brow look of concentration you have when studying. Besides forming the framework of the face, the facial bones protect and provide support for the entrances to the digestive and respiratory systems. Together, the cranial and facial bones protect and support the delicate special sense organs for vision, taste, smell, hearing, and equilibrium.

Other than the auditory ossicles (see Figure 16.19), which are involved in hearing and are located within the temporal bones, the mandible is the only movable bone of the skull. Most of the skull bones are held together by immovable joints, called sutures, which are especially noticeable on the outer surface of the skull.

TABLE 7.3

Bones of the Adult Skull

CRANIAL BONES	FACIAL BONES
Frontal (1)*	Nasal (2)
Parietal (2)	Maxillae (2)
Temporal (2)	Zygomatic (2)
Occipital (1)	Mandible (1)
Sphenoid (1)	Lacrimal (2)
Ethmoid (1)	Palatine (2)
	Inferior nasal conchae (2)
	Vomer (1)

*The numbers in parentheses indicate how many of each bone are present.

The skull also has many surface markings such as foramina and fissures through which blood vessels, nerves, and ligaments pass. We will mention most of the **foramina** of the skull in the descriptions of the cranial and facial bones that they penetrate. Structures that pass through the foramina will be described in *WileyPLUS*.

Checkpoint

4. What are the functions of the cranial bones? The facial bones?
5. What do the auditory ossicles and mandible have in common?
6. What function is served by foramina?

7.3 The cranial bones include the frontal, parietal, temporal, occipital, sphenoid, and ethmoid bones.

Frontal Bone

The **frontal** bone forms the forehead (the anterior part of the skull), the roofs of the *orbits* (eye sockets), and most of the anterior part of the cranial floor (Figure 7.2). Soon after birth the left and right sides of the frontal bone are united by the *metopic suture*, which usually disappears between the ages of six and eight. The frontal bone gradually slopes inferiorly from the coronal suture on top of the skull, then angles abruptly and becomes almost vertical. Superior to the orbits the frontal bone thickens, forming the

Figure 7.2 Anterior view of the skull.

Anterior view

🗝 The skull consists of cranial bones and facial bones.

supraorbital margin (*supra-* = above; *-orbital* = wheel rut). From this margin the frontal bone extends posteriorly to form the roof of the orbit and part of the floor of the cranial cavity. Within the supraorbital margin is the *supraorbital foramen*, a passageway for nerves of the eyebrow and eyelid. The *frontal sinuses* lie deep to the frontal bone.

Parietal Bones

The two **parietal bones** (pa-RĪ-e-tal; *pariet-* = wall) form the greater portion of the sides and roof of the cranial cavity (Figure 7.3). The internal surfaces of the parietal bones contain many protrusions and depressions that accommodate the blood vessels supplying the dura mater, the superficial membrane covering the brain.

Temporal Bones

The two **temporal bones** (*tempor-* = temple) form the inferior lateral sides of the skull and part of the cranial floor (Figure 7.3). Projecting from the inferior portion of the temporal bone is the *zygomatic process*, which articulates (forms a joint) with the temporal process of the zygomatic (cheek) bone. Together, the zygomatic process of the temporal bone and the temporal process of the zygomatic bone form

Figure 7.3   Right lateral view of the skull.

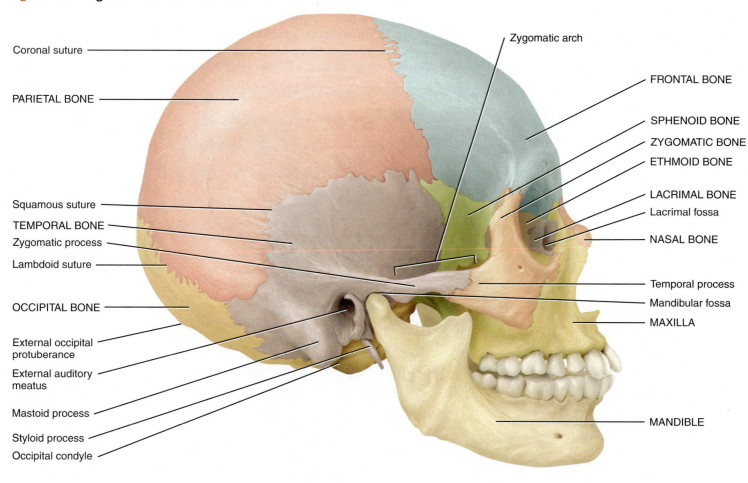

Coronal suture

PARIETAL BONE

Squamous suture

TEMPORAL BONE

Zygomatic process

Lambdoid suture

OCCIPITAL BONE

External occipital protuberance

External auditory meatus

Mastoid process

Styloid process

Occipital condyle

Zygomatic arch

FRONTAL BONE

SPHENOID BONE

ZYGOMATIC BONE

ETHMOID BONE

LACRIMAL BONE

Lacrimal fossa

NASAL BONE

Temporal process

Mandibular fossa

MAXILLA

MANDIBLE

Right lateral view

The zygomatic arch is formed by the zygomatic process of the temporal bone and the temporal process of the zygomatic bone.

the *zygomatic arch*. On the inferior posterior surface of the zygomatic process of the temporal bone is a socket called the *mandibular fossa*. The mandibular fossa articulates with the mandible (lower jawbone) to form the *temporomandibular joint (TMJ)*.

Inferior to the zygomatic process is an external opening in the temporal bone, the *external auditory meatus* (meatus = passageway), or ear canal, which directs sound waves into the ear (Figure 7.3). The *mastoid process* is a rounded projection of the temporal bone posterior to the external auditory meatus. It serves as a point of attachment for several neck muscles. The *styloid process* (*styl-* = stake or pole) projects from the inferior surface of the temporal bone and serves as a point of attachment for muscles and ligaments of the tongue and neck (Figures 7.3 and 7.4). Between the styloid process and the mastoid process is the *stylomastoid foramen* (Figure 7.4) through which a nerve passes that controls movement of facial muscles.

Figure 7.4 **Inferior view of the skull.** The mandible (lower jawbone) has been removed.

Inferior view

🔑 The occipital condyles of the occipital bone articulate with the first cervical vertebra to allow you to nod your head.

At the floor of the cranial cavity (see Figure 7.7) is the *petrous portion* (*petrous* = rock) of the temporal bone. The petrous portion is triangular and located at the base of the skull between the sphenoid and occipital bones. The petrous portion houses the internal ear and the middle ear, structures involved in hearing and equilibrium (balance). The *internal auditory meatus* (see Figures 7.5 and 7.7) is the opening in the petrous portion through which nerve impulses for hearing and equililrium are carried to the brain. The petrous portion also contains the *carotid foramen*, through which the internal carotid artery carries blood to the brain (Figure 7.4). Located in the suture between the petrous portion and the occipital bone is the *jugular foramen*, a passageway for the internal jugular vein carrying blood away from the brain.

Occipital Bone

The **occipital bone** (ok-SIP-i-tal; *occipit-* = back of head) forms the posterior part and most of the base of the skull (Figures 7.3, 7.4, and 7.6). The *foramen magnum* (= large hole) is in the inferior part of the bone (Figure 7.4). Within this foramen, the medulla oblongata (inferior part of the brain) connects with the spinal cord. The vertebral and spinal arteries also pass through this foramen. The *occipital condyles* are oval processes with convex surfaces, one on either side of the foramen magnum (Figure 7.4). They articulate with the first cervical vertebra (atlas), which allows you to nod your head "yes." Superior to each occipital condyle on the inferior surface of the skull is the *hypoglossal canal* (*hypo-* = under; *-glossal* = tongue), a passageway for nerve impulses that control movement of the tongue (see Figure 7.5).

Figure 7.5　Sagittal section of the skull. Although the hyoid bone is not part of the skull, it is included in the illustration for reference.

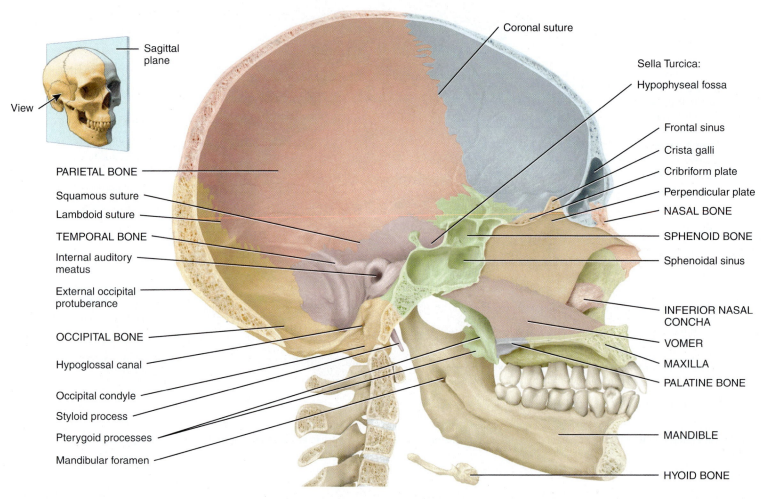

(a) Medial view of sagittal section

The cranial bones are the frontal, parietal, temporal, occipital, sphenoid, and ethmoid bones. The facial bones are the nasal bone, maxillae, zygomatic bones, lacrimal bones, palatine bones, mandible, and vomer.

Figure 7.6 **Posterior view of the skull.** The sutures are exaggerated for emphasis.

Posteroinferior view

🔑 **The occipital bone forms most of the posterior and inferior portions of the skull.**

The *external occipital protuberance* is a prominent midline projection on the posterior surface of the occipital bone just superior to the foramen magnum. You may be able to feel this structure as a definite bump on the back of your head, just above your neck (Figure 7.3). Extending laterally from the protuberance are two curved lines, the *superior nuchal lines*, and below these are two *inferior nuchal lines*, which are areas of muscle attachment (Figure 7.6).

Sphenoid Bone

The **sphenoid bone** (SFĒ-noyd = wedge-shaped) lies at the middle of the base of the skull (Figures 7.4 and 7.7). This bone is the keystone of the cranial floor because it articulates with all the other cranial bones, holding them together. Viewing the cranial floor from above (see Figure 7.7a), note the sphenoid bone joins anteriorly with the frontal bone, laterally with the parietal and temporal bones, and posteriorly with the occipital bone. It lies posterior and slightly superior to the nasal cavity and forms part of the floor, side walls, and rear wall of the orbit (see Figure 7.11).

The shape of the sphenoid bone resembles a butterfly with outstretched wings (Figure 7.7b). The *body* of the sphenoid bone is the hollowed cubelike medial portion between the ethmoid and occipital bones. The space inside the body is the *sphenoidal sinus*, which drains into the nasal cavity (see Figure 7.12). The *sella turcica* (SEL-a TUR-si-ka; *sella* = saddle; *turcica* = Turkish) is a bony saddle-shaped structure on the superior surface of the body of the sphenoid bone (Figure 7.7a). The seat of the saddle is a depression, the **hypophyseal fossa** (hī-pō-FIZ-ē-al), which contains the pituitary gland.

Figure 7.7   Sphenoid bone.

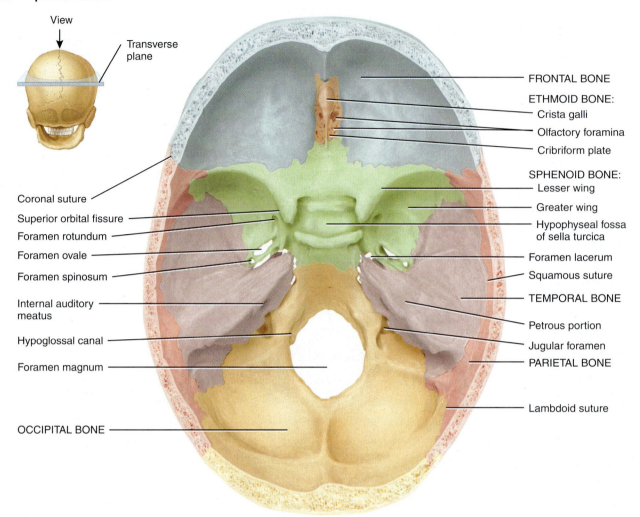

View

Transverse plane

FRONTAL BONE

ETHMOID BONE:
Crista galli
Olfactory foramina
Cribriform plate

SPHENOID BONE:
Lesser wing

Greater wing

Hypophyseal fossa of sella turcica

Foramen lacerum

Squamous suture

TEMPORAL BONE

Petrous portion

Jugular foramen

PARIETAL BONE

Lambdoid suture

Coronal suture

Superior orbital fissure

Foramen rotundum

Foramen ovale

Foramen spinosum

Internal auditory meatus

Hypoglossal canal

Foramen magnum

OCCIPITAL BONE

(a) Superior view of sphenoid bone in floor of cranium

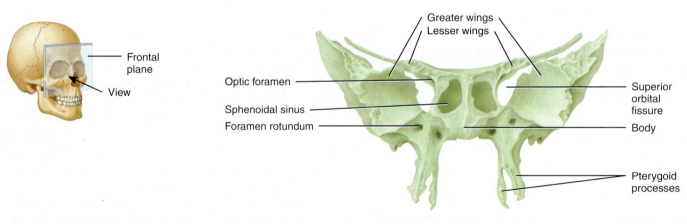

Greater wings
Lesser wings

Frontal plane

View

Optic foramen

Sphenoidal sinus

Foramen rotundum

Superior orbital fissure

Body

Pterygoid processes

(b) Anterior view of sphenoid bone

🔑 **The sphenoid bone is called the keystone of the cranial floor because it articulates with all other cranial bones, holding them together.**

The *greater wings* of the sphenoid bone project laterally from the body, forming the anterolateral cranial floor (Figure 7.7a) and part of the lateral wall of the skull just anterior to the temporal bone (see Figure 7.3). The *lesser wings*, which are smaller than the greater wings, form a ridge of bone anterior and superior to the greater wings. They form part of the cranial floor (Figure 7.7a) and the posterior part of the orbit of the eye (see Figure 7.11). In Figures 7.4 and 7.7b, you can see the *pterygoid processes* (TER-i-goyd = winglike) extending inferiorly from the sphenoid bone. Some of the muscles that move the mandible attach to the pterygoid processes.

Between the body and lesser wing just anterior to the sella turcica is the *optic foramen* (*optic* = eye), through which nerve impulses of vision are carried to the brain (Figure 7.7b and see Figure 7.11). Lateral to the body between the greater and lesser wings is a triangular slit called the *superior orbital fissure*, through which nerve impulses are transmitted that control eyeball movement (Figure 7.7b and see Figure 7.11). In the greater wing, three

foramina, from anterior to posterior, are the *foramen rotundum* (= round), through which passes a nerve responsible for facial sensations, the *foramen ovale* (= oval hole), through which passes a nerve that controls movement of the mandible, and the *foramen spinosum* (= resembling a spine), through which a blood vessel carries blood to the membranes covering the brain (Figure 7.7a). The *foramen lacerum* (= lacerated) is a jagged passageway between the sphenoid and occipital bones for a blood vessel supplying the brain (Figures 7.4 and 7.7a).

Ethmoid Bone

The **ethmoid bone** (ETH-moyd = like a sieve) is spongelike in appearance and is located in the anterior part of the cranial floor medial to the orbits (Figure 7.8). It is anterior to the sphenoid and posterior to the nasal bones. The ethmoid bone forms (1) part of the anterior cranial floor; (2) the medial wall of the orbits; (3) the

Figure 7.8 **Ethmoid bone.**

Sagittal plane

View

Frontal sinus

ETHMOID BONE:
Crista galli
Olfactory foramen
Cribriform plate
Superior nasal concha
Middle nasal concha
Inferior nasal concha
Palatine bone
Maxilla

Sphenoidal sinus

(a) Medial view of sagittal section

(continues)

Figure 7.8 (continued)

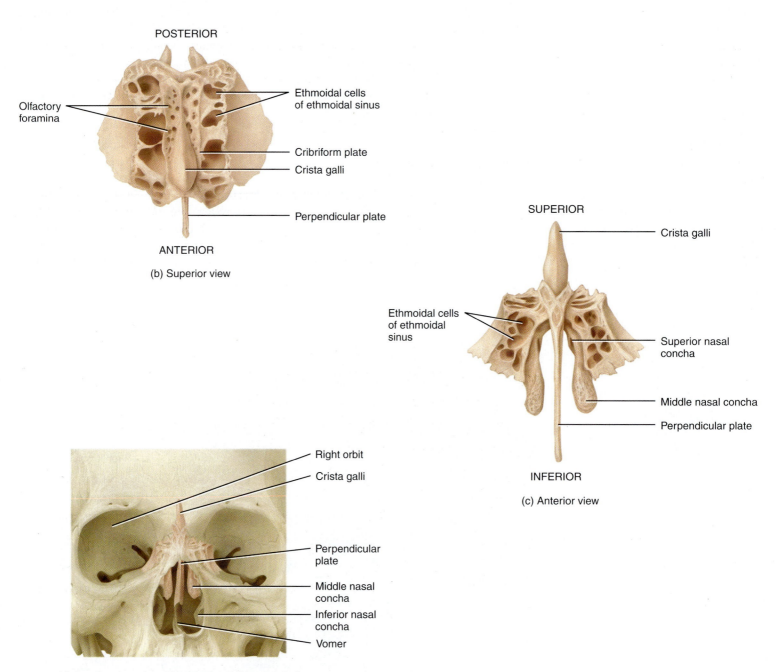

POSTERIOR

Olfactory
foramina

Ethmoidal cells
of ethmoidal sinus

Cribriform plate

Crista galli

Perpendicular plate

ANTERIOR

(b) Superior view

SUPERIOR

Crista galli

Ethmoidal cells
of ethmoidal
sinus

Superior nasal
concha

Middle nasal concha

Perpendicular plate

INFERIOR

(c) Anterior view

Right orbit

Crista galli

Perpendicular
plate

Middle nasal
concha

Inferior nasal
concha

Vomer

(d) Anterior view of position of ethmoid bone in skull

The ethmoid bone forms part of the anterior portion of the cranial floor, the medial wall of the orbits, the superior portions of the nasal septum, and most of the sidewalls of the nasal cavity.

superior portions of the nasal septum, which divides the nasal cavity into right and left sides; and (4) most of the superior sidewalls of the nasal cavity.

The *cribriform plate* (*cribri-* = sieve) of the ethmoid bone lies in the anterior cranial floor and forms the roof of the nasal cavity (Figures 7.5 and 7.7a). The cribriform plate contains the *olfactory foramina* (*olfact-* = to smell) through which smell sensations are transmitted to the brain. Projecting superiorly from the cribriform plate is a triangular process called the *crista galli* (*crista* = crest; *galli* = cock) (Figures 7.5 and 7.8). This structure serves as a point of attachment for the membranes that separate the two sides of the brain. Projecting inferiorly from the cribriform plate is the *perpendicular plate*, which forms the superior portion of the nasal septum (see Figure 7.9).

The ethmoid bone contains 3 to 18 air spaces, or "cells," that give this bone a sievelike appearance. The ethmoidal cells together form the *ethmoidal sinuses* (see Figure 7.12). Also part of the ethmoid bone are two thin, scroll-shaped projections on either side of the nasal septum (Figure 7.8). These are the *superior nasal concha* (KONG-ka = shell) and the *middle nasal concha*. The plural term is *conchae* (KONG-kē). A third pair of conchae, the inferior nasal conchae, are separate bones (discussed shortly). The conchae increase the vascular and mucous membrane surface area in the nasal cavity, which warms and moistens inhaled air before it passes into the lungs. The conchae also cause inhaled air to swirl, and the result is that many inhaled particles become trapped in the mucus that lines the nasal cavity. This action of the conchae helps cleanse inhaled air before it passes into the rest of the respiratory tract. The superior nasal conchae aid the sense of smell because the mucous membrane covering them contains sensory receptors for olfaction (smell).

Checkpoint

7. List the cranial bones, and indicate which ones are single and which ones are paired.
8. With which bones does the parietal bone articulate? With which bones does the temporal bone articulate?
9. What are the components of the zygomatic arch? The temporomandibular joint?
10. Which marking of the skull articulates with the first cervical vertebra? Which one forms a bump on the back of your head?
11. Starting at the ethmoid bone and going in a clockwise direction, list the bones that articulate with the sphenoid bone.
12. What functions are served by the internal auditory meatus and external auditory meatus? On which cranial bone are these markings located?
13. What functions are served by nasal conchae? Which nasal conchae are part of the ethmoid bone?

7.4 Facial bones include the nasal bones, maxillae, zygomatic bones, lacrimal bones, palatine bones, inferior nasal conchae, vomer, and mandible.

The shape of the face changes dramatically during the first two years after birth. The brain and cranial bones expand, the teeth form and erupt (emerge), and the paranasal sinuses increase in size. Growth of the face ceases at about 16 years of age. The 14 facial bones include two nasal bones, two maxillae, two zygomatic bones, the mandible, two lacrimal bones, two palatine bones, two inferior nasal conchae, and the vomer.

Nasal Bones

The paired **nasal bones** meet at the midline (see Figure 7.2) and form the bridge of the nose. The rest of the supporting tissue of the nose consists of cartilage (Figure 7.9).

Maxillae

The paired **maxillae** (mak-SIL-ē = jawbones; singular is *maxilla*) unite to form the upper jawbone. They articulate with every bone of the face except the mandible, or lower jawbone (see Figures 7.2, 7.3, and 7.4). The maxillae form part of the floors of the orbits, part of the lateral walls and floor of the nasal cavity, and most of the hard palate. The *hard palate* is the bony roof of the mouth and is formed by the palatine processes of the maxillae and horizontal plates of the palatine bones. The hard palate separates the nasal cavity from the oral cavity.

Each maxilla contains a large *maxillary sinus* that empties into the nasal cavity (see Figure 7.12). The *alveolar process* (al-VĒ-ō-lar; *alveol-* = small cavity) of the maxilla is an arch that contains

Figure 7.9   Nasal cavity.

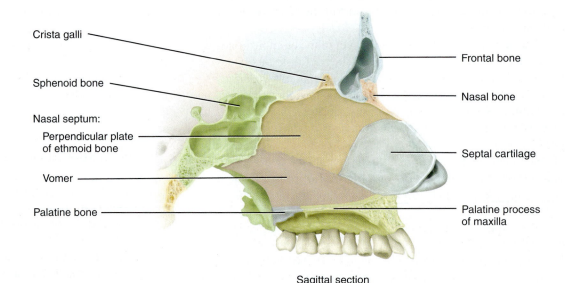

Sagittal section

🔑 **The structures that form the nasal septum are the perpendicular plate of the ethmoid bone, the vomer, and the septal cartilage.**

the alveoli (sockets) for the maxillary (upper) teeth (see Figure 7.2). The *palatine process* is a horizontal projection of the maxilla that forms the anterior hard palate (see Figures 7.4 and 7.6). The union and fusion of the maxillary bones normally is completed before birth. If the fusion fails, this condition is referred to as a cleft palate.

The *infraorbital foramen* (*infra-* = below), which can be seen in the anterior view of the skull in Figure 7.2, is a passageway in the maxilla below the orbit for a nerve to the face. Another prominent foramen, the *incisive foramen* (= incisor teeth), seen in Figure 7.4, is a passageway in the hard palate just posterior to the incisor teeth for nerve supply to the hard palate. Nerves and blood vessels to the face pass through the *inferior orbital fissure*, located within the orbit between the maxilla and sphenoid bone (see Figure 7.11).

Zygomatic Bones

The two **zygomatic bones** (*zygo-* = like a yoke), commonly called cheekbones, form the prominences of the cheeks and part of the lateral wall and floor of each orbit (see Figures 7.2, 7.3, and 7.11).

They articulate with the maxilla and the frontal, sphenoid, and temporal bones. The *temporal process* of the zygomatic bone projects posteriorly and articulates with the zygomatic process of the temporal bone to form the *zygomatic arch* (see Figure 7.3).

Lacrimal Bones

The paired **lacrimal bones** (LAK-ri-mal; *lacrim-* = teardrops) roughly resemble a fingernail in size and shape (see Figures 7.2, 7.3, and 7.11). These bones, the smallest bones of the face, are posterior and lateral to the nasal bones and form a part of the medial wall of each orbit. The lacrimal bones each contain a *lacrimal fossa*, a vertical tunnel formed with the maxilla, through which tears pass into the nasal cavity (see Figures 7.3 and 7.11).

Palatine Bones

The two L-shaped **palatine bones** (PAL-a-tīn) form the posterior hard palate, part of the floor and lateral wall of the nasal cavity, and a small portion of the floors of the orbits (see Figures 7.4 and 7.9).

Inferior Nasal Conchae

The two **inferior nasal conchae** are inferior to the middle nasal conchae of the ethmoid bone and are not part of the ethmoid bone (see Figures 7.2 and 7.8a). These scroll-like bones form a part of the inferior lateral wall of the nasal cavity and project into the nasal cavity. All three pairs of nasal conchae (superior, middle, inferior) increase the surface area of the nasal cavity and help swirl and filter air before it passes into the lungs.

Vomer

The **vomer** (VŌ-mer = plowshare) is a roughly triangular bone on the floor of the nasal cavity that articulates superiorly with the perpendicular plate of the ethmoid bone and inferiorly with both the maxillae and palatine bones (see Figures 7.2, 7.4, and 7.9). It forms the inferior portion of the nasal septum.

Figure 7.10   Mandible.

Right lateral view

 The mandible is the largest and strongest facial bone.

Mandible

The **mandible** (*mand-* = to chew), or lower jawbone, is the largest, strongest facial bone (see Figures 7.2, 7.3, and 7.10). It is the only freely movable skull bone (other than the auditory ossicles). In the lateral view, you can see that the mandible consists of a curved, horizontal portion, the *body*, and two perpendicular portions, the *rami* (RĀ-mī = branches; singular is *ramus*). The *angle* of the mandible is the area where each *ramus* meets the body. Each ramus has a posterior *condylar process* (KON-di-lar) that articulates with the mandibular fossa of the temporal bone (see Figure 7.3) to form the *temporomandibular joint (TMJ)*. It also has an anterior *coronoid process* (KOR-ō-noyd) to which the temporalis muscle attaches (Figure 7.10). The depression between the coronoid and condylar processes is called the *mandibular notch*. The *alveolar process* is an arch containing the *alveoli* (sockets) for the mandibular (lower) teeth.

The *mental foramen* (*ment-* = chin) is a hole in the mandible that dentists use when injecting anesthetics. Another foramen associated with the mandible is the *mandibular foramen* on the medial surface of each ramus, another site often used by dentists to inject anesthetics. Through the mandibular foramen pass nerves and blood vessels that supply the mandibular teeth.

Checkpoint

14. On which bones do eyeglasses normally rest?
15. What bones form the roof of the mouth? What is the anatomical name for the bony roof of the mouth?
16. Through which bones do tears travel from the eyes to the nasal cavity?
17. What anatomical structure do the maxillae and mandible have in common?
18. Which bone forms the inferior part of the nasal septum?
19. Which bone attaches to other skull bones by freely movable joints?
20. What is the anatomical name for the lateral corner of your jaw?

Hassan and Baru quickly climbed down into the excavation pit and rushed over to the pile of skeletal remnants. "There may be men, women, and even children here. We need to assess the surface features for anatomic differences." Baru pointed to a small smooth-domed skull. "See this one? The smoothness of the frontal bone above the eye sockets tells me that it is likely a female skull."

Baru and Hassan mobilized as many students as they could. The media had been alerted, and Baru explained to a reporter, "There is evidence of a mass slaughter. Many skulls show signs of blows to the back of the head and unhealed fractures. But these remains tell us much more than how they died."

"There's this poor woman." Baru picked up a skull. The orbit of the left side was enlarged, with rough, bony bumps above the eye socket. "This deformity shows the presence of a large tumor. And this is how it was treated." Baru turned the skull delicately in his hands. A fairly large hole had been cut into the parietal bone. "Trepanation. The hole would have been cut with stone or bronze surgical tools, without anesthesia. The bone has healed around the wound, suggesting a successful procedure. We have skeletal remains from many different social classes—slaves, workers, priests, soldiers. The data we gather will help us to understand not only how they died but who they were, what their health status was, and how they lived."

Hassan walked over to a nearby table where Liu Han, a postdoctoral student, was working. "What do you have there, Liu?" asked Hassan.

She replied, "Have a look at this frontal bone. Adult male, late twenties to early thirties, I'd guess. There's a metopic suture present. That's not the only one we've found. We have three other adult skulls with metopic sutures. Even in populations with a high incidence, they are found in only 5 to 10% of the people."

"What happened to this poor soul?" asked Hassan, pointing to a skull. "It looks like an injury to the orbit. The orbit was crushed laterally and the nasal bone on that side was broken, but they appear to have healed."

Liu examined the skull. "A lot of the smaller skeletal structures become damaged and worn, eroding with time. A bone inside the nasal cavity has worn or broken away. But we can still find gross changes such as the damage from this fracture," she said.

A. Baru is using surface markings to identify the gender of a skull. What two major types of surface markings do bones have?

B. Why are Liu and Hassan surprised to find a metopic suture on an adult skull? In which skull bone do metopic sutures occur?

C. What delicate skeletal structures are found inside the nasal cavity that might be missing from an excavated skull?

7.5 Unique features of the skull include the nasal septum, orbits, sutures, paranasal sinuses, and fontanels.

The skull exhibits several unique features and a structural organization not seen in other bones of the body. These include the nasal septum, orbits, paranasal sinuses, sutures, and fontanels.

Nasal Septum

The inside of the nose, called the nasal cavity, is divided into right and left sides by a vertical partition called the **nasal septum**. The three components of the nasal septum are the vomer, septal cartilage, and perpendicular plate of the ethmoid bone (see Figure 7.9). The anterior border of the vomer articulates with the septal cartilage, which is hyaline cartilage, to form the anterior portion of the nasal septum. The superior border of the vomer articulates with the perpendicular plate of the ethmoid bone to form the remainder of the nasal septum. The term *broken nose*, in most cases, refers to damage to the septal cartilage rather than the nasal bones themselves.

Orbits

Seven bones of the skull join to form each **orbit** (eye socket) (Figure 7.11). The three cranial bones of the orbit are the frontal, sphenoid, and ethmoid; the four facial bones are the palatine, zygomatic, lacrimal, and maxilla. Each orbit contains an eyeball and its associated structures.

Associated with each orbit are five openings:

• The *supraorbital foramen* is found on the supraorbital margin of the frontal bone.

• The *optic foramen* is in the sphenoid bone.

• The *superior orbital fissure* is in the sphenoid bone.

• The *inferior orbital fissure* is located between the sphenoid bone, zygomatic bone, and maxilla.

• The *lacrimal fossa* is in the lacrimal bone.

Figure 7.11 **Details of the orbit.**

FRONTAL BONE

Supraorbital margin

SPHENOID BONE

Optic foramen

Superior orbital fissure

Zygomaticofacial foramen

ZYGOMATIC BONE

Infraorbital foramen

Supraorbital foramen

NASAL BONE

LACRIMAL BONE

ETHMOID BONE

Lacrimal fossa

MAXILLA

Inferior orbital fissure

🔑 **The orbit is a cavity that contains the eyeball and associated structures.**

Paranasal Sinuses

The **paranasal sinuses** (*para-* = beside), mucous membrane-lined cavities near the nasal cavity, are found within the frontal, sphenoid, ethmoid, and maxillary bones (Figure 7.12). The paranasal sinuses lighten the mass of the skull and increase the surface area of the nasal mucosa to help moisten and cleanse inhaled air. In addition, the paranasal sinuses serve as resonating (echo) chambers that intensify and prolong sounds, thereby enhancing the quality of the voice. When you have a cold, excess mucus blocks the sinus passageways through which sound travels, changing the quality of your voice.

Figure 7.12 **Paranasal sinuses.**

Frontal sinus

Ethmoidal cells of ethmoidal sinus

Sphenoidal sinus

Maxillary sinus

(a) Anterior view

(b) Right lateral view

🔑 **Paranasal sinuses are mucous membrane-lined spaces in the frontal, sphenoid, ethmoid, and maxillary bones that connect to the nasal cavity.**

Sutures

Sutures (SOO-churs = seams), found only between skull bones, hold most skull bones together. Sutures in the skulls of infants and children often are movable, but those in an adult usually form immovable joints. Of the many sutures found in the skull, we will identify only four prominent ones:

- The **coronal suture** (kō-RŌ-nal; *coron-* = crown) unites the frontal bone and both parietal bones (see Figure 7.3).

- The **sagittal suture** (SAJ-i-tal; *sagitt-* = arrow) unites the two parietal bones on the superior midline of the skull (see Figure 7.6). The sagittal suture is so named because in the infant, before the bones of the skull are firmly united, the suture and the fontanels (soft spots) associated with it resemble an arrow.

- The **lambdoid suture** (LAM-doyd) unites the two parietal bones to the occipital bone. This suture is so named because of its resemblance to the Greek letter lambda (λ), as can be seen in Figure 7.6. The sagittal and lambdoid sutures may contain small bones called sutural bones (SOO-chur-al; *sutur-* = seam).

- The **squamous sutures** (SKWĀ-mus; *squam-* = flat) unite the parietal and temporal bones on the lateral sides of the skull (see Figure 7.3).

Figure 7.13   Fontanels at birth.

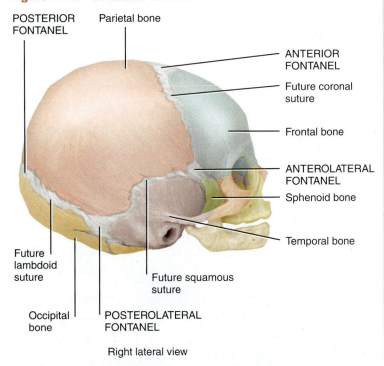

Right lateral view

🔑 Fontanels are mesenchyme-filled spaces between cranial bones that are present at birth.

Fontanels

The skeleton of an embryo consists of cartilage or mesenchyme arranged in sheetlike layers that resemble membranes shaped like the bones they will become. Gradually, osseous tissue replaces the cartilage and mesenchyme. At birth, mesenchyme-filled spaces called **fontanels** (fon-ta-NELZ = little fountains) are present between the cranial bones (Figure 7.13). Commonly called "soft spots," fontanels are areas of unossified mesenchyme. Eventually, they will be replaced with osseous tissue by intramembranous ossification to become sutures. Functionally, fontanels provide some flexibility to the fetal skull, allowing the skull to change shape as it passes through the birth canal and later permitting rapid growth of the brain during infancy. Although an infant may have many fontanels at birth, the form and location of six are fairly constant:

- The unpaired **anterior fontanel**, located at the midline between the two parietal bones and the frontal bone, is the largest fontanel. It usually closes 18 to 24 months after birth.

- The unpaired **posterior fontanel** is located at the midline between the two parietal bones and the occipital bone. It generally closes about 2 months after birth.

- The paired **anterolateral fontanels**, located laterally between the frontal, parietal, temporal, and sphenoid bones, are small. Normally, they close about 3 months after birth.

- The paired **posterolateral fontanels** are located laterally between the parietal, occipital, and temporal bones. They begin to close 1 to 2 months after birth, but closure is generally not complete until 12 months.

The amount of closure in fontanels helps a physician gauge the degree of brain development. In addition, the anterior fontanel serves as a landmark for withdrawal of blood for analysis from the superior sagittal sinus (a large vein on the midline surface of the brain).

Checkpoint

21. What structures make up the nasal septum? What kind of tissue composes the anterior portion of the nasal septum?
22. What are the locations and functions of the paranasal sinuses?
23. What bones are joined by the squamous suture? The lambdoid suture? The coronal suture? The sagittal suture?
24. Which fontanel is bordered by four different skull bones?
25. Following the birth of a baby, how many months are needed before all fontanels are ossified into sutures?

7.7   The vertebral column protects the spinal cord, supports the head, and is a point of attachment for the ribs, bones of the lower limbs, and back muscles.

The **vertebral column**, also called the *spine*, makes up about two-fifths of your total height and is composed of a series of bones called **vertebrae** (VER-te-brē; singular is *vertebra*) and connective tissue. The vertebral column, sternum, and ribs form the skeleton of the trunk of the body. The length of the column is about 71 cm (28 in.) in an average adult male and about 61 cm (24 in.) in an average adult female. The vertebral column functions as a strong, flexible rod with elements that can rotate and move forward, backward, and sideways. In addition to enclosing and protecting the spinal cord, the vertebral column supports the head, and serves as a point of attachment for the ribs, pelvic girdle, and muscles of the back.

The total number of vertebrae during early development is 33. Then, several vertebrae in the sacral and coccygeal regions fuse. As a result, the vertebrae of the adult vertebral column typically are distributed as follows:

7 **cervical vertebrae** (*cervic-* = neck) are in the neck region.

12 **thoracic vertebrae** (*thorax* = chest) are posterior to the thoracic cavity.

5 **lumbar vertebrae** (*lumb-* = loin) support the lower back.

1 **sacrum** (SĀ-krum = sacred bone) consists of five fused **sacral vertebrae**.

1 **coccyx** (KOK-siks = cuckoo, because the shape resembles the bill of a cuckoo bird), consists of four fused **coccygeal vertebrae** (kok-SIJ-ē-al).

The cervical, thoracic, and lumbar vertebrae are movable, but the sacrum and coccyx are not. We discuss each of these regions in detail shortly.

Normal Curves of the Vertebral Column

When viewed from the side, the adult vertebral column shows four slight bends called **normal curves** (Figure 7.15b). Relative to the front of the body, the *cervical* and *lumbar curves* are convex (bulging anteriorly), and the *thoracic* and *sacral curves* are concave (curving posteriorly). The curves of the vertebral column increase its strength, help maintain balance in the upright position, absorb shocks during walking, and help protect the vertebrae from fracture.

Intervertebral Discs

Between the bodies of adjacent vertebrae from the second cervical vertebra to the sacrum are **intervertebral discs** (Figure 7.15c). Each disc has an outer fibrous ring consisting of fibrocartilage called the *annulus fibrosus* (*annulus* = ring or ringlike) and an inner soft, pulpy, highly elastic substance called the *nucleus pulposus* (*pulposus* = pulplike). The discs form strong joints, permit various movements of the vertebral column, and absorb vertical shock. Under compression, they flatten and broaden. During the course of a day the discs compress so that we are a bit shorter at night. While we are sleeping there is less compression so that we are taller when we awaken in the morning. As we age, the nucleus pulposus hardens and becomes less elastic. Narrowing of the discs and compression of vertebrae result in a decrease in height with age.

Since intervertebral discs are avascular, the annulus fibrosus and nucleus pulposus rely on blood vessels from the bodies of vertebrae to obtain oxygen and nutrients and remove wastes. Certain stretching exercises, such as yoga, decompress discs and increase blood circulation, both of which speed up the uptake of oxygen and nutrients by discs and the removal of wastes.

Checkpoint

27. What are the functions of the vertebral column?

28. Name the five types of vertebrae and state the number of each type. How are the vertebrae of the sacrum and coccyx functionally different from those of the rest of the vertebral column?

29. What are the functions of the normal curves of the vertebral column? Which curves of the adult vertebral column are concave (relative to the anterior side of the body)?

30. What are the functions of the intervertebral discs?

7.6 The hyoid bone supports the tongue and attaches to tongue, pharynx, and larynx muscles.

The single **hyoid bone** (= U-shaped) is a unique component of the axial skeleton because it does not articulate with any other bone (see Figure 7.5). Rather, it is suspended from the styloid processes of the temporal bones by ligaments and muscles (see Figure 11.6). Located in the anterior neck between the mandible and larynx (Figure 7.14a), the hyoid bone supports the tongue and provides attachment sites for some tongue muscles and for muscles of the pharynx and larynx. The hyoid bone also helps to keep the larynx (voice box) open. The hyoid bone consists of a *body* and paired projections called the *lesser horns* and the *greater horns*

(Figure 7.14b,c). Muscles and ligaments attach to these projections.

The hyoid bone and cartilages of the larynx and trach often fractured during strangulation. Thus, they are carefu amined at autopsy when strangulation is suspected.

Checkpoint

26. In what way is the hyoid bone different from all the other bones of the skeleton?

Figure 7.14   Hyoid bone.

(a) Position of hyoid

Mandible

Hyoid bone

Larynx

(b) Anterior view

Gre
ho

Le
ho

B

(c) Right lateral view

Greater horn

Lesser horn

Body

🔑 **The hyoid bone supports the tongue and provides attachment sites for muscles of the tongue, neck, and pharynx.**

Figure 7.15 **Vertebral column.** The numbers in parentheses in (a) indicate the number of vertebrae in each region. In (c), the relative size of the disc has been enlarged for emphasis.

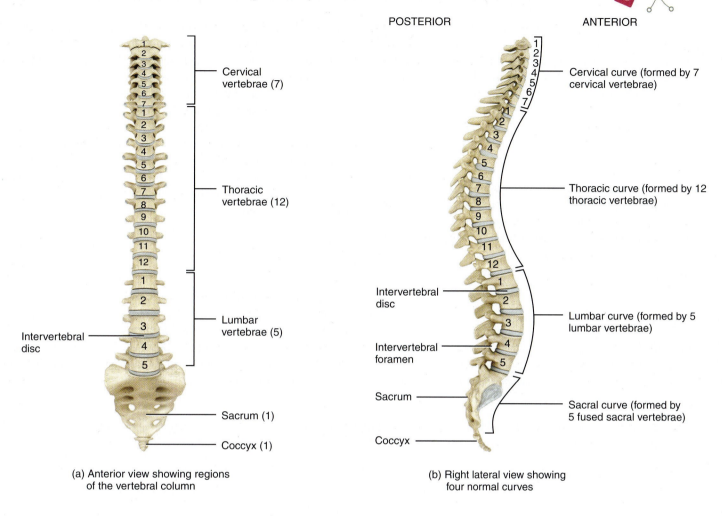

POSTERIOR · ANTERIOR

Cervical vertebrae (7)

Thoracic vertebrae (12)

Intervertebral disc

Lumbar vertebrae (5)

Sacrum (1)

Coccyx (1)

(a) Anterior view showing regions of the vertebral column

Cervical curve (formed by 7 cervical vertebrae)

Thoracic curve (formed by 12 thoracic vertebrae)

Intervertebral disc

Lumbar curve (formed by 5 lumbar vertebrae)

Intervertebral foramen

Sacrum

Sacral curve (formed by 5 fused sacral vertebrae)

Coccyx

(b) Right lateral view showing four normal curves

Intervertebral foramen

Vertebral body

Annulus fibrosus

Normal intervertebral disc

Compressed intervertebral disc in a weight-bearing situation

(c) Intervertebral disc

🔑 **The adult vertebral column typically contains 26 vertebrae.**

7.8   A vertebra usually consists of a body, a vertebral arch, and several processes.

Even though vertebrae in different regions of the vertebral column vary in size, shape, and detail, they are similar enough that we can discuss the structures (and the functions) of a typical vertebra (Figure 7.16a). Vertebrae typically consist of a body, a vertebral arch, and several processes.

Body

The **body** is the thick, disc-shaped anterior portion that is the weight-bearing part of a vertebra. Its superior and inferior surfaces are roughened for the attachment of intervertebral discs. The anterior and lateral surfaces contain nutrient foramina through which blood vessels supply the osseous tissue.

Vertebral Arch

The **vertebral arch** extends posteriorly from the body of the vertebra and together with the body of the vertebra surrounds the spinal cord. Two short, thick processes, the *pedicles* (PED-i-kuls = little feet), form the lateral portions of the vertebral arch. The pedicles project posteriorly from the body to unite with the laminae. The *laminae* (LAM-i-nē = thin layers) are the flat parts that join to form the posterior portion of the vertebral arch. The *vertebral foramen* lies between the vertebral arch and body and contains the spinal cord, adipose tissue, areolar connective tissue, and blood vessels. Collectively, the vertebral foramina of all vertebrae form the **vertebral canal**. When the vertebrae are stacked on top of one another, there is an opening between adjoining vertebrae on both sides of the column (Figure 7.15). Each opening, called an *intervertebral foramen*, permits the passage of a single spinal nerve to a specific region of the body.

Processes

Seven **processes** arise from the vertebral arch (Figure 7.16). At the point where a lamina and pedicle join, a *transverse process* extends laterally on each side. A single *spinous process* projects posteriorly from the junction of the laminae. These three processes serve as points of attachment for muscles. The remaining four processes form joints with other vertebrae above or below. The two

Figure 7.16   Structure of a typical vertebra, as illustrated by a thoracic vertebra. In (b), only one spinal nerve has been included, and it has been extended beyond the intervertebral foramen for clarity.

(a) Superior view

(b) Right posterolateral view of articulated vertebrae

🔑 **A vertebra consists of a body, a vertebral arch, and several processes.**

superior articular processes of a vertebra articulate (form joints) with the two inferior articular processes of the vertebra immediately above them. In turn, the two *inferior articular processes* of that vertebra articulate with the two superior articular processes of the vertebra immediately below them. The articulating surfaces of the articular processes are referred to as *facets* (FAS-et or fa-SET = little faces) and are covered with hyaline cartilage.

Checkpoint

31. What part of a vertebra is most responsible for supporting a person's weight?
32. What are the components of the vertebral arch?
33. Through what foramen does the spinal cord pass?
34. To which processes do muscles attach? Which processes attach to other vertebrae? Which parts of processes physically articulate with other bones?

Return to Hassan's Story

When they did finally find it, the process of uncovering the tomb entrance was slow and meticulous. A sarcophagus was discovered immediately inside the entrance, with the image of a young woman holding a baby painted on its surface. Hassan looked more closely.

"There is a royal seal here on the front. But who are the woman and child?"

"Professor, what do these symbols say?" asked Liu, motioning to a series of hieroglyphs on the lid of the sarcophagus.

"It will take time to decipher all the writings. From what I can make out, this may be the King's wife and child."

Hassan motioned for his men to place their steel pry bars beneath the lid. He stepped back as the men struggled to lift the lid slowly and carefully.

The men stood back as the three scientists moved forward, peering into the stone box that now lay open before them. Inside were two skeletons, a small, delicately framed woman with arms folded around what appeared to be a child.

"They're not mummified!" exclaimed Liu.

Hassan shined his light up and down the interior of the sarcophagus. "Yes, but the preservation is still remarkable considering the age of the tomb. Look at their heads. There is still hair on the woman's skull, and fontanels are visible on the child's head. Baru, what do you think? Mother and child?"

Baru scrutinized the contents of the sarcophagus closely. "We will need to X-ray the remains to estimate the age of the woman, and a DNA analysis will tell if they are related. Look here." Baru gently touched one of the woman's thoracic vertebrae. "These lesions suggest bone tuberculosis, which is most common in children. We have found several examples of tubercular lesions in vertebrae and long bones in the remains outside the tomb. The vertebral bodies and intervertebral discs collapse as the bacteria invade the osseous tissue. Notice that the body of this vertebra shows significant porosity, the articular processes are damaged, and the articulating surfaces look arthritic."

"But this is not a child—a teenage girl perhaps, but not a child," stated Liu.

"Baru, look at this," said Liu. "The sacrum is partially fused, yet I would say this girl is perhaps only 16 or 17 years old."

Baru turned his attention to the skeletons again. "Liu, make sure you get good pictures of the skulls."

Liu noticed a small fracture in the frontal bone. The curve of the child's neck was unnatural, she thought.

D. *How would Hassan and his team be able to tell the ages of the skeletal remains of the woman and the baby?*

E. *What features of the vertebral column would the larger skeleton in the sarcophagus show to indicate it was female?*

F. *What bone in the neck region other than cervical vertebrae may be damaged during strangulation or neck trauma?*

7.9 Vertebrae in the different regions of the vertebral column vary in size, shape, and detail.

We turn now to the five regions of the vertebral column, beginning superiorly and moving inferiorly. The regions are the cervical, thoracic, lumbar, sacral, and coccygeal. Note that vertebrae in each region are numbered in sequence, from superior to inferior. When you actually view the bones of the vertebral column, you will notice that the transition from one region to the next is not abrupt but gradual, a feature that helps vertebrae fit together.

Cervical Region

The bodies of **cervical vertebrae** (C1–C7) are smaller than all other vertebrae except those that form the coccyx (Figure 7.17a). Typical cervical vertebrae have three foramina: one vertebral foramen and two transverse foramina (Figure 7.17b,d). The vertebral foramina of cervical vertebrae are the largest in the spinal column because they house the cervical enlargement of the spinal cord. The *transverse foramina* serve as a passageway for the vertebral artery that supplies blood to the brain. The spinous processes of C2 through C6 are often *bifid*—that is, split into two parts (Figure 7.17a,d).

The first two cervical vertebrae differ considerably from the others. Named after the mythological Atlas, who supported the world on his shoulders, the first cervical vertebra (C1), the **atlas**, supports the head (Figure 7.17a,b). The atlas lacks a body and a spinous process. Instead, it consists of a ring of bone with *anterior* and *posterior arches*. The superior, lateral surfaces contain *superior articular facets* that articulate with the occipital condyles of the occipital bone. These articulations permit the movement seen when moving the head to signify "yes." The inferior lateral surfaces contain the *inferior articular facets* that articulate with the second cervical vertebra.

The second cervical vertebra (C2), called the **axis** (Figure 7.17a,c), does have a body. A peglike process on the body, the *dens* (= tooth), projects superiorly through the anterior portion of the vertebral foramen of the atlas. The dens makes a pivot on which the atlas and head rotate. This articulation between the anterior arch of the atlas and dens of the axis permits side-to-side movement of the head, as when you move your head to signify "no." In some instances of trauma, the dens of the axis may be driven into the medulla oblongata of the brain. This type of injury is the usual cause of death from whiplash injuries.

Figure 7.17   Cervical vertebrae.

Location of cervical vertebrae

(a) Posterior view of articulated cervical vertebrae

Posterior arch

Superior articular
facet

Vertebral foramen

Transverse foramen

Articular surface
for dens of axis

Anterior arch

Transverse
process

ANTERIOR

(b) Superior view of atlas (C1)

POSTERIOR

Spinous process

Lamina

Vertebral
foramen

Dens

Inferior articular
facet

Transverse process

Superior articular
facet

Body

ANTERIOR

(c) Superior view of axis (C2)

POSTERIOR

Bifid spinous
process

Lamina

Vertebral
foramen

Transverse
process

Transverse
foramen

Superior articular
facet

Pedicle

Body

ANTERIOR

(d) Superior view of a typical cervical vertebra

The cervical vertebrae are found in the neck region.

The third through sixth cervical vertebrae (C3–C6), represented by the vertebra in Figure 7.17d, correspond to the structural pattern of the typical cervical vertebra previously described. The seventh cervical vertebra (C7), called the *vertebra prominens*, has a single large spinous process that may be seen and felt at the base of the posterior neck (Figure 7.17a).

Thoracic Region

Thoracic vertebrae (T1–T12; Figures 7.16 and 7.18) are considerably larger and stronger than cervical vertebrae. In addition, the spinous processes on T1 through T10 are long, laterally flat-tened, and directed inferiorly. In contrast, the spinous processes on T11 and T12 are shorter, broader, and directed more posteriorly. Compared to cervical vertebrae, thoracic vertebrae have longer transverse processes (see Figure 7.15a).

The most distinguishing feature of thoracic vertebrae is that they articulate with the ribs. As you can see in Figure 7.18, the bodies of thoracic vertebrae have either facets or demifacets (half facets) for articulation with the *heads* of the ribs (see Figure 7.22c). The transverse processes of most thoracic vertebrae also have facets for articulating with the *tubercles* of the ribs (see Figure 7.22c). Movements of the thoracic region are limited by the attachment of the ribs to the sternum.

Figure 7.18 Thoracic vertebrae.

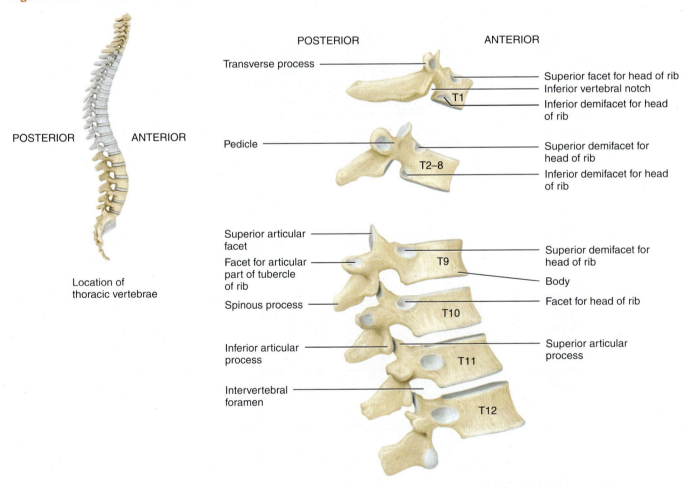

(a) Right lateral view of several articulated thoracic vertebrae

Lumbar Region

The **lumbar vertebrae** (L1–L5) are the largest and strongest of the unfused bones in the vertebral column (Figure 7.19a-c) because the amount of body weight supported by the vertebrae increases toward the inferior end of the spine. Their processes are short and thick. The superior articular processes are directed medially instead of superiorly, and the inferior articular processes are directed laterally instead of inferiorly. The spinous processes are thick and broad, and project posteriorly. The spinous processes are well adapted for the attachment of the large back muscles.

A summary of the major structural differences among cervical, thoracic, and lumbar vertebrae is presented in Table 7.4.

Sacrum

The **sacrum** is a triangular bone formed by the union of the five sacral vertebrae (Figure 7.20). The fusion of the sacral vertebrae begins between ages 16 and 18 and is usually completed by age

30. Positioned at the posterior portion of the pelvic cavity medial to the two hip bones, the sacrum serves as a strong foundation for the pelvic girdle. To accommodate pregnancy and childbirth, the female sacrum is shorter, wider, and more curved than the male sacrum.

The concave, anterior side of the sacrum faces the pelvic cavity. It is smooth and contains four *transverse lines* that mark the joining of the sacral vertebral bodies (Figure 7.20a). At the ends of these lines are four pairs of *sacral foramina* that open to both the anterior and posterior surfaces and through which nerves and blood vessels pass. Extending laterally on either side from the superior surface is a winglike *sacral ala* (= wing).

The convex, posterior surface of the sacrum contains a *median sacral crest*, which is the fused spinous processes of sacral vertebrae, and a *lateral sacral crest*, which is the fused transverse processes of sacral vertebrae (Figure 7.20b). The *sacral canal* is continuous with the vertebral canal. The laminae of the fifth sacral vertebra, and sometimes the fourth, fail to meet. This leaves an inferior entrance to the vertebral canal called the *sacral hiatus* (hī-Ā-tus = opening).

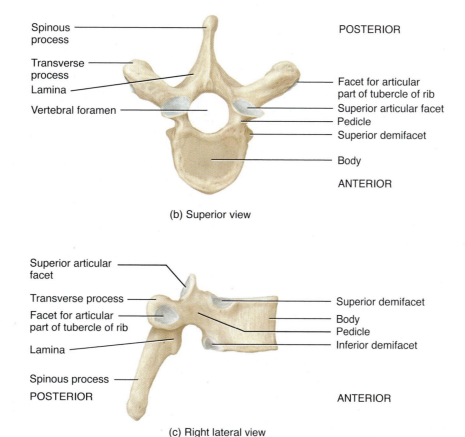

(b) Superior view

(c) Right lateral view

🔑 **The thoracic vertebrae are found in the chest region and articulate with the ribs.**

Figure 7.19   Lumbar vertebrae.

POSTERIOR   ANTERIOR

Intervertebral foramen

Intervertebral disc

Superior articular process

Transverse process

Body

Spinous process

Inferior articular facet

POSTERIOR   ANTERIOR

Location of
lumbar vertebrae

(a) Right lateral view of articulated lumbar vertebrae

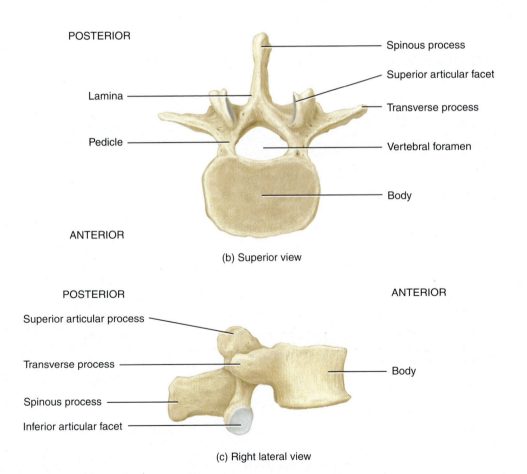

POSTERIOR

Spinous process

Superior articular facet

Lamina

Transverse process

Pedicle

Vertebral foramen

Body

ANTERIOR

(b) Superior view

POSTERIOR   ANTERIOR

Superior articular process

Transverse process

Body

Spinous process

Inferior articular facet

(c) Right lateral view

Lumbar vertebrae are found in the lower back.

TABLE 7.4

Comparison of Major Structural Features of Cervical, Thoracic, and Lumbar Vertebrae

CHARACTERISTIC	CERVICAL	THORACIC	LUMBAR
Overall structure	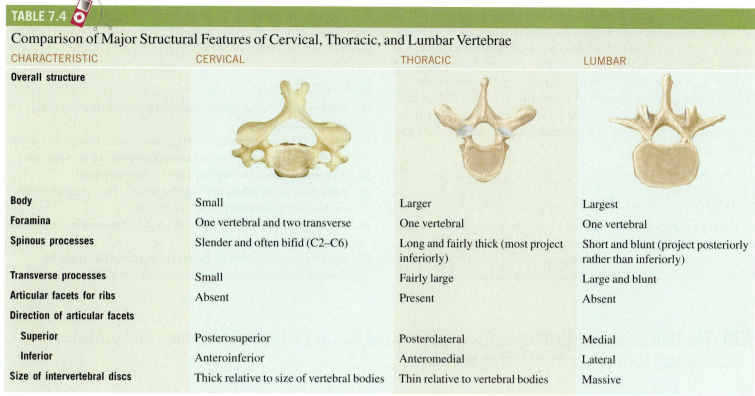		
Body	Small	Larger	Largest
Foramina	One vertebral and two transverse	One vertebral	One vertebral
Spinous processes	Slender and often bifid (C2–C6)	Long and fairly thick (most project inferiorly)	Short and blunt (project posteriorly rather than inferiorly)
Transverse processes	Small	Fairly large	Large and blunt
Articular facets for ribs	Absent	Present	Absent
Direction of articular facets			
Superior	Posterosuperior	Posterolateral	Medial
Inferior	Anteroinferior	Anteromedial	Lateral
Size of intervertebral discs	Thick relative to size of vertebral bodies	Thin relative to vertebral bodies	Massive

Figure 7.20   Sacrum and coccyx.

(a) Anterior view

(b) Posterior view

The sacrum is formed by the union of five sacral vertebrae, and the coccyx is formed by the union of four coccygeal vertebrae.

The anteriorly projecting border of the superior sacrum, called the *sacral promontory* (PROM-on-tō-rē), is a landmark used for measuring the pelvis. On both lateral surfaces, the sacrum has a large *auricular surface* (*auricular* = ear) that articulates with the ilium of each hip bone to form the sacroiliac joint, which contains depressions for the attachment of ligaments. The auricular surface articulates with the hip bones. The *superior articular processes* of the sacrum articulate with the inferior articular processes of the fifth lumbar vertebra.

Coccyx

The **coccyx** is also triangular and is formed by the fusion of usually four coccygeal vertebrae, indicated in Figure 7.20 as Co1–Co4. The coccygeal vertebrae fuse when a person is between 20 and 30 years of age. The *superior articular processes* of the first coc-cygeal vertebra connect by ligaments to the sacrum. In females, the coccyx points inferiorly to allow the passage of a baby during birth; in males, it points anteriorly.

Checkpoint

35. Which vertebrae contain transverse foramina? What function is served by transverse foramina?
36. How are the spinous processes different among cervical, thoracic, and lumbar vertebrae? Which cervical vertebra has the most prominent spinous process?
37. What are the names of the first and second cervical vertebrae?
38. Which bones permit nodding of the head to indicate "yes"? Which permit the rotational movement of the head to indicate "no"?
39. Which vertebrae articulate with ribs? Which parts of those vertebrae articulate with the ribs?
40. Name the superior and inferior openings of the sacrum that extend the vertebral column. Which sacral structures unite with the hip bones?

7.10 The thoracic cage protects vital organs in the thorax and upper abdomen and provides support for the bones of the upper limbs.

The term **thorax** refers to the entire chest. The skeletal part of the thorax, the **thoracic cage**, is a bony enclosure formed by the sternum, ribs and their costal cartilages, and the bodies of the thoracic vertebrae (Figure 7.21). The thoracic cage is narrower at its supe-rior end and broader at its inferior end. It encloses and protects the organs in the thoracic and superior abdominal cavities and provides support for the bones of the upper limbs.

Figure 7.21   Skeleton of the thorax.

(a) Anterior view of sternum

(b) Anterior view of skeleton of thorax

🔑 The bones of the thorax enclose and protect organs in the thoracic cavity and upper abdominal cavity.

Sternum

The **sternum**, or breastbone, is a flat, narrow bone located in the center of the anterior thoracic wall that measures about 15 cm (6 in.) in length and consists of three parts (Figure 7.21). The superior part is the **manubrium** (ma-NOO-brē-um = handlelike); the middle and largest part is the **body**; and the inferior, smallest part is the **xiphoid process** (ZĪ-foyd = sword-shaped). The parts of the sternum typically fuse by age 25, and the points of fusion are marked by transverse ridges.

The junction of the manubrium and body forms the *sternal angle*. The sternal angle is a common clinical landmark for identifying the second rib, a starting point for counting ribs to assist in locating underlying structures such as heart valves. The manubrium has a depression on its superior surface, the *suprasternal notch*. Lateral to the suprasternal notch are *clavicular notches* that articulate with the medial ends of the clavicles. The manubrium also articulates with the costal cartilages of the first and second ribs.

The body of the sternum articulates directly or indirectly with the costal cartilages of the second through tenth ribs. The xiphoid process consists of hyaline cartilage during infancy and childhood and does not ossify completely until about age 40. No ribs are attached to it, but the xiphoid process provides attachment for some abdominal muscles. Incorrect positioning of the hands of a rescuer during cardiopulmonary resuscitation (CPR) may fracture the xiphoid process, driving it into internal organs.

Ribs

Twelve pairs of **ribs** give structural support to the sides of the thoracic cavity (Figure 7.21b). The ribs increase in length from the first through seventh, then decrease in length to the twelfth rib. Each articulates posteriorly with its corresponding thoracic vertebra.

The first through seventh pairs of ribs have a direct anterior attachment to the sternum by a strip of hyaline cartilage called *costal cartilage* (*cost-* = rib). The costal cartilages contribute to the elasticity of the thoracic cage to allow breathing and to prevent various blows to the chest from fracturing the sternum and/or ribs. The ribs that have costal cartilages and attach directly to the sternum are called *true ribs*. The remaining five pairs of ribs are termed *false ribs,* because their costal cartilages either attach indirectly to the sternum or do not attach to the sternum at all. The cartilages of the eighth, ninth, and tenth pairs of false ribs attach to one another and then to the cartilages of the seventh pair of ribs. The eleventh and twelfth false ribs are also known as as *floating ribs* because the costal cartilage at their anterior ends does not attach to the sternum at all. Floating ribs attach only posteriorly to the thoracic vertebrae.

Figure 7.22a shows the parts of a typical (third through ninth) rib. The *head* is a projection at the posterior end of the rib that contains *superior* and *inferior facets*.

Figure 7.22 **The structure of ribs.**

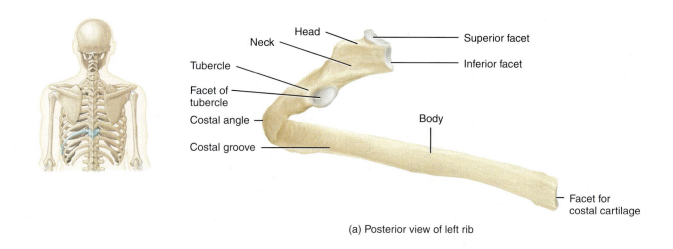

(a) Posterior view of left rib

(continues)

Figure 7.22 (continued)

(b) Posterior view of left ribs articulated with thoracic vertebrae and sternum

(c) Superior view of left rib articulated with thoracic vertebra

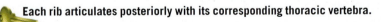

🔑 **Each rib articulates posteriorly with its corresponding thoracic vertebra.**

The *neck* is a constricted portion just lateral to the head. A knoblike structure on the posterior surface, where the neck joins the body, is called the *tubercle* (TOO-ber-kul). The *body* is the main shaft of the rib. A short distance beyond the tubercle, an abrupt change in the curvature of the body occurs. This point is called the *costal angle*. The inner surface of the rib has a *costal groove* that protects blood vessels and a small nerve.

The posterior portion of the rib connects to a thoracic vertebra by its head and the tubercle. The facet of the head fits into a facet on the body of a single vertebra or into the demifacets of two adjoining vertebrae. The facet of the tubercle articulates with the facet of a transverse process of the vertebra (Figure 7.22c).

Spaces between ribs, called *intercostal spaces*, are occupied by intercostal muscles, blood vessels, and nerves. Surgical access to the lungs or other structures in the thoracic cavity is commonly obtained through an intercostal space. Special rib retractors are used to create a wide separation between ribs. The costal cartilages are sufficiently elastic in younger individuals to permit considerable bending without breaking.

Checkpoint

41. From superior to inferior, what are the three components of the sternum?
42. What bones articulate with the manubrium? What structures articulate with the body?
43. What is the function of the xiphoid process?
44. What component of the thoracic cage allows the thorax to expand and contract during breathing?
45. How does a rib articulate with a thoracic vertebra? What is the space between two ribs called, and what is its clinical importance?

Hassan's Story: Epilogue and Discussion

▶▶▶ "It looks as if the child's neck may have been broken," Liu announced as she snapped another picture. "Baru, look at the transverse processes in the mid-cervical region. They appear crushed."

Hassan had been looking at the pictographs on the walls surrounding the sarcophagus. "I think you may be right, Liu. These scenes depict some kind of tragedy. The land had been plagued by drought, famine, and disease, and the death of the princess is repeated through the scenes. To appease the gods, the King offered up his firstborn child."

Liu put her camera down. "That's horrible."

Hassan responded, "Those were desperate times."

Our intrepid explorers have uncovered the skeletal remains of a group of ancient people. As we have seen, knowledge of the skeleton is important for many disciplines. Bones are living tissue. They are constantly remodeled and shaped by a person's activities, health, and diet. In the case of Hassan, Baru, and Liu, their

knowledge of skeletal anatomy and physiology has yielded many clues. However, their work has only just begun. In the next chapter we will continue their investigation. The bones will reveal additional clues to help them solve the mystery of the people of the unearthed tomb. As you conclude your study of this chapter, consider the following questions. ◀◀◀

G. If the bones of a person found at the excavation site were mixed up and out of sequence, how could the anthropologists determine which vertebrae were cervical, lumbar, or thoracic?

H. Why would bones, with their hard structure, be subject to and show signs of the disease that destroyed this community?

I. What structure passes through the transverse foramina of cervical vertebrae that would cause death if damaged as in the child's skeleton?

Concept and Resource Summary

Concept	Resources

Introduction

1. The bones, muscles, and joints form an integrated system called the *musculoskeletal system.*

Concept 7.1 Bones of the axial skeleton and appendicular skeleton have characteristic surface markings.

1. The 206 bones of the adult human skeleton can be categorized into two divisions: the **axial skeleton** and the **appendicular skeleton.**
2. The axial skeleton consists of bones around the body's longitudinal axis including skull bones, ribs, sternum, vertebral column, auditory ossicles, and hyoid bone.
3. The appendicular skeleton is composed of the bones of the upper and lower limbs, and the shoulder and hip girdles.
4. Bone **surface markings** are important anatomical features that indicate sites of muscle attachment, ligaments, joints, and passage of blood vessels and nerves.

See
- **Anatomy Overview – The Axial Skeleton**

Concept 7.2 The skull provides attachment sites for muscles and membranes, and protects and supports the brain and sense organs.

1. The **skull** contains 22 bones, 8 **cranial bones** and 14 **facial bones.**
2. The cranial bones protect the brain and sense organs, and allow attachment of membranes and muscles.
3. The facial bones form the face, allow muscle attachment, and protect and support the entrances to the digestive and respiratory tracts.
4. The immovable joints in the skull are called **sutures**, whereas the openings for blood vessels and nerves are called **foramina**.

Concept 7.3 The cranial bones include the frontal, parietal, temporal, occipital, sphenoid, and ethmoid bones.

1. The **frontal** bone forms the forehead and anterior cranial floor, helps form the orbits, and houses the frontal sinuses.
2. The two **parietal bones** make up a large portion of the sides and roof of the cranial cavity.
3. The two **temporal bones** form the inferior lateral sides of the skull and cranial floor. The *mandibular fossa* is a socket that articulates with the mandible, forming the *temporomandibular joint (TMJ).*
4. The **occipital bone** forms the posterior part and most of the base of the skull. Inferiorly, the *foramen magnum* allows the brain to connect with the spinal cord.

Concept   **Resources**

5. The **sphenoid bone** lies in the middle of the base of the skull and articulates with all other cranial bones. It contains sphenoidal sinuses.

6. The **ethmoid bone** is in the anterior part of the cranial floor, between the orbits; the ethmoid bone forms the anterior cranial floor, part of the orbits, part of the nasal septum, and most of the superior side walls of the nasal cavity.

Concept 7.4 Facial bones include the nasal bones, maxillae, zygomatic bones, lacrimal bones, palatine bones, inferior nasal conchae, vomer, and mandible.

1. The two **nasal bones** unite medially to form the bridge of the nose.
2. Paired **maxillae** form the upper jawbone, articulate with every facial bone except the mandible, and contribute to the orbits, nasal cavity, and hard palate.
3. The two **zygomatic bones** form the cheeks and part of the orbits.
4. The smallest facial bones, the paired **lacrimal bones**, contribute to the orbits; each contains the *lacrimal fossa* through which tears pass to the nasal cavity.
5. Paired **palatine bones** help form the hard palate, nasal cavity, and orbits.
6. Scroll-like **inferior nasal conchae** are inferior to the conchae of the ethmoid bone.
7. The **vomer** helps to form the nasal septum.
8. The strongest facial bone, and only freely movable skull bone, is the **mandible**, or lower jawbone.

See
• **Anatomy Overview – Facial Bones**

Concept 7.5 Unique features of the skull include the nasal septum, orbits, sutures, paranasal sinuses, and fontanels.

1. The nasal cavity has a vertical partition called the **nasal septum** formed by the vomer, the perpendicular plate of the ethmoid, and septal cartilage.
2. Each **orbit** houses the eyeball and comprises parts of seven bones: frontal, sphenoid, ethmoid, palatine, zygomatic, lacrimal, and maxilla.
3. **Paranasal sinuses** are cavities lined by mucous membranes found in the frontal, sphenoid, ethmoid, and maxillary bones.
4. The immovable joints between most skull bones are called **sutures**. Four principal sutures are the **coronal suture**, the **sagittal suture**, the **lambdoid suture**, and the **squamous sutures**.
5. **Fontanels** are mesenchyme-filled spaces between skull bones that are not yet fully ossified in the fetus and infant. They allow for skull flexibility during delivery and accommodate brain growth.

See
• **Anatomy Overview – Fibrous Joints and Sutures**

Hear
• **Figure 7.11 – Details of the Orbit**

Concept 7.6 The hyoid bone supports the tongue and attaches to tongue, pharynx, and larynx muscles.

1. The **hyoid bone** in the anterior neck does not articulate with any other bones. It is suspended by ligaments and muscles, and attaches the tongue and muscles of the neck and pharynx.

See
• **Anatomy Overview – The Hyoid Bone**

Concept 7.7 The vertebral column protects the spinal cord, supports the head, and is a point of attachment for the ribs, bones of the lower limbs, and back muscles.

1. The **vertebral column** is composed of a series of **vertebrae** that surround and protect the spinal cord, support the head, and serve as a point of attachment for the ribs, pelvic girdle, and back muscles.
2. The vertebral column includes 7 cervical vertebrae, 12 thoracic vertebrae, 5 lumbar vertebrae, a sacrum, and a coccyx.
3. Except for the sacrum and coccyx, the vertebrae are freely movable.
4. The adult vertebral column has four **normal curves** that increase strength, absorb shock, help maintain balance, and protect against vertebral fracture.
5. **Intervertebral** discs occur between bodies of adjacent vertebrae; they allow vertebral column movement and absorb shock.

See
• **Anatomy Overview – The Vertebral Column**

Hear
• **Figure 7.15 – Vertebral Column**

Concept 7.8 A vertebra usually consists of a body, a vertebral arch, and several processes.

1. The **body** is the weight-bearing portion of the vertebra.
2. The **vertebral arch** has two *pedicles* and two *laminae*.
3. The body and vertebral arch enclose an opening called the *vertebral foramen*. The foramina from all the vertebrae form the **vertebral canal** for the spinal cord.
4. Lateral openings, called *intervertebral foramina*, occur between articulating vertebrae on both sides of the column to allow passage of spinal nerves.
5. The vertebral **processes** include two transverse processes, a spinous process, two superior articular processes, and two inferior articular processes.

Concept	Resources

Concept 7.9 Vertebrae in the different regions of the vertebral column vary in size, shape, and detail.

1. Each transverse process of most **cervical vertebrae** has a *transverse foramen* through which the vertebral arteries pass to supply blood to the brain.
2. The **atlas** (C1) articulates with the occipital condyles of the skull; lacks a body and spinous process; and has modifications that allow head movement to signify "yes."
3. The **axis** (C2) has the *dens,* which projects superiorly to act as a pivot for rotation of the atlas to signify "no."
4. The unique feature of the **thoracic vertebrae** is their articulation with ribs. Their bodies and transverse processes have facets and demifacets for articulation with ribs.
5. The **lumbar vertebrae** have short, thick processes and are the largest, strongest vertebrae.
6. The **sacrum** is formed by the fusion of the five sacral vertebrae. It articulates laterally with the two hip bones and sits posteriorly in the pelvis.
7. The **coccyx**, formed by the fusion of four coccygeal vertebrae, points inferiorly in females and anteriorly in males.

See

- **Anatomy Overview – The Vertebral Column**

Hear

- **Figure 7.20 –** Sacrum and Coccyx
- **Table 7.4 – Comparison of Major Structural Features of Cervical, Thoracic, and Lumbar Vertebrae**

Concept 7.10 The thoracic cage protects vital organs in the thorax and upper abdomen and provides support for the bones of the upper limbs.

1. The **thoracic cage** is composed of the sternum, ribs and their costal cartilages, and bodies of thoracic vertebrae.
2. The **sternum** is in the midline of the anterior thoracic wall. It includes three parts: **manubrium**, **body**, and **xiphoid process**.
3. Twelve pairs of **ribs** articulate posteriorly with the thoracic vertebrae: There are seven pairs of *true ribs* and five pairs of *false ribs*, with the last two pairs being called *floating ribs*.

See

- **Anatomy Overview – The Thorax**

Hear

- **Figure 7.21 –** Skeleton of the Thorax

Do

- **Exercise – Bones, Bones, Bones**

Other Chapter Resources

For additional information on clinical applications, medical disorders, and medical terminology, see Chapter 4 in the *Clinical Connections* manual packaged with your text.

WILEY PLUS

Within *WileyPLUS*

- **Chapter Quiz**
- **Visual Anatomy**
- **Mastering Vocabulary**
- **Audio Glossary**
- **Web Links**

• Understanding the Concepts

1. What components of the nervous system join together within the largest foramen in the skull? Where is it located?

2. Which opening in the skull is a passageway for nerve impulses related to the sense of vision? For nerve impulses related to eyeball movement? In which cranial bone is each of these openings located?

3. What part of the ethmoid bone forms the superior part of the nasal septum? Forms the roof of the nasal cavity? Is a passageway for nerve impulses related to the sense of smell? Is an attachment site for membranes separating the two sides of the brain?

4. Crying often produces a runny nose as tears fill the nasal cavity. Which opening in the skull allows tears to reach the nose? In which facial bone is this opening located?

5. Which foramina would you want your dentist to know the exact location of prior to initiating potentially painful dental work? Why? In which facial bone can these foramina be found?

6. How does the tissue composition of fontanels justify their common name, "soft spots"? What is the functional significance of fontanels?

7. If the body of a vertebra were to degenerate and collapse, as occurs in some musculoskeletal diseases, what foramen would compress the spinal nerve? What would be the effect on the rest of the body?

8. Why must the lumbar vertebrae be the largest and strongest in the vertebral column?

9. What are the components of the thoracic cage? What functions does it serve?

10. Why are "kidney punches" (blows to the lowermost ribs in boxing) considered dangerous and illegal?

8

The Appendicular Skeleton

Stefan groaned as he hoisted the large wooden crate onto his workbench, first one end, then the other. "Never anyone around to help when you need them," he muttered under his breath. The crate was dusty, marked with bold, black lettering indicating the university name and "Attention: Department of Archaeology." Hassan, director of the Ishtamira archaeological dig site, had been sending such boxes for months now. For Stefan, it had meant unpacking, categorizing, photographing, and sorting the important or interesting finds. Stefan was not a professional archaeologist, but he had worked with Hassan for almost 15 years now, and he had learned a great deal from him.

When Stefan last spoke to him, Hassan had been very excited about a particular box, and here it was, the crate that held the remains of the princess. He picked up a large metal pry bar from his workbench and began removing the lid. It wasn't easy, but the nails finally gave way with a screeching sound. Inside the box were bones carefully and neatly wrapped in soft cotton padding. Stefan removed the cotton and lifted out the bones one at a time: a humerus, a radius, an ulna, and several small cube-shaped bones. These were followed by shorter bones that looked like those left over after a chicken dinner.

Very neatly, Stefan began arranging the bones on the table. Stefan knew his anatomy, and, slowly, a skeletal upper limb began to appear. Next he arranged the vertebrae by size. The atlas and axis were easy to place, then the thoracic and lumbar vertebrae, next the sacrum.

As we saw from Hassan's story in Chapter 7, skeletal anatomy gives us many clues as to who an individual was and how he or she may have lived. As we explore the appendicular components of the skeleton, we will continue the story of Hassan and his colleagues, and learn how Hassan and his team used their knowledge of the skeletal system to solve the ancient riddle they had uncovered.

INTRODUCTION

As noted in Chapter 7, the two main divisions of the skeletal system are the axial skeleton and the appendicular skeleton. The general function of the axial skeleton is the protection of internal organs; the primary function of the appendicular skeleton, the focus of this chapter, is movement. The appendicular skeleton includes the bones that make up the

The appendicular skeleton includes the bones that make up the upper and lower limbs as well as the bones, arranged in formations called girdles, that attach the limbs to the axial skeleton.

upper and lower limbs as well as the bones, arranged in formations called girdles, that at-

tach the limbs to the axial skeleton. As you progress through this chapter, you will see how the bones of the appendicular skeleton are connected with one another and with skeletal muscles, making possible a wide array of movements. This arrangement permits you to do things such as walk, write, use a computer, dance, swim, and play a musical instrument.

8.1 Each pectoral girdle, which consists of a clavicle and scapula, attaches an upper limb to the axial skeleton.

The **pectoral** (PEK-tō-ral) or **shoulder girdles** attach the bones of the upper limbs to the axial skeleton (Figure 8.1). Each of the two pectoral girdles consists of a clavicle and a scapula. The clavicle, the anterior bone, articulates with the manubrium of the sternum; the scapula, the posterior bone, articulates with the clavicle and with the humerus. The pectoral girdles do not articulate (form joints) with the vertebral column and are held in position and stabilized by muscles that extend from the vertebral column and ribs to the scapula.

Clavicle

Each slender, S-shaped **clavicle** (KLAV-i-kul = key), or *collarbone*, is a long, slender bone that lies horizontally across the anterior part of the thorax superior to the first rib (Figure 8.2). The medial half of the clavicle is convex anteriorly, and the lateral half is concave anteriorly. It is rougher and more curved in males. The medial end of the clavicle, the *sternal end*, is rounded and articulates with the manubrium of the sternum, whereas the broad, flat, lateral end, the *acromial end* (a-KRŌ-mē-al), articulates with the acromion of the scapula (Figure 8.1).

Figure 8.1   Right pectoral girdle.

Pectoral girdle:
Clavicle
Scapula

CLAVICLE
Sternum

SCAPULA

Rib

Humerus

(a) Anterior view of pectoral girdle

CLAVICLE

SCAPULA

Rib

Humerus

Vertebrae

(b) Posterior view of pectoral girdle

The clavicle is the anterior bone of the pectoral girdle and the scapula is the posterior bone.

Figure 8.2 Right clavicle.

LATERAL

MEDIAL

POSTERIOR

ANTERIOR

(a) Superior view

View

Clavicle

Acromial
end

Sternal
end

Clavicle

ANTERIOR

View

POSTERIOR

(b) Inferior view

🔑 **The clavicle articulates medially with the manubrium of the sternum and laterally with the acromion of the scapula.**

Scapula

Each **scapula** (SCAP-ū-la; plural is *scapulae*), or *shoulder blade*, is a large, triangular, flat bone situated in the superior part of the posterior thorax (Figure 8.3). A prominent ridge called the *spine* runs diagonally across the posterior surface of the scapula (Figure 8.3b). The lateral end of the spine projects as a flattened, expanded process called the *acromion* (a-KRŌ-mē-on; *acrom-* = topmost; *omos* = shoulder), easily felt as the high point of the shoulder. Tailors measure the length of the upper limb from the acromion. The acromion articulates with the acromial end of the clavicle. Inferior to the acromion is a shallow depression, the *glenoid cavity*, that articulates with the head of the humerus (arm bone).

The thin edge of the scapula closer to the vertebral column is called the *medial border* (Figure 8.3). The thick edge closer to the arm is called the *lateral border*. The medial and lateral borders join at the *inferior angle*. The superior edge of the scapula, called the *superior border*, joins the medial border at the *superior angle*.

At the lateral end of the superior border of the scapula is a projection of the anterior surface called the *coracoid process* (KOR-a-koyd = like a crow's beak), to which the tendons of muscles and ligaments attach. Superior and inferior to the spine are two fossae: the *supraspinous fossa* (sū-pra-SPĪ-nus) and the *infraspinous fossa* (in-fra-SPĪ-nus), respectively. Both serve as a surface of attachment for a shoulder muscle. On the anterior surface is a slightly hollowed-out area called the *subscapular fossa*, also a surface of attachment for a shoulder muscle.

Figure 8.3   Right scapula.

(a) Anterior view

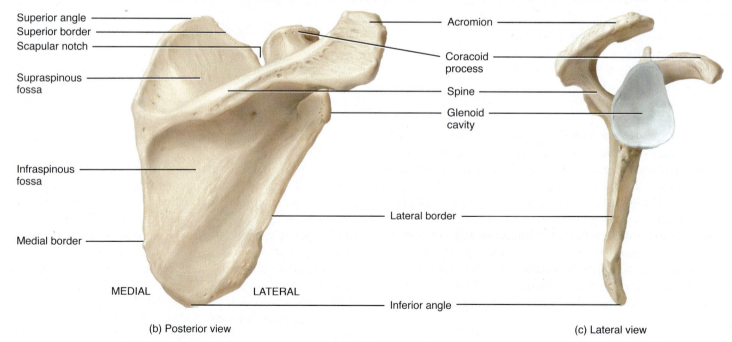

(b) Posterior view

(c) Lateral view

🔑 **The glenoid cavity of the scapula articulates with the head of the humerus to form the shoulder joint.**

Checkpoint

1. Which parts of the clavicle and sternum articulate? Which parts of the clavicle and scapula articulate?
2. Which parts of the scapula and humerus articulate?
3. Which two fossae are separated by the spine of the scapula?

8.2 The bones of each upper limb include the humerus, ulna, radius, carpals, metacarpals, and phalanges.

Each **upper** limb consists of 30 bones: a humerus in each arm, ulna and radius in each forearm, 8 carpals in each wrist, 5 metacarpals in each palm, and 14 phalanges in the fingers of each hand (Figure 8.4).

Humerus

The **humerus** (HŪ-mer-us), or arm bone, is the longest and largest bone of the upper limb (Figure 8.5). It articulates proximally with the scapula and distally at the elbow joint with both the ulna and the radius.

The proximal end of the humerus features a rounded *head* that articulates with the glenoid cavity of the scapula to form the shoulder joint (Figure 8.4). Distal to the head is the *anatomical neck*, which is visible as an oblique groove. The *greater tubercle* is a lateral projection distal to the anatomical neck. It is the most laterally palpable bony landmark of the shoulder region. The *lesser tubercle* projects anteriorly. Between both tubercles runs an *intertubercular sulcus*. The *surgical neck* is a constriction in the humerus just distal to the tubercles, where the head tapers to the shaft; it is so named because fractures often occur here.

The *body (shaft)* of the humerus contains a roughened, V-shaped area called the *deltoid tuberosity*. This area serves as a point of attachment for the tendons of the deltoid muscle.

Several prominent features are evident at the distal end of the humerus. The *capitulum* (ka-PIT-ū-lum; *capit-* = head) is a rounded knob on the lateral aspect of the bone that articulates with the head of the radius. The *radial fossa* is an anterior depression above the capitulum that receives the head of the radius when the elbow joint is flexed (bent). The *trochlea* (TRŌK-lē-a = pulley), located medial to the capitulum, is a spool-shaped surface that articulates with the ulna. The *coronoid fossa* (KOR-ō-noyd = crown-shaped) is an anterior depression that receives the coronoid process of the ulna when the elbow joint is flexed. The *olecranon fossa* (ō-LEK-ra-non = elbow) is a posterior depression that receives the olecranon of the ulna when the elbow joint is extended (straightened). The *medial epicondyle* and *lateral epicondyle* are rough projections on either side of the distal end of the humerus to which most muscles of the forearm are attached. The ulnar nerve, the superficial nerve that makes you feel a very severe pain when you hit your elbow, may be palpated by rolling a finger over the skin surface above the posterior surface of the medial epicondyle.

Figure 8.4   Right upper limb.

Anterior view of upper limb

🔑 **Each upper limb consists of a humerus, ulna, radius, carpals, metacarpals, and phalanges.**

Figure 8.5   Right humerus in relation to the scapula, ulna, and radius.

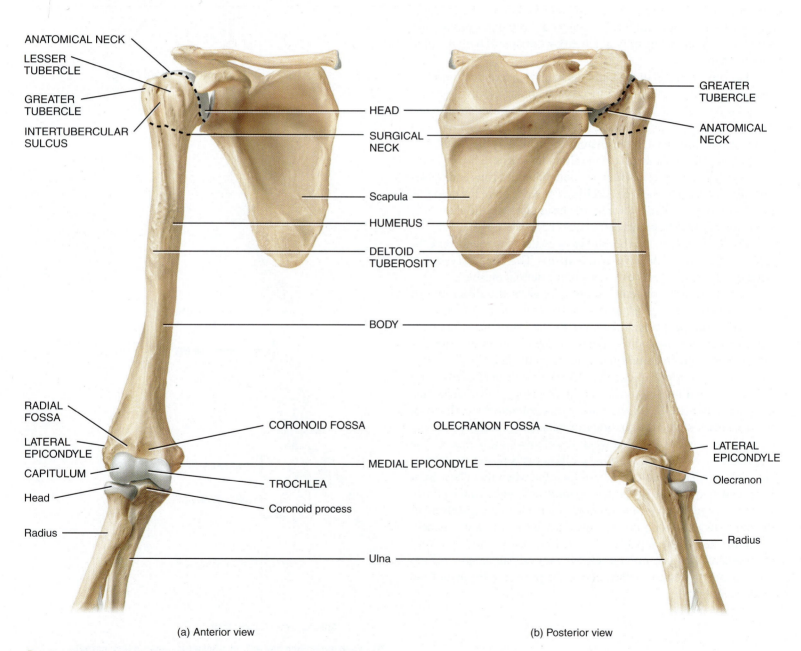

(a) Anterior view

(b) Posterior view

The humerus is the longest and largest bone of the upper limb.

Ulna and Radius

The **ulna** is located on the medial aspect (the little-finger side) of the forearm and is longer than the radius (Figure 8.6). It is sometimes convenient to use an aid to learn new or unfamiliar informa-
tion. Such an aid is called a *mnemonic* (nē-MON-ik = memory). One such mnemonic to help you remember the location of the ulna in relation to the hand is "p.u." (the **p**inky is on the **u**lna side). At the proximal end of the ulna (Figures 8.6b and 8.7b) is the *olecranon*, which forms the prominence of the elbow. The *coronoid*

Figure 8.6 **Right ulna and radius in relation to the humerus and carpals.**

(a) Anterior view   (b) Posterior view

In the forearm, the ulna is on the medial side, and the radius is on the lateral side.

process (Figures 8.6a and 8.7b) is an anterior projection distal to the olecranon. The *trochlear notch* (Figure 8.7b) is a large curved area between the olecranon and coronoid process that articulates with the trochlea of the humerus at the elbow joint. The *radial notch* (Figure 8.7b) is a depression that is lateral and inferior to the trochlear notch that articulates with the head of the radius. The distal end of the ulna consists of a *head* on the anterior side and a *styloid process* on the posterior side (Figure 8.6). The styloid process provides attachment for the ulnar collateral ligament to the wrist.

The **radius** is located on the lateral aspect (thumb side) of the forearm (Figure 8.6). The proximal end of the radius has a disc-shaped *head* that articulates with the capitulum of the humerus and the radial notch of the ulna (Figure 8.6a). Inferior to the head is the constricted *neck*. A roughened area inferior to the neck on the medial side, called the *radial tuberosity*, is a point of attachment for the biceps brachii muscle. The shaft of the radius widens distally where it contains a depression, the *ulnar notch*, on the medial side and a *styloid process* on the lateral side, which can be felt proximal to the thumb. The styloid process provides attachment for the brachioradialis muscle and wrist bones. Fracture of the distal end of the radius is the most common fracture in adults older than 50 years.

The ulna and radius articulate with the humerus at the elbow joint. The articulation occurs in two places: where the head of the radius articulates with the capitulum of the humerus, and where the trochlear notch of the ulna receives the trochlea of the humerus (Figure 8.7).

The ulna and the radius connect with one another at three sites. First, a broad, flat, fibrous connective tissue called the *interosseous membrane* (in´-ter-OS-ē-us; *inter-* = between, *-osseous* = bone) joins the shafts of the two bones. The ulna and radius also articulate at their proximal and distal ends. Proximally, the head of the radius articulates with the ulna's *radial notch* (Figure 8.7b). Distally, the head of the ulna articulates with the *ulnar notch* of the radius (Figure 8.7c). Finally, the distal end of the radius articulates with three bones of the wrist—the lunate, the scaphoid, and the triquetrum—to form the *wrist joint*.

Carpals, Metacarpals, and Phalanges

The **carpus** (wrist) is the proximal region of the hand and consists of eight small bones, the **carpals**, joined to one another by ligaments (Figure 8.8). The carpals are arranged in two transverse rows of four bones each. Their names reflect their shapes. The carpals in the proximal row, from lateral to medial, are the **scaphoid**

(SKAF-oyd = boatlike), **lunate** (LOO-nāt = moon-shaped), **triquetrum** (trī-KWĒ-trum = three-cornered), and **pisiform** (PIS-i-form = pea-shaped). The proximal row of carpals articulates with the distal ends of the ulna and radius to form the *wrist joint*. The carpals in the distal row, from lateral to medial, are the **trapezium** (tra-PĒ-zē-um = four-sided figure with no two sides parallel), **trapezoid** (TRAP-e-zoyd = four-sided figure with two sides parallel), **capitate** (KAP-i-tāt = head-shaped), and **hamate** (HAM-āt = hooked). The capitate is the largest carpal bone; its rounded projection, the head, articulates with the lunate. The hamate is named for a large hook-shaped projection on its anterior surface. In about 70% of carpal fractures, only the scaphoid is broken. This is because the force of a fall on an outstretched hand is transmitted from the capitate through the scaphoid to the radius.

The anterior concave space formed by the pisiform and hamate (on the ulnar side), and the scaphoid and trapezium (on the radial side), with the roof-like covering of the *flexor retinaculum* (fibrous bands of fascia) is the **carpal tunnel**. Tendons for muscles to the digits and thumb and the median nerve pass through the carpal tunnel. Narrowing of the carpal tunnel, due to such factors as inflammation, may give rise to a condition called carpal tunnel syndrome.

A mnemonic for learning the names of the carpal bones is shown in Figure 8.8. The first letters of the carpal bones from lateral to medial (proximal row, then distal row) correspond to the first letter of each word in the mnemonic.

The **metacarpus** (*meta-* = beyond), or palm, is the intermediate region of the hand and consists of five bones called **metacarpals**. The metacarpals are numbered I to V (or 1–5), starting with the thumb, from lateral to medial (Figure 8.8). The proximal ends articulate with the distal row of carpal bones whereas the distal ends articulate with the proximal phalanges. The distal ends of the metacarpals are commonly called knuckles and are readily visible in a clenched fist.

The **phalanges** (fa-LAN-jēz; *phalanx* = a battle line), or bones of the fingers, make up the distal region of the hand. There are 14 phalanges in each hand and, like the metacarpals, the fingers are numbered I to V (or 1–5), beginning with the thumb, from lateral to medial. A single bone of a finger or toe is referred to as a *phalanx* (FĀ-lanks). There are two phalanges (proximal and distal) in the thumb and three phalanges (proximal, middle, and distal) in each of the other four fingers. In order from the thumb, these other four fingers are commonly referred to as the index finger, middle finger, ring finger, and little finger. The *proximal* phalanges articulate with the metacarpal bones and middle phalanges. The *middle* phalanges articulate with the proximal row and *distal* phalanges. The distal phalanges articulate with the middle phalanges.

Figure 8.7 **Articulations formed by the right ulna and radius.** (a) Right elbow joint. (b) Joint surfaces at proximal end of the ulna. (c) Joint surfaces at distal ends of radius and ulna.

(a) Medial view in relation to humerus

(b) Lateral view of proximal end of ulna

(c) Inferior view of distal ends of radius and ulna

🗝️ The elbow joint is formed by two articulations: (1) the trochlear notch of the ulna with the trochlea of the humerus and (2) the head of the radius with the capitulum of the humerus.

Figure 8.8   Right wrist and hand in relation to the ulna and radius.

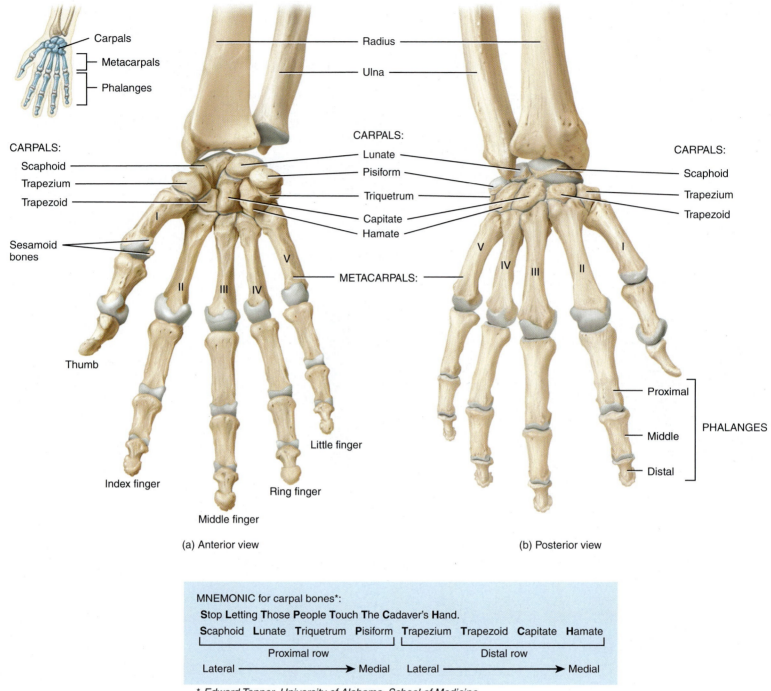

(a) Anterior view

(b) Posterior view

MNEMONIC for carpal bones*:
Stop **L**etting **T**hose **P**eople **T**ouch **T**he **C**adaver's **H**and.
Scaphoid **L**unate **T**riquetrum **P**isiform **T**rapezium **T**rapezoid **C**apitate **H**amate

Proximal row	Distal row
Lateral ⟶ Medial	Lateral ⟶ Medial

* *Edward Tanner, University of Alabama, School of Medicine*

🔑 **The skeleton of the hand consists of the proximal carpals, the intermediate metacarpals, and the distal phalanges.**

Checkpoint

4.   How many bones make up the upper limb?

5.   Which part of which bone forms the most lateral projection of the shoulder?

6.   Which part of the ulna makes up the projecting point of the elbow?

7.   Which parts of the humerus, ulna, and radius articulate at the elbow joint? Which depression of the humerus and projection of the ulna articulate when the elbow is flexed? When the elbow is straight?

8.   Which carpal bones form a concave space that is part of the carpal tunnel?

9.   Which bones form the palm of the hand? Are commonly called the knuckles?

As Stefan slowly unwrapped the cotton, a flat, triangular-shaped bone was revealed. "Right scapula," he thought, and placed it above the now-articulated right upper limb. He aligned the socket of the shoulder blade with the head of the humerus. Immediately under the right scapula in the shipping crate was the left scapula. "Where are your collarbones, Princess?" he mused, as he searched carefully through the packing material in the crate.

Stefan always enjoyed this part of his job. Rearticulating the skeleton was like piecing together an ancient puzzle. "Ah, here we go." Stefan pulled out two bundles, each containing a narrow, S-shaped bone. Turning them over in his hands, he quickly decided which was right and which left, then placed each clavicle by its neighboring scapula.

Arranging the right and left humerus, radius, and ulna was easy for Stefan using their obvious landmarks—the head, deltoid tuberosity, and olecranon fossa of the humerus, the olecranon and pointed styloid process of the ulna, and the circular head and wide styloid process of the radius. The deltoid tuberosity on the right humerus was somewhat larger than on the left. "Hmm, must have been right-handed," Stefan noted.

The wrist and hand bones were more difficult to put back in order. Stefan unwrapped each individual bone, identifying them first as phalanges, metacarpals, or carpals. Unlike the larger bones of the upper limb, these small bones would have required quite a bit more time to determine right from left. Luckily, the field researchers had kept them separate. Stefan focused on the wrist bones. "Pisiform goes here, scaphoid, and… let's see here, this is…".

Stefan's quiet contemplation was shattered by the thud of the metal door banging against the wall. "Hello, Stefan!" A tall, thin young man stood in the doorway.

Stefan grimaced. "Octavio, don't you ever knock?"

"Oh, Stefan, relax. Did I disturb you and your new girlfriend? Don't worry; I don't think she can run away too fast." Octavio grinned.

Stefan didn't care much for Octavio. But he had to admit Octavio was as good if not better at paleoanthropology than Liu or Professor Baru.

"I just came to see how you were doing. You should stay calm, Stefan, like your friend here. Nothing gets under her skin."

"Octavio, please go. I will call you when I have finished. You can do your analysis later." Still laughing, Octavio just shook his head and left the room.

A. Which clue would tell Stefan which scapular surface was anterior and which was posterior? What is the name of the shallow, oval socket of the scapula that Stefan placed next to the humerus?

B. Which bone is Stefan referring to as the "collarbone"?

C. Which surface markings could Stefan use to distinguish the right humerus from the left? How could he identify the right and left radius and ulna?

D. Why would Stefan think that an enlarged right deltoid tuberosity might indicate right-handedness?

8.3 The pelvic girdle supports the vertebral column and pelvic viscera and attaches the lower limbs to the axial skeleton.

The **pelvic (hip) girdle** consists of the two **hip bones**, also called **coxal bones** (KOK-sal; *cox-* = hip) (Figure 8.9). The hip bones unite anteriorly at a joint called the **pubic symphysis** (PŪ-bik SIM-fi-sis). They unite posteriorly with the sacrum at the *sacroiliac joints*. The complete ring composed of the hip bones, pubic symphysis, and sacrum forms a deep, basinlike structure called the **pelvis** (*pelv-* = basin). Functionally, the pelvis provides a strong and stable support for the vertebral column and protects pelvic and lower abdominal organs. The pelvic girdle of the pelvis also connects the bones of the lower limbs to the axial skeleton.

There are some significant differences between the pectoral girdle and pelvic girdle. The pectoral girdle does not directly articulate with the vertebral column, but the pelvic girdle does so via the sacroiliac joint. The pectoral girdle sockets (glenoid cavities) that articulate with the upper limbs are shallow and maximize movement, in contrast to the pelvic girdle sockets (acetabula) that articulate with the lower limbs, which are deep and allow less movement. Overall, the structure of the pectoral girdle offers more mobility than strength, and that of the pelvic girdle offers more strength than mobility.

Each of the two hip bones of a newborn consists of three bones separated by cartilage: a superior *ilium*, an inferior and anterior *pubis*, and an inferior and posterior *ischium*. By age 23, the three separate bones fuse together (see Figure 8.10a). Although the hip bones function as single bones, anatomists commonly discuss each hip bone as though they still consisted of three bones.

Figure 8.9   Pelvis. Shown here is the female pelvis.

Pelvic (hip) girdle

Hip bone

Sacrum

Coccyx

Pubic symphysis

Sacroiliac joint

Sacral promontory

Pelvic brim

Acetabulum

Obturator foramen

🔑 **The hip bones are united anteriorly at the pubic symphysis and posteriorly at the sacrum to form the pelvis.**

Ilium

The **ilium** (IL-ē-um = flank) is the largest of the three components of the hip bone (Figure 8.10). It is comprised of a superior *ala* (= wing) and an inferior *body*. Its superior border, the *iliac crest*, ends anteriorly in a blunt *anterior superior iliac spine* (Figure 8.10b,c). Below this spine is the *anterior inferior iliac spine*. Posteriorly, the iliac crest ends in a sharp *posterior superior iliac spine*. Below this spine is the *posterior inferior iliac spine*. The spines serve as points of attachment for the muscles of the trunk, hip, and thighs. Below the posterior inferior iliac spine is the *greater sciatic notch* (sī-AT-ik), through which the sciatic nerve, the longest nerve in the body, passes.

The medial surface of the ilium contains a concavity, the *iliac fossa*, a site of muscle attachment. Posterior to this fossa is the *auricular surface* (*auric-* = ear-shaped), which articulates with the sacrum to form the *sacroiliac joint* (Figure 8.9). Projecting anteriorly and inferiorly from the auricular surface of the ilium is a ridge called the *arcuate line* (AR-kū-āt; *arc-* = bow) (Figure 8.10c).

The other conspicuous markings of the ilium are three arched lines on its lateral surface called the *posterior gluteal line* (*glut-* = buttock), the *anterior gluteal line*, and the *inferior gluteal line*. The gluteal muscles attach to the ilium between these three lines.

Figure 8.10   Right hip bone. The lines of fusion of the ilium, ischium, and pubis depicted in (a) are not always visible in an adult.

(a) Lateral view showing parts of hip bone

(b) Detailed lateral view

(c) Detailed medial view

The acetabulum is the socket formed where the three parts of the hip bone converge.

Ischium

The **ischium** (IS-kē-um = hip), the inferior, posterior portion of the hip bone (Figure 8.10), comprises a superior *body* and an inferior *ramus* (ram- = branch; plural is *rami*). The ramus is the portion of the ischium that fuses with the pubis. Features of the ischium include the prominent *ischial spine*, a *lesser sciatic notch* below the spine, and a rough and thickened *ischial tuberosity*. This prominent tuberosity may hurt someone's thigh when you sit on his or her lap. Together, the ischium and the pubis surround the *obturator foramen* (OB-too-rā-tōr; *obtur-* = closed up), the largest foramen in the skeleton. The foramen is so named because, even though blood vessels and nerves pass through it, it is nearly completely closed by the fibrous *obturator membrane*.

Pubis

The **pubis**, or **pubic bone**, is the anterior and inferior part of the hip bone (Figure 8.10b,c). A *superior ramus*, an *inferior ramus*, and a *body* between the rami comprise the pubis. The anterior, superior border of the body is the *pubic crest*, and at its lateral end is a projection called the *pubic tubercle*. The *pubic symphysis* is the joint between the two hip bones (Figure 8.9). It consists of a disc of fibrocartilage. Inferior to this joint, the inferior rami of the two pubic bones converge to form the *pubic arch*. In the later stages of pregnancy, the hormone relaxin increases the flexibility of the pubic symphysis to ease delivery of the baby. Weakening of the joint, together with an already compromised center of gravity due to an enlarged uterus, also changes the gait during pregnancy.

Acetabulum

The *acetabulum* (as-e-TAB-ū-lum = vinegar cup) is a deep fossa (depression) formed by the ilium, ischium, and pubis. It functions as the socket that articulates with the rounded head of the femur to form the *hip joint*.

False and True Pelves

The pelvis is divided into superior and inferior portions by a boundary called the *pelvic brim* (Figure 8.11a). Beginning posteriorly at the sacral promontory of the sacrum, the pelvic brim is a ridge that runs laterally along the arcuate lines, then the pubic crest to the superior portion of the pubic symphysis. Because of the tilt of the pelvis, the pelvic brim is higher in the back than in the front (Figure 8.11b).

The portion of the pelvis superior to the pelvic brim is the **false pelvis** (Figure 8.11b). It is bordered by the lumbar vertebrae posteriorly, the upper portions of the hip bones laterally, and the abdominal wall anteriorly. The space enclosed by the false pelvis is part of the abdomen and does not contain pelvic organs, except for the urinary bladder (when it is full) and the uterus during pregnancy.

The portion of the pelvis inferior to the pelvic brim is the **true pelvis** (Figure 8.11b). It is bounded by the sacrum and coccyx posteriorly, inferior portions of the ilium and ischium laterally, and the pubic bones anteriorly. The true pelvis surrounds the pelvic cavity (see Figure 1.9). The superior opening of the true pelvis, bordered by the pelvic brim, is called the *pelvic inlet*; the inferior opening of the true pelvis is the *pelvic outlet*. The *pelvic axis* is an imaginary line that curves through the true pelvis from the center of the pelvic inlet to the center of the pelvic outlet. During childbirth the pelvic axis is the route taken by the baby's head as it descends through the pelvis.

Checkpoint

10. What are the functions of the pelvic girdle and the pelvis?
11. Distinguish among a hip bone, pelvic girdle, and pelvis.
12. What are the four spines of the ilium? What is their function?
13. Which part of the hip bone articulates with the sacrum? With the femur?
14. What is the significance of the pelvic axis?

Figure 8.11 **True and false pelves.** Shown here is the female pelvis. For simplicity, in (a) the landmarks of the pelvic brim are shown only on the left side of the body, and the outline of the pelvic brim is shown only on the right side. The entire pelvic brim is shown in **Table 8.1**.

Hip bone

Sacrum

Pelvic brim

Pelvic brim landmarks:

Sacral promontory

Arcuate line

Pubic crest

Pubic symphysis

(a) Anterosuperior view of pelvic girdle

Midsagittal plane

POSTERIOR

ANTERIOR

Sacral canal

Sacrum

TRUE PELVIS

Coccyx

Plane of pelvic outlet

Sacral promontory

FALSE PELVIS

Plane of pelvic brim

Pelvic axis

Pubic symphysis

(b) Midsagittal section indicating locations of true and false pelves

🔑 **The true and false pelves are separated by the pelvic brim.**

TABLE 8.1

Comparison of Female and Male Pelves

POINT OF COMPARISON	FEMALE	MALE
General structure	Light and thin.	Heavy and thick.
False pelvis	Shallow.	Deep.
Pelvic brim (inlet)	Larger and more oval.	Smaller and heart-shaped.
Acetabulum	Small and faces anteriorly.	Large and faces laterally.
Obturator foramen	Oval.	Round.
Pubic arch	Greater than 90° angle.	Less than 90° angle.

Anterior views

Iliac crest	Less curved.	More curved.
Ilium	Less vertical.	More vertical.
Greater sciatic notch	Wide.	Narrow.
Coccyx	More movable and more curved anteriorly.	Less movable and less curved anteriorly.
Sacrum	Shorter, wider (see anterior views), and less curved anteriorly.	Longer, narrower (see anterior views), and more curved anteriorly.

Right lateral views

Pelvic outlet	Wider.	Narrower.
Ischial tuberosity	Shorter, farther apart, and more medially projecting.	Longer, closer together, and more laterally projecting.

Inferior views

8.4 Male pelves are generally larger, heavier, and have more prominent markings; female pelves are generally wider and shallower.

In the following discussion it is assumed that the male and female are comparable in age and physical stature. Generally, the bones of a male are larger and heavier than those of a female and have larger surface markings (Table 8.1). Sex-related differences in the features of bones are readily apparent when comparing the female and male pelves. Most of the structural differences in the pelves are adaptations to the requirements of pregnancy and childbirth. Because the female's pelvis is wider and shallower than the male's, there is more space in the true pelvis of the female, especially in the pelvic inlet and pelvic outlet, which accommodate the passage of the infant's head at birth. Other significant structural differences between pelves of females and males are listed and illustrated in Table 8.1.

Checkpoint

15. Why are there structural differences between female and male pelves?
16. What adaptations of the female pelvis assist the infant's head exiting at birth?

Return to Stefan's Story

Stefan kept at his work, without further interruption, until he had the remains completely articulated. He had made some notes of his observations and decided it was time to telephone Professor Hassan.

"Hello, Professor Hassan?"

"Yes," came the reply.

"This is Stefan. I have the remains of the princess here and Octavio will be looking at them tonight or tomorrow. They are remarkably preserved."

"I agree. We are most fortunate. What do you think, Stefan? Are they mother and child?"

Stefan shifted in his chair; his back hurt from hours of leaning over the remains. "The lack of brow ridge development and pelvic appearance support the gender of both as female. The adult is possibly the mother; the pubic symphyses had defined, discrete lesions that could have occurred by pulling of the cartilage of the pubic symphysis during childbirth."

Hassan was silent for a moment. "Have Octavio prepare his report as soon as he can, then have him call me. Baru thought that the child was probably about two to three years old at the time of death. It could be her child."

"Professor, we should be getting some of the results from the genetic testing soon. What do you think happened there?"

Hassan paused again, then replied. "We have found some evidence suggesting intentional homicide, but many of the remains do not give such an indication. There could have been a plague, but most diseases produce soft tissue injury and act quickly, leaving no evidence in bones. I have sent several samples directly to the genetics lab to test for some possible suspects. There may be spores left in the soil, or perhaps on the bones of some of the remains. Be careful, Stefan. No food or drink in the lab."

Stefan felt his stomach tighten. "Of course, Professor. Uh, you don't think whatever killed them could still be contagious after thousands of years, do you?"

Hassan laughed. "Curse of the mummy, Stefan? Like King Tut? Don't worry, my friend. I must go now. Goodbye."

E. What is the location of the pubic symphysis Stefan refers to in the story?

F. Which adaptations would have taken place in the pubic symphysis of the female skeleton during the later stages of her pregnancy in preparation for the birthing process?

G. What clues could Stefan have used to identify the gender of the pelvis?

H. How would Stefan have distinguished between the right and the left coxal bones?

I. Would the bones of the pelvis be fused in the female child's skeleton?

8.5 The bones of each lower limb include the femur, patella, tibia, fibula, tarsals, metatarsals, and phalanges.

Each **lower limb** is composed of 30 bones: the femur in the thigh, the patella (kneecap), the tibia and fibula in the leg, 7 tarsals in the ankle, 5 metatarsals in the foot, and 14 phalanges in the toes (Figure 8.12).

Femur

The **femur**, or thigh bone, is the longest, heaviest, and strongest bone in the body (Figure 8.13). Its proximal end articulates with the acetabulum of the hip bone. Its distal end articulates with the tibia and patella. The *body (shaft)* of the femur angles medially; as a result, the knee joints are closer to the midline. The angle is greater in females because the female pelvis is broader.

The proximal end of the femur consists of a rounded *head* that articulates with the acetabulum of the hip bone to form the hip joint. The head contains a small centered depression called the *fovea capitis* (FŌ-vē-a CAP-i-tis; *fovea* = pit; *capitis* = of the head). The ligament of the head of the femur connects the fovea capitis of the femur to the acetabulum of the hip bone. The *neck* of the femur is a constricted region distal to the head. A "broken hip" is more often associated with a fracture in the neck of the femur than fractures of the hip bones. The *greater trochanter* (trō-KAN-ter) and *lesser trochanter* are projections of the femur that serve as points of attachment for thigh and buttock muscles. The greater trochanter is the prominence felt and seen anterior to the hollow on the side of the hip. It is a landmark commonly used to locate the site for intramuscular injections into the lateral surface of the thigh. The lesser trochanter is inferior and medial to the greater trochanter. Between the trochanters is an anterior *intertrochanteric line* (Figure 8.13a) and a posterior *intertrochanteric crest* (Figure 8.13b).

On the posterior surface of the body of the femur is a vertical ridge called the *gluteal tuberosity* that blends into another vertical ridge called the *linea aspera* (LIN-ē-a AS-per-a; *asper-* = rough). Both of these ridges serve as attachment points for several thigh muscles.

The distal end of the femur expands to include the *medial condyle* (= knuckle) and the *lateral condyle*. These articulate with the medial and lateral condyles of the tibia to form the *knee joint*. Superior to the condyles are the *medial epicondyle* and the *lateral epicondyle*. A depressed area between the condyles on the posterior surface is called the *intercondylar fossa* (in-ter-KON-di-lar). The *patellar surface*, located anteriorly between the condyles, is a smooth surface over which the patella glides.

Figure 8.12   Right lower limb.

Hip bone

Sacrum

FEMUR

PATELLA

TIBIA

FIBULA

TARSALS

METATARSALS

PHALANGES

Anterior view of lower limb

Each lower limb consists of a femur, patella, tibia, fibula, tarsals, metatarsals, and phalanges.

223

Figure 8.13 **Right femur in relation to the hip bone, patella, tibia, and fibula.**

Femur

Hip bone

GREATER TROCHANTER

HEAD

NECK

INTERTROCHANTERIC LINE   CREST

LESSER TROCHANTER

FEMUR

BODY

GREATER TROCHANTER

GLUTEAL TUBEROSITY

LINEA ASPERA

LATERAL EPICONDYLE

LATERAL CONDYLE

Fibula

MEDIAL EPICONDYLE

MEDIAL CONDYLE

Patella

Tibia

LATERAL EPICONDYLE

INTERCONDYLAR FOSSA

LATERAL CONDYLE

Fibula

(a) Anterior view

(b) Posterior view

(continues)

Figure 8.13 (continued)

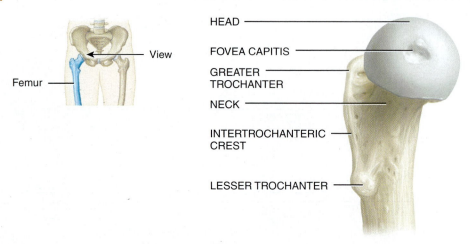

HEAD

FOVEA CAPITIS

GREATER
TROCHANTER

NECK

INTERTROCHANTERIC
CREST

LESSER TROCHANTER

View

Femur

(c) Medial view of proximal end of femur

The acetabulum of the hip bone and head of the femur articulate to form the hip joint.

Patella

The **patella** (= little dish), or kneecap, is a small, triangular sesamoid bone located anterior to the knee joint (Figure 8.14). The broad superior end of the patella is called the *base* and the pointed inferior end is the *apex*. The posterior surface contains two *articular facets*, one for each condyle of the femur. During normal flexion and extension of the knee, the patella glides up and down in the groove between the two femoral condyles. The patella is enclosed superiorly within the tendon of the quadriceps femoris (an anterior thigh muscle) and inferiorly is attached to the tibial tuberosity of the tibia by the patellar ligament. The patella increases the leverage of the quadriceps femoris, maintains the position of the tendon when the knee is flexed (bent), and protects the knee joint.

Tibia and Fibula

The **tibia**, or shin bone, is the larger, medial, weight-bearing bone of the leg (Figure 8.15). The tibia articulates at its proximal end with the femur and fibula and at its distal end with the fibula and the talus of the ankle. The tibia and fibula, like the ulna and radius, are connected by an interosseous membrane.

The proximal end of the tibia is expanded into a *lateral condyle* and a *medial condyle*, which articulate with the condyles of the femur to form the knee joint. The inferior surface of the lateral condyle articulates with the head of the fibula. The slightly concave condyles are separated by an upward projection called the *intercondylar eminence* (Figure 8.15b). The *tibial tuberosity* on the anterior surface is a point of attachment for the patellar ligament.

Figure 8.14   Right patella.

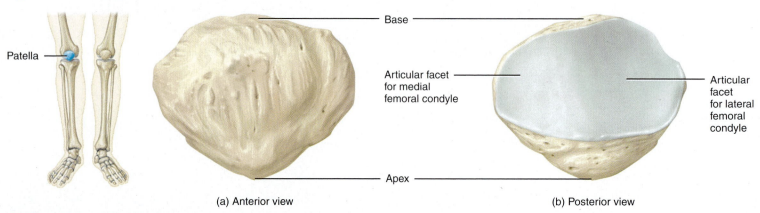

Patella

Base

Articular facet
for medial
femoral condyle

Articular
facet
for lateral
femoral
condyle

Apex

(a) Anterior view

(b) Posterior view

The patella articulates with the lateral and medial condyles of the femur.

Figure 8.15 Right tibia and fibula in relation to the femur, patella, and talus.

Tibia

Fibula

MNEMONIC for location of tibia and fibula:
The fibu**LA** is **LA**teral.

Femur

Patella

INTERCONDYLAR
EMINENCE

LATERAL
CONDYLE

MEDIAL CONDYLE

LATERAL
CONDYLE

HEAD

TIBIAL TUBEROSITY

HEAD

FIBULA

TIBIA

Interosseous membrane

FIBULA

ANTERIOR BORDER

MEDIAL MALLEOLUS

LATERAL
MALLEOLUS

Talus

Calcaneus

LATERAL
MALLEOLUS

(a) Anterior view

(b) Posterior view

Tibia

View

POSTERIOR

ANTERIOR

FIBULAR
NOTCH

MEDIAL
MALLEOLUS

(c) Lateral view of distal end of tibia

The tibia articulates with the femur and fibula proximally, and with the fibula and talus distally.

Inferior to and continuous with the tibial tuberosity is a sharp ridge that can be felt below the skin, known as the *anterior border* or shin.

The medial surface of the distal end of the tibia forms the *medial malleolus* (mal-LĒ -ō-lus = hammer). This structure articulates with the talus of the ankle and forms the prominence that can be felt on the medial surface of the ankle. The *fibular notch* (Figure 8.15c) articulates with the distal end of the fibula.

The **fibula** is parallel and lateral to the tibia, but is considerably smaller. (See Figure 8.15 for a mnemonic describing the relative positions of the tibia and fibula.) The *head* of the fibula, the proximal end, articulates with the inferior surface of the lateral condyle of the tibia below the level of the knee joint. The distal end has a projection, the *lateral malleolus*, that articulates with the talus of the ankle. The lateral malleolus is the prominence on the lateral surface of the ankle. As noted previously, the fibula also articulates with the tibia at the fibular notch.

Tarsals, Metatarsals, and Phalanges

The **tarsus** (ankle) is the proximal region of the foot and consists of seven **tarsal bones** (Figure 8.16). They include the **talus** (TĀ-lus = ankle bone) and **calcaneus** (kal-KĀ-nē-us = heel), located

Figure 8.16   Right foot.

(a) Superior view   (b) Inferior view

MNEMONIC for tarsal bones:						
Tall	**Centers**	**Never**	**Take**	**Shots**	**From**	**Corners.**
Talus	Calcaneus	Navicular	Third cuneiform	Second cuneiform	First cuneiform	Cuboid

🔑 **The skeleton of the foot consists of the proximal tarsals, the intermediate metatarsals, and the distal phalanges.**

in the posterior part of the foot. The calcaneus is the largest and strongest tarsal bone. The anterior tarsal bones are the **navicular** (= like a little boat), three **cuneiform bones** (= wedge-shaped) called the **lateral**, **intermediate**, and **medial cuneiforms**, and the **cuboid** (= cube-shaped). The talus, the most superior tarsal bone, is the only bone of the foot that articulates with the fibula and tibia. It articulates on one side with the medial malleolus of the tibia and on the other side with the lateral malleolus of the fibula to form the *ankle joint*. During walking, the talus transmits about half the weight of the body to the calcaneus. The remainder is transmitted to the other tarsal bones.

The **metatarsus** is the intermediate region of the foot and consists of five **metatarsal bones** numbered I to V (or 1–5) from the medial to lateral position (Figure 8.16). The metatarsals articulate proximally with the cuneiform bones and with the cuboid. Distally, they articulate with the proximal row of phalanges. The first metatarsal is thicker than the others because it bears more weight.

The **phalanges** comprise the distal component of the foot and resemble those of the hand both in number and arrangement. The toes are numbered I to V (or 1–5) beginning with the great toe, from medial to lateral. The great or big toe has two large, heavy phalanges called proximal and distal phalanges. The other four toes each have three phalanges—proximal, middle, and distal.

Checkpoint

17. Which bone is the strongest bone in the body?
18. Which part or parts of the femur articulate with the acetabulum at the hip joint? Are attachment sites for buttock muscles? Is a lateral landmark for intramuscular injections?
19. The patella is classified as which type of bone? Why? What are its functions?
20. The part of which bone forms the bony prominences found on the medial surface of the ankle? On the lateral surface of the ankle?
21. Which tarsal bone articulates with the tibia and fibula? Is the bony component of the heel of the foot?

Stefan's Story: Epilogue and Discussion

▶▶▶ "Professor Hassan, this is Octavio."

"Hello, Octavio," Hassan replied. "What have you found?"

"Based on the teeth of the female, I estimate an age of 15 to 17 years. The epiphyseal plates of the phalanges of the foot had not yet closed. The lack of arthritic growths and generally youthful appearance of the calcaneus and talus facet rims would also seem to support this age. The child was probably about two to three years old at the time of death. The results from the DNA lab came back this morning. It is her child. The fractures in the child's cervical vertebrae appear to be postmortem. The bones likely were damaged after the soft tissue had decayed. I don't think her neck was intentionally broken.

"There was evidence of periostitis in the princess. There are bilateral lesions on both tibia. It's likely that the tuberculosis you suspected had become a systemic infection. It's hard to say if tuberculosis killed her. I think it's more likely that these people succumbed to a virulent, infectious disease that killed them before a bony response could take place—something that spread quickly, and that affected people of all different ages. The lab results have confirmed the presence of *Mycobacterium tuberculosis* DNA. We have also found high concentrations of lead in many of the bone samples."

After a brief pause Hassan replied, "Interesting findings. They seem to fit with our translations of the hieroglyphs. The sarcophagus inscriptions indicate that this is the princess of Ishtamira and the child was her daughter. Something happened, a plague of some sort. The princess and her child were 'taken by the gods.' So many were dying that the tomb intended to be the final resting place of the king became a mass grave for his people."

"Something puzzles me, Professor. I understand that they would have needed to bury the bodies of the dead, but why the sacrifices? I didn't think that human sacrifice was endemic to this region, even that long ago."

"Difficult to say, Octavio. Many of the cooking vessels, cups, and utensils we recovered had lead in them, which is a potent neurotoxin. Lead ingestion may have caused a low level of lead poisoning. Perhaps the king went mad and ordered the sacrifices. We will never know for sure."

You have seen that the anatomy and physiology of bone reveal much about who we are. Nutritional status, activity level, exposure to certain diseases, gender, age, and many other pieces of information can be deduced from skeletal remains. In thinking about the skeleton and the stories from Chapters 7 and 8, see if you can answer the following questions: ◀◀◀

J. How do infants' and children's skeletons differ from those of adults? How do male and female skeletons differ overall?
K. Where in bone is DNA located?
L. What information do the surface markings of bones reveal about an individual's activities during life?
M. How would the mineral content of bone reflect an individual's health status?
N. Octavio mentions that the phalangeal epiphyseal plates were unclosed, an observation he used to help him estimate the age of the remains. What are the epiphyseal plates, and why are they useful for determining age?

Concept and Resource Summary

Concept   **Resources**

Introduction

1. The appendicular skeleton is composed of the upper and lower limbs, and the pectoral and pelvic girdles.

Concept 8.1 Each pectoral girdle, which consists of a clavicle and scapula, attaches an upper limb to the axial skeleton.

1. Each of the **pectoral girdles** is composed of a clavicle and scapula and attaches an upper limb to the axial skeleton.
2. Each **clavicle**, or collarbone, articulates medially with the sternum and laterally with the scapula.
3. Each **scapula** has an *acromion* that articulates with a clavicle. The scapula's *glenoid cavity* articulates with the head of the humerus, and the large surfaces of each scapula serve as attachment points for shoulder muscles.

See

- Anatomy Overview – The Appendicular Skeleton —The Pectoral Girdle

Hear

- Figure 8.1 – Right Pectoral Girdle

Concept 8.2 The bones of each upper limb include the humerus, ulna, radius, carpals, metacarpals, and phalanges.

1. The 30 bones of each upper limb include the following: a humerus in the arm; an ulna and radius in the forearm; 8 carpals in the wrist; 5 metacarpals in the palm; and 14 phalanges in the fingers.
2. The **humerus** articulates proximally by its *head* with the glenoid cavity of the scapula. Distally, at the *elbow joint* the *capitulum* of the humerus articulates with the head of the radius, and the *trochlea* of the humerus articulates with the ulna. The distal posterior surface has an *olecranon fossa* that receives the olecranon of the ulna when the elbow is extended.
3. The forearm bones, the medial **ulna** and the lateral **radius**, articulate proximally, at the elbow, with the humerus; distally, the radius articulates with three of the carpals. The ulna's *olecranon* forms the prominence of the elbow.
4. The **carpus** contains eight carpals. The concave space formed by several carpals plus the *flexor retinaculum* is the **carpal tunnel**. Tendons of muscles to the digits and thumb, and the median nerve, pass through this tunnel.
5. The **metacarpus**, or palm, consists of five **metacarpals** that articulate proximally with the carpals and distally with the phalanges.
6. The **phalanges** are the bones of the fingers. There are three phalanges in each finger, except the thumb, which has two phalanges.

See

- Anatomy Overview – The Upper Limb

Hear

- Figure 8.5 – Right Humerus in Relation to the Scapula, Ulna, and Radius

---

Concept 8.3 The pelvic girdle supports the vertebral column and pelvic viscera and attaches the lower limbs to the axial skeleton.

1. The **pelvic girdle** is composed of two **hip bones** that unite anteriorly at the **pubic symphysis** and articulate posteriorly with the sacrum. The hip bones, pubic symphysis, and sacrum form a basinlike structure called the **pelvis**, which gives strong support for the vertebral column, and protects lower abdominal and pelvic viscera. The pelvic girdle connects the lower limbs to the axial skeleton. Each hip bone is composed of three fused bones: the **ilium**, **ischium**, and **pubis**.
2. The ilium, the largest of the three components of the hip bone, comprises a superior *ala* and an inferior *body*.
3. The ischium, the inferior, posterior portion of the hip bone, comprises a superior *body* and an inferior *ramus*, which joins to the pubis.
4. The parts of the pubis include the *superior ramus, inferior ramus,* and *body*.
5. The point where these three bones fuse on the lateral surface of each hip bone is a socket, called the *acetabulum*, that accepts the head of the femur.
6. The pelvis is divided by the *pelvic brim* into a superior **false pelvis** and an inferior **true pelvis**. The pelvic brim is a ridge that runs from the sacral promontory to the arcuate lines to the superior region of the pubic symphysis.
7. The superior opening of the true pelvis, bordered by the pelvic brim, is called the *pelvic inlet*; its inferior opening is called the *pelvic outlet*.

See

- Anatomy Overview – The Pelvic Girdle

Concept	Resources

Concept 8.4 Male pelves are generally larger, heavier, and have more prominent markings; female pelves are generally wider and shallower.

1. The male pelvis is larger, heavier, and narrow compared to the female pelvis. Most structural differences in the female are to accommodate pregnancy and childbirth. The female pelvis is wide and shallow with more space in the true pelvis (see Table 8.1).

Concept 8.5 The bones of each lower limb include the femur, patella, tibia, fibula, tarsals, metatarsals, and phalanges.

1. The 30 bones of each **lower limb** include the femur in the thigh, the patella, the tibia and fibula in the leg, 7 tarsals in the ankle, 5 metatarsals in the foot, and 14 phalanges in the toes.
2. The **femur** is the longest, strongest bone in the body. Its head articulates with the acetabulum and its distal end articulates with the tibia and patella.
3. The **patella** is a sesamoid bone, enclosed in the tendon of the quadriceps, and located anterior to the *knee joint*.
4. The **tibia** is the more medial and weight-bearing leg bone. It articulates proximally with the femur to form the knee joint, and articulates distally with the talus of the ankle.
5. The **fibula** is lateral and parallel to the tibia. Its proximal end articulates with the tibia below the level of the knee joint, and the fibula articulates distally with the talus of the ankle.
6. The **tarsus** consists of seven tarsal bones.
7. There are five **metatarsal bones** of the **metatarsus** of the foot.
8. There are fourteen **phalanges**, three in each toe except for the big toe, which has two phalanges.

See
- Anatomy Overview – The Lower Limb

Hear
- **Figure 8.13 – Right Femur in Relation to the Hip Bone, Patella, Tibia, and Fibula**

Do
- Exercise – Bones, Bones, Bones

Other Chapter Resources

For additional information on clinical applications, medical disorders, and medical terminology, see Chapter 4 in the *Clinical Connections* manual packaged with your text.

 Within *WileyPLUS*
- Chapter Quiz
- Visual Anatomy
- Mastering Vocabulary
- Audio Glossary
- Web Links

• Understanding the Concepts

1. Why is the acromion used as a landmark in obtaining accurate measurements of sleeve length?

2. The skeletal component projecting from the medial side of our elbows is commonly called the funny bone because, when we accidentally hit it, we feel a strange, tingling pain due to stimulation of the ulnar nerve. Which part of which bone is the funny bone? With what other bone or bones does it articulate?

3. When raising your arm to ask a question in class, your deltoid muscle lifts your arm by pulling which part of the humerus? When flexing your elbow to raise a cup to your mouth, your biceps brachii muscle lifts your forearm by pulling which part of the radius?

4. How many carpals do you have in each wrist? Name each one.

5. On the way to class you slip on ice, landing hard on your right buttock. Which part of which hip bone is likely to take the brunt of the fall?

6. How would you distinguish between the false and true pelves? Between the pelvic inlet and pelvic outlet?

7. Some bicycle seats are designed for gender comfort; seats for female riders have two depressions that are farther apart than seats for male riders. Why?

8. Which part of which leg bone is injured when you accidentally bang your shin on a low table? Which of the two leg bones bears the weight of the body?

9. Name the bones that form the upper and lower limbs, from proximal to distal.

9

Articulations

The alarm went off at 5 A.M. Virtjean didn't hear it; her hearing wasn't that great anymore. But the flashing of the red digital clock's large numbers blinked in her face insistently. As she reached for the alarm clock, her hand, gnarled from arthritis, glowed red in the light of the clock's display.

Normally, Virtjean didn't need an alarm clock to wake her up. She had always been an early riser. "Early to bed and early to rise. Isn't that how the saying goes?" she would joke with her granddaughter. But today she was having visitors and wanted extra time to get ready. The occupational therapist at her new "home" said Virtjean could use the kitchen facility to make special goodies for her great-grandchildren. Virtjean knew her way around the kitchen. Seven children and a lifetime on a farm had made it a requirement as the matriarch of her clan.

But first she would have to get up. Every morning it was hard. She had often remarked to her friends that it was now just a part of life. Virtjean could no longer remember a time when her joints did not ache. She had first started noticing intermittent signs of arthritis when she was in her fifties. Now, at age 84, her hands were almost unusable. But she had learned to adapt and she was not going to let anything stop her today. She didn't get visits very often and she wanted this time to be special.

Virtjean has rheumatoid arthritis, a painful, chronic joint disease. It is an autoimmune disorder in which the immune system attacks the cartilage and linings of freely movable joints (synovial joints). Disorders that affect joints can limit our mobility and reduce our quality of life. As we will see in this chapter, proper joint functioning is important for movement of the body and manipulation of our environment.

INTRODUCTION

Bones are too rigid to bend without being damaged. Fortunately, flexible connective tissues form joints that hold bones together while still permitting, in most cases, some degree of movement. A **joint**, also called an **articulation**, is a point of contact between two bones, between bone and cartilage, or between bone and teeth. When we say one bone *articulates* with another bone, we mean that the bones

> *Flexible connective tissues form joints that hold bones together while still permitting, in most cases, some degree of movement.*

form a joint. Because most movements of the body occur at joints, you can appreciate their importance if you imagine how a cast over your knee joint makes walking difficult, or how a splint on a finger limits your ability to manipulate small objects. In this chapter we will explore the different types of joints, their structures, and the way in which joints allow the myriad movements that the human body can produce.

9.1 Joints are classified structurally as fibrous, cartilaginous, or synovial; they are classified functionally as synarthroses, amphiarthroses, or diarthroses.

Joints are classified structurally based on their anatomical characteristics and functionally based on the type of movement they permit. The structural classification of joints is based on (1) the presence or absence of a space between the articulating bones, called a synovial cavity, and (2) the type of connective tissue that holds the bones together. Structurally, joints are classified as one of the following types:

- **Fibrous joints** (FĪ-brus): The bones are held together by dense connective tissue that is rich in collagen fibers, and there is no synovial cavity.

- **Cartilaginous joints** (kar-ti-LAJ-i-nus): The bones are held together by cartilage, and there is no synovial cavity.

- **Synovial joints** (si-NŌ-vē-al; *syn-* = together): There is a synovial cavity, and the bones are united by the dense connective tissue of an articular capsule and often by accessory ligaments.

The functional classification of joints relates to the degree of movement they permit. Functionally, joints are classified as one of the following types:

- **Synarthrosis** (sin´-ar-THRŌ-sis): An immovable joint. The plural is *synarthroses*.

- **Amphiarthrosis** (am´-fē-ar-THRŌ-sis; *amphi-* = on both sides): A slightly movable joint. The plural is *amphiarthroses*.

- **Diarthrosis** (dī-ar-THRŌ-sis = movable joint): A freely movable joint. The plural is *diarthroses*. All diarthroses are synovial joints. They have a variety of shapes and permit several different types of movements.

The following sections present the joints of the body according to their structural classifications. As we examine the structure of each type of joint, we will also explore its functional attributes.

Checkpoint

1. On what basis are joints classified?
2. Which kind of joint is united by cartilage? Which kind is immovable? Which kind is highly movable?

9.2 Fibrous joints lack a synovial cavity and are held together by dense connective tissue.

Fibrous joints lack a synovial cavity, and the articulating bones are held together very closely by dense connective tissue. These joints, which permit little or no movement, include sutures, syndesmoses, and gomphoses.

Sutures

A **suture** (SOO-chur; *sutur*i- = seam) is a fibrous joint composed of a thin layer of dense connective tissue. Such joints are found only in the skull. An example is the coronal suture between the parietal and frontal bones (Figure 9.1a). The irregular, interlocking edges of sutures give them added strength and decrease their chance of fracturing. Because a suture is immovable, it is classified functionally as a synarthrosis.

Some sutures, although present during childhood, are replaced by osseous tissue in the adult. Such a suture is called a **synostosis** (sin´-os-TŌ-sis; *os-* = bone), or bony joint—a joint in which there is a complete fusion of osseous tissue across the suture line. An example is the *metopic suture* (me-TŌ-pik; *metopon* = forehead) between the left and right sides of the frontal bone that begins to

fuse during infancy. A synostosis is classified functionally as a synarthrosis.

Syndesmoses

A **syndesmosis** (sin´-dez-MŌ-sis; plural is *syndesmoses*; *syndesmo-* = band or ligament) is a fibrous joint in which there is a greater distance between the articulating bones and dense connective tissue than in a suture. The dense connective tissue can be arranged either as a bundle (ligament) or as a sheet (interosseous membrane) (Figure 9.1b). One example of a syndesmosis is the interosseous membrane between the parallel borders of the tibia and fibula. Because it permits slight movement, a syndesmosis is classified functionally as an amphiarthrosis.

Another example of a syndesmosis is a **gomphosis** (gom-FŌ-sis; plural is *gomphoses*; *gompho-* = a bolt or nail in which a cone-shaped peg fits into a socket). The only examples of gomphoses are the articulations between the roots of the teeth and their sockets (alveoli) in the maxillae and mandible (Figure 9.1b). The dense connective tissue between a tooth and its socket is the periodontal ligament. A gomphosis is classified functionally as a synarthrosis, an immovable joint.

Figure 9.1 **Fibrous joints.**

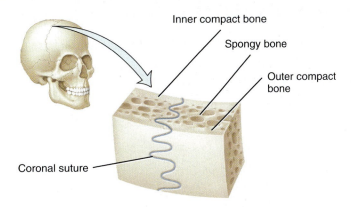

Inner compact bone

Spongy bone

Outer compact bone

Coronal suture

(a) Suture between skull bones

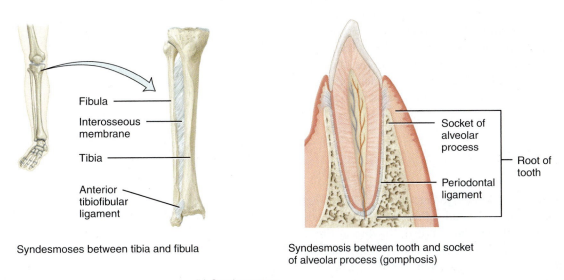

Fibula

Interosseous membrane

Tibia

Anterior tibiofibular ligament

Socket of alveolar process

Root of tooth

Periodontal ligament

Syndesmoses between tibia and fibula

Syndesmosis between tooth and socket of alveolar process (gomphosis)

(b) Syndesmoses

At a fibrous joint the bones are held together by dense connective tissue.

Checkpoint

3. Where in the body can sutures be found? Where can gomphoses be found?
4. Functionally, why are sutures classified as synarthroses and syndesmoses as amphiarthroses?

9.3 Cartilaginous joints lack a synovial cavity and are held together by cartilage.

Like a fibrous joint, a **cartilaginous joint** lacks a synovial cavity and allows little or no movement. Here the articulating bones are tightly connected, either by hyaline cartilage or by fibrocartilage. The two types of cartilaginous joints are synchondroses and symphyses.

Synchondroses

A **synchondrosis** (sin´-kon-DRŌ-sis; plural is *synchondroses*; *chondro-* = cartilage) is a cartilaginous joint in which the connecting material is hyaline cartilage. An example of a synchondrosis is the epiphyseal plate that connects the epiphysis and diaphysis of a growing bone (Figure 9.2a). A photomicrograph of the epiphyseal plate is shown in Figure 6.8a. Functionally, a synchondrosis is a synarthrosis, an immovable joint. When bone elongation ceases, osseous tissue replaces the hyaline cartilage, and the synchondrosis becomes a synostosis, a bony joint. Another example of a synchondrosis is the joint between the first rib and the manubrium of the sternum, which also ossifies during adult life and becomes an immovable synostosis.

Symphyses

A **symphysis** (SIM-fi-sis = growing together; plural is *symphyses*) is a cartilaginous joint in which the ends of the articulating bones are covered with hyaline cartilage, but a broad, flat disc of fibrocartilage connects the bones. All symphyses occur in the midline of the body. The pubic symphysis between the anterior surfaces of the hip bones is one example of a symphysis (Figure 9.2b). This type of joint is also found at the junction of the manubrium and body of the sternum (see Figure 7.21) and at the intervertebral joints between the bodies of vertebrae (see Figure 7.19a). A portion of the intervertebral disc is fibrocartilage. A symphysis is classified functionally as an amphiarthrosis, a slightly movable joint.

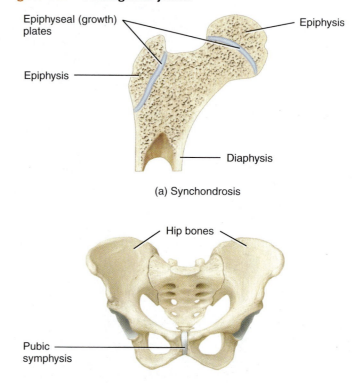

Figure 9.2   Cartilaginous joints.

Epiphyseal (growth) plates

Epiphysis

Epiphysis

Diaphysis

(a) Synchondrosis

Hip bones

Pubic symphysis

(b) Symphysis

At a cartilaginous joint the bones are held together by cartilage.

Checkpoint

5. How does the connective tissue in synchondroses and symphyses differ?
6. The epiphyseal plate in a young child is an example of which type of joint? When the epiphyseal plate ossifies, which kind of joint does it become?
7. Which types of cartilaginous joints are synarthroses? Which are amphiarthroses?

9.4 Articulating surfaces of bones at a synovial joint are covered with articular cartilage and enclosed within a synovial cavity.

Synovial joints have certain characteristics that distinguish them from other joints. The unique characteristic of a synovial joint is the presence of a space called a **synovial (joint) cavity** between the articulating bones (Figure 9.3). The synovial cavity allows the joint to be freely movable. Hence, all synovial joints are classified functionally as diarthroses. **Articular cartilage** covers the articulating surfaces of the bones at a synovial joint and provides a smooth, slippery surface that reduces friction between the bones in the joint during movement and also helps to absorb shock.

Articular Capsule

A sleevelike **articular capsule** surrounds the synovial joint, enclosing the synovial cavity and uniting the articulating ends of the bone. The articular capsule is composed of two layers, an outer fibrous membrane and an inner synovial membrane (Figure 9.3).

Figure 9.3 **Structure of a typical synovial joint.** Note the two layers of the articular capsule—the fibrous membrane and the synovial membrane.

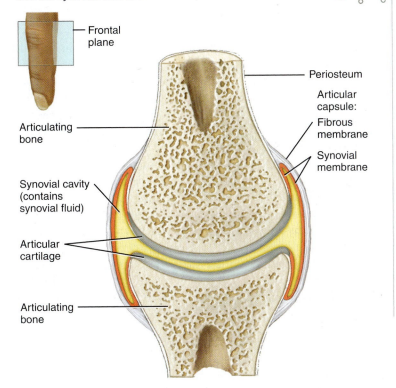

(a) Frontal section

The **fibrous membrane** usually consists of dense connective tissue (mostly collagen fibers) that attaches to the periosteum of the articulating bones. The flexibility of the fibrous membrane permits considerable movement at a joint while its great tensile strength (resistance to stretching) helps prevent the bones from dislocating. The fibers of some fibrous membranes are arranged in parallel bundles that are highly adapted for resisting strains. Such fiber bundles, called **ligaments** (*liga-* = bound or tied), are one of the principal mechanical factors that hold bones close together in a synovial joint. The inner layer of the articular capsule, the **synovial membrane**, is composed of areolar connective tissue. Many synovial joints include cushioning adipose tissue called **articular fat pads**. An example is the *infrapatellar fat pad* in the knee (see Figure 9.14c).

Synovial Fluid

The synovial membrane secretes **synovial fluid** (*ov-* = egg), which forms a thin film over the surfaces within the articular capsule. This viscous, clear fluid was named for its similarity in appearance and consistency to uncooked egg white (albumin). Synovial fluid consists of hyaluronic acid secreted by fibroblast-like cells in the synovial membrane and interstitial fluid filtered from blood plasma. Its functions include reducing friction by lubrication of the joint, absorbing shocks, and supplying oxygen and nutrients to and removing carbon dioxide and metabolic wastes from the chondrocytes within articular cartilage. (Recall that cartilage is an avascular tissue, so it does not have blood vessels to perform the latter function.) Synovial fluid also contains phagocytic cells that

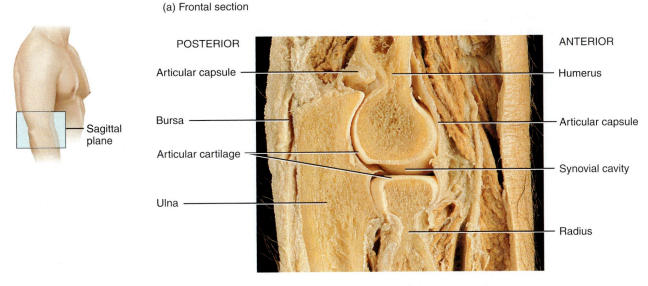

(b) Sagittal section of right elbow joint

The distinguishing feature of a synovial joint is the synovial cavity between the articulating bones.

remove microbes and the debris that results from normal wear and tear in the joint. When a synovial joint is immobile for a time, the fluid becomes quite viscous (gel-like), but as joint movement increases, the fluid thins, becoming less viscous. One of the benefits of warming up before exercise is that it stimulates the production and secretion of synovial fluid; more fluid means less stress on the joints during exercise.

We are all familiar with the cracking sounds heard as certain joints move, or the popping sounds that arise when a person pulls on the fingers to "crack" the knuckles. According to one theory, when the synovial cavity expands, the pressure of the synovial fluid decreases, creating a partial vacuum. This suction draws carbon dioxide and oxygen out of blood vessels in the synovial membrane, forming bubbles in the fluid. When the bubbles burst, as when the fingers are flexed (bent), the cracking or popping sound is heard.

Accessory Ligaments and Articular Menisci

Many synovial joints also contain **accessory ligaments** that lie outside and inside the articular capsule. Examples of accessory ligaments outside the articular capsule are the fibular and tibial collateral ligaments of the knee joint (see Figure 9.14d). Examples of accessory ligaments inside the articular capsule are the anterior and posterior cruciate ligaments of the knee joint (see Figure 9.14d).

Inside some synovial joints, such as the knee, pads of fibrocartilage lie between the articular surfaces of the bones and are attached to the fibrous membrane. These pads are called **articular menisci** (me-NIS-sī or me-NIS-kī; singular is *meniscus*). Figure 9.14d depicts the lateral and medial menisci in the knee joint. The menisci usually subdivide the synovial cavity into two separate spaces. This separation can allow separate movements to occur in each space. By modifying the shape of the joint surfaces of the articulating bones, menisci allow two bones of different shapes to fit more tightly together. Menisci also help to maintain the stability of the joint and direct the flow of synovial fluid to areas of greatest friction.

Bursae and Tendon Sheaths

The various movements of the body create friction between moving parts. Saclike structures called **bursae** (BER-sē = purses; singular is *bursa*) are strategically situated to alleviate friction in some joints, such as the shoulder and knee joints (see Figures 9.11a and 9.14c). Bursae are not strictly parts of synovial joints, but they do resemble joint capsules because their walls consist of connective tissue lined by a synovial membrane. They are also filled with a small amount of fluid similar to synovial fluid. Bursae are located between the skin and bone, tendons and bones, muscles and bones, and ligaments and bones. The fluid-filled bursal sacs cushion the movement of these body parts over one another.

Structures called tendon sheaths also reduce friction at joints. **Tendon sheaths** are tubelike bursae, which wrap around tendons that experience considerable friction. This occurs where tendons pass through synovial cavities, such as the tendon of the biceps brachii muscle at the shoulder joint (see Figure 9.11c). Tendon sheaths are also found in the fingers and toes, where there is a great deal of movement (see Figure 11.17a).

Checkpoint

8. Nutritional supplements that claim to rebuild knee cartilage are most likely referring to helping which component of the synovial joint?

9. What are the functions of synovial fluid?

10. What is the functional classification of synovial joints?

11. As you may have noticed in the chapters on the skeleton, the ends of long bones are often rounded. What structures add stability to synovial joints and compensate for this anatomical phenomenon?

12. How are bursae similar to and different from joint capsules? What is the function of bursae?

Virtjean got out of bed slowly, turning first onto her side, and then pushing herself up with her arms. As she bent forward to retrieve her slippers, she gasped in response to a sharp twinge of pain in her lower back. She paused for a moment, then reached for her slippers with a small gnarled hand. Her fingers were bent at odd angles, the knuckles enlarged and swollen.

Janey, the occupational therapist at Shady Acres, had spent quite a bit of time showing Virtjean how to use her remaining strength and mobility to manipulate her environment, such as using built-up handles on her toothbrush and eating utensils and getting dressed using a special "grabber" tool. Today Janey was going to help Virtjean in the kitchen so she could make cookies for her great-grandchildren.

Virtjean walked down the hall, passing several open doors. "Good morning, Gertrude," she called as she passed by.

"Good morning, Virj. Need any help in the kitchen?" Gertrude asked.

"I wouldn't mind the company."

Virtjean proceeded to get out the spoons and measuring cups, and Gertrude retrieved the flour and sugar from the cabinets. Virtjean was able to open the utensil drawer by hooking her middle and index fingers through the handle. She grasped a wooden spoon with a large-diameter foam handle, and struggled to pull apart a set of measuring cups, finally spilling them onto the counter with a clatter.

"I just can't get these old hands to cooperate," Virtjean sputtered.

Gertrude put an arm around her. "Honey, you're going to make cookies for your great-grandchildren today, arthritis or not."

A. Would the sutures in Virtjean's skull be affected by her rheumatoid arthritis? Why or why not?
B. What functional classification of joints is found at Virtjean's knuckles?
C. What properties of synovial joints allow healthy interphalangeal joints to move freely? What damage could have occurred inside Virtjean's interphalangeal joints to reduce their mobility?
D. The articular capsules of Virtjean's metacarpophalangeal joints (knuckles) are deformed and damaged. Describe the articular capsule, and explain its role in the synovial joint.

9.5   Synovial joints are described as planar, hinge, pivot, condyloid, saddle, or ball-and-socket.

Although all synovial joints are similar in structure, the shapes of the articulating surfaces vary, allowing a variety of movements. Accordingly, synovial joints are divided into six types: planar, hinge, pivot, condyloid, saddle, and ball-and-socket.

Planar Joints

The articulating surfaces of bones in a **planar joint** are flat or slightly curved (Figure 9.4a). Planar joints primarily permit side-to-side and back-and-forth gliding movements between flat surfaces of bones. Examples of planar joints are the intercarpal joints (between carpal bones at the wrist), intertarsal joints (between tarsal bones at the ankle), sternoclavicular joints (between the manubrium of the sternum and the clavicle), acromioclavicular joints (between the acromion of the scapula and the clavicle), and vertebrocostal joints (between the ribs and transverse processes of thoracic vertebrae).

Hinge Joints

In a **hinge joint**, the convex surface of one bone fits into the concave surface of another bone (Figure 9.4b). As the name implies, hinge joints produce an angular, opening-and-closing motion like that of a hinged door. In most joint movements, one bone remains in a fixed position while the other moves around an axis. An axis is a straight line around which a bone moves. Hinge joints allow motion along a single axis. Examples of hinge joints are the knee, elbow, ankle, and interphalangeal joints (between the phalanges of the fingers and toes).

Pivot Joints

In a **pivot joint**, the rounded or pointed surface of one bone articulates with a ring formed partly by another bone and partly by a ligament (Figure 9.4c). A pivot joint allows rotation only around its own axis. Examples of pivot joints are the atlanto-axial joint, in which the atlas rotates around the axis and permits the head to turn from side-to-side as in signifying "no" (see Figure 9.9a), and the radioulnar joints that enable the palms to turn anteriorly and posteriorly (see Figure 9.10h).

Condyloid Joints

In a **condyloid joint** (KON-di-loyd; *condyl-* = knuckle), the convex oval-shaped projection of one bone fits into the oval-shaped depression of another bone (Figure 9.4d). A condyloid joint permits movement around two axes, both up-and-down and from side-to-side. Examples are the wrist and the metacarpophalangeal joints (between the metacarpals and phalanges) of the second through fifth digits.

Figure 9.4   **Types of synovial joints.** For each type, a drawing of the actual joint and a simplified diagram are shown.

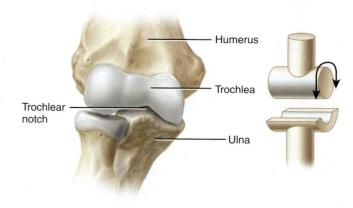

(a) Planar joint between navicular and second and third cuneiforms of tarsus in foot

(b) Hinge joint between trochlea of humerus and trochlear notch of ulna at the elbow

(c) Pivot joint between head of radius and radial notch of ulna

(d) Condyloid joint between radius and scaphoid and lunate bones of the wrist

(e) Saddle joint between trapezium of the wrist and metacarpal of the thumb

(f) Ball-and-socket joint between head of femur and acetabulum of hip bone

🔑 **Synovial joints are classified on the basis of the shapes of the articulating bone surfaces.**

Saddle Joints

In a **saddle joint**, the articular surface of one bone is saddle-shaped (the concave surface); the articular surface of the other bone fits into the saddle like a rider sitting on a horse (the convex surface) (Figure 9.4e). Saddle joints permit movement around three axes: side-to-side, up-and-down, and rotation. An example of a saddle joint is the carpometacarpal joint between the trapezium of the carpus and metacarpal of the thumb.

Ball-and-Socket Joints

A **ball-and-socket joint** consists of the ball-like surface of one bone fitting into a cuplike depression of another bone (Figure 9.4f). Such joints permit movement around three axes plus all directions in between. Examples of functional ball-and-socket joints are the shoulder and hip joints. At the shoulder joint the head of the humerus fits into the glenoid cavity of the scapula. At the hip joint the head of the femur fits into the acetabulum of the hip bone.

Table 9.1 summarizes the structural and functional categories of joints.

TABLE 9.1

Structural and Functional Classifications of Joints

STRUCTURAL CLASSIFICATION	DESCRIPTION	FUNCTIONAL CLASSIFICATION	EXAMPLE
Fibrous No synovial cavity; articulating bones held together by dense connective tissue.			
Suture	Articulating bones united by a thin layer of dense connective tissue, found between bones of the skull. With age, some sutures are replaced by a synostosis, in which separate cranial bones fuse into a single bone.	Synarthrosis (immovable).	Coronal suture.
Syndesmosis	Articulating bones united by dense connective tissue, either a ligament or an interosseous membrane.	Amphiarthrosis (slightly movable).	Distal tibiofibular joint.
Gomphosis	Articulating bones united by periodontal ligament; cone-shaped peg fits into a socket.	Synarthrosis.	At roots of teeth in alveoli (sockets) of maxillae and mandible.
Cartilaginous No synovial cavity; articulating bones united by cartilage.			
Synchondrosis	Connecting material is hyaline cartilage; becomes a synostosis when bone elongation ceases.	Synarthrosis.	Epiphyseal plate between the diaphysis and epiphysis of a long bone.
Symphysis	Connecting material is a broad, flat disc of fibrocartilage.	Amphiarthrosis.	Intervertebral joints and pubic symphysis.
Synovial Characterized by a synovial cavity, articular cartilage, and an articular capsule; may contain accessory ligaments, articular menisci, and bursae.			
Planar	Articulated surfaces are flat or slightly curved.	Diarthroses (freely movable); gliding motion.	Intercarpal, intertarsal, sternocostal (between sternum and the second to seventh pairs of ribs), and vertebrocostal joints.
Hinge	Convex surface fits into a concave surface.	Monaxial diarthrosis; flexion and extension.	Elbow, ankle, and interphalangeal joints.
Pivot	Rounded or pointed surface fits into a ring formed partly by bone and partly by a ligament.	Monaxial diarthrosis; rotation.	Atlanto-axial and radioulnar joints.
Condyloid	Oval-shaped projection fits into an oval-shaped depression.	Biaxial diarthrosis; flexion, extension, abduction, adduction, and circumduction.	Radiocarpal and metacarpophalangeal joints.
Saddle	Articular surface of one bone is saddle-shaped, and the articular surface of the other bone "sits" in the saddle.	Triaxial diarthrosis; flexion, extension, abduction, adduction, and circumduction.	Carpometarcarpal joint between trapezium and thumb.
Ball-and-socket	Ball-like surface fits into a cuplike depression.	Multiaxial diarthrosis; flexion, extension, abduction, adduction, circumduction, and rotation.	Shoulder and hip joints.

Checkpoint

13. Skip is asked by his instructor if he knows what a pivot joint is and he vigorously shakes his head side-to-side, indicating "no." His instructor responds "Good job. Nice example." Skip is confused—why is the instructor happy?
14. Why are condyloid and saddle joints important to many musicians?
15. Which types of synovial joints allow movement around one axis, two axes, three axes, no axes?

9.6 Synovial joint movement terminology indicates the direction of movement or the relationships of body parts during movement.

Anatomists, physical and occupational therapists, and kinesiologists use specific terminology to designate movements that can occur at synovial joints. These precise terms may indicate the form of motion, the direction of movement, or the relationship of one body part to another during movement. Movements at synovial joints are grouped into four main categories: (1) gliding, (2) angular movements, (3) rotation, and (4) special movements.

Gliding

Gliding is a simple movement in which relatively flat bone surfaces move back-and-forth and from side-to-side with respect to one another (Figure 9.5). There is no significant alteration of the angle between the bones. Gliding movements are limited in range due to the structure of the articular capsule and associated ligaments and bones. Gliding occurs at planar joints (see Figure 9.4a).

Angular Movements

In **angular movements**, there is an increase or a decrease in the angle between articulating bones. The principal angular move-

Figure 9.5   Gliding movements at synovial joints.

Gliding

Intercarpal joints

Gliding motion consists of side-to-side and back-and-forth movements.

ments are flexion, extension, lateral flexion, hyperextension, abduction, adduction, and circumduction. These movements are always discussed with respect to the body in the anatomical position.

Flexion, Extension, Lateral Flexion, and Hyperextension

Flexion and extension are opposite movements. In **flexion** (FLEK-shun; *flex-* = to bend) there is a decrease in the angle between articulating bones; in **extension** (eks-TEN-shun; *exten-* = to stretch out) there is an increase in the angle between articulating bones, often to restore a part of the body to the anatomical position after it has been flexed (Figure 9.6). Hinge, condyloid, saddle, and ball-and-socket joints all permit flexion and extension (Figure 9.4b,d–f).

All of the following are examples of flexion:

- Bending the head toward the chest at the atlanto-occipital joint between the atlas (the first vertebra) and the occipital bone of the skull, and at the intervertebral joints between the cervical vertebrae (Figure 9.6a).

- Bending the trunk forward at the intervertebral joints, as occurs when bending over to touch your toes.

- Moving the humerus forward at the shoulder joint, as in swinging the arms forward while walking (Figure 9.6b).

- Moving the forearm toward the arm at the elbow joint between the humerus, ulna, and radius (Figure 9.6c).

- Moving the palm toward the forearm at the wrist joint (Figure 9.6d).

- Bending at the interphalangeal joints between phalanges, as when making a fist of the fingers or when curling the toes.

- Moving the femur forward at the hip joint, as in walking (Figure 9.6e).

- Moving the leg toward the thigh at the knee joint, as occurs when bending the knee (Figure 9.6f).

Flexion and extension usually occur along the sagittal plane, although there are a few exceptions. An example is movement of the trunk sideways to the right or left at the waist as in a side bend. This movement, which occurs along the frontal plane and involves the intervertebral joints, is called **lateral flexion** (Figure 9.6g).

Figure 9.6 Angular movements at synovial joints.

(a) Atlanto-occipital and cervical intervertebral joints

(b) Shoulder joint

(c) Elbow joint

(d) Wrist joint

(e) Hip joint

(f) Knee joint

(g) Intervertebral joints

🔑 In angular movements, there is an increase or decrease in the angle between articulating bones.

The movement that returns body parts to the anatomical position is extension (Figure 9.6a–f). Continuation of extension beyond the anatomical position is called **hyperextension** (*hyper-* = beyond or excessive). Examples of hyperextension include the following:

- Bending the head backward at the atlanto-occipital and cervical intervertebral joints (Figure 9.6a).

- Bending the trunk backward at the intervertebral joints, as in a back bend.

- Moving the humerus backward at the shoulder joint, as in swinging the arms backward while walking (Figure 9.6b).

- Moving the palm backward at the wrist joint (Figure 9.6d).

- Moving the femur backward at the hip joint, as in walking (Figure 9.6e).

Hyperextension of hinge joints, such as the elbow, interphalangeal, and knee joints, is usually prevented by the arrangement of ligaments and the anatomical alignment of the bones.

Abduction, Adduction, and Circumduction

Abduction (ab-DUK-shun; *ab-* = away; *-duct-* = to lead) is the movement of a bone away from the midline; **adduction** (ad-DUK-shun; *ad-* = toward) is the movement of a bone toward the midline. Both movements usually occur along the frontal plane. Condyloid, saddle, and ball-and-socket

Figure 9.7   Angular movements at synovial joints—abduction and adduction.

(a) Shoulder joint
(b) Wrist joint
(c) Hip joint

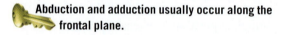

(d) Metacarpophalangeal joints of the fingers (not the thumb)

🔑 **Abduction and adduction usually occur along the frontal plane.**

joints permit abduction and adduction. Examples of abduction include moving the humerus laterally at the shoulder joint, moving the palm laterally (away from the body) at the wrist joint, and moving the femur laterally at the hip joint (Figure 9.7a–c). The movement that returns each of these body parts to the anatomical position is adduction (Figure 9.7a–c).

With respect to the digits, the midline of the body is not used as a point of reference for abduction and adduction. In abduction of the fingers, an imaginary line is drawn through the longitudinal axis of the middle (longest) finger, and the fingers move away (spread out) from the middle finger (Figure 9.7d). Abduction of the toes is relative to

an imaginary line drawn through the second toe. Adduction of the fingers and toes involves returning them to the anatomical position.

Circumduction (ser-kum-DUK-shun; *circ-* = circle) is movement of the distal end of a body part in a circle (Figure 9.8). Condyloid, saddle, and ball-and-socket joints permit circumduction. Circumduction occurs as a result of a continuous sequence of flexion, abduction, extension, and adduction. Examples of circumduction are moving the humerus in a circle at the shoulder joint (Figure 9.8a), moving the hand in a circle at the wrist joint, moving the thumb in a circle at the carpometacarpal joint (between the carpal and metacarpal), moving a finger in a circle at the metacarpophalangeal joint (between the metacarpal and proximal phalanx), and moving the femur in a circle at the hip joint (Figure 9.8b). Both the shoulder and hip joints permit circumduction, but circumduction is more limited at the hip due to greater tension on the ligaments and muscles.

Figure 9.8   Angular movements at synovial joints—circumduction.

(a) Shoulder joint
(b) Hip joint

🔑 **Circumduction is the movement of the distal end of a body part in a circle.**

Rotation

In **rotation** (rō-TA-shun; *rota-* = revolve), a bone revolves around its own longitudinal axis. Pivot and ball-and-socket joints permit rotation. One example is turning the head from side to side at the atlanto-axial joint, as in signifying "no" (Figure 9.9a). Another is turning the trunk from side to side at the intervertebral joints while keeping the hips and lower limbs in the anatomical position.

In the limbs, rotation is defined relative to the midline, and specific qualifying terms are used. If the anterior surface of a bone of the limb is turned toward the midline, the movement is called *medial rotation*. You can medially rotate the humerus at the shoulder joint as follows: Starting in the anatomical position, flex your elbow and then move your palm across the chest (Figure 9.9b). You can medially rotate the femur at the hip joint as follows: Lie on your back, bend your knee, and then move your leg and foot laterally from the midline. Although you are moving your leg and foot laterally, the femur is rotating medially (Figure 9.9c). If the anterior surface of the bone of a limb is turned away from the midline, the movement is called *lateral rotation* (Figure 9.9b,c).

Special Movements

Special movements are movements that occur only at certain joints. They include elevation, depression, protraction, retraction, inversion, eversion, dorsiflexion, plantar flexion, supination, pronation, and opposition (Figure 9.10):

- **Elevation** (el-e-VĀ-shun = to lift up) is an upward movement of a part of the body, such as closing the mouth at the temporomandibular joint (between the mandible and temporal bone) to elevate the mandible (Figure 9.10a) or shrugging the shoulders at the acromioclavicular joint (between the acromion of the scapula and the clavicle) to elevate the scapula.

- **Depression** (de-PRESH-un = to press down) is a downward movement of a part of the body, such as opening the mouth to depress the mandible (Figure 9.10b) or returning shrugged shoulders to the anatomical position to depress the scapula.

- **Protraction** (prō-TRAK-shun = to draw forth) is a movement of a part of the body anteriorly in the transverse plane. You can protract your mandible at the temporomandibular joint by thrusting it outward (Figure 9.10c).

- **Retraction** (rē-TRAK-shun = to draw back) is a movement of a protracted part of the body back to the anatomical position (Figure 9.10d).

- **Inversion** (in-VER-zhun = to turn inward) is movement of the sole of the foot medially at the intertarsal joints (between the tarsals) so that it faces the other foot (Figure 9.10e). Physical therapists also refer to inversion of the feet as *supination*.

- **Eversion** (ē-VER-zhun = to turn outward) is a movement of the sole laterally at the intertarsal joints so that it faces away from the other foot (Figure 9.10f). Physical therapists also refer to eversion of the feet as *pronation.*

- **Dorsiflexion** (dor-si-FLEK-shun) refers to bending of the foot at the ankle in the direction of the dorsum (superior surface) (Figure 9.10g). Dorsiflexion occurs when you stand on your heels.

Figure 9.9 Rotation at synovial joints.

(a) Atlanto-axial joint

(b) Shoulder joint

(c) Hip joint

 In rotation, a bone revolves around its own axis.

Figure 9.10 **Special movements at synovial joints.**

(a)   Temporomandibular joint   (b)   (c)   Temporomandibular joint   (d)

(e)   Intertarsal joints   (f)   (g) Ankle joint

(h) Radioulnar joint   (i) Carpometacarpal joint

🔑 Special movements occur only at certain synovial joints.

- **Plantar flexion** involves bending of the foot at the ankle joint in the direction of the plantar or inferior surface (Figure 9.10g), as when you elevate your body by standing on your toes.

- **Supination** (soo-pi-NĀ-shun) is a movement of the forearm at the proximal and distal radioulnar joints in which the palm is turned anteriorly, or forward (Figure 9.10h). This position of the palms is one of the defining features of the anatomical position.

- **Pronation** (prō-NĀ-shun) is a movement of the forearm so that the distal end of the radius crosses over the ulna and the palm is turned posteriorly, or backward (Figure 9.10h).

- **Opposition** (op-ō-ZISH-un) is the movement of the thumb at the carpometacarpal joint (between the trapezium and metacarpal of the thumb) in which the thumb moves across the palm to touch the tips of the fingers on the same hand (Figure 9.10i).

A summary of the movements that occur at synovial joints is presented in Table 9.2.

Checkpoint

16. Which type of angular movement describes what is happening at a joint when someone:
    a. Straightens the knees to move from a squatting position to a standing position?
    b. Draws a big circle during art class?
    c. Uses the hip joint to move the lower limb so that the toes point laterally?
    d. Brings the other fingers toward the middle finger to slip his hand through the sleeve of a coat?
    e. Closes her mouth?
17. In what way is considering adduction as "adding your limb to your trunk" an effective learning device?
18. A young bird is trying to fly. Which two movements is it producing as it flaps its wings up and down?
19. A common test of dexterity is to have someone touch, as fast as he can, his fingertips, one at a time, to the thumb on the same hand. Which type of movement is this?

TABLE 9.2

Movements at Synovial Joints

MOVEMENT	DESCRIPTION
Gliding	Movement of relatively flat bone surfaces back-and-forth and from side-to-side over one another; little change in the angle between bones.
Angular	Increase or decrease in the angle between bones.
Flexion	Decrease in the angle between articulating bones, usually in the sagittal plane.
Lateral flexion	Movement of the trunk in the frontal plane.
Extension	Increase in the angle between articulating bones, usually in the sagittal plane.
Hyperextension	Extension beyond the anatomical position.
Abduction	Movement of a bone away from the midline, usually in the frontal plane.
Adduction	Movement of a bone toward the midline, usually in the frontal plane.
Circumduction	Flexion, abduction, extension, and adduction in succession, in which the distal end of a body part moves in a circle.
Rotation	Movement of a bone around its longitudinal axis; in the limbs, it may be medial (toward midline) or lateral (away from midline).
Special	Occurs at specific joints.
Elevation	Superior movement of a body part.
Depression	Inferior movement of a body part.
Protraction	Anterior movement of a body part in the transverse plane.
Retraction	Posterior movement of a body part in the transverse plane.
Eversion	Lateral movement of the sole so that it faces away from the other foot.
Inversion	Medial movement of the sole so that it faces the other foot.
Dorsiflexion	Bending the foot in the direction of the dorsum (superior surface).
Plantar flexion	Bending the foot in the direction of the plantar surface (sole).
Supination	Movement of the forearm that turns the palm anteriorly.
Pronation	Movement of the forearm that turns the palm posteriorly.
Opposition	Movement of the thumb across the palm to touch fingertips on the same hand.

Return to Virtjean's Story

Janey walked into the kitchen. "I was hoping that you could show me some of the special kitchen appliances you told me about. I'm so clumsy with these old hands," Virtjean sighed.

Janey smiled. "From what I can see, it looks like you're off to a good start. There's an electric mixer for mixing your dough, and you and Gertrude can probably find most of what you need on your own. I would like to just watch how you're doing, Virtjean. That way, I can assess any difficulties, come up with a treatment plan, and think about other adaptive equipment."

Janey took some notes: *84-year-old female, history of rheumatoid arthritis, currently on pain medication (check with nursing staff on meds), has stage 3 R.A., deformity and limited mobility of fingers. Check arm strength, appears to have difficulty in extension of forearm at elbow, metacarpophalangeal joints of index and middle fingers may be subluxed (partially dislocated).*

Virtjean used a wooden spoon with a built-up handle to spoon cookie dough onto the sheet. Janey noticed that she remembered her instructions, hyperextending her hand at the wrist so that her fingers naturally flexed palmward. This simple trick of the arm's anatomy allowed someone with a weak grip like Virtjean to close their fingers to their palm.

Virtjean had trouble abducting her arms as well, so the women worked at a table that was adjusted to a lower than normal height. She stood with an awkward posture; her knees were flexed, her thighs somewhat medially rotated at the hips, and her feet everted. Janey wrote on her notepad: *New shoes to address postural concerns?*

E. Which type of synovial joint is the metacarpophalangeal joint? Which type of movement does it normally permit that Virtjean would now have difficulty doing?

F. Which movements would Virtjean normally have to make in order to spoon dough from a bowl onto a cookie sheet?

G. When Virtjean's knees are flexed and her thighs are rotated inward at the hips, are her knees abducted or adducted?

H. Virtjean has trouble abducting her arms. Explain what this means in terms that she could understand.

9.7 The shoulder, elbow, hip, and knee joints provide examples of synovial joint components, classifications, and movements.

In Chapters 7 and 8, we discussed the major bones and their markings. In this chapter we have examined how joints are classified according to their structure and function, and we have introduced the movements that occur at joints. Table 9.3 (selected joints of the axial skeleton) and Table 9.4 (selected joints of the appendicular skeleton) will help you integrate the information you learned about bones with the information about joint classifications and movements. These tables list some of the major joints of the body according to their articular components (the bones that enter into their formation), their structural and functional classifications, and the type(s) of movement that occur(s) at each joint.

Next we will examine four selected synovial joints in more detail. The joints described are the shoulder (glenohumeral) joint, elbow joint, hip (coxal) joint, and knee joint. Because these joints are described in detail in the text, they are not included in Tables 9.3 and 9.4. The temporomandibular and ankle joints are described in detail in *WileyPLUS* and are, therefore, also omitted from Tables 9.3 and 9.4. Each discussion of a selected synovial joint is accompanied by figures that illustrate that joint. Structures that are labeled in all capital letters are specifically referred to in the joint discussion. Additional related structures are also labeled as a reference source.

TABLE 9.3

Selected Joints of the Axial Skeleton

JOINT	ARTICULAR COMPONENTS	CLASSIFICATION	MOVEMENTS
Suture	Between skull bones.	Structural: fibrous. Functional: synarthrosis.	None.
Atlanto-occipital	Between superior articular facets of atlas and occipital condyles of occipital bone.	Structural: synovial (condyloid). Functional: diarthrosis.	Flexion, extension, and hyperextension of head and slight lateral flexion of head to either side.
Atlanto-axial	(1) Between dens of axis and anterior arch of atlas and (2) between lateral masses of atlas and axis.	Structural: synovial (pivot) between dens and anterior arch, and synovial (planar) between lateral masses. Functional: diarthrosis.	Rotation of head.
Intervertebral	(1) Between vertebral bodies and (2) between vertebral arches.	Structural: cartilaginous (symphysis) between vertebral bodies, and synovial (planar) between vertebral arches. Functional: amphiarthrosis between vertebral bodies, and diarthrosis between vertebral arches.	Flexion, extension, lateral flexion, and rotation of vertebral column.
Vertebrocostal	(1) Between facets of heads of ribs and facets of bodies of adjacent thoracic vertebrae and intervertebral discs between them and (2) between articular part of tubercles of ribs and facets of transverse processes of thoracic vertebrae.	Structural: synovial (planar). Functional: diarthrosis.	Slight gliding.
Sternocostal	Between sternum and first seven pairs of ribs.	Structural: cartilaginous (synchondrosis) between sternum and first pair of ribs, and synovial (planar) between sternum and second through seventh pairs of ribs. Functional: synarthrosis between sternum and first pair of ribs, and diarthrosis between sternum and second through seventh pairs of ribs.	None between sternum and first pair of ribs; slight gliding between sternum and second through seventh pairs of ribs.
Lumbosacral	(1) Between body of fifth lumbar vertebra and base of sacrum and (2) between inferior articular facets of fifth lumbar vertebra and superior articular facets of first vertebra of sacrum.	Structural: cartilaginous (symphysis) between body and base, and synovial (planar) between articular facets. Functional: amphiarthrosis between body and base, and diarthrosis between articular facets.	Flexion, extension, lateral flexion, and rotation of vertebral column.

TABLE 9.4

Selected Joints of the Appendicular Skeleton

JOINT	ARTICULAR COMPONENTS	CLASSIFICATION	MOVEMENTS
Sternoclavicular	Between sternal end of clavicle, manubrium of sternum, and first costal cartilage.	Structural: synovial (planar and pivot). Functional: diarthrosis.	Gliding, with limited movements in nearly every direction.
Acromioclavicular	Between acromion of scapula and acromial end of clavicle.	Structural: synovial (planar). Functional: diarthrosis.	Gliding and rotation of scapula on clavicle.
Radioulnar	Proximal radioulnar joint between head of radius and radial notch of ulna; distal radioulnar joint between ulnar notch of radius and head of ulna.	Structural: synovial (pivot). Functional: diarthrosis.	Supination and pronation of forearm.
Wrist (radiocarpal)	Between distal end of radius and scaphoid, lunate, and triquetrum of carpus.	Structural: synovial (condyloid). Functional: diarthrosis.	Flexion, extension, abduction, adduction, circumduction, and slight hyperextension of wrist.
Intercarpal	Between proximal row of carpal bones, distal row of carpal bones, and between both rows of carpal bones (midcarpal joints).	Structural: synovial (planar), except for hamate, scaphoid, and lunate (midcarpal) joint, which is synovial (saddle). Functional: diarthrosis.	Gliding plus flexion, extension, abduction, adduction, and slight rotation at midcarpal joints.
Carpometacarpal	Carpometacarpal joint of thumb between trapezium of carpus and first metacarpal; carpometacarpal joints of remaining digits formed between carpus and second through fifth metacarpals.	Structural: synovial (saddle) at thumb and synovial (planar) at remaining digits. Functional: diarthrosis.	Flexion, extension, abduction, adduction, and circumduction at thumb, and opposition.
Metacarpophalangeal and metatarsophalangeal	Between heads of metacarpals (or metatarsals) and bases of proximal phalanges.	Structural: synovial (condyloid). Functional: diarthrosis.	Flexion, extension, abduction, adduction, and circumduction of phalanges.
Interphalangeal	Between heads of phalanges and bases of more distal phalanges.	Structural: synovial (hinge). Functional: diarthrosis.	Flexion and extension of phalanges.
Sacroiliac	Between auricular surfaces of sacrum and ilia of hip bones.	Structural: synovial (planar). Functional: diarthrosis.	Slight gliding (even more so during pregnancy).
Pubic symphysis	Between anterior surfaces of hip bones.	Structural: cartilaginous (symphysis). Functional: amphiarthrosis.	Slight movements (even more so during pregnancy).
Tibiofibular	Proximal tibiofibular joint between lateral condyle of tibia and head of fibula; distal tibiofibular joint between distal end of fibula and fibular notch of tibia.	Structural: synovial (planar) at proximal joint, and fibrous (syndesmosis) at distal joint. Functional: diarthrosis at proximal joint, and amphiarthrosis at distal joint.	Slight gliding at proximal joint, and slight rotation of fibula during dorsiflexion of foot.
Ankle (talocrural)	(1) Between distal end of tibia and its medial malleolus and talus and (2) between lateral malleolus of fibula and talus.	Structural: synovial (hinge). Functional: diarthrosis.	Dorsiflexion and plantar flexion of foot.
Intertarsal	Subtalar joint between talus and calcaneus of tarsus; talocalcaneonavicular joint between talus and calcaneus and navicular of tarsus; calcaneocuboid joint between calcaneus and cuboid of tarsus.	Structural: synovial (planar) at subtalar and calcaneocuboid joints, and synovial at talocalcaneonavicular joint. Functional: diarthrosis.	Inversion and eversion of foot.
Tarsometatarsal	Between three cuneiforms of tarsus and bases of five metatarsal bones.	Structural: synovial (planar). Functional: diarthrosis.	Slight gliding.

The Shoulder Joint

Anatomical Components

The **shoulder joint** is a ball-and-socket joint formed by the head of the humerus and the glenoid cavity of the scapula (Figure 9.11c). It also is referred to as the *glenohumeral joint*. The **articular capsule** completely envelops the joint and extends from the glenoid cavity to the anatomical neck of the humerus (Figure 9.11a,c). The articular capsule is fairly thin and loose, allowing a free range of movement about the shoulder joint. Reinforcing ligaments of the shoulder joint are located primarily on the anterior and superior aspects of the joint. The **coracohumeral ligament** strengthens the superior part of the articular capsule and extends from the coracoid process of the scapula to the greater tubercle of the humerus (Figure 9.11a). The **glenohumeral ligaments** extend over the anterior surface of the joint from the glenoid cavity to the lesser tubercle and anatomical neck of the humerus. These ligaments are often indistinct or absent and provide only minimal strength (Figure 9.11a,b). The **glenoid labrum** forms a narrow rim of fibrocartilage around the edge of the glenoid cavity that slightly deepens and enlarges the socket (Figure 9.11b,c). Four **bursae** help reduce friction where body parts move across the shoulder joint. They are the *subscap-*

ular bursa (Figure 9.11a), *subdeltoid bursa, subacromial bursa* (Figure 9.11a–c), and *subcoracoid bursa.*

Movements

The shoulder joint allows flexion, extension, abduction, adduction, medial rotation, lateral rotation, and circumduction of the arm (see Figures 9.6–9.9). It has more freedom of movement than any other joint of the body. This freedom results from the looseness of the articular capsule and shallowness of the glenoid cavity in relation to the large size of the head of the humerus. The loose fit of the shoulder joint certainly allows mobility, but at the sacrifice of stability; hence, shoulder dislocations are common.

Although the ligaments of the shoulder joint strengthen it to some extent, most of the strength results from the muscles that surround the joint, especially the *rotator cuff muscles*. These muscles (supraspinatus, infraspinatus, teres minor, and subscapularis) join the scapula to the humerus (see also Figure 11.14). The tendons of the rotator cuff muscles encircle the joint and fuse with the articular capsule. The rotator cuff muscles work as a group to hold the head of the humerus in the glenoid cavity.

Figure 9.11   Right shoulder joint.

Clavicle

Acromioclavicular ligament

Coracoacromial ligament

Acromion of scapula

SUBACROMIAL BURSA

CORACOHUMERAL LIGAMENT

GLENOHUMERAL LIGAMENTS

TRANSVERSE HUMERAL LIGAMENT

Tendon of subscapularis muscle

Humerus

Coracoclavicular ligament:

Conoid ligament

Trapezoid ligament

Superior transverse scapular ligament

Coracoid process of scapula

SUBSCAPULAR BURSA

ARTICULAR CAPSULE

Scapula

Tendon of biceps brachii muscle (long head)

(a) Anterior view

View

SUPERIOR

Acromion of scapula ——————————————— Coracoacromial ligament

SUBACROMIAL BURSA ——————————— Tendon of supraspinatus muscle

Tendon of biceps brachii muscle (long head) ————————— CORACOHUMERAL LIGAMENT

CORACOID PROCESS OF SCAPULA

Tendon of infraspinatus muscle ————————— Tendon of subscapularis muscle

GLENOID CAVITY —————————————

ARTICULAR CAPSULE ————————————— GLENOHUMERAL LIGAMENTS

Tendon of teres minor muscle ————————— GLENOID LABRUM

POSTERIOR ANTERIOR

(b) Lateral view (opened with synovial membrane removed)

Frontal plane

Acromioclavicular ligament

Clavicle

Acromion of scapula ——————————— Tendon of supraspinatus muscle

Coracoacromial ligament —————————

SUBACROMIAL BURSA ————————— GLENOID LABRUM

ARTICULAR CAPSULE ————————————

Head of humerus ——————————— Scapula

GLENOID CAVITY

Tendon sheath ——————————— Articular cartilage

GLENOID LABRUM

ARTICULAR CAPSULE:

Tendon of biceps brachii muscle (long head) —————————

Synovial membrane

Fibrous membrane

Humerus

(c) Frontal section

Most of the stability of the shoulder joints results from the arrangement of the rotator cuff muscles.

The Elbow Joint

Anatomical Components

The **elbow joint** is a hinge joint formed by the trochlea of the humerus, the trochlear notch of the ulna, and the head of the radius (see Figure 8.7). The **articular capsule** extends from the humerus to the ulna and the annular ligament that encircles the head of the radius (Figure 9.12a,b). The **ulnar collateral ligament** extends from the medial epicondyle of the humerus to the coronoid process and olecranon of the ulna (Figure 9.12a). The **radial collat-** **eral ligament** extends from the lateral epicondyle of the humerus to the annular ligament of the radius (Figure 9.12b).

Movements

The elbow joint allows flexion and extension of the forearm (see Figure 9.6c). The strong ulnar and radial collateral ligaments restrict side-to-side movements at the elbow. The annular ligament connects the head of the radius to the ulna, while allowing the head to rotate within the ligament during forearm supination and pronation.

Figure 9.12   Right elbow joint. (See also Figure 9.3b.)

(a) Medial aspect

(b) Lateral aspect

The elbow joint is formed by parts of three bones: humerus, ulna, and radius.

The Hip Joint

Anatomical Components

The **hip joint** *(coxal joint)* is a ball-and-socket joint formed by the head of the femur (see Figure 8.13) and the acetabulum of the hip bone (see Figure 8.10b). The **articular capsule** of the hip joint is very dense and strong and extends from the rim of the acetabulum to the neck of the femur (Figure 9.13c). This capsule, which is one of the strongest structures of the body, consists of overlapping circular and longitudinal fibers. The **acetabular labrum** forms a fibrocartilage rim attached to the margin of the acetabulum that increases the depth of the socket (Figure 9.13c). Because the acetabulum provides such a deep socket and the diameter of the acetabular rim is smaller than that of the head of the femur, dislocation of the femur is rare.

Reinforcing the capsule are several strong accessory ligaments. The **iliofemoral ligament** extends from the ilium of the hip bone to the intertrochanteric line of the femur (Figure 9.13a,b). The **pubofemoral ligament** extends from the pubic wall of the acetabulum to the neck of the femur (Figure 9.13a). The **ischiofe-** **moral ligament** extends from the ischial wall of the acetabulum to the neck of the femur (Figure 9.13b). The **ligament of the head of the femur** extends from the fossa of the acetabulum to the fovea capitis, a small depression in the center of the head of the femur (Figure 9.13c). Also present is the **transverse ligament of the acetabulum**, which supports the acetabular labrum by crossing over the inferior acetabulum to connect with the ligament of the head of the femur and the articular capsule (Figure 9.13c).

Movements

The hip joint allows flexion, extension, abduction, adduction, circumduction, medial rotation, and lateral rotation of the thigh (see Figures 9.6–9.9). The extreme stability of the hip joint is related to the very strong articular capsule and its accessory ligaments, the manner in which the femur fits into the acetabulum, and the muscles surrounding the joint. Although the shoulder and hip joints are both ball-and-socket joints, the hip joint does not have so wide a range of motion due to tension of the hip joint's strong ligaments and the deep socket of the acetabulum.

Figure 9.13   Right hip joint.

Greater trochanter of femur

ILIOFEMORAL LIGAMENT

Lesser trochanter of femur

Femur

Tendon of rectus femoris muscle

PUBOFEMORAL LIGAMENT

Obturator canal

Obturator membrane

Hip bone

(a) Anterior view

(continues)

Figure 9.13 (continued)

ILIOFEMORAL LIGAMENT

Hip bone

Reflected tendon of rectus femoris muscle

Greater trochanter of femur

ISCHIOFEMORAL LIGAMENT

Zona orbicularis

Lesser trochanter of femur

Femur

(b) Posterior view

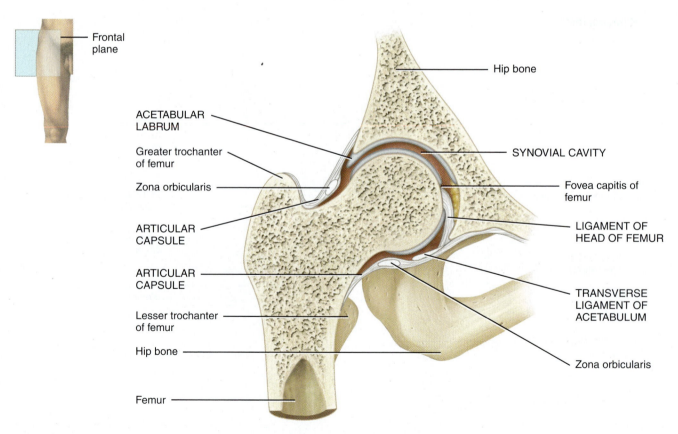

Frontal plane

ACETABULAR LABRUM

Greater trochanter of femur

Zona orbicularis

ARTICULAR CAPSULE

ARTICULAR CAPSULE

Lesser trochanter of femur

Hip bone

Femur

Hip bone

SYNOVIAL CAVITY

Fovea capitis of femur

LIGAMENT OF HEAD OF FEMUR

TRANSVERSE LIGAMENT OF ACETABULUM

Zona orbicularis

(c) Frontal section

🔑 The articular capsule of the hip joint is one of the strongest structures in the body.

The Knee Joint

Anatomical Components

The **knee joint**, the largest and most complex joint of the body, actually consists of three joints within a single synovial cavity (Figure 9.14). The lateral tibiofemoral joint is formed by the lateral condyle of the femur, lateral meniscus, and lateral condyle of the tibia; it is a modified hinge joint. The medial tibiofemoral joint is formed between the medial condyle of the femur, medial meniscus, and medial condyle of the tibia; it is also a modified hinge joint. The third joint, the patellofemoral joint, is located between the patella and the patellar surface of the femur and is a planar joint. The **medial** and **lateral menisci**, two C-shaped **articular** discs of fibrocartilage located between the tibial and femoral condyles, help compensate for the irregular shapes of the articulating bones (Figure 9.14c,d).

The **articular capsule** of the knee joint is weak and incomplete, but it is strengthened by muscle tendons and ligaments associated with the joint (Figure 9.14a,b). The tendon of the anterior thigh muscle (quadriceps femoris) continues across the anterior surface of the knee and inserts on the anterior surface of the tibia as the **medial** and **lateral patellar retinacula** and the patellar ligament (Figure 9.14a,c). The **patellar ligament** extends from the patella to the tibial tuberosity (Figure 9.14d) and is separated from the synovial membrane of the joint by an **infrapatellar fat pad** (Figure 9.14c). Together, the medial and lateral patellar retinacula and the patellar ligament strengthen the anterior surface of the joint.

The **oblique popliteal ligament** extends from the intercondylar fossa (between the condyles) of the femur to the head of the tibia (Figure 9.14b). The **arcuate popliteal ligament** extends from the lateral condyle of the femur to the head of the fibula (Figure 9.14b). The two popliteal ligaments reinforce the posterior surface of the joint.

The **tibial collateral ligament** extends from the medial condyle of the femur to the medial condyle of the tibia and strengthens the medial surface of the joint (Figure 9.14a,b,d). Because the tibial collateral ligament is firmly attached to the medial meniscus, tearing of the ligament frequently results in tearing of the meniscus and damage to the anterior cruciate ligament, described shortly. The **fibular collateral ligament** extends from the lateral condyle of the femur to the head of the fibula (Figure 9.14a,b,d). It strengthens the lateral aspect of the joint.

Figure 9.14 Right knee joint.

(a) Anterior superficial view

Femur

Vastus lateralis muscle

Tendon of quadriceps femoris muscle

Patella

LATERAL PATELLAR RETINACULUM

FIBULAR COLLATERAL LIGAMENT

Head of fibula

INFRAPATELLAR BURSA

Fibula

SUPRAPATELLAR BURSA

Vastus medialis muscle

MEDIAL PATELLAR RETINACULUM

Infrapatellar fat pad

PATELLAR LIGAMENT

TIBIAL COLLATERAL LIGAMENT

ARTICULAR CAPSULE

PATELLAR LIGAMENT

Tibia

(b) Posterior deep view

Femur

Tendon of adductor magnus muscle

Medial head of gastrocnemius muscle

TIBIAL COLLATERAL LIGAMENT

Popliteus muscle

Semimembranosus tendon

Tibia

ARTICULAR CAPSULE

Lateral head of gastrocnemius muscle

OBLIQUE POPLITEAL LIGAMENT

ARCUATE POPLITEAL LIGAMENT

FIBULAR COLLATERAL LIGAMENT

Posterior ligament of head of fibula

Fibula

(continues)

Figure 9.14 (continued)

(c) Sagittal section

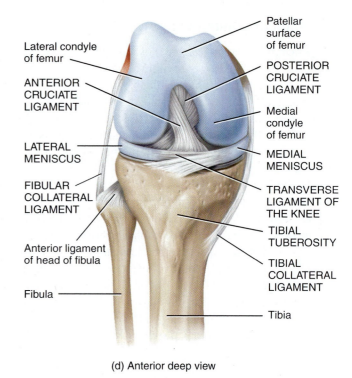

(d) Anterior deep view

🔑 **The knee joint is the largest and most complex joint in the body.**

Within the articular capsule, two ligaments connect the tibia and femur. The anterior and posterior cruciate ligaments (KROO-shē-āt = shaped like a cross) are named based on their attachment sites on the tibia and because they cross one another on their way to their destinations on the femur. The **anterior cruciate ligament (ACL)** extends posteriorly and laterally from a point *anterior* to the intercondylar area of the tibia to the medial surface of the lateral condyle of the femur (Figure 9.14d). The ACL limits hyperextension of the knee and prevents the anterior sliding of the tibia on the femur. This ligament is stretched or torn in about 70% of all serious knee injuries. The **posterior cruciate ligament (PCL)** extends anteriorly and medially from the *posterior* intercondylar area of the tibia to the lateral surface of the medial condyle of the femur (Figure 9.14d). The PCL prevents the posterior sliding of the tibia on the femur, especially when the knee is flexed. This is very important when walking down stairs or a steep incline.

Several important bursae are associated with the knee to help relieve friction at the knee joint. For example, the **prepatellar bursa**, located between the patella and skin, is often damaged when we bump the anterior knee (Figure 9.14c).

Movements

The knee joint allows flexion, extension, and slight medial rotation, and lateral rotation of the leg in the flexed position (see Figures 9.6f and 9.9c).

Checkpoint

20. At the shoulder joint, which ligaments connect the scapula and humerus? Glenoid cavity and humerus?
21. Which structures provide the shoulder joint with stability?
22. At the elbow joint, which ligaments connect the humerus and ulna? Humerus and radius?
23. What movements can occur in the elbow joint?
24. At the hip joint, which ligaments connect the ilium and femur? Pubis and femur?
25. At the knee joint, which ligaments connect the anterior thigh muscle tendon and tibia? Lateral condyle of femur and fibula?

Virtjean's Story: Epilogue and Discussion

▶▶▶The sweet smell of baking cookies drifted out of the kitchen and down the hallways of Shady Acres. Two small children burst through the door of the kitchen, calling out, "Granny! Granny!" Virtjean's great-granddaughter stood in the doorway.

"Hi, Granny. We're finally here. You kids be careful. They were so excited to see you." She tipped her head back and inhaled deeply. "Mmm, you made cookies. Are they ready?"

"Yeah, are they ready, Granny? I love your cookies!" chimed in the children.

"Let's take a peek!" replied Virtjean, smiling as she turned toward the oven.

In this chapter we have seen that joints include any point of contact between two bones. These articulations are classified structurally and functionally, based on the type of movement they permit. The aging process, stress, and daily wear and tear can all have negative effects on joint function. Joint mobility is also reduced by

diseases such as Virtjean's rheumatoid arthritis, which attacks the synovial membrane and articular cartilage, or osteoarthritis, which produces a progressive loss of articular cartilage and formation of new osseous tissue into obstructive bone spurs. Any number of conditions can alter joint flexibility and significantly impact our daily lives. An understanding of the structure and function of joints can assist in determining treatments and making adaptations to improve the quality of life. Perhaps in the future we will gain the knowledge to restore Virtjean's hands.◀◀◀

I.   How are synovial joints different from fibrous or cartilaginous joints? Which type of joints are the most seriously affected by Virtjean's rheumatoid arthritis?

J.   Virtjean has rheumatoid arthritis; the articular cartilage in her joints has been damaged and the synovial cavity is reduced. Explain how these conditions affect her joints.

Concept and Resource Summary

Concept	Resources
Introduction 1. A **joint**, or **articulation**, is the area where two or more bones meet, or where bone and cartilage, or bone and teeth, meet. 2. Most movements of the body occur at joints.	
Concept 9.1 Joints are classified structurally as fibrous, cartilaginous, or synovial; they are classified functionally as synarthroses, amphiarthroses, or diarthroses. 1. *Structural* classification of joints is based on (1) presence or absence of a synovial cavity and (2) the type of connective tissue binding the bones together. 2. There are three types of joints based on structure. **Fibrous joints** are held together by dense connective tissue and do not have a synovial cavity. **Cartilaginous joints** are held together by cartilage and do not have a synovial cavity. **Synovial joints** are held together by dense connective tissue of an articular capsule and possess a synovial cavity. 3. *Functional* classification of joints is based on the degree of movement that they permit. There are three functional types of joints: immovable **synarthoses**; slightly movable **amphiarthroses**; and freely movable **diarthroses**.	See • Anatomy Overview – Joints
Concept 9.2 Fibrous joints lack a synovial cavity and are held together by dense connective tissue. 1. **Fibrous joints** include sutures, syndesmoses, and gomphoses. 2. A **suture** is composed of thin dense connective tissue. Sutures are synarthroses and are only found in the skull. 3. Some sutures, called **synostoses**, are present during childhood but eventually get replaced by osseous tissue in adulthood. 4. In a **syndesmosis**, the dense connective tissue of a ligament or interosseous membrane unites the bones and permits slight movement. 5. A **gomphosis** is a peg-in-a-socket joint composed of a bone held in its bony socket by a periodontal ligament. It is a synarthrosis; the only examples in the human body are the teeth.	See • Anatomy Overview – Fibrous Joints
Cncept 9.3 Cartilaginous joints lack a synovial cavity and are held together by cartilage. 1. **Cartilaginous joints** include synchondroses and symphyses. 2. A **synchondrosis** is composed of hyaline cartilage and is a synarthrosis. 3. A **symphysis** has fibrocartilage connecting the bones at their articular cartilage surfaces; it is an amphiarthrosis.	See • Anatomy Overview – Cartilaginous Joints

Concept

Concept 9.4   Articulating surfaces of bones at a synovial joint are covered with articular cartilage and enclosed within a synovial cavity.

1. All **synovial joints** have a **synovial cavity** that allows the joint to function as a diarthrosis. The bones have a covering of **articular cartilage** on their articulating surface to reduce friction and absorb shock.
2. The synovial cavity is enclosed by a sleevelike **articular capsule** composed of an outer **fibrous membrane** that helps stabilize the joint and an inner **synovial membrane**.
3. The synovial membrane secretes a viscous, clear **synovial fluid** that forms a thin film over the articular capsule surfaces to reduce friction, absorb shock, supply oxygen and nutrients, and remove wastes.
4. **Accessory ligaments** located inside and outside the articular capsule are often found at synovial joints.
5. Certain synovial joints, like the knee, have pads of fibrocartilage, called **articular menisci**, lying between the articular surfaces of the bones and attached to the fibrous membrane. These discs stabilize the joint and direct the flow of synovial fluid to areas of greatest friction.
6. **Bursae** are sacs lined with a synovial membrane and filled with synovial fluid. They cushion adjacent body parts at certain joints and help alleviate friction between them. Tubelike bursae called **tendon sheaths** wrap around muscle tendons to prevent friction.

See

- **Anatomy Overview – Synovial Joints**

Hear

- **Figure 9.3 –** Structure of a Typical Synovial Joint

Do

- **Exercise – Judge the Joint Concepts and Connections – Joint Classifications**

Concept 9.5   Synovial joints are described as planar, hinge, pivot, condyloid, saddle, or ball-and-socket.

1. The shapes of the articulating surfaces of bones vary in synovial joints and govern their range of motion. The six types of synovial joints include planar, hinge, pivot, condyloid, saddle, and ball-and-socket.
2. Bones in a **planar joint** have flat articular surfaces permitting gliding movements that do not involve rotation around an axis.
3. In a **hinge joint**, the convex surface of one bone fits into the concave surface of another bone. An opening-and-closing motion, like a hinged door, occurs at this joint.
4. A **pivot joint** has the rounded surface of one bone articulating with a ring-shaped bone or ligament to allow rotation around its own axis.
5. A **condyloid joint** has a convex oval-shaped projection of one bone fitting into an oval-shaped depression of another bone. It allows both up-and-down and side-to-side movements.
6. A **saddle joint** involves a concave articular surface of one bone that fits into the convex surface, or "saddle," of another bone. This joint allows side-to-side, up-and-down movements, and rotation.
7. A **ball-and-socket joint** has the ball-like articular surface of one bone fitting into a cuplike depression of another bone. This joint permits movement around three axes and all directions in between. Of all the joints, this type is the most freely movable.

See

- **Anatomy Overview – Structural Subtypes of Synovial Joints**
- **Anatomy Overview – Specific Synovial Joints**

Concept 9.6   Synovial joint movement terminology indicates the direction of movement or the relationships of body parts during movement.

1. Movements at synovial joints are grouped into four categories: (1) gliding, (2) angular movements, (3) rotation, and (4) special movements.
2. **Gliding** occurs at planar joints where flat bone surfaces move back-and-forth and side-to-side in relation to one another without alteration of the angles between the bones.
3. **Angular movements** occur when the angle between articulating bones increases or decreases during movement. Angular movements include flexion, extension, lateral flexion, hyperextension, abduction, adduction, and circumduction.
4. **Flexion** is a decrease in the angle between articulating bones; **extension** is an increase in the angle between articulating bones. When flexion occurs laterally along the frontal plane, rather than along the sagittal plane, it is **lateral flexion**. When body parts at a joint undergo extension beyond their anatomical position, it is **hyperextension**.
5. **Abduction** is the movement of a bone away from the midline; **adduction** is the movement of a bone toward the midline.
6. **Circumduction** is movement of the distal end of a body part in a circle. It results from a sequence of movements: flexion, abduction, extension, and adduction.
7. **Rotation** involves a bone revolving around its own axis. In *medial rotation* a limb rotates toward the midline; in *lateral rotation* a limb rotates away from the midline.
8. **Special movements** occur only at certain joints. These include elevation, depression, protraction, retraction, inversion, eversion, dorsiflexion, plantar flexion, supination, pronation, and opposition.
9. Upward movement of a body part is **elevation**, whereas downward movement of a body part is **depression**.
10. Movement of a body part anteriorly in the transverse plane is **protraction**, whereas movement of a protracted body part back to the anatomical position is **retraction**.

See

- **Anatomy Overview – Movements Produced at Synovial Joints**

Hear

- **Figure 9.6 –** Angular Movements at Synovial Joints
- **Figure 9.7 –** Angular Movements at Synovial Joints—Abduction and Adduction
- **Figure 9.8 –** Angular Movements at Synovial Joints—Circumduction
- **Figure 9.9 –** Rotation at Synovial Joints

Do

- **Exercise – Manage Movements**

Concept	Resources

11. Moving the soles of the feet medially is **inversion**; moving the soles laterally is **eversion**.

12. Bending of the foot at the ankle in the direction of its superior surface is termed **dorsiflexion**; bending of the foot at the ankle in the direction of its inferior surface is **plantar flexion**.

13. **Supination** is movement of the forearm resulting in the palm facing anteriorly; **pronation** is the movement of the forearm resulting in the palm facing posteriorly.

14. Movement of the thumb at the carpometacarpal joint across the palm to touch the fingertips is **opposition**.

Concept 9.7 The shoulder, elbow, hip, and knee joints provide examples of synovial joint components, classifications, and movements.

1. The **shoulder joint** is formed by the head of the humerus and the glenoid cavity of the scapula. Anterior and superior ligaments help reinforce the joint, and four bursae help reduce friction. Most joint stability comes from the *rotator cuff muscles* and their tendons that encircle the joint. Rotation and every type of angular movement are permitted at this joint.

2. The **elbow joint** is a hinge joint formed by the trochlea of the humerus, the trochlear notch of the ulna, and the head of the radius. It is stabilized by strong ligaments. This joint permits flexion and extension of the forearm.

3. The **hip joint** is a ball-and-socket joint formed by the head of the femur and the acetabulum of the coxal bone. Its articular capsule is dense and strong, and reinforced by several strong ligaments. The acetabulum is a deep socket that secures articulation with the femur. Rotation and every angular movement are permitted; however, the hip joint does not have so wide a range of motion compared to the shoulder joint because of its strong ligaments and deep acetabulum.

4. The **knee joint** is the body's largest and most complex joint, consisting of three joints within one synovial cavity: (1) the lateral tibiofemoral joint, (2) the medial tibiofemoral joint, and (3) the patellofemoral joint. The weak, incomplete articular capsule is strengthened by the muscle tendon of the anterior thigh muscle and other tendons and ligaments. The tibial and fibular collateral ligaments give the joint stability.

See

- **Anatomy Overview – Specific Synovial Diarthrotic Joints**

Hear

- **Figure 9.14 – Right Knee Joint**

Other Chapter Resources

 For additional information on clinical applications, medical disorders, and medical terminology, see Chapter 5 in the *Clinical Connections* manual packaged with your text.

Within *WileyPLUS*

- Chapter Quiz
- Visual Anatomy
- Mastering Vocabulary
- Audio Glossary
- Homeostatic Imbalance: The Case of the Painful Hands
- Web Links

• Understanding the Concepts

1. What are the functional classifications of joints? Give an example of each type.

2. Maria is walking to class when she trips and falls, dislocating her shoulder. Which structures in the articular capsule have failed Maria?

3. Jamar likes to run but hates to stretch. Why will this eventually lead to an early retirement from regular running?

4. Jill likes to "crack" her knuckles, but her mother hates it and tells Jill that it can't be good for Jill's knuckles when she makes her bones snap into each other. What does cause knuckles to crack? Is it damaging the articulating ends of Jill's bones?

5. Which type of joint is found between vertebrae? How movable are these joints normally? How do they become painful, as in Virtjean's case?

6. After standing for several hours at a grocery store cash register, a checker's feet are sore and tired. While continuing to stand and check groceries, she attempts to relieve her foot discomfort. Which movements is the checker using when she raises her heels while rocking onto her toes? Raises her feet by rocking onto her heels?

7. Why does the shoulder joint have more freedom of movement than any other joint of the body?

8. Why is the hip joint so stable? List all of the structures that contribute to its stability.

9. Describe the movement thiat occurs at the knee joint when the anterior thigh muscles contract.

10. People often require hip replacement surgery following a broken hip. In which component of the hip joint would hip fractures most likely occur? Which type of synovial joint is the hip? Which structures normally stabilize the hip joint?

10

Muscle Tissue

Jesse moved down the basketball court with a determined look. His mind focused on his next move as he spun and drove forward to put the ball up. Jesse felt his feet leave the ground as he leaped upward. He grasped the basketball with two hands and brought them up, putting just the right spin on the ball with his fingers. But then a strange thing happened; he couldn't complete the move. His left arm would not obey his commands and the ball shot up past the backboard. "Oh man, what are you doing? What was that?" Jesse's friend Ronnie looked incredulously at Jesse. "What's the matter with you, man? If you can't make a shot like that, pass me the ball."

Jesse shot back an angry glance. "I was fouled. Look, if you're gonna get in my face, you can find someone else to play." Jesse turned and strode off the concrete playground court, not even slowing down as he grabbed his jacket and gym bag. He didn't stop jogging until he was several blocks away. He knew he wasn't really mad at Ronnie, he was mad at himself. He had always been good at sports; he played football, ran for the track team, and he loved basketball. But he had noticed lately that something wasn't right. He couldn't raise his left arm up over his head. It felt weaker the higher he tried to raise it. It had slowly been getting worse, and even increasing his weight lifting at the gym didn't seem to help. He had noticed, too, that his shoulder blade felt funny when he made certain movements, like it was coming loose. Even though he didn't want to admit it, he knew something was wrong. He wanted to find out what was happening; he knew that he couldn't make the varsity team in the fall if this kept getting worse. He would have to tell his mother.

When you think about muscles, what do you think of? Strength? Power? Professional athletes? We all have muscles that help us perform a variety of tasks, not just moving our bones to help sink a basket. In the next two chapters we will learn about what muscles are, what they do, and, in Jesse's case, the effects on daily life if they do not function properly.

Although bones provide leverage and form the framework of the body, they cannot move body parts by themselves. Motion results from the alternating contraction and relaxation of muscles, which constitute 40–50% of total body weight. Your muscle strength reflects

The prime function of muscle is the transformation of chemical energy into mechanical energy to generate force, perform work, and produce movement.

the prime function of muscle— the transformation of chemical energy into mechanical energy to generate force, perform work, and produce movement. In addition, muscle tissue stabilizes the body's position, regulates organ volume, generates heat, and propels fluids and food through various body systems.

10.1 Skeletal, cardiac, and smooth muscle tissues differ in location, structure, and function.

There are three types of **muscle tissue:** skeletal, cardiac, and smooth (Figure 10.1). Although the three types of muscle tissue share some properties, they differ from one another in their microscopic anatomy, location, and control by the nervous and endocrine systems.

Skeletal muscle tissue is so named because the function of most skeletal muscles is to move bones of the skeleton. (A few skeletal muscles attach to and move the skin or other skeletal muscles.) Skeletal muscle tissue is referred to as *striated* because alternating light and dark bands (*striations*) are visible when the tissue is examined under a microscope (see Figure 10.6). Skeletal muscle tissue works primarily in a *voluntary* manner; that is, its activity can be consciously (voluntarily) controlled by the nervous system. Most skeletal muscles also are controlled subconsciously to some extent. For instance, your diaphragm continues to alternately contract and relax while you are asleep so that you don't stop breathing. Also, you do not need to consciously think about contracting the skeletal muscles that maintain your posture or stabilize body positions.

Cardiac muscle tissue is found only in the heart, where it forms most of the heart wall. Like skeletal muscle, cardiac muscle is *striated*, but its action is *involuntary*—its alternating contraction and relaxation cannot be consciously controlled. Rather, the heart beats because it has a pacemaker that initiates each contraction; this built-in (intrinsic) rhythm is called *autorhythmicity*.

Smooth muscle tissue is located in the walls of hollow internal structures, such as blood vessels, airways, and most organs in the abdominopelvic cavity. It is also attached to hair follicles in the skin. Smooth muscle tissue gets its name because, under a microscope, this tissue lacks striations; hence, it appears *nonstriated* or *smooth*. The action of smooth muscle is usually *involuntary*, and, like cardiac muscle, some smooth muscle tissue, such as the muscles that propel food through your gastrointestinal tract, has autorhythmicity. Both cardiac muscle and smooth muscle are regulated by the autonomic (involuntary) division of the nervous system and by hormones released by endocrine glands.

Checkpoint

1. Which types of muscle tissue can be consciously controlled? Which cannot be consciously controlled?

2. How does the appearance of smooth muscle differ from that of cardiac or skeletal muscle?

Figure 10.1   Muscle tissue.

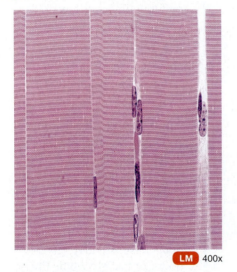

LM 400x

(a) Longitudinal section of skeletal muscle tissue

LM 600x

(b) Longitudinal section of cardiac muscle tissue

LM 350x

(c) Longitudinal section of smooth muscle tissue

🔑 Skeletal muscle tissue is attached to bones by tendons, cardiac muscle tissue forms the walls of the heart, and smooth muscle tissue is found in the walls of hollow organs.

10.2 Muscle tissue performs four functions and possesses four properties.

Functions of Muscle Tissue

Through sustained contraction or alternating contraction and relaxation, muscle tissue has four key functions:

- *Producing body movements.* Movements of the whole body, such as walking and running, and localized movements, such as grasping a pencil or nodding the head, rely on the integrated functioning of skeletal muscles, bones, and joints.

- *Stabilizing body positions.* Skeletal muscle contractions stabilize joints and help maintain body positions, such as standing or sitting. Postural muscles contract continuously when a person is awake; for example, sustained contractions in neck muscles hold the head upright.

- *Storing and moving substances within the body.* Sustained contractions of ringlike bands of smooth muscles called *sphincters* prevent outflow of the contents of a hollow organ. Temporary storage of food in the stomach or urine in the urinary bladder is possible because smooth muscle sphincters close off the outlets of these organs. Cardiac muscle contractions pump blood through the body's blood vessels. Contraction and relaxation of smooth muscle in the walls of blood vessels help adjust their diameter and thus regulate the rate of blood flow. Smooth muscle contractions also move food and substances such as bile and enzymes through the gastrointestinal tract, push gametes (sperm and oocytes) through the passageways of the reproductive systems, and propel urine through the urinary system. Skeletal muscle contractions promote the flow of lymph and aid the return of blood to the heart.

- *Generating heat.* As muscle tissue contracts, it also produces heat. Much of the heat released by muscles is used to maintain normal body temperature. Involuntary contractions of skeletal muscle, known as *shivering*, can dramatically increase the rate of heat production.

Properties of Muscle Tissue

Muscle tissue has four special properties that enable it to function and contribute to homeostasis:

- **Electrical excitability**, a property of both muscle cells and neurons, is the ability to respond to certain stimuli by producing electrical signals called **action potentials** or **nerve impulses**. Nerve impulses propagate (travel) along the plasma membrane due to the presence of specific ion channels. The stimuli that trigger action potentials may be electrical signals arising in the muscle tissue itself, such as occurs in the heart's pacemaker, or chemical stimuli, such as neurotransmitters released by neurons, hormones distributed by the blood, or even local changes in pH.

- **Contractility** is the ability of muscle tissue to contract forcefully when stimulated by an action potential. When a muscle contracts, it generates *tension* (force of contraction) while pulling on its attachment points. If the tension generated is great enough to overcome the resistance of the object to be moved, the muscle shortens and movement occurs.

- **Extensibility** is the ability of muscle tissue to stretch without being damaged. Normally, smooth muscle is subject to the greatest amount of stretching. For example, each time your stomach fills with food, the muscle in the wall is stretched. Cardiac muscle also is stretched each time the heart fills with blood.

- **Elasticity** is the ability of muscle tissue to return to its original length and shape after contraction or extension.

Skeletal muscle tissue is the focus of much of this chapter. Cardiac muscle tissue and smooth muscle tissue are described briefly. Cardiac muscle tissue will be discussed in more detail in Chapter 19; smooth muscle tissue will be examined in detail in Chapter 14, as well as in discussions of the various organs containing smooth muscle.

Checkpoint

3. What are the general functions of muscle tissue?
4. What properties do all muscle tissues have in common?
5. Which property do muscle tissues and neurons share?

10.3 Skeletal muscles are surrounded by connective tissues and are well supplied with nerves and blood vessels.

Each of your skeletal muscles is a separate organ composed of hundreds to thousands of cells. The cells are called *muscle fibers* because of their elongated shapes. Connective tissues surround muscle fibers and whole muscles, and blood vessels and nerves penetrate skeletal muscles to exert their effects on individual muscle fibers (Figure 10.2). To understand skeletal muscle contraction, you first need to understand its gross and microscopic anatomy.

Figure 10.2 **Organization of skeletal muscle and its connective tissue coverings.**

Transverse plane

Bone

Transverse sections

Fascicle

Periosteum

Tendon

Skeletal muscle

Perimysium

Epimysium

Fascicle

Perimysium

Muscle fiber

Myofibril

Perimysium

Endomysium

Motor neuron

Blood capillary

Endomysium

Nucleus

Muscle fiber

Striations

Sarcoplasm

Sarcolemma

Myofibril

Filament

Functions of Muscle Tissue
1. Produces body motions.
2. Stabilizes body positions.
3. Stores and moves substances within the body.
4. Produces heat.

🔑 **A skeletal muscle consists of individual muscle fibers bundled into fascicles and surrounded by three connective tissue layers that are extensions of the fascia.**

Connective Tissue Components

Connective tissue surrounds and protects muscle tissue. The **hypodermis**, or **subcutaneous layer**, separates muscle from skin (see Figure 11.9c). It is composed of areolar connective tissue and adipose tissue, provides a pathway for nerves and blood and lymphatic vessels to enter and exit muscles, and protects muscles from physical trauma. **Fascia** (FASH-ç-a = bandage) is a sheet or broad band of dense connective tissue that supports and surrounds muscles and other organs of the body. Fascia holds muscles with similar functions together; allows free movement of muscles; carries nerves, blood vessels, and lymphatic vessels, and fills spaces between muscles.

Three layers of connective tissue extend from the fascia to further protect and strengthen skeletal muscle (Figure 10.2). The outermost layer, which encircles the entire muscle, is the **epimysium** (ep-i-MĪZ-ē-um; *epi-* = upon). **Perimysium** (per-i-MĪZ-ē-um; *peri-* = around) surrounds groups of 10 to 100 or more individual muscle fibers, separating them into bundles called **fascicles** (FAS-i-kuls = little bundles). Many fascicles are large enough to be seen with the naked eye. They give a cut of meat its characteristic "grain"; if you tear a piece of meat, it rips apart along the fascicles. Both epimysium and perimysium are dense irregular connective tissue. Penetrating the interior of each fascicle and separating individual muscle fibers from one another is **endomysium** (en´-dō-MĪZ-ē-um; *endo-* = within), a thin sheath of areolar connective tissue.

Epimysium, perimysium, and endomysium may extend together beyond the muscle fibers to form a **tendon**—a cord of dense regular connective tissue composed of parallel bundles of collagen fibers that attach a muscle to the periosteum of a bone. An example is the calcaneal (Achilles) tendon of the gastrocnemius (calf) muscle, which attaches the muscle to the calcaneus (see Figure 11.20d). When the connective tissue extends as a broad, flat layer, the tendon is called an **aponeurosis** (ap-ō-noo-RŌ-sis; *apo-* = from; *-neur-* = a sinew). An example of an aponeurosis is the epicranial aponeurosis on top of the skull between the occipital and frontal bellies of the occipitofrontalis muscle (shown in Figure 11.3 a,c).

Certain tendons, especially those of the wrist and ankle, are enclosed by tubes of fibrous connective tissue called *tendon sheaths*, which are similar in structure to bursae (see Figure 11.17a). Tendon sheaths contain a film of synovial fluid that reduces friction as tendons slide back and forth within the sheaths.

Nerve and Blood Supply

Skeletal muscles are well supplied with nerves and blood vessels. Generally, an artery and one or two veins accompany each nerve that penetrates a skeletal muscle. The neurons that stimulate skeletal muscle fibers to contract are called **motor neurons**. Each motor neuron has a threadlike extension, called an **axon**, that extends from the brain or spinal cord to a group of skeletal muscle fibers (see Figure 10.3). The axon of a motor neuron typically branches many times. Each axon collateral (branch) extends to a different skeletal muscle fiber. The ends of the axon terminals enlarge into swellings known as *synaptic end bulbs*.

Microscopic blood vessels called capillaries are plentiful in muscle tissue; each muscle fiber is in close contact with one or more capillaries (Figure 10.3). The blood capillaries bring in oxygen and nutrients and remove heat and the waste products of muscle metabolism. Especially during contraction, a muscle fiber synthesizes and uses considerable ATP (adenosine triphosphate). These reactions require oxygen, glucose, fatty acids, and other substances that are supplied in the blood.

Checkpoint

6. What functions does fascia serve?
7. Which connective tissue coat surrounds groups of muscle fibers, separating them into bundles? What are these bundles called?
8. What are the structural and functional differences among tendons, aponeuroses, and tendon sheaths?

Figure 10.3 **Blood and nerve supply to skeletal muscle.** The site of synaptic contact between a motor neuron and a skeletal muscle fiber is the neuromuscular junction (NMJ). Two NMJs are present in this view.

Blood capillary

Axon collateral

Motor neuron

Axon collateral

Synaptic end bulbs

Axon terminal

Synaptic end bulbs

Skeletal muscle fiber

SEM 1650x

Branches of one motor neuron typically contact many muscle fibers.

10.4 Each skeletal muscle fiber is covered by a sarcolemma; each of its myofibrils is surrounded by sarcoplasmic reticulum and contains sarcomeres.

The most important components of a skeletal muscle are the muscle fibers themselves. Mature skeletal muscle fibers range from 10 to 100 μm in diameter.* The typical length of a mature skeletal muscle fiber is about 10 cm (4 in.), although some are as long as 30 cm (12 in.). During embryonic development, each skeletal muscle fiber arises from the fusion of a hundred or more small mesoder-mal cells called *myoblasts* (Figure 10.4a). Hence, each mature skeletal muscle fiber is a single cell with a hundred or more nuclei. Once fusion has occurred, the muscle fiber loses its ability to undergo cell division. Thus, the number of skeletal muscle fibers is set before you are born, and most of these cells last a lifetime.

*One micrometer (μm) is 10^{-6} meter (1/25,000 in.).

Figure 10.4   Microscopic organization of skeletal muscle. (a) During embryonic development, many myoblasts fuse to form one skeletal muscle fiber. (b-d) The sarcolemma of the fiber encloses sarcoplasm and myofibrils. Sarcoplasmic reticulum wraps around each myofibril. Thousands of tranverse tubules invaginate from the sarcolemma toward the center of the muscle fiber.

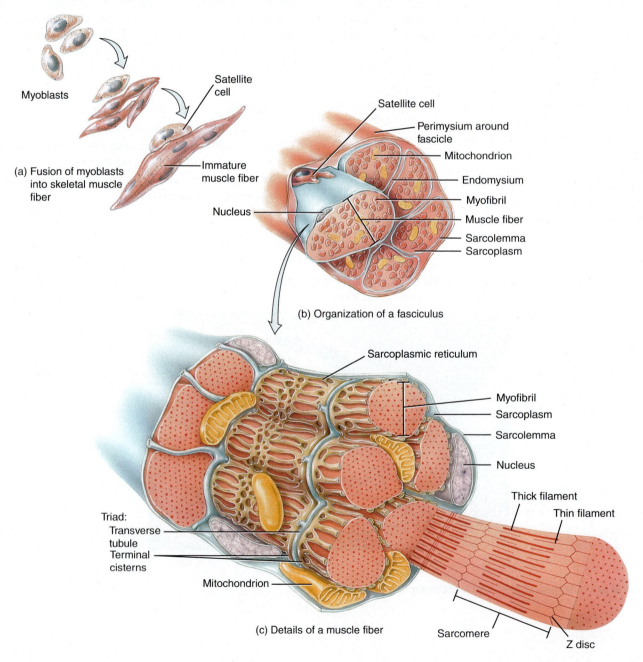

Myoblasts

Satellite cell

(a) Fusion of myoblasts into skeletal muscle fiber

Immature muscle fiber

Nucleus

Satellite cell

Perimysium around fascicle

Mitochondrion

Endomysium

Myofibril

Muscle fiber

Sarcolemma

Sarcoplasm

(b) Organization of a fasciculus

Sarcoplasmic reticulum

Myofibril

Sarcoplasm

Sarcolemma

Nucleus

Thick filament

Thin filament

Triad:
Transverse tubule
Terminal cisterns

Mitochondrion

(c) Details of a muscle fiber

Sarcomere

Z disc

The dramatic muscle growth that occurs after birth occurs by enlargement of existing muscle fibers, called **hypertrophy** (hī-PER-trō-fē; *hyper-* = above or excessive; *-trophy* = nourishment), rather than an increase in the number of fibers, which is referred to as **hyperplasia** (hī-per-PLĀ-zē-a; *-plasis* = molding). During childhood, human growth hormone and other hormones stimulate an increase in the size of skeletal muscle fibers. The hormone testosterone (from the testes in males and in small amounts from other tissues, such as the ovaries in females) promotes further enlargement of muscle fibers. If muscle tissue is not actively used, it loses size and strength, a process called **atrophy** (A-trō-fē; *a-* = without, *-trophy* = nourishment).

A few myoblasts do persist in mature skeletal muscle as *satellite cells* (Figure 10.4a). These cells retain the capacity to fuse with one another or with damaged muscle fibers to regenerate functional muscle fibers. However, the number of new skeletal muscle fibers formed is not enough to compensate for significant skeletal muscle damage. In such cases, skeletal muscle tissue undergoes **fibrosis**, the replacement of muscle fibers by fibrous scar tissue. For this reason, skeletal muscle tissue can regenerate only to a limited extent.

(d) Simplistic representation of a muscle fiber

The contractile elements of muscle fibers, the myofibrils, contain overlapping thick and thin filaments.

Sarcolemma, Transverse Tubules, and Sarcoplasm

The multiple nuclei of a skeletal muscle fiber are located just beneath the **sarcolemma** (*sarc-* = flesh; *-lemma* = sheath), the plasma membrane of a muscle fiber (Figure 10.4b, c). Thousands of tiny tunnel-like extensions of the sarcolemma, called **transverse (T) tubules**, extend toward the center of each muscle fiber. T tubules are open to the outside of the fiber and thus are filled with interstitial fluid. Muscle action potentials propagate along the sarcolemma and through the T tubules, quickly spreading throughout the muscle fiber. This arrangement ensures that all the superficial and deep parts of the muscle fiber become excited by an action potential at essentially the same time.

The sarcolemma surrounds the **sarcoplasm**, the cytoplasm of a muscle fiber. Sarcoplasm includes a substantial amount of glycogen; glycogen can be split into glucose, which is used for synthesis of ATP. In addition, the sarcoplasm contains a red-colored protein called **myoglobin** (mī-ō-GLŌB-in). This protein, found only in muscle, binds oxygen molecules that diffuse into muscle fibers from interstitial fluid. Myoglobin releases oxygen when mitochondria need it for ATP production. The mitochondria lie in rows throughout the muscle fiber, strategically close to the muscle proteins that use ATP during contraction (Figure 10.4c).

Myofibrils and Sarcoplasmic Reticulum

At high magnification the sarcoplasm appears stuffed with little parallel threads. These small structures are the contractile elements of skeletal muscle, the **myofibrils** (Figure 10.4c). Myofibrils are about 2 μm in diameter and extend the entire length of the muscle fiber. Their prominent striations make the whole muscle fiber appear striped.

A fluid-filled system of membranous sacs called the **sarcoplasmic reticulum** (sar´-kō-PLAZ-mik re-tik-ū-lum) or **SR** wraps around each myofibril (Figure 10.4c). This elaborate system is similar to smooth endoplasmic reticulum in nonmuscle cells. Dilated end sacs of the sarcoplasmic reticulum called **terminal cisterns** (= reservoirs) butt against the T tubules from both sides. One T tubule and the two terminal cisterns

on either side of it form a **triad** (*tri-* = three). In a relaxed muscle fiber, the sarcoplasmic reticulum stores calcium ions (Ca^{2+}). Release of Ca^{2+} from the terminal cisterns into the sarcoplasm triggers muscle contraction.

Filaments and the Sarcomere

Within myofibrils are smaller structures, called *filaments*, which are directly involved in the contractile process (Figure 10.4c,d). The diameter of the **thin filaments** is about 8 nm*, and that of the **thick filaments** is about 16 nm. In the regions of the sarcomere where filaments overlap, each thick filament is surrounded by six thin filaments. The filaments inside a myofibril do not extend the entire length of a muscle fiber. Instead, they are arranged in compartments called **sarcomeres** (*-mere* = part), the basic functional units of a myofibril (Figure 10.5a). Narrow, zigzagging zones of dense protein material called **Z discs** separate one sarcomere from the next. A sarcomere extends from one Z disc to the next.

*One nanometer (nm) is 10^{-9} meter (0.001 μm).

Figure 10.5   Arrangement of filaments within a sarcomere.

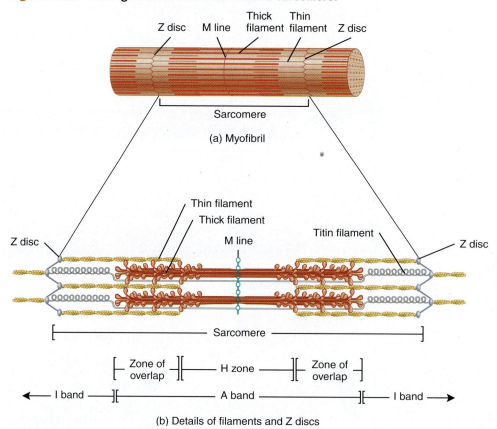

Myofibrils contain two types of contractile filaments: thick filaments and thin filaments.

Figure 10.6 Transmission electron micrograph showing the characteristic zones and bands of a sarcomere.

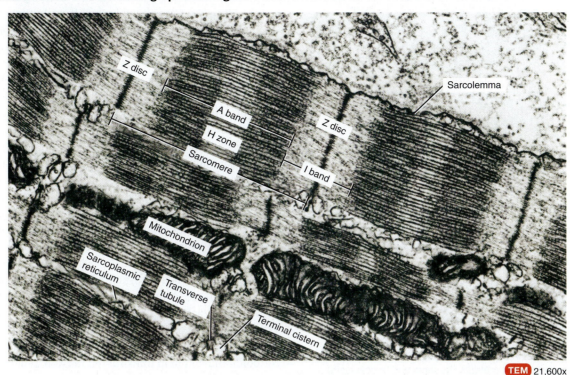

TEM 21,600x

The striations of skeletal muscle are alternating darker A bands and lighter I bands.

The thick and thin filaments overlap one another to a greater or lesser extent, depending on whether the muscle is contracted, relaxed, or stretched. The pattern of their overlap, consisting of a variety of zones and bands (Figure 10.5b), creates the striations that can be seen both in single myofibrils and in whole muscle fibers. The darker middle part of the sarcomere is the **A band**, which extends the entire length of the thick filaments (Figure 10.5b). A narrow **H zone** in the center of each A band contains thick filaments but no thin filaments. Toward each end of the A band is a *zone of overlap*, where the thick and thin filaments lie side by side. The **I band** is a lighter, less dense area that contains thin filaments but no thick filaments (Figure 10.5b). A Z disc passes through the center of each I band. Supporting proteins that hold the thick filaments together at the center of the H zone form the *M line*, so named because it is at the *middle* of the sarcomere. Figure 10.6 shows the relations of the zones, bands, and lines as seen in a transmission electron micrograph.

Muscle Proteins

Myofibrils contain *contractile proteins*, which generate force during contraction. The two contractile proteins in muscle are myosin and actin, which are the main components of thick and thin filaments, respectively. **Myosin** functions as a *motor protein* in all three types of muscle tissue. Motor proteins push or pull various cellular structures to achieve movement by converting the chemical energy in ATP to the mechanical energy of motion or the production of force. In skeletal muscle, about 300 molecules of myosin form a single thick filament. Each myosin molecule is shaped like two golf clubs twisted together (Figure 10.7a). The *myosin tail* (twisted golf club handles) points toward the M line in the center of the sarcomere. Tails of neighboring myosin molecules lie parallel to one another, forming the shaft of the thick filament. The two projections of each myosin molecule (golf club heads) are called *myosin heads*. The heads project outward from the thick filament

Figure 10.7   Structure of thick and thin filaments. (a) A thick filament contains myosin molecules (see enlargement). (b) Thin filaments contain actin, troponin, and tropomyosin.

Thick filament

Myosin tail   Myosin heads

(a) One thick filament (above) and a myosin molecule (below)

Actin   Troponin   Tropomyosin

Myosin-binding site (covered by tropomyosin)

(b) Portion of a thin filament

🔑 **Contractile proteins (myosin and actin) generate force during contraction; regulatory proteins (troponin and tropomyosin) help switch contraction on and off.**

in a spiraling fashion, each head extending toward one of the six thin filaments that surround each thick filament.

Thin filaments extend from the Z discs (see Figure 10.5b). Their main component is the protein **actin**. Individual actin molecules join to form an actin filament that is twisted into a helix (Figure 10.7b). On each actin molecule is a *myosin-binding site*, where a myosin head can attach. Thin filaments also contain smaller amounts of two *regulatory proteins*, **tropomyosin** and

troponin, which help switch the contraction process on and off. In relaxed muscle, myosin is blocked from binding to actin because strands of tropomyosin cover actin's myosin-binding site. The tropomyosin strand, in turn, is held in place by troponin.

Besides contractile and regulatory proteins, muscle contains about a dozen *structural proteins* that contribute to the alignment, stability, elasticity, and extensibility of myofibrils. One key structural protein is **titin** (*titan* = gigantic), the third most plentiful protein in skeletal muscle (after actin and myosin). Each titin filament connects a Z disc to an M line, thereby helping stabilize the position of the thick filament (see Figure 10.5b). Because it can stretch to at least four times its resting length and then spring back unharmed, titin accounts for much of the elasticity and extensibility of myofibrils. Titin probably helps the sarcomeres return to their resting length after a muscle has contracted or been stretched, may help prevent overextension of sarcomeres, and maintains the central location of the A bands.

Dystrophin is a cytoskeletal protein that links thin filaments of the sarcomere to integral membrane proteins of the sarcolemma, which are attached in turn to proteins in the connective tissue extracellular matrix that surrounds muscle fibers (see Figure 10.4d). Dystrophin and its associated proteins are thought to reinforce the sarcolemma and help transmit the tension generated by the sarcomeres to muscle tendons.

Checkpoint

9. What are the functions of T tubules, myoglobin, and sarcoplasmic reticulum, which are all unique to muscle fibers?

10. Arrange the following elements from smallest to largest: muscle fiber, thick filament, myofibril.

11. List the components of a myofibril found within the I band, the A band, and the H zone. Which components attach to the M line? To the Z disc?

12. What type of protein composes thick filaments? Thin filaments?

10.5   The neuromuscular junction is where a muscle action potential is initiated.

A muscle fiber contracts in response to muscle action potentials propagating along its sarcolemma and through its system of T tubules. As noted earlier in the chapter, the neurons that stimulate skeletal muscle fibers to contract are called **motor neurons**. Muscle action potentials arise at the **neuromuscular junction (NMJ)**, the structural point of synaptic contact and functional site of communication between a motor neuron and a skeletal muscle fiber (Figure 10.8a). A **synapse** is a region where communication occurs between a neuron and another cell—in

this case, between a motor neuron and a skeletal muscle fiber. A small gap, called the *synaptic cleft*, separates the two cells. Because the cells do not physically touch, the action potential from the motor neuron cannot "jump the gap" to directly excite the muscle fiber. Instead, the motor neuron communicates with the muscle fiber indirectly, by releasing a chemical called a **neurotransmitter**.

As you saw in Figure 10.3, each motor neuron axon terminal divides into a cluster of synaptic end bulbs (Figure 10.8a).

Suspended in the cytosol within each synaptic end bulb are hundreds of membrane-enclosed sacs called **synaptic vesicles**. Inside each synaptic vesicle are thousands of molecules of **acetylcholine** (as´-ē-til-KŌ-lēn), abbreviated **ACh**, the neurotransmitter released at the NMJ.

The region of the sarcolemma of the muscle fiber that is opposite the synaptic end bulb, called the **motor end plate**, is the muscle fiber part of the NMJ (Figure 10.8b,c). Within each motor end plate are 30 to 40 million **acetylcholine receptors** that bind specifically to ACh. A neuromuscular junction thus includes the synaptic end bulb on one side of the synaptic cleft, plus the motor end plate of the muscle fiber on the other side.

A nerve impulse excites a skeletal muscle fiber in the following way (Figure 10.8b,c):

1 *Release of acetylcholine.* A nerve impulse travels from the brain or spinal cord along a motor neuron to the muscle fiber (Figure 10.8b). Arrival of the nerve impulse at the synaptic end bulb causes many synaptic vesicles to undergo exocytosis (Figure 10.8c). During exocytosis, the synaptic vesicles fuse with the motor neuron's plasma membrane, which liberates ACh into the synaptic cleft. The ACh then diffuses across the synaptic cleft between the motor neuron and the motor end plate.

2 *Activation of ACh receptors.* Binding of two molecules of ACh to the receptor opens the ion channel part of the ACh receptor. Once the channel is open, small cations, most importantly Na⁺, can flow across the sarcolemma into the muscle fiber.

Figure 10.8 **Structure of the neuromuscular junction, the synapse between a motor neuron and a skeletal muscle fiber.**

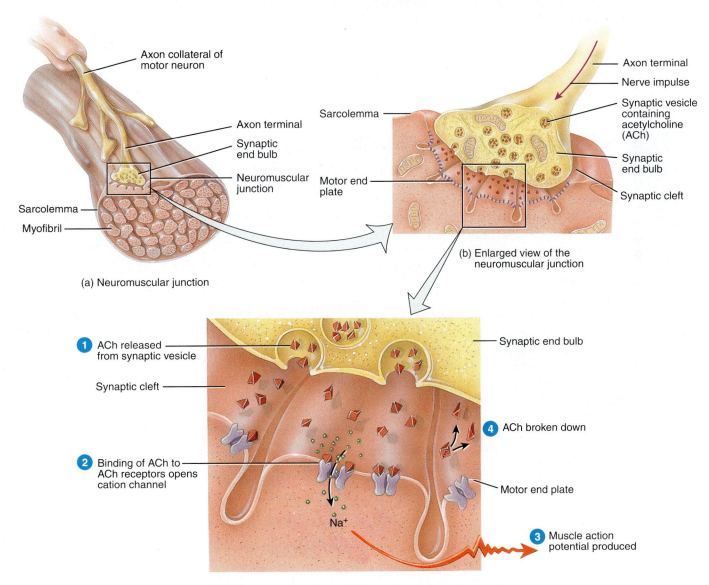

(a) Neuromuscular junction

(b) Enlarged view of the neuromuscular junction

(c) Binding of acetylcholine to ACh receptors in the motor end plate

Synaptic end bulbs at the tips of axon terminals contain synaptic vesicles filled with acetylcholine.

③ Generation of muscle action potential. The inflow of Na⁺ (down its electrochemical gradient) generates a muscle action potential. The details of action potential generation are discussed in Chapter 12. The muscle action potential then propagates along the sarcolemma and into the T tubules (see red arrows in Figure 10.12), stimulating the contraction process.

④ Termination of ACh activity. The effect of ACh binding lasts only briefly because ACh is rapidly broken down by an enzyme in the synaptic cleft called **acetylcholinesterase (AChE).** AChE breaks down ACh into products that cannot activate the ACh receptor.

Each nerve impulse normally elicits one muscle action potential. If another nerve impulse releases more acetylcholine, then steps 2 and 3 repeat. When nerve impulses cease in the motor neuron, ACh release stops, and AChE rapidly breaks down the ACh already present in the synaptic cleft. The ion channels in the sarcolemma close, and action potentials are no longer generated in the muscle fiber.

Because skeletal muscle fibers often are very long cells, the NMJ usually is located near the midpoint of a skeletal muscle fiber. Muscle action potentials arise at the NMJ and then propagate toward both ends of the fiber. This arrangement permits nearly simultaneous activation (and thus contraction) of all parts of the muscle fiber.

Checkpoint

13. Which portion of the sarcolemma contains acetylcholine receptors?
14. What prevents an action potential arriving at the synaptic end bulbs of a motor neuron from being transferred directly to a muscle fiber?
15. Describe the steps that occur for an action potential arriving at a synaptic end bulb to result in an action potential in a muscle fiber.
16. What happens to acetylcholine once it has bound to its receptor?

Return to Jesse's Story

As Jesse helped his mother unpack the groceries, he grabbed a large can of peaches. He tried to put it in the cupboard, but couldn't lift his arm high enough. Jesse set the heavy can back down. "Momma, I think I hurt my shoulder. I can't lift my arm up any higher than this."

As he waited in the doctor's examining room, Jesse thought about the night before, when he overheard his parents talking about his Uncle Walter, who had something called muscular dystrophy. "Come in," Jesse said in response to a knock on the door.

After introducing himself, Dr. Fitts had Jesse lower his arms as he pushed on the shoulder muscles with his fingers, checking the muscle tone. He then had Jesse raise his arms and bring them around in front. "Hmm, you have some winging of the scapula. Your shoulder is somewhat unstable. It's supposed to stay flat against the ribs, but sometimes the muscles weaken and it doesn't stay where it's supposed to." When Dr. Fitts asked Jesse to whistle, he could not pucker his lips. "I'm going to order some tests—a blood test, an electromyogram (EMG), a nerve conduction study, and a muscle biopsy."

As Jesse got dressed, Dr. Fitts spoke to his mother, indicating that Jesse's muscle weakness might be related to Uncle Walter's muscular dystrophy, because it is a genetic disorder. He did his best to reassure her, explaining that Jesse's symptoms indicated a less severe form of the disease.

A. How could a reduced blood supply to the muscles of Jesse's shoulder explain his muscle weakness?

B. Jesse will have a biopsy (removal of a sample of living tissue for microscopic analysis) performed to assess the microscopic structure of his muscle; describe the cells and tissues that will be seen.

C. The pathology report on Jesse's muscle indicated the presence of "isolated small atrophic fibers intermixed with hypertrophic muscle fibers." What is muscle hypertrophy? What is atrophy?

D. The pathology report also indicated that the sarcomeres "appear regular and show no signs of disarray or degeneration." What is a sarcomere, and what makes skeletal muscle tissue appear to have regular repeating striations?

10.6 An action potential releases calcium ions that allow thick filaments to bind to and pull thin filaments toward the center of the sarcomere.

Once an action potential is propagated along the sarcolemma and T tubules, muscle contraction begins. The model describing the contraction of muscle is known as the **sliding filament mechanism**. Muscle contraction occurs because myosin heads attach to and "walk" along the thin filaments at both ends of a sarcomere, progressively pulling the thin filaments toward the M line (Figure 10.9). As a result, the thin filaments *slide* inward and meet at the center of a sarcomere. They may even move so far inward that their ends overlap (Figure 10.9c). As the thin filaments slide inward, the Z discs come closer together, and the sarcomere shortens. Note that the lengths of the individual thick and thin filaments do not change; it is the amount of overlap between them that changes. Shortening of the individual sarcomeres causes shortening of the whole muscle fiber, and ultimately shortening of the entire muscle.

Excitation–Contraction Coupling

Excitation–contraction coupling refers to the steps that connect excitation (a muscle action potential propagating along the sarcolemma and into the T tubules) to contraction (sliding of the filaments). An increase in Ca^{2+} concentration in the sarcoplasm starts muscle contraction; a decrease stops it. When a muscle fiber is relaxed, the concentration of Ca^{2+} in its sarcoplasm is very low. However, a huge amount of Ca^{2+} is stored inside the sarcoplasmic reticulum (Figure 10.10a). When a muscle action potential propagates along the sarcolemma and into the T tubules, it opens **Ca^{2+} release channels** in the SR membrane (Figure 10.10b). As these channels open, Ca^{2+} flows from the SR into the sarcoplasm around the thick and thin filaments. The released Ca^{2+} combines with troponin, moving the troponin–tropomyosin complex away from the myosin-binding sites on actin. Once these binding sites are exposed, myosin heads bind to them, and the contraction cycle begins.

Figure 10.9 Sliding filament mechanism of muscle contraction, as it occurs in two adjacent sarcomeres.

(a) Relaxed muscle

(b) Partially contracted muscle

(c) Maximally contracted muscle

During muscle contractions, thin filaments move toward the M line of each sarcomere.

Figure 10.10   The role of Ca²⁺ in the regulation of contraction by troponin and tropomyosin. (a) During relaxation, the level of Ca²⁺ in the sarcoplasm is low because active transport pumps force Ca²⁺ into the sarcoplasmic reticulum (SR). (b) A muscle action potential propagating along a T tubule opens Ca²⁺ release channels in the SR, calcium ions flow into the sarcoplasm, and contraction begins.

Troponin holds tropomyosin in position to block myosin-binding sites on actin.

(a) Relaxation

Key:
○ = Ca²⁺
⊗ = Ca²⁺ active transport pumps
< = Ca²⁺ release channels

Ca²⁺ binds to troponin, which uncovers the myosin-binding sites on actin.

(b) Contraction

🔑 **An increase in the Ca⁺² level in the sarcoplasm starts the sliding of thin filaments; when the level of Ca⁺² in the sarcoplasm declines, sliding stops.**

The Contraction Cycle

Calcium released from the sarcoplasmic reticulum into the sarcoplasm causes the troponin–tropomyosin complexes to move away from the myosin-binding sites on actin. Once the binding sites are uncovered, the contraction cycle begins. The **contraction cycle**—a repeating sequence of events that causes the filaments to slide—consists of four steps (Figure 10.11):

① *ATP splits.* The myosin head includes an ATP-binding site and ATPase, an enzyme that splits ATP into ADP (adenosine diphosphate) and P (a phosphate group). This splitting reaction reorients and energizes the myosin head. Notice that ADP and a phosphate group are still attached to the myosin head.

② *Myosin attaches to actin.* The energized myosin head attaches to the myosin-binding site on actin and releases the phosphate group.

③ *Power stroke occurs.* Binding of the myosin head to actin triggers the power stroke of contraction. During the power stroke, the myosin head rotates or swivels and releases the ADP. The myosin head generates force as it rotates toward the center of the sarcomere, sliding the thin filament past the thick filament toward the M line.

④ *Myosin detaches from actin.* At the end of the power stroke, the myosin head remains firmly attached to actin until it binds another molecule of ATP. As ATP binds to the ATP-binding pocket on the myosin head, the myosin head detaches from actin.

The contraction cycle repeats for as long as ATP and Ca²⁺ are available in the sarcoplasm. Splitting of ATP by the myosin ATPase again reorients the myosin head and transfers energy from ATP to the myosin head. Once reoriented and energized, the myosin head combines with the next myosin-binding site farther along the thin filament. The myosin heads keep rotating back and forth

Figure 10.11 **The contraction cycle.**

Key:
= Ca²⁺

1 Myosin heads split ATP and become reoriented and energized

ADP P

2 Myosin heads bind to actin

ATP

Contraction cycle continues if ATP is available and Ca²⁺ level in the sarcoplasm is high

ADP

4 As myosin heads bind ATP, they detach from actin

ATP

ADP

3 Myosin heads rotate toward center of the sarcomere (power stroke)

During the power stroke of contraction, myosin heads rotate and move the thin filaments past the thick filaments toward the center of the sarcomere.

with each power stroke. Each power stroke pulls the thin filaments closer to the M line. Each of the 600 myosin heads in one thick filament repeatedly attach to actin, rotate, and detach about five times per second. At any one instant, some of the myosin heads are attached to actin and producing force, and others are detached and preparing to bind again.

The contraction cycle is analogous to walking on a foot-powered, nonmotorized treadmill. One foot (myosin head) strikes the treadmill belt (thin filament) and pushes it backward (toward the M line). Then the other foot comes down and imparts a second push. The belt (thin filament) moves smoothly while the walker (thick filament) remains stationary. Each myosin head progressively "walks" along a thin filament, coming closer to the Z disc with each "step," while the thin filament moves closer to the M line. Like the legs of a walker, the myosin head needs a constant supply of energy to keep going—one molecule of ATP for each contraction cycle.

As the contraction cycle continues, movement of myosin heads applies the force that draws the Z discs toward each other, and the sarcomere shortens. During a maximal muscle contraction, the sarcomere can shorten by as much as half the resting length. The Z discs, in turn, pull on neighboring sarcomeres, shortening the whole muscle fiber, which ultimately leads to shortening of the entire muscle. As the fibers of a skeletal muscle start to shorten,

they first pull on their connective tissue coverings (endomysium, perimysium, and epimysium) and tendons. The coverings and tendons become taut, and the tension passed through the tendons pulls on the bones to which they are attached. The result is movement of a part of the body.

Relaxation

Two changes permit a muscle fiber to relax after it has contracted. First, when nerve impulses cease being transmitted by the motor neuron, ACh release stops and AChE rapidly breaks down the ACh already present in the synaptic cleft (see Figure 10.8). As ion channels close, muscle action potentials are no longer generated and propagated throughout the T tubules, and the Ca²⁺ release channels in the sarcoplasmic reticulum membrane close.

Second, the sarcoplasmic reticulum membrane also contains **Ca²⁺ active transport pumps** that use ATP to constantly move Ca²⁺ from the sarcoplasm into the SR (Figure 10.10a). With Ca²⁺ release channels closed and the pumps returning Ca²⁺ back into the SR, the concentration of calcium ions in the sarcoplasm quickly decreases. Inside the SR, molecules of a calcium-binding protein, appropriately called *calsequestrin*, bind to the Ca²⁺, enabling even

Figure 10.12   Summary of the events of contraction and relaxation in a skeletal muscle fiber.

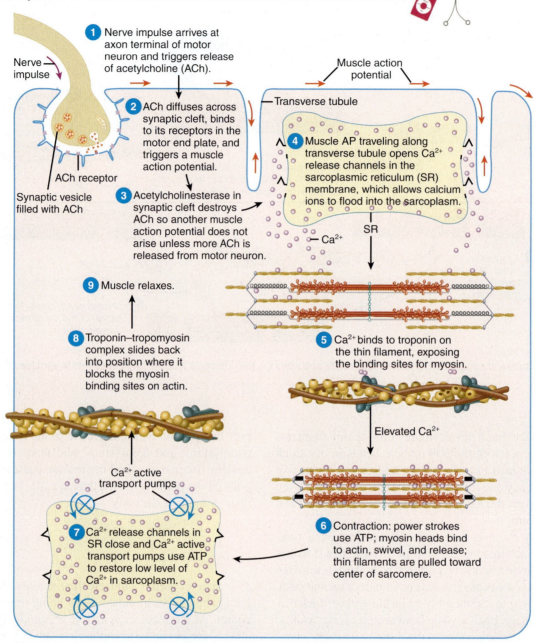

1 Nerve impulse arrives at axon terminal of motor neuron and triggers release of acetylcholine (ACh).

Nerve impulse

2 ACh diffuses across synaptic cleft, binds to its receptors in the motor end plate, and triggers a muscle action potential.

ACh receptor

Synaptic vesicle filled with ACh

3 Acetylcholinesterase in synaptic cleft destroys ACh so another muscle action potential does not arise unless more ACh is released from motor neuron.

Muscle action potential

Transverse tubule

4 Muscle AP traveling along transverse tubule opens Ca²⁺ release channels in the sarcoplasmic reticulum (SR) membrane, which allows calcium ions to flood into the sarcoplasm.

Ca²⁺

SR

9 Muscle relaxes.

8 Troponin–tropomyosin complex slides back into position where it blocks the myosin binding sites on actin.

5 Ca²⁺ binds to troponin on the thin filament, exposing the binding sites for myosin.

Elevated Ca²⁺

Ca²⁺ active transport pumps

7 Ca²⁺ release channels in SR close and Ca²⁺ active transport pumps use ATP to restore low level of Ca²⁺ in sarcoplasm.

6 Contraction: power strokes use ATP; myosin heads bind to actin, swivel, and release; thin filaments are pulled toward center of sarcomere.

Acetylcholine released at the neuromuscular junction triggers a muscle action potential, which leads to muscle contraction.

more Ca²⁺ to be stored or sequestered within the SR. As the Ca²⁺ level in the sarcoplasm drops, the troponin–tropomyosin complexes slide back over and cover the myosin-binding sites on actin. No longer attached to myosin heads, the thin filaments slip back to their relaxed positions, and the muscle fiber relaxes. As the fibers of a skeletal muscle lengthen, their connective tissue coverings and tendons become slack, and tendons decrease their pull on the bones to which they are attached.

Figure 10.12 summarizes the events of contraction and relaxation in a skeletal muscle fiber.

Checkpoint

17. What happens to the distance between neighboring Z discs when thin filaments in a sarcomere slide toward the M line? When thin filaments slide away from the M line?

18. As each myosin head detaches from a thin filament, what stops the thin filament from sliding back to its original, relaxed position?

19. How does sarcomere shortening result in shortening of an entire muscle and movement of a bone?

20. How does a skeletal muscle fiber relax?

21. What are three functions of ATP in muscle contraction and relaxation?

10.7 Muscle tension is controlled by stimulation frequency and motor unit recruitment.

A single nerve impulse in a motor neuron elicits a single muscle action potential in all muscle fibers with which it forms synapses. The contraction that results from a single muscle action potential has significantly smaller force than the maximum force or tension the fiber is capable of producing. The total tension that a single fiber can produce depends on the rate at which nerve impulses arrive at the neuromuscular junction. The number of impulses per second is the *frequency of stimulation*. When considering the contraction of a whole muscle, the total tension it can produce depends on the number of fibers that are contracting in unison.

Motor Units

Even though each skeletal muscle fiber has only a single neuromuscular junction, the axon of a motor neuron branches out and forms neuromuscular junctions with many different muscle fibers. A **motor unit** consists of one motor neuron plus all the skeletal muscle fibers it stimulates (Figure 10.13). A single motor neuron makes contact with an average of 150 muscle fibers within a muscle, and all muscle fibers in one motor unit contract in unison. Typically, the muscle fibers of a motor unit are dispersed throughout a muscle rather than clustered together.

Muscles that control small, precise movements consist of many small motor units. For instance, muscles of the larynx (voice box) that control voice production have as few as two or three muscle fibers per motor unit, and muscles controlling eye movements may have 10–20 muscle fibers per motor unit. In contrast, skeletal muscles responsible for large, powerful movements, such as the biceps brachii muscle in the arm and the gastrocnemius muscle in the calf of the leg, have fewer motor units, with as many as 2000–3000 muscle fibers in each. Stimulation of one motor neuron causes all the muscle fibers in that motor unit to contract at the same time. Accordingly, the total strength of a muscle contraction depends, in part, on the size of the motor units and on the number of motor units that are activated at the same time.

Twitch Contraction

A **twitch contraction** is the brief contraction of all the muscle fibers in a motor unit in response to a single nerve impulse in its motor neuron. In the laboratory, a twitch can be produced by direct electrical stimulation of a motor neuron or its muscle fibers. The record of a muscle contraction, called a **myogram**, is shown in Figure 10.14.

Figure 10.13 Motor units. Shown are two motor neurons, each supplying the muscle fibers of its motor unit.

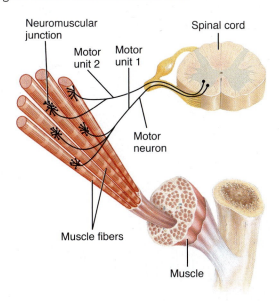

Figure 10.14 Myogram of a twitch contraction. The arrow indicates the time at which the stimulus occurred.

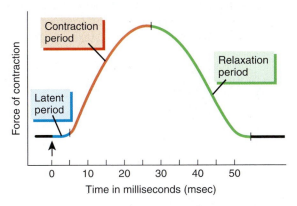

A myogram is a record of a muscle contraction.

A motor unit consists of a motor neuron plus all the muscle fibers it stimulates.

Note that a brief delay, called the **latent period**, occurs between application of the stimulus (time zero on the graph) and the beginning of contraction. During the latent period, the muscle action potential sweeps over the sarcolemma, and calcium ions are released from the SR. During the second phase, the **contraction period** (upward tracing), calcium ions bind to troponin, myosin-binding sites on actin are exposed, myosin heads attach to actin, and the resulting power strokes generate the force of contraction. During the third phase, the **relaxation period**, calcium ions are actively transported back into the SR; the level of Ca^{2+} in the sarcoplasm decreases, myosin-binding sites are covered by tropomyosin, myosin heads detach from actin, power strokes cease, and the muscle fiber relaxes. The actual duration of these periods depends on the type of skeletal muscle fiber. Some fibers, such as the fast-twitch fibers that move the eyes (described shortly), have contraction periods as brief as 10 msec and equally brief relaxation periods. Others, such as the slow-twitch fibers that move the legs, have contraction and relaxation periods of about 100 msec each.

Frequency of Stimulation

If a second stimulus occurs before a muscle fiber has relaxed, the second contraction will actually be stronger than the first (Figure 10.15b). This phenomenon, in which stimuli arriving in close sequence cause larger contractions, is called **wave summation**. When a skeletal muscle is stimulated at a rate of 20 to 30 times per second, it can only partially relax between stimuli. The result is a sustained but wavering contraction called **unfused tetanus** (*tetan-* = rigid, tense; Figure 10.15c). When a skeletal muscle fiber is stimulated at a higher rate of 80 to 100 times per second, it does not relax at all. The result is **fused tetanus**, a sustained contraction in which individual twitches cannot be detected (Figure 10.15d).

Wave summation and both kinds of tetanus result from the release of additional Ca^{2+} from the sarcoplasmic reticulum by subsequent stimuli; the extra Ca^{2+} adds to the Ca^{2+} still in the sarcoplasm from the previous stimulus. Because of the Ca^{2+} buildup in the sarcoplasm, contractions generated during fused tetanus are 5 to 10 times more forceful than those produced during a single twitch. Even so, smooth, sustained voluntary muscle contractions are achieved mainly by out-of-synchrony unfused tetanus in different motor units.

Motor Unit Recruitment

The force of a muscle contraction is greater as more motor units are activated. The process by which the number of contracting motor units is increased is called **motor unit recruitment**. Normally, the various motor units of a whole muscle are stimulated to contract *asynchronously* (at different times): While some motor units are contracting, others are relaxed. Because alternately contracting motor units relieve one another, muscle fatigue is delayed and contraction of a whole muscle can be sustained for long periods.

Figure 10.15   Myograms showing the effects of different frequencies of stimulation. (a) Single twitch. (b) When a second stimulus occurs before the muscle has relaxed, wave summation occurs, and the second contraction is stronger than the first. (The dashed line indicates the force of contraction expected in a single twitch.) (c) In unfused tetanus, the curve looks jagged due to partial relaxation of the muscle between stimuli. (d) In fused tetanus, the contraction force is steady, sustained, and strong.

(a) Single twitch   (b) Wave summation   (c) Unfused tetanus   (d) Fused tetanus

Due to wave summation, the tension produced during a sustained contraction is greater than during a single twitch.

Recruitment is one factor responsible for producing smooth movements rather than a series of jerky movements. As mentioned, the number of muscle fibers innervated by one motor neuron varies greatly. Precise movements, as occurs in muscles of the eye, are brought about by small changes in muscle contraction. Typically, the muscles that produce precise movements are composed of small motor units. In this way, when one motor unit is recruited or turned off, only slight changes occur in muscle tension. On the other hand, large motor units are active where large tension is needed and precision is less important, as occurs in muscles of the thigh and leg during locomotion.

Muscle Tone

Even at rest, a muscle exhibits **muscle tone** (*tonos* = tension), a small amount of tautness or tension in the muscle due to weak, involuntary contractions of its motor units. To sustain muscle tone, small groups of motor units are alternately active and inactive in a constantly shifting pattern. Muscle tone keeps skeletal muscles firm, but it does not result in a contraction strong enough to produce movement. For example, the tone of muscles in the back of the neck keep the head upright and prevent it from slumping forward on the chest, but they do not generate enough force to draw the head backward into hyperextension. Muscle tone also is important in smooth muscle tissue such as that found in the gastro-intestinal tract, where the walls of the digestive organs maintain a steady pressure on their contents. The tone of smooth muscle tissue in the walls of blood vessels plays a crucial role in maintaining blood pressure.

Recall that a skeletal muscle contracts only after it is stimulated by motor neurons. Muscle tone is established by the brain and spinal cord that excite the muscle's motor neurons. When the motor neurons serving a skeletal muscle are damaged, the muscle becomes **flaccid** (FLAS-sid = flabby), a state of limpness in which muscle tone is lost.

Isotonic and Isometric Contractions

In an **isotonic contraction** (*iso-* = equal; *-tonic* = tension), there is a change in the length of a muscle but no change in its tension. Isotonic contractions are used for body movements and for moving objects. The two types of isotonic contractions are concentric and eccentric. In a **concentric isotonic contraction**, a muscle shortens and pulls on another structure, such as a tendon, to produce movement and to reduce the angle at a joint. During a concentric contraction, the tension generated by the myosin heads is great enough to overcome the *load* (the resistance of the object to being moved), so the muscle shortens. Picking up a book from a table involves concentric isotonic contractions of the biceps brachii muscle in the arm (Figure 10.16a).

Figure 10.16 **Comparison between isotonic (concentric and eccentric) and isometric contractions.** Parts (a) and (b) show isotonic contraction of the biceps brachii muscle in the arm; part (c) shows isometric contraction of shoulder and arm muscles.

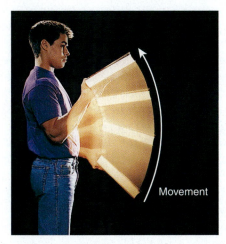

(a) Concentric contraction while picking up a book

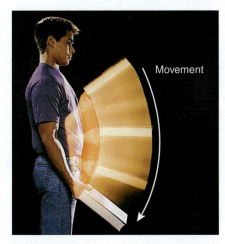

(b) Eccentric contraction while lowering a book

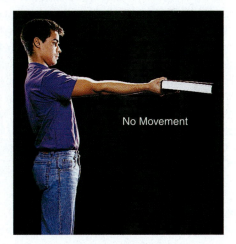

(c) Isometric contraction while holding a book steady

In an isotonic contraction, tension remains constant as muscle length decreases or increases; in an isometric contraction, tension increases greatly without a change in muscle length.

By contrast, as you lower the book to place it back on the table, the previously shortened biceps lengthens in a controlled manner while it continues to contract. When the length of a muscle increases during a contraction, the contraction is an **eccentric isotonic contraction** (Figure 10.16b). During an eccentric contraction, the tension exerted by the myosin head activity resists contraction of another muscle or the pull of gravity (the weight of the book, in this case) and slows the lengthening process. For reasons that are not well understood, repeated eccentric isotonic contractions produce more muscle damage and more delayed-onset muscle soreness than do concentric isotonic contractions.

In an **isometric contraction** (-*metric* = measure), the myosin head activity generates considerable tension, but not enough to pull the thin filaments inward (toward the M line). As a result, the muscle does not shorten. Isometric contractions occur when the load equals or exceeds the muscle tension. An example would be holding a book steady using an outstretched arm (Figure 10.16c). As the book pulls the arm downward, stretching the shoulder and arm muscles, isometric contraction of the shoulder and arm muscles counteracts the stretch. Isometric contractions are important for maintaining posture, for supporting objects in a fixed position, and for stabilizing some joints as other joints are moved. Most activities include both isotonic and isometric contractions.

Checkpoint

22. What is a motor unit?
23. Which events occur during the latent period of a twitch contraction? During the relaxation period?
24. What is the difference between unfused tetanus and fused tentanus? Which type of stimulation can allow you to lift a heavier object?
25. How are calcium ions related to tetanus?
26. In what ways does motor unit recruitment assist muscle function?

Return to Jesse's Story

"Hi, Jesse, I'm Lisa. I'll be doing your nerve conduction and EMG studies today. We're going to measure two things—the electrical signals from your nerves that talk to your muscles and the electrical activity of the muscles themselves that make the muscles contract. I'm going to give you a small stimulus and record the response of your nerves. It's kind of like a static electricity shock.

"Okay, Jesse you did great. We're going to do the electromyogram next," said Lisa. "The EMG is for measuring the function of your muscles directly. The muscle receives an electrical signal from the nerve, then produces its own electrical signal which allows it to contract. These tests today help us figure out if it's your nerves or your muscles that aren't working properly. I need you to make the muscles contract. Raise your arm up. That's good. We have a pretty good signal."

Dr. Fitts read the results of Jesse's tests. His blood tests showed a normal serum creatine kinase but the left deltoid muscle biopsy revealed findings consistent with facioscapulohumeral muscular dystrophy; isolated small atrophic fibers were present. His nerve conduction study and EMG reveal a myopathic process, most likely FSHD. Fitts sat back in his chair and picked up the phone.

"Hello, Mrs. Beatty?" he said.

"Yes," she replied.

"This is Dr. Fitts. I have the results of Jesse's tests. The good news is that Jesse's condition is not life-threatening. He has what is called *facioscapulohumeral muscular dystrophy*. As I explained before, there are different types of muscular dystrophy. FSHD is generally less severe than the Duchenne muscular dystrophy that his uncle has. It may affect Jesse later in life, but he has a chance to be quite active. I think we can even help him with his shoulder weakness. We have a surgical procedure that will stabilize his scapula and allow a more normal range of motion."

Jesse didn't really understand all the talk about "autosomal dominant disorders" and "phenotypic expression." What he did understand was that his shoulder could be fixed with surgery, at least for now. They scheduled the surgery during the middle of his first semester.

E. *What is the name for the electrical signals that Lisa is measuring? What happens when the electrical signal from the nerve reaches the muscle fibers in Jesse's shoulder? How would the lack of proper electrical signals from the nervous system cause muscle weakness in Jesse's shoulder?*

F. *Lisa is measuring the electrical signals in both the nerve and the muscle; how are these two signals linked to initiate muscle contraction? As Jesse raises his arms, his muscles are actively contracting; which process couples muscle action potential to muscle contraction?*

G. *Based on the events of muscle stimulation and contraction that you have read about so far, with which events might problems occur that could lead to Jesse's muscle weakness? Could Jesse's muscles contract if the nerves that innervate those muscles were not functioning?*

H. *Jesse's nerve conduction test results reveal that the nerve impulses are traveling normally to the muscle. How does his loss of muscle fibers due to the muscular dystrophy affect the force generated by motor units in his atrophied muscles?*

10.8 Muscle fibers produce ATP from creatine phosphate, by anaerobic cellular respiration, and by aerobic cellular respiration.

Production of ATP in Muscle Fibers

Unlike most cells of the body, skeletal muscle fibers often switch between virtual inactivity, when they are relaxed and using only a modest amount of ATP, and great activity, when they are contracting and using ATP at a rapid pace. A huge amount of ATP is needed to power the contraction cycle, to pump Ca^{2+} back into the SR to achieve muscle relaxation, and for other metabolic reactions involved in muscle contraction. However, the ATP present inside muscle fibers is enough to power contraction for only a few sec-

onds. If strenuous exercise continues for more than a few seconds, additional ATP must be synthesized.

ATP is the only molecule within muscle fibers that can directly transfer its stored energy to the contraction cycle. Fortunately, muscle fibers have three ways to produce ATP: (1) from creatine phosphate, (2) by anaerobic cellular respiration, and (3) by aerobic cellular respiration (Figure 10.17). All body cells make ATP through anaerobic and aerobic cellular respiration, but the use of creatine phosphate for ATP production is unique to muscle fibers. We consider the events of cellular respiration briefly here and then in detail in Chapter 23.

Figure 10.17 **Production of ATP for muscle contraction.** (a) Creatine phosphate is formed from ATP while the muscle is relaxed, then releases ATP as the muscle contracts. (b) Muscle glycogen is broken down into glucose, which glycolysis converts to pyruvic acid and ATP. Without sufficient oxygen, pyruvic acid is converted to lactic acid. (c) If adequate oxygen is available, mitochondria use pyruvic acid, fatty acids, and amino acids to produce additional ATP.

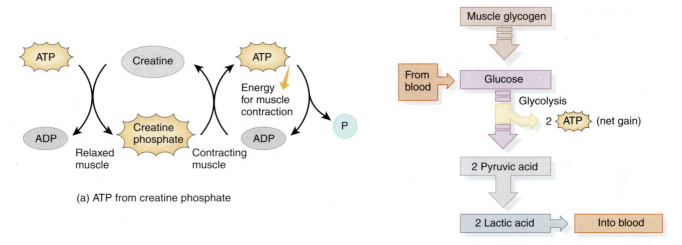

(a) ATP from creatine phosphate

(b) ATP from anaerobic cellular respiration

(c) ATP from aerobic cellular respiration

🔑 **During a long-term event such as a marathon race, most ATP is produced aerobically.**

Creatine Phosphate

While they are relaxed, muscle fibers produce more ATP than they need for resting metabolism. The excess ATP is used to synthesize **creatine phosphate**, an energy-rich molecule that is found only in muscle fibers (Figure 10.17a). The enzyme *creatine kinase* catalyzes the transfer of one high-energy phosphate group from ATP to creatine, forming creatine phosphate and ADP. Creatine phosphate is three to six times more plentiful than ATP in the sarcoplasm of a relaxed muscle fiber. When contraction begins and the ADP level starts to rise, the high-energy phosphate group can be transferred from creatine phosphate back to ADP, quickly forming new ATP molecules. Together, creatine phosphate and ATP provide enough energy for muscles to contract maximally for about 15 seconds. This amount of energy is sufficient for short bursts of activity, for example, to run a 100-meter dash.

Anaerobic Cellular Respiration

Anaerobic cellular respiration is a series of ATP-producing reactions that do not require oxygen. When muscle activity continues past the 15-second mark, the supply of creatine phosphate is depleted, and glucose is catabolized to generate ATP. Glucose passes easily from the blood into contracting muscle fibers and is also produced within muscle fibers by the breakdown of glycogen (Figure 10.17b). Then, a series of reactions known as *glycolysis* quickly breaks down each glucose molecule into two molecules of pyruvic acid. Glycolysis uses two ATP but produces four ATP for a net gain of two.

Ordinarily, the pyruvic acid formed by glycolysis in the sarcoplasm enters mitochondria. There, it enters a series of oxygen-requiring reactions called aerobic respiration (described next) that produce a large amount of ATP. During some activities, however, not enough oxygen is available. In such situations, anaerobic reactions convert most of the pyruvic acid to lactic acid. Most of the lactic acid diffuses out of the skeletal muscle fibers into the blood. Liver cells can convert some of the lactic acid back to glucose. This conversion has two benefits: providing new glucose molecules and reducing blood acidity. Glycolysis can provide enough energy for about 30 to 40 seconds of maximal muscle activity. Together, creatine phosphate and glycolysis can provide enough ATP to run a 400-meter race.

Aerobic Cellular Respiration

Muscle activity that lasts longer than half a minute depends increasingly on **aerobic cellular respiration**, a series of oxygen-requiring reactions that produce ATP in mitochondria. If enough oxygen is present, pyruvic acid from glycolysis enters the mitochondria, where it is completely oxidized in reactions that generate ATP, carbon dioxide, water, and heat (Figure 10.17c). Although aerobic cellular respiration is slower than glycolysis alone, it yields much more ATP, about 36 molecules of ATP from each glucose molecule that entered glycolysis.

Muscle fibers have two sources of oxygen: (1) oxygen that diffuses into them from the blood and (2) oxygen released by myoglobin in the sarcoplasm. Both myoglobin (found only in muscle cells) and hemoglobin (found only in red blood cells) bind oxygen when it is plentiful and release oxygen when it is scarce.

Aerobic cellular respiration provides enough ATP for prolonged activity as long as sufficient oxygen and nutrients are available. Besides pyruvic acid obtained from glycolysis of glucose, these nutrients include fatty acids from the breakdown of triglycerides in adipose cells and amino acids from the breakdown of proteins. In activities that last more than 10 minutes, aerobic cellular respiration provides most of the needed ATP. At the end of an endurance event, such as a marathon race, nearly 100% of the ATP is being produced by aerobic cellular respiration.

Muscle Fatigue

The inability of a muscle to contract forcefully after prolonged activity is called **muscle fatigue**. Even before actual muscle fatigue occurs, a person may have feelings of tiredness and the desire to cease activity. This response, termed *central fatigue*, is caused by changes in the central nervous system (brain and spinal cord) and may be a protective mechanism to stop a person from exercising before muscle becomes damaged. Although the precise mechanisms that cause muscle fatigue are still unclear, several factors are thought to contribute. One is inadequate release of calcium ions from the SR, resulting in a decline of Ca^{2+} level in the sarcoplasm. Other factors that contribute to muscle fatigue include depletion of creatine phosphate, insufficient oxygen, depletion of glycogen and other nutrients, buildup of lactic acid and ADP, and failure of nerve impulses in the motor neuron to release enough acetylcholine.

Oxygen Consumption after Exercise

During prolonged periods of muscle contraction, increases in breathing rate and blood flow enhance oxygen delivery to muscle tissue. After muscle contraction has stopped, heavy breathing continues for a time, and oxygen consumption remains above the resting level. Depending on the intensity of the exercise, the recovery period may be just a few minutes or it may last as long as several hours. The term **oxygen debt** refers to the added oxygen, over and above the oxygen consumed at rest, that is taken into the body after exercise. This extra oxygen is used to "pay back" or restore metabolic conditions to the resting level in three ways: (1) to convert lactic acid back into glycogen stores in the liver, (2) to resynthesize creatine phosphate and ATP in muscle fibers, and (3) to replace the oxygen removed from myoglobin.

The metabolic changes that occur *during exercise* can account for only some of the extra oxygen used *after exercise*. Only a small amount of glycogen is resynthesized from lactic acid. Instead, most glycogen stores are replenished much later from dietary carbohydrates. Much of the lactic acid that remains after exercise is converted back to pyruvic acid and used for ATP production via aerobic cellular respiration. Ongoing changes after exercise also boost oxygen use. First, the elevated body temperature after strenuous exercise increases the rate of chemical reactions throughout the body. Faster reactions use ATP more rapidly, and more oxygen is needed to produce ATP. Second, the heart and muscles used in breathing are still working harder than they were at rest, and thus they consume more ATP. Third, tissue repair processes are occurring at an increased pace. For these reasons, **recovery oxygen uptake** is a better term than oxygen debt for the elevated use of oxygen after exercise.

Checkpoint

27. Which molecule is the only direct source of contraction energy for a muscle fiber?
28. Which ATP-producing reactions provide ATP during a 1000-meter run? Which occur within mitochondria?
29. What is muscle fatigue?
30. Why is the term recovery oxygen uptake more accurate than oxygen debt?

10.9 Skeletal muscle fibers are classified as slow oxidative fibers, fast oxidative–glycolytic fibers, or fast glycolytic fibers.

Skeletal muscle fibers vary structurally in their content of myoglobin, the red protein that binds oxygen in muscle fibers. Those with a low myoglobin content appear pale and are called *white muscle fibers*. White muscle fibers are prevalent in the "white meat" in chicken breasts. Skeletal muscle fibers with a high myoglobin content and many blood capillaries have a dark, reddish appearance and are called *red muscle fibers*. Red muscle fibers are in the "dark meat" in chicken legs and thighs.

Skeletal muscle fibers contract and relax at different speeds; they also vary in which metabolic reactions are used to generate ATP and how quickly they fatigue. For example, a fiber is categorized as either slow or fast depending on how rapidly the ATPase in its myosin heads splits ATP. Based on these structural and functional characteristics, skeletal muscle fibers are classified into three main categories: (1) slow oxidative fibers, (2) fast oxidative–glycolytic fibers, and (3) fast glycolytic fibers.

Slow Oxidative Fibers

Slow oxidative (SO) fibers are smallest in diameter and thus are the least powerful type of skeletal muscle fibers. They appear dark red because they contain large amounts of myoglobin and many blood capillaries. Because they have many mitochondria, SO fibers generate ATP mainly by aerobic cellular respiration, which is why they are called oxidative fibers. These fibers are said to be "slow" because the ATPase in the myosin heads splits ATP relatively slowly; thus, SO fibers have a slow speed of contraction. However, slow fibers are very resistant to fatigue and are capable of prolonged, sustained contractions for many hours. These slow-twitch, fatigue-resistant fibers are adapted for maintaining posture and for aerobic, endurance-type activities such as running a marathon.

Fast Oxidative–Glycolytic Fibers

Fast oxidative–glycolytic (FOG) fibers are intermediate in diameter between the other two types of fibers. Like slow oxidative fibers, they contain large amounts of myoglobin and many capillaries. FOG fibers can generate considerable ATP by aerobic cellular respiration, which gives them a moderately high resistance to fatigue. Because their glycogen level is high, they also generate ATP by glycolysis. FOG fibers are "fast" because the ATPase in their myosin heads splits ATP faster than the myosin ATPase in SO fibers, which makes their speed of contraction faster. FOG fibers contribute to activities such as walking and sprinting.

Fast Glycolytic Fibers

Fast glycolytic (FG) fibers are largest in diameter and contain the most myofibrils. Hence, they can generate the most powerful contractions. FG fibers are white fibers that have low myoglobin content and relatively few blood capillaries, but they contain large amounts of glycogen and generate ATP mainly by glycolysis. Due to their large size and their ability to use ATP at a fast rate, FG fibers contract strongly and quickly. These fast-twitch fibers are adapted for intense anaerobic movements of short duration, such as weight lifting or throwing a ball, but they fatigue quickly. Strength training programs that engage a person in activities requiring great strength for short periods increase the size, strength, and glycogen content of FG fibers. The FG fibers of a weight lifter may be 50% larger than those of a sedentary person or an endurance athlete. The increase in size is due to increased synthesis of muscle proteins. The result is muscle enlargement due to hypertrophy of the FG fibers.

Distribution and Recruitment of Different Types of Fibers

Most skeletal muscles are a mixture of all three types of skeletal muscle fibers. The proportions vary somewhat, depending on the action of the muscle, the person's training regimen, and genetic factors. For example, the continually active postural muscles of the neck, back, and legs have a high proportion of SO fibers. Muscles of the shoulders and arms, in contrast, are not constantly active but are used briefly now and then to produce large amounts of tension, such as in lifting and throwing. These muscles have a high proportion of FG fibers. Leg muscles, which not only support the body but are also used for walking and running, have large numbers of both SO and FOG fibers.

Even though most skeletal muscles are a mixture of all three types of skeletal muscle fibers, the skeletal muscle fibers of any given motor unit are all of the same type. The different motor units in a muscle are recruited in a specific order, depending on need. For example, if weak contractions suffice to perform a task, only SO motor units are activated. If more force is needed, the motor units of FOG fibers are also recruited. Finally, if maximal force is required, motor units of FG fibers are also called into action. Activation of various motor units is controlled by the brain and spinal cord.

Table 10.1 summarizes the characteristics of the three types of skeletal muscle fibers.

Checkpoint

31. Why are some skeletal muscle fibers classified as "fast" and others are said to be "slow"?

32. In what order are the various types of skeletal muscle fibers recruited when you sprint to make it to the bus stop?

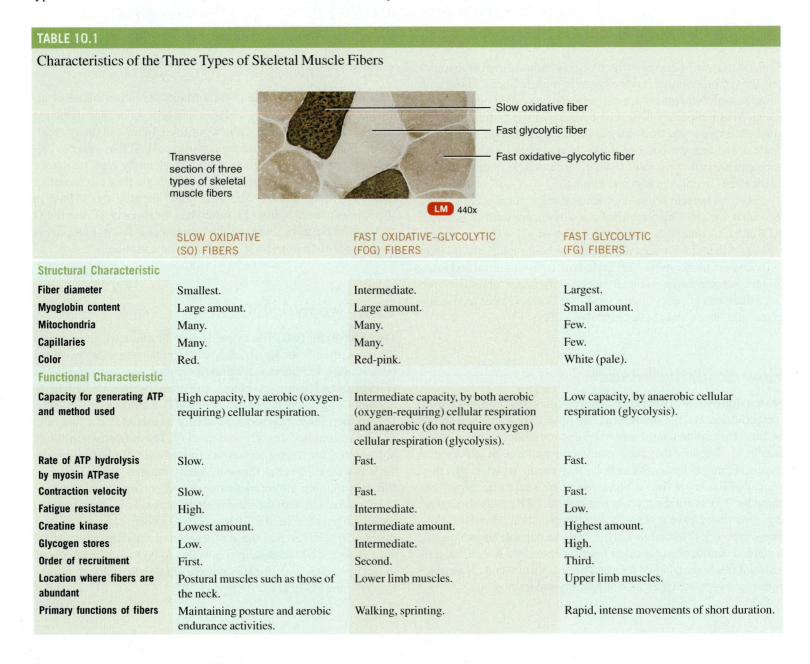

TABLE 10.1

Characteristics of the Three Types of Skeletal Muscle Fibers

Transverse section of three types of skeletal muscle fibers

— Slow oxidative fiber
— Fast glycolytic fiber
— Fast oxidative–glycolytic fiber

LM 440x

	SLOW OXIDATIVE (SO) FIBERS	FAST OXIDATIVE–GLYCOLYTIC (FOG) FIBERS	FAST GLYCOLYTIC (FG) FIBERS
Structural Characteristic			
Fiber diameter	Smallest.	Intermediate.	Largest.
Myoglobin content	Large amount.	Large amount.	Small amount.
Mitochondria	Many.	Many.	Few.
Capillaries	Many.	Many.	Few.
Color	Red.	Red-pink.	White (pale).
Functional Characteristic			
Capacity for generating ATP and method used	High capacity, by aerobic (oxygen-requiring) cellular respiration.	Intermediate capacity, by both aerobic (oxygen-requiring) cellular respiration and anaerobic (do not require oxygen) cellular respiration (glycolysis).	Low capacity, by anaerobic cellular respiration (glycolysis).
Rate of ATP hydrolysis by myosin ATPase	Slow.	Fast.	Fast.
Contraction velocity	Slow.	Fast.	Fast.
Fatigue resistance	High.	Intermediate.	Low.
Creatine kinase	Lowest amount.	Intermediate amount.	Highest amount.
Glycogen stores	Low.	Intermediate.	High.
Order of recruitment	First.	Second.	Third.
Location where fibers are abundant	Postural muscles such as those of the neck.	Lower limb muscles.	Upper limb muscles.
Primary functions of fibers	Maintaining posture and aerobic endurance activities.	Walking, sprinting.	Rapid, intense movements of short duration.

10.10 Cardiac muscle tissue is found in the heart, and smooth muscle tissue is found in hollow internal structures.

Cardiac Muscle Tissue

The principal tissue in the heart wall is **cardiac muscle tissue**. Compared with skeletal muscle fibers, cardiac muscle fibers are shorter in length and less circular in transverse section (Figure 10.18a). They also exhibit branching, which gives individual cardiac muscle fibers a "stair-step" appearance. Usually, one centrally located nucleus is present. Cardiac muscle fibers interconnect with

Figure 10.18 **Histology of cardiac muscle.** A photomicrograph of cardiac muscle tissue is shown in Figure 10.1b.

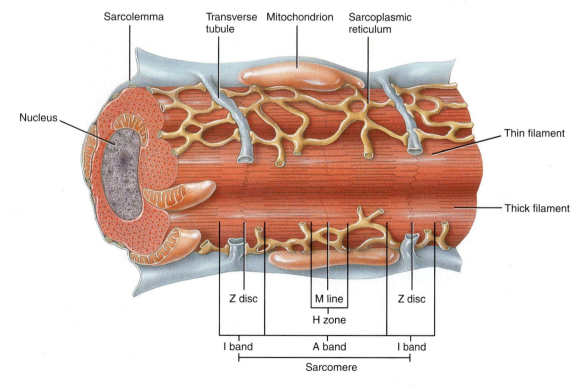

(a) Cardiac muscle fibers

(b) Arrangement of components in a cardiac muscle fiber

Intercalated discs hold the fibers together and allow muscle action potentials to spread quickly from one cardiac muscle fiber to another.

one another by irregular transverse thickenings of the sarcolemma called **intercalated discs** (in-TER-ka-lāt-ed = inserted between). The discs contain cell junctions: **desmosomes** hold the fibers together, and **gap junctions** allow muscle action potentials to spread quickly from one cardiac muscle fiber to another.

Cardiac muscle fibers are striated like skeletal muscle fibers with the same arrangement of actin and myosin, and the same bands, zones, and Z discs (Figure 10.18b). The sarcoplasmic reticulum of cardiac muscle fibers is somewhat smaller than the SR of skeletal muscle fibers.

In response to a single action potential, cardiac muscle tissue remains contracted 10 to 15 times longer than skeletal muscle tissue. The long contraction is due to prolonged delivery of Ca^{2+} into the sarcoplasm. In cardiac muscle fibers, Ca^{2+} enters the sarcoplasm both from the sarcoplasmic reticulum and from the interstitial fluid that bathes the fibers. Because channels in the sarcolemma that allow inflow of Ca^{2+} from interstitial fluid stay open for a relatively long time, a cardiac muscle contraction lasts longer than a skeletal muscle twitch.

We have seen that skeletal muscle tissue contracts only when stimulated by acetylcholine released by a nerve impulse in a motor neuron. In contrast, cardiac muscle tissue contracts when stimulated by its own autorhythmic muscle fibers. Under normal resting conditions, cardiac muscle tissue contracts and relaxes about 75 times per minute. This continuous, rhythmic activity is a major physiological difference between cardiac and skeletal muscle tissue. Thus, cardiac muscle tissue requires a constant supply of oxygen and nutrients. The mitochondria in cardiac muscle fibers are larger and more numerous than in skeletal muscle fibers and produce most of the needed ATP via aerobic cellular respiration. In addition, cardiac muscle fibers can use lactic acid produced by skeletal muscle fibers to make ATP, a benefit during exercise.

Smooth Muscle Tissue

Of the two types of smooth muscle tissue, the more common type is **visceral (single-unit) smooth muscle tissue** (Figure 10.19a). It is found in sheets that form part of the walls of small arteries and veins and walls of hollow organs such as the stomach, intestines, uterus, and urinary bladder. Like cardiac muscle, visceral smooth muscle is autorhythmic and the fibers connect to one another by gap junctions, forming a network through which muscle action potentials can spread. When a neurotransmitter, hormone, or autorhythmic signal stimulates one fiber, the muscle action potential is transmitted to neighboring fibers, which then contract in unison, as a single unit.

Figure 10.19   Histology of smooth muscle tissue. In (a), one autonomic motor neuron synapses with several visceral smooth muscle fibers. In (b), three autonomic motor neurons synapse with individual multiunit smooth muscle fibers. In (c), a smooth muscle fiber is shown in the relaxed state and the contracted state. A photomicrograph of smooth muscle tissue is shown in Figure 10.1c.

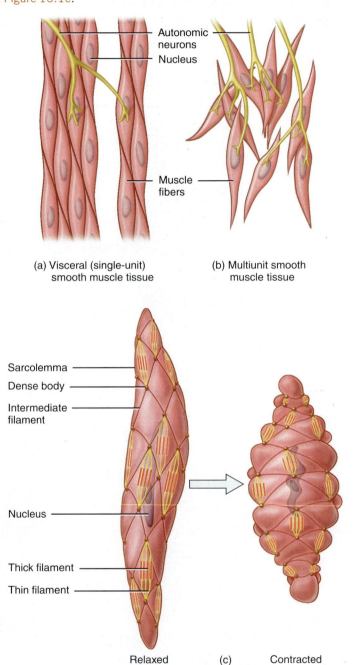

(a) Visceral (single-unit) smooth muscle tissue

(b) Multiunit smooth muscle tissue

Relaxed   (c)   Contracted

Smooth muscle fibers have thick and thin filaments but no transverse tubules and little sarcoplasmic reticulum.

The second type of smooth muscle tissue is **multiunit smooth muscle tissue** (Figure 10.19b). It consists of individual fibers, each with its own motor neuron terminals and with few gap junctions between neighboring fibers. Stimulation of one visceral muscle fiber causes contraction of many adjacent fibers, but stimulation of one multiunit fiber causes contraction of that fiber only. Multiunit smooth muscle tissue is found in the walls of large arteries, in airways to the lungs, in the arrector pili muscles that attach to hair follicles, and in the internal eye muscles.

Smooth muscle fibers are considerably smaller in length and diameter than skeletal muscle fibers and are tapered at both ends (Figure 10.19c). Within each fiber is a single, centrally located nucleus. The sarcoplasm of smooth muscle fibers contains both *thick filaments* and *thin filaments*, but they are not arranged in orderly sarcomeres as in striated muscle. Smooth muscle fibers also contain *intermediate filaments*, elements of the cytoskeleton. Because the various filaments have no regular pattern of overlap, smooth muscle lacks alternating dark and light bands and thus has a smooth appearance. Bundles of intermediate filaments stretch between structures called *dense bodies*, which are similar to the Z discs found in striated muscle (Figure 10.19c). Smooth muscle fibers also lack transverse tubules and have only scanty sarcoplasmic reticulum for storage of Ca^{2+}.

Although the principles of contraction are similar in all three types of muscle tissue, smooth muscle tissue exhibits some important physiological differences. During contraction, the sliding filament mechanism involving thick and thin filaments generates tension that is transmitted to intermediate filaments. These, in turn, pull on the dense bodies attached to the sarcolemma, causing a lengthwise shortening of the muscle fiber. As a smooth muscle fiber contracts, it rotates as a corkscrew turns. The fiber twists in a helix as it contracts, and rotates in the opposite direction as it relaxes (Figure 10.19c).

Compared with contraction in a skeletal muscle fiber, contraction in smooth muscle fiber starts more slowly and lasts much longer. In addition, smooth muscle can both shorten and stretch to a greater extent than other muscle types.

An increase in the concentration of Ca^{2+} in the sarcoplasm of a smooth muscle initiates contraction, just as in striated muscle.

However, the regulatory protein *calmodulin*, rather than troponin, activates myosin head cycling in smooth muscle. Calcium ions flow into smooth muscle sarcoplasm from both the interstitial fluid and sarcoplasmic reticulum, but because there are no transverse tubules in smooth muscle fibers, it takes longer for Ca^{2+} to reach the filaments in the center of the fiber and trigger the contractile process. This accounts, in part, for the slow onset of contraction of smooth muscle.

Calcium ions also move out of the muscle fiber slowly, which delays relaxation. The prolonged presence of Ca^{2+} in the cytosol provides for *smooth muscle tone*, a state of continued partial contraction. Smooth muscle tissue can thus sustain long-term tone, which is important in the walls of blood vessels that maintain a steady pressure on blood, and in the walls of gastrointestinal tract organs that maintain a steady pressure on their contents.

Like cardiac muscle tissue, smooth muscle tissue is usually activated involuntarily. Most smooth muscle fibers contract or relax in response to action potentials from the autonomic nervous system. In addition, many smooth muscle fibers contract or relax in response to stretching, hormones, or local factors such as changes in pH, oxygen and carbon dioxide levels, temperature, and ion concentrations. For example, the hormone epinephrine causes relaxation of smooth muscle in the airways and in some blood vessel walls.

Table 10.2 summarizes the major characteristics of the three types of muscle tissue.

Checkpoint

33. Of the three types of muscle tissue, which have contractile proteins organized into sarcomeres? Contain transverse tubules? Contain intermediate filaments? Contain dense bodies?

34. What are the functions of intercalated discs? In which muscle tissue type are they found?

35. What is the source of calcium ions for each of the three types of muscle tissue?

36. Which type of smooth muscle is found in the walls of hollow organs? Which is found in large blood vessels?

37. How does the speed of onset and duration of contraction in a smooth muscle fiber compare with that in a skeletal muscle fiber?

TABLE 10.2

Major Features of the Three Types of Muscle Tissue

CHARACTERISTIC	SKELETAL MUSCLE	CARDIAC MUSCLE	SMOOTH MUSCLE
Microscopic appearance and features	Long cylindrical fiber with many peripherally located nuclei; striated.	Branched cylindrical fiber with one centrally located nucleus; intercalated discs join neighboring fibers; striated.	Fiber is thickest in the middle, tapered at each end, and has one centrally positioned nucleus; not striated.
Location	Attached primarily to bones by tendons.	Heart.	Walls of hollow viscera, airways, blood vessels, iris and ciliary body of eye, arrector pili of hair follicles.
Fiber diameter	Very large (10–100 μm).	Large (10–20 μm).	Small (5–10 μm).
Connective tissue components	Endomysium, perimysium, and epimysium.	Endomysium.	Endomysium.
Fiber length	100 μm–30 cm.	50–100 μm.	20–500 μm.
Contractile proteins organized into sarcomeres	Yes.	Yes.	No.
Sarcoplasmic reticulum	Abundant.	Some.	Scanty.
Transverse tubules present	Yes, aligned with each A–I band junction.	Yes, aligned with each Z disc.	No.
Junctions between fibers	None.	Intercalated discs contain gap junctions and desmosomes.	Gap junctions in visceral smooth muscle; none in multiunit smooth muscle.
Autorhythmicity	No.	Yes.	Yes, in visceral smooth muscle.
Source of Ca^{2+} for contraction	Sarcoplasmic reticulum.	Sarcoplasmic reticulum and interstitial fluid.	Sarcoplasmic reticulum and interstitial fluid.
Speed of contraction	Fast.	Moderate.	Slow.
Nervous control	Mostly voluntary.	Involuntary.	Involuntary.
Contraction regulated by:	Acetylcholine released by somatic motor neurons.	Acetylcholine and norepinephrine released by autonomic motor neurons; several hormones.	Acetylcholine and norepinephrine released by autonomic motor neurons; several hormones; local chemical changes; stretching.
Capacity for regeneration	Limited, via satellite cells.	Limited, under certain conditions.	Considerable (compared with other muscle tissues, but limited compared with epithelium), via pericytes.

Jesse's Story: Epilogue and Discussion

▶▶▶ Jesse, a 16-year-old male athlete, presents to his sports medicine doctor complaining of arm weakness and shoulder discomfort. Upon examination, Jesse's doctor finds that Jesse has a winging scapula, indicating a weakening of the shoulder joint, and cannot raise his left arm above shoulder level. In addition, Jesse has facial weakness and cannot pucker his lips or whistle. Laboratory findings indicate that Jesse's nerves are functioning normally, but his shoulder muscles have some pathological changes including atrophied muscle fibers. There is a family history of muscular dystrophy.

Based on the physical examination and laboratory results, the doctor diagnoses *facioscapulohumeral muscular dystrophy (FSHD)*. This condition is characterized by muscle degeneration and weakness, like all muscular dystrophies. However, there are many types of muscular dystrophy. FSHD weakens muscles in the face, around the shoulder blades, and in the upper arm. The disease occurs with a frequency of 1 in 20,000, and symptoms usually appear during the first or second decade of life. The severity of the symptoms is worse the earlier the disease manifests itself. Unlike Duchenne muscular dystrophy, a more common and severe form of muscular dystrophy, FSHD does not affect life span. However, the severity of the disorder can vary from individual to individual. In Jesse's case, stabilizing the scapula by surgically fixing its medial borders to the ribs will probably result in improved abduction and flexion of the arms and improve range of motion in spite of the muscle weakness. ◀◀◀

I.   What are the three types of muscle tissues found in the human body? Which of these was not functioning properly in Jesse's body? How were the functions of Jesse's muscle tissue impaired by his disorder?

J.   Jesse had a nerve conduction study and an EMG. What was being tested?

K.   Why was the doctor interested in checking nerve function if it was the muscles in Jesse's shoulder and arm that were weak?

L.   Jesse plans on lifting weights and strength training as part of his recuperation; how will strengthening his shoulder help his quality of life?

M.   Why is surgical fixation and stabilization of the scapula necessary in Jesse's case? How would Jesse's condition and its associated processes affect Jesse if they were left untreated?

Concept and Resource Summary

Concept   **Resources**

Introduction

1. The principal function of muscle is to change chemical energy into mechanical energy to generate force, perform work, and produce movement.

Concept 10.1 Skeletal, cardiac, and smooth muscle tissues differ in location, structure, and function.

See
- Anatomy Overview – Muscle Tissue

1. There are three types of **muscle tissue**: skeletal, cardiac, and smooth. **Skeletal muscle tissue** is *striated*, under *involuntary* control, and functions to move bones of the skeleton. **Cardiac muscle tissue** is striated, under **involuntary** control, has **autorhythmicity**, and is only found in the heart. **Smooth muscle tissue** is **nonstriated**, under **involuntary** control, and is located in the walls of hollow internal structures.

Concept 10.2 Muscle tissue performs four functions and possesses four properties.

1. Muscle tissue has four important functions: *producing body movements, stabilizing body positions, storing and moving substances within the body*, and *producing heat*.
2. Muscle tissue has four special properties: **electrical excitability**, **contractility**, **extensibility** and **elasticity**.

Concept

Concept 10.3 Skeletal muscles are surrounded by connective tissues and are well supplied with nerves and blood vessels.

1. A skeletal muscle is an organ composed of elongated **muscle fibers** plus associated nerves, blood vessels, and connective tissues.
2. **Hypodermis** separates muscle from skin. **Fascia** unites muscles with similar functions, carries nerves and vessels, and fills spaces between muscles.
3. Three layers of connective tissue protect and strengthen skeletal muscle: the outermost layer encircles the muscle, and is the **epimysium**; **perimysium** bundles together groups of muscle fibers into **fascicles**; and the innermost **endomysium** surrounds each muscle fiber. These three layers may extend beyond the muscle to form a **tendon**, or a broad flat **aponeurosis**. Tendons and aponeuroses attach a muscle to bone. **Tendon sheaths** contain synovial fluid and enclose certain tendons to reduce friction.
4. One artery and one or two veins supply each skeletal muscle.
5. **Motor neurons** stimulate skeletal muscle and extend an **axon**, with axon collaterals, to muscle fibers that it controls. Each muscle fiber meets a *synaptic end bulb* of an axon terminal at the **neuromuscular junction**.

See

- Anatomy Overview – Cross Section of Skeletal Muscle
- Animation – Contraction and Movement

Concept 10.4 Each skeletal muscle fiber is covered by a sarcolemma; each of its myofibrils is surrounded by sarcoplasmic reticulum and contains sarcomeres.

1. Embryonic development of skeletal muscle fibers arises from fusion of *myoblasts* into one elongated, multinucleate, amitotic muscle fiber. Some myoblasts persist as *satellite cells* in mature skeletal muscle that can fuse and regenerate muscle fibers in damaged tissue. Extensive damage involves **fibrosis**, the replacement of muscle fibers with scar tissue.
2. After birth, muscle growth occurs from enlargement of existing muscle fibers, termed **hypertrophy**, stimulated by human growth hormone, testosterone, and other hormones.
3. The **sarcolemma** has many invaginated, tunnel-like extensions called **transverse (T) tubules** that are also open to the cell's exterior.
4. The **sarcoplasm** contains glycogen, and a red protein, called **myoglobin**, that binds oxygen molecules for use in ATP synthesis.
5. The sarcoplasm is full of long contractile elements, called **myofibrils**, that give the fiber its striated appearance.
6. The **sarcoplasmic reticulum (SR)** wraps around each myofibril. The SR stores and releases Ca^{2+}. The dilated end sacs of the SR, called **terminal cisterns**, lie on each side of one T tubule, forming a **triad**.
7. Myofibrils contain **thin filaments** and **thick filaments** arranged in basic functional units called **sarcomeres**. Sarcomeres in a myofibril are separated from one another by **Z discs**.
8. A darker middle part of the sarcomere containing thick and thin overlapping filaments is the **A band**, at the center of which is a narrow **H zone** containing only thick myofilaments. **I bands** contain only thin filaments and are located near the Z discs at the ends of a sarcomere.
9. The contractile proteins, myosin and actin, are located inside myofibrils. **Myosin** converts chemical energy in ATP to the mechanical energy of motion. About 300 myosin molecules form one thick filament with myosin heads projecting toward the surrounding thin filaments. **Actin** is the main component of thin filaments. Actin has a *myosin-binding site* where the myosin head can attach. Two regulatory proteins, **tropomyosin** and **troponin**, are also part of the thin filament, and help switch contraction on and off.
10. In addition to contractile and regulatory proteins, muscle contains structural proteins such as **titin** and **dystrophin**. Titin helps anchor a Z disc to the M line, and **dystrophin** links thin filaments of the sarcomere to the sarcolemma.

See

- Anatomy Overview – Skeletal Muscle
- Animation – Muscle Cell Structures

Concept	Resources

Concept 10.5 The neuromuscular junction is where a muscle action potential is initiated.

1. Muscle action potentials arise at the **neuromuscular junction (NMJ)**, the **synapse** between a motor neuron and a skeletal muscle fiber.
2. The synaptic end bulbs have **synaptic vesicles** containing the neurotransmitter **acetylcholine (ACh)**. The motor neuron releases ACh into the synaptic cleft; ACh binds to **acetylcholine receptors** on the **motor end plate** region of the sarcolemma of the muscle fiber.
3. Events in the excitation of a skeletal muscle fiber include the following: release of acetylcholine as a nerve impulse from the brain or spinal cord reaches the synaptic end bulbs; activation of ACh receptors when ACh binds to a motor end plate ACh receptor, which opens ion channels; generation of a muscle action potential that propagates along the sarcolemma and T tubules, stimulating a contraction; and breakdown of ACh by the enzyme **acetylcholinesterase (AChE)** in the synaptic cleft.

See
- Animation – Neuromuscular Junctions

Do
- Concepts and Connections – Events at the Neuromuscular Junction

Concept 10.6 An action potential releases calcium ions that allow thick filaments to bind to and pull thin filaments toward the center of the sarcomere.

1. The **sliding filament mechanism** of muscle contraction involves thin filaments at both ends of the sarcomere being pulled to the center of the sarcomere by myosin head activity. Z discs come closer together and the sarcomere shortens.
2. When a muscle action potential propagates along the sarcolemma and T tubules, it opens Ca^{2+} **release channels** in the SR membrane. Ca^{2+} flows into the sarcoplasm and combines with troponin, moving the troponin–tropomyosin complex away from myosin-binding sites on actin. Myosin heads then bind to actin.
3. The **contraction cycle** consists of four steps: ATP splits in the myosin head to reorient and energize it; myosin attaches to actin when the energized myosin head attaches to the myosin-binding site on actin; the power stroke occurs when the cross-bridge releases the phosphate group from the myosin head, triggering ADP release and myosin rotation, which slides the thin filament toward the M line; and myosin detaches from actin as ATP binds to the myosin head.
4. ATPase splits ATP on the myosin head, and energy is transferred to the myosin head as it is energized and reoriented in position. The contraction cycle repeats as successive power strokes result in shortening of the sarcomeres. As Z discs pull on adjacent sarcomeres, the whole muscle fiber shortens, and, ultimately, the entire muscle shortens.
5. Cessation of nerve impulses stops ACh release, and AChE breaks down any ACh in the synaptic cleft. Ion channels close, action potentials stop, and Ca^{2+} **active transport pumps** use ATP to move Ca^{2+} back into the SR. The troponin–tropomyosin complexes slide back to cover myosin-binding sites of actin, and thin filaments return to their relaxed positions.

See
- Animation – Contraction of a Sarcomere

Hear
- Figure 10.9 – Sliding Filament Mechanism of Muscle Contraction
- Figure 10.11 – The Contraction Cycle
- Figure 10.12 – Summary of the Events of Contraction and Relaxation in a Skeletal Muscle Fiber

Do
- Exercise – Contraction Connections
- Exercise – Muscle Car
- Concepts and Connections – Excitation of Skeletal Muscle
- Concepts and Connections – Skeletal Muscle Contraction Cycle

Concept 10.7 Muscle tension is controlled by stimulation frequency and motor unit recruitment.

1. A single nerve impulse in a motor neuron elicits a single muscle twitch contraction in all the muscle fibers that it innervates. The frequency of stimulation governs the total tension that can be produced by a single muscle fiber. The total tension produced in a whole muscle depends on the number of fibers contracting in unison.
2. A **motor unit** is one motor neuron and all the skeletal muscle fibers the motor neuron stimulates.
3. Muscles controlling small, precise movements are innervated by many small motor units; muscles controlling large, powerful movements have fewer large motor units. The size of a muscle's motor units and the number of motor units activated contribute to the contraction strength.

See
- Animation – Control of Muscle Tension

Do
- Exercise – Increase Muscle Tension

Concept

4. The response of a motor unit to a single nerve impulse in its motor neuron is a **twitch contraction**. The three phases of a twitch are the **latent period** of cell events leading up to contraction; the **contraction period** of power strokes generating tension; and the **relaxation period**, during which the muscle is allowed to resume its original length.

5. Multiple stimuli that arrive before the muscle fiber has fully relaxed lead to **wave summation**. When the frequency of stimulation allows partial relaxation it is called **unfused tetanus**; rapid frequency of stimulation and sustained contraction is called **fused tetanus**.

6. **Motor unit recruitment** is the process of increasing the number of contracting motor units.

7. A muscle, at rest, exhibits **muscle tone**, a small amount of tension due to involuntary alternating contractions of a small number of its motor units that does not produce movement. Damaged motor neurons that cannot maintain muscle tone cause a muscle to become **flaccid**.

8. Isotonic contractions involve a change in muscle length without a change in its tension. There are two types: **concentric isotonic contractions** occur when the muscle shortens; **eccentric isotonic contractions** occur when the muscle lengthens.

9. An **isometric contraction** occurs when the load equals or exceeds the muscle tension, and the muscle does not lengthen or shorten.

Concept 10.8 Muscle fibers produce ATP from creatine phosphate, by anaerobic cellular respiration, and by aerobic cellular respiration.

1. ATP is the only direct source of energy for muscle contraction. Muscle fibers have three ways to produce ATP: from creatine phosphate, by anaerobic cellular respiration, and by aerobic cellular respiration.

2. Muscle fibers break down excess ATP and transfer a phosphate group to creatine, forming **creatine phosphate** and ADP. During contraction, muscle fibers transfer the phosphate group from creatine phosphate to ADP, forming ATP.

3. A muscle at peak activity quickly depletes available ATP and creatine phosphate, and it will then catabolize glucose molecules from glycogen. *Glycolysis* is the initial pathway in glucose breakdown and yields two ATP molecules and two pyruvic acid molecules. When oxygen is unavailable, **anaerobic** reactions convert pyruvic acid to lactic acid. Blood removes lactic acid from skeletal muscle and carries much of it to the liver for reconversion to glucose.

4. When oxygen is available, the pyruvic acid molecules from glycolysis enter the mitochondria, where **aerobic cellular respiration** completely oxidizes glucose for a total of 36 ATP molecules from one glucose molecule that had entered glycolysis.

5. **Muscle fatigue** is the inability of muscle to contract forcefully after prolonged activity.

6. Heavy breathing after prolonged muscle activity helps to repay the **oxygen debt**, or **recovery oxygen uptake**. This increased oxygen intake helps to restore metabolic conditions to the resting level by conversion of lactic acid back to glycogen, resynthesis of creatine phosphate and ATP, and replacement of oxygen in muscle fibers.

See
- Animation – Muscle Metabolism

Do
- Exercise – Fueling Contraction and Recovery
- Concepts and Connections – Muscle Metabolism

Concept 10.9 Skeletal muscle fibers are classified as slow oxidative fibers, fast oxidative–glycolytic fibers, or fast glycolytic fibers.

1. Skeletal muscle fibers with low myoglobin content appear pale and are called *white muscle fibers*; skeletal muscle fibers with high myoglobin content have a dark, reddish appearance and are called *red muscle fibers*.

2. Skeletal muscle fibers are classified as slow oxidative, fast oxidative–glycolytic, and fast glycolytic.

3. *Slow oxidative fibers* use aerobic cellular respiration, have a slow speed of contraction, and are fatigue resistant. These fibers are beneficial for posture and endurance activities.

4. **Fast oxidative–glycolytic fibers** use aerobic cellular respiration and glycolysis, have fast speeds of contraction, and are moderately fatigue resistant. These fibers are used for walking and sprinting.

5. **Fast glycolytic fibers** mainly use glycolysis, contract strongly and rapidly, and are adapted for intense bursts of anaerobic movements, but fatigue rapidly.

6. Most skeletal muscles in the body are a mixture of all three types of muscle fibers. Training, genetics, and muscle action can slightly alter proportions of the fiber types.

Concept	Resources

Concept 10.10 Cardiac muscle tissue is found in the heart, and smooth muscle tissue is found in internal hollow structures.

1. **Cardiac muscle fibers** are found in the wall of the heart. These fibers have one nucleus and are branched and striated. Cardiac muscle fibers interconnect by **intercalated discs** containing **desmosomes** and **gap junctions**. Long, sustained contractions are supported by inflow of Ca^{2+} into the sarcoplasm.
2. Specialized cardiac muscle fibers have **autorhythmicity**, the ability to repeatedly generate spontaneous action potentials. A constant supply of oxygen and nutrients is needed for the continuous contraction–relaxation that occurs in cardiac muscle fibers.
3. There are two types of smooth muscle tissue: (1) **visceral (single-unit) smooth muscle tissue** is autorhythmic, and the fibers are connected by gap junctions allowing action potentials to spread throughout the network so that cells contract as a single unit; and (2) **multiunit smooth muscle tissue** acts independently, has few gap junctions, and lacks autorhythmicity.
4. Smooth muscle fibers have tapered ends, one central nucleus, are nonstriated, and lack sarcomeres. Intermediate filaments form bundles that stretch between dense bodies.
5. Thin and thick filaments of smooth muscle have a sliding mechanism that generates tension, resulting in lengthwise shortening of the fiber. Smooth muscle contraction starts slowly and lasts an extended time.
6. Smooth muscle is involuntary and responds to autonomic nervous system impulses, hormones, and local factors.

See
- **Anatomy Overview – Cardiac Muscle**
- **Anatomy Overview – Smooth Muscle Tissue**

Hear
- **Figure 10.18 – Histology of Cardiac Muscle**
- **Figure 10.19 – Histology of Smooth Muscle Tissue**

Other Chapter Resources

For additional information on clinical applications, medical disorders, and medical terminology, see Chapter 6 in the *Clinical Connections* manual packaged with your text.

Within *WileyPLUS*
- Chapter Quiz
- Visual Anatomy
- Mastering Vocabulary
- Audio Glossary
- Homeostatic Imbalance: The Case of Muscles and More
- Homeostatic Imbalance: The Case of the Weakened Teenager
- Web Links

• Understanding the Concepts

1. Describe the functions of titin and dystrophin.
2. How is the muscle found in your heart different from the muscle that moves your arms?
3. Why does your mother take your temperature when she thinks you're sick? What is the relationship between skeletal muscles and body temperature?
4. When a biopsy of a muscle is taken, that area of muscle may form scar tissue. Explain the process that is involved when skeletal muscle forms scar tissue.
5. If the lengths of the thick and thin filaments do not change, how does a skeletal muscle fiber shorten during muscle contraction?
6. Which of the numbered steps in Figure 10.12 are part of excitation–contraction coupling?
7. Would muscles of the fingers likely contain many small or a few large motor units? Is this also true of lower back muscles? Explain your answers.
8. Which type of contraction can allow you to make a sudden small jerk of your arm when startled? Maintain your neck's position while you are walking? Drag a chair to a table? Slowly sit down in the chair?
9. The label on your energy drink claims it contains "creatine for strong muscle, reduced fatigue, and building muscle mass." Why might increasing creatine levels in muscle increase the muscle's ability to maintain force? How could an energy drink containing carbohydrates possibly enhance athletic performance?
10. What are the major functional differences between cardiac and skeletal muscle tissue? Between smooth and skeletal muscle tissue?

11

The Muscular System

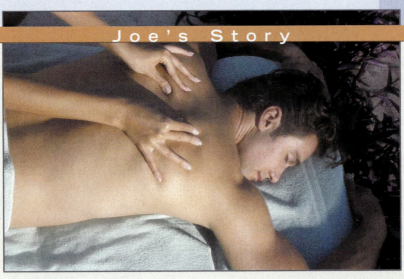

Joe's friend Pat had given him a gift certificate for a 1-hour massage at the Healing Touch Therapy Center. Pat was always trying to get him to do new things, most of which he resisted. But after starting his new exercise program, he was feeling so sore and tired he thought it couldn't hurt. As he walked through the door, he was hit by the strong scent of incense. The lighting was soft and music played quietly in the background. "Hi, I'm Joe, I have a 5:30 appointment."

A young girl with dreadlocks looked up from an appointment book. "Hi, Joe, I'm Jasmine. You're right on time. Have you been here before?"

Joe cleared his throat and replied, "No, I got this gift card from my friend, so I thought I'd give it a try. You're licensed, aren't you?"

Jasmine replied, "As a matter of fact we are. All of our therapists are licensed through the state medical board. You haven't had massage therapy before, have you?"

"Well, no. I started going back to the gym recently. My doctor said I needed to lose a few pounds, and I'm feeling my age." Joe grabbed his upper arm, flexing and extending it up and down. "Man, I'm so sore I can hardly move."

"Well, Joe, you've come to the right place. Massage therapy has a long history as one of the healing arts. Our therapists have all attended accredited training programs."

Joe asked, "What kind of training?"

Jasmine smiled. "Well I had to take anatomy, physiology, pathology, medical ethics, kinesiology, psychology, and even a course in business principles in my program." Jasmine was enjoying the surprised look on Joe's face.

"Well, that makes sense, I guess. This is a business. I guess I never thought that massage therapists had to take all those classes. I thought you just rubbed people's backs."

"Well, not quite, Joe. Doing a massage properly requires a solid foundation in anatomy. Plus there are many different techniques that can be used depending on the person's concerns or the doctor's recommendations."

Joe was really flabbergasted now. "You work with doctors?"

Jasmine nodded. "Sometimes we do. Come on in and we'll get started."

A solid understanding of the anatomy and physiology of the human body is essential for any number of professions, including those not usually associated with what we think of as traditional medicine. As in all other body systems, the anatomy of the muscular system is intimately linked with muscle function.

INTRODUCTION

The voluntarily controlled skeletal muscle tissues of your body make up the **muscular system**. Almost 700 individual skeletal muscles, containing both muscle and connective tissues, are found in this important system. The primary function of most muscles is to produce movements of body parts. A few muscles function mainly to stabilize bones so that other skeletal muscles can execute a movement more effectively. This chapter presents many of the major muscles in the body. Most of them are found **bilaterally**, meaning on both the right and left sides of

> Developing a working knowledge of key aspects of skeletal muscle anatomy will help you to understand how normal movements occur.

the body. While learning the names of skeletal muscles, we will identify their attachment sites and the movements that they produce. Developing a working knowledge of these key aspects of skeletal muscle anatomy will help you to understand how normal movements occur. This knowledge is especially crucial for professionals, such as those in the allied health and physical rehabilitation fields, who work with patients whose normal patterns of movement and physical mobility have been disrupted by physical trauma, surgery, or muscular paralysis.

CONCEPTS

- **11.1** Skeletal muscles produce movement when the insertion is pulled toward the origin.

- **11.2** Skeletal muscles are named based on size, shape, action, location, or attachments.

- **11.3** Muscles of the head produce facial expressions, eyeball movement, and assist in biting, chewing, swallowing, and speech.

- **11.4** Muscles of the neck assist in swallowing and speech, and allow balance and movement of the head.

- **11.5** Muscles of the abdomen protect the abdominal viscera, move the vertebral column, and assist breathing.

- **11.6** Muscles of the pelvic floor and perineum support the pelvic viscera, function as sphincters, and assist in urination, erection, ejaculation, and defecation.

- **11.7** Muscles inserting on the upper limb move and stabilize the pectoral girdle, and move the arm, forearm, and hand.

- **11.8** Deep muscles of the back move the head and vertebral column.

- **11.9** Muscles originating on the pelvic girdle or lower limb move the femur, leg, and foot.

11.1 Skeletal muscles produce movement when the insertion is pulled toward the origin.

Muscle Attachment Sites: Origin and Insertion

As we have already mentioned, not all skeletal muscles produce movements. Those skeletal muscles that produce movements do so by exerting force on tendons, which in turn pull on bones or other structures (such as skin). Most muscles cross at least one joint and are usually attached to the articulating bones that form the joint (Figure 11.1a).

When a skeletal muscle contracts, it pulls one of the articulating bones toward the other. The two articulating bones usually do not move equally in response to contraction. One bone remains stationary or near its original position; the attachment of a muscle's tendon to the stationary bone is called the **origin**. The attachment of the muscle's other tendon to the movable bone is called the **insertion**. A useful rule of thumb is that the origin is usually proximal and the insertion distal, especially in the limbs; the insertion is usually pulled toward the origin. The fleshy portion of the muscle between the tendons is called the **belly**. The **actions** of a muscle are the main movements that occur when the muscle contracts. A good analogy is a spring on a door. The part of the spring attached to the frame represents the origin, the part attached to the door is the insertion, the coils of the spring are the belly, and closing the door would be the action.

Figure 11.1 **Relationship of skeletal muscles to bones.** (a) Muscles are attached to bones by tendons at their origin and insertion. (b) Skeletal muscles produce movements by pulling on bones. Bones serve as levers, and joints act as fulcrums for the levers. Here the lever–fulcrum principle is illustrated by the movement of the forearm. Note where the load (resistance) and effort are applied in this example.

(a) Origin and insertion of a skeletal muscle

(b) Movement of the forearm lifting a weight

In the limbs, the origin of a muscle is usually proximal and the insertion is usually distal.

Muscles that move a body part often do not cover the moving part. Figure 11.1b shows that although one of the functions of the biceps brachii is to move the forearm, the belly of the muscle lies over the humerus, not over the forearm. You will also see that muscles that cross two joints, such as the rectus femoris and sartorius of the thigh, have more complex actions than muscles that cross only one joint.

Lever Systems and Leverage

A **lever** is a rigid structure that moves around a fixed point called the **fulcrum**. In producing movement, bones act as levers, and joints function as the fulcrums of these levers. A lever is acted on by two different forces: the **effort**, which causes movement, and the **load**, which resists movement. The effort is the force exerted by muscle contraction; the load is typically the weight of the body part that is moved plus any external object your body is moving. Motion occurs when the effort applied to the bone at the insertion exceeds the load. Consider the biceps brachii flexing the forearm at the elbow as an object is lifted (Figure 11.1b). When the forearm is raised, the elbow is the fulcrum. The weight of the forearm plus the weight of the object in the hand is the load. The force of contraction of the biceps brachii pulling the forearm up is the effort.

A lever operates at a *mechanical advantage* when the load is closer to the fulcrum and the effort is applied farther from the fulcrum; less effort is required to move the load. Conversely, a lever operates at a *mechanical disadvantage* when the load is farther from the fulcrum and the effort is closer to the fulcrum; more effort is required to move the load. For example, it is much easier to crush a piece of hard food (the load) with the teeth in the back of your mouth than with your front teeth because your back teeth are closer to the fulcrum (your jaw, or temporomandibular joint). Here is one more example you can try. Straighten out a paper clip. Now try to cut the paper clip with the tip of a pair of scissors (mechanical disadvantage) versus near the pivot of the scissors (mechanical advantage).

Effects of Fascicle Arrangement

Recall from Chapter 10 that the skeletal muscle fibers (cells) within a muscle are arranged in bundles known as **fascicles**. Within a fascicle, all muscle fibers are parallel to one another. The fascicles, however, may form one of five patterns with respect to the tendons: parallel, fusiform (spindle-shaped, narrow toward the ends and wide in the middle), circular, triangular, or pennate (shaped like a feather) (Table 11.1).

Fascicular arrangement affects a muscle's power and range of motion. Range of motion refers to the range, measured in degrees of a circle, through which the bones of a joint can be moved. The longer and more parallel the fibers in a muscle, the greater is the range of motion it can produce. However, the contraction power of a muscle depends not on length but on its total cross-sectional area. Thicker muscles contain a greater number of fascicles, hence have more muscle fibers to produce a more forceful contraction. Fascicular arrangement often represents a compromise between power and range of motion. Pennate muscles, for instance, have a large number of short-fibered fascicles distributed over their tendons, giving them greater power but a smaller range of motion. In contrast, parallel muscles have comparatively few fascicles, with long fibers resulting in less power, yet because the fascicles extend the length of the muscle, they have a greater range of motion.

Coordination among Muscles

Movements often are the result of several skeletal muscles acting as a group rather than individually. Most skeletal muscles are arranged in opposing pairs at joints—that is, flexors–extensors, abductors–adductors, and so on. A muscle that causes a desired action is referred to as the **prime mover** or **agonist** (= leader). Often, another muscle, called the **antagonist** (*ant-* = against), relaxes while the prime mover contracts. The antagonist has an effect opposite to that of the prime mover; that is, the antagonist stretches and yields to the movement of the prime mover. When you flex your elbow, the biceps brachii is the prime mover. While the biceps brachii is contracting, the triceps brachii (the antagonist), is relaxing (see Figure 11.1a). The antagonist and prime mover are usually located on opposite sides of the bone or joint, as is the case in this example.

With an opposing pair of muscles, the roles of the prime mover and antagonist can switch for different movements. For example, when extending the forearm at the elbow (i.e., lowering the load shown in Figure 11.1b), the triceps brachii becomes the prime mover and the biceps brachii is the antagonist. If a prime mover and its antagonist contract at the same time with equal force, there will be no movement.

TABLE 11.1

Arrangement of Fascicles

Parallel Fascicles parallel to longitudinal axis of muscle; terminate at either end in flat tendons.

Example: Stylohyoid muscle (see Figure 11.7)

Fusiform Fascicles nearly parallel to longitudinal axis of muscle; terminate in flat tendons; muscle tapers toward tendons, where diameter is less than at belly.

Example: Digastric muscle (see Figure 11.7)

Circular Fascicles in concentric circular arrangements form sphincter muscles that enclose an orifice (opening).

Example: Orbicularis oculi muscle (see Figure 11.3)

Triangular Fascicles spread over broad area converge at thick central tendon; give muscle a triangular appearance.

Example: Pectoralis major muscle (see Figure 11.2a)

Pennate Short fascicles in relation to total muscle length; tendon extends nearly entire length of muscle.

Unipennate Fascicles are arranged on only one side of tendon.	**Bipennate** Fascicles are arranged on both sides of centrally positioned tendons.	**Multipennate** Fascicles attach obliquely from many directions to several tendons.

Example: Extensor digitorum longus muscle (see Figure 11.20a,b)

Example: Rectus femoris muscle (see Figure 11.19a)

Example: Deltoid muscle (see Figure 11.9a)

Sometimes a prime mover crosses other joints before it reaches the joint at which its primary action occurs. The biceps brachii, for example, spans both the shoulder and elbow joints but has its primary action at the elbow joint. To prevent unwanted movements at intermediate joints or to otherwise aid the movement of the prime mover, muscles called **synergists** (SIN-er-gists; *syn-* = together; *-ergon* = work) contract and stabilize the intermediate joints. As an example, muscles that flex the fingers (prime movers) cross the carpal joints (intermediate joints). If movement at these intermediate joints was unrestrained, you would not be able to flex your fingers without flexing the wrist at the same time. Synergistic contraction of the wrist extensor muscles stabilizes the wrist joint and prevents unwanted movements, while the flexor muscles of the fingers contract to bring about the primary action, flexion of the fingers.

Some muscles in a group also act as **fixators**, stabilizing the origin of the prime mover so that the prime mover can act more efficiently. Fixators steady the proximal end of a limb while movements occur at the distal end. For example, the scapula in the pectoral (shoulder) girdle is a freely movable bone that serves as the origin for several muscles that move the arm. When the arm muscles contract, the scapula must be held steady. In abduction of the arm, the deltoid serves as the prime mover. As the deltoid pulls on the humerus to abduct the arm, fixators (pectoralis minor, trapezius, subclavius, serratus anterior, and others) hold the scapula firmly against the back of the chest (see

Figure 11.14). Depending on the movement needed at the moment, muscles may act as prime movers, antagonists, synergists, or fixators.

Checkpoint

1. How do skeletal muscles produce body movements by pulling on bones? *(Hint: Use the terms origin, insertion, and belly.)*
2. When you flex your knee, where are the fulcrum and load?
3. Would a narrow, long muscle likely have a large range of motion or a highly forceful contraction?

11.2 Skeletal muscles are named based on size, shape, action, location, or attachments.

The figures and tables that follow will assist you in learning the names of the principal skeletal muscles in various regions of the body. (Remember that there are nearly 700 individual muscles in the human body; we have not included them all here.) The selected skeletal muscles have been divided into groups according to the part of the body on which they act.

In certain figures, superficial muscles cover deeper muscles. In other figures, superficial muscles have been removed or reflected (cut or pulled aside) to expose the deeper muscles. Within the figures, muscles that are named in all capital letters are specifically referred to in the corresponding tables. Additional related structures are also labeled in the figures for reference. Figure 11.2 provides an overview of principal superficial skeletal muscles. As you study groups of muscles, refer to Figure 11.2 to see how each group is related to the others.

The names of most of the skeletal muscles contain combinations of the word roots of their distinctive features. Learning the terms that refer to these features will help you remember the names of muscles. Such muscle features include the pattern of the muscle's fascicles; the size, shape, action, number of origins, and location of the muscle; and the sites of origin and insertion of the muscle. On the other hand, knowing the names of the muscles will give you clues to their features. Study Table 11.2 to become familiar with the terms used in muscle names.

To make it easier for you to learn and understand how skeletal muscles are named and to say their names, the tables include word roots and phonetic pronunciations. As you review each muscle in

the table, notice the information provided next to its name. Once you have mastered the naming of the muscles, their actions will have more meaning and be easier to remember. The tables also provide the origin, insertion, actions, and innervation of each selected muscle. By learning the origin and insertion of a muscle, we can better understand its action. Employing common sense is often easier than memorization. As you think of a muscle's attachments, you can often reason through the movements, or the "actions," that muscle could produce when it contracts.

The innervation section of each table lists the nerve or nerves that cause contraction of each muscle. In general, cranial nerves, which arise from the lower parts of the brain, serve muscles in the head region. Spinal nerves, which arise from the spinal cord within the vertebral column, innervate muscles in the rest of the body. Cranial nerves are designated by both a name and a Roman numeral—for example, the facial (VII) nerve. Spinal nerves are numbered in groups according to the part of the spinal cord from which they arise: C = cervical (neck region), T = thoracic (chest region), L = lumbar (lower back region), and S = sacral (buttocks region). An example is T1, the first thoracic spinal nerve.

Checkpoint

4. Using Figure 11.2, give an example of a muscle named for each of the following characteristics: direction of fibers, shape, action, size, origin and insertion, location, and number of tendons of origin.

Figure 11.2   Principal superficial skeletal muscles.

Occipitofrontalis (frontal belly)
Nasalis
Orbicularis oris
Depressor anguli oris
Omohyoid
Sternohyoid
Latissimus dorsi
Rectus abdominis
Brachioradialis
External oblique
Tensor fasciae latae
Iliacus
Extensor pollicis longus
Psoas major
Pectineus
Adductor longus
Sartorius
Adductor magnus
Gracilis
Vastus lateralis
Rectus femoris
Vastus medialis
Tendon of quadriceps femoris
Patella
Gastrocnemius
Soleus
Tibia

Epicranial aponeurosis
Temporalis
Orbicularis oculi
Masseter
Platysma (cut)
Sternocleidomastoid
Trapezius
Scalenes
Deltoid
Pectoralis major
Serratus anterior
Biceps brachii
Brachialis
Triceps brachii
Pronator teres
Brachioradialis
Flexor carpi radialis
Flexor digitorum superficialis
Flexor carpi ulnaris
Abductor pollicis longus
Thenar muscles
Hypothenar muscles
Iliotibial tract
Patellar ligament
Tibialis anterior
Fibularis longus
Tibia

(a) Anterior view

Epicranial aponeurosis

Occipitofrontalis (occipital belly)

Splenius capitis

Trapezius

Deltoid

Brachialis

Triceps brachii

Brachioradialis

Anconeus

Extensor carpi radialis brevis

Extensor digitorum

Extensor carpi ulnaris

Flexor carpi ulnaris

Abductor pollicis longus

Extensor pollicis brevis

Temporalis

Sternocleidomastoid

Infraspinatus

Teres minor

Teres major

Latissimus dorsi

External oblique

Gluteus medius

Flexor carpi ulnaris

Brachioradialis

Tensor fasciae latae

Gluteus maximus

Gracilis

Adductor magnus

Semitendinosus

Biceps femoris

Iliotibial tract

Semimembranosus

Popliteal fossa

Sartorius

Plantaris

Gastrocnemius

Soleus

Soleus

Flexor digitorum longus

Calcaneal tendon

(b) Posterior view

Most movements require several skeletal muscles acting in groups rather than individually.

TABLE 11.2

Characteristics Used to Name Muscles

NAME	MEANING	EXAMPLE	FIGURE
Direction: Orientation of muscle fascicles relative to the body's midline			
Rectus	Parallel to midline	Rectus abdominis	11.9b
Transverse	Perpendicular to midline	Transversus abdominis	11.9b
Oblique	Diagonal to midline	External oblique	11.9c
Size: Relative size of the muscle			
Maximus	Largest	Gluteus maximus	11.2b
Medius	Intermediate	Gluteus medius	11.19c
Minimus	Smallest	Gluteus minimus	11.19c
Longus	Long	Adductor longus	11.19a
Brevis	Short	Adductor brevis	11.19b
Latissimus	Widest	Latissimus dorsi	11.14b
Longissimus	Longest	Longissimus capitis	11.18a
Magnus	Large	Adductor magnus	11.19a
Major	Larger	Pectoralis major	11.9a
Minor	Smaller	Pectoralis minor	11.13a
Vastus	Huge	Vastus lateralis	11.19a
Shape: Relative shape of the muscle			
Deltoid	Triangular	Deltoid	11.9a
Trapezius	Trapezoid	Trapezius	11.2b
Serratus	Saw-toothed	Serratus anterior	11.13b
Rhomboid	Diamond-shaped	Rhomboid major	11.14c
Orbicularis	Circular	Orbicularis oculi	11.3a
Pectinate	Comblike	Pectineus	11.19a
Piriformis	Pear-shaped	Piriformis	11.19c
Platys	Flat	Platysma	11.3c
Quadratus	Square, four-sided	Quadratus femoris	11.19c
Gracilis	Slender	Gracilis	11.19a
Action: Principal action of the muscle			
Flexor	Decreases a joint angle	Flexor carpi radialis	11.16a
Extensor	Increases a joint angle	Extensor carpi ulnaris	11.16c
Abductor	Moves a bone away from the midline	Abductor pollicis longus	11.16c
Adductor	Moves a bone closer to the midline	Adductor longus	11.19a
Levator	Raises or elevates a body part	Levator scapulae	11.13a
Depressor	Lowers or depresses a body part	Depressor labii inferioris	11.3b
Supinator	Turns palm anteriorly	Supinator	11.16b
Pronator	Turns palm posteriorly	Pronator teres	11.16a
Sphincter	Decreases the size of an opening	External anal sphincter	11.11
Tensor	Makes a body part rigid	Tensor fasciae latae	11.19a
Rotator	Rotates a bone around its longitudinal axis	Rotatore	11.18a
Number of Origins: Number of tendons of origin			
Biceps	Two origins	Biceps brachii	11.15a
Triceps	Three origins	Triceps brachii	11.15b
Quadriceps	Four origins	Quadriceps femoris	11.19a
Location: Structure near which a muscle is found			
Example: Temporalis, a muscle near the temporal bone			11.3c
Origin and Insertion: Sites where muscle originates and inserts			
Example: Sternocleidomastoid, originating on the sternum and clavicle and inserting onto the mastoid process of the temporal bone			11.2a

11.3 Muscles of the head produce facial expressions, eyeball movement, and assist in biting, chewing, swallowing, and speech.

Muscles of Facial Expression

The muscles of facial expression provide us with the ability to express a wide variety of emotions, including displeasure, surprise, fear, and happiness (Figure 11.3; Table 11.3). The muscles themselves lie within the hypodermis (connective tissue beneath the skin). They usually originate in the fascia or bones of the skull and insert into skin or other muscles. Because of their insertions, the muscles of facial expression move the skin rather than a joint when they contract.

Figure 11.3   Muscles of facial expression.

(a) Anterior superficial view   (b) Anterior deep view

(continues)

Figure 11.3 (continued)

(c) Right lateral superficial view

🔑 **When they contract, muscles of facial expression move the skin rather than a joint.**

TABLE 11.3				
Muscles of Facial Expression				
MUSCLE	ORIGIN	INSERTION	ACTION	INNERVATION
Scalp Muscles				
Occipitofrontalis (ok-sip´-i-tō-frun-TĀ-lis)				
Frontal belly	Epicranial aponeurosis.	Skin superior to supraorbital margin.	Draws scalp anteriorly, raises eyebrows, and wrinkles skin of forehead horizontally as in a look of surprise.	Facial (VII) nerve.
Occipital belly (*occipit-* = back of the head)	Occipital bone and mastoid process of temporal bone.	Epicranial aponeurosis.	Draws scalp posteriorly.	Facial (VII) nerve.

TABLE 11.3 (continued)

Muscles of Facial Expression

MUSCLE	ORIGIN	INSERTION	ACTION	INNERVATION
Mouth Muscles				
Orbicularis oris (or-bi´-kū-LAR-is OR-is; *orb-* = circular; *oris* = of the mouth)	Muscle fibers surrounding opening of mouth.	Skin at corner of mouth.	Closes and protrudes lips, as in kissing; compresses lips against teeth; and shapes lips during speech.	Facial (VII) nerve.
Zygomaticus major (zī-gō-MA-ti-kus; *zygomatic* = cheek bone; *major* = greater)	Zygomatic bone.	Skin at angle of mouth and orbicularis oris.	Draws angle of mouth superiorly and laterally, as in smiling.	Facial (VII) nerve.
Zygomaticus minor (*minor* = lesser)	Zygomatic bone.	Upper lip.	Raises (elevates) upper lip, exposing maxillary teeth.	Facial (VII) nerve.
Levator labii superioris (le-VĀ-tor LĀ-bē-ī soo-per´-ē-OR-is; *levator* = raises or elevates; *labii* = lip; *superioris* = upper)	Superior to infraorbital foramen of maxilla.	Skin at angle of mouth and orbicularis oris.	Raises upper lip.	Facial (VII) nerve.
Depressor labii inferioris (de-PRE-sor = depresses or lowers; *inferioris* = lower)	Mandible.	Skin of lower lip.	Depresses (lowers) lower lip.	Facial (VII) nerve.
Depressor anguli oris (*angul-* = angle or corner)	Mandible.	Angle of mouth.	Draws angle of mouth laterally and inferiorly, as in opening mouth.	Facial (VII) nerve.
Levator anguli oris	Inferior to infraorbital foramen.	Skin of lower lip and orbicularis oris.	Draws angle of mouth laterally and superiorly.	Facial (VII) nerve.
Buccinator (BUK-si-nā´-tor; *bucc-* = cheek)	Alveolar processes of maxilla and mandible and pterygomandibular raphe (fibrous band extending from the pterygoid process to the sphenoid bone to the mandible).	Orbicularis oris.	Presses cheeks against teeth and lips, as in whistling, blowing, and sucking; draws corner of mouth laterally; and assists in mastication (chewing) by keeping food between the teeth (and not between teeth and cheeks).	Facial (VII) nerve.
Risorius (ri-ZOR-ē-us; *risor-* = laughter)	Fascia over parotid (salivary) gland.	Skin at angle of mouth.	Draws angle of mouth laterally, as in grimacing.	Facial (VII) nerve.
Mentalis (men-TĀ-lis; *ment-* = the chin)	Mandible.	Skin of chin.	Elevates and protrudes lower lip and pulls skin of chin up, as in pouting.	Facial (VII) nerve.
Neck Muscle				
Platysma (pla-TIZ-ma; *platys-* = flat, broad)	Fascia over deltoid and pectoralis major muscles.	Mandible, muscle around angle of mouth, and skin of lower face.	Draws outer part of lower lip inferiorly and posteriorly, as in pouting; depresses mandible.	Facial (VII) nerve.
Orbit and Eyebrow Muscles				
Orbicularis oculi (OK-ū-lī = of the eye)	Medial wall of orbit.	Circular path around orbit.	Closes eye.	Facial (VII) nerve.
Corrugator supercilii (KOR-a-gā´-tor soo-per-SI-lē-ī; *corrugat-* = wrinkle; *supercilii* = of the eyebrow)	Medial end of superciliary arch of frontal bone.	Skin of eyebrow.	Draws eyebrow inferiorly and wrinkles skin of forehead vertically as in frowning.	Facial (VII) nerve.
Levator palpebrae superioris (PAL-pe-brē = eyelids) (see also Figure 11.4a)	Roof of orbit (lesser wing of sphenoid bone).	Skin of upper eyelid.	Elevates upper eyelid (opens eye).	Oculomotor (III) nerve.

The **occipitofrontalis** is an unusual muscle in this group because it is made up of two parts: an anterior part called the **frontal belly**, which is superficial to the frontal bone, and a posterior part called the **occipital belly**, which is superficial to the occipital bone. The two muscular portions are held together by a strong aponeurosis (sheetlike tendon), the **epicranial aponeurosis** (ep-i-KRĀ-nē-al ap´-ō-noo-RŌ-sis), which covers the superior and lateral surfaces of the skull.

The **buccinator** forms the major muscular portion of the cheek. The buccinator is so named because it compresses the cheeks (*bucc-* = cheek) during blowing—for example, when a musician plays a wind instrument such as a trumpet. It functions in whistling, blowing, and sucking, and assists in chewing.

Among the noteworthy muscles in this group are those surrounding the orifices (openings) of the head, such as the eyes, nose, and mouth. These muscles function as **sphincters** (SFINGK-ters), which close the orifices, and *dilators,* which open the orifices. For example, the **orbicularis oculi** closes the eye, and the **levator palpebrae superioris** opens it.

Muscles That Move the Eyeballs

Muscles that move the eyeballs are called **extrinsic eye muscles** because they originate outside the eyeballs (in the orbit) and insert on its outer surface, the sclera ("white of the eye") (Figure 11.4; Table 11.4). The extrinsic eye muscles are some of the fastest contracting and most precisely controlled skeletal muscles in the body.

Figure 11.4   Extrinsic muscles of the eyeball.

(a) Lateral view of right eyeball

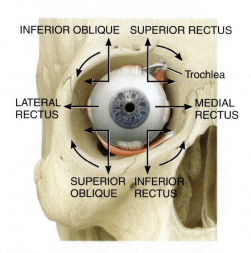

(b) Movements of right eyeball in response to contraction of extrinsic muscles

The extrinsic muscles of the eyeball are among the fastest contracting and most precisely controlled skeletal muscles in the body.

TABLE 11.4

Muscles That Move the Eyeballs—Extrinsic Eye Muscles

MUSCLE	ORIGIN	INSERTION	ACTION	INNERVATION
Superior rectus (*rectus* = fascicles parallel to midline)	Common tendinous ring (attached to orbit around optic foramen).	Superior and central part of eyeball.	Moves eyeball superiorly (elevation) and medially (adduction), and rotates it medially.	Oculomotor (III) nerve.
Inferior rectus	Same as above.	Inferior and central part of eyeball.	Moves eyeball inferiorly (depression) and medially (adduction), and rotates it medially.	Oculomotor (III) nerve.
Lateral rectus	Same as above.	Lateral side of eyeball.	Moves eyeball laterally (abduction).	Abducens (VI) nerve.
Medial rectus	Same as above.	Medial side of eyeball.	Moves eyeball medially (adduction).	Oculomotor (III) nerve.
Superior oblique (*oblique* = fascicles diagonal to midline)	Sphenoid bone, superior and medial to the tendinous ring in the orbit.	Eyeball between superior and lateral recti. The muscle inserts into the superior and lateral surfaces of the eyeball via a tendon that passes through the trochlea.	Moves eyeball inferiorly (depression) and laterally (abduction), and rotates it medially.	Trochlear (IV) nerve.
Inferior oblique	Maxilla in floor of orbit.	Eyeball between inferior and lateral recti.	Moves eyeball superiorly (elevation) and laterally (abduction), and rotates it laterally.	Oculomotor (III) nerve.

Three pairs of extrinsic eye muscles control movements of the eyeballs: (1) superior and inferior recti, (2) lateral and medial recti, and (3) superior and inferior obliques. The four recti muscles (superior, inferior, lateral, and medial) arise from a common tendinous ring at the back of the orbit and insert into the sclera of the eyeball. The names of the four recti muscles clearly imply their actions. When the **superior rectus** pulls the eyeball, the eye looks up; as the **inferior rectus** pulls, the eye looks down. Contraction of the **lateral rectus** makes the eye rotate laterally, and when the **medial rectus** pulls, the eye moves medially.

The oblique muscles—superior and inferior—rotate the eyeball on its axis. The **superior oblique** originates posteriorly near the tendinous ring, passes anteriorly, and ends in a round tendon that extends through a pulleylike loop called the *trochlea* (= pulley), where it turns and inserts on the superior, lateral eyeball. When the superior oblique pulls the eyeball, the eye looks down and laterally. The **inferior oblique** originates on the maxilla at the anterior, medial floor of the orbit. It then passes posteriorly and laterally and inserts on the posterior, lateral eyeball. Because of this arrangement, contraction of the inferior oblique results in the eye looking up and laterally.

Muscles That Move the Mandible

The muscles that move the mandible (lower jawbone) at the temporomandibular joint (TMJ) are known as the muscles of mastication (chewing) (Figure 11.5; Table 11.5). These muscles also assist in speech. Of the four pairs of muscles involved in mastication, three are powerful closers of the jaw and account for the strength of the bite: **masseter**, **temporalis**, and **medial pterygoid**. Of these, the masseter is the strongest muscle of mastication. The **medial** and **lateral pterygoid** assist in mastication by moving the mandible from side to side to help grind food. Additionally, these muscles protract the mandible (thrust it forward).

Figure 11.5   Muscles that move the mandible.

Parietal bone

TEMPORALIS

Zygomatic arch (cut)

Occipital bone

Temporomandibular joint

MEDIAL PTERYGOID (deep to mandible)

Ramus of mandible (cut and transparent)

Frontal bone

Nasal bone

Zygomatic bone (cut)

LATERAL PTERYGOID

Maxilla

Buccinator

Orbicularis oris

Body of mandible

Right lateral superficial view

The muscles that move the mandible are also known as muscles of mastication.

TABLE 11.5

Muscles That Move the Mandible

MUSCLE	ORIGIN	INSERTION	ACTION	INNERVATION
Masseter (MA-se-ter = a chewer) (see Figure 11.3c)	Maxilla and zygomatic arch.	Angle and ramus of mandible.	Elevates mandible, as in closing mouth.	Mandibular division of trigeminal (V) nerve.
Temporalis (tem´-pō-RĀ-lis; *tempor-* = time or temples)	Temporal bone.	Coronoid process and ramus of mandible.	Elevates and retracts mandible.	Mandibular division of trigeminal (V) nerve.
Medial pterygoid (TER-i-goyd; *medial* = closer to midline; *pterygoid* = like a wing)	Medial surface of lateral portion of pterygoid process of sphenoid bone; maxilla.	Angle and ramus of mandible.	Elevates and protracts (protrudes) mandible and moves mandible from side to side.	Mandibular division of trigeminal (V) nerve.
Lateral pterygoid (*lateral* = farther from midline)	Greater wing and lateral surface of lateral portion of pterygoid process of sphenoid bone.	Condyle of mandible; temporomandibular joint (TMJ).	Protracts mandible, depresses mandible as in opening mouth, and moves mandible from side to side.	Mandibular division of trigeminal (V) nerve.

Muscles That Move the Tongue

The tongue is a highly mobile structure that is vital to digestive functions such as mastication, perception of taste, and deglutition (swallowing). It is also important in speech. The tongue's mobility is greatly aided by its suspension from the mandible, styloid process of the temporal bone, and hyoid bone.

The tongue is divided into lateral halves by a median fibrous septum. The septum extends throughout the length of the tongue. Inferiorly, the septum attaches to the hyoid bone. Muscles of the tongue are of two principal types: extrinsic and intrinsic (Figure 11.6; Table 11.6). **Extrinsic tongue muscles** originate outside the tongue and insert into it. They move the entire tongue in various directions, such as anteriorly, posteriorly, and laterally. **Intrinsic tongue muscles** originate and insert within the tongue. These muscles alter the shape of the tongue rather than moving the entire tongue. The extrinsic and intrinsic muscles of the tongue insert into both lateral halves of the tongue. Only the extrinsic tongue muscles are considered in Table 11.6.

When you study the extrinsic tongue muscles, you will notice that all of their names end in -*glossus*, meaning tongue. You will notice that the actions of the muscles are obvious, considering the

Figure 11.6 Muscles that move the tongue.

Superior constrictor
Styloid process of temporal bone
Mastoid process of temporal bone
Digastric (posterior belly–cut)
Middle constrictor
Stylohyoid
Stylopharyngeus
HYOGLOSSUS
Hyoid bone
Inferior constrictor
Thyroid cartilage of larynx

STYLOGLOSSUS
PALATOGLOSSUS
Hard palate (cut)
Tongue
GENIOGLOSSUS
Mandible (cut)
GENIOHYOID
Mylohyoid
Intermediate tendon of digastric (cut)
Fibrous loop for intermediate tendon of digastric
Thyrohyoid membrane (connects hyoid bone to larynx)

Right side deep view

🔑 The extrinsic and intrinsic muscles of the tongue insert into both lateral halves of the tongue.

TABLE 11.6				
Muscles That Move the Tongue—Extrinsic Tongue Muscles				
MUSCLE	ORIGIN	INSERTION	ACTION	INNERVATION
Genioglossus (jē´-nē-ō-GLOS-us; *genio-* = the chin; *-glossus* = tongue)	Mandible.	Undersurface of tongue and hyoid bone.	Depresses tongue and thrusts it anteriorly (protraction).	Hypoglossal (XII) nerve.
Styloglossus (stī´-lō-GLOS-us; *stylō-* = stake or poles for styloid process of temporal bone)	Styloid process of temporal bone.	Side and undersurface of tongue.	Elevates tongue and draws it posteriorly (retraction).	Hypoglossal (XII) nerve.
Palatoglossus (pal´-a-tō-GLOS-us; *palato-* = the roof of the mouth or palate)	Anterior surface of soft palate.	Side of tongue.	Elevates posterior portion of tongue and draws soft palate down on tongue.	Pharyngeal plexus, which contains axons from both the vagus (X) and accessory (XI) nerves.
Hyoglossus (hī´-ō-GLOS-us)	Greater horn and body of hyoid bone.	Side of tongue.	Depresses tongue and draws down its sides.	Hypoglossal (XII) nerve.

positions of the mandible, styloid process, hyoid bone, and soft palate, which serve as origins for these muscles. For example, the **genioglossus** (origin: the mandible) pulls the tongue downward and forward, the **styloglossus** (origin: the styloid process) pulls the tongue upward and backward, the **hyoglossus** (origin: the hyoid bone) pulls the tongue downward and flattens it, and the **palatoglossus** (origin: the soft palate) raises the back portion of the tongue.

Checkpoint

5.   Why do the muscles of facial expression move the skin rather than a joint?
6.   Which muscles of facial expression would you use to blow up a balloon? To squint? To pucker your lips? To frown?
7.   Which extrinsic eye muscles contract as you gaze to your left without moving your head? Which relax?
8.   Which extrinsic eye muscles cause elevation of the anterior eyeball?
9.   What are the functions of the tongue?

11.4 Muscles of the neck assist in swallowing and speech, and allow balance and movement of the head.

Muscles That Move the Hyoid Bone and Larynx

Two groups of muscles are associated with the anterior aspect of the neck (Figure 11.7; Table 11.7): the **suprahyoid muscles**, so called because they are located superior to the hyoid bone, and the **infrahyoid muscles**, named for their position inferior to the hyoid bone. Both groups of muscles stabilize the hyoid bone, allowing it to serve as a firm base on which the tongue can move.

As a group, the suprahyoid muscles elevate the hyoid bone, floor of the oral cavity, and tongue during swallowing. As its name suggests, the **digastric** (*di-* = two) has two bellies, anterior and posterior, united by an intermediate tendon that is held in position by a fibrous loop. This muscle elevates the hyoid bone and lar-

ynx (voice box) during swallowing and speech and depresses the mandible. Together, the **stylohyoid**, **mylohyoid**, and **geniohyoid** elevate the hyoid bone during swallowing.

Most of the infrahyoid muscles depress the hyoid bone, and some move the thyroid cartilage (Adam's apple) of the larynx during swallowing and speech. The **omohyoid**, like the digastric, is composed of two bellies connected by an intermediate tendon. In this case, however, the two bellies are referred to as *superior* and *inferior,* rather than anterior and posterior. Together, the omohyoid, **sternohyoid**, and **thyrohyoid** depress the hyoid bone. In addition, the **sternothyroid** depresses the thyroid cartilage (Adam's apple) of the larynx to produce low sounds; the thyrohyoid elevates the thyroid cartilage to produce high sounds.

Figure 11.7   Muscles of the anterior neck.

Parotid gland
DIGASTRIC:
    Anterior belly
    Posterior belly
STYLOHYOID
STERNOHYOID
OMOHYOID
Sternocleidomastoid

Mandible
Masseter
MYLOHYOID
Intermediate tendon of digastric
Fibrous loop for intermediate tendon
Hyoid bone
Levator scapulae
THYROHYOID
Thyroid cartilage of larynx
STERNOTHYROID
Scalene muscles

(a) Anterior superficial view   (b) Anterior deep view

TABLE 11.7

Muscles That Move the Hyoid Bone and Larynx

MUSCLE	ORIGIN	INSERTION	ACTION	INNERVATION
Suprahyoid Muscles				
Digastric (dī´-GAS-trik; *di-* = two; *-gastr-* = belly)	Anterior belly from inner side of inferior border of mandible; posterior belly from temporal bone.	Body of hyoid bone via an intermediate tendon.	Elevates hyoid bone and depresses mandible, as in opening the mouth.	Anterior belly: mandibular division of trigeminal (V) nerve. Posterior belly: facial (VII) nerve.
Stylohyoid (stī´-lō-HĪ-oyd; *stylo-* = stake or pole, for styloid process of temporal bone; *-hyo-* = U-shaped, pertaining to hyoid bone)	Styloid process of temporal bone.	Body of hyoid bone.	Elevates hyoid bone and draws it posteriorly.	Facial (VII) nerve.
Mylohyoid (mī´-lō-HĪ-oyd) (*mylo-* = mill)	Inner surface of mandible.	Body of hyoid bone.	Elevates hyoid bone and floor of mouth and depresses mandible.	Mandibular division of trigeminal (V) nerve.
Geniohyoid (jē´-nē-ō-HĪ-oyd; *genio-* = chin) (see Figure 11.6)	Inner surface of mandible.	Body of hyoid bone.	Elevates hyoid bone, draws hyoid bone and tongue anteriorly, and depresses mandible.	First cervical spinal nerve.
Infrahyoid Muscles				
Omohyoid (ō-mō-HĪ-oyd; *omo-* = relationship to the shoulder)	Superior border of scapula and superior transverse ligament.	Body of hyoid bone.	Depresses hyoid bone.	Branches of spinal nerves C1–C3.
Sternohyoid (ster´-nō-HĪ-oyd; *sterno-* = sternum)	Medial end of clavicle and manubrium of sternum.	Body of hyoid bone.	Depresses hyoid bone.	Branches of spinal nerves C1–C3.
Sternothyroid (ster´-nō-THĪ-royd; *-thyro-* = thyroid gland)	Manubrium of sternum.	Thyroid cartilage of larynx.	Depresses thyroid cartilage of larynx.	Branches of spinal nerves C1–C3.
Thyrohyoid (thī´-rō-HĪ-oyd)	Thyroid cartilage of larynx.	Greater horn of hyoid bone.	Elevates thyroid cartilage and depresses hyoid bone.	Branches of spinal nerves C1–C2 and descending hypoglossal (XII) nerve.

(c) Anterior superficial view   (d) Anterior deep view

 The suprahyoid muscles elevate the hyoid bone, the floor of the oral cavity, and the tongue during swallowing.

Muscles That Move the Head

Balance and movement of the head on the vertebral column involve the action of several neck muscles (Figure 11.8; Table 11.8). Acting together (bilaterally), contraction of the **sternocleidomastoid** flexes the head at the neck. Bilateral contraction of each pair of **semispinalis capitis**, **splenius capitis**, and **longissimus capitis** extends the head at the neck. However, when the above muscles contract unilaterally, their actions are quite different, involving primarily rotation of the head. For example, acting singly (unilaterally), each sternocleidomastoid laterally flexes the neck and rotates the head toward the opposite side.

The sternocleidomastoid is an important landmark that divides the neck into an **anterior triangle** and **posterior triangle**. The triangles are important anatomically and surgically because of the structures that lie within their boundaries.

Figure 11.8 **Selected muscles of the lateral neck.**

SEMISPINALIS CAPITIS

SPLENIUS CAPITIS

STERNOCLEIDOMASTOID

Levator scapulae

Splenius cervicis

Rhomboid minor

Rhomboid major

LONGISSIMUS CAPITIS

Splenius cervicis

Levator scapulae

Middle scalene

Posterior scalene

Longissimus cervicis

Iliocostalis cervicis

Longissimus thoracis

(a) Posterior superficial view (left) and deep view (right)

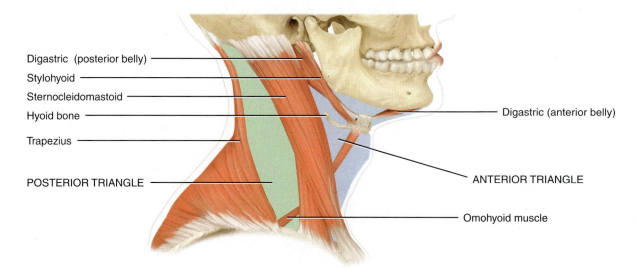

Digastric (posterior belly)

Stylohyoid

Sternocleidomastoid

Hyoid bone

Trapezius

POSTERIOR TRIANGLE

Digastric (anterior belly)

ANTERIOR TRIANGLE

Omohyoid muscle

(b) Right lateral view of triangles of neck

Contraction of both sternocleidomastoids flexes the cervical curve of the vertebral column, which brings the anterior head toward the chest (see Figure 9.6a).

TABLE 11.8

Muscles That Move the Head

MUSCLE	ORIGIN	INSERTION	ACTION	INNERVATION
Sternocleidomastoid (ster´-nō-klī´-dō-MAS-toid; *sterno-* = breastbone; *cleido-* = clavicle; *mastoid* = -mastoid process of temporal bone)	Sternum and clavicle.	Mastoid process of temporal bone.	Acting together (bilaterally), flex cervical portion of vertebral column, flex head, and elevate sternum during forced inhalation; acting singly (unilaterally), laterally extend and rotate head to side opposite contracting muscle.	Accessory (XI) nerve.
Semispinalis capitis (se´-mē-spi-NĀ-lis KAP-i-tis; *semi-* = half; *spine* = spinous process; *capit-* = head)	Transverse processes of first six or seven thoracic vertebrae and seventh cervical vertebra, and articular processes of fourth, fifth, and sixth cervical vertebrae.	Occipital bone between superior and inferior nuchal lines.	Acting together, extend head; acting singly, rotate head to side opposite contracting muscle.	Cervical spinal nerves.
Splenius capitis (SPLĒ-nē-us; *splenion-* = bandage)	Ligamentum nuchae and spinous processes of seventh cervical vertebra and first three or four thoracic vertebrae.	Occipital bone and mastoid process of temporal bone.	Acting together, extend head; acting singly, laterally flex and rotate head to same side as contracting muscle.	Cervical spinal nerves.
Longissimus capitis (lon-JIS-i-mus = longest)	Transverse processes of upper four thoracic vertebrae and articular processes of last four cervical vertebrae.	Mastoid process of temporal bone.	Acting together, extend head; acting singly, laterally flex and rotate head to same side as contracting muscle.	Cervical spinal nerves.

Checkpoint

10. What is the combined action of the suprahyoid and infrahyoid muscles?

11. What should happen to the hyoid bone when the suprahyoid muscles contract? When the infrahyoid muscles contract?

12. Which muscles do you contract to signify "yes"? Which do you contract to signify "no"?

Return to Joe's Story

Jasmine led Joe to a small, quiet room with a padded table. "Joe, there are sheets in the cabinet. I'll check back with you in a few minutes."

Jasmine returned after Joe had undressed and covered up with one of the sheets. "Are there any areas that are bothering you in particular?" she asked.

"My chest, arms, and back mostly. My neck is always sore, too; I spend a lot of time at my desk."

Jasmine replied, "We'll start with a light relaxation massage. If we find any trouble spots I can do some deep muscle massage."

After pulling the sheet from his shoulders, Jasmine poured oil on her hands, rubbed them together, and evenly distributed the oil on Joe's skin. With long gliding strokes, she moved her hands down Joe's spine, then back up and lightly over his shoulders. "Wow, that's sore," said Joe, "but it sure feels good."

Jasmine replied, "If you feel any discomfort at any point let me know. You're carrying a lot of tension in your shoulders; your trapezius, the muscle on the top of your shoulder, is really tight. It's attached to the head, the shoulder blade, and the spine. It's a common place where stress is concentrated."

Jasmine used light gliding movements, working from Joe's shoulders to the base of his skull. She applied pressure to the back of his head with the tips of her thumbs at the inferior nuchal line. "Whatever you're doing there feels great," he said.

"I'm putting some pressure on the origin of the trapezius. It helps the muscle relax," she said softly. "Do you get tension headaches?"

"I get these headaches in my temples sometimes," he replied. Jasmine released the pressure slowly, then, using her thumbs, began making circular motions on the back of Joe's scalp. She massaged Joe's temples, the line of his jaw, even the side of his neck. Joe felt his entire body relaxing.

"Joe, I need you to turn over onto your back so I can work your neck some more." Jasmine worked the front and top of Joe's shoulders, moving her fingertips from the point of each shoulder toward his neck, then back. She also massaged the front of his neck, his chin, and his face. Jasmine then slid her hands under Joe's neck, using his own body weight to increase the pressure she was applying to the back of his neck. Joe felt twinges and sore spots where he didn't even know he had muscles.

A. As Jasmine begins her work, she will initially focus on working muscle groups known as the prime movers; what are prime movers? What is the prime mover for flexion of the forearm? What is that muscle's antagonist?

B. What is the body of the trapezius that Jasmine is working on? How does it differ from the muscle's origin and insertion?

C. Which muscle on Joe's face is primarily responsible for chewing or clenching his teeth? Which muscle located on Joe's temporal bone is receiving relief from the massage?

D. Joe spends a considerable amount of time at a computer and he doesn't have very good posture; which neck muscles must he use to stabilize his head in this position?

11.5 Muscles of the abdomen protect the abdominal viscera, move the vertebral column, and assist breathing.

Muscles That Protect Abdominal Viscera and Move the Vertebral Column

The anterolateral abdominal wall includes four pairs of muscles: external oblique, internal oblique, transversus abdominis, and rectus abdominis (Figure 11.9; Table 11.9). From superficial to deep, the external oblique, internal oblique, and transversus abdominis form three layers of muscle around the abdomen. In each layer, the muscle fascicles extend in a different direction. This is a structural arrangement that affords considerable protection to the abdominal viscera, especially when the muscles have good tone. The **external oblique** is the superficial muscle. Its fascicles extend inferiorly and medially. The **internal oblique** is the intermediate

muscle, with fascicles that extend at right angles to those of the external oblique. The **transversus abdominis** is the deep muscle, with most of its fascicles directed transversely around the abdominal wall.

The **rectus abdominis** is a long muscle that extends the entire length of the anterior abdominal wall, from the pubic crest and pubic symphysis to the cartilages of ribs 5–7 and the xiphoid process of the sternum. The anterior surface of the muscle is interrupted by three transverse fibrous bands of tissue called *tendinous intersections.*

Recall that an *aponeurosis* (ap´-ō-noo-RŌ-sis) is a broad, flat tendon. The aponeuroses of the external oblique, internal oblique, and transversus abdominis form the *rectus sheaths,* which enclose the rectus abdominis. The sheaths meet at the midline to form the

Figure 11.9   Muscles of the anterolateral abdominal wall.

(a) Anterior superficial view   (b) Anterior deep view

TABLE 11.9

Muscles That Protect Abdominal Viscera and Move the Vertebral Column

MUSCLE	ORIGIN	INSERTION	ACTION	INNERVATION
Rectus abdominis (REK-tus ab-DOM-in-is; *rectus* = fascicles parallel to midline; *abdomin* = abdomen)	Pubic crest and pubic symphysis.	Cartilage of fifth to seventh ribs and xiphoid process.	Flexes vertebral column, especially lumbar portion, and compresses abdomen to aid in defecation, urination, forced exhalation, and childbirth.	Thoracic spinal nerves T7–T12.
External oblique (ō-BLĒK; *external* = closer to surface; *oblique* = fascicles diagonal to midline)	Inferior eight ribs.	Iliac crest and linea alba.	Acting together (bilaterally), compress abdomen and flex vertebral column; acting singly (unilaterally), laterally flex vertebral column, especially lumbar portion, and rotate vertebral column.	Thoracic spinal nerves T7–T12 and the iliohypogastric nerve.
Internal oblique (*internal* = farther from surface)	Iliac crest, inguinal ligament, and thoracolumbar fascia.	Cartilage of last three or four ribs and linea alba.	Acting together, compress abdomen and flex vertebral column; acting singly, laterally flex vertebral column, especially lumbar portion, and rotate vertebral column.	Thoracic spinal nerves T8–T12, iliohypogastric nerve, and ilioinguinal nerve.
Transversus abdominis (tranz-VER-sus = fascicles perpendicular to midline)	Iliac crest, inguinal ligament, lumbar fascia, and cartilages of inferior six ribs.	Xiphoid process, linea alba, and pubis.	Compresses abdomen.	Thoracic spinal nerves T8–T12, iliohypogastric nerve, and ilioinguinal nerve.
Quadratus lumborum (kwod-RĀ-tus lum-BOR-um; *quad-* = four; *lumbo-* = lumbar region) (see Figure 11.10b)	Iliac crest and iliolumbar ligament.	Inferior border of twelfth rib and first four lumbar vertebrae.	Acting together, pull twelfth ribs inferiorly during forced exhalation, fix twelfth ribs to prevent their elevation during deep inhalation, and help extend lumbar portion of vertebral column; acting singly, laterally flex vertebral column, especially lumbar portion.	Thoracic spinal nerve T12 and lumbar spinal nerves L1–L3 or L1–L4.

linea alba (= white line), a tough, fibrous band that extends from the xiphoid process of the sternum to the pubic symphysis.

As a group, the muscles of the anterolateral abdominal wall help contain and protect the abdominal viscera; flex, laterally flex, and rotate the vertebral column at the intervertebral joints; compress the abdomen during forced exhalation; and produce the force required for defecation, urination, and childbirth.

The posterior abdominal wall is formed by the lumbar vertebrae, parts of the ilia of the hip bones, quadratus lumborum (see Figure 11.10b), psoas major, and iliacus (see Figure 11.19a).

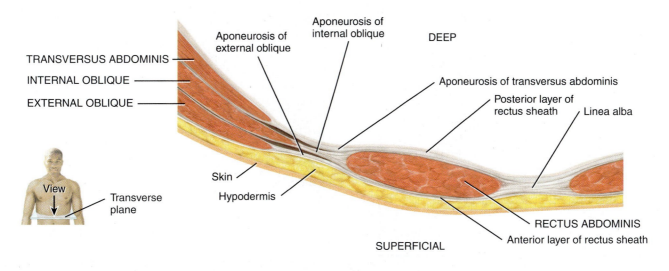

(c) Transverse section of anterior abdominal wall superior to umbilicus (navel)

The anterolateral abdominal muscles protect the abdominal viscera, move the vertebral column, and assist in forced expiration, defecation, urination, and childbirth.

Muscles Used in Breathing

The muscles described here alter the size of the thoracic cavity so that breathing can occur (Figure 11.10; Table 11.10). Inhalation (breathing in) occurs when the thoracic cavity increases in size, and exhalation (breathing out) occurs when the thoracic cavity decreases in size.

The dome-shaped **diaphragm** separates the thoracic and abdominal cavities. The diaphragm has a convex superior surface that forms the floor of the thoracic cavity (Figure 11.10c) and a concave, inferior surface that forms the roof of the abdominal cavity (Figure 11.10d). The muscular portion of the diaphragm is around the periphery of the muscle. The fibers of the muscular portion converge and insert into the *central tendon,* a strong aponeurosis located near the center of the muscle (Figure 11.10c,d). The central tendon fuses with the inferior surface of the pericardium (covering of the heart) and the pleurae (coverings of the lungs). The diaphragm has three major openings through which the aorta, esophagus, and inferior vena cava pass between the thorax and abdomen (Figure 11.10c,d).

Figure 11.10   Muscles used in breathing.

Clavicle

INTERNAL INTERCOSTALS

EXTERNAL INTERCOSTALS

Pectoralis minor (cut)

Ribs

External oblique (cut)

Rectus abdominis (cut)

Transversus abdominis and aponeurosis

Rectus abdominis (covered by anterior layer of rectus sheath [cut])

Linea alba

Internal oblique

Anterior superior iliac spine

Aponeurosis of internal oblique

Inguinal ligament

Spermatic cord

Ribs

EXTERNAL INTERCOSTALS

INTERNAL INTERCOSTALS

INNERMOST INTERCOSTALS

Sternum

Central tendon

DIAPHRAGM

QUADRATUS LUMBORUM

Transversus abdominis

Fourth lumbar vertebra

Iliac crest

Sacrum

Pubis

Pubic symphysis

(a) Anterior superficial view

(b) Anterior deep view

Sternum

Fifth costal cartilage

Pectoralis major

Hypodermis

Skin

Pericardium covering central tendon

Pleura (cut)

Inferior vena cava in caval opening

Vagus nerve

Esophagus

Central tendon

INNERMOST INTERCOSTAL

DIAPHRAGM

Body of T10

Fifth rib

Pleura (cut)

Serratus anterior

DIAPHRAGM

Sixth rib

Central tendon

EXTERNAL INTERCOSTAL

Seventh rib

INTERNAL INTERCOSTAL

Eighth rib

Latissimus dorsi

Ninth rib

Erector spinae

Spinal cord

Aorta

Thoracic duct

Azygos vein

(c) Superior view of diaphragm

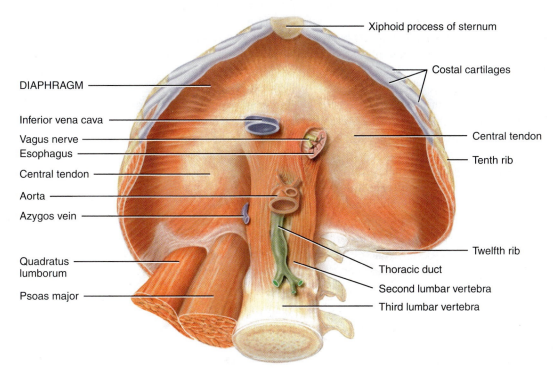

Xiphoid process of sternum

Costal cartilages

DIAPHRAGM

Inferior vena cava

Vagus nerve

Esophagus

Central tendon

Aorta

Azygos vein

Quadratus lumborum

Psoas major

Central tendon

Tenth rib

Twelfth rib

Thoracic duct

Second lumbar vertebra

Third lumbar vertebra

(d) Inferior view of diaphragm

The muscles used in breathing alter the size of the thoracic cavity.

TABLE 11.10

Muscles Used in Breathing

MUSCLE	ORIGIN	INSERTION	ACTION	INNERVATION
Diaphragm (DĪ-a-fram; *dia-* = across; *-phragm* = wall)	Xiphoid process of the sternum, costal cartilages and adjacent portions of the inferior six ribs, lumbar vertebrae and their intervertebral discs, and the twelfth rib.	Central tendon.	Contraction of the diaphragm causes it to flatten and increases the vertical dimension of the thoracic cavity, resulting in inhalation; relaxation of the diaphragm causes it to move superiorly and decreases the vertical dimension of the thoracic cavity, resulting in exhalation.	Phrenic nerve, which contains axons from cervical spinal nerves (C3–C5).
External intercostals (in´-ter-KOS-tals; *external* = closer to surface; *inter-* = between; *-costa* = rib)	Inferior border of rib above.	Superior border of rib below.	Contraction elevates the ribs and increases the anteroposterior and lateral dimensions of the thoracic cavity, resulting in inhalation; relaxation depresses the ribs and decreases the anteroposterior and lateral dimensions of the thoracic cavity, resulting in exhalation.	Thoracic spinal nerves T2–T12.
Internal intercostals (*internal* = further from surface)	Superior border of rib below.	Inferior border of rib above.	Contraction draws adjacent ribs together to further decrease the anteroposterior and lateral dimensions of the thoracic cavity during forced exhalation.	Thoracic spinal nerves T2–T12.

The diaphragm is the most important muscle that powers breathing. During contraction, it depresses into a flatter shape, increasing thoracic cavity volume, which results in inhalation. As the diaphragm relaxes, it elevates back to the dome shape, decreasing thoracic cavity volume to produce exhalation. Movements of the diaphragm also help return blood from the abdomen to the heart. Together, the diaphragm and anterolateral abdominal muscles can be voluntarily contracted to help increase pressure in the abdomen to evacuate the pelvic contents during defecation, urination, and childbirth. The increase in intra-abdominal pressure will also help support the vertebral column and prevent flexion during weight lifting. This greatly assists the back muscles in lifting a heavy weight.

Other muscles involved in breathing, called intercostal muscles, span the spaces between ribs. The **external intercostals** occupy the superficial layer, and their fibers run obliquely and anteriorly from the rib above to the rib below. They elevate the ribs during inhalation to help expand the thoracic cavity. The **internal intercostals** lie deep to the external intercostals. The fibers of these muscles are at right angles to fibers of the external intercos-tals and run obliquely and posteriorly from the rib below to the rib above. They draw adjacent ribs together during forced exhalation to help decrease the size of the thoracic cavity. The deepest muscle layer is the paired **innermost intercostal muscles**. These muscles extend in the same direction as the internal intercostals and may have the same role.

As you will see in Chapter 22, the diaphragm and external intercostals are used during quiet inhalation and exhalation. However, during deep, forceful inhalation, as occurs during exercise or playing a wind instrument, the sternocleidomastoid, scalene, and pectoralis minor are also used, and during deep, forceful exhalation, the external oblique, internal oblique, transverse abdominis, rectus abdominis, and internal intercostals are also used.

Checkpoint

13. State the direction of the fascicles for the external oblique, internal oblique, transversus abdominis, and rectus abdominis.
14. Which abdominal muscle aids in urination?
15. Name the structures that pass through the diaphragm.

11.6 Muscles of the pelvic floor and perineum support the pelvic viscera, function as sphincters, and assist in urination, erection, ejaculation, and defecation.

Muscles of the Pelvic Floor

The pelvic floor includes the coccygeus and the levator ani muscle group (Figure 11.11; Table 11.11). These muscles and their fascia are referred to as the **pelvic diaphragm**, which stretches from the pubis anteriorly, to the coccyx posteriorly, and from one lateral wall of the pelvis to the other. The anal canal and urethra pass through the pelvic diaphragm in both sexes, and the vagina goes through it in females.

The **levator ani** muscle group supports the pelvic viscera and resists the inferior thrust that accompanies increases in intra-abdominal pressure during functions such as forced exhalation, coughing, vomiting, urination, and defecation. The muscle also functions as a sphincter for the anal canal, urethra, and vagina. In addition to assisting the levator ani, the ischiococcygeus pulls the coccyx anteriorly after it has been pushed posteriorly during defecation or childbirth. Figure 11.11 shows these muscles in the female and Figure 11.12 illustrates them in the male.

Muscles of the Perineum

The **perineum** is the region of the trunk inferior to the pelvic diaphragm. It is a diamond-shaped area that extends from the pubic symphysis anteriorly, to the coccyx posteriorly, and to the ischial tuberosities laterally. The female and the male perineums may be compared in Figures 11.11 and 11.12, respectively. A transverse line drawn between the ischial tuberosities divides the perineum into an anterior *urogenital triangle* that contains the external genitals and a posterior *anal triangle* that contains the anus (see Figure 25.21). Clinically, the perineum is very important to physicians who care for women during pregnancy and who treat disorders related to the female genital tract, urogenital organs, and the anorectal region.

Figure 11.11   Muscles of the female pelvic floor.

EXTERNAL URETHRAL SPHINCTER

COMPRESSOR URETHRAE

SPHINCTER URETHRO-VAGINALIS

Perineal body

Obturator internus

Anus

Anococcygeal raphe

ISCHIOCOCCYGEUS

Clitoris

Inferior public ramus

Urethral orifice

ISCHIOCAVERNOSUS

Ischiopubic ramus

BULBOSPONGIOSUS

Vagina

Perineal membrane

SUPERFICIAL TRANSVERSE PERINEAL

Ischial tuberosity

EXTERNAL ANAL SPHINCTER

LEVATOR ANI:
 PUBOCOCCYGEUS
 ILIOCOCCYGEUS

Coccyx

Gluteus maximus

Inferior superficial view

The pelvic diaphragm supports the pelvic viscera.

TABLE 11.11

Muscles of the Pelvic Floor

MUSCLE	ORIGIN	INSERTION	ACTION	INNERVATION
Levator ani (le-VĀ-tor Ā-nē; *levator* = raises; *ani* = anus) This muscle is divisible into two parts, the pubococcygeus muscle and the iliococcygeus muscle.				
Pubococcygeus (pū′-bō-kok-SIJ-ē-us; *pubo-* = pubis; *-coccygeus* = coccyx)	Pubis.	Coccyx, urethra, anal canal, perineal body of the perineum (a wedge-shaped mass of fibrous tissue in the center of the perineum), and anococcygeal ligament (narrow fibrous band that extends from anus to coccyx).	Supports and maintains position of pelvic viscera; resists increase in intra-abdominal pressure during forced exhalation, coughing, vomiting, urination, and defecation; constricts anus, urethra, and vagina.	Sacral spinal nerves S2–S4.
Iliococcygeus (il′-ē-ō-kok-SIJ-ē-us; *ilio-* = ilium)	Ischial spine.	Coccyx.	As above.	Sacral spinal nerves S2–S4.
Ischiococcygeus (is′-kē-ō-kok-SIJ-ē-us)	Ischial spine.	Lower sacrum and upper coccyx.	Supports and maintains position of pelvic viscera; resists increase in intra-abdominal pressure during forced exhalation, coughing, vomiting, urination, and defecation; and pulls coccyx anteriorly following defecation or childbirth.	Sacral spinal nerves S4–S5.

Figure 11.12   **Muscles of the male perineum.**

Inferior superficial view

🔑 **The muscles of the perineum assist in urination and help strengthen the pelvic floor in both females and males, and assist in ejaculation in males.**

The superficial muscles of the perineum help maintain erection of the penis in males and clitoris in females, and facilitate ejaculation in males (Table 11.12). The deep muscles of the perineum assist in urination and closing the vagina in females and urination and ejaculation in males. The **external anal sphincter** keeps the anal canal and anus closed except during defecation.

Checkpoint

16. What are the borders of the pelvic diaphragm?
17. Which muscles help expel urine and semen? Which maintain erection of the penis and clitoris?
18. What are the borders of the perineum?
19. What is contained within the borders of the urogenital triangle and the anal triangle?

TABLE 11.12

Muscles of the Perineum

MUSCLE	ORIGIN	INSERTION	ACTION	INNERVATION
Superficial Perineal Muscles				
Superficial transverse perineal (per-i-NĒ-al; *superficial* = closer to surface; *transverse* = across; *perineus* = perineum)	Ischial tuberosity.	Perineal body of perineum.	Stabilizes perineal body of perineum.	Perineal branch of pudendal nerve of sacral plexus.
Bulbospongiosus (bul´-bō-spon´-jē-Ō-sus; *bulb* = a bulb; *spongio-* = sponge)	Perineal body of perineum.	Perineal membrane of deep muscles of perineum, corpus spongiosum of penis, and deep fascia on dorsum of penis in male; pubic arch and root and dorsum of clitoris in female.	Helps expel urine during urination, helps propel semen along urethra, assists in erection of the penis in male; constricts vaginal orifice and assists in erection of clitoris in female.	Perineal branch of pudendal nerve of sacral plexus.
Ischiocavernosus (is´-kē-ō-ka´-ver-NŌ-sus; *ischio-* = the hip)	Ischial tuberosity and ischial and pubic rami.	Corpus cavernosum of penis in male and clitoris in female.	Maintains erection of penis in male and clitoris in female.	Perineal branch of pudendal nerve of sacral plexus.
Deep Perineal Muscles				
Deep transverse perineal (*deep* = farther from surface)	Ischial rami.	Perineal body of perineum.	Helps expel last drops of urine and semen in male and urine in female.	Perineal branch of pudendal nerve of sacral plexus.
External urethral sphincter (ū-RĒ-thral SFINGK-ter) (see Figure 24.21)	Ischial and pubic rami.	Median raphe in male and vaginal wall in female.	Helps expel last drops of urine and semen in male and urine in female.	Sacral spinal nerve S4 and inferior rectal branch of pudendal nerve.
Compressor urethrae (ū-RĒ-thrē)	Ischiopubic ramus.	Blends with partner on other side anterior to urethra.	Serves as accessory sphincter of the urethra.	Perineal branch of the pudendal nerve of the sacral plexus.
Sphincter urethrovaginalis (ū-RĒ-thrō-vaj-i-NAL-is)	Perineal body.	Blends with partner on other side anterior to urethra.	Serves as a sphincter of urethra and facilitates closing of vagina.	Perineal branch of the pudendal nerve of the sacral plexus.
External anal sphincter (Ā-nal)	Anococcygeal ligament.	Perineal body of perineum.	Keeps anal canal and anus closed.	Sacral spinal nerve S4 and inferior rectal branch of pudendal nerve.

11.7   Muscles inserting on the upper limb move and stabilize the pectoral girdle, and move the arm, forearm, and hand.

Muscles That Move the Pectoral Girdle

The main action of the muscles that move the pectoral (shoulder) girdle (Figure 11.13; Table 11.13) is to stabilize the scapula so it can function as a steady origin for most of the muscles that move the humerus. Because scapular movements usually accompany humeral movements, the muscles also move the scapula to increase the range of motion of the humerus. For example, it would not be possible to raise your hand in class, which requires abducting your humerus past the horizontal, if the scapula did not follow the humerus by rotating upward.

Muscles that move the pectoral girdle can be classified into two groups based on their location in the thorax: anterior and posterior thoracic muscles. The **anterior thoracic muscles** are the subclavius, pectoralis minor, and serratus anterior. The **subclavius** extends from the clavicle to the first rib and steadies the clavicle during movements of the pectoral girdle. The **pectoralis minor** lies deep to the pectoralis major. In addition to its role in movements of the scapula, the pectoralis minor assists in forced inhalation. The **serratus anterior**, a large, fan-shaped muscle between the ribs and scapula, is named for the saw-toothed appearance of its origins on the ribs.

Figure 11.13   **Muscles that move the pectoral girdle.**

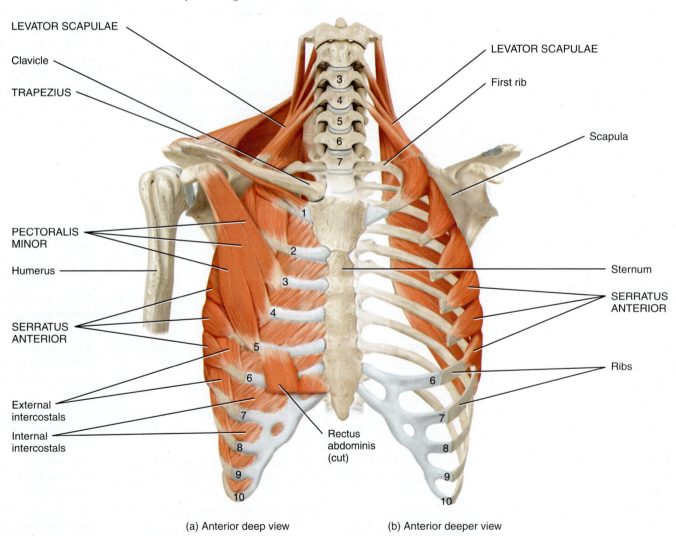

(a) Anterior deep view   (b) Anterior deeper view

Muscles that move the pectoral girdle originate on the axial skeleton and insert on the clavicle or scapula.

The **posterior thoracic muscles** are the trapezius, levator scapulae, rhomboid major, and rhomboid minor. The most superficial back muscle, the **trapezius**, extends from the skull and vertebral column medially to the pectoral girdle laterally. The two trapezius muscles form a trapezoid (diamond-shaped quadrangle)—hence its name. The **levator scapulae** lies in the posterior neck deep to the sternocleidomastoid and trapezoid. As its name suggests, one of its actions is to elevate the scapula. The **rhomboid major** and **rhomboid minor** lie deep to the trapezius and pass from the vertebrae to the scapula. They are named based on their shape—that is, a rhomboid (an oblique parallelogram). Both muscles are used when forcibly lowering the raised upper limbs, as in driving a stake with a sledgehammer.

TABLE 11.13

Muscles That Move the Pectoral Girdle

MUSCLE	ORIGIN	INSERTION	ACTION	INNERVATION
Anterior Thoracic Muscles				
Subclavius (sub-KLĀ-vē-us; *sub-* = under; *-clavius* = clavicle	First rib.	Clavicle.	Depresses and moves clavicle anteriorly and helps stabilize pectoral girdle.	Subclavian nerve.
Pectoralis minor (pek´-tō-RĀ-lis; *pector-* = the breast, chest, thorax; *minor* = lesser)	Second through fifth, third through fifth, or second through fourth ribs.	Coracoid process of scapula.	Abducts scapula and rotates it downward; elevates third through fifth ribs during forced inhalation when scapula is fixed.	Medial pectoral nerve.
Serratus anterior (ser-Ā-tus = saw-toothed; *anterior* = front)	Superior eight or nine ribs.	Vertebral border and inferior angle of scapula.	Abducts scapula and rotates it upward; elevates ribs when scapula is stabilized; known as "boxer's muscle" because it is important in horizontal arm movements such as punching and pushing.	Long thoracic nerve.
Posterior Thoracic Muscles				
Trapezius (tra-PĒ-zē-us; = trapezoid-shaped)	Superior nuchal line of occipital bone, ligamentum nuchae, and spines of seventh cervical and all thoracic vertebrae.	Clavicle and acromion and spine of scapula.	Superior fibers elevate scapula and can help extend head; middle fibers adduct scapula; inferior fibers depress scapula; superior and inferior fibers together rotate scapula upward; stabilizes scapula.	Accessory (XI) nerve and cervical spinal nerves C3–C5.
Levator scapulae (le-VĀ-tor SKA-pū-lē; *levator* = raises; *scapulae* = of the scapula)	Superior four or five cervical vertebrae.	Superior vertebral border of scapula.	Elevates scapula and rotates it downward.	Dorsal scapular nerve and cervical spinal nerves C3–C5.
Rhomboid major (rom-BOYD = rhomboid or diamond-shaped) (see Figure 11.14c)	Spines of second to fifth thoracic vertebrae.	Vertebral border of scapula inferior to spine.	Elevates and adducts scapula and rotates it downward; stabilizes scapula.	Dorsal scapular nerve.
Rhomboid minor (see Figure 11.14c)	Spines of seventh cervical and first thoracic vertebrae.	Vertebral border of scapula superior to spine.	Elevates and adducts scapula and rotates it downward; stabilizes scapula.	Dorsal scapular nerve.

Muscles That Move the Humerus

Of the nine muscles that cross the shoulder joint, all except the pectoralis major and latissimus dorsi originate on the scapula (Figure 11.14; Table 11.14). These two muscles originate on the axial skeleton. The **pectoralis major** is a large, fan-shaped muscle that covers the superior part of the thorax. The **latissimus dorsi** is a broad, triangular muscle located on the inferior part of the back. It is commonly called the "swimmer's muscle" because its many actions are used while swimming; consequently, many competitive swimmers have well-developed "lats." The **deltoid** is a thick, powerful muscle that covers the shoulder joint and forms the rounded contour of the shoulder. This muscle is a frequent site

of intramuscular injections. As you study the deltoid, note that its fascicles originate from three different points and that each group of fascicles moves the humerus differently.

The strength and stability of the shoulder joint are not provided by the shape of the articulating bones or its ligaments. Instead, the **subscapularis**, **supraspinatus**, **infraspinatus**, and **teres minor** strengthen and stabilize the shoulder joint as they join the scapula to the humerus. Their tendons fuse together to form the *rotator cuff,* a nearly complete circle of tendons around the shoulder joint, like the cuff on a shirtsleeve. The supraspinatus is especially predisposed to wear and tear because of its location between the head of the humerus and acromion of the scapula, which compresses its tendon during shoulder movements, especially abduction of the arm.

Figure 11.14   Muscles that move the humerus.

(a) Anterior deep view (the intact pectoralis major muscle is shown in Figure 11.9a)

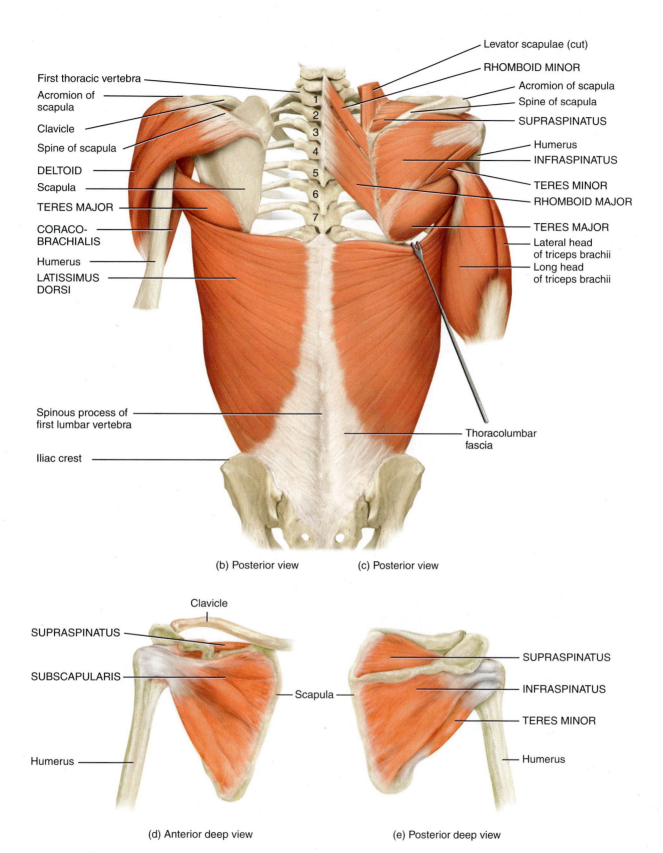

Levator scapulae (cut)

RHOMBOID MINOR

First thoracic vertebra

Acromion of scapula

Acromion of scapula

Spine of scapula

Clavicle

SUPRASPINATUS

Spine of scapula

Humerus

DELTOID

INFRASPINATUS

Scapula

TERES MINOR

TERES MAJOR

RHOMBOID MAJOR

CORACO-BRACHIALIS

TERES MAJOR

Lateral head of triceps brachii

Humerus

Long head of triceps brachii

LATISSIMUS DORSI

Spinous process of first lumbar vertebra

Thoracolumbar fascia

Iliac crest

(b) Posterior view   (c) Posterior view

Clavicle

SUPRASPINATUS

SUPRASPINATUS

SUBSCAPULARIS

INFRASPINATUS

Scapula

TERES MINOR

Humerus

Humerus

(d) Anterior deep view   (e) Posterior deep view

The strength and stability of the shoulder joint are provided by the tendons that form the rotator cuff.

TABLE 11.14

Muscles That Move the Humerus

MUSCLE	ORIGIN	INSERTION	ACTION	INNERVATION
Axial Muscles That Move the Humerus				
Pectoralis major (pek´-tō-RĀ-lis; *pector-* = chest; *major* = larger) (see also Figure 11.9a)	Clavicle (clavicular head), sternum, and costal cartilages of second to sixth ribs and sometimes first to seventh ribs (sternocostal head).	Greater tubercle and lateral lip of the intertubercular sulcus of humerus.	As a whole, adducts and medially rotates arm at shoulder joint; clavicular head flexes arm, and sternocostal head extends the flexed arm to side of trunk.	Medial and lateral pectoral nerves.
Latissimus dorsi (la-TIS-i-mus DOR-sī; *latissimus* = widest; *dorsi* = of the back)	Spines of inferior six thoracic vertebrae, lumbar vertebrae, crests of sacrum and ilium, inferior four ribs.	Intertubercular sulcus of humerus.	Extends, adducts, and medially rotates arm at shoulder joint; draws arm inferiorly and posteriorly.	Thoracodorsal nerve.
Scapular Muscles That Move the Humerus				
Deltoid (DEL-toyd = triangularly shaped)	Acromial extremity of clavicle (anterior fibers), acromion of scapula (lateral fibers), and spine of scapula (posterior fibers).	Deltoid tuberosity of humerus.	Lateral fibers abduct arm at shoulder joint; anterior fibers flex and medially rotate arm at shoulder joint; posterior fibers extend and laterally rotate arm at shoulder joint.	Axillary nerve.
Subscapularis (sub-scap´-ū-LĀ-ris; *sub-* = below; -*scapularis* = scapula)	Subscapular fossa of scapula.	Lesser tubercle of humerus.	Medially rotates arm at shoulder joint.	Upper and lower subscapular nerve.
Supraspinatus (soo-pra-spī-NĀ-tus; *supra-* = above; -*spina-* = spine of the scapula)	Supraspinous fossa of scapula.	Greater tubercle of humerus.	Assists deltoid muscle in abducting arm at shoulder joint.	Suprascapular nerve.
Infraspinatus (in´-fra-spī-NĀ-tus; *infra-* = below)	Infraspinous fossa of scapula.	Greater tubercle of humerus.	Laterally rotates and adducts arm at shoulder joint.	Suprascapular nerve.
Teres major (TE-rēz; *teres* = long and round)	Inferior angle of scapula.	Medial lip of intertubercular sulcus of humerus.	Extends arm at shoulder joint and assists in adduction and medial rotation of arm at shoulder joint.	Lower subscapular nerve.
Teres minor	Inferior lateral border of scapula.	Greater tubercle of humerus.	Laterally rotates, extends, and adducts arm at shoulder joint.	Axillary nerve.
Coracobrachialis (kor´-a-kō-brā-kē-Ā-lis; *coraco-* = coracoid process of the scapula; -*brachi-* = arm)	Coracoid process of scapula.	Middle of medial surface of shaft of humerus.	Flexes and adducts arm at shoulder joint.	Musculocutaneous nerve.

Muscles That Move the Radius and Ulna

Most of the muscles that move the radius and ulna (Figure 11.15; Table 11.15) are divided into **forearm flexors** and **forearm extensors**. Recall that the elbow joint is a hinge joint, capable only of flexion and extension. The biceps brachii, brachialis, and brachioradialis are flexors of the elbow joint; the triceps brachii and the anconeus are extensors.

The **biceps brachii** is the large muscle located on the anterior surface of the arm. As indicated by its name (*bi-* = two), it has two heads of origin (long and short), both from the scapula. Because the muscle spans both the shoulder and elbow joints, it flexes the arm at the shoulder joint, flexes the forearm at the elbow joint, and supinates the forearm at the radioulnar joints. The **brachialis** is deep to the biceps brachii. It is the most powerful flexor of the forearm at the elbow joint. For this reason, it is called the "workhorse" of the elbow flexors. The **brachioradialis** also flexes the forearm at the elbow joint, especially when a quick movement is required or when a weight is lifted slowly during flexion of the forearm.

Figure 11.15   Muscles that move the radius and ulna.

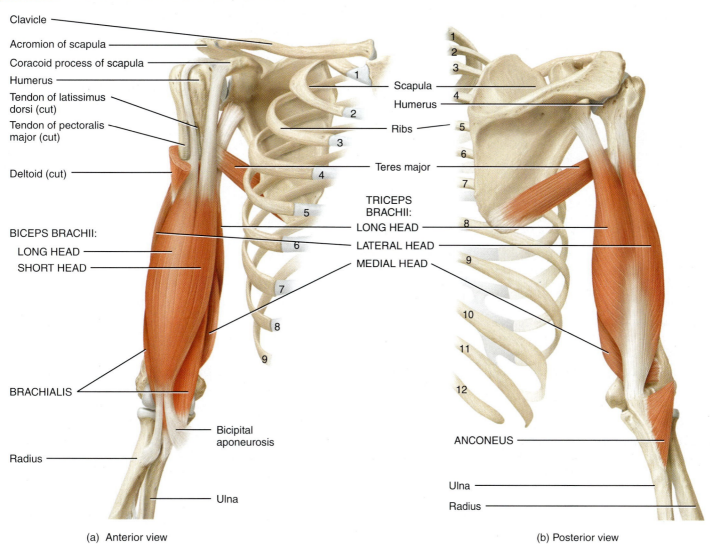

(a) Anterior view (b) Posterior view

🔑 The anterior arm muscles flex the forearm; the posterior arm muscles extend it.

TABLE 11.15

Muscles That Move the Radius and Ulna

MUSCLE	ORIGIN	INSERTION	ACTION	INNERVATION
Forearm Flexors				
Biceps brachii (BĪ-ceps BRĀ-kē-ī; *biceps* = two heads of origin; *brachii* = arm)	Long head: tubercle above glenoid cavity of scapula (supraglenoid tubercle). Short head: coracoid process of scapula.	Radial tuberosity of radius and bicipital aponeurosis.*	Flexes forearm at elbow joint, supinates forearm at radioulnar joints, and flexes arm at shoulder joint.	Musculocutaneous nerve.
Brachialis (brā-kē-Ā-lis)	Distal, anterior surface of humerus.	Ulnar tuberosity and coronoid process of ulna.	Flexes forearm at elbow joint.	Musculocutaneous and radial nerves.
Brachioradialis (brā´-kē-ō-rā-dē-Ā-lis; *-radi* = radius) (see Figure 11.16a)	Lateral border of distal end of humerus.	Superior to styloid process of radius.	Flexes forearm at elbow joint; supinates and pronates forearm at radioulnar joints to neutral position.	Radial nerve.
Forearm Extensors				
Triceps brachii (TRĪ-ceps = three heads of origin)	Long head: infraglenoid tubercle, a projection inferior to glenoid cavity of scapula. Lateral head: lateral and posterior surface of humerus superior to radial groove. Medial head: entire posterior surface of humerus inferior to groove for radial nerve.	Olecranon of ulna.	Extends forearm at elbow joint and extends arm at shoulder joint.	Radial nerve.
Anconeus (an-KŌ-nē-us; *ancon-* = the elbow) (see also Figure 11.15b)	Lateral epicondyle of humerus.	Olecranon and superior portion of shaft of ulna.	Extends forearm at elbow joint.	Radial nerve.
Forearm Pronators (see Figure 11.16a)				
Pronator teres (PRŌ-nā-tor TE-rēz; *pronator* = turns palm posteriorly)	Medial epicondyle of humerus and coronoid process of ulna.	Midlateral surface of radius.	Pronates forearm at radioulnar joints and weakly flexes forearm at elbow joint.	Median nerve.
Pronator quadratus (kwod-RĀ-tus = square, four-sided)	Distal portion of shaft of ulna.	Distal portion of shaft of radius.	Pronates forearm at radioulnar joints.	Median nerve.
Forearm Supinator (see Figure 11.16b)				
Supinator (SOO-pi-nā-tor = turns palm anteriorly)	Lateral epicondyle of humerus and ridge near radial notch of ulna (supinator crest).	Lateral surface of proximal one-third of radius.	Supinates forearm at radioulnar joints.	Deep radial nerve.

*The **bicipital aponeurosis** is a broad aponeurosis from the tendon of insertion of the biceps brachii muscle that descends medially across the brachial artery and fuses with deep fascia over the forearm flexor muscles.

The **triceps brachii** is the large muscle located on the posterior surface of the arm. It is the more powerful of the extensors of the forearm at the elbow joint. As its name implies (*tri-* = three), it has three heads of origin, one from the scapula (long head) and two from the humerus (lateral and medial heads). The **anconeus** is a small muscle located on the lateral part of the posterior elbow that assists the triceps brachii in extending the forearm at the elbow joint.

Some muscles that move the radius and ulna are concerned with supination and pronation of the forearm. The pronators, as suggested by their names, are the **pronator teres** and **pronator quadratus**. The supinator of the forearm is aptly named the **supinator**. You use the powerful action of the supinator when you twist a corkscrew or turn a screw with a screwdriver.

Muscles That Move the Wrist, Hand, and Fingers

Muscles that move the wrist, hand, and fingers are located on the forearm and are many and varied (Figure 11.16; Table 11.16).

Those muscles in this group that move the fingers in various ways are known as **extrinsic muscles of the hand** because they originate *outside* the hand and insert within it. As you will see, the names for the muscles give some indication of their origin, insertion, or action.

Figure 11.16   Muscles that move the wrist, hand, and fingers.

Biceps brachii
Brachialis
Medial epicondyle of humerus
Lateral epicondyle of humerus
Tendon of biceps brachii
Bicipital aponeurosis
PRONATOR TERES
BRACHIORADIALIS
PALMARIS LONGUS
FLEXOR CARPI RADIALIS
FLEXOR CARPI ULNARIS
EXTENSOR CARPI RADIALIS LONGUS
FLEXOR DIGITORUM SUPERFICIALIS
FLEXOR POLLICIS LONGUS
PRONATOR QUADRATUS
Radius
Flexor retinaculum
Palmar aponeurosis
Metacarpal
Tendon of flexor digitorum superficialis
Tendon of flexor digitorum profundus

Humerus
Supinator
FLEXOR DIGITORUM PROFUNDUS
FLEXOR POLLICIS LONGUS
Pronator quadratus
Tendon of flexor pollicis longus
Tendon of flexor digitorum profundus

(a) Anterior superficial view

(b) Anterior intermediate view

(c) Anterior deep view

(continues)

Figure 11.16 (continued)

Triceps brachii

Humerus

BRACHIORADIALIS

EXTENSOR CARPI RADIALIS
LONGUS

Medial epicondyle of humerus

Lateral epicondyle of humerus

Olecranon of ulna

ANCONEUS

Supinator

EXTENSOR CARPI ULNARIS

EXTENSOR DIGITORUM

Ulna

EXTENSOR CARPI RADIALIS
BREVIS

FLEXOR CARPI ULNARIS

ABDUCTOR POLLICIS LONGUS

EXTENSOR POLLICIS LONGUS

EXTENSOR INDICIS

EXTENSOR DIGITI MINIMI

EXTENSOR POLLICIS BREVIS

Tendon of extensor carpi ulnaris

Tendon of extensor digiti minimi

Tendon of extensor indicis

Dorsal interossei

Tendons of extensor digitorum

(d) Posterior superficial view

(e) Posterior deep view

The anterior forearm muscles function as flexors, and the posterior forearm muscles function as extensors.

The anterior forearm muscles originate on the humerus, typically insert on the carpals, metacarpals, and phalanges, and function as flexors. The superficial anterior muscles in the forearm are arranged in the following order from lateral to medial: **flexor carpi radialis**, **palmaris longus**, and **flexor carpi ulnaris**. The **flexor digitorum superficialis** is deep to the other three muscles. The deep anterior muscles are arranged in the following order from lateral to medial: **flexor pollicis longus** (the only flexor of the distal phalanx of the thumb) and **flexor digitorum profundus** (ends in four tendons that insert into the distal phalanges of the fingers).

The posterior forearm muscles originate on the humerus, insert on the metacarpals and phalanges, and function as extensors. The superficial posterior muscles are arranged in the following order from lateral to medial: **extensor carpi radialis longus**, **extensor carpi radialis brevis**, **extensor digitorum** (divides into four tendons that insert into the middle and distal phalanges of the fingers), **extensor digiti minimi**, and the **extensor carpi ulnaris**. The deep posterior muscles are arranged in the following order from lateral to medial: **abductor pollicis longus**, **extensor pollicis brevis**, **extensor pollicis longus**, and **extensor indicis**.

TABLE 11.16

Muscles That Move the Wrist, Hand, and Fingers

MUSCLE	ORIGIN	INSERTION	ACTION	INNERVATION
Superficial Anterior (Flexor) Compartment of the Forearm				
Flexor carpi radialis (FLEK-sor KAR-pē rā´-dē-Ā-lis; *flexor* = decreases angle at joint; *carpi* = of the wrist; *radi-* = radius)	Medial epicondyle of humerus.	Second and third metacarpals.	Flexes and abducts hand (radial deviation) at wrist joint.	Median nerve.
Palmaris longus (pal-MA-ris LON-gus; *palma-* = palm; *longus* = long)	Medial epicondyle of humerus.	Flexor retinaculum and palmar aponeurosis (fascia in center of palm).	Weakly flexes hand at wrist joint.	Median nerve.
Flexor carpi ulnaris (ul-NAR-is = ulna)	Medial epicondyle of humerus and superior posterior border of ulna.	Pisiform, hamate, and base of fifth metacarpal.	Flexes and adducts hand (ulnar deviation) at wrist joint.	Ulnar nerve.
Flexor digitorum superficialis (di-ji-TOR-um soo´-per-fish´-ē-Ā-lis; *digit* = finger or toe; *superficialis* = closer to surface)	Medial epicondyle of humerus, coronoid process of ulna, and a ridge along lateral margin of anterior surface (anterior oblique line) of radius.	Middle phalanx of each finger.*	Flexes middle phalanx of each finger at proximal interphalangeal joint, proximal phalanx of each finger at metacarpophalangeal joint, and hand at wrist joint.	Median nerve.
Deep Anterior (Flexor) Compartment of the Forearm				
Flexor pollicis longus (POL-li-sis = thumb)	Anterior surface of radius and interosseous membrane (sheet of fibrous tissue that holds shafts of ulna and radius together).	Base of distal phalanx of thumb.	Flexes distal phalanx of thumb at interphalangeal joint.	Median nerve.
Flexor digitorum profundus (prō-FUN-dus = deep)	Anterior medial surface of body of ulna.	Base of distal phalanx of each finger.	Flexes distal and middle phalanges of each finger at interphalangeal joints, proximal phalanx of each finger at metacarpophalangeal joint, and hand at wrist joint.	Median and ulnar nerves.

(continues)

TABLE 11.16 (continued)

Muscles That Move the Wrist, Hand, and Fingers

MUSCLE	ORIGIN	INSERTION	ACTION	INNERVATION
Superficial Posterior (Extensor) Compartment of the Forearm				
Extensor carpi radialis longus (eks-TEN-sor = increases angle at joint)	Lateral supracondylar ridge of humerus.	Second metacarpal.	Extends and abducts hand at wrist joint.	Radial nerve.
Extensor carpi radialis brevis (BREV-is = short)	Lateral epicondyle of humerus.	Third metacarpal.	Extends and abducts hand at wrist joint.	Radial nerve.
Extensor digitorum	Lateral epicondyle of humerus.	Distal and middle phalanges of each finger.	Extends distal and middle phalanges of each finger at interphalangeal joints, proximal phalanx of each finger at metacarpophalangeal joint, and hand at wrist joint.	Radial nerve.
Extensor digiti minimi (DIJ-i-tē MIN-i-mē; *digit* = finger or toe; *minimi* = smallest)	Lateral epicondyle of humerus.	Tendon of extensor digitorum on fifth phalanx.	Extends proximal phalanx of little finger at metacarpophalangeal joint and hand at wrist joint.	Deep radial nerve.
Extensor carpi ulnaris	Lateral epicondyle of humerus and posterior border of ulna.	Fifth metacarpal.	Extends and adducts hand at wrist joint.	Deep radial nerve.
Deep Posterior (Extensor) Compartment of the Forearm				
Abductor pollicis longus (ab-DUK-tor = moves part away from midline)	Posterior surface of middle of radius and ulna and interosseous membrane.	First metacarpal.	Abducts and extends thumb at carpometacarpal joint and abducts hand at wrist joint.	Deep radial nerve.
Extensor pollicis brevis	Posterior surface of middle of radius and interosseous membrane.	Base of proximal phalanx of thumb.	Extends proximal phalanx of thumb at metacarpophalangeal joint, first metacarpal of thumb at carpometacarpal joint, and hand at wrist joint.	Deep radial nerve.
Extensor pollicis longus	Posterior surface of middle of ulna and interosseous membrane.	Base of distal phalanx of thumb.	Extends distal phalanx of thumb at interphalangeal joint, first metacarpal of thumb at carpometacarpal joint, and abducts hand at wrist joint.	Deep radial nerve.
Extensor indicis (IN-di-kis; = index)	Posterior surface of ulna.	Tendon of extensor digitorum of index finger.	Extends distal and middle phalanges of index finger at interphalangeal joints, proximal phalanx of index finger at metacarpophalangeal joint, and hand at wrist joint.	Deep radial nerve.

*Reminder: The thumb or pollex is the first finger and has two phalanges: proximal and distal. The remaining fingers are numbered II–V (2–5), and each has three phalanges: proximal, middle, and distal.

The tendons of the muscles of the forearm, which attach to the wrist or continue into the hand with blood vessels and nerves, are held close to bones by fascia. The tendons are also surrounded by tendon sheaths. At the wrist, the fascia is thickened into fibrous bands called *retinacula* (*retinacul-* = a holdfast). The *flexor reti-* *naculum,* located over the palmar surface of the carpal bones, secures the tendons of the flexors of the fingers and wrist. The *extensor retinaculum,* located over the dorsal surface of the carpal bones, stabilizes tendons of the extensors of the wrist and fingers, which pass deep to it.

Intrinsic Muscles of the Hand

The **intrinsic muscles of the hand** originate and insert *within* the hand (Figure 11.17; Table 11.17). Whereas the extrinsic hand muscles create powerful but crude movements of the fingers, the intrinsic muscles produce the weak but intricate and precise movements of the fingers that characterize the human hand.

Figure 11.17 **Intrinsic muscles of the hand.**

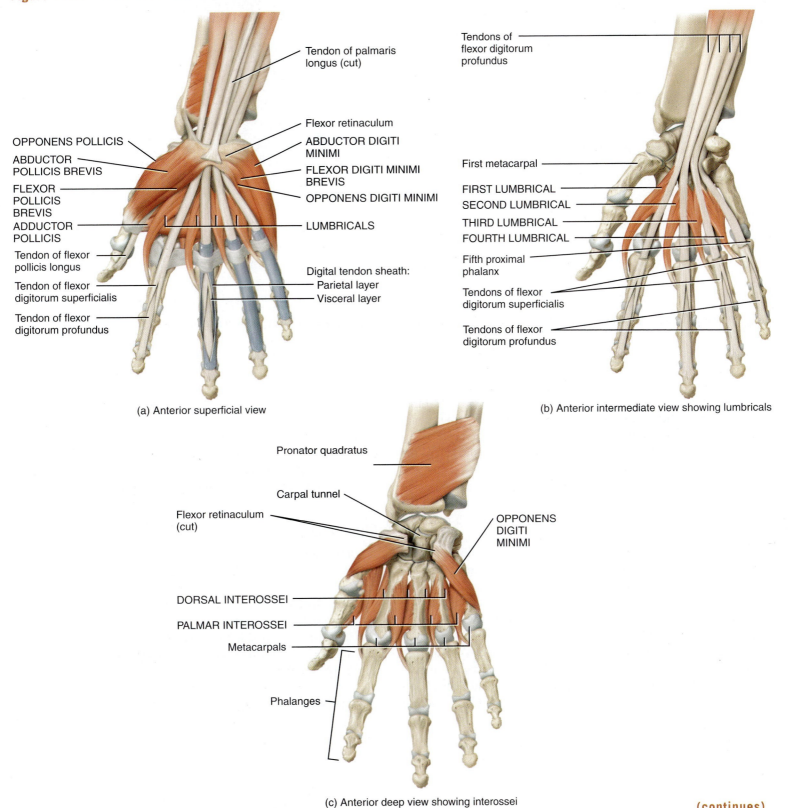

(a) Anterior superficial view

(b) Anterior intermediate view showing lumbricals

(c) Anterior deep view showing interossei

(continues)

Figure 11.17 (continued)

PALMAR INTEROSSEI

Fifth
metacarpal

Fifth proximal
phalanx

(d) Anterior deep view of palmar interossei

DORSAL INTEROSSEI

Fifth
metacarpal

Fifth
proximal
phalanx

(e) Anterior deep view of dorsal interossei

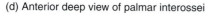 **The intrinsic muscles of the hand produce the intricate and precise movements of the fingers that characterize the human hand.**

The intrinsic muscles of the hand are divided into three groups: **thenar**, **hypothenar**, and **intermediate**. The four thenar muscles act on the thumb and form the **thenar eminence**, the lateral rounded contour on the palm that is also called the ball of the thumb. The three hypothenar muscles act on the little finger and form the **hypothenar eminence**, the medial rounded contour on the palm that is also called the ball of the little finger. One of the hypothenar muscles, the **abductor digiti minimi**, is a powerful muscle that plays an important role in grasping an object with outspread fingers. The 11 intermediate muscles act on all the fingers except the thumb. Two intermediate muscles, the **palmar interossei** and **dorsal interossei**, are located between the metacarpals and are important in abduction, adduction, flexion, and extension of the fingers, and in movements involved in skilled activities such as writing, typing, and playing a piano.

The functional importance of the hand is readily apparent when you consider that certain hand injuries can result in permanent disability. Most of the dexterity of the hand depends on movements of the thumb. The general activities of the hand are free motion, power grip (forcible movement of the fingers and thumb against the palm, as in squeezing), precision handling (a change in position of a handled object that requires exact control of finger and thumb positions, as in winding a watch or threading a needle), and pinch (compression between the thumb and index finger or between the thumb and first two fingers).

Checkpoint

20. Name the muscles that flex, extend, and adduct your arm at the shoulder joint.
21. The tendons of which muscles make up the rotator cuff?
22. Flex your arm at the elbow joint. Which muscles are contracting? Which muscles must relax so that you can flex your arm?
23. Name the muscles that supinate and pronate your forearm.
24. State the functions of the flexor retinaculum and extensor retinaculum.
25. Which extrinsic and intrinsic muscles of the hand are used to flex the thumb? To abduct the thumb?
26. Name the extrinsic and intrinsic muscles of the hand that flex the fingers and those that are used to adduct the fingers.

TABLE 11.17

Intrinsic Muscles of the Hand

MUSCLE	ORIGIN	INSERTION	ACTION	INNERVATION
Thenar (Lateral Aspect of Palm)				
Abductor pollicis brevis (ab-DUK-tor POL-li-sis BREV-is; *abductor* = moves part away from middle; *pollic-* = the thumb; *brevis* = short)	Flexor retinaculum, scaphoid, and trapezium.	Lateral side of proximal phalanx of thumb.	Abducts thumb at carpometacarpal joint.	Median nerve.
Opponens pollicis (op-PŌ-nenz = opposes)	Flexor retinaculum and trapezium.	Lateral side of first metacarpal (thumb).	Moves thumb across palm to meet little finger (opposition) at the carpometacarpal joint.	Median nerve.
Flexor pollicis brevis (FLEK-sor = decreases angle at joint)	Flexor retinaculum, trapezium, capitate, and trapezoid.	Lateral side of proximal phalanx of thumb.	Flexes thumb at carpometacarpal and metacarpophalangeal joints.	Median and ulnar nerves.
Adductor pollicis (ad-DUK-tor = moves part toward midline)	Oblique head: capitate and second and third metacarpals; transverse head: third metacarpal.	Medial side of proximal phalanx of thumb by a tendon containing a sesamoid bone.	Adducts thumb at carpometacarpal and metacarpophalangeal joints.	Ulnar nerve.
Hypothenar (Medial Aspect of Palm)				
Abductor digiti minimi (DIJ-i-tē MIN-i-mē; *digit-* = finger or toe; *minimi* = little)	Pisiform and tendon of flexor carpi ulnaris.	Medial side of proximal phalanx of little finger.	Abducts and flexes little finger at metacarpophalangeal joint.	Ulnar nerve.
Flexor digiti minimi brevis	Flexor retinaculum and hamate.	Medial side of proximal phalanx of little finger.	Flexes little finger at carpometacarpal and metacarpophalangeal joints.	Ulnar nerve.
Opponens digiti minimi	Flexor retinaculum and hamate.	Medial side of fifth metacarpal (little finger).	Moves little finger across palm to meet thumb (opposition) at the carpometacarpal joint.	Ulnar nerve.
Intermediate (Midpalmar)				
Lumbricals (LUM-bri-kals; *lumbric-* = earthworm) (four muscles)	Lateral sides of tendons and flexor digitorum profundus of each finger.	Lateral sides of tendons of extensor digitorum on proximal phalanges of each finger.	Flex each finger at metacarpophalangeal joints and extend each finger at interphalangeal joints.	Median and ulnar nerves.
Palmar interossei (PAL-mer in´-ter-OS-ē-ī *palmar* = palm; *inter-* = between; *-ossei* = bones) (three muscles)	Sides of shafts of metacarpals of all fingers (except the middle one).	Sides of bases of proximal phalanges of all fingers (except the middle one).	Adduct each finger at metacarpophalangeal joints; flex each finger at metacarpophalangeal joints.	Ulnar nerve.
Dorsal interossei (DOR-sal = back surface) (four muscles)	Adjacent sides of metacarpals.	Proximal phalanx of each finger.	Abduct fingers 2–4 at metacarpophalangeal joints; flex fingers 2–4 at metacarpophalangeal joints; and extend each finger at interphalangeal joints.	Ulnar nerve.

11.8 Deep muscles of the back move the head and vertebral column.

The muscles that move the vertebral column (spine) (Figure 11.18; Table 11.18) are quite complex because they have multiple origins and insertions and there is considerable overlap among them. The **splenius** muscles are attached to the sides and back of the neck. The two muscles in this group are named on the basis of their insertions: **splenius capitis** (head region) and **splenius cervicis** (cervical region). They extend, laterally flex, and rotate the head at the neck.

Figure 11.18   Muscles that move the vertebral column.

SPINALIS CAPITIS

LONGISSIMUS CAPITIS

SPINALIS CERVICIS

LONGISSIMUS CERVICIS

ILIOCOSTALIS THORACIS

SPINALIS THORACIS

ILIOCOSTALIS LUMBORUM

SEMISPINALIS CAPITIS

Ligamentum nuchae

SPLENIUS CAPITIS

SPLENIUS CERVICIS

ILIOCOSTALIS CERVICIS

SEMISPINALIS CERVICIS

LONGISSIMUS THORACIS

SEMISPINALIS THORACIS

INTERTRANSVERSARIUS

ROTATORE

MULTIFIDUS

(a) Posterior view

The **erector spinae** is the largest muscle mass of the back, forming a prominent bulge on either side of the vertebral column (Figure 11.18a). It is the chief extensor of the vertebral column. It is also important in controlling flexion, lateral flexion, and rotation of the vertebral column and in maintaining the lumbar curve, because the main mass of the muscle is in the lumbar region. The erector spinae consists of three groups of muscles: **iliocostalis group** (il´-ē-ō-kos-TĀ-lis), **longissimus group** (lon-JI-si-mus), and **spinalis group** (spi-NĀ-lis). The iliocostalis group is the most lateral, the longissimus group is intermediate, and the spinalis group is the most medial.

The **transversospinales**, so named because their fibers run from the transverse processes to the spinous processes of the vertebrae, extend, flex, and rotate the vertebral column, and rotate the head. The **segmental** muscles unite the spinous and transverse processes of consecutive vertebrae, and function primarily in stabilizing the vertebral column during its movements (Figure 11.18b). Because the **scalene** muscles assist in moving the vertebral column, they are also included in Table 11.18. The scalenes flex and rotate the head, and assist in deep inhalation (Figure 11.18c). Note in Table 11.9 that the rectus abdominis, external oblique, internal oblique, and quadratus lumborum also play roles in moving the vertebral column.

(b) Posterolateral view

(c) Anterior view

🔑 The erector spinae group (iliocostalis, longissimus, and spinalis muscles) is the chief extensor of the vertebral column.

TABLE 11.18

Muscles That Move the Vertebral Column

MUSCLE	ORIGIN	INSERTION	ACTION	INNERVATION
Splenius (SPLĒ-nē-us)				
Splenius capitis (KAP-i-tis; *splenium* = bandage; *capit-* = head)	Ligamentum nuchae and spinous processes of seventh cervical vertebra and first three or four thoracic vertebrae.	Occipital bone and mastoid process of temporal bone.	Acting together (bilaterally), extend head; acting singly (unilaterally), laterally flex and rotate head to same side as contracting muscle.	Middle cervical spinal nerves.
Splenius cervicis (SER-vi-kis; *cervic-* = neck)	Spinous processes of third through sixth thoracic vertebrae.	Transverse processes of first two or four cervical vertebrae.	Acting together, extend head; acting singly, laterally flex and rotate head to same side as contracting muscle.	Inferior cervical spinal nerves.
Erector Spinae (e-REK-tor SPI-nē) Consists of iliocostalis muscles (lateral), longissimus muscles (intermediate), and spinalis muscles (medial).				
Iliocostalis Group (Lateral)				
Iliocostalis cervicis (il´-ē-ō-kos-TĀL-is; *ilio-* = flank; *-costa-* = rib)	Superior six ribs.	Transverse processes of fourth to sixth cervical vertebrae.	Acting together, muscles of each region (cervical, thoracic, and lumbar) extend and maintain erect posture of vertebral column of their respective regions; acting singly, laterally flex vertebral column of their respective regions.	Cervical and thoracic spinal nerves.
Iliocostalis thoracis (thō-RĀ-sis; = chest)	Inferior six ribs.	Superior six ribs.		Thoracic spinal nerves.
Iliocostalis lumborum (lum-BOR-um)	Iliac crest.	Inferior six ribs.		Lumbar spinal nerves.
Longissimus Group (Intermediate)				
Longissimus capitis (lon-JIS-i-mus = longest)	Transverse processes of superior four thoracic vertebrae and articular processes of inferior four cervical vertebrae.	Mastoid process of temporal bone.	Acting together, both longissimus capitis muscles extend head; acting singly, rotate head to same side as contracting muscle.	Middle and inferior cervical and spinal nerves.
Longissimus cervicis	Transverse processes of fourth and fifth thoracic vertebrae.	Transverse processes of second to sixth cervical vertebrae.	Acting together, longissimus cervicis and both longissimus thoracis muscles extend vertebral column of their respective regions; acting singly, laterally flex vertebral column of their respective regions.	Cervical and superior thoracic spinal nerves.
Longissimus thoracis	Transverse processes of lumbar vertebrae.	Transverse processes of all thoracic and superior lumbar vertebrae and ninth and tenth ribs.		Thoracic and lumbar spinal nerves.
Spinalis Group (Medial)				
Spinalis capitis (spi-NĀ-lis = vertebral column)	Arises with semispinalis capitis.	Occipital bone.	Acting together, muscles of each region (cervical, thoracic, and lumbar) extend vertebral column of their respective regions.	Cervical and superior thoracic spinal nerves.
Spinalis cervicis	Ligamentum nuchae and spinous process of seventh cervical vertebra.	Spinous process of axis.		Inferior cervical and thoracic spinal nerves.
Spinalis thoracis	Spinous processes of superior lumbar and inferior thoracic vertebrae.	Spinous processes of superior thoracic vertebrae.		Thoracic spinal nerves.

MUSCLE	ORIGIN	INSERTION	ACTION	INNERVATION
Transversospinales (trans-ver-sō-spi-NĀ-lēz)				
Semispinalis capitis (sem´-ē-spi-NĀ-lis; *semi-* = partially or one half)	Transverse processes of first six or seven thoracic vertebrae and seventh cervical vertebra, and articular processes of fourth, fifth, and sixth cervical vertebrae.	Occipital bone.	Acting together, extend head; acting singly, rotate head to side opposite contracting muscle.	Cervical and thoracic spinal nerves.
Semispinalis cervicis	Transverse processes of superior five or six thoracic vertebrae.	Spinous processes of first to fifth cervical vertebrae.	Acting together, both semispinalis cervicis and both semispinalis thoracis muscles extend vertebral column of their respective regions; acting singly, rotate head to side opposite contracting muscle.	Cervical and thoracic spinal nerves.
Semispinalis thoracis	Transverse processes of sixth to tenth thoracic vertebrae.	Spinous processes of superior four thoracic and last two cervical vertebrae.		Thoracic spinal nerves.
Multifidus (mul-TIF-i-dus; *multi-* = many; *-fid-* = segmented)	Sacrum, ilium, transverse processes of lumbar, thoracic, and inferior four cervical vertebrae.	Spinous process of a more superior vertebra.	Acting together, extend vertebral column; acting singly, laterally flex vertebral column and rotate head to side opposite contracting muscle.	Cervical, thoracic, and lumbar spinal nerves.
Rotatores (rō-ta-TŌ-rēz; singular is *rotatore* = to rotate)	Transverse processes of all vertebrae.	Spinous process of vertebra superior to the one of origin.	Acting together, extend vertebral column; acting singly, rotate vertebral column to side opposite contracting muscle.	Cervical, thoracic, and lumbar spinal nerves.
Segmental (seg-MEN-tal)				
Interspinales (in-ter-spī-NĀ-lēz; *inter-* = between)	Superior surface of all spinous processes.	Inferior surface of spinous process of vertebra superior to the one of origin.	Acting together, extend vertebral column; acting singly, stabilize vertebral column during movement.	Cervical, thoracic, and lumbar spinal nerves.
Intertransversarii (in´-ter-trans-vers-AR-ē-ī; singular is *intertransversarius*)	Transverse processes of all vertebrae.	Transverse process of vertebra superior to the one of origin.	Acting together, extend vertebral column; acting singly, laterally flex vertebral column and stabilize it during movements.	Cervical, thoracic, and lumbar spinal nerves.
Scalenes (SKĀ-lēnz)				
Anterior scalene (SKĀ-lēn; *anterior* = front; *scalene* = uneven)	Transverse processes of third through sixth cervical vertebrae.	First rib.	Acting together, right and left anterior scalene and middle scalene muscles flex head and elevate first ribs during deep inhalation; acting singly, laterally flex head and rotate head to side opposite contracting muscle.	Cervical spinal nerves C5–C6.
Middle scalene	Transverse processes of inferior six cervical vertebrae.	First rib.		Cervical spinal nerves C3–C8.
Posterior scalene	Transverse processes of fourth through sixth cervical vertebrae.	Second rib.	Acting together, flex head and elevate second ribs during deep inhalation; acting singly, laterally flex head and rotate head to side opposite contracting muscle.	Cervical spinal nerves C6-C8.

Checkpoint

27.   Which muscle groups constitute the erector spinae?

28.   Name the muscle groups that laterally flex the head and those that laterally flex the vertebral column.

Return to Joe's Story

Joe noticed that his sides and stomach were sore from his new workout routine. The bottom of his ribs hurt, too, which he thought was odd. As he lay there, he tried to breathe deeply, and eventually fell into a regular rhythm.

"Joe, why don't you turn back over onto your stomach." Starting at his wrist, Jasmine oiled his arm up to his shoulder, then applied light pressure as her hands glided down his arm. Jasmine could feel the ropelike tension in Joe's forearm; the muscles moved under his skin as she felt for each one, then worked them in a feathering motion toward his elbow. She noticed that Joe was holding his arm up, not completely relaxing it. "Joe, relax your arm, and let me take the weight." She glided her hands down and began working the palm of his hand with her thumbs. Jasmine then raised Joe's elbow up with one hand and bent his arm at the elbow so his hand lay on his back. Then she began working her fingers along the medial border of his scapula, between his vertebral column and shoulder blade.

"Wow, is that sore," he said. She worked the muscles above and below the spine of the scapula, moving her thumbs along the body of the muscles toward Joe's heart.

After massaging both of his arms, Jasmine moved back to his neck. She drew her hands up along the sides of Joe's spine from his lower back up to his neck and back down. Just below Joe's shoulder blades she worked from the vertebrae toward Joe's armpits with the palms of her hands, sweeping them up toward each side. She repeated this motion several times, her hands gliding and increasing pressure as she did so. Jasmine also alternated using the point of her elbow and the heel of her palm as her hands began to tire. Using her thumbs, she made small overlapping circles over the sacrum, gradually moving up the sides of the spine. She oiled his legs starting at the right ankle. As she kneaded the calf of Joe's leg, working toward the back of his knee, she carefully avoided putting pressure there. At the thigh, she again made her strokes in motions toward Joe's hip. Joe's calves were particularly sore because he had been jogging the day before.

E. Jasmine has moved Joe's arm to elevate the scapula and finds that his "rhomboids" are tight; what are the rhomboids, and what do they do?

F. Joe experienced great relief when Jasmine rubbed his hands; what are the intrinsic muscles of the hand, and how are they different from the extrinsic muscles of the hand?

G. Jasmine might be experiencing pain in her thenar eminence from performing the massage; which muscles make up the thenar eminence?

H. Which prominent muscle group is Jasmine massaging that extends Joe's vertebral column?

11.9 Muscles originating on the pelvic girdle or lower limb move the femur, leg, and foot.

Muscles That Move the Femur

While upper limb muscles are characterized by versatility of movement, lower limb muscles are larger and more powerful to provide stability, locomotion, and maintenance of posture. In addition, muscles of the lower limbs often cross two joints and act equally on both.

The majority of muscles that move the femur (thigh bone) (Figure 11.19; Table 11.19) originate on the pelvic girdle and insert on the femur. The **psoas major** and **iliacus** have a common insertion (lesser trochanter of the femur) and together are known as the **iliopsoas** (il´-ē-ō-SŌ-as). There are three gluteal muscles: gluteus maximus, gluteus medius, and gluteus minimus. The **gluteus maximus** is the largest and heaviest of the three muscles and is one of the largest muscles in the body. It is the chief extensor of the femur at the hip joint. The **gluteus medius** is mostly deep to the gluteus maximus and is a powerful abductor of the femur at the hip joint. It is a common site for an intramuscular injection. The **gluteus minimus** is the smallest of the gluteal muscles and lies deep to the gluteus medius. The **tensor fasciae latae** is located on the lateral surface of the thigh. Together, the tendons of the tensor fasciae latae and gluteus maximus form a structure called the *iliotibial tract* that inserts into the lateral condyle of the tibia. Three muscles on the medial thigh, the **adductor longus**, **adductor brevis**, and **adductor magnus**, adduct, flex, and medially rotate the femur at the hip joint.

The *femoral triangle* is a space between inguinal ligament, adductor longus, and sartorius. Within the femoral triangle are the femoral nerve, femoral artery, femoral vein, and inguinal lymph nodes.

Figure 11.19 **Muscles that move the femur.**

Twelfth rib

Quadratus lumborum — Psoas minor

— PSOAS MAJOR

Iliac crest

ILIACUS

Anterior superior iliac spine — Sacrum

Femoral triangle

TENSOR FASCIAE LATAE — Inguinal ligament

SARTORIUS

QUADRICEPS FEMORIS: — Pubic tubercle

RECTUS FEMORIS (cut) — PECTINEUS

VASTUS LATERALIS — ADDUCTOR LONGUS

VASTUS INTERMEDIUS

— GRACILIS

VASTUS MEDIALIS

RECTUS FEMORIS (cut)

Iliotibial tract

Section of fascia lata (cut)

Tendon of quadriceps femoris

— Patella

Patellar ligament

(a) Anterior superficial view (the femoral triangle is indicated by a dashed line)

(continues)

Figure 11.19 (continued)

TENSOR FASCIAE LATAE (cut)

SARTORIUS (cut)

RECTUS FEMORIS (cut)

Iliofemoral ligament of hip joint

Inguinal ligament

PECTINEUS (cut)

Pubis

OBTURATOR EXTERNUS

ADDUCTOR LONGUS (cut)

PECTINEUS (cut)

ADDUCTOR BREVIS

ADDUCTOR MAGNUS

ADDUCTOR LONGUS (cut)

GRACILIS

Femur

SARTORIUS (cut)

Patella

GLUTEUS MEDIUS

GLUTEUS MAXIMUS

TENSOR FASCIAE LATAE

Illiotibial tract

VASTUS LATERALIS

GRACILIS

ADDUCTOR MAGNUS

SEMITENDINOSUS

SEMIMEMBRANOSUS

BICEPS FEMORIS:

LONG HEAD

SHORT HEAD

SARTORIUS

Gastrocnemius

(b) Anterior deep view

(c) Posterior superficial view

Iliac crest

Sacrum

GLUTEUS MAXIMUS (cut)

OBTURATOR INTERNUS

Coccyx

Ischial tuberosity

Sciatic nerve

GRACILIS

SARTORIUS

GLUTEUS MEDIUS (cut)

GLUTEUS MINIMUS

PIRIFORMIS

SUPERIOR GEMELLUS

Greater trochanter

INFERIOR GEMELLUS

QUADRATUS FEMORIS

GLUTEUS MAXIMUS (cut)

Femur

ADDUCTOR MAGNUS

HAMSTRINGS:

SEMITENDINOSUS

BICEPS FEMORIS

SEMIMEMBRANOSUS

VASTUS LATERALIS

Tendon of biceps femoris

Popliteal fossa

Gastrocnemius

(d) Posterior superficial and deep view

(continues)

Figure 11.19 (continued)

Psoas minor
Iliac crest
ILIACUS
PSOAS MAJOR
PECTINEUS
ADDUCTOR BREVIS
Femur
ADDUCTOR LONGUS
GRACILIS
ADDUCTOR MAGNUS

(e) Anterior deep view

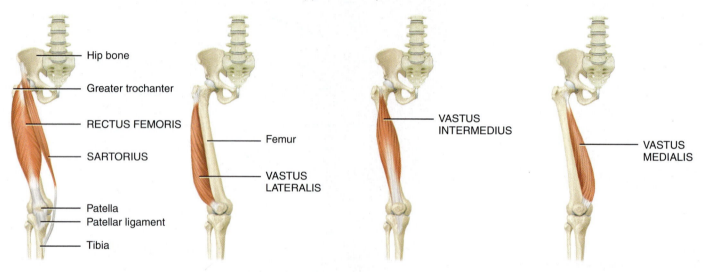

Hip bone
Greater trochanter
RECTUS FEMORIS
SARTORIUS
Patella
Patellar ligament
Tibia

Femur
VASTUS LATERALIS

VASTUS INTERMEDIUS

VASTUS MEDIALIS

(f) Anterior views

SEMITENDINOSUS
BICEPS FEMORIS, LONG HEAD
Tibia

Ischial tuberosity
Femur
SEMIMEMBRANOSUS
BICEPS FEMORIS, SHORT HEAD
Fibula

(g) Posterior deep views

Most muscles that move the femur originate on the pelvic girdle and insert on the femur.

TABLE 11.19

Muscles That Move the Femur

MUSCLE	ORIGIN	INSERTION	ACTION	INNERVATION
Iliopsoas				
Psoas major (SŌ-as; *psoa-* = a muscle of the loin)	Transverse processes and bodies of lumbar vertebrae.	With iliacus into lesser trochanter of femur.	Psoas major and iliacus muscles acting together flex thigh at hip joint, rotate thigh laterally, and flex trunk on hip as in sitting up from supine position.	Lumbar spinal nerves L2–L3.
Iliacus (il´-ē-A-cus; *iliac-* = ilium)	Iliac fossa and sacrum.	With psoas major into lesser trochanter of femur.		Femoral nerve.
Gluteus maximus (GLOO-tē-us MAK-si-mus; *glute-* = rump or buttock; *maximus* = largest)	Iliac crest, sacrum, coccyx, and aponeurosis of sacrospinalis.	Iliotibial tract of fascia lata and lateral part of linea aspera (gluteal tuberosity) under greater trochanter of femur.	Extends thigh at hip joint and laterally rotates thigh.	Inferior gluteal nerve.
Gluteus medius (MĒ-de-us = middle)	Ilium.	Greater trochanter of femur.	Abducts thigh at hip joint and medially rotates thigh.	Superior gluteal nerve.
Gluteus minimus (MIN-i-mus = smallest)	Ilium.	Greater trochanter of femur.	Abducts thigh at hip joint and medially rotates thigh.	Superior gluteal nerve.
Tensor fasciae latae (TENsor FA-shē-ē LĀ-tē; *tensor* = makes tense; *fasciae* = of the band; *lat-* = wide)	Iliac crest.	Tibia by way of the iliotibial tract.	Flexes and abducts thigh at hip joint.	Superior gluteal nerve.
Piriformis (pir-i-FOR-mis; *piri-* = pear; *-form-* = shape)	Anterior sacrum.	Superior border of greater trochanter of femur.	Laterally rotates and abducts thigh at hip joint.	Sacral spinal nerves S1 or S2, mainly S1.
Obturator internus (OB-too-rā´-tor in-TER-nus; *obturator* = obturator foramen; *intern-* = inside)	Inner surface of obturator foramen, pubis, and ischium.	Medial surface of greater trochanter of femur.	Laterally rotates and abducts thigh at hip joint.	Nerve to obturator internus.
Obturator externus (ex-TER-nus = outside)	Outer surface of obturator membrane.	Deep depression inferior to greater trochanter (trochanteric fossa) of femur.	Laterally rotates and abducts thigh at hip joint.	Obturator nerve.
Superior gemellus (jem-EL-lus; *superior* = above; *gemell-* = twins)	Ischial spine.	Medial surface of greater trochanter of femur.	Laterally rotates and abducts thigh at hip joint.	Nerve to obturator internus.
Inferior gemellus (*inferior* = below)	Ischial tuberosity.	Medial surface of greater trochanter of femur.	Laterally rotates and abducts thigh at hip joint.	Nerve to quadratus femoris.
Quadratus femoris (kwod-RĀ-tus FEM-or-is; *quad-* = square, four-sided; *femoris* = femur)	Ischial tuberosity.	Elevation superior to mid-portion of intertrochanteric crest (quadrate tubercle) on posterior femur.	Laterally rotates and stabilizes hip joint.	Nerve to quadratus femoris.
Adductor longus (LONG-us; *adductor* = moves part closer to midline; *longus* = long)	Pubic crest and pubic symphysis.	Linea aspera of femur.	Adducts and flexes thigh at hip joint and laterally rotates thigh.	Obturator nerve.
Adductor brevis (BREV-is; = short)	Inferior ramus of pubis.	Superior half of linea aspera of femur.	Adducts and flexes thigh at hip joint and medially rotates thigh.	Obturator nerve.
Adductor magnus (MAG-nus = large)	Inferior ramus of pubis and ischium to ischial tuberosity.	Linea aspera of femur.	Adducts thigh at hip joint and laterally rotates thigh; anterior part flexes thigh at hip joint, and posterior part extends thigh at hip joint.	Obturator and sciatic nerves.
Pectineus (pek-TIN-ē-us; = a comb)	Superior ramus of pubis.	Pectineal line of femur, between lesser trochanter and linea aspera.	Flexes and adducts thigh at hip joint.	Femoral nerve.

Muscles That Move the Femur, Tibia, and Fibula

Medial muscles of the thigh adduct the femur at the hip joint (Figure 11.19; Table 11.20). The medial thigh muscles, the adductor magnus, adductor longus, adductor brevis, and pectineus, are included in Table 11.19 because they act on the femur. The **gracilis**, a long, straplike muscle on the medial aspect of the thigh and knee, not only adducts the thigh, but also flexes the leg at the knee joint. For this reason, it is listed in Table 11.20.

The muscles of the anterior aspect of the thigh extend the leg at the knee joint, and some also flex the thigh at the hip joint. The **quadriceps femoris** is the largest muscle in the body, covering most of the anterior surface and sides of the thigh. It has four distinct parts, usually described as four separate muscles: **rectus femoris** on the anterior aspect of the thigh, **vastus lateralis** on the lateral aspect of the thigh, **vastus medialis** on the medial aspect of the thigh, and **vastus intermedius** located deep to the rectus femoris between the vastus lateralis and vastus medialis. The common tendon for the four muscles is the *quadriceps tendon,* which inserts on the patella. The tendon continues below the patella as the *patellar ligament* and attaches to the tibial tuberosity. The **sartorius** is a long, narrow muscle that forms a band across the thigh from the ilium of the hip bone to the medial side of the tibia. The various movements it produces help effect the cross-legged sitting position in which the heel of one limb is placed on the knee of the opposite limb. It is known as the tailor's muscle because tailors often assume this cross-legged sitting position.

The muscles of the posterior thigh flex the leg and extend the thigh. The three posterior thigh muscles are collectively called the **hamstrings**: the **biceps femoris**, **semitendinosus**, and **semimembranosus**. The hamstrings are so named because butchers can hang hams for smoking by the tendons of these muscles, which are long and stringlike in the popliteal area. Because the hamstrings span two joints (hip and knee), they are able to both extend the thigh and flex the leg. The *popliteal fossa* is a diamond-shaped space on the posterior aspect of the knee bordered laterally by the tendons of the biceps femoris and medially by the tendons of the semitendinosus and semimembranosus.

TABLE 11.20

Muscles That Move the Femur, Tibia, and Fibula

MUSCLE	ORIGIN	INSERTION	ACTION	INNERVATION
Medial (Adductor) Compartment of the Thigh				
Adductor magnus (MAG-nus)	See **Table 11.19**			
Adductor longus (LONG-us)				
Adductor brevis (BREV-is)				
Pectineus (pek-TIN-ē-us)				
Gracilis (gra-SIL-is = slender)	Body and inferior ramus of pubis.	Medial surface of body of tibia.	Adducts thigh at hip joint, medially rotates thigh, and flexes leg at knee joint.	Obturator nerve.

TABLE 11.20 (continued)

Muscles That Move the Femur, Tibia, and Fibula

MUSCLE	ORIGIN	INSERTION	ACTION	INNERVATION
Anterior (Extensor) Compartment of the Thigh				
Quadriceps femoris (KWOD-ri-ceps FEM-or-is; *quadriceps* = four heads of origin; *femoris* = femur)				
Rectus femoris (REK-tus = fascicles parallel to midline)	Anterior inferior iliac spine.	Patella via quadriceps tendon and then tibial tuberosity via patellar ligament.	All four heads extend leg at knee joint; rectus femoris muscle acting alone also flexes thigh at hip joint.	Femoral nerve.
Vastus lateralis (VAS-tus lat´-e-RĀ-lis; *vast* = huge; *lateralis* = lateral)	Greater trochanter and linea aspera of femur.			
Vastus medialis (*medialis* = medial)	Linea aspera of femur.			
Vastus intermedius (*intermedius* = middle)	Anterior and lateral surfaces of body of femur.			
Sartorius (sar-TOR-ē-us; *sartor* = tailor; longest muscle in body)	Anterior superior iliac spine.	Medial surface of body of tibia.	Flexes leg at knee joint; flexes, abducts, and laterally rotates thigh at hip joint, thus crossing leg.	Femoral nerve.
Posterior (Flexor) Compartment of the Thigh				
Hamstrings A collective designation for three separate muscles.				
Biceps femoris (BĪ-ceps = two heads of origin)	Long head: ischial tuberosity. Short head: linea aspera of femur.	Head of fibula and lateral condyle of tibia.	Flexes leg at knee joint and extends thigh at hip joint.	Tibial and common peroneal nerves from sciatic nerve.
Semitendinosus (sem´-ē-ten-di-NŌ-sus; *semi-* = half; *-tendo-* = tendon)	Ischial tuberosity.	Proximal part of medial surface of shaft of tibia.	Flexes leg at knee joint and extends thigh at hip joint.	Tibial nerve from sciatic nerve.
Semimembranosus (sem´-ē-mem-bra-NŌ-sus; *membran-* = membrane)	Ischial tuberosity.	Medial condyle of tibia.	Flexes leg at knee joint and extends thigh at hip joint.	Tibial nerve from sciatic nerve.

Muscles That Move the Foot and Toes

Muscles that move the foot and toes (Figure 11.20; Table 11.21) are located in the leg. Muscles of the anterior aspect of the leg dorsiflex the foot at the ankle joint. Besides dorsiflexion, the **tibialis anterior**, easily palpated against the lateral surface of the tibia, inverts the foot, while the **fibularis tertius** everts the foot. The **extensor hallucis longus** and **extensor digitorum longus** dorsiflex the ankle and extend the toes. Analogous to the wrist, the tendons of anterior leg muscles are held firmly to the ankle by thickenings of deep fascia called the *superior extensor retinaculum* and *inferior extensor retinaculum*.

The lateral leg contains two muscles that plantar flex and evert the foot at the ankle joint: the **fibularis longus** and **fibularis brevis**.

The superficial muscles of the posterior aspect of the leg are the gastrocnemius, soleus, and **plantaris**—the so-called calf muscles. The superficial muscles and most of the deep posterior leg muscles plantar flex the foot at the ankle joint. The large size of these muscles is directly related to the characteristic upright stance of humans.

Figure 11.20 Muscles that move the foot and toes.

Vastus lateralis

Vastus medialis

Biceps femoris

Tendon of quadriceps femoris

Iliotibial tract

Patella

Tendon of sartorius

Fibula

Patellar ligament

Tibia

TIBIALIS ANTERIOR

GASTROCNEMIUS

FIBULARIS LONGUS

SOLEUS

EXTENSOR DIGITORUM LONGUS

FLEXOR DIGITORUM LONGUS

Fibularis brevis

Calcaneal tendon

Fibula

Patellar ligament

(a) Anterior superficial view

(b) Right lateral superficial view

The **gastrocnemius** is the most superficial muscle and forms the prominence of the calf. The broad, flat **soleus**, which lies deep to the gastrocnemius, derives its name from its resemblance to a flat fish (sole). The superficial posterior leg muscles share a common tendon of insertion, the *calcaneal (Achilles) tendon*, the strongest tendon of the body. It inserts into the calcaneus, the heel bone of the ankle.

The deep muscles of the posterior leg are the popliteus, tibialis posterior, flexor digitorum longus, and flexor hallucis longus. The **popliteus** forms the floor of the popliteal fossa. The **tibialis posterior** is the deepest muscle in the posterior compartment. It lies between the **flexor digitorum longus** and **flexor hallucis longus**, muscles that flex the toes.

Sartorius

Semitendinosus

Semimembranosus

Tibia

Biceps femoris

Femur

PLANTARIS

GASTROCNEMIUS (cut)

Tibia

PLANTARIS

POPLITEUS

GASTROCNEMIUS

SOLEUS (cut)

Fibula

SOLEUS

TIBIALIS POSTERIOR

FIBULARIS LONGUS

FLEXOR DIGITORUM LONGUS

FLEXOR HALLUCIS LONGUS

Fibularis brevis

Calcaneal tendon (cut)

Fibula

(c) Posterior superficial view

(d) Posterior deep view

(continues)

Figure 11.20 (continued)

Anterior views

Right lateral view

Posterior deep views

(e) Anterior, right lateral, and posterior views

🔑 The superficial muscles of the posterior leg share a common tendon of insertion, the calcaneal (Achilles) tendon, that inserts into the calcaneus of the ankle.

TABLE 11.21

Muscles That Move the Foot and Toes

MUSCLE	ORIGIN	INSERTION	ACTION	INNERVATION
Anterior Compartment of the Leg				
Tibialis anterior (tib´-ē-Ā-lis = tibia; *anterior* = front)	Lateral condyle and body of tibia and interosseous membrane (sheet of fibrous tissue that holds shafts of tibia and fibula together).	First metatarsal and first (medial) cuneiform.	Dorsiflexes foot at ankle joint and inverts foot at intertarsal joints.	Deep fibular (peroneal) nerve.
Extensor hallucis longus (eks-TEN-sor HAL-ū-sis LON-gus; *extensor* = increases angle at joint; *halluc-* = hallux, or great toe; *longus* = long)	Anterior surface of fibula and interosseous membrane.	Distal phalanx of great toe.	Dorsiflexes foot at ankle joint and extends proximal phalanx of great toe at metatarsophalangeal joint.	Deep fibular (peroneal) nerve.
Extensor digitorum longus (di´-ji-TOR-um)	Lateral condyle of tibia, anterior surface of fibula, and interosseous membrane.	Middle and distal phalanges of toes 2–5.*	Dorsiflexes foot at ankle joint and extends distal and middle phalanges of each toe at interphalangeal joints and proximal phalanx of each toe at metatarsophalangeal joint.	Deep fibular (peroneal) nerve.
Fibularis (peroneus) tertius (fib-ū-LĀ-ris TER-shus; *peron-* = fibula; *tertius* = third)	Distal third of fibula and interosseous membrane.	Base of fifth metatarsal.	Dorsiflexes foot at ankle joint and everts foot at intertarsal joints.	Deep fibular (peroneal) nerve.
Lateral (Fibular) Compartment of the Leg				
Fibularis (peroneus) longus	Head and body of fibula and lateral condyle of tibia.	First metatarsal and first cuneiform.	Plantar flexes foot at ankle joint and everts foot at intertarsal joints.	Superficial fibular (peroneal) nerve.
Fibularis (peroneus) brevis (BREV-is = short)	Body of fibula.	Base of fifth metatarsal.	Plantar flexes foot at ankle joint and everts foot at intertarsal joints.	Superficial fibular (peroneal) nerve.
Superficial Posterior Compartment of the Leg				
Gastrocnemius (gas´-trok-NĒ-mē-us; *gastro -* = belly; *-cnem-* = leg)	Lateral and medial condyles of femur and capsule of knee.	Calcaneus by way of calcaneal (Achilles) tendon.	Plantar flexes foot at ankle joint and flexes leg at knee joint.	Tibial nerve.
Soleus (SŌ-lē-us; *sole* = a type of flat fish)	Head of fibula and medial border of tibia.	Calcaneus by way of calcaneal (Achilles) tendon.	Plantar flexes foot at ankle joint.	Tibial nerve.
Plantaris (plan-TĀR-is = the sole)	Femur superior to lateral condyle.	Calcaneus by way of calcaneal (Achilles) tendon.	Plantar flexes foot at ankle joint and flexes leg at knee joint.	Tibial nerve.
Deep Posterior Compartment of the Leg				
Popliteus (pop-LIT-ē-us; *poplit-* = back of knee)	Lateral condyle of femur.	Proximal tibia.	Flexes leg at knee joint and medially rotates tibia to unlock the extended knee.	Tibial nerve.
Tibialis posterior (*posterior* = back)	Tibia, fibula, and interosseous membrane.	Second, third, and fourth metatarsals; navicular; all three cuneiforms; and cuboid.	Plantar flexes foot at ankle joint and inverts foot at intertarsal joints.	Tibial nerve.
Flexor digitorum longus (FLEK-sor = decreases angle at joint)	Posterior surface of tibia.	Distal phalanges of toes 2–5.	Plantar flexes foot at ankle joint; flexes distal and middle phalanges of each toe at interphalangeal joints and proximal phalanx of each toe at metatarsophalangeal joint.	Tibial nerve.
Flexor hallucis longus	Inferior two-thirds of fibula.	Distal phalanx of great toe.	Plantar flexes foot at ankle joint; flexes distal phalanx of great toe at interphalangeal joint and proximal phalanx of great toe at metatarsophalangeal joint.	Tibial nerve.

*Reminder: The great toe, or hallux, is the first toe and has two phalanges: proximal and distal. The remaining toes are numbered II–V (2–5), and each has three phalanges: proximal, middle, and distal.

Intrinsic Muscles of the Foot

The muscles in Table 11.22 are termed **intrinsic foot muscles** because they originate and insert *within* the foot (Figure 11.21). The muscles of the hand are specialized for precise and intricate movements, but those of the foot are limited to support and locomotion.

The intrinsic muscles of the foot primarily originate on the tarsal and metatarsal bones. They move the toes and contribute to the longitudinal arch of the foot. The fascia of the foot forms the *plantar aponeurosis* that extends from the calcaneus to the phalanges of the toes. The plantar aponeurosis supports the longitudinal arch of the foot and encloses the flexor tendons of the foot.

Figure 11.21   Intrinsic muscles of the foot.

Tendon of flexor hallucis longus

Tendons of flexor digitorum brevis (cut)

ADDUCTOR HALLUCIS

LUMBRICALS

FLEXOR HALLUCIS BREVIS

PLANTAR INTEROSSEI

FLEXOR DIGITI MINIMI BREVIS

FLEXOR DIGITORUM BREVIS

ABDUCTOR HALLUCIS

ABDUCTOR DIGITI MINIMI

Calcaneus

Tendon of flexor hallucis longus

Tendons of flexor digitorum longus

FLEXOR HALLUCIS BREVIS

Navicular

QUADRATUS PLANTAE

Tendon of tibialis posterior

Tendon of flexor hallucis longus

(a) Plantar superficial view

(b) Plantar intermediate view

The muscles of the hand are specialized for precise and intricate movements; those of the foot are limited to support and movement.

TABLE 11.22

Intrinsic Muscles of the Foot

MUSCLE	ORIGIN	INSERTION	ACTION	INNERVATION
Dorsal				
Extensor digitorum brevis (eks-TEN-sor di-ji-TOR-um BREV-is; *extensor* = increases angle at joint; *digit* = finger or toe; *brevis* = short) (see Figure 11.20a,b)	Calcaneus and inferior extensor retinaculum.	Tendons of extensor digitorum longus on toes 2–4 and proximal phalanx of great toe.*	Extensor hallucis brevis extends great toe at metatarsophalangeal joint and extensor digitorum brevis extends toes 2–4 at interphalangeal joints.	Deep fibular (peroneal) nerve.
Plantar				
First Layer (most superficial)				
Abductor hallucis (ab-DUK-tor HAL-ū-sis; *abductor* = moves part away from midline; *hallucis* = hallux, or great toe)	Calcaneus, plantar aponeurosis, and flexor retinaculum.	Medial side of proximal phalanx of great toe with the tendon of the flexor hallucis brevis.	Abducts and flexes great toe at metatarsophalangeal joint.	Medial plantar nerve.
Flexor digitorum brevis (FLEK-sor = decreases angle at joint)	Calcaneus and plantar aponeurosis.	Sides of middle phalanx of toes 2–5.	Flexes toes 2–5 at proximal interphalangeal and metatarsophalangeal joints.	Medial plantar nerve.
Abductor digiti minimi (DIJ-i-tē MIN-i-mē; *minimi* = little)	Calcaneus and plantar aponeurosis.	Lateral side of proximal phalanx of little toe with the tendon of the flexor digiti minimi brevis.	Abducts and flexes little toe at metatarsophalangeal joint.	Lateral plantar nerve.
Second Layer				
Quadratus plantae (kwod-RĀ-tus plan-TĒ; *quad-* = square, four-sided; *planta* = the sole)	Calcaneus.	Tendon of flexor digitorum longus.	Assists flexor digitorum longus to flex toes 2–5 at interphalangeal and metatarsophalangeal joints.	Lateral plantar nerve.
Lumbricals (LUM-bri-kals; *lumbric-* = earthworm)	Tendons of flexor digitorum longus.	Tendons of extensor digitorum longus on proximal phalanges of toes 2–5.	Extend toes 2–5 at interphalangeal joints and flex toes 2–5 at metatarsophalangeal joints.	Medial and lateral plantar nerves.
Third Layer				
Flexor hallucis brevis	Cuboid and third (lateral) cuneiform.	Medial and lateral sides of proximal phalanx of great toe via a tendon containing a sesamoid bone.	Flexes great toe at metatarsophalangeal joint.	Medial plantar nerve.
Adductor hallucis (ad-DUK-tor = moves part toward midline)	Metatarsals 2–4, ligaments of 3–5 metatarsophalangeal joints, and tendon of peroneus longus.	Lateral side of proximal phalanx of great toe.	Adducts and flexes great toe at metatarsophalangeal joint.	Lateral plantar nerve.
Flexor digiti minimi brevis	Metatarsal 5 and tendon of peroneus longus.	Lateral side of proximal phalanx of little toe.	Flexes little toe at metatarsophalangeal joint.	Lateral plantar nerve.
Fourth Layer (deepest)				
Dorsal interossei (DOR-sal in-ter-OS-ē-ī; *dorsal* = back surface; *inter-* = between; *-ossei* = bones) (not illustrated)	Adjacent side of metatarsals.	Proximal phalanges: both sides of toe 2 and lateral side of toes 3 and 4.	Abduct and flex toes 2–4 at metatarsophalangeal joints and extend toes at interphalangeal joints.	Lateral plantar nerve.
Plantar interossei (PLAN-tar = relating to sole of foot)	Metatarsals 3–5.	Medial side of proximal phalanges of toes 3–5.	Adduct and flex proximal metatarsophalangeal joints and extend toes at interphalangeal joints.	Lateral plantar nerve.

*The tendon that inserts into the proximal phalanx of the great toe, together with its belly, is often described as a separate muscle, the extensor hallucis brevis.

Checkpoint

29. What are the principal differences between the muscles of the upper and lower limbs?

30. Which muscles extend the leg at the knee joint?

31. Name the muscles that are part of the iliopsoas, quadriceps femoris, and hamstrings. Where is each muscle group found?

32. Which muscle tendons form the medial and lateral borders of the popliteal fossa?

33. Name the muscles that dorsiflex the foot at the ankle joint and those that plantar flex it.

34. Which muscles flex the toes? Which extend the toes?

35. How do the intrinsic muscles of the hand and foot differ in function?

Joe's Story: Epilogue and Discussion

►►►The hour was over before Joe even realized it. "Joe, I'm done now. You relax here as long as you like. When you're ready, I'll meet you up front." As she spoke, Jasmine finished wiping the last of the oil from Joe's back with a clean white towel. Joe lay on the table for a moment, feeling the beat of his heart. His skin felt warm, and he was as relaxed as he could ever remember being. When he finally arose, he felt a bit light-headed. It seemed to take him a bit longer than usual to change into his clothes. But by the time he had finished and was getting ready to leave, he had decided that this was definitely worth doing again.

Massage is one of the oldest and simplest forms of therapy. Using a system of stroking, pressing, and kneading, different areas of the body are manipulated to relieve pain, relax muscles, and stimulate the body. Massage does more than create a pleasant sensation; it manipulates muscles, connective tissues, tendons, and ligaments. Although massage largely affects those muscles just under the skin, its benefits may also reach the deeper layers of muscle, depending on the type of massage therapy employed. Massage also stimulates blood circulation and assists the lymphatic system by increasing the return of venous blood and lymphatic fluids to the heart. Massage therapy is most effective when performed by someone with a good knowledge of muscle anatomy. Understanding the location, shape, size, and action of muscles provides the therapist with a basis for manipulating the muscle using the most effective techniques. ◄◄◄

I.   What is Joe's prime mover for flexing the neck, abducting the arm at the shoulder joint, extending the forearm at the elbow joint, extending the vertebral column, and extending the femur at the hip joint?

J.   As Jasmine massaged Joe's muscles, she also stretched the tendons linking each muscle to its origin and insertion. Explain the terms origin and insertion in your own words.

K.   State the function of the gluteus medius and minimus muscles of Joe's thigh. Would you expect these muscles to be sore from jogging?

L.   Name the calf muscles that are causing Joe to experience pain.

M.   Why would Jasmine massage the muscle along the muscle fibers rather than across them?

Concept and Resource Summary

Concept	Resources

Introduction

1. The **muscular system** is made up of the voluntarily controlled skeletal muscles of the body.

2. Most skeletal muscles produce movement of body parts; a few skeletal muscles stabilize bones during their movement.

Concept 11.1 Skeletal muscles produce movement when the insertion is pulled toward the origin.

1. The contraction of skeletal muscles pulls on bones to cause movement. A muscle's **insertion** is its attachment on the movable bone that is typically moved toward its **origin**, which is its attachment on the immovable bone.

2. A **lever** is a rigid structure that moves around a fixed point called the **fulcrum**. **Effort** applied to the lever moves the **load**, or resistance. Bones act as levers, joints are the fulcrums, and skeletal muscles supply the effort.

See
• Anatomy Overview – Skeletal Muscle

Hear
• Figure 11.1 – Relationship of Skeletal Muscles to Bones

3. When a smaller effort can move a heavier load, the lever operates at a *mechanical advantage*. A lever operates at a *mechanical disadvantage* when a larger effort moves a lighter load.

4. The arrangements of **fascicles** include parallel, fusiform, circular, triangular, and pennate.

5. The muscle that accomplishes a desired action is termed the **prime mover** (**agonist**), and the muscle that accomplishes the opposing action is the **antagonist**. **Synergists** assist the prime mover and stabilize intermediate joints to prevent unwanted movements. **Fixators** stabilize the origin of the prime mover to increase efficiency.

Concept 11.2 Skeletal muscles are named based on size, shape, action, location, or attachments.

1. Skeletal muscle names are derived from specific features such as size, action, and attachments.

Hear

- **Figure 11.2 –**
  Principal Superficial
  Skeletal Muscles

Concept 11.3 Muscles of the head produce facial expressions, eyeball movement, and assist in biting, chewing, swallowing, and speech.

1. Muscles of facial expression move the skin or muscles rather than bone when they contract. Around the eyes, nose, and mouth are *sphincter* muscles that encircle and close the orifices; *dilator* muscles open the orifices.

2. The **extrinsic eye muscles** are among the fastest contracting and most precisely controlled skeletal muscles in the body. They permit you to elevate, depress, abduct, adduct, and medially and laterally rotate the eyeballs.

3. Muscles for mastication (chewing) move the mandible. The **masseter**, **temporalis**, and **medial pterygoid** elevate the mandible for powerful closing of the jaw. The medial and **lateral pterygoids** move the mandible from side to side and protract it.

4. Tongue movements are important for mastication, swallowing, and speech. *Extrinsic tongue muscles* originate outside the tongue and insert into it to move it in various directions. *Intrinsic tongue muscles* originate and insert within the tongue to change its shape.

See

- Anatomy Overview – Muscles of Facial Expression
- Anatomy Overview – Muscles Moving Eyeballs
- Anatomy Overview – Muscles for Speech, Swallowing, and Chewing

---

Concept 11.4 Muscles of the neck assist in swallowing and speech, and allow balance and movement of the head.

1. Two groups of muscles stabilize and move the hyoid bone, enabling it to serve as a firm base for attachment and action of the tongue. The **suprahyoid muscles** elevate the hyoid bone, oral cavity floor, and tongue during swallowing; the **infrahyoid muscles** depress the hyoid bone, and some move the larynx during swallowing and speech.

2. Several neck muscles balance and move the head on the vertebral column. Bilateral action of the sternocleidomastoids result in flexion of the head; unilateral action results in head rotation.

See

- Anatomy Overview – Muscles that Move the Head
- Anatomy Overview – Muscles for Speech, Swallowing, and Chewing

---

Concept 11.5 Muscles of the abdomen protect the abdominal viscera, move the vertebral column, and assist breathing.

1. Muscles that act on the anterolateral abdominal wall help contain and protect the abdominal viscera, move the vertebral column, compress the abdomen, and produce the force required for defecation, urination, vomiting, and childbirth.

2. The **diaphragm**, a large, dome-shaped muscle that separates the thoracic and abdominal cavities, is the most important muscle for breathing. Contraction results in inhalation; relaxation results in exhalation.

3. The **external intercostals** elevate the ribs during inhalation. The **internal intercostals** help decrease the thoracic cavity volume during forced exhalation.

See

- Anatomy Overview – Muscles of the Torso
- Anatomy Overview – Muscles for Breathing

Hear

- **Figure 11.10 –**
  Muscles Used in Breathing

Concept 11.6 Muscles of the pelvic floor and perineum support the pelvic viscera, function as sphincters, and assist in urination, erection, ejaculation, and defecation.

1. The **diaphragm** supports pelvic viscera, resists increases in intra-abdominal pressure, and acts as a sphincter for defecation, urination, and vaginal control.
2. The **perineum**, a diamond-shaped region inferior to the pelvic diaphragm, is important during childbirth. Muscles of the perineum also assist in erection of the penis and clitoris, ejaculation and urination, and defecation.

Concept 11.7 Muscles inserting on the upper limb move and stabilize the pectoral girdle, and move the arm, forearm, and hand.

1. Muscles that move the pectoral girdle stabilize the scapula and facilitate its function as the origin for most of the muscles that move the humerus.
2. Seven of the nine muscles that cross the shoulder joint originate on the scapula; two originate on the axial skeleton. Tendons of several shoulder muscles form the *rotator cuff,* which encircles the shoulder joint to give it strength and stability.
3. The flexors of the elbow joint are the **biceps brachii**, **brachialis**, and **brachioradialis**. The extensors of the elbow joint are the **triceps brachii** and **anconeus**. The **pronator teres** and **pronator quadratus** pronate the forearm; the **supinator** allows supination of the forearm.
4. The numerous muscles that move the wrist, hand, and fingers are located on the forearm. The anterior forearm muscles act as flexors, whereas the posterior forearm muscles act as extensors. The fascia at the wrist is thickened into fibrous bands called the *flexor retinaculum* and *extensor retinaculum*, which secure the tendons of certain forearm muscles.
5. **Intrinsic muscles of the hand** originate within the hand and provide us with the ability to grasp and manipulate objects precisely.

See

- **Anatomy Overview – Muscles That Move the Pectoral Girdle**
- **Anatomy Overview – Muscles That Move the Arm**
- **Anatomy Overview – Muscles That Move the Forearm**
- **Anatomy Overview – Muscles That Move the Hand**

Hear

- **Figure 11.16 – Muscles That Move the Wrist, Hand, and Fingers**

Do

- **Exercise – Shoulder Movement Target Practice**

Concept 11.8 Deep muscles of the back move the head and vertebral column.

1. The **splenius** muscles attach to the sides and back of the neck for head extension, flexion, and rotation. The largest muscle mass of the back, the **erector spinae** group, consists of the **ilocostalis group**, **longissimus group**, and **spinalis group**. The erector spinae muscles are prime movers of the vertebral column, allowing back extension. The **transversospinales** and **segmental** muscles also function in vertebral column movements.

See

- **Anatomy Overview – Muscles of the Torso**

Concept 11.9 Muscles originating on the pelvic girdle or lower limb move the femur, leg, and foot.

1. Most muscles that move the thigh at the hip originate on the pelvic girdle and insert on the femur. The **iliopsoas** flexes the thigh. The **adductor longus**, **adductor brevis**, and **adductor magnus** adduct, medially rotate, and flex the thigh. The **gluteus medius** and **gluteus minimus** abduct the thigh at the hip. The **tensor fascia latae** flexes and abducts the thigh.
2. The **gracilis** is a long, straplike muscle that adducts the thigh and flexes the leg at the knee. The **quadriceps femoris** muscle group is a powerful extensor of the leg at the knee. The **hamstrings** muscle group on the posterior thigh flexes the leg at the knee and extends the thigh at the hip.
3. Muscles that move the foot at the ankle and toes are located in the leg. The **tibialis anterior**, **fibularis tertius**, **extensor hallucis longus**, and **extensor digitorum longus** dorsiflex the foot. The **tibialis posterior**, **fibularis longus**, and **fibularis brevis** plantar flex and evert the foot. The **gastrocnemius**, **soleus**, and **plantaris** also plantar flex the foot. The **flexor digitorum longus** and **flexor hallucis longus** flex the toes.
4. **Intrinsic foot muscles** originate and insert within the foot to move the toes and contribute to the longitudinal arch of the foot.

See

- **Anatomy Overview – Muscles That Move the Thigh**
- **Anatomy Overview – Muscles of the Foot**

Hear

- **Figure 11.19 – Muscles That Move the Femur**

Do

- **Exercise – Lower Limb Movement Target Practice**
- **Exercise – Manage Those Muscles**

Other Chapter Resources

For additional information on clinical applications, medical disorders, and medical terminology, see Chapter 6 in the *Clinical Connections* manual packaged with your text.

Within *WileyPLUS*

- Chapter Quiz
- Audio Glossary
- Visual Anatomy
- Mastering Vocabulary
- Web Links

• Understanding the Concepts

1. A soccer player kicks a soccer ball down the field, flexing his thigh forward at the hip joint while keeping his knee joint locked in extension. What is the lever? The fulcrum? The effort? The load?

2. Two girls sit down on a playground seesaw, or teeter-totter. Both children weigh the same amount. Maria sits at the end of her side of the seesaw while Tonya sits halfway between the end and the fulcrum. Which girl will exert more force and be lowered to the ground?

3. What would happen if an injury resulted in loss of nerve stimulation (innervation) to your masseter and temporalis?

4. When your physician says "Open your mouth, stick out your tongue, and say *ahh*" so she can examine the inside of your mouth for possible signs of infection, which muscles do you contract?

5. Which muscles do you contract when you "suck in your tummy," thereby compressing the anterior abdominal wall? Which do you contract when you rotate the vertebral column?

6. Which muscles do you contract to increase the dimension of your thoracic cavity during quiet, normal inhalation? When you exhale?

7. Name the muscles you use to raise your shoulders and those you use to lower your shoulders.

8. Extend your hand at the wrist. Which muscles are contracting? Which muscles must relax so that you can extend your hand?

9. Which muscles do you exercise when performing sit-ups? Why might your ribs become sore from doing stomach exercises?

10. Which busy muscles flex, adduct, and medially rotate the femur at the hip joint as you rush to class? Which one flexes, abducts, and laterally rotates the thigh at the hip joint and flexes the leg at the knee joint?

12

Introduction to the Nervous System

Mary fumbled through her purse looking for her keys. "Not again, I'm late already," she said to herself. In the back seat of her car, a little face peered at her from the car's safety seat. "There they are," Mary said, reaching for the keys. Thrusting one of them in the ignition, she turned it and felt a sharp pain shoot through the palm of her hand. "Ow!" She shook her hand as the car rumbled to life. Mary had decided to continue working at home after her daughter's birth and had thought the time away from her office might help her relax and enjoy her new baby. But nothing seemed to have gone that way. Mary wanted to keep working; she needed her job, but it had meant more typing and computer work, and the baby had been a handful these first few months. Mary had first noticed the numbness and tingling sensations in her hands when she had been working on her laptop for several hours straight, with her hands contorted to fit the small keyboard. The sensations kept getting worse, and she thought her hands seemed weaker, too. Now, several months later, she was finding it hard to grasp the baby's bottle during feedings, and simply holding a pen was difficult. The problem was getting worse and she knew it was time to see someone.

Mary is experiencing signs and symptoms that indicate an imbalance in the nervous system. Although Mary's muscle weakness could be due to problems with the muscles, you know now that the nervous system makes these muscles contract. The nervous system also senses our environment, processes that information, and allows us to respond to the sensory information. To understand Mary's case, knowledge of the basic anatomy and physiology of the nervous system is required. She is sensing numbness and tingling; What is causing this sensation? She is experiencing weakness; Why? In this chapter we will study the anatomy and physiology of the central nervous system to help us understand Mary's problem.

INTRODUCTION

The nervous system and endocrine system share the responsibilities of maintaining the internal homeostasis of the human body. Both of these systems must sense changes in normal physiological set points (such as temperature maintenance or blood pressure), integrate the information they are receiving, and respond by making changes that will maintain homeostasis. The nervous system uses this three-step approach in generating both sensory and motor output.

Very simply put, all you see, feel, think, and do is controlled by your nervous system. A simple example of information processing by the nervous system is the "knee-jerk reflex," used in the course of many medical exams. The doctor taps your knee with a hammer, a stretch receptor in your muscle tendons at-tached to a sensory nerve cell (**sensory neuron**) is activated, and a response is sent into your spinal cord where special nerve cells (**interneurons**) process that information. The interneurons act as integrating centers, processing incoming sensory information. If the sensory stimulus is great enough, the interneurons excite a third type of nerve cell, the **motor neuron**. The motor neuron, when stimulated by the interneurons, will send signals to the appropriate muscle groups within the leg, causing them to contract and resulting in extension of the leg.

> *The nervous system and endocrine system share the responsibilities of maintaining the internal homeostasis of the human body.*

The simple reflex just described is only one way in which information can be perceived, processed, and responded to by your nervous system. All behaviors are controlled by the nervous system in some way, including how you squint when your roommate turns on the light in the middle of the night, the way you jump in your seat when a library book falls off the shelf, or even the "butterflies in your stomach" caused by your attraction to that certain special someone. The differences among these behaviors include the type of information being perceived and the amount of information processing that must take place. Your thoughts, memories, and emotions are processed in your brain by groups of interneurons specialized for their specific tasks. How these groups of cells communicate with one another, process information, and generate responses to your environment and to the people around you is only now beginning to be understood.

With a mass of only 2 kg (4.5 lb), about 3% of total body weight, the nervous system is one of the smallest and yet the most complex of the 11 body systems. The nervous system consists of two main subdivisions: the **central nervous system (CNS)**, which consists of the brain and spinal cord, and the **peripheral** (pe-RIF-er-al) **nervous system (PNS)**, which includes all nervous tissue outside the CNS. **Neurology** is the medical field that concerns itself with the study and treatment of diseases of the nervous system.

12.1 The nervous system maintains homeostasis and integrates all body activities.

Structures of the Nervous System

The **nervous system** is a complex, highly organized network of billions of neurons and even more neuroglia. The structures that make up the nervous system include the brain, spinal cord, nerves, ganglia, enteric plexuses, and sensory receptors (Figure 12.1). The skull encloses the **brain**. The **spinal cord** connects to the brain through the foramen magnum of the skull and is encircled by the bones of the vertebral column. **Peripheral nerves** are located outside the brain and spinal cord, and are classified by their origin. **Cranial nerves** emerge from the base of the brain; **spinal nerves** emerge from the spinal cord. Each nerve follows a defined path and serves a specific region of the body. For example, the right cranial nerve I carries signals for the sense of smell from the right side of the nose to the brain. **Ganglia** (GANG-lē-a = swelling or knot; singular is *ganglion*) are small masses of nervous tissue located outside the brain and spinal cord and are closely associated with cranial and spinal nerves. Located within the walls of the gastrointestinal tract are extensive networks of neurons called **enteric plexuses** that help regulate the digestive system. **Sensory re-**

ceptors may be dendrites of sensory neurons (described shortly) as well as separate, specialized cells that monitor changes in the internal or external environment, such as the sudden sound of a honking horn.

Functions of the Nervous System

The nervous system is responsible for all of your perceptions, behaviors, memories, and movements. To accomplish these functions, it carries out a complex array of tasks, such as sensing various smells, producing speech, remembering, providing signals that control body movements, and regulating the operation of internal organs. These activities can be divided into three groups:

1. *Sensory function.* Sensory receptors *detect* internal stimuli, such as an increase in blood acidity, and external stimuli, such as a raindrop landing on your arm. This **sensory** information is then carried into the brain and spinal cord through cranial and spinal nerves.

Figure 12.1   Major structures of the nervous system.

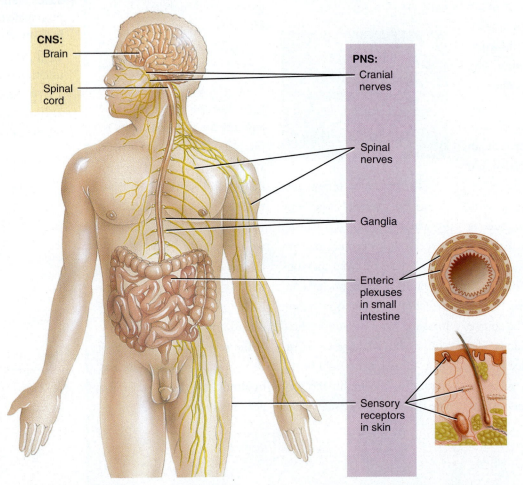

The central nervous system (CNS) consists of the brain and spinal cord; the peripheral nervous system (PNS) consists of all nervous tissue outside the CNS.

2. Integrative function. The nervous system *integrates* (processes) sensory information received from afferent neurons by analyzing and storing some of it, and by making decisions for appropriate responses. An important integrative function is **perception**, the conscious awareness of sensory stimuli. Perception occurs in the brain.

3. Motor function. Once sensory information is integrated, the nervous system may elicit an appropriate motor response by activating **effectors** (muscles and glands) through cranial and spinal nerves. Stimulation of the effectors causes muscles to contract and glands to secrete.

12.2 The nervous system is organized into the central and peripheral nervous systems.

Organization of the Nervous System

Recall that the nervous system consists of two main subdivisions: the central nervous system and the peripheral nervous system. The CNS (brain and spinal cord) processes many different kinds of incoming sensory information and is the source of thoughts, emotions, and memories. Most nerve impulses that stimulate muscles to contract and glands to secrete originate in the CNS and have projections to their targets that are part of the PNS. The PNS includes cranial nerves and their branches, spinal nerves and their branches, ganglia, and sensory receptors. The PNS may be subdivided further into a **somatic nervous system (SNS)** (*somat-* = body), an **autonomic nervous system (ANS)** (*auto-* = self; *-nomic* = law), and an **enteric nervous system (ENS)** (*enter-* = intestines) (Figure 12.2).

The SNS consists of sensory neurons that convey information to the CNS from somatic receptors in the head, body wall, and limbs and from receptors for the special senses of vision, hearing, taste, and smell. The SNS also includes motor neurons that conduct impulses from the CNS to *skeletal muscles* only. Because these motor responses can be consciously controlled, the action of this part of the PNS is *voluntary*.

The ANS consists of sensory neurons that convey information to the CNS from autonomic sensory receptors, located primarily in visceral organs such as the stomach and lungs. The ANS also consists of motor neurons that conduct nerve impulses from the CNS to *smooth muscle, cardiac muscle*, and *glands*. Because its motor responses are not normally under conscious control, the action of the ANS is *involuntary*. The ANS consists of two branches, the **sympathetic division** and the **parasympathetic division**. With a few exceptions, effectors are innervated by both the sympathetic and parasympathetic divisions, usually with opposing actions. For example, sympathetic neurons increase heart rate, and parasympathetic neurons slow it down.

Figure 12.2 Organization of the nervous system. Blue boxes represent sensory components of the peripheral nervous system (PNS); red boxes represent motor components of the PNS; and green boxes represent effectors (muscles and glands).

The SNS is under voluntary control; the ANS and ENS are under involuntary control.

The ENS consists of approximately 100 million neurons in enteric plexuses that extend the length of the gastrointestinal (GI) tract. Enteric sensory neurons monitor chemical changes within the GI tract and the stretching of its walls. Enteric motor neurons govern contraction of GI tract smooth muscle to propel food through the GI tract, secretions of the GI tract organs such as acid secretion by the stomach, and activity of GI tract endocrine cells, which secrete hormones. The operation of the ENS, the "brain of the gut," is involuntary. Many neurons of the enteric plexuses function independently of the CNS to some extent, although they may communicate with the CNS via sympathetic and parasympathetic neurons.

Checkpoint

1. State three functions of the nervous system.
2. Would a receptor more likely be located in your eyes, brain, or legs? In which of these locations would you expect to find an effector?
3. List the components of the CNS and the PNS.
4. What are the components and functions of the SNS, ANS, and ENS? Which subdivisions have voluntary actions? Involuntary actions?

12.3 Neurons are responsible for most of the unique functions of the nervous system.

Nervous tissue consists of two types of cells: neurons and neuroglia. **Neurons** (nerve cells) carry out most of the unique functions of the nervous system, such as sensing, thinking, remembering, controlling muscle activity, and regulating glandular secretions. Like muscle cells, neurons have the property of **electrical excitability**, the ability to produce **action potentials** in response to stimuli. A **stimulus** is any change in the environment that is strong enough to initiate an action potential. Action potentials, also called **nerve impulses**, are electrical signals that propagate (travel) along the surface of a neuron. **Neuroglia** support, nourish, and protect neurons, and maintain homeostasis in the interstitial fluid that bathes them.

Some neurons are tiny and propagate impulses over a short distance (less than 1 mm) within the CNS. Others are the longest cells in the body; for example, sensory neurons that allow you to feel the position of a feather tickling your wiggling toes stretch all the way from your foot to the lower portion of your brain.

Parts of a Neuron

Most neurons have three parts: (1) a cell body, (2) dendrites, and (3) an axon (Figure 12.3a). The **cell body** contains a nucleus surrounded by cytoplasm that contains typical organelles such as lysosomes, mitochondria, and a Golgi complex. Neuronal cell bodies also contain prominent clusters of rough endoplasmic reticulum, termed *Nissl bodies*, where protein synthesis occurs (Figure 12.3b). Newly synthesized proteins produced by Nissl bodies are used to replace cellular components, as material for growth of neurons, and to regenerate damaged axons in the PNS. The cytoskeleton includes both *neurofibrils*, composed of bundles of intermediate filaments that provide the cell shape and support, and *microtubules*, which assist in moving materials between the cell body and axon. Most neurons cannot divide (are amitotic) because they lack the centrioles that are essential for mitosis to occur.

Two kinds of processes (extensions) emerge from the cell body of a neuron: multiple dendrites and a single axon. **Dendrites** (= little trees) are the receiving or input portions of a neuron. They usually are short, tapering, and highly branched. In many neurons the dendrites form a tree-shaped array of processes extending from the cell body.

The single **axon** (= axis) of a neuron propagates nerve impulses toward another neuron, a muscle fiber, or a gland cell. An axon is a long, thin, cylindrical projection that often joins the cell body at a cone-shaped elevation called the **axon hillock** (= small hill). The portion of the axon closest to the axon hillock is the **initial segment**. In most neurons, impulses arise at the junction of the axon hillock and the initial segment, called the **trigger zone**, from which impulses are conducted toward the distal end of the axon. The cytoplasm of an axon, called **axoplasm**, is surrounded by a plasma membrane known as the **axolemma** (*lemma* = sheath or husk). Along the length of an axon, side branches called **axon collaterals** may branch off, typically at a right angle to the axon. The axon and its collaterals end by dividing into many fine processes called **axon terminals**. The tips of some axon terminals swell into bulb-shaped structures called **synaptic end bulbs**, which are the sites of communication between the neuron and another neuron or an effector cell.

The cell body is the portion of a neuron that synthesizes new cell products or recycles old ones. Newly synthesized substances are conveyed in the axoplasm from the cell body toward the axon terminals (**anterograde transport**) for developing or regenerating axons and to replenish axoplasm in mature axons. Some substances are transported through axons by "motor" proteins that move substances along the surfaces of microtubules. Materials that form the membranes of the axolemma, synaptic end bulbs, and synaptic vesicles utilize microtubule transport. Some materials are transported from the axon terminal back to the cell body (**retrograde transport**) to be degraded or recycled; others influence neuronal growth. Damage to an axon can impede the flow of axoplasm, leading to altered transmission of chemical and electrical signals.

Figure 12.3 **Structure of a typical motor neuron and Schwann cells.** Arrows indicate the direction of information flow: dendrites → cell body → axon → axon terminals.

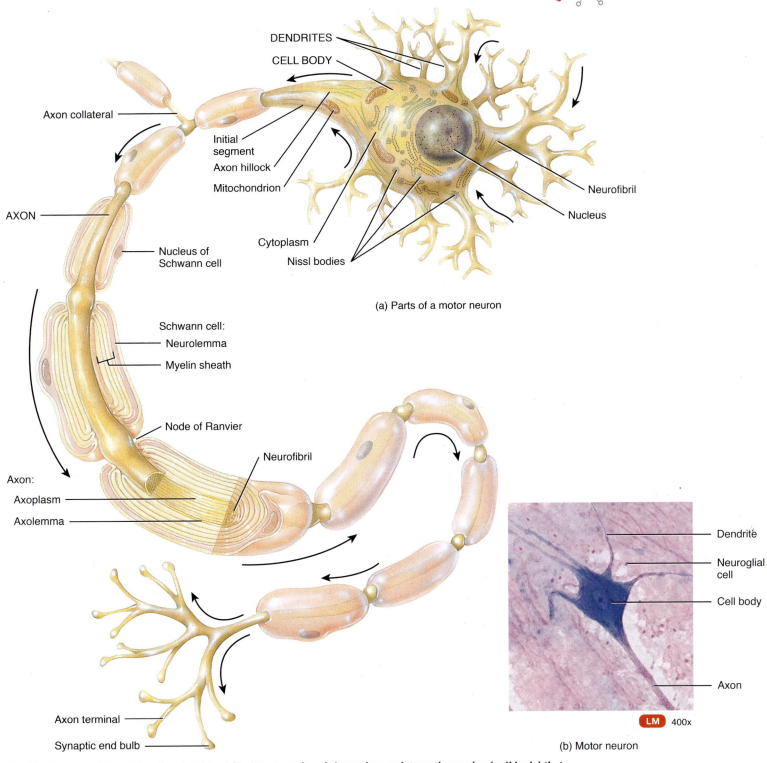

(a) Parts of a motor neuron

DENDRITES

CELL BODY

Axon collateral

Initial segment

Axon hillock

Mitochondrion

AXON

Nucleus of Schwann cell

Cytoplasm

Nissl bodies

Neurofibril

Nucleus

Schwann cell:

Neurolemma

Myelin sheath

Node of Ranvier

Neurofibril

Axon:

Axoplasm

Axolemma

Axon terminal

Synaptic end bulb

LM 400x

(b) Motor neuron

Dendrite

Neuroglial cell

Cell body

Axon

Most neurons have an input region (dendrites) that receives information, an integration region (cell body) that analyzes the information, and an output region (axon) that transmits information toward another cell.

Structural Diversity and Classification of Neurons

Neurons display great diversity in size and shape. Their cell bodies range in diameter from 5 micrometers (μm) (slightly smaller than a red blood cell) up to 135 μm (barely large enough to see with the unaided eye). The pattern of dendritic branching is varied and distinctive for neurons in different parts of the nervous system. A few small neurons lack an axon, and many others have very short axons. The longest axons, however, are almost as long as a person is tall, extending from the toes to the lowest part of the brain.

Both structural and functional features are used to classify the various neurons in the body. Structurally, neurons are classified according to the number of processes extending from the cell body (Figure 12.4).

- **Multipolar neurons** usually have several dendrites and one axon (Figure 12.4a). Most neurons in the brain and spinal cord are of this type.

- **Bipolar neurons** have one main dendrite and one axon (Figure 12.4b). They are found in the retina of the eye, in the inner ear, and in the olfactory (*olfact-* = to smell) area of the brain.

- **Unipolar neurons** have dendrites and one axon that are fused together to form a continuous process that emerges from the cell body (Figure 12.4c). The dendrites of most unipolar neurons function as sensory receptors that detect a sensory stimulus such as touch, pressure, pain, or thermal stimuli (see Figure 12.9).

The trigger zone for nerve impulses in a unipolar neuron is at the junction of the dendrites and axon. The impulses then propagate toward the synaptic end bulbs. The cell bodies of most unipolar neurons are located in the ganglia of spinal and cranial nerves.

Functionally, neurons are classified according to the direction in which the nerve impulse (action potential) is conveyed with respect to the CNS.

- **Sensory** or **afferent neurons** (AF-e-rent NOO-ronz; *af-* = toward; *-ferrent* = carried) either contain sensory receptors at their distal ends (dendrites) (see Figure 12.9) or are located just after sensory receptors that are separate cells. Once an appropriate stimulus activates a sensory receptor, the sensory neuron forms an action potential in its axon and the action potential is conveyed *into* the CNS through cranial or spinal nerves. Most sensory neurons are unipolar in structure.

- **Motor** or **efferent neurons** (EF-er-ent; *ef-* = away from) convey action potentials away from the CNS to **effectors** (muscles and glands) in the periphery (PNS) through cranial or spinal nerves (see Figure 12.9). Most motor neurons are multipolar in structure.

- **Interneurons** are mainly located within the CNS between sensory and motor neurons (see Figure 12.9). Interneurons integrate (process) incoming sensory information from sensory neurons and then elicit a motor response by activating the appropriate motor neurons. Most interneurons are multipolar in structure.

Figure 12.4   Structural classification of neurons. Breaks indicate that axons are longer than shown.

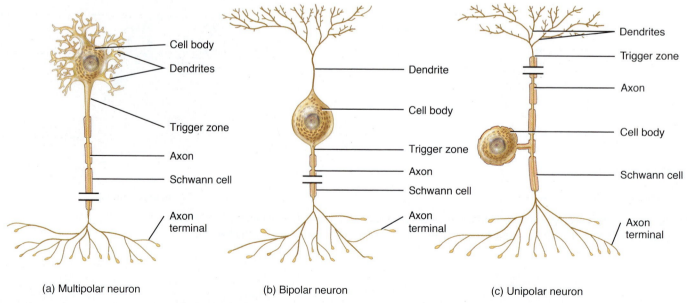

(a) Multipolar neuron   (b) Bipolar neuron   (c) Unipolar neuron

A multipolar neuron has many processes extending from the cell body, a bipolar neuron has two, and a unipolar neuron has one.

Checkpoint

5. Describe the parts of a neuron and the functions of each.
6. Name the location where a neuron and muscle cell interact.
7. What roles do the dendrites, cell body, and axon play in communication of signals?

8. Describe three structural types of neurons.
9. What kind of neuron carries input toward your CNS? What kind of neuron carries input toward your PNS?

12.4 Neuroglia support, nourish, and protect neurons and maintain homeostasis.

Neuroglia (noo-RŌG-lē-a; *-glia* = glue), or **glia**, constitute about half the volume of the CNS (Figure 12.5). Their name derives from the idea of early histologists that they were the "glue" that held nervous tissue together. We now know that neuroglia are not merely passive bystanders but rather active participants in the operation of nervous tissue. Generally, neuroglia are smaller than neurons, and they are 5 to 50 times more numerous (see Figure 12.3b). In contrast to neurons, glia do not generate or propagate action potentials, and they can readily multiply and divide. In cases of injury or disease, neuroglia multiply to fill in the spaces formerly occupied

Figure 12.5   Neuroglia of the central nervous system.

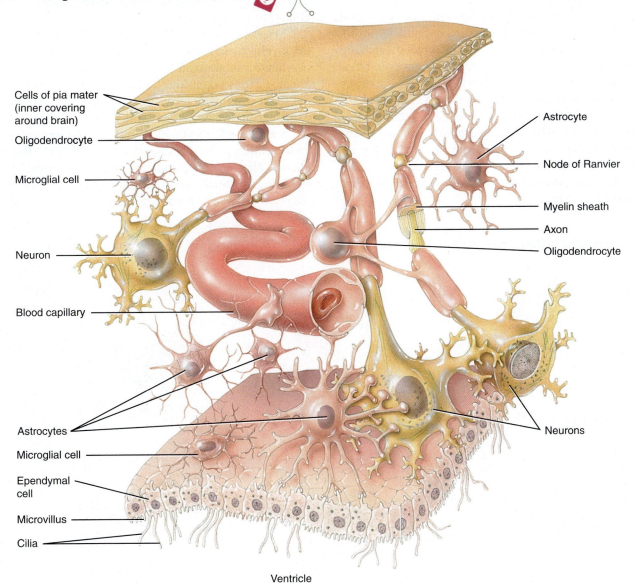

Cells of pia mater (inner covering around brain)
Oligodendrocyte
Microglial cell
Neuron
Blood capillary
Astrocytes
Microglial cell
Ependymal cell
Microvillus
Cilia
Astrocyte
Node of Ranvier
Myelin sheath
Axon
Oligodendrocyte
Neurons
Ventricle

Neuroglia of the CNS can be distinguished by size and the arrangement of their processes.

by neurons. Brain tumors derived from glia, called **gliomas**, tend to be highly malignant and to grow rapidly. Four of the six types of neuroglia—astrocytes, oligodendrocytes, microglia, and ependymal cells—are found only in the CNS. The remaining two types—Schwann cells and satellite cells—are present in the PNS.

Neuroglia of the CNS

Neuroglia of the CNS include astrocytes, oligodendrocytes, microglia, and ependymal cells (Figure 12.5).

Astrocytes (AS-trō-sīts; *astro-* = star; *-cyte* = cell) These star-shaped cells have many processes and are the largest and most numerous of the neuroglia. The processes of astrocytes make contact with blood capillaries, neurons, and the pia mater (a thin membrane around the brain and spinal cord). Astrocytes cling to and support neurons. Astrocytes help to maintain the appropriate chemical environment for the generation of nerve impulses by providing nutrients to neurons, removing excess neurotransmitters (described shortly), and regulating the concentration of important ions. Astrocyte processes wrap around blood capillaries to inhibit movement of potentially harmful substances in blood, creating a *blood–brain barrier* that restricts the movement of substances between the blood and neurons of the CNS. Details of the blood–brain barrier are discussed in Chapter 13.

Oligodendrocytes (OL-i-gō-den′-drō-sīts; *oligo-* = few; *-dendro-* = tree) These resemble astrocytes but are smaller and contain fewer processes. Oligodendrocyte processes are responsible for forming and maintaining the myelin sheath (described shortly) around CNS axons.

Microglia (mī-KROG-lē-a; *micro-* = small) These neuroglia are small cells with slender processes that give off numerous spinelike projections. Microglia phagocytize microbes and damaged nervous tissue.

Ependymal cells (ep-EN-de-mal; *epen-* = above; *-dym-* = garment) These are cuboidal to columnar cells arranged in a single layer that have microvilli and cilia. These cells line the *ventricles* of the brain and central canal of the spinal cord (spaces filled with cerebrospinal fluid, which protects and nourishes the brain and spinal cord). Ependymal cells produce and assist in the circulation of cerebrospinal fluid. They also form the blood–cerebrospinal fluid barrier, which is discussed in Chapter 13.

Neuroglia of the PNS

Neuroglia of the PNS completely surround axons and cell bodies. The two types of neuroglia in the PNS are Schwann cells and satellite cells (Figure 12.6).

Schwann cells (SCHVON or SCHWON) These cells form the myelin sheath around axons in the PNS. A Schwann cell can myelinate a single axon (Figure 12.6a) or enclose multiple unmyelinated axons (axons that lack a myelin sheath) (Figure 12.6b). Schwann cells participate in axon regeneration, which is more easily accomplished in the PNS than in the CNS.

Satellite cells (SAT-i-līt) These flat cells surround the cell bodies of neurons within PNS ganglia (Figure 12.6c). (Recall that ganglia are collections of neuronal cell bodies outside the CNS.) Besides providing structural support, satellite cells regulate the exchanges of materials between neuronal cell bodies and interstitial fluid.

Figure 12.6   Neuroglia of the peripheral nervous system.

Neuron cell body in a ganglion

Satellite cell

Node of Ranvier

Schwann cell

Myelin sheath

Axon

Schwann cell

Unmyelinated axons

Schwann cell

Axon

(a)   (b)   (c)

Neuroglia of the PNS completely surround axons and cell bodies of neurons.

Myelination

The axons of most neurons are surrounded by a multilayered lipid and protein covering called the **myelin sheath** that electrically insulates them and increases the speed of nerve impulse conduction. Axons with a myelin sheath are referred to as **myelinated** (Figure 12.7a) whereas axons without such a covering are said to be **unmyelinated** (Figure 12.7b). Schwann cells produce myelin sheaths around axons in the PNS; oligodendrocytes produce myelin sheaths around axons in the CNS.

Schwann cells begin to form myelin sheaths around axons during fetal development. Each Schwann cell wraps about 1 millimeter (1 mm = 0.04 in.) of a single axon's length by spiraling many times around the axon (Figure 12.7a). Eventually, multiple

Figure 12.7 **Myelinated and unmyelinated axons.**

(a) Transverse sections of stages in the formation of a myelin sheath

(b) Transverse section of unmyelinated axons

(c) Transverse section of myelinated axon

(d) Transverse section of unmyelinated axons

Myelinated axons are surrounded by a myelin sheath produced by Schwann cells in the PNS or by oligodendrocytes in the CNS.

layers of Schwann cell plasma membrane surround the axon, with the Schwann cell's cytoplasm and nucleus forming the outermost layer. The inner portion, consisting of up to 100 layers of Schwann cell plasma membrane, is the myelin sheath. The outer, nucleated cytoplasmic layer of the Schwann cell enclosing the myelin sheath is the **neurolemma**. When an axon is injured, the neurolemma aids regeneration by forming a regeneration tube that guides and stimulates regrowth of the axon. Gaps in the myelin sheath, called **nodes of Ranvier** (RON-vē-ā), appear at intervals along the axon between adjacent Schwann cells (see Figure 12.3).

In the CNS, an oligodendrocyte myelinates parts of several axons. Each oligodendrocyte extends about 15 broad, flat processes that spiral around CNS axons, forming a myelin sheath. A neurolemma is not present, however, because the oligodendrocyte cell body and nucleus do not envelop the axon. Nodes of Ranvier are present, but they are fewer in number than seen with PNS neurons. Axons in the CNS display little regrowth after injury. This is thought to be due, in part to the absence of a neurolemma and in part to an inhibitory influence exerted by the oligodendrocytes on axon regrowth.

The amount of myelin increases from birth to maturity, and its presence greatly increases the speed of nerve impulse conduc- tion. An infant's responses to stimuli are neither as rapid nor as coordinated as those of an older child or an adult, in part because myelination is still in progress during infancy.

Gray and White Matter

In a freshly dissected section of the brain or spinal cord, some regions look white and glistening, and others appear gray (Figure 12.8). **White matter** consists primarily of the myelinated axons. The whitish color of myelin gives white matter its name. **Gray matter** contains neuronal cell bodies, dendrites, unmyelinated axons, axon terminals, and neuroglia. It looks grayish, rather than white, because the Nissl bodies impart a gray color and there is little or no myelin in these areas. In the spinal cord, the white matter surrounds an inner core of gray matter that, depending on how imaginative you are, is shaped like a butterfly or the letter H. In the brain, a thin outer shell of gray matter covers the deeper white matter. The arrangement of gray matter and white matter in the spinal cord and brain is discussed more extensively in Chapter 13.

Figure 12.8   Distribution of gray and white matter in the spinal cord and brain.

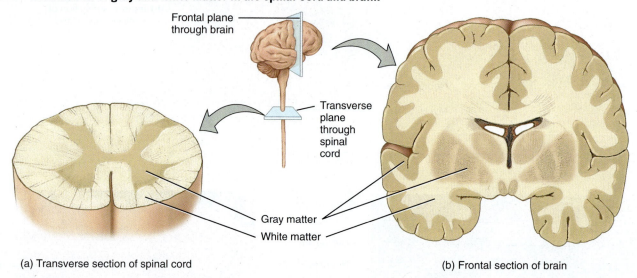

Frontal plane through brain

Transverse plane through spinal cord

Gray matter

White matter

(a) Transverse section of spinal cord

(b) Frontal section of brain

🔑 **White matter primarily consists of myelinated axons of many neurons. Gray matter consists of neuron cell bodies, dendrites, axon terminals, unmyelinated axons, and neuroglia.**

Checkpoint

10. Which neuroglia could fight infection by phagocytizing bacterial cells and dead cells? Are these glial cells located within the CNS or PNS?

11. Which neuroglia help ensure a healthy environment for the nervous system? Are these glial cells located within the CNS or PNS?

12. Which neuroglia help move cerebrospinal fluid through the CNS?

13. What is the functional advantage of the myelin sheath?

14. Name two neuroglia that produce myelin sheaths. Are each of these glial cells located within the CNS or PNS?

15. Differentiate between the myelin sheath and neurolemma.

16. How do white matter and gray matter differ in structure?

Return to Mary's Story

"Hello Mary. How is the baby doing?"

"Hello, Dr. Addison. She's a handful, but she's doing great. We're not due for her checkup for another month or so; I'm really here because of my hands. I started getting these shooting pains in the thumbs and first two fingers of both of my hands a couple of months into the pregnancy. They came and went, so I thought it was maybe part of being pregnant, you know with all the new sensations and changes. I guess I thought it would go away after I had some rest. But it's gotten worse, and my hands are really weak, especially my right hand. I can hardly hold a can of soup anymore. I'm kind of afraid I could drop Caitlin." She shifted the baby on her knee. "My hand will just give way sometimes and I lose strength."

Dr. Addison replied, "Mary, numbness and tingling are fairly common. The weakness does concern me, though. I'm going to refer you to a neurologist for an evaluation. I have an idea of what might be causing the problem, especially with all the typing and computer work you do. I am thinking it may be a repetitive motion type injury."

A. What kind of information is carried along nerves in Mary's peripheral nervous system?

B. How would damage to the peripheral nerves explain Mary's signs and symptoms?

C. Consider the structure of the neuron and peripheral nerves. Repetitive motion injuries damage peripheral nerves. What neuronal structures are going to be involved in Mary's problem? How do these structures relate to the function of Mary's nervous system?

D. What type of supporting cells may be involved in Mary's problem? Based on your understanding of Concept 12.4, will Mary's signs and symptoms resolve if these supporting cells are involved in her problem?

E. Think about the structures of the nervous system you have learned about so far. Make a list of structures that could be involved in causing Mary's signs or symptoms and develop a preliminary diagnosis.

12.5 Neurons communicate with other cells.

Like muscle fibers, neurons are electrically excitable cells. They communicate with one another using two types of electrical signals: (1) *Graded potentials* are used for short-distance communication only. (2) *Action potentials* allow communication over long distances within the body. An action potential in a muscle fiber is called a *muscle action potential*. When an action potential occurs in a neuron, it is called a *nerve action potential* (*nerve impulse*). To understand the functions of graded potentials and action potentials, consider how the nervous system allows you to feel the smooth surface of a pen that you have picked up from a table (Figure 12.9):

1 As you touch the pen, a graded potential develops in a sensory receptor in the skin of the fingers.

2 The graded potential triggers the axon of the sensory neuron to form a nerve action potential, which travels along its axon into the CNS and ultimately causes the release of neurotransmitter at its synapse with an interneuron. A neurotransmitter is a chemical released by a neuron that excites or inhibits other neurons or effector cells.

3 The neurotransmitter stimulates the interneuron to form a graded potential in its dendrites and cell body.

4 In response to the graded potential, the axon of the interneuron forms a nerve action potential. The nerve action potential travels along the axon, which results in neurotransmitter release at the next synapse with another interneuron.

5 This process of neurotransmitter release at a synapse followed by the formation of a graded potential, and then a nerve action potential, occurs over and over as interneurons in higher parts of the brain (such as the thalamus and cerebral cortex) are activated. Once interneurons in the cerebral cortex, the outer part of the brain, are activated, perception occurs and you are now aware of the smooth surface of the pen against your fingers.

Suppose that you want to use the pen to write a letter. The nervous system would respond in the following way (Figure 12.9):

6 A stimulus in the brain causes a graded potential to form in the dendrites and cell body of an upper motor neuron, a motor neuron that synapses with a lower motor neuron farther down in the CNS. The graded potential subsequently causes a nerve action potential to occur in the axon of the upper motor neuron, leading to neurotransmitter release.

7 The neurotransmitter generates a graded potential in a **lower motor neuron**, a motor neuron that directly supplies skeletal muscle fibers. The graded potential triggers a nerve action potential and then release of the neurotransmitter at neuromuscular junctions formed with skeletal muscle fibers that control movements of the fingers.

8 The neurotransmitter stimulates the formation of muscle action potentials in muscle fibers that control finger movements. The muscle action potentials cause these muscle fibers to contract, allowing you to write with the pen.

The production of graded potentials and action potentials depends on two basic features of the plasma membrane of excitable cells: the existence of a resting membrane potential and the presence of specific types of ion channels. Like most other cells in the body, the plasma membrane of excitable cells exhibits a **membrane potential**, an electrical potential difference (voltage) across the plasma membrane. In excitable cells, this voltage is termed the **resting membrane potential**. The membrane potential is like voltage stored in a battery. If you connect the positive and negative terminals of a battery with a piece of wire, electrons will flow along the wire. This flow of charged particles is called **current**. In living cells, the flow of ions (rather than electrons) constitutes the electrical current.

Figure 12.9 Overview of communication within the nervous system.

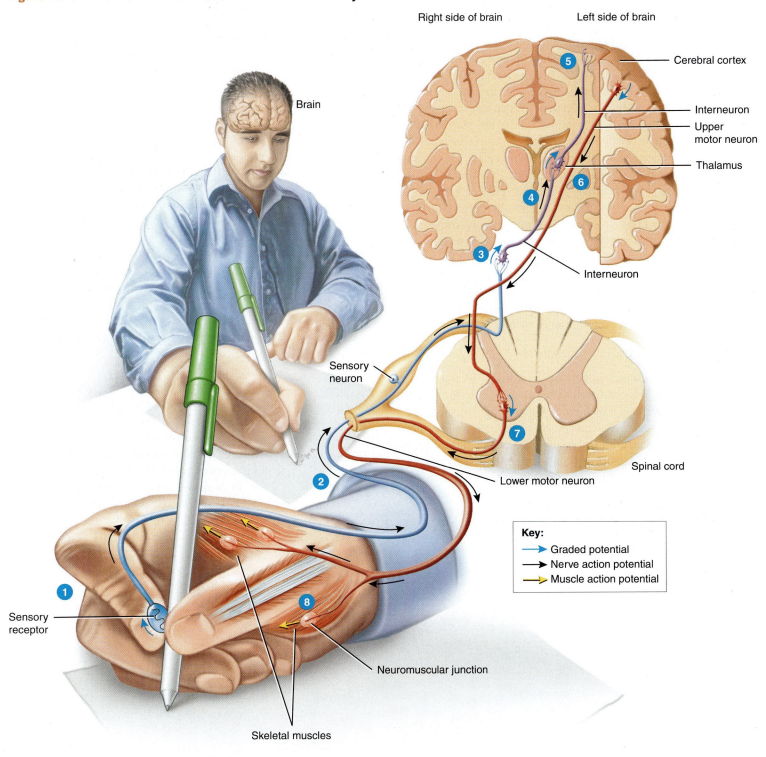

Right side of brain
Left side of brain

Cerebral cortex

Interneuron

Upper motor neuron

Thalamus

Interneuron

Brain

Sensory neuron

Spinal cord

Lower motor neuron

Key:
- → Graded potential
- → Nerve action potential
- → Muscle action potential

Sensory receptor

Neuromuscular junction

Skeletal muscles

🔑 Graded potentials and nerve and muscle action potentials are involved in the relay of sensory stimuli, integrative functions such as perception, and motor activities.

Ion Channels

Graded potentials and action potentials occur because the plasma membranes of neurons contain many different kinds of ion channels (Figure 12.10) that open or close in response to specific stimuli. Because the lipid bilayer of the plasma membrane is a good electrical insulator, the main paths for current to flow across the membrane are through the ion channels. When ion channels are open, they allow specific ions to move across the plasma membrane, down their electrochemical gradients—a concentration (chemical) difference plus an electrical difference. Recall that ions move from where they are more concentrated to where they are less concentrated. Also, positively charged ions (cations) move toward a negatively charged area, and negatively charged ions (anions) move toward a positively charged area. As ions move, they constitute a flow of electrical current that can change the membrane potential.

Some ion channels contain "gates," part of the channel protein that can seal the channel pore shut or move aside to open the pore (see Figure 3.6). The electrical signals produced by neurons and muscle fibers rely on four types of ion channels:

- Leakage channels are always open to ions (Figure 12.10a).

- A **ligand-gated channel** opens and closes in response to a specific **ligand**, a molecule that binds to a receptor (Figure 12.10b). A wide variety of ligands—including neurotransmitters, hormones, and ions—can open or close ligand-gated channels. The neurotransmitter acetylcholine, for example, opens cation channels that allow sodium ions (Na^+) and calcium ions (Ca^{2+}) to diffuse inward and potassium ions (K^+) to diffuse outward.

- A **mechanically gated channel** opens or closes in response to mechanical stimulation such as sound waves, pressure, or tis-

sue stretching (Figure 12.10c). The force distorts the channel, opening the gate.

- A **voltage-gated channel** opens in response to a change in membrane potential (voltage) (Figure 12.10d). Voltage-gated channels participate in the generation and conduction of action potentials.

Resting Membrane Potential

The resting membrane potential exists because of a small buildup of negative ions in the cytosol along the inside of the membrane, and a buildup of positive ions in the extracellular fluid along the outside surface of the membrane (Figure 12.11a). Such a separation of positive and negative electrical charges is a form of potential energy, which is measured in volts or millivolts (1 mV = 0.001 V). The greater the difference in charge across the membrane, the larger is the membrane potential (voltage). Notice in Figure 12.11a that the buildup of charge occurs only very close to either side of the membrane. The cytosol or extracellular fluid elsewhere contains equal numbers of positive and negative charges and is electrically neutral.

The resting membrane potential of a cell can be measured in the following way: The tip of a recording microelectrode is inserted inside the cell, and a reference electrode is placed outside the cell in the extracellular fluid. Electrodes are devices that conduct electrical charges. The recording microelectrode and the reference electrode are connected to an instrument known as a voltmeter, which detects the electrical difference (voltage) across the plasma membrane (Figure 12.11b). In neurons, the typical resting membrane potential is approximately –70 mV. The minus sign indicates that the cytosol side of the membrane is negative relative to the outside. A cell that exhibits a membrane potential is said to be polarized.

Figure 12.10 **Ion channels in the plasma membrane.** (a) Leakage channels randomly open and close. (b) A chemical stimulus—here, the neurotransmitter acetylcholine—opens a ligand-gated channel. (c) A mechanical stimulus opens a mechanically-gated channel. (d) A change in membrane potential opens voltage-gated K⁺ channels during an action potential.

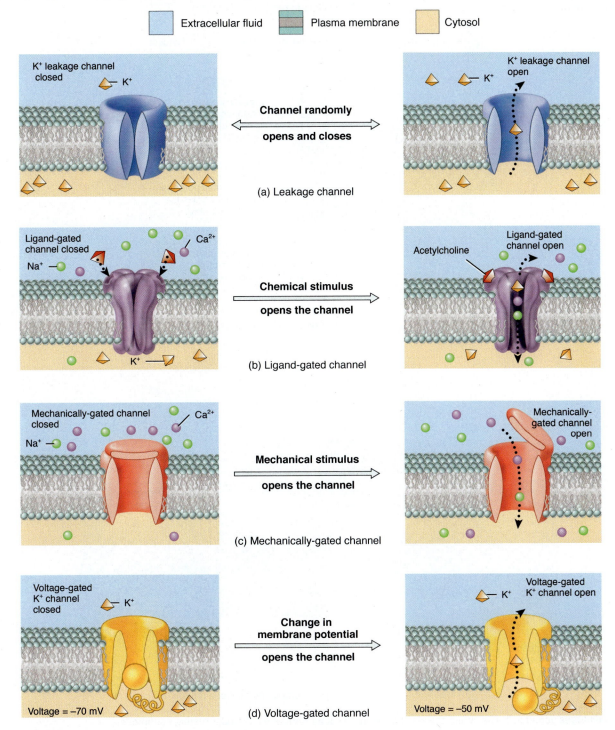

When ion channels are open, specific ions can move across the plasma membrane down their electrochemical gradients.

Figure 12.11 **Resting membrane potential.** (a) Distribution of charges that produce the resting membrane potential of a neuron. (b) Measurement of the resting membrane potential of a neuron. The tip of the recording microelectrode is inserted inside the neuron, and the reference electrode is placed in the extracellular fluid (ECF). The electrodes are connected to a voltmeter that measures the difference in charge across the plasma membrane (in this case −70 mV inside relative to outside).

(a) Distribution of charges that produce the resting membrane potential of a neuron.

(b) Measurement of the resting membrane potential of a neuron.

🔑 **The resting membrane potential exists across the plasma membrane of an excitable cell under resting conditions.**

The resting membrane potential arises from three major factors:

1. **Unequal distribution of ions across the plasma membrane.** A major factor that contributes to the resting membrane potential is the unequal distributions of various ions in extracellular fluid and cytosol (Figure 12.12). Extracellular fluid is rich in Na^+ and Cl^- while cytosol contains large concentrations of K^+ and anions such as phosphates and amino acids. Because the plasma membrane typically has more K^+ leakage channels than Na^+ leakage channels, the number of potassium ions that diffuse down their concentration gradient out of the cell is greater than the number of sodium ions that diffuse down their concentration gradient into the cell. As more and more positive potassium ions exit, the cytosol side of the membrane becomes increasingly negative and the outside of the membrane becomes increasingly positive.

2. **Inability of most anions to leave the cell.** Another factor contributes to the relative negativity of the cell interior: Most anions inside the cell are not free to leave because they are attached to nondiffusible molecules such as ATP and large molecules (Figure 12.12).

3. **Electrogenic nature of the sodium–potassium pump.** The small inward Na^+ leak and outward K^+ leak through leakage channels are offset by sodium–potassium pumps (Figure 12.12). These pumps help maintain the resting membrane potential by pumping out Na^+ as fast as it leaks in and, at the same time, returning K^+ to the cell interior. The sodium–potassium pumps expel three Na^+ for each two K^+ imported. Since these pumps remove more positive charges from the cell than they bring into the cell, they are *electrogenic*, which means they contribute to the relative negativity of the cell interior.

Checkpoint

17. How does the cell membrane of a neuron allow the cell to store electrical charges?

18. What are the three stimuli that result in ion channels opening or closing? What happens when these channels open?

19. How are leakage channels different from gated channels?

20. What two conditions allow maintenance of the resting membrane potential?

Figure 12.12   Factors that contribute to the resting membrane potential. The cell interior is relatively negative and the extracellular fluid is relatively positive because (1) the plasma membrane has more K^+ leakage channels (blue) than Na^+ leakage channels (red), resulting in more K^+ ions leaving the cell than Na^+ ions enter the cell; (2) most anions are unable to leave the cell; (3) the electrogenic Na^+/K^+ pumps (purple) expel more positive ions (three Na^+) than they bring into the cell (two K^+).

The resting membrane potential exists because the cytosol side of the plasma membrane has relatively more negative ions while the extracellular fluid has relatively more positive ions.

12.6 Graded potentials are the first response of a neuron to stimulation.

A **graded potential** is a small deviation from the membrane potential that makes the membrane either more polarized (inside more negative) or less polarized (inside less negative). When the response makes the membrane more polarized (increases the difference in the charge between inside and outside), it is termed a **hyperpolarizing graded potential** (Figure 12.13a). When the response makes the membrane less polarized, it is termed a **depolarizing graded potential** (Figure 12.13b).

Figure 12.13   Graded potentials. Most graded potentials occur in the dendrites and cell body (areas colored blue in the inset).

(a) Hyperpolarizing graded potential

(b) Depolarizing graded potential

During a hyperpolarizing graded potential, the membrane polarization is more negative than the resting level; during a depolarizing graded potential, the membrane polarization is less negative than the resting level.

To say that these electrical signals are *graded* means that they vary in amplitude (size), depending on the strength of the stimulus (Figure 12.14). They are larger or smaller depending on how many ligand-gated or mechanically gated channels have opened (or closed) and how long each remains open. The opening or closing of these channels alters the flow of specific ions across the membrane, producing a flow of current that is localized, which means that it spreads along the plasma membrane for a short distance and then dies out. Because they die out within a few millimeters of their point of origin, graded potentials are useful for short-distance communication only. Although graded potentials may only travel a short distance, they can generate another type of membrane potential, an action potential (described shortly) that travels the entire length of an axon.

A graded potential occurs when a stimulus causes ligand-gated or mechanically gated channels to open or close in an excitable cell's plasma membrane (Figure 12.15). Typically, ligand-gated and mechanically gated channels are present only in the dendrites

Figure 12.14   The graded nature of graded potentials. As the stimulus strength increases, the amplitude (size) of each resulting depolarizing graded potential increases. Although not shown, a similar relationship exists between stimulus strength and the amplitude of a hyperpolarizing graded potential.

The amplitude of a graded potential depends on the strength of the stimulus.

Figure 12.15 **Generation of graded potentials.** (a) A mechanical stimulus opens a mechanically-gated channel that allows cations to enter the cell; a depolarizing graded potential occurs as the cell interior becomes less negative than at rest. (b) A ligand stimulus (the neurotransmitter acetylcholine) opens a ligand-gated channel that allows the exchange of ions, and a depolarizing graded potential occurs as the cell interior becomes less negative than at rest. (c) A ligand stimulus (the neurotransmitter glycine) opens a ligand-gated channel that allows Cl⁻ ions into the cell, and a hyperpolarizing graded potential occurs as the cell interior becomes more negative than at rest.

(a) Depolarizing graded potential caused by pressure, a mechanical stimulus

(b) Depolarizing graded potential caused by the neurotransmitter acetylcholine, a ligand stimulus

(c) Hyperpolarizing graded potential caused by the neurotransmitter glycine, a ligand stimulus

A graded potential forms in response to the opening of mechanically-gated or ligand-gated channels.

and cell bodies of neurons. Hence, graded potentials occur mainly in the dendrites and cell body of a neuron.

Graded potentials become stronger and last longer by **summation**, the process by which graded potentials add together. If two depolarizing graded potentials summate, the net result is a larger depolarizing graded potential (Figure 12.16). If two hy-

Figure 12.16   Summation of graded potentials. Summation of two depolarizing graded potentials happens when two stimuli occur very close together in time. The dotted lines represent the individual depolarizing graded potentials that would form if summation did not occur.

🔑 **Summation occurs when two or more graded potentials add together to become larger in amplitude.**

perpolarizing graded potentials summate, the net result is a larger hyperpolarizing graded potential. You will learn more about the process of summation later in this chapter.

Graded potentials have different names depending on where they occur. For example, graded potentials resulting from a neurotransmitter binding to the dendrites or cell body of a neuron are called *postsynaptic potentials*, and graded potentials occurring in sensory receptors are termed *receptor potentials* (explained in Chapter 15).

Checkpoint

21. As more and more channels are opened, what happens to the magnitude of a graded potential?
22. What do we mean when we say graded potentials are "localized"?
23. In what neuronal structures do most graded potentials occur?
24. What types of ion channels produce graded potentials?

12.7   The action potential is an all-or-none electrical signal.

An **action potential**, or **nerve impulse**, is a sequence of rapidly occurring events that reverse the membrane potential and then eventually restore it to the resting state. An action potential has two main phases (Figure 12.17). During the **depolarizing phase**, the negative membrane potential becomes less negative, reaches zero, and then becomes positive. Then the **repolarizing phase** restores the membrane potential to the resting state of −70 mV.

Two types of voltage-gated channels open and then close during an action potential. These channels are present mainly in the plasma membrane of the axon and axon terminals. The first channels to open, voltage-gated Na^+ channels, allow Na^+ to rush into the cell, resulting in the depolarizing phase. Then voltage-gated K^+ channels open, allowing K^+ to flow out, producing the repolarizing phase.

The generation of an action potential depends on whether a stimulus is able to bring the membrane potential to a certain level termed the **threshold** (about −55 mV in many neurons) (Figure 12.18). An action potential will not occur in response to a **subthreshold stimulus**, a stimulus that is a weak depolarization that cannot bring the membrane potential to threshold. However, an action potential will occur in response to a **threshold stimulus**, a stimulus that is just strong enough to depolarize the membrane to

threshold. Several action potentials will form in response to a **suprathreshold stimulus**, a stimulus that is strong enough to depolarize the membrane *above* threshold. Each of the action potentials caused by a suprathreshold stimulus has the same amplitude (size) as an action potential caused by a threshold stimulus. Therefore, once an action potential is generated, the amplitude of an action potential is always the same and does not depend on stimulus intensity. Instead, the greater the stimulus strength above threshold, the greater is the frequency of action potentials.

As you have just learned, an action potential is generated in response to a threshold stimulus but does not form when there is a subthreshold stimulus. In other words, an action potential either occurs completely or it does not occur at all. This characteristic of an action potential is known as the **all-or-none principle**. The all-or-none principle of the action potential is similar to pushing the first domino in a long row of standing dominoes. When the push on the first domino is strong enough (when depolarization reaches threshold), that domino falls against the second domino, and the entire row topples (an action potential occurs). Stronger pushes on the first domino produce the identical effect—toppling of the entire row. Thus, pushing on the first domino produces an all-or-none event: The dominoes all fall or none fall.

Figure 12.17 **Action potential.** When a stimulus depolarizes the membrane to threshold (−55 mV), an action potential is generated. The action potential arises at the trigger zone, the junction of the axon hillock and the initial segment (area colored green in the inset).

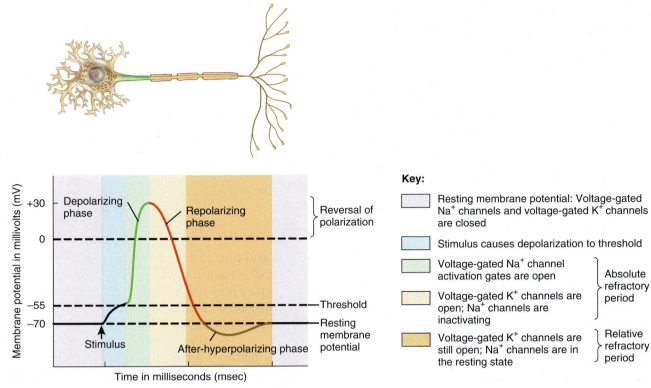

Key:

Resting membrane potential: Voltage-gated Na⁺ channels and voltage-gated K⁺ channels are closed

Stimulus causes depolarization to threshold

Voltage-gated Na⁺ channel activation gates are open ⎱ Absolute refractory period

Voltage-gated K⁺ channels are open; Na⁺ channels are inactivating ⎰

Voltage-gated K⁺ channels are still open; Na⁺ channels are in the resting state ⎱ Relative refractory period

An action potential consists of depolarizing and repolarizing phases.

Figure 12.18 **Stimulus strength and action potential generation.** A subthreshold stimulus does not result in an action potential because it does not bring the membrane potential to threshold. An action potential does occur in response to a threshold stimulus, a stimulus just strong enough to depolarize the membrane to threshold. Several action potentials form in response to a suprathreshold stimulus, which depolarizes the membrane above threshold.

An action potential will only occur when the membrane potential reaches threshold.

Depolarizing Phase

Depolarizing graded potentials originate in the dendrites or cell body of a neuron and then travel to the trigger zone. If the graded potential is able to depolarize the membrane to threshold, voltage-gated Na^+ channels open quickly. As voltage-gated Na^+ channels open, electrical and chemical gradients favor inward movement of Na^+. The resulting influx of Na^+ produces the depolarizing phase of the action potential as the membrane potential rises from the threshold level of −55 mV to +30 mV (see Figure 12.18). At the peak of the action potential, the cytosol side of the membrane is 30 mV more positive than the outside.

Each voltage-gated Na^+ channel has an *activation gate* and an *inactivation gate*. In the *resting state* of a voltage-gated Na^+ channel, the inactivation gate is open, but the activation gate is closed (step 1 in Figure 12.19). As a result, Na^+ cannot move into the cell through these channels and the membrane potential is at −70 mV.

During the depolarizing phase, the resting membrane potential rises to threshold, both the activation and inactivation gates in the Na^+ channel are open, and Na^+ influx begins (step 2 in Figure 12.19). As Na^+ rushes through the channels into the neuron, the buildup of positive charges on the cytosol side of the membrane increases the membrane potential to +30 mV.

Although the voltage-gated Na^+ channels are open for only a few ten-thousandths of a second, about 20,000 Na^+ ions flow across the membrane and can change the membrane potential considerably. Because only a few thousand Na^+ enter during a single action potential, the sodium–potassium pumps easily bail them out and maintain the low concentration of Na^+ inside the cell.

Repolarizing Phase

Shortly after the activation gates open, the inactivation gates close and Na^+ stops entering the neuron (step 3 in Figure 12.19). Besides opening voltage-gated Na^+ channels, a threshold-level depolarization also opens voltage-gated K^+ channels (steps 3 and 4 in Figure 12.19). Because voltage-gated K^+ channels open slowly, their opening occurs at about the same time the voltage-gated Na^+ channels are closing. The slower opening of voltage-gated K^+ channels and the closing of previously opened Na^+ channels produce the re-

polarizing phase of the action potential. With K^+ channels open, K^+ flows out of the neuron, negative charges build up on the cytosol side of the membrane, and the membrane potential changes from +30 mV to −70 mV, restoring the resting membrane potential.

After-Hyperpolarizing Phase

While the voltage-gated K^+ channels are open, outflow of K^+ may be large enough to cause an **after-hyperpolarizing phase** of the action potential (see Figure 12.17). During this phase, voltage-gated K^+ channels remain open and the membrane potential becomes even more negative than the resting level of −70 mV. As the voltage-gated K^+ channels close, the membrane potential returns to the resting level.

Refractory Period

The period of time after an action potential begins, during which an excitable cell cannot generate another action potential, is called the **refractory period** (see key in Figure 12.19). The **absolute refractory period** occurs from the time the Na^+ channel activation gates open to when the Na^+ channel inactivation gates close (steps 2–3 in Figure 12.19). During this period, even a very strong stimulus cannot initiate a second action potential because inactivated Na^+ channels must return to the resting state before they can re-open (step 1 in Figure 12.19).

The **relative refractory period** is the period of time during which a second action potential can be initiated, but only by a larger-than-normal stimulus. It coincides with the period when the voltage-gated K^+ channels are still open after inactivated Na^+ channels have returned to their resting state (see key in Figure 12.17).

Checkpoint

25. How are graded potentials and the action potential threshold related?

26. What is the all-or-none principle? Are graded potentials all-or-none?

27. What ions and ion channels are responsible for the depolarizing and repolarizing phases of the action potential? (Which channels are open during depolarization? During repolarization?)

28. What is the refractory period of an action potential? How are the absolute and relative refractory periods different?

12.8 Action potentials propagate from the trigger zone to axon terminals.

To communicate information from one part of the body to another, action potentials must travel from where they arise at a trigger zone to the axon terminals. This mode of travel is called **propagation**, and it depends on positive feedback. As sodium ions flow into the neuron, voltage-gated Na^+ channels in adjacent segments of the

membrane open. Thus, the action potential self-propagates along the membrane, rather like toppling that long row of dominoes by pushing over the first one in the line. In actuality, it is not the same action potential that propagates along the entire axon. Instead, the action potential regenerates over and over at adjacent regions of

Figure 12.19 **Changes in ion flow through voltage-gated channels during the depolarizing and repolarizing phases of an action potential.** Leakage channels and sodium–potassium pumps are not shown.

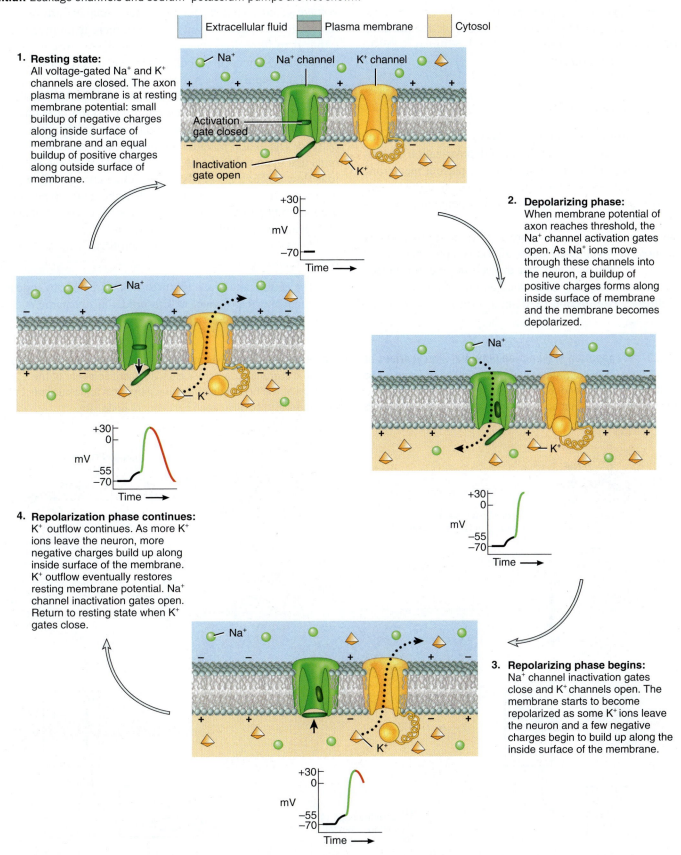

Extracellular fluid   Plasma membrane   Cytosol

1. **Resting state:**
All voltage-gated Na⁺ and K⁺ channels are closed. The axon plasma membrane is at resting membrane potential: small buildup of negative charges along inside surface of membrane and an equal buildup of positive charges along outside surface of membrane.

2. **Depolarizing phase:**
When membrane potential of axon reaches threshold, the Na⁺ channel activation gates open. As Na⁺ ions move through these channels into the neuron, a buildup of positive charges forms along inside surface of membrane and the membrane becomes depolarized.

3. **Repolarizing phase begins:**
Na⁺ channel inactivation gates close and K⁺ channels open. The membrane starts to become repolarized as some K⁺ ions leave the neuron and a few negative charges begin to build up along the inside surface of the membrane.

4. **Repolarization phase continues:**
K⁺ outflow continues. As more K⁺ ions leave the neuron, more negative charges build up along inside surface of the membrane. K⁺ outflow eventually restores resting membrane potential. Na⁺ channel inactivation gates open. Return to resting state when K⁺ gates close.

Inflow of sodium ions (Na⁺) causes the depolarizing phase and outflow of potassium ions (K⁺) causes the repolarizing phase of an action potential.

membrane from the trigger zone to the axon terminals. An action potential propagates in only one direction, toward the axon terminals (see leading edge of action potential in Figure 12.20), because any region of the axon that has just undergone an action potential must recover (experience the refractory period) before it is able to generate another action potential.

Continuous and Saltatory Conduction

The type of impulse propagation described so far is **continuous conduction**, which involves step-by-step depolarization and re-polarization of each adjacent segment of the plasma membrane as ions flow through each voltage-gated channel along the membrane. Unmyelinated axons propagate action potentials by continuous conduction.

Action potentials propagate more rapidly along myelinated axons than along unmyelinated axons. If you compare parts a and b in Figure 12.20, you will see that the action potential propagates much farther along the myelinated axon in the same period of time. **Saltatory conduction** (SAL-ta-tō-rē; *saltat-* = leaping),

the special mode of action potential propagation that occurs along myelinated axons, occurs because of the uneven distribution of voltage-gated channels. Voltage-gated channels are present primarily at the nodes of Ranvier (where there is no myelin sheath) rather than in regions where a myelin sheath covers the plasma membrane. Hence, current carried by Na^+ and K^+ flows across the membrane mainly at the nodes of Ranvier.

When an action potential propagates along a myelinated axon, an electric current flows from one node to the next through the extracellular fluid surrounding the myelin sheath and through the cytosol. The action potential at the first node generates ionic currents that open voltage-gated Na^+ channels at the second node. As ions flow through the newly opened channels, an action potential is generated at the second node. Then the action potential at the second node generates an ionic current that opens voltage-gated Na^+ channels at the third node, and so on.

The flow of current across the membrane only at the nodes of Ranvier has two consequences:

1.   The action potential appears to leap from node to node as each nodal area depolarizes to threshold, thus the name "saltatory."

Figure 12.20   Propagation of an action potential after it arises at the trigger zone. Dotted lines indicate ionic current flow. Insert shows path of current flow. (a) In continuous conduction along an unmyelinated axon, ionic currents flow across each adjacent segment of the membrane. (b) In saltatory conduction along a myelinated axon, the action potential at the first node of Ranvier generates ionic currents in the cytosol and interstitial fluid that open voltage-gated Na^+ channels at the second node, and so on at each subsequent node.

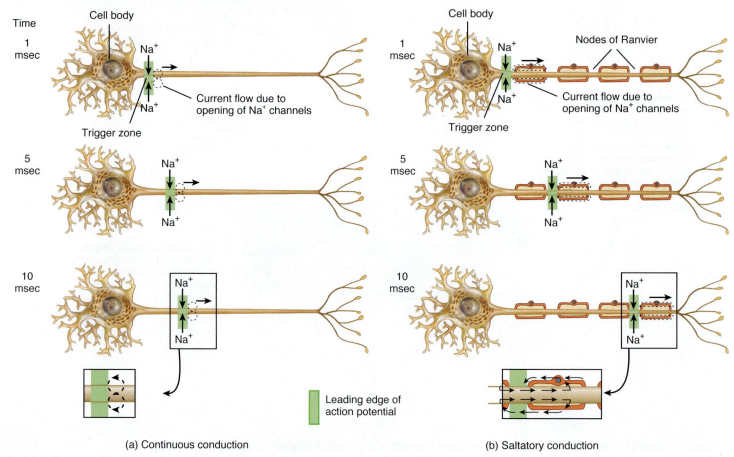

(a) Continuous conduction

(b) Saltatory conduction

🔑 **Unmyelinated axons exhibit continuous conduction; myelinated axons exhibit saltatory conduction.**

Because an action potential "leaps" across long segments of the myelinated axon as current flows from one node to the next, saltatory conduction moves action potentials much faster than continuous conduction in an unmyelinated axon of the same diameter.

2. Opening a smaller number of channels only at the nodes, rather than many channels in each adjacent segment of plasma membrane, results in minimal inflow of Na^+ and outflow of K^+ each time an action potential occurs. Thus, less ATP is used by sodium–potassium pumps to maintain the low intracellular concentration of Na^+ and the low extracellular concentration of K^+.

Factors That Affect the Speed of Propagation

The speed of propagation of an action potential is affected by three major factors: amount of myelination, axon diameter, and temperature.

- *Amount of myelination.* As you have just learned, action potentials propagate more rapidly along myelinated axons than along unmyelinated axons.

- *Axon diameter.* Larger-diameter axons propagate action potentials faster than smaller ones due to their larger surface areas. The largest-diameter axons (about 5–20 μm) propagate action potentials at speeds of 12 to 130 m/sec (27–280 mi/hr), medium-diameter axons (about 2–3 μm) propagate at speeds speeds up to 15 m/sec (32 mi/hr), and action potentials transmitted in the smallest-diameter axons (0.5–1.5 μm) travel only at a speed of 0.5 to 22 m/sec (1–44 mi/hr).

- *Temperature.* Axons propagate action potentials at lower speeds when cooled.

Encoding of Stimulus Intensity

How can your sensory systems detect stimuli of differing intensities if all nerve impulses are the same size? Why does a light touch feel different from firmer pressure? The main answer to such questions is the *frequency of action potentials*—how often they are generated at the trigger zone. A light touch generates a low frequency of action potentials. By contrast, a firmer grip elicits action potentials that pass down the axon at a higher frequency. In addition to this "frequency code" for stimulus intensity, a second factor is the number of sensory neurons recruited (activated) by the stimulus. For example, a firm grip stimulates more pressure-sensitive neurons than a light touch.

Comparison of Electrical Signals Produced by Excitable Cells

We have seen that excitable cells—neurons and muscle fibers—produce two types of electrical signals: graded potentials and action potentials (impulses). One obvious difference between them is that graded potentials function only in short-distance communication, but the propagation of action potentials permits communication over long distances. Table 12.1 presents a summary of the differences between graded potentials and action potentials.

Checkpoint

29. Why is saltatory conduction faster than continuous conduction of an action potential?
30. Why do large-diameter myelinated axons conduct action potentials more quickly than small-diameter unmyelinated axons?
31. How are stimuli of differing intensity detected as being different, such as warm soup versus hot soup?
32. Differentiate between graded potentials and action potentials.

TABLE 12.1		
Comparison of Graded Potentials and Action Potentials in Neurons		
CHARACTERISTIC	GRADED POTENTIALS	ACTION POTENTIALS
Origin	Arise mainly in dendrites and cell body.	Arise at trigger zones and propagate along the axon.
Types of channels	Ligand-gated or mechanically gated ion channels.	Voltage-gated channels for Na^+ and K^+.
Conduction	Not propagated; permit communication over short distances.	Propagated; permit communication over longer distances.
Amplitude (size)	Depending on strength of stimulus, varies from less than 1 mV to more than 50 mV.	All-or-none; typically about 100 mV.
Duration	Typically longer, ranging from several milliseconds to several minutes.	Shorter, ranging from 0.5 to 2 msec.
Polarity	May be hyperpolarizing (inhibitory to generation of an action potential) or depolarizing (excitatory to generation of an action potential).	Always consists of depolarizing phase followed by repolarizing phase and return to resting membrane potential.
Refractory period	Not present, thus summation can occur.	Present, thus summation cannot occur.

12.9   The synapse is a special junction between neurons.

The site of communication between two neurons or between a neuron and an effector cell is called a **synapse** (SIN-aps). In Chapter 10 we described the events occurring at one type of synapse, the neuromuscular junction. Our focus in this chapter is on synaptic communication among the billions of neurons in the nervous system. Neurons filter, integrate, and process information using synaptic communication. During learning, the structure and function of particular synapses change. For example, the changes in your synapses from studying will determine how well you do on your anatomy and physiology tests! Some diseases and neurological disorders result from disruptions of synaptic communication. Synapses also are the sites of action for many therapeutic and addictive chemicals.

At a synapse between neurons, the neuron sending the signal is called the **presynaptic neuron**, and the neuron receiving the message is called the **postsynaptic neuron** (Figure 12.21). Most synapses are **axoaxonic** (from axon to axon), **axodendritic** (from axon to dendrite), or **axosomatic** (from axon to cell body). The two types of synapses—electrical and chemical—differ both structurally and functionally.

Figure 12.21   Examples of synapses. Arrows indicate the direction of information flow: presynaptic neuron → postsynaptic neuron. Presynaptic neurons usually synapse on the axon (axoaxonic; red), a dendrite (axodendritic; blue), or the cell body (axosomatic; green).

Dendrites

Axon

Cell body

Axons of presynaptic neurons

Synapses

Postsynaptic neuron

LM 2500x

(a) Examples of synapses

(b) Synapses between neurons

🔑 **Neurons communicate with other neurons at synapses, which are junctions between the axon of one neuron and the dendrite, cell body, or axon of a second neuron.**

Chemical Synapses

Unlike the gap junctions in electrical synapses, the plasma membranes of presynaptic and postsynaptic neurons in a **chemical synapse** are close, but they are not connected directly (Figure 12.22). They are separated by the **synaptic cleft**, a space of 20–50 nm that is filled with interstitial fluid. Because action potentials cannot conduct across the synaptic cleft, an alternate, indirect form of communication occurs. In response to a nerve impulse, the presynaptic neuron releases a neurotransmitter that diffuses through the fluid in the synaptic cleft and binds to receptors in the plasma membrane of the postsynaptic neuron. The postsynaptic neuron receives the chemical signal and, in turn, produces a **postsynaptic potential**, a type of graded potential. Thus, the presynaptic neuron converts an electrical signal (nerve impulse) into a chemical signal (released neurotransmitter). The postsynaptic neuron receives the chemical signal and, in turn, generates an electrical signal (postsynaptic potential). The time required for these processes at a chemical synapse, a synaptic delay of about 0.5 msec, is the reason that chemical synapses relay signals more slowly than electrical synapses.

A typical chemical synapse transmits a signal as follows (Figure 12.22):

1 A nerve impulse arrives at the synaptic end bulb of a presynaptic axon.

2 The depolarizing phase of the nerve impulse opens **voltage-gated Ca^{2+} channels** in the membrane of synaptic end bulbs. Because calcium ions are more concentrated in the extracellular fluid, Ca^{2+} flows inward through the opened channels.

3 An increase in the concentration of Ca^{2+} inside the presynaptic neuron triggers exocytosis of some of the synaptic vesicles. As vesicle membranes merge with the plasma membrane, neurotransmitter molecules within the synaptic vesicles are released into the synaptic cleft. Each synaptic vesicle contains several thousand molecules of neurotransmitter.

4 The neurotransmitter molecules diffuse across the synaptic cleft and bind to **neurotransmitter receptors** in the postsynaptic neuron's plasma membrane. The receptor shown in Figure 12.22 is part of a ligand-gated channel (see Figure 12.10b).

5 Binding of neurotransmitter molecules to their receptors on ligand-gated channels opens the channels and allows particular ions to flow across the membrane.

6 As ions flow through the opened channels, the voltage across the membrane changes.

This change in membrane voltage is a **postsynaptic potential**. Depending on which ions the channels admit, the postsynaptic potential may be a depolarization or hyperpolarization. For example, opening of Na^+ channels allows inflow of Na^+, which causes depolarization. Opening of K^+ channels allows K^+ to move out, producing hyperpolarization.

7 When a depolarizing postsynaptic potential reaches threshold, it triggers an action potential in the postsynaptic neuron.

At most chemical synapses, only *one-way information transfer* can occur—from a presynaptic neuron to a postsynaptic neuron or to an effector, such as a muscle fiber or a gland cell. For example, synaptic transmission at a neuromuscular junction proceeds from a motor neuron to a skeletal muscle fiber (but not in the opposite direction). Only synaptic end bulbs of presynaptic neurons can release neurotransmitter, and only the postsynaptic neuron's membrane has the receptor proteins that can recognize and bind that neurotransmitter. As a result, action potentials move in one direction.

Figure 12.22 **Signal transmission at a chemical synapse.** Through exocytosis of synaptic vesicles, a presynaptic neuron releases neurotransmitter molecules. After diffusing across the synaptic cleft, the neurotransmitter binds to receptors in the plasma membrane of the postsynaptic neuron and produces a postsynaptic potential.

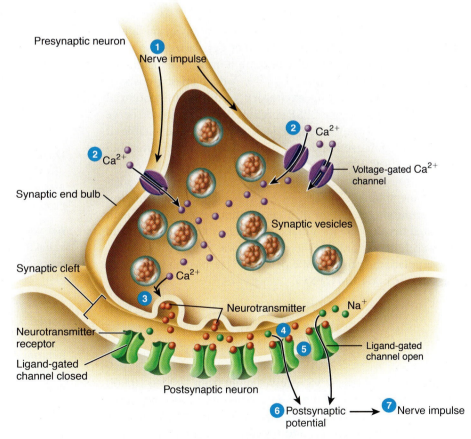

At a chemical synapse, a presynaptic neuron converts an electrical signal (nerve impulse) into a chemical signal (neurotransmitter release). The postsynaptic neuron then converts the chemical signal back into an electrical signal (postsynaptic potential).

Electrical Synapses

At an **electrical synapse**, action potentials conduct directly between adjacent cells through **gap junctions**, which act like tunnels to connect the cytosol of the two cells. As ions flow from one cell to the next, the action potential spreads from cell to cell. Gap junctions are common in visceral smooth muscle, cardiac muscle, and the CNS.

Electrical synapses have two main advantages:

• *Faster communication.* At an electrical synapse, the action potential passes directly from the presynaptic cell to the postsynaptic cell. Because action potentials conduct directly through gap junctions, electrical synapses are faster than chemical synapses.

• *Synchronization.* Electrical synapses can synchronize (coordinate) the activity of a group of neurons or muscle fibers. Large numbers of neurons can produce action potentials in unison if they are connected by gap junctions. Synchronizing action potentials in the heart or in visceral smooth muscle coordinates contractions that produce a heartbeat or move food through the gastrointestinal tract.

Excitatory and Inhibitory Postsynaptic Potentials

A neurotransmitter causes either an excitatory or an inhibitory graded potential. A neurotransmitter that depolarizes the postsynaptic membrane is excitatory because it brings the membrane closer to threshold (see Figure 12.13b). A depolarizing postsynaptic potential is called an **excitatory postsynaptic potential (EPSP)**. Although a single EPSP normally does not initiate a nerve impulse, the postsynaptic neuron does become more excitable. It is partially depolarized and thus more likely to reach threshold when the next EPSP occurs.

A neurotransmitter that causes *hyperpolarization* of the postsynaptic membrane (see Figure 12.13a) is inhibitory. During hyperpolarization, generation of a nerve impulse is more difficult than usual because the membrane potential is more negative and thus even farther from threshold than it was in its resting state. A hyperpolarizing postsynaptic potential is termed an **inhibitory postsynaptic potential (IPSP)**.

Summation of Postsynaptic Potentials

A typical neuron in the CNS receives input from 1000 to 10,000 synapses. Integration of these inputs involves **summation** of postsynaptic potentials. Recall that summation is the process by which graded potentials add together. The greater the sum-

mation of EPSPs, the greater is the chance that threshold will be reached. At threshold, one or more nerve impulses (action potentials) arise. A single postsynaptic neuron receives input from many presynaptic neurons, some of which release excitatory neurotransmitters and some of which release inhibitory neurotransmitters (Figure 12.23). The sum of all the excitatory and inhibitory effects at any given time determines the effect on the postsynaptic neuron, which may respond in one of the following ways:

• *EPSP.* If the total excitatory effects are greater than the total inhibitory effects but less than the threshold level of stimulation, the result is an EPSP that does not reach threshold. Following an EPSP, subsequent stimuli can more easily generate an action potential through summation because the neuron is partially depolarized.

• *Action potential(s).* If the total excitatory effects are greater than the total inhibitory effects and the threshold level of stimulation is reached, one or more action potentials will be triggered.

• *IPSP.* If the total inhibitory effects are greater than the excitatory effects, the membrane hyperpolarizes (IPSP). The result is inhibition of the postsynaptic neuron and an inability to generate an action potential.

Removal of Neurotransmitter

Removal of the neurotransmitter from the synaptic cleft is essential for normal synaptic function. If a neurotransmitter could linger in the synaptic cleft, it would influence the postsynaptic neuron, muscle fiber, or gland cell indefinitely. Neurotransmitter is removed in three ways:

• *Diffusion.* Some of the released neurotransmitter molecules diffuse away from the synaptic cleft. Once a neurotransmitter molecule is out of reach of its receptors, it can no longer exert an effect.

• *Enzymatic degradation.* Certain neurotransmitters are inactivated through enzymatic degradation. For example, the enzyme acetylcholinesterase breaks down acetylcholine in the synaptic cleft.

• *Uptake by cells.* Many neurotransmitters are actively transported back into the neuron that released them (reuptake). Others are transported into neighboring neuroglia (uptake). The neurons that release the neurotransmitter norepinephrine, for example, rapidly take up the norepinephrine and recycle it into new synaptic vesicles.

Table 12.2 summarizes the structural and functional elements of a neuron.

Figure 12.23 **Summation of postsynaptic potentials at the trigger zone of a postsynaptic neuron.** Presynaptic neurons 1, 3, and 5 release excitatory neurotransmitters (red dots) that generate excitatory postsynaptic potentials (EPSPs) (red arrows) in the membrane of a postsynaptic neuron. Presynaptic neurons 2 and 4 release inhibitory neurotransmitters (purple dots) that generate inhibitory postsynaptic potentials (IPSPs) (purple arrows) in the membrane of the postsynaptic neuron. The net summation of these EPSPs and IPSPs determines whether an action potential will be generated.

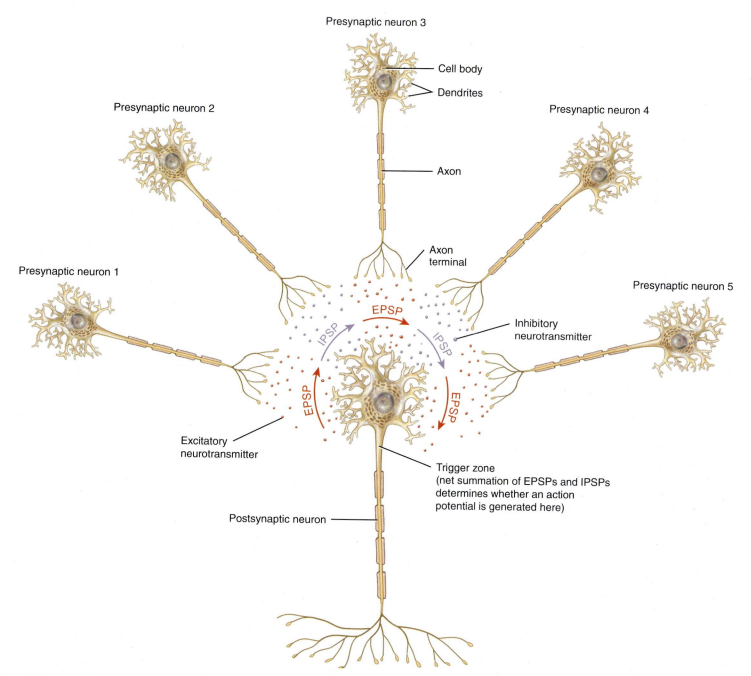

If the sum of all excitatory and inhibitory postsynaptic potentials is a depolarization that reaches threshold, then an action potential will occur at the trigger zone of a postsynaptic neuron.

TABLE 12.2

Neuronal Structure and Function

STRUCTURE	FUNCTIONS
Dendrites	Receive stimuli through activation of ligand-gated or mechanically gated ion channels; in sensory neurons, produce generator or receptor potentials; in motor neurons and interneurons, produce excitatory and inhibitory postsynaptic potentials (EPSPs and IPSPs).
Cell body	Receives stimuli and produces EPSPs and IPSPs through activation of ligand-gated ion channels.
Junction of axon hillock and initial segment of axon	Trigger zone in many neurons; integrates EPSPs and IPSPs and, if sum is a depolarization that reaches threshold, initiates action potential (nerve impulse).
Axon	Propagates nerve impulses from initial segment (or from dendrites of sensory neurons) to axon terminals in a self-regenerating manner; impulse amplitude does not change as it propagates along the axon.
Axon terminals and synaptic end bulbs (or varicosities)	Inflow of Ca^{2+} caused by depolarizing phase of nerve impulse triggers exocytosis of neurotransmitter from synaptic vesicles.

Key:

▬ Plasma membrane includes chemically gated channels

▬ Plasma membrane includes voltage-gated Na^+ and K^+ channels

▬ Plasma membrane includes voltage-gated Ca^{2+} channels

Checkpoint

33. How do electrical and chemical synapses differ?

34. What sequence of events must occur for the release of neurotransmitter into the synaptic cleft?

35. Why is calcium required for the release of neurotransmitter at the synapse?

36. Explain the effects of excitatory and inhibitory postsynaptic potentials on the postsynaptic neuron.

Return to Mary's Story

When Mary finally gets in to see the neurologist, she finds the examination to be quite different from her visits with Dr. Addison. In addition to taking an extensive medical history from Mary, the neurologist performs a neurological examination, checking Mary's cranial nerves, her reflexes, posture, gait, and balance. All are normal. However, the neurologist finds significant weakness in both hands and decreased sensation in the thumb and first three fingers. An electromyogram and a nerve conduction study are ordered to determine if the peripheral muscles or nerves are functioning properly. Both tests reveal that Mary's median nerve (the nerve that passes from the arm to the hand through the carpal tunnel of the wrist) is functioning abnormally, just as Dr. Addison had suspected. Based on Mary's history and the neurological exam, the neurologist diagnoses carpal tunnel syndrome.

Mary looks up as the neurologist steps back into the exam room. "Mary, there are several options for treatment of carpal tunnel syndrome, depending on the severity of the symptoms. I think in your case we should consider a lightweight wrist splint and possibly a vitamin B supplement. You can use an analgesic like acetaminophen or ibuprofen as you need it."

Mary leans forward in her chair and asks, "It will get better, though, won't it?"

The neurologist replies, "Well, if we can address the cause of the inflammation, perhaps reconsider your use of a laptop, and make other ergonomic corrections to your work space, you may not need surgery. But if you continue to work the way you have been, the symptoms could get worse."

F. Based on what you learned in Concepts 12.5 and 12.6, how could damage to the peripheral nerve and loss of myelination, or compression, cause a tingling sensation in Mary's hands?

G. How would damage to the motor axons in Mary's nerves cause weakness? Consider the coverage of propagation and conduction and the role of myelin in Concept 12.8 in your answer.

H. In view of the fact that Mary's neurological exam was otherwise normal, why might the neurologist be interested in Mary's electrolyte balance (i.e., Na^+ and K^+ distribution in her body)? How might changing Na^+ and K^+ levels alter action potential conduction?

I. Mary has been diagnosed with carpal tunnel syndrome. How does the neurologist rule out a central nervous system disorder or other conditions that may be causing peripheral nerve damage or demyelination (loss of myelin)?

12.10 PNS neurons have a greater capacity for repair and regeneration than CNS neurons.

Throughout your life, your nervous system exhibits **plasticity**, the capability to change based on experience. At the level of individual neurons, the changes that can occur include the sprouting of new dendrites, synthesis of new proteins, and changes in synaptic contacts with other neurons. Undoubtedly, both chemical and electrical signals drive the changes that occur. Despite plasticity, however, mammalian neurons have very limited powers of repair or **regeneration**, the capability to replace destroyed cells.

Damage and Repair in the PNS

In the PNS, damage to dendrites and myelinated axons may be repaired if the cell body remains intact and if the Schwann cells that produce myelination remain active (Figure 12.24). Most nerves in the PNS consist of processes that are covered with a neurolemma. A person who injures axons of a nerve in an upper limb, for example, has a good chance of regaining nerve function.

When there is damage to an axon, changes usually occur in both the cell body and axon of the affected neuron. Following injury, the Nissl bodies break up into fine granular masses (Figure 12.24b). This alteration is called **chromatolysis** (krō-ma-TOL-i-sis; *chromato-* = color; *-lysis* = destruction). Within days the part of the axon distal to the damaged region breaks up into fragments and the myelin sheath deteriorates (Figure 12.24b). Degeneration of the distal portion of the axon and myelin sheath is called **Wallerian degeneration**.

Following chromatolysis, signs of recovery in the cell body become evident. Macrophages phagocytize the debris. Schwann cells on either side of the injured site multiply by mitosis, grow toward each other, and may form a **regeneration tube** across the injured area (Figure 12.24c). The tube guides growth of a new axon from the proximal area, across the injured area, into the distal area occupied by the original axon, and eventually toward receptors and effectors previously contacted by the neuron. Thus, some sensory and motor connections are reestablished and some functions restored. In time, the Schwann cells form a new myelin sheath.

Damage and Repair in the CNS

Little or no repair of damage to neurons occurs in the brain and spinal cord. Axons in the CNS are myelinated by oligodendrocytes rather than Schwann cells, and this CNS myelin is one of the factors inhibiting regeneration of neurons. Also, after axonal damage, nearby astrocytes proliferate rapidly, forming scar tissue that is a physical barrier to regeneration. Thus, injury of the brain or spinal cord usually is permanent. Ongoing research

Figure 12.24 Damage and repair of a neuron in the PNS.

(a) Normal neuron

(b) Chromatolysis and Wallerian degeneration

(c) Regeneration

🔑 Myelinated axons in the peripheral nervous system may be repaired if the cell body remains intact and if Schwann cells remain active.

seeks ways to improve the environment for existing spinal cord axons to bridge the injury gap. Scientists also are trying to find ways to replace neurons lost through damage or disease and to develop tissue-cultured neurons that can be used for transplantation purposes.

Checkpoint

37. Describe the repair process for a damaged PNS neuron.
38. How are Schwann cells involved in regeneration?
39. How is repair of nervous tissue in the PNS different from what occurs in the CNS?

Mary's Story: Epilogue and Discussion

▶▶▶Mary, a 28-year-old administrative assistant in good health, recently postpartum, visits the doctor complaining of intermittent numbness, tingling, and weakness in both her hands. She has no family history of neurological disorders, and her job requires her to use her hands in a repetitive manner. A neurological exam reveals a nerve conduction abnormality in the median nerves of both wrists. Mary is tentatively diagnosed with carpal tunnel syndrome. As you have learned in this chapter, the nervous system receives sensory information, processes this information in the central nervous system, and allows a response to sensory stimuli through the peripheral nervous system. Along with the endocrine system, the nervous system is primarily responsible for maintaining homeostasis. The information you have learned in this chapter will be crucial to your understanding of neu-

rological illness. Hopefully, you have seen that Mary's problem could be diagnosed and understood fairly well in the context of what you have learned so far about the nervous system. ◀◀◀

J.   Why would reducing movement of Mary's wrist (with the splints) and swelling around the median nerve (with adjustment of her work habits) reduce Mary's symptoms?

K.   Would hand-strengthening exercises be appropriate for Mary?

L.   How likely is it that Mary will recover full function of her hands and have reduced numbness and tingling if she follows her neurologist's advice?

Concept and Resource Summary

Concept	Resources

Introduction

1. The nervous system, together with the endocrine system, governs homeostasis by sensing and responding to changes in normal physiological body set points.
2. The nervous system controls all body activities. The field of medicine that studies and treats nervous system diseases is **neurology**.

Concept 12.1 The nervous system maintains homeostasis and integrates all body activities.

1. **Neurons** and **neuroglia** are the cells that form nervous tissue.
2. Nervous system structures include the **brain**, **spinal cord**, and peripheral **cranial nerves** arising from the brain, and spinal nerves originating from the spinal cord.
3. Small collections of neurons outside the brain and spinal cord, called **ganglia**, plus the **sensory receptors**, are also part of the nervous system.
4. The **sensory function** involves detection of internal changes by **sensory receptors** that relay this information to **sensory** or **afferent neurons** of peripheral nerves that carry the information to the spinal cord and brain.
5. The **integrative function** involves the processing of sensory information by **interneurons** of the brain and spinal cord.
6. The **motor function** occurs when **motor** or **efferent neurons** carry motor commands from the brain to the spinal cord and then to peripheral nerves that innervate body **effectors** such as muscles and glands.

See
- **Anatomy Overview – The Nervous System: Overview**
- **Animation – Neuron Structure and Function**

Concept 12.2 The nervous system is organized into the central and peripheral nervous systems.

1. The **central nervous system (CNS)** consists of the **brain** and **spinal cord**. The CNS is responsible for integration of incoming sensory information, generation of motor commands, and thoughts and memories.
2. The **peripheral nervous system (PNS)** is composed of **cranial** and **spinal nerves**, **ganglia**, and **sensory receptors**. Subdivisions of the PNS include the **somatic nervous system (SNS)**, **autonomic nervous system (ANS)**, and **enteric nervous system (ENS)**.
3. The SNS consists of **sensory neurons** that conduct impulses from somatic and special sense receptors to the CNS. The SNS also has **motor neurons** that innervate skeletal muscles under **voluntary control.**
4. The ANS consists of sensory neurons that conduct impulses from **visceral organs** to the CNS and motor neurons that convey impulses from the CNS to smooth muscle tissue, cardiac muscle tissue, and glands.
5. The ENS consists of neurons in **enteric plexuses** in the gastrointestinal (GI) tract that function somewhat independently of the ANS and CNS. The ENS monitors sensory changes in and controls operation of the GI tract.

See
- **Animation – Structure and Function of the Nervous System: System Organization**

| Concept | Resources **WILEY PLUS** |

Concept 12.3 Neurons are responsible for most of the unique functions of the nervous system.

1. There are three parts to most neurons: cell body, dendrites, and axon.
2. The **cell body** houses the nucleus surrounded by cytoplasm containing organelles such as clusters of rough endoplasmic reticulum called Nissl bodies, lysosomes, mitochondria, Golgi complexes, and a cytoskeleton of neurofibrils and microtubules.
3. Multiple **dendrites** are processes that receive incoming impulses.
4. A single long, thin **axon** conducts impulses away from the cell body toward another neuron, muscle fiber, or gland cell.
5. The **axon hillock** is where the axon joins the cell body. **Axon collaterals** are side branches of an axon. The **axoplasm** is the axon's cytoplasm.
6. **Axon terminals** are numerous, fine processes at the ends of an axon and its collaterals.
7. Synaptic end bulbs are swollen tips of axon terminals where communication occurs between the neuron and other cells.
8. Newly synthesized cell products move by **anterograde transport** from the cell body toward the axon. Materials for recycling and degradation move from the axon to the cell body by **retrograde transport**.
9. Neurons can be classified structurally as **multipolar** (several dendrites and one axon), **bipolar** (one dendrite and one axon), or **unipolar** (one process arising from the cell body that branches into two axon-like processes).
10. Multipolar neurons are common in the CNS; bipolar neurons are found in the retina, the inner ear, and the brain's olfactory area; and unipolar neurons are sensory neurons.

See
- Anatomy Overview – Nerve
- Anatomy Overview – Nervous Tissue: Neuron
- Animation – Neuron Structure and Function

Hear
- **Figure 12.3 –** Structure of a Typical Motor Neuron and Schwann Cells

Do
- Exercise – Paint a Neuron

Concept 12.4 Neuroglia support, nourish, and protect neurons and maintain homeostasis.

1. Neuroglia are smaller and more numerous than neurons. They can divide but cannot conduct impulses.
2. There are six types of neuroglia, each of which has one or more unique functions: astrocytes, oligodendrocytes, microglia, ependymal cells, Schwann cells, and satellite cells.
3. The **myelin sheath** is a multilayered lipid and protein covering produced by layers of plasma membrane from neuroglia that wrap around myelinated axons.
4. In the PNS, Schwann cells form myelin sheaths; in the CNS, oligodendrocytes form myelin sheaths.
5. The unmyelinated gaps occurring at intervals along an axon are called the **nodes of Ranvier**.
6. The **neurolemma** is the outermost, nucleated layer of a Schwann cell that encloses the myelin sheath and functions in regeneration of damaged PNS axons.
7. The myelin sheath of CNS oligodendrocytes lacks a neurolemma. CNS axons show little regeneration.
8. Myelination increases from birth to maturity and functions to increase speed of impulse conduction along an axon.
9. Nervous system **white matter** appears white due to the presence of myelinated axons; the darker **gray matter** is composed of unmyelinated axons, cell bodies, and neuroglia.

See
- Anatomy Overview – Nerve
- Anatomy Overview – Nervous Tissue: Neuroglia
- Animation – Factors That Affect Conduction Rates: Myelination

Hear
- **Figure 12.5 –** Neuroglia of the Central Nervous System

Concept 12.5 Neurons communicate with other cells.

1. Neurons are electrically excitable cells that communicate by electrical signals called **graded potentials** over short distances or by **action potentials** over longer distances.
2. **Nerve action potentials** (*nerve impulses*) occur in neurons, whereas muscle action potentials occur in muscle fibers.
3. Graded potentials typically occur in sensory receptors, dendrites, and cell bodies, which allow axons to form nerve action potentials resulting in release of **neurotransmitters**.
4. Excitable cell plasma membranes have a **resting membrane potential** due to an electrical voltage difference across the membrane established by specific **ion channels**.
5. Gated ion channels open or close in response to specific stimuli. When the channels open, the movement of cations and anions results in a flow of electrical current that can change the membrane potential.
6. There are four types of ion channels: **leakage channels** have alternating open and closed gates; **voltage-gated channels** open when the membrane potential changes and participate in action potentials; **ligand-gated channels** open and close due to a chemical stimulus in which ligand binds to a membrane receptor; and **mechanically gated channels** open and close due to mechanical stimulation.
7. The **resting membrane potential** is more negative along the inside of the membrane and more positive along the outside membrane surface. The separation of these charges forms potential energy.
8. A cell with a membrane potential, such as a neuron, is **polarized**; neurons typically have a resting membrane potential of −70 mV.
9. Resting membrane potentials result from differences in distribution of charged ions across the membrane and from a greater permeability to K^+ and Cl^- compared to Na^+.

See
- Animation – Introduction to Membrane Potentials
- Animation – Membrane Transport Proteins
- Animation – Resting Membrane Potential

Hear
- **Figure 12.11 –** Resting Membrane Potential

Do
- Exercise – Keep the Resting Potential

Concept

Concept 12.6 Graded potentials are the first response of a neuron to stimulation.

1. The opening or closing of ligand-gated and mechanically gated channels in response to a stimulus produces graded potentials.
2. **Depolarizing graded potentials** result in a less polarized membrane.
3. **Hyperpolarizing graded potentials** result in an even more polarized membrane.
4. The size of a graded potential varies with the strength of the stimulus.
5. Opening or closing of ion channels in graded potentials causes a *localized* flow of current along the membrane.

See
- Animation – Graded Potentials

Concept 12.7 The action potential is an all-or-none electrical signal.

1. **Action potentials (nerve impulses)** are rapid electrical events occurring in two phases: the depolarizing phase and the repolarizing phase.
2. Action potentials follow the **all-or-none principle**; once **threshold** depolarization occurs, voltage-gated channels open, and an action potential that is always the same size occurs.
3. During the **depolarizing phase**, voltage-gated Na^+ channels quickly open, Na^+ rushes into the cell, and the membrane potential becomes positive.
4. During the **repolarizing phase**, voltage-gated K^+ channels slowly open, K^+ flows out of the cell, and the membrane is repolarized.
5. Following an action potential there is a **refractory period** during which an excitable cell cannot generate another action potential.
6. The **absolute refractory period**, which is briefer in large-diameter axons than smaller ones, is the time when even a very strong stimulus cannot generate a second action potential.
7. During a **relative refractory period**, a very strong stimulus can initiate a second action potential.

See
- Animation – Action Potentials

Hear
- Figure 12.17 –
  Action Potential

Do
- Exercise – Ruling the Gated Channels
- Exercise – Define Stages of the Action Potential
- Concepts and Connections – Membrane Potentials

Concept 12.8 Action potentials propagate from the trigger zone to axon terminals.

1. When Na^+ flows into open channels in one area of the membrane, **positive feedback** causes depolarization along adjacent regions, resulting in **propagation** of the nerve impulse toward the axon terminals.
2. **Continuous conduction** occurs in unmyelinated axons and muscle fibers.
3. Myelinated axons undergo a more rapid **saltatory conduction** in which Na^+ and K^+ flow across the membrane only at the unmyelinated nodes of Ranvier as the impulse leaps from one node to another.
4. Impulses are conducted at a faster pace along larger-diameter axons.
5. Greater stimulus intensity, in addition to a greater number of recruited sensory neurons, leads to higher frequency of impulses and the ability to differentiate between stimuli.
6. Action potentials permit communication over long distances because they are propagated; graded potentials function only in short-distance communication because they are not propagated. Their origins, amplitudes, duration, and types of channels also differ.

See
- Animation – Propagation of Nerve Impulses

Do
- Exercise – Investigate Types of Impulse Conduction
- Exercise – Speed Up Nerve Impulse Propagation

Concept 12.9 The synapse is a functional junction between neurons.

1. The **synapse** is the junction where communication occurs between two neurons or a neuron and an effector.
2. **Electrical synapses** conduct impulses from a neuron to another cell through **gap junctions**. Electrical synapses allow faster communication and synchronization of activity of cell groups. They are common in visceral smooth muscle and cardiac muscle.
3. **Chemical synapses** conduct impulses from a neuron to another cell through a **synaptic cleft**. A nerve impulse in a presynaptic axon opens Ca^{2+} channels in synaptic end bulbs, leading to release of neurotransmitter that crosses the synaptic cleft and binds to postsynaptic neuron receptors, causing a **postsynaptic potential**.
4. Some neurotransmitters will either depolarize the postsynaptic neuron's membrane to produce an **excitatory postsynaptic potential (EPSP)**, bringing the membrane potential closer to threshold. Other neurotransmitters will hyperpolarize the postsynaptic neuron's membrane to produce an **inhibitory postsynaptic potential (IPSP)**, moving the membrane potential farther from threshold.
5. **Summation** of postsynaptic potentials determines whether the postsynaptic neuron will generate an action potential.
6. Neurotransmitter is removed from the synaptic cleft by **diffusion**, **enzymatic degradation**, and **cellular uptake**.

See
- Animation – Events at the Synapse

Do
- Exercise – Are Your Synapses Working?
- Exercise – Summation Smash
- Concepts and Connections – Synaptic Summation

Concept 12.10 PNS neurons have a greater capacity for repair and regeneration than CNS neurons.

1. Repair of PNS neurons begins with **Wallerian degeneration** of the distal region of the axon and myelin sheath. Debris is phagocytized, and neuronal RNA and protein synthesis increases.
2. A **regeneration tube** formed by Schwann cells allows entry of the regenerating axon.
3. CNS neurons have little to no repair capacity.

Other Chapter Resources

For additional information on clinical applications, medical disorders, and medical terminology, see Chapter 7 in the *Clinical Connections* manual packaged with your text.

Within *WileyPLUS*

- Chapter Quiz
- Visual Anatomy
- Mastering Vocabulary
- Audio Glossary
- Homeostatic Imbalance: The Case of the Girl with On and Off Symptoms
- Web Links

• Understanding the Concepts

1. What types of problems would result if your sensory neurons were damaged? If your interneurons were damaged? If your motor neurons were damaged?

2. Which of the following are voluntary? Involuntary? Both? Explain each of your answers.
   - Moving your eyes across this page.
   - Blinking your eyes.
   - The tearing of your eyes when sand is blown into them.

3. What structural type of neuron would your eyes likely use to receive images of words on this page? What structural type of neuron would your brain likely use to determine what the words mean? Explain the steps involved in each process.

4. Which component of the nervous system would assist the repair of severed axons following a cut to your finger? Explain how this repair would occur.

5. Your toothpaste tube is almost empty. Starting with the seam end, you tightly wrap the tube around a pencil until the last of the toothpaste is collected in the cap region of the tube. Assuming the toothpaste tube and pencil are analogous to components of nervous tissue, what structure would be represented by each of the following?
   - The cap region of the toothpaste tube containing the remaining toothpaste.
   - The tightly rolled component of the tube that is now empty of toothpaste.
   - The pencil.

6. What would happen to the membrane potential of a membrane with mechanically gated channels if it were subjected to continuous pressure?

7. Why may electrical synapses work in two directions, but chemical synapses can transmit a signal in one direction only?

8. How would the arrival of hyperpolarizing graded potentials at the trigger zone of a neuron affect the generation of action potentials?

9. If the myelin sheath were destroyed or damaged, would you expect action potential propagation to speed up or slow down? Why?

13

The Central Nervous System

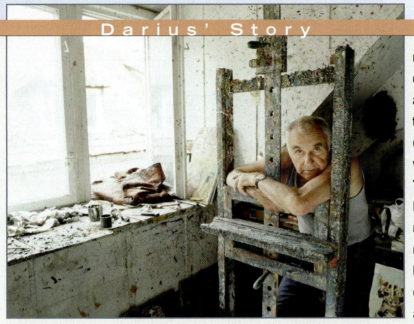

Darius, a 45-year-old artist, has been working all day in his studio completing a new portrait commissioned by the 15th Duke of Buccleuch, when suddenly he drops his brush and cries out in pain. Paolo, Darius' young apprentice, comes running in from the next room.

"Dar, what's wrong?" asks Paolo urgently.

"I don't know, my head, it feels as if I've been hit by lightning, like it's exploding, the worst headache of my life, I can't…stand…the pain…"

Suddenly, Darius staggers forward and vomits. Paolo rushes toward Darius and manages to catch him just as he loses consciousness and collapses to the floor. After repeated attempts to wake him, Paolo makes a hasty call to the 911 operator.

The paramedics arrive to find Darius unconscious but still breathing. His heartbeat is strong but his pulse is racing (115) and his blood pressure is elevated to 200/105. Darius still does not respond to attempts to rouse him; his pupils are dilated but responsive to light. Paolo describes what happened and tells the paramedics that Darius has a history of high blood pressure for which he takes medication. The paramedics administer IV saline and oxygen, and rush him to the emergency room. The ER physician, on hearing the description of his collapse, immediately orders a CT scan of Darius' head. The scan confirms the doctor's suspicion: Darius has suffered a subarachnoid hemorrhage, a rupture of a blood vessel within the membranes surrounding his brain, and the hemorrhage is bleeding into the space between his skull and brain.

In this chapter we will explore the structure of the brain and spinal cord, how they are protected and nourished, and the functions of their various regions. Through an understanding of Darius's case, you will begin to appreciate the complexity of the central nervous system and understand how it functions to control your higher thought processes, consciousness, and basic bodily functions.

INTRODUCTION

The central nervous system (CNS) consists of the brain and the spinal cord. The brain is the control center for your thoughts, emotions, creativity, wisdom, memories, and all the activities and behaviors that make you who you are. Made up of about 100 billion neurons and 10–50 trillion neuroglia, the adult brain has a mass of only about 1300 g (almost 3 lb).

Humans have realized the importance of the brain for thousands of years. The Egyptians are given credit for the first anatomical descriptions of the brain, which date back as far as 3000 B.C. Egyptian records describe paralysis, loss of cognitive function, and personality changes associated with skull fracture and brain damage. As far back as the fifth century B.C., the Greek physician

The central nervous system consists of the brain and spinal cord.

Hippocrates realized the importance of the brain in mediating personality, emotion, and thought. Some South American cultures were even known to perform a primitive form of brain surgery known as *trepa-nation* (trep-uh-NĀ-shun), in which a hole was drilled into the skull with bronze surgical tools (see Chapter 7 case story). Although our understanding of the human brain has improved dramatically since ancient times, especially over the last 150 years, we have only begun to scratch the surface.

The spinal cord controls some of your most rapid reactions to environmental changes. If you pick up something hot, you may drop the object even before the sensation of extreme heat or pain reaches your conscious perception. The spinal cord is the pathway for sensory nerve impulses traveling to the brain and motor nerve impulses traveling from the brain toward skeletal muscles and other effectors.

13.1 The CNS consists of the brain, the spinal cord, and several protective structures.

The brain consists of four major parts: cerebrum, diencephalon, brain stem, and cerebellum (Figure 13.1). The **cerebrum** (se-RĒ-brum = brain) is the largest part of the brain and is divided into two sides, the left and right **cerebral hemispheres** (Figure 13.1b). Immediately below the cerebrum is the **diencephalon** (dī-en-SEF-a-lon; *di-* = through; *-encephalon* = brain), consisting mainly

Figure 13.1   The brain. The pituitary gland is discussed with the endocrine system in Chapter 17.

(a) Sagittal section, medial view

(b) Anterior

CEREBRUM

DIENCEPHALON:
Thalamus

Hypothalamus

BRAIN STEM:
Midbrain

CEREBELLUM

Pons

Medulla oblongata

Spinal cord

(c) Sagittal section, medial view

🔑 **The four principal parts of the brain are the cerebrum, diencephalon, brain stem, and cerebellum.**

of the thalamus and hypothalamus. The diencephalon provides a structural connection between the cerebrum and the **brain stem**. The brain stem consists of the midbrain, pons, and medulla oblongata (Figure 13.1c). Posterior to the brain stem is the **cerebellum** (ser´-e-BEL-um = little brain). The **spinal cord** extends inferiorly from the medulla oblongata to the level of the first or second lumbar vertebra (Figure 13.2).

Two types of connective tissue coverings protect the central nervous system: (1) the bony skull and vertebrae and (2) the cranial and spinal meninges. Additionally, a cushion of cerebrospinal fluid (produced in the brain) surrounds and protects the delicate nervous tissue of the brain and spinal cord.

Skeletal Protection

The brain is located within the cranial cavity of the skull, formed by the securely interlocked cranial bones (Figure 13.3). The spinal cord is located within the vertebral canal of the vertebral column. The vertebral foramina of all the vertebrae, stacked one on top of the other, form the vertebral canal. The surrounding vertebrae provide a sturdy shelter to protect the enclosed spinal cord against trauma (see Figure 13.4c).

Meninges

The **meninges** (me-NIN-jēz); singular is *meninx* (MEN-inks) are three connective tissue coverings that encircle the brain and spinal cord. The **cranial meninges** surround and protect the brain and

are continuous with the **spinal meninges**, which surround and protect the spinal cord. The cranial meninges and the spinal meninges have the same basic organizational structure and bear the same anatomical names: the outer dura mater, the middle arachnoid mater, and the inner pia mater (Figures 13.3 and 13.4).

The most superficial and strongest of the three meninges is the **dura mater** (DOO-ra MĀ-ter = tough mother), which is composed of dense, irregular connective tissue. Three extensions of the dura mater separate portions of the brain. (1) The **falx cerebri** (FALKS SER-e-brē; *falx* = sickle-shaped) separates the two hemispheres (sides) of the cerebrum. (2) The **falx cerebelli** (cer-e-BEL-li) separates the two hemispheres of the cerebellum. (3) The **tentorium cerebelli** (ten-TŌ-rē-um = tent) separates the cerebrum from the cerebellum. The dura mater of the brain adheres directly to the periosteum of the interior surface of cranial bones. Within the vertebral canal, an **epidural space** exists between the dura mater and the wall of the vertebral canal. The spinal cord is protected not only by the strong dura mater, but also by a cushion of fat and connective tissue located within the epidural space. The dura mater of the brain contains **dural sinuses**, spaces within the dura mater itself such as the **superior sagittal sinus** (see Figure 13.3a), which drain blood from the brain and deliver it to the internal jugular veins of the neck.

Deep to the dura mater is the **arachnoid mater** (a-RAK-noyd; *arachn-* = spider; *-oid* = similar to), an avascular covering so named because of its spider's web arrangement of delicate collagen fibers and some elastic fibers. The arachnoid mater surrounds both the brain and spinal cord. Between the dura mater and the arachnoid mater is a thin **subdural space**, which contains interstitial fluid.

Figure 13.2   **External anatomy of the spinal cord and the spinal nerves.**

Medulla oblongata

Atlas (first cervical vertebra)

Cervical enlargement

First thoracic vertebra

Lumbar enlargement

First lumbar vertebra

Conus medullaris

Cauda equina

Ilium

Sacrum

Filum terminale

Posterior view of entire spinal cord and portions of spinal nerves

The spinal cord extends from the medulla oblongata of the brain to the superior border of the second lumbar vertebra.

Figure 13.3 **The protective coverings of the brain.**

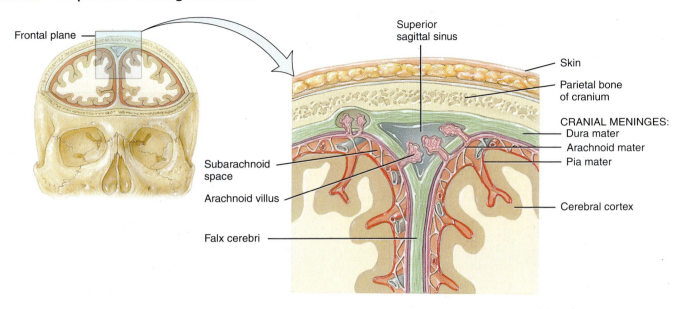

(a) Anterior view of frontal section through skull showing the cranial meninges

(b) Sagittal section of extensions of the dura mater

🔑 Cranial bones and the cranial meninges protect the brain.

The innermost membrane is the **pia mater** (PĒ-a MĀ-ter; *pia* = delicate), a thin, transparent connective tissue layer that adheres tightly to the surface of the brain and spinal cord. It consists of interlacing bundles of collagen fibers and some fine elastic fibers. Between the arachnoid mater and the pia mater is the **subarachnoid space**, which contains cerebrospinal fluid. As blood ves-sels pass along the surface of the brain and spinal cord, they are sheathed by a loose-fitting sleeve of pia mater.

All three spinal meninges cover the spinal nerves up to the point of exit from the spinal column through the intervertebral foramina. Triangular-shaped membranous extensions of the pia mater, called **denticulate ligaments** (den-TIK-ū-lāt = small

Figure 13.4   Gross anatomy of the spinal cord. The spinal meninges are evident in both views.

SPINAL CORD:
Gray matter
White matter

Posterior
median sulcus
Central canal
Anterior
median fissure

SPINAL
MENINGES:
Pia mater
(inner)

Spinal nerve

Denticulate
ligament

Arachnoid
mater
(middle)

Subarachnoid
space

Dura mater
(outer)

Subdural
space

(a) Anterior view and transverse section through spinal cord

SUPERIOR

Fourth ventricle

Cerebellum of
brain (cut)

Occipital bone
(cut)

Posterior median
sulcus

Denticulate
ligament

Dura mater
and arachnoid (cut)

Posterior
rootlets of
spinal nerve

INFERIOR

(b) Posterior view of cervical region of spinal cord

View

Transverse
plane

POSTERIOR

Spinous process of
vertebra

Subarachnoid space

Posterior root of spinal
nerve

Denticulate ligament

Anterior root
of spinal nerve

Transverse foramen

Body of vertebra

Dura mater and
arachnoid mater

Spinal cord

Pia mater

Epidural space

Superior articular
facet of vertebra

Posterior ramus
of spinal nerve

Spinal nerve

Anterior ramus
of spinal nerve

Vertebral artery in
transverse foramen

ANTERIOR

(c) Transverse section of the spinal cord within a cervical vertebra

🔑 **Meninges are connective tissue coverings that surround the spinal cord and brain.**

tooth), suspend the spinal cord in the middle of its dural sheath. They project laterally and fuse with the arachnoid mater and inner surface of the dura mater between the anterior and posterior nerve roots of spinal nerves on either side. Extending all along the length of the spinal cord, the denticulate ligaments protect the spinal cord against sudden displacement that could result in shock.

Checkpoint

1.   Which of the four principal parts of the brain is the largest? Which part attaches to the spinal cord?
2.   Where are the spinal meninges located? What are the three layers of the meninges, from superficial to deep?
3.   Where are the epidural, subdural, and subarachnoid spaces located?
4.   In which meningeal space does cerebrospinal fluid circulate?

13.2 The CNS is nourished and protected by blood and cerebrospinal fluid.

Blood Flow to the Brain and Spinal Cord

Blood supplies oxygen and nutrients to the CNS. It flows to the brain mainly via the internal carotid and vertebral arteries (see Figure 20.18). Blood leaving the brain flows into the dural sinuses, which empty into the internal jugular veins (see Figure 20.23). Blood is supplied to the spinal cord by the posterior intercostal and lumbar arteries (see Figure 20.19). Blood leaves the spinal cord in the posterior intercostal and lumbar veins (see Figure 20.25).

In an adult, the brain represents only 2% of total body weight, but it consumes about 20% of the oxygen and glucose used even at rest. Neurons synthesize ATP almost exclusively from glucose via reactions that use oxygen (oxidative phosphorylation in mitochondria). When the activity of neurons and neuroglia increases in a particular region of the brain, blood flow to that area also increases.

Even a brief slowing of brain blood flow may cause unconsciousness. Typically, an interruption in blood flow for 1 or 2 minutes impairs neuronal function, and total deprivation of oxygen for about 4 minutes causes permanent injury. Because virtually no glucose is stored in the brain, the supply of glucose also must be continuous. If blood entering the brain has a low level of glucose, mental confusion, dizziness, convulsions, and loss of consciousness may occur. If the arterial supply to the brain is blocked, the resulting loss of blood flow can damage the brain, producing a cerebrovascular accident (CVA) or stroke. CVAs are the most common brain disorder. They affect 500,000 people each year in the United States and represent the third leading cause of death, behind heart attacks and cancer.

The **blood–brain barrier (BBB)** protects the CNS from harmful substances and pathogens by preventing passage of these substances from blood into the interstitial fluid of brain tissue. Once inside brain tissue, the cerebral arteries quickly divide into capillaries. Tight cell junctions seal together the endothelial cells of brain capillaries, which also are surrounded by a thick basement membrane. In addition, the processes of many astrocytes (one type of neuroglia) press up against the capillaries. The astrocytic processes secrete chemicals that maintain the permeability characteristics of the endothelial cell junctions, allowing selective passage of some substances from blood into brain tissue while inhibiting the passage of others. A few water-soluble substances (for example, glucose) cross the BBB by active transport. Other substances, such as creatinine, urea, and most ions, cross the BBB very slowly. Still other substances—proteins and most antibiotic drugs—do not pass at all from the blood into brain tissue. However, lipid-soluble substances, such as oxygen, carbon dioxide, alcohol, and most anesthetic agents, easily cross the blood–brain barrier. Trauma, certain toxins, and inflammation can cause a breakdown of the blood–brain barrier.

Cerebrospinal Fluid

Cerebrospinal fluid (CSF) is a clear, colorless liquid that protects the brain and spinal cord against chemical and physical injuries. It also carries oxygen, glucose, and other needed chemicals from the blood to neurons and neuroglia. CSF continuously circulates through cavities in the brain and spinal cord and around the brain and spinal cord in the subarachnoid space (between the arachnoid mater and pia mater).

The total volume of CSF is 80 to 150 milliliters (3 to 5 oz) in an adult. CSF contains glucose, proteins, lactic acid, urea, cations (Na^+, K^+, Ca^{2+}, Mg^{2+}), and anions (Cl^- and HCO_3^-); it also contains some white blood cells. The CSF contributes to homeostasis in three main ways:

- **Mechanical protection**. CSF serves as a shock-absorbing medium that protects the delicate tissue of the brain and spinal cord from jolts that would otherwise cause them to hit the bony walls of the cranial cavity and vertebral canal. The fluid also buoys the brain so that it "floats" in the cranial cavity.

- **Chemical protection**. CSF provides an optimal chemical environment for accurate neuronal signaling. Even slight changes in the ionic composition of CSF within the brain can seriously disrupt production of action potentials and postsynaptic potentials.

- **Circulation**. CSF is a medium for exchange of nutrients and waste products between the blood and nervous tissue.

Formation of CSF in the Ventricles

Figure 13.5 shows the four CSF-filled cavities within the brain, which are called **ventricles** (VEN-tri-kuls = little cavities). One **lateral ventricle** is located in each hemisphere of the cerebrum. Anteriorly, the lateral ventricles are separated by a thin membrane, the **septum pellucidum** (SEP-tum pe-LOO-si-dum; *pellucid* = transparent; see Figures 13.5b and 13.6a). The **third ventricle** is a narrow cavity along the midline superior to the hypothalamus and between the right and left halves of the thalamus. The **fourth ventricle** lies between the brain stem and the cerebellum.

Figure 13.5   Locations of ventricles within a "transparent" brain. One interventricular foramen on each side connects a lateral ventricle to the third ventricle, and the cerebral aqueduct connects the third ventricle to the fourth ventricle.

POSTERIOR

ANTERIOR

Cerebrum

LATERAL VENTRICLES

INTERVENTRICULAR FORAMEN

FOURTH VENTRICLE

THIRD VENTRICLE

LATERAL APERTURE

Cerebellum

CEREBRAL AQUEDUCT

MEDIAN APERTURE

Pons

Medulla oblongata

CENTRAL CANAL

Spinal cord

(a) Right lateral view of brain

Lateral
ventricle

Septum
pellucidum

Interventricular
foramen

Third
ventricle

Cerebral
aqueduct

Fourth
ventricle

Lateral
aperture

Central
canal

(b) Anterior view of brain

Ventricles are cavities within the brain that are filled with cerebrospinal fluid.

The sites of CSF production are the **choroid plexuses** (KŌ-royd = membranelike), networks of capillaries (microscopic blood vessels) in the walls of the ventricles. The capillaries are covered by ependymal cells (a type of neuroglia) that form cerebrospinal fluid from blood plasma by filtration and secretion. Because the ependymal cells are tightly joined by cell junctions, materials entering CSF from choroid capillaries cannot leak between these cells; instead, they must pass through the ependymal cells. This **blood–cerebrospinal fluid barrier** permits certain substances to enter the CSF but excludes others, protecting the brain and spinal cord from potentially harmful blood-borne substances. A point of clarification: The blood–brain barrier controls passage of substances from blood into interstitial fluid of neural tissue, whereas the blood–cerebrospinal fluid barrier controls passage of substances from blood into CSF.

Circulation of CSF

Ependymal cells line the ventricles of the brain and central canal of the spinal cord. Movement of their cilia assists in the flow of CSF. The CSF formed in the choroid plexuses of each lateral ventricle flows into the third ventricle through two narrow, oval openings, the **interventricular foramina** (singular is *foramen*; Figure 13.6). Additional CSF is added by the choroid plexus in the roof of the third ventricle. The fluid then flows through the **cerebral aqueduct**, which passes through the midbrain, into the fourth ventricle. The choroid plexus of the fourth ventricle contributes more fluid. CSF enters the subarachnoid space through three openings in the roof of the fourth ventricle: a **median aperture** and the paired **lateral apertures**, one on each side. CSF then circulates in the central canal of the spinal cord and in the subarachnoid space around the surface of the brain and spinal cord.

Figure 13.6 **Pathways of circulating cerebrospinal fluid.**

(a) Superior view of transverse section of brain showing choroid plexuses

(continues)

Figure 13.6 (continued)

POSTERIOR

ANTERIOR

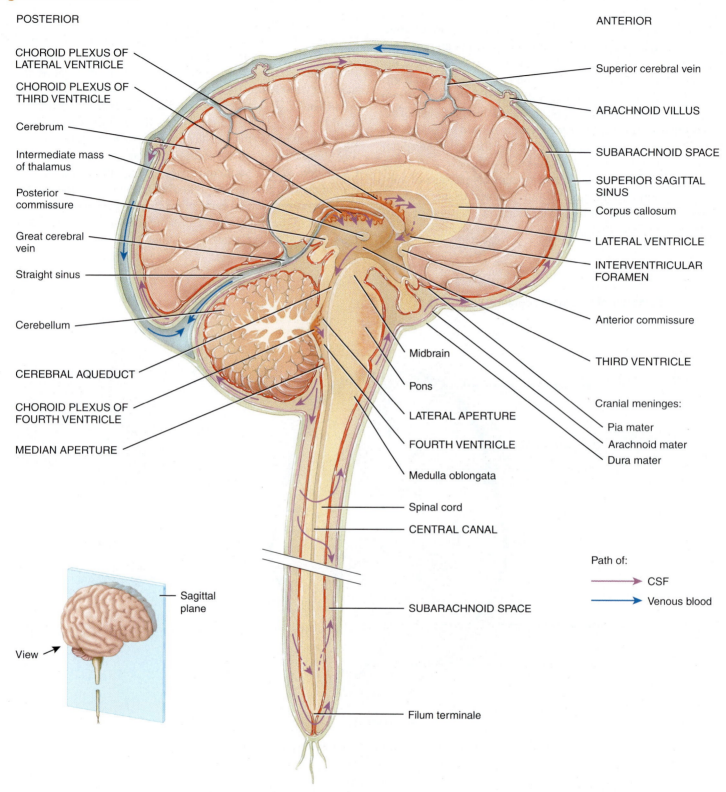

CHOROID PLEXUS OF
LATERAL VENTRICLE

CHOROID PLEXUS OF
THIRD VENTRICLE

Cerebrum

Intermediate mass
of thalamus

Posterior
commissure

Great cerebral
vein

Straight sinus

Cerebellum

CEREBRAL AQUEDUCT

CHOROID PLEXUS OF
FOURTH VENTRICLE

MEDIAN APERTURE

Superior cerebral vein

ARACHNOID VILLUS

SUBARACHNOID SPACE

SUPERIOR SAGITTAL
SINUS

Corpus callosum

LATERAL VENTRICLE

INTERVENTRICULAR
FORAMEN

Anterior commissure

THIRD VENTRICLE

Cranial meninges:

Pia mater

Arachnoid mater

Dura mater

Midbrain

Pons

LATERAL APERTURE

FOURTH VENTRICLE

Medulla oblongata

Spinal cord

CENTRAL CANAL

SUBARACHNOID SPACE

Filum terminale

Sagittal
plane

View

Path of:

CSF

Venous blood

(b) Sagittal section of brain and spinal cord

Superior sagittal sinus

ARACHNOID VILLUS

Falx cerebri

LATERAL VENTRICLE

CHOROID PLEXUS

Cerebrum

CEREBRAL AQUEDUCT

Cerebellum

FOURTH VENTRICLE

Corpus callosum
Septum pellucidum

THIRD VENTRICLE

SUBARACHNOID SPACE
(surrounding brain)

Tentorium cerebelli

LATERAL APERTURE

MEDIAN APERTURE

Frontal
plane

View

SPINAL CORD

SUBARACHNOID SPACE
(surrounding spinal cord)

(c) Frontal section of brain and spinal cord

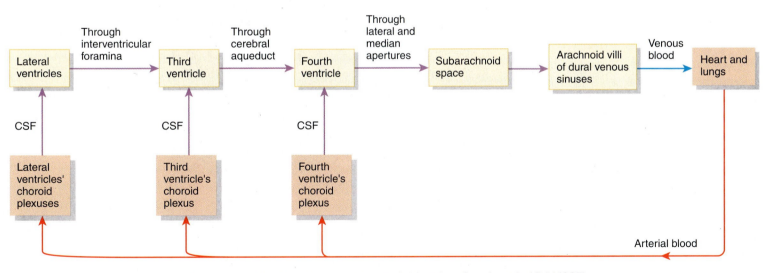

(d) Summary of the formation, circulation, and absorption of cerebrospinal fluid (CSF)

CSF is formed by ependymal cells that cover the choroid plexuses of the ventricles.

CSF is gradually reabsorbed from the subarachnoid space into the blood through **arachnoid villi**, fingerlike extensions of the arachnoid mater that project into the dural sinuses, especially the superior sagittal sinus (see Figure 13.3a). (A sinus is similar to a vein but has thinner walls.) Normally, CSF is reabsorbed as rapidly as it is formed by the choroid plexuses, at a rate of about 20 mL/hr (480 mL/day). Because the rates of formation and reabsorption are the same, the pressure of CSF normally is constant. Figure 13.6d summarizes the production and flow of CSF.

Checkpoint

5. What vessels return blood from the head to the heart?
6. How does CSF protect the brain and spinal cord?
7. Which brain region is anterior to the fourth ventricle? Which is posterior to it?
8. What structures are the sites of CSF production, and where are they located?
9. Where is CSF reabsorbed into the blood?

13.3 The cerebrum interprets sensory impulses, controls muscular movements, and functions in intellectual processes.

The cerebrum is the "seat of intelligence." It provides us with the ability to read, write, and speak; to make calculations and compose music; and to remember the past, plan for the future, and imagine things that have never before existed.

The cerebrum consists of an outer cerebral cortex, an internal region of cerebral white matter, and nuclei, masses of gray matter deep within the white matter. Gray matter, where neurons form synapses with one another, receives and integrates incoming and outgoing information. The **cerebral cortex** (*cortex* = rind or bark) is a region of gray matter that forms the outer rim of the cerebrum (Figure 13.7). Although only 2–4 mm (0.08–0.16 in.) thick, the cerebral cortex contains billions of neurons.

Figure 13.7   Cerebrum. Because the insula cannot be seen externally, it has been projected to the surface in (b).

Gyrus
Sulcus
Cerebral cortex
Cerebral white matter
Fissure

Details of a gyrus, sulcus, and fissure

ANTERIOR

Longitudinal fissure
Precentral gyrus
Central sulcus
Postcentral gyrus

Frontal lobe
Parietal lobe
Occipital lobe

Left hemisphere   Right hemisphere

POSTERIOR

(a) Superior view

Postcentral gyrus

Parietal lobe

Parieto-occipital sulcus

Occipital lobe

Transverse fissure

Cerebellum

Central sulcus

Precentral gyrus

Frontal lobe

Insula (projected to surface)

Lateral cerebral sulcus

Temporal lobe

(b) Right lateral view

POSTERIOR

ANTERIOR

Central sulcus

Postcentral gyrus

Parietal lobe

Occipital lobe

Cerebellum

Precentral gyrus

Frontal lobe

Insula

Temporal lobe (cut)

Medulla oblongata

Spinal cord

(c) Right lateral view with temporal lobe cut away

The cerebrum provides us with the ability to read, write, and speak; to make calculations and compose music; to remember the past and plan for the future; and to create.

During embryonic development, when brain size increases rapidly, the gray matter of the cortex enlarges much faster than the deeper white matter. As a result, the cortical region rolls and folds upon itself. The folds are called **gyri** (JĪ -rī = circles). The deepest grooves between folds are known as **fissures**; the shallower grooves between folds are termed **sulci** (SUL-sī = grooves). (The singular terms are *gyrus* and *sulcus*.) The most prominent fissure, the **longitudinal fissure**, separates the cerebrum into right and left halves called **cerebral hemispheres**. The cerebral hemispheres are connected internally by the **corpus callosum** (kal-LŌ-sum; *corpus* = body; *callosum* = hard), a broad band of white matter containing axons that extend between the hemispheres (see Figure 13.6).

Lobes of the Cerebrum

Each cerebral hemisphere can be further subdivided into four lobes. The lobes are named after the bones that cover them: frontal, parietal, temporal, and occipital lobes (Figure 13.7). The **central sulcus** (SUL-kus) separates the **frontal lobe** from the **parietal lobe**. The **precentral gyrus** is located immediately anterior to the central sulcus, and the **postcentral gyrus** is located immediately posterior to the central sulcus. The **lateral cerebral sulcus** separates the **frontal lobe** from the **temporal lobe**. The **parieto-occipital sulcus** separates the **parietal lobe** from the **occipital lobe**. A fifth part of the cerebrum, the **insula**, cannot be seen at the surface

of the brain because it lies within the lateral cerebral sulcus, deep to the parietal, frontal, and temporal lobes (Figure 13.7b,c).

Cerebral White Matter

The **cerebral white mattter** consists primarily of **tracts**, highways of myelinated axons that propagate impulses throughout the CNS to allow communication between its regions. Whereas tracts are bundles of axons in the CNS, recall that nerves are bundles of axons in the PNS. The three types of tracts are as follows (Figure 13.8):

- *Association tracts* contain axons that conduct nerve impulses between gyri in the same hemisphere.

- *Commissural tracts* contain axons that conduct nerve impulses from gyri in one cerebral hemisphere to corresponding gyri in the other cerebral hemisphere. Three important groups of commissural tracts are the **corpus callosum**, **anterior commissure**, and **posterior commissure.**

- *Projection tracts* contain axons that conduct nerve impulses from the cerebrum to lower parts of the CNS (thalamus, brain stem, or spinal cord) or from lower parts of the CNS to the cerebrum. An example is the **internal capsule**, a thick band of white matter that contains both ascending and descending axons (see Figure 13.9b).

Figure 13.8 **Organization of white matter tracts of the left cerebral hemisphere.**

Medial view of tracts revealed by removing gray matter from a midsagittal section

Association tracts, commissural tracts, and projection tracts form white matter areas in the cerebral hemispheres.

Basal Nuclei

Deep within each cerebral hemisphere are three nuclei (masses of neuronal cell bodies) that are collectively termed the **basal nuclei** (**basal ganglia**) (Figure 13.9). Two of the basal nuclei are side by side, just lateral to the thalamus. They are the **globus pal-**lidus (GLŌ-bus PAL-i-dus; *globus* = ball; *pallidus* = pale), which is closer to the thalamus, and the **putamen** (pū-TĀ-men = shell), which is closer to the cerebral cortex. The third member of the basal nuclei is the **caudate nucleus** (*caud-* = tail), which has a large "head" connected to a smaller "tail" by a long, comma-shaped "body."

Figure 13.9   Basal nuclei. The basal nuclei are shown in purple. In (a) the basal nuclei have been projected to the surface.

(a) Lateral view of right side of brain

POSTERIOR   ANTERIOR

Lateral ventricle — Body of caudate nucleus
Thalamus — Frontal lobe of cerebrum
Tail of caudate nucleus — Putamen
Occipital lobe of cerebrum — Head of caudate nucleus

(b) Anterior view of frontal section

Frontal plane
View

Longitudinal fissure — Cerebrum
Septum pellucidum — Corpus callosum
Internal capsule — Lateral ventricle
Insula — Caudate nucleus
Thalamus — Putamen — } Basal nuclei
Hypothalamus and associated nuclei — Globus pallidus
— Third ventricle
— Optic tract

The basal nuclei help initiate and terminate movements, suppress unwanted movements, and regulate muscle tone.

The basal nuclei receive input from the cerebral cortex and provide output to motor portions of the cerebral cortex via nuclei of the thalamus. In addition, basal nuclei have extensive connections with one another. A major function of the basal nuclei is to help initiate and terminate movements. The basal nuclei also suppress unwanted movements and regulate muscle tone. In addition, the basal nuclei influence many aspects of cortical function, including sensory, limbic, cognitive, and linguistic functions. Basal nuclei functions are described in more detail in Chapter 15.

See Table 13.2 for a summary of the functions of the cerebrum.

Checkpoint

10. During development, does the gray matter or white matter enlarge more rapidly?
11. What are the brain folds, shallow grooves, and deep grooves called?
12. What deep groove separates the right and left cerebral hemispheres? What shallow grooves separate each lobe of the cerebrum from the next?
13. Which tracts carry impulses between gyri of the same hemisphere? Between gyri in opposite hemispheres? Between the cerebrum, thalamus, brain stem, and spinal cord?

13.4 The limbic system controls emotions, behavior, and memory.

Encircling the upper part of the brain stem and the corpus callosum is a ring of structures on the inner border of the cerebrum and floor of the diencephalon that constitutes the **limbic system** (*limbic* = border; Figure 13.10). The limbic system is sometimes called the "emotional brain" because it plays a primary role in a range of emotions, including pleasure, pain, docility, affection, fear, and anger.

Although behavior is a function of the entire nervous system, the limbic system controls most of its involuntary aspects related to survival. Animal experiments suggest that it has a major role in controlling overall patterns of behavior. When different areas of the limbic system are stimulated, the animals' reactions indicate extreme pleasure or intense pain. Stimulation of other limbic system areas produces tameness and signs of affection. Stimulation of a cat's **amygdala** (a-MIG-da-la), one of the basal nuclei of the cerebrum, produces fear and a behavioral pattern called rage—the cat extends its claws, raises its tail, opens its eyes wide, hisses, and spits. By contrast, removal of the amygdala produces an animal that lacks fear and aggression. Likewise, a person whose amygdala is damaged fails to recognize fearful expressions in others or fails to express fear in appropriate situations.

Together with other parts of the cerebrum, the limbic system also functions in memory. People with damage to certain limbic system structures forget recent events and cannot commit anything to memory.

Checkpoint

14. What are the functions of the limbic system?

Figure 13.10   The limbic system. Components of the limbic system are shown in green.

Sagittal plane

View

The limbic system governs emotional aspects of behavior.

13.5 The cerebral cortex can be divided functionally into sensory areas, motor areas, and association areas.

Specific types of sensory, motor, and integrative signals are processed in certain regions of the cerebral cortex (Figure 13.11). Generally, *sensory areas* receive sensory impulses, *motor areas* initiate voluntary movements, and *association areas* deal with more complex integrative functions such as memory, emotions, reasoning, will, judgment, personality traits, and intelligence.

Sensory Areas

Sensory impulses arrive mainly in the posterior half of both cerebral hemispheres, in regions behind the central sulci. In the cerebral cortex, primary sensory areas receive information from peripheral sensory receptors.

The following are some important sensory areas (Figure 13.11):

- The **primary somatosensory area** is located in each parietal lobe in the postcentral gyrus directly posterior to the central sulcus. It extends from the lateral cerebral sulcus, along the lateral surface of the parietal lobe to the longitudinal fissure, and then along the medial surface of the parietal lobe within the longitudinal fissure. The primary somatosensory area receives nerve impulses for touch, proprioception (joint and muscle position), pain, itch, tickle, and temperature (coldness and warmth). A "map" of the entire body is present in the primary somatosensory area: Each point within the area receives impulses from a specific part of the body (see Figure 15.5a). The size of the cortical area receiving impulses from a particular part of the body depends on the number of receptors present rather than on the size of the body part. For example, the region of the somatosensory area that receives impulses from the lips and fingertips is larger than the area that receives impulses from the thorax or hip. The major function of the primary somatosensory area is to localize the exact points of the body where somatic sensations originate, so you know exactly where on your body to swat that mosquito.

- The **primary visual area** is located at the posterior tip of the occipital lobe, mainly on the medial surface (next to the longitudinal fissure). It receives impulses that convey information

Figure 13.11 **Functional areas of the cerebrum.** Broca's speech area is in the left cerebral hemisphere of most people; it is shown here to indicate its relative location.

Central sulcus

PRIMARY SOMATOSENSORY AREA (postcentral gyrus)

SOMATOSENSORY ASSOCIATION AREA

Parietal lobe

COMMON INTEGRATIVE AREA

WERNICKE'S AREA

VISUAL ASSOCIATION AREA

PRIMARY VISUAL AREA

Occipital lobe

Temporal lobe

PRIMARY MOTOR AREA (precentral gyrus)

PREMOTOR AREA

PRIMARY GUSTATORY AREA

FRONTAL EYE FIELD AREA

Frontal lobe

BROCA'S AREA

PREFRONTAL CORTEX

Lateral cerebral sulcus

PRIMARY AUDITORY AREA

AUDITORY ASSOCIATION AREA

POSTERIOR

ANTERIOR

Lateral view of right cerebral hemisphere

🔑 **Particular areas of the cerebral cortex process sensory, motor, and integrative signals.**

for vision. Axons of neurons in the eye carry visual information to the thalamus. From the thalamus, neurons project to the primary visual area, carrying information concerning shape, color, and movement of visual stimuli.

• The **primary auditory area** is located in the superior part of the temporal lobe near the lateral cerebral sulcus. It receives information concerning the basic characteristics of sound, such as pitch and rhythm.

• The **primary gustatory** area is located at the base of the postcentral gyrus superior to the lateral cerebral sulcus in the parietal cortex. It receives impulses for taste.

• The **primary olfactory area** is located in the temporal lobe on the medial aspect (and thus is not visible in Figure 13.11). It receives impulses for smell.

Motor Areas

Motor output from the cerebral cortex flows mainly from the anterior part of each hemisphere. Among the most important motor areas are the following (Figure 13.11):

• The **primary motor area** is located in each frontal lobe in the precentral gyrus directly anterior to the central sulcus. It extends along the lateral surface of the frontal lobe to the longitudinal fissure, and then along the medial surface of the frontal lobe within the longitudinal fissure. Each region in the primary motor area controls voluntary contractions of specific muscles or groups of muscles (see Figure 15.5b). Electrical stimulation of any point in the primary motor area causes contraction of specific skeletal muscle fibers on the opposite side of the body. As is true for representation in the primary somatosensory area, different muscles are represented unequally in the primary motor area. More cortical area is devoted to those muscles involved in skilled, complex, or delicate movement. For instance, the cortical region devoted to muscles that move the fingers is much larger than the region for muscles that move the toes, which is why most pianists play with their fingers.

• **Broca's area** (BRŌ-kaz) is located in the frontal lobe close to the lateral cerebral sulcus. Speaking and understanding language are complex activities that involve several sensory, association, and motor areas of the cortex. The planning and production of speech occurs in Broca's area, located in one frontal lobe—the *left* frontal lobe in 97% of the population. From Broca's area, nerve impulses pass to the premotor area to control the muscles of the larynx, pharynx, and mouth for the specific, coordinated contractions that enable you to speak. Simultaneously, impulses propagate from Broca's area to the primary motor area to control the breathing muscles and to regulate the proper flow of air past the vocal cords.

Association Areas

The association areas consist of large areas of all lobes of the cerebral cortex. Association areas often are adjacent to the primary sensory areas. They usually receive input both from the primary sensory areas and from other brain regions. Association areas integrate sensory experiences to generate meaningful patterns of recognition and awareness. A person with damage in the *primary* visual area would be blind in at least part of his visual field; a person with damage to a visual *association* area might see normally yet be unable to recognize a friend. Association areas are connected with one another by association tracts and include the following (Figure 13.11):

• The **somatosensory association area** is just posterior to and receives input from the primary somatosensory area, as well as from the thalamus and other parts of the brain. It integrates and interprets sensations. This area permits you to determine the exact shape and texture of an object by feeling it, to determine the orientation of one object with respect to another as they are felt, and to sense the relationship of one body part to another. Another role of the somatosensory association area is the storage of memories of past somatic sensory experiences, enabling you to compare current sensations with previous experiences. For example, the somatosensory association area allows you to recognize objects such as a pencil and a paper clip simply by touching them.

• The **prefrontal cortex** is an extensive area in the anterior portion of the frontal lobe that is well developed in primates, especially humans. This area has numerous connections with other areas of the cerebral cortex, thalamus, hypothalamus, limbic system, and cerebellum. The prefrontal cortex is concerned with the makeup of a person's personality, intellect, complex learning abilities, recall of information, initiative, judgment, foresight, reasoning, conscience, intuition, mood, planning for the future, and development of abstract ideas. A person with bilateral damage to the prefrontal cortices typically becomes rude, inconsiderate, incapable of accepting advice, moody, inattentive, less creative, unable to plan for the future, and incapable of anticipating the consequences of rash or reckless words or behavior.

• The **visual association area** is located in the occipital lobe anterior to the primary visual area. It receives sensory impulses from the primary visual area and the thalamus. It relates present and past visual experiences and is essential for recognizing and evaluating what is seen. For example, the visual association area allows you to recognize an object such as a spoon simply by looking at it.

• The **auditory association area** is located inferior and posterior to the primary auditory area in the temporal cortex. It allows you to realize whether a sound is speech, music, or noise.

- **Wernicke's area** (VER-ni-kēz) is a broad region in the left temporal and parietal lobes. This area interprets the meaning of speech by recognizing spoken words. It is active as you translate words into thoughts. The regions in the *right* hemisphere that correspond to the locations of Broca's and Wernicke's areas in the left hemisphere also contribute to verbal communication by adding emotional content, such as anger or joy, to spoken words.

- The **common integrative area** is bordered by the somatosensory, visual, and auditory association areas. It integrates sensory interpretations from all sensory association areas, allowing the formation of thoughts based on a variety of sensory inputs. This area then transmits signals to other parts of the brain to cause the appropriate response to the sensory signals it has interpreted.

- The **premotor area** is a motor association area that is immediately anterior to the primary motor area. The premotor area controls learned, skilled motor activities of a complex and sequential nature. It generates nerve impulses that cause specific groups of muscles to contract in a specific sequence. Through repetition, the premotor area serves as a memory bank to store specific patterns of movements. For instance, while the primary motor area controls general movements of muscles of the hand, the premotor area directs muscles of the hand to tie a bow.

- The **frontal eye field area** in the frontal cortex controls voluntary scanning movements of the eyes—like those you just used in reading this sentence.

Hemispheric Lateralization

Although the brain is almost symmetrical on its right and left sides, subtle anatomical differences between the two hemispheres exist. This asymmetry appears in the human fetus at about 30 weeks of gestation. Furthermore, although the two hemispheres share performance of many functions, each hemisphere also specializes in performing certain unique functions. This functional asymmetry is termed **hemispheric lateralization**.

In the most obvious example of hemispheric lateralization, the left hemisphere receives somatic sensory signals from and controls muscles on the right side of the body, and the right hemisphere receives sensory signals from and controls muscles on the left side of the body. In most people the left hemisphere is more important for reasoning, numerical and scientific skills, spoken and written language, and the ability to use and understand sign language. For example, patients with damage in the left hemisphere may be unable to properly articulate or form words. Conversely, the right hemisphere is more specialized for musical and artistic awareness; spatial and pattern perception; recognition of faces and emotional content of language; discrimination of different smells; and generating mental images of sight, sound, touch, taste, and smell to compare relationships among them. Patients with damage in the right hemisphere regions speak in a monotonous voice, having lost the ability to impart emotional inflection to what they say.

Despite some dramatic differences in functions of the two hemispheres, there is considerable variation from one person to another. Lateralization seems less pronounced in females than in males, both for language (left hemisphere) and for visual and spatial skills (right hemisphere). A possibly related observation is that females have, on average, a 12% larger anterior commissure and a broader posterior portion of the corpus callosum than males. Recall that both the anterior commissure and the corpus callosum are commissural tracts that provide communication between the two hemispheres.

Table 13.1 summarizes some of the distinctive functions that are more likely to reside in the left or the right hemisphere.

TABLE 13.1	
Functional Differences between the Two Cerebral Hemispheres	
LEFT HEMISPHERE FUNCTIONS	RIGHT HEMISPHERE FUNCTIONS
Receives somatic sensory signals from and controls muscles on right side of body.	Receives somatic sensory signals from and controls muscles on left side of body.
Reasoning.	Musical and artistic awareness.
Numerical and scientific skills.	Space and pattern perception.
Ability to use and understand sign language.	Recognition of faces and emotional content of facial expressions.
Spoken and written language.	Generating emotional content of language.
	Generating mental images to compare spatial relationships.
	Identifying and discriminating among odors.

Checkpoint

15. What does the term *hemispheric lateralization* mean?
16. Which areas of the cerebrum contain neurons that demonstrate hemispheric lateralization for speech?

Return to Darius' Story

Paolo paces nervously in the waiting room. The neurosurgeon comes in and explains that a cerebral angiogram has shown that one of the arteries that supply blood to Darius' brain has weakened. The artery formed a balloonlike structure called an aneurysm, which has ruptured. "We need to perform surgery immediately to clamp the ruptured blood vessel. He has a fifty–fifty chance of coming through, even if the surgery is successful," states the surgeon.

Once Paolo regains his composure, he begins making calls to Darius' friends and family. After several hours, the surgeon finally returns.

"Paolo, Darius is doing very well. We were able to clamp the aneurysm. It was located below the brain at the anterior communicating artery. We'll be most concerned with keeping his blood pressure down and monitoring his recovery for the next few hours in the ICU. Darius is awake, he can move all his extremities, and he is responding well to verbal commands. All we can do now is keep a close watch on him."

Paolo has been allowed in to the ICU to see Darius. When Darius has difficulty staying awake, Paolo calls the nurse.

"Darius, open your eyes. Darius, squeeze my hand if you can hear me!" calls out the nurse. The nurse turns to Paolo, saying, "He's not responding. You'll have to leave," and calls the neurologist.

After ordering another CT scan, the neurologist learns that Darius has developed acute hydrocephalus, or elevated CSF pressure. She informs Paolo that blood has leaked into Darius' brain, increasing the volume of fluid in the space around Darius' brain. "We need to insert a tube, called a shunt, into his head to drain the excess fluid so that his brain is not permanently damaged."

Following surgery, Darius wakes up but has severe weakness on his right side. He can obviously hear, yet is unable to respond to verbal commands from the ICU staff. The neurologist orders another CT scan and finds that Darius has suffered a left middle cerebral artery stroke. She consults again with Paolo. "Your friend has suffered a stroke on the left side of his brain. The blood that has entered the space around the brain irritated the blood vessels, causing them to constrict. In Darius' case, the areas of the brain that control the understanding and initiation of speech have been damaged by the lack of blood flow. He has what we call aphasia. At this time, there's no way to tell if he will recover his ability to speak."

A. Between which of Darius' membranes has bleeding occurred?

B. What would be the result of the loss of blood flow to Darius' brain?

C. Which damaged areas of Darius' brain prevent him from speaking and from understanding speech?

D. If the stroke has occurred on the left side of the brain, why does Darius have weakness on his right side?

13.6 The diencephalon includes the thalamus, hypothalamus, and pineal gland.

The **diencephalon** extends from the brain stem to the cerebrum and surrounds the third ventricle. Major regions of the diencephalon include the thalamus, hypothalamus, and pineal gland.

Thalamus

The **thalamus** (THAL-a-mus = inner chamber) measures about 3 cm (1.2 in.) in length and makes up 80% of the diencephalon (Figure 13.12). It consists of paired oval masses of gray matter, organized into nuclei, with interspersed tracts of white matter (Figures 13.1 and 13.9b). In most people, a bridge of gray matter called the **intermediate mass** crosses the third ventricle, joining the right and left halves of the thalamus (Figure 13.12). A vertical Y-shaped sheet of white matter called the **internal medullary lamina** divides the gray matter of the right and left sides of the thalamus. It consists of myelinated axons that enter and leave the various thalamic nuclei. Axons that connect the thalamus and cerebral cortex pass through the **internal capsule**, a thick band of white matter lateral to the thalamus (see Figure 13.9b).

The thalamus relays and processes ascending sensory and descending motor information. The thalamus is the major relay station for most sensory impulses that reach the primary sensory areas of the cerebral cortex from the spinal cord and brain stem. In addition, the thalamus contributes to motor functions by transmitting information from the cerebellum and basal nuclei to the primary motor area of the cerebral cortex. It also relays nerve impulses between different areas of the cerebrum.

In addition to serving as a relay station, thalamic nuclei function along with other parts of the brain in movement control, emotions, maintenance of consciousness, pain perception, learning, memory, and cognition (thinking and knowing).

Figure 13.12 **Thalamus.** Note the position of the thalamus in (a) the lateral view, and in (b) the medial view. The various thalamic nuclei shown in (c) are correlated by color to the cortical regions in (a) and (b) to which they project.

(a) Lateral view of right cerebral hemisphere

(b) Medial view of left cerebral hemisphere

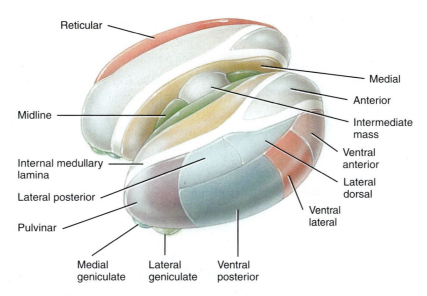

(c) Superolateral view of thalamus showing locations of thalamic nuclei (reticular nucleus is shown on the left side only; all other nuclei are shown on the right side)

The thalamus is the main relay station for sensory impulses that reach the cerebral cortex from other parts of the brain and the spinal cord.

Hypothalamus

The **hypothalamus** (*hypo-* = under) is a small part of the diencephalon located inferior to the thalamus (see Figures 13.1 and 13.9b). It is composed of a dozen or so nuclei, as shown in Figure 13.13. Among these are the **mammillary bodies**, small, rounded nuclei projecting from the hypothalamus that serve as relay stations for reflexes related to the sense of smell. The stalklike **infundibulum** connects the pituitary gland to the hypothalamus.

The hypothalamus controls many body activities and is one of the major regulators of homeostasis. Sensory impulses related to both somatic and visceral senses arrive at the hypothalamus, as do impulses from receptors for vision, taste, and smell. Other receptors within the hypothalamus itself continually monitor conditions within the blood, including osmotic pressure, glucose level, certain hormone concentrations, and temperature. The hypothalamus has several very important connections with the pituitary gland and produces a variety of hormones. Some functions can be attributed to specific hypothalamic nuclei, but others are not so precisely localized. Important functions of the hypothalamus include the following:

- **Control of the ANS**. The hypothalamus controls and integrates activities of the autonomic nervous system (ANS), which regulates contraction of smooth muscle and cardiac muscle and the secretions of many glands. Axons extend from the hypothalamus to sympathetic and parasympathetic nuclei in the brain stem and spinal cord. Through the ANS, the hypothalamus is a major regulator of visceral activities, including heart rate, movement of food through the gastrointestinal tract, and contraction of the urinary bladder.

- **Production of hormones**. The hypothalamus produces several hormones and has two types of important connections with the pituitary gland, an endocrine gland located inferior to the hypothalamus (see Figures 13.1 and 13.13). First, hypothalamic hormones travel in the bloodstream through the infundibulum to the pituitary, where they stimulate or inhibit secretion of pituitary hormones. Second, axons extend from the hypothalamus through the infundibulum into the pituitary. The cell bodies of these neurons make one of two hormones (oxytocin or antidiuretic hormone). Their axons transport the hormones to the pituitary, where they are released.

- **Regulation of emotional and behavioral patterns**. Together with the limbic system, the hypothalamus participates in expressions of rage, aggression, pain, and pleasure, and the behavioral patterns related to sexual arousal.

Figure 13.13   Hypothalamus. Shown is a three-dimensional representation of hypothalamic nuclei located within the wall of the third ventricle.

Sagittal section of brain showing hypothalamic nuclei

The hypothalamus controls many body activities and is an important regulator of homeostasis.

- *Regulation of eating and drinking*. The hypothalamus regulates food intake. It contains a **feeding center**, which promotes eating, and a **satiety center**, which causes a sensation of fullness and cessation of eating. The hypothalamus also contains a **thirst center**. When certain cells in the hypothalamus are stimulated by rising osmotic pressure of the extracellular fluid, they cause the sensation of thirst. The intake of water by drinking restores the osmotic pressure to normal, removing the stimulation and relieving the thirst.

- *Control of body temperature*. The hypothalamus also functions as the body's *thermostat*. If the temperature of blood flowing through the hypothalamus is above normal, the hypothalamus directs the autonomic nervous system to stimulate activities that promote heat loss. When blood temperature is below normal, the hypothalamus generates impulses that promote heat production and retention.

- *Regulation of circadian rhythms and states of consciousness*. The hypothalamus serves as the body's biological clock by establishing patterns of awakening and sleep that occur on a circadian (daily) schedule.

Pineal Gland

The **pineal gland** (PĪN-ē-al = pinecone like) is about the size of a small pea and protrudes from the posterior midline of the third ventricle (Figure 13.13). Because the pineal gland secretes the hormone **melatonin**, it is considered part of the endocrine system. Melatonin appears to contribute to the setting of the body's biological clock. Melatonin is also thought to promote sleepiness because more melatonin is liberated during darkness than in light.

The functions of the diencephalon are summarized in Table 13.2.

Checkpoint

17. What structure usually connects the right and left halves of the thalamus?
18. Which part of the diencephalon contains the mammillary bodies? What is the function of the mammillary bodies?
19. What are the important functions of the hypothalamus?
20. What are the proposed functions of the pineal gland?

13.7 The midbrain, pons, and medulla oblongata of the brain stem serve as a relay station and control center.

The **brain stem** (Figure 13.14) is the part of the brain between the spinal cord and the diencephalon. It consists of three structurally and functionally connected regions: (1) midbrain, (2) pons, and (3) medulla oblongata, or more simply the medulla. Extending through the brain stem is the reticular formation, a netlike region of interspersed gray and white matter. The midbrain, pons, and

Figure 13.14 Inferior aspect of brain showing components of the brain stem.

Longitudinal fissure

Optic nerve

Mammillary body

Pons

Pyramids of medulla oblongata

Cerebellum

Frontal lobe

Midbrain

Temporal lobe

Spinal cord

The brain stem consists of the midbrain, pons, and medulla oblongata.

medulla contain both tracts and nuclei. Collectively, the structures of the brain stem act as relay centers for processing and controlling involuntary reflexes related to visual and auditory processing, eye movement, and regulation of autonomic functions including respiration, heart rate, blood pressure, and digestion.

Midbrain

The **midbrain** extends from the pons to the diencephalon (see Figures 13.1, 13.6b, and 13.14) and is about 2.5 cm (1 in.) long. The cerebral aqueduct passes through the midbrain, connecting the third ventricle above with the fourth ventricle below.

The anterior part of the midbrain contains paired bundles of axons (tracts) known as the **cerebral peduncles** (pe-DUNG-kuls = little feet; see Figure 13.15). The cerebral peduncles consist of axons of motor (descending) tracts that conduct nerve impulses from motor areas in the cerebral cortex to the medulla, pons, and spinal cord.

The posterior part of the midbrain contains four rounded elevations known as the **corpora quadrigemina** (kor-POR-a kwod-ri-JEM-i-nah). The two superior elevations are known as the **superior colliculi** (ko-LIK-ū-lī = little hills; singular is *colliculus*; Figure 13.15). They serve as reflex centers for certain visual activities. Through neural circuits from the retina of the eye to the superior colliculi to the extrinsic eye muscles, visual stimuli elicit eye movements for tracking moving images (such as a moving car) and for scanning stationary images (as you are doing to read this sentence). The superior colliculi are also responsible for reflexes that govern movements of the head, eyes, and trunk in response to visual stimuli. The two inferior elevations, the **inferior colliculi**, are part of the auditory pathway, relaying impulses from the receptors for hearing in the ear to the brain. They also are reflex centers for the *startle reflex,* sudden movements of the head, eyes, and trunk that occur when you are surprised by a loud noise such as a gunshot.

The midbrain contains several nuclei. Neurons of nuclei that release dopamine extend to the basal nuclei to help control subconscious muscle activities. Loss of these neurons is associated with Parkinson disease. Axons from the cerebellum and cerebral cortex form synapses in midbrain nuclei that help control voluntary movements of the limbs.

Figure 13.15 **Midbrain.** Cranial nerves, shown in yellow, will be discussed in Chapter 14.

(a) Posterior view of midbrain in relation to brain stem

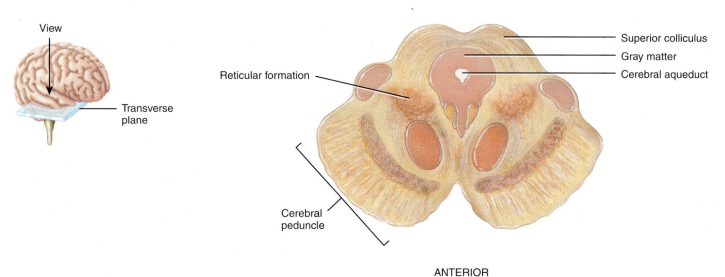

POSTERIOR

View

Transverse plane

Reticular formation

Superior colliculus

Gray matter

Cerebral aqueduct

Cerebral peduncle

ANTERIOR

(b) Transverse section of midbrain

Superior colliculus

Inferior colliculus

Cerebral aqueduct

ARBOR VITAE

CEREBELLAR CORTEX

POSTERIOR

Cerebellum

Midsagittal plane

View

Pineal gland

Cerebral peduncle

Mammillary body

Pons

Fourth ventricle

Medulla oblongata

Central canal of spinal cord

ANTERIOR

(c) Midsagittal section of cerebellum and brain stem

The midbrain connects the pons to the diencephalon.

Pons

The **pons** (= bridge) lies directly inferior to the midbrain and anterior to the cerebellum, and is about 2.5 cm (1 in.) long (see Figures 13.1, 13.14, and 13.15c). As its name implies, the pons is a bridge that connects parts of the brain with one another. These connections are provided by bundles of axons. Like the midbrain, the pons consists of nuclei, sensory tracts, and motor tracts.

Signals for voluntary movements are relayed from the motor areas of the cerebral cortex to the cerebellum. The **pneumotaxic area** (noo-mō-TAK-sik) and the **apneustic area** (ap-NOO-stik), shown in Figure 22.23, are nuclei in the pons that work with the medullary rhythmicity area of the medulla to help control breathing.

Medulla Oblongata

The **medulla oblongata** (me-DOOL-a ob-long-GA-ta), or more simply the **medulla**, forms the inferior part of the brain stem and is a continuation of the spinal cord (see Figures 13.1, 13.14, and 13.15c). The medulla begins at the inferior border of the pons and extends to the foramen magnum, a distance of about 3 cm (1.2 in.).

Within the medulla's white matter are all sensory (ascending) tracts and motor (descending) tracts extending between the spinal cord and other parts of the brain. Some of the white matter forms bulges on the anterior aspect of the medulla. These protrusions are the **pyramids** (see Figures 13.14 and 13.16), formed by the large motor tracts that pass from the cerebrum to the spinal cord. Just superior to the junction of the medulla with the spinal cord, 90% of the axons in the left pyramid cross to the right side, and 90% of the axons in the right pyramid cross to the left side. This crossing is called the **decussation of pyramids** (deh-ku-SĀ-shun; *decuss-* = crossing) and explains why each side of the brain controls movements on the opposite side of the body.

The medulla also contains several nuclei that control vital body functions. The **cardiovascular center**, shown in Figure 20.11, regulates the rate and force of the heartbeat and the diameter of blood vessels. The **medullary rhythmicity area**, shown in Figure 22.23, adjusts the basic rhythm of breathing. Other nuclei in the medulla control reflexes for vomiting, coughing, sneezing, and hiccupping. Just lateral to each pyramid is an oval-shaped swelling called an **olive** (Figure 13.16). Neurons within the olives relay impulses to the cerebellum that help the cerebellum make adjustments to muscle activities as you learn new motor skills.

Nuclei associated with sensations of touch, conscious proprioception, pressure, and vibration are also located in the medulla. Many ascending sensory axons form synapses in these nuclei, and postsynaptic neurons then relay the sensory information to the thalamus on the opposite side of the brain (see Figure 15.6a).

Reticular Formation

In addition to the well-defined nuclei already described, much of the brain stem consists of small clusters of neuronal cell bodies (gray matter) interspersed among small bundles of myelinated axons (white matter). The broad region where white matter and gray matter exhibit a netlike arrangement is known as the **reticular formation** (*ret-* = net; Figure 13.15b). It extends from the lower part of the diencephalon, throughout the brain stem, and into the upper part of the spinal cord. Neurons within the reticular formation have both ascending (sensory) and descending (motor) functions. Part of the reticular formation, called the **reticular activating system (RAS)**, consists of sensory axons that project to the cerebral cortex (see Figure 15.9). The RAS helps maintain consciousness and is active during awakening from sleep. For example, we awaken to the sound of an alarm clock, to a flash of lightning, or to a painful pinch because of RAS activity that arouses the cerebral cortex. The reticular formation also filters incoming sensory messages so as not to overload the brain. Familiar or repeated impulses are ignored, while new or unusual impulses are passed on to the cerebral cortex. As you walk through a forest you may not notice the sound of a breeze stirring the overhead canopy of leaves, but you would be aware of a branch snapping behind you. The reticular formation's main descending functions are to help regulate posture and *muscle tone*, the slight degree of contraction in normal resting muscles.

The functions of the brain stem are summarized in Table 13.2.

Checkpoint

21. Where are the midbrain, pons, and medulla located relative to one another?
22. Which part of the brain stem contains the cerebral peduncles? What is their importance?
23. In which regions of the brain stem are the nuclei that help control breathing?
24. Define decussation of pyramids. What is the functional consequence of decussation of the pyramids?
25. What are important functions of the reticular formation?

Figure 13.16 **Internal anatomy of the medulla oblongata.**

View

Transverse plane

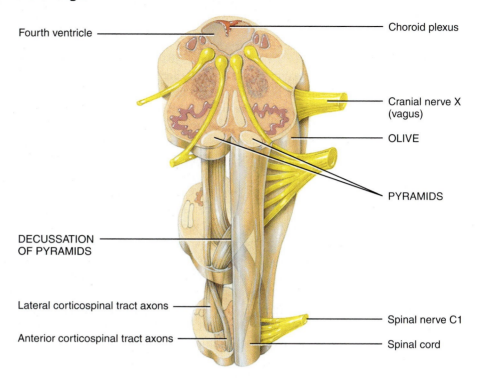

Fourth ventricle

Choroid plexus

Cranial nerve X (vagus)

OLIVE

PYRAMIDS

DECUSSATION OF PYRAMIDS

Lateral corticospinal tract axons

Anterior corticospinal tract axons

Spinal nerve C1

Spinal cord

Transverse section and anterior surface of medulla oblongata

🔑 **The pyramids of the medulla contain large motor tracts that run from the cerebrum to the spinal cord.**

Return to Darius' Story

There days have passed. Darius' parents and his sister have arrived, and Paolo has told them what has happened so far. Each day passes with hope but concern for Darius. The morning of the third day, the neurologist sits down with the family and Paolo. "Darius' condition has gotten worse; the fluid in his brain is being drained off, but the swelling in his brain is severe. The brain itself is swelling and pushing down on parts of the brain below the cerebrum. From what we can tell, his brain stem structures have been damaged. Darius' pupils are not responding to light, he has lost reflexive eye movements, and he is unresponsive to our attempts to rouse him. I am afraid the outlook is not good. We'll do our best. But you must understand; Darius may not wake up again."

E. *What body functions do the nuclei in Darius' brain stem control?*

F. *Darius has been on a ventilator to assist with his breathing. What parts of his brain stem would normally maintain breathing?*

G. *Darius' eyes are not following a moving pen light. What part of the brain stem could be damaged?*

H. *Darius' body is no longer regulating its temperature. What additional area of the brain that controls temperature regulation might be damaged?*

13.8 The cerebellum coordinates movements and helps maintain normal muscle tone, posture, and balance.

The **cerebellum**, the second-largest part of the brain, occupies the inferior and posterior aspects of the cranial cavity. It accounts for about a tenth of the brain mass yet contains nearly half of the neurons in the brain. The cerebellum is posterior to the medulla and pons and inferior to the posterior portion of the cerebrum (see Figures 13.1 and 13.14). In superior or inferior views, the shape of the cerebellum is somewhat like a butterfly. The central constricted area is the **vermis** (= worm), and the lateral "wings" or lobes are the **cerebellar hemispheres** (Figure 13.17). Each hemisphere consists of lobes separated by deep and distinct fissures. The **anterior lobe** and **posterior lobe** govern subconscious aspects of skeletal muscle movements. The **flocculonodular lobe** (*flocculo-* = wool-like tuft) on the inferior surface contributes to equilibrium and balance.

The superficial layer of the cerebellum, called the **cerebellar cortex**, consists of gray matter in a series of slender, parallel ridges. Deep to the gray matter are tracts of white matter called **arbor vitae** (= tree of life; see Figure 13.15c) that resemble branches of a tree. Even deeper, within the white matter, are the **cerebellar nuclei**, regions of gray matter that give rise to axons carrying impulses from the cerebellum to other brain centers and to the spinal cord.

Three paired **cerebellar peduncles** attach the cerebellum to the brain stem (see Figures 13.15a and 13.17b). These bundles of white matter consist of tracts that conduct impulses between the cerebellum and other parts of the CNS. The **superior cerebellar peduncles** carry communication from the cerebellum to the mid-brain and thalamus; the **middle cerebellar peduncles** carry commands for voluntary movements (those that originate in the motor areas of the cerebral cortex) from the pons into the cerebellum; and the **inferior cerebellar peduncles** carry sensory information into the cerebellum from the medulla, pons, and spinal cord.

The primary function of the cerebellum is to evaluate how well movements initiated by motor areas in the cerebrum are actually being carried out. When movements initiated by the cerebral motor areas are not being carried out correctly, the cerebellum detects the discrepancies. It then sends feedback signals to motor areas of the cerebral cortex via the cerebellar peduncles. The feedback signals help correct the errors, smooth the movements, and coordinate complex sequences of skeletal muscle contractions. Besides coordinating skilled movements, the cerebellum is the main brain region that regulates posture and balance. These aspects of cerebellar function make possible all skilled muscular activities, from catching a baseball to dancing to speaking.

The functions of the cerebellum are summarized in Table 13.2.

Checkpoint

26. How is the cerebellum externally similar to the cerebrum?
27. What are the functions of the three lobes of the cerebellum?
28. Which structures contain the axons that carry information into and out of the cerebellum?

Figure 13.17   Cerebellum.

(a) Superior view

(b) Inferior view

🔑 **The cerebellum coordinates skilled movements and regulates posture and balance.**

13.9 The spinal cord receives sensory input and provides motor output through spinal nerves.

External Anatomy of the Spinal Cord

The **spinal cord**, although roughly cylindrical, is flattened slightly anteriorly and posteriorly. The length of the adult spinal cord ranges from 42 to 45 cm (16–18 in.) (see Figure 13.2). Its diameter is about 2 cm (0.75 in.) in the midthoracic region, somewhat larger in the cervical and lumbar regions, and smallest at the inferior tip. When the spinal cord is viewed externally, two conspicuous enlargements can be seen. The superior enlargement, called the **cervical enlargement**, supplies nerves to and from the upper limbs. The inferior enlargement, called the **lumbar enlargement**, provides nerves to and from the lower limbs. Inferior to the lumbar enlargement, the spinal cord tapers to a conical portion referred to as the **conus medullaris** (KŌ-nus med-ū-LAR-is; *conus* = cone). Arising from the conus medullaris is the **filum terminale** (FĪ-lum ter-mi-NAL-ē = terminal filament), an extension of the pia mater that extends inferiorly and anchors the spinal cord to the coccyx.

In adults, the spinal cord extends from the medulla oblongata, the inferior part of the brain, to the level of the intervertebral disc between the first or second lumbar vertebra (see Figure 13.2). During early childhood, both the spinal cord and the vertebral column grow longer as part of overall body growth. Around age 4 or 5, however, elongation of the spinal cord stops. Because the vertebral column continues to elongate, the spinal cord does not extend the entire length of the adult vertebral column.

Spinal nerves are the paths of communication between the spinal cord and specific regions of the body. The spinal cord appears to be segmented because the 31 pairs of spinal nerves emerge at regular intervals through intervertebral foramina of the vertebral column (see Figures 13.2 and 13.4c). For ease of discussion, each pair of spinal nerves is said to arise from a *spinal segment*. However, within the spinal cord there is no obvious segmentation.

Two bundles of axons, called **roots**, connect each spinal nerve to a segment of the cord (see Figures 13.4c and 13.18). The **posterior root** contains only sensory axons, which conduct nerve impulses from sensory receptors in the skin, muscles, and internal organs into the central nervous system. Each posterior root also has a swelling, the **posterior root ganglion**, which contains the cell bodies of sensory neurons. The **anterior root** contains axons of motor neurons, which conduct nerve impulses from the CNS to effector organs and cells.

Because the spinal cord does not extend the full length of the vertebral column, nerves that arise from the inferior part of the spinal cord do not leave the vertebral column at the same level as they exit from the cord. The roots of these nerves angle inferiorly in the vertebral canal from the end of the spinal cord like wisps of hair.

Appropriately, the roots of these nerves are collectively named the **cauda equina** (KAW-da ē-KWĪ-na), meaning "horse's tail."

Internal Anatomy of the Spinal Cord

The spinal cord white matter surrounds an inner core of gray matter (Figure 13.18). The white matter of the spinal cord consists primarily of bundles of myelinated axons of neurons. Two grooves penetrate the white matter and divide it into right and left sides. The **anterior median fissure** is a wide groove on the anterior side. The **posterior median sulcus** is a narrow furrow on the posterior side. The gray matter of the spinal cord is shaped like the letter H or a butterfly; it consists of cell bodies of neurons, unmyelinated axons, dendrites, and neuroglia. The **gray commissure** (KOM-mishur) forms the crossbar of the H, connecting the gray matter on the right and left sides. In the center of the gray commissure is the **central canal**, a passageway filled with cerebrospinal fluid that extends the length of the spinal cord. At its superior end, the central canal is continuous with the fourth ventricle (a space that contains cerebrospinal fluid) in the medulla oblongata of the brain. Anterior to the gray commissure, the **anterior white commissure** connects the white matter of the right and left sides of the spinal cord.

The gray matter on each side of the spinal cord is subdivided into regions called **horns**. The **posterior gray horns** contain interneurons as well as axons of incoming sensory neurons. The **anterior gray horns** contain cell bodies of somatic motor neurons that provide nerve impulses for skeletal muscle contractions. Between the posterior and anterior gray horns are the **lateral gray horns**. The lateral gray horns contain the cell bodies of autonomic motor neurons that regulate the activity of cardiac muscle, smooth muscle, and glands.

The white matter of the spinal cord, like the gray matter, is organized into regions. The anterior and posterior gray horns divide the white matter on each side into three broad areas called **columns:** (1) **anterior white columns**, (2) **posterior white columns**, and (3) **lateral white columns**. Within each column are tracts containing distinct bundles of axons having a common origin or destination and carrying similar information. These tracts may extend long distances up or down the spinal cord. **Sensory (ascending) tracts** consist of axons that conduct nerve impulses toward the brain. Tracts consisting of axons that carry nerve impulses down the spinal cord are called **motor (descending) tracts**. Sensory and motor tracts of the spinal cord are continuous with sensory and motor tracts in the brain.

Figure 13.18   Internal anatomy of the spinal cord: the organization of gray matter and white matter. For simplicity, dendrites are not shown. Blue, red, and green arrows in (a) indicate the direction of nerve impulse propagation.

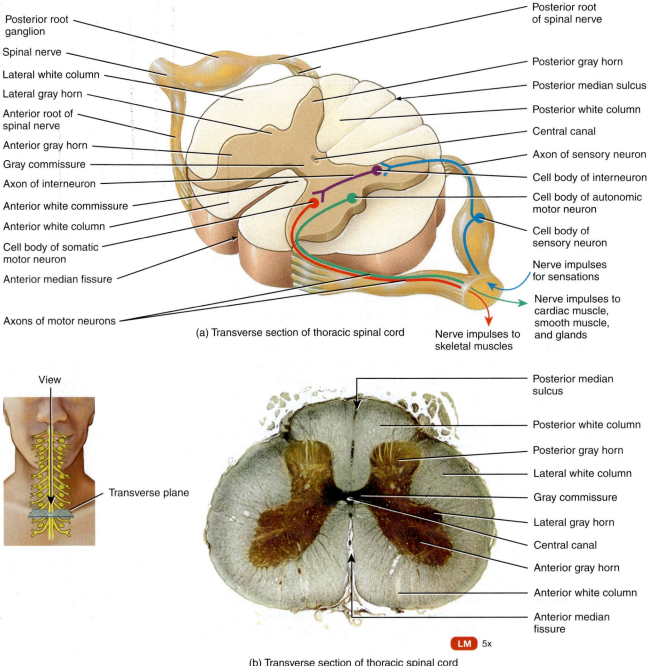

Posterior root ganglion

Spinal nerve

Lateral white column

Lateral gray horn

Anterior root of spinal nerve

Anterior gray horn

Gray commissure

Axon of interneuron

Anterior white commissure

Anterior white column

Cell body of somatic motor neuron

Anterior median fissure

Axons of motor neurons

Posterior root of spinal nerve

Posterior gray horn

Posterior median sulcus

Posterior white column

Central canal

Axon of sensory neuron

Cell body of interneuron

Cell body of autonomic motor neuron

Cell body of sensory neuron

Nerve impulses for sensations

Nerve impulses to cardiac muscle, smooth muscle, and glands

Nerve impulses to skeletal muscles

(a) Transverse section of thoracic spinal cord

View

Transverse plane

Posterior median sulcus

Posterior white column

Posterior gray horn

Lateral white column

Gray commissure

Lateral gray horn

Central canal

Anterior gray horn

Anterior white column

Anterior median fissure

LM 5x

(b) Transverse section of thoracic spinal cord

🔑 **In the spinal cord, white matter surrounds the gray matter.**

Checkpoint

29. What portion of the spinal cord connects with nerves of the upper limbs? With nerves of the lower limbs?

30. What are the conus medullaris, filum terminale, and cauda equina?

31. What structures are found in the posterior roots? The posterior root ganglia? The anterior roots?

32. What connects the gray matter on the two sides of the spinal cord? The white matter on the two sides of the spinal cord?

33. Describe the composition of the posterior, anterior, and lateral gray horns. What is the difference between a horn and a column in the spinal cord?

34. Where are ascending and descending tracts located in the spinal cord?

13.10 The spinal cord conducts nerve impulses between spinal nerves and the brain, and contains reflex pathways.

The spinal cord has two principal functions in maintaining homeostasis: nerve impulse propagation and information integration. Tracts within the *white matter* of the spinal cord serve as highways for the propagation of sensory input toward the brain and motor output from the brain toward skeletal muscles and other effector tissues. The *gray matter* of the spinal cord receives and integrates incoming and outgoing information.

Sensory and Motor Tracts

As noted previously, the spinal cord conducts nerve impulses along tracts. Often, the name of a tract indicates its position in the white matter and where it begins and ends. For example, the anterior corticospinal tract is located in the *anterior* white column; it begins in the *cerebral cortex* and ends in the *spinal cord*. Notice that the location of the axon terminals comes last in the name. This regularity in naming allows you to determine the direction of information flow along any tract named according to this convention. Thus, because the anterior corticospinal tract conveys nerve impulses from the brain toward the spinal cord, it is a motor (descending) tract.

Figure 13.19 highlights the major sensory and motor tracts in the spinal cord. Sensory and motor tracts are described in more detail in Chapter 15.

Nerve impulses from sensory receptors propagate up the spinal cord to the brain along two main routes on each side: the spinothalamic tract and the posterior column. The **spinothalamic tract** conveys nerve impulses for sensing pain, warmth, coolness, itching, and tickling. The **posterior column** consists of tracts that convey nerve impulses for touch, pressure, vibration, and conscious proprioception (the awareness of the positions and movements of muscles, tendons, and joints).

Sensory input keeps the CNS informed of changes in the external and internal environments. As sensory information is conveyed to the spinal cord and brain, it becomes part of a large pool of sensory input. Interneurons in the CNS integrate (process) each piece of incoming information with all the other information arriving from sensory neurons. Responses to the integrative decisions are brought about by motor activities (muscular contractions and glandular secretions). The cerebral cortex plays a major role in controlling precise voluntary muscular movements. Other brain regions provide important integration for regulation of automatic movements.

Figure 13.19   Locations of major sensory and motor tracts, shown in a transverse section of the spinal cord. Sensory tracts are indicated on one half and motor tracts on the other half of the cord, but actually all tracts are present on both sides.

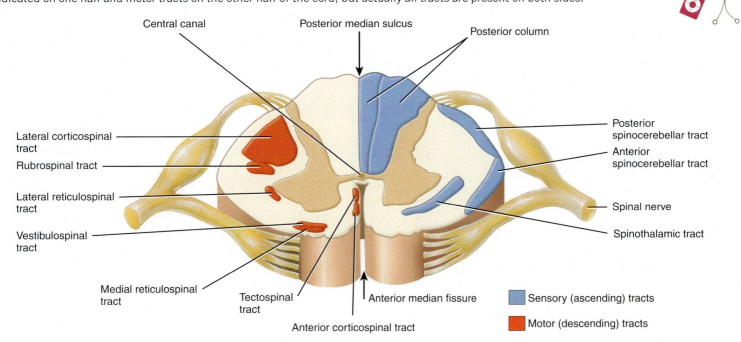

The name of a tract often indicates its location in the white matter and where it begins and ends.

Motor output to skeletal muscles travels down the spinal cord in two types of descending pathways: direct and indirect. The **direct pathways** convey nerve impulses without synapsing from the cerebral cortex and are destined to cause *voluntary* movements of skeletal muscles.

Indirect pathways convey nerve impulses from the brain stem to cause *automatic movements* that regulate muscle tone, posture, balance, and orientation of the head and body.

Spinal Reflexes

The spinal cord also promotes homeostasis by serving as an integrating center for some reflexes. Reflexes are fast, involuntary responses to stimuli. When integration takes place in the spinal cord gray matter, the reflex is a **spinal reflex**. By contrast, if integration occurs in the brain rather than the spinal cord, the reflex is a **cranial reflex**. Reflexes will be described in more detail in Chapter 14.

The functions of the spinal cord are summarized in Table 13.2.

Checkpoint

35. Based on its name, what are the position in the spinal cord, origin, and destination of the anterior spinocerebellar tract? Is this a sensory or a motor tract?

36. Which component of the spinal cord white matter transmits information to your brain that someone has touched your shoulder? Which components transmit information to your brain that your body feels cold?

37. Which kind of pathway would transmit motor instructions to consciously pick up a book? To maintain an upright posture while you sit in class?

Darius' Story: Epilogue and Discussion

▶▶▶ "I'm afraid Darius is gone…" the neurologist said as he stood in front of Paolo and Darius' family.

"But the surgeon said he was okay after they stopped the bleeding," said Paolo.

"This type of event is very unpredictable. The bleeding was just the beginning of Darius' ordeal. Escaping blood irritated brain tissue, the brain swelled, and permanent brain damage resulted," said the neurologist.

Darius' sister stepped forward. "But we just saw him. He looks alive. What do you mean he's dead?"

"Well," explained the neurologist, "according to the electroencephalogram, Darius has no brain activity. The part of his brain stem that keeps him breathing is damaged, so he can't breathe on his own. A blood flow study conducted on Darius' brain indicates that no blood is flowing to his brain. Darius is what we consider brain-dead. You need to think about Darius' wishes. Paolo has indicated that Darius wanted to be an organ donor, but in these cases we always consult with the family first."

Although the doctor's news appears unexpected to his hopeful family members and friends, perhaps it is not so to you. From your study of the CNS and review of the case you should be able to determine what most likely led to Darius' death. Let's summarize what we know up to this point.

- The brain CT scan revealed a subarachnoid hemorrhage.
- A cerebral angiogram identified the location of the bleed and the bleeding was stopped. This should have reduced elevated fluid volume in the subarachnoid space and pressure on the brain.

- Darius did recover consciousness, was able to move all his extremities, and retained cognitive function. No brain damage was initially evident.
- Darius began to have difficulty maintaining consciousness. He developed acute hydrocephalus, so a shunt was put in to relieve the pressure.
- When Darius woke up, the ICU staff learned that a stroke had affected his right side.
- Brain swelling occurred secondary to the initial damage.
- Darius lost his pupillary reflexes, became unconscious and unresponsive, and could not breathe on his own.
- The last tests revealed no blood flow to his brain and no electrical activity in the cerebral cortex. ◀◀◀

I.   *Would you expect Darius' cranial reflexes to be functioning? Why or why not?*

J.   *Would you expect Darius to exhibit normal spinal reflexes? Why or why not?*

K.   *Damage to what area of the brain would prevent Darius from "waking up"?*

L.   *Darius suffered a series of catastrophic events—a hemorrhage, acute hydrocephalus, a stroke, and brain stem damage. Which of these, based on your study of the CNS and the case, was most likely directly responsible for his death?*

M.   *How would you explain Darius' death in terms of brain stem function?*

TABLE 13.2

Functions of Principal Parts of the Central Nervous System

PART	FUNCTIONS	PART	FUNCTIONS
Cerebrum Cerebrum	Sensory areas interpret sensory impulses, motor areas control voluntary movement, and association areas function in emotional and intellectual processes. Basal nuclei coordinate automatic muscle movements and regulate muscle tone. Limbic system functions in emotional aspects of behavior related to survival.	**Brain stem** Midbrain	*Midbrain:* Relays motor impulses from the cerebral cortex to the pons and sensory impulses from the spinal cord to the thalamus. Superior colliculi coordinate movements of the head, eyes, and trunk in response to visual stimuli, and the inferior colliculi coordinate movements of the head, eyes, and trunk in response to auditory stimuli. Contributes to control of movements.
Diencephalon Pineal gland Thalamus Hypothalamus	*Thalamus:* Relays almost all sensory input to the cerebral cortex. Contributes to motor functions by transmitting information from the cerebellum and basal ganglia to the primary motor area of the cerebral cortex. Also plays a role in the maintenance of consciousness. *Hypothalamus:* Controls and integrates activities of the autonomic nervous system and pituitary gland. Regulates emotional and behavioral patterns. Controls body temperature and regulates eating and drinking behavior. Helps maintain the waking state and establishes patterns of sleep. Produces hormones. *Pineal gland:* Secretes melatonin, sets the body's biological clock.		*Pons:* Relays impulses between cerebral cortex and cerebellum and between the medulla and midbrain. Pneumotaxic and apneustic area, together with the medulla oblongata, help control breathing. Pons
			Medulla oblongata: Relays motor and sensory impulses between other parts of the brain and the spinal cord. Vital centers regulate heartbeat, breathing (together with pneumotaxic and apneustic area of pons), and blood vessel diameter. Other centers coordinate swallowing, vomiting, coughing, sneezing, and hiccuping. Medulla oblongata
Cerebellum Cerebellum	Compares intended movements with what is actually happening to smooth and coordinate complex, skilled movements. Regulates posture and balance. May have a role in cognition and language processing.		*Reticular formation:* Helps maintain consciousness, causes awakening from sleep, filters repetitive sensory input, and contributes to regulation of muscle tone. Reticular formation
		Spinal cord Spinal cord	Conducts sensory nerve impulses toward the brain and motor nerve impulses from the brain toward skeletal muscles and other effector tissues. Integrates spinal reflexes.

Concept and Resource Summary

Concept   **Resources**   WILEY **PLUS**

Introduction

1. The central nervous system consists of the brain, the control center for thoughts, memories, and behaviors; and the spinal cord, the pathway for sensory nerve impulses traveling to the brain and motor impulses traveling from the brain toward skeletal muscles and other effectors.

Concept 13.1 The CNS consists of the brain, the spinal cord, and several protective structures.

1. The four major parts of the brain are the **cerebrum**, the **diencephalon**, the **brain stem**, and the **cerebellum**.
2. The **spinal cord** extends from the medulla oblongata to the second lumbar vertebra.
3. The skull, vertebrae, and meninges protect the CNS. The brain produces cerebrospinal fluid that also protects the brain and spinal cord.
4. The **meninges** are three connective tissue coverings layered from superficial to deep as the dura mater, arachnoid mater, and pia mater. The **dura mater**, the strongest meninx, adheres to the periosteum of the cranial bones, but it is separated from the vertebrae around the spinal cord by the **epidural space**. The middle **arachnoid mater** contains a web of delicate collagen and elastic fibers, and is avascular. The **pia mater** is transparent and adheres tightly to the surface of the brain and spinal cord and protects against shock and displacement of the spinal cord.
5. The **subarachnoid space** between the arachnoid mater and the pia mater contains cerebrospinal fluid.

See
* **Anatomy Overview – The Brain**
* **Anatomy Overview – The Spinal Cord**

Hear
* **Figure 13.1 – The Brain**
* **Figure 13.2 – External Anatomy of the Spinal Cord and the Spinal Nerves**
* **Figure 13.4 – Gross Anatomy of the Spinal Cord**

Concept 13.2 The CNS is nourished and protected by blood and cerebrospinal fluid.

1. Blood from the internal carotid and vertebral arteries carries oxygen and nutrients to the brain; blood leaves through dural sinuses to the internal jugular veins.
2. Blood is carried to the spinal cord by the posterior intercostal and lumbar arteries; blood leaves by the posterior intercostal and lumbar veins.
3. The **blood–brain barrier** is created by tightly connected endothelial cells of CNS capillaries and astrocyte processes that press up against the capillaries. Glucose, oxygen, carbon dioxide, most anesthetics, and several other substances can cross the blood–brain barrier. Toxins, most antibiotics, and proteins cannot cross it.
4. **Cerebrospinal fluid** is formed in the choroid plexuses and circulates through the **lateral ventricles**, **third ventricle**, **fourth ventricle**, **subarachnoid space**, and **central canal**. Ependymal cells use their cilia to maintain cerebrospinal fluid flow. Cerebrospinal fluid is returned to the blood across **arachnoid villi** that project into the dural sinuses.
5. Cerebrospinal fluid provides mechanical protection, chemical protection, and circulation of nutrients to the brain and spinal cord.

See
* **Anatomy Overview – The Nervous System: Overview**

Concept 13.3 The cerebrum interprets sensory impulses, controls muscular movements, and functions in intellectual processes.

1. The cerebrum consists of an outer rim of gray matter called the **cerebral cortex**, deeper cerebral white matter, and gray matter nuclei. It has folds called **gyri**; grooves called **sulci**; and deep grooves called **fissures**. The **longitudinal fissure** separates the cerebrum into left and right **cerebral hemispheres** connected by the **corpus callosum**.
2. Each cerebral hemisphere is subdivided by sulci into **four lobes**: **frontal lobe**, **parietal lobe**, **temporal lobe**, and **occipital lobe**. A fifth deep region of the cerebrum, the **insula**, cannot be seen at the surface.
3. The **cerebral white matter** contains **tracts** of myelinated axons that propagate impulses for communication between CNS regions. There are three types of tracts: **association tracts** conduct impulses between gyri of the same hemisphere; **commisural tracts** conduct impulses between gyri of different hemispheres; and **projection tracts** conduct impulses between the cerebrum and lower CNS regions.
4. **Basal nuclei** located deep in each cerebral hemisphere include: the **globus pallidus**, the **putamen**, and the **caudate nucleus**. Basal nuclei help initiate and terminate movements, suppress unwanted movements, and regulate muscle tone.

See
* **Anatomy Overview – Cerebrum**

Hear
* **Figure 13.7 – Cerebrum**

Do
* **Exercise – Paint the Functional Areas of the Cerebral Cortex**

Concept

Concept 13.4 The limbic system controls emotions, behavior, and memory.

1. The **limbic system** encircles the upper brain stem and corpus callosum. It plays a primary role in the emotional aspects of behavior and memory.

Concept 13.5 The cerebral cortex can be divided functionally into sensory areas, motor areas, and association areas.

1. The *sensory areas* of the cerebral cortex allow perception of sensory information. The *motor areas* control the execution of voluntary movements. The *association areas* are concerned with more complex integrative functions such as memory, personality traits, and intelligence.

2. The **primary somatosensory area** receives nerve impulses from somatic sensory receptors for touch, pressure, vibration, itch, tickle, temperature, pain, and proprioception and is involved in the perception of these sensations. The primary somatosensory area localizes the part of the body from which a sensation originates. The **primary visual area** receives visual information, the **primary auditory area** receives sound information, the **primary gustatory area** receives taste information, and the **primary olfactory area** receives smell information.

3. Motor areas include the **primary motor area**, which controls voluntary contractions of muscles, and **Broca's area**, which controls production of speech.

4. The **somatosensory association area** permits you to evaluate an object by touch and to sense the relationship of one body part to another. It also stores memories of past somatic sensory experiences.

5. The **prefrontal cortex** is concerned with personality, intellect, complex learning abilities, judgment, reasoning, conscience, intuition, and development of abstract ideas.

6. The **visual association area** utilizes past visual experiences for recognizing and evaluating what is seen.

7. The **auditory association area** allows you to recognize a particular sound.

8. **Wernicke's area** interprets the meaning of speech by translating words into thoughts.

9. The **common integrative area** integrates sensory interpretations from the association areas, allowing thoughts based on sensory inputs.

10. The **premoter area** controls sequential muscle contractions. It also serves as a memory bank for complex movements.

11. The **frontal eye field area** controls voluntary scanning movements of the eyes.

12. Each hemisphere has unique functions in a division of labor called **hemispheric lateralization.** Each hemisphere receives sensory signals from and controls movements of the opposite side of the body. The left hemisphere is more important for language, numerical and scientific skills, and reasoning. The right hemisphere is more important for musical and artistic awareness, spatial and pattern perception, recognition of faces, emotional content of language, identifying odors, and generating mental images of sight, sound, touch, taste and smell.

Concept 13.6 The diencephalon includes the thalamus, hypothalamus, and pineal gland.

1. The thalamus, hypothalamus, and pineal gland are regions of the diencephalon.

2. The **thalamus** consists of paired oval masses of gray matter organized into nuclei that, in most people, are connected by gray matter called the **intermediate mass**.

3. The thalamus relays and processes sensory impulses to the primary sensory areas of the cerebral cortex. It transmits input from the cerebellum to the primary motor area of the cerebral cortex. Together with other brain areas, it functions in movement control, emotions, consciousness, pain perception, learning, memory, and cognition.

4. The **hypothalamus** consists of many nuclei, including the **mammillary bodies**, which are relay stations for olfaction. The **infundibulum** connects the pituitary gland to the hypothalamus.

5. The hypothalamus is responsible for many body activities, including control of the autonomic nervous system, production of hormones, regulation of emotions and behavior, regulation of eating and drinking, control of body temperature, and regulation of circadian rhythms.

6. The **pineal gland** secretes the hormone **melatonin**, which is thought to promote sleep.

See

* **Anatomy Overview– Diencephalon**

Hear

* **Figure 13.13 –** **Hypothalarmus**

Concept

Concept 13.7 The midbrain, pons, and medulla oblongata of the brain stem serve as a relay station and control center.

1. The three functional regions of the brain stem are the midbrain, the pons, and the medulla oblongata.
2. The tracts and nuclei of brain stem structures act as relay centers for processing and controlling involuntary reflexes for vision and hearing, and govern reflexes vital to life.
3. The **midbrain** contains anterior tracts called the **cerebral peduncles**; posteriorly, the **corpora quadrigemina** has **superior colliculi** for visual reflexes and **inferior colliculi** for auditory reflexes. Nuclei of the midbrain release dopamine for control of subconscious muscle activities.
4. The **pons** relays signals for voluntary movements from the cerebral cortex to the cerebellum. The **pneumotaxic area** and the **apneustic area** are pons nuclei that function with the medulla to control breathing.
5. The **medulla oblongata** forms the inferior part of the brain stem and is a continuation of the spinal cord. Some of its white matter forms anterior bulges called the **pyramids** where there is a crossing of motor axons (**decussation of pyramids**) to the opposite side before continuing to the spinal cord. Nuclei of the medulla control vital body functions of heart rate, blood pressure, and the basic rhythm of breathing. Olives, lateral to each pyramid, relay motor impulses to the cerebellum while learning new motor skills.
6. The broad region of the brain stem where white matter and gray matter form a netlike arrangement is called the **reticular formation**. Part of the reticular formation, called the **reticular activating system**, helps arouse the body from sleep and maintain consciousness. Its motor function helps regulate posture and muscle tone.

See
- **Anatomy Overview – Brain Stem**

Concept 13.8 The cerebellum coordinates movements and helps maintain normal muscle tone, posture, and balance.

1. The **cerebellum** has a central area called the **vermis** and two lateral **cerebellar hemispheres**. The superficial **cerebellar cortex** consists of gray matter, whereas deeper, branching white matter is the **arbor vitae**.
2. The cerebellum smoothes and coordinates the contractions of skeletal muscles. It also maintains posture and balance.

See
- **Anatomy Overview – Cerebellum**

Concept 13.9 The spinal cord receives sensory input and provides motor output through spinal nerves.

1. The spinal cord extends from the medulla oblongata to the second lumbar vertebra. It has **cervical** and **lumbar enlargements**, and an inferior **conus medullaris**. The **filum terminale** extends from the conus medullaris to anchor the spinal cord to the coccyx.
2. There are 31 pairs of spinal nerves, which function in communication between the spinal cord and peripheral nerves. Each spinal nerve connects to a segment of the spinal cord by two **roots**. The **posterior root** contains sensory axons and has a swelling, the **posterior root ganglion**, containing the cell bodies of sensory neurons. The **anterior root** contains axons of motor neurons.
3. The **cauda equina** is a collection of spinal nerve roots that hang inferiorly from the spinal cord in the vertebral canal.
4. The H-shaped gray matter has a **central canal** that contains cerebrospinal fluid and is subdivided into **horns**. The **posterior gray horns** contain interneurons and axons of sensory neurons; the **anterior gray horns** contain cell bodies of somatic motor neurons; and the **lateral gray horns** contain cell bodies of autonomic motor neurons.
5. The white matter of the spinal cord is organized into **anterior white columns**, **posterior white columns**, and **lateral white columns**. The columns contain **sensory (ascending) tracts** that conduct impulses to the brain and **motor (descending) tracts** that conduct impulses from the brain down the spinal cord.

See
- **Anatomy Overview – Spinal Nerves**

Concept	Resources

Concept 13.10 The spinal cord conducts nerve impulses between spinal nerves and the brain, and contains reflex pathways.

1. The spinal cord conducts impulses along sensory and motor tracts. Ascending nerve impulses from sensory receptors travel along both sides of the spinal cord to the brain in the **spinothalamic tract** and the **posterior column**.
2. **Direct pathways** transmit nerve impulses from the cerebral cortex to skeletal muscles for voluntary movements. **Indirect pathways** transmit nerve impulses from the brain stem *for automatic movements* that regulate muscle tone, balance, and posture.
3. The spinal cord functions as an integration center for **spinal reflexes**; the brain integrates **cranial reflexes**.

See
- **Animation – Somatic Sensory and Motor Pathways**

Hear
- **Figure 13.19 –** **Locations of Major Sensory and Motor Tracts**

Do
- **Exercise – Concentrate on Your Brain**

Other Chapter Resources

For additional information on clinical applications, medical disorders, and medical terminology, see Chapter 7 in the *Clinical Connections* manual packaged with your text.

Within *WileyPLUS*
- **Chapter Quiz**
- **Visual Anatomy**
- **Mastering Vocabulary**
- **Audio Glossary**
- **Homeostatic Imbalance: The Case of the Forgetful Father**
- **Homeostatic Imbalance: The Case of the Man with the Tingling Arm**
- **Web Links**

• Understanding the Concepts

1. How does the skeleton protect the spinal cord?
2. How does the blood–brain barrier protect the central nervous system?
3. What is the functional difference between the blood–brain barrier and the blood–cerebrospinal fluid barrier?
4. What structures must cerebrospinal fluid pass through to enter the dural sinuses? Are red blood cells and large proteins normally able to pass through these structures? Why or why not?
5. Describe the location, composition, and functions of the basal nuclei.
6. What area(s) of the cerebral cortex receives information about the colors in a flower? Controls flexion of your fingers? Instructs your fingers to play a familiar song on the piano? Distinguishes a baby's cry from a dog's bark? Interprets directions for installing software in a computer? Recognizes an object flying by as a basketball?
7. Why is the thalamus considered a "relay station" in the brain?
8. How does the hypothalamus influence the autonomic nervous system? Why is it considered part of both the nervous system and the endocrine system?
9. What functions are carried out by the superior colliculi? By the inferior colliculi?
10. How is the spinal cord partially divided into right and left sides?

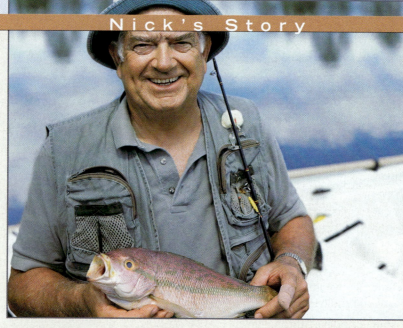

Nick's Story

The door to the exam room opened slowly. "Well, hello, Nick, I haven't seen you in quite a while," said the doctor.

"Hi, Doc. Yeah, it's been a while; changed jobs again, and with my new insurance company's rules, well, I was afraid they wouldn't cover the cost of the visit." Nick shifted in his chair and began to stand up. He rose slowly, steadying himself on the arms of the chair.

"Steady there, Nick, let me help you over to the exam table."

"Sorry, Doc. I'm only fifty-six, but I feel like a hundred and fifty-six. I've been trying to keep the diabetes under control, but there's been a lot of stress in the last two years, what with my divorce and the company laying off its machinists. I've been feeling worse and worse. I got these weird pains in my feet, like burning and prickling, and I've been so clumsy lately. I get real dizzy when I sit down or stand up, which isn't too safe working in the shop, you know. I nearly broke my neck when I bent down to grab a tool the other day!"

The doctor quickly flipped through Nick's medical chart. "Hmm, it's been almost two years since you were here last, Nick.

You know you really need to come in more often than that. Diabetes can lead to a lot of other health problems if you don't watch your blood sugars and get regular checkups."

"Oh, you docs are all the same, you just want our money," Nick said, winking at the doctor.

"Nick, you always get right to the point. I think you're just afraid of doctors!" They both laughed, but Dr. Jerouen was more concerned than he was ready to admit to Nick.

Nick is a Type 2 diabetic; his body is not able to regulate its blood sugar levels properly. This pathology causes very high blood sugar levels that can starve cells of the energy they need, damage the kidneys, and, in Nick's case, damage the nerves. As we follow Nick's case we will see that his neurological symptoms resulted from damage to his peripheral nerves. In order to understand his case, we need to know more about the peripheral nervous system. What are nerves? What information do they carry? How does damage to the peripheral nervous system affect the body?

INTRODUCTION

The **peripheral nervous system (PNS)** consists of nervous tissue found outside the brain and spinal cord, including nerves that exit the central nervous system and ganglia found throughout the body cavities. (**Ganglia** are collections of neuronal cell bodies outside the central nervous system.) The peripheral nervous system can be subdivided into two functional systems, the **somatic nervous system (SNS)** and the **autonomic nervous system (ANS)**.

The somatic nervous system consists of afferent, sensory cranial and spinal nerves that connect somatic receptors to the central nervous system (CNS) and efferent, motor cranial and spinal nerves that connect the CNS to skeletal muscles. Functionally, the so-

The peripheral nervous system can be subdivided into two functional systems, the somatic nervous system and the autonomic nervous system.

matic nervous system is associated with voluntary or conscious activities.

The autonomic nervous system includes autonomic sensory neurons, integrating centers within the CNS, and autonomic motor neurons. The ANS usually operates without conscious control to regulate the activity of smooth muscle, cardiac muscle, and glands.

In this chapter, we explore the structure and function of the peripheral nervous system and its regulation by the CNS.

14.1 Nerves have three protective connective tissue coverings.

A **nerve** is a bundle of hundreds to thousands of axons plus associated connective tissue and blood vessels that lie outside the brain and spinal cord. Nerves emerging from the brain are called **cranial nerves**; those emerging from the spinal cord are called **spinal nerves**. Each cranial nerve and spinal nerve contains layers of protective connective tissue coverings (Figure 14.1). Individual axons, whether myelinated or unmyelinated, are wrapped in **endoneurium** (en´-dō-NOO-rē-um; *endo-* = within or inner; *-neurium* = nerve). Groups of axons with their endoneurium are arranged in bundles called **fascicles**, each of which is wrapped in **perineurium** (per´-i-NOO-rē-um; *peri-* = around). The superficial covering over the entire nerve is the **epineurium** (ep´-i- NOO-rē-um; *epi-* = over). The dura mater of the CNS meninges fuses with the epineurium as the nerve passes through foramina of the skull or vertebral column. Note the presence of many blood vessels, which nourish nerves, within the perineurium and epineurium (Figure 14.1b). You may recall from Chapter 10 that the connective tissue coverings of skeletal muscles—endomysium, perimysium, and epimysium—are similar in organization to those of nerves.

Checkpoint

1. *What is a nerve?*
2. *What protective connective tissue coverings are associated with nerves? Which part of the nerve does each connective tissue covering surround?*

Figure 14.1   Organization and connective tissue coverings of a spinal nerve. [Part (b) from Dr. Richard Kessel and Randy Kardon/Tissues & Organs/Visuals Unlimited. Reprinted by permission.]

(a) Transverse section showing the coverings of a spinal nerve

(b) Transverse section of 12 nerve fascicles

Three layers of connective tissue wrappings protect axons: endoneurium surrounds individual axons, perineurium surrounds bundles of axons (fascicles), and epineurium surrounds an entire nerve.

14.2 Twelve pairs of cranial nerves distribute primarily to regions of the head and neck.

The 12 pairs of **cranial nerves** are so named because they arise from the brain inside the cranial cavity and pass through various foramina in the cranial bones. The cranial nerves and their branches are part of the peripheral nervous system. Each cranial nerve has both a number, designated by a Roman numeral, and a name (Figure 14.2). The numbers indicate the order, from anterior to posterior, in which the nerves arise from the brain. The names designate a nerve's distribution or function.

Three cranial nerves, I (olfactory), II (optic), and VIII (vestibulocochlear), contain only sensory axons and thus are called **sensory nerves**. The rest are **mixed nerves** because they contain axons of both sensory and motor neurons. The cell bodies of sensory neurons are located in ganglia outside the brain; the cell bodies of motor neurons lie in nuclei within the brain. Cranial nerves III (oculomotor), VII (facial), IX (glossopharyngeal), and X (vagus) include both somatic and autonomic motor

Figure 14.2   Origins of cranial nerves.

ANTERIOR

Inferior aspect of brain

There are 12 cranial nerves located on the inferior surface of the brain.

axons. The somatic axons innervate skeletal muscles, and the autonomic axons (part of the parasympathetic division) innervate glands, smooth muscle, and cardiac muscle. Although the cranial nerves are mentioned singly in the following descriptions of their type, location, and function, remember that they are paired structures.

Olfactory (I) Nerve

The **olfactory nerve** (ol-FAK-tō-rē; *olfact-* = to smell) is entirely sensory; it contains axons that conduct nerve impulses for the sense of smell, or olfaction (Figure 14.3). The olfactory epithelium occupies the superior part of the nasal cavity, covering the inferior surface of the cribriform plate and extending down along the superior nasal concha. The olfactory receptors are located in the olfactory epithelium. Each has a single odor-sensitive dendrite projecting from one side of the cell body and an axon extending from the other side. Bundles of axons of olfactory receptors extend superiorly through the olfactory foramina in the cribriform plate of the ethmoid bone. On each side of the nose, these bundles of axons collectively form the right and left olfactory nerves.

Olfactory nerves end in the brain in paired masses of gray matter called the **olfactory bulbs**, two extensions of the brain that rest on the cribriform plate. Within the olfactory bulbs, the axon terminals of olfactory receptors form synapses with neurons in the olfactory pathway. The axons of these neurons make up the **olfactory tracts**, which extend posteriorly from the olfactory bulbs (see Figure 14.2). Axons in the olfactory tracts end in the primary olfactory area in the temporal lobe of the cerebral cortex.

Figure 14.3 Olfactory (I) nerve.

Olfactory tract

Olfactory bulb

Bundle of axons of olfactory receptors

Cribriform plate

Olfactory epithelium

Olfactory receptor

Anterior

Olfactory bulb

Cribriform plate

Olfactory epithelium

Olfactory (I) nerve

Olfactory bulb

Olfactory tract

Posterior

Inferior surface of brain

🔑 The olfactory epithelium is located on the inferior surface of the cribriform plate and superior nasal conchae.

Optic (II) Nerve

The **optic nerve** (OP-tik; = eye, vision) is entirely sensory; it contains axons that conduct nerve impulses for vision (Figure 14.4). In the retina of each eye, rods and cones initiate visual signals that are relayed via the optic nerves to the visual processing centers of the brain. The two optic nerves pass through the optic foramen of the orbit (see Figure 7.11), then merge to form the **optic chiasm** (KĪ-azm = cross, as in the letter X). Within the chiasm, visual information from each eye may either continue on to the same side of the brain or cross to the opposite side for visual–spatial information processing. Posterior to the chiasm, the regrouped axons form the **optic tracts**. Axons from the optic tracts project to the primary visual area of the occipital lobe in the cerebral cortex (see Figure 13.11).

Figure 14.4 Optic (II) nerve.

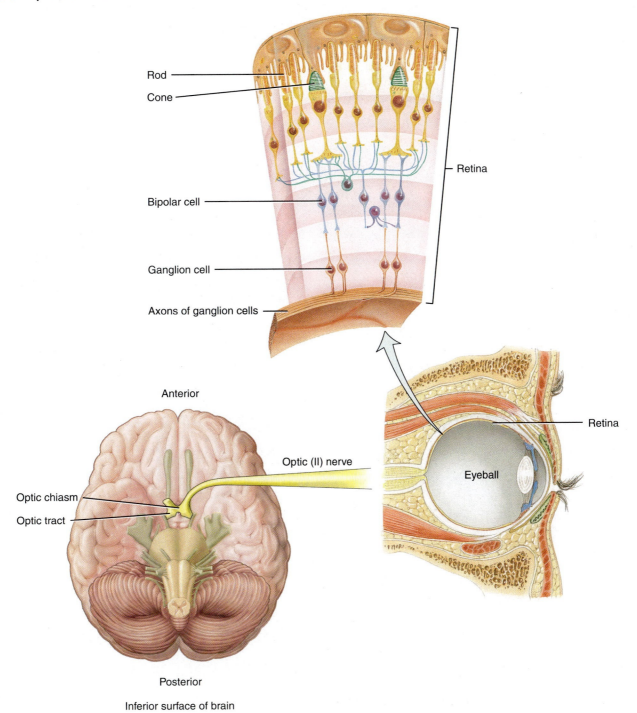

Rod
Cone
Retina
Bipolar cell
Ganglion cell
Axons of ganglion cells
Anterior
Retina
Optic (II) nerve
Eyeball
Optic chiasm
Optic tract
Posterior
Inferior surface of brain

🔑 **Visual signals are relayed from rods and cones in the retina to the primary visual area of the cerebral cortex.**

Oculomotor (III) Nerve

The **oculomotor nerve** (ok´-ū-lō-MŌ-tor; *oculo-* = eye; *-motor* = a mover) is a mixed cranial nerve (Figure 14.5). The oculomotor nerve passes through the superior orbital fissure into the orbit (see Figure 7.11). Sensory axons from proprioceptors in the extrinsic eyeball muscles extend to the midbrain. These axons convey nerve impulses for **proprioception**, the nonvisual perception of the movements and position of the body. Axons of somatic motor neurons innervate extrinsic eyeball muscles for movement of the eyeball, and the upper eyelid muscle (the levator palpebrae superioris) for raising the upper eyelid. Autonomic impulses propagate from the midbrain to the ciliary muscle that adjusts the shape of the lens (see Figure 16.8), and the iris to adjust the size of the pupil (see Figure 16.9).

Trochlear (IV) Nerve

The **trochlear nerve** (TRŌK-lē-ar = a pulley) is a mixed cranial nerve with axons passing through the superior orbital fissure of the orbit. Sensory axons convey nerve impulses for proprioception from an extrinsic eyeball muscle (the superior oblique) to the midbrain. Motor neurons originate in the midbrain and extend axons that innervate the superior oblique to help control movement of the eyeball (Figure 14.5b).

Figure 14.5   Oculomotor (III), trochlear (IV), and abducens (VI) nerves.

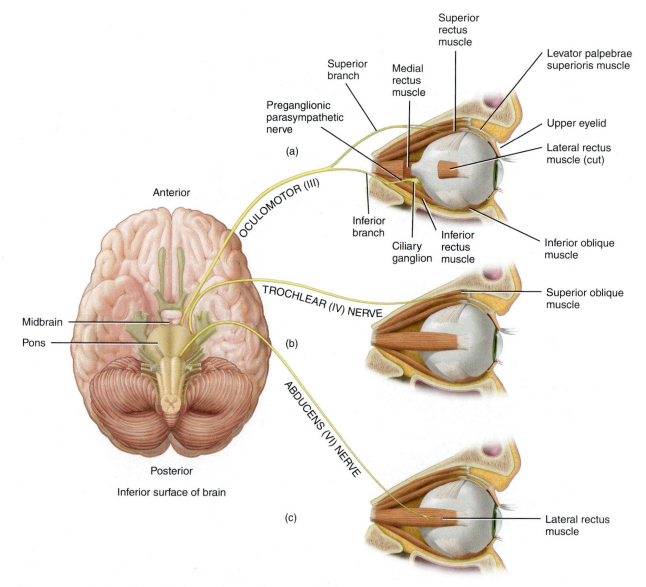

The oculomotor nerve has the widest distribution to extrinsic eye muscles.

Trigeminal (V) Nerve

The **trigeminal nerve** (trī-JEM-i-nal = triple, for its three branches), the largest of the cranial nerves, is a mixed cranial nerve that emerges on the lateral surface of the pons (Figure 14.6). As indicated by its name, the trigeminal nerve has three branches: ophthalmic, maxillary, and mandibular. The **ophthalmic nerve** (of-THAL-mik = eye) enters the orbit via the superior orbital fissure; the **maxillary nerve** (*maxilla* = upper jawbone) passes through the foramen rotundum; and the **mandibular nerve** (*mandibula* = lower jawbone) exits through the foramen ovale (see Figure 7.7a). Sensory axons in the trigeminal nerve carry nerve impulses for touch, pain, and temperature sensations to the pons. The ophthalmic nerve contains sensory axons from the upper eyelid, eyeball, lacrimal glands (which secrete tears), nasal cavity, nose, forehead,

and anterior scalp. The maxillary nerve includes sensory axons from the nose, palate, upper mouth, and lower eyelid. The mandibular nerve contains sensory axons from the anterior tongue (but not those for taste), cheek, skin over the jaw and side of the head, and lower mouth. The trigeminal nerve also contains sensory axons from proprioceptors located in the muscles of mastication. Somatic motor axons of the mandibular nerve supply muscles of mastication that control chewing movements.

Abducens (VI) Nerve

The **abducens nerve** (ab-DOO-senz; *ab-* = away; *-ducens* = to lead) is a mixed cranial nerve that originates in the pons (Figure 14.5c). Sensory axons extend from proprioceptors in an extrin-

Figure 14.6 **Trigeminal (V) nerve.**

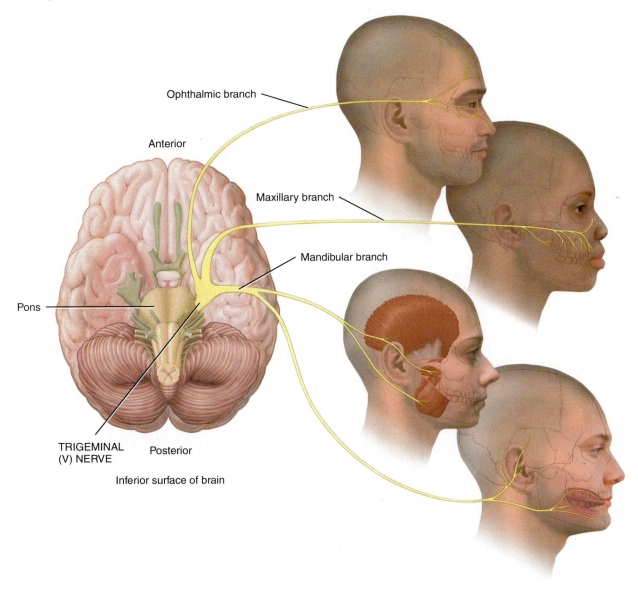

Ophthalmic branch

Anterior

Maxillary branch

Mandibular branch

Pons

TRIGEMINAL (V) NERVE   Posterior

Inferior surface of brain

The three branches of the trigeminal nerve leave the skull through the superior orbital fissure, foramen rotundum, and foramen ovale.

sic eyeball muscle (the lateral rectus) to the pons. Somatic motor axons extend from the pons to the lateral rectus through the superior orbital fissure of the orbit. The abducens nerve is so named because nerve impulses cause abduction of the eyeball (lateral rotation).

Facial (VII) Nerve

The **facial nerve** (FĀ-shal = face) is a mixed cranial nerve (Figure 14.7). Its sensory axons extend from the taste buds of the anterior tongue, pass through the stylomastoid foramen (see Figure 7.4) to the pons, and then continue on to the primary gustatory area in the parietal lobe of the cerebral cortex (see Figure 13.11). The sensory portion of the facial nerve also contains axons from proprioceptors in muscles of the face and scalp. Axons of somatic motor neurons arise in the pons and innervate facial, scalp, and neck muscles for facial expression (Figure 14.7). Axons of autonomic motor neurons innervate lacrimal glands and salivary glands.

Vestibulocochlear (VIII) Nerve

The **vestibulocochlear nerve** (ve-sti´-bū-lō-KŌ-klē-ar; *vestibulo-* = small cavity; *-cochlear* = spiral, snail-like) is entirely sensory and has two branches, the vestibular branch and the cochlear branch (Figure 14.8). The **vestibular branch** carries impulses for equilibrium from the semicircular canals, saccule, and utricle of the inner ear (see Figure 16.20) to the pons and cerebellum. The **cochlear branch** carries impulses for hearing from the spiral organ of the inner ear (see Figure 16.21b) to the primary auditory area in the temporal lobe of the cerebral cortex (see Figure 13.11).

Glossopharyngeal (IX) Nerve

The **glossopharyngeal nerve** (glo´-sō-fa-RIN-jē-al; *glosso-* = tongue; *-pharyngeal* = throat) is a mixed cranial nerve (Figure 14.9). Axons of the glossopharyngeal nerve arise in the medulla

Figure 14.7   Facial (VII) nerve.

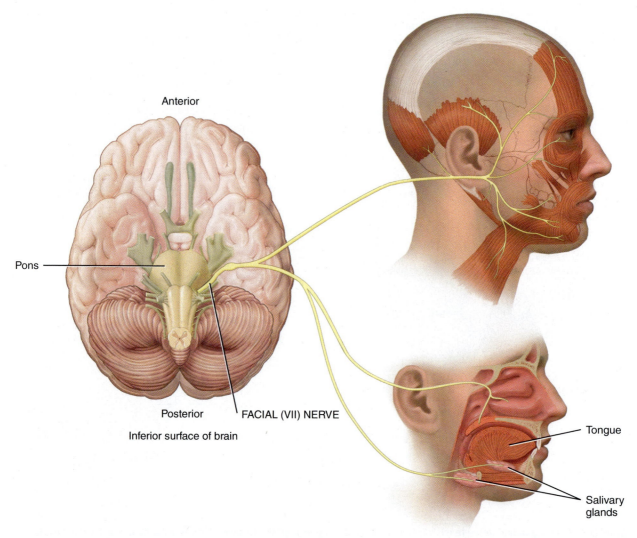

Anterior

Pons

Posterior

FACIAL (VII) NERVE

Inferior surface of brain

Tongue

Salivary glands

The facial nerve causes contraction of the muscles of facial expression.

Figure 14.8 **Vestibulocochlear (VIII) nerve.**

Inferior surface of brain

🔑 The vestibular branch carries impulses for equilibrium, and the cochlear branch carries impulses for hearing.

Figure 14.9 **Glossopharyngeal (IX) nerve.**

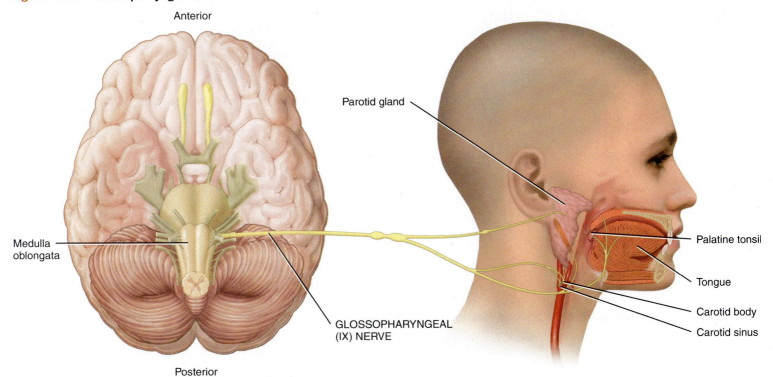

Inferior surface of brain

🔑 Sensory axons in the glossopharyngeal nerve carry signals from the taste buds.

and pass through the jugular foramen (see Figure 7.7a). Sensory axons arise from taste buds and sensory receptors on the posterior tongue, from proprioceptors in swallowing muscles, from baroreceptors of the carotid sinuses in the neck that monitor blood pressure, and from chemoreceptors of the carotid bodies in the neck that monitor blood gas levels. Somatic motor neurons innervate a muscle of the pharynx used in swallowing. Autonomic motor neurons stimulate the parotid gland to secrete saliva.

Vagus (X) Nerve

The **vagus nerve** (VĀ-gus = vagrant or wandering) is a mixed cranial nerve that is distributed from the head and neck into the thorax and abdomen and is the only cranial nerve that extends beyond the head and neck region (Figure 14.10). The nerve derives its name from its wide distribution. Sensory axons arise from taste buds in the epiglottis and pharynx, and from proprioceptors

Figure 14.10   Vagus (X) nerve.

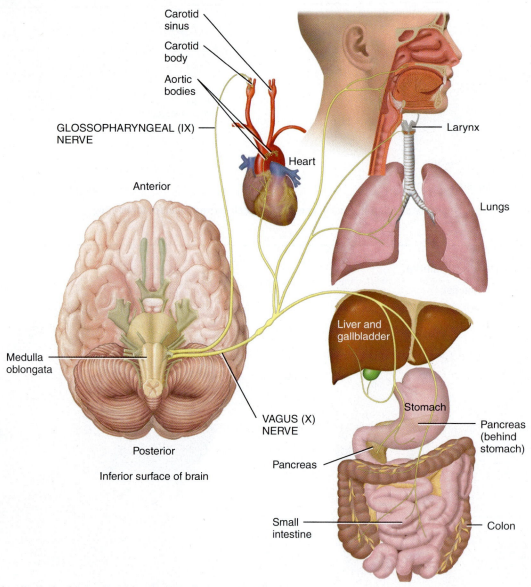

The vagus nerve is widely distributed in the head, neck, thorax, and abdomen.

in muscles of the neck and throat. Sensory axons from receptors in the carotid sinus monitor blood pressure, and those from the carotid body and the aortic bodies (at the arch of the aorta) monitor respiratory function. Sensory axons also come from most organs of the thoracic and abdominal cavities. Sensory axons in the vagus nerve pass through the jugular foramen and end in the medulla and pons. The somatic motor neurons arise in the medulla and supply muscles of the pharynx and larynx that are used in speech and swallowing. Autonomic motor neurons innervate the lungs, heart, and smooth muscle and glands of the respiratory passageways and gastrointestinal tract.

Accessory (XI) Nerve

The **accessory nerve** (ak-SES-ō-rē = assisting) is a mixed cranial nerve. It differs from all other cranial nerves because it originates from *both* the brain stem and the spinal cord (Figure 14.11). The cranial branch arises from the medulla, passes through the jugular foramen, and stimulates muscles of the pharynx, larynx, and soft palate that are used in swallowing. The spinal branch arises in the anterior gray horn of the cervical portion of the spinal cord, enters the skull through the foramen magnum, and then exits through the jugular foramen (see Figure 7.7a). Sensory axons in the spinal branch originate from proprioceptors in neck muscles. The spinal branch conveys sensory impulses from proprioceptors in neck muscles and motor impulses to neck muscles to coordinate head movements.

Figure 14.11 Accessory (XI) nerve.

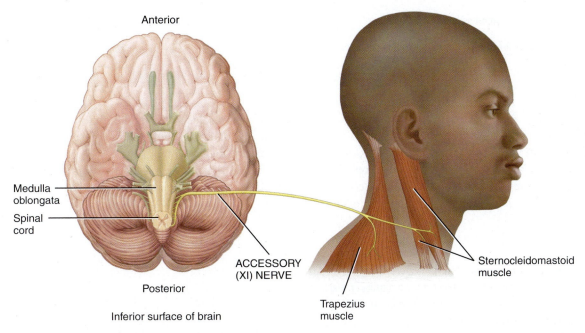

Inferior surface of brain

The accessory nerve exits the skull through the jugular foramen.

Figure 14.12   Hypoglossal (XII) nerve.

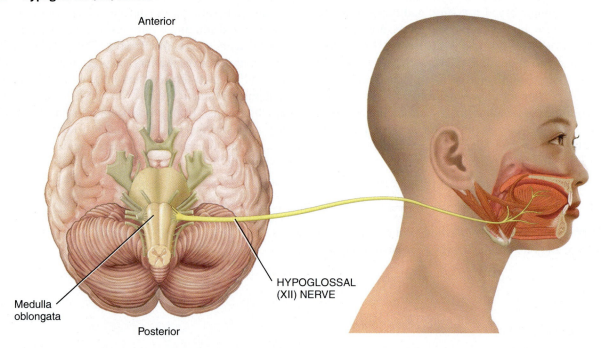

Anterior

Medulla
oblongata

Posterior

HYPOGLOSSAL
(XII) NERVE

Inferior surface of brain

🔑 The hypoglossal nerve exits the skull through the hypoglossal canal.

Hypoglossal (XII) Nerve

The **hypoglossal nerve** (hī´-pō-GLO-sal; *hypo-* = below; *-glossal* = tongue) is a mixed cranial nerve (Figure 14.12). Axons of the hypoglossal nerve pass through the hypoglossal canal (see Figure 7.5). Sensory axons conduct nerve impulses from proprioceptors in tongue muscles to the medulla. Somatic motor axons originate in the medulla and supply tongue muscles with nerve impulses for speech and swallowing.

Table 14.1 presents a summary of the cranial nerves, including illustrations of their individual exits from the CNS and discussions of their locations and functions.

Checkpoint

3.   How are cranial nerves named and numbered? How many are there?
4.   What is the difference between a mixed cranial nerve and a sensory cranial nerve?
5.   How are somatic motor axons different from autonomic motor axons?
6.   Where do the motor axons of the facial nerve originate?
7.   Through which foramen does the glossopharyngeal nerve exit the skull?

TABLE 14.1

Cranial Nerves

NUMBER AND NAME*	TYPE AND LOCATION	FUNCTION
Olfactory (I) nerve Olfactory bulb Olfactory nerve Olfactory tract	**Sensory** Arises in the olfactory mucosa, passes through the olfactory bulb and then the olfactory tract, which extends to the primary olfactory area of the cerebral cortex.	*Function*: Smell.
Optic (II) nerve Optic nerve Optic tract	**Sensory** Arises in the retina of the eye, forms the optic chiasm and then the optic tracts, and terminates in the primary visual area of the cerebral cortex.	*Function*: Vision.
Oculomotor (III) nerve Oculomotor nerve	**Mixed** *Sensory portion*: Consists of axons from proprioceptors in extrinsic eyeball muscles that terminate in the midbrain. *Motor portion*: Originates in the midbrain. Axons of somatic motor neurons innervate the upper eyelid muscle and extrinsic eyeball muscles. Parasympathetic axons innervate intrinsic eyeball muscles.	*Sensory function*: Proprioception. *Somatic motor function*: Movement of upper eyelid and eyeball. *Autonomic motor function:* Alters lens shape and pupil size.
Trochlear (IV) nerve Trochlear nerve	**Mixed** *Sensory portion*: Consists of axons from proprioceptors in an extrinsic eyeball muscle, which terminate in the midbrain. *Motor portion*: Originates in the midbrain and innervates an extrinsic eyeball muscle.	*Sensory function*: Proprioception. *Somatic motor function*: Movement of the eyeball.

(continues)

TABLE 14.1

Cranial Nerves (continued)

NUMBER AND NAME*	TYPE AND LOCATION	FUNCTION
Trigeminal (V) nerve — Trigeminal nerve	**Mixed** *Sensory portion*: Consists of three branches, all of which end in the pons. The **ophthalmic nerve** (*ophthalm-* = the eye) contains axons from the scalp, forehead, upper eyelid, eyeball, lacrimal glands, and nose. The **maxillary nerve** (*maxilla* = upper jawbone) contains axons from the lower eyelid, nose, and mouth. The **mandibular nerve** (*mandibula* = lower jaw bone) contains axons from the cheek, tongue, lower mouth, mandible, side of head, and lower side of face. *Motor portion*: Originates in the pons and innervates muscles of mastication.	*Sensory function*: Conveys impulses for touch, pain, and temperature sensations and proprioception. *Somatic motor function*: Chewing.
Abducens (VI) nerve — Abducens nerve	**Mixed** *Sensory portion*: Consists of axons from proprioceptors in an extrinsic eyeball muscle that end in the pons. *Motor portion*: Originates in the pons and innervates an extrinsic eyeball muscle.	*Sensory function*: Proprioception. *Somatic motor function*: Movement of the eyeball.
Facial (VII) nerve — Facial nerve	**Mixed** *Sensory portion*: Arises from taste buds on the tongue, enters the pons, and then passes to the primary gustatory area of the cerebral cortex. Also contains axons from proprioceptors in muscles of the face and scalp. *Motor portion*: Originates in the pons. Axons of somatic motor neurons innervate facial, scalp, and neck muscles. Parasympathetic axons innervate lacrimal and salivary glands.	*Sensory function*: Taste and proprioception. *Somatic motor function*: Facial expression. *Autonomic motor function*: Secretion of tears and saliva.
Vestibulocochlear (VIII) nerve — Vestibulocochlear nerve	**Sensory** The **vestibular nerve** arises in the organs of equilibrium in the inner ear and ends in the pons and cerebellum. The **cochlear nerve** arises in the organ of hearing, in the inner ear, and relays impulses to the primary auditory area of the cerebral cortex.	*Vestibular branch function*: Conveys impulses related to equilibrium. *Cochlear branch function*: Conveys impulses for hearing.

NUMBER AND NAME*	TYPE AND LOCATION	FUNCTION

Glossopharyngeal (IX) nerve

Glossopharyngeal nerve

Mixed

Sensory portion: Consists of axons from taste buds and sensory receptors on the tongue, from proprioceptors in swallowing muscles, and from receptors in carotid sinus and carotid body. Axons end in the medulla oblongata.

Motor portion: Originates in the medulla oblongata. Axons of somatic motor neurons innervate a muscle of the pharynx used in swallowing. Parasympathetic axons innervate the parotid gland.

Sensory function: Taste and sensations (touch, pain, temperature) from the tongue; proprioception in swallowing muscles; monitoring blood pressure and blood gases.

Somatic motor function: Swallowing and speech.

Autonomic motor function: Stimulates secretion of saliva.

Vagus (X) nerve

Vagus nerve

Mixed

Sensory portion: Consists of axons from taste buds in the epiglottis and pharynx, proprioceptors in muscles of the neck and throat, receptors in the carotid sinus, carotid body, aortic body, and thoracic and abdominal cavity organs. Axons end in the medulla oblongata and pons.

Motor portion: Originates in medulla oblongata. Axons of somatic motor neurons innervate muscles in the throat and neck.

Autonomic axons innervate the lungs, heart, and smooth muscle and glands of the respiratory passageways and gastrointestinal (GI) tract.

Sensory function: Taste and sensations (touch, pain, temperature, and proprioception) from epiglottis and pharynx; monitoring of blood pressure and respiratory function, sensations from visceral organs in thorax and abdomen.

Somatic motor function: Swallowing and speech.

Autonomic motor function: Muscle contractions in organs of the respiratory and GI tract, and secretion of digestive fluids.

Accessory (XI) nerve

Accessory nerve

Mixed

Sensory portion: Consists of axons from proprioceptors in muscles of the pharynx, larynx, and soft palate and ends in the medulla oblongata.

Motor portion: Supplies muscles of the pharynx, larynx, soft palate, and neck muscles.

Sensory function: Proprioception.

Somatic motor function: Swallowing and head movement.

Hypoglossal (XII) nerve

Hypoglossal nerve

Mixed

Sensory portion: Consists of axons from proprioceptors in tongue muscles and ends in the medulla oblongata.

Motor portion: Originates in the medulla oblongata and supplies muscles of the tongue.

Sensory function: Proprioception.

Motor function: Movement of tongue during speech and swallowing.

---

*MNEMONIC for cranial nerves:

Oh	**O**h	**O**h	**T**o	**T**ouch	**A**nd	**F**eel	**V**ery	**G**reen	**V**egetables	**AH!**	
Olfactory	**O**ptic	**O**culomotor	**T**rochlear	**T**rigeminal	**A**bducens	**F**acial	**V**estibulocochlear	**G**lossopharyngeal	**V**agus	**A**ccessory	**H**ypoglossal

14.3   Each spinal nerve branches into a posterior ramus, anterior ramus, meningeal branch, and rami communicantes.

Like the 12 pairs of cranial nerves, spinal nerves and the nerves that branch from them (Figure 14.13) are part of the peripheral nervous system. Spinal nerves connect the CNS to sensory receptors, muscles, and glands in all parts of the body. The 31 pairs of spinal nerves are named and numbered according to the region and level of the vertebral column from which they emerge (Figure 14.13). There are 8 pairs of *cervical nerves* (represented as C1–C8), 12 pairs of *thoracic nerves* (T1–T12), 5 pairs of *lumbar*

Figure 14.13   External anatomy of the spinal cord and spinal nerves.

CERVICAL PLEXUS (C1–C5):
 Lesser occipital nerve
 Ansa cervicalis
 Transverse cervical nerve
 Supraclavicular nerve
 Phrenic nerve

BRACHIAL PLEXUS (C5–T1):
 Musculocutaneous nerve
 Axillary nerve
 Median nerve
 Radial nerve
 Ulnar nerve

Intercostal nerves

LUMBAR PLEXUS (L1–L4):
 Iliohypogastric nerve
 Ilioinguinal nerve
 Genitofemoral nerve
 Lateral cutaneous nerve
 Femoral nerve
 Obturator nerve

SACRAL PLEXUS (L4–S4):
 Superior gluteal nerve
 Inferior gluteal nerve
 Sciatic nerve:
  Common fibular nerve
  Tibial nerve
 Posterior cutaneous nerve of thigh
 Pudendal nerve

Medulla oblongata
Atlas (first cervical vertebra)
CERVICAL NERVES (8 pairs)
Cervical enlargement
First thoracic vertebra
THORACIC NERVES (12 pairs)
Lumbar enlargement
First lumbar vertebra
Conus medullaris
LUMBAR NERVES (5 pairs)
Cauda equina
Ilium of hip bone
Sacrum
SACRAL NERVES (5 pairs)
COCCYGEAL NERVES (1 pair)
Filum terminale

C1, C2, C3, C4, C5, C6, C7, C8, T1, T2, T3, T4, T5, T6, T7, T8, T9, T10, T11, T12, L1, L2, L3, L4, L5, S1, S2, S3, S4, S5

Posterior view of entire spinal cord and portions of spinal nerves

The spinal nerves are named and numbered according to the region of the vertebral column from which they emerge.

nerves (L1–L5), 5 pairs of *sacral nerves* (S1–S5), and 1 pair of *coccygeal nerves* (Co1). The first cervical pair (C1) emerges between the occipital bone and the atlas (first cervical vertebra). All other spinal nerves emerge from the vertebral column through the intervertebral foramina between adjoining vertebrae.

Not all spinal cord segments are aligned with their corresponding vertebrae. Recall that the spinal cord ends near the level of the superior border of the second lumbar vertebra, and that the roots of the lower lumbar, sacral, and coccygeal nerves descend at an angle to reach their respective foramina before emerging from the vertebral column. This arrangement constitutes the cauda equina.

As noted in Chapter 13, a typical **spinal nerve** has two connections to the cord: a **posterior root** and an **anterior root** (see Figure 13.18). The posterior and anterior roots unite to form a spi-nal nerve at the intervertebral foramen. Spinal nerves are mixed nerves because they contain *both* sensory axons and motor axons. The posterior root contains a **posterior root ganglion** in which cell bodies of sensory neurons are located.

Distribution of Spinal Nerves

Branches

A short distance after passing through its intervertebral foramen, a spinal nerve divides into several branches (Figure 14.14). These branches are known as **rami** (RĀ-mī = branches). The **posterior ramus** (RĀ-mus; singular form) serves the muscles and skin of

Figure 14.14 Branches of a typical spinal nerve, shown in transverse section through the thoracic portion of the spinal cord (see also Figure 13.4b).

(a) Transverse section of thoracic spinal cord

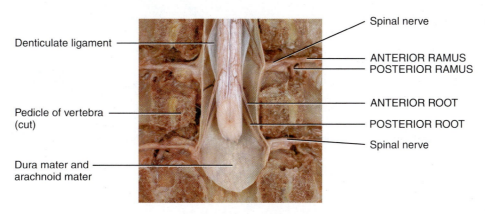

(b) Anterior view and oblique section of spinal cord

The branches of a spinal nerve are the posterior ramus, anterior ramus, meningeal branch, and rami communicantes.

the posterior trunk of the body. The **anterior ramus** serves the muscles and structures of the upper and lower limbs and the skin of the lateral and anterior trunk. In addition to posterior and anterior rami, spinal nerves also give off a **meningeal branch** that reenters the vertebral canal through the intervertebral foramen and supplies the vertebrae, spinal cord, and meninges. Other branches of a spinal nerve are the rami communicantes components of the autonomic nervous system, which will be discussed shortly.

Plexuses

Axons from the anterior rami of spinal nerves, except for thoracic nerves T2–T12, do not go directly to the body structures they supply. Instead, they form networks on both the left and right sides of the body by joining with various numbers of axons from anterior rami of adjacent nerves. Such a network of axons is called a **plexus** (= braid or network). Because axons of anterior rami recombine within each plexus, nerves emerging from a plexus contain axons of several spinal nerves. Because nerves to peripheral structures are supplied by several spinal nerves, total paralysis of a muscle is less likely when one spinal cord segment is damaged. The principal plexuses are the cervical plexus, brachial plexus, lumbar plexus, and sacral plexus (see Figure 14.13). The nerves branching from the plexuses bear names that are often descriptive of the general regions they serve or the course they take. Each of the nerves, in turn, may have several branches named for the specific structures they innervate.

The **cervical plexus** (SER-vi-kul) is formed by the first four cervical nerves (C1–C4), with contributions from C5 (Figure 14.15). The cervical plexus supplies the skin and muscles of the head, neck, and superior part of the shoulders and chest. The **phrenic nerve** is an important nerve that arises from the cervical plexus. The phrenic nerve stimulates the diaphragm, a major respiratory muscle, to contract. Table 14.2 summarizes the major nerves of the cervical plexus.

Spinal nerves C5–C8 and T1 form the **brachial plexus** (BRĀ-kē-al; Figure 14.16a). It passes above the first rib posterior to the

Figure 14.15   Cervical plexus in anterior view.

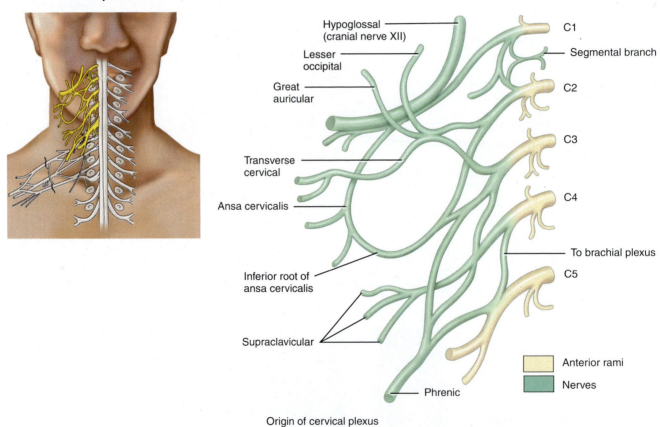

Hypoglossal (cranial nerve XII)
Lesser occipital
Great auricular
Transverse cervical
Ansa cervicalis
Inferior root of ansa cervicalis
Supraclavicular
Phrenic

C1
Segmental branch
C2
C3
C4
To brachial plexus
C5

Anterior rami
Nerves

Origin of cervical plexus

The cervical plexus innervates the skin and muscles of the head, neck, superior portion of the shoulders and chest, and diaphragm.

Table 14.2

Cervical Plexus

NERVE	ORIGIN	DISTRIBUTION
Superficial (Sensory) Branches		
Lesser occipital	C2	Skin of scalp posterior and superior to ear.
Great auricular (aw-RIK-ū-lar)	C2–C3	Skin over ear and over parotid glands.
Transverse cervical	C2–C3	Skin over anterior neck.
Supraclavicular	C3–C4	Skin over superior chest and shoulder.
Deep (Largely Motor) Branches		
Ansa cervicalis (AN-sa ser-vi-KAL-is) (superior and inferior roots)	C1–C3	Infrahyoid and geniohyoid muscles of neck.
Phrenic (FREN-ik)	C3–C5	Diaphragm.
Segmental branches	C1–C5	Deep muscles of neck, levator scapulae, and middle scalene muscles.

clavicle and then enters the axilla (Figure 14.16b). The brachial plexus provides the entire nerve supply of the shoulders and upper limbs. Since the brachial plexus is so complex, an explanation of its various parts is helpful. The anterior rami of several spinal nerves in the plexus unite to form *trunks*; the trunks subdivide into *divisions*, which regroup into *cords*. Five important nerves branch

Figure 14.16 Brachial plexus in anterior view.

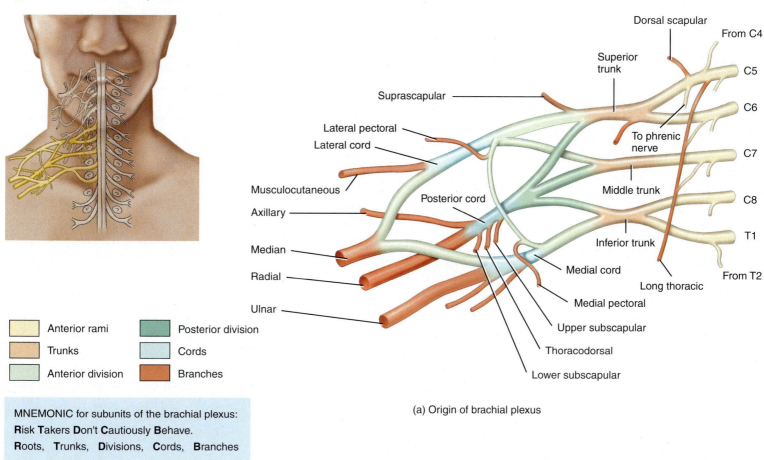

Legend:
- Anterior rami
- Trunks
- Anterior division
- Posterior division
- Cords
- Branches

MNEMONIC for subunits of the brachial plexus:
Risk **T**akers **D**on't **C**autiously **B**ehave.
Roots, **T**runks, **D**ivisions, **C**ords, **B**ranches

(a) Origin of brachial plexus

(continues)

Figure 14.16 (continued)

(b) Distribution of nerves from the brachial plexus

The brachial plexus innervates the shoulders and upper limbs.

TABLE 14.3

Brachial Plexus

NERVE	ORIGIN	DISTRIBUTION
Dorsal scapular (SKAP-ū-lar)	C5	Levator scapulae and rhomboid muscles.
Long thoracic (thor-AS-ik)	C5–C7	Serratus anterior muscle.
Suprascapular	C5–C6	Supraspinatus and infraspinatus muscles.
Musculocutaneous (mus´-kū-lo-kū-TĀN-ē-us)	C5–C7	Flexors of arm (coracobrachialis, biceps brachii, and brachialis muscles); skin over lateral forearm.
Pectoral (PEK-to-ral)(medial and lateral)	C5–T1	Pectoralis major and pectoralis minor muscles.
Subscapular (upper and lower)	C5–C6	Subscapularis and teres major muscles.
Thoracodorsal (thor´-ak-ō-DŌR-sal)	C6–C8	Latissimus dorsi muscle.
Axillary(AK-si-lar-ē)	C5–C6	Deltoid and teres minor muscles; skin over deltoid and superior posterior arm.
Median	C5–T1	Flexors of forearm, except flexor carpi ulnaris and some muscles of the hand; skin of lateral two-thirds of palm of hand and fingers.
Radial	C5–T1	Triceps brachii and other extensor muscles of arm and extensor muscles of forearm; skin of posterior arm and forearm, lateral two-thirds of dorsum of hand, and fingers over proximal and middle phalanges.
Ulnar	C8–T1	Flexor carpi ulnaris, flexor digitorum profundus, and most muscles of the hand; skin over medial hand.

from the cords of the brachial plexus: (1) The **axillary nerve** innervates the shoulder. (2) The **musculocutaneous nerve** innervates the anterior arm and lateral forearm. (3) The **radial nerve** innervates the posterior arm and forearm. (4) The **median nerve** supplies the anterior forearm and some of the hand. (5) The **ulnar nerve** innervates the medial forearm and most of the hand. Table 14.3 summarizes the major nerves of the brachial plexus.

Spinal nerves L1–L4 form the **lumbar plexus** (LUM-bar) (Figure 14.17a). The lumbar plexus supplies the anterior and lateral abdominal wall, external genitals, and part of the lower limbs (Figure 14.17b). Some important nerves arising from this plexus are the **femoral nerve** innervating the anterior and medial thigh and medial leg, and the **obturator nerve** innervating the medial thigh. Table 14.4 summarizes the major nerves of the lumbar plexus.

Figure 14.17 **Lumbar plexus in anterior view.**

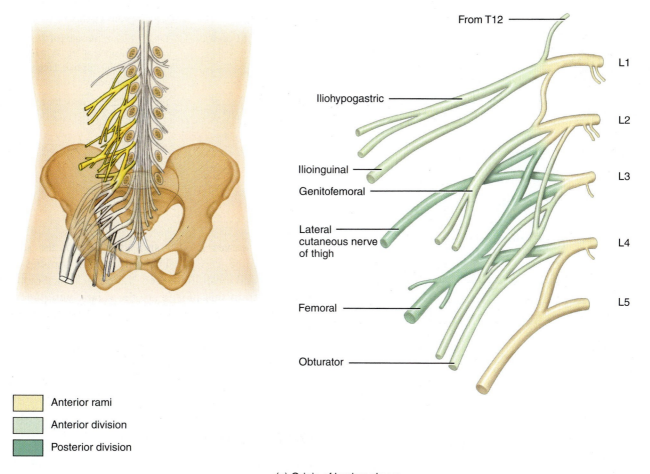

Iliohypogastric

Ilioinguinal

Genitofemoral

Lateral cutaneous nerve of thigh

Femoral

Obturator

From T12

L1

L2

L3

L4

L5

Anterior rami

Anterior division

Posterior division

(a) Origin of lumbar plexus

(continues)

Figure 14.17 (continued)

Anterior view   Posterior view

(b) Distribution of nerves from the lumbar and sacral plexuses

🔑 **The lumbar plexus innervates the anterolateral abdominal wall, external genitals, and part of the lower limbs.**

TABLE 14.4		
Lumbar Plexus		
NERVE	ORIGIN	DISTRIBUTION
Iliohypogastric (il´-ē-ō-hī-pō-GAS-trik)	L1	Abdominal muscles; skin over inferior abdomen and buttocks.
Ilioinguinal (il´-ē-ō-IN-gwi-nal)	L1	Abdominal muscles (with iliohypogastric); skin of superior medial thigh, penis and scrotum in male, and labia majora and mons pubis in female.
Genitofemoral (jen´-i-tō-FEM-or-al)	L1–L2	Cremaster muscle; skin over middle anterior thigh, scrotum in male, and labia majora in female.
Lateral cutaneous nerve of thigh	L2–L3	Skin over lateral, anterior, and posterior thigh.
Femoral (FEM-or-al)	L2–L4	Flexor muscles of thigh and extensor muscles of leg; skin over anterior and medial thigh and medial side of leg and foot.
Obturator (OB-too-rā-tor)	L2–L4	Adductor muscles of thigh; skin over medial thigh.

Spinal nerves L4–L5 and S1–S4 form the **sacral plexus** (SĀ-kral) (Figure 14.18). The sacral plexus supplies the buttocks, perineum, and lower limbs (see Figure 14.17b). Among the nerves that arise from this plexus are the sciatic and pudendal nerves. The **sciatic nerve**—the longest and thickest nerve in the body—innervates most of the lower limb, except for the anterior and medial thigh. The **pudendal nerve** innervates the perineum. Table 14.5 summarizes the major nerves of the sacral plexus.

Intercostal Nerves

The anterior rami of spinal nerves T2–T12 do not enter into the formation of plexuses and are known as **intercostal nerves** (see Figure 14.13). These nerves and their branches directly innervate the structures they supply: the intercostal muscles between ribs, and the muscles and skin of the thorax and abdominal wall.

Figure 14.18 **Sacral and coccygeal plexuses in anterior view.** The distribution of the nerves of the sacral plexus is shown in Figure 14.17b.

Anterior rami

Anterior division

Posterior division

L4 contribution to femoral nerve

L4

L5

S1

Superior gluteal

Inferior gluteal

S2

S3

S4

S5

Tibial

Common fibular

Pudendal

Sciatic

Posterior cutaneous nerve of thigh

Origin of sacral plexus

🔑 **The sacral plexus innervates the buttocks, perineum, and lower limbs.**

TABLE 14.5		
Sacral Plexus		
NERVE	ORIGIN	DISTRIBUTION
Superior gluteal (GLOO-tē-al)	L4–L5 and S1	Gluteus minimus and gluteus medius muscles, and tensor fasciae latae.
Inferior gluteal	L5–S2	Gluteus maximus muscle.
Posterior cutaneous nerve of thigh	S1–S3	Skin over anal region, inferior lateral aspect of buttocks, superior posterior aspect of thigh, superior part of calf, scrotum in male, and labia majora in female.
Sciatic (sī-AT-ik)	L4–S3	Hamstring and adductor magnus muscles. Composed of two nerves bound together by common sheath that splits, usually at the knee.
Tibial (TIB-ē-al) (including medial plantar branches)	L4–S3	Gastrocnemius, plantaris, soleus, popliteus, tibialis posterior muscles, flexors of toes, intrinsic muscles of the foot; skin over posterior leg and plantar surface of foot.
Common fibular (FIB-ū-lar) (including superficial and deep fibular branches)	L4–S2	Fibularis muscles (brevis, longus, tertius), tibialis anterior muscles, and extensor muscles of toes; skin over anterior leg and dorsum of foot.
Pudendal (pū-DEN-dal)	S2–S4	Muscles of perineum; skin of penis and scrotum in male and clitoris, labia majora, labia minora, and vagina in female.

Checkpoint

8. How are spinal nerves named and numbered?
9. How do spinal nerves connect to the spinal cord?
10. Why are all spinal nerves classified as mixed nerves?

11. Where do the posterior and anterior rami originate and what regions of the body does each serve?
12. Which regions of the body are served by each plexus?
13. Which body regions are served by intercostal nerves?

14.4 A reflex is produced by a reflex arc in response to a particular stimulus.

Reflexes and Reflex Arcs

A **reflex** is a fast, involuntary, unplanned sequence of actions that occurs in response to a particular stimulus. Some reflexes are present from birth; for example, the reflex involved in pulling your hand away from a painful stimulus is intrinsic to your nervous system. Other reflexes are learned or acquired. For instance, you learn many reflexes while acquiring driving expertise. Slamming on the brakes in an emergency is one example.

The brain stem and spinal cord serve as integrating centers for reflexive information. When integration occurs in the gray matter of the brain stem, the reflex is a **cranial reflex**. An example of a cranial reflex is the tracking movements of your eyes as you read this sentence. If the spinal cord gray matter is the site of integration, the reflex is a **spinal reflex**. For example, jerking your bare foot up from hot pavement is a spinal reflex.

You are probably most aware of **somatic reflexes**, which involve contraction of skeletal muscles. Equally important, however, are the **autonomic reflexes**, which generally are not consciously perceived. They involve responses of smooth muscle, cardiac muscle, and glands. Body functions such as heart rate, digestion, urination, and defecation are controlled by the autonomic nervous system through autonomic reflexes.

Nerve impulses propagating into, through, and out of the CNS follow specific pathways, depending on the kind of information, its origin, and its destination. The pathway followed by nerve impulses that produce a reflex is called a **reflex arc**. A reflex arc includes the following five functional components (Figure 14.19):

1 **Sensory receptor**. The distal end (dendrite) of a sensory neuron or an associated sensory structure serves as a sensory receptor. It responds to a specific **stimulus**—a change in the internal or external environment—by triggering one or more nerve impulses in the sensory neuron.

2 **Sensory neuron**. The nerve impulses propagate from the sensory receptor along the sensory neuron to the axon terminals in the gray matter of the spinal cord or brain stem.

3 **Integrating center**. One or more regions of gray matter within the CNS act as an integrating center. A reflex pathway having only one synapse in the integrating center is termed a **monosynaptic reflex arc** (*mono-* = one). More often, the integrating center includes one or more interneurons, which may relay impulses to other interneurons as well as to a motor neuron. A **polysynaptic reflex arc** (*poly-* = many) involves more than two types of neurons and more than one integrating center synapse.

4 **Motor neuron**. Impulses triggered by the integrating center propagate out of the CNS along a motor neuron to the part of the body that will respond.

5 **Effector**. The part of the body that responds to the motor nerve impulse is the effector. Its action is called a reflex. If the effector is skeletal muscle, the reflex is a somatic reflex. If the effector is smooth muscle, cardiac muscle, or a gland, the reflex is an autonomic reflex.

Figure 14.19   General components of a reflex arc. The arrows show the direction of nerve impulse propagation.

Interneuron

2 SENSORY NEURON (axon conducts impulses from receptor to integrating center)

1 SENSORY RECEPTOR (responds to a stimulus by producing a generator or receptor potential)

3 INTEGRATING CENTER (one or more regions within the CNS that relay impulses from sensory to motor neurons)

4 MOTOR NEURON (axon conducts impulses from integrating center to effector)

5 EFFECTOR (muscle or gland that responds to motor nerve impulses)

A reflex is a fast, predictable sequence of involuntary actions that occur in response to certain changes in the environment.

Because reflexes are normally so predictable, they provide useful information about the health of the nervous system and can greatly aid diagnosis of disease. Damage or disease anywhere along its reflex arc can cause a reflex to be absent or abnormal.

For example, tapping the patellar ligament normally causes reflex extension of the knee joint. Absence of the patellar reflex could indicate damage of the sensory or motor neurons, or a spinal cord injury in the lumbar region. Somatic reflexes generally can be tested simply by tapping or stroking the body surface. In contrast, most autonomic reflexes are not practical diagnostic tools because it is difficult to stimulate visceral receptors, which are deep inside the body. An exception is the pupillary light reflex, in which the pupils of both eyes decrease in diameter when either eye is exposed to light. Because the reflex arc includes synapses in the brain stem, the *absence of a normal pupillary light reflex* may indicate brain damage or injury.

The Stretch Reflex

A **stretch reflex** causes contraction of a skeletal muscle (the effector) in response to stretching of the muscle. This type of reflex is a monosynaptic reflex arc with activation of a single sensory neuron that synapses in the CNS with a single motor neuron. Stretch reflexes can be elicited by tapping on tendons attached to muscles at the elbow, wrist, knee, and ankle joints.

A stretch reflex operates as follows (Figure 14.20):

1 Slight stretching of a muscle stimulates sensory receptors in the muscle called **muscle spindles** (shown in more detail in Figure 15.4). Muscle spindles monitor changes in the length of the muscle.

2 In response to being stretched, a muscle spindle generates one or more nerve impulses that propagate along a somatic

Figure 14.20   Stretch reflex. This monosynaptic reflex arc has only one synapse in the CNS—between the sensory neuron and motor neuron. Also illustrated is a polysynaptic reflex arc, with two synapses in the CNS and one interneuron, that innervates antagonistic muscles. Plus signs (+) indicate excitatory synapses; the minus sign (–) indicates an inhibitory synapse.

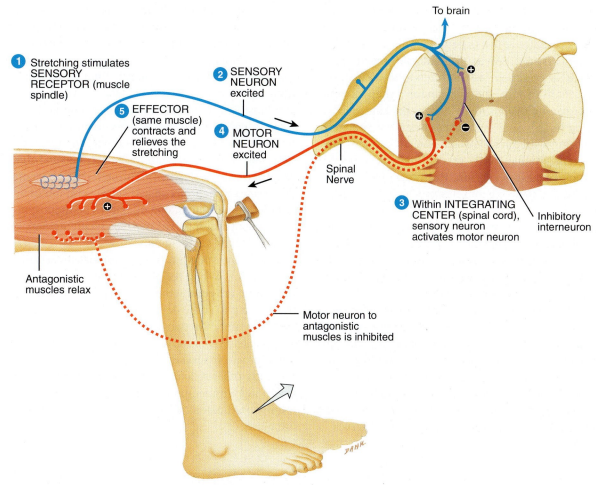

🔑 **The stretch reflex causes contraction of a muscle that has been stretched.**

sensory neuron that extends through the posterior root of the spinal nerve into the spinal cord.

3 In the spinal cord gray matter (integrating center), the sensory neuron synapses with and activates a motor neuron.

4 Nerve impulses arise in the motor neuron and propagate along its axon from the spinal cord into the anterior root to and on the stimulated muscle.

5 Acetylcholine is released at the neuromuscular junctions between the axon terminals of the motor neuron and skeletal muscle fibers of the stretched muscle. Acetylcholine triggers action potentials in the stretched muscle (effector), and the muscle contracts. Thus, muscle stretch is followed by muscle contraction, which relieves the stretching.

In the reflex arc just described, sensory nerve impulses enter the spinal cord on the same side from which motor nerve impulses leave it. This arrangement is called an **ipsilateral reflex** (ip´-si-LAT-er-al = same side). All monosynaptic reflexes are ipsilateral.

By adjusting how vigorously a muscle spindle responds to stretching, the brain sets an overall level of **muscle tone**, which is the small degree of contraction present while the muscle is at rest. Because the stimulus for the stretch reflex is stretching of muscle, this reflex helps avert injury by preventing overstretching of muscles.

Although the stretch reflex pathway itself is monosynaptic (just two neurons and one synapse), a polysynaptic reflex arc to the antagonistic muscles operates at the same time. This arc involves three neurons and two synapses. An axon collateral (branch) from the sensory neuron also synapses with an inhibitory interneuron in the integrating center. In turn, the interneuron synapses with and inhibits a motor neuron that normally excites the antagonistic muscles (dotted line in Figure 14.20). Thus, when the stretched muscle contracts during a stretch reflex, antagonistic muscles that oppose the contraction simultaneously relax. This type of neural circuit is termed **reciprocal innervation**. Reciprocal innervation prevents conflict between opposing muscles and is vital in coordinating body movements.

Axon collaterals of the sensory neuron also relay nerve impulses to the brain about the state of stretch or contraction of skeletal muscles, enabling the brain to coordinate muscular movements and posture. The nerve impulses that pass to the brain also allow conscious awareness that the reflex has occurred.

The patellar reflex (knee jerk) is an example of a stretch reflex. The patellar reflex occurs when the leg extends at the knee joint due to contraction of the quadriceps femoris muscle in response to tapping the patellar ligament (Figure 14.20). This reflex is blocked by damage to either nerves supplying the quadriceps femoris muscle or the integrating centers in the lumbar segments of the spinal cord. The patellar reflex is often absent in people with chronic diabetes mellitus or neurosyphilis, both of which cause degeneration of nerves.

The Flexor Reflex

A polysynaptic reflex arc results from a painful stimulus, such as stepping on a tack. In response to such a painful stimulus, you immediately withdraw your leg. This reflex, called the **flexor** or **withdrawal reflex**, operates as follows (Figure 14.21):

1 Stepping on a tack stimulates a pain-sensitive sensory neuron (pain receptor).

2 This stimulated sensory neuron generates nerve impulses, which propagate into the spinal cord.

3 Within the spinal cord (integrating center), the sensory neuron activates interneurons that extend to several spinal cord segments.

4 The interneurons activate several motor neurons. The stimulated motor neurons generate nerve impulses, which propagate toward the axon terminals.

5 Acetylcholine released by the motor neurons causes the flexor muscles in the thigh (effectors) to contract, producing withdrawal of the leg. This reflex is protective because contraction of flexor muscles moves a limb away from the source of a possibly damaging stimulus.

The flexor reflex, like the stretch reflex, is ipsilateral—the incoming and outgoing impulses propagate into and out of the same side of the spinal cord. The flexor reflex also illustrates another feature of polysynaptic reflex arcs. Moving your entire lower or upper limb away from a painful stimulus involves contraction of more than one muscle group. Hence, several motor neurons must simultaneously convey impulses to several limb muscles. Because nerve impulses from one sensory neuron ascend and descend in the spinal cord and activate interneurons in several segments of the spinal cord, this type of reflex is called an **intersegmental reflex arc** (*inter-* = between). Through intersegmental reflex arcs, a single sensory neuron can activate several motor neurons, thereby stimulating more than one effector. The monosynaptic stretch reflex, in contrast, involves muscles receiving nerve impulses from one spinal cord segment only.

As in the stretch reflex, reciprocal innervation occurs in the flexor reflex. When the flexor muscles of a painfully stimulated lower limb are contracting, the extensor muscles of the same limb are relaxing to some degree. If both sets of muscles contracted at the same time, the two sets of muscles would pull on the bones in opposite directions, which might immobilize the limb. Because of reciprocal innervation, however, one set of muscles contracts while the other relaxes. Reflexes involving intersegmental reflex arcs along with reciprocal innervation of antagonistic muscles result in coordinated movements of the body away from dangerous situations.

Figure 14.21 **Flexor (withdrawal) reflex.** This reflex arc is polysynaptic and ipsilateral. Plus signs (+) indicate excitatory synapses.

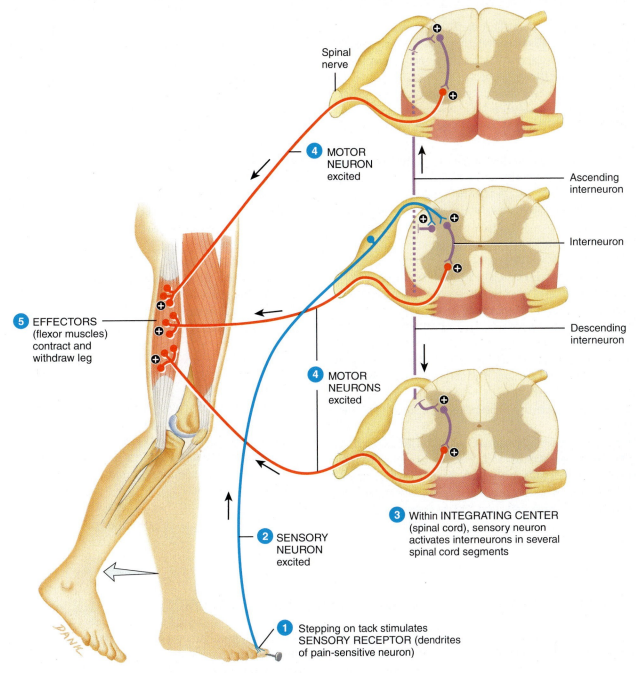

Spinal nerve

4 MOTOR NEURON excited

Ascending interneuron

Interneuron

Descending interneuron

5 EFFECTORS (flexor muscles) contract and withdraw leg

4 MOTOR NEURONS excited

3 Within INTEGRATING CENTER (spinal cord), sensory neuron activates interneurons in several spinal cord segments

2 SENSORY NEURON excited

1 Stepping on tack stimulates SENSORY RECEPTOR (dendrites of pain-sensitive neuron)

The flexor reflex causes withdrawal of a part of the body in response to a painful stimulus.

Checkpoint

14. How are cranial and spinal reflexes similar? How do they differ?

15. How are somatic and autonomic reflexes similar? How are they different?

16. List the components of a reflex arc in order.

17. Give the meanings of each of the following terms in relation to reflex arcs: monosynaptic, polysynaptic, reciprocal innervation, and intersegmental.

"**W**ell, Nick," said Dr. Jerouen. "You were diagnosed with Type 2 diabetes mellitus eight years ago. You've made pretty regular visits up until now, with no complaints. When did these problems start?"

"I've been having these prickling sensations, in both my feet, since March or April. I lose my balance sometimes when I'm getting in and out of my boat. And my eyes get funny once in a while."

Dr. Jerouen interrupted. "So this has been progressing for about a year. Have you been monitoring your blood glucose levels regularly?"

"I haven't been real good about it," admitted Nick.

"You may have developed diabetic neuropathy, nerve damage from not controlling your blood sugar. I'm going to run some blood tests and refer you to a neurologist to check your nerve function."

About a week later, Nick went to the neurologist.

"Hi, Nick, I'm Dr. Meyerson. Tell me a little bit about what's been happening." As Nick began recounting his symptoms, Dr. Meyerson worked quickly and efficiently. She closely observed Nick's face, looking for symmetry, noting his facial muscle tone, and listening to the sound of his voice and the articulation of his words. "Smile. Now stick out your tongue. Wrinkle your forehead. Good. I'm going to shine a light into your eyes. Hmmm. Follow the end of this pen with your eyes as I move it around in front of you. Good. Have you had any loss of hearing or problems with your vision?" she asked.

"I have trouble focusing sometimes, and my hearing is always bad," said Nick in a slightly hoarse voice.

"Nick, I'm going to tap you with this hammer to check your reflexes," said Dr. Meyerson. "So, Nick? Do you have any hobbies?" Dr. Meyerson was only half listening to him talk as she briskly tapped Nick's patellar tendon several times and observed the response. She noticed that Nick's knee reflex was somewhat weak. Moving up to Nick's arms, the doctor tapped the back of his arm, then the inside of the elbow.

Using the blunt end of the reflex hammer, Dr. Meyerson ran the tool up the bottom of Nick's foot from the outside of the heel to the ball of his foot. She observed his toes. "Did you feel that?" she asked Nick.

"Not really. Can't feel much in the feet these days," admitted Nick.

"Nick, I'm going to put pressure on your foot. You resist me as hard as you can." Placing her hand on top of Nick's foot, Dr. Meyerson pushed down. "Nick, hold your foot up as I try to push it down. Hmm, you've lost quite a bit of muscle strength and sensation, and your reflexes are somewhat weak. Your tests suggest some peripheral nerve damage. There's not much we can do for the neuropathy except try to get your diabetes under control, to prevent further nerve damage. If your pain increases, there are a few drugs we can try. Come back and see me in six months. But if your symptoms get worse or you begin having trouble with indigestion, vomiting, or constipation, you should come in sooner. It could be a sign of autonomic involvement."

"Auto *what*?" asked Nick.

"Your autonomic nervous system automatically controls things like digestion and blood pressure. But the nerves that carry those signals can be damaged just like your somatic nervous system, the part of your nervous system that voluntarily directs your muscles to move your arms and legs."

A.   Which symptoms that Nick has described so far are relevant to the nervous system? Are his symptoms sensory, motor, or both?

B.   Do you think the symptoms Nick describes are likely caused by peripheral nerve damage? Could they be caused by damage to the central nervous system?

C.   Diabetic neuropathies damage peripheral nerves. Which component of the reflex arc is most likely to be damaged in Nick's situation?

14.5 The autonomic nervous system produces involuntary movements.

Like the somatic nervous system, the **autonomic nervous system (ANS)** operates via reflex arcs. A continual flow of nerve impulses propagates from autonomic sensory neurons in visceral organs and blood vessels into integrating centers in the CNS. Then, impulses in autonomic motor neurons propagate to various effector tissues, thereby regulating the activity of smooth muscle, cardiac muscle, and glands.

The somatic nervous system includes both sensory and motor neurons. Sensory neurons convey input from receptors for the **special senses** (vision, hearing, taste, smell, and equilibrium) and from receptors for **somatic senses** (pain, thermal, tactile, and proprioceptive sensations). All these sensations normally are consciously perceived. In turn, somatic motor neurons innervate skeletal muscle—the effector tissue of the somatic nervous system—and produce voluntary movements. When a somatic motor neuron stimulates a skeletal muscle, the muscle contracts; the effect is always excitation. If somatic motor neurons cease to stimulate a muscle, the result is a paralyzed, limp muscle that has no muscle tone. In addition, even though we are generally not conscious of breathing, the muscles that generate respiratory movements are skeletal muscles controlled by somatic motor neurons. If the respiratory motor neurons become inactive, breathing stops.

The main input to the ANS comes from **autonomic sensory neurons**. Mostly, these neurons are associated with **interoceptors** (IN-ter-ō-sep´-tors), sensory receptors that monitor the internal environment of the body, such as chemoreceptors that monitor blood CO_2 level and mechanoreceptors that detect the degree of stretch in the walls of organs or blood vessels. Unlike sensory signals triggered by a flower's perfume, a beautiful painting, or a delicious meal, sensory signals from interoceptors are not consciously perceived most of the time. On the other hand, intense activation of interoceptors may produce conscious sensations. Two examples of perceived visceral sensations are pain from damaged viscera and angina pectoris (chest pain) from inadequate blood flow to the heart. The ANS can be influenced by sensory input from somatic sensory and special sensory neurons. For example, somatic pain can produce dramatic changes in some autonomic activities.

Autonomic motor neurons regulate visceral activities by either increasing (exciting) or decreasing (inhibiting) ongoing activities in their effector tissues (cardiac muscle, smooth muscle, and glands). Changes in the diameter of the pupils, dilation and constriction of blood vessels, and adjustment of the rate and force of the heartbeat are examples of autonomic motor responses. Unlike skeletal muscle, tissues innervated by the ANS often function to some extent even if their nerve supply is damaged. The heart continues to beat when it is removed for transplantation into another person, smooth muscle in the lining of the gastrointestinal tract contracts rhythmically on its own, and glands produce some secretions in the absence of ANS control.

The ANS usually operates without conscious control and was originally named *autonomic* because it was thought to function autonomously, or in a self-governing manner, without control by the CNS. However, we now know that the hypothalamus and brain stem do regulate ANS activity. Most autonomic responses cannot be consciously altered to any great degree. You probably cannot voluntarily slow your heartbeat to half its normal rate. For this reason, some autonomic responses are the basis for polygraph ("lie detector") tests. However, practitioners of yoga or other techniques of meditation may learn how to regulate at least some of their autonomic activities through long practice. **Biofeedback**, in which monitoring devices display information about a body function such as heart rate or blood pressure, enhances the ability to learn such conscious control. Signals from the general somatic and special senses, acting via the limbic system, also influence responses of autonomic motor neurons. Seeing a bike about to hit you, hearing squealing brakes of a nearby car, or being grabbed by an attacker would all increase the rate and force of your heartbeat.

The axon of a single, myelinated somatic motor neuron extends from the CNS all the way to the skeletal muscle fibers in its motor unit (Figure 14.22a). By contrast, most autonomic motor pathways consist of two motor neurons in series, that is, one following the other (Figure 14.22b). The first autonomic motor neuron has its cell body in the CNS; its myelinated axon extends from the CNS to an **autonomic ganglion**. (Recall that a ganglion is a collection of neuronal cell bodies in the PNS.) The cell body of the second neuron is also in that same autonomic ganglion; its unmyelinated axon extends directly from the ganglion to the effector (smooth muscle, cardiac muscle, or a gland). All somatic motor neurons release only acetylcholine as their neurotransmitter, but autonomic motor neurons may release acetylcholine or norepinephrine.

The motor portion of the ANS has two principal branches: the **sympathetic division** and the **parasympathetic division**. Most organs receive **dual innervation**; that is, they receive impulses from both sympathetic and parasympathetic motor neurons. In general, nerve impulses from one division of the ANS stimulate the organ to increase its activity (excitation), and impulses from the other division decrease the organ's activity (inhibition). For example, an increased rate of nerve impulses from the *sympathetic* division increases heart rate, and an increased rate of nerve impulses from the *parasympathetic* division decreases heart rate.

Table 14.6 compares the somatic and autonomic nervous systems.

Figure 14.22   Motor neuron pathways in the (a) somatic nervous system and (b) autonomic nervous system. Note that somatic motor neurons release acetylcholine (ACh); autonomic motor neurons release either ACh or norepinephrine (NE).

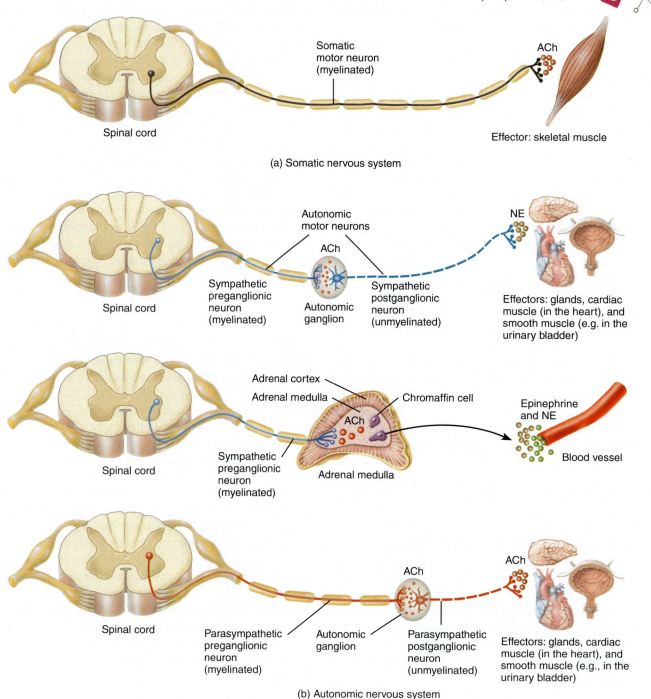

(a) Somatic nervous system

(b) Autonomic nervous system

Somatic nervous system stimulation always excites its effectors (skeletal muscle fibers); stimulation by the autonomic nervous system either excites or inhibits visceral effectors.

TABLE 14.6

Comparison of Somatic and Autonomic Nervous Systems

	SOMATIC NERVOUS SYSTEM	AUTONOMIC NERVOUS SYSTEM
Sensory input	Special senses and somatic senses.	Mainly from interoceptors; some from somatic senses and special senses.
Control of motor output	Voluntary control from cerebral cortex, with contributions from basal nuclei, cerebellum, brain stem, and spinal cord.	Involuntary control from hypothalamus, brain stem, and spinal cord with limited contribution from cerebral cortex.
Motor neuron pathway	One-neuron pathway: Somatic motor neurons extending from CNS synapse directly with effector.	Two-neuron pathway: Preganglionic neurons extending from CNS synapse with postganglionic neurons in an autonomic ganglion. Postganglionic neurons extending from ganglion synapse with a visceral effector.
Neurotransmitters and hormones	Somatic motor neurons release acetylcholine (ACh).	Preganglionic neurons release ACh. Most sympathetic postganglionic neurons release norepinephrine; those of sweat glands release ACh. Parasympathetic postganglionic neurons release ACh.
Effectors	Skeletal muscle.	Smooth muscle, cardiac muscle, and glands.
Responses	Contraction of skeletal muscle.	Contraction or relaxation of smooth muscle; increased or decreased rate and force of contraction of cardiac muscle; increased or decreased secretions of glands.

Checkpoint

18. What are the main input and output components of the autonomic nervous system?
19. What are the effectors of the somatic nervous system? The ANS?
20. How is the motor neuron pathway of the ANS different from the motor neuron pathway of the somatic nervous system?
21. What is dual innervation?

14.6 The ANS includes preganglionic neurons, autonomic ganglia and plexuses, and postganglionic neurons.

The first of the two motor neurons in any autonomic motor pathway is called a **preganglionic neuron** (see Figure 14.22b). Its cell body is in the brain stem or spinal cord, and its axon exits the CNS as part of a cranial or spinal nerve and extends to an autonomic ganglion. There it synapses with a **postganglionic neuron**, the second neuron in the autonomic motor pathway (Figure 14.22b). Notice that the postganglionic neuron lies entirely outside the CNS. Its cell body and dendrites are located in an autonomic ganglion, where it forms synapses with one or more preganglionic axons, then extends to a visceral effector. Thus, preganglionic neurons convey nerve impulses from the CNS to autonomic ganglia, and postganglionic neurons relay the impulses from autonomic ganglia to visceral effectors.

Preganglionic Neurons

In the sympathetic division, cell bodies of preganglionic neurons are located in the gray matter in the thoracic and first two or three lumbar segments of the spinal cord (Figure 14.23). Cell bodies of preganglionic neurons of the parasympathetic division are located in the brain stem and sacral segments of the spinal cord (Figure 14.24).

Autonomic Ganglia

There are two major groups of autonomic ganglia: (1) sympathetic ganglia, which are components of the sympathetic division of the ANS, and (2) parasympathetic ganglia, which are components of the parasympathetic division of the ANS.

Figure 14.23   Structure of the sympathetic division of the autonomic nervous system. Solid lines represent preganglionic axons; dashed lines represent postganglionic axons. Although the innervated structures are shown only for one side of the body for diagrammatic purposes, the sympathetic division actually innervates tissues and organs on both sides.

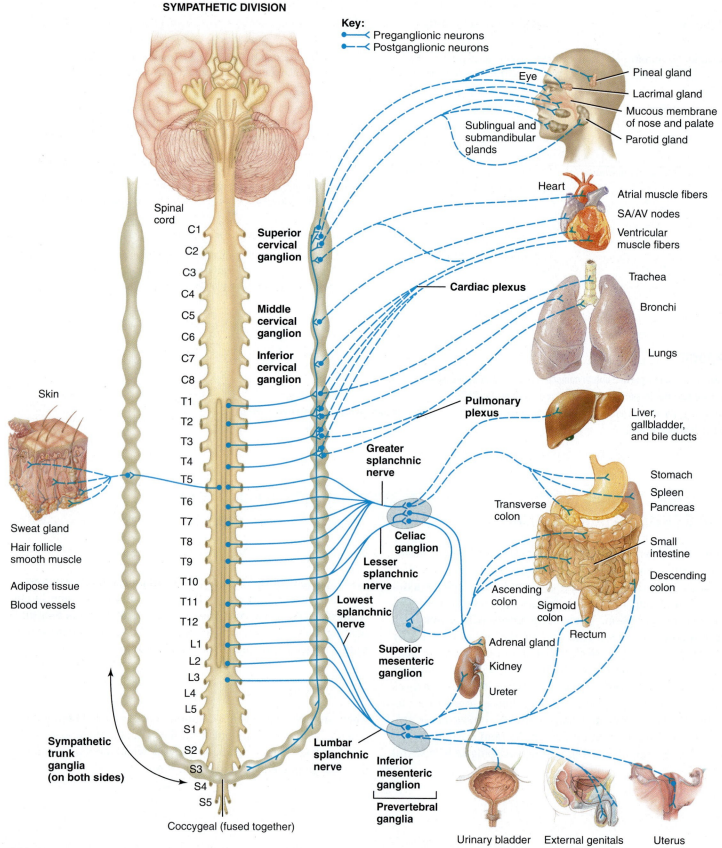

SYMPATHETIC DIVISION

Key:
● ─< Preganglionic neurons
● ─< Postganglionic neurons

Spinal cord

C1
C2
C3
C4
C5
C6
C7
C8
T1
T2
T3
T4
T5
T6
T7
T8
T9
T10
T11
T12
L1
L2
L3
L4
L5
S1
S2
S3
S4
S5

Coccygeal (fused together)

Superior cervical ganglion

Middle cervical ganglion

Inferior cervical ganglion

Cardiac plexus

Pulmonary plexus

Greater splanchnic nerve

Celiac ganglion

Lesser splanchnic nerve

Lowest splanchnic nerve

Superior mesenteric ganglion

Lumbar splanchnic nerve

Inferior mesenteric ganglion

Prevertebral ganglia

Sympathetic trunk ganglia (on both sides)

Skin

Sweat gland

Hair follicle smooth muscle

Adipose tissue

Blood vessels

Eye
Pineal gland
Lacrimal gland
Mucous membrane of nose and palate
Sublingual and submandibular glands
Parotid gland

Heart
Atrial muscle fibers
SA/AV nodes
Ventricular muscle fibers

Trachea
Bronchi
Lungs

Liver, gallbladder, and bile ducts

Stomach
Spleen
Pancreas
Transverse colon
Small intestine
Ascending colon
Descending colon
Sigmoid colon
Rectum

Adrenal gland
Kidney
Ureter

Urinary bladder   External genitals   Uterus

🔑 **Cell bodies of sympathetic preganglionic neurons are located in the gray matter in the thoracic and lumbar segments of the spinal cord.**

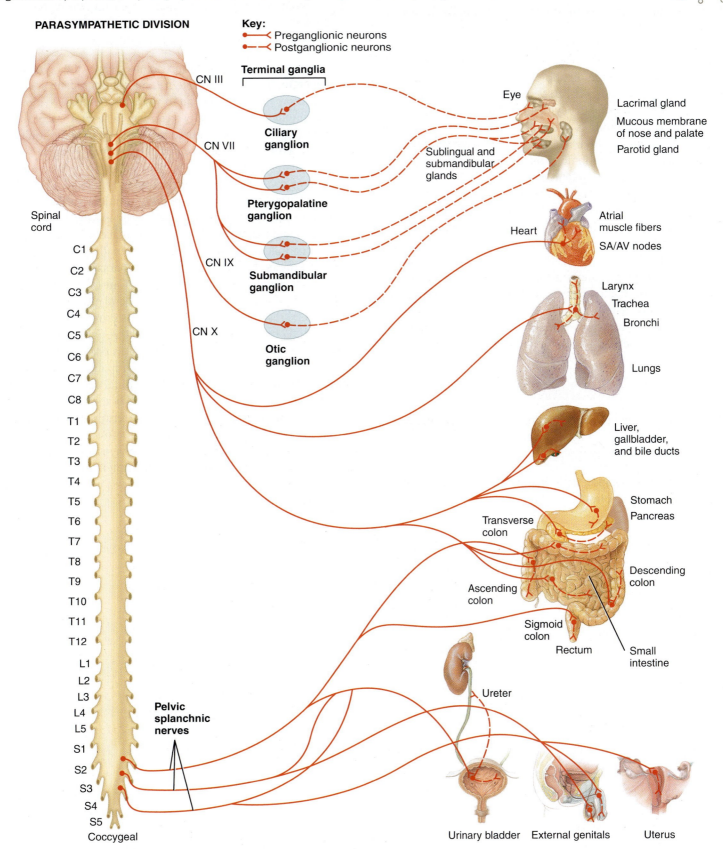

Figure 14.24 **Structure of the parasympathetic division of the autonomic nervous system.** Solid lines represent preganglionic axons; dashed lines represent postganglionic axons. Although the innervated structures are shown only for one side of the body for diagrammatic purposes, the parasympathetic division actually innervates tissues and organs on both sides.

PARASYMPATHETIC DIVISION

Key:
- ●—< Preganglionic neurons
- ●--< Postganglionic neurons

Terminal ganglia

CN III

CN VII

CN IX

CN X

Ciliary ganglion

Pterygopalatine ganglion

Submandibular ganglion

Otic ganglion

Spinal cord

C1
C2
C3
C4
C5
C6
C7
C8
T1
T2
T3
T4
T5
T6
T7
T8
T9
T10
T11
T12
L1
L2
L3
L4
L5
S1
S2
S3
S4
S5
Coccygeal

Pelvic splanchnic nerves

Eye

Lacrimal gland

Mucous membrane of nose and palate

Parotid gland

Sublingual and submandibular glands

Heart

Atrial muscle fibers

SA/AV nodes

Larynx

Trachea

Bronchi

Lungs

Liver, gallbladder, and bile ducts

Stomach

Pancreas

Transverse colon

Descending colon

Ascending colon

Sigmoid colon

Rectum

Small intestine

Ureter

Urinary bladder   External genitals   Uterus

Cell bodies of parasympathetic preganglionic neurons are located in the brain stem and in the gray matter in the sacral segments of the spinal cord.

Sympathetic ganglia are the sites of synapses between sympathetic preganglionic and postganglionic neurons. There are two major types of sympathetic ganglia: sympathetic trunk ganglia and prevertebral ganglia. **Sympathetic trunk ganglia** lie in a vertical row on either side of the vertebral column (see Figure 14.23). These ganglia extend from the base of the skull to the coccyx. Postganglionic axons from sympathetic trunk ganglia primarily innervate organs above the diaphragm. Because the sympathetic trunk ganglia are near the spinal cord, most sympathetic preganglionic axons are short and most sympathetic postganglionic axons are long. The second group of sympathetic ganglia, the **prevertebral ganglia**, lies in the abdomen and anterior to the vertebral column. In general, postganglionic axons from prevertebral ganglia innervate organs below the diaphragm.

Sympathetic preganglionic axons leave the spinal cord through the anterior roots of spinal nerves and enter short pathways, the spinal nerve branches called rami communicantes (kō-mū-ni-KAN-tē-z), to arrive at the sympathetic trunk ganglia (Figure 14.25).

Parasympathetic preganglionic neurons exiting the brain stem extend outward on cranial nerves to **terminal ganglia** that are located close to or within their target organ (see Figure 14.24). Therefore, parasympathetic preganglionic axons are long, in contrast to parasympathetic postganglionic axons, which are short.

Postganglionic Neurons

Once axons of sympathetic preganglionic neurons pass to sympathetic trunk ganglia, they may connect with postganglionic neurons in one of the following ways (Figure 14.25):

❶ An axon may synapse with postganglionic neurons in the ganglion it first reaches.

❷ An axon may ascend or descend to a higher or lower ganglion before synapsing with postganglionic neurons.

❸ An axon may continue, without synapsing, through the sympathetic trunk ganglion to end at a prevertebral ganglion and synapse with postganglionic neurons there.

Sympathetic postganglionic axons leave the sympathetic trunk ganglia by passing through rami communicantes to spinal nerves, where they continue on to visceral effectors. Sympathetic postganglionic axons typically terminate in several visceral effectors (see Figure 14.23). A single sympathetic preganglionic fiber has many axon collaterals (branches) and may synapse with 20 or more postganglionic neurons. This pattern of projection is an example of divergence and helps explain why many sympathetic responses affect almost the entire body simultaneously.

Parasympathetic preganglionic axons exiting from both the brain stem and sacral segments of the spinal cord extend to terminal ganglia that are located close to or actually within the wall of a visceral organ. In a terminal ganglion, parasympathetic presynaptic axons usually synapses with only four or five postsynaptic neurons, all of which supply a single visceral effector. Thus, parasympathetic responses can be localized to a single effector.

Cranial preganglionic axons extend through cranial nerves to terminal ganglia, where they synapse with postsynaptic neurons innervating structures in the head. Parasympathetic preganglionic neurons leaving sacral segments of the spinal cord extend through the anterior roots of spinal nerves, then branch off to form **pelvic splanchnic nerves** that continue to terminal ganglia near or within the walls of visceral effectors (Figure 14.26). In the terminal ganglia, pelvic splanchnic nerves synapse with parasympathetic postganglionic neurons, which innervate smooth muscle and glands in the walls of the colon, ureters, urinary bladder, and reproductive organs.

Figure 14.25 **Types of connections between ganglia and postganglionic neurons in the sympathetic division of the ANS.** Also illustrated are the rami communicantes.

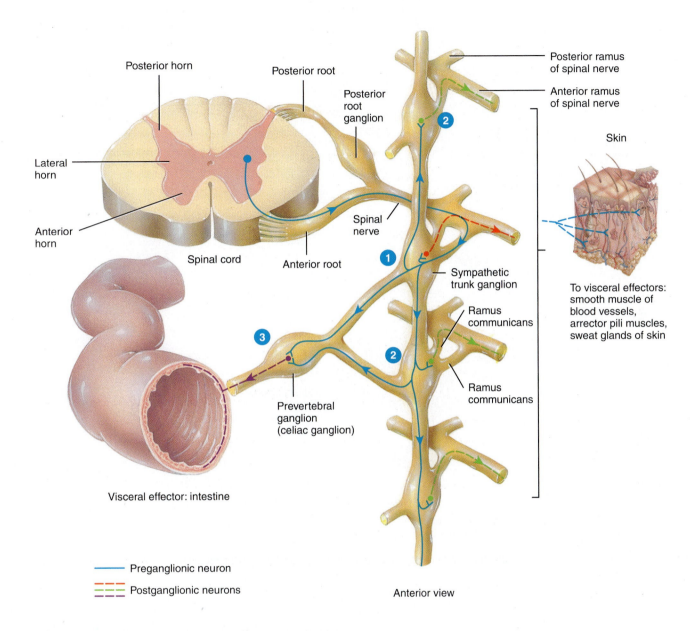

Posterior horn

Posterior root

Posterior root ganglion

Posterior ramus of spinal nerve

Anterior ramus of spinal nerve

Lateral horn

Skin

Anterior horn

Spinal nerve

Spinal cord

Anterior root

Sympathetic trunk ganglion

To visceral effectors: smooth muscle of blood vessels, arrector pili muscles, sweat glands of skin

Ramus communicans

Ramus communicans

Prevertebral ganglion (celiac ganglion)

Visceral effector: intestine

—— Preganglionic neuron
= = = Postganglionic neurons

Anterior view

🔑 Sympathetic ganglia lie in two chains on either side of the vertebral column (sympathetic trunk ganglia) and anterior to the vertebral column (prevertebral ganglia).

Figure 14.26   Pelvic splanchnic nerves.

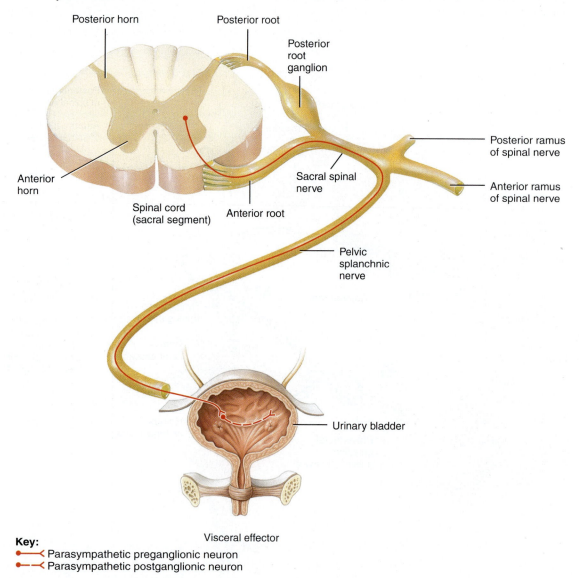

Posterior horn

Posterior root

Posterior root ganglion

Anterior horn

Spinal cord (sacral segment)

Anterior root

Sacral spinal nerve

Posterior ramus of spinal nerve

Anterior ramus of spinal nerve

Pelvic splanchnic nerve

Urinary bladder

Visceral effector

Key:
•──< Parasympathetic preganglionic neuron
•── ─< Parasympathetic postganglionic neuron

🔑 **Parasympathetic preganglionic axons extend through pelvic splanchnic nerves to parasympathetic postganglionic neurons in terminal ganglia within the walls of pelvic viscera.**

Autonomic Plexuses

In the thorax, abdomen, and pelvis, axons of both sympathetic and parasympathetic neurons form tangled networks called **autonomic plexuses**, many of which lie along major arteries (Figure 14.27). The autonomic plexuses may also contain sympathetic ganglia. The major plexuses in the thorax are the **cardiac plexus**, which supplies the heart, and the **pulmonary plexus**, which supplies the bronchial tree. Located along the abdominal aorta are several autonomic plexuses; often the plexuses that innervate abdominal viscera are named after the artery along which they are distributed (e.g., *celiac, superior mesenteric, inferior mesenteric, hypogastric,* and *renal*).

Figure 14.27 **Autonomic plexuses in the thorax, abdomen, and pelvis.**

Trachea

Right vagus (X) nerve

Arch of aorta

Left vagus (X) nerve

CARDIAC PLEXUS

PULMONARY PLEXUS

Right primary bronchus

Esophagus

Right sympathetic
trunk ganglion

Thoracic aorta

ESOPHAGEAL PLEXUS

Inferior vena cava (cut)

Diaphragm

Celiac trunk
(artery)

CELIAC GANGLION AND PLEXUS

SUPERIOR MESENTERIC
GANGLION AND PLEXUS

Superior mesenteric artery

RENAL GANGLION
AND RENAL PLEXUS

Right kidney

INFERIOR MESENTERIC
GANGLION AND PLEXUS

Inferior mesenteric artery

Right sympathetic
trunk ganglion

HYPOGASTRIC PLEXUS

🔑 An autonomic plexus is a network of sympathetic and parasympathetic axons and may also include sympathetic ganglia.

Checkpoint

22. What is the functional difference between preganglionic and postganglionic neurons?
23. Contrast the locations of sympathetic trunk ganglia, prevertebral ganglia, and terminal ganglia.
24. Which ganglia are associated with the sympathetic division? Parasympathetic division?
25. Which division, sympathetic or parasympathetic, has longer preganglionic axons? Why?

14.7 ANS neurons release acetylcholine or norepinephrine, resulting in excitation or inhibition.

Autonomic neurons release neurotransmitters at synapses between neurons (preganglionic to postganglionic) and at synapses with autonomic effectors. Based on the neurotransmitter they produce and release, autonomic neurons are classified as either cholinergic or adrenergic. Autonomic neurotransmitters exert their effects by binding to specific receptors located in the plasma membrane of a postsynaptic neuron or effector cell.

Cholinergic Neurons and Receptors

Cholinergic neurons (kō-lin-ER-jik) release the neurotransmitter **acetylcholine (ACh)**. Cholinergic neurons include (1) all sympathetic and parasympathetic preganglionic neurons, (2) sympathetic postganglionic neurons that innervate most sweat glands, and (3) all parasympathetic postganglionic neurons (Figure 14.28).

ACh is stored in synaptic vesicles and released by exocytosis. It then diffuses across the synaptic cleft and binds with specific **cholinergic receptors** on the *postsynaptic* plasma membrane. The two types of cholinergic receptors are nicotinic receptors and muscarinic receptors. **Nicotinic receptors** are present on dendrites and cell bodies of both sympathetic and parasympathetic postganglionic neurons (Figure 14.28c). They are so named because nicotine mimics the action of ACh by binding to these receptors. (Nicotine, a natural substance in tobacco leaves, is not a naturally occurring substance in humans and is not normally present in nonsmokers.) **Muscarinic receptors** are present on all effectors (smooth muscle, cardiac muscle, and glands) innervated by parasympathetic postganglionic axons and on most sweat glands (Figure 14.28b,c). These receptors are so named because a mushroom poison called muscarine mimics the actions of ACh by binding to them. Nicotine does not activate muscarinic receptors, and muscarine does not activate nicotinic receptors, but ACh does activate both types of cholinergic receptors.

Activation of nicotinic receptors by ACh causes depolarization and thus excitation of the postsynaptic cell, which can be a postganglionic neuron, an autonomic effector, or a skeletal muscle fiber. Activation of muscarinic receptors by ACh sometimes causes depolarization (excitation) and sometimes causes hyperpolarization (inhibition), depending on the target cell. For example, binding of ACh to muscarinic receptors inhibits (relaxes) smooth muscle sphincters in the gastrointestinal tract but excites (contracts) smooth muscle fibers in the the iris of the eye. Because acetylcholine is quickly inactivated by the enzyme **acetylcholinesterase**, effects triggered by cholinergic neurons are brief.

Adrenergic Neurons and Receptors

Adrenergic neurons (ad´-ren-ER-jik) release **norepinephrine (NE)**, also known as noradrenaline (Figure 14.28a). Most sym-

Figure 14.28   Cholinergic neurons and adrenergic neurons in the sympathetic and parasympathetic divisions.

(a) Sympathetic division—innervation to most effector tissues

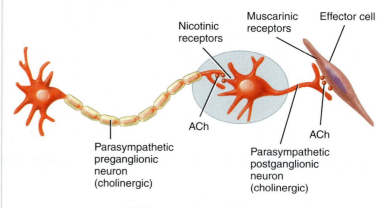

(b) Sympathetic division—innervation to most sweat glands

(c) Parasympathetic division

🔑 **Cholinergic neurons release acetylcholine; adrenergic neurons release norepinephrine.**

pathetic postganglionic neurons are adrenergic. Like ACh, NE is synthesized and stored in synaptic vesicles and released by exocytosis. Molecules of NE diffuse across the synaptic cleft and bind to specific adrenergic receptors on the postsynaptic membrane, causing either excitation or inhibition of the effector cell.

Adrenergic receptors bind both norepinephrine and epinephrine. Norepinephrine can be released either as a neurotransmitter by sympathetic postganglionic neurons, or as a hormone into the blood by the adrenal medullae; epinephrine is released as a hormone. The two main types of adrenergic receptors are **alpha (α) receptors** and **beta (β) receptors**, which are found on visceral effectors innervated by most sympathetic postganglionic axons. These receptors are further classified into subtypes—α_1, α_2, β_1, β_2, and β_3—based on the specific responses they elicit and their selective binding of drugs that activate or block them.

The activity of norepinephrine at a synapse is terminated either when NE is taken up by the axon that released it or when NE is inactivated by enzymes. Compared to ACh, norepinephrine lingers in the synaptic cleft for a longer time. Thus, effects triggered by adrenergic neurons typically are longer lasting than those triggered by cholinergic neurons.

Table 14.7 describes the locations of cholinergic and adrenergic receptors and summarizes the responses that occur when each type of receptor is activated.

TABLE 14.7

Locations and Responses of Adrenergic and Cholinergic Receptors

TYPE OF RECEPTOR	MAJOR LOCATIONS	ACTIVATING NEUROTRANSMITTER	EFFECTS OF RECEPTOR ACTIVATION
Cholinergic			
Nicotinic	Sympathetic and parasympathetic postganglionic neurons.	Acetylcholine.	Impulses in postsynaptic neurons.
	Skeletal muscles.		Contraction.
Muscarinic	Effectors innervated by parasympathetic postganglionic neurons.	Acetylcholine.	Excitation or inhibition of postsynaptic cell.
	Sweat glands.		Increased sweating.
Adrenergic			
α_1	Blood vessels serving salivary glands, skin, mucosa, kidneys, and abdominal viscera.	Norepinephrine and epinephrine.	Vasoconstriction.
	Sphincter muscles of stomach and urinary bladder.		Closing of sphincters.
	Salivary glands.		Increased secretion.
α_2	Blood vessels.	Norepinephrine and epinephrine.	Vasodilation.
	Pancreas.		Decreased secretion of insulin and digestive enzymes.
	Platelets in blood.		Aggregation to promote blood clotting.
β_1	Heart.	Norepinephrine and epinephrine.	Increased force and rate of contraction.
	Kidneys.		Secretion of renin.
	Adipose cells.		Breakdown of triglycerides to fatty acids.
β_2	Lungs.	Norepinephrine and epinephrine.	Dilation of airways.
	Blood vessels serving heart, skeletal muscle, adipose tissue, and liver.		Vasodilation.
	Digestive and urinary organs.		Relaxation of organ walls.
	Liver.		Breakdown of glycogen to glucose.
β_3	Adipose tissue.	Norepinephrine and epinephrine.	Heat production.

Checkpoint

26. Define cholinergic and adrenergic neurons and receptors.
27. Which neurotransmitters bind to cholinergic receptors and which bind to adrenergic receptors?
28. Which autonomic neurons produce a shorter effect, cholinergic or adrenergic? Why?

Return to Nick's Story

As Nick stood up he swayed a bit and placed one hand on the table and the other on his head. "Hey, Nick, are you feeling alright?" his friend Tony asked.

"I get a little dizzy sometimes when I stand up. It's nothing. Let's just get to the lake. I want to catch some fish," said Nick.

"Alright," agreed Tony. "Just don't pass out on me. You're too big to carry back to town on my shoulders."

As they drove down the narrow dirt road, Tony asked, "So, Nick, you been feeling alright?"

Nick shook his head. "My stomach's been giving me an awful time. I've been throwing up, can't keep anything down. What I do keep down, well, it doesn't come out. I've been stopped up for weeks. I should probably go see the doctor. I was supposed to go about 3 months ago, but I just didn't have time. Here we are. Let's get the boat in the water."

As Nick got out of the truck his head felt light, and it felt like someone was grabbing his scalp. "Tony, something bad's going to happen." As Tony turned, he saw Nick fall to the ground.

Nick's eyes opened slowly. "Nick, wake up. Can you hear me?" asked Dr. Meyerson.

"Where am I?" Nick said hoarsely.

"You're in the intensive care unit. You've been unconscious for several hours." The doctor shined a light in Nick's eyes. His pupils constricted slowly. "Hmm," she said. She turned his head quickly to the right and then to the left, noting the movement of Nick's eyes. Dr. Meyerson pinched Nick's index finger hard. "Can you feel that?" she asked Nick.

"I can feel you squeezing my finger," he said.

Pulling out her reflex hammer, the doctor tapped Nick's biceps tendon, then his patellar tendon.

"What happened to me, anyway?" asked Nick.

"You passed out. Your blood sugars were sky high. You're lucky your friend got you here fast. We can talk more tomorrow, Nick. Just rest for now," advised Dr. Meyerson.

The doctor stopped by the nurse's station. "How's he doing?" asked the nurse.

"He's got some substantial neuropathy; pupillary reflex is slow, and pain sensation is pretty limited. Were you here when he came in yesterday?"

Yolanda sighed. "Yes, and it wasn't pretty. He was sweating buckets—rapid heart rate, elevated blood pressure, rapid respiration. He responded quickly to treatment. I guess he hasn't been taking care of his diabetes. Blood sugars were really high, and he's been having a lot of gastrointestinal problems—vomiting, constipation, things like that."

D.   Which division of the autonomic nervous system would be affected and would be causing Nick's GI tract symptoms?

E.   Nick's light-headedness is caused by a condition known as orthostatic hypotension, a rapid drop in blood pressure upon standing up. Based on what you have learned so far, how does the autonomic nervous system control blood pressure?

F.   After becoming comatose, Nick was sweating profusely, and had rapid heart and respiratory rates and elevated blood pressure. Which area of the brain interacts with the autonomic nervous system during physical stress to initiate these responses?

G.   Nick has digestive symptoms indicating reduced gastrointestinal mobility. What autonomic receptors regulate closing of sphincters and relaxation of organ walls?

14.8   The sympathetic division supports vigorous physical activity; the parasympathetic division conserves body energy.

As noted earlier, most body organs receive innervation from both divisions of the ANS, which typically work in opposition to one another. The balance between sympathetic and parasympathetic activity or "autonomic tone" is regulated by the hypothalamus. Typically, the hypothalamus turns up sympathetic tone at the same time it turns down parasympathetic tone, and vice versa. The two divisions can affect body organs differently because their postganglionic neurons release different neurotransmitters and because the effector organs possess different adrenergic and cholinergic receptors. A few structures receive only sympathetic innervation—sweat glands, arrector pili muscles attached to hair follicles in the skin, the kidneys, the spleen, most blood vessels, and the adrenal medullae (see Figure 14.23). In these structures there is no opposition from the parasympathetic division. Still, an increase in sympathetic tone has one effect, and a decrease in sympathetic tone produces the opposite effect. Although both the sympathetic and parasympathetic divisions are concerned with maintaining health, they do so in dramatically different ways.

Sympathetic Responses

During physical activity or emotional stress, the sympathetic division dominates the parasympathetic division. High sympathetic tone favors body functions that can support vigorous physical activity and rapid production of ATP. At the same time, the sympathetic division reduces body functions that favor the storage of energy. Besides physical exertion, various emotions—such as fear, embarrassment, or rage—stimulate the sympathetic division. Visualizing body changes that occur during "E situations" such as exercise, emergency, excitement, and embarrassment will help you remember most of the sympathetic responses. Sympathetic activities result in increased alertness and metabolic activities in order to prepare the body for an emergency situation. Activation of the sympathetic division and release of hormones by the adrenal medullae set in motion a series of physiological responses collectively called the **fight-or-flight response,** which includes the following effects:

- Pupils of the eyes dilate.

- Heart rate, force of heart contraction, and blood pressure increase.

- Airways dilate, allowing faster movement of air into and out of the lungs.

- Blood vessels supplying the kidneys and gastrointestinal tract constrict, which decreases blood flow through these tissues. The result is a slowing of urine formation and digestive activities, which are not essential during exercise.

- Blood vessels supplying organs involved in exercise or fighting off danger—skeletal muscles, cardiac muscle, liver, and adipose tissue—dilate, allowing greater blood flow through these tissues.

- The liver increases breakdown of glycogen to glucose and adipose tissue increases breakdown of triglycerides to fatty acids and glycerol, raising blood levels of these molecules for greater ATP production.

- Processes that are not essential for meeting the stressful situation are inhibited. For example, muscular movements of the gastrointestinal tract and digestive secretions slow down or even stop.

Parasympathetic Responses

In contrast to the *fight-or-flight* activities of the sympathetic division, the parasympathetic division enhances *rest-and-digest* activities. Parasympathetic responses support body functions that conserve and restore body energy during times of rest and recovery. In the quiet intervals between periods of exercise, parasympathetic impulses to the digestive glands and the smooth muscle of the gastrointestinal tract prevail over sympathetic impulses. This allows energy-supplying food to be digested and absorbed. At the same time, parasympathetic responses reduce body functions that support physical activity.

The acronym *SLUDD* can be helpful in remembering five parasympathetic responses. It stands for salivation (S), lacrimation (L), urination (U), digestion (D), and defecation (D). All of these activities are stimulated mainly by the parasympathetic division. Besides the increasing SLUDD responses, other important parasympathetic responses are "three decreases": decreased heart rate, decreased diameter of airways, and decreased diameter (constriction) of the pupils.

Table 14.8 compares the structural and functional features of the sympathetic and parasympathetic divisions of the ANS.

Table 14.9 lists the responses of glands, cardiac muscle, and smooth muscle to stimulation by the sympathetic and parasympathetic divisions of the ANS.

TABLE 14.8

Comparison of Sympathetic and Parasympathetic Divisions of the ANS

	SYMPATHETIC	PARASYMPATHETIC
Origin of preganglionic neuron	Spinal cord segments T1–L2.	Cranial nerves exiting brain stem (III, VII, IX, and X) and spinal cord segments S2–S4.
Associated ganglia	Sympathetic trunk ganglia and prevertebral ganglia.	Terminal ganglia.
Ganglia locations	Close to CNS and distant from visceral effectors.	Typically near or within wall of visceral effectors.
Axon length and divergence	Short preganglionic neuron axons; long postganglionic neuron axons that pass to many visceral effectors.	Long preganglionic neuron axons; short postganglionic neuron axons that pass to a single visceral effector.
Rami communicantes	Present.	Absent.
Neurotransmitters	Preganglionic neurons release acetylcholine (ACh); most postganglionic neurons release norepinephrine (NE); postganglionic neurons that innervate most sweat glands and some blood vessels in skeletal muscle release ACh.	Preganglionic neurons and postganglionic neurons release ACh.
Physiological effects	Fight-or-flight responses to prepare body for intense physical activity.	Rest-and-digest responses to conserve and restore energy.

TABLE 14.9

Effects of Sympathetic and Parasympathetic Divisions of the ANS

VISCERAL EFFECTOR	EFFECT OF SYMPATHETIC STIMULATION	EFFECT OF PARASYMPATHETIC STIMULATION
Glands		
Adipose tissue*	Release of fatty acids into blood.	No known effect.
Adrenal medullae	Secretion of epinephrine and norepinephrine.	No known effect.
Gastric glands	Inhibits secretion of gastric juice.	Stimulates secretion of gastric juice.
Intestinal glands	Inhibits secretion of intestinal juice.	Stimulates secretion of intestinal juice.
Kidneys*	Secretion of renin.	No known effect.
Lacrimal glands	No known effect.	Stimulates secretion of tears.
Liver*	Glucose secretion; decreased bile secretion.	Glycogen synthesis; increased bile secretion.
Pancreas	Inhibits secretion of digestive enzymes and insulin; stimulates secretion of glucagon.	Stimulates secretion of digestive enzymes and insulin.
Pineal gland	Increases secretion of melatonin.	No known effect.
Posterior pituitary	Secretion of antidiuretic hormone (ADH).	No known effect.
Salivary glands	Inhibits secretion of saliva.	Stimulates secretion of saliva.
Sweat glands	Increases sweating.	No known effect.
Cardiac (Heart) Muscle	Increased heart rate and force of contractions.	Decreased heart rate and force of contractions.
Smooth Muscle		
Ciliary muscle of eye	Relaxation for distant vision.	Contraction for close vision.
Gallbladder and ducts	Storage of bile.	Release of bile into small intestine.
Hair follicles, arrector pili muscle	Erection of hairs.	No known effect.
Iris of eye	Dilation of pupil.	Constriction of pupil.
Lungs, bronchial muscle	Airway dilation.	Airway constriction.
Sex organs	Ejaculation of semen (males).	Erection of clitoris (females) and penis (males).
Spleen	Discharge of stored blood into general circulation.	No known effect.
Stomach and intestines	Decreased motility; contraction of sphincters.	Increased motility; relaxation of sphincters.
Urinary bladder	Relaxation of muscular wall; contraction of sphincter.	Contraction of muscular wall; relaxation of sphincter.
Uterus	Inhibits contraction in nonpregnant woman; promotes contraction in pregnant woman.	Minimal effect.
Vascular Smooth Muscle		
Abdominal viscera arterioles	Vasoconstriction.	No known effect.
Coronary (heart) arterioles	Vasodilation.	Minimal effect.
Kidney arterioles	Decreased production of urine.	No known effect.
Skeletal muscle arterioles	Vasodilation.	No known effect.

*Grouped with glands because they release substances into the blood.

Checkpoint

29. Which part of the brain controls the balance between sympathetic and parasympathetic tone?

30. What happens during the fight-or-flight response? Which division of the ANS is responsible?

31. What happens during the rest-and-digest response of the parasympathetic division of the ANS?

32. Describe the sympathetic response in a frightening situation for each of the following body parts: iris of eye, lungs, urinary bladder, stomach, intestines, heart, arterioles of the abdominal viscera, and arterioles of skeletal muscles.

33. Give three examples of the opposite effects of the sympathetic and parasympathetic divisions of the autonomic nervous system.

14.9 Autonomic reflexes regulate controlled body conditions and are primarily integrated by the hypothalamus.

Autonomic Reflexes

Autonomic reflexes are responses that occur when nerve impulses pass through an autonomic reflex arc. These reflexes play a key role in regulating controlled conditions in the body such as *blood pressure*, by adjusting heart rate, force of ventricular contraction, and blood vessel diameter; *digestion*, by adjusting the motility (movement) and muscle tone of the gastrointestinal tract; and *defecation and urination*, by regulating the opening and closing of sphincters.

The components of an autonomic reflex arc are as follows:

- *Receptor*. Like the receptor in a somatic reflex arc (see Figure 14.19), the receptor in an autonomic reflex arc is the distal end of a sensory neuron, which responds to a stimulus and produces a change that will ultimately trigger nerve impulses.

- *Sensory neuron*. Conducts nerve impulses from receptors to the CNS.

- *Integrating center*. Interneurons within the CNS relay signals from sensory neurons to motor neurons. Integrating centers for most autonomic reflexes are located in the hypothalamus and brain stem. Some autonomic reflexes, such as those for urination and defecation, have integrating centers in the spinal cord.

- *Motor neurons*. Nerve impulses triggered by the integrating center propagate out of the CNS along motor neurons to an effector. In an autonomic reflex arc, two motor neurons connect the CNS to an effector: The preganglionic neuron conducts motor impulses from the CNS to an autonomic ganglion, and the postganglionic neuron conducts motor impulses from an autonomic ganglion to an effector (see Figure 14.22).

- *Effector*. In an autonomic reflex arc, the effectors are smooth muscle, cardiac muscle, and glands, and the reflex is called an autonomic reflex.

Autonomic Control by Higher Centers

Normally, we are not aware of muscular contractions of our digestive organs, heartbeat, changes in the diameter of our blood vessels, and pupil dilation and constriction because the integrating centers for these autonomic responses are in the spinal cord or brain stem. Somatic or autonomic sensory neurons deliver input to these centers, and autonomic motor neurons provide output that adjusts activity in the visceral effector, usually without our conscious perception.

The hypothalamus is the major control and integration center of the ANS. The hypothalamus receives sensory input related to visceral functions, olfaction (smell), and gustation (taste), as well as input related to changes in temperature, osmotic pressure, and levels of various substances in blood. It also receives input relating to emotions from the limbic system. Output from the hypothalamus influences autonomic centers both in the brain stem (such as the cardiovascular, salivation, swallowing, and vomiting centers) and in the spinal cord (such as the defecation and urination reflex centers).

Anatomically, the hypothalamus is connected to both the sympathetic and parasympathetic divisions of the ANS by axons that form tracts from the hypothalamus to sympathetic and parasympathetic neurons in the brain stem and spinal cord through relays in the reticular formation. The posterior and lateral parts of the hypothalamus control the sympathetic division. Stimulation of these areas produces an increase in heart rate and force of contraction, a rise in blood pressure due to constriction of blood vessels, an increase in body temperature, dilation of the pupils, and inhibition of the gastrointestinal tract. In contrast, the anterior and medial parts of the hypothalamus control the parasympathetic division. Stimulation of these areas results in a decrease in heart rate, lowering of blood pressure, constriction of the pupils, and increased secretion and motility of the gastrointestinal tract.

Checkpoint

34. Give three examples of controlled conditions in the body that are kept in homeostatic balance by autonomic reflexes.

35. How does an autonomic reflex arc differ from a somatic reflex arc?

36. How is the hypothalamus anatomically connected to the autonomic nervous system?

Nick's Story: Epilogue and Discussion

▶▶▶From Nick's story and the information in this chapter, you have learned that the peripheral nervous system consists of somatic and autonomic divisions, both of which can be affected by disturbances in body homeostasis. In this case, Nick's diabetes mellitus led to a chronic condition of elevated blood glucose (regulation of blood glucose by the endocrine system will be discussed further in Chapter 17). Chronically elevated glucose levels damage peripheral nerves. Small unmyelinated fibers in the distal lower extremities are usually affected first, with a resulting loss of pain and temperature sensation. Large myelinated fibers can also be affected, causing further sensory and proprioceptive sensation loss.

Typically, symptoms like Nick's do not resolve themselves but can be controlled as the blood sugar levels are normalized. In extreme cases of uncontrolled diabetes mellitus, significant somatic and autonomic involvement and even coma can occur. There is no cure for diabetic polyneuropathy, but it is treatable by controlling blood glucose levels and by continued monitoring of the diabetic condition. It is important to rule out other possible causes of peripheral neuropathy before diagnosing diabetic polyneuropathy, since treatment methods may vary. ◀◀◀

H.   Why would the term polyneuropathy be appropriate for the symptoms Nick was experiencing? (Hint: Poly means "many.")

I.   What symptoms noted by Nick's primary care physician indicated a polyneuropathy?

J.   Why are Nick's generalized symptoms more indicative of a peripheral polyneuropathy than a central nervous system lesion to the brain or spinal cord?

K.   Which of Nick's symptoms were related to somatic reflexes? Which were related to autonomic reflexes?

Concept and Resource Summary

Concept	Resources

Introduction

1. The **peripheral nervous system** consists of nervous tissue outside the brain and spinal cord, and is subdivided into the **somatic nervous system** and the **autonomic nervous system**.
2. The somatic nervous system is involved in voluntary skeletal muscle activities.
3. The autonomic nervous system regulates the activities of smooth muscle, cardiac muscle, and glands.

Concept 14.1 Nerves have three protective tissue coverings.

1. A **nerve** is a bundle of axons outside the CNS, along with its associated connective tissues and blood vessels. **Cranial nerves** arise from the brain and **spinal nerves** arise from the spinal cord. Each axon in a nerve is wrapped in **endoneurium**. Groups of axons are then wrapped by **perineurium** into bundles called **fascicles**. The entire nerve is covered with **epineurium**.

See
- **Anatomy Overview – Nerves**

Concept 14.2 Twelve pairs of cranial nerves distribute primarily to regions of the head and neck.

1. The 12 pairs of **cranial nerves** are numbered from anterior to posterior in order of their emergence from the brain and spinal cord.
2. Most of the cranial nerves are **mixed nerves** containing axons of both sensory and motor neurons. A few cranial nerves are solely **sensory nerves**.
3. The **olfactory (I) nerve** is entirely sensory and functions in the sense of smell.
4. The **optic (II) nerve** is entirely sensory and conducts visual impulses. The paired nerves form the **optic chiasm**, in which visual information may continue to the same side of the brain or cross to the opposite side.
5. The **oculomotor (III) nerve** is a mixed nerve stimulating eyeball and eyelid movement, and change in lens shape and pupil size.
6. The **trochlear (IV) nerve** is a mixed nerve that transmits sensory impulses for proprioception and motor impulses that stimulate eyeball movement.

Concept

7. The **trigeminal (V) nerve** has three branches that carry sensory impulses for touch, pain, temperature sensations, and proprioception. Motor impulses stimulate chewing movements.

8. The **abducens (VI) nerve** is a mixed nerve that transmits sensory impulses for proprioception and motor impulses that stimulate eyeball movement.

9. The **facial (VII) nerve** is a mixed nerve that transmits sensory impulses for taste from taste buds of the tongue and for proprioception from the face and scalp. Motor impulses stimulate facial expression and secretion by lacrimal glands and salivary glands.

10. The **vestibulocochlear (VIII) nerve** is a sensory nerve with two branches serving the inner ear. One branch conveys impulses for equilibrium; the other branch conveys impulses for hearing.

11. The **glossopharyngeal (IX) nerve** is a mixed nerve that transmits sensory impulses from the tongue for taste and somatic sensations, blood pressure and blood gas levels from sensory receptors, and proprioception from swallowing muscles. Motor impulses stimulate the pharynx for swallowing and speech.

12. The **vagus (X) nerve** is a mixed nerve that transmits sensory impulses for taste from taste buds of the epiglottis and pharynx, proprioception from muscles of the neck and throat, blood pressure and respiratory functions from sensory receptors, and sensations from most thoracic and abdominal organs. Motor impulses stimulate speech, swallowing, smooth muscle contraction of respiratory and abdominal organs, and secretion of digestive fluids.

13. The **accessory (XI) nerve** is a mixed nerve that transmits sensory impulses for proprioception from muscles of the throat, and motor impulses for swallowing and head movement.

14. The **hypoglossal (XII) nerve** is a mixed nerve that transmits sensory impulses for proprioception from the tongue, and motor impulses for speech and swallowing.

See

- Anatomy Overview – Cranial Nerves

Hear

- **Figure 14.2 – Origins of Cranial Nerves**

Do

- **Exercise – Cranial Nerve Target Practice – Sensory Functions**
- **Exercise – Cranial Nerve Target Practice – Motor Functions**

Concept 14.3 Each spinal nerve branches into a posterior ramus, anterior ramus, meningeal branch, and rami communicantes.

1. The 31 pairs of spinal nerves are numbered according to where they emerge from the vertebral column, beginning with the first cervical pair that emerge between the atlas and the occipital bone. There are eight pairs of *cervical nerves,* twelve pairs of *thoracic nerves,* five pairs of *lumbar nerves,* five pairs of *sacral nerves,* and one pair of *coccygeal nerves.*

2. Spinal nerves connect the CNS to sensory receptors, muscles, and glands. Because **spinal nerves** arise from the union of **dorsal** and **ventral roots** of the spinal cord, all spinal nerves are mixed nerves; the dorsal root contains sensory axons conducting impulses from sensory receptors into the CNS, and the ventral root contains motor axons conducting impulses from the CNS out to effectors.

3. After each short spinal nerve passes through its intervertebral foramen, the nerve divides into branches. Branches include the **posterior rami**, serving muscles and skin of the posterior trunk; the **anterior rami**, serving the muscles and structures of limbs and the skin of the lateral and anterior trunk; **meningeal branches**, which supply the vertebrae, spinal cord, and meninges; and the **rami communicantes** of the autonomic nervous system.

4. Except for T2–T12, ventral rami do not go directly to body structures, but instead form **plexuses** with axons of adjacent ventral rami. Nerves emerge from each plexus.

5. The **cervical plexus** serves the skin and muscles of the head, neck, and superior shoulders and chest. An important nerve of this plexus is the **phrenic nerve**, which innervates the diaphragm.

6. The **brachial plexus**, innervates the shoulders and upper limbs. The important nerves arising from this plexus are the **axillary nerve**, the **musculocutaneous nerve**, the **radial nerve**, the **median nerve**, and the **ulnar nerve**.

7. The **lumbar plexus** innervates the anterior and lateral abdominal wall, external genitals, and part of the lower limbs. The **femoral nerve** and **obturator nerve** arise from this plexus.

8. The **sacral plexus** supplies the buttocks, perineum, and lower limbs. One of its nerves, the long, thick **sciatic nerve**, innervates most of the lower limb. The **pudendal nerve** innervates the perineum.

9. The ventral rami of T2–T12 do not form plexuses. They are the **intercostal nerves** that innervate intercostal muscles, and muscles and skin of the thorax and abdominal wall.

See

- Anatomy Overview – Spinal Nerves

Hear

- **Figure 14.13 – External Anatomy of the Spinal Cord and Spinal Nerves**

Concept

Concept 14.4 A reflex is produced by a reflex arc in response to a particular stimulus.

1. A **reflex** is a fast, involuntary response to a stimulus.
2. Integrating centers for reflexive information are the gray matter of the brain stem in a **cranial reflex** and spinal cord gray matter in a **spinal reflex**.
3. **Somatic reflexes** involve motor responses by skeletal muscles; **autonomic reflexes** involve smooth muscle, cardiac muscle, and glands as the effectors.
4. A **reflex arc** has five functional components in the sequential pathway of nerve impulse propagation in response to a **stimulus**: (1) **sensory receptor**, (2) **sensory neuron**, (3) **integrating center**, (4) **motor neuron**, and (5) **effector**.
5. A **monosynaptic reflex arc** has one synapse in the CNS, and a **polysynaptic reflex arc** has more than one CNS synapse.
6. A **stretch reflex** begins when stretching of a muscle stimulates **muscle spindles**, and the monosynaptic reflex response is contraction of the stretched muscle. There is simultaneous relaxation of antagonist muscles by a polysynaptic reflex arc governed by **reciprocal innervation**. Stretch reflexes function in maintaining muscle tone.
7. The stretch reflex is an example of an **ipsilateral reflex** because sensory impulses enter the spinal cord on the same side as motor impulses leave it.
8. The **flexor** or **withdrawal reflex** is a polysynaptic arc initiated by a painful stimulus, in which the response of flexor leg muscle contraction causes withdrawal of the leg. Several muscle groups respond through activation of interneurons in several segments of the spinal cord. This is termed an **intersegmental reflex arc**.

See
- Animation – Reflex Arcs
- Animation – Reflexes

Hear
- **Figure 14.19** – General Components of a Reflex Arc

Do
- Exercise – Assemble an Arc
- Exercise – Stretch Reflex

Concept 14.5 The autonomic nervous system produces involuntary movements.

1. The somatic nervous system is under conscious control. It has sensory neurons conveying input from receptors for **special senses** and **somatic senses**. Somatic motor neurons innervate skeletal muscle effectors, which always results in excitation for muscle contraction.
2. The **autonomic nervous system (ANS)** involves **autonomic sensory neurons** conveying input from interoceptors in viscera and blood vessels. **Autonomic motor neurons** carry motor impulses to cardiac muscle, smooth muscle, and glands.
3. The ANS usually operates without conscious control. The hypothalamus and brain stem regulate ANS activity.
4. The ANS has two principal branches: the **sympathetic division** and the **parasympathetic division**.
5. The somatic nervous system has one motor neuron from the CNS to the skeletal muscle. The sympathetic and parasympathetic ANS pathways have two motor neurons in a chain. The first motor neuron has its cell body in the CNS, and its myelinated axon extends from the CNS to an autonomic ganglion that contains the cell body of the second motor neuron. The unmyelinated axon of the second motor neuron extends to an effector.
6. All somatic motor neurons release acetylcholine; autonomic motor neurons release acetylcholine or norepinephrine.
7. Most organs receive **dual innervation** from the sympathetic and parasympathetic divisions of the ANS. Generally, these two divisions transmit opposing signals to the organ that they innervate.

See
- Anatomy Overview – Visceral Receptors
- Anatomy Overview – Visceral Effectors

Hear
- **Figure 14.22** – Motor Neuron Pathways in the Somatic Nervous System and Autonomic Nervous System

Concept	Resources

Concept 14.6 The ANS includes preganglionic neurons, autonomic ganglia and plexuses, and postganglionic neurons.

1. The ANS pathway involves a **preganglionic neuron** with its cell body in the CNS and a myelinated axon extending to the autonomic ganglion, where it synapses with the **postganglionic neuron**. The cell body and dendrites of this second motor neuron are in the autonomic ganglion, and its unmyelinated axon extends to a visceral effector.
2. In the sympathetic division, cell bodies of preganglionic neurons are in thoracic and lumbar spinal cord segments; in the parasympathetic division, preganglionic neuron cell bodies are in the brain stem and sacral spinal cord segments.
3. **Sympathetic trunk ganglia** extend on each side of the vertebral column, and their postganglionic axons typically innervate organs above the diaphragm. **Prevertebral ganglia** lie anterior to the vertebral column, and their postganglionic axons typically innervate organs below the diaphragm.
4. Preganglionic axons of the parasympathetic division extend from the CNS to synapse with postganglionic neurons in **terminal ganglia** located close to visceral organs.
5. Sympathetic preganglionic axons leave the spinal cord, travel through rami communicantes, enter sympathetic trunk ganglia, and connect to postganglionic neurons, then continue in spinal nerves to effectors.
6. Parasympathetic preganglionic axons leave the brain stem and spinal cord, enter terminal ganglia located near or in visceral organs, connect to postganglionic neurons supplying local effectors. Sacral parasympathetic preganglionic neurons travel through **pelvic splanchnic nerves** to reach terminal ganglia.
7. In the thorax, abdomen, and pelvis, sympathetic and parasympathetic neurons form **autonomic plexuses** adjacent to major arteries. These plexuses may also contain sympathetic ganglia.

See
- **Anatomy Overview – Organization of the ANS**
- **Animation – ANS: Motor Pathways**

Hear

- **Figure 14.23 –** Structure of the Sympathetic Division of the Autonomic Nervous System
- **Figure 14.24 –** Structure of the Parasympathetic Division of the Autonomic Nervous System

Do
- **Exercise – Assemble the Structures of the ANS**

Concept 14.7 ANS neurons release acetylcholine or norepinephrine, resulting in excitation or inhibition.

1. In an autonomic pathway, neurotransmitters are released at synapses between preganglionic and postganglionic neurons, and between postganglionic neurons and effectors.
2. **Cholinergic neurons** release **acetylcholine (ACh)**. Cholinergic neurons include all sympathetic and parasympathetic preganglionic neurons, sympathetic postganglionic neurons that innervate sweat glands, and all parasympathetic postganglionic neurons.
3. ACh binds to **cholinergic receptors** on postsynaptic plasma membranes. There are two types of cholinergic receptors: **nicotinic receptors** on dendrites and cell bodies of sympathetic and parasympathetic postganglionic neurons, and in the motor end plate at the neuromuscular junction; and **muscarinic receptors** on all effectors innervated by parasympathetic postganglionic axons, and most sweat glands.
4. ANS **adrenergic neurons** release **norepinephrine (NE)**. Most sympathetic postganglionic neurons are adrenergic.
5. Adrenergic receptors bind both norepinephrine, released by sympathetic postganglionic neurons, and epinephrine, an adrenal hormone. The two adrenergic receptors are **alpha (α) receptors** and **beta (β) receptors** found on visceral effectors.

See
- **Anatomy Overview – Neurotransmitters**
- **Animation – The ANS: Types of Neurotransmitters and Neurons**

Hear
- **Figure 14.28 –** Cholinergic Neurons and Adrenergic Neurons in the Sympathetic and Parasympathetic Divisions

Concept

Concept 14.8 The sympathetic division supports vigorous physical activity; the parasympathetic division conserves body energy.

1. The balance between sympathetic and parasympathetic activity is regulated by the hypothalamus.
2. Although most body structures have dual innervation by the ANS, some structures receive only sympathetic innervation. A decrease in sympathetic tone produces an opposing effect on these structures.
3. High sympathetic tone supports vigorous physical activity, rapid ATP production, and decreased energy storage. Activation of the sympathetic division gives rise to a series of physiological responses called the **fight-or-flight response**.
4. Parasympathetic responses support rest-and-digest activities such as conservation and restoration of body energy during rest and recovery periods, and increased gastrointestinal tract activity. The acronym SLUDD represents parasympathetic responses: salivation, lacrimation, urination, digestion, and defecation. Parasympathetic responses also include decreased heart rate, decreased diameter of airways, and decreased diameter of pupils.

See
- Anatomy Overview – Effectors
- Animation – Physiological Effects of the ANS
- Animation – The Alarm Reaction

Do
- Exercise – Sort ANS Functions
- Exercise – What Is Your ANS Status?

Concept 14.9 Autonomic reflexes regulate controlled body conditions and are primarily integrated by the hypothalamus.

1. **Autonomic reflexes** regulate controlled body conditions including blood pressure, digestion, defecation, and urination.
2. The components of a typical autonomic reflex arc are: **receptor**, **sensory neuron**, **integrating center** in the hypothalamus and brain stem, **motor neurons** arranged in a two-motor-neuron chain of the ANS, and a visceral **effector** such as smooth muscle, cardiac muscle, and glands.
3. The hypothalamus is the major control and integration center of the ANS. It is connected to both the sympathetic and the parasympathetic divisions by tracts that relay through the reticular formation of the brain stem.

See
- Anatomy Overview – The ANS Control Centers

Do
- Concepts and Connections – Nervous System Negative Feedback Loop

---

Other Chapter Resources

For additional information on clinical applications, medical disorders, and medical terminology, see Chapter 7 in the *Clinical Connections* manual packaged with your text.

Within *WileyPLUS*

- Chapter Quiz
- Visual Anatomy
- Mastering Vocabulary
- Audio Glossary

- Homeostatic Imbalance: The Case of the Forgetful Father
- Homeostatic Imbalance: The Case of the Man with the Tingling Arm
- Web Links

Understanding the Concepts

1. How does the accessory nerve differ from the other cranial nerves?

2. What important loss of function could reveal damage to each cranial nerve?

3. Why could severing of the spinal cord in the cervical region cause respiratory arrest?

4. Injury of which nerve could cause paralysis of the biceps brachii muscle? The triceps brachii muscle? From which plexus do they branch?

5. Injury to which nerve could cause pain that extends down the anterior lower limb? Down the posterior lower limb?

6. How are the intercostal nerves different from other spinal nerves?

7. Which branch of the nervous system includes all integrating centers for reflexes?

8. How is the response of ANS effectors different from the response of somatic nervous system effectors?

9. Why can the sympathetic division produce simultaneous effects throughout the body, but parasympathetic effects typically are localized to specific organs?

10. Why is the parasympathetic division of the ANS sometimes referred to as an energy conservation or restoration system?

15

Sensory, Motor, and Integrative Systems

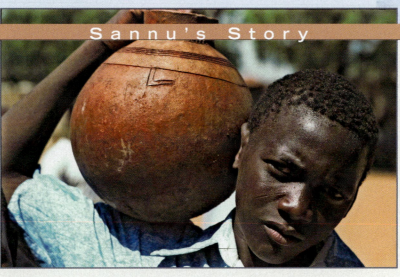

Sannu's Story

As Sannu looked up toward the sun, he wiped the back of his hand over his face, feeling the rough sensation of sand and grit against the sweat of his forehead. Pausing, he lightly touched the bumps on his face. He wondered what they looked like as he bent down to pick up the jug of water he would carry on his head for 3 miles back to his village. He had been taking this trail for 10 years now, since he was 6 years old. His feet were well worn, thickly calloused like those of all the villagers; he felt little of the earth under his feet. Sharp rocks and pebbles scraped at his feet as he quickly moved along the trail. Again he glanced toward the setting sun as his neck made quick, muscular adjustments to balance the heavy jug of water on his head.

Then it happened. A brief moment of distraction and his foot came down hard against a sharp rock. He steadied himself and kept walking. The sun was going down and it was wise to reach the village before dark. The African savanna was not a place for a young man to be out alone at night.

As he finally approached the village, his sister Rolande came running to greet him. "Sannu!" she exclaimed, "you are bleeding. Your foot—what happened?" she asked.

"I must have stepped on something on the trail. It is no matter. It does not hurt. My feet are always worn and bruised," he said. Sannu placed his water jug down and sat on a neatly made grass mat with an intricately woven pattern of red, green, and yellow. He looked at his foot. "Perhaps tomorrow I will see Mama Bingel for medicines, although they never seem to work very well," he thought to himself. The nearest clinic was over 150 miles away. With the rainy season coming, it wouldn't be a trip he would likely make soon.

Clean water, sufficient food, and access to basic medical care are luxuries available to most of us. For Sannu, however, they are difficult to come by. He lives a hard life, with limited access to medical care. Sannu has contracted leprosy, which is a treatable and curable disease for those with the opportunity and the resources to get medical attention. Sannu's equilibrium, balance, and muscular strength are all intact; however, the disease has caused significant sensory loss. Sannu is also malnourished and has chronic infections. As we see in this chapter, the nervous system depends on peripheral receptors that sense our environment. Diseases that affect these receptors or cause damage to the receptors themselves reduce our ability to respond to our environment.

INTRODUCTION

In the three previous chapters we have explored how the nervous system processes information and responds to the environment. In order for such responses to occur, the body must be able to sense the internal and external environment and convey these sensations to the appropriate processing areas of the central nervous system (CNS). As sensory impulses reach the CNS, they become part of a large pool of sensory input. However, not every bit of input to the CNS elicits a response. Rather, each piece of incoming information is combined with other arriving and previously stored in-formation in a process called *integration*. Integration occurs at many sites along pathways

> *Sensory information is perceived, modified, and integrated to produce appropriate motor responses.*

in the CNS, such as the spinal cord, brain stem, cerebellum, basal nuclei, and cerebral cortex. The motor responses that govern muscle contraction are modified at several of these CNS levels to produce appropriate motor responses to sensory input.

Disruption of any of the sensory, motor, or integrative structures or pathways can cause significant disturbances in homeostasis. By learning about the functional role of each of these components, it will be easier to understand the disease processes associated with them.

In this chapter we learn more about how information is sensed, the pathways along which information travels, and how sensory information is perceived, modified, and integrated to produce appropriate motor responses. We also introduce two complex integrative functions of the brain: (1) wakefulness and sleep and (2) learning and memory.

15.1 Sensations arise as a result of stimulation, transduction, generation, and integration.

In its broadest definition, **sensation** is the conscious or subconscious awareness of changes in the external or internal environment. The nature of the sensation and the type of reaction generated vary according to the ultimate destination of nerve impulses that convey sensory information to the CNS. Sensory impulses relayed to the spinal cord may serve as input for spinal reflexes, such as the stretch reflex you learned about in Chapter 14. Sensory impulses that reach the lower brain stem elicit more complex reflexes, such as changes in heart rate or breathing rate. When sensory impulses reach the cerebral cortex, we become consciously aware of the sensory stimuli and can precisely locate and identify specific sensations such as touch, pain, hearing, or taste.

Perception, the conscious awareness and interpretation of sensations, is primarily a function of the cerebral cortex. We have no perception of some sensory information because it never reaches the cerebral cortex. For example, certain sensory receptors constantly monitor the pressure of blood in blood vessels. Because the nerve impulses conveying blood pressure information propagate to the cardiovascular center in the medulla oblongata rather than to the cerebral cortex, blood pressure is not consciously perceived.

Sensory Modalities

Each unique type of sensation—such as touch, pain, vision, or hearing—is called a **sensory modality** (mō-DAL-i-tē). A given sensory neuron carries information for only one sensory modality. Neurons relaying impulses for touch to the somatosensory area of the cerebral cortex do not transmit impulses for pain. Likewise, nerve impulses from the eyes are perceived as sight, and those from the ears are perceived as sounds.

The different sensory modalities can be grouped into two classes: general senses and special senses.

- The **general senses** include both **somatic senses** and **visceral senses**. Somatic senses (*somat-* = of the body) include tactile sensations (touch, pressure, vibration, itch, and tickle), thermal sensations (warm and cold), pain sensations, and proprioceptive sensations. Proprioceptive sensations allow perception of the positions and movements of our head and limbs. Visceral senses provide information about conditions within internal organs.

- The **special senses** include the sensory modalities of smell, taste, vision, hearing, and equilibrium or balance.

In this chapter we discuss the somatic senses and visceral pain. The special senses will be the focus of Chapter 16. Visceral senses were introduced in Chapter 14 and will be discussed in association with individual organs in later chapters.

The Process of Sensation

The process of sensation begins in a **sensory receptor**, which can be either a specialized cell or the dendrites of a sensory neuron. A given sensory receptor responds to only one particular kind of **stimulus**, a change in the environment that can activate certain sensory receptors. This characteristic of sensory receptors is known as **selectivity**.

For a sensation to arise, the following four events typically occur:

1. **Stimulation of the sensory receptor**. An appropriate stimulus must occur within the sensory receptor's *receptive field*, that is, the body region where stimulation activates the receptor and produces a response.

2. **Transduction of the stimulus**. A sensory receptor *transduces* (converts) energy from a stimulus into a graded potential. For example, air molecules that we can smell contain chemical energy that is transduced by olfactory (smell) receptors in our nose into electrical energy in the form of a graded potential. Recall that graded potentials vary in amplitude (size), depending on the strength of the stimulus that causes them, and are not propagated. (See Concept 12.8 to review the differences between graded potentials and action potentials.) Each type of sensory receptor exhibits selectivity; it can transduce only one kind of stimulus.

3. **Generation of nerve impulses**. When a graded potential in a sensory neuron reaches threshold, it triggers one or more nerve impulses, which then propagate toward the CNS. Sensory neurons that conduct impulses from the peripheral nervous system (PNS) into the CNS are called **first-order neurons**.

4. **Integration of sensory input**. A particular region of the CNS receives and integrates the sensory nerve impulses. Conscious sensations or perceptions are integrated in the cerebral cortex. You seem to see with your eyes, hear with your ears, and feel pain in an injured part of your body because sensory impulses from each part of the body arrive in a specific region of the cerebral cortex, which interprets the sensation as coming from the stimulated sensory receptors.

Checkpoint

1. What is a sensory modality?
2. Distinguish between general senses and special senses.
3. What is receptor selectivity?
4. What four events must happen for a sensation to occur?
5. What is transduction?

15.2 Sensory receptors can be classified structurally, functionally, or by the type of stimulus detected.

Types of Sensory Receptors

Several structural and functional characteristics of sensory receptors can be used to group them into different classes. Structurally, sensory receptors may be free nerve endings or encapsulated nerve endings of first-order sensory neurons, or separate cells that synapse with first-order sensory neurons (Figure 15.1).

- **Free nerve endings** are bare dendrites; they lack any structural specializations that can be seen under a light microscope (Figure 15.1a). Receptors for pain, temperature, tickle, itch, and some touch sensations are free nerve endings.

- **Encapsulated nerve endings** are dendrites enclosed in a connective tissue capsule (Figure 15.1b). Receptors for pressure, vibration, and some touch sensations are encapsulated nerve endings.

- **Separate cells** may synapse with sensory neurons (Figure 15.1c). Sensory receptors for some special senses are specialized, separate cells. These include *hair cells* for hearing and equilibrium in the inner ear, *gustatory receptors* in taste buds, and *photoreceptors* in the retina of the eye for vision. You will learn more about separate receptor cells for the special senses in Chapter 16.

Figure 15.1   Types of sensory receptors and their relationship to first-order sensory neurons. (a) Free nerve endings, in this case a cold-sensitive receptor, are bare dendrites of first-order neurons. (b) An encapsulated nerve ending, in this case a pressure-sensitive receptor, are enclosed dendrites of first-order neurons. (c) A separate receptor cell—here, a gustatory (taste) receptor—synapses with a first-order neuron.

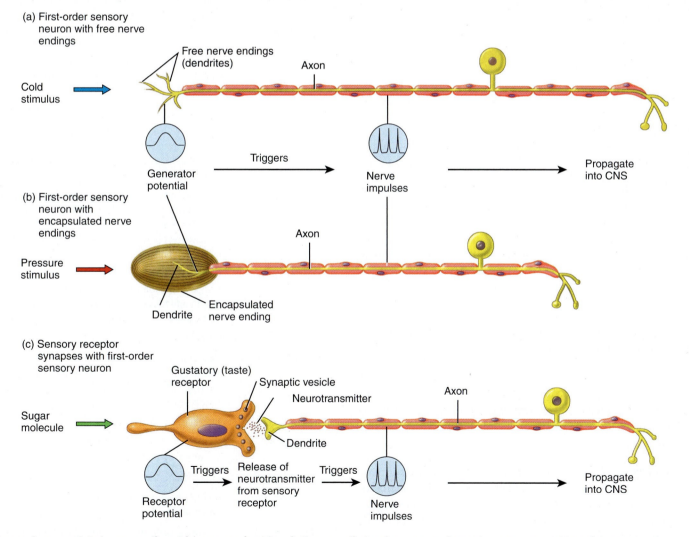

Free and encapsulated nerve endings trigger nerve impulses in the same first-order neurons. Separate sensory receptors release a neurotransmitter that triggers nerve impulses in a first-order neuron.

Sensory receptors produce two kinds of graded potentials—generator potentials and receptor potentials—in response to a stimulus. When stimulated, the dendrites of free nerve endings and encapsulated nerve endings produce **generator potentials** (Figure 15.1a,b). When a generator potential is large enough to reach threshold, it triggers one or more action potentials in the axon of a first-order sensory neuron. The resulting nerve impulse propagates along the axon into the CNS.

By contrast, sensory receptors that are separate cells produce graded potentials termed **receptor potentials**. Receptor potentials trigger release of neurotransmitter (Figure 15.1c). The neurotransmitter molecules produce a postsynaptic potential (PSP) in the first-order neuron. The PSP may trigger one or more action potentials, which propagate as a nerve impulse along the first-order neuron into the CNS.

The amplitude of both generator potentials and receptor potentials varies; an intense stimulus produces a large potential, and a weak stimulus elicits a small one. Similarly, large generator potentials or receptor potentials trigger nerve impulses at higher frequencies (more impulses) in the first-order neuron, in contrast to small generator potentials or receptor potentials, which trigger nerve impulses at lower frequencies (fewer impulses).

Another way to group sensory receptors is based on the location of the receptors and the origin of the stimuli that activate them.

- **Exteroceptors** (EKS-ter-ō-sep´-tors) are located at or near the external surface of the body. They are sensitive to stimuli originating outside the body and provide information about the *external* environment. The sensations of hearing, vision, smell, taste, touch, pressure, vibration, temperature, and pain are conveyed by exteroceptors.

- **Interoceptors** (IN-ter-ō-sep´-tors) are located in blood vessels, visceral organs, muscles, and the nervous system. Interoceptors monitor conditions in the *internal* environment. The nerve impulses produced by interoceptors usually are not consciously perceived.

- **Proprioceptors** (PRŌ-prē-ō-sep´-tors; *proprio-* = one's own) are located in muscles, tendons, joints, and the inner ear. They provide information about body position, muscle length and tension, and movement of your joints.

A third way to group sensory receptors is according to the type of stimulus they detect. Stimuli may be in the form of electromagnetic energy, such as light or heat; mechanical energy, such as sound waves or pressure changes; or chemical energy, such as in a molecule of glucose.

- **Mechanoreceptors** are sensitive to mechanical stimuli such as the deformation, stretching, or bending of cells. Mechanoreceptors provide sensations of touch, pressure, vibration, proprioception, hearing, and equilibrium. They also monitor the stretching of blood vessels and internal organs.

- **Thermoreceptors** detect changes in temperature.

- **Nociceptors** respond to painful stimuli resulting from physical or chemical damage to tissue.

- **Photoreceptors** detect light that strikes the retina of the eye.

- **Chemoreceptors** detect chemicals in the mouth (taste), nose (smell), and body fluids.

- **Osmoreceptors** detect the osmotic pressure of body fluids.

Table 15.1 summarizes the classification of sensory receptors.

TABLE 15.1	
Classification of Sensory Receptors	
BASIS OF CLASSIFICATION	DESCRIPTION
Microscopic Features	
Free nerve endings	Bare dendrites associated with pain, thermal, tickle, itch, and some touch sensations.
Encapsulated nerve endings	Dendrites enclosed in a connective tissue capsule for pressure, vibration, and some touch sensations.
Separate cells	Receptor cells synapse with first-order sensory neurons; located in the retina of the eye (photoreceptors), inner ear (hair cells), and taste buds of the tongue (gustatory receptor cells).
Receptor Location and Activating Stimuli	
Exteroceptors	Located at or near body surface; sensitive to stimuli originating outside body; provide information about external environment; convey visual, smell, taste, touch, pressure, vibration, thermal, and pain sensations.
Interoceptors	Located in blood vessels, visceral organs, and nervous system; provide information about internal environment; impulses produced usually are not consciously perceived but occasionally may be felt as pain or pressure.
Proprioceptors	Located in muscles, tendons, joints, and inner ear; provide information about body position, muscle length and tension, position and motion of joints, and equilibrium (balance).
Type of Stimulus Detected	
Mechanoreceptors	Detect mechanical stimuli; provide sensations of touch, pressure, vibration, proprioception, and hearing and equilibrium; also monitor stretching of blood vessels and internal organs.
Thermoreceptors	Detect changes in temperature.
Nociceptors	Respond to painful stimuli resulting from physical or chemical damage to tissue.
Photoreceptors	Detect light that strikes the retina of the eye.
Chemoreceptors	Detect chemicals in mouth (taste), nose (smell), and body fluids.
Osmoreceptors	Sense the osmotic pressure of body fluids.

Adaptation in Sensory Receptors

A characteristic of most sensory receptors is **adaptation**, in which the generator potential or receptor potential decreases in amplitude during a maintained, constant stimulus. As a result, the frequency of nerve impulses in the first-order neuron decreases. Because of adaptation, the perception of a sensation may fade or disappear even though the stimulus persists. For example, when you first step into a hot shower, the water may feel very hot, but soon the sensation decreases to one of comfortable warmth even though the stimulus (the high temperature of the water) does not change.

Receptors vary in how quickly they adapt. **Rapidly adapting receptors** adapt very quickly. They are specialized for signaling *changes* in a stimulus. Receptors associated with pressure, touch, and smell are rapidly adapting. **Slowly adapting receptors**, by contrast, adapt slowly and continue to trigger nerve impulses as long as the stimulus persists. Slowly adapting receptors monitor stimuli associated with pain, body position, and chemical composition of the blood.

Checkpoint

6. How do sensory receptors differ structurally?
7. What is adaptation? How does it occur?
8. What is the difference between rapidly adapting and slowly adapting receptors?

Return to Sannu's Story

"**S**annu, wake up. The doctor from the mission is here!" exclaimed Rolande.

"Hello, Sannu. We talked about your condition last time I was here. You need treatment. Why haven't you come to the clinic?" asked Dianna.

"I have my brothers and sisters to care for, and the clinic is expensive, Dr. Dianna," he replied.

"Let's take a look at you. Your skin lesions are worse," noted Dianna as she examined his hands. "Let me see your feet, Sannu. Have you been keeping your wounds clean and bandaging them like I showed you?" she asked. Dianna noticed his feet showed signs of burns and one had a large gash. "Do these wounds hurt much?" she asked.

"No. I do not feel pain in my feet or in my hands," responded Sannu.

Dianna felt the enlarged peripheral nerves under the skin of Sannu's calf. She removed her reflex hammer and tapped his calcaneal tendon, then stroked the sole of his foot, trying to elicit a reflex response. "No response on the Achilles or Babinski reflexes. Not a good sign," she thought to herself. "Sannu, I am going to hold your foot. Push against my hand as hard as you can. Hmm, you've lost muscle strength. You're at risk of more injury. You must come to the clinic," advised Dianna.

Sannu traveled for 5 days to get to the clinic. As he stopped to take water from a small dirty stream, Sannu noticed the wound on his foot was dirty and infected. Sannu washed it with water from the stream and filled his water skin. By nightfall he reached the clinic, where he found a space in the large open-air room provided for patients and their families.

The next morning was one that Sannu would remember for the rest of his life. Dr. Dianna looked grim. "Sannu, your foot is terribly infected. I am afraid it's gangrene. We must remove it before the infection gets any worse. It could kill you," she warned.

"Is there no other way?" asked Sannu, as tears welled up in his eyes.

"Once you've recovered, I will arrange for transport to the leprosarium where you can receive treatment," explained Dianna.

A. Leprosy, caused by the bacterium Mycobacterium leprae, is a leading cause of peripheral neuropathy worldwide; although it is treatable, the damage caused by the disease is irreversible. In Sannu's case, why is there both sensory loss and muscle weakness?

B. Leprosy can affect the skin and Schwann cells of peripheral nerves. Which events of sensation processing are most likely not functioning properly in Sannu's leprosy?

C. Were somatic, visceral, or special senses being investigated when Dianna evaluated Sannu's Achilles and Babinski reflex activity?

D. Sannu has lost sensations of pain, temperature, light touch, and pressure. What types of receptor endings mediate the detection of these sensations?

E. Leprosy infects the body via the skin, typically in the cooler regions of the body. Based on this information, would you expect Sannu's interoceptors to be affected?

15.3 Somatic sensations include tactile sensations, thermal sensations, pain, and proprioception.

Somatic sensations arise from stimulation of sensory receptors embedded in the skin or subcutaneous layer; in mucous membranes of the mouth, vagina, and anus; in muscles, tendons, and joints; and in the inner ear. The sensory receptors for somatic sensations are distributed unevenly—some parts of the body surface are densely populated with receptors, and others contain only a few. The areas with the highest density of somatic sensory receptors are the tip of the tongue, the lips, and the fingertips. Somatic sensations that arise from stimulating the skin surface are **cutaneous sensations** (kū-TĀ-nē-us; *cutane-* = skin). There are four modalities of somatic sensations: tactile, thermal, pain, and proprioceptive.

Tactile Sensations

The **tactile sensations** (TAK-tĭl; *tact-* = touch) include touch, pressure, vibration, itch, and tickle. Encapsulated mechanoreceptors mediate sensations of touch, pressure, and vibration. Other touch sensations, as well as itch and tickle sensations, are detected by free nerve endings. Tactile receptors in the skin or subcutaneous layer include corpuscles of touch, hair root plexuses, type I and II cutaneous mechanoreceptors, lamellated corpuscles, and free nerve endings (Figure 15.2).

Figure 15.2 **Structure and location of sensory receptors in the skin and subcutaneous layer.**

The skin contains sensory receptors for somatic sensations of touch, pressure, vibration, warmth, cold, and pain.

Touch

Sensations of **touch** generally result from stimulation of tactile receptors in the skin or subcutaneous layer. There are two types of rapidly adapting touch receptors. **Corpuscles of touch**, or **Meissner corpuscles** (MĪS-ner), are receptors located in the dermal papillae of hairless skin. Each corpuscle is an egg-shaped mass of dendrites enclosed by a capsule of connective tissue. Because corpuscles of touch are rapidly adapting receptors, they generate nerve impulses mainly at the onset of a touch. They are abundant in the fingertips, hands, eyelids, tip of the tongue, lips, nipples, soles, clitoris, and tip of the penis. **Hair root plexuses** are rapidly adapting touch receptors found in hairy skin; they consist of free nerve endings wrapped around hair follicles. Hair root plexuses detect movements on the skin surface that disturb hairs. For example, an insect landing on a hair causes movement of the hair shaft, which in turn stimulates the free nerve endings.

There are also two types of slowly adapting touch receptors. **Type I cutaneous mechanoreceptors**, also known as **Merkel discs**, are saucer-shaped, flattened free nerve endings that make contact with Merkel cells of the stratum basale (see Figure 5.2). These mechanoreceptors are plentiful in the fingertips, hands, lips, and external genitalia. **Type II cutaneous mechanoreceptors**, or **Ruffini corpuscles**, are elongated, encapsulated receptors located deep in the dermis, and in ligaments and tendons. Present in the hands and abundant on the soles, they are most sensitive to stretching that occurs as digits or limbs are moved.

Pressure

Pressure is a sustained sensation that is felt over a larger area than touch. It occurs when deeper tissues are compressed. Receptors that contribute to sensations of pressure include corpuscles of touch, type I cutaneous mechanoreceptors, and lamellated corpuscles. **Lamellated**, or **pacinian**, **corpuscles** (pa-SIN-ē-an) are large oval structures composed of a multilayered connective tissue capsule that encloses a dendrite. Lamellated corpuscles adapt rapidly. They are widely distributed in the body: in the dermis and subcutaneous layer; in tissues that underlie mucous and serous membranes; around joints, tendons, and muscles; in the periosteum; and in the mammary glands, external genitalia, and some viscera.

Vibration

Sensations of **vibration** result from rapidly repetitive sensory signals from tactile receptors. The receptors for vibration sensations are corpuscles of touch and lamellated corpuscles. Corpuscles of touch can detect lower frequency vibrations; lamellated corpuscles detect higher frequency vibrations.

Itch

The **itch** sensation results from stimulation of free nerve endings by certain chemicals, such as bradykinin, often because of a local inflammatory response.

Tickle

Free nerve endings are thought to mediate the **tickle** sensation. This intriguing sensation typically arises only when someone else touches you, not when you touch yourself. The explanation of this puzzle seems to lie in the impulses that conduct to and from the cerebellum when you are moving your fingers and touching yourself, which don't occur when someone else is tickling you.

Thermal Sensations

Two distinct **thermal sensations**—coldness and warmth—are mediated by different free nerve ending **thermoreceptors**. Temperatures between 10°C and 40°C (50–105°F) activate **cold receptors** located in the epidermis. Temperatures between 32°C and 48°C (90–118°F) are stimulated by **warm receptors** located in the dermis. Cold and warm receptors both adapt rapidly at the onset of a stimulus but continue to generate impulses at a lower frequency throughout a prolonged stimulus. Temperatures below 10°C and above 48°C primarily stimulate nociceptors (pain receptors) rather than thermoreceptors.

Pain Sensations

Pain is indispensable for survival. It serves a protective function by signaling the presence of harmful, tissue-damaging conditions. From a medical standpoint, describing the type and location of pain may help pinpoint the underlying cause of disease.

Nociceptors (nō′-sī-SEP-tor; *noci*- = harmful), the receptors for pain, are free nerve endings found in every tissue of the body except the brain (see Figure 15.2). Intense thermal, mechanical, or chemical stimuli can activate nociceptors. Tissue irritation or injury releases chemicals such as prostaglandins, kinins, and potassium ions that stimulate nociceptors. Conditions that elicit pain include excessive stretching of a structure, prolonged muscular contractions, muscle spasms, or ischemia (inadequate blood flow to an organ). Pain may persist even after a pain-producing stimulus is removed because pain-mediating chemicals linger and because nociceptors exhibit very little adaptation.

Types of Pain

There are two types of pain: fast and slow. The perception of **fast pain** occurs very rapidly, usually within 0.1 second after a stimulus is applied. This type of pain is also known as acute, sharp, or pricking pain. The pain felt from a needle puncture or knife cut to the skin is fast pain. The perception of **slow pain**, by contrast, begins a second or more after a stimulus is applied. It then gradually increases in intensity over a period of several seconds or minutes. This type of pain is also referred to as chronic, burning, aching, or throbbing pain, like that associated with a toothache. You can perceive the difference in these two types of pain when you stub your toe; first you feel the sharp sensation of fast pain, followed by the slower, aching sensation of slow pain.

Pain that arises from stimulation of receptors in the skin is called **superficial somatic pain**. Stimulation of receptors in skeletal muscles, joints, tendons, and fascia causes **deep somatic pain**. **Visceral pain** results from stimulation of nociceptors in visceral organs. Diffuse stimulation of visceral nociceptors might result from distention or ischemia of an internal organ. For example, a kidney stone might cause severe pain by obstructing and distending a ureter.

Localization of Pain

Fast pain is very precisely localized to the stimulated area. For example, if someone pricks you with a pin, you know exactly which part of your body was stimulated. Slow pain has more diffuse localization (involves large areas); it usually appears to come from a larger area of the skin. In some instances of visceral slow pain, the affected area is where the pain is felt. However, in many instances of visceral pain, the pain is felt in or just deep to the skin that overlies the stimulated organ or in a surface area far from the stimulated organ. This phenomenon is called **referred pain**. Figure 15.3 shows skin regions to which visceral pain may be referred. In general, the visceral organ involved and the area to which the pain is referred are served by the same segment of the spinal cord. For example, sensory fibers from the heart, the skin over the heart, and the skin along the medial aspect of the left arm enter spinal cord segments T1 to T5. Thus, the pain of a heart attack typically is felt in the skin over the heart and along the left arm.

Proprioceptive Sensations

Proprioceptive sensations allow us to know where our head and limbs are located and how they are moving even if we are not looking at them, so that we can walk, type, or dress without using our eyes. Proprioceptive sensations arise in receptors called **proprioceptors**. Proprioceptors embedded in muscles and tendons inform us of the degree to which muscles are contracted, the amount of tension on tendons, and the positions of joints. Hair cells of the inner ear are proprioceptors for balance and equilibrium that monitor the orientation of the head relative to gravity and head position during movements. Because proprioceptors adapt only slightly, the brain continually receives nerve impulses related to the position of different body parts to ensure coordination.

Figure 15.3   Distribution of referred pain. The colored parts of the diagrams indicate skin areas to which visceral pain is referred.

(a) Anterior view   (b) Posterior view

🔑 **Nociceptors are present in almost every tissue of the body.**

Proprioceptive sensations also allow us to estimate the weight of objects and determine the muscular effort necessary to perform a task. For example, as you pick up an object you quickly realize how heavy it is, and you then exert the correct amount of effort needed to lift it. Three types of proprioceptors include muscle spindles, tendon organs, and joint kinesthetic receptors.

Muscle Spindles

Muscle spindles are the proprioceptors in skeletal muscles that monitor changes in the length of skeletal muscles and participate in stretch reflexes (shown in Figure 14.20). By adjusting how vigorously a muscle spindle responds to stretching of a skeletal muscle, the brain sets an overall level of **muscle tone**, the small degree of contraction that is present while the muscle is at rest.

Each **muscle spindle** contains several slowly adapting sensory nerve endings that wrap around specialized muscle fibers called **intrafusal muscle fibers** (*intrafusal* = within a spindle; Figure 15.4). Muscle spindles are interspersed among most skeletal muscle fibers and aligned parallel to them. In muscles that produce finely controlled movements, such as those of the fingers, muscle spindles are plentiful. Muscles involved in coarser but more forceful movements, like the hamstrings of the thigh, have fewer muscle spindles.

Figure 15.4 **Two types of proprioceptors: a muscle spindle and a tendon organ.** In muscle spindles, which monitor changes in skeletal muscle length, sensory nerve endings wrap around intrafusal muscle fibers. In tendon organs, which monitor the force of muscle contraction, sensory nerve endings are activated by increasing tension on a tendon.

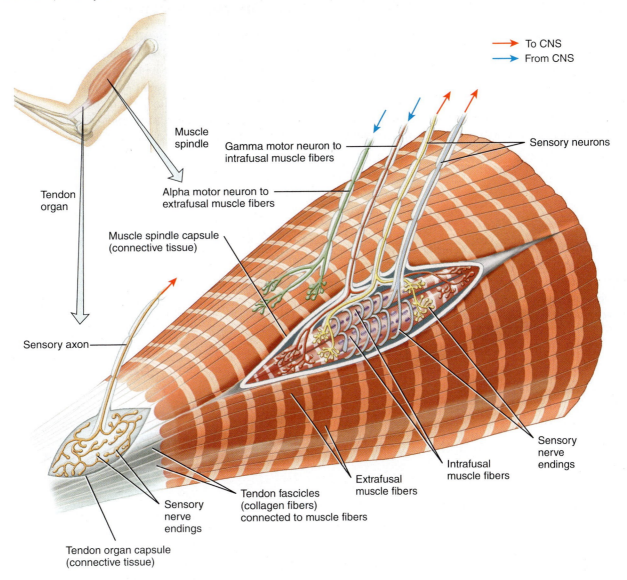

Proprioceptors provide information about body position and movement.

The main function of muscle spindles is to measure *muscle length*—how much a muscle is being stretched. Stretching of the intrafusal muscle fibers stimulates the sensory nerve endings. The resulting nerve impulses propagate into the CNS. Information from muscle spindles arrives at the somatic sensory areas of the cerebral cortex, allowing conscious perception of limb positions and movements. At the same time, impulses from muscle spindles pass to the cerebellum, where the input is used to coordinate muscle contractions.

In addition to their sensory nerve endings that wrap around the central portion of intrafusal fibers, muscle spindles contain motor neurons, called **gamma motor neurons**, that terminate near both ends of the intrafusal fibers. Gamma motor neurons adjust the tension in a muscle spindle to variations in the length of the muscle. For example, when a muscle shortens, gamma motor neurons stimulate the intrafusal fibers to contract slightly. This keeps the intrafusal fibers taut and maintains the sensitivity of the muscle spindle to stretching of the muscle. As the frequency of impulses in its gamma motor neuron increases, a muscle spindle becomes more sensitive to stretching.

Surrounding muscle spindles are ordinary skeletal muscle fibers, called **extrafusal muscle fibers** (*extrafusal* = outside a spindle), that are innervated by **alpha motor neurons**. During the stretch reflex, impulses in muscle spindle sensory axons propagate into the spinal cord and brain stem, where they activate alpha motor neurons that are connected to extrafusal muscle fibers in the same muscle. In this way, activation of its muscle spindles causes contraction of a skeletal muscle, which relieves the stretching.

Tendon Organs

Tendon organs are located at the junction of a tendon and a muscle. By initiating reflexes that cause muscle relaxation before a muscle contracts too forcefully, tendon organs protect tendons and their associated muscles from damage. Each **tendon organ** contains a few collagen fiber bundles called **tendon fascicles** (see Figure 15.4). Sensory nerve endings entwine among and around the collagen fibers of the tendon organ. As a muscle contracts, the tendon organs generate nerve impulses that propagate into the CNS, providing information about changes in muscle tension. Reflex activity decreases muscle tension by causing muscle relaxation.

Joint Kinesthetic Receptors

Several types of **joint kinesthetic receptors** (kin´-es-THET-ik; *kin-* = motion) are present within and around the articular capsules of synovial joints to sense joint position and movement. Joint kinesthetic receptors respond to joint pressure and acceleration and deceleration of joints during movement.

Table 15.2 summarizes the types of somatic sensory receptors and the sensations they convey.

Checkpoint

9. Which somatic sensory receptors adapt slowly, and which adapt rapidly?
10. Why do we need pain receptors? How are they activated?
11. Compare superficial somatic, deep somatic, and visceral pain. How are they similar? How do they differ?
12. What aspects of muscle function are monitored by muscle spindles and tendon organs?
13. What is the function of joint kinesthetic receptors?

TABLE 15.2

Receptors for Somatic Sensations

RECEPTOR TYPE	RECEPTOR STRUCTURE AND LOCATION	SENSATIONS	ADAPTATION RATE
Tactile Receptors			
Corpuscles of touch (Meissner corpuscles)	Capsule surrounds mass of dendrites in dermal papillae of hairless skin.	Touch, pressure, and slow vibrations.	Rapid.
Hair root plexuses	Free nerve endings wrapped around hair follicles in skin.	Touch.	Rapid.
Type I cutaneous mechanoreceptor (Merkel disc)	Saucer-shaped free nerve endings make contact with Merkel cells in epidermis.	Touch and pressure.	Slow.
Type II cutaneous mechanoreceptors (Ruffini corpuscles)	Elongated capsule surrounds dendrites deep in dermis and in ligaments and tendons.	Stretching of skin.	Slow.
Lamellated (pacinian) corpuscles	Oval, layered capsule surrounds dendrites; present in dermis and subcutaneous layer, submucosal tissues, joints, periosteum, and some viscera.	Pressure and fast vibrations.	Rapid.
Itch and tickle receptors	Free nerve endings in skin and mucous membranes.	Itching and tickling.	Both slow and rapid.
Thermoreceptors			
Warm receptors and cold receptors	Free nerve endings in skin and mucous membranes of mouth, vagina, and anus.	Warmth or cold.	Initially rapid, then slow.
Pain Receptors			
Nociceptors	Free nerve endings in every tissue of the body except the brain.	Pain.	Slow.
Proprioceptors			
Muscle spindles	Sensory nerve endings wrap around central area of encapsulated intrafusal muscle fibers within most skeletal muscles.	Muscle length.	Slow.
Tendon organs	Capsule encloses collagen fibers and sensory nerve endings at junction of tendon and muscle.	Muscle tension.	Slow.
Joint kinesthetic receptors	Pacinian corpuscles, Ruffini corpuscles, tendon organs, and free nerve endings.	Joint position and movement.	Rapid.

15.4   The somatosensory and primary motor areas of the cerebral cortex unequally serve different body regions.

Specific areas of the cerebral cortex receive somatic sensory input from particular parts of the body. Other areas of the cerebral cortex provide output in the form of instructions for movement of particular parts of the body. The *somatic sensory map* and the *somatic motor map* relate body parts to these cortical areas.

Precise localization of somatic sensations occurs when nerve impulses arrive at the **primary somatosensory area** (see Figure 13.11), which occupies the postcentral gyri of the parietal lobes of the cerebral cortex. Each region in this area receives sensory input from a different part of the body. Figure 15.5a maps the destination of somatic sensory signals coming from different parts of the left side of the body to the somatosensory area of the right cerebral

hemisphere. The left cerebral hemisphere has a similar primary somatosensory area that receives sensory input from the right side of the body.

Note that some parts of the body—chiefly the lips, face, tongue, and hand—provide input to large regions in the somatosensory area. Other parts of the body, such as the trunk and lower limbs, project to much smaller cortical regions. The relative sizes of these regions in the somatosensory area are proportional to the sensitivity, based on the number of specialized sensory receptors, of the corresponding body part. For example, there are many sensory receptors in the skin of the lips but few in the skin of the trunk.

Figure 15.5   Somatic sensory and somatic motor maps in the cerebral cortex. (a) Primary somatosensory area (postcentral gyrus) and (b) primary motor area (precentral gyrus) of the right cerebral hemisphere. The left hemisphere has similar representation. (After Penfield and Rasmussen.)

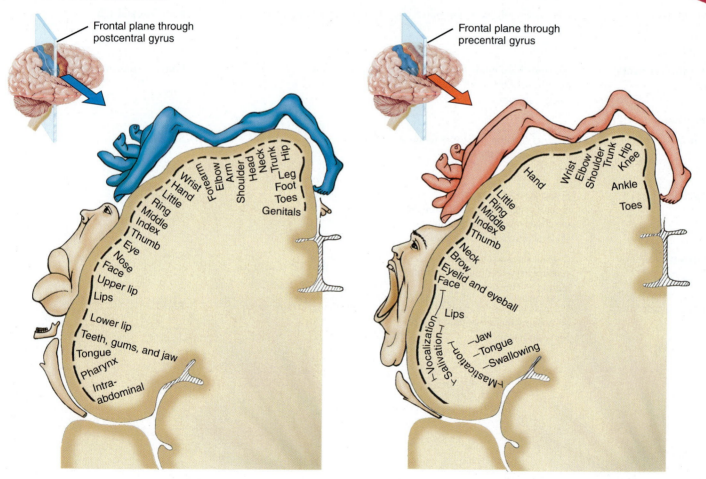

(a) Frontal section of primary somatosensory area in right cerebral hemisphere

(b) Frontal section of primary motor area in right cerebral hemisphere

Each point on the body surface maps to a specific region in both the primary somatosensory area and the primary motor area.

Control of body movements occurs via neural circuits in several regions of the brain. The **primary motor area** (see Figure 13.11), located in the precentral gyrus of the frontal lobe, is a major control region for the execution of voluntary movements. The adjacent **premotor area** also influences voluntary movements. As is true for somatic sensory representation in the somatosensory area, different muscles are represented unequally in the primary motor area (Figure 15.5b). More cortical area is devoted to those muscles involved in skilled, complex, or delicate movements. Muscles in the hand, lips, tongue, and vocal cords have large representations; the trunk has a much smaller representation. By comparing Figures 15.5a and b, you can see that somatosensory and somatic motor representations are similar but not identical for most parts of the body.

Checkpoint

14. Which body parts have the largest representation in the primary somatosensory areas?
15. Which body parts have the largest representation in the primary motor areas?
16. Why are certain body regions represented by larger portions of the primary motor area than others?

Return to Sannu's Story

Rolande lightly stroked Sannu's forehead. "Wake up, Brother," she said. "Today we go to Mbe, to the leprosarium." She squeezed his hand, but Sannu did not stir. She poked him hard in the ribs, then shook him. Sannu's eyes opened, and he smiled.

"You will make me fall from the bed with all this shaking, Sister," he joked.

"How do you feel today?" Rolande asked.

"My leg itches and it hurts again, but no cramps today," said Sannu.

"How can you feel a leg that is not there, Brother?" asked Rolande, looking down at Sannu's stump.

Dianna approached Rolande and Sannu. "Good morning. How are you feeling today, Sannu?"

"My leg burns and itches, but the cramps are gone," Sannu said.

Dianna kneeled down next to Sannu's bed. "Phantom limb sensations are not uncommon after amputation, even in patients with leprosy. They should diminish with time. I will check in with you after you have settled in at the leprosarium. You must get used to moving around with crutches until I can get you a prosthesis. And you must take the leprosy medicines, and take care of your wounds so that you do not get more infections. The bacteria that cause leprosy damage your nerves, particularly the ones closest to the surface of the body. The protective covering of your nerves, called myelin, is damaged so you are unable to feel things like you normally would. You have to be very careful, Sannu, so something like this doesn't happen to you again. You won't recover any sensation that you have lost, but drug therapy can stop the leprosy from doing more damage."

After living at the leprosarium for 3 months, Sannu still had not received a prosthetic leg but was progressing well. Rolande stayed with him, and his brother and two aunts had moved in with him. The days were filled with tending the fields, gathering water, and the fullness of village life. Sannu missed his own village, but life was good here.

"It's not so bad here," offered Eric.

Sannu looked at the old man next to him and asked, "How long have you been here?"

Sighing heavily, Eric sat back against a large tree in the shade. "Since the mission started this place. Must be 40 years now. How are you feeling? Is your missing leg still giving you pain?" asked Eric.

"Yes, especially at night. It feels like my foot and calf are all cramped up," complained Sannu.

"Well, this also happened with me. Those feelings went away after many years, but now it is getting harder to walk, even with the fake leg they gave me. I am old. I must take more care now," said Eric. Sannu had noticed that the old man did indeed move around awkwardly and hesitantly. Sannu hoped his leg would stop hurting. He wanted to go home.

F. Sannu describes pain in his leg. What receptors mediate sensations of pain? What type of pain is Sannu's "burning" pain, fast or slow?

G. What type of peripheral receptors would be activated in a normal, intact limb to produce the sensations Sannu is experiencing in his "phantom limb"?

H. Sannu has experienced two devastating events: loss of myelination of peripheral nerves and below-the-knee amputation of a leg. Sannu has been experiencing sensations in a limb that had no sensation prior to the amputation. How might this be possible?

I. Since Sannu has lost his leg, and since sensory input to the somatosensory cortex of the brain will be nonexistent, what might happen to the representation of his amputated leg on the somatosensory map?

J. Could the loss of input from proprioceptors in the amputated leg affect Sannu's sense of balance and equilibrium?

15.5   Somatic sensory pathways relay information from sensory receptors to the cerebral cortex and cerebellum.

The sensory and motor pathways of the body provide routes for input into the brain and spinal cord and output to targeted organs for responses, such as muscle contraction. **Somatic sensory pathways** relay information from somatic sensory receptors to the primary somatosensory area in the cerebral cortex and to the cerebellum. These ascending pathways consist of thousands of sets of three neurons: a first-order neuron, a second-order neuron, and a third-order neuron.

1. **First-order neurons** conduct impulses from somatic receptors into the brain stem or spinal cord. From the face, mouth, teeth, and eyes, somatic sensory impulses propagate along *cranial nerves* into the brain stem. From the neck, trunk, limbs, and posterior head, somatic sensory impulses propagate along *spinal nerves* into the spinal cord.

2. **Second-order neurons** conduct impulses from the brain stem and spinal cord to the thalamus. Axons of second-order neurons *decussate* (cross over to the opposite side) in the brain stem or spinal cord before ascending to the thalamus. Thus, all somatic sensory information from one side of the body reaches the thalamus on the opposite side.

3. **Third-order neurons** conduct impulses from the thalamus to the primary somatosensory area of the cerebral cortex on the same side.

Regions within the CNS where neurons synapse with other neurons that are a part of a particular pathway are known as **relay stations** because neural signals are being relayed from one region of the CNS to another. For example, the neurons of many sensory pathways synapse with neurons in the thalamus; therefore the thalamus functions as a relay station. In addition to the thalamus, many other regions of the CNS, including the spinal cord and brain stem, can function as relay stations.

Somatic sensory impulses may ascend to the cerebral cortex through two general pathways: (1) the posterior column–medial lemniscus pathway and (2) the anterolateral pathways. Somatic sensory impulses reach the cerebellum via the spinocerebellar tracts.

Posterior Column–Medial Lemniscus Pathway to the Cerebral Cortex

Nerve impulses for most touch sensations, vibration, and proprioception ascend to the cerebral cortex along the **posterior column–medial lemniscus pathway** (Figure 15.6a). The name of

the pathway comes from the names of two white matter tracts, the posterior column of the spinal cord and the medial lemniscus of the brain stem.

First-order neurons extend from sensory receptors in the limbs and trunk into the spinal cord and ascend to the medulla oblongata on the same side of the body. In the spinal cord, their axons form the **posterior columns** (white matter located on the posterior side of the spinal cord). The axons synapse with second-order neurons in nuclei of the medulla oblongata. Axons of second-order neurons cross to the opposite side of the medulla and enter the **medial lemniscus** (lem-NIS-kus = ribbon), a thin, ribbonlike projection tract that extends from the medulla and travels superiorly through the brain stem to the thalamus. In the thalamus, the axons of second-order neurons synapse with third-order neurons, which project their axons to the primary somatosensory area of the cerebral cortex.

Anterolateral Pathway to the Cerebral Cortex

Nerve impulses for pain, temperature, itch, and tickle ascend to the cerebral cortex along the **anterolateral** or **spinothalamic pathway** (spī-nō-tha-LAM-ik). Like the posterior column–medial lemniscus pathway, the anterolateral pathway is composed of three-neuron sets (Figure 15.6b). First-order neurons connect receptors in the limbs and trunk with the spinal cord. Axons of the first-order neuron synapse with second-order neurons in the posterior gray horn of the spinal cord. Axons of second-order neurons cross to the opposite side of the spinal cord, then ascend upward to the brain stem as the **spinothalamic tract** to the thalamus, where they synapse with third-order neurons. Axons of third-order neurons project to the primary somatosensory area of the cerebral cortex on the same side as the thalamus.

Somatic Sensory Pathways to the Cerebellum

Two tracts in the spinal cord—the **posterior spinocerebellar tract** (spī-nō-ser-e-BEL-ar) and the **anterior spinocerebellar tract**—are the major routes proprioceptive impulses take to reach the cerebellum (see Figure 13.19). Although they are not consciously perceived, sensory impulses conveyed to the cerebellum along these two pathways are critical for posture, balance, and coordination of skilled movements.

Figure 15.6 **Somatic sensory pathways.**

(a) Posterior column–medial lemniscus pathway

(b) Anterolateral (spinothalamic) pathways

🔑 **Nerve impulses propagate along sets of first-order, second-order, and third-order neurons to the primary somatosensory area of the cerebral cortex.**

Checkpoint

17. Which sensations could be lost if the posterior column–medial lemniscus pathway is damaged?

18. State three differences between the posterior column–medial lemniscus pathway and the anterolateral pathway.

19. What type of sensory information is carried in the spinocerebellar pathways, and what is its usefulness?

15.6 Somatic motor pathways carry impulses from the brain to effectors.

Neurons in the brain and spinal cord orchestrate all voluntary and involuntary movements. Ultimately, all excitatory and inhibitory signals that control movement converge on the neurons known as **lower motor neurons (LMNs)**, which innervate skeletal muscles. From the brain stem, axons of LMNs extend though cranial nerves to innervate skeletal muscles of the head. From the spinal cord, axons of LMNs extend through spinal nerves to innervate skeletal muscles of the limbs and trunk.

Other neurons in the brain and spinal cord participate in control of movement by providing input to lower motor neurons:

- *Local circuit neurons*. Nearby interneurons termed **local circuit neurons** help coordinate rhythmic activity in specific muscle groups, such as alternating flexion and extension of the lower limbs during walking.

- *Upper motor neurons*. Local circuit neurons and lower motor neurons receive input from **upper motor neurons (UMNs)**. UMNs from the cerebral cortex execute voluntary movements of the body. UMNs originating in the brain stem help maintain muscle tone, posture, and balance.

- *Basal nuclei neurons*. Neurons of the basal nuclei assist movement by providing input to upper motor neurons. Basal nuclei neurons help initiate and terminate movements, suppress unwanted movements, and establish a normal level of muscle tone.

- *Cerebellar neurons*. Neurons of the cerebellum also aid movement by controlling the activity of upper motor neurons. A prime function of the cerebellum is to monitor differences between intended movements and movements actually performed. Then it issues commands to upper motor neurons to reduce errors in movement. The cerebellum thus coordinates body movements and helps maintain normal posture and balance.

Organization of Upper Motor Neuron Pathways

The axons of upper motor neurons extend from the brain to lower motor neurons via two types of descending somatic motor pathways—direct and indirect. Direct motor pathways provide input to lower motor neurons via axons that extend directly from the cerebral cortex. Indirect motor pathways provide input to lower motor neurons from motor centers in the brain stem. Direct and indirect pathways both govern generation of nerve impulses in the lower motor neurons, the neurons that stimulate contraction of skeletal muscles.

Direct Motor Pathways

Nerve impulses for voluntary movements propagate from the cerebral cortex to lower motor neurons via the direct motor pathways

(Figure 15.7). The **direct motor pathways** consists of axons that descend from upper motor neurons located in the primary motor area of the cerebral cortex.

Figure 15.7   Somatic motor pathways. Shown here are the two most direct pathways whereby signals initiated by the primary motor area in one hemisphere control skeletal muscles on the opposite side the body.

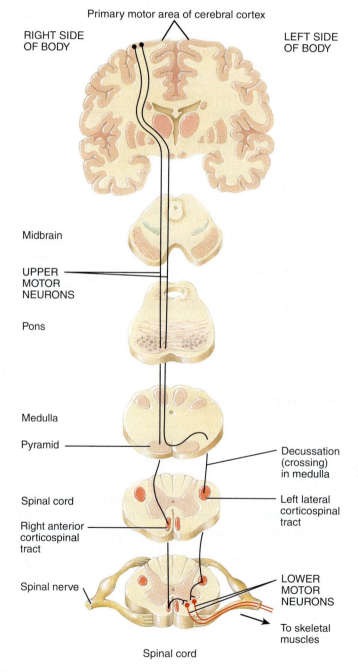

Direct motor pathways convey impulses that result in precise, voluntary movements.

The **corticospinal tracts** (kor´-ti-kō-SPĪ-nal) are major tracts that conduct impulses for the control of muscles of the limbs and trunk. While descending from the primary motor area, corticospinal axons cross to the opposite side in the medulla oblongata or in the spinal cord and then synapse with a lower motor neuron. Thus, the right cerebral cortex controls muscles on the left side of the body, and the left cerebral cortex controls muscles on the right side of the body.

Corticospinal axons that cross in the medulla form the **lateral corticospinal tract** in the lateral white columns of the spinal cord (Figure 15.7). These axons synapse with lower motor neurons in the spinal cord. Axons of these lower motor neurons exit the cord in the anterior roots of spinal nerves and terminate in skeletal muscles that control movements for precise, agile, and highly skilled movements of the hands and feet. Examples include the movements needed to button a shirt or play the piano.

Corticospinal axons that do not cross in the medulla oblongata form the **anterior corticospinal tract** in the anterior white columns of the spinal cord (Figure 15.7). At each spinal cord level, some of these axons cross via the anterior white commissure. Then they synapse with lower motor neurons in the anterior gray horn. Axons of these lower motor neurons exit the spinal cord in the anterior roots of spinal nerves and terminate in skeletal muscles that control movements of the trunk and proximal parts of the limbs.

In addition to the corticospinal tracts, another direct motor pathway is the **corticobulbar tract** (kor´-ti-kō-BUL-bar), which conducts impulses to skeletal muscles in the head. Some of the axons of the corticobulbar tract decussate, and others do not, as they descend from the cerebral cortex and then synapse with lower motor neurons of cranial nerves exiting the brain stem. These lower motor neurons convey impulses that control precise, voluntary movements of the eyes, tongue, and neck, plus chewing, facial expression, and speech.

Indirect Motor Pathways

The **indirect motor pathways** include all somatic motor tracts other than the corticospinal and corticobulbar tracts. Axons of upper motor neurons that give rise to the indirect pathways descend from the brain stem into spinal cord tracts that terminate on local circuit neurons or lower motor neurons.

Modulation of Movement by the Basal Nuclei

As previously noted, the basal nuclei and cerebellum influence movement through their effects on upper motor neurons. The basal nuclei play a major role in the initiation and termination of movements. They receive input from sensory, association, and motor areas of the cerebral cortex, then send feedback signals to the motor areas of the cerebral cortex by way of the thalamus. This circuit—from cortex to basal nuclei to thalamus to cortex—appears to function in initiating and terminating movements. The basal nuclei suppress unwanted movements by their inhibitory effects on the thalamus and midbrain. The basal nuclei also influence muscle tone by sending impulses into the brain stem that reduce muscle tone.

Modulation of Movement by the Cerebellum

In addition to maintaining proper posture and balance, the cerebellum is active in both learning and performing rapid, coordinated, highly skilled movements such as hitting a golf ball, speaking, and swimming. Cerebellar function involves four activities (Figure 15.8):

1 The cerebellum *monitors intentions for movement* by receiving impulses from the motor areas of the cerebral cortex and basal nuclei via the pons regarding what movements are planned (red arrows).

2 The cerebellum *monitors actual movement* by receiving input from proprioceptors in joints and muscles that reveals what is actually happening (blue arrows). Nerve impulses from the vestibular (equilibrium-sensing) apparatus in the inner ear and from the eyes also enter the cerebellum.

Figure 15.8   Input to and output from the cerebellum.

Sagittal plane

Motor areas of cerebral cortex

Corrective feedback

Thalamus

Cortex of cerebellum

Motor centers in brain stem

Pons

Direct pathways

Indirect pathways

Sensory signals from proprioceptors in muscles and joints, vestibular apparatus, and eyes

Signals to lower motor neurons

Sagittal section through brain and spinal cord

🔑 **The cerebellum coordinates and smoothes contractions of skeletal muscles during skilled movements.**

❸ The cerebellum *compares the command signals* (intentions for movement) *with sensory information* (actual movement performed).

❹ If there is a discrepancy between intended and actual movement, the cerebellum *sends out corrective feedback* to upper motor neurons. This information travels via the thalamus to the cerebral cortex and goes directly to upper motor neurons in brain stem motor centers (green arrows). As movements occur, the cerebellum continuously provides error correction to upper motor neurons, which decreases errors and smoothes motions. The cerebellum also contributes over longer periods to the learning of new motor skills.

Skilled activities such as volleyball provide good examples of the contribution of the cerebellum to movement. To block a spike,

you must bring your arms forward just far enough to make solid contact. How do you stop at exactly the right point? Before you even hit the ball, the cerebellum has sent nerve impulses to the cerebral cortex and basal nuclei informing them where your swing must stop. In response to impulses from the cerebellum, the cortex and basal nuclei transmit motor impulses to opposing body muscles to stop the swing.

Checkpoint

20. What is the function of lower motor neurons?

21. Summarize the functions of neurons that provide input to lower motor neurons.

22. Why are the two main somatic motor pathways named "direct" and "indirect"?

15.7 Wakefulness and memory are integrative functions of the brain.

We turn now to a fascinating, though incompletely understood, function of the cerebrum: integration, the processing of sensory information by analyzing and storing it and making decisions for various responses. The **integrative functions** include cerebral activities such as sleep and wakefulness, learning and memory, and emotional responses.

Wakefulness and Sleep

Humans sleep and awaken in a 24-hour cycle called a **circadian rhythm** (ser-KĀ-dē-an; *circa-* = about; *-dia-* = a day) that is established by the hypothalamus. A person who is awake is in a state of readiness and is able to react consciously to various stimuli. Electroencephalogram (EEG) recordings show that the cerebral cortex is very active during wakefulness; fewer impulses arise during most stages of sleep.

How does your nervous system make the transition between the two states? Stimulation of some parts of the reticular formation in the brain stem (see Concept 13.7) increases activity of the cerebral cortex to maintain consciousness. Thus, a portion of the reticular formation is known as the **reticular activating system (RAS)** (Figure 15.9). When this area is active, many nerve impulses are transmitted to widespread areas of the cerebral cortex. The effect is a generalized increase in cerebral cortex activity.

Arousal, or awakening from sleep, also involves increased activity in the RAS. For arousal to occur, the RAS must be stimulated. Many sensory stimuli can activate the RAS: painful stimuli detected by nociceptors, touch and pressure on the skin, movement of the limbs, bright light, or the buzz of an alarm clock. Once the RAS is activated, the cerebral cortex is also activated, and arousal occurs. The result is a state of wakefulness called **consciousness**. Notice in Figure 15.9 that even though the RAS receives input from somatic sensory receptors, the eyes, and the ears, there is no

Figure 15.9 The reticular activating system (RAS) consists of neurons whose axons project from the reticular formation through the thalamus to the cerebral cortex.

Sagittal plane

Thalamus

Cerebral cortex

RAS projections to cerebral cortex

Cerebellum

Visual impulses from eyes

Pons

Reticular formation

Medulla oblongata

Spinal cord

Auditory and vestibular impulses from ears

Somatic sensory impulses (from nociceptors, proprioceptors, and touch receptors)

Sagittal section through brain and spinal cord

🔑 Increased activity of the RAS causes awakening from sleep (arousal).

input from olfactory receptors; even strong odors may fail to cause arousal. People who die in house fires usually succumb to smoke inhalation without awakening. For this reason, all sleeping areas should have a nearby smoke detector that emits a loud alarm. A vibrating pillow or flashing light can serve the same purpose for those who are hearing impaired.

Sleep is a state of altered consciousness or partial unconsciousness from which an individual can be aroused. Although it is essential, the exact functions of sleep are still unclear. Sleep deprivation impairs attention, learning, and performance. Several physiological changes occur during sleep. One apparent sleep-inducing chemical is adenosine, which accumulates during periods of high use of ATP (adenosine triphosphate) by the nervous system. Adenosine binds to specific receptors on neurons of the RAS that participate in arousal. Thus, activity in the RAS during sleep is low due to the inhibitory effect of adenosine. Caffeine (in coffee) and theophylline (in black tea)—substances known for their ability to maintain wakefulness—bind to and block the adenosine receptors, preventing adenosine from binding and inducing sleep. During sleep, activity in the parasympathetic division of the autonomic nervous system increases while sympathetic activity decreases. For example, heart rate and blood pressure decrease during sleep.

Learning and Memory

Without memory, we would repeat mistakes and be unable to learn. Similarly, we would not be able to repeat our successes or accomplishments, except by chance. Although both learning and memory have been studied extensively, we still have no completely satisfactory explanation for how we recall information or how we remember events.

Learning is the ability to acquire new information or skills through instruction or experience. **Memory** is the process by which information acquired through learning is stored and retrieved. For an experience to become part of memory, it must produce structural and functional changes in the brain. This capability for change involves changes in individual neurons—for example, synthesis of different proteins or sprouting of new dendrites—as well as changes in the strengths of synaptic connections among neurons. The parts of the brain known to be involved with memory include the association areas of the cerebral cortex, thalamus and hypothalamus, and parts of the limbic system. If a particular body part is used more intensively or in a newly learned activity, such as reading Braille, the cortical areas devoted to that body part gradually expand.

Memory occurs in stages over a period of time. **Immediate memory** is the ability to recall ongoing experiences for a few sec-

onds. It provides a perspective to the present time that allows us to know where we are and what we are doing. **Short-term memory** is the temporary ability to recall a few pieces of information for seconds to minutes. One example is when you look up an unfamiliar telephone number, cross the room to the phone, and then dial the new number. If the number has no special significance, it is usually forgotten within a few seconds. Some evidence supports the notion that short-term memory depends more on electrical and chemical events in the brain than on structural changes, such as the formation of new synapses.

Information in short-term memory may later be transformed into a more permanent type of memory, called **long-term memory**, which lasts from days to years. If you use that new telephone number often enough, it becomes part of long-term memory. Information in long-term memory usually can be retrieved for use whenever needed. Long-term memories for information that can be expressed by language, such as a telephone number, apparently are stored in wide regions of the cerebral cortex. Memories for motor skills, such as how to serve a tennis ball, are stored in the basal ganglia and cerebellum as well as in the cerebral cortex.

Although the brain receives many stimuli, we pay attention to only a few of them at a time. It has been estimated that only 1% of all the information that comes to our consciousness is stored as long-term memory. Moreover, much of what goes into long-term memory is eventually forgotten. Memory does not record every detail as if it were magnetic tape. Even when details are lost, we can often explain the idea or concept using our own words and ways of viewing things.

Several conditions that inhibit the electrical activity of the brain, such as anesthesia, coma, electroconvulsive therapy, and ischemia of the brain, disrupt retention of recently acquired information without altering previously established long-term memories. People who suffer retrograde amnesia cannot remember anything that occurred during the 30 minutes or so before the amnesia developed. As a person recovers from amnesia, the most recent memories return last.

Anatomical changes occur in neurons when they are stimulated. For example, electron micrographs of neurons subjected to prolonged, intense activity reveal an increase in the number of axon terminals and enlargement of synaptic end bulbs in presynaptic neurons, as well as an increase in the number of dendritic branches in postsynaptic neurons. Opposite changes occur when neurons are inactive. For example, the cerebral cortex in the visual area of animals that have lost their eyesight becomes thinner.

Checkpoint

23. How is wakefulness related to the reticular activating system?
24. Define memory. What are the three kinds of memory?

▶▶▶Sannu is a young African man who lives a hard life. He has contracted leprosy. Leprosy is a leading cause of peripheral nerve disease worldwide but is treatable with appropriate drugs. Sannu was unable to obtain treatment during the early stages of his disease. Geographic and monetary limitations often reduce the likelihood of obtaining treatment in developing countries. Because of damage to the myelination of his peripheral nerves, Sannu lost peripheral sensation and was unaware of an injury to his foot and a subsequent infection. This resulted in the amputation of his foot as the infection became gangrenous. Following the amputation, Sannu experienced phantom limb sensations and pain in the missing appendage. Sannu finally received treatment for his leprosy, but he is concerned about developing further complications from his previous health problems. Consider the following questions about Sannu's case: ◀◀◀

K. Was the sensory loss that Sannu experienced initially due to interference of central nervous system processing or sensory pathway transmission?

L. What damage was done by bacterial infection of the nervous system that led to Sannu's symptoms?

M. What type of receptor should have initially sensed the injury to Sannu's foot?

N. What part of his brain was mediating the odd sensations Sannu felt following the amputation of his foot?

O. Is Sannu at any higher risk than anyone else in his village for developing Parkinson disease (which involves changes in the basal nuclei) because of his leprosy infection and symptoms or the amputation?

Concept and Resource Summary

Concept

Resources

Introduction

1. The nervous system must sense the internal and external environments, and convey the sensations to the CNS for modification and integration with other incoming or stored information in order to determine appropriate motor responses.

Concept 15.1 Sensations arise as a result of stimulation, transduction, generation, and integration.

1. **Sensation** is the conscious or subconscious detection of changes in the external or internal environment.
2. **Perception** is the conscious awareness and interpretation of sensations.
3. A **sensory modality** is one unique type of sensation. A sensory receptor has **selectivity**, in that it responds vigorously to only one kind of **stimulus**.
4. Sensory modalities are grouped into two classes: general senses and special senses. **General senses** include **somatic senses** and **visceral senses**. Somatic sensory modalities include tactile, thermal, pain, and proprioceptive sensations. Visceral sensations mediated by the autonomic nervous system provide information about conditions within internal organs. **Special senses** include the sensory modalities of smell, taste, vision, hearing, and equilibrium.
5. The process of sensation involves a sequence of four events: (1) stimulation of a sensory receptor, (2) transduction of the stimulus, (3) generation of nerve impulses, and (4) integration of sensory input.

Concept 15.2 Sensory receptors can be classified structurally, functionally, or by the type of stimulus detected.

1. There are three types of sensory receptors based on structure: **free nerve endings** are bare dendrites; **encapsulated nerve endings** are dendrites enclosed in connective tissue; and **separate cells**.
2. Graded potentials have amplitudes that vary with the intensity of the stimulus. Two types are generator potentials and receptor potentials. Free nerve endings and encapsulated nerve endings produce **generator potentials**. Separate cells produce **receptor potentials**.
3. Sensory receptors can be grouped based on location of receptors and origin of the stimuli. **Exteroceptors** monitor the external environment. **Interoceptors** monitor the internal environment. **Proprioceptors** provide information about body position and movement.
4. Sensory receptors can also be distinguished by the type of stimulus they detect, including electromagnetic energy, mechanical energy, and chemical energy.
5. Adaptation is a decrease in sensitivity during a constant stimulus. There are **rapidly adapting receptors** and **slowly adapting receptors**.

See

- Anatomy Overview – Chemoreceptors
- Anatomy Overview – Baroreceptors
- Anatomy Overview – Proprioceptors

Concept	Resources

Concept 15.3 Somatic sensations include tactile sensations, thermal sensations, pain, and proprioception.

1. Somatic sensations arise from stimulation of sensory receptors in skin, certain mucous membranes, muscles, tendons, joints, and the inner ear. The four modalities of somatic sensations are tactile, thermal, pain, and proprioceptive sensations.
2. **Tactile sensations** arise by activation of encapsulated mechanoreceptors for touch, pressure, and vibration; and by activation of free nerve endings for itch and tickle sensations.
3. Rapidly adapting **touch** receptors include **Meissner corpuscles** in hairless skin and **hair root plexuses** in hairy skin. Slowly adapting touch receptors include **type I cutaneous mechanoreceptors** in the epidermis of the skin and **type II cutaneous mechanorecepters** in the dermis of the skin and in ligaments and tendons.
4. **Pressure receptors** are widely distributed in the body, and include corpuscles of touch, type I cutaneous mechanoreceptors, and **lamellated corpuscles**.
5. **Vibration** receptors are corpuscles of touch and lamellated corpuscles.
6. **Itch** and **tickle** sensations result from stimulation of free nerve endings.
7. **Thermoreceptors** are free nerve endings on the skin that detect coldness and warmth. **Cold receptors** are in the epidermis, and **warm receptors** are in the dermis.
8. Receptors for pain are free nerve endings, called **nociceptors**, and are found in every body tissue except the brain. **Fast pain** is known as acute, sharp, or pricking pain. **Slow pain** is known as chronic, burning, aching, or throbbing pain. Depending on the location of the receptors, pain can be **superficial somatic pain**, **deep somatic pain**, or **visceral pain**. **Referred pain**, in a surface away from the affected organ, can also be experienced with visceral pain.
9. **Proprioceptive sensations** allow us to be aware of the movements and position of the head and limbs. **Proprioceptors** embedded in muscles and tendons convey to the brain the amount of tension and stretch in muscles and joints. **Muscle spindles**, composed of specialized **intrafusal muscles fibers** and nerve endings, are proprioceptors that facilitate the setting of overall **muscle tone** by the brain. The function of muscle spindles is to measure muscle length.
10. **Tendon organs** give information about changes in muscle tension.
11. **Joint kinesthetic receptors** sense joint position and movement.

See
- **Anatomy Overview – Sensory Receptors of the Skin**

Hear

- **Figure 15.3 –** **Distribution of Referred Pain**
- **Figure 15.4 – Two Types of Proprioceptors: a Muscle Spindle and a Tendon Organ**

Concept 15.4 The somatosensory and primary motor areas of the cerebral cortex unequally serve different body regions.

1. Precise localization of somatic sensations occurs when nerve impulses arrive at the **primary somatosensory area** in the parietal lobes. Each region of this area receives sensory input from different body areas.
2. Body movements are controlled via neural circuits principally in the **primary motor area** in the frontal lobe. The size of the cortical area governing a muscle is proportional to the number of motor units in the muscle.

See
- **Anatomy Overview – Cerebrum**

Hear

- **Figure 15.5 –** **Somatic Sensory and Somatic Motor Maps in the Cerebral Cortex**

Concept 15.5 Somatic sensory pathways relay information from sensory receptors to the cerebral cortex and cerebellum.

1. **Somatic sensory pathways** relay information from somatic sensory receptors to the primary somatosensory area in the cerebral cortex and to the cerebellum. **First-order neurons** conduct impulses from somatic receptors via nerves into the brain stem or spinal cord. **Second-order neurons** conduct impulses from where they *decussate* in the brain stem or spinal cord to the thalamus. **Third-order neurons** conduct impulses from the thalamus to the primary somatosensory area of the cortex on the same side.
2. The **posterior column–medial lemniscus pathway** conducts nerve impulses for conscious proprioception, touch, and vibratory sensations to the primary somatosensory area of the cerebral cortex.
3. The **anterolateral** or **spinothalamic pathway** relays impulses for pain, temperature, tickle, and itch sensations to the primary somatosensory area of the cerebral cortex.
4. The **posterior** and **anterior spinocerebellar tracts** of the spinal cord are the major routes for proprioceptive impulses to travel to the cerebellum. These are critical for posture, balance, and coordination of skilled movements.

See
- **Animation – Somatic Sensory Pathways**

Hear
- **Figure 15.6 –** **Somatic Sensory Pathways**

Concept	Resources

Concept 15.6 Somatic motor pathways carry impulses from the brain to effectors.

1. Axons of **lower motor neurons (LMNs)** extend from cranial and spinal nerves to skeletal muscles.

2. Neurons in the brain and spinal cord give input to LMNs. **Local circuit neurons** help coordinate rhythmic activity in specific muscle groups. **Upper motor neurons (UMNs)** from the cerebral cortex give input for voluntary movements. **Basal nuclei neurons** help initiate and terminate movements, suppress unwanted movements, and establish normal muscle tone. **Cerebellar neurons** control activity of UMNs to reduce movement errors.

3. **Direct motor pathways** have axons that extend from the cerebral cortex to LMNs. **Indirect motor pathways** provide input to LMNs from motor centers in the brain stem.

4. Three tracts contain axons of upper motor neurons of direct motor pathways: **lateral corticospinal tracts** control precise, skilled movements of the hands and feet; **anterior corticospinal tracts** control movements of the trunk and proximal limbs; and **corticobulbar tracts** control skeletal muscles of the head.

5. **Indirect motor pathways** include all somatic motor pathways other than the corticospinal and corticobulbar tracts. Axons from UMNs carrying impulses from indirect pathways terminate on LMNs.

6. Basal nuclei receive input from sensory, association, and motor areas of the cerebral cortex, then respond with feedback signals to the motor cortex. Basal nuclei suppress unwanted movements and influence muscle tone.

7. The cerebellum monitors intentions for movements, monitors actual movements, compares them and sends out corrective signals to LMNs.

See

- Animation – Somatic Motor Pathways

Hear

- **Figure 15.7 –** Somatic Motor Pathways

Concept 15.7 Wakefulness and memory are integrative functions of the brain.

1. **Wakefulness** and **sleep** are integrative functions that are controlled by the **reticular activating system (RAS)** in the brain stem. The reticular activating system sends nerve impulses to the cerebral cortex, both directly and via the thalamus, resulting in increased cortical activity. **Arousal** from sleep involves increased RAS activity. The RAS can be activated by many different sensory stimuli, which then activate the cerebral cortex, resulting in **consciousness**.

2. **Learning** is the ability to acquire new information or skills.

3. **Memory**, the ability to store and recall information or skills, involves structural changes in neurons and synapses within the brain. The three types are immediate, short-term, and long-term memory.

Other Chapter Resources

For additional information on clinical applications, medical disorders, and medical terminology, see Chapter 7 in the *Clinical Connections* manual packaged with your text.

Within *WileyPLUS*

- Chapter Quiz
- Visual Anatomy
- Mastering Vocabulary
- Audio Glossary
- Web Links

• Understanding the Concepts

1. How is a sensation different from a perception?

2. How are generator potentials and receptor potentials similar? How do they differ?

3. Which body parts would be more sensitive, those with a higher density of somatic sensory receptors or those with a lower density? Explain your answer.

4. What is referred pain, and how is it useful in diagnosing internal disorders?

5. Why would you expect the face to be represented by a larger portion of the somatosensory area of the cerebral cortex than a physically larger area such as the trunk?

6. What is the significance of the genitals being represented in the primary somatosensory area but not in the primary motor area of the cerebral cortex? (*Hint*: Genital tissues contain smooth muscle but no skeletal muscle tissue.)

7. Compare representation of the hand in the somatosensory and motor areas of the cerebral cortex.

8. What types of sensory deficits could be produced by damage to the right lateral spinothalamic tract?

9. What movement-related differences would you expect to see with damage to the basal nuclei versus damage to the cerebellum?

10. Why should every sleeping room have a smoke detector?

16

The Special Senses

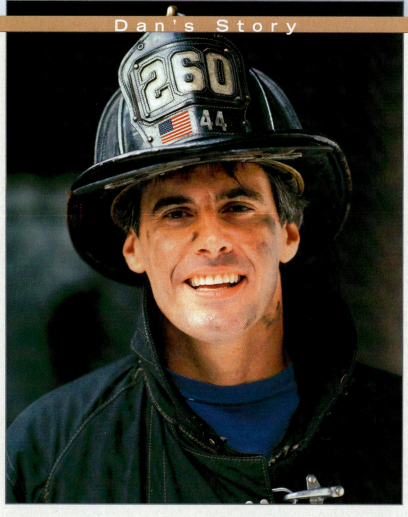

Dan rolled over in bed. The muscles in his back ached and he could feel the pain in his knees again, but it had been worth it. He played the events of the last run over in his head: the old wooden building, the steep second story stairs; the screams of the terrified elderly woman. He didn't think about the danger when it was happening, he just went in. He was proud of being a volunteer firefighter.

He heard a soft giggle and felt tickling on his bare back. He knew it was little Danny tickling him, so he rolled over and grabbed the toddler, pulling him into bed. Dan could smell the sweet odor of dried milk and sugar-coated cereal on Danny's pajamas and fresh coffee brewing in the kitchen. He thought, "I'm lucky to smell anything at all after breathing in all that smoke last night."

"Dan! We have to get going to the beach before it gets too crowded."

"Go tell Mommy I'll be there in a minute," said Dan.

Dan swung his feet over the edge of the bed. He reached for the window shade, but his hand slipped. The spring-loaded roll shade flung open with a snap, flooding the room with the bright morning sun. "No clouds today, I guess," he thought to himself. He reached for a pair of sunglasses on the nightstand. Dan walked into the kitchen with the glasses on.

The phone rang suddenly. Dan picked it up. "Three alarms, huh? Alright. Bye." Dan said, "There's been a fire at the chemical plant down by the county line; they need everyone they can get. If that place explodes, it will take out half the town. I'm sorry," said Dan. "We'll get to the beach another day, but I have to go now."

Consider the sensations experienced by Dan. Like many of us, he takes his special senses for granted. As we will learn, the special senses are not only important for sensing our environment, but also for enjoying the richness of life.

INTRODUCTION

The **special senses** include smell, taste, vision, hearing, and equilibrium. In this chapter we discuss how the special senses differ from the general senses in how they receive stimuli and how the stimuli are converted into signals that can be understood by the nervous system. Familiar sensations that you take for granted, such as pleasant smells, the sound of laughter, the taste of your favorite food, and the ability to see the face of someone you love

The special senses include smell, taste, vision, hearing, and equilibrium.

or the beauty of a sunrise are mediated by the special senses. Recall from Chapter 15 that the general senses include the somatic senses (tactile, thermal, pain, proprioceptive) and visceral senses. As you learned in that chapter, receptors for the general senses are scattered throughout the body and are rel-

atively simple in structure. The receptors for the special senses are more complex in structure and are concentrated in sensory organs of the head, such as the eyes and ears. Neural pathways for the special senses are also more complex than those of the general senses.

In this chapter we examine the structure and function of the special sense organs and the pathways involved in conveying their information to the central nervous system (CNS).

16.1 Impulses for smell propagate along the olfactory nerve to the brain.

Both smell and taste are chemical senses; the sensations arise from the interaction of molecules with smell or taste receptors. Because impulses for smell and taste propagate to the limbic system (and to higher cortical areas as well), certain odors and tastes can evoke strong emotional responses or a flood of memories.

Anatomy of the Olfactory Epithelium

The nose contains between 10 and 100 million receptors for the sense of smell, or **olfaction** (ol-FAK-shun; *olfact-* = smell). The total area of the **olfactory epithelium** is only 5 cm² (a little less than 1 in.²). It occupies the superior part of the nasal cavity, covering the inferior surface of the cribriform plate of the ethmoid bone and extending along the superior nasal concha (Figure 16.1a). The olfactory epithelium consists of three types of cells: olfactory receptor cells, supporting cells, and basal cells (Figure 16.1b,c).

Olfactory receptor cells are the first-order neurons of the olfactory pathway. They are bipolar neurons, with one dendrite and one axon. The dendrite of each olfactory receptor cell projects to the surface of the olfactory epithelium. From the base of each olfactory receptor cell, a single axon projects through a foramen of the cribriform plate of the ethmoid bone, then into the olfactory bulb. The sites of the olfactory receptor cell that respond to inhaled chemicals are the **olfactory hairs**, nonmotile cilia that project from the dendrite. **Odorants** are chemicals that can stimulate the olfactory hairs and, therefore, be detected as odors. Receptors on the olfactory hairs respond to the chemical stimulation of an odorant molecule by producing a generator potential, thus initiating the olfactory response.

Figure 16.1 **Olfactory epithelium and olfactory receptor cells.** (a) Location of olfactory epithelium in the nasal cavity. (b) Axons of olfactory receptor cells extend through the cribriform plate and terminate in the olfactory bulb. (c) Histology of the olfactory epithelium.

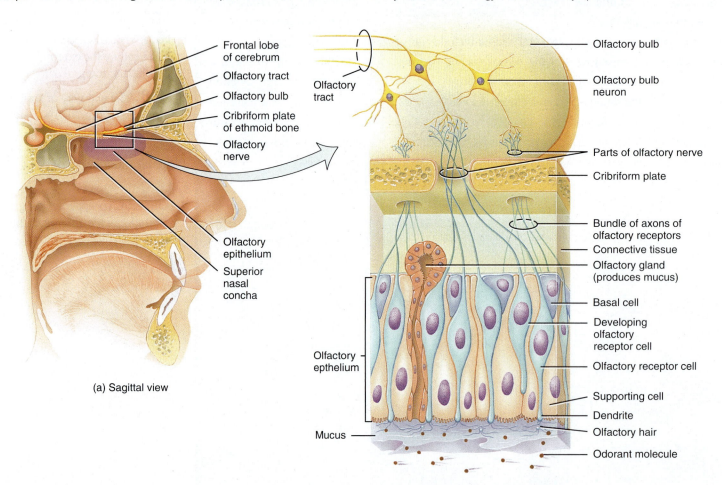

(a) Sagittal view

Frontal lobe of cerebrum
Olfactory tract
Olfactory bulb
Cribriform plate of ethmoid bone
Olfactory nerve
Olfactory epithelium
Superior nasal concha

Olfactory tract
Olfactory bulb
Olfactory bulb neuron
Parts of olfactory nerve
Cribriform plate
Bundle of axons of olfactory receptors
Connective tissue
Olfactory gland (produces mucus)
Basal cell
Developing olfactory receptor cell
Olfactory receptor cell
Supporting cell
Dendrite
Olfactory hair
Odorant molecule

Olfactory epthelium
Mucus

(b) Enlarged aspect of olfactory epithelium

Supporting cells are columnar epithelial cells of the mucous membrane lining the nose. They provide physical support, nourishment, and electrical insulation for the olfactory receptor cells, and they help detoxify chemicals that come in contact with the olfactory epithelium. **Basal cells** are stem cells that lie between the bases of the supporting cells and continually undergo cell division to produce new olfactory receptor cells, which live for only a month or so before being replaced. This process is remarkable considering that olfactory receptors are neurons, and, as you have already learned, mature neurons are generally not replaced.

Within the connective tissue that supports the olfactory epithelium are **olfactory glands**, which produce mucus that is carried to the surface of the epithelium by ducts. The secretion moistens the surface of the olfactory epithelium and dissolves odorants. Both supporting cells of the nasal epithelium and olfactory glands are innervated by branches of the facial nerve, which can be stimulated by certain chemicals. Impulses in these branches of the facial nerve in turn stimulate the lacrimal glands in the eyes and nasal mucous glands. The result is tears and a runny nose after inhaling substances such as pepper, onion, or the vapors of household ammonia.

Physiology of Olfaction

Many attempts have been made to distinguish among and classify "primary" sensations of smell. Genetic evidence now suggests the existence of hundreds of primary odors. Our ability to recognize about 10,000 different odors probably depends on patterns of activity in the brain that arise from activation of many different combinations of olfactory receptor cells.

Olfactory receptor cells react to odorant molecules in the same way that most sensory receptors react to their specific stimuli: A

Connective tissue
Basal cell
Olfactory gland
Olfactory receptor cell

Supporting cell
Olfactory hairs

LM 300x

(c) Histology of olfactory epithelium

The olfactory epithelium consists of olfactory receptor cells, supporting cells, and basal cells.

Figure 16.2 Olfactory transduction. Binding of the odorant molecule to the plasma protein receptor activates adenylate cyclase, resulting in the production of cAMP. Cyclic AMP opens sodium channels and Na$^+$ enters the olfactory receptor cell. The resulting depolarization may generate an action potential, which is transmitted to the axon terminals.

Odorants can produce depolarizing generator potentials, which can lead to action potentials.

generator potential develops and triggers one or more action potentials. How the generator potential arises in olfactory receptors is known in some cases. Many odorants bind to receptor proteins, called *odorant binding proteins*, on the plasma membrane and activate the enzyme adenylate cyclase (Figure 16.2). The result is the following chain of events: production of cyclic adenosine monophosphate (cAMP) → opening of sodium ion (Na$^+$) channels → inflow of Na$^+$ → generation of depolarizing generator potential → triggering of action potential → propagation of nerve impulses along axon of olfactory receptor cell.

Odor Thresholds and Adaptation

Olfaction, like all the special senses, has a low threshold. Only a few molecules of certain substances need be present in air to be perceived as an odor. A good example is the chemical methyl mercaptan, which has a foul odor and can be detected in concentrations as low as 1/25 billionth of a milligram per milliliter of air. Because the natural gas used for cooking and heating is odorless but lethal and potentially explosive if it accumulates, a small amount of methyl mercaptan is added to it to provide olfactory warning of natural gas leaks.

Adaptation (decreasing sensitivity) to odors occurs rapidly. Olfactory receptors adapt by about 50% in the first second or so after stimulation but adapt very slowly thereafter. Still, complete insensitivity to certain strong odors occurs in about a minute after exposure. Apparently, reduced sensitivity involves an adaptation process in the central nervous system as well.

The Olfactory Pathway

On each side of the nose, bundles of the slender, unmyelinated axons of olfactory receptor cells extend through about 20 olfactory foramina in the cribriform plate of the ethmoid bone (Figure 16.1b). These 40 or so bundles of axons collectively form the right and left **olfactory nerves**. The olfactory nerves terminate in the brain in paired masses of gray matter called the **olfactory bulbs**, which are located below the frontal lobes of the cerebrum. Within the olfactory bulbs, the axon terminals of olfactory receptor cells form synapses with the dendrites and cell bodies of olfactory bulb neurons.

Axons of olfactory bulb neurons extend posteriorly and form the **olfactory tract** (Figure 16.1b). These axons project to the primary olfactory area of the cerebral cortex, where conscious awareness of smells begins (Figure 16.3). Other axons of the olfactory tract project to the limbic system and hypothalamus; these connections account for our emotional and memory-evoked responses to odors. Examples include sexual excitement upon smelling a certain perfume, nausea upon smelling a food that once made you violently ill, or an odor-evoked memory of a childhood experience.

From the primary olfactory area, pathways also project to the frontal lobe. Odor identification and discrimination occur in the frontal lobe. People who suffer damage to this area have been reported to have difficulty identifying different odors.

Checkpoint

1. What is the function of the supporting cells of the nasal epithelium?
2. How do basal cells contribute to olfaction?
3. Olfactory receptor cells are unique among neurons because they can undergo what process?
4. Why must the olfactory epithelium have a coating of mucus?
5. How long would it take for your olfactory receptor cells to adapt to the smell of something very rotten?

Figure 16.3   The olfactory pathway.

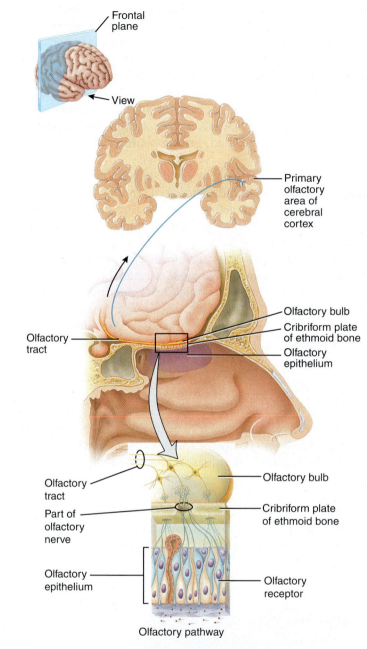

Olfactory pathway

🔑 From the olfactory epithelium, olfactory nerve impulses are propagated along the olfactory bulb and tract to the primary olfactory area of the cerebral cortex.

16.2 Impulses for taste propagate along the facial, glossopharyngeal, and vagus nerves to the brain.

Like olfaction, taste is a chemical sense; to be detected by either sense, stimulating molecules must be dissolved. Taste, or **gustation** (gus-TĀ-shun; *gust-* = taste), is much simpler than olfaction in that only five primary tastes can be distinguished: bitter, sour, salty, sweet, and umami (ū-MAM-ē). The umami taste, described by Japanese scientists, is detected as "meaty" or "savory." Umami is believed to arise from taste receptors that are stimulated by monosodium glutamate (MSG), a substance naturally present in many foods and added to others as a flavor enhancer. All other flavors, such as chocolate, pepper, and coffee, are combinations of the five primary tastes, plus accompanying olfactory and tactile (touch) sensations. Odors from food can pass upward from the mouth into the nasal cavity, where they stimulate olfactory receptor cells. Because olfaction is much more sensitive than taste, a given concentration of a food substance may stimulate the olfactory system thousands of times more strongly than it stimulates the gustatory system. When people with colds or allergies complain that they cannot taste their food, they are actually reporting blockage of olfaction, not of taste.

Anatomy of Taste Buds and Papillae

The receptors for sensations of taste are located in the taste buds (Figure 16.4). The nearly 10,000 taste buds of a young adult are mainly on the tongue, but they are also found on the soft palate (posterior portion of roof of mouth), pharynx (throat), and epiglottis (cartilage lid over voice box). The number of taste buds declines with age.

Figure 16.4 **The relationship of gustatory receptor cells in taste buds to tongue papillae.**

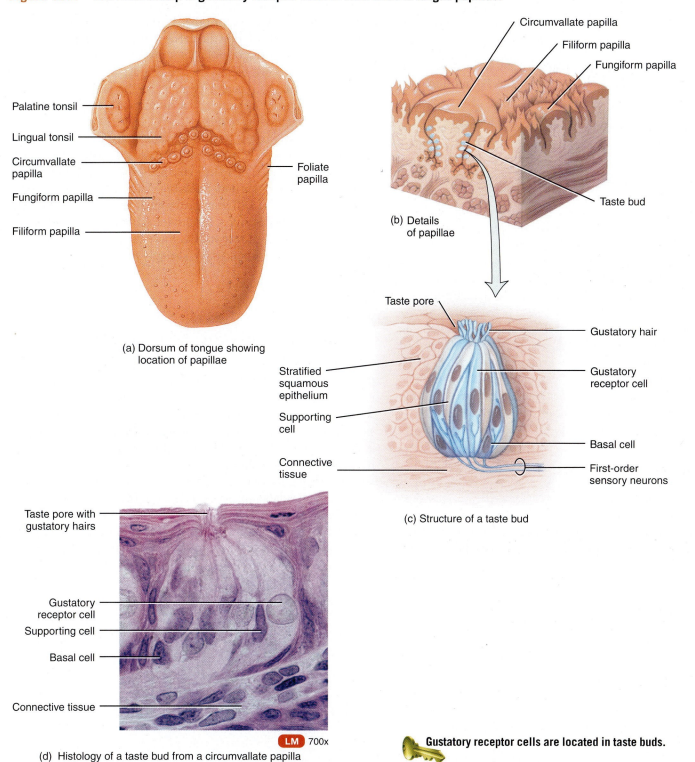

Palatine tonsil

Lingual tonsil

Circumvallate papilla

Fungiform papilla

Filiform papilla

Foliate papilla

(a) Dorsum of tongue showing location of papillae

Circumvallate papilla

Filiform papilla

Fungiform papilla

Taste bud

(b) Details of papillae

Taste pore

Gustatory hair

Stratified squamous epithelium

Supporting cell

Gustatory receptor cell

Connective tissue

Basal cell

First-order sensory neurons

(c) Structure of a taste bud

Taste pore with gustatory hairs

Gustatory receptor cell

Supporting cell

Basal cell

Connective tissue

LM 700x

(d) Histology of a taste bud from a circumvallate papilla

Gustatory receptor cells are located in taste buds.

Taste buds are found on the tongue in elevations called **papillae** (pa-PIL-ē), which provide a rough texture to the upper surface of the tongue (Figure 16.4a,b). Three types of papillae contain taste buds.

- About 12 very large, circular **circumvallate papillae** (ser-kum-VAL-āt = wall-like) form an inverted V-shaped row at the back of the tongue. Each of these papillae houses 100–300 taste buds.

- **Fungiform papillae** (FUN-ji-form = mushroomlike) are mushroom-shaped elevations scattered over the entire surface of the tongue that contain about five taste buds each.

- **Foliate papillae** (FŌ-lē-āt = leaflike) are located in small trenches on the lateral margins of the tongue, but most of their taste buds degenerate in early childhood.

In addition, the entire surface of the tongue has **filiform papillae** (FIL-i-form = threadlike). These pointed, threadlike structures contain tactile receptors but no taste buds. They increase friction between the tongue and food, making it easier for the tongue to move food in the oral cavity.

Each **taste bud** is an oval body consisting of three kinds of epithelial cells: supporting cells, gustatory receptor cells, and basal cells (Figure 16.4c,d). The **supporting cells** surround the **gustatory receptor cells**. A single, long microvillus, called a **gustatory hair**, projects from each gustatory receptor cell to the external surface through the **taste pore**, an opening in the taste bud. **Basal cells**, stem cells found at the periphery of the taste bud near the connective tissue layer, produce supporting cells, which then develop into gustatory receptor cells with a life span of only about 10 days. At their base, the gustatory receptor cells synapse with dendrites of the first-order neurons that form the first part of the gustatory pathway. Each first-order neuron has many dendrites, which receive input from many gustatory receptor cells in several taste buds.

Physiology of Gustation

Chemicals that stimulate gustatory receptor cells are known as **tastants**. Once a tastant is dissolved in saliva, it can make contact with the plasma membrane of the gustatory hairs, which are the sites on gustatory receptor cells that respond to tastants. The result is a receptor potential that stimulates release of neurotransmitter molecules from the gustatory receptor cell. In turn, the liberated neurotransmitter molecules trigger action potentials in the first-order neurons that synapse with gustatory receptor cells.

The receptor potential arises differently for different tastants. The sodium ions (Na^+) in a salty food enter gustatory receptor cells via Na^+ channels in the plasma membrane. The accumulation of Na^+ inside causes depolarization, which leads to release of neurotransmitter. The hydrogen ions (H^+) in sour tastants may flow into gustatory receptor cells via H^+ channels. They also influence opening and closing of other types of ion channels. Again, the result is a depolarization that leads to release of neurotransmitter. Other tastants, responsible for stimulating sweet, bitter, and umami tastes, do not themselves enter gustatory receptor cells. Rather, they bind to receptor proteins on the plasma membrane that activate second messengers, resulting in depolarization with the release of neurotransmitter.

If all tastants cause release of neurotransmitter from many gustatory receptor cells, why do foods taste different? The answer to this question is thought to lie in the patterns of nerve impulses in groups of first-order gustatory neurons that synapse with the gustatory receptor cells. Different tastes elicit activation of different groups of gustatory neurons. In addition, although each individual gustatory receptor cell responds to more than one of the five primary tastes, it may respond more strongly to some tastants than to others.

Taste Thresholds and Adaptation

The threshold for taste varies for each of the primary tastes. The threshold for bitter substances, such as quinine, is lowest. Because poisonous substances often are bitter, the low threshold (or high sensitivity) may have a protective function. The threshold for sour substances, such as lemon, as measured by using hydrochloric acid, is somewhat higher. The thresholds for salty substances, represented by sodium chloride, and for sweet substances, such as sucrose, are higher than those for bitter or sour substances.

Complete adaptation to a specific taste can occur in 1–5 minutes of continuous stimulation. Taste adaptation is due to changes that occur in the gustatory receptor cells, in olfactory receptor cells, and in neurons of the gustatory pathway in the CNS.

The Gustatory Pathway

Three cranial nerves contain axons of gustatory neurons from taste buds. The facial nerve serves the anterior two-thirds of the tongue, the glossopharyngeal nerve serves the posterior one-third of the tongue, and the vagus nerve serves the throat and epiglottis (Figure 16.5). From taste buds, nerve impulses propagate along these cranial nerves to the medulla oblongata. From the medulla, some axons carrying taste signals project to the limbic system and the hypothalamus; others project to the thalamus. Taste signals that project from the thalamus to the primary gustatory area in the parietal lobe of the cerebral cortex give rise to the conscious perception of taste.

Checkpoint

6. For each of the primary tastes, give an example of a food that strongly represents that taste.
7. Where on the tongue are each of the four types of papillae located?
8. What is the function of supporting cells in taste buds?
9. Which type of receptors are olfactory and gustatory receptor cells (see Concept 15.2)?
10. What is the sequence of events from the binding of a tastant molecule to a gustatory hair to the generation of an action potential in a first-order gustatory neuron?

Figure 16.5   **The gustatory pathway.**

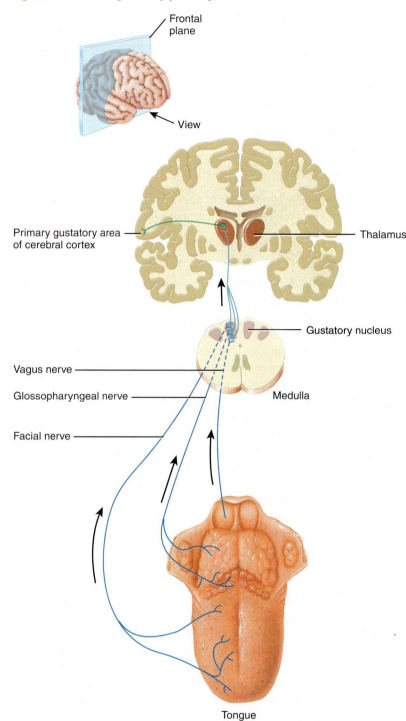

When gustatory nerve impulses arrive at the primary gustatory area of the cerebral cortex, we consciously perceive taste.

16.3   The eye is protected by eyelids, eyelashes, eyebrows, and a lacrimal apparatus.

More than half the sensory receptors in the human body are located in the eyes, and a large part of the cerebral cortex is devoted to processing visual information. We begin our exploration of the eyes by introducing the accessory structures that protect or assist their function.

Accessory Structures of the Eye

The **accessory structures** of the eye include the eyelids, eyelashes, eyebrows, lacrimal apparatus, and extrinsic eye muscles.

Eyelids

The upper and lower **eyelids**, or **palpebrae** (PAL-pe-brē), shade the eyes during sleep, protect the eyes from excessive light and foreign objects, and spread lubricating secretions over the eyeballs (Figure 16.6). The upper eyelid is more movable than the lower and contains in its superior region the **levator palpebrae superioris**, which raises the upper eyelid (Figure 16.7a). The space between the upper and lower eyelids that exposes the eyeball is the **palpebral fissure** (PAL-pe-bral). Its angles are known as the **lateral commissure** (KOM-i-shur), which is closer to the temporal bone, and the **medial commissure**, which is nearer the nasal bone. In the medial commissure is a small, reddish elevation, the **lacrimal caruncle** (KAR-ung-kul), which contains sebaceous (oil) glands and sudoriferous (sweat) glands. The whitish material that sometimes collects in the medial commissure comes from these glands.

From superficial to deep, each eyelid consists of skin overlying the orbicularis oculi, Meibomian glands, and conjunctiva (Figure 16.7a). The **orbicularis oculi** closes the eyelid. Embedded in each eyelid is a row of modified sebaceous glands, known as **Meibomian glands** (mī-BŌ-mō-an), which secrete a fluid that helps keep the eyelids from adhering to each other. The **conjunctiva** (kon´-junk-TĪ-va) is a thin, protective mucous membrane. The **palpebral conjunctiva** lines the inner aspect of each eyelid. The **bulbar conjunctiva** passes from the eyelids onto the anterior surface of the eyeball, where it covers the sclera (the "white" of the eye) but not the cornea, which is a transparent region that forms the outer anterior surface of the eyeball. Both the sclera and the cornea are discussed in more detail shortly. Dilation and congestion of the blood vessels of the bulbar conjunctiva due to local irritation or infection are the cause of **bloodshot eyes**.

Figure 16.6   Surface anatomy of the right eye.

🔑 **The eye is protected by the eyebrow, eyelashes, and eyelids.**

Eyelashes and Eyebrows

The **eyelashes** project from the border of each eyelid, and the **eyebrows** arch transversely above the upper eyelids. Eyelashes and eyebrows help protect the eyeballs from foreign objects, perspiration, and the direct rays of the sun.

The Lacrimal Apparatus

The **lacrimal apparatus** (*lacrim-* = tears) is a group of structures that produces and drains **lacrimal fluid**, or **tears**. The **lacrimal glands**, each about the size and shape of an almond, secrete lacrimal fluid, which drains into **lacrimal ducts** that empty tears onto the surface of the conjunctiva of the upper lid (Figure 16.7b). From there, the tears pass medially over the anterior surface of the eyeball to enter two small openings called **lacrimal puncta**. Tears then pass into two ducts, the **lacrimal canals**, which lead into the **lacrimal sac** and then into the **nasolacrimal duct**. This duct carries the lacrimal fluid into the nasal cavity.

Figure 16.7 **Accessory structures of the eye.**

Sagittal plane

Levator palpebrae superioris

Superior rectus

Orbicularis oculi

Eyebrow

Bulbar conjunctiva

Palpebral conjunctiva

Scleral venous sinus

Cornea

Pupil
Lens
Iris

Eyelashes

Optic nerve

Meibomian glands

Inferior oblique

Orbicularis oculi

Inferior rectus

(a) Sagittal section of eye and its accessory structures

FLOW OF TEARS

Lacrimal gland

Lacrimal duct

Superior lacrimal canal

Lacrimal punctum

Lacrimal sac

Inferior lacrimal canal

Nasolacrimal duct

Nasal cavity

Lacrimal gland
↓
Lacrimal ducts
↓
Superior or inferior lacrimal canal
↓
Lacrimal sac
↓
Nasolacrimal duct
↓
Nasal cavity

(b) Anterior view of the lacrimal apparatus

Accessory structures of the eye include the eyelids, eyelashes, eyebrows, the lacrimal apparatus, and extrinsic eye muscles.

Lacrimal fluid is a watery solution containing salts, some mucus, and **lysozyme**, a protective bactericidal enzyme. The fluid protects, cleans, lubricates, and moistens the eyeball. After being secreted, lacrimal fluid is spread medially over the surface of the eyeball by the blinking of the eyelids. Each gland produces about 1 mL of lacrimal fluid per day.

Normally, tears are cleared away as fast as they are produced, either by evaporation or by passing into the lacrimal canals and then into the nasal cavity. If an irritating substance makes contact with the conjunctiva, however, the lacrimal glands are stimulated to oversecrete, and tears accumulate (watery eyes) as the tears dilute and wash away the irritating substance. Watery eyes also occur when an inflammation of the nasal mucosa, such as occurs with a cold, obstructs the nasolacrimal ducts and blocks drainage of tears. Humans are unique in expressing emotions, both happiness and sadness, by **crying**. In response to parasympathetic stimulation, the lacrimal glands produce excessive lacrimal fluid that may spill over the edges of the eyelids and even fill the nasal cavity with fluid. Thus, crying sometimes produces a runny nose.

Extrinsic Eye Muscles

The extrinsic eye muscles extend from the orbit of the skull to the outer surface of the eyeball. Six extrinsic eye muscles move each eyeball: the **superior rectus**, **inferior rectus**, **lateral rectus**, **medial rectus**, **superior oblique**, and **inferior oblique** (see Figures 16.7a and 16.8). The motor units in these muscles are small and serve only two or three muscle fibers—fewer than in any other part of the body except the larynx (voice box). Such small motor units permit smooth, precise, and rapid movements of the eyes. As indicated in Figure 11.4b, the extrinsic eye muscles move the eyeball laterally, medially, superiorly, and inferiorly. For example, looking to the right requires simultaneous contraction of the right eye's lateral rectus and left eye's medial rectus and relaxation of the left eye's lateral rectus and right eye's medial rectus. The oblique muscles preserve rotational stability of the eyeball. Neural circuits in the brain stem and cerebellum coordinate and synchronize the movements of the eyes.

Checkpoint

11. Give the name and function of each muscle of the eyelids.
12. Why do eyelids of healthy eyes not stick together?
13. Which structure shown in Figure 16.7 is continuous with the inner lining of the eyelids?
14. What is lacrimal fluid, and what are its functions?
15. Why does your nose run when you cry?

16.4   The eye is constructed of three layers and two chambers.

The adult **eyeball** measures about 2.5 cm (1 in.) in diameter. Of its total surface area, only the anterior one-sixth is exposed; the remainder is recessed and protected by the orbit, into which it fits. Anatomically, the wall of the eyeball consists of three layers: fibrous tunic, vascular tunic, and retina.

Fibrous Tunic

The **fibrous tunic**, the superficial layer of the eyeball, is avascular and consists of the anterior cornea and posterior sclera (Figure 16.8). The **cornea** (KOR-nē-a) is a transparent coat that covers the colored iris. Because it is curved, the cornea helps focus light onto the retina. The cornea is composed primarily of dense connective tissue, but the many layers of collagen fibers are arranged to allow transmission of light. Since the central part of the cornea receives oxygen from the outside air, contact lenses that are worn for long periods of time must be permeable to permit oxygen to pass through them. The **sclera** (SKLE-ra; *scler-* = hard), the "white" of the eye, is composed of dense connective tissue. The sclera covers the entire eyeball except the cornea; it gives shape to the eyeball, makes it more rigid, protects its inner parts, and is an attachment site for extrinsic eye muscles. At the junction of the sclera and cornea is an opening known as the **scleral venous sinus** that drains a fluid called aqueous humor (see Figure 16.11).

Vascular Tunic

The **vascular tunic** is the middle layer of the eyeball and has three parts: choroid, ciliary body, and iris (Figure 16.8). The highly vascularized **choroid** (KŌ-royd), which is the posterior portion of the vascular tunic, lines most of the internal surface of the sclera. It provides nutrients to the retina and, because of its dark pigmentation, absorbs stray light rays.

In the anterior portion of the vascular tunic, the choroid is continuous with the **ciliary body** (SIL-ē-ar´-ē). It extends from the **ora serrata** (Ō-ra ser-RĀ-ta), the jagged anterior margin of the retina, to a point just posterior to the junction of the sclera and cornea. The ciliary body consists of the ciliary processes and the ciliary muscle. The **ciliary processes** are protrusions or folds on the internal surface of the ciliary body. They contain capillaries that secrete aqueous humor. Extending from the ciliary process are **suspensory ligaments** that attach to the lens. The **ciliary muscle** is a circular band of smooth muscle that alters the shape of the lens, adapting it for near or far vision.

The **iris**, the colored portion of the eyeball, has a flat, disc shape with a hole in the center, the pupil. The iris is suspended between the cornea and the lens and is attached at its outer margin to the ciliary processes. It consists of melanin-producing melanocytes and circular and radial smooth muscle fibers that regulate the diameter of the pupil. The amount of melanin in the iris

TABLE 16.1

The Structures of the Eyeball

STRUCTURE	FUNCTION	STRUCTURE	FUNCTION
Fibrous tunic Cornea / Sclera	*Cornea:* Admits and refracts (bends) light. *Sclera:* Provides shape and protects inner parts.	**Lens** Lens	Focuses light on the retina.
Vascular tunic Iris / Ciliary body / Choroid	*Iris:* Regulates amount of light that enters eyeball. *Ciliary body:* Secretes aqueous humor and alters the shape of the lens for near or far vision (accommodation). *Choroid:* Provides blood supply and absorbs scattered light.	**Anterior cavity** Anterior cavity	Contains aqueous humor that helps maintain the shape of the eyeball and supplies oxygen and nutrients to the lens and the cornea.
Retina Retina	Receives light and converts it into nerve impulses. Provides output to the brain via axons of ganglion cells, which form the optic nerve.	**Vitreous chamber** Vitreous chamber	Contains the vitreous body that helps maintain the shape of the eyeball and keeps the retina attached to the choroid.

The pressure in the eye, called **intraocular pressure**, is produced mainly by the aqueous humor and partly by the vitreous body; normally it is about 16 mm Hg (millimeters of mercury). The intraocular pressure maintains the shape of the eyeball and prevents it from collapsing.

Table 16.1 summarizes the structures associated with the eyeball.

Checkpoint

16. What are the components of the fibrous tunic and vascular tunic?
17. What produces differences in pupil color?
18. How does melanin assist vision?
19. Which part of the retina produces the sharpest vision when light falls on it?
20. What is the function of the aqueous humor? The vitreous body?
21. What separates the anterior and posterior chambers of the eyeball? The anterior cavity from the vitreous chamber?

As Dan swung in behind the parked fire engine, he noticed that the putrid metallic chemical odor seemed less intense, even though he was closer to the fire. Dan quickly put on his gear, then grabbed his breathing apparatus as he moved toward the building. As he entered the building, Dan felt resistance as he tried to inhale; probably a sticky valve on his respirator. He stopped, took off the mask, and quickly checked it again. His nose began to run and his eyes began to tear from the acrid smoke. Just as he went to put his mask back on, he smelled the foul odor of a natural gas leak. He fumbled with the straps on the head gear, trying to get his respirator back on. Then a large bright flash blinded him. Dan felt a moment of pain, then nothing.

Dan woke with a start. He felt dizzy and couldn't see. Most disturbing of all, he couldn't hear anything. He felt something in his throat, a hard plastic tube. His mouth was dry. He couldn't taste anything, but he thought he smelled a faint odor of hospital antiseptic cleaner.

Dan's wife Pam grasped his arm. He reached up and felt the bandages on his eyes.

"Someone come quickly, please! He's waking up!" shouted Pam.

The nurse came in and checked Dan's IV line and the tubes that seemed to be sticking out of every orifice of his body. "I'll call the doctor," she said as she left the room.

A moment later Dr. Burke entered the room.

"Dr. Burke, can he hear us?" asked Pam.

"Well, Dan was lucky. His equipment spared his face and body from the shrapnel of the blast. He may have some temporary blindness from the flash of the explosion. It's hard to say with his hearing, though. Both his tympanic membranes were ruptured. But I'm most concerned about the internal injuries to his lungs and some of his internal organs. We'll keep a close watch on him."

A few days later, Dr. Burke returned with some good news. "Hi, Dan. How are you doing?" he asked. "We're going to take the bandages off today and check your eyes."

"Well, my hearing's better today, but thankfully, I can't taste the food," joked Dan in a raspy voice. As the bandages came off, Dan felt a rush of relief. He could see—not well, but he could see! His eyes felt sore, puffy, and watery.

"Dan, how many fingers am I holding up? Two? Good. Now follow my finger with your eyes, that's good. How's the hearing? Are you still feeling dizzy?" Dan was only half listening to the doctor's questions as he scanned the room.

Later, Dan sat in his hospital room trying to watch the local news. He had been in the hospital for over a week now and was feeling better in spite of the constant pain. His internal injuries hadn't been serious. The television appeared blurry, and he had trouble focusing on the screen. Dan rubbed his eyes and looked up. He saw what looked like a curtain falling over the television. He blinked and rubbed his eyes again, trying to clear his vision, but it was still there. He reached for the call button to summon the nurse.

A. Dan has suffered a blast injury. Although rare, these are devastating events involving injury to multiple organ systems. The odor of which chemical was detected by Dan prior to the explosion? Why did the foul odor cause a strong sense of fear in Dan?

B. If Dan has damage to his olfactory receptor cells, can these receptor cells be replaced? Will he be able to taste normally? Explain your answers.

C. If Dan's olfactory epithelia are damaged, will he be able to taste normally?

D. Dan's eyes were protected during the explosion, so it's not likely that debris has damaged them. What might be causing his blurred vision? Why might his eyes hurt? Which membranes are inflamed?

E. Dan is experiencing a rapid onset of visual loss. Which structures could be involved in blocking the path of light to his retina? Is it likely that such a rapid loss of vision is due to blockage of light entering the eye, or is something wrong with the retina?

16.5 Image formation involves refraction of light rays, change in lens shape, and constriction of the pupil.

In some ways the eye is like a camera: Its optical elements focus an image of some object on a light-sensitive "film"—the retina—while ensuring the correct amount of light to make the proper "exposure." To understand how the eye forms clear images of objects on the retina, we must examine three processes: (1) the refraction, or bending, of light by the lens and cornea; (2) accommodation, the change in shape of the lens; and (3) constriction, or narrowing, of the pupil.

Refraction of Light Rays

When light rays traveling through a transparent substance (such as air) pass into a second transparent substance with a different density (such as water), they bend at the junction between the two substances. This bending is called **refraction** (Figure 16.12a). As light rays enter the eye, they are refracted at the anterior and posterior surfaces of the cornea. Both surfaces of the lens of the eye further refract the light rays so they come into exact focus on the retina.

Images focused on the retina are inverted, or upside-down (Figure 16.12b,c). They also undergo right-to-left reversal; that is, light from the right side of an object strikes the left side of the retina, and vice versa. The reason the world does not look inverted and reversed is that the brain "learns" early in life to coordinate visual images with the orientations of objects. The brain stores the inverted and reversed images we acquired when we first reached for and touched objects and interprets those visual images as being correctly oriented in space.

About 75% of the total refraction of light occurs at the cornea. The lens provides the remaining 25% of focusing power and also changes the focus to view near or distant objects. When an object is 6 m (20 ft) or more away from the viewer, the light rays reflected from the object are nearly parallel to one another (Figure 16.12b). These parallel rays must be bent enough to fall exactly focused on the central fovea, where vision is sharpest. Because light rays that are reflected from objects closer than 6 m (20 ft) are divergent rather than parallel (Figure 16.12c), the rays must be refracted more if they are to be focused on the retina. This additional refraction is accomplished in the process called accommodation.

Accommodation and the Near Point of Vision

A surface that curves outward, like the surface of a ball, is said to be *convex*. When the surface of a lens is convex, that lens will refract incoming light rays toward each other, so that they eventually intersect. If the surface of a lens curves inward, like the inside of a hollow ball, the lens is said to be *concave* and causes light rays to refract away from one another. The lens of the eye is convex on both its anterior and posterior surfaces, and its focusing power increases as its curvature becomes greater. When the eye is focusing on a close object, the lens becomes more curved and refracts the light rays more. This increase in the curvature of the lens for near vision is called **accommodation** (Figure 16.12c). The **near point of vision** is the minimum distance from the eye that an object can be clearly focused with maximum accommodation. This distance is about 10 cm (4 in.) in a young adult.

How does accommodation occur? When you are viewing distant objects, the ciliary muscle is relaxed and the lens is flatter be-

Figure 16.12 Refraction of light rays. (a) Refraction is the bending of light rays at the junction of two transparent substances with different densities. (b) The cornea and lens refract light rays from distant objects so the image is focused on the retina. (c) In accommodation, the lens becomes more spherical, which increases the refraction of light.

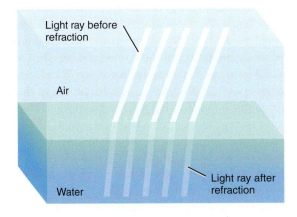

(a) Refraction of light rays

(b) Viewing distant object

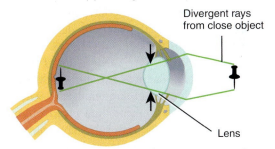

(c) Accommodation

🔑 **Images focused on the retina are inverted and left-to-right reversed.**

cause it is stretched in all directions by taut suspensory ligaments (Figure 16.12b). When you view a close object, the ciliary muscle contracts. As the muscle contracts, the ciliary muscle is pulled toward the lens, which pulls the ciliary process forward toward the lens (Figure 16.12c). This action releases tension on the lens and suspensory ligaments. Because it is elastic, the lens becomes more spherical (more convex), which increases its focusing power and causes greater convergence of the light rays.

Refraction Abnormalities

The normal eye, known as an **emmetropic eye** (em´-e-TROP-ik), can sufficiently refract light rays from an object 6 m (20 ft) away so that a clear image is focused on the retina (Figure 16.13a). Many people, however, lack this ability because of refraction abnormalities. Among these abnormalities are **myopia** (mī-Ō-pē-a), or nearsightedness, in which the eye is able to focus on near objects but not distant objects. Myopia occurs when the eyeball is too long relative to the focusing power of the cornea and lens, or when the lens is thicker than normal, so an image converges in front of the retina. In **hyperopia** (hī-per-Ō-pē-a), or farsightedness, the eye is able to focus on distant objects but not nearby objects. Hyperopia occurs because the eyeball length is short relative to the focusing power of the cornea and lens, or the lens is thinner than normal, so an image converges behind the retina. Figure 16.13 illustrates these conditions and explains how they are corrected. Another refraction abnormality is **astigmatism** (a-STIG-ma-tizm), in which either the cornea or the lens has an irregular curvature. As a result, parts of the image are out of focus, blurring or distorting vision.

Most errors of vision can be corrected by eyeglasses, contact lenses, or surgical procedures. A contact lens floats on a film of tears over the cornea. The anterior outer surface of the contact lens corrects the visual defect, and its posterior surface matches the curvature of the cornea. LASIK involves reshaping the cornea to permanently correct refraction abnormalities.

Constriction of the Pupil

The circular muscle fibers of the iris also have a role in the formation of clear retinal images. Part of the accommodation mechanism consists of the contraction of the circular muscles of the iris to constrict the pupil (see Figure 16.9). **Constriction of the pupil** is a narrowing of the diameter of the hole through which light enters the eye. This autonomic reflex occurs simultaneously with accommodation for near objects and prevents light rays from entering the eye through the periphery of the lens. Light rays entering at the periphery would not be brought to focus on the retina and would result in blurred vision. The pupil, as noted earlier, also constricts in bright light.

Convergence

Because of the position of their eyes in their heads, many animals see one set of objects off to the left through one eye and an entirely different set of objects off to the right through the other. In humans, both eyes focus on only one set of objects—a characteristic called **binocular vision**. This feature of our visual system allows the perception of depth and an appreciation of the three-dimensional nature of objects. Binocular vision occurs when light

Figure 16.13   Refraction abnormalities in the eyeball and their correction. (a) Normal (emmetropic) eye. (b) In the nearsighted or myopic eye, the image is focused in front of the retina. The condition may result from an elongated eyeball or thickened lens. (c) Myopia can be corrected by use of a concave lens that diverges entering light rays so that they come into focus directly on the retina. (d) In the farsighted or hyperopic eye, the image is focused behind the retina. The condition results from a shortened eyeball or a thin lens. (e) Hyperopia can be corrected by a convex lens that converges entering light rays so that they focus directly on the retina.

(a) Normal (emmetropic) eye

(b) Nearsighted (myopic) eye, uncorrected   (c) Nearsighted (myopic) eye, corrected

(d) Farsighted (hyperopic) eye, uncorrected   (e) Farsighted (hyperopic) eye, corrected

🔑 **In myopia (nearsightedness), distant objects can't be seen clearly; in hyperopia (farsightedness), nearby objects can't be seen clearly.**

rays from an object strike corresponding points on the two retinas. When we stare straight ahead at a distant object, the incoming light rays are aimed directly at both pupils and are refracted to comparable spots on the retinas of both eyes. As we move closer to an object, however, the eyes must rotate medially if the light rays from the object are to strike the same points on both retinas. The term **convergence** refers to this medial movement of the two eyeballs so that both are directed toward the object being viewed. The nearer the object, the greater the degree of convergence needed to maintain binocular vision. The coordinated action of the extrinsic eye muscles brings about convergence.

Checkpoint

22. What is refraction? Which components of the eye are primarily responsible for refracting light?

23. If you look at the horizon to determine where you are, then look down to read a map, what process must your eyes accomplish to keep your vision focused?

24. Which sequence of events occurs when you look at a distant object? A close object?

25. Which refraction abnormality do you likely have if both near and far objects are out of focus?

26. What is convergence? Why is it important for human vision?

16.6 The neural pathway of light is photoreceptors → bipolar cells → ganglion cells → optic nerve → primary visual cortex.

Photoreceptors and Photopigments

Rods and cones were named for differences in the appearance of the *outer segment*—the distal end of the photoreceptor next to the pigmented layer. The outer segments of rods are cylindrical or rod-shaped; those of cones are tapered or cone-shaped (Figure 16.14). Transduction of light energy into a receptor potential occurs in the outer segment. The photopigments are proteins in the plasma membrane of the outer segment. In cones the plasma membrane folds back and forth in a pleated fashion, but in rods the pleats pinch off from the plasma membrane to form discs. The outer segment of each rod contains a stack of about 1000 discs, piled up like coins inside a wrapper. Photoreceptor outer segments are renewed at an astonishingly rapid pace. In rods, one to three new discs are added to the base of the outer segment every hour while old discs slough off at the tip and are phagocytized by epithelial cells of the pigmented layer of the retina.

The *inner segment* of a rod or cone contains the cell nucleus, Golgi complex, and many mitochondria. At its proximal end, the photoreceptor expands into bulblike synaptic terminals filled with synaptic vesicles.

The first step in visual transduction is absorption of light by a **photopigment**, a colored protein that undergoes structural changes when it absorbs light, in the outer segment of a photoreceptor. Light absorption initiates the events that lead to the production of a receptor potential. The photopigment in rods is **rhodopsin** (*rhod-* = rose; *-opsin* = related to vision). Three different **cone photopigments** are present in the retina, one in each of three types of cones. Color vision results from different colors of light selectively activating the different cone photopigments.

Photopigments contain two parts: a glycoprotein known as **opsin** and a derivative of vitamin A called **retinal**. Vitamin A derivatives are formed from carotenoids, the plant pigments that give carrots their orange color. Good vision depends on adequate dietary intake of carotenoid-rich vegetables such as carrots, spinach, broccoli, and yellow squash, or other foods that contain vitamin A, such as liver. Retinal is the light-absorbing part of photopigments. In the human retina, there are four different opsins, one for each cone photopigment and another for rhodopsin. Small molecular variations of the different opsins permit the rods and cones to absorb different colors (wavelengths) of incoming light.

Figure 16.14 Structure of rod and cone photoreceptors. The inner segments contain the metabolic machinery for synthesis of photopigments. The photopigments are embedded in the membrane discs or folds of the outer segments. New discs in rods and new folds in cones form at the junction of the outer and inner segments. Epithelial cells of the pigmented retinal layer phagocytize the old discs and folds that slough off the distal tip of the outer segments.

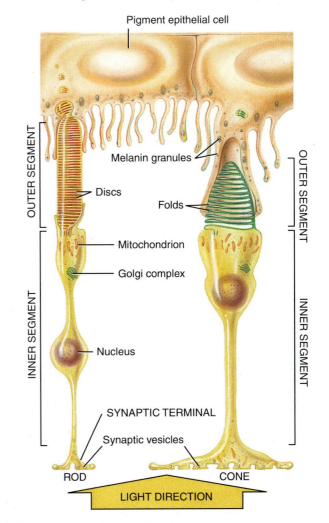

Transduction of light energy into receptor potentials occurs in the outer segments of rods and cones.

Photopigments respond to light in the following cyclical process (Figure 16.15):

1 In darkness, retinal has a bent shape, called *cis*-retinal, which fits snugly into the opsin portion of the photopigment. When *cis*-retinal absorbs a photon of light, it straightens out to a shape called *trans*-retinal. This *cis*-to-*trans* conversion is called **isomerization** and is the first step in visual transduction. After retinal isomerizes, several chemical changes occur that generate a receptor potential.

2 In about a minute, *trans*-retinal completely separates from opsin. The final products look colorless, so this part of the cycle is termed **bleaching** of photopigment.

3 An enzyme called **retinal isomerase** converts *trans*-retinal back to *cis*-retinal.

4 The *cis*-retinal binds to opsin, reforming a functional photopigment. This part of the cycle—resynthesis of a photopigment—is called **regeneration**.

Figure 16.15   The cyclical bleaching and regeneration of photopigment. Yellow section of cycle indicates light exposure. Blue arrows indicate bleaching steps; black arrows indicate regeneration steps.

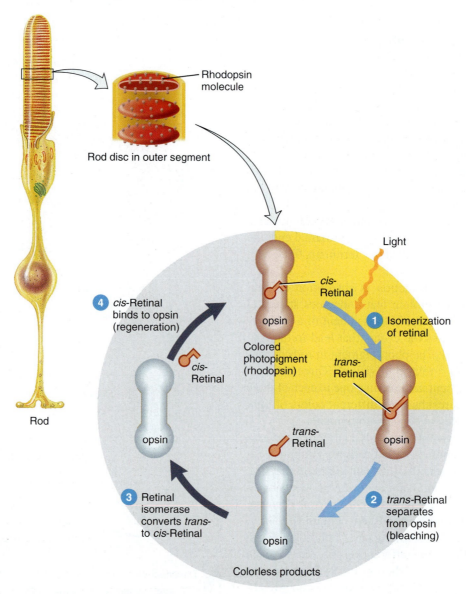

Rhodopsin molecule

Rod disc in outer segment

Rod

Light

cis-Retinal

4 *cis*-Retinal binds to opsin (regeneration)

cis-Retinal

Colored photopigment (rhodopsin)

opsin

1 Isomerization of retinal

trans-Retinal

opsin

opsin

trans-Retinal

3 Retinal isomerase converts *trans*- to *cis*-Retinal

trans-Retinal

opsin

Colorless products

2 *trans*-Retinal separates from opsin (bleaching)

Retinal, a derivative of vitamin A, is the light-absorbing component of photopigments.

Cone photopigments regenerate much more quickly than rhodopsin. After complete bleaching, regeneration of half of the rhodopsin takes 5 minutes; half of the cone photopigments regenerate in only 90 seconds. Full regeneration of bleached rhodopsin takes 30 to 40 minutes. Bleaching of photopigment occurs when you continue to see an image after a camera flashbulb has flashed at you. When your retinas are exposed to a bright light, the photopigments can't respond to further light stimulation until regeneration has occurred. Until the photoreceptors recover, your retinas respond to an afterimage, a "ghost" image of the bright light.

Light and Dark Adaptation

When you emerge from dark surroundings (say, a tunnel) into the sunshine, **light adaptation** occurs—your visual system adjusts in seconds to the brighter environment by decreasing its sensitivity. On the other hand, when you enter a darkened room such as a movie theater, your visual system undergoes **dark adaptation**—its sensitivity increases slowly over many minutes.

Bleaching and regeneration of the photopigments account for some (but not all) of the sensitivity changes during light and dark adaptation. As the light level increases, more and more photopigment is bleached, assisting light adaptation. As the light level decreases, more and more photopigment regenerates to assist your adaptation to darkness. In complete darkness, full regeneration of cone photopigments occurs during the first 8 minutes of dark adaptation. Rhodopsin regenerates more slowly, and our visual sensitivity increases until

even a single photon (the smallest unit of light) can be detected. In that situation, barely perceptible light appears gray-white, regardless of its color. At very low light levels, such as starlight, objects appear as shades of gray because only the rods are functioning.

Release of Neurotransmitter by Photoreceptors

The absorption of light and isomerization of retinal initiates chemical changes in the photoreceptor that lead to production of a receptor potential. To understand how the receptor potential arises, however, we first need to examine the operation of photoreceptors in the *absence* of light. In darkness, sodium ions (Na^+) flow into photoreceptor outer segments through Na^+ channels that are held open by **cyclic GMP (guanosine monophosphate)** or **cGMP** (Figure 16.16a). The inflow of Na^+ partially depolarizes the photoreceptor. As a result, in darkness the membrane potential of a photoreceptor is about -30 mV. This is much closer to zero than a typical neuron's resting membrane potential of -70 mV. The depolarization during darkness triggers continual release of neurotransmitter at the synaptic terminals. The neurotransmitter in rods, and perhaps in cones, is the amino acid glutamate. At synapses between rods and bipolar cells, glutamate is an inhibitory neurotransmitter. Glutamate triggers inhibitory postsynaptic potentials (IPSPs) that hyperpolarize the bipolar cells and prevent them from generating and transmitting receptor potentials to the ganglion cells.

Figure 16.16 **Operation of rod photoreceptors.**

(a) In darkness

cGMP-gated Na⁺ channels open
↓
Inflow of Na⁺
↓
Membrane potential of −30 mV
↓
Glutamate released at synaptic terminals inhibits bipolar cell

cGMP
Na⁺
Glutamate

(b) In light

Activation of enzyme that breaks down cGMP
↓
cGMP-gated Na⁺ channels close
↓
Inflow of Na⁺ decreases
↓
Hyperpolarizing receptor potential
↓
Glutamate release turned off, which excites bipolar cell

Light causes a hyperpolarizing receptor potential in photoreceptors, which decreases release of an inhibitory neurotransmitter (glutamate).

When light strikes the retina, enzymes are activated that break down cGMP. As a result, some cGMP-gated Na^+ channels close, Na^+ inflow decreases, and the membrane potential becomes more negative, approaching -70 mV (Figure 16.16b). This sequence of events produces a hyperpolarizing receptor potential in the photoreceptor that decreases the release of glutamate. Dim lights cause small and brief receptor potentials that partially turn off glutamate release; brighter lights elicit larger and longer receptor potentials that more completely shut down neurotransmitter release. As less inhibitory neurotransmitter is released by the photoreceptors, the bipolar cells generate more receptor potentials. Thus, light excites the bipolar cells that synapse with rods by turning off the release of an inhibitory neurotransmitter.

The Visual Pathway

As bipolar cells transmit receptor potentials to ganglion cells, the ganglion cells generate action potentials. After considerable processing of visual signals in the retina at synapses among the various types of neurons (see Figure 16.10 and related discussion), the axons of the ganglion cells provide output from the retina to the brain. They exit the eyeball as the **optic nerve**.

Processing of Visual Input in the Retina

Light stimulation of rods and cones induces receptor potentials in bipolar cells. Between 6 and 600 rods synapse with a single bipolar cell in the outer synaptic layer. A cone more often synapses with just one bipolar cell. The convergence of many rods onto a single bipolar cell increases the light sensitivity of rod vision but slightly blurs the image that is perceived. Cone vision, although less sensitive, has higher acuity because of the one-to-one synapses between cones and their bipolar cells.

Brain Pathway and Visual Fields

The axons of the optic nerve pass through the **optic chiasm** (KĪ-a-zm = a crossover, as in the letter X), a crossing point of the optic nerves (Figure 16.17a,b). Some axons cross to the opposite side, and others remain uncrossed. After passing through the optic chiasm, the axons, now part of the **optic tract**, enter the brain and terminate in the thalamus. Here they synapse with neurons having axons that form the **optic radiations**. The optic radiations project to the primary visual areas in the occipital lobes of the cerebral cortex (see Figure 13.11).

Everything that can be seen by one eye is that eye's **visual field**. Because our eyes are located anteriorly in the head, the visual fields of the two eyes overlap considerably (Figure 16.17b). We have binocular vision due to the large region where the visual fields of the two eyes overlap. The visual field of each eye is divided into two regions: the **nasal half** and the temporal half (Figure 16.17c,d). For each eye, light rays from an object in the nasal half of the visual field fall on the temporal half of the retina, and light rays from an object in the temporal half of the visual field fall on the nasal half of the retina. To further complicate matters, visual information from the *right* half of each visual field is conveyed to the *left* side of the brain, and visual information from the *left* half of each visual field is conveyed to the *right* side of the brain, as follows (Figure 16.17c,d):

1. The axons of ganglion cells in one eye exit the eyeball at the optic disc and form the optic nerve on that side.

2. At the optic chiasm, axons from the temporal half of each retina do not cross but continue directly to the thalamus on the same side.

3. In contrast, axons from the nasal half of each retina cross and continue to the opposite side of the thalamus.

4. Each optic tract consists of crossed and uncrossed axons that project from the optic chiasm to the thalamus on one side.

5. Axon collaterals (branches) of the ganglion cells project to the midbrain, where they participate in neural circuits that govern constriction of the pupils in response to light and coordination of head and eye movements. Collaterals also extend to the hypothalamus, which establishes patterns of sleep and other activities that occur on a circadian or daily schedule in response to intervals of light and darkness.

6. The axons of thalamic neurons form the optic radiations as they project from the thalamus to the primary visual area of the cerebral cortex on the same side.

Checkpoint

27. What are the functional similarities between rods and cones?
28. What is the conversion of *cis*-retinal to *trans*-retinal called? What is the conversion of *trans*-retinal to *cis*-retinal called?
29. How do photopigments respond to light and recover in darkness?
30. Why does a decreased release of glutamate by photoreceptors generate a receptor potential in bipolar cells?
31. What is the composition of the optic nerve?
32. Light rays from an object in the temporal half of the visual field strike which half of the retina?

Figure 16.17 **The visual pathway.** (a) Partial dissection of the brain reveals the optic radiations (axons extending from the thalamus to the occipital lobe). (b) An object in the binocular visual field can be seen with both eyes. In (c) and (d), note that information from the right side of the visual field of each eye projects to the left side of the brain, and information from the left side of the visual field of each eye projects to the right side of the brain.

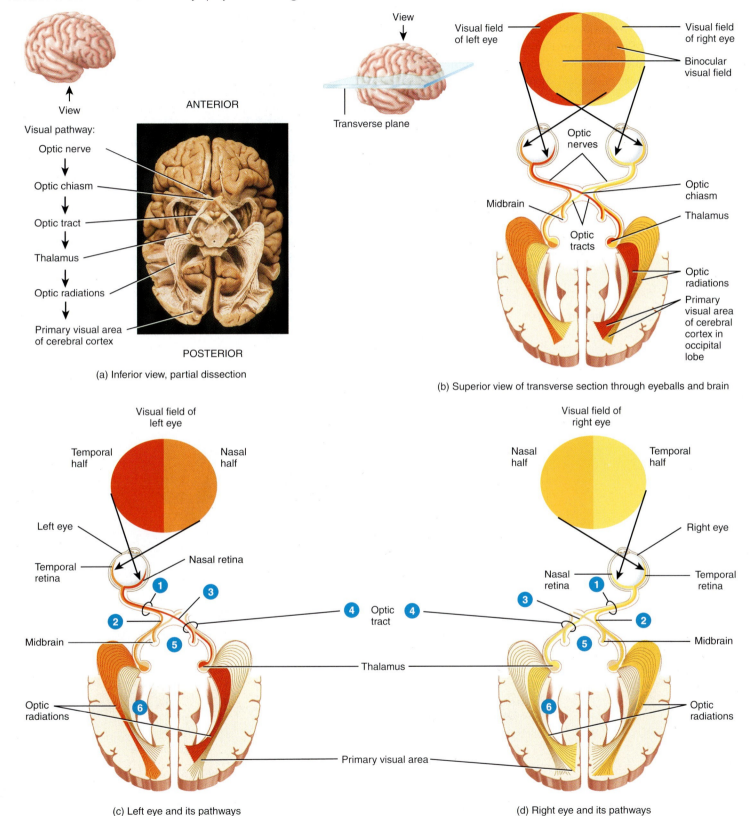

(a) Inferior view, partial dissection

(b) Superior view of transverse section through eyeballs and brain

(c) Left eye and its pathways

(d) Right eye and its pathways

🔑 The axons of ganglion cells in the temporal half of each retina extend to the thalamus on the same side; the axons of ganglion cells in the nasal half of each retina extend to the thalamus on the opposite side.

16.7 The three main regions of the ear are the external, middle, and internal ear.

The ear is an engineering marvel because its sensory receptors can transduce sound vibrations with amplitudes as small as the diameter of an atom of gold (0.3 nm) into electrical signals 1000 times faster than photoreceptors can respond to light. Besides receptors for sound waves, the ear also contains receptors for equilibrium.

Anatomy of the Ear

The ear is divided into three main regions: the external ear, which collects sound waves and channels them inward; the middle ear, which conveys sound vibrations to the oval window; and the internal ear, which houses the receptors for hearing and equilibrium (Figure 16.18).

External (Outer) Ear

The **external (outer) ear** consists of the auricle, external auditory canal, and tympanic membrane. The **auricle** is a flap of elastic cartilage shaped like the flared end of a trumpet and covered by skin. The **external auditory canal** (*audit-* = hearing) is a curved tube about 2.5 cm (1 in.) long that lies in the temporal bone and leads from the auricle to the tympanic membrane. The **tympanic membrane** (tim-PAN-ik; *tympan-* = a drum), or **eardrum**, separates the external auditory canal from the middle ear. The tympanic membrane is a thin, semitransparent connective tissue membrane covered by skin on the side facing the external auditory canal and mucus membrane on the internal surface. Tearing of the tympanic membrane is called a **perforated eardrum**. It may be due to pressure from a cotton swab, trauma, or a middle ear infection. A perforated eardrum usually heals within a month.

Figure 16.18   Anatomy of the ear.

- ■ External ear
- ■ Middle ear
- ■ Internal ear

Frontal section through the right side of the skull showing the three principal regions of the ear

🔑 The external ear collects sound waves, which are transmitted through the middle ear to the internal ear, where the receptors for hearing and equilibrium are located.

Near the exterior opening, the external auditory canal contains a few hairs and specialized sebaceous (oil) glands called **ceruminous glands** (se-ROO-mi-nus) that secrete earwax or **cerumen** (se-ROO-men). The combination of hairs and cerumen helps prevent dust and foreign objects from entering the ear. Cerumen also protects the delicate skin of the external ear canal from water and insects. Cerumen usually dries up and falls out of the ear canal. Some people, however, produce a large amount of cerumen, which can muffle incoming sounds. The treatment for *impacted cerumen* is usually periodic ear irrigation or removal of wax by trained medical personnel.

Middle Ear

The **middle ear** is a small, air-filled cavity in the petrous portion of the temporal bone (Figure 16.19). It is separated from the internal ear by a thin bony partition that contains two small membrane-covered openings: the oval window and the round window. Extending across the middle ear and attached to it by ligaments are the three smallest bones in the body, the **auditory ossicles** (OS-si-

kuls). The bones, named for their shapes, are the malleus, incus, and stapes—commonly called the hammer, anvil, and stirrup, respectively. The auditory ossicles are connected to one another by synovial joints. The "handle" of the **malleus** (MAL-ē-us) attaches to the internal surface of the tympanic membrane. The head of the malleus articulates with the body of the incus. The **incus** (ING-kus), the middle bone in the series, articulates with the head of the stapes. The base or footplate of the **stapes** (STĀ-pēz) fits into the **oval window**. Directly below the oval window is another membrane-covered opening, the **round window**.

Besides the ligaments, two tiny skeletal muscles also attach to the auditory ossicles. The **tensor tympani** (TIM-pan-ē) limits movement and increases tension on the tympanic membrane to prevent damage to the inner ear from loud noises. The **stapedius** (sta-PĒ-de-us) is the smallest of all skeletal muscles. By dampening large vibrations of the stapes due to loud noises, it protects the oval window. Because it takes a fraction of a second for the tensor tympani and stapedius muscles to contract, they help protect the inner ear from prolonged loud noises, but not from sudden, brief ones such as a gunshot or explosion.

Figure 16.19 **The right middle ear containing the auditory ossicles.**

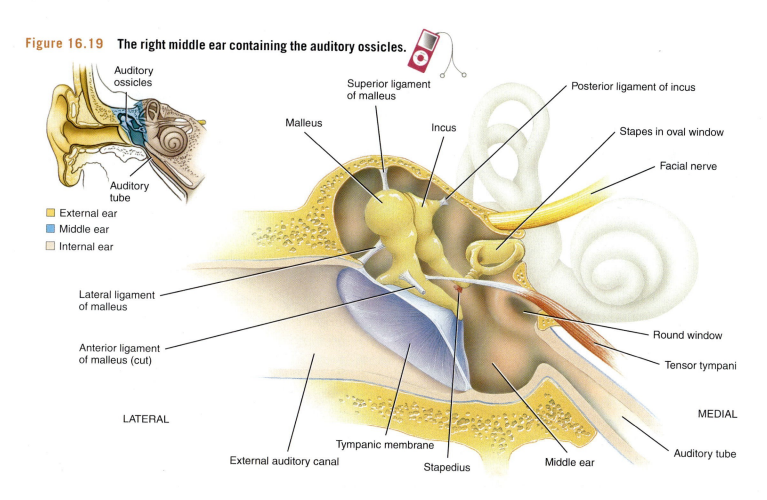

Frontal section showing location of auditory ossicles

The malleus, incus, and stapes are attached to one another by synovial joints and suspended in the middle ear by ligaments.

The anterior wall of the middle ear contains the opening of the **auditory tube**, commonly known as the **eustachian tube**. The auditory tube connects the middle ear with the nasopharynx (superior portion of the throat). It is normally closed at its medial (pharyngeal) end. During swallowing and yawning, it opens, allowing air to enter or leave the middle ear until the pressure in the middle ear equals the atmospheric pressure. Most of us have experienced our ears popping as the pressures equalize. When the pressures are balanced, the tympanic membrane vibrates freely as sound waves strike it. If the pressure is not equalized, intense pain, hearing impairment, ringing in the ears, and vertigo could develop. The auditory tube also is a route by which pathogens may travel from the nose and throat to the middle ear, causing an ear infection.

Internal (Inner) Ear

The **internal (inner) ear** is a complicated series of canals (Figure 16.20). Structurally, it consists of two main divisions: an outer bony labyrinth that encloses an inner membranous labyrinth. The **bony labyrinth** is a series of cavities in the petrous portion of the temporal bone divided into three areas: (1) the semicircular canals and (2) the vestibule, both of which contain receptors for equilibrium, and (3) the cochlea, which contains receptors for hearing. The bony labyrinth contains **perilymph**. This fluid, which is chemically similar to cerebrospinal fluid, surrounds the **membranous labyrinth**, a series of sacs and tubes inside the bony labyrinth and having the same general shape. The membranous labyrinth contains **endolymph**.

The **vestibule** (VES-ti-būl) is the oval central portion of the bony labyrinth. The membranous labyrinth in the vestibule consists of two sacs called the **utricle** (Ū-tri-kul = little bag) and the **saccule** (SAK-ūl = little sac), which are connected by a small duct. Projecting superiorly and posteriorly from the vestibule are the three bony **semicircular canals**, each of which lies at approximately right angles to the other two. Based on their positions, they are named the anterior, posterior, and lateral semicircular canals. The anterior and posterior semicircular canals are vertically oriented; the lateral canal is horizontally oriented. At one end of each canal is a swollen enlargement called the **ampulla** (am-PUL-la =

Figure 16.20   The right internal ear. The outer, cream-colored area is part of the bony labyrinth; the inner, pink-colored and orange-colored fluid-filled chambers compose the membranous labyrinth.

External ear
Middle ear
Internal ear

Semicircular canals (contain semicircular ducts):

Anterior

Posterior

Lateral

LATERAL

Ampulla of semicircular duct

Oval window

Round window

Bony labyrinth (contains perilymph)

Membranous labyrinth (contains endolymph)

Ampulla of semicircular canal

Utricle

Vestibule

Saccule

Cochlea

Cochlear duct

MEDIAL

The bony labyrinth contains perilymph, and the membranous labyrinth contains endolymph.

saclike duct). The portions of the membranous labyrinth that lie inside the bony semicircular canals are called the **semicircular ducts**. These structures communicate with the utricle of the vestibule. The vestibular branch of the vestibulocochlear nerve transmits impulses for equilibrium from the vestibule and semicircular canals. Cell bodies of the sensory neurons are located in the **vestibular ganglia** (see Figure 16.21b).

Anterior to the vestibule is the **cochlea** (KOK-lē-a = snail-shaped), a bony spiral canal (Figure 16.21a) that resembles a snail's shell and makes almost three turns around a central bony hub. Sections through the cochlea reveal that it is divided into three channels: cochlear duct, scala vestibuli, and scala tympani (Figure 16.21a–c). The **cochlear duct** is a continuation of the membranous labyrinth into the cochlea; it is filled with endolymph. The channel above the bony partition is the **scala vestibuli**, which ends at the oval window. The channel below is the **scala tympani**, which ends at the round window. Both the scala vestibuli and scala tympani are part of the bony labyrinth of the cochlea; therefore, these chambers are filled with perilymph. The scala vestibuli and scala tympani are completely separated by the cochlear duct, except for an opening at the apex of the cochlea, the **helicotrema** (hel-i-kō-TRĒ-ma; see Figure 16.22). The cochlea adjoins the wall of the vestibule, into which the scala vestibuli opens. The perilymph in the vestibule is continuous with that of the scala vestibuli.

The **vestibular membrane** separates the cochlear duct from the scala vestibuli, and the **basilar membrane** separates the cochlear duct from the scala tympani. Resting on the basilar membrane is the **spiral organ** (Figure 16.21c,d). The spiral organ is a coiled sheet of epithelial cells, including supporting cells, and about 16,000 **hair cells**, which are the receptors for hearing. At the apical tip of each hair cell is a **hair bundle**, consisting of 40–80 long, hairlike microvilli that extend into the endolymph of the cochlear duct and are arranged in several rows of graded height. Projecting over and in contact with the hair cells of the spiral organ is the **tectorial membrane** (*tector-* = covering), a flexible gelatinous membrane. At their basal ends, hair cells synapse with first-order sensory neurons from the cochlear branch of the vestibulocochlear nerve. Cell bodies of the sensory neurons are located in the **spiral ganglion** (Figure 16.21b,c).

Figure 16.21   Semicircular canals, vestibule, and cochlea of the right ear. Note that the cochlea makes nearly three complete turns.

- ☐ External ear
- ☐ Middle ear
- ☐ Internal ear

LATERAL

MEDIAL

Utricle

Stapes in oval window

Saccule

Scala vestibuli

Cochlea

Scala tympani

Cochlear duct

Scala vestibuli

Vestibular membrane

Cochlear duct

Basilar membrane

Round window

Scala tympani

Transmission of sound waves from scala vestibuli to scala tympani by way of helicotrema

(a) Sections through the cochlea

(continues)

Figure 16.21 (continued)

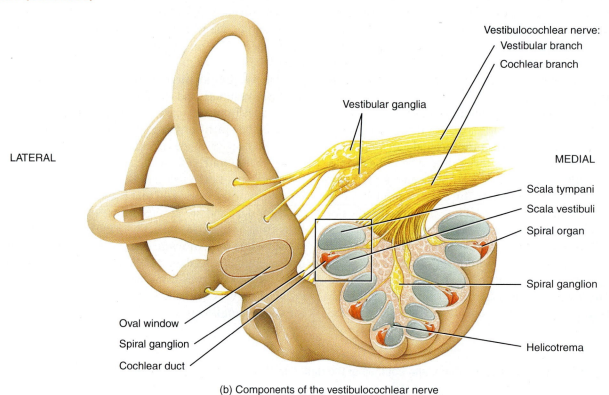

LATERAL

MEDIAL

Vestibulocochlear nerve:
Vestibular branch
Cochlear branch

Vestibular ganglia

Scala tympani
Scala vestibuli
Spiral organ

Spiral ganglion

Helicotrema

Oval window
Spiral ganglion
Cochlear duct

(b) Components of the vestibulocochlear nerve

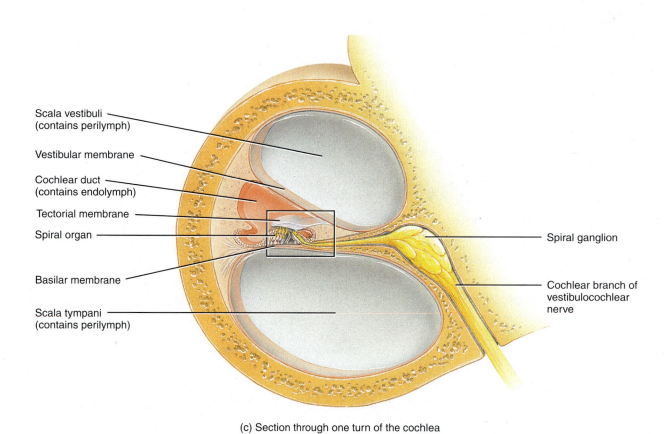

Scala vestibuli
(contains perilymph)

Vestibular membrane

Cochlear duct
(contains endolymph)

Tectorial membrane

Spiral organ

Basilar membrane

Scala tympani
(contains perilymph)

Spiral ganglion

Cochlear branch of
vestibulocochlear
nerve

(c) Section through one turn of the cochlea

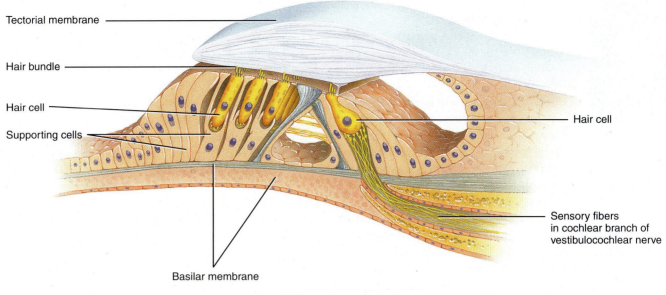

Tectorial membrane

Hair bundle

Hair cell

Supporting cells

Hair cell

Sensory fibers in cochlear branch of vestibulocochlear nerve

Basilar membrane

(d) Enlargement of spiral organ

Cochlear duct

Tectorial membrane

Hair cells

Hair cell

Supporting cells

Basilar membrane

Scala tympani

LM 140x

(e) Histology of the spiral organ

🔑 **The three channels in the cochlea are the scala vestibuli, the scala tympani, and the cochlear duct.**

Checkpoint

33. If you lost your auricles (common in severe burns of the head), how would your hearing be affected?
34. Which structures separate the middle ear from the external ear and from the internal ear?
35. List the auditory ossicles in order of sound wave transmission through the middle ear. For each ossicle, name two structures to which it is attached.
36. What are the three subdivisions of the bony labyrinth?
37. Which component of the cochlea contains the receptors for hearing?
38. Which nerve innervates the inner ear? Which branch of that nerve transmits equilibrium signals? Which branch transmits auditory signals?

Return to Dan's Story

Dr. Burke held a penlight in front of Dan, watching the pupils constrict as he swept the light toward Dan's eyes.

"Stay focused on the end of this pen as I move it," requested Dr. Burke. He brought the pen in toward Dan's face, then pulled away, then back in.

"It's most likely you have a detached retina, probably from the impact of the explosion. A layer of your retina has pulled away from the inner lining of the eye. It might have happened anyway, but these things usually don't occur in healthy 30-year-old men. The ophthalmologist said you had some abnormalities with your eyes: high intraocular pressure, some small hemorrhages in the vasculature of your retinas, poor accommodation. But this problem won't just heal by itself," said Dr. Burke.

Dr. Burke consulted with an eye surgeon, who visited Dan shortly thereafter. Reattaching the retina required delicate laser surgery. The detachment would disrupt the supply of blood to the retina, which could lead to permanent blindness if the retina wasn't reattached. During the laser procedure, his eye was numbed, then a gas bubble was injected into the eye. After the procedure Dan could see, but his vision was blurry in the eye that had been operated on. It was explained to Dan that the retina had only detached at the perimeter. It then slid down over his central fovea, the part of his retina needed for good vision.

Pam sat next to Dan's hospital bed. "Okay, now that they repaired your eye, what did they say about the hearing loss?" she asked.

"Well, I can hear better out of the right ear, but I am pretty much deaf in the left. The audiologist said my eardrum wasn't quite as damaged on the right side. They said my eardrum might grow back, but I might need surgery on it, too."

Pam searched Dan's face. "But will you be able to hear like before? Is there any long-term damage?" she asked.

"The inner ear might be damaged. The doc said that was more serious, less likely to heal," explained Dan.

"Hi, Pam. Hi, Dan, how you doing today? I heard they're letting you go soon," Captain Donald stood in the doorway.

"Hey, Cap!" Dan replied.

"Are you still dizzy all the time?" asked Captain Donald.

"It comes and goes," responded Dan.

"How are the ears?" asked Donald.

"What, what?" Dan joked, laughing. "Well, 50% loss in the right, 75% in the left. But the eardrums are healing."

"Dan, you know I can't let you back on the fire department if you're not at 100%. Someone could get hurt."

Dan looked at Captain Donald. He knew Cap was right, but it was hard to hear.

F. A "detached retina" occurs when the neural layer of the retina separates from the pigmented layer. The usual cause is a sudden impact to the eye. If this occurred at the central fovea, would Dan lose all of his vision from that eye? Why or why not?

G. Which of Dan's middle ear structures would have served a protective function during the explosion?

H. Why would damage to Dan's inner ear be less likely to heal than damage to his tympanic membrane? Which components of his inner ear may be damaged?

I. Why would a firefighter need good vision, sight, and hearing in addition to a high level of fitness?

16.8 The pathway of sound is tympanic membrane → ossicles → oval window → cochlea → vestibulocochlear nerve → primary auditory cortex.

The Nature of Sound Waves

Sound waves are alternating high- and low-pressure regions traveling in the same direction through some medium (such as air). They originate from a vibrating object in much the same way that ripples arise and travel over the surface of a pond when you toss a stone into it. The number of pressure waves that arrive in a given time is defined as **frequency** of a sound and is measured in *hertz* (Hz; 1 Hz = 1 cycle per second). For example, the entire audible range of sounds heard by the human ear extends from 20 to 20,000 Hz. We interpret different sound frequencies as differences in *pitch*. The higher the frequency of vibration, the higher is the pitch.

The harder an object, such as a tuning fork, is struck, the louder is the sound coming from that vibrating object. Sound volume is defined as its **intensity** (size or amplitude). The larger the intensity of the vibration, the *louder* is the sound. Sound intensity is measured in units called *decibels (dB)*. The hearing threshold—the point at which an average young adult can just distinguish sound from silence—is defined as 0 dB. Rustling leaves have a decibel level of 15; whispered speech, 30; normal conversation, 60; a vacuum cleaner, 75; shouting, 80; and a nearby motorcycle or jackhammer, 90. Sound becomes uncomfortable to a normal ear at about 120 dB, and painful above 140 dB.

Physiology of Hearing

The following events are involved in hearing (Figure 16.22):

1 The auricle directs sound waves into the external auditory canal.

2 When sound waves strike the tympanic membrane, the alternating high and low pressure of the air causes the tympanic membrane to vibrate back and forth. The distance it moves, which is very small, depends on the intensity and frequency of the sound waves. The tympanic membrane vibrates slowly in response to low-frequency (low-pitched) sounds and rapidly in response to high-frequency (high-pitched) sounds. It vibrates more forcefully in response to higher intensity sounds, more gently in response to lower intensity sounds.

3 The central area of the tympanic membrane connects to the malleus, which also starts to vibrate. The vibration is transmitted from the malleus to the incus and then to the stapes.

4 As the stapes moves back and forth, it pushes the membrane of the oval window in and out. The oval window vibrates about 20 times more vigorously than the eardrum because the auditory ossicles efficiently transmit small vibrations, spread over a large surface area (tympanic membrane), into larger vibrations of a smaller surface (oval window).

5 The movement of the oval window sets up fluid pressure waves in the cochlea by pushing on the perilymph of the scala vestibuli.

6 Pressure waves are transmitted from the scala vestibuli to the scala tympani and eventually to the round window, where the pressure waves are finally absorbed as the flexible membrane of the round window bulges outward into the middle ear. (See **9** in the figure.)

7 As the pressure waves deform the walls of the scala vestibuli and scala tympani, they also push the vestibular membrane back and forth, creating pressure waves in the endolymph inside the cochlear duct.

8 The pressure waves in the endolymph cause the basilar membrane to vibrate, which moves the hair cells of the spiral organ against the tectorial membrane. This leads to bending of the hair cell microvilli, which produces receptor potentials that ultimately lead to the generation of nerve impulses.

Figure 16.22 **Events in the stimulation of auditory receptors in the right ear.** The cochlea has been uncoiled to more easily visualize the transmission of sound waves and their distortion of the vestibular and basilar membranes of the cochlear duct.

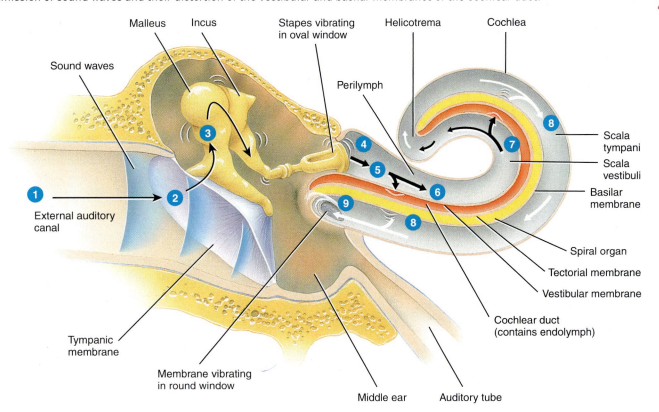

🔑 **Hair cells of the spiral organ convert a mechanical vibration (stimulus) into an electrical signal (receptor potential).**

Sound waves of various frequencies cause certain regions of the basilar membrane to vibrate more intensely than other regions. Each segment of the basilar membrane is "tuned" for a particular pitch. Because the membrane is narrower and stiffer at the base of the cochlea (portion closer to the oval window), high-frequency (high-pitched) sounds near 20,000 Hz induce maximal vibrations in this region. Toward the apex of the cochlea near the helicotrema, the basilar membrane is wider and more flexible; low-frequency (low-pitched) sounds near 20 Hz cause maximal vibration of the basilar membrane there. As noted previously, loudness is determined by the intensity of sound waves. High-intensity sound waves cause larger vibrations of the basilar membrane, which leads to a higher frequency of nerve impulses reaching the brain.

The hair cells transduce mechanical vibrations into electrical signals. As the basilar membrane vibrates, the hair bundles at the apex of the hair cell bend back and forth. Bending of the hair bundles stimulates the hair cells to release a neurotransmitter, which generates nerve impulses in the sensory neurons that innervate the hair cells. As more neurotransmitter is re-leased, the frequency of nerve impulses in the sensory neurons increases.

The Auditory Pathway

Sensory neurons that innervate the hair cells form the cochlear branch of the vestibulocochlear nerve (Figure 16.23). These axons synapse in nuclei in the medulla oblongata on the same side. From there, axons ascend directly to the midbrain or cross to the opposite side and terminate either in nuclei of the pons or in the midbrain. Slight differences in the timing of impulses arriving from your two ears at the pons allow you to locate the source of a sound. From the pons, axons ascend to the midbrain in a tract called the **lateral meniscus**. From the midbrain, nerve impulses are conveyed to nuclei in the thalamus, and finally to the primary auditory area in the temporal lobe of the cerebral cortex. Because many auditory axons cross over in the medulla oblongata while others remain on the same side, the right and left primary auditory areas receive nerve impulses from both ears.

Figure 16.23 The auditory pathway.

Nucleus in thalamus

Cochlear branch of vestibulocochlear nerve

Nuclei in medulla oblongata

Primary auditory area in cerebral cortex

Midbrain

Lateral menisci

Nucleus in pons

Cerebellum

🔑 From the cochlea, auditory nerve impulses are propagated along the cochlear branch of the vestibulocochlear nerve and then to the brain stem, thalamus, and cerebral cortex.

Checkpoint

39. Would strongly blowing through a whistle produce sound waves of high or low frequency? High or low intensity?
40. How are sound waves transmitted from the auricles to the spiral organ?
41. How does the tympanic membrane respond to quieter sounds? To low-pitched sounds?
42. If fluid waves in the cochlea bounced back and forth, sounds would echo inside your cochleas. What stops fluid waves from traveling more than once through the cochlea?
43. How is the human ear able to identify the origin of a sound?
44. What is the pathway for auditory impulses from the cochlea to the cerebral cortex?

16.9 Impulses for equilibrium propagate along the vestibulocochlear nerve to the brain.

Physiology of Equilibrium

There are two types of **equilibrium** (balance). **Static equilibrium** refers to the maintenance of the position of the body (mainly the head) relative to the force of gravity. Body movements that stimulate the receptors for static equilibrium include tilting the head and *linear* acceleration or deceleration, such as when the body is being moved in an elevator or in a car that is speeding up or slowing down. **Dynamic equilibrium** is the maintenance of body position (mainly the head) in response to *rotational* acceleration or deceleration. Collectively, the receptor organs for equilibrium are called the **vestibular apparatus** (ves-TIB-ū-lar); these include the saccule, utricle, and semicircular ducts.

Otolithic Organs: Saccule and Utricle

The walls of both the utricle and the saccule contain a small, thickened region called a **macula** (MAK-ū-la; Figure 16.24). The two maculae (plural), which are perpendicular to one another, are the receptors for static equilibrium. They provide sensory information on the position of the head in space, essential to maintaining appropriate posture and balance. The maculae also detect linear acceleration and deceleration.

The maculae consist of two kinds of cells: **hair cells**, which are the sensory receptors, and **supporting cells**. Hair cells feature **hair bundles** that consist of 40–80 microvilli. Scattered among the hair cells are columnar supporting cells that probably secrete the thick, gelatinous layer, called the **otolithic membrane**, that rests on the hair cells. A layer of calcium carbonate crystals, called **otoliths** (*oto-* = ear; *-liths* = stones), extends over the entire surface of the otolithic membrane.

The densely packed otoliths add weight to the otolithic membrane, amplifying the pull of gravity during movements. Because the otolithic membrane sits on top of the macula, when you tilt your head forward, the otolithic membrane and the otoliths are pulled by gravity and slide downhill over the hair cells in the direction of the tilt, bending the hair bundles. The movement of the hair bundles produces receptor potentials releasing neurotransmitter that stimulates nerve impulses in first-order sensory neurons in the vestibular branch of the vestibulocochlear nerve (see Figure 16.21b).

Semicircular Ducts

The three semicircular ducts function in dynamic equilibrium. The ducts lie at right angles to one another in three planes of space (Figure 16.25): The two vertical ducts are the anterior and posterior semicircular ducts, and the horizontal one is the lateral semicircular duct. This positioning permits detection of rotational acceleration or deceleration. In the **ampulla**, the dilated portion of each duct, is a small elevation called the **crista** (KRIS-ta). Each crista contains a group of **hair cells** and **supporting cells**. Covering the crista is a mass of gelatinous material called the **cupula** (KŪ-pū-la). Hair bundles project from the hair cells into the cupula. When the head moves along the plane of one of the semicircular ducts, the enclosed endolymph moves through the semicircular ducts, pushing on the cupula and bending the hair bundles. The three different planes of the three semicircular ducts in each ear are able to respond to virtually any rotational movement of the head. When you nod "yes," fluid waves stimulate hair cells in the anterior semicircular duct. As you shake your head "no," hair cells are stimulated in the lateral semicircular duct. Tilting your head from shoulder to shoulder generates fluid waves that stimulate hair cells in the posterior semicircular duct. Bending of the hair bundles produces receptor potentials. In turn, the receptor potentials lead to nerve impulses that pass along the vestibular branch of the vestibulocochlear nerve (see Figure 16.21b).

Figure 16.24   Location and structure of receptors in the maculae of the right ear.

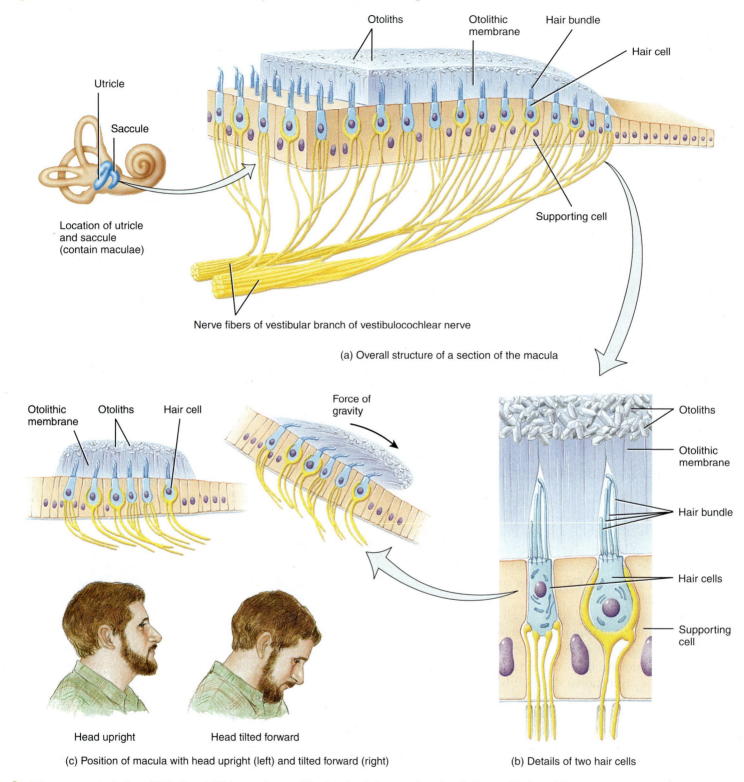

Utricle

Saccule

Otoliths

Otolithic membrane

Hair bundle

Hair cell

Location of utricle and saccule (contain maculae)

Supporting cell

Nerve fibers of vestibular branch of vestibulocochlear nerve

(a) Overall structure of a section of the macula

Otolithic membrane

Otoliths

Hair cell

Force of gravity

Otoliths

Otolithic membrane

Hair bundle

Hair cells

Supporting cell

Head upright   Head tilted forward

(c) Position of macula with head upright (left) and tilted forward (right)

(b) Details of two hair cells

The movement of microvilli by the otolithic membrane ultimately leads to nerve impulses in the vestibulocochlear nerve.

Figure 16.25 Location and structure of the semicircular ducts of the right ear.

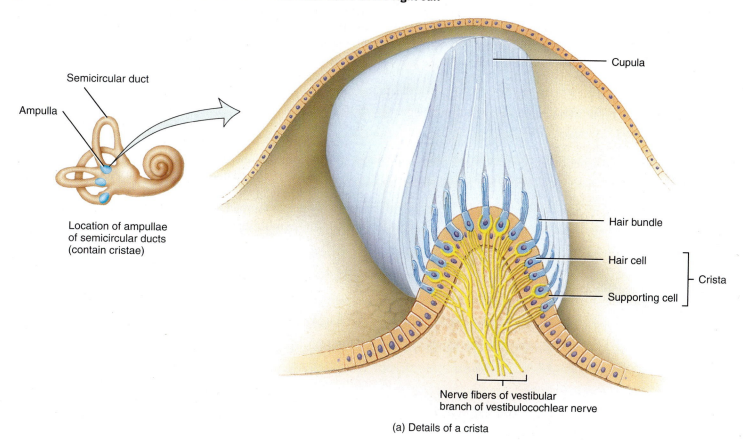

Semicircular duct

Ampulla

Location of ampullae of semicircular ducts (contain cristae)

Cupula

Hair bundle

Hair cell

Supporting cell

Crista

Nerve fibers of vestibular branch of vestibulocochlear nerve

(a) Details of a crista

Cupula

Ampulla

Cupula sensing movement and direction of flow of endolymph

Head in still position

Head rotating

(b) Position of a cupula with the head in the still position (left) and when the head rotates (right)

The positions of the semicircular ducts permit detection of rotational movements.

Equilibrium Pathways

Bending of hair cells in the utricle, saccule, or semicircular ducts causes the release of a neurotransmitter, which generates nerve impulses in the sensory neurons that innervate the hair cells. Sensory neurons that innervate the hair cells form the vestibular branch of the vestibulocochlear nerve (Figure 16.26). Most of these axons synapse with sensory neurons in nuclei in the medulla oblongata and pons. The remaining axons terminate in the cerebellum.

The medulla and pons integrate information arriving from the utricle, saccule, semicircular ducts, and eyes and then send commands to: (1) the oculomotor, trochlear, and abducens nerves that control coupled movements of the eyes with those of the head to help maintain focus on the visual field; (2) the accessory nerves that help control head and neck movements to help maintain equi-librium; (3) the vestibulospinal tract that conveys impulses down the spinal cord to maintain muscle tone in skeletal muscles to help maintain equilibrium; and (4) the thalamus and then to the vestibular area of the cerebral cortex (which is part of the primary somatosensory area; see Figure 13.11) to provide us with the conscious awareness of the position and movements of the head.

The cerebellum plays a key role in maintaining equilibrium. In response to input from the utricle, saccule, and semicircular ducts, the cerebellum continuously sends nerve impulses to the motor areas of the cerebrum. This feedback allows correction of signals from the motor cortex to specific skeletal muscles to smooth movements and coordinate complex sequences of muscle contractions to help maintain equilibrium.

Table 16.2 summarizes the structures of the ear related to hearing and equilibrium.

Figure 16.26   The equilibrium pathway.

Nucleus in thalamus

Vestibular branch of vestibulocochlear nerve

Nuclei in medulla oblongata and pons

Spinal cord

Vestibular area in cerebral cortex

Oculomotor nerve nucleus

Trochlear nerve motor nucleus

Abducens nerve motor nucleus

Cerebellum

Accessory nerve nucleus

Vestibulospinal tract

🔑 **From the semicircular ducts, utricle, and saccule, equilibrium nerve impulses are propagated along the vestibular branch of the vestibulocochlear nerve to the brain stem, cerebellum, thalamus, and cerebral cortex.**

TABLE 16.2

Structures of the Ear

REGIONS OF THE EAR AND KEY STRUCTURES	FUNCTION
External (outer) ear External auditory canal, Auricle, Tympanic membrane	*Auricle:* Collects sound waves. *External auditory canal:* Directs sound waves to the tympanic membrane. *Tympanic membrane:* Sound waves cause it to vibrate, which in turn causes the malleus to vibrate.
Middle ear 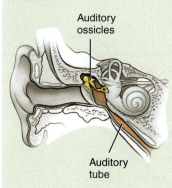 Auditory ossicles, Auditory tube	*Auditory ossicles:* Transmit and amplify vibrations from tympanic membrane to oval window. *Auditory tube:* Equalizes air pressure on both sides of the tympanic membrane.

Internal (inner) ear 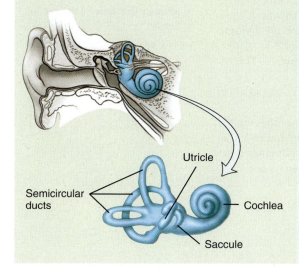 Utricle, Semicircular ducts, Cochlea, Saccule	*Cochlea:* Contains a series of fluids, channels, and membranes that transmit vibrations to the spiral organ, the organ of hearing; hair cells in the spiral organ produce reactions that elicit nerve impulses in the cochlear branch of the vestibulocochlear nerve.
	 Vestibular apparatus: Includes semicircular ducts, utricle, and saccule, which generate nerve impulses that propagate along the vestibular branch of the vestibulocochlear nerve.
	    Semicircular ducts: Contain cristae, sites of hair cells for dynamic equilibrium.
	    Utricle: Contains maculae, sites of hair cells for static and dynamic equilibrium.
	    Saccule: Contains maculae, sites of hair cells for static and dynamic equilibrium.

Checkpoint

45. What is the difference between static and dynamic equilibrium?
46. How are the receptors for hearing and equilibrium structurally similar? How do they differ?
47. What function is served by the otoliths?
48. Why do we need three, rather than one or two, semicircular canals?
49. With which type of equilibrium are the semicircular ducts, the utricle, and the saccule associated?
50. Where do axons in the vestibular branch of the vestibulocochlear nerve terminate? What purposes are served by the transmission of equilibrium input to these locations?

Dan's Story: Epilogue and Discussion

▶▶▶ **D**an is a healthy 30-year-old male who has suffered a blast injury from a natural gas explosion. The explosion has caused significant damage to his special senses. Dan initially had swelling and bruising, and possibly a concussion from the blast. His vision has been disrupted, and he suffered a detached retina, which was surgically repaired. Immediately after his accident, his sense of smell was abnormal, and he had a reduced ability to taste. Hearing loss is significant due to damage to middle ear components that conduct sound to the inner ear, and possibly due to damage to the hair cells within the cochlea. Dan is also experiencing chronic vertigo, a feeling of dizziness, due to blast shock damage to the vestibular apparatus.

Dan was fortunate that he was wearing protective equipment when he was exposed to the explosion. Typically, the predominant post-explosion injuries suffered in such a blast involve penetrating injuries and blunt trauma. ◀◀◀

J. Which structures of Dan's inner ear could have been damaged to produce his continued sensations of dizziness?

K. How could the explosion have damaged hair cells in Dan's vestibular apparatus?

L. Is it likely that Dan's senses of taste and smell will return? Why or why not?

M. During Dan's entry into the burning building, how did his special senses warn him of impending danger?

Concept and Resource Summary

Concept	Resources

Introduction

1. The **special senses** include smell, taste, vision, hearing, and equilibrium. Receptors for these senses are structurally complex and located in sensory organs of the head.

Concept 16.1 Impulses for smell propagate along the olfactory nerve to the brain.

1. Sensations of smell and taste arise from molecules interacting with smell or taste receptors that propagate impulses to the limbic system and higher cortical areas.

2. The **olfactory epithelium**, in the superior area of the nasal cavity, contains receptors for **olfaction**. **Olfactory receptor cells** are first-order neurons; their axons project into the olfactory bulb; their **olfactory hairs** respond to inhaled chemicals, or **odorants**. **Supporting cells** detoxify chemicals, give electrical insulation of receptors, and provide support and nourishment. **Basal cells** continually divide to produce new olfactory receptor cells. **Olfactory glands** produce mucus to dissolve odorants.

3. Olfactory receptor cells are excited by odorant molecules binding to *odorant binding proteins*. This results in production of a generator potential, which may lead to an action potential.

4. Olfaction has a low threshold; only a few molecules are needed to be perceived as an odor. Adaptation to odors occurs rapidly.

5. Bundles of axons of olfactory receptor cells pass through foramina in the cribriform plate of the ethmoid bone and form the left and right **olfactory nerves**. The nerves terminate in the **olfactory bulbs**. Within the olfactory bulbs, receptor cell axons synapse on olfactory bulb neurons, forming the **olfactory tract** that projects to the primary olfactory area of the cerebral cortex, where awareness of smells begins.

See

- Anatomy Overview – **Chemoreceptors**
- Anatomy Overview – **Olfactory Nerve**
- Anatomy Overview – **The Special Senses**

Concept	Resources

Concept 16.2 Impulses for taste propagate along the facial, glossopharyngeal, and vagus nerves to the brain.

1. **Gustation** involves distinguishing five tastes: bitter, sour, salty, sweet, and umami.
2. Receptors for taste are located in taste buds. There are three types of taste buds, which are in elevated **papillae** on the tongue: **circumvallate**, **fungiform**, and **foliate**. The tongue surface also has **filiform papillae** that increase friction with food to help manipulate it. Each **taste bud** consists of **supporting cells**, **basal cells**, and **gustatory receptor cells** that have a **gustatory hair** projecting from the cell through a **taste pore** to the surface.
3. **Tastants** dissolve in saliva and bind to gustatory hairs, stimulating the receptor cells to have a receptor potential, releasing neurotransmitter that triggers action potentials in first-order neurons.
4. Na^+ from salty foods and H^+ from sour foods enter gustatory receptor cells, causing depolarization that leads to the release of neurotransmither. Sweet, bitter, and umami tastes bind to receptors that stimulate second messengers to release neurotransmitter.
5. The threshold for each of the primary tastes varies. Complete adaptation occurs quickly.
6. Three cranial nerves propagate nerve impulses to the medulla; some then project to the limbic system and hypothalamus, whereas others project to the thalamus and continue to the primary gustatory area in the parietal lobe of the cerebral cortex.

See

- Anatomy Overview – Chemoreceptors
- Anatomy Overview – Facial Nerve
- Anatomy Overview – Glossopharyngeal Nerve
- Anatomy Overview – Vagus Nerve

Concept 16.3 The eye is protected by eyelids, eyelashes, eyebrows, and a lacrimal apparatus.

1. The **accessory structures** of the eye include eyelids, eyelashes, eyebrows, lacrimal apparatus, and extrinsic eye muscles.
2. The **eyelids** shade and protect the eyes, and spread lubricants over the eyeballs. Each eyelid contains the **orbicularis oculi**, **Meibomian glands**, and **conjunctiva**. The orbicularis oculi closes the eyelid. The conjunctiva is mucous membrane lining each eyelid and covering the anterior eyeball surface. The upper eyelid contains the **levator palpebrae superioris muscle** that raises the upper eyelid. The medial **lacrimal caruncle** contains sebaceous and sweat glands.
3. The **eyelashes** and **eyebrows** help protect the eyeballs.
4. The **lacrimal apparatus** includes the **lacrimal glands**, **lacrimal ducts**, **lacrimal puncta**, **lacrimal canals**, **lacrimal sac**, and **nasolacrimal duct**. This group of structures produce and drain **lacrimal fluid**. Tears are a watery solution with salts, mucus, and the bactericidal enzyme **lysozyme**. Tears protect, clean, and lubricate the eyeball.
5. Six extrinsic muscles move each eyeball: the **superior rectus**, **inferior rectus**, **lateral rectus**, **medial rectus**, **superior oblique**, and **inferior oblique**.

See

- Anatomy Overview – Oculomotor Nerve
- Anatomy Overview – Trochlear Nerve
- Anatomy Overview – Abducens Nerve

Concept 16.4 The eye is constructed of three layers and two chambers.

1. The wall of the eyeball has three layers: the fibrous tunic, vascular tunic, and retina.
2. The superficial **fibrous tunic** is avascular: It consists of the posterior, tough, protective, white **sclera**; and the anterior, curved, transparent **cornea**. The **scleral venous sinus** is at the junction of the sclera and cornea, and functions to drain aqueous humor.
3. The **middle vascular tunic** has three parts: the vascular **choroid** that supplies nutrients to the retina and absorbs scattered light rays, the **ciliary body** consisting of the **ciliary processes** and **ciliary muscle** that secretes aqueous humor and controls the shape of the lens, and the **iris**, which regulates the diameter of its central opening, the **pupil**.
4. The **retina** is the innermost tunic. The outer **pigmented layer** is composed of melanin-containing epithelial cells that prevent scattering of light rays. The **neural layer** contains three types of retinal neurons: **photoreceptors**, **bipolar cells**, and **ganglion cells**. The **rods** are photoreceptors that are stimulated by even low light and allow night or dim vision. **Cones** are photoreceptors that are stimulated by bright light and produce color vision.

Hear

- **Figure 16.8 –** Anatomy of the eyeball

5. Visual information passes from photoreceptors to bipolar cells to ganglion cells. Axons of ganglion cells extend to the **optic disc** and exit the eyeball as the optic nerve. The optic disc is called the **blind spot** because it contains no rods or cones.

6. The central, posterior portion of the retina contains only cones. This area, the **central fovea**, is the area of highest **visual acuity** or **resolution**.

7. The **lens** is posterior to the pupil and iris, and suspended in the eyeball cavity by the suspensory ligaments attached to the ciliary processes. It is transparent and contains proteins called **crystallins**. The lens functions to focus images on the retina.

8. The lens divides the interior of the eyeball into the anterior cavity and the vitreous chamber. The **anterior cavity**, the space anterior to the lens, contains **anterior** and **posterior chambers** filled with **aqueous humor**. Aqueous humor is produced by the ciliary processes and drained by the scleral venous sinus. The larger **vitreous chamber**, between the lens and the retina, contains the **vitreous body**. Aqueous humor and the vitreous body contribute to **intraocular pressure**, which prevents the eyeball from collapsing.

Concept 16.5 Image formation involves refraction of light rays, change in lens shape, and constriction of the pupil.

1. Light rays entering the eye undergo **refraction**, the bending of light rays traveling from one transparent substance into another. This occurs at each surface of the cornea and the lens; finally, the refracted light rays are focused on the retina.

2. The anterior and posterior lens surfaces are convex, and increasing the lens curvature increases its focusing power. **Accommodation** is an increase in the curvature of the lens for near vision.

3. An **emmetropic eye** is a normal eye that can refract light rays so that a clear image is focused on the retina. **Myopia** (nearsightedness) is the inability to focus properly on distant objects. **Hyperopia** (farsightedness) results in the inability to focus on nearby objects. **Astigmatism** occurs when there is an irregular curvature to the cornea or lens that causes parts of an image to be out of focus or distorted.

4. **Constriction of the pupil** is an autonomic reflex occurring simultaneously with accommodation for near objects and bright light. It prevents light rays from entering at the periphery of the lens, which causes blurred vision.

5. Human eyes have **binocular vision**; both eyes focus on only one set of objects as light rays strike corresponding points on both retinas. This allows depth perception. When objects move closer, the eyes must move medially and undergo **convergence**, so that light rays from the object can strike the same points on both eyes.

Concept 16.6 The neural pathway of light is photoreceptors → bipolar cells → ganglion cells → optic nerve → occipital lobe.

1. In the outer segment of photoreceptors, plasma membrane proteins, called **photopigments**, transduce light energy into a receptor potential. Rods have the photopigment **rhodopsin**; cones have three different **cone photopigments** that absorb blue, green, and red light.

2. Photopigments contain two parts: the glycoprotein **opsin**, and **retinal**, which is a derivative of vitamin A and is the light-absorbing part of photopigments. When retinal absorbs light, it changes shape from *cis* to *trans* in a process called **isomerization**. The *trans*-retinal separates from opsin during **bleaching**, and then **retinal isomerase** converts *trans*-retinal back to *cis*-retinal. The photopigment finally undergoes **regeneration** when *cis*-retinal binds to opsin.

See

• **Anatomy Overview – Optic Nerve**

Hear

• **Figure 16.17 –**
The Visual Pathway

3. **Light adaptation** occurs when photoreceptors adjust to a brighter environment by decreasing their sensitivity as increasing amounts of photopigment are bleached.

4. **Dark adaptation** occurs when photoreceptors increase sensitivity by increasing photopigment regeneration. Only rods function at low light intensity.

5. Light striking the retina activates enzymes that lead to a receptor potential in the photoreceptors, then receptor potentials in bipolar cells that synapse on photoreceptors.

6. Bipolar cells transmit receptor potentials to ganglion cells, which generate action potentials. Ganglion cell axons exit the eyeball as the **optic nerve**. When optic nerve axons pass through the **optic chiasm**, they either cross to the opposite side or continue straight ahead, forming the **optic tract** that enters the thalamus. **Optic radiations** allow for projection to the primary visual areas of the cerebral cortex.

7. The eye's **visual field** is everything that can be seen by one eye.

Concept 16.7 The three main regions of the ear are the external, middle, and internal ear.

1. The ear contains receptors for sound waves and for equilibrium.

2. The external ear collects and channels sound waves inward; the middle ear conveys sound vibrations to the oval window; and the inner ear has the receptors for hearing and equilibrium.

3. The **external ear** consists of three strucures: the **auricle** collects sound waves; the **external auditory canal** carries sound waves from the auricle to tympanic membrane; and the **tympanic membrane** separate the external auditory canal from the middle ear.

4. The **middle ear** is an air-filled chamber in the temporal bone. Three small bones, the **auditory ossicles**, extend across the middle ear: the **malleus**, attached to the eardrum; the **incus**, which is the middle bone; and the **stapes**, which fits into the **oval window**. The **tensor tympani** and the **stapedius** attach to the ossicles and protect against damaging vibrations to the inner ear. The middle ear is connected to the nasopharynx by the **auditory tube**, which equalizes atmospheric pressure.

5. The **internal ear** consists of two principal divisions: the **bony labyrinth** and the **membranous labyrinth**. The bony labyrinth is made up of cavities in the temporal bone divided into the semicircular canals and the vestibule, with receptors for equilibrium; and the cochlea, which contains receptors for hearing.

6. The **vestibule** contains two sacs, the **utricle** and **saccule**. Three bony **semicircular canals** project from the vestibule and contain the **semicircular ducts** that communicate with the utricle. The vestibule and semicircular canals transmit nerve impulses to the vestibular branch of the vestibulocochlear nerve.

7. The spiral-shaped **cochlea** has three interior channels: the **scala vestibuli**, **scala tympani**, and **cochlear duct**. The **spiral organ** contains **hair cells**, the receptors for hearing. The basal ends of hair cells synapse with first-order sensory neurons from the cochlear branch of the vestibulocochlear nerve.

Hear

- **Figure 16.19** – **The Right Middle Ear Containing the Auditory Ossicles**
- **Figure 16.20** – **The Right Internal Ear**

Concept

Resources

WILEY PLUS

Concept 16.8 The pathway of sound is tympanic membrane → ossicles → oval window → cochlea → vestibulocochlear nerve → primary auditory cortex.

1. **Sound waves** originate from a vibrating object and are alternating high- and low-pressure regions traveling in a medium.
2. The number of pressure waves that arrive in a given time is **frequency** of a sound, measured in hertz (Hz); different sound frequencies are interpreted as differences in *pitch*. Sound volume is its **intensity** (size or amplitude), measured in *decibels (dB)*.
3. Sound waves directed by the auricle into the external auditory canal cause vibrations of the tympanic membrane. The ossicles vibrate beginning with the malleus and ending with the stapes striking the oval window.
4. Fluid pressure waves occur in the cochlear perilymph and then in endolymph, causing vibrations of the basilar membrane that move the hair cells against the tectorial membrane. Segments of the basilar membrane are tuned for a particular pitch; low-frequency sounds cause vibrations toward the apex of the cochlea, and high-frequency sounds induce vibrations of the base of the cochlea. Louder sounds lead to a higher frequency of nerve impulses leaving the spiral organ.
5. As the microvilli of hair cells bend against the tectorial membrane, they transduce mechanical vibrations to electrical signals and generate nerve impulses in sensory neurons in the cochlear branch of the vestibulocochlear nerve.
6. Cochlear branch neurons terminate in the medulla oblongata. Axons carrying auditory signals then project to the pons, midbrain, thalamus, and finally to the primary auditory area in the cerebral cortex.

See

• Anatomy Overview –
  Vestibulocochlear Nerve

Hear

• **Figure 16.22** –
  **Events in the Stimulation
  of Auditory Receptors
  in the Right Ear**

Concept 16.9 Impulses for equilibrium propagate along the vestibulocochlear nerve to the brain.

1. **Static equilibrium** is the maintenance of the position of the body relative to the force of gravity. **Dynamic equilibrium** is the maintenance of body position in response to rotational movements.
2. The receptor organs for equilibrium, the **vestibular apparatus**, consist of the saccule, utricle, and semicircular ducts.
3. The **maculae** of the utricle and saccule are the receptors for static equilibrium. The maculae consist of **hair cells**, which act as the sensory receptors, and **supporting cells**.
4. Hair cells have **hair bundles** consisting of microvilli. Resting on the hair cells is the **otolithic membrane** covered by a layer of **otoliths**, which amplify the pull of gravity during movements.
5. The three semicircular ducts function in dynamic equilibrium. Each duct has cristae that contain **hair cells**. When the head moves, endolymph moves through the ducts, bending the hair bundles and producing receptor potentials in the hair cells. The receptor potentials result in nerve impulses in the vestibulocochlear nerve.
6. Most vestibular branch axons terminate in the medulla and pons; remaining axons terminate in the cerebellum.
7. The cerebellum receives sensory information from all structures of the vestibular apparatus and plays a key role in maintaining static and dynamic equilibrium by making corrective adjustments.
8. Bending of hair cells in the utricle, saccule, or semicircular ducts generates nerve impulses in sensory neurons of the vestibular branch of the vestibulocochlear nerve. Most vestibular branch axons of the vestibulocochlear nerve enter the brain stem and terminate in the medulla and pons; other axons enter the cerebellum.

See

• Anatomy Overview –
  Vestibulocochlear Nerve

Other Chapter Resources

For additional information on clinical applications, medical disorders, and medical terminology, see Chapter 7 in the *Clinical Connections* manual packaged with your text.

Within *WileyPLUS*

- Chapter Quiz
- Visual Anatomy
- Mastering Vocabulary
- Audio Glossary
- Web Links

Understanding the Concepts

1. Describe the sequence of events from the binding of an odorant molecule to an olfactory hair to the arrival of a nerve impulse in an olfactory bulb.

2. What is the difference between the olfactory nerves, olfactory bulbs, and olfactory tract in the olfactory pathway?

3. How do the receptor cells for olfaction and gustation differ in structure and function? How do the olfactory and gustatory pathways differ?

4. Why is the low threshold of bitter substances a survival advantage?

5. What protective functions would be compromised if a person lost his or her eyelashes and eyebrows?

6. Which division of the autonomic nervous system causes constriction of pupils? Which causes dilation of pupils?

7. Through which structures does light pass as it travels through the neural layer of the retina? How does visual information pass from photoreceptors to the optic nerve?

8. Where is aqueous humor produced, what is its circulation path, and where does it drain from the eyeball?

9. Which characteristics allow the lens to transmit light? Why are we unable to see an image that strikes the blind spot?

10. Describe the pathway of nerve impulses triggered by an object in the nasal half of the visual field of the left eye to the primary visual area of the cortex.

11. Why do small particles, such as dust from a dirt trail, normally not travel to the interior of your external auditory canal?

12. An infection of the throat can lead to a secondary infection of the middle ear (otitis media). How are these structures connected?

13. Why might loud noises cause dizziness?

14. Which part of the basilar membrane vibrates most vigorously in response to banging on a bass drum (low-frequency sounds)? How would the basilar membrane respond to someone quietly whispering (low-intensity sounds)?

15. If you were blindfolded, would maculae or cristae sense that you were hanging upside down? Where are maculae and cristae found?

17

The Endocrine System

The hot Oklahoma wind blew through the windows of Karen's small, single-wide trailer home as she talked on the phone with her daughter Amanda. "I've made some chicken for dinner tonight and I'm making a flower arrangement for my neighbor, Sara. These fresh flowers are just going to curl up and die before her wedding tomorrow with this heat. How are the kids, Am?"

Amanda was always amazed at her mother's constant stream of conversation. "We're enjoying the summer, but I'll be glad when the kids are back in school. How are you doing, Mom? Are you still so tired all the time?"

Karen sighed, "I keep busy. But by the afternoon, I'm just so tired I need a nap. On this TV talk show the other day, they were talking about some new disease everyone's getting, called fibromyalgia. This lady had symptoms just like mine."

"Mom, every time you see a show about some disease you think you have it. Remember when Ronald Reagan was diagnosed with Alzheimer disease? You thought you had that, too," said Amanda.

"Oh, I know. But I have all the symptoms. I'm tired all the time, weak, puffy, and swollen. And you know how I can't lose weight. I've tried all the diets," said Karen as she reached for a handful of chocolate-covered peanuts that sat next to two half-empty cans of soda pop.

Amanda thought to herself, *Here we go again*. "Mom, it's probably just your asthma. You really ought to think about moving out west. The mountain air would do wonders for your health."

"Now, Am. You know I like it here. This is where my friends are, and your brothers and you are all within three hours' drive. I can't move." Karen got up from the table as she talked. "I need to go, Amanda. I can't find my floral wire. I think it's in the shed."

Karen went outside. When she opened the storage shed, a hot gust of air hit her face. The air was hard to breathe, and the heat made her dizzy. Karen looked for the wire, hoping to find it quickly. Her chest felt tighter, and it was becoming progressively harder to breathe. She felt like she couldn't exhale. Her heart was pounding. Karen stumbled out of the shed. She knew well how an asthma attack felt, and this was a bad one. She reached for the phone and dialed 911.

The effects of hormones on the body can be subtle or dramatic. As we shall see in Karen's case, hormones can significantly influence overall short-term and long-term health in subtle and not-so-subtle ways.

INTRODUCTION

Around age 12, boys and girls enter puberty and start to develop striking differences in physical appearance and behavior. Perhaps no other period in life so dramatically shows the impact of the endocrine system in directing development and regulating body functions. The endocrine system controls body activities by releasing chemical mediators called **hormones**. In girls, estrogens promote accumulation of adipose tissue in the breasts and hips, sculpting a feminine shape. At the same time (or a little later in boys), testosterone begins to help build muscle mass and enlarge

The endocrine system controls body activities by releasing chemical mediators called hormones.

the vocal cords, producing a lower-pitched voice. These changes are just a few examples of the powerful influence of endocrine secretions, or hormones. Less dramatically, perhaps, multitudes of hormones help maintain homeostasis on a daily basis. They regulate the activity of smooth and cardiac muscle, alter metabolism, influence reproductive processes, promote growth and development, and participate in the establishment of circadian (daily) rhythms established by the hypothalamus.

In the nervous system chapters you learned that the nervous system is a control and integration center for your organ systems. In this chapter, we focus on the major endocrine glands and explore how their hormones interact with the nervous system to govern body activities.

17.1   The endocrine system works more slowly than the nervous system, releasing hormones that regulate virtually all body cells.

The nervous and endocrine systems act together to coordinate functions of all body systems. Recall that the nervous system acts through nerve impulses (action potentials) conducted along axons of neurons. At synapses, nerve impulses trigger the release of mediator molecules called *neurotransmitters* (shown in Figure 12.22). The endocrine system controls body activities by releasing mediator molecules called *hormones*, but the means of control of the two systems are very different.

A **hormone** (*hormon* = to excite or get moving) is a molecule that is released in one part of the body but regulates the activity of cells in other parts of the body. Most hormones enter interstitial fluid and then the bloodstream. The circulating blood delivers hormones to cells throughout the body. Both neurotransmitters and hormones exert their effects by binding to receptors on or in their "target" cells. Several mediator molecules act as both neurotransmitters and hormones. One example is norepinephrine, which is released as a neurotransmitter by sympathetic postganglionic neurons and as a hormone by cells of the adrenal medullae.

Responses produced by the endocrine system often are slower than responses initiated by the nervous system. Although some hormones act within seconds, most take several minutes or more to cause a response. When glucose levels in your blood rise after ingesting a sugary treat, insulin is secreted from the pancreas, travels through the bloodstream, and stimulates uptake of glucose by body cells; after several minutes, blood glucose levels return to normal. In contrast, reflex activity of your nervous system moves your hand away from a hot pan within milliseconds. The effects of nervous system activation are generally briefer than those of the endocrine system. The nervous system acts on specific muscles to contract or relax and on glands to secrete more or less of their product. The influence of the endocrine system is much broader; it helps regulate virtually all types of body cells.

The nervous and endocrine systems function together as an interlocking supersystem. For example, certain parts of the nervous system stimulate or inhibit the release of hormones, which in turn may promote or inhibit the generation of nerve impulses.

Table 17.1 compares the characteristics of the nervous and endocrine systems.

Endocrine Glands

Recall from Chapter 4 that the body contains two kinds of glands: exocrine glands and endocrine glands. **Exocrine glands** (*exo-* = outside) secrete their products into ducts that carry the secretions into body cavities, into the lumen of an organ, or to the outer surface of the body. Exocrine glands include sudoriferous (sweat), sebaceous (oil), mucous, and digestive glands. **Endocrine glands** secrete their products (hormones) into the interstitial fluid surrounding the secretory cells rather than into ducts. From the interstitial fluid, hormones diffuse into capillaries and blood carries them to target cells throughout the body. As they approach the target cells, hormones diffuse from blood into the interstitial fluid, where they interact with their target cells. Most hormones are required in very small amounts, so circulating levels typically are low.

The endocrine glands include the pituitary, thyroid, parathyroid, adrenal, and pineal glands (Figure 17.1). In addition, several organs and tissues do not function exclusively as endocrine glands but contain cells that secrete hormones. These include the hypothalamus, thymus, pancreas, ovaries, testes, kidneys, stomach, liver, small intestine, skin, heart, adipose tissue, and placenta. Taken together, all endocrine glands and hormone-secreting cells constitute the *endocrine system*.

Checkpoint

1. Why is the release of acetylcholine at the neuromuscular junction not considered a "hormonal event"?
2. Which system produces a quicker response, the nervous system or the endocrine system? Which system produces a longer-acting response?
3. How does a sudoriferous (sweat) gland differ from an adrenal gland?
4. Why are organs such as the kidneys, stomach, heart, and skin considered to be part of the endocrine system?

TABLE 17.1		
Comparison of Control by the Nervous and Endocrine Systems		
CHARACTERISTIC	NERVOUS SYSTEM	ENDOCRINE SYSTEM
Mediator molecules	Neurotransmitters released locally in response to nerve impulses.	Hormones delivered to tissues throughout the body by the blood.
Site of mediator action	Close to site of release, at a synapse; binds to receptors in postsynaptic membrane.	Far from site of release (usually); binds to receptors on or in target cells.
Types of target cells	Muscle (smooth, cardiac, and skeletal) cells, gland cells, other neurons.	Cells throughout the body.
Time to onset of action	Typically within milliseconds (thousandths of a second).	Seconds to hours or days.
Duration of action	Generally briefer (milliseconds).	Generally longer (seconds to days).

Figure 17.1 **Location of many endocrine glands.** Also shown are other organs that contain endocrine cells and associated structures.

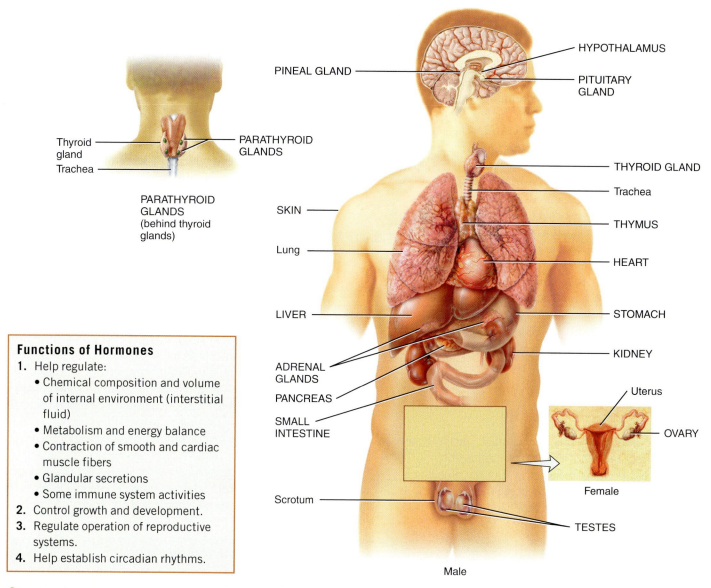

PINEAL GLAND

HYPOTHALAMUS

PITUITARY GLAND

Thyroid gland

PARATHYROID GLANDS

Trachea

PARATHYROID GLANDS (behind thyroid glands)

SKIN

THYROID GLAND

Trachea

THYMUS

Lung

HEART

LIVER

STOMACH

Functions of Hormones
1. Help regulate:
   - Chemical composition and volume of internal environment (interstitial fluid)
   - Metabolism and energy balance
   - Contraction of smooth and cardiac muscle fibers
   - Glandular secretions
   - Some immune system activities
2. Control growth and development.
3. Regulate operation of reproductive systems.
4. Help establish circadian rhythms.

ADRENAL GLANDS

PANCREAS

SMALL INTESTINE

KIDNEY

Uterus

OVARY

Female

Scrotum

TESTES

Male

Endocrine glands secrete hormones, which circulating blood delivers to target tissues.

17.2 The secretion of hormones is regulated by the nervous system, chemical changes in the blood, and other hormones.

The Role of Hormone Receptors

As a rule, most of the body's 50 or so hormones affect only a few types of cells. As a given hormone travels throughout the body in the blood, it affects only certain **target cells**. Hormones, like neurotransmitters, influence their target cells by chemically binding to specific protein **receptors**. Only the target cells for a given hormone have receptors that bind to and recognize that specific hormone. For example, thyroid-stimulating hormone (TSH) binds to receptors on cells of the thyroid gland, but it does not bind to cells of the ovaries because ovarian cells do not have TSH receptors.

Receptors, like other cellular proteins, are constantly being synthesized and broken down. Generally, a target cell has 2000 to 100,000 receptors for a particular hormone. When a hormone is present in excess, the number of receptors on target cells may decrease, making the target cell *less sensitive* to the hormone. For example, when certain cells of the testes are exposed to a high concentration of luteinizing hormone (LH), the number of LH receptors decreases. This decreases the responsiveness of target cells to the hormone. In contrast, when a hormone is deficient, the number of receptors may increase to make the target tissue more sensitive to the hormone.

Chemical Classes of Hormones

Chemically, hormones can be divided into two broad classes: those that are soluble in lipids (fats), and those that are soluble in water. This chemical classification is also useful functionally because the two classes exert their effects on target cells differently.

The lipid-soluble hormones include steroid hormones, thyroid hormones, and nitric oxide. **Steroid hormones** are derived from cholesterol. The two **thyroid hormones** (T_3 and T_4) are synthesized by attaching iodine to the amino acid tyrosine. The gas **nitric oxide (NO)** is both a hormone and a neurotransmitter.

Most water-soluble hormones are amine, peptide, and protein hormones. **Amine hormones** are synthesized by modifying certain amino acids. They are called amines because they retain an amino group $—NH_3^+$. For instance, histamine is synthesized from the amino acid histidine, whereas serotonin and melatonin are derived from the amino acid tryptophan. **Peptide hormones** consist of chains of 3 to 49 amino acids. Examples of peptide hormones are antidiuretic hormone (ADH) and oxytocin. The larger **protein hormones** include 50 to 200 amino acids. Protein hormones include human growth hormone and insulin.

Table 17.2 summarizes the classes of lipid-soluble and water-soluble hormones. It also provides an overview of the major hormones and their sites of secretion.

Hormone Transport in the Blood

Most water-soluble hormone molecules circulate in the watery blood plasma in a "free" form (not attached to plasma proteins), but most lipid-soluble hormone molecules travel in the bloodstream bound to **transport proteins**. The transport proteins, which are synthesized by cells in the liver, have three functions:

- They make lipid-soluble hormones temporarily water-soluble, thus increasing their solubility in blood.

- They retard passage of small hormone molecules through the filtering mechanism in the kidneys, thus slowing the rate of hormone loss in the urine.

- They provide a ready reserve of hormone in the bloodstream.

In general, 0.1–10% of the molecules of a lipid-soluble hormone are not bound to a transport protein. This **free fraction** diffuses out of capillaries, binds to receptors, and triggers hormonal responses. As free hormone molecules leave the blood and bind to their receptors, transport proteins release new ones to replenish the free fraction.

Mechanism of Hormone Action

The response to a hormone depends on both the hormone and the target cell. Various target cells respond differently to the same hormone. Insulin, for example, stimulates synthesis of glycogen in liver cells but synthesis of triglycerides in adipose cells.

The response to a hormone is not always the synthesis of new molecules. Other hormonal effects include changing the permeability of the plasma membrane, stimulating transport of a substance into or out of the target cells, altering the rate of specific metabolic reactions, or causing contraction of smooth muscle or cardiac muscle. In part, these varied effects of hormones are possible because a single hormone can set in motion several different cellular responses. First, however, a hormone must "announce its arrival" to a target cell by binding to its receptors. The receptors for lipid-soluble hormones are located inside target cells; the receptors for water-soluble hormones are part of the plasma membrane of target cells.

TABLE 17.2

Hormones by Chemical Class

CHEMICAL CLASS	HORMONES	SITE OF SECRETION
Lipid-Soluble		
Steroid hormones Aldosterone	Aldosterone, cortisol, and androgens.	Adrenal cortex.
	Calcitriol.	Kidneys.
	Testosterone.	Testes.
	Estrogens and progesterone.	Ovaries.
Thyroid hormones Triiodothyronine (T_3)	T_3 (triiodothyronine) and T_4 (thyroxine).	Thyroid gland (follicular cells).
Gas	Nitric oxide (NO).	Endothelial cells lining blood vessels.

Action of Lipid-Soluble Hormones

Lipid-soluble hormones diffuse through the lipid bilayer of the plasma membrane and bind to their receptors *within* target cells. They exert their effect in the following ways (Figure 17.2):

1 A lipid-soluble hormone molecule detaches from its transport protein in the bloodstream. Then, the free hormone diffuses from blood into interstitial fluid, and through the plasma membrane into a cell.

2 If the cell is a target cell, the hormone binds to and activates receptors located within the cell. The activated receptor–hormone complex then alters gene expression by turning specific genes on or off.

3 As the DNA is transcribed, new messenger RNA (mRNA) forms, leaves the nucleus, and enters the cytosol. There, it directs synthesis of a new protein, often an enzyme, on the ribosomes.

4 The new proteins alter the cell's activity and cause the responses typical of that specific hormone.

Figure 17.2 Mechanism of action of lipid-soluble hormones.

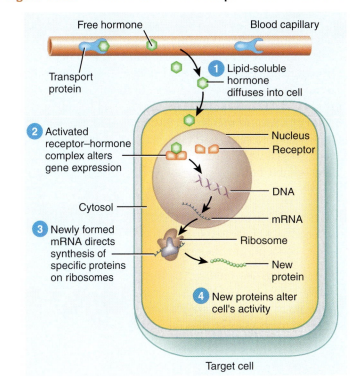

Lipid-soluble hormones bind to receptors inside target cells.

TABLE 17.2 (continued)		
Hormones by Chemical Class		
CHEMICAL CLASS	HORMONES	SITE OF SECRETION
Water-soluble		
Amines	Epinephrine and norepinephrine (catecholamines).	Adrenal medulla.
	Melatonin.	Pineal gland.
	Histamine.	Mast cells in connective tissues.
Norepinephrine	Serotonin.	Platelets in blood.
Peptides and proteins	All hypothalamic releasing and inhibiting hormones.	Hypothalamus.
	Oxytocin, antidiuretic hormone.	Posterior pituitary.
	Human growth hormone, thyroid-stimulating hormone, adrenocorticotropic hormone, follicle-stimulating hormone, luteinizing hormone, prolactin, melanocyte-stimulating hormone.	Anterior pituitary.
	Insulin, glucagon, somatostatin, pancreatic polypeptide.	Pancreas.
	Parathyroid hormone.	Parathyroid glands.
	Calcitonin.	Thyroid gland (parafollicular cells).
Oxytocin	Gastrin, secretin, cholecystokinin, GIP (glucose-dependent insulinotropic peptide).	Stomach and small intestine (enteroendocrine cells).
	Erythropoietin.	Kidneys.
	Leptin.	Adipose tissue.
Eicosanoids	Prostaglandins, leukotrienes.	All cells except red blood cells.
A leukotriene (LTB$_4$)		

Action of Water-soluble Hormones

Because most amino acid–based hormones are not lipid-soluble, they cannot diffuse through the lipid bilayer of the plasma membrane. Instead, water-soluble hormones bind to receptors that protrude from the target cell surface. When a water-soluble hormone binds to its receptor at the outer surface of the plasma membrane, it acts as the **first messenger**. The first messenger then causes production of a **second messenger** inside the cell, where specific hormone-stimulated responses take place. One common second messenger is **cyclic AMP (cAMP)**.

Water-soluble hormones exert their effects as follows (Figure 17.3):

1 A water-soluble hormone (the first messenger) diffuses from the blood and binds to its receptor on the exterior surface of a target cell's plasma membrane.

2 As a result of the binding, the enzyme **adenylate cyclase** converts ATP into cAMP. Because the enzyme's active site is on the inner surface of the plasma membrane, this reaction occurs in the cytosol of the cell.

3 Cyclic AMP (the second messenger) causes the activation of several enzymes.

4 Activated enzymes catalyze reactions that produce physiological responses.

Figure 17.3   Mechanism of action of water-soluble hormones.

Water-soluble hormones bind to receptors embedded in the plasma membrane of target cells.

5 After a brief period, cAMP is inactivated. Thus, the cell's response is turned off unless new hormone molecules continue to bind to receptors in the plasma membrane.

Hormones that bind to plasma membrane receptors can induce their effects at very low concentrations because they initiate a cascade or chain reaction, each step of which multiplies or amplifies the initial effect. For example, the binding of a single molecule of epinephrine (first messenger) to its receptor on a liver cell may activate a hundred or so adenylate cyclase molecules. If each adenylate cyclase produces even 1000 cyclic AMP, then 100,000 of these second messengers will be liberated inside the cell. Each cyclic AMP may activate an enzyme molecule, which in turn can act on hundreds or thousands of substrate molecules. In the breakdown of glycogen, binding of a single molecule of epinephrine to its receptor results in the breakdown of millions of glycogen molecules into glucose monomers.

Hormone Interactions

The responsiveness of a target cell to a hormone depends on (1) the hormone's concentration, (2) the abundance of the target cell's hormone receptors, and (3) influences exerted by other hormones. A target cell responds more vigorously when the level of a hormone rises or when the target cell has more receptors. In addition, the actions of some hormones on target cells require a simultaneous or recent exposure to a second hormone to exert their full effects. In such cases, the second hormone is said to have a **permissive effect**. For example, epinephrine alone only weakly stimulates lipolysis (the breakdown of triglycerides). Small amounts of thyroid hormones (T_3 and T_4) must be present for epinephrine to powerfully stimulate lipolysis. Sometimes the permissive hormone increases receptors for the other hormone, and sometimes it promotes the synthesis of an enzyme required for the expression of the other hormone's effects.

The **synergistic effect** occurs when more than one hormone produces the same effects and their combined effects are greater or more extensive than the sum of each hormone acting alone. For example, normal development of oocytes in the ovaries requires both follicle-stimulating hormone from the anterior pituitary and estrogens from the ovaries. Neither hormone alone is sufficient.

When one hormone opposes the actions of another hormone, the two hormones are said to have **antagonistic effects**. An example of an antagonistic pair of hormones is insulin, which promotes synthesis of glycogen by liver cells, and glucagon, which stimulates breakdown of glycogen in the liver.

Control of Hormone Secretion

The release of most hormones occurs in short bursts, with little or no secretion between bursts. When stimulation increases, an endocrine gland will release its hormone in more frequent bursts, thus elevating the hormone's overall blood concentration. In the absence of stimulation, bursts are inhibited, and the blood level of the hormone decreases. Regulation of hormone secretion normally maintains homeostasis and prevents overproduction or underproduction of any given hormone.

Hormone secretion is regulated by (1) signals from the nervous system, (2) chemical changes in the blood, and (3) other hormones. For example, nerve impulses to the adrenal medullae regulate the release of epinephrine; blood Ca^{2+} levels regulate the secretion of parathyroid hormone; and a hormone from the anterior pituitary (adrenocorticotropic hormone) stimulates the release of cortisol by the adrenal cortex. Most systems that regulate secretion of hormones work by negative feedback (see Figure 1.3), but a few operate by positive feedback (see Figure 1.4). For example, during childbirth, the hormone oxytocin stimulates contractions of the uterus, and uterine contractions, in turn, stimulate more oxytocin release, a positive feedback effect.

Now that you have a general understanding of the roles of hormones in the endocrine system, we next discuss the various endocrine glands and the hormones they secrete.

Checkpoint

5. Why would follicle-stimulating hormone act on the ovaries but not on the thyroid?
6. Identify the chemical classes of hormones.
7. Where do lipid-soluble hormones bind to their target cells? Where do water-soluble hormones bind to their target cells?
8. Why is cyclic AMP considered a "second messenger"?
9. What factors determine the responsiveness of a target cell to a hormone?
10. What are the three influences over hormone secretion?

17.3 The hypothalamus regulates anterior pituitary hormone secretion of seven important hormones.

The Hypothalamus

For many years, the **pituitary gland** (pi-TOO-i-tār-ē), or **hypophysis** (hī-POF-i-sis), was called the master endocrine gland because it secretes several hormones that control other endocrine glands. We now know that the pituitary gland itself has a master—the **hypothalamus**. This small region of the brain inferior to the thalamus is the major integrating link between the nervous and endocrine systems. Cells in the hypothalamus synthesize at least nine different hormones, and the pituitary gland secretes seven. Hormones secreted by the hypothalamus and the pituitary gland play important roles in the regulation of virtually all aspects of growth, development, metabolism, and homeostasis.

The Pituitary Gland

The pituitary gland is a pea-shaped structure measuring 1–1.5 cm (0.5 in.) in diameter that lies in the hypophyseal fossa of the sella turcica of the sphenoid bone and attaches to the hypothalamus by a stalk, the **infundibulum** (= a funnel; Figure 17.4a). The pituitary gland has two anatomically and functionally separate portions. The **anterior pituitary** accounts for about 75% of the total weight of the gland; it produces and secretes several hormones that control a range of organs. The **posterior pituitary** contains axons of neurons with cell bodies located in the hypothalamus. As we will see shortly, the posterior pituitary does not produce any hormones, but it does release hormones that are synthesized by the hypothalamus.

Anterior Pituitary

The **anterior pituitary**, or **adenohypophysis** (ad´-e-nō-hī-POF-i-sis; *adeno-* = gland; *-hypophysis* = undergrowth), secretes hormones that regulate a wide range of bodily activities. Release of anterior pituitary hormones is stimulated by **releasing hormones** and suppressed by **inhibiting hormones** from the hypothalamus (Figure 17.4b). These hypothalamic hormones are an important link between the nervous and endocrine systems.

Hypothalamic hormones reach the anterior pituitary through a portal system. Usually, blood passes from the heart through an artery to a capillary to a vein and back to the heart. In a portal system, blood flows from one capillary network into a portal vein, and then into a second capillary network without passing through the heart. In the **hypophyseal portal system** (hī´-pō-FIZ-ē-al; Figure 17.4c), blood flows from capillaries in the hypothalamus into portal veins that carry blood to capillaries of the anterior pituitary. The **superior hypophyseal arteries**, branches of the internal carotid arteries, bring blood to the infundibulum, where the arteries divide into a capillary network called the **primary plexus**. From the primary plexus, blood drains into the **hypophyseal portal veins** that pass down the infundibulum, to the anterior pituitary, where the veins divide again and form another capillary network called the **secondary plexus**.

Figure 17.4   Hypothalamus and pituitary gland, and their blood supply. Part (b) indicates that releasing and inhibiting hormones synthesized by hypothalamic neurons are transported within axons and released from the axon terminals. The hormones diffuse into capillaries of the primary plexus of the hypophyseal portal system and are carried by the hypophyseal portal veins to the secondary plexus of the hypophyseal portal system for distribution to target cells in the anterior pituitary.

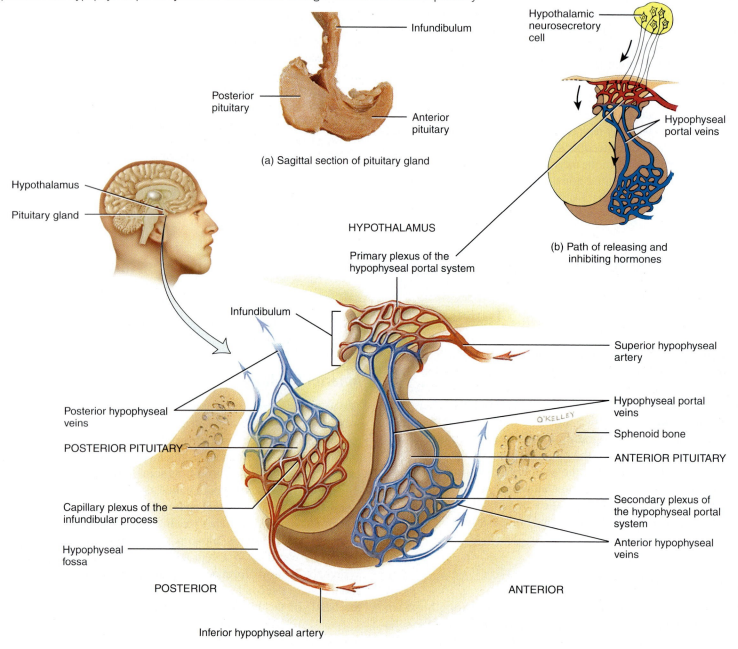

(a) Sagittal section of pituitary gland

(b) Path of releasing and inhibiting hormones

(c) Relationship of the hypothalamus to the pituitary gland

🔑 **Hypothalamic hormones are an important link between the nervous and endocrine systems.**

The hypothalamus contains clusters of specialized neurons called **neurosecretory cells**. They synthesize the hypothalamic releasing and inhibiting hormones in their cell bodies and package the hormones inside vesicles, which reach the axon terminals by axonal transport. When nerve impulses stimulate the axon terminals, the vesicles undergo exocytosis. The hormones then diffuse into the primary plexus of the hypophyseal portal system.

Quickly, the hypothalamic hormones flow with the blood through the hypophyseal portal veins and into the secondary plexus. This direct route permits hypothalamic hormones to act immediately on anterior pituitary cells, before the hormones are diluted or destroyed in the general circulation. Hormones secreted by anterior pituitary cells pass into the secondary plexus capillaries, which drain into the anterior hypophyseal veins. Now in the general circulation, anterior pituitary hormones travel to target tissues throughout the body.

Those anterior pituitary hormones that influence another endocrine gland are called **tropic hormones** or **tropins**. Several of the anterior pituitary hormones are tropins. The two gonadotropins, follicle-stimulating hormone and luteinizing hormone, regulate the functions of the gonads (ovaries and testes). Thyrotropin stimulates the thyroid gland, and corticotropin acts on the cortex of the adrenal gland.

Control of Secretion by the Anterior Pituitary

Secretion of anterior pituitary hormones is regulated in two ways. First, neurosecretory cells in the hypothalamus secrete releasing hormones, which stimulate secretion of anterior pituitary hormones, and inhibiting hormones, which suppress secretion of anterior pituitary hormones (Table 17.3). Second, negative feedback, in the form of rising blood levels of hormones released by target glands, decreases secretions of some anterior pituitary cells (Figure 17.5). For example, corticotropin-releasing hormone (CRH) stimulates secretion of corticotropin, or adrenocorticotropic hormone (ACTH), by the anterior pituitary. ACTH stimulates the cortex of the adrenal gland to secrete glucocorticoids, mainly cortisol. In turn, an elevated blood level of cortisol decreases secretion of both corticotropin and corticotropin-releasing hormone due to negative feedback suppression of the anterior pituitary and hypothalamus.

Human Growth Hormone and Insulinlike Growth Factors

Human growth hormone (hGH), also known as **somatotropin**, is the most abundant anterior pituitary hormone. The main function of hGH is to promote synthesis and secretion of small protein hormones called **insulinlike growth factors (IGFs)**. IGFs are so named because some of their actions are similar to those of insulin. In response to human growth hormone, cells in the liver, skeletal muscle, cartilage, bones, and other tissues secrete IGFs, which may either enter the bloodstream or act locally. IGFs cause cells to grow and multiply by increasing uptake of amino acids into cells and accelerating protein synthesis. IGFs also decrease the breakdown of proteins and the use of amino acids for ATP production.

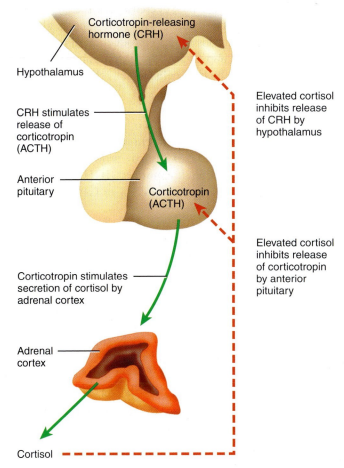

Figure 17.5 **Negative feedback regulation of hypothalamic and anterior pituitary secretion.** Solid green arrows show stimulation of secretions; dashed red lines show inhibition of secretion by negative feedback.

Cortisol secreted by the adrenal cortex suppresses secretion of CRH and ACTH.

Due to these effects of the IGFs, human growth hormone increases the growth rate of the skeleton and skeletal muscles during childhood and the teenage years. In adults, human growth hormone and IGFs help maintain the mass of muscles and bones and promote healing of injuries and tissue repair.

TABLE 17.3

Control of Secretion by the Anterior Pituitary

HORMONE	SECRETED BY	RELEASING HORMONE (STIMULATES SECRETION)	INHIBITING HORMONE (SUPPRESSES SECRETION)
Human growth hormone (hGH)	Somatotrophs.	Growth hormone–releasing hormone (GHRH).	Growth hormone–inhibiting hormone (GHIH).
Thyroid-stimulating hormone (TSH)	Thyrotrophs.	Thyrotropin-releasing hormone (TRH).	Growth hormone–inhibiting hormone (GHIH).
Follicle-stimulating hormone (FSH)	Gonadotrophs.	Gonadotropin-releasing hormone (GnRH)	—
Luteinizing hormone (LH)	Gonadotrophs.	Gonadotropin-releasing hormone (GnRH).	—
Prolactin (PRL)	Lactotrophs.	Prolactin-releasing hormone (PRH).	Prolactin-inhibiting hormone (PIH).
Adrenocorticotropic hormone (ACTH)	Corticotrophs.	Corticotropin-releasing hormone (CRH).	—
Melanocyte-stimulating hormone (MSH)	Corticotrophs.	Corticotropin-releasing hormone (CRH).	Dopamine.

Insulinlike growth factors elevate the level of glucose in blood. IGFs stimulate the breakdown of triglycerides (fats), releasing fatty acids for ATP production by body cells. Human growth hormone and IGFs influence carbohydrate metabolism by decreasing glucose uptake, which decreases the use of glucose for ATP production by most body cells. This action spares glucose so that it is available to neurons for ATP production in times of glucose scarcity. IGFs and human growth hormone also stimulate the breakdown of glycogen by liver cells, which releases glucose into the blood.

The anterior pituitary releases hGH in bursts that occur every few hours, especially during sleep. Two hypothalamic hormones control secretion of hGH: **growth hormone–releasing hormone (GHRH)** promotes secretion of human growth hormone, and **growth hormone–inhibiting hormone (GHIH)** suppresses it. Blood glucose levels are a major regulator of GHRH and GHIH secretion (Figure 17.6):

① **Hypoglycemia** (hī´-po-glī-SĒ-mē-a), abnormally low blood glucose levels, stimulates the hypothalamus to secrete GHRH, which flows through the hypophyseal portal veins to the anterior pituitary.

② GHRH stimulates the anterior pituitary to release human growth hormone.

③ Human growth hormone stimulates secretion of insulinlike growth factors, which speed up breakdown of liver glycogen into glucose, causing glucose to enter the blood more rapidly.

④ As a result, blood glucose rises to the normal level (about 90 mg/100 mL of blood plasma).

⑤ An increase in blood glucose above the normal level inhibits release of GHRH.

⑥ **Hyperglycemia** (hī´-per-glī-SĒ-mē-a), an abnormally high blood glucose concentration, stimulates the hypothalamus to secrete GHIH.

⑦ On reaching the anterior pituitary in portal blood, GHIH inhibits secretion of human growth hormone.

⑧ A low level of human growth hormone and IGFs slows breakdown of glycogen in the liver, and glucose is released into the blood more slowly.

⑨ Blood glucose falls to the normal level.

⑩ A decrease in blood glucose below the normal level (hypoglycemia) inhibits release of GHIH.

Other stimuli that promote secretion of human growth hormone include decreased fatty acids and increased amino acids in the blood; deep sleep; increased activity of the sympathetic division of the autonomic nervous system, such as might occur with

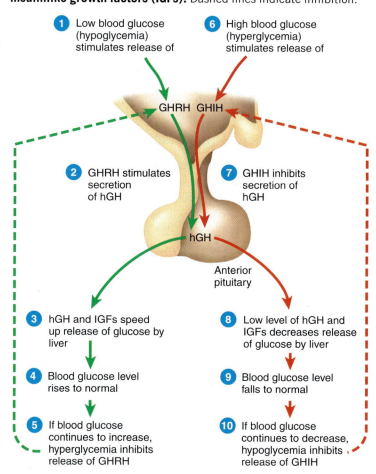

Figure 17.6   Effects of human growth hormone (hGH) and insulinlike growth factors (IGFs). Dashed lines indicate inhibition.

1 Low blood glucose (hypoglycemia) stimulates release of

6 High blood glucose (hyperglycemia) stimulates release of

GHRH   GHIH

2 GHRH stimulates secretion of hGH

7 GHIH inhibits secretion of hGH

hGH

Anterior pituitary

3 hGH and IGFs speed up release of glucose by liver

8 Low level of hGH and IGFs decreases release of glucose by liver

4 Blood glucose level rises to normal

9 Blood glucose level falls to normal

5 If blood glucose continues to increase, hyperglycemia inhibits release of GHRH

10 If blood glucose continues to decrease, hypoglycemia inhibits release of GHIH

🔑 **Secretion of hGH is stimulated by growth hormone–releasing hormone (GHRH) and inhibited by growth hormone–inhibiting hormone (GHIH).**

stress or vigorous physical exercise; and other hormones. Factors that inhibit human growth hormone secretion include increased levels of fatty acids and decreased levels of amino acids in the blood, light sleep, emotional deprivation, obesity, and human growth hormone itself (through negative feedback).

Thyroid-Stimulating Hormone

Thyroid-stimulating hormone (TSH), or **thyrotropin**, stimulates the synthesis and secretion of the two thyroid hormones, triiodothyronine (T_3) and thyroxine (T_4), by the thyroid gland. **Thyrotropin-releasing hormone (TRH)** from the hypothalamus controls TSH secretion. Release of TRH, in turn, depends on blood levels of T_3 and T_4, which inhibit secretion of TRH via negative feedback. There is no thyrotropin-inhibiting hormone. The release of TRH is explained later in the chapter (see Figure 17.11).

Follicle-stimulating Hormone

In females, the ovaries are the targets for **follicle-stimulating hormone (FSH)**. Each month FSH initiates the development of several ovarian follicles, saclike arrangements of secretory cells that surround a developing oocyte (future ovum). In males, FSH stimulates sperm production in the testes. **Gonadotropin-releasing hormone (GnRH)** from the hypothalamus stimulates FSH release. Release of GnRH and FSH is suppressed by estrogens (female sex hormones) in females and by testosterone (the principal male sex hormone) in males through negative feedback systems. There is no gonadotropin-inhibiting hormone.

Luteinizing Hormone

In females, **luteinizing hormone (LH)** triggers **ovulation**, the release of an oocyte by an ovary. After ovulation, LH stimulates formation of the corpus luteum in the ovary and the secretion of progesterone (another female sex hormone) by the corpus luteum. Together, FSH and LH also stimulate ovarian follicles to secrete estrogens (additional female sex hormones). Estrogens and progesterone prepare the uterus for implantation of a fertilized ovum and help prepare the mammary glands for milk secretion. In males, LH stimulates the testes to secrete testosterone. Secretion of LH, like that of FSH, is stimulated by gonadotropin-releasing hormone (GnRH).

Prolactin

Prolactin (PRL), together with other hormones, initiates and maintains milk secretion by the mammary glands. Only after the mammary glands have been primed by other hormones exerting permissive effects (estrogens, progesterone, glucocorticoids, human growth hormone, thyroxine, and insulin) does PRL bring about milk secretion. Ejection of milk from the mammary glands depends on the hormone oxytocin, which is released from the posterior pituitary. These two hormones act synergistically. Together, milk secretion and ejection constitute *lactation*.

The hypothalamus secretes both excitatory and inhibitory hormones that regulate prolactin secretion. In females, **prolactin-inhibiting hormone (PIH)** inhibits the release of prolactin from the anterior pituitary most of the time. Each month, just before menstruation begins, the secretion of PIH diminishes and the blood level of prolactin rises, but not enough to stimulate milk production. Breast tenderness just before menstruation may be caused by elevated prolactin. As the menstrual cycle begins anew, PIH is again secreted and the prolactin level drops. During pregnancy, the prolactin level rises, stimulated by **prolactin-releasing hormone (PRH)** from the hypothalamus. The sucking action of a nursing infant causes a reduction in hypothalamic secretion of PIH, and therefore an increase in prolactin secretion.

The function of prolactin is not known in males, but its hypersecretion causes erectile dysfunction (impotence, the inability to have an erection of the penis).

Adrenocorticotropic Hormone

Adrenocorticotropic hormone (ACTH), or **corticotropin**, controls the production and secretion of glucocorticoids, primarily cortisol, by the cortex (outer portion) of the adrenal glands. **Corticotropin-releasing hormone (CRH)** from the hypothalamus stimulates secretion of ACTH. There is no corticotropin-inhibiting hormone. Stress-related stimuli, such as low blood glucose or physical trauma, and interleukin-1, a substance produced by macrophages, also stimulate release of ACTH. Glucocorticoids inhibit both CRH and ACTH release via negative feedback (see Figure 17.16).

Melanocyte-stimulating Hormone

There is little circulating **melanocyte-stimulating hormone (MSH)** in humans. MSH increases skin pigmentation in amphibians by stimulating the dispersion of melanin granules in melanocytes. Its exact role in humans is unknown, but the presence of MSH receptors in the brain suggests it may influence brain activity. Excessive levels of corticotropin-releasing hormone (CRH) can stimulate MSH release, and prolactin-inhibiting hormone (PIH) inhibits MSH release.

Table 17.4 summarizes the principal actions of the anterior pituitary hormones.

Checkpoint

11. What is the functional importance of the hypophyseal portal veins?
12. Which anterior pituitary hormones stimulate a second endocrine gland to release a hormone?
13. You have noticed your neighbor's 8-year-old child is the same height as her 5-year-old cousin. This child could be deficient in which hormone?
14. If a person has a pituitary tumor that secretes a large amount of hGH and the tumor cells are not responsive to regulation by GHRH and GHIH, will hyperglycemia or hypoglycemia be more likely? Explain your answer.
15. Explain how blood levels of T_3/T_4, TSH, and TRH would change in a laboratory animal that has undergone a thyroidectomy (complete removal of its thyroid gland).

TABLE 17.4

The Principal Actions of Anterior Pituitary Hormones

HORMONE AND TARGET TISSUES	PRINCIPAL ACTIONS
Human growth hormone (hGH) Liver	Stimulates liver, muscle, cartilage, bone, and other tissues to secrete insulinlike growth factors (IGFs); IGFs promote growth of body cells, protein synthesis, tissue repair, lipolysis, and elevation of blood glucose concentration.
Thyroid-stimulating hormone (TSH) Thyroid gland	Stimulates secretion of thyroid hormones by thyroid gland.
Follicle-stimulating hormone (FSH) Ovary Testis	In females, initiates development of oocytes and induces ovarian secretion of estrogens. In males, stimulates testes to produce sperm.
Luteinizing hormone (LH) Ovary Testis	In females, stimulates secretion of estrogens and progesterone, ovulation, and formation of corpus luteum. In males, stimulates interstitial cells in testes to secrete testosterone.
Protactin (PRL) Mammary glands	Together with other hormones, promotes milk secretion by the mammary glands.
Adrenocorticotropic hormone (ACTH) Adrenal cortex	Stimulates secretion of glucocorticoids (mainly cortisol) by adrenal cortex.
Melanocyte-stimulating hormone (MSH) Brain	Exact role in humans is unknown, but may influence brain activity; when present in excess, can cause darkening of skin.

17.4 Oxytocin and antidiuretic hormone originate in the hypothalamus and are stored in the posterior pituitary.

Although the **posterior pituitary**, or **neurohypophysis**, does not *synthesize* hormones, it does *store* and *release* two hormones. The posterior pituitary contains axon terminals of neurosecretory cells with cell bodies located in the hypothalamus. Hypothalamic neurosecretory cell axons begin in the hypothalamus and end near blood capillaries in the posterior pituitary gland (Figure 17.7).

Different neurosecretory cells produce two hormones: **oxytocin (OT)** (ok´-sē-TŌ-sin; *oxytoc-* = quick birth) and **antidiuretic hormone (ADH)**, also called **vasopressin** (vā-sō-PRES-in; *vaso-* = blood vessel; *-pressus* = to press).

After their production in the hypothalamic cell bodies of neurosecretory cells, oxytocin and antidiuretic hormone are packed

Figure 17.7 **Axons of hypothalamic neurosecretory cells extend from the hypothalamus to the posterior pituitary.** Hormone molecules synthesized in the cell body of a neurosecretory cell move down to the axon terminals. Nerve impulses trigger exocytosis of the vesicles and release the hormone.

Oxytocin and antidiuretic hormone are synthesized in the hypothalamus and released into capillaries in the posterior pituitary.

into vesicles and transported down the axons to the axon terminals in the posterior pituitary, where they are stored until nerve impulses trigger exocytosis and release of the hormone. As the vesicles undergo exocytosis, the hormone within them is released and diffuses into the capillaries of the posterior pituitary.

Blood is supplied to the posterior pituitary by the **inferior hypophyseal arteries** (see Figure 17.4), which branch from the internal carotid arteries. In the posterior pituitary, the inferior hypophyseal arteries drain into the **capillary plexus** of the **infundibular process**, a capillary network that receives oxytocin and antidiuretic hormone secreted from the neurosecretory cells of the hypothalamus. From this plexus, hormones pass into the **posterior hypophyseal veins** for distribution to target cells.

Oxytocin

During and after delivery of a baby, oxytocin affects two target tissues: the mother's uterus and breasts. During delivery, oxytocin enhances contraction of smooth muscle cells in the wall of the uterus. Initial uterine contractions push the baby's head into the cervix, the lowest part of the uterus. Stretching of the cervix sends nerve impulses to the hypothalamic neurosecretory cells, stimulating the release of OT from the posterior pituitary. A positive feedback loop occurs as OT causes even greater contraction of the uterus, pushing the baby farther down to the cervix. As the cervix stretches further, nerve impulses arriving at the hypothalamus stimulate the release of even more OT, and even greater contraction of the uterus occurs (see Figure 1.4).

After delivery, OT stimulates milk ejection ("letdown") from the mammary glands in response to the mechanical stimulus provided by a suckling infant. OT secretion operates by positive feedback. Nerve impulses from a suckled nipple are transmitted to the hypothalamus, stimulating hypothalamic neurosecretory cells to secrete OT from the posterior pituitary, and additional breast milk is ejected.

The function of oxytocin in males and in nonpregnant females is unclear. Experiments with animals have suggested that its actions within the brain foster parental caretaking behavior toward young offspring. OT may also be responsible, in part, for the feelings of sexual pleasure during and after intercourse.

Antidiuretic Hormone

As its name implies, an **antidiuretic** (an´-ti-dī-ū-RET-ik; *anti-* = against; *-diuretic* = increases urine production) is a substance that decreases urine production. ADH causes the kidneys to return more water to the blood, thus decreasing urine volume. In the absence of ADH, urine output increases more than tenfold, from the normal 1 to 2 liters to about 20 liters a day. ADH also decreases the water lost through sweating and causes constriction of arterioles, which increases blood pressure. This hormone's other name, vasopressin, reflects its effect on blood pressure.

The amount of ADH secreted varies with blood osmotic pressure and blood volume. Blood osmotic pressure is proportional to the concentration of solutes in the blood plasma. Figure 17.8 shows regulation of ADH secretion and the actions of ADH.

❶ High blood osmotic pressure—due to dehydration or a decline in blood volume because of hemorrhage, diarrhea, or excessive sweating—stimulates **osmoreceptors**, neurons in the hypothalamus that monitor blood osmotic pressure.

❷ Osmoreceptors activate the hypothalamic neurosecretory cells that synthesize and release ADH.

❸ When neurosecretory cells receive excitatory input from the osmoreceptors, they generate nerve impulses that travel down the axons of neurosecretory cells to the axon terminals in the posterior pituitary, where they cause exocytosis of ADH-containing vesicles. This liberates ADH, which diffuses into blood capillaries of the posterior pituitary.

❹ The blood carries ADH to three target tissues: the kidneys, sudoriferous (sweat) glands, and smooth muscle in blood vessel walls. The kidneys respond by retaining more water, which decreases urine output. Secretory activity of sweat glands decreases, which lowers the rate of water loss by perspiration from the skin. Smooth muscle in the walls of arterioles (small arteries) contracts in response to high levels of ADH, which constricts (narrows) the lumen of these blood vessels and increases blood pressure.

❺ Low blood osmotic pressure (or increased blood volume) inhibits the osmoreceptors.

❻ Inhibition of osmoreceptors reduces or stops ADH secretion. The kidneys then retain less water by forming a larger volume of urine, secretory activity of sweat glands increases, and arterioles dilate. The blood volume and osmotic pressure of body fluids return to normal.

Secretion of ADH can also be altered in other ways. Pain, stress, trauma, anxiety, acetylcholine, nicotine, and drugs such as morphine, tranquilizers, and some anesthetics stimulate ADH secretion, leading to decreased urination. Drinking alcohol often causes frequent and copious urination because the alcohol inhibits secretion of ADH. The resulting dehydration may cause both the thirst and the headache typical of a hangover.

Figure 17.8   Regulation of secretion and actions of antidiuretic hormone (ADH).

❶ High blood osmotic pressure stimulates hypothalamic osmoreceptors

❷ Osmoreceptors activate the neurosecretory cells that synthesize and release ADH

Osmoreceptors

❺ Low blood osmotic pressure inhibits hypothalamic osmoreceptors

❻ Inhibition of osmoreceptors reduces or stops ADH secretion

Hypothalamus

❸ Nerve impulses liberate ADH from axon terminals in the posterior pituitary into the bloodstream

ADH

Target tissues

❹ Kidneys retain more water, which decreases urine output

Sudoriferous (sweat) glands decrease water loss by perspiration from the skin

Arterioles constrict, which increases blood pressure

🔑 **ADH acts to retain body water and increase blood pressure.**

Table 17.5 lists the posterior pituitary hormones, control of their secretion, and their principal actions.

Checkpoint

16. Where are the hormones secreted by the posterior pituitary produced? How do OT and ADH get from their production site to the capillaries of the posterior pituitary?

17. Functionally, how are the axons of hypothalamic neurosecretory cells and the hypophyseal portal veins similar?

18. Is OT regulated by positive or negative feedback? What feedback stimuli influence the release of OT?

19. Sometimes during difficult deliveries or following long labor a synthetic form of oxytocin is given to the mother. Why?

20. What effect would drinking a large volume of water have on the osmotic pressure of your blood? The level of ADH?

TABLE 17.5

Posterior Pituitary Hormones

HORMONE AND TARGET TISSUES	CONTROL OF SECRETION	PRINCIPAL ACTIONS
Oxytocin (OT) Uterus Mammary glands	Neurosecretory cells of hypothalamus secrete OT in response to uterine distention and stimulation of nipples.	Stimulates contraction of smooth muscle cells of uterus during childbirth; stimulates milk ejection from mammary glands.
Antidiuretic hormone (ADH) or vasopressin Kidneys Sudoriferous (sweat) glands Arterioles	Neurosecretory cells of hypothalamus secrete ADH in response to elevated blood osmotic pressure, dehydration, loss of blood volume, pain, or stress; low blood osmotic pressure, high blood volume, and alcohol inhibit ADH secretion.	Conserves body water by decreasing urine volume; decreases water loss through perspiration; raises blood pressure by constricting arterioles.

Return to Karen's Story

Karen told the emergency room doctor that she had been taking asthma medication, some sort of steroid she could never remember the name of. "Dr. B, I just know something's wrong with me. I'm tired all the time, I'm constipated, and I've been gaining this weight no matter how much I diet." Karen went on with her life history, telling the doctor about all of her children, and about how after every one of her pregnancies she had nearly died, what with all the bleeding, what her doctor had called "postpartum hemorrhaging." She had never been able to breast-feed. After her last child was born, she had stopped menstruating and had never had another baby. She was 54 now, with seven grandchildren.

"Your medical history and symptoms suggest endocrine system involvement. Because of your symptoms—fatigue, weakness, intolerance to cold, and weight gain—I'm going to order a thyroid function test," Dr. Beaumont said.

When Dr. Beaumont returned with the test results, Karen asked, "What did you find out?"

"You're slightly hypothyroid. That, along with your history, is suggestive of a form of hypopituitarism called Sheehan's syndrome," replied the doctor. "It's not life-threatening, but we should consider hormone therapy. It will help with the symptoms you've been having."

"You mean there's a pill I could have been taking all along?" asked Karen.

"It's a bit more complicated than that," Dr. Beaumont explained. "The master gland, the pituitary gland, controls all the other hormones in your body, and it isn't functioning properly. When a woman gives birth, sometimes excessive bleeding can cause the blood pressure to become too low, reducing blood flow to the pituitary gland and damaging it. The pituitary gland regulates your menstrual cycle, your metabolism, and lots of other things. It's likely the reason you stopped menstruating and couldn't breast-feed, and feel fatigued, weak, and have gained weight over the years since. The endocrinologist will explain the therapy."

"Hi, Dr. Allen. This is Dr. Beaumont. I have a patient I'd like you to see. She presented with an acute asthma exacerbation, which we treated successfully. However, she did have a history of amenorrhea, excessive postpartum bleeding, and an inability to lactate after each of her five pregnancies. I suspected she might have some sort of pituitary insufficiency. We ran some tests; the thyroid function tests showed she had low T_3 and T_4 resin uptake values, but normal thyrotropin levels. I'm thinking she may have secondary hypothyroidism caused by a primary hypopituitarism, maybe Sheehan's syndrome. What do you think?"

"Sheehan's syndrome—that's so rare!" exclaimed Dr. Allen. "But with the history of postpartum bleeding, it is possible she did have postpartum necrosis of the pituitary. Was her thyroid enlarged?" asked Dr. Allen.

"No," answered Dr. Beaumont.

Karen went to see the endocrinologist. "Hello, Karen. We have the results back from your ACTH stimulation and thyroid function tests. The lab tests show that your thyroid gland is not functioning properly, most likely due to a problem with your pituitary gland. The hypothyroidism is a secondary symptom of the pituitary problem. The pituitary was probably damaged when you were giving birth, years ago. The ACTH stimulation test helped us confirm this. I'd like to start you on hormone therapy."

A. If Karen's glands were not producing hormones, how would that affect the number of target cell receptors?

B. Are Karen's thyroid hormones lipid-soluble or water-soluble molecules? How are thyroid hormones carried in the blood? How would this affect measurement of their levels in the blood?

C. Could any of Karen's symptoms be related to the function of the anterior pituitary? What questions would you have asked Karen to discover whether there was damage to the posterior pituitary? Which anterior pituitary hormone and posterior pituitary hormones control lactation and could have been missing in Karen postpartum (following childbirth)?

17.5 The thyroid gland secretes thyroxine, triiodothyronine, and calcitonin.

The butterfly-shaped **thyroid gland** is located just inferior to the larynx (voice box); the right and left **lobes** lie on either side of the trachea (Figure 17.9a,d). Connecting the lobes is a mass of tissue called an **isthmus** (IS-mus) that lies anterior to the trachea. The gland usually weighs about 30 g (1 oz) and has a rich blood supply, receiving 80–120 mL of blood per minute.

Figure 17.9   The thyroid gland.

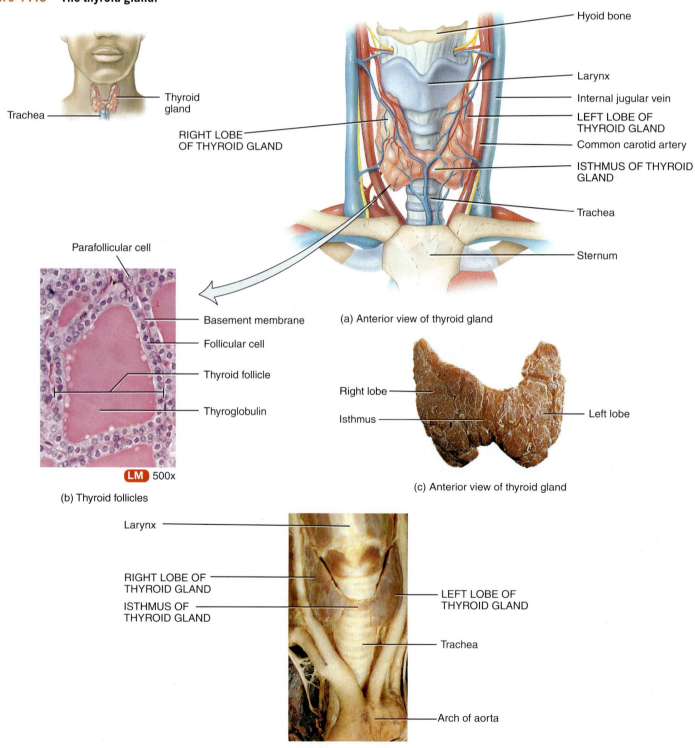

Trachea — Thyroid gland

RIGHT LOBE OF THYROID GLAND

Hyoid bone

Larynx

Internal jugular vein

LEFT LOBE OF THYROID GLAND

Common carotid artery

ISTHMUS OF THYROID GLAND

Trachea

Sternum

(a) Anterior view of thyroid gland

Parafollicular cell

Basement membrane

Follicular cell

Thyroid follicle

Thyroglobulin

LM 500x

(b) Thyroid follicles

Right lobe

Isthmus

Left lobe

(c) Anterior view of thyroid gland

Larynx

RIGHT LOBE OF THYROID GLAND

ISTHMUS OF THYROID GLAND

LEFT LOBE OF THYROID GLAND

Trachea

Arch of aorta

(d) Anterior view

Thyroid hormones regulate oxygen use and basal metabolic rate, cellular metabolism, and growth and development.

Microscopic spherical sacs called **thyroid follicles** (Figure 17.9b) make up most of the thyroid gland. The wall of each follicle consists primarily of cells called **follicular cells** that surround the lumen (internal space) of the follicle. The follicular cells produce two hormones: **triiodothyronine** (trī´-ī-ō-dō-THĪ -rō-nēn), which is also called T_3, because it contains three atoms of iodine, and **thyroxine** (thī-ROK-sēn), or T_4, which contains four atoms of iodine. T_3 and T_4 together are also known as **thyroid hormones**. A few cells called **parafollicular cells** lie between follicles (Figure 17.9b). They produce the hormone **calcitonin** (kal-si-TŌ -nin), which helps regulate calcium homeostasis.

Formation, Storage, and Release of Thyroid Hormones

The thyroid gland is the only endocrine gland that stores its secretory product in large quantity—normally about a 100-day supply. Synthesis and secretion of T_3 and T_4 occur as follows (Figure 17.10):

1 *Iodide trapping.* Thyroid follicular cells trap iodide ions (I^-) by actively transporting them from the blood into the cytosol. As a result, the thyroid gland normally contains most of the iodide in the body.

2 *Synthesis of thyroglobulin.* While the follicular cells are trapping I^-, they are also synthesizing **thyroglobulin (TGB)**. TGB, a large glycoprotein, is secreted into the lumen of the follicle. The material that accumulates in the lumen is termed **colloid**.

3 *Oxidation of iodide.* Some of the amino acids in TGB are tyrosines that must bind with iodine. However, negatively charged iodide ions cannot bind to tyrosine until they undergo oxidation (removal of electrons), transforming them into iodine molecules ($2I^- \rightarrow I_2$). As iodine molecules form, they pass out of the follicular cells into the colloid.

4 *Iodination of tyrosine.* I_2 binds with the tyrosine components of thyroglobulin (TGB) molecules. Binding of TGB with one iodine atom from an iodine molecule yields monoiodotyrosine (T_1), and a second iodination produces diiodotyrosine (T_2).

5 *Coupling of T_1 and T_2.* During the last step in the synthesis of thyroid hormone, one T_1 and one T_2 join to form T_3 or two T_2 molecules join to form T_4.

6 *Pinocytosis and digestion of colloid.* Droplets of colloid reenter follicular cells by pinocytosis and merge with lysosomes. Digestive enzymes break down TGB, cleaving off separate molecules of T_3 and T_4.

7 *Secretion of thyroid hormones.* Because T_3 and T_4 are lipid-soluble, they diffuse through the plasma membrane and enter the bloodstream. T_4 normally is secreted in greater quantity than T_3, but T_3 is several times more potent. In addition, after

Figure 17.10 **Steps in the synthesis and secretion of thyroid hormones.**

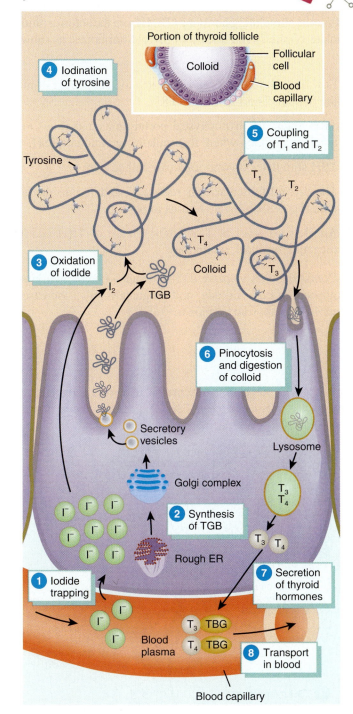

Thyroid hormones are synthesized by attaching iodine atoms to tyrosine. Refer to text for explanation of abbreviations.

T_4 enters a body cell, most of it is converted to T_3 by removal of one iodine.

8 *Transport in the blood.* More than 99% of both T_3 and T_4 combine with transport proteins in the blood, mainly **thyroxine-binding globulin (TBG)**.

Control of Thyroid Hormone Secretion

Thyrotropin-releasing hormone (TRH) from the hypothalamus and thyroid-stimulating hormone (TSH) from the anterior pituitary stimulate synthesis and release of thyroid hormones, as shown in Figure 17.11:

1 Low blood levels of T_3 and T_4 or low metabolic rate stimulate the hypothalamus to secrete TRH.

2 TRH enters the hypophyseal portal veins and flows to the anterior pituitary, where it stimulates secretion of TSH.

3 TSH stimulates thyroid follicular cell activity.

4 The thyroid follicular cells release T_3 and T_4 into the blood until the metabolic rate returns to normal.

5 An elevated level of T_3 inhibits release of TRH and TSH (negative feedback inhibition).

Conditions that increase ATP demand—a cold environment, hypoglycemia, high altitude, and pregnancy—also increase the secretion of the thyroid hormones.

Figure 17.11   Regulation of secretion of thyroid hormones. TRH = thyrotropin-releasing hormone, TSH = thyroid-stimulating hormone, T_3 = triiodothyronine, and T_4 = thyroxine.

1 Low blood levels of T_3 and T_4 or low metabolic rate stimulate release of

TRH

Hypothalamus

2 TRH, carried by hypophyseal portal veins to anterior pituitary, stimulates release of TSH

5 Elevated T_3 inhibits release of TRH and TSH (negative feedback)

3 TSH released into blood stimulates thyroid follicular cells

TSH

Anterior pituitary

4 T_3 and T_4 released into blood by follicular cells

Thyroid follicle

TSH promotes release of thyroid hormones (T_3 and T_4) by the thyroid gland.

Actions of Thyroid Hormones

Because most body cells have receptors for thyroid hormones, T_3 and T_4 exert their effects throughout the body.

- Thyroid hormones increase **basal metabolic rate (BMR)**, the rate of oxygen consumption under standard or basal conditions (awake, at rest, and fasting), by stimulating the use of cellular oxygen to produce ATP. When the basal metabolic rate increases, cellular metabolism of carbohydrates, lipids, and proteins also increases.

- A second major effect of thyroid hormones is to stimulate synthesis of additional sodium–potassium pumps (Na^+/K^+ ATPases), which use large amounts of ATP to continually eject sodium ions (Na^+) from the cytosol into the extracellular fluid and potassium ions (K^+) from the extracellular fluid into the cytosol. As cells use more ATP, more heat is given off, and body temperature rises. This phenomenon is called the **calorigenic effect**. In this way, thyroid hormones play an important role in the maintenance of normal body temperature. Normal mammals can survive in freezing temperatures, but those whose thyroid glands have been removed cannot.

- Thyroid hormones stimulate protein synthesis and increase the use of glucose and fatty acids for ATP production. They also increase the breakdown of triglycerides and enhance cholesterol excretion, thus reducing blood cholesterol level.

- The thyroid hormones enhance some actions of norepinephrine and epinephrine. For this reason, symptoms of hyperthyroidism include increased heart rate, more forceful heartbeats, and increased blood pressure.

- Thyroid hormones stimulate body growth, particularly the growth and development of nervous tissue. Deficiency of thyroid hormones during fetal development, infancy, or childhood causes severe mental retardation and stunted bone growth.

Calcitonin

The hormone produced by the **parafollicular cells** of the thyroid gland (see Figure 17.9b) is **calcitonin (CT)**. CT can inhibit the action of osteoclasts, the cells that break down bone matrix. The secretion of CT is controlled by a negative feedback system (see Figure 17.13).

When its blood level is high, calcitonin lowers the amount of blood calcium and phosphates by inhibiting bone resorption (breakdown of bone extracellular matrix) by osteoclasts and by accelerating uptake of calcium and phosphates into bone extracellular matrix. Miacalcin, a calcitonin extract derived from salmon that is 10 times more potent than human calcitonin, is prescribed to treat osteoporosis.

Table 17.6 summarizes the hormones produced by the thyroid gland, control of their secretion, and their principal actions.

TABLE 17.6

Thyroid Gland Hormones

HORMONE AND SOURCE	CONTROL OF SECRETION	PRINCIPAL ACTIONS
T_3 (triiodothyronine) and T_4 (thyroxine) or **thyroid hormones** from follicular cells	Secretion is increased by TRH from hypothalamus, which stimulates release of TSH from anterior pituitary in response to low thyroid hormone levels, low metabolic rate, cold, pregnancy, and high altitudes; TRH and TSH secretion are inhibited in response to high thyroid hormone levels.	Increase basal metabolic rate, stimulate synthesis of proteins, increase use of glucose and fatty acids for ATP production, increase breakdown of triglycerides, enhance cholesterol excretion, stimulate body growth, and contribute to development of the nervous system.
Calcitonin (CT) from parafollicular cells	High blood Ca^{2+} levels stimulate secretion: low blood Ca^{2+} levels inhibit secretion.	Lowers blood levels of Ca^{2+} and phosphates by inhibiting bone resorption by osteoclasts and uptake of calcium and phosphates into bone extracellular matrix.

Checkpoint

21. Which cells of the thyroid gland secrete T_3 and T_4? Which secrete calcitonin? Which of these hormones are also called thyroid hormones?
22. How are the thyroid hormones synthesized, stored, and transported in the bloodstream?
23. How is the secretion of T_3 and T_4 regulated?
24. Why might hypothyroidism (low thyroid hormone levels) in an infant result in mental retardation?
25. What is the primary target tissue for calcitonin? What is the action of calcitonin?

17.6 The parathyroid glands secrete parathyroid hormone, which regulates calcium, magnesium, and phosphate ion levels.

Partially embedded in the posterior surface of the lateral lobes of the thyroid gland are several small, round masses of tissue called the **parathyroid glands** (*para-* = beside). Each parathyroid gland has a mass of about 40 mg (0.04 g). Usually, one superior and one inferior parathyroid gland are attached to each thyroid lobe (Figure 17.12a,d).

Microscopically, the parathyroid glands contain two kinds of epithelial cells (Figure 17.12b,c). The more numerous cells, called **chief cells**, produce **parathyroid hormone (PTH)**. The function of the other kind of cell, called an *oxyphil cell*, is not known.

Parathyroid Hormone

Parathyroid hormone is the major regulator of the levels of calcium (Ca^{2+}), magnesium (Mg^{2+}), and phosphate (HPO_4^{2-}) ions in the blood. Blood calcium levels directly control the secretion of both calcitonin and parathyroid hormone through negative feedback, and the two hormones have opposite effects on blood Ca^{2+} levels (Figure 17.13):

❶ A higher-than-normal level of calcium ions (Ca^{2+}) in the blood stimulates parafollicular cells of the thyroid gland to release more calcitonin.

Figure 17.12   The parathyroid glands.

Parathyroid glands (behind thyroid gland)

Trachea

Right internal jugular vein

Right common carotid artery

Thyroid gland

LEFT SUPERIOR PARATHYROID GLAND

Esophagus

LEFT INFERIOR PARATHYROID GLAND

RIGHT SUPERIOR PARATHYROID GLAND

RIGHT INFERIOR PARATHYROID GLAND

Trachea

O'KELLEY

(a) Posterior view

Chief cell

Blood vessel

Oxyphil cell

LM 325x

(b) Parathyroid gland

Parathyroid

Thyroid

Chief cell

Oxyphil cell

Parathyroid gland

Follicular cell

Parafollicular cell

Thyroid gland

Blood vessel

(c) Portion of the thyroid gland (left) and parathyroid gland (right)

Parathyroid gland

Thyroid gland

Parathyroid gland

(d) Posterior view of parathyroid glands

🔑 **The parathyroid glands, normally four in number, are embedded in the posterior surface of the thyroid gland.**

Figure 17.13 **The roles of calcitonin (green arrows), parathyroid hormone (blue arrows), and calcitriol (orange arrows) in homeostasis of blood calcium levels.**

1 High level of Ca²⁺ in blood stimulates thyroid gland parafollicular cells to release more calcitonin.

6 CALCITRIOL stimulates increased absorption of Ca²⁺ from foods, which increases blood Ca²⁺ level.

5 PTH also stimulates the kidneys to release CALCITRIOL.

4 PTH stimulates osteoclasts and retards loss of Ca²⁺ in urine, increasing blood Ca²⁺ level.

2 CALCITONIN inhibits osteoclasts, decreasing blood Ca²⁺ level.

3 Low level of Ca²⁺ in blood stimulates parathyroid gland chief cells to release more PARATHYROID HORMONE (PTH).

Calcitonin and PTH have opposite effects on the level of calcium ions (Ca²⁺) in the blood.

2 Calcitonin inhibits the activity of osteoclasts, decreasing the release of Ca²⁺ into the blood from bone extracellular matrix and resulting in decreased blood Ca²⁺ levels.

3 A lower-than-normal level of Ca²⁺ in the blood stimulates chief cells of the parathyroid gland to release more PTH.

4 PTH promotes resorption of bone extracellular matrix by osteoclasts, which releases Ca²⁺ into the blood, and slows loss of Ca²⁺ in the urine, raising the blood level of Ca²⁺.

5 PTH also stimulates the kidneys to synthesize **calcitriol**, the active form of vitamin D.

6 Calcitriol stimulates increased absorption of Ca²⁺ from foods in the gastrointestinal tract, which helps increase the blood level of Ca²⁺.

PTH increases the number and activity of osteoclasts, which releases calcium (Ca²⁺) and phosphates (HPO₄²⁻) in bone to the blood. PTH also acts on the kidneys. First, it slows the rate at which Ca²⁺ and Mg²⁺ are lost from blood into the urine. Second, it increases loss of HPO₄²⁻ from blood into the urine. Because more HPO₄²⁻ is lost in the urine than is gained from the bones, PTH decreases blood HPO₄²⁻ levels and increases blood Ca²⁺ and Mg²⁺ levels. A third effect of PTH on the kidneys is to promote formation of the hormone calcitriol, the active form of vitamin D. Calcitriol acts on the gastrointestinal tract to increase the rate of Ca²⁺, HPO₄²⁻, and Mg²⁺ absorption from food within the gastrointestinal tract into the blood.

Table 17.7 summarizes control of secretion and the principal actions of parathyroid hormone.

Checkpoint

26. What are the primary target tissues for PTH and calcitriol?
27. How is secretion of parathyroid hormone regulated?
28. In which ways are the actions of PTH and calcitriol similar and different?
29. What is the benefit of adding vitamin D to various food products such as milk?
30. Would hypopituitarism affect calcium balance in the body?

TABLE 17.7		
Parathyroid Gland Hormone		
HORMONE AND SOURCE	CONTROL OF SECRETION	PRINCIPAL ACTIONS
Parathyroid hormone (PTH) from chief cells Chief cell	Low blood Ca²⁺ levels stimulate secretion. High blood Ca²⁺ levels inhibit secretion.	Increases blood Ca²⁺ and Mg²⁺ levels and decreases blood phosphate level; increases bone resorption by osteoclasts; increases Ca²⁺ reabsorption and phosphate excretion by kidneys; and promotes formation of calcitriol (active form of vitamin D), which increases rate of dietary Ca²⁺ and Mg²⁺ absorption.

17.7 The adrenal cortex secretes mineralocorticoids, glucocorticoids, and androgens; the adrenal medulla secretes epinephrine and norepinephrine.

The paired **adrenal glands**, or **suprarenal glands** (*supra-* = above; *-renal* = kidney), one of which lies superior to each kidney, have a flattened pyramidal shape (Figure 17.14a,c). In an adult, each adrenal gland is 3–5 cm in height, 2–3 cm in width, and a little less than 1 cm thick; it weighs 3.5–5 g, only half its weight at birth. Each adrenal gland has two structurally and functionally distinct regions: a large, peripherally located **adrenal cortex**, comprising 80–90% of the gland, and a small, centrally located

Figure 17.14 The adrenal (suprarenal) glands.

(a) Anterior view

(b) Section through left adrenal gland

(c) Anterior view of adrenal gland and kidney

(d) Subdivisions of the adrenal gland

LM 50x

The adrenal cortex secretes steroid hormones that are essential for life; the adrenal medulla secretes norepinephrine and epinephrine.

adrenal medulla (Figure 17.14b). A connective tissue capsule covers the gland. The adrenal glands, like the thyroid gland, are highly vascularized.

The adrenal cortex produces steroid hormones that are essential for life. Complete loss of adrenocortical hormones leads to death due to dehydration and electrolyte imbalances in a few days to a week, unless hormone replacement therapy begins promptly. The adrenal medulla produces two hormones: norepinephrine and epinephrine.

Adrenal Cortex

The adrenal cortex is subdivided into three zones, each of which secretes different hormones (Figure 17.14d). The outer zone, just deep to the connective tissue capsule, secretes hormones called **mineralocorticoids** (min´-er-al-ō-KOR-ti-koyds) because they affect mineral homeostasis. The middle zone secretes mainly **glucocorticoids** (gloo´-kō-KOR-ti-koyds), so named because they affect glucose homeostasis. The inner zone synthesizes small amounts of weak **androgens** (andro- = a man), steroid hormones that have masculinizing effects.

Mineralocorticoids

Aldosterone (al-DOS-ter-ōn) is the major mineralocorticoid. It regulates homeostasis of two mineral ions, sodium ions (Na$^+$) and potassium ions (K$^+$). Aldosterone increases reabsorption of Na$^+$ from the urine into the blood, and it stimulates excretion of K$^+$ into the urine. Aldosterone also helps adjust blood pressure and blood volume.

Secretion of aldosterone is controlled by the **renin–angiotensin–aldosterone pathway** (RĒ -nin an´-jē-ō-TEN-sin; Figure 17.15):

1 The renin–angiotensin–aldosterone pathway is activated by stimuli that include dehydration, low blood sodium, or hemorrhage. These conditions decrease blood volume, which leads to decreased blood pressure.

2 Decreased blood pressure stimulates the kidney to secrete the enzyme **renin** into the blood.

3 Renin converts **angiotensinogen**, a plasma protein produced by the liver, into **angiotensin I**.

4 As blood rich in angiotensin I flows through the lungs, another enzyme called **angiotensin-converting enzyme** converts inactive angiotensin I into the active hormone **angiotensin II**.

5 Angiotensin II stimulates contraction of smooth muscle in the walls of arterioles. The resulting vasoconstriction of the arterioles helps raise blood pressure to normal.

6 Angiotensin II also stimulates the adrenal cortex to secrete aldosterone.

7 Blood containing increased levels of aldosterone acts on the kidneys to promote the movement of Na$^+$ and water from

Figure 17.15   The renin–angiotensin–aldosterone pathway.

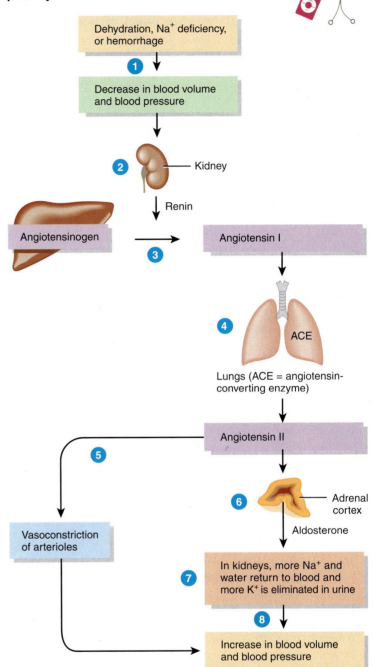

Aldosterone helps regulate blood volume, blood pressure, and levels of Na$^+$ and K$^+$ in the blood.

urine to the blood. Aldosterone also stimulates the kidneys to increase secretion of K$^+$ into the urine.

8 As more water returns to the blood (and less is lost in the urine), blood volume increases. As blood volume increases and arterioles constrict, blood pressure increases to normal.

Therefore, blood pressure increases to normal due to the aldosterone-influenced increase in blood volume and the angiotensin II-influenced vasoconstriction of arterioles.

Glucocorticoids

The glucocorticoids, which regulate metabolism and resistance to stress, include **cortisol** (**hydrocortisone**), **corticosterone**, and **cortisone**. Of the three, cortisol is the most abundant, accounting for about 95% of glucocorticoid activity.

Control of glucocorticoid secretion is by a typical negative feedback system (Figure 17.16; see also Figure 17.5):

❶ Low blood levels of glucocorticoids, mainly cortisol, stimulate neurosecretory cells in the hypothalamus to secrete **corticotropin-releasing hormone (CRH)**.

❷ CRH travels through the infundibulum to the anterior pituitary, where it promotes the release of ACTH from the anterior pituitary.

❸ ACTH flows in the blood to the adrenal cortex, where it stimulates glucocorticoid secretion.

❹ As the level of glucocorticoids rises, it exerts negative feedback inhibition both on the anterior pituitary to reduce release of ACTH and on the hypothalamus to reduce release of CRH.

Glucocorticoids have the following effects:

- **Protein breakdown.** Glucocorticoids increase the rate of protein breakdown, mainly in muscle fibers, and thus increase amino acid concentrations in the blood. These amino acids can be used by body cells for synthesis of new proteins and for ATP production.

- **Glucose formation.** On stimulation by glucocorticoids, liver cells may convert certain amino acids or lactic acid to glucose, which cells can use for ATP production. Such conversion of a substance other than glycogen or another monosaccharide into glucose is called **gluconeogenesis** (gloo´-ko-nē-ō-JEN-e-sis).

- **Breakdown of triglycerides.** Glucocorticoids stimulate the breakdown of triglycerides in adipose tissue. The fatty acids released into the blood can be used for ATP production by many body cells.

- **Resistance to stress.** Glucocorticoids work in many ways to provide resistance to stress. The additional amino acids, glucose, and fatty acids provide tissues with an increased source of ATP to combat a range of stresses, including exercise, fasting, fright, temperature extremes, high altitude, bleeding, infection, surgery, trauma, and disease.

- **Anti-inflammatory effects.** Glucocorticoids inhibit white blood cells that participate in inflammatory responses. Unfortunately, glucocorticoids also retard tissue repair, which slows wound healing. Although high doses can cause severe mental disturbances, synthetic glucocorticoids are very useful in the treatment of chronic inflammatory disorders such as rheumatoid arthritis.

Figure 17.16 Negative feedback regulation of glucocorticoid secretion.

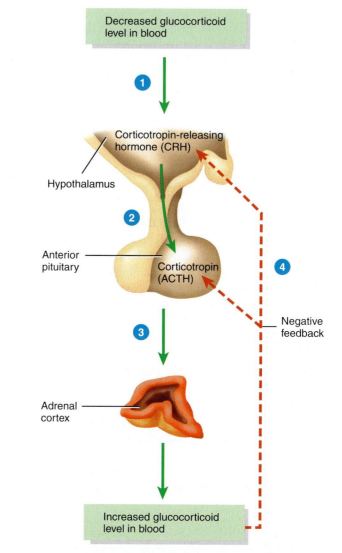

Low levels of glucocorticoids stimulate the release of ACTH; ACTH in turn stimulates glucocorticoid secretion by the adrenal cortex.

- **Depression of immune responses.** High doses of glucocorticoids depress immune responses. For this reason, glucocorticoids are prescribed for organ transplant recipients to retard tissue rejection by the immune system.

Androgens

In both males and females, the adrenal cortex secretes small amounts of weak androgens. After puberty in males, the androgen testosterone is also released in much greater quantity by the testes. Thus, the relative amount of androgens secreted by the adrenal gland in males is usually so low that their effects are insignificant. In females, however, adrenal androgens play important roles. They contribute to libido (sex drive) and are converted into estrogens (feminizing sex steroids) by other body tissues. After menopause, when ovarian secretion of estrogens ceases, all female estrogens come from conversion of adrenal androgens. Adrenal androgens also stimulate growth of axillary and pubic hair in boys and girls and contribute to growth spurts before puberty. Although control of adrenal androgen secretion is not fully understood, the main hormone that stimulates its secretion is ACTH.

Adrenal Medulla

The inner region of the adrenal gland, the **adrenal medulla**, is a modified sympathetic ganglion of the autonomic nervous system (ANS). It develops from the same embryonic tissue as all other sympathetic ganglia, but its cells, which lack axons, form clusters around large blood vessels. Rather than releasing a neurotransmitter, the cells of the adrenal medulla secrete hormones. Because the ANS exerts direct control over the adrenal medulla through innervation from sympathetic preganglionic neurons, hormone release can occur very quickly.

The two major hormones synthesized by the adrenal medulla are **epinephrine** and **norepinephrine (NE)**, also called **adrenaline** and **noradrenaline**. In stressful situations and during exercise, impulses from the hypothalamus stimulate sympathetic pre-ganglionic neurons, which in turn stimulate the cells of the adrenal medulla to secrete epinephrine and norepinephrine. These two hormones greatly augment the fight-or-flight response that was described in Chapter 14 (see Concept 14.8). By increasing heart rate and force of contraction, epinephrine and norepinephrine increase the pumping output of the heart, which increases blood pressure. They also increase blood flow to the heart, liver, skeletal muscles, and adipose tissue; dilate airways to the lungs; and increase blood levels of glucose and fatty acids.

Table 17.8 summarizes the hormones produced by the adrenal glands, control of their secretion, and their principal actions.

Checkpoint

31. In which two ways can angiotensin II increase blood pressure? What are its target tissues for each mechanism?
32. Which stimulus initiates the renin–angiotensin–aldosterone pathway? Which stimulus turns the pathway off? Is this a negative or positive feedback loop?
33. How do the hypothalamus and anterior pituitary influence glucocorticoid secretion?
34. How do glucocorticoids influence ATP production? How does this influence your resistance to stress?
35. Which chemicals do the adrenal medulla and the autonomic nervous system both produce?
36. Complete removal of the adrenal glands would eliminate the release of which hormones?

TABLE 17.8

Adrenal Gland Hormones

HORMONE AND SOURCES	CONTROL OF SECRETION	PRINCIPAL ACTIONS
Adrenal Cortex Hormones		
Mineralocorticoids (mainly aldosterone)	Increased blood K⁺ level and angiotensin II stimulate secretion.	Increase blood levels of Na⁺ and water and decrease blood level of K⁺.
Glucocorticoids (mainly cortisol)	ACTH stimulates release; corticotropin-releasing hormone (CRH) promotes ACTH secretion in response to stress and low blood levels of glucocorticoids.	Increase protein breakdown (except in liver), stimulate gluconeogenesis and lipolysis, provide resistance to stress, dampen inflammation, and depress immune responses.
Androgens	ACTH stimulates secretion.	Assist in early growth of axillary and pubic hair in both sexes; in females, contribute to libido and are source of estrogens after menopause.
Adrenal Medulla Hormones		
Epinephrine and norepinephrine	Sympathetic preganglionic neurons release acetylcholine, which stimulates secretion.	Produce effects that enhance those of the sympathetic division of the autonomic nervous system during stress.

17.8 The pancreatic islets regulate blood glucose levels by secreting glucagon and insulin.

The **pancreas** (*pan-* = all; *-creas* = flesh) is both an endocrine gland and an exocrine gland (Figure 17.17). We discuss its endocrine functions in this chapter and its exocrine functions in Chapter 23 in the coverage of the digestive system. The pancreas is a flattened organ that measures about 12–15 cm (4.5–6 in.) in length. It is located in the curve of the duodenum, the first part of the small intestine

Figure 17.17   The pancreas.

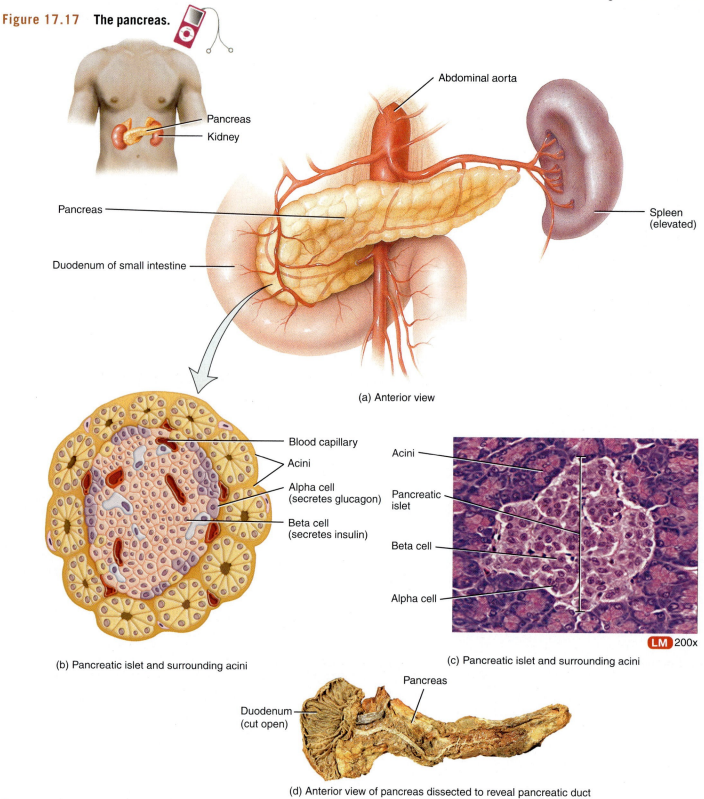

(a) Anterior view

(b) Pancreatic islet and surrounding acini

(c) Pancreatic islet and surrounding acini

LM 200x

(d) Anterior view of pancreas dissected to reveal pancreatic duct

Pancreatic hormones regulate blood glucose level.

(Figure 17.17a). Roughly 99% of the pancreas consists of clusters of exocrine cells called **acini**. The acini produce digestive enzymes, which flow into the gastrointestinal tract through a network of ducts. Scattered among the acini are 1–2 million tiny clusters of endocrine tissue called **pancreatic islets** or **islets of Langerhans** (LAHNG-er-hanz; Figure 17.17b,c). Abundant capillaries serve both the exocrine and endocrine portions of the pancreas.

Regulation of Glucagon and Insulin Secretion

Some of the pancreatic islet cells, the **alpha cells**, secrete the hormone **glucagon** (GLOO-ka-gon). Other neighboring islet cells, the **beta cells**, secrete **insulin** (IN-soo-lin). The principal action of glucagon is to increase blood glucose levels when they fall below normal. Insulin, on the other hand, helps lower blood glucose levels when they are too high. The level of blood glucose controls secretion of glucagon and insulin via negative feedback. Figure 17.18 shows the conditions that stimulate the pancreatic islets to secrete their hormones, the ways in which glucagons and insulin produce their effects on blood glucose levels, and negative feedback control of hormone secretion.

1 Low blood glucose levels (hypoglycemia) stimulate secretion of glucagon from alpha cells of the pancreatic islets.

2 Glucagon acts on liver cells to promote breakdown of glycogen into glucose and formation of glucose from lactic acid and certain amino acids.

3 As a result, the liver releases glucose into the blood more rapidly, and blood glucose levels rise.

4 If blood glucose continues to rise, high blood glucose levels (hyperglycemia) inhibit release of glucagon by alpha cells (negative feedback).

5 At the same time, high blood glucose (hyperglycemia) stimulates secretion of insulin by beta cells of the pancreatic islets.

6 Insulin acts on various cells in the body to increase facilitated diffusion of glucose into cells to speed synthesis of glycogen from glucose; to increase uptake of amino acids by cells; to increase protein synthesis; and to increase synthesis of fatty acids. Because insulin increases the synthesis of proteins and fatty acids within cells, it is an important hormone when tissues are developing, growing, or being repaired.

7 As glucose enters and is utilized by cells, blood glucose levels fall.

8 If blood glucose levels drop below normal, low blood glucose inhibits release of insulin by beta cells (negative feedback) and stimulates release of glucagon.

Figure 17.18 Regulation of blood glucose level by negative feedback systems involving glucagon (blue arrows) and insulin (orange arrows).

1 Low blood glucose (hypoglycemia) stimulates alpha cells to secrete

5 High blood glucose (hyperglycemia) stimulates beta cells to secrete

GLUCAGON

INSULIN

2 Glucagon acts on liver cells to:
- convert glycogen into glucose form glucose from
- lactic acid and certain amino acids

3 Glucose released by liver cells raises blood glucose level to normal

4 If blood glucose continues to rise, hyperglycemia inhibits release of glucagon

6 Insulin acts on various body cells to:
- accelerate facilitated diffusion of glucose into cells
- speed conversion of glucose into glycogen
- increase uptake of amino acids and increase protein synthesis
- speed synthesis of fatty acids

7 Blood glucose level falls

8 If blood glucose continues to fall, hypoglycemia inhibits release of insulin

🔑 **Low blood glucose stimulates secretion of glucagon; high blood glucose stimulates secretion of insulin.**

Although blood glucose levels are the most important regulator of glucagon and insulin, these hormones are also regulated by other hormones, the autonomic nervous system, and your diet. Human growth hormone and adrenocorticotropic hormone stimulate insulin secretion because they elevate blood glucose levels. The sympathetic division of the ANS stimulates secretion of glucagon during exercise. High blood levels of amino acids following a meal containing mainly proteins also stimulate glucagon secretion. The parasympathetic division of the ANS, by contrast, stimulates secretion of insulin following a meal rich in carbohydrates.

Table 17.9 summarizes the control of secretion and principal actions of insulin and glucagon.

TABLE 17.9

Selected Pancreatic Islet Hormones

HORMONE AND SOURCE	CONTROL OF SECRETION	PRINCIPAL ACTIONS
Glucagon from alpha cells of pancreatic islets Alpha cell	Decreased blood levels of glucose, stimulation by sympathetic division of ANS during exercise, and mainly protein meals stimulate secretion. Insulin inhibits secretion.	Raises blood glucose levels by accelerating breakdown of glycogen into glucose in liver, converting other nutrients into glucose in liver, and releasing glucose into the blood.
Insulin from beta cells of pancreatic islets Beta cell	Increased blood levels of glucose, stimulation by parasympathetic division of ANS following high-carbohydrate meals, hGH, and ACTH stimulate secretion.	Lowers blood glucose levels by accelerating transport of glucose into cells, converting glucose into glycogen, and stimulating protein and fatty acid synthesis.

Checkpoint

37. Why is the pancreas classified as both an exocrine gland and an endocrine gland?
38. Why are the digestive enzymes secreted by the pancreas not considered hormones?
39. You are reading through a medical journal and encounter an article entitled "Clinical Implications of Beta-cell Tumor Hypersecretion." What is the article about?
40. You've just eaten a big meal and are feeling very relaxed. Which division of the autonomic nervous system interacts with the pancreas to regulate glucose levels after the meal? The secretion of which pancreatic islet hormone is stimulated?
41. You're swimming laps in the school pool. Which division of the autonomic nervous system interacts with the pancreas to regulate glucose levels while you swim? The secretion of which pancreatic islet hormone is stimulated?

Return to Karen's Story

Karen understood the basics; something was wrong with her hormones. "It's the *change*, isn't it—menopause?"

The doctor replied, "No, Karen, you haven't been menstruating for almost 25 years. Your pituitary gland controls the release of hormones that regulate the menstrual cycle. During menopause the ovaries become unresponsive to the pituitary hormones that stimulate the release of estrogens. Basically, you skipped menopause or went through it earlier in life without realizing it. But we might still consider putting you on estrogen–progesterone replacement. I'm going to prescribe a glucocorticoid, T₄ replacement, and an estrogen–progesterone combination. We'll need to schedule a follow-up visit to check your progress."

"Hi, Anne. It's Karen." Karen was seated at a small wooden table just outside her home. She had been making calls all morning. "Yes, I'm going to be fine. The doctor said I needed some medications. My hormones are all off or something. I don't see how it could have anything to do with my babies. The doctor said something about my blood pressure being low because all the bleeding damaged some gland or something. Anyway, there's a craft fair this weekend over in Midville. We should go."

After being on hormone replacement therapy for several weeks, Karen was feeling better. She and her friend Anne were in line for tickets to the craft fair. "Karen, I am about to die from this heat. How are you doing?" asked Anne.

"It's so hot, I think I'm going to melt. I can't take too much heat, or cold for that matter. You know, I stopped taking those estrogen pills the doctor gave me. I was having these spells; I'd feel all hot and uncomfortable; I felt tired all the time, too. I thought it must be those pills because I saw this article in the paper that said that estrogens were bad for you now. Are you still taking your Premarin?"

Anne replied, "Oh, yes. I feel a lot better on it, but I saw that article, too, and I'm not sure I should keep taking it. My doctor says I should stay on it if it helps."

D. Karen's thyroid gland appeared normal according to the doctor's examination. Where would he have palpated to check her thyroid gland? Which hormone levels would you measure to test Karen's thyroid function?

E. The hormone therapy Dr. Beaumont has prescribed includes T₄ therapy. Which of Karen's symptoms would be resolved by taking this drug? Is T₄ replacement therapy likely to improve Karen's calcium homeostasis?

F. Why would the lack of thyroid hormones make Karen cold-intolerant?

G. How do insulin and glucocorticoids affect Karen's blood glucose levels? Which pancreatic hormone is most like the glucocorticoid cortisol in its function?

17.9 The ovaries produce estrogens, progesterone, and inhibin; the testes produce testosterone and inhibin.

Gonads are the organs that produce gametes—oocytes in females and sperm in males. In addition to their reproductive function, the gonads secrete hormones.

The female gonads, the **ovaries**, paired organs located in the pelvic cavity, produce the female sex hormones **estrogens** and **progesterone**. Along with FSH and LH from the anterior pituitary, the female sex hormones regulate the female reproductive cycle, maintain pregnancy, and prepare the mammary glands for lactation. They also help establish and maintain feminine secondary sex characteristics, such as breast development and widening of the hips. The ovaries produce **inhibin**, a hormone that inhibits secretion of FSH. During pregnancy, the ovaries and placenta produce a hormone called **relaxin**, which increases the flexibility of the pubic symphysis during pregnancy and

helps dilate the uterine cervix during labor and delivery. These actions help ease the baby's passage by enlarging the birth canal.

The male gonads, or **testes**, lie in the scrotum and produce and secrete **testosterone**, the primary **androgen** or male sex hormone. Testosterone stimulates descent of the testes before birth, stimulates development and maintenance of masculine secondary sex characteristics such as beard growth and deepening of the voice, and regulates production of sperm. The testes also produce inhibin, which inhibits FSH secretion.

The detailed structure of the ovaries and testes and the specific roles of sex hormones will be discussed in Chapter 25.

Table 17.10 summarizes the hormones produced by the ovaries and testes and their principal actions.

TABLE 17.10	
Hormones of the Ovaries and Testis	
HORMONE	PRINCIPAL ACTIONS
Ovarian Hormones	
Estrogens and progesterone Ovary	Together with gonadotropic hormones of the anterior pituitary gland, regulate the female reproductive cycle, maintain pregnancy, prepare the mammary glands for lactation, and promote development and maintenance of female secondary sex characteristics.
Relaxin	Increases flexibility of pubic symphysis during pregnancy and helps dilate uterine cervix during labor and delivery.
Inhibin	Inhibits secretion of FSH from anterior pituitary.
Testicular Hormones	
Testosterone Testis	Stimulates descent of testes before birth, regulates spermatogenesis, and promotes development and maintenance of male secondary sex characteristics.
Inhibin	Inhibits secretion of FSH from anterior pituitary.

Checkpoint

42. Which two hormones are involved in labor and delivery? (*Hint:* One is a posterior pituitary hormone, the other is an ovarian hormone.)
43. What two endocrine glands produce androgens?
44. Where do androgens in females come from?
45. What is the target tissue of inhibin?

17.10 The pineal gland secretes melatonin, which contributes to setting the body's biological clock.

The **pineal gland** (PĪ N-ē-al = pinecone shape) is a small endocrine gland attached to the roof of the third ventricle of the brain at the midline (see Figure 17.1). It is positioned between the two superior colliculi of the midbrain and has a mass of 0.1–0.2 g. The gland consists of masses of neuroglia and secretory cells.

The pineal gland secretes the hormone **melatonin** that appears to contribute to the setting of the body's biological clock, which is controlled by the hypothalamus. As more melatonin is liberated during darkness than in light, this hormone is thought to promote sleepiness. In response to visual input from the eyes (retina), the hypothalamus stimulates sympathetic neurons, which in turn stimulate the pineal gland to secrete melatonin in a rhythmic pattern, with low levels of melatonin secreted during the day and significantly higher levels secreted at night.

During sleep, levels of melatonin in the bloodstream increase tenfold and then decline to a low level again before awakening. Small doses of melatonin given orally can induce sleep and reset daily rhythms, which might benefit workers whose shifts alternate between daylight and nighttime hours.

In animals that breed during specific seasons, melatonin inhibits reproductive functions outside the breeding season, but it is unclear whether melatonin influences human reproductive function. Melatonin levels are higher in children and decline with age into adulthood, but there is no evidence that changes in melatonin secretion correlate with the onset of puberty and sexual maturation. Nevertheless, because melatonin causes atrophy of the gonads in several animal species, the possibility of adverse effects on human reproduction must be studied before its use to reset daily rhythms can be recommended.

Checkpoint

46. Why might world travelers flying across several time zones be tempted to try small doses of melatonin supplements?

47. How is the autonomic nervous system related to the pineal gland?

Karen's Story: Epilogue and Discussion

▶▶▶ **K**aren is an asthmatic 54-year-old female. She presents to the emergency room with an asthma exacerbation induced by heat and mild exertion. The physician noted that she had experienced excessive hemorrhaging during her pregnancies decades earlier, had an inability to breast-feed postpartum, and had amenorrhea (absence of menstruation) since her late twenties. She is overweight, and has complained of progressive weight gain over the course of many years. Other complaints include fatigue, generalized weakness, cold intolerance, puffiness and swelling, and constipation. Testing indicated secondary hypothyroidism due to hypopituitarism, most likely due to Sheehan's syndrome, a rare disorder occurring with postpartum pituitary necrosis. ◀◀◀

H. One of Karen's complaints is not being able to sleep at night. Which hormone regulates your biological clock?

I. Karen's weight gain is most likely due to insufficient levels of which hormone? Why?

J. What was likely the hormonal cause of Karen's inability to breast-feed after giving birth to her children?

K. Amenorrhea can occur when hormone insufficiency prevents ovulation. Deficiencies of which hormones might have caused Karen's amenorrhea?

L. How would hyperpituitarism have manifested itself if Karen had a pituitary tumor rather than pituitary necrosis? (Hint: What would happen if you increased the levels of releasing hormone?

Concept and Resource Summary

Concept	Resources

Introduction

1. The endocrine system releases hormones to control body activities and help maintain homeostasis.

Concept 17.1 The endocrine system works more slowly than the nervous system, releasing hormones that regulate virtually all body cells.

1. A **hormone** is a molecule that is released in one part of the body but regulates the activity of cells in other parts of the body.
2. In contrast to the rapid communication and control associated with the nervous system, the endocrine system responses are slower and more sustained.
3. **Exocrine gland** secretions enter ducts that carry the secretions to body surfaces or into cavities. **Endocrine glands** secrete hormones into interstitial fluid to diffuse into capillaries and circulate in the blood to target tissues. Several organs are not exclusively endocrine glands but do contain cells that secrete hormones.
4. The *endocrine system* comprises all the endocrine glands and hormone-secreting cells.

See

- Anatomy Overview – The Endocrine System

Concept 17.2 The secretion of hormones is regulated by the nervous system, chemical changes in the blood, and other hormones.

1. **Target cells** have specific **receptors** for a hormone to bind to and affect cell activities.
2. There are two classes of hormones: lipid-soluble and water-soluble.
3. Water-soluble hormones circulate in blood plasma unattached to plasma proteins; most lipid-soluble hormones circulate attached to transport proteins. **Transport proteins** increase blood solubility of lipid-soluble hormones, prevent loss of small hormone molecules to urine, and provide a ready reserve of hormones in the blood.
4. The response to a hormone depends on both the particular hormone and the target cell. Hormones can stimulate synthesis of molecules, alterations in membrane permeability, stimulation of membrane transport, alterations in metabolic rate, and contraction of smooth and cardiac muscle.
5. Three factors influence the responsiveness of a target cell to a hormone: the hormone's concentration, abundance of target cell receptors, and influences exerted by other hormones.
6. Hormone secretion is regulated by: nervous system signals, chemical changes in the blood, and other hormones.

See

- Anatomy Overview – Lipid-soluble Hormones
- Anatomy Overview – Water-soluble Hormones
- Anatomy Overview – Local Hormones
- Animation – Introduction to Hormonal Regulation, Secretion, and Concentration
- Animation – Mechanisms of Hormone Action
- Animation – Introduction to Hormone Feedback Loops

Concept 17.3 The hypothalamus regulates anterior pituitary secretion of seven important hormones.

1. The **hypothalamus** is the major integrating link between the nervous and endocrine systems. Although the **pituitary gland** secretes several hormones that control other endocrine glands, it is also controlled by the hypothalamus.
2. The pituitary gland sits in the sella turcica of the sphenoid bone and is attached by a stalk, the **infundibulum**, to the hypothalamus. The pituitary gland contains two functionally distinct regions: the anterior pituitary and the posterior pituitary.
3. The **anterior pituitary** secretes hormones when stimulated by **releasing hormones** and stops secreting hormones when inhibited by **inhibiting hormones**. These hypothalamic hormones reach the anterior pituitary through the **hypophyseal portal system** that links the hypothalamus and the anterior pituitary.
4. Several anterior pituitary hormones are **tropic hormones**, hormones that influence the secretion of other endocrine glands.
5. Anterior pituitary hormone secretion is regulated by two factors: releasing and inhibiting hormones of the hypothalamus, and negative feedback from rising blood levels of target gland hormones.
6. **Human growth hormone (hGH)** promotes synthesis and secretion of protein hormones called **insulinlike growth factors (IGFs)**. IGFs stimulate body growth and repair. **Growth hormone–releasing hormone (GHRH)** and **growth hormone–inhibiting hormone (GHIH)** control hGH secretion. Blood glucose level is a major regulator of GHRH and GHIH secretion.

See

- Anatomy Overview – The Hypothalamus and Pituitary Gland
- Anatomy Overview – Hormones of the Anterior Pituitary Gland
- Anatomy Overview – Anterior Pituitary Reproductive Hormones
- Animation – GHRH/hGH
- Animation – ACTH/Cortisol—Glycogenolysis
- Animation – THR/TSH—Production
- Animation – hGH—Growth and Development
- Animation – hGH—Glycogenolysis and Lipolysis

Hear

- Figure 17.4 – Hypothalamus and Pituitary Gland, and Their Blood Supply

Concept

7. **Thyroid-stimulating hormone (TSH)** stimulates the thyroid gland to secrete **triiodothyronine** (T_3) and **thyroxine** (T_4). Blood levels of T_3 and T_4 influence the hypothalamus to release **thyrotropin-releasing hormone (TRH)** to control TSH secretion by the anterior pituitary.

8. **Follicle-stimulating hormone (FSH)** targets the ovaries for monthly development of several ovarian follicles that surround a developing oocyte. FSH also stimulates the testes to produce sperm. **Gonadotropin-releasing hormone (GnRH)** from the hypothalamus stimulates FSH release by the anterior pituitary. Estrogens and testosterone suppress release of GnRH and FSH.

9. **Luteinizing hormone (LH)** triggers **ovulation** and formation of the corpus luteum and its secretion of progesterone. LH and FSH also stimulate secretion of estrogens by the ovaries. In males, LH stimulates secretion of testosterone by the testes. GnRH stimulates the anterior pituitary to secrete LH.

10. **Prolactin (PRL)** initiates milk secretion by mammary glands. Secretion of PRL is regulated by **prolactin-releasing hormone** and **prolactin-inhibiting hormone (PIH)** of the hypothalamus. The function of prolactin in males is unknown, but its hypersecretion causes erectile dysfunction.

11. **Adrenocorticotropic hormone (ACTH)** controls the secretion of glucocorticoids such as cortisol from the adrenal cortex. **Corticotropin-releasing hormone (CRH)** from the hypothalamus stimulates secretion of ACTH. Glucocorticoids cause inhibition of CRH and ACTH release.

12. Although there are MSH receptors in the brain, the function of normal levels of **melanocyte-stimulating hormone (MSH)** in humans is unknown. CRH stimulates its release and PIH inhibits secretion.

Do

- Exercise – Hormone Actions

Concept 17.4 Oxytocin and antidiuretic hormone originate in the hypothalalmus and are stored in the posterior pituitary.

1. The **posterior pituitary** stores and releases two hypothalamic hormones, oxytocin and antidiuretic hormone. After oxytocin and ADH are produced in the neurosecretory cells of the hypothalamus, they are transported down to the posterior pituitary. Nerve impulses trigger their release.

2. **Oxytocin (OT)** enhances smooth muscle contractions in the uterine wall to facilitate labor and delivery, and stimulates milk ejection from mammary glands after delivery. Release of OT is governed by a positive feedback mechanism triggered by uterine stretching and suckling.

3. **Antidiuretic hormone (ADH)** causes the kidneys to return more water to the blood while decreasing urine volume. ADH decreases the water lost through sweating, and causes constriction of arterioles to increase blood pressure. Blood osmotic pressure and blood volume affect secretion of ADH.

See

- Anatomy Overview – The Hypothalamus
- Anatomy Overview – Hormones of the Hypothalamus
- Anatomy Overview – Hypothalamic Reproductive Hormones
- Animation – Antidiuretic Hormone

Concept 17.5 The thyroid gland secretes thyroxine, triiodothyronine, and calcitonin.

1. The **thyroid gland** lies inferior to the larynx and anterior to the trachea. Internally, the gland is made up of sacs called **thyroid follicles** with **follicular cells** that produce two **thyroid hormones**: triiodothyronine (T_3) and thyroxine (T_4). **Parafollicular cells** produce **calcitonin**, which participates in calcium homeostasis.

2. The thyroid gland is the only endocrine gland that stores its secretory products in large supply.

3. Thyroid hormones are synthesized from iodine and tyrosine within **thyroglobulin**. They are transported in the blood bound to plasma proteins, mostly as **thyroxine-binding globulin**.

4. Secretion is controlled by TRH from the hypothalamus and thyroid-stimulating hormone from the anterior pituitary.

5. Thyroid hormones regulate oxygen use and metabolic rate, cellular metabolism, and growth and development.

6. **Calcitonin (CT)** lowers the blood levels of calcium ions (Ca^{2+}) and promote deposition of Ca^{2+} into bone matrix. Secretion of CT is controlled by the Ca^{2+} level in the blood.

See

- Anatomy Overview – The Thyroid
- Anatomy Overview – Thyroid Hormones
- Animation – TRS/TSH
- Animation – Calcitonin
- Animation – Thyroid Hormones and Glucose and Lipid Catabolism

Hear

- Figure 17.10 – Steps in the Synthesis and Secretion of Thyroid Hormones

Do

- Exercise – Calcium Homeostasis

Concept	Resources

Concept 17.6 The parathyroid glands secrete parathyroid hormone, which regulates calcium, magnesium, and phosphate ion levels.

1. The **parathyroid glands** are embedded in the posterior surface of the thyroid gland. The **chief cells** produce **parathyroid hormone (PTH)**.
2. PTH plays a major role in regulating blood levels of calcium (Ca^{2+}), magnesium (Mg^{2+}), and phosphate (HPO_4^{2-}) ions.
3. Low blood calcium levels stimulate PTH secretion, causing increased Ca^{2+} and HPO_4^{2-} release into the blood from bones. Calcitonin from the thyroid gland has an opposite effect.
4. PTH also stimulates synthesis of **calcitriol** by the kidneys. Calcitriol increases absorption of Ca^{2+}, HPO_4^{2-}, and Mg^{2+} from ingested foods into the blood.
5. PTH also decreases the loss of Ca^{2+} and Mg^{2+} from the blood to urine and increases the loss of HPO_4^{2-} from blood to urine.

See
- **Anatomy Overview – Parathyroid Glands**
- **Anatomy Overview – Hormones of the Parathyroid**
- **Animation – Parathyroid Hormone**

Do
- **Exercise – Calcium Homeostasis**

Concept 17.7 The adrenal cortex secretes mineralocorticoids, glucocorticoids, and androgens; the adrenal medulla secretes epinephrine and norepinephrine.

1. The paired **adrenal glands** lie superior to each kidney. Each adrenal gland has an outer **adrenal cortex** and inner **adrenal medulla**.
2. The adrenal cortex is subdivided into three zones: the outer zone secretes **mineralocorticoids**, the middle zone secretes **glucocorticoids**, and the inner zone secretes weak **androgens**.
3. The major mineralocorticoid, **aldosterone**, increases reabsorption of Na^+ by the kidneys into the blood, stimulates excretion of K^+ into the urine, and helps adjust blood pressure and blood volume. Secretion of aldosterone is controlled by the **renin–angiotensin–aldosterone pathway**.
4. The glucocorticoids (mainly **cortisol**) regulate metabolism and resistance to stress. Secretion of glucocorticoids is stimulated by **corticotropin-releasing hormone** of the hypothalamus, which stimulates release of ACTH from the anterior pituitary. Glucocorticoids increase protein breakdown, enhance glucose formation, break down triglycerides, inhibit inflammation, and depress immune system responses.
5. The adrenal cortex secretes androgens that stimulate growth of axillary and pubic hair, aid prepubertal growth spurts, and contribute to libido in females.
6. The adrenal medulla secretes **epinephrine** and **norepinephrine**. These hormones augment the fight-or-flight response.

See
- **Anatomy Overview – The Adrenal Glands**
- **Anatomy Overview – Hormones of the Adrenal Glands**
- **Animation – ACTH/Cortisol**
- **Animation – Epinephrine/NE**
- **Animation – Cortisol**

Hear
- **Figure 17.14 –** **Location, blood supply, and histology of the adrenal glands**
- **Figure 17.15 – The renin–angiotensin–aldosterone pathway**

Concept 17.8 The pancreatic islets regulate blood glucose levels by secreting glucagon and insulin.

1. The **pancreas**, which is located in the abdomen near the duodenum of the small intestine, is both an endocrine and an exocrine gland. Exocrine cells produce digestive enzymes. The pancreatic endocrine tissue is referred to as **pancreatic islets** and is composed of **alpha cells** that secrete the hormone **glucagon** and **beta cells** that secrete the hormone **insulin**.
2. Low blood glucose levels cause secretion of glucagon, which acts on liver cells to release glucose into the blood. High blood glucose levels cause secretion of insulin, which acts on cells to increase diffusion of glucose into those cells. Insulin also increases protein and fatty acid synthesis.
3. In addition to blood glucose levels, the autonomic nervous system, diet, and other hormones can also stimulate glucagon and insulin release.

See
- **Anatomy Overview – The Pancreas**
- **Anatomy Overview – Hormones of the Pancreas**
- **Animation – Glucagon**
- **Animation – Insulin**

Hear

- **Figure 17.17 –** **The Pancreas**
- **Figure 17.18 – Regulation of Blood Glucose Level by Negative Feedback Systems Involving Glucagon and Insulin**

Do
- **Exercise – Glucose Regulation Feedback Loop**
- **Concepts and Connections – Blood Glucose Regulation**

Concept	Resources

Concept 17.9 The ovaries produce estrogens, progesterone, and inhibin; the testes produce testosterone and inhibin.

1. The paired **ovaries** are located in the pelvic cavity and produce **estrogens** and **progesterone**. Along with FSH and LH, these hormones regulate the female reproductive cycle, maintain pregnancy, and prepare the mammary glands for lactation. They also help develop and maintain female secondary sex characteristics.
2. The ovaries also produce **inhibin**, a hormone that inhibits FSH secretion.
3. During pregnancy, the ovaries and placenta produce the hormone **relaxin**, which increases the flexibility of the pubic symphysis and helps dilate the cervix.
4. The paired **testes** are located in the scrotum and produce **testosterone**, the primary male sex hormone or **androgen**, which regulates sperm production and stimulates development and maintenance of masculine secondary sex characteristics.
5. The testes also produce **inhibin**, which inhibits FSH secretion.

See

- Anatomy Overview – The Ovaries
- Anatomy Overview – Ovarian Hormones
- Anatomy Overview – The Testes
- Anatomy Overview – Testicular Hormones
- Animation – Hormonal Control of Male Reproductive Function
- Animation – Hormonal Regulation of Female Reproductive System

Do

- Exercise – Match the Male Hormones
- Exercise – Match the Female Hormones

Concept 17.10 The pineal gland secretes melatonin, which contributes to setting the body's biological clock.

1. The **pineal gland** is attached to the roof of the third ventricle in the brain. It secretes the hormone **melatonin**, which contributes to setting the body's biological clock.
2. During sleep, blood levels of melatonin rise.

See

- Anatomy Overview – The Endocrine System
- Anatomy Overview – The Pineal Gland
- Animation – The General Adaptation Syndrome

Do

- Exercise – Concentrate on the Endocrine System
- Exercise – Produce That Hormone
- Exercise – Hormone Actions
- Concepts and Connections – Endocrine System Negative Feedback Loop

Other Chapter Resources

For additional information on clinical applications, medical disorders, and medical terminology, see Chapter 8 in the *Clinical Connections* manual packaged with your text.

Within *WileyPLUS*

- Chapter Quiz
- Visual Anatomy
- Mastering Vocabulary
- Audio Glossary
- Homeostatic Imbalance: The Case of the Girl with Fruity Breath
- Homeostatic Imbalance: The Case of the Growing Girl
- Web Links

• Understanding the Concepts

1. A report on a blood hormone concentration states that the "free fraction" levels were 0.6 micrograms per deciliter of blood. Is this level an accurate reflection of the actual total concentration of the hormone in the blood? Why or why not?

2. How do lipid-soluble hormones influence target cells to produce new molecules? Which type of molecule is produced?

3. What are the differences among permissive effects, synergistic effects, and antagonistic effects of hormones?

4. How do hypothalamic releasing and inhibiting hormones influence secretions of the anterior pituitary?

5. Iodized salt is common throughout the United States. Why might adding iodine to salt be a good idea?

6. A fellow student with hyperthyroidism complains of feeling hot all the time; explain why.

7. If you received a thyroidectomy, should you be concerned about your skeletal system? Why or why not?

8. Would a drug that blocks the action of angiotensin-converting enzyme be used to raise or lower blood pressure? Where does this enzyme function?

9. Why are high-dose glucocorticoids sometimes used following tissue transplant surgery?

10. If a female has an adrenal tumor that causes masculinization, is the tumor located in the adrenal cortex or adrenal medulla? Which hormone class is being produced in excess?

11. You have fallen asleep in class and wake up in time to hear the instructor say that the adrenal gland is composed of nervous tissue. You initially are confused because you thought the lecture was about the endocrine system, but then you realize that the instructor is partly right. Why?

12. How is the release of glucagon and insulin regulated if no hypothalamic releasing or inhibiting factors are present in the body?

18

The Cardiovascular System: The Blood

Ed leaned back in his chair, reaching for his margarita. A warm Gulf breeze, the first of the early evening, moved through his hair. He had been coming to Belize for over 10 years now and his work with the government park service had been very successful. Documenting the local flora and fauna had been a daunting but satisfying task for the ecologist, who at 46 had finally settled into his mid-career stride. As he sat enjoying the sunset, he swatted at the occasional insect. Tomorrow he and his team of students would be heading back to the States. It had been a particularly hot, wet season, and the mosquitoes had been intense in the national park where they had been working. He felt wiped out, which was not unusual for having finished a month of intensive fieldwork with students. As he sat, his eyelids felt heavy and his eyes began to close. The high-pitched whine of the mosquito's wings didn't disturb Ed as he dozed. The *Anopheles* mosquito lit on his arm and began its work. It probed Ed's skin, finding just the right spot with its sharp-ended proboscis, then pierced through the epithelial barrier. The mosquito began pumping salivary enzymes into the dermis to prevent the blood from coagulating. Her meal of blood would require only seconds to complete, yet it was still enough time for the protozoan hitchhikers living in her salivary glands to be injected into Ed's arm. The malarial sporozoites of *Plasmodium falciparum* squirmed inside the dermis, penetrating the superficial capillaries of Ed's skin to begin their journey to his liver.

Malaria is a very serious and sometimes fatal parasitic disease affecting the blood. Although it is rare in the United States and Europe, malaria is a leading cause of death worldwide and is endemic to (constantly present in) certain regions of the world. The vast majority of malaria cases in the United States (about 1200 each year) occur in individuals who have traveled to high-risk areas and in immigrants from such regions. We will use Ed's experience to understand several concepts related to the function of blood and will see that blood is a complex tissue that is intimately involved in maintaining the body's homeostasis.

INTRODUCTION

The **cardiovascular system** (*cardio-* = heart; *-vascular* = blood vessels) consists of three interrelated components: blood, the heart, and blood vessels. The focus of this chapter is blood; the next two chapters will explore the heart and blood vessels, respectively. Blood transports various substances, helps regulate several life processes, and affords protection against disease. For all of its similarities in origin, composition, and functions, blood is as unique from one person to another as are skin, bone, and hair. Despite these differences, blood is the most easily and widely shared of human tissues, saving many thousands of lives every year through blood transfusions. Health-care professionals routinely examine and analyze the blood through various tests when trying to determine the cause of disease.

Most cells of the body cannot move around to obtain oxygen and nutrients and

Blood transports various substances, helps regulate several life processes, and affords protection against disease.

get rid of carbon dioxide and other wastes. Instead, two fluids meet these needs: blood and interstitial fluid. **Blood** is a connective tissue composed of a liquid extracellular matrix called plasma that suspends various cells and cell fragments. **Interstitial fluid** is the fluid that bathes body cells and is constantly renewed by the blood. Blood transports oxygen from the lungs and nutrients from the gastrointestinal tract to cells throughout the body. The oxygen and nutrients diffuse from the blood into the interstitial fluid, the fluid found between body cells, and then into body cells. Carbon dioxide and other wastes move in the reverse direction from the body cells into the interstitial fluid and then into the blood. Blood then transports the wastes to various organs— the lungs, kidneys, skin, and digestive system—for elimination from the body.

In order for blood to reach all the cells, it must be moved throughout the body. The *heart* is the pump that circulates the blood. *Blood vessels* convey blood from the heart to body cells (via arteries) and from body cells back to the heart (via veins).

18.1   Blood contains plasma and formed elements and transports essential substances through the body.

Functions of Blood

Blood, a liquid connective tissue, has three general functions:

- **Transportation.** Blood transports oxygen from the lungs to cells throughout the body and carbon dioxide from the cells to the lungs. It also carries nutrients from the gastrointestinal tract to body cells, heat and waste products away from cells, and hormones from endocrine glands to other body cells.

- **Regulation.** Circulating blood helps maintain homeostasis in all body fluids. Blood helps regulate pH through the use of buffers. It also helps adjust body temperature through the heat-absorbing and coolant properties of the water in blood and blood's variable rate of flow through the skin, where excess heat can be lost from the blood to the environment. Also, through interactions of dissolved ions and proteins, blood's osmotic pressure influences the water content of cells.

- **Protection.** Blood can clot (become gel-like) in response to an injury, which protects against its excessive loss from the cardiovascular system. In addition, white blood cells protect against disease by carrying on phagocytosis; several blood proteins, including antibodies, interferons, and complement, also help protect against disease.

Physical Characteristics of Blood

Blood is denser and more viscous (thicker) than water, which is part of the reason it flows more slowly than water. The temperature of blood is 38°C (100.4°F), about 1°C higher than oral or rectal body temperature, and it has a slightly alkaline pH ranging from 7.35 to 7.45. The color of blood varies with its oxygen content. When it has a high oxygen content, it is bright red. When it has a low oxygen content, it is dark red. Blood constitutes about 8% of the total body weight. The blood volume is 5 to 6 liters (1.5 gal) in an average-sized adult male and 4 to 5 liters (1.2 gal) in an average-sized adult female. The difference in volume is due to differences in body size. Several hormonal negative feedback systems ensure that blood volume and osmotic pressure remain relatively constant.

Components of Blood

Whole blood is composed of plasma, a watery liquid that contains dissolved substances, and formed elements, which are cells and cell fragments. If a sample of blood is centrifuged (spun at high speed) in a small glass tube, the cells sink to the bottom of the tube while the lighter-weight plasma forms a layer on top (Figure 18.1a). Blood is about 45% formed elements and about 55% plasma. Normally, more than 99% of the formed elements are red blood cells (RBCs). Pale, colorless white blood cells (WBCs) and platelets occupy less than 1% of the formed elements. Because they are less dense than red blood cells but more dense than plasma, they form a very thin layer, called the *buffy coat*, between the packed RBCs and plasma in centrifuged blood. Figure 18.1b shows the composition of plasma and the numbers of the various types of formed elements in blood.

Plasma

When the formed elements are removed from blood, a straw-colored liquid called **plasma** remains. Plasma is about 91.5% water, 7% proteins, and 1.5% solutes other than proteins. Proteins in the blood, the **plasma proteins**, are synthesized mainly by hepatocytes (liver cells). The most plentiful plasma proteins are the **albumins**, which account for about 54% of all plasma proteins. Among other functions, albumins help maintain proper blood osmotic pressure, which is an important factor in the exchange of fluids across capillary walls. **Globulins**, which compose 38% of plasma proteins, include **antibodies**, or **immunoglobulins**, defensive proteins produced during certain immune responses. **Fibrinogen** makes up about 7% of plasma proteins and is a key protein in formation of blood clots. Other solutes in plasma include electrolytes, nutrients, gases, regulatory substances such as enzymes and hormones, vitamins, and waste products.

Table 18.1 describes the chemical composition of plasma.

Figure 18.1 **Components of blood in a normal adult.**

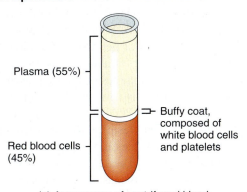

Plasma (55%)

Buffy coat, composed of white blood cells and platelets

Red blood cells (45%)

(a) Appearance of centrifuged blood

Functions of Blood
1. Transports oxygen, carbon dioxide, nutrients, hormones, heat, and wastes.
2. Regulates pH, body temperature, and water content of cells.
3. Protects against blood loss through clotting and against disease through phagocytic white blood cells and antibodies.

Whole blood 8%

Other fluids and tissues 92%

Plasma 55%

Proteins 7%

Water 91.5%

Other solutes 1.5%

Albumins 54%

Globulins 38%

Fibrinogen 7%

All others 1%

Electrolytes
Nutrients
Gases
Regulatory substances
Waste products

PLASMA (weight)

SOLUTES

Formed elements 45%

Platelets 150,000–400,000

White blood cells 5,000–10,000

Red blood cells 4.8–5.4 million

Neutrophils 60–70%

Lymphocytes 20–25%

Monocytes 3–8%

Eosinophils 2–4%

Basophils 0.5–1.0%

BODY WEIGHT

VOLUME

FORMED ELEMENTS (number per μL)

WHITE BLOOD CELLS

(b) Components of blood

Blood is a connective tissue that consists of plasma (liquid) plus formed elements (red blood cells, white blood cells, and platelets).

TABLE 18.1	
Substances in Plasma	
CONSTITUENT	**DESCRIPTION**
Water (91.5%)	Liquid portion of blood. Acts as solvent and suspending medium for components of blood; absorbs, transports, and releases heat.
Plasma Proteins (7.0%)	Exert coloid osmotic pressure, which helps maintain water balance between blood and tissues and regulates blood volume.
Albumins	Smallest and most numerous plasma proteins: produced by liver. Function as transport proteins for several steroid hormones and for fatty acids.
Globulins	Produced by liver and by plasma cells, which develop from B lymphocytes. Antibodies (immunoglobulins) help attack viruses and bacteria. Alpha and beta globulins transport iron, lipids, and fat-soluble vitamis.
Fibrinogen	Produced by liver. Plays essential role in blood clotting.
Other Solutes (1.5%)	
Electrolytes	Inorganic salts. Positively charged ions (cations) include Na^+, K^+, Ca^{2+}, Mg^{2+}; negatively charged ions (anions) include Cl^-, HPO_4^{2-}, SO_4^{2-} and HCO_3^-. Help maintain osmotic pressure and play essential roles in the function of cells.
Nutrients	Products of digestion pass into blood for distribution to all body cells. Include amino acids (from proteins), glucose (from carbohydrates), fatty acids and glycerol (from triglycerides), vitamins, and minerals.
Gases	Oxygen (O_2), carbon dixode (CO_2), and nitrogen (N_2). More O_2 is associated with hemoglobin inside red blood cells, but more CO_2 is dissolved in plasma. N_2 is present but has no known function in the body.
Regulatory substances	Enzymes, produced by body cells, catalyze chemical reactions. Hormones, produced by endocrine glands, regulate metabolism, growth, and development. Vitamins are cofactors for enzymatic reactions.
Waste products	Most are breakdown products of protein metabolism and are carried by blood to organs of excretion. Include urea, uric acid, creatine, creatinine, bilirubin, and ammonia.

Formed Elements

The **formed elements** of the blood include red blood cells (RBCs), white blood cells (WBCs), and platelets (Figure 18.2). RBCs and WBCs are whole cells, but platelets are cell fragments. There are several distinct types of WBCs—neutrophils, lymphocytes, monocytes, eosinophils, and basophils—each with a unique microscopic appearance. WBCs have a larger number of functions than RBCs and platelets. The roles of each type of WBC are discussed later in this chapter.

The percentage of total blood volume occupied by RBCs is called the **hematocrit** (hē-MAT-ō-krit). For example, a hematocrit of 40% means that 40% of the volume of blood is composed of RBCs. The normal range of hematocrit for adult females is about 38–46% (average = 42%); for adult males it is about 40–54% (average = 47%). The hormone testosterone, present in much higher concentration in males than in females, stimulates synthesis of **erythropoietin** (e-rith´-rō-POY-e-tin), the hormone that stimulates production of RBCs. Thus, testosterone contributes to higher hematocrits in males. Lower values in women during their reproductive years also may be due to excessive loss of blood during menstruation. A significant drop in the hematocrit indicates anemia, a lower-than-normal number of RBCs. In **polycythemia** the percentage of RBCs is abnormally high, and the hematocrit may be 65% or higher. Polycythemia may be caused by conditions such as an abnormal increase in RBC production, tissue hypoxia (low oxygen levels), dehydration, and blood doping or the use of EPO by athletes.

Figure 18.2 **Scanning electron micrograph and photomicrograph of the formed elements of blood.**

White blood cell

Platelet

Red blood cell

SEM 3500x

(a)

White blood cell (neutrophil)

Blood plasma

Red blood cell

Platelet

White blood cell (monocyte)

LM 400x

(b)

🔑 The formed elements of blood are red blood cells, white blood cells, and platelets.

Checkpoint

1. List the substances that blood transports.
2. Which processes does blood help regulate?
3. Which formed elements of blood are most numerous?

4. What is the buffy coat?
5. List the functions of plasma proteins.
6. Which formed elements of the blood are cell fragments?

18.2 Hemopoiesis is the production of formed elements.

Although some lymphocytes have a lifetime measured in years, most formed elements of the blood (Figure 18.3) are continually dying and being replaced within hours, days, or weeks. Negative feedback systems regulate the total number of RBCs and plate- lets in circulation, and their numbers normally remain steady. The abundance of the different types of WBCs, however, varies in response to challenges by invading pathogens and other for- eign antigens.

Figure 18.3 **Origin, development, and structure of formed elements.** Some of the generations of some cell lines have been omitted.

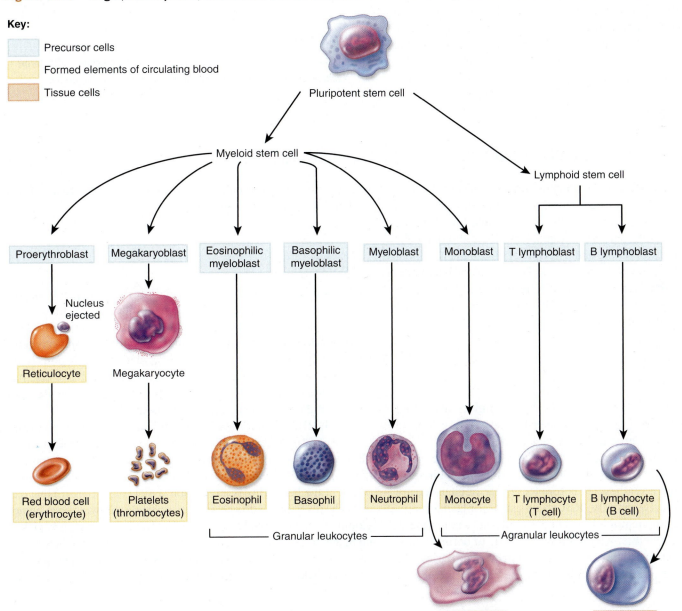

Key:
- Precursor cells
- Formed elements of circulating blood
- Tissue cells

🔑 Formed element production is called hemopoiesis and occurs mainly in red bone marrow after birth.

The process by which the formed elements of blood develop is called **hemopoiesis** (hē-mō-poy-Ē-sis; *-poiesis* = making) or **hematopoiesis**. Before birth, hemopoiesis occurs in the liver, spleen, thymus, and lymph nodes of a fetus. Red bone marrow becomes the primary site of hemopoiesis in the last 3 months before birth, and continues as the source of formed elements after birth and throughout life.

Red bone marrow is a highly vascularized connective tissue located in the microscopic spaces between trabeculae of spongy bone tissue. It is present chiefly in bones of the axial skeleton, pectoral and pelvic girdles, and the proximal epiphyses of the humerus and femur. About 0.05–0.1% of red bone marrow cells are derived from mesenchyme and are called **pluripotent stem cells** (ploo-RIP-ō-tent; *pluri-* = several). Pluripotent stem cells have the capacity to develop into many different types of cells (Figure 18.3).

In newborns all bone marrow is red and thus active in blood cell production. As an individual grows and in adulthood, the rate of blood cell formation decreases; the red bone marrow in the medullary (marrow) cavity of long bones becomes inactive and is replaced by yellow bone marrow, which is largely fat cells. Under certain conditions, such as severe bleeding, yellow bone marrow can revert to red bone marrow to assist in blood cell production.

In order to form blood cells, pluripotent stem cells generate two further types of stem cells, *myeloid stem cells* and *lymphoid stem cells*, which have the capacity to develop into several types of cells (Figure 18.3). Myeloid stem cells begin and complete their development in red bone marrow and give rise to red blood cells, platelets, eosinophils, basophils, neutrophils, and monocytes. Lymphoid stem cells begin their development in red bone marrow but complete it in lympatic tissues; they give rise to lymphocytes.

Myeloid and lymphoid stem cells give rise to **precursor cells**. Precursor cells are committed to a specific formed element pathway. Over several cell divisions they develop into the actual formed elements of blood. For example, monoblasts develop into monocytes, eosinophilic myeloblasts develop into eosinophils, and so on.

Checkpoint

7. What is the primary site for the development of formed elements after birth?
8. From which connective tissue cells do pluripotent stem cells develop?

18.3 Mature red blood cells are biconcave cells containing hemoglobin.

Red blood cells (RBCs) or **erythrocytes** (e-RITH-rō-sīts; *erythro-* = red; *-cyte* = cell) contain the oxygen-carrying protein **hemoglobin**, which is a pigment that gives whole blood its red color. A healthy adult male has about 5.4 million red blood cells per microliter (μL) of blood, and a healthy adult female has about 4.8 million. (One drop of blood is about 50 μL.) To maintain normal numbers of RBCs, new mature cells must enter the circulation at the astonishing rate of at least 2 million per second, a pace that balances the equally high rate of RBC destruction.

RBC Anatomy

RBCs are biconcave (concave on both sides) discs averaging about 8 micrometers (μm) in diameter (Figure 18.4a). Mature RBCs have a simple structure; they lack a nucleus and other organelles and can neither reproduce nor carry on extensive metabolic activities. The cytosol of RBCs contains hemoglobin molecules, which were synthesized before loss of the nucleus during RBC production. Essentially, RBCs consist of a plasma membrane, cytosol, and hemoglobin. The RBC plasma membrane is both strong and flexible, which allows the cells to deform without rupturing as they squeeze through narrow capillaries. As you will see later, certain glycolipids in the plasma membrane are antigens that account for the various blood groups such as the ABO and Rh groups.

RBC Physiology

Red blood cells are highly specialized for their oxygen transport function. Because mature RBCs have no nucleus, all their internal space is available for oxygen transport. Moreover, RBCs lack mitochondria and generate ATP anaerobically (without oxygen). Hence, they do not use up any of the oxygen they transport. Even the shape of an RBC facilitates its function. A biconcave disc has a much greater surface area for the diffusion of gas molecules into and out of the RBC than, say, a sphere or a cube.

Each RBC contains about 280 million hemoglobin molecules (Figure 18.4b). A hemoglobin molecule consists of a protein called **globin**, composed of four polypeptide chains; a ringlike nonprotein pigment called a **heme** is bound to each of the four chains. At the center of the heme ring is an iron ion (Fe^{2+}) (Figure 18.4c) that can combine reversibly with one oxygen molecule, allowing each hemoglobin molecule to bind four oxygen molecules. The oxygen picked up in the lungs is transported bound to the iron ion of the heme ring. As blood flows through tissue capillaries, hemoglobin releases oxygen, which diffuses first into the interstitial fluid and then into cells.

Figure 18.4 **Shapes of a red blood cell and a hemoglobin molecule, and structure of a heme group.** In (b), note that each of the four polypetide chains (blue) of a hemoglobin molecule has one heme group (gold), which contains an iron ion, Fe^{2+} (shown in red).

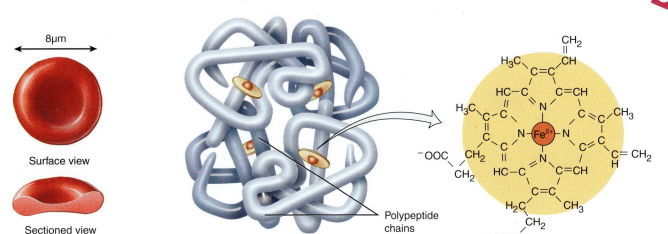

8µm

Surface view

Sectioned view

(a) RBC shape

(b) Hemoglobin molecule

Polypeptide chains

(c) Iron-containing heme

The iron portion of a heme group binds oxygen for transport by hemoglobin.

Hemoglobin also transports about 13% of the total carbon dioxide, a waste product of metabolism. Blood flowing through tissue capillaries picks up carbon dioxide, some of which combines with amino acids in the globin part of hemoglobin. As blood flows through the lungs, the carbon dioxide is released from hemoglobin and then exhaled.

Checkpoint

9. Which characteristics of red blood cells facilitate oxygen transport?
10. How many molecules of O_2 can one hemoglobin molecule transport?
11. Which part of the hemoglobin molecule carries oxygen? Which part carries carbon dioxide?

18.4 Red blood cells have a life cycle of 120 days.

Red blood cells live only about 120 days because of the wear and tear their plasma membranes undergo as they squeeze through blood capillaries. Without a nucleus and other organelles, RBCs cannot synthesize new components to replace damaged ones. The plasma membrane becomes more fragile with age, and the cells are more likely to burst, especially as they squeeze through narrow channels in the spleen. Ruptured red blood cells are removed from circulation and destroyed by phagocytic macrophages in the spleen and liver, and the breakdown products are recycled, as follows (Figure 18.5):

1 Macrophages in the spleen, liver, and red bone marrow phagocytize ruptured and worn-out red blood cells, splitting apart the globin and heme portions of hemoglobin.

2 Globin is broken down into amino acids, which can be reused by body cells to synthesize other proteins.

3 Iron removed from the heme portion associates with the plasma protein **transferrin** (trans-FER-in; *trans-* = across; *-ferr-* =iron), a transporter for iron in the bloodstream.

4 The iron–transferrin complex is then carried to red bone marrow, where RBC precursor cells use it in hemoglobin synthesis. Iron is needed for the heme portion of the hemoglobin molecule, and amino acids are needed for the globin portion. Vitamin B_{12} is also needed for synthesis of hemoglobin.

5 Within red bone marrow, erythropoiesis (red blood cell production) releases red blood cells into the circulation.

6 When iron is removed from heme, the non-iron portion of heme is converted to **biliverdin** (bil´-i-VER-din), a green pigment, and then into **bilirubin** (bil´-i-ROO-bin), a yellow-orange pigment. Bilirubin enters the blood and is transported to the liver. Within the liver, bilirubin is secreted by hepatocytes into bile, which passes into the small intestine and then into the large intestine.

Figure 18.5   Formation and destruction of red blood cells and recycling of hemoglobin components.

🔑 **The rate of RBC formation by red bone marrow normally equals the rate of RBC destruction by macrophages.**

7 In the large intestine, bacteria convert bilirubin into **urobilinogen** (ūr-ō-bī-LIN-ō-jen). Some urobilinogen is absorbed back into the blood, where it travels to the kidneys, is converted to a yellow pigment called **urobilin** (ūr-ō-BĪ-lin), and excreted in urine. Most urobilinogen is eliminated in feces in the form of a brown pigment called **stercobilin** (ster′-kō-BĪ-lin), which gives feces its characteristic color.

Checkpoint

12. Why do red blood cells live for only about 120 days?

13. As red blood cells die, how are the components of hemoglobin recycled or excreted?

Return to Ed's Story

As Ed sat on the plane, he had no idea that his immune system was on alert. It had been 24 hours since the intruder had entered his bloodstream. The microscopic sporozoites had hitched a ride on the stream of Ed's life-giving blood. As they traveled, they collided with red blood cells, white blood cells, and proteins. The sporozoite surfaces were covered with alien proteins that Ed's body didn't recognize. A few were destroyed by white blood cells, but most found their way to the liver. Proteolytic enzymes released by the sporozoites started eating away at the cell surface while microscopic molecular motors inside the parasites pushed them into Ed's hepatocytes. Once hidden inside the cells of Ed's own body, the sporozoites would transform and multiply in relative safety.

Ed had been home now for 4 weeks. He had gone to bed early, feeling a bit nauseous and achy. By 3 o'clock the next morning, he was tossing and turning in his bed, and sweat soaked the sheets. Inside his body the next stage in the protozoa's life cycle had begun. Thousands of transformed *Plasmodium* merozoites now burst forth from the liver and traveled into the bloodstream. The merozoites bound to the surface of red blood cells and invaded the interior. Once

inside, the merozoites hijacked the RBCs' protein-manufacturing centers, robbing them of vital nutrients and masking the surface of the infected cells with rogue proteins to prevent detection by Ed's immune system. As the parasites grew and multiplied, the infected RBCs swelled and finally ruptured, releasing toxins and cellular debris into Ed's bloodstream. Each ruptured RBC meant more potential merozoites to infect more RBCs. Ed awoke sweating and gasping for air. His muscles ached as if he had run a race.

A. How would Ed's blood help protect him from a foreign invader such as the one now in his system?

B. The sporozoites have traveled to Ed's liver. How is the function of hepatocytes related to blood?

C. Ed's red blood cells are being destroyed by the parasite that infects him. What is the process of blood cell formation that replenishes lost blood cells? Where in his body would new red blood cells be produced?

D. Why would Ed's RBCs be a good host for a parasitic protozoan?

18.5 Erythropoiesis is the process of red blood cell formation.

As you have already learned, the production of the formed elements of blood is called hemopoiesis. The specific production of RBCs is termed **erythropoiesis** (e-rith´-rō-poy-Ē-sis). Erythropoiesis starts in the red bone marrow with a precursor cell called a pro-erythroblast (see Figure 18.3). The proerythroblast divides several times, producing cells that begin to synthesize hemoglobin. Ultimately, a cell near the end of the development sequence ejects its nucleus and becomes a **reticulocyte** (re-TIK-ū-lō-sit). Loss of the nucleus causes the center of the cell to indent, producing the RBC's distinctive biconcave shape. Reticulocytes, which are about 34% hemoglobin and retain some mitochondria, ribosomes, and endoplasmic reticulum, pass from red bone marrow into the bloodstream. Reticulocytes develop into erythrocytes (mature RBCs) within 1 or 2 days after their release from bone marrow.

Normally, erythropoiesis and red blood cell destruction proceed at roughly the same pace. If the oxygen-carrying capacity of the blood falls because erythropoiesis is not keeping up with RBC destruction, a negative feedback system increases RBC production (Figure 18.6). The controlled condition in this particular feedback loop is the amount of oxygen delivered to the kidneys (and thus to body tissues in general). Cellular oxygen deficiency, called **hypoxia** (hī-POKS-ē-a), may occur if too little oxygen enters the blood. For example, the reduced oxygen content of air at high altitudes lowers the level of oxygen in the blood. Oxygen delivery may also decrease due to anemia, which has many causes: Lack of iron, lack of certain amino acids, and lack of vitamin B_{12} are but a few. Circulatory problems that reduce blood flow to tissues may also reduce oxygen delivery. Whatever the cause, hypoxia stimulates the kidneys to increase the release of erythropoietin. This hormone circulates through the blood to the red bone marrow, where it speeds the development of proerythroblasts into reticulocytes. As the number of circulating RBCs increases, more oxygen can be delivered to body tissues.

Checkpoint

14. Which forms of red blood cells are found in circulating blood?

15. A hematocrit indicates the percentage of the blood volume occupied by red blood cells. How might your hematocrit change if you moved from a town at sea level to a high mountain village where the air contains less oxygen than at sea level?

Figure 18.6 **Negative feedback regulation of erythropoiesis (red blood cell formation).**

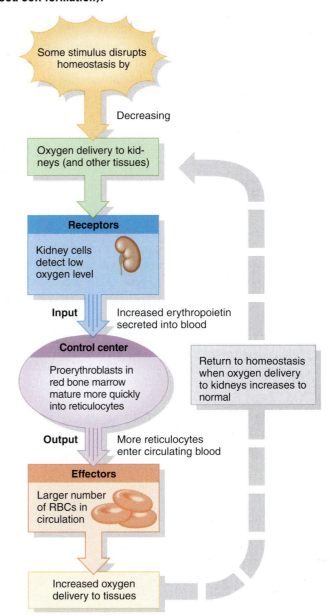

Some stimulus disrupts homeostasis by

Decreasing

Oxygen delivery to kidneys (and other tissues)

Receptors
Kidney cells detect low oxygen level

Input — Increased erythropoietin secreted into blood

Control center
Proerythroblasts in red bone marrow mature more quickly into reticulocytes

Return to homeostasis when oxygen delivery to kidneys increases to normal

Output — More reticulocytes enter circulating blood

Effectors
Larger number of RBCs in circulation

Increased oxygen delivery to tissues

The main stimulus for erythropoiesis is when kidney cells detect low oxygen levels in the blood.

18.6   Blood is categorized into groups based on surface antigens.

The surfaces of red blood cells contain a genetically determined assortment of **antigens** composed of glycoproteins and glycolipids. These antigens, called **agglutinogens** (ag´-loo-TIN-ō-jens) occur in characteristic combinations. Based on the presence or absence of various antigens, blood is categorized into different **blood groups**. There are more than 100 antigens that can be detected on the surface of red blood cells, and at least 24 currently recognized blood groups. Within a given blood group there may be two or more different **blood types**. Here we discuss two major blood groups: ABO and Rh.

ABO Blood Group

The **ABO blood group** is based on two antigens called A and B (Figure 18.7). People whose RBCs display only antigen A have **type A** blood. Those who have only antigen B are **type B**. Individuals who have both A and B antigens are **type AB**, and those who have neither antigen A nor antigen B are **type O**. In about 80% of the population, soluble antigens of the ABO type appear in saliva and other body fluids, in which case blood type can be identified from a sample of saliva. The incidence of ABO blood types varies among different population groups, as indicated in Table 18.2.

In addition to antigens on the surfaces of RBCs, plasma usually contains antibodies or *agglutinins* (a-GLOO-ti-nins) that react with RBC surface antigens if the two are mixed. Regarding the ABO blood group, **anti-A antibody** reacts with antigen A, and **anti-B antibody** reacts with antigen B. The antibodies present in each of the four ABO blood types are also shown in Figure 18.7. You do not have antibodies that react with your own antigens, but most likely you do have antibodies for any antigens of the ABO group that your RBCs lack. For example, if you have type A blood, you have A antigens on the surfaces of your RBCs, and you have anti-B antibodies in your plasma. If you had anti-A antibodies in your plasma, they would attack your RBCs. Because anti-A and anti-B antibodies are large molecules that do not cross the placenta, ABO incompatibility between a mother and her fetus rarely causes problems.

TABLE 18.2

Blood Types in the United States

POPULATION GROUP	BLOOD TYPE (PERCENTAGE)				
	O	A	B	AB	RH+
European-American	45	40	11	4	85
African-American	49	27	20	4	95
Korean-American	32	28	30	10	100
Japanese-American	31	38	21	10	100
Chinese-American	42	27	25	6	100
Native American	79	16	4	1	100

Figure 18.7   Antigens and antibodies involved in the ABO blood groups.

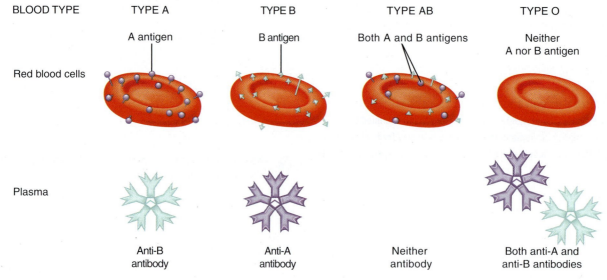

BLOOD TYPE	TYPE A	TYPE B	TYPE AB	TYPE O
	A antigen	B antigen	Both A and B antigens	Neither A nor B antigen
Red blood cells				
Plasma	Anti-B antibody	Anti-A antibody	Neither antibody	Both anti-A and anti-B antibodies

The antibodies in your plasma do not react with the antigens on your red blood cells.

Rh Blood Group

The **Rh blood group** is so named because the antigen was discovered in the blood of the *Rhesus* monkey. People whose RBCs have the Rh antigen are designated Rh$^+$ (Rh positive); those who lack the Rh antigen are designated Rh$^-$ (Rh negative). The percentages of Rh$^+$ and Rh$^-$ individuals in various populations are shown in Table 18.2. Normally, plasma does not contain anti-Rh antibodies. If an Rh$^-$ person receives an Rh$^+$ blood transfusion, however, the immune system starts to make anti-Rh antibodies that do remain in the blood. If a second transfusion of Rh$^+$ blood is given later, the previously formed anti-Rh antibodies will cause clumping and **hemolysis** (rupture) of the RBCs in the donated blood, and a severe reaction may occur.

Transfusions

Despite the differences in RBC antigens, blood is the most easily shared of human tissues, saving many thousands of lives every year through transfusions. A **transfusion** (trans-FŪ-zhun) is the transfer of whole blood or blood components (red blood cells or plasma, for example) into the bloodstream. Most often a transfusion is given to alleviate anemia, to increase blood volume (for example, after a severe *hemorrhage*, the loss of a large amount of blood), or to improve immunity. However, the normal components of one person's RBC plasma membrane can trigger damaging antigen–antibody responses in a transfusion recipient.

In an incompatible blood transfusion, antibodies in the recipient's plasma bind to the antigens on the donated RBCs, which causes **agglutination** (a-gloo-ti-NĀ-shun), or clumping, of the RBCs. Agglutination is an antigen–antibody response in which RBCs become cross-linked to one another. (Note that agglutination is not the same as blood clotting.) Formation of these antigen–antibody complexes causes hemolysis and the release of hemoglobin into the plasma. The liberated hemoglobin may cause kidney damage by clogging the filtration membranes.

As an example of an incompatible blood transfusion, consider what happens if a person with type A blood receives a transfusion of type B blood. The recipient's blood (type A) contains A antigens on the red blood cells and anti-B antibodies in the plasma. The donor's blood (type B) contains B antigens and anti-A antibodies. In this situation, two things can happen. First, the anti-B antibodies in the recipient's plasma can bind to the B antigens on the donor's RBCs, causing hemolysis of the donated RBCs. Second, the anti-A antibodies in the donor's plasma can bind to the A antigens on the recipient's RBCs. The second reaction is usually not serious because the donor's anti-A antibodies become so diluted in the recipient's plasma that they do not cause any significant hemolysis of the recipient's RBCs. Table 18.3 summarizes the interactions of the four blood types of the ABO system.

People with type AB blood do not have anti-A or anti-B antibodies in their plasma. They are sometimes called "universal recipients" because theoretically they can receive blood from donors of all four ABO blood types. They have no antibodies to attack antigens on donated RBCs (Table 18.3). People with type O blood have neither A nor B antigens on their RBCs and are sometimes called "universal donors." Because there are no antigens on their RBCs for antibodies to attack, type O people, theoretically, can donate blood to all four ABO blood types. Type O persons requiring blood may receive only type O blood, as they have antibodies to both A and B antigens in their plasma (Table 18.3). In practice, use of the terms *universal recipient* and *universal donor* is misleading and dangerous. Blood contains antigens and antibodies other than those associated with the ABO system that can also cause transfusion problems. Thus, blood should always be carefully matched before transfusion.

TABLE 18.3

ABO Blood Group Interactions

CHARACTERISTIC	BLOOD TYPE			
	A	B	AB	O
Antigen (agglutinogen) on RBC	A	B	Both A and B	Neither A nor B
Antibody (agglutinin) in plasma	Anti-B	Anti-A	Neither anti-A nor anti-B	Both anti-A and anti-B
Compatible donor blood types (no hemolysis)	A, O	B, O	A, B, AB, O	O
Incompatible donor blood types (hemolysis)	B, AB	A, AB	—	A, B, AB

Checkpoint

16. Which ABO antigens are found in type AB blood? In type O blood?
17. Where in blood are antibodies to the A or B antigens found?
18. Which antibodies are found in type A blood? In type AB blood? In type O blood?
19. Which blood type can be given to people with type AB blood?
20. What is hemolysis, and how can it occur after a mismatched Rh blood transfusion?

18.7 White blood cells combat inflammation and infection.

WBC Anatomy and Types

Unlike red blood cells, **white blood cells**, or **leukocytes** (LOO-kō-sīts; *leuko-* = white), have a nucleus and do not contain hemoglobin (Figure 18.8). WBCs are classified as either granular or agranular, depending on whether they contain cytoplasmic granules (vesicles) that are made visible by staining when viewed through a light microscope. *Granular leukocytes* include neutrophils, eosinophils, and basophils; *agranular leukocytes* include lymphocytes and monocytes. As shown in Figure 18.3, granular leukocytes and monocytes develop from a myeloid stem cell, and lymphocytes develop from a lymphoid stem cell.

Granular Leukocytes

After staining, each of the three types of granular leukocytes displays conspicuous granules with distinctive coloration that can be recognized under a light microscope. Granular leukocytes can be distinguished as follows:

- **Eosinophil** (ē-ō-SIN-ō-fil). The large, uniform-sized granules within an eosinophil are *eosinophilic* (= eosin-loving)—they stain red-orange with acidic dyes (Figure 18.8a). The granules usually do not cover or obscure the nucleus, which most often has two or three lobes connected by a thin strand of nuclear material.

- **Basophil** (BĀ-sō-fil). The round, variable-sized granules of a basophil are *basophilic* (= basic-loving)—they stain blue-purple with basic dyes (Figure 18.8b). The granules commonly obscure the nucleus, which has two lobes.

- **Neutrophil** (NOO-trō-fil). The granules of a neutrophil are smaller, evenly distributed, and pale lilac in color (Figure 18.8c). Because the granules do not strongly attract either the acidic (red) or basic (blue) stain, these WBCs are *neutrophilic* (= neutral-loving). The nucleus has two to five lobes, connected by very thin strands of nuclear material. As the cells age, the number of nuclear lobes increases. Because older neutrophils thus have several differently shaped nuclear lobes, they are often called *polymorphonuclear leukocytes (PMNs)*, or "polys."

Agranular Leukocytes

Even though so-called agranular leukocytes possess cytoplasmic granules, the granules are not visible under a light microscope because of their small size and poor staining qualities. Characteristics of agranular leukocytes are as follows:

- **Lymphocyte** (LIM-fō-sīt). The nucleus of a lymphocyte stains dark and is round or slightly indented (Figure 18.8d). The cytoplasm stains sky blue and forms a rim around the nucleus. The larger the cell, the more cytoplasm is visible. Lymphocytes may be as small as 6–9 μm in diameter or as large as 10–14 μm in diameter. Although the functional significance of the size difference between small and large lymphocytes is unclear, the distinction is still clinically useful because an increase in the number of large lymphocytes has diagnostic significance in acute viral infections and in some immunodeficiency diseases.

- **Monocyte** (MON-ō-sīt´). The nucleus of a monocyte is usually kidney-shaped or horseshoe-shaped, and the cytoplasm is blue-gray and has a foamy appearance (Figure 18.8e). The cytoplasm's color and appearance are due to very fine *azurophilic granules* (az´-ū-rō-FIL-ik; *azur* = blue; *philos* = loving), which are lysosomes. Blood is merely a conduit for monocytes, which migrate from the blood into the tissues, where they enlarge and differentiate into **macrophages** (MAK-rō-fāj = large eaters). Some become **fixed macrophages**, which means they reside in a particular tissue; examples are alveolar macrophages in the lungs or macrophages in the spleen. Others become **wandering macrophages**, which roam the tissues and gather at sites of infection or inflammation.

Figure 18.8   Types of white blood cells.

LM all 1600x

(a) Eosinophil   (b) Basophil   (c) Neutrophil   (d) Lymphocyte   (e) Monocyte

🔑 **White blood cells are distinguished from one another by the shapes of their nuclei and the staining properties of their cytoplasmic granules.**

WBC Functions

The skin and mucous membranes of the body are continuously exposed to microscopic organisms, such as bacteria, some of which are capable of invading deeper tissues and causing disease. Once pathogens enter the body, the general function of white blood cells is to combat them by phagocytosis or immune responses. To accomplish these tasks, many WBCs leave the bloodstream and collect at points of pathogen invasion or inflammation. Once granular leukocytes and monocytes leave the bloodstream to fight injury or infection, they never return to it. Lymphocytes, on the other hand, continually recirculate—from blood to interstitial spaces of tissues to lymphatic fluid and back to blood. Only 2% of the total lymphocyte population is circulating in the blood at any given time; the rest are in lymphatic fluid and organs such as skin, lungs, lymph nodes, and spleen.

RBCs are contained within the bloodstream, but WBCs are able to cross capillary walls by a process termed **emigration** (em´-i-GRĀ-shun; *e-* = out; *-migra-* = wander), formerly called *diapedesis*. During emigration, WBCs roll along the endothelium that forms capillary walls, stick to it, and then squeeze between the endothelial cells (Figure 18.9). The precise signals that stimulate emigration through a particular capillary vary for the different types of WBCs. Molecules known as **adhesion molecules** help WBCs stick to the endothelium. For example, in response to nearby injury and inflammation, endothelial cells display adhesion molecules called *selectins*. Selectins stick to the surface of neutrophils, causing them to slow down and roll along the endothelial surface. On the neutrophil surface are other adhesion molecules called *integrins*, which anchor neutrophils to the endothelium and assist their movement through the capillary wall and into the interstitial fluid of the injured tissue.

Neutrophils and macrophages are active in **phagocytosis**; they can ingest bacteria and dispose of dead matter. Several different chemicals released by pathogens and inflamed tissues attract phagocytes, a phenomenon called **chemotaxis**.

Among WBCs, neutrophils respond most quickly to tissue destruction by bacteria. After engulfing a pathogen during phagocytosis, a neutrophil unleashes several destructive chemicals to destroy the ingested pathogen. These chemicals include the enzyme **lysozyme**, which destroys certain bacteria, and strong **oxidants**, such as hydrogen peroxide (H_2O_2) and the hypochlorite anion (OCl^-), which is similar to household bleach. Neutrophils also contain **defensins**, proteins that exhibit a broad range of antibiotic activity against bacteria and fungi. Within a neutrophil, granules containing defensins form peptide "spears" that poke holes in microbe membranes; the resulting loss of cellular contents kills the invader.

Monocytes take longer to reach a site of infection than do neutrophils, but they arrive in large numbers. Upon their arrival, monocytes enlarge and differentiate into wandering macrophages, which phagocytize many more microbes than can neutrophils. They also clean up cellular debris following an infection.

At sites of inflammation, basophils leave capillaries, enter tissues, and release heparin and histamine. These substances intensify the inflammatory reaction and are involved in hypersensitivity (allergic) reactions.

Eosinophils leave the capillaries and enter tissue fluid. They are believed to release enzymes, such as histaminase, that combat the effects of histamine and other mediators of inflammation in allergic reactions. Eosinophils also phagocytize antigen–antibody complexes and are effective against certain parasitic worms. A high eosinophil count often indicates an allergic condition or a parasitic infection.

Lymphocytes are the major combatants in immune responses. Most lymphocytes continually move among lymphoid tissues, lymph, and blood, spending only a few hours at a time in blood. Thus, only a small proportion of the total lymphocytes are present in the blood at any given time. There are three types of lymphocytes: T lymphocytes (T cells), B lymphocytes (B cells), and natural killer (NK) cells. B cells are particularly effective in destroying bacteria and inactivating their toxins. T cells attack viruses, fungi, transplanted cells, cancer cells, and some bacteria. Immune responses carried out by B cells and T cells help combat infection and provide protection against some diseases. NK cells attack a wide variety of infectious microbes and certain tumor cells.

Figure 18.9 **Emigration of white blood cells.**

Interstitial fluid

Blood flow

Neutrophil

Endothelial cell

Rolling

Sticking

Squeezing between endothelial cells

Key:

Selectins on endothelial cells

Integrins on neutrophil

Adhesion molecules (selectins and integrins) assist the emigration of WBCs from the bloodstream into interstitial fluid.

WBC Life Span

Microbes have continuous access to the body through the mouth, nose, and pores of the skin. Furthermore, many cells age and die daily, and their remains must he removed. However, a WBC can phagocytize only a certain amount of material before the ingested material interferes with the WBC's own metabolic activities. Thus, the life span of most WBCs is only a few days. During a period of infection, phagocytic WBCs may live only a few hours. However, some lymphocytes can live for several months or years.

RBCs outnumber white blood cells by about 700:1. There are normally about 5000–10,000 WBCs per microliter of blood. *Leukocytosis* (loo´-kō-sī-TŌ-sis), an increase in the number of WBCs (above 10,000/μL), is a normal, protective response to stresses such as invading microbes, strenuous exercise, anesthesia, and surgery. Leukocytosis usually indicates an inflammation or infection. An abnormally low level of white blood cells (below 5000/μL) is termed *leukopenia* (loo-kō-PĒ-nē-a). It is never beneficial and may be caused by radiation, shock, and certain chemotherapeutic agents.

A physician may order a **differential white blood cell count**, a count of each of the five types of WBCs, to detect infection or inflammation, determine the effects of possible poisoning by chemicals or drugs, monitor blood disorders (such as leukemia) and the effects of chemotherapy, or detect allergic reactions and parasitic infections. Differential counts measure the number of each kind of WBC in a sample of 100 WBCs. Because each type of white blood cell plays a different role, determining the *percentage* of each type in the blood assists in diagnosing the condition.

Table 18.4 lists the significance of both elevated and depressed WBC counts.

TABLE 18.4		
Significance of High and Low White Blood Cell Counts		
WBC TYPE	HIGH COUNT MAY INDICATE	LOW COUNT MAY INDICATE
Neutrophils	Bacterial infection, burns, stress, inflammation.	Radiation exposure, drug toxicity, vitamin B_{12} deficiency, systemic lupus erythematosus (SLE).
Lymphocytes	Viral infections, some leukemias.	Prolonged illness, immunosuppression, treatment with cortisol.
Monocytes	Viral or fungal infections, tuberculosis, some leukemias, other chronic diseases.	Bone marrow suppression, treatment with cortisol.
Eosinophils	Allergic reactions, parasitic infections, autoimmune diseases.	Drug toxicity, stress.
Basophils	Allergic reactions, leukemias, cancers, hypothyroidism.	Pregnancy, ovulation, stress, hyperthyroidism.

Checkpoint

21. Which WBCs are called agranular leukocytes? Why?
22. In which way is the "traffic pattern" of lymphocytes in the body different from those of other WBCs?
23. What is the importance of emigration, chemotaxis, and phagocytosis in fighting bacterial invaders?
24. How do neutrophils utilize chemicals to kill ingested pathogens?
25. What is a differential white blood cell count?

18.8 Platelets reduce blood loss from damaged vessels.

In addition to developing into RBCs and WBCs, hemopoietic stem cells also differentiate into cells that produce platelets. Under the influence of the hormone **thrombopoietin**, myeloid stem cells develop into precursor cells called megakaryoblasts (see Figure 18.3). Megakaryoblasts transform into megakaryocytes, huge cells that splinter into 2000–3000 fragments in the red bone marrow and then enter the bloodstream. Each fragment, enclosed by a piece of megakaryocyte cell membrane, is a **platelet**, or **thrombocyte**. Between 150,000 and 400,000 platelets are present in each microliter of blood. Platelets are disc-shaped, have a diameter of 2–4 μm, and exhibit many vesicles but no nucleus. When blood vessels are damaged, platelets help stop blood loss by forming a platelet plug that fills the gap in the blood vessel wall. Their vesicles also contain chemicals that promote blood clotting. After their short life span of just 5–9 days, platelets are removed from circulation by macrophages in the spleen and liver.

Table 18.5 summarizes the formed elements in blood.

Checkpoint

26. Why are platelets not called blood cells?
27. What functions are served by platelets?

TABLE 18.5

Formed Elements in Blood

NAME AND APPEARANCE	NUMBER	CHARACTERISTICS[*]	FUNCTIONS
Red Blood Cells (RBCs) or Erythrocytes	4.8 million/μL in females; 5.4 million/μL in males	7–8 μm diameter, biconcave discs, without nuclei; live for about 120 days.	Hemoglobin within RBCs transports most of the oxygen and part of the carbon dioxide in the blood.
White Blood Cells (WBCs) or Leukocytes	5000–10,000/μL	Most live for a few hours to a few days.[†]	Combat pathogens and other foreign substances that enter the body.
Granular leukocytes **Neutrophils**	60–70% of all WBCs	10–12 μm diameter; nucleus has 2–5 lobes connected by thin strands of chromatin; cytoplasm has very fine, pale lilac granules.	Phagocytosis. Destruction of bacteria with lysozyme, defensins, and strong oxidants, such as superoxide anion, hydrogen peroxide, and hypochlorite anion.
Eosinophils	2–4% of all WBCs	10–12 μm diameter; nucleus usually has 2 lobes connected by thick strand of chromatin; large, red-orange granules fill the cytoplasm.	Combat the effects of histamine in allergic reactions, phagocytize antigen–antibody complexes, and destroy certain parasitic worms.
Basophils	0.5–1% of all WBCs	8–10 μm diameter; nucleus has 2 lobes; large cytoplasmic granules appear deep blue-purple.	Liberate heparin, histamine, and serotonin in allergic reactions that intensify the overall inflammatory response.
Agranular leukocytes **Lymphocytes** **(T cells, B cells,** and **natural killer cells)**	20–25% of all WBCs	Small lymphocytes are 6–9 μm in diameter; large lymphocytes are 10–14 μm in diameter; nucleus is round or slightly indented; cytoplasm forms a rim around the nucleus that looks sky blue; the larger the cell, the more cytoplasm is visible.	Mediate immune responses, including antigen–antibody reactions. B cells develop into plasma cells, which secrete antibodies. T cells attack invading viruses, cancer cells, and transplanted tissue cells, Natural killer cells attack a wide variety of infectious microbes and certain spontaneously arising tumor cells.
Monocytes	3–8% of all WBCs	12–20 μm diameter; nucleus is kidney-shaped or horseshoe-shaped; cytoplasm is blue-gray and has foamy appearance.	Phagocytosis (after transforming into fixed or wandering macrophages).
Platelets (Thrombocytes)	150,000–400,000/μL	2–4 μm diameter cell fragments that live for 5–9 days; contain many vesicles but no nucleus.	Form platelet plug in hemostasis; release chemicals that promote vascular spasm and blood clotting.

[*]Colors are those seen when using Wright's stain.

[†]Some lymphocytes, called T and B memory cells, can live for many years once they are established.

Return to Ed's Story

"**S**o, what seems to be troubling you, Ed?" asked the doctor as he reviewed Ed's chart.

"Jerry, I'm sore and achy all over, I've been having headaches, hot and cold spells, night sweats; I wake up soaking wet! I just got back from Central America about 4 weeks ago. I usually get a cold or mild flu when I travel, but this is the worst I've had. On top of everything else, I've been having diarrhea, too."

Jerry nodded and asked, "Were you up-to-date on your shots before you left?"

"I've been going to Belize with my students for 10 years, so I keep up-to-date on the shots and I got chloroquine for malaria prophylaxis. I was supposed to take 300 milligrams weekly, but I got busy. I might have missed a few doses."

"Well, you are definitely a candidate for a malarial infection. We'd better check to be sure. Malaria is nothing to take lightly; it can cause severe anemia, kidney and liver damage, and seizures. We'll do a blood smear and I'll also draw some more blood for lab analysis."

Jerry returned once the lab work was back. "Ed, your blood smear was positive for malarial parisitemia. The blood work suggests your kidneys have sustained some damage; hematocrit is low, bilirubin levels are elevated, blood glucose is low. I'd like to admit you to the hospital right away. We'll need to start treatment immediately."

"I thought it was just the flu," Ed said weakly, then reached for his cell phone to call his wife.

Jerry left the room and walked to the front desk. "Linda, we need to get this guy admitted right away. He's got malaria. His kidneys are shutting down and he may need dialysis. We'll need to get him started right away on IV quinidine, for the malaria. And try to get hold of nephrology. We're going to need a consult on this one."

As he lay in the hospital bed, Ed heard two voices in a hushed but serious conversation just outside his door. "I don't think quinidine is a good choice. His platelet count is already low and quinidine will cause further thrombocytopenia. We could try quinine instead to address the malaria, without increasing his risk of internal bleeding. At any rate, we'll have to get him on dialysis right away. His sodium levels are way down and his potassium and bilirubin levels are sky high."

Jerry replied, "I'll start him on 10 mg/kg quinine dihydrochloride in 5% dextrose. I don't think he could handle oral medication right now. He's been vomiting quite a bit and says he can't keep food or drink down. The dextrose should also help with his hypoglycemia."

Jerry entered Ed's room. "Hi, Ed. How are you feeling?" he asked.

"I'm feeling okay, but I'm nauseous and stiff, and it feels warm in here. Is everything okay? I heard you talking in the hallway."

"The nephrologist and I were just consulting on your case. Your platelet count is quite low and your kidneys have temporarily shut down. We'll get you on medication for the malaria and we'll have to put you on dialysis for a while until your kidneys begin to work on their own again."

Ed pushed himself up in the bed and asked, "Shouldn't the chloroquine have prevented this, Doc? I only missed a few doses."

"Well, *Plasmodium falciparum* is chloroquine-resistant, Ed. Just like the excessive use of antibiotics has led to antibiotic-resistant microbes, certain species of *Plasmodium* have developed resistance. There's still not a lot known about how *Plasmodium* is so successful in fighting off the human immune system, but it does. Your white blood cell count actually was high, but this bug has a knack for hiding in cells and altering the recognition of infected red cells by the immune system."

E.   If erythropoiesis cannot keep up with the pace of destruction of RBCs, what will happen to the oxygen-carrying capacity of Ed's blood?

F.   Jaundice is a condition characterized by a yellowish color to the skin. Which pigments are produced from the breakdown of blood that might cause Ed to appear jaundiced?

G.   Why would Ed potentially need a transfusion of blood?

H.   Ed's kidneys have been damaged as they receive the released hemoglobin and RBC fragments. How is the kidney involved in blood cell production?

I.   Which type of WBC phagocytizes damaged tissues and pathogens and may have been active early in Ed's infection (at the site of the bite where the inflammation was occurring)?

18.9   Hemostasis is the sequence of events that stops bleeding from a damaged blood vessel.

Hemostasis (hē´-mō-STĀ-sis; -*stasis* = standing still) is a sequence of responses that stops bleeding when blood vessels are injured. (A word of caution: Be sure not to confuse the terms *hemostasis* and *homeostasis*.) When blood vessels are damaged or ruptured, the hemostatic response must be quick, localized to the region of damage, and carefully controlled. Three mechanisms can reduce loss of blood from blood vessels: (1) vascular spasm, (2) platelet plug formation, and (3) blood clotting (coagulation). When successful, hemostasis prevents **hemorrhage** (HEM-or-ij; -*rhage* = burst forth), the loss of a large amount of blood from the vessels. Hemostasis can prevent hemorrhage from smaller blood vessels, but extensive hemorrhage from larger vessels usually requires medical intervention.

Vascular Spasm

When a blood vessel is damaged, the smooth muscle in its wall contracts immediately, a response called **vascular spasm**. Vascular spasm reduces blood loss for several minutes to several hours, during which time the other hemostatic mechanisms begin to operate. The spasm is probably caused by damage to the smooth muscle, by substances released from activated platelets, and by reflexes initiated by pain receptors.

Platelet Plug Formation

When platelets come into contact with parts of a damaged blood vessel, they quickly form a platelet plug that helps fill the gap in the injured blood vessel wall. Platelet plug formation occurs as follows (Figure 18.10):

1 Initially, platelets contact and stick to parts of a damaged blood vessel, such as collagen fibers of the connective tissue underlying the damaged endothelial cells. This process is called **platelet adhesion**.

2 Due to adhesion, the platelets become activated, and their characteristics change dramatically. They extend many projections that enable them to contact and interact with one another, and they begin to liberate the chemicals contained in their vesicles. This phase is called the **platelet release reaction**. Liberated chemicals activate nearby platelets and sustain the vascular spasm, which decreases blood flow through the injured vessel.

3 The release of platelet chemicals makes other platelets in the area sticky, and the stickiness of the newly recruited and activated platelets causes them to stick to the originally activated platelets. This gathering of platelets is called **platelet aggregation**. Eventually, the accumulation and attachment of large numbers of platelets form a mass called a **platelet plug**.

A platelet plug is very effective in preventing blood loss in a small vessel. Although initially the platelet plug is loose, it becomes quite tight when reinforced by fibrin threads formed during clotting (Figure 18.11; see following section). A platelet plug can stop blood loss completely if the hole in a blood vessel is not too large.

Blood Clotting

Normally, blood remains in its liquid form as long as it stays within its vessels. If it is withdrawn from the body, however, it thickens and forms a gel. Eventually, the gel separates from the liquid. The

Figure 18.10 Platelet plug formation.

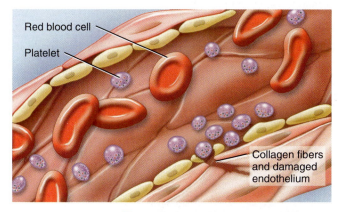

Red blood cell

Platelet

Collagen fibers and damaged endothelium

1 Platelet adhesion

Chemicals liberated from platelets

2 Platelet release reaction

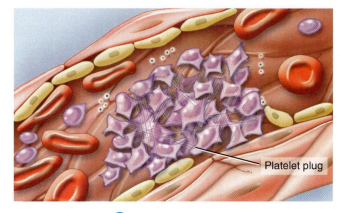

Platelet plug

3 Platelet aggregation

🔑 **A platelet plug can stop blood loss completely if the hole in a blood vessel is small.**

straw-colored liquid, called **serum**, is simply plasma minus the clotting proteins. The gel is called a **clot** and consists of a network of insoluble protein fibers called **fibrin** in which the formed elements of blood are trapped (Figure 18.11).

Figure 18.11   Blood clot formation. Notice the platelet and red blood cells entrapped in fibrin threads.

SEM   900x

(a) Early stage

SEM   900x

(b) Intermediate stage

SEM   900x

(c) Late stage

Red blood cell

Fibrin threads

SEM   1600x

(d) Red blood cells trapped in fibrin threads

Platelet

Fibrin threads

🔑 **A blood clot is a gel that contains formed elements of the blood entangled in fibrin threads.**

The process of gel formation, called **clotting** or **coagulation**, is a series of chemical reactions that culminates in the formation of fibrin threads. If blood clots too easily, the result can be **thrombosis**—clotting in an undamaged blood vessel. If the blood takes too long to clot, hemorrhage can result.

Clotting is a complex cascade of enzymatic reactions in which various chemicals known as **clotting factors** (calcium ions, enzymes, and molecules associated with platelets or damaged tissues) activate one another in a fixed sequence. Finally, the series of chemical reactions forms a network of insoluble protein fibers. Clotting can be divided into three stages (Figure 18.12):

1. Several reactions of the clotting sequence end in formation of **prothrombinase**. Two pathways, called the extrinsic pathway (Figure 18.12a) and the intrinsic pathway (Figure 18.12b), produce prothrombinase.

2. Prothrombinase converts **prothrombin** (a plasma protein formed by the liver) into the enzyme **thrombin**.

3. Thrombin converts soluble **fibrinogen** (another plasma protein formed by the liver) into insoluble fibrin. Fibrin forms the threads of the clot.

The Extrinsic Pathway

The **extrinsic pathway** of blood clotting occurs rapidly, within a matter of seconds if trauma is severe. It is so named because a tissue protein called **tissue factor** leaks into the blood from damaged tissue cells *outside (extrinsic to)* blood vessels (Figure 18.12a). Tissue factor begins a sequence of reactions requiring calcium ions (Ca^{2+}) and several clotting factors that eventually form the active enzyme prothrombinase. This completes the extrinsic pathway.

The Intrinsic Pathway

The **intrinsic pathway** of blood clotting (Figure 18.12b) is more complex than the extrinsic pathway, and it occurs more slowly, usually requiring several minutes. The intrinsic pathway is so named because its activators are either in direct contact with blood or contained *within (intrinsic to)* the blood; outside tissue damage is not needed. If endothelial cells lining the blood vessels become roughened or damaged, blood can come in contact with collagen fibers in the adjacent connective tissue. Such contact activates clotting factors. In addition, trauma to endothelial cells activates platelets, causing them to release phospholipids that can also activate certain clotting factors. After several additional reactions that require Ca^{2+} and several clotting factors, prothrombinase is formed, completing the intrinsic pathway.

The Common Pathway

Once prothrombinase is formed, the next two stages of clotting are the same for both the extrinsic and intrinsic pathways. Together these two stages are referred to as the **common pathway** (Figure 18.12c). In the second stage of blood clotting, prothrombinase and Ca^{2+} catalyze the conversion of prothrombin to thrombin. In the third stage, thrombin, in the presence of Ca^{2+}, converts fibrinogen, which is soluble, to loose fibrin threads, which are insoluble. Thrombin also activates a clotting factor that strengthens and stabilizes the fibrin threads into a sturdy clot.

Figure 18.12 Blood clotting.

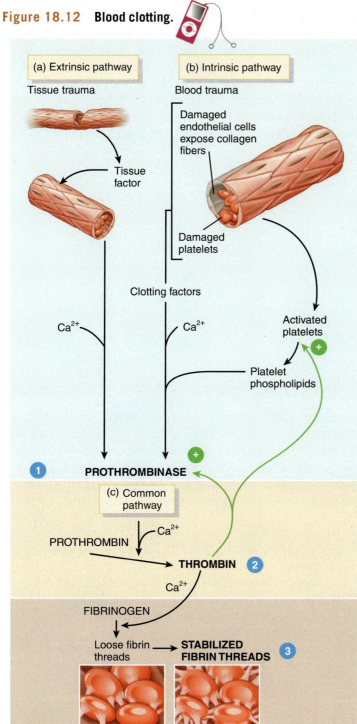

During blood clotting, the clotting factors activate one another, resulting in a cascade of reactions that includes positive feedback cycles.

As thrombin forms, a positive feedback cycle begins (see green ➕ arrows in Figure 18.12). Thrombin activates more platelets, increasing their release of platelet phospholipids, and therefore increasing formation of prothrombinase. Also, thrombin directly accelerates the formation of prothrombinase. The addi-

tional prothrombinase, in turn, accelerates the production of more thrombin, and so on. As the positive feedback loop continues, the fibrin clot grows.

Clot Retraction and Blood Vessel Repair

Once a clot is formed, it plugs the ruptured area of the blood vessel and thus stops blood loss. **Clot retraction** is the consolidation or tightening of the fibrin clot. Platelets trapped around the fibrin threads contract, pulling on the fibrin threads and tightening the clot. Because the fibrin threads are attached to the damaged surfaces of the blood vessel, as the clot retracts, it pulls the edges of the damaged blood vessel closer together, decreasing the risk of further damage. Permanent repair of the blood vessel can then take place. In time, fibroblasts form connective tissue in the ruptured area, and new endothelial cells repair the blood vessel lining.

Hemostatic Control Mechanisms

Clot formation normally remains localized at the site of blood vessel damage. A clot does not extend beyond the wound site into the general circulation, in part because fibrin absorbs thrombin into the clot. This helps stop the spread of thrombin into the blood and thus inhibits clotting except at the wound site. Another reason for localized clot formation is that some clotting factors are dispersed by the blood; therefore, their concentrations are not high enough to bring about widespread clotting.

Even when blood vessels are not damaged, small clots begin to form many times a day, often at a site of minor roughness inside a blood vessel. Because blood clotting involves positive feedback cycles, a clot has a tendency to enlarge, creating the potential for impairment of blood flow through undamaged vessels. Usually, small, inappropriate clots dissolve in a process called **fibrinolysis** (fī´-bri-NOL-i-sis). When a clot is formed, an inactive plasma enzyme called **plasminogen** is incorporated into the clot. Both body tissues and blood contain substances that can activate plasminogen to **plasmin**, an active plasma enzyme. Once plasmin is formed, it can dissolve the clot by digesting fibrin threads and inactivating substances such as fibrinogen and prothrombin.

In addition, substances that inhibit clotting, called **anticoagulants**, are present in blood. Most block the action of clotting factors. An example is *heparin*, an anticoagulant produced by mast cells and basophils that helps block the action of thrombin.

Clotting in Blood Vessels

Despite fibrinolysis and the action of anticoagulants, blood clots sometimes form within blood vessels. The endothelial surfaces of a blood vessel may be roughened as a result of *atherosclerosis* (accumulation of fatty substances on arterial walls), trauma, or infection. These conditions induce adhesion of platelets. Clots may also form in blood vessels when blood flows too slowly, allowing clotting factors to accumulate in high enough concentrations to initiate a clot.

Clotting in an unbroken blood vessel is called **thrombosis** (*thromb-* = clot; *-osis* = a condition of). The clot itself, called a **thrombus**, may dissolve spontaneously. If it remains intact, however, the thrombus may become dislodged and be swept away in the blood. A blood clot, bubble of air, fat from broken bones, or a piece of debris transported by the bloodstream is called an **embolus** (*em-* = in; *-bolus* = a mass; plural is *emboli*). An embolus that breaks away from an arterial wall may lodge in a smaller-diameter artery downstream. If it blocks blood flow to the brain, kidney, or heart, the embolus can cause a stroke, kidney failure, or heart attack, respectively. When an embolus lodges in the lungs, the condition is called a **pulmonary embolism**. Massive emboli in the lungs may result in heart failure and death in a few minutes or hours.

Checkpoint

28. Describe how vascular spasm and platelet plug formation occur.
29. What is the outcome of the first stage of clotting?
30. Why is the clotting mechanism described as a positive feedback cycle?
31. Why is adequate dietary calcium important to hemostasis?
32. How does clot retraction assist in blood vessel repair?

Ed's Story: Epilogue and Discussion

▶▶▶ Ed, a 46-year-old ecologist, returns from a field research excursion to Belize. Approximately 4 weeks after his return, he seeks medical attention for flulike symptoms: fever, chills, sweating, nausea and vomiting, and abdominal pain. He admits to poor compliance with his malaria chemoprophylaxis (a chemical that prevents disease). A blood smear at the time of his exam indicates the presence of *Plasmodium falciparum*. His laboratory blood tests reveal he has thrombocytopenia (low platelet count), leukocytosis (high WBC count), hypoglycemia (low blood glucose level), hyponatremia (low blood sodium level), hyperkalemia (elevated potassium level), and elevated creatinine (waste product of muscle metabolism). A diagnosis of acute malaria with hemolytic anemia and acute renal failure requiring dialysis is made. Ed is fortunate that his physician did not misdiagnose this illness as a common flu. The most revealing clue was that Ed had traveled to a region of the world where malaria is endemic. The malarial parasite is one of four species of *Plasmodium*, a protozoan with a multistage life cycle. The majority of the pest's time in humans is spent multiply-

ing inside RBCs; when the number of creatures is great enough, the RBCs lyse, discharging organisms that infect more RBCs. Since the organisms live inside RBCs and multiply rapidly, WBCs have difficulty mounting a significant immune response. A regular pattern of fevers and chills following a 48-hour cycle is observed in most patients with this condition. The fever spikes correlate with the release of toxins into the bloodstream from massive RBC hemolysis. ◀◀◀

J.   Which organ systems were involved in Ed's case?
K   How would the hemolysis of Ed's RBCs affect the function of his blood?
L.   Why would you expect Ed to have a high eosinophil count?
M.   Why would Ed's low platelet count concern his physicians? What are the three hemostatic mechanisms that normally occur in a healthy person?
N.   If Ed's platelet count is low, which hormone could be administered to stimulate platelet production?

Concept and Resource Summary

Concept	Resources

Introduction

1. The **cardiovascular system** consists of three interrelated components: blood, the heart, and blood vessels. **Blood** is a connective tissue composed of plasma (liquid portion) and formed elements (cells and cell fragments).

Concept 18.1 Blood contains plasma and formed elements and transports essential substances through the body.

1. **Blood** *transports* oxygen, nutrients, and hormones to body cells, and transports carbon dioxide, heat, and metabolic wastes away from cells; *regulates* homeostasis in all body fluids; and *protects* through the clotting mechanism and white blood cell response to pathogens and disease.
2. Blood is viscous and slightly alkaline, ranging from pH 7.35 to 7.45 with a temperature of 38°C (100.4°F). Its volume is 4–6 liters.
3. Whole blood is composed of plasma, a watery liquid containing dissolved substances, and formed elements, which are cells and cell fragments.
4. **Plasma** is composed of 91.5% water, 7% proteins, and 1.5% solutes other than proteins. *Plasma proteins* include **albumins**, **globulins**, **antibodies**, and **fibrinogen**. Solutes, other than plasma proteins, include electrolytes, nutrients, gases, enzymes, hormones, vitamins, and waste products.
5. The **formed elements** include red blood cells (RBCs), white blood cells (WBCs), and platelets.
6. A **hematocrit** is the percentage of total blood volume occupied by RBCs.

See
- Anatomy Overview – Blood
- Animation – Erythropoietin

Do
- Concepts and Connections – Blood

Concept 18.2 Hemopoiesis is the production of formed elements.

1. Hemopoiesis is the process by which formed elements of the blood develop.
2. The **pluripotent stem cells** of the red bone marrow give rise to all formed elements. Hormonal stimulation causes pluripotent stem cells to generate *lymphoid stem cells* and *myeloid stem cells*. Myeloid stem cells differentiate in red bone marrow into all the formed elements except lymphocytes. Lymphoid stem cells differentiate into lymphocytes in red bone marrow, and then in lymphatic tissues.

See
- Anatomy Overview – Erythrocytes

Hear
- Figure 18.3 – Origin, Development, and Structure of Formed Elements

Concept 18.3 Mature red blood cells are biconcave cells containing hemoglobin.

1. **Red blood cells** contain the oxygen-carrying protein pigment **hemoglobin**, which gives RBCs their red color. An adult male has about 5.4 million RBCs per microliter (μL) of blood, and an adult female has about 4.8 million/μL of blood.
2. RBCs are biconcave discs. Mature RBCs lack a nucleus.
3. A hemoglobin molecule consists of the protein **globin** composed of four polypeptide chains; bound to each chain is a ringlike, nonprotein pigment called **heme**. Heme contains iron that binds to oxygen. About 13% of the carbon dioxide transported by the blood combines with globin.

See
- Anatomy Overview • Erythrocytes

Hear
- Figure 18.4 – Shapes of a Red Blood Cell and a Hemoglobin Molecule, and Structure of a Heme Group

Concept 18.4 Red blood cells have a life cycle of 120 days.

1. The life span of a red blood cell is about 120 days. Fragile, old, or damaged RBCs are destroyed by phagocytic macrophages.
2. As macrophages phagocytize RBCs, their hemoglobin is broken down and recycled. Amino acids from globin are used to make proteins. Iron from heme is used to synthesize hemoglobin for new RBCs. The non-iron portion of heme is eventually converted to **bilirubin** and secreted into bile that passes into the intestines.

Hear
- Figure 18.5 – Formation and Destruction of Red Blood Cells and Recycling of Hemoglobin Components

Concept 18.5 Erythropoiesis is the process of red blood cell formation.

1. The production of red blood cells, termed **erythropoiesis**, begins in the red bone marrow. **Reticulocytes** enter the bloodstream where, within 1 to 2 days, they become mature erythrocytes.
2. Normally, erythropoiesis occurs at the same pace as red blood cell destruction. Erythropoiesis is stimulated by hypoxia, which results in the release of erythropoietin by the kidneys.

See
- Animation – Erythropoietin

Concept 18.6 Blood is categorized into groups based on surface antigens.

1. Genetically determined **agglutinogens** are found on the surfaces of RBCs. These surface antigens are the basis for different **blood groups** and **blood types**.
2. The **ABO blood group** is based on the presence or absence of antigen A and antigen B on the RBC surface. Plasma typically contains **anti-A antibodies** or **anti-B antibodies**, called *agglutinins*, directed against A or B antigens, respectively.
3. The **Rh blood group** is based on the Rh antigen. RBCs with Rh antigen are termed Rh$^+$, and those without the Rh antigen are termed Rh$^-$.
4. The transfer of whole blood or blood components into the bloodstream is called a **transfusion**. In an incompatible transfusion, the recipient's agglutinins recognize foreign donor cell antigens, which cause RBC clumping.

Concept 18.7 White blood cells combat inflammation and infection.

1. **White blood cells** have a nucleus, do not have hemoglobin, and are classified as either granular or agranular.
2. The three types of granular leukocytes can be distinguished under a light microscope. **Eosinophils** have large, uniform granules stained red-orange. The nucleus has two or three lobes. **Basophils** have variable-sized granules stained blue-purple that obscure the two-lobed nucleus. **Neutrophils** have small, pale lilac granules. The nucleus has two to five lobes.
3. The two types of agranular leukocytes can be distinguished. **Lymphocytes** have a sky blue cytoplasm with a dark, round, slightly indented nucleus. **Monocytes** have a blue-gray cytoplasm with a foamy appearance and a kidney-shaped nucleus. Monocytes that migrate into tissues enlarge into **macrophages**.
4. WBCs can leave the bloodstream by **emigration** to move to tissue sites of pathogen invasion or inflammation.
5. **Chemotaxis** occurs when pathogens and inflamed tissues release chemicals that attract phagocytic neutrophils and macrophages. These cells ingest and dispose of pathogens and dead matter during **phagocytosis**. Neutrophils respond most rapidly to an infection site. Monocytes arrive later at an infection site but arrive in large numbers.
6. Basophils release heparin, histamine, and serotonin that intensify the inflammatory reaction.
7. Eosinophils stop the effects of histamine and other mediators of inflammation in allergic reactions. They also attack parasitic worms.
8. Lymphocytes function in immune responses. B cells destroy bacteria and inactivate their toxins. T cells combat viruses, fungi, transplanted cells, cancer cells, and some bacteria. NK cells attack various microbes and tumor cells.
9. Most WBCs live only a few hours to a few days. Normally there are 5000–10,000 WBCs per microliter of blood. A **differential white blood cell count** measures the number of each type of WBC in a sample of 100 WBCs.

See
- Anatomy Overview – White Blood Cells

Hear
- Figure 18.8 – Types of White Blood Cells

Do
- Concepts and Connections – Blood

Concept 18.8 Platelets reduce blood loss from damaged vessels.

1. The hormone **thrombopoietin** stimulates myeloid stem cells to develop into megakaryocytes. These huge cells splinter into 2000–3000 fragments and enter the bloodstream. Each fragment is a **platelet**.
2. Platelets lack a nucleus and have chemicals that promote blood clotting. Platelets stop blood loss in damaged vessels by forming a platelet plug in the vessel wall.

See
- Anatomy Overview – Platelets

Do
- Concepts and Connections – Blood

Concept

Concept 18.9 Hemostasis is the sequence of events that stops bleeding from a damaged blood vessel.

1. **Hemostasis** is a sequence of responses to stop blood loss from a damaged blood vessel. Three mechanisms are involved: vascular spasm, platelet plug formation, and blood clotting.
2. **Vascular spasms** of the smooth muscle in the wall of a damaged vessel help reduce blood loss. Platelets that accumulate at the site release chemicals that enhance vasoconstriction.
3. Platelets that come into contact with damaged blood vessels undergo **platelet plug** formation.
4. Blood withdrawn from the body will thicken and form a gel called a **clot** consisting of protein fibers called **fibrin** and trapped formed elements. **Clotting** is a cascade of reactions involving **clotting factors** that activate one another. Stages of clotting include the following: formation of **prothrombinase**, conversion by prothrombinase of the plasma protein **prothrombin** into **thrombin**, and conversion by thrombin of the plasma protein **fibrinogen** into insoluble protein threads called **fibrin**.
5. In the **extrinsic pathway**, damaged tissue cells leak a protein called **tissue factor** into the blood from outside blood vessels. Activators of the **intrinsic pathway** are within the blood. The intrinsic and extrinsic pathways both culminate in the formation of prothrombinase. Because the next two stages of clotting are the same for both pathways, they are collectively called the **common pathway**.
6. The clot plugs the damaged area of the blood vessel and undergoes **clot retraction**, which pulls the blood vessel edges closer together.
7. Clotting in an unbroken blood vessel is called **thrombosis**; the clot is called a **thrombus**. A dislodged thrombus can be swept away in the blood and is then called an **embolus**.

Hear

• **Figure 18.12** – **Blood Clotting**

Do

• **Concepts and Connections – Blood**

Other Chapter Resources

For additional information on clinical applications, medical disorders, and medical terminology, see Chapter 9 in the *Clinical Connections* manual packaged with your text.

Within *WileyPLUS*

WILEY PLUS
• **Chapter Quiz**
• **Visual Anatomy**
• **Mastering Vocabulary**
• **Audio Glossary**

• **Homeostatic Imbalance: The Case of the Man with Yellow Eyes**
• **Web Links**
• **Crossword Puzzle**

• Understanding the Concepts

1. How does the volume of blood in your body compare to the volume of fluid in a 2-liter bottle of soda?

2. What is the significance of a lower-than-normal hematocrit? A higher-than-normal hematocrit? What is the effect of a bacterial infection on the hematocrit?

3. Compare the development of lymphocytes with the development of the other formed elements.

4. How are red blood cells able to squeeze through capillaries that are smaller in diameter than they are?

5. What is erythropoiesis? Which factors speed up and slow down erythropoiesis?

6. What would happen if a person with type B blood were given a transfusion of type O blood?

7. During an anatomy and physiology exam you are asked to view white blood cells in prepared slides of standard human blood smears. Based on the observations below, what is the name and function of each WBC?

   • WBC has a round nucleus surrounded by a blue halo of cytoplasm with no visible granules.

   • WBC contains dense blue-purple granules that hide the nucleus.

   • WBC has a U-shaped nucleus and a bluish, foamy cytoplasm with no visible granules.

   • WBC contains small, pale lilac granules and a four-lobed nucleus.

   • WBC contains red-orange granules and a two-lobed nucleus.

8. Why would the level of leukocytes be higher in an individual who has been infected with a parasitic disease?

9. In regions where malaria is endemic, some people build up immune resistance to the malaria pathogen. Which WBCs are responsible for the immune response against pathogens? How do they function?

10. What is the function of prothrombinase and thrombin in clotting? Explain how the extrinsic and intrinsic pathways of blood clotting differ.

19

The Cardiovascular System: The Heart

Prakesh's Story

It was a warm summer morning as Prakesh awoke. Saturday was his day to cut the lawn, and although he had felt somewhat out of breath for the last few days, he thought the exercise might do him good. He had suffered a heart attack three years earlier and had undergone coronary artery bypass surgery. Although it had taken him a while to recover, he hadn't felt poorly until just the last few days. He had been very scared that day three years ago when he had felt the crushing pain of the heart attack in his chest. But he had been fortunate enough to have been taken to the hospital emergency department quickly. Ever since the event, he had tried to take better care of himself by sticking to a low-fat diet, reduced salt intake, and exercise. He had never missed taking his one-mile daily walk, but recently he had noticed it was getting harder to finish. He felt short of breath, and his ankles and feet swelled. He had mentioned it in passing to his wife, who suggested he see his cardiologist.

Prakesh is a 57-year-old man with a history of heart disease. As we will learn in this chapter, the heart pumps blood to the organs and tissues of the body, including cardiac tissue. When the heart can't pump enough blood to the body's organs, the entire body's homeostasis is affected.

INTRODUCTION

In the last chapter we examined the composition and functions of blood. For blood to reach body cells and exchange materials with them, it must constantly be pumped by the heart through the body's blood vessels. The heart beats about 100,000 times every day, which adds up to about 35 million beats in a year and about 2.5 billion times in an average lifetime. The left side of the heart pumps blood through an estimated 100,000 km (60,000 mi) of blood vessels. The right side of the heart pumps blood through the lungs, so that it can pick up oxygen and unload carbon dioxide. Even while you are sleeping, your heart pumps 30 times its own weight each minute, which amounts to about 5 liters (5.3 qt) to the lungs and the same volume to the rest of the body.

> The heart beats about 100,000 times every day, which adds up to about 35 million beats in a year and about 2.5 billion times in an average lifetime.

At this rate, the heart pumps more than 14,000 liters (3600 gal) of blood in a day or 10 million liters (2.6 million gal) in a year. Because you don't spend all your time sleeping, and your heart pumps more vigorously when you are active, the actual blood volume the heart pumps in a single day is even larger.

The scientific study of the normal heart and the diseases associated with it is cardiology (kar´-dē-OL-ō-jē; *cardio-* = heart; *-logy* = study of). This chapter explores the structure of the heart and the unique properties that permit it to pump for a lifetime without a moment of rest.

19.1 The heart is located in the mediastinum and has a muscular wall covered by pericardium.

Location of the Heart

For all its might, the heart is relatively small, roughly the same size as a closed fist—about 12 cm (5 in.) long, 9 cm (3.5 in.) wide at its broadest point, and 6 cm (2.5 in.) thick. Its mass averages 250 g (8 oz) in adult females and 300 g (10 oz) in adult males. The heart rests on the diaphragm, near the midline of the thoracic cavity in the mediastinum (mē´-dē-a-STĪ-num), an anatomical region between the lungs that extends from the sternum to the vertebral column (Figure 19.1a,b).

About two-thirds of the mass of the heart lies to the left of the body's midline. You can visualize the heart as a cone lying on its side. The pointed **apex** is formed by the tip of the left ventricle (a lower chamber of the heart) and rests on the diaphragm. The apex is directed anteriorly, inferiorly, and to the left. The **base** of the heart, the broad superior portion opposite the apex, is formed by the atria (upper chambers of the heart). The major blood vessels of the heart enter and exit at the base.

Figure 19.1   Position of the heart and associated structures in the mediastinum (dashed outline). In this and subsequent illustrations, blood vessels that carry oxygenated blood (which looks bright red) are colored red, whereas those that carry deoxygenated blood (which looks dark red) are colored blue.

(a) Inferior view of transverse section of thoracic cavity showing the heart in the mediastinum

Pericardium

The membrane that surrounds and protects the heart is the **peri-cardium** (*peri-* = around). It confines the heart to its position in the mediastinum, while allowing sufficient freedom of movement for vigorous and rapid contraction. The pericardium consists of two principal parts: the fibrous pericardium and the serous peri-cardium (Figure 19.2a). The superficial **fibrous pericardium** is composed of tough, inelastic, dense irregular connective tissue. It prevents overstretching of the heart, provides protection, and an-chors the heart in the mediastinum.

The deeper **serous pericardium** is a thinner, more delicate membrane that forms a double layer around the heart (Figure 19.2a,b). The outer **parietal layer** is fused to the fibrous pericar-dium, and the inner **visceral layer**, also called the **epicardium** (*epi-* = on top of), adheres tightly to the surface of the heart. Between the parietal and visceral layers is a **pericardial cavity** containing **pericardial fluid**, a slippery, lubricating secretion of the pericardial membranes that reduces friction between the se-rous pericardial membranes as the heart moves.

Layers of the Heart Wall

The wall of the heart (Figure 19.2a) is composed of three layers: epicardium (external layer), myocardium (middle layer), and en-docardium (inner layer). As noted earlier, the **epicardium**, the thin, outer layer of the heart wall, is also known as the visceral layer of serous pericardium. It is composed of mesothelium and connec-tive tissue that imparts a smooth, slippery texture to the outermost surface of the heart. The **myocardium** (*myo-* = muscle) consists of cardiac muscle tissue, which constitutes about 95% of the heart and is responsible for its pumping action. Although it is striated like skeletal muscle, cardiac muscle is involuntary like smooth muscle (see Chapter 4 for a discussion of muscle histology). The cardiac muscle fibers swirl diagonally around the heart in interlac-ing bundles (Figure 19.2c). The **endocardium** (*endo-* = within) is a thin layer of endothelium and connective tissue that provides a smooth lining for the heart chambers and covers the heart valves (Figure 19.2a). The endocardium is continuous with the endothe-lial lining of the large blood vessels attached to the heart.

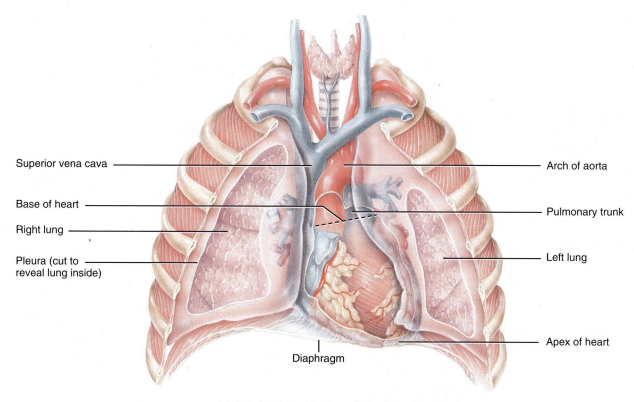

Superior vena cava

Base of heart

Right lung

Pleura (cut to reveal lung inside)

Diaphragm

Arch of aorta

Pulmonary trunk

Left lung

Apex of heart

(b) Anterior view of the heart in the thoracic cavity

The heart is located between the lungs, with about two-thirds of its mass to the left of the midline.

Figure 19.2   Pericardium and heart wall.

Pericardium
Epicardium
Myocardium
Endocardium

PERICARDIUM
Heart wall

FIBROUS PERICARDIUM
PARIETAL LAYER OF
SEROUS PERICARDIUM

Pericardial cavity

VISCERAL LAYER OF
SEROUS PERICARDIUM

ENDOCARDIUM

Trabeculae carneae

Coronary blood vessels

MYOCARDIUM

(a) Portion of pericardium and right ventricular heart wall showing the
divisions of the pericardium and layers of the heart wall

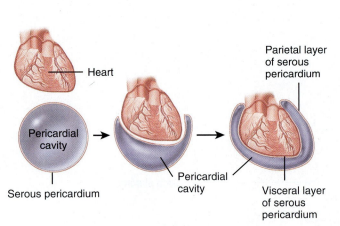

Heart

Pericardial
cavity

Serous pericardium

Pericardial
cavity

Parietal layer
of serous
pericardium

Visceral layer
of serous
pericardium

(b) Simplified relationship of the serous pericardium to the heart

Superior
vena cava

Deep
muscle
bundle in
ventricle

Aorta

Pulmonary
trunk

Superficial
muscle
bundles
in atria

Superficial
muscle
bundles in
ventricles

(c) Cardiac muscle bundles of the myocardium

🔑 The pericardium is a triple-layered sac that surrounds and protects the heart.

Checkpoint

1.   How is the heart positioned in the mediastinum?
2.   Which layer of the pericardium is both a part of the pericardium and a part of
the heart wall?

3.   What is the function of pericardial fluid?
4.   From most superficial to deepest, what are the layers of the pericardium and
heart wall?

19.2   The heart has four chambers, two upper atria and two lower ventricles.

The heart contains four chambers. The two upper receiving chambers are the atria (= entry halls or chambers), and the two lower pumping chambers are the ventricles (= little bellies). On the anterior surface of each atrium is a wrinkled pouchlike structure called an auricle (OR-i-kul; *auri-* = ear), so named because of its resemblance to a dog's ear (Figure 19.3). Each auricle slightly increases the capacity of an atrium so that it can hold a greater volume of blood. Also on the surface of the heart are a series of grooves,

Figure 19.3 **Structure of the heart: surface anatomy.**

Brachiocephalic trunk

Superior vena cava

Ascending aorta

Right pulmonary artery

Fibrous pericardium (cut)

Right pulmonary veins

RIGHT AURICLE OF RIGHT ATRIUM

Right coronary artery

RIGHT ATRIUM

CORONARY SULCUS

RIGHT VENTRICLE

Inferior vena cava

Left common carotid artery

Left subclavian artery

Arch of aorta

Ligamentum arteriosum

Left pulmonary artery

Pulmonary trunk

Left pulmonary veins

LEFT AURICLE OF LEFT ATRIUM

Branch of left coronary artery

LEFT VENTRICLE

ANTERIOR INTERVENTRICULAR SULCUS

Descending aorta

(a) Anterior external view showing surface features

Brachiocephalic trunk

Superior vena cava

Ascending aorta

Right pulmonary veins

RIGHT AURICLE OF
RIGHT ATRIUM

RIGHT ATRIUM

CORONARY SULCUS

RIGHT VENTRICLE

Left subclavian artery

Left common carotid artery

Arch of aorta

Ligamentum arteriosum

Left pulmonary artery

Left pulmonary veins

Pulmonary trunk

LEFT AURICLE OF LEFT ATRIUM

LEFT VENTRICLE

ANTERIOR INTERVENTRICULAR
SULCUS

(b) Anterior external view

(continues)

Figure 19.3 (continued)

Left common carotid artery

Left subclavian artery

Arch of aorta

Descending aorta

Left pulmonary artery

AURICLE OF LEFT ATRIUM

Left pulmonary veins

LEFT ATRIUM

Coronary sinus
(in the coronary sulcus)

LEFT VENTRICLE

POSTERIOR
INTERVENTRICULAR SULCUS

Brachiocephalic trunk

Superior vena cava

Ascending aorta

Right pulmonary artery

Right pulmonary veins

RIGHT ATRIUM

Right coronary artery

Inferior vena cava

Middle cardiac vein

RIGHT VENTRICLE

(c) Posterior external view showing surface features

Sulci are grooves that contain blood vessels and fat and mark the boundaries between the various chambers.

called sulci (SUL-sī), that contain coronary blood vessels and a variable amount of fat. Each sulcus (SUL-kus) marks the external boundary between two chambers of the heart. The **coronary sulcus** (*coron-* = resembling a crown) encircles most of the heart and marks the external boundary between the superior atria and inferior ventricles. The **anterior interventricular sulcus** marks the external boundary between the right and left ventricles. This sulcus continues around to the posterior surface of the heart as the **posterior interventricular sulcus**, which marks the external boundary between the ventricles on the posterior aspect of the heart.

Right Atrium

The **right atrium** receives blood from three veins: *superior vena cava, inferior vena cava,* and *coronary sinus* (Figure 19.4). The posterior wall of the right atrium is smooth, but the anterior wall is rough due to muscular ridges called **pectinate muscles** (*pectin* = comb), which also extend into the auricle (Figure 19.4b). Between the right atrium and left atrium is a thin partition called the **interatrial septum** (*inter-* = between; *-septum* = a dividing wall or partition). A prominent feature of this septum is an oval depression called the **fossa ovalis**, the remnant of the *foramen ovale*, an

opening in the interatrial septum of the fetal heart that normally closes soon after birth. Blood passes from the right atrium into the right ventricle through the **tricuspid valve** (trī-KUS-pid; *tri-* = three; *-cuspid* = point), so named because it consists of three cusps or leaflets (Figure 19.4a).

Right Ventricle

The **right ventricle** forms most of the anterior surface of the heart (Figure 19.3a). The wall of the right ventricle contains a series of ridges formed by raised bundles of cardiac muscle fibers called **trabeculae carneae** (tra-BEK-ū-lē KAR-nē-ē; *trabeculae* = little beams; *carneae* = fleshy; Figure 19.4). The cusps of the tricuspid valve are connected to tendon-like cords, the **chordae tendineae** (KOR-dē ten-DIN-ē-ē; *chord-* = cord; *tend-* = tendon), which in turn are connected to cone-shaped trabeculae carneae called **papillary muscles** (*papill-* = nipple). The right ventricle is separated from the left ventricle by a partition called the **interventricular septum**. Blood passes from the right ventricle through the **pulmonary valve** into a large artery, the *pulmonary trunk*, which carries blood to the lungs for oxygenation. Arteries always take blood away from the heart.

Figure 19.4 **Structure of the heart: internal anatomy.**

Frontal plane

Left common carotid artery
Left subclavian artery
Brachiocephalic trunk

Arch of aorta
Ligamentum arteriosum

Ascending aorta
Superior vena cava
Right pulmonary artery
PULMONARY VALVE

Left pulmonary artery
Pulmonary trunk

Left pulmonary veins
LEFT ATRIUM
AORTIC VALVE
BICUSPID VALVE

Right pulmonary veins

CHORDAE TENDINEAE
LEFT VENTRICLE
INTERVENTRICULAR SEPTUM

Opening of superior vena cava

Fossa ovalis
RIGHT ATRIUM
Opening of coronary sinus
Opening of inferior vena cava

PAPILLARY MUSCLE
TRABECULAE CARNEAE

TRICUSPID VALVE

RIGHT VENTRICLE

Inferior vena cava

Descending aorta

(a) Anterior view of frontal section showing internal anatomy

Brachiocephalic trunk

Left subclavian artery
Left common carotid artery

Superior vena cava

Arch of aorta

Ligamentum arteriosum

Right pulmonary vein

Ascending aorta

RIGHT AURICLE (cut open)

Pulmonary trunk

Pectinate muscles

Left pulmonary vein
LEFT AURICLE

RIGHT ATRIUM

Cusp of tricuspid valve

Chordae tendineae

LEFT VENTRICLE

Papillary muscle

INTERVENTRICULAR SEPTUM

RIGHT VENTRICLE

TRABECULAE CARNEAE

(b) Anterior view of partially sectioned heart

(continues)

Figure 19.4 (continued)

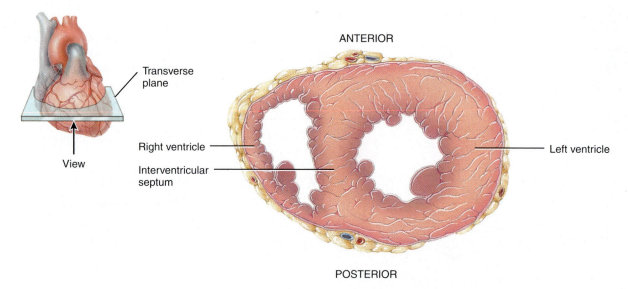

(c) Inferior view of transverse section showing differences
in thickness of ventricular walls

The thickness of the four chambers varies according to their functions.

Left Atrium

The **left atrium** receives blood from the lungs through four *pulmonary veins* (Figure 19.4). Like the right atrium, the inside of the left atrium has a smooth posterior wall. Because the ridged pectinate muscles are confined to the auricle of the left atrium, the anterior wall of the left atrium is smooth. Blood passes from the left atrium into the left ventricle through the **bicuspid (mitral) valve** (*bi-* = two), which, as its name implies, has two cusps. The term *mitral* refers to its resemblance to a bishop's miter (hat), which is two-sided.

Left Ventricle

The **left ventricle** is the thickest chamber of the heart, averaging 10–15 mm (0.4–0.6 in.), and forms the apex of the heart (see Figure 19.1b). Like the right ventricle, the left ventricle contains trabeculae carneae and has chordae tendinae that anchor the cusps of the bicuspid valve to papillary muscles (Figure 19.4). Blood passes from the left ventricle through the **aortic valve** into the largest artery of the body, the *ascending aorta* (*aorte* = to suspend, because the aorta once was believed to lift up the heart). Some of the blood in the aorta flows into the *coronary arteries*, which branch from the ascending aorta and carry blood to the heart wall; the remainder of the blood passes into the *arch of the aorta* and *descending aorta*. Branches of the arch of the aorta and descending aorta carry blood throughout the body.

During fetal life, a temporary blood vessel, called the *ductus arteriosus*, shunts blood from the pulmonary trunk into the aorta

to bypass the nonfunctioning fetal lungs. The ductus arteriosus normally closes shortly after birth, leaving a remnant known as the **ligamentum arteriosum**, which connects the arch of the aorta and pulmonary trunk (Figure 19.4a).

Myocardial Thickness and Function

The thickness of the myocardium of the four chambers varies according to the amount of work each chamber has to perform. The walls of the atria are thin compared to those of the ventricles because the atria need only enough cardiac muscle tissue to deliver blood into the ventricles (Figure 19.4a). Since the ventricles pump blood greater distances, their walls are thicker. Although the right and left ventricles act as two separate pumps that simultaneously eject equal volumes of blood, the right ventricle pumps blood only a short distance to the lungs (pulmonary circulation); the left ventricle pumps blood to all other parts of the body (systemic circulation). The left ventricle works much harder than the right ventricle to maintain the same rate of blood flow; therefore, the muscular wall of the left ventricle is considerably thicker than the wall of the right ventricle (Figure 19.4c).

Checkpoint

5.   What function do the auricles serve?

6.   The coronary sulcus forms a boundary between which chambers of the heart?

7.   The anterior interventricular sulcus forms a boundary between which chambers of the heart?

8.   What structure separates the atria? The ventricles?

19.3 Heart valves ensure one-way flow of blood.

Blood flows through the heart from areas of higher blood pressure to areas of lower blood pressure. As the walls of each chamber contract and relax, resulting pressure differences across the heart valves force valves to open and close. Each of the four valves helps to ensure one-way flow of blood by opening to let blood through and then closing to prevent its backflow.

Operation of the Atrioventricular Valves

Because they are located between an atrium and a ventricle, the tricuspid and bicuspid valves are also called **atrioventricular (AV)**

valves. When an AV valve is open, the rounded ends of the cusps project into the ventricle (Figure 19.5a,d). When the ventricles are relaxed, the papillary muscles are relaxed, the chordae tendineae are slack, and blood moves from a higher pressure in the atria to a lower pressure in the ventricles through open AV valves. When a ventricle contracts, the pressure of the ventricular blood drives the cusps upward until their edges meet and close the valve opening (Figure 19.5b,e). At the same time, the papillary muscles contract, which pulls on and tightens the chordae tendineae, preventing the valve cusps from pushing up into the atria in response to the high ventricular pressure. If the AV valves or chordae tendineae were to become damaged, blood might flow back into the atria when the ventricles contract.

Figure 19.5 **Responses of the valves to the pumping of the heart.**

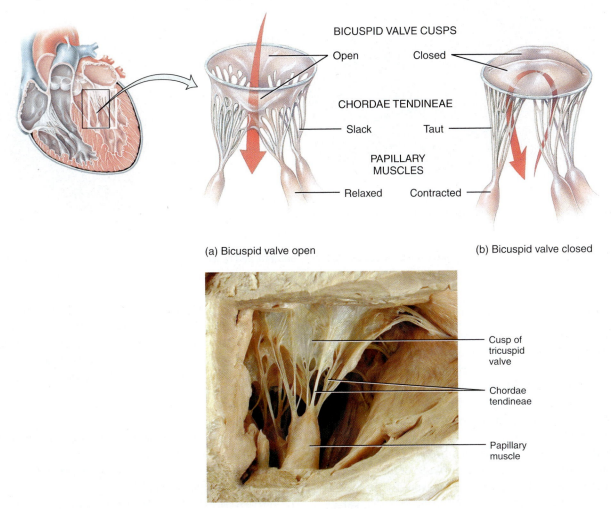

(a) Bicuspid valve open

(b) Bicuspid valve closed

(c) Tricuspid valve open

(continues)

Figure 19.5 (continued)

ANTERIOR

Pulmonary valve (closed)
Aortic valve (closed)
Left coronary artery
Right coronary artery
Bicuspid valve (open)
Tricuspid valve (open)

POSTERIOR

(d) Superior view with atria removed: pulmonary and aortic valves closed, bicuspid and tricuspid valves open

ANTERIOR

Pulmonary valve (open)
Aortic valve (open)
Bicuspid valve (closed)
Tricuspid valve (closed)

POSTERIOR

(e) Superior view with atria removed: pulmonary and aortic valves open, bicuspid and tricuspid valves closed

ANTERIOR

Pulmonary trunk
Ascending aorta
PULMONARY VALVE
Right coronary artery
Pectinate muscle of left atrium
Pectinate muscle of right atrium
Left coronary artery
AORTIC VALVE
BICUSPID VALVE
TRICUSPID VALVE
Coronary sinus

POSTERIOR

(f) Superior view of atrioventricular and semilunar valves

Semilunar cusp

(g) Superior view of aortic valve

🔑 Heart valves open and close in response to pressure changes as the heart contracts and relaxes.

Operation of the Semilunar Valves

Near the origin of the pulmonary trunk and aorta are the two **semilunar (SL) valves** (*semi-* = half; *-lunar* = moon-shaped), the pulmonary and aortic valves, that prevent arterial blood from flowing back into the ventricles. Each SL valve consists of three crescent-moon-shaped cusps that attach to an artery wall by their outer margins (Figure 19.5d). The free margins of the cusps project into the lumen of the artery. Like the atrioventricular valves, the semilunar valves permit blood to flow in one direction only—in this case, from ventricles into arteries. When the walls of the ventricles contract, the increasing ventricular blood pressure forces the SL valves to open, permitting ejection of blood through the SL valves into the pulmonary trunk and aorta (Figure 19.5e). As the ventricles relax, blood starts to flow back toward the heart. This backflowing blood fills the cusps, which causes the semilunar valves to close (Figure 19.5d).

Surprisingly perhaps, there are no valves between the venae cavae and the right atrium or between the pulmonary veins and the left atrium. As the atria contract, a small amount of blood does flow backward into these vessels. However, backflow is minimized because contracting atria compress the venous entry points.

Checkpoint

9. What is the main force that causes blood to flow through the heart and the heart valves to open and to close?

10. As the ventricles contract, what prevents the atrioventricular valves from swinging upward into the atria?

11. Why does blood ejected into the pulmonary trunk or aorta normally not flow back through the semilunar valve?

19.4 The heart pumps blood to the lungs for oxygenation, then pumps oxygen-rich blood throughout the body.

Systemic and Pulmonary Circulations

With each beat, the heart pumps blood into two circuits—the **pulmonary circulation** (*pulmon-* = lung) carrying blood to the air sacs (alveoli) of the lungs and the **systemic circulation** carrying blood to the rest of the body. The two circuits are arranged in series so that the output of one becomes the input of the other, as would happen if you attached two garden hoses (see Figure 20.14). Figure 19.6 shows the route of blood flow through the heart and pulmonary and systemic circulations.

The right side of the heart is the pump for pulmonary circulation. The right atrium receives deoxygenated (oxygen-poor) blood returning from systemic circulation and then delivers it into the right ventricle, which pumps the deoxygenated blood into the *pulmonary trunk*. The pulmonary trunk divides into *right* and *left pulmonary arteries* that carry blood to the corresponding lung. In each lung, blood flows through arteries (blood vessels that carry blood away from the heart—artery = *a*way) to smaller-diameter *pulmonary arterioles*, and finally into extensive beds of *pulmonary capillaries*. In the thin-walled pulmonary capillaries, deoxygenated blood unloads CO_2 (carbon dioxide) and picks up O_2 (oxygen). This oxygenated blood flows into *pulmonary venules* and eventually into larger *pulmonary veins*.

The left side of the heart is the pump for systemic circulation. Oxygenated blood from pulmonary veins enters the left atrium then passes into the left ventricle. The left ventricle ejects blood into the *aorta* (see Figure 19.4a), which branches into progressively smaller *systemic arteries* that carry the oxygenated blood to all organs of the body—except for the air sacs (alveoli) of the lungs, which are supplied by pulmonary circulation. Arteries give rise to *systemic arterioles*, which finally lead into *systemic capillaries*. Exchange of nutrients and gases occurs across the thin capillary walls. Blood unloads O_2 and picks up CO_2. Blood flows from capillaries into *systemic venules* that carry deoxygenated blood away from tissues and merge to form larger *systemic veins*. Ultimately the blood flows back to the right atrium through three veins (blood vessels that return blood to the heart), the superior vena cava, inferior vena cava, and coronary sinus (see Figures 19.3c and 19.4a).

Coronary Circulation

Nutrients are not able to diffuse quickly enough from blood within the chambers of the heart to supply all the layers of cells that make up the heart wall. For this reason, the wall of the heart has its own blood vessels. The flow of blood through the many vessels that pierce the myocardium is called the **coronary (cardiac) circulation** (*coron-* = crown) because these arteries encircle the heart like a crown encircles the head (see Figure 19.7a).

Figure 19.6 **Systemic and pulmonary circulations.**

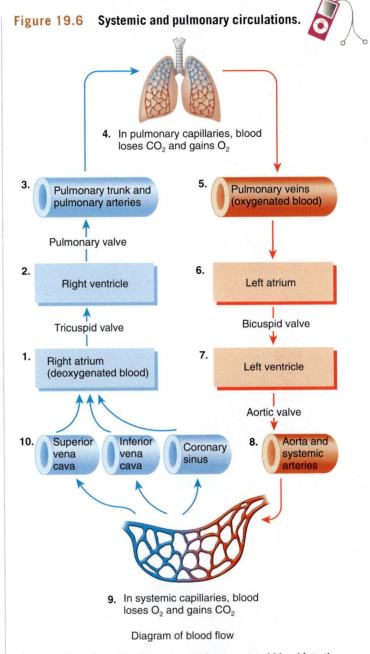

4. In pulmonary capillaries, blood loses CO_2 and gains O_2

3. Pulmonary trunk and pulmonary arteries

5. Pulmonary veins (oxygenated blood)

Pulmonary valve

2. Right ventricle

6. Left atrium

Tricuspid valve

Bicuspid valve

1. Right atrium (deoxygenated blood)

7. Left ventricle

Aortic valve

10. Superior vena cava | Inferior vena cava | Coronary sinus

8. Aorta and systemic arteries

9. In systemic capillaries, blood loses O_2 and gains CO_2

Diagram of blood flow

🔑 The right side of the heart pumps deoxygenated blood into the pulmonary circulation to the air sacs (alveoli) of the lungs; the left side of the heart pumps freshly oxygenated blood into the systemic circulation to all tissues of the body except the air sacs of the lungs.

Coronary Arteries

The principal coronary vessels, the **left** and **right coronary arteries**, originate as branches of the ascending aorta and supply oxygenated blood to the myocardium (Figure 19.7a). The **left coronary artery** passes inferior to the left auricle and divides into the anterior interventricular and circumflex branches. The **anterior interventricular branch**, or *left anterior descending (LAD) artery*, is in the anterior interventricular sulcus and supplies oxygenated blood to the walls of both ventricles. The **circumflex branch** lies in the coronary sulcus and distributes oxygenated blood to the walls of the left atrium and left ventricle.

The **right coronary artery** divides into small branches that supply the right atrium. It continues inferior to the right auricle and ultimately divides into the posterior interventricular and marginal branches (Figure 19.7a). The **posterior interventricular branch** follows the posterior interventricular sulcus and supplies the walls of the two ventricles with oxygenated blood. The **marginal branch** in the coronary sulcus transports oxygenated blood to the myocardium of the right ventricle.

Most parts of the body receive blood from branches of more than one artery, and where two or more arteries supply the same body region, they usually connect. These connections, called **anastomoses** (a-nas´-tō-MŌ-sēs), provide alternate routes for blood to reach a particular organ or tissue. The myocardium contains many anastomoses that connect branches of coronary arteries to provide detours for arterial blood if a main route becomes obstructed. Thus, cardiac muscle tissue may receive sufficient oxygen even if one of the coronary arteries is partially blocked.

Figure 19.7   Coronary circulation. The views of the heart from the anterior aspect in (a) and (b) are drawn as if the heart were transparent to reveal blood vessels on the posterior aspect.

(a) Anterior view of coronary arteries

(b) Anterior view of coronary veins

Coronary Veins

After blood passes through the coronary arteries, it flows into capillaries, where it delivers oxygen and nutrients to cardiac muscle tissue and collects carbon dioxide and wastes, and then moves into coronary veins. Most of the deoxygenated blood from the myocardium drains into the **coronary sinus** on the posterior surface of the heart (Figure 19.7b). (A *vascular sinus* is a thin-walled vein.) The deoxygenated blood in the coronary sinus empties into the right atrium. The principal tributaries carrying blood into the coronary sinus are the following (Figure 19.7b):

- **Great cardiac vein** in the anterior interventricular sulcus, which drains the left atrium and left and right ventricles.

- **Middle cardiac vein** in the posterior interventricular sulcus, which drains the left and right ventricles.

- **Small cardiac vein** in the coronary sulcus, which drains the right atrium and right ventricle.

- **Anterior cardiac veins**, which drain the right ventricle.

Checkpoint

12. From which vessels do each of the atria receive blood? Into which vessels do the ventricles eject blood?
13. Which blood vessels that enter and exit the heart carry oxygenated blood? Which carry deoxygenated blood?
14. Which blood vessel supplies blood to the coronary arteries?
15. Which coronary blood vessel delivers oxygenated blood to the myocardium of the left atrium and left ventricle? Which drains deoxygenated blood from the myocardium of the left atrium and left ventricle? Which drains deoxygenated blood from the myocardium of the right atrium and right ventricle?

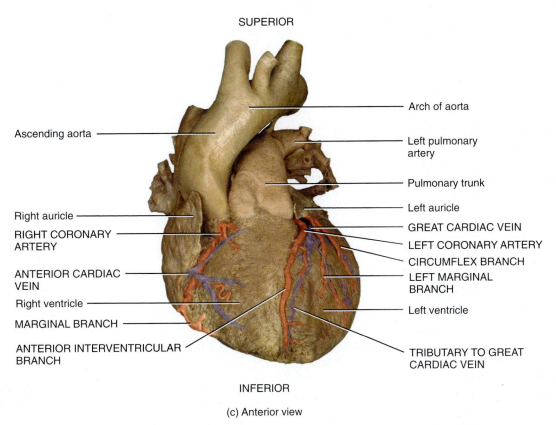

SUPERIOR

Ascending aorta

Right auricle

RIGHT CORONARY ARTERY

ANTERIOR CARDIAC VEIN

Right ventricle

MARGINAL BRANCH

ANTERIOR INTERVENTRICULAR BRANCH

Arch of aorta

Left pulmonary artery

Pulmonary trunk

Left auricle

GREAT CARDIAC VEIN

LEFT CORONARY ARTERY

CIRCUMFLEX BRANCH

LEFT MARGINAL BRANCH

Left ventricle

TRIBUTARY TO GREAT CARDIAC VEIN

INFERIOR

(c) Anterior view

The right and left coronary arteries deliver blood to the heart; the coronary veins drain blood from the heart into the coronary sinus.

Return To Prakesh's Story

Prakesh sat up in his bed Sunday morning feeling a bit light-headed. He turned to sit on the edge of the bed and glanced around the floor for his socks. As he bent to pick them up, he noticed that his feet and ankles were swollen, as they had been since the previous morning.

"Prakesh, are you all right?" Startled, he turned and saw his wife Manjula.

"Yes, dear, I couldn't sleep last night. Terrible dreams. I'm a bit out of sorts, you know, a bit short of breath this morning. It's nothing."

Manjula frowned. "It's nothing? What about your heart? We should go to the clinic this morning!"

"I have no chest pain," said Prakesh.

Manjula wagged her finger at him and said, "Get dressed! We are going in. That's final."

At the clinic, Prakesh waited anxiously. When the nurse came in, she calmly went about her business. "Your blood pressure is 110 over 52; your pulse is a bit fast, though, at 130. Let's get your temperature; open up. Mrs. Singh, how long ago was your husband's last visit?"

Manjula thought for a moment and replied, "Last year about this time. He doesn't like coming to the doctor. He's been exercising and eating well, though; and he quit smoking two years ago!"

"Well, that's great," said the nurse. "Temp is 98.9°F. Dr. Abbas will be right in."

The nurse's exit was followed quickly by a brisk knock on the door. "Hello, Mr. Singh. I'm Dr. Abbas. This is your wife, I presume?"

Prakesh shook the doctor's hand and said, "Doctor, please tell my wife this is nothing. I have been a good boy. I exercise, I quit smoking, I could still lose some weight...but...I'm just a bit tired is all."

Dr. Abbas took a deep breath and glanced down at his notes. "Well, looking at your vitals here, and based on your complaints, I'm not ready to send you home yet. I'd like to get an electrocardiogram and a chest X-ray first. With your history of heart disease, the swelling in your feet and your shortness of breath may be signs of something more serious than fatigue."

A. Develop a patient history for Prakesh including his age, gen-der, and health history. What are his vital signs? His primary symptoms?

B. Based on what you know about blood flow, what might be causing Prakesh's swelling? Is this related to his shortness of breath?

C. What are Prakesh's risk factors for cardiac disease?

D. How might Prakesh's heart valves be related to his symptoms?

E. Why would the doctor think the problem lies in Prakesh's heart rather than his lungs? Identify all organ systems that could be involved.

19.5 The cardiac conduction system coordinates heart contractions for effective pumping.

An inherent and rhythmical electrical activity is the reason for the heart's lifelong beat. The source of this electrical activity is a net-work of specialized cardiac muscle fibers called **autorhythmic fibers** (*auto-* = self) because they are self-excitable; they repeat-edly and rhythmically generate action potentials that trigger heart contractions. Autorhythmic fibers continue to stimulate a heart to beat even after it is removed from the body—for example, to be transplanted into another person—and all of its nerves have been cut. The remaining "working" myocardial fibers, **contractile fi-bers**, provide the powerful contractions that propel blood.

Autorhythmic cells have two important functions. They pro-duce the electrical excitation that triggers heart contraction. They also form a **conduction system** that delivers action potentials throughout the heart. The conduction system ensures that car-diac chambers are stimulated to contract in a coordinated manner, which makes the heart an effective pump.

Cardiac action potentials propagate through the conduction system in the following sequence (Figure 19.8):

1 Cardiac excitation normally begins in the **sinoatrial (SA) node**, located in the right atrial wall just inferior to the open-ing of the superior vena cava. SA node cells do not have a stable resting potential. Rather, they repeatedly depolarize to threshold spontaneously, which triggers an action potential.

Each action potential from the SA node quickly propagates throughout both atria, from one atrial fiber to the next, via gap junctions (see Figure 10.18a). Following the action po-tential, the atria contract.

2 By propagating along atrial muscle fibers, the action poten-tial reaches the **atrioventricular (AV) node**, located in the interatrial septum, just anterior to the opening of the coro-nary sinus.

3 From the AV node, the action potential enters the **atrioven-tricular (AV) bundle** (also known as the **bundle of His**), the only site where action potentials can be transferred from the atria to the ventricles.

4 After propagating along the AV bundle, the action potential enters both the **right** and **left bundle branches** that course through the interventricular septum toward the apex of the heart.

5 Finally, **Purkinje fibers** conduct the action potential, first to the apex of the heart and then upward to the remainder of the ventricular myocardium. As the ventricles contract supe-riorly from the apex, blood is pushed toward the semilunar valves.

Figure 19.8 **The conduction system of the heart.** Autorhythmic fibers in the sinoatrial node act as the heart's pacemaker to initiate cardiac action potentials that cause contraction of the heart's chambers.

Frontal plane

Right atrium

1 SINOATRIAL NODE

2 ATRIOVENTRICULAR (AV) NODE

3 ATRIOVENTRICULAR BUNDLE

4 RIGHT AND LEFT BUNDLE BRANCHES

Right ventricle

5 PURKINJE FIBERS

Left atrium

Left ventricle

Anterior view of frontal section

The conduction system ensures that cardiac chambers contract in a coordinated manner.

The SA node initiates action potentials about 100 times per minute, faster than those of any other autorhythmic fibers. Action potentials from the SA node spread through the conduction system and stimulate other regions before those regions are able to generate an action potential at their own, slower rate. Thus, the SA node sets the rhythm for contraction of the heart—it is the natural **pacemaker** of the heart. Nerve impulses from the autonomic nervous system (ANS) and blood-borne hormones (such as epinephrine) *modify the timing and strength* of each heartbeat, but they *do not establish the fundamental rhythm*. In a person at rest, for example, acetylcholine released by the parasympathetic division of the ANS typically slows SA node pacing to about 75 action potentials per minute, causing 75 heartbeats per minute.

Checkpoint

16. What are autorhythmic cells? What are their two functions?
17. Which component of the conduction system provides the only route for action potentials to conduct between the atria and the ventricles?
18. What is the path of an action potential through the conduction system of the heart?
19. Which component of the conduction system sets the pace of contraction in a normal heart?

19.6 The electrocardiogram is a record of electrical activity associated with each heartbeat.

Electrocardiogram

As action potentials propagate through the heart, they generate electric currents that can be detected by electrodes placed on the body's surface. A recording of the electrical changes that accompany the heartbeat is called an **electrocardiogram** (e-lek´-trō-KAR-dē-ō-gram), which is abbreviated as either **ECG** or **EKG** (from the German word *electrokardiogram*). The ECG is a composite of action potentials, graphed as a series of up-and-down waves, pro-

duced by cardiac muscle fibers during each heartbeat. The instrument used to record the changes is an **electrocardiograph**.

In clinical practice, electrodes are positioned on the arms and legs (limb leads) and at six positions on the chest (chest leads) to record the ECG. The electrocardiograph amplifies the heart's electrical signals and produces 12 different tracings from different combinations of limb and chest leads. Each limb and chest electrode records slightly different electrical activity because it is in a different position relative to the heart.

In a typical record, three clearly recognizable waves accompany each heartbeat (Figure 19.9) The first, called the **P wave**, is a small upward deflection on the ECG; it represents **atrial depolarization**, the depolarizing (electrical excitation) phase of the cardiac action potential as it spreads from the SA node through contractile fibers in both atria leading to their contraction. The second wave, called the **QRS complex**, begins as a downward deflection (Q); continues as a large, upright, triangular wave (R); and ends as a downward wave (S). The **QRS complex** represents the onset of **ventricular depolarization**, as the cardiac action potential spreads through ventricular contractile fibers. Shortly after the QRS complex begins, the ventricles start to contract. The third wave is an upward deflection called the **T wave**; it indicates **ventricular repolarization** and occurs just before the ventricles start to relax.

Variations in the size and duration of waves of an ECG are useful in diagnosing abnormal cardiac rhythms and conduction patterns and in following the course of recovery from a heart attack. An ECG can also detect the presence of a living fetus.

Figure 19.9   Normal electrocardiogram (ECG) of a single heartbeat (Lead II). P wave = atrial depolarization; QRS complex = onset of ventricular depolarization; T wave = ventricular repolarization.

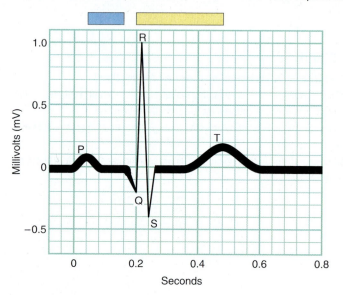

Key:

�in blue	Atrial contraction
▬ yellow	Ventricular contraction

🔑 **An electrocardiogram is a recording of the electrical activity that initiates each heartbeat.**

Correlation of ECG Waves with Heart Activity

As we have seen, the atria and ventricles depolarize and then contract at different times because the conduction system routes cardiac action potentials along a specific pathway. The term **systole** (SIS-tō-lē = contraction) refers to the phase of contraction; the phase of relaxation is **diastole** (dī-AS-tō-lē = dilation or expansion). The ECG waves predict the timing of atrial and ventricular systole and diastole. The sequence of systole and diastole, as they relate to the waves of an ECG, is as follows (Figure 19.10):

❶ A cardiac action potential arises in the SA node. It propagates throughout the atrial muscle and down to the AV node. As the atrial contractile fibers depolarize, the P wave appears in the ECG.

❷ Contraction of atrial contractile fibers (atrial systole) begins after the P wave appears.

❸ The action potential enters the AV bundle, where it is propagated through the bundle branches, Purkinje fibers, and the entire ventricular myocardium. Depolarization progresses down the interventricular septum, then upward and outward from the apex, producing the QRS complex. At the same time, atrial repolarization is occurring, but it is not usually evident in an ECG because the larger QRS complex masks it. As the atria repolarize, they relax (atrial diastole).

❹ Contraction of ventricular contractile fibers (ventricular systole) begins after the QRS complex appears. As contraction proceeds from the apex toward the base of the heart, blood is squeezed superiorly toward the semilunar valves.

❺ Repolarization of ventricular contractile fibers begins at the apex and spreads throughout the ventricular myocardium. This produces the T wave in the ECG.

❻ After the T wave begins, the ventricles relax (ventricular diastole).

Briefly, contractile fibers in both the atria and ventricles are relaxed. Then the P wave appears again in the ECG, the atria begin to contract, and the cycle repeats. Events in the heart occur in cycles that repeat for as long as you live. Next, we will see how the pressure changes associated with relaxation and contraction of the heart chambers allow the heart to alternately fill with blood and then eject blood into the aorta and pulmonary trunk.

Checkpoint

20. What is an electrocardiogram?
21. How does each ECG wave relate to the phases of an action potential?
22. What is the diagnostic significance of the ECG?
23. How does each ECG wave relate to contraction and relaxation of the atria and ventricles?

Figure 19.10 **Timing and route of action potential depolarization through the heart.** Green indicates depolarization and red indicates repolarization.

1 Depolarization of atrial contractile fibers produces P wave

P

Action potential in SA node

0   0.2
Seconds

6 Ventricular diastole (relaxation)

P

0   0.2   0.4   0.6   0.8
Seconds

2 Atrial systole (contraction)

P

0   0.2
Seconds

5 Repolarization of ventricular contractile fibers produces T wave

P   T

0   0.2   0.4   0.6
Seconds

3 Depolarization of ventricular contractile fibers produces QRS complex

R

P

Q   S

0   0.2   0.4
Seconds

4 Ventricular systole (contraction)

P

0   0.2   0.4
Seconds

🔑 **Depolarization causes contraction and repolarization causes relaxation of cardiac muscle fibers.**

19.7 The cardiac cycle represents all the events associated with one heartbeat.

The **cardiac cycle** comprises all the events associated with one heartbeat. During the cardiac cycle, the two atria contract (atrial systole) while the two ventricles relax (ventricular diastole). Then while the two ventricles contract (ventricular systole), the two atria relax (atrial diastole).

Pressure and Volume Changes during the Cardiac Cycle

In each cardiac cycle, contraction of the atria and then ventricles forces blood from areas of higher pressure to areas of lower pressure. As each chamber of the heart contracts, blood pressure within it increases, pushing blood into a ventricle or out of the heart into an artery. Figure 19.11 shows the relation between the heart's electrical signals (ECG) and changes in atrial pressure, ventricular pressure, aortic pressure, and ventricular volume during the cardiac cycle. The pressures given in Figure 19.11 apply to the left side of the heart; pressures on the right side are considerably lower. Each ventricle, however, expels the same volume of blood per beat, and the same pattern exists for both pumping chambers. To examine and correlate the events taking place during a cardiac cycle, we will begin with atrial systole.

Atrial Systole

During **atrial systole**, the atria are contracting. At the same time, the ventricles are relaxed.

1 An action potential from the SA node results in atrial depolarization, marked by the P wave in an ECG.

2 Atrial depolarization produces atrial systole. As the atria contract, the blood within is forced through the open AV valves into the ventricles.

3 Because the SL valves are still closed, the ventricles fill with blood. Atrial systole contributes a final 25% of the blood in each ventricle. At the end of ventricular diastole (relaxation), each ventricle contains about 130 mL of blood. This blood volume is the **end-diastolic volume (EDV)**. At the end of ventricular diastole, the AV valves are still open and the SL valves are still closed.

Ventricular Systole

During **ventricular systole**, the ventricles are contracting. At the same time, the atria are relaxed, in **atrial diastole**.

4 The QRS complex in the ECG marks the onset of ventricular depolarization.

5 Ventricular depolarization leads to **ventricular systole** (contraction of the ventricles). Ventricular contraction pushes the blood within against the AV valves, forcing them shut. For a split second, both the SL and AV valves are closed. This is the period of **isovolumetric contraction** (*iso-* = same). During this interval, the ventricles contract, yet because all four valves are closed, blood volume in the ventricles remains the same (isovolumic).

6 As the ventricles continue to contract, pressure inside the closed chambers rises sharply. When ventricular pressure rises above the pressure in the aorta and pulmonary trunk, both SL valves are forced open, and blood is ejected from the heart. The period when the SL valves are open is termed **ventricular ejection**. During this period, the ventricles are contracting so strongly that pressure within the chambers continues to rise even as blood is ejected through the SL valves.

7 In the resting body, the volume of blood ejected from each ventricle during ventricular systole is about 70 mL (a little more than 2 oz). The volume of blood remaining in each ventricle at the end of systole, about 60 mL, is the **end-systolic volume (ESV)**.

Relaxation Period

During the **relaxation period**, all four chambers are in diastole.

8 The T wave in the ECG marks the onset of ventricular repolarization.

9 Ventricular repolarization causes **ventricular diastole**. As the ventricles relax, pressure within the chambers falls, and blood starts to flow backward from the pulmonary trunk and aorta toward the regions of lower pressure in the ventricles. As this backflowing blood becomes trapped in the semilunar cusps, the semilunar valves close. Rebound of blood off the closed cusps of the aortic valve produces the **dicrotic wave** on the aortic pressure curve. After the SL valves close, there is a brief interval when ventricular blood volume does not change because all four valves are closed. This is the period of **isovolumetric relaxation**.

10 As the ventricles continue to relax, ventricular pressure falls quickly. When ventricular pressure drops below atrial pressure, the AV valves are forced open, and **ventricular filling** begins. Blood that has been flowing into and building up in the atria during ventricular systole rushes rapidly into the ventricles. About 75% of the ventricular filling occurs after the AV valves open and before the atria contract. Another cardiac cycle begins as atrial depolarization, signaled by the P wave in the ECG, causes atrial systole.

Figure 19.11 **Cardiac cycle.** (a) ECG. (b) Changes in left atrial pressure (green line), left ventricular pressure (blue line), and aortic pressure (red line) as they relate to the opening and closing of heart valves. (c) Heart sounds. (d) Changes in left ventricular volume. (e) Phases of the cardiac cycle with green indicating atrial depolarization.

(a) ECG

	0.1 sec	0.3 sec	0.4 sec
	Atrial systole	Ventricular systole	Relaxation period

(b) Pressure (mm Hg)

- 9 Aortic valve closes
- Dicrotic wave
- Aortic pressure
- 6 Aortic valve opens
- Left ventricular pressure
- 5 Bicuspid valve closes
- 10 Bicuspid valve opens
- Left atrial pressure

(c) Heart sounds

S1   S2 S3   S4

(d) Volume in ventricle (mL)

- 3 End-diastolic volume
- Stroke volume
- 7 End-systolic volume

(e) Phases of the cardiac cycle

Atrial contraction | Isovolumetric contraction | Ventricular ejection | Isovolumetric relaxation | Ventricular filling | Atrial contraction

A cardiac cycle is composed of all the events associated with one heartbeat.

Heart Sounds

Auscultation (aws-kul-TĀ-shun; *ausculta-* = listening) is the act of listening to sounds within the body, and it is usually done with a stethoscope. **Heart sounds** come primarily from turbulence in blood flow created by the closure of the heart valves. Smoothly flowing blood is silent. Recall the sounds made by whitewater rapids or a waterfall as compared with the silence of a smoothly flowing river. During each cardiac cycle, there are four heart sounds, but in a normal heart only the first and second heart sounds are loud enough to be heard by listening through a stethoscope. Figure 19.11c shows the timing of heart sounds relative to other events in the cardiac cycle.

The first sound (S1), which can be described as a **lubb** sound, is louder and a bit longer than the second sound, and is from the AV valves closing after ventricular systole begins. The second sound (S2), which is shorter and not as loud as the first, can be described as a **dupp** sound, and is from the SL valves closing at the end of ventricular systole. The cardiac cycle is heard as lubb, dupp, pause; lubb, dupp, pause; lubb, dupp, pause with the pause signifying the relaxation period.

Although S1 and S2 are associated with the closure of valves, they are best heard at the surface of the chest in locations that are slightly different from the locations of the valves (Figure 19.12) because the sound is carried away from the valves by blood flow.

Checkpoint

24. Which term is used for the contraction phase of the cardiac cycle? The relaxation phase?
25. Which phase of the cardiac cycle are the atria in during ventricular systole?
26. What is the volume of blood called that remains in each ventricle at the end of ventricular diastole? What is the volume of blood called that remains in each ventricle at the end of ventricular systole?
27. Which heart events can usually be heard through a stethoscope?

Figure 19.12   Heart sounds. Location of valves (purple) and ausculation sites (red) for heart sounds.

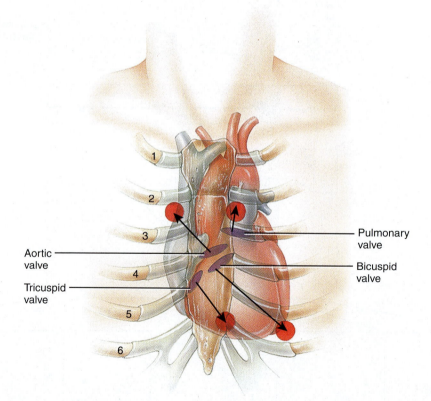

Aortic valve

Tricuspid valve

Pulmonary valve

Bicuspid valve

Anterior view of heart valve locations and auscultation sites

Listening to sounds within the body is called ausculation; it is usually done with a stethoscope.

Return to Prakesh's Story

The chest X-ray didn't take long. Afterward Prakesh and Manjula were escorted to another room for the electrocardiogram. Looking at the X-ray, Dr. Abbas noticed that Prakesh's heart was enlarged. "Hmm, cardiomegaly with pulmonary edema. Could be congestive heart failure; might be worth trying a diuretic," he said to himself. The diuretic drug would rid Prakesh of the excess fluids that were congesting the tissues around his heart. The results of the electrocardiogram would provide more information to help Dr. Abbas determine the cause of the congestive heart failure.

"Prakesh, I have the results of the ECG and the chest X-rays. There are a couple of things going on we need to address. You've got some fluid built up around the heart that's making it difficult for you to breathe and causing the shortness of breath. The fluid buildup is most likely caused by the heart not beating properly. Because the heart muscle isn't receiving the signals from the pacemaker of the heart that tell it to contract, your heart isn't pumping hard enough to send blood out to your body. Your kidneys are responding by retaining water and salt, which is the cause of the excess fluid retention and swelling. I would like to prescribe two medications. One, called a diuretic, will rid your body of the excess water, and the second drug will increase the force of your heart."

F. If the left bundle branch of the heart was damaged during Prakesh's previous heart attack, how might the function of his heart be affected?

G. Which ECG wave is most likely to be abnormal on Prakesh's ECG if he has a left bundle branch block?

H. Dr. Abbas suspects that Prakesh's end-systolic volume is higher than it should be. Why? (Hint: Remember that his heart is pumping with less force.)

19.8 Cardiac output is the blood volume ejected by a ventricle each minute.

Although the heart's autorhythmic fibers enable it to beat independently, its operation is governed by events occurring throughout the body. All body cells must receive a certain amount of oxygenated blood each minute to maintain health and life. When cells are metabolically active, as during exercise, they take up even more oxygen from the blood and the heart is stimulated to work harder. During rest periods, cellular metabolic need is reduced, and the workload of the heart decreases.

Cardiac output is the volume of blood ejected by a ventricle per minute. Cardiac output equals the **stroke volume**, the volume of blood ejected by the ventricle during each beat (contraction), multiplied by the **heart rate**, the number of beats per minute:

$$\text{Cardiac output (mL/min)} = \text{stroke volume (mL/beat)} \times \text{heart rate (beats/min)}$$

In a typical resting adult, stroke volume averages 70 mL/beat, and heart rate is about 75 beats/min. Thus the average cardiac output in a resting adult is 70 mL/beat × 75 beats/min, which equals 5250 mL/min (5.25 liters/min). This volume is close to the total blood volume, which is about 5 liters in a typical adult. Thus, your entire blood volume flows through the pulmonary and systemic circulations each minute. When body tissues use more or less oxygen, cardiac output changes to meet the need. Factors that increase stroke volume or make the heart beat faster, such as exercise, increase cardiac output. Conversely, factors that decrease stroke volume or heart rate decrease cardiac output.

Regulation of Stroke Volume

Although some blood is always left in the ventricles at the end of their contraction, a healthy heart pumps out the entire volume of blood that has entered its chambers during the previous diastole. In other words, the more blood that returns to the heart during diastole, the more blood that is ejected during the next systole. Three factors regulate stroke volume and ensure that the left and right ventricles pump equal volumes of blood:

1. **Preload**. The **preload** is the degree to which the heart is stretched before it contracts. Preload can be compared to the stretching of a rubber band. The more the rubber band is stretched, the more forcefully it will snap back. Within limits, the more the heart is stretched as it fills during diastole, the greater is the force of contraction during systole, a relationship known as the **Frank–Starling law of the heart**. The Frank–Starling law of the heart equalizes the output of the right and left ventricles and keeps the same volume of blood flowing to both the systemic and pulmonary circulations. If the left side of the heart pumps a little more blood than the right side, for example, a larger volume of blood returns to the right ventricle. On the next beat the right ventricle contracts more forcefully, and the two sides are again in balance.

2. **Contractility**. The forcefulness of contraction of individual ventricular muscle fibers is called their **contractility**. Even at a constant degree of stretch, the heart can contract more or less

forcefully when certain substances are present. Stimulation by the sympathetic division of the autonomic nervous system (ANS), by hormones such as epinephrine and norepinephrine, by increased Ca^{2+} level in the interstitial fluid, and by the drug digitalis, all increase the force of contraction of cardiac muscle fibers. In contrast, inhibition of the sympathetic division of the ANS, some anesthetics, and increased K^+ levels in the extracellular fluid decrease contraction force.

3. *Afterload*. Ejection of blood from the heart begins when the semilunar valves open due to greater pressure in the right ventricle than the pulmonary trunk and greater pressure in the left ventricle than the aorta. The amount of pressure the contracting ventricles must produce to force open the SL valves is termed the **afterload**. When the required pressure is higher than normal, the valves open later than normal, stroke volume decreases, and more blood remains in the ventricles at the end of systole. Conditions that can increase afterload include hypertension (elevated blood pressure) and narrowing of arteries by atherosclerosis.

Regulation of Heart Rate

As we have seen, cardiac output depends on heart rate as well as stroke volume. Adjustments in heart rate are important in the short-term control of cardiac output and blood pressure. If left to itself, the sinoatrial node would set a constant heart rate of about 100 beats/min. However, tissues require different volumes of blood flow under different conditions. During exercise, for example, cardiac output rises to supply working tissues with increased amounts of oxygen and nutrients. Stroke volume may fall if the ventricular

myocardium is damaged or if bleeding reduces blood volume. In these cases, adequate cardiac output is maintained by increasing the heart rate. Among the several factors that contribute to regulation of heart rate, the most important are the autonomic nervous system and hormones.

Autonomic Regulation of Heart Rate

Nervous system regulation of the heart originates in the **cardiovascular center** in the medulla oblongata. This region of the brain stem receives input from a variety of sensory receptors and from higher brain centers, such as the limbic system and cerebral cortex. In response, the cardiovascular center increases or decreases the frequency of nerve impulses sent to the heart through both the sympathetic and parasympathetic branches of the ANS (Figure 19.13).

Even before physical activity begins, especially in competitive situations, heart rate may climb. This anticipatory increase occurs because the limbic system sends nerve impulses to the cardiovascular center in the medulla oblongata. As physical activity begins, **proprioceptors** that are monitoring the position of limbs and muscles send an increased frequency of nerve impulses to the cardiovascular center. Proprioceptor input is a major stimulus for the quick rise in heart rate that occurs at the onset of physical activity. Other sensory receptors that provide input to the cardiovascular center include chemoreceptors and baroreceptors. **Chemoreceptors** monitor chemical changes in the blood, and **baroreceptors** detect changes in blood pressure by monitoring the stretching of major arteries and veins caused by the pressure of the blood flowing through them.

Figure 19.13   Nervous system control of the heart.

> The cardiovascular center in the medulla oblongata controls both sympathetic and parasympathetic nerves that innervate the heart.

Sympathetic neurons extend from the cardiovascular center to the heart in **cardiac accelerator nerves**. They innervate the SA node, AV node, and most portions of the myocardium. Impulses in the cardiac accelerator nerves trigger the release of norepinephrine, which increases heart rate and force of contraction. As a result, a greater volume of blood is ejected during systole.

Also arising from the cardiovascular center are parasympathetic neurons that reach the heart via the right and left **vagus nerves**. These parasympathetic neurons terminate in the SA node, AV node, and atrial myocardium. The neurotransmitter they release—acetylcholine—decreases the heart rate.

A continually shifting balance exists between sympathetic and parasympathetic stimulation of the heart. At rest, parasympathetic stimulation predominates. The resting heart rate—about 75 beats/min—is usually lower than the autorhythmic rate of the SA node (about 100 beats/min).

Chemical Regulation of Heart Rate

Certain chemicals influence both the basic physiology of cardiac muscle and heart rate. Several hormones and cations have major effects on the heart:

1. *Hormones*. Epinephrine and norepinephrine (from the adrenal medullae) enhance the heart's pumping effectiveness by increasing both heart rate and contractility. Exercise, stress, and excitement cause the adrenal medullae to release more hormones. Thyroid hormones also enhance cardiac contractility and increase heart rate. One sign of hyperthyroidism (excessive thyroid hormone) is **tachycardia**, an elevated resting heart rate.

2. *Cations*. Given that differences between intracellular and extracellular concentrations of several cations are crucial for the production of action potentials in all nerve and muscle fibers, it is not surprising that ionic imbalances can quickly compromise the pumping effectiveness of the heart. A moderate increase in Ca^{2+} level increases heart rate and contraction force. Elevated blood levels of Na^+ or K^+ decrease heart rate and contractility. Excess Na^+ blocks Ca^{2+} inflow into cardiac muscle fibers, thereby decreasing the force of contraction. Excess K^+ decreases heart rate by blocking generation of action potentials.

Other Factors in Heart Rate Regulation

Age, gender, physical fitness, and body temperature also influence resting heart rate. A newborn baby is likely to have a resting heart rate over 120 beats/min; the rate then declines throughout life. Adult females generally have slightly higher resting heart rates than adult males, although regular exercise tends to bring resting heart rate down in both sexes. A physically fit person may even exhibit **bradycardia**, a resting heart rate under 50 beats/min. This is a beneficial effect of endurance-type training because a slowly beating heart is more energy efficient than one that beats more rapidly.

Increased body temperature, such as occurs during fever or strenuous exercise, increases heart rate, whereas decreased body temperature decreases heart rate and force of contraction. During surgical repair of certain heart abnormalities, it is helpful to slow a patient's heart rate by hypothermia (hī´-pō-THER-mē-a), deliberate cooling of the body to a low core temperature. Hypothermia also slows metabolism, which reduces the oxygen needs of the tissues, allowing the heart and brain to withstand short periods of interrupted or reduced blood flow during the procedure.

Checkpoint

28. What is cardiac output and how is it calculated?
29. What is stroke volume? Which factors regulate stroke volume?
30. Where in the central nervous system is heart rate adjusted?
31. Which sensory receptors provide feedback to the cardiovascular center? Which specific input does each receptor provide?
32. How do hormones alter heart activity?

Prakesh's Story: Epilogue and Discussion

▶▶▶ Prakesh is a late middle-aged man with a history of myocardial infarction. He underwent cardiac bypass surgery three years ago and recently had been noticing shortness of breath and swelling of his feet and ankles. He had been following a daily exercise regimen, walking one mile a day. Prakesh is overweight but claims to eat a healthy diet. He awoke with breathlessness during the last couple of nights. His blood pressure was 110/52, pulse 130, and irregular. His temperature was normal. Prakesh was found by Dr. Abbas to be very anxious but exhibited normal cognitive function. A chest X-ray revealed cardiomegaly (an enlarged heart) and pulmonary edema. Electrocardiogram testing revealed Prakesh has a left bundle branch block. The doctor suggested that Prakesh consult with his cardiologist and also recommended several treatment options including drugs that would increase the contractility of the heart, reduce peripheral vascular resistance, and decrease the fluid retention and swelling. ◀◀◀

I. What is Prakesh's primary problem; that is, the main cause of all the signs and symptoms? Consider the role of his previous heart attack.

J. Dr. Abbas has told Prakesh that his heart is enlarged. Why wouldn't increased filling of the heart cause Prakesh's heart to function more efficiently? Use the Frank–Starling law of the heart in your answer.

K. Prakesh's heart was beating very quickly. What is this condition called?

L. Why would Prakesh's heart be increasing its rate of contraction? Consider the relationship of cardiac output to stroke volume and heart rate.

Concept and Resource Summary

| Concept | Resources |

Introduction

1. The heart functions to pump blood to all body cells.
2. The study of the heart and its diseases is **cardiology**.

Concept 19.1 The heart is located in the mediastinum and has a muscular wall covered by pericardium.

See

- **Anatomy Overview – Cardiac Muscle**

1. The heart is about the size of a closed fist and is located in the mediastinum of the thoracic cavity, where it rests on the diaphragm. It is situated obliquely, with the **apex** (pointed end) directed inferiorly to the left and the **base** positioned superiorly to the right.
2. The heart is surrounded by a protective membrane, the **pericardium**, consisting of a tough superficial **fibrous pericardium** and a deeper more delicate **serous pericardium** that forms a double layer around the heart, an outer **parietal layer** and an inner **visceral layer**. A thin, slippery, lubricating **pericardial fluid** is found in the **pericardial cavity** between the parietal and visceral layers of the serous pericardium.
3. The three layers of the heart wall from superficial to deep are the **epicardium** (visceral layer of serous pericardium), the **myocardium** (cardiac muscle tissue), and the **endocardium** (endothelium overlying connective tissue). The endocardium also covers the valves of the heart.

Concept 19.2 The heart has four chambers, two upper atria and two lower ventricles.

See

- **Anatomy Overview – Heart Structures**

Hear

- **Figure 19.3 –**
  Structure of the Heart: Surface Anatomy

Do

- **Exercise – Paint the Heart**

1. There is a wrinkled, pouchlike auricle on the anterior surface of each atrium that increases blood volume capacity.
2. The heart surface has sulci (grooves) that contain coronary blood vessels and fat. The coronary sulcus encircles the heart between the superior atria and inferior ventricles. The anterior interventricular sulcus marks the anterior boundary between the right and left ventricles.
3. The **right atrium** receives blood from three veins: *superior vena cava, inferior vena cava,* and *coronary sinus.*
4. The anterior wall has muscular ridges called **pectinate muscles**. The **interatrial septum** separates the right and left atria, and has the **fossa ovalis**, a shallow depression that is a remnant of the fetal *foramen ovale*. The **tricuspid valve** lies between the right atrium and the right ventricle, where blood passes as it flows from the atrium to the ventricle.
5. The inner wall of the right ventricle has raised ridges of bundles of cardiac muscle fibers called **trabeculae carneae**. The cusps of the tricuspid valve connect to **chordae tendineae** that in turn are connected to cone-shaped **papillary muscles** of the ventricular wall. The right ventricle is separated from the left ventricle by the **interventricular septum**. Blood exiting the right ventricle passes through the **pulmonary valve** into the large pulmonary trunk that carries blood to the lungs.
6. The **left atrium** receives blood from the lungs through four *pulmonary veins*. Blood from the left atrium passes through the **bicuspid (mitral) valve** as it flows into the left ventricle.
7. The **left ventricle** forms the apex of the heart, and contains trabeculae carneae and chordae tendineae that anchor the cusps of the bicuspid valve to papillary muscles. Blood exiting the left ventricle passes through the **aortic valve** and into the *ascending aorta*. Some blood in the aorta flows into the *coronary arteries* that carry blood to the heart wall; remaining blood passes into the *arch of the aorta* and *descending aorta* that carry blood throughout the body.
8. In the fetus, the *ductus arteriosus* shunts blood from the pulmonary trunk into the aorta to bypass the nonfunctioning lungs. This vessel closes after birth, leaving a remnant called the **ligamentum arteriosum** that connects the arch of the aorta and pulmonary trunk.
9. Atrial walls are thinner than ventricular walls because their contraction only pumps blood into the ventricles. Although the right and left ventricles have thicker walls than the atria, the right ventricle, needing only to pump blood to the nearby lungs, has a thinner wall compared to the left ventricle. The muscular wall of the left ventricle is very thick because it must generate contractions that pump blood to all parts of the body.

Concept	Resources

Concept 19.3 Heart valves ensure one-way flow of blood.

1. Blood flows through the heart from areas of higher pressure to areas of lower pressure. Contraction of the atrial and ventricular walls causes increased pressure of the blood within the chambers.
2. Heart valves open and close in response to pressure differences across the valves.
3. The location of the bicuspid and tricuspid valves between the atria and ventricles has given them the name **atrioventricular (AV) valves**.
4. Contraction of the walls of an atrium results in increased pressure of the blood within it, causing blood to push open the AV valve and flow into the ventricle.
5. Contraction of the walls of a ventricle causes an increase in the pressure of ventricular blood, driving the cusps of the AV valve upward until they meet and close the valve. Contraction of the papillary muscles tightens the chordae tendineae, preventing the cusps from pushing up into the atrium. Damage to the AV valves or chordae tendineae can result in backflow of blood into the atria when the ventricles contract.
6. The **semilunar (SL) valves** are located at the origin of the pulmonary trunk and aorta. These valves permit flow of blood from the ventricles into the arteries and prevent backflow of blood into the ventricles during ventricular relaxation.

See
- Anatomy Overview – Heart Structures

Do
- Exercise – Paint the Heart

Concept 19.4 The heart pumps blood to the lungs for oxygenation, then pumps oxygen-rich blood throughout the body.

1. The heart pumps blood into two circuits with each heartbeat: the **pulmonary circulation** and the **systemic circulation**.
2. The right side of the heart is the pump for the pulmonary circulation. **Deoxygenated blood** returning from the systemic circulation enters the right atrium through the two vena cavas and the coronary sinus. Blood from the right atrium enters the right ventricle, which pumps blood into the pulmonary trunk to the lungs. Deoxygenated blood flows through pulmonary capillaries, where it unloads CO_2 and picks up O_2.
3. The left side of the heart is the pump for the systemic circulation. Freshly **oxygenated blood** from pulmonary veins enters the left atrium, then the left ventricle. Ventricular contraction ejects blood into the aorta and its branches, which carry oxygenated blood to all parts of the body except for the air sacs of the lungs. In the systemic circulation, blood flows through arteries to smaller arterioles and then into *systemic capillaries* where nutrients, gases, and wastes are exchanged. Venules carry deoxygenated blood away into larger veins and back to the right atrium of the heart.
4. The heart itself must receive blood enriched with nutrients and gases, and have wastes removed. The wall of the heart has its own blood vessels called the **coronary (cardiac) circulation** that delivers blood to and from the myocardium.
5. The principal coronary vessels are the **left** and **right coronary arteries** that arise from the ascending aorta. The **left coronary artery** divides into the **anterior interventricular branch** serving both ventricles, and the **circumflex branch** serving the left atrium and ventricle. The **right coronary artery** branches to supply the right atrium. Branches of the right coronary artery form the **posterior interventricular branch** serving walls of both ventricles, and the **marginal branch** supplying the right ventricle.
6. Arteries of the coronary circulation have many **anastomoses** (connections) that allow detours for arterial blood if a main route becomes obstructed.
7. Deoxygenated blood from the myocardium drains into the posterior **coronary sinus** that empties into the right atrium. The **greater**, **middle**, **small**, and **anterior cardiac veins** carry blood into the coronary sinus.

See
- Anatomy Overview – Heart Structures
- Anatomy Overview – Pulmonary Circulation

Hear
- Figure 19.6 – Systemic and Pulmonary Circulations

Do
- Exercise – Drag and Drop Blood Flow

Concept 19.5 The cardiac conduction system coordinates heart contractions for effective pumping.

1. The heart has specialized cardiac muscle fibers, called **autorhythmic fibers**, that are self-excitable (have inherent rhythmical electrical activity). The powerful contractions of **contractile fibers**, which comprise the majority of myocardial fibers, propel blood throughout the body.
2. The autorhythmic cells act as a **pacemaker** and form the heart's **conduction system**, which coordinates and delivers action potentials throughout the heart muscle.
3. Conduction system action potentials occur as follows: (1) Excitation is initiated at the heart's pacemaker, the **sinoatrial (SA) node**, in the right atrial wall, and action potentials propagate throughout both atria; (2) the action potential reaches the **atrioventricular (AV) node** in the interatrial septum; (3) the action potential next enters the **atrioventricular (AV) bundle**; (4) after conducting along the AV bundle, the action potential enters the **right** and **left bundle branches** through the interventricular septum; and (5) **Purkinje fibers** conduct the action potential to the apex and then upward to the remainder of the ventricular myocardium.

See
- Anatomy Overview – Heart Structures
- Animation – Cardiac Conduction

Hear
- Figure 19.8 – The Conduction System of the Heart

Do
- Exercise – Sequence Cardiac Conduction

Concept	Resources

4. The SA node initiates action potentials faster than any other autorhythmic fibers and sets the rhythm for contraction of the heart. ANS nerve impulses speed or slow pacing by the SA node but they do not establish the fundamental rhythm.

Concept 19.6 The electrocardiogram is a record of electrical activity associated with each heartbeat.

1. An **electrocardiogram** (ECG or EKG) is a recording of the electrical changes that accompany each heartbeat. An **electrocardiograph** is the instrument used to amplify the electrical activity for the ECG. Electrodes positioned on the arms, legs, and chest produce 12 tracings of electrical activity.
2. The ECG is a composite of all the action potentials produced by cardiac muscle fibers during each heartbeat and graphed as a series of three waves: (1) The **P wave** is a small upward deflection that represents depolarization; (2) the **QRS complex** begins as a downward deflection (Q), continues as a large upward triangular wave (R), and ends as a downward wave (S), representing ventricular depolarization; (3) the **T wave** represents ventricular repolarization.
3. ECG waves correlate to atrial and ventricular relaxation, or **diastole**, and contraction, or **systole**. The sequence of systole and diastole related to the ECG is as follows: (1) An action potential arises in the SA node and propagates throughout atrial cardiac muscle, causing depolarization and the P wave; (2) atrial systole occurs; (3) an action potential propagates through the AV bundle, bundle branches, Purkinje fibers, and entire ventricular myocardium, causing ventricular depolarization and the QRS complex; (4) ventricular systole begins shortly after the QRS complex appears; (5) repolarization of the ventricles begins at the apex and spreads to produce the T wave; and (6) ventricular diastole occurs.

See

• **Animation – Cardiac Cycle and ECG**

Do

• **Exercise – ECG Jigsaw Puzzle**

Concept 19.7 The cardiac cycle represents all the events associated with one heartbeat.

1. The heart's electrical signals cause alternating contraction and relaxation of the atria and ventricles. During each cardiac cycle, contraction of the heart chambers causes an increase in the blood pressure within the chambers, which pushes blood to areas of lower pressure (ventricles, aorta, pulmonary trunk).
2. Although pressures are higher on the left side of the heart compared to the right side, each ventricle ejects the same volume of blood per beat.
3. The events during a **cardiac cycle** include the following: (1) depolarization of SA node, atrial depolarization, and P wave; (2) atrial systole, with increased pressure of chamber blood, forcing it across open AV valves into relaxed ventricles; (3) filling of each ventricle with an **end-diastole volume (EDV)** of 130 mL of blood while AV valves are open and SL valves are closed; (4) QRS complex, which signals the onset of ventricular depolarization; (5) **atrial diastole** and **ventricular systole**, which push blood against the AV valves so they close while SL valves briefly remain closed in a period of **isovolumetric contraction**; (6) increased pressure inside the chambers, opening of both SL valves, and **ventricular ejection**; (7) (at rest) ejection of 70 mL of blood from each ventricle; (8) ventricular repolarization and T wave; (9) **ventricular diastole** with a pressure decrease that causes a backflow of blood from the pulmonary trunk and aorta, trapping of blood in the semilunar cusps that rebounds to cause the **dicrotic wave**, and closing of SL valves during **isovolumetric relaxation**; and (10) ventricular diastole, which causes a pressure drop, AV valves to open, and **ventricular filling** to begin.
4. Listening to sounds within the body, usually with a stethoscope, is called **auscultation**.
5. Using a stethoscope, the first two of the four **heart sounds** of a cardiac cycle are loud enough to be heard. The first sound (S1) is heard as a **lubb** when the AV valves close after ventricular systole begins. The second sound (S2) is heard as a **dupp** when the SL valves close at the end of ventricular systole. The heart sounds arise primarily from turbulence in blood flow created by closure of the valves.

See

• **Animation – Cardiac Cycle**

Do

• **Concepts and Connections – Cardiac Cycle**
• **Exercise – Cardiac Cycle**

Concept 19.8 Cardiac output is the blood volume ejected by a ventricle each minute.

1. **Cardiac output** is the volume of blood ejected each minute from the left ventricle into the aorta; the same amount of blood is also ejected by the right ventricle into the pulmonary trunk.
2. **Stroke volume** is the volume of blood ejected by the ventricle with each contraction (beat). **Heart rate** is the number of heartbeats per minute. Cardiac output = stroke volume × heart rate. The average cardiac output in a resting adult is about 5 L per minute. Exercise increases stroke volume, heart rate, and cardiac output.

See

• **Animation – Cardiac Output Factors**

Do

• **Exercise – Cardiac Ouput**
• **Concepts and Connections – Cardiac Output**

3. Three factors regulate stroke volume and ensure that the left and right ventricles pump equal volumes of blood: (1) **Preload**, or the degree to which the heart is stretched before it contracts; greater stretch in diastole leads to greater force of contraction during systole—a relationship known as the **Frank–Starling law of the heart**; (2) **contractility**, or the forcefulness of contraction of individual ventricular muscle fibers, is regulated by the sympathetic ANS division, hormones (epinephrine and norepinephrine), ions (Ca^{2+}, K^+), the drug digitalis, and some anesthetics; and (3) **afterload**, or the amount of pressure the contracting ventricles must produce to force open the SL valves and eject blood. Conditions such as hypertension and atherosclerosis increase afterload.

4. The **cardiovascular center** in the medulla oblongata is the origin of nervous system regulation of the heart. This region receives input from **proprioceptors**, **chemoreceptors**, **baroreceptors**, and from the limbic system and cerebral cortex.

5. Sympathetic neurons arise from the cardiovascular center and reach the heart via **cardiac accelerator nerves** that innervate the conduction system, atria, and ventricles. They release norepinephrine to increase heart rate.

6. Parasympathetic neurons arise from the cardiovascular center and reach the heart via the right and left **vagus nerves**, which innervate the conduction system and the atria, releasing acetylcholine to decrease heart rate.

7. Chemicals that influence cardiac muscle and its contraction include: (1) the **hormones** epinephrine, norepinephrine, and thyroid hormones that increase heart rate and contractility; and (2) **cations** that modify heart rate and contractility. Elevated Na^+ or K^+ decreases heart rate and contractility, while elevated Ca^{2+} increases heart rate and contractility.

8. Other factors that influence heart rate include age, gender, physical fitness, and body temperature.

Other Chapter Resources

For additional information on clinical applications, medical disorders, and medical terminology, see Chapter 9 in the *Clinical Connections* manual packaged with your text.

Within *WileyPLUS*

- Chapter Quiz
- Visual Anatomy
- Mastering Vocabulary
- Audio Glossary
- Homeostatic Imbalance: The Case of the Man with a Pain in his Jaw
- Homeostatic Imbalance: The Case of the Man with Labored Breathing
- Web Links

• Understanding the Concepts

1. Which chamber of the heart has the thickest wall? Why?

2. Describe the sequence of heart chambers, heart valves, and blood vessels a drop of blood encounters from the time it flows into the right atrium until it reaches the aorta. When during its journey does the drop of blood contain the most oxygen? The least oxygen?

3. Why is blood flowing through the chambers within the heart unable to supply sufficient oxygen or remove enough carbon dioxide from the myocardium? What is the body's solution to this problem?

4. Describe the status of each of the four heart valves during ventricular filling. How would a defect in the atrioventricular valve affect ventricular filling?

5. Why must ventricular pressure be greater than arterial pressure during ventricular ejection?

6. Which events cause the AV valves to open and shut? Which events make the SL valves open and shut?

7. Will stroke volume increase or decrease in each of the following situations? Explain each of your answers.
   - Your blood pressure rises when you're angry.
   - An anesthetic decreases the strength of contraction of your ventricles.
   - When you are exercising, contraction of skeletal muscles returns more blood to the heart.

8. What mechanism does the cardiovascular center use to speed up or slow down heart rate?

20

The Cardiovascular System: Blood Vessels

Sandy got up early to get ready for school. As she made coffee in the small kitchenette of the apartment where she and her grandfather lived, she heard him coughing. "Grandpa David, it's time to get up," she called as she finished putting water in the coffeepot.

"I'm just going to stay in here a while. I've got a headache," replied David.

Sandy thought it was odd. Her grandfather usually got up early, much earlier than she did. She walked down the hallway and knocked on his door. "Grandpa, are you all right?" she asked as she entered his room.

"Just leave me alone. I'm fine, just a little under the weather today," he answered.

She walked over to him and put her hand on his head. His pale skin was cool and clammy, and a strange mottled rash covered his face and neck. She noticed that he was breathing rapidly, and felt his pulse racing in the artery that ran along his temple. "Grandpa David, I think you're really sick. Maybe you should go to the doctor."

David grabbed his blanket and turned away from her. "It's just the flu. I threw up last night. I'll be fine. I just need some rest. Now go to school. Tell your Grandma to come in here. I want to talk to her."

Sandy stared at him and said, "Grandpa, she's not here." Sandy wasn't sure what to do. Her grandmother had been dead for 10 years. She felt scared. Something was not right—Grandpa David was sick, and confused. "Grandpa David, I'm going to see if Mrs. Mahoney can drive us to the clinic."

The cardiovascular system contributes to homeostasis of other body systems by transporting and distributing blood throughout the body, delivering oxygen and nutrients, and carrying away wastes. Defects in the functioning of the cardiovascular system can have profound effects on the overall health of the body. Such defects can take many forms, and, as we will see in David's case, failure of the cardiovascular system to deliver enough oxygen and nutrients to the body can have significant consequences.

INTRODUCTION

The cardiovascular system transports and distributes blood throughout the body to deliver materials (such as oxygen, nutrients, and hormones) and carry away wastes. Blood vessels form a closed system of tubes for the flow of blood to and from the heart and the exchange of nutrients and wastes in tissues.

They also play an important role in adjusting the velocity

Blood vessels form a closed system of tubes for the flow of blood.

and volume of blood flow. In Chapters 18 and 19 we discussed the composition and functions of blood and the structure and function of the heart. In this chapter we examine the structure and functions of the various types of blood vessels (arteries, arterioles, capillaries, venules, and veins); the forces involved in circulating blood throughout the body; and the blood vessels that constitute the major circulatory routes.

20.1   Most blood vessel walls have three distinct tissue layers.

The wall of a generalized blood vessel consists of three layers, or tunics, of different tissues. The three structural layers from innermost to outermost are the *tunica interna*, *tunica media*, and *tunica externa* (Figure 20.1). Modifications of this basic design account for the structural and functional differences among the various blood vessel types. These structural variations correlate to the differences in function that occur throughout the cardiovascular system.

Tunica Interna

The **tunica interna** (*tunic* = garment; *interna* = innermost) forms the thin inner lining of a blood vessel and is in direct contact with blood as it flows through the **lumen**, or interior opening, of the vessel (Figure 20.1a,b). Its innermost layer, the **endothelium**, is a single layer of flattened cells forming a simple squamous epithelium that lines the inner surface of the entire cardiovascular system (heart and blood vessels). Endothelial cells are active participants in a variety of vessel-related activities, including influencing blood flow, secreting chemical mediators that influence vessel contraction, and assisting with capillary permeability. Deep to the endothelium is a *basement membrane*. It anchors the endothelium to the underlying connective tissue and provides a physical support base that imparts significant tensile strength and resilience for stretching and recoil. The outermost part of the tunica interna is the *internal elastic lamina* (*lamina* = thin plate), a thin sheet of elastic fibers with windowlike openings that give it the look of Swiss cheese. These openings facilitate diffusion of materials through the tunica interna to the thicker tunica media.

Figure 20.1   Comparative structure of blood vessels. The capillary in (c) is enlarged relative to the structures shown in parts (a) and (b).

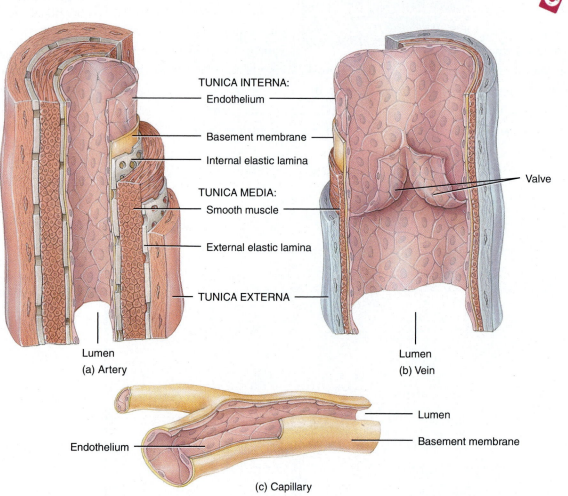

TUNICA INTERNA:
Endothelium
Basement membrane
Internal elastic lamina

Valve

TUNICA MEDIA:
Smooth muscle

External elastic lamina

TUNICA EXTERNA

Lumen
(a) Artery

Lumen
(b) Vein

Lumen
Endothelium
Basement membrane

(c) Capillary

Tunica Media

The **tunica media** (*media* = middle) is a relatively thick layer composed mainly of smooth muscle cells and substantial amounts of elastic fibers (Figure 20.1a,b). The primary role of the smooth muscle cells, which extend circularly around the lumen like a ring encircles your finger, is to regulate the diameter of the lumen. As you will learn in more detail shortly, smooth muscle contraction of the tunica media in some blood vessels regulates blood flow through different parts of the body and monitors blood pressure. In addition to regulating blood flow and blood pressure, smooth muscle in the tunica media contracts when vessels are damaged to help limit loss of blood through the injured vessel. The tunica media is the most variable of the tunics. As you study the different types of blood vessels, you will see that the structural differences in this layer account for the many variations in function among the different vessel types.

Tunica Externa

The outer covering of a blood vessel, the **tunica externa** (*externa* = outermost), consists of elastic and collagen fibers (Figure 20.1a,b). Separating the tunica externa from the tunica media is a network of elastic fibers called the *external elastic lamina*, which is part of the tunica media. The tunica externa contains numerous nerves and, especially in larger vessels tiny blood vessels, that supply blood to the tissue of the vessel wall. The tunica externa also helps anchor the vessel to surrounding tissues.

Next we investigate the types of blood vessels: arterioles, capillaries, venules, and veins.

LM 200x

(d) Transverse section through a muscular artery

Internal elastic lamina

External elastic lamina

Tunica externa

Lumen with blood cells

Tunica interna

Tunica media

Connective tissue

Connective tissue

Red blood cell

Capillary endothelial cells

LM 600x

(e) Red blood cells passing through a capillary

🔑 Arteries carry blood from the heart to tissues; veins carry blood from tissues back to the heart.

20.2 Blood ejected from the heart flows through elastic arteries, muscular arteries, and then arterioles.

Arteries (AR-ter-ēz) carry *blood away from the heart* (arteries = *a*way) to other organs. Because arteries (*ar-* = air; *-ter-* = to carry) were found empty at death, in ancient times they were thought to contain only air. The wall of an artery has the three layers of a typical blood vessel, but has a thick tunica media (see Figure 20.1a).

Sympathetic axons of the autonomic nervous system (ANS) innervate the smooth muscle of arteries. An increase in sympathetic stimulation typically stimulates the smooth muscle to contract, squeezing the arterial wall and narrowing the lumen. Such a decrease in the diameter of the lumen of a blood vessel is called **vasoconstriction**. In contrast, when sympathetic stimulation decreases, or in the presence of certain chemicals, smooth muscle fibers relax. The resulting increase in lumen diameter is called **vasodilation**. Additionally, when an artery or arteriole is damaged, its smooth muscle contracts, producing vascular spasm of the vessel called a **vasospasm**. Vasospasm limits blood flow through the damaged vessel and helps reduce blood loss if the vessel is small.

Elastic Arteries

Elastic arteries are the largest diameter arteries in the body, ranging from the garden hose–sized aorta and pulmonary trunk to the finger-sized branches of the aorta. Their vessel walls (approximately 10% of the vessel's total diameter) are relatively thin compared to the overall size of the vessel. These vessels are characterized by a thick tunica media that is dominated by elastic fibers. Elastic arteries leave the heart and perform an important function:

They help propel blood onward while the ventricles are relaxing. As blood is ejected from the heart into elastic arteries, their highly elastic walls stretch, easily accommodating the surge of blood. As they stretch, the elastic fibers momentarily store mechanical energy, functioning as a **pressure reservoir** (Figure 20.2a). Then, while the ventricles are relaxing, the elastic fibers in the artery walls recoil, forcing blood onward toward the smaller arteries (Figure 20.2b). Because they conduct blood from the heart to muscular arteries, elastic arteries also are called *conducting arteries*. Elastic arteries include the two major trunks that exit the heart (the aorta and the pulmonary trunk), along with the aorta's major initial branches, such as the brachiocephalic, subclavian, common carotid, and common iliac arteries (see Figure 20.16a).

Muscular Arteries

Large elastic arteries branch into medium-sized **muscular arteries**. The tunica media of muscular arteries contains more smooth muscle and fewer elastic fibers than elastic arteries. The large amount of smooth muscle, approximately three-quarters of the total mass, makes the walls of muscular arteries relatively thick. Thus, muscular arteries are capable of greater vasoconstriction and vasodilation to adjust the rate of blood flow. Muscular arteries span a range of sizes from the pencil-sized femoral and axillary arteries to string-sized arteries that enter organs. Compared to elastic arteries, the walls of muscular arteries comprise a larger percentage (approximately 25%) of the total vessel diameter. Muscular arteries branch out into the various regions of the body.

Figure 20.2   Pressure reservoir function of elastic arteries.

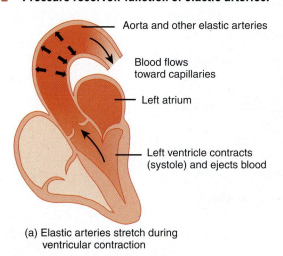

(a) Elastic arteries stretch during ventricular contraction

- Aorta and other elastic arteries
- Blood flows toward capillaries
- Left atrium
- Left ventricle contracts (systole) and ejects blood

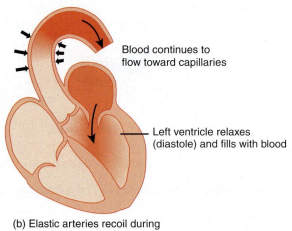

(b) Elastic arteries recoil during ventricular relaxation

- Blood continues to flow toward capillaries
- Left ventricle relaxes (diastole) and fills with blood

Recoil of elastic arteries keeps blood flowing during ventricular relaxation (diastole).

Because the muscular arteries distribute blood to each of the various organs, they are called *distributing arteries*. Examples include the brachial artery in the arm and radial artery in the forearm (see Figure 20.18b).

Anastomoses

Most tissues of the body receive blood from more than one artery. The union of the branches of two or more arteries supplying the same body region is called an **anastomosis** (a-nas-tō-MŌ-sis = connecting; plural is *anastomoses*). Anastomoses between arteries provide alternative routes for blood to reach a tissue or organ. If blood flow stops for a short time when normal movements compress a vessel, or if a vessel is blocked by disease, injury, or surgery, then circulation to a part of the body is not necessarily stopped. The alternative circulatory route of blood flow to a body part through an anastomosis is known as **collateral circulation**. Anastomoses may also occur between veins and between arterioles and venules. Arteries that do not anastomose are known as **end arteries**. Obstruction of an end artery interrupts the blood supply to a whole segment of an organ, producing necrosis (death) of that segment. Alternative blood routes may also be provided by nonanastomosing vessels that supply the same region of the body.

Arterioles

Muscular arteries branch into small arteries, which in turn divide into still smaller arteries called **arterioles** (ar-TĒR-ē-ōls). Literally meaning small arteries, arterioles are abundant microscopic vessels that regulate the flow of blood into the capillary networks of the body's tissues (Figure 20.3). The wall thickness of arterioles is about one-half of the total vessel diameter. Arterioles have a thin tunica interna with a thin internal elastic lamina that disappears at the terminal end. Arterioles have a tunica media that consists of one to two layers of smooth muscle cells having a circular orientation in the vessel wall. The terminal end of the arteriole, the region called the **metarteriole** (*meta-* = after), tapers toward the capillary junction. At the metarteriole–capillary junction, the distalmost muscle cell forms the **precapillary sphincter** (= to bind tight), which monitors the blood flow into the capillary. The tunica externa of the arteriole contains abundant sympathetic nerves. These nerves, along with the actions of local chemical mediators, can alter the diameter of arterioles to vary the flow of blood through these vessels.

Arterioles play a key role in regulating blood flow from arteries into capillaries by regulating *resistance*, the opposition to blood flow. In a blood vessel, resistance is due mainly to friction between blood and the inner walls of blood vessels. When the blood

Figure 20.3 Arterioles, capillaries, and venules. Precapillary sphincters regulate the flow of blood through capillary beds.

(a) Sphincters relaxed: blood flowing through capillary bed

(b) Sphincters contracted: blood flowing through thoroughfare channel

🔑 **In capillaries, nutrients, gases, and wastes are exchanged between the blood and interstitial fluid.**

vessel diameter is smaller, the friction is greater. Contraction of the smooth muscle of an arteriole causes vasoconstriction, which increases resistance and decreases blood flow into capillaries supplied by that arteriole. By contrast, relaxation of the smooth muscle of an arteriole causes vasodilation, which decreases resistance and increases blood flow into capillaries. A change in diameter of arterioles can also affect blood pressure: vasoconstriction of arterioles increases blood pressure, and vasodilation of arterioles decreases blood pressure.

Checkpoint

1. What is the function of elastic fibers and smooth muscle in the tunica media of arteries?
2. How do elastic arteries and muscular arteries differ in structure?
3. Contrast the functions of elastic arteries, muscular arteries, and arterioles.
4. What is the relationship between anastomoses and collateral circulation?

20.3 Capillaries are microscopic blood vessels that function in exchange between blood and interstitial fluid.

As arterioles enter a tissue, they branch into numerous tiny vessels called **capillaries** (KAP-i-lar´-ēz; *capillar-* = hairlike). Capillaries, the smallest of blood vessels, are microscopic vessels that form the U-turns that connect arterial outflow from the heart to venous return (see Figure 20.3). Capillaries form an extensive network of branched, interconnecting blood vessels that course among individual body cells. The primary function of capillaries is the exchange of substances between the blood and the interstitial fluid bathing body cells. Capillaries are found near almost every cell in the body, but their number varies with the metabolic activity of the tissue they serve. Body tissues with high metabolic requirements, such as muscles, the liver, the kidneys, and the nervous system, use more oxygen and nutrients and thus have extensive capillary networks. Tissues with lower metabolic requirements, such as tendons and ligaments, contain fewer capillaries. Capillaries are absent in a few tissues, for instance, the cornea and lens of the eye, and cartilage.

can flow through a capillary network from an arteriole into a venule as follows:

1. *Capillaries.* In this route, blood flows from an arteriole into capillaries and then into venules. As noted earlier, at the junctions between the metarteriole and the capillaries are precapillary sphincters that control the flow of blood through the capillaries. When the precapillary sphincters are relaxed (open), blood flows into the capillaries (see Figure 20.3a); when precapillary sphincters contract (close), blood flow through the capillaries stops or decreases (see Figure 20.3b).

2. *Thoroughfare channel.* The proximal end of a metarteriole is surrounded by scattered smooth muscle fibers whose contraction and relaxation help regulate blood flow. The distal end of the vessel has no smooth muscle and is called a **thoroughfare channel**. Such a channel provides a direct route for blood from an arteriole to a venule, thus bypassing capillaries.

Structure of Capillaries

The primary function of capillaries is the exchange of substances between the blood and interstitial fluid. The thin walls of capillaries are well suited to this function. Capillaries have no tunica media or tunica externa (see Figure 20.1c). Capillary walls are composed of only a single layer of endothelial cells and a basement membrane. Exchange of materials occurs only through the walls of capillaries and the beginning of venules; the walls of arteries, arterioles, most venules, and veins present too thick a barrier. Capillaries form extensive branching networks that increase the surface area available for rapid exchange of materials.

Throughout the body, capillaries function as part of a **capillary bed** (see Figure 20.3a), a network of 10–100 capillaries that arises from a single metarteriole. In most parts of the body, blood

Autoregulation of Capillary Blood Flow

In each capillary bed, local changes can regulate vasodilation and vasoconstriction. When vasodilators are released by tissue cells, they cause dilation of nearby arterioles and relaxation of precapillary sphincters. Then, blood flow into the capillary networks increases, and O_2 delivery to the tissue rises. Vasoconstrictors have the opposite effect. The ability of a tissue to automatically adjust its blood flow to match its metabolic demands is called **autoregulation**. In tissues such as the heart and skeletal muscle, where the demand for O_2 and nutrients and for the removal of wastes can increase as much as tenfold during physical activity, autoregulation is an important contributor to increased blood flow through the tissue. Autoregulation also controls regional blood flow in the brain; blood distribution to various parts of the brain changes dramatically for different mental and physical activities. During a conversation, for example, blood flow increases to your motor speech areas when you are talking and increases to the auditory areas when you are listening.

Types of Capillaries

There are three different types of capillaries: continuous capillaries, fenestrated capillaries, and sinusoids (Figure 20.4). Most capillaries are **continuous capillaries**, in which the plasma membranes of endothelial cells form a continuous tube that is interrupted only by *intercellular clefts*, gaps between neighboring endothelial cells (Figure 20.4a). Most areas of the brain contain continuous capillaries, which allow only a few substances to move across their walls. Because of the "tight" nature of these capillary walls, the resulting blockade to movement of materials into and out of brain capillaries is known as the *blood–brain barrier* (see Concept 13.2).

Some capillaries are **fenestrated capillaries** (*fenestr-* = window). The plasma membranes of the endothelial cells in these capillaries contain many *fenestrations*, small pores (holes) (Figure 20.4b). The many fenestrations allow these capillaries greater exchange of materials. Fenestrated capillaries are found in the kidneys, where filtration of blood actively occurs.

Sinusoids have endothelial cells with large fenestrations, an incomplete or absent basement membrane, and large intercellular clefts (Figure 20.4c). Sinusoid walls are so leaky that large molecules and even blood cells can pass from a tissue into the bloodstream. For example, newly formed blood cells enter the bloodstream through the sinusoids of red bone marrow.

Usually blood passes from the heart and then in sequence through arteries, arterioles, capillaries, venules, and veins and then back to the heart. In some parts of the body, however, blood passes from one capillary network into another through a vein called a portal vein. Such a circulation of blood is called a **portal system**.

Capillary Exchange

The mission of the entire cardiovascular system is to keep blood flowing through capillaries to allow **capillary exchange**, the movement of substances between blood and interstitial fluid. The 7% of the blood in systemic capillaries at any given time is continually exchanging materials with interstitial fluid.

Diffusion

Most capillary exchange occurs by simple diffusion. Many substances, such as oxygen (O_2) and carbon dioxide (CO_2), enter and leave capillaries by simple diffusion. Because O_2 and nutrients are normally present in higher concentrations in blood, they diffuse down their concentration gradients into interstitial fluid and then

Figure 20.4 Types of capillaries.

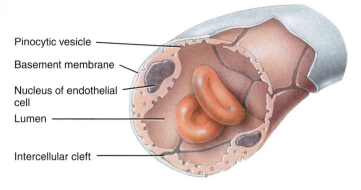

(a) Continuous capillary formed by endothelial cells

(b) Fenestrated capillary

(c) Sinusoid

🔑 **Capillaries are microscopic blood vessels that connect arterioles and venules.**

into body cells. CO_2 and other wastes released by body cells are present in higher concentrations in interstitial fluid, so they diffuse into blood.

Water-soluble substances diffuse across capillary walls through intercellular clefts or fenestrations (Figure 20.4). Lipid-soluble materials diffuse across capillary walls directly through the lipid bilayer of endothelial cell plasma membranes. Most plasma proteins and red blood cells cannot pass through capillary walls of continuous and fenestrated capillaries because they are too large to fit through the intercellular clefts and fenestrations. In sinusoids, however, the intercellular clefts are so large that they allow even proteins and blood cells to pass through their walls.

Transcytosis

A small quantity of material crosses capillary walls by **transcytosis** (*trans-* = across; *-cyt-* = cell; *-osis* = process). In this process, substances in blood plasma become enclosed within tiny pinocytic vesicles that first enter endothelial cells by endocytosis, then move across the cell and exit on the other side by exocytosis. This method of transport is important mainly for large, lipid-insoluble molecules that cannot cross capillary walls in any other way. For example, certain antibodies and proteins pass from the maternal circulation into the fetal circulation by transcytosis.

Influences on Capillary Exchange

The exchange of gases, nutrients, hormones, and metabolic waste products between capillary blood and interstitial fluid occurs due to pressure differences. **Capillary blood pressure**, the pressure of blood against the walls of capillaries, "pushes" fluid out of capillaries into interstitial fluid. An opposing pressure, termed **colloid osmotic pressure**, "pulls" fluid into capillaries. (Recall that an osmotic pressure is the pressure of a fluid due to its solute concentration. The higher the solute concentration, the greater is the osmotic pressure.) Most solutes exchange freely and so are present in nearly equal concentrations in blood and interstitial fluid. Plasma proteins, however, are too large to pass through either fenestrations or gaps between endothelial cells. The presence of proteins in plasma and their virtual absence in interstitial fluid gives blood the higher osmotic pressure. Colloid osmotic pressure is osmotic pressure due mainly to plasma proteins.

Whether fluids leave or enter capillaries depends on the balance of pressures. If the pressures that push fluid out of capillar-

Figure 20.5   Capillary exchange.

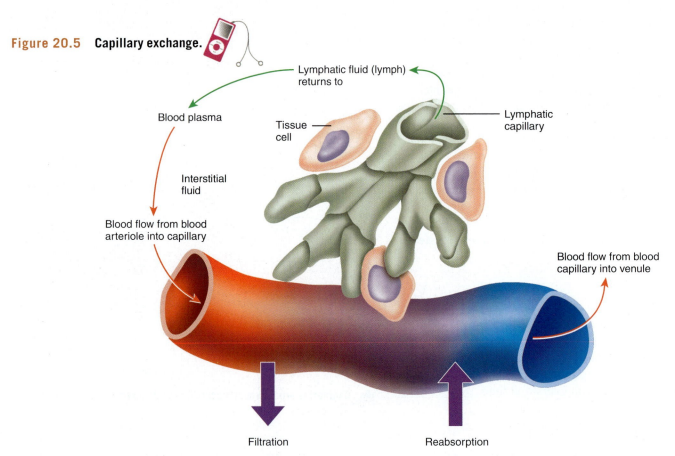

Filtration   Reabsorption

🔑 Capillary blood pressure pushes fluid out of capillaries (filtration); blood colloid osmotic pressure pulls fluid into capillaries (reabsorption).

ies exceed the pressures that pull fluid into capillaries, fluid will move *from* capillaries *into* interstitial spaces, a movement called **filtration** (Figure 20.5). If, however, the pressures that push fluid out of interstitial spaces into capillaries exceed the pressures that pull fluid out of capillaries, then fluid will move *from* interstitial spaces *into* capillaries, a process termed **reabsorption**. At the arterial end of a capillary, capillary blood pressure is higher than blood osmotic pressure. Thus, water and solutes are filtered out of the blood capillary into the surrounding interstitial fluid. Capillary blood pressure decreases progressively as blood flows along a capillary. At the venous end of a capillary, blood pressure drops below colloid osmotic pressure. Then, water and solutes are reabsorbed from interstitial fluid into the blood capillary. On average, about 85% of the fluid filtered out of capillaries is reabsorbed back into capillaries. The excess filtered fluid and the few plasma proteins

that do escape from blood into interstitial fluid enter lymphatic capillaries (Figure 20.5) and eventually are returned by the lymphatic system to the cardiovascular system.

Checkpoint

5. How do materials move through capillary walls?
6. How is blood flow directed through a capillary bed?
7. What is autoregulation?
8. Contrast the structure and function of the three types of capillaries.
9. State the ways substances enter and leave blood plasma.
10. What happens to excess filtered fluid and proteins that are not reabsorbed back into capillaries?
11. How do capillary blood pressure and colloid osmotic pressure determine fluid movement across the walls of capillaries?

20.4 Venules and veins return blood to the heart.

Venules

Groups of capillaries within a tissue reunite to form small vessels called **venules** (VEN-ūls = little veins). Unlike their thick-walled arterial counterparts, venules and veins have thin walls that do not readily maintain their shape. Venules drain the capillary blood and begin the return flow of blood back toward the heart (see Figure 20.3). The smallest venules, those closest to the capillaries, have loosely organized intercellular junctions (the weakest endothelial contacts encountered along the entire vascular tree) and thus are very porous. Venules function as sites of exchange of nutrients and wastes and movement of white blood cells from the bloodstream into an inflamed or infected tissue. As venules continue to enlarge, they acquire thicker walls across which exchanges with the interstitial fluid can no longer occur.

Veins

Veins (VĀNZ) are the blood vessels that convey blood from the tissues *back to the heart*. Veins have very thin walls relative to their total diameter (average thickness is less than one-tenth of the vessel diameter). Although veins are composed of essentially the same three coats as arteries, the relative thicknesses of the layers are different (see Figure 20.1b). The tunica interna and tunica media of veins is much thinner than in arteries. The tunica externa of veins is the thickest layer and consists of collagen and elastic fibers. Because they lack the external or internal elastic laminae found in arteries, veins are distensible enough to adapt to variations in the volume and pressure of blood passing through them, but they are not structured to withstand high pressure. The lumen

of a vein is larger than that of a comparable artery, and their thinner walls often result in veins appearing collapsed (flattened) when sectioned.

By the time blood leaves capillary beds and moves into veins, it has lost a great deal of pressure. Therefore, the average blood pressure in veins is considerably lower than in arteries. The difference in pressure can be noticed when blood flows from a cut vessel. Blood leaves a cut vein in an even, slow flow but spurts rapidly from a cut artery. Most of the structural differences between arteries and veins reflect this pressure difference. For example, the walls of veins are not as strong as those of arteries. Many veins, especially those in the limbs, contain **valves**, thin folds of tunica interna that form flaplike cusps (see Figure 20.1b). The valve cusps project into the lumen, pointing toward the heart (Figure 20.6). The low blood pressure in veins allows blood returning to the heart to slow and even back up; the valves aid in venous return by preventing the backflow of blood.

A **venous sinus** is a vein with a thin endothelial wall that has no smooth muscle to alter its diameter. In a venous sinus, the surrounding dense connective tissue replaces the tunica media and tunica externa in providing support. For example, the coronary sinus of the heart is a venous sinus that conveys deoxygenated blood from the heart wall to the right atrium (see Figure 19.3c).

Venous Return

Venous return, the volume of blood flowing back to the heart through the systemic veins, is primarily assisted by blood pressure generated by contraction of the heart's ventricles. Blood pressure is measured in millimeters of mercury and abbreviated

Figure 20.6 **Venous valves.**

Photographs of a valve in a vein

🔑 Valves in veins allow blood to flow in one direction only—toward the heart.

mm Hg. Although small, the blood pressure difference from venules (averaging about 16 mm Hg) to the right atrium (near 0 mm Hg) normally is sufficient to cause venous return to the heart. For example, when you stand up at the end of an anatomy and physiology lecture, the pressure pushing blood up the veins in your lower limbs is barely enough to overcome the force of gravity pushing it back down. Besides the remaining blood pressure generated by the heart, two other mechanisms help boost venous return by "pumping" blood from the lower body back to the heart: (1) the skeletal muscle pump and (2) the respiratory pump. Both pumps depend on the presence of valves in veins.

The **skeletal muscle pump** operates as follows (Figure 20.7):

1 While you are standing at rest, both the venous valve closer to the heart and the one farther from the heart in this part of the leg are open, and blood flows upward toward the heart.

2 Contraction of leg muscles, such as when you stand on tiptoes or take a step, compresses the vein. The compression pushes blood through the valve closer to the heart, an action called *milking*. At the same time, the valve farther from the

heart in the uncompressed segment of the vein closes as some blood is pushed back against it. People who are immobilized through injury or disease lack these contractions of leg muscles. As a result, their venous return is slower and they may develop circulation problems.

3 Just after muscle relaxation, pressure falls in the previously compressed section of vein, which causes the valve closer to the heart to close. The valve farther from the heart now opens because blood pressure in the foot is higher than in the leg, and the vein fills with blood from the foot.

The **respiratory pump** is also based on alternating compression and decompression of veins. During inhalation the diaphragm moves downward, which causes a decrease in pressure in the thoracic cavity and an increase in pressure in the abdominal cavity. As a result, abdominal veins are compressed, and a greater volume of blood moves from the compressed abdominal veins into the decompressed thoracic veins and then into the right atrium. When the pressures reverse during exhalation, the valves in the veins prevent backflow of blood from the thoracic veins to the abdominal veins.

A summary of the distinguishing features of blood vessels is presented in Table 20.1.

Figure 20.7 **Action of the skeletal muscle pump in returning blood to the heart.** Steps are described in the text.

🔑 Milking refers to skeletal muscle contractions that drive venous blood toward the heart.

TABLE 20.1

Distinguishing Features of Blood Vessels

	DIAMETER	TUNICA INTERNA	TUNICA MEDIA	TUNICA EXTERNA	FUNCTION
Elastic arteries	Greater than 1 cm	Endothelium, basement membrane, and incomplete internal elastic lamina.	Smooth muscle and higher proportion of elastic fibers and thin external elastic lamina.	Collagen and elastic fibers.	Conduct blood from the heart to muscular arteries.
Muscular arteries	0.1–10 mm	Endothelium, basement membrane, and thin internal elastic lamina.	Higher proportion of smooth muscle, fewer elastic fibers, and prominent external elastic lamina.	Collagen and elastic fibers.	Distribute blood to arterioles.
Arterioles (near arteries from which they branch)	10–100 μm	Endothelium, basement membrane, and internal elastic lamina.	Smooth muscle and very few elastic fibers.	Collagen and elastic fibers.	Deliver blood to capillaries and help regulate blood flow.
Capillaries	4–10 μm	Endothelium and basement membrane.	None.	None.	Permit exchange of nutrients and wastes between blood and interstitial fluid.
Venules (closer to convergence with veins)	10–100 μm	Endothelium and basement membrane.	Smooth muscle.	Collagen and elastic fibers.	Collect blood from capillaries and pass it on to veins.
Veins	0.1 mm–greater than 1 mm	Endothelium and basement membrane; contains valves.	Smooth muscle and elastic fibers.	Collagen and elastic fibers.	Return blood to the heart, facilitated by valves in veins in limbs.

Blood Distribution

The largest portion of your blood volume at rest—about 64%—is in systemic veins and venules (Figure 20.8). Systemic arteries and arterioles hold about 13% of the blood volume, pulmonary blood vessels hold about 9%, systemic capillaries hold about 7%, and the heart chambers hold about 7%. Because systemic veins and venules contain a large percentage of the blood volume, they function as **blood reservoirs** from which blood can be diverted quickly if the need arises. For example, when there is increased muscular activity, the cardiovascular center in the brain stem sends more sympathetic impulses to veins. As a result, veins constrict, reducing the volume of blood in reservoirs and allowing a greater blood volume to flow to skeletal muscles, where it is needed most. A similar mechanism operates in cases of hemorrhage; when blood volume and pressure decrease, veins constrict to help counteract the drop in blood pressure. Among the principal blood reservoirs are the veins of the abdominal organs (especially the liver and spleen) and the veins of the skin.

Figure 20.8 Blood distribution in the cardiovascular system at rest.

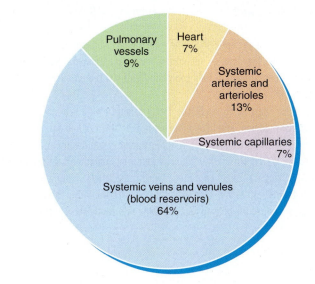

Because systemic veins and venules contain more than half the total blood volume, they are called blood reservoirs.

Checkpoint

12.   Which vessel—the femoral artery or the femoral vein—has a thicker wall? Which has a wider lumen?

13.   What are the primary structural and functional differences between arteries, capillaries, and veins?

14.   Explain how venous blood is returned to the heart.

15.   Referencing Figure 20.8, if your total blood volume is 5 liters, what volume is in your venules and veins right now? In your capillaries?

Return to David's Story

Sandy left her grandfather to get her neighbor, Mrs. Mahoney. Mrs. Mahoney knocked on the door. "David, you don't look so hot. How long has this been going on?" She felt his forehead. He felt cool, he looked pale, and his hands felt cold. "I think Sandy is right. You need to get to the clinic. Better to be safe than sorry."

Reluctantly, David went to the clinic with Sandy and Mrs. Mahoney. The nurse found David's blood pressure to be quite low, only 90/60. His pulse was a rapid 130, and his respirations were rapid. "David, how long have you been feeling poorly?" she asked.

"About the last week or so I've been really tired. I've been helping a friend tear down his garage."

"Is that how you got those scrapes on your arms?" the nurse asked. She looked at David's arm; one of the wounds was red and puffy, possibly infected. "How long ago did that happen, David?"

"About two weeks ago."

She made a note of the accident on his chart. "David, have you been having any trouble urinating in the last few weeks?"

"Yeah," he said, looking down at his hands. "But at my age, that's nothing new. And I did have a fever about a week ago, but it went away."

Dr. Rolandi saw the nurse exit the exam room. "So who's in room three, Gail?"

"David, possible septic shock; he's got a nasty wound on his arm he didn't take care of. He's hypotensive, respirations are rapid, temperature is below normal, hasn't been urinating. He's complaining of nausea, vomiting, and had a fever last week."

"Thanks," said Dr. Rolandi. The doctor asked David a few more questions, palpated his abdomen, and listened to his heart. "David, I'd like to admit you to the hospital for treatment. All your symptoms point to septicemia and impending septic shock, probably from that injury on your arm. Sometimes, if the conditions are right, a bacterial infection can overwhelm the body's immune system. The infection itself doesn't even have to be that bad. It's the toxins the bacteria produce that can cause the body to react by going into septic shock. Your capillaries get leaky, you begin to lose fluid into your tissues, and your blood volume drops, which makes your blood pressure drop. Your heart has to work harder. If the condition is left untreated, your lungs and kidneys can fail."

The doctor left the exam room to find the nurse. "Gail, we need to admit David, in exam room three. Let's get him started on IV saline and oxygen right away, and I'll be ordering an antibiotic. Put a rush on the labs. We'll need arterial blood gases, complete blood cell count, and a blood culture. Get a culture of that arm wound, too."

Gail nodded. "Septicemia?" she asked.

"Most likely. We want to avoid septic shock," replied Dr. Rolandi. "His capillary refill times were delayed, and he has some of the symptoms. He's lucky his granddaughter brought him in."

Sandy and Mrs. Mahoney approached. "Doctor, how's Grandpa?"

"Not well, I'm sorry to say. He must be admitted to the hospital. His infected wound may have caused septicemia. Bacterial toxins have produced a systemic inflammatory response. We need to get his blood volume up, which will increase his blood pressure. Right now, his blood is pooling in his capillary beds and he's losing plasma proteins as the capillaries become more leaky. This has led to what we call circulatory insufficiency; there's not enough blood pressure to keep his blood moving throughout his body as it should."

A.   David's skin is cool to the touch, pale, and blotchy. Which blood vessels regulate blood flow to the capillaries of the skin? What changes in the peripheral vasculature of the skin might cause the changes in its appearance?

B.   David's blood pressure is low. Which blood vessels regulate blood pressure?

C.   What structures control the flow of David's blood into his capillary beds?

D.   How would David's blood reservoirs respond to his low systemic blood pressure?

E.   How would the loss of plasma proteins affect capillary exchange in David's tissues?

20.5 Blood flow occurs from regions of higher pressure to those of lower pressure.

Blood flow is the volume of blood that flows through any tissue in a given time (in mL/min). *Total blood flow* is cardiac output (CO), the volume of blood that circulates through systemic (or pulmonary) blood vessels each minute. Cardiac output depends on heart rate and stroke volume. How the cardiac output becomes distributed into circulatory routes that serve various body tissues depends on two factors: (1) the *pressure difference* that drives the blood flow through a tissue and (2) the *resistance* to blood flow in specific blood vessels.

Blood Pressure

Blood flows from regions of higher pressure to regions of lower pressure; the greater the pressure difference, the greater is the blood flow. Contraction of the ventricles generates **blood pressure (BP)**, the pressure exerted by blood on the walls of a blood vessel. BP is highest in the aorta and large systemic arteries, where in a resting young adult it rises to about 110 mm Hg during systole (ventricular contraction) and drops to about 70 mm Hg during diastole (ventricular relaxation). **Systolic blood pressure** is the highest pressure attained in arteries during systole, and **diastolic blood pressure** is the lowest arterial pressure during diastole (Figure 20.9). As blood leaves the aorta and flows through the systemic circulation, its pressure falls progressively as the distance from the left ventricle increases. Blood pressure decreases to about 35 mm Hg as blood passes from systemic arteries through systemic arterioles and into capillaries, where the pressure fluctuations disappear. At the venous end of capillaries, blood pressure drops to about 16 mm Hg. Blood pressure continues to drop as blood enters systemic venules and then veins, because these vessels are farthest from the left ventricle. Finally, blood pressure reaches 0 mm Hg as blood returns to the right atrium.

Blood pressure depends in part on the total volume of blood in the cardiovascular system. The normal volume of blood in an adult is about 5 liters (5.3 qt). Any decrease in this volume, as from hemorrhage, decreases the amount of blood that is circulated through the arteries. A modest decrease can be compensated for by homeostatic mechanisms that help maintain blood pressure, but if the decrease in blood volume is greater than 10% of total blood volume, blood pressure drops. Conversely, anything that increases blood volume, such as water retention in the body, tends to increase blood pressure.

Vascular Resistance

Vascular resistance is the opposition to blood flow due to friction between blood and the walls of blood vessels. As vascular resistance increases, blood flow decreases. An increase in vascular resistance increases blood pressure; a decrease in vascular resistance has the opposite effect. Vascular resistance depends on the size of the blood vessel lumen, blood viscosity, and the total blood vessel length.

- *Size of the lumen.* The smaller the lumen of a blood vessel, the greater is its resistance to blood flow. Vasoconstriction narrows the lumen, and vasodilation widens it. Normally, moment-to-moment fluctuations in blood flow through a given tissue are due to vasoconstriction and vasodilation of the tissue's arterioles. As arterioles dilate, resistance decreases, and blood pressure falls. As arterioles constrict, resistance increases, and blood pressure rises.

- *Blood viscosity.* The viscosity (thickness) of blood depends mostly on the ratio of red blood cells to plasma (fluid) volume, and to a smaller extent on the concentration of proteins in plasma. The higher the blood's viscosity, the higher the resistance. Any condition that increases the viscosity of blood, such as dehydration or polycythemia (an unusually high number of red blood cells), thus increases blood pressure. A depletion of red blood cells or plasma proteins, due to anemia or hemorrhage, decreases viscosity and thus decreases blood pressure.

Figure 20.9 Blood pressure in various parts of the cardiovascular system. The dashed line is the mean (average) blood pressure in the aorta, arteries, and arterioles.

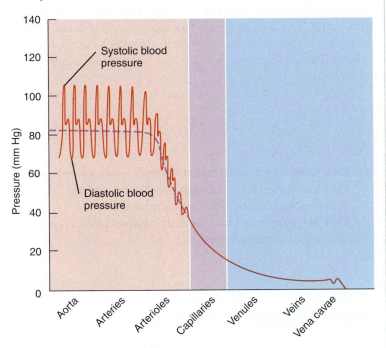

Blood pressure falls progressively as blood flows from systemic arteries through capillaries and back to the right atrium.

Figure 20.10   Summary of factors that increase blood pressure. Changes noted within green boxes increase cardiac output; changes noted within blue boxes increase systemic vascular resistance.

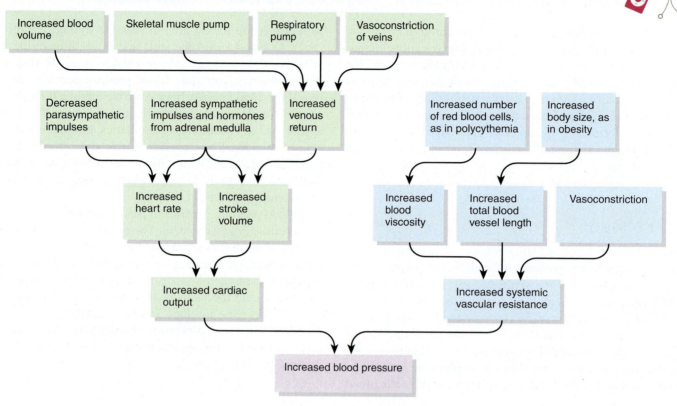

Increases in cardiac output and increases in systemic vascular resistance will increase blood pressure.

• *Total blood vessel length.* Resistance to blood flow through a vessel is directly proportional to the length of the blood vessel. The longer a blood vessel, the greater is the contact between the vessel wall and the blood. The greater the contact between the vessel wall and the blood, the greater is the friction against blood flow. An estimated 200 miles of additional blood vessels develop for each extra pound of fat, one reason why overweight individuals may have hypertension (elevated blood pressure).

Systemic vascular resistance (SVR), also known as *total peripheral resistance,* refers to all the vascular resistances offered by systemic blood vessels. The diameters of arteries and veins are large, so their resistance is small because most of the blood does not come into physical contact with the walls of the blood vessel. The smallest vessels—arterioles, capillaries, and venules—

contribute the most resistance. A major function of arterioles is to control SVR—and therefore blood pressure and blood flow to particular tissues—by changing their diameters. Arterioles need to vasodilate or vasoconstrict only slightly to have a large effect on SVR.

Figure 20.10 summarizes several factors that increase blood pressure.

Checkpoint

16. Which activity is occurring in the heart to produce systolic blood pressure? Which heart activity is occurring during diastolic blood pressure?
17. What is vascular resistance? What factors contribute to it?
18. Which type of blood vessel exerts the most control over systemic vascular resistance, and how is it achieved?

20.6   Blood pressure is regulated by neural and hormonal negative feedback systems.

Several interconnected negative feedback systems control blood pressure by adjusting heart rate, stroke volume, systemic vascular resistance, and blood volume. Some systems allow rapid adjustments to cope with sudden changes, such as the drop in blood pressure in the brain that occurs when you get out of bed; others act more slowly to provide long-term regulation of blood pressure.

Role of the Cardiovascular Center

In Chapter 19, we noted how the **cardiovascular center** in the medulla oblongata helps regulate heart rate and stroke volume. The cardiovascular center also controls the neural and hormonal negative feedback systems that regulate blood pressure and blood flow to specific tissues.

The cardiovascular center receives input both from higher brain regions and from sensory receptors (Figure 20.11). Nerve impulses descend from the cerebral cortex, limbic system, and hypothalamus to affect the cardiovascular center. For example, even before you start to run a race, your heart rate may increase due to nerve impulses conveyed from the limbic system to the cardiovascular center. If your body temperature rises during a race, the hypothalamus sends nerve impulses to the cardiovascular center. The resulting vasodilation of skin blood vessels allows heat to dissipate more rapidly from the surface of the skin. The three main types of sensory receptors that provide input to the cardiovascular center are proprioceptors, baroreceptors, and chemoreceptors. *Proprioceptors* monitor movements of joints and muscles, and provide input to the cardiovascular center during physical activity. Their activity accounts for the rapid increase in heart rate at the beginning of exercise. *Baroreceptors* monitor changes in pressure and stretch in the walls of blood vessels, and *chemoreceptors* monitor the concentration of various chemicals in the blood.

Output from the cardiovascular center flows along sympathetic and parasympathetic neurons of the autonomic nervous system (Figure 20.11). Opposing sympathetic (stimulatory) and parasympathetic (inhibitory) influences control the heart. Parasympathetic stimulation, conveyed along the **vagus nerves**, decreases heart rate; sympathetic impulses in the **cardiac accelerator nerves** increase both the rate and force of contraction of the heart. The cardiovascular center also continually sends impulses to smooth muscle in blood vessel walls throughout the body via **vasomotor nerves**. The result is a moderate state of vasoconstriction, called **vasomotor tone**, that sets the resting level of systemic vascular resistance. Sympathetic stimulation of most veins results in their constriction and movement of blood out of venous blood reservoirs, which increases blood pressure.

Neural Regulation of Blood Pressure

The nervous system regulates blood pressure via negative feedback loops that occur as two types of reflexes: baroreceptor reflexes and chemoreceptor reflexes.

Baroreceptor Reflexes

Baroreceptors (pressure receptors) are located in the aorta, internal carotid arteries (arteries in the neck that supply blood to the brain), and other large arteries in the neck and chest. They send impulses to the cardiovascular center to help regulate blood pressure.

Figure 20.11   The cardiovascular center. The cardiovascular center receives input from higher brain centers, proprioceptors, baroreceptors, and chemoreceptors. Then, it provides output to the sympathetic and parasympathetic divisions of the autonomic nervous system.

INPUT TO CARDIOVASCULAR CENTER (nerve impulses)

From higher brain centers: cerebral cortex, limbic system, and hypothalamus

From proprioceptors: monitor joint movements

From baroreceptors: monitor blood pressure

From chemoreceptors: monitor blood acidity (H^+), CO_2, and O_2

OUTPUT TO EFFECTORS (increased frequency of nerve impulses)

Heart: decreased rate

Heart: increased rate and contractility

Blood vessels: vasoconstriction

Vagus nerves (parasympathetic)

Cardiac accelerator nerves (sympathetic)

Vasomotor nerves (sympathetic)

Cardiovascular center

The cardiovascular center is the main region for nervous system regulation of the heart and blood vessels.

When blood pressure falls, the baroreceptors are stretched less, and they send nerve impulses at a slower rate to the cardiovascular center (Figure 20.12). In response, the cardiovascular center decreases parasympathetic stimulation of the heart via the vagus nerves and increases sympathetic stimulation of the heart via cardiac accelerator nerves. Another consequence of increased sympathetic stimulation is increased secretion of epinephrine and norepinephrine by the adrenal medulla, which increase heart rate and force of contraction. As the heart beats faster and more forcefully, cardiac output increases. Vasoconstriction increases systemic vascular resistance. Both increased cardiac output and increased systemic vascular resistance raise blood pressure back to the normal level.

Conversely, when an increase in pressure is detected, the baroreceptors send impulses at a faster rate. The cardiovascular center responds by increasing parasympathetic stimulation and decreasing sympathetic stimulation. The resulting decreases in heart rate and force of contraction reduce the cardiac output. The cardiovascular center also slows the rate at which it sends sympathetic impulses along vasomotor neurons that normally cause vasoconstriction. The resulting vasodilation lowers systemic vascular resistance. Decreased cardiac output and decreased systemic vascular resistance both lower blood pressure to the normal level.

Chemoreceptor Reflexes

Chemoreceptors, sensory receptors that monitor the chemical composition of blood, are located close to the baroreceptors of the carotid sinus and arch of the aorta in small structures called **carotid bodies** and **aortic bodies**, respectively. These chemoreceptors detect changes in blood levels of O_2, CO_2, and H^+. *Hypoxia* (lowered O_2 availability), *acidosis* (an increase in H^+ concentration), or *hypercapnia* (excess CO_2) stimulates the chemoreceptors to send impulses to the cardiovascular center. In response, the cardiovascular center increases sympathetic stimulation to arterioles and veins, producing vasoconstriction and an increase in blood pressure.

Hormonal Regulation of Blood Pressure

Several hormones help regulate blood pressure and blood flow by altering cardiac output, changing systemic vascular resistance, or adjusting the total blood volume:

• *Renin–angiotensin–aldosterone system*. When blood volume falls or blood flow to the kidneys decreases, the kidneys secrete **renin** into the bloodstream. Together, renin and angiotensin-converting enzyme produce the active hormone **angiotensin II**, which raises blood pressure in two ways. Angiotensin II causes vasoconstriction; it raises blood pressure by increasing systemic vascular resistance. Second, it increases secretion of **aldosterone**, which increases reabsorption of sodium ions (Na^+) and water by the kidneys. The water reabsorption increases total blood volume, which increases blood pressure.

• *Epinephrine and norepinephrine*. In response to sympathetic stimulation, the adrenal medulla releases epinephrine and norepinephrine. These hormones increase cardiac output by increasing the rate and force of heart contractions, and cause

Figure 20.12 Negative feedback regulation of blood pressure via baroreceptor reflexes.

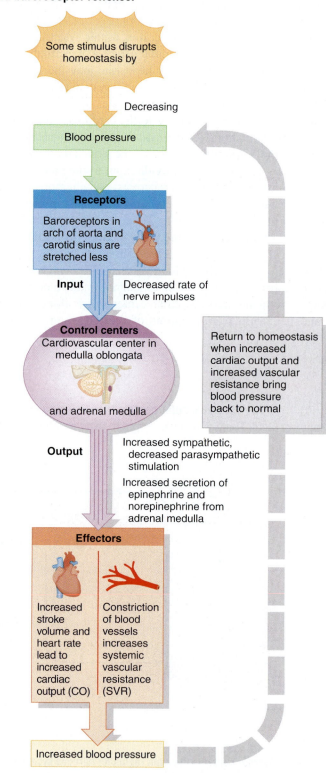

When blood pressure decreases, heart rate increases.

vasoconstriction of some arterioles and veins. The resulting increase in cardiac output and vasoconstriction raises blood pressure. Epinephrine also produces vasodilation of arterioles in skeletal muscles, which helps increase blood flow to muscles during exercise.

- **Antidiuretic hormone.** Antidiuretic hormone is produced by the hypothalamus and released from the posterior pituitary in response to dehydration or decreased blood volume. Among other actions, antidiuretic hormone causes vasoconstriction, which increases blood pressure.

- **Atrial natriuretic peptide.** Released by cells in the atria of the heart, atrial natriuretic peptide lowers blood pressure by causing vasodilation and by promoting the loss of salt and water in the urine, which reduces blood volume.

Table 20.2 summarizes the regulation of blood pressure by hormones.

Checkpoint

19. What are the principal inputs to and outputs from the cardiovascular center?
20. What types of effector tissues are regulated by the cardiovascular center?
21. Following a drop in blood pressure, how does the baroreceptor reflex return blood pressure to normal levels?
22. What is the role of chemoreceptors in the regulation of blood pressure?
23. How do hormones regulate blood pressure?

TABLE 20.2

Blood Pressure Regulation by Hormones

FACTOR INFLUENCING BLOOD PRESSURE	HORMONE	EFFECT ON BLOOD PRESSURE
Cardiac Output		
Increased heart rate and contractility	Norepinephrine	Increase
	Epinephrine	
Systemic Vascular Resistance		
Vasoconstriction	Angiotensin II	Increase
	Antidiuretic hormone (vasopressin)	
	Norepinephrine*	
	Epinephrine†	
Vasodilation	Atrial natriuretic peptide	Decrease
	Epinephrine†	
	Nitric oxide	
Blood Volume		
Blood volume increase	Aldosterone	Increase
	Antidiuretic hormone	
Blood volume decrease	Atrial natriuretic peptide	Decrease

*Acts at α_1 receptors in arterioles of abdomen and skin.
†Acts at β_2 receptors in arterioles of cardiac and skeletal muscle; norepinephrine has a much smaller vasodilating effect.

20.7 Measurement of the pulse and blood pressure help assess cardiovascular system function.

Pulse

The alternate expansion and recoil of elastic arteries after each contraction of the left ventricle creates a traveling pressure wave that is called the **pulse**. The pulse is strongest in the arteries closest to the heart, becomes weaker in arterioles, and disappears altogether in capillaries. The pulse may be felt in any artery that lies near the surface of the body that can be compressed against a bone or other firm structure. Table 20.3 depicts some common pulse points.

The pulse rate normally is the same as the heart rate, about 70 to 80 beats per minute at rest. **Tachycardia** (tak´-i-KAR-dē-a; *tachy-* = fast) is a rapid resting heart or pulse rate over 100 beats/min. **Bradycardia** (brād´-i-KAR-dē-a; *brady-* = slow) is a slow resting heart or pulse rate under 50 beats/min. Endurance-trained athletes normally exhibit bradycardia.

Measuring Blood Pressure

In clinical use, the term **blood pressure** usually refers to the pressure in arteries generated by the left ventricle during systole and the pressure remaining in the arteries when the ventricle is in diastole.

Blood pressure is usually measured in the brachial artery in the left arm (Table 20.3). The device used to measure blood pressure is a **sphygmomanometer** (sfig´-mō-ma-NOM-e-ter; *sphygmo-* = pulse; *-manometer* = instrument used to measure pressure). It consists of a rubber cuff connected to a rubber bulb that is used to inflate the cuff and a meter that registers the pressure in the cuff. With the arm resting on a table so that it is about the same level as the heart, the cuff of the sphygmomanometer is wrapped around a bared arm. The cuff is inflated by squeezing the bulb until the brachial artery is compressed and blood flow stops, about 30 mm Hg higher than the person's usual systolic pressure. The technician places a stethoscope below the cuff on the brachial artery and slowly deflates the cuff. When the cuff is deflated enough to allow the artery to open, a spurt of blood passes through, resulting in the first sound heard through the stethoscope. This sound corresponds to **systolic blood pressure**—the force with which blood is pushing on arterial walls just after ventricular contraction (Figure 20.13). As the cuff is deflated further, the sounds suddenly become too faint to be heard through the stethoscope. This level, called the **diastolic blood pressure**, represents the force exerted by the blood remaining in arteries during ventricular relaxation. At pressures below diastolic blood pressure, sounds disappear altogether.

TABLE 20.3

Pulse Points

STRUCTURE	LOCATION	STRUCTURE	LOCATION
Superficial temporal artery	Lateral to orbit of eye.	Femoral artery	Inferior to inguinal ligament.
Facial artery	Mandible (lower jawbone) on a line with the corners of the mouth.	Popliteal artery	Posterior to knee.
Common carotid artery	Lateral to larynx (voice box).	Radial artery	Distal aspect of wrist.
Brachial artery	Medial side of biceps brachii muscle.	Dorsal artery of the foot (dorsalis pedis artery)	Superior to instep of foot.

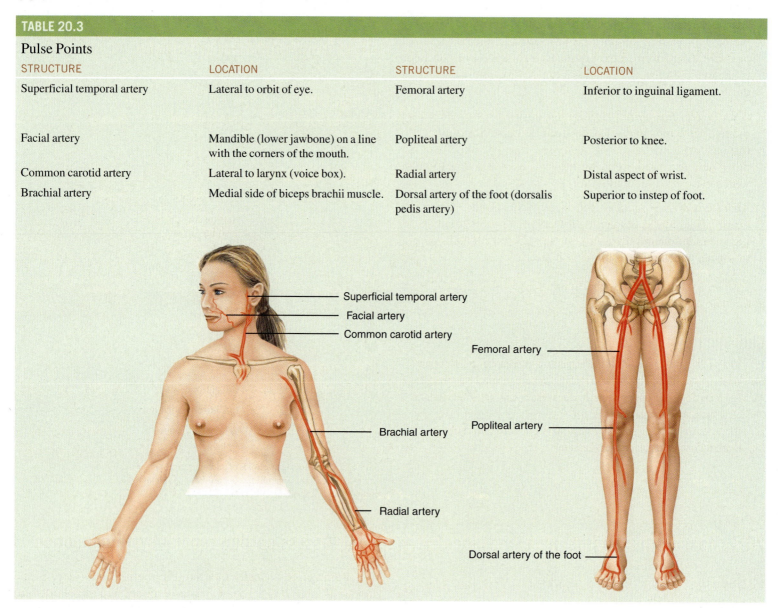

The various sounds that are heard while taking blood pressure are called **Korotkoff sounds** (kō-ROT-kof).

The normal blood pressure of an adult male is less than 120 mm Hg systolic and less than 80 mm Hg diastolic. For example, "110 over 70" (written as 110/70) is a normal blood pressure. In young adult females, the pressures are 8 to 10 mm Hg less. People who exercise regularly and are in good physical condition may have even lower blood pressures. Thus, blood pressure slightly lower than 120/80 may be a sign of good health and fitness.

The difference between systolic and diastolic pressure is called **pulse pressure**. This pressure, normally about 40 mm Hg, provides information about the condition of the cardiovascular system. For example, conditions such as atherosclerosis greatly increase pulse pressure.

Figure 20.13   Relationship of blood pressure changes to cuff pressure.

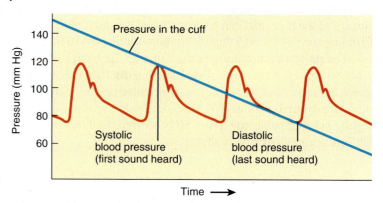

As the cuff is deflated, sounds first occur at the systolic blood pressure; the sounds suddenly become faint at the diastolic blood pressure.

Checkpoint

24. What causes the pulse?
25. Explain the terms tachycardia and bradycardia.
26. How are systolic and diastolic blood pressures measured with a sphygmomanometer?
27. If a blood pressure is reported as "142 over 95," what are the diastolic and systolic pressures?

20.8 The two main circulatory routes are the pulmonary circulation and the systemic circulation.

Blood vessels are organized into **circulatory routes** that carry blood to specific organs in the body (Figure 20.14). The routes are parallel—in most cases a portion of the cardiac output flows separately to each tissue of the body. Thus, each organ receives its own supply of freshly oxygenated blood. The two main circulatory routes are the pulmonary circulation and the systemic circulation.

Figure 20.14   Circulatory routes.
Short black arrows indicate the pulmonary circulation (detailed in Figure 20.15), long black arrows show the systemic circulation (detailed in Figures 20.16–20.27), and red arrows outline the hepatic portal circulation (detailed in Figure 20.28).

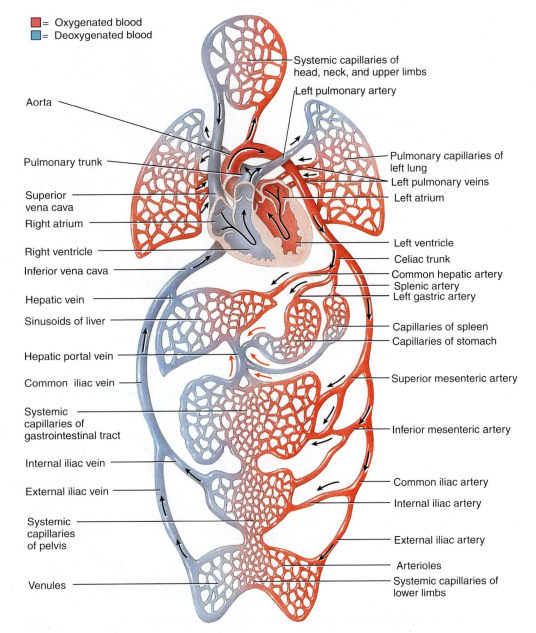

■ = Oxygenated blood
■ = Deoxygenated blood

Systemic capillaries of head, neck, and upper limbs
Left pulmonary artery
Aorta
Pulmonary trunk
Pulmonary capillaries of left lung
Left pulmonary veins
Left atrium
Superior vena cava
Right atrium
Left ventricle
Right ventricle
Celiac trunk
Inferior vena cava
Common hepatic artery
Splenic artery
Left gastric artery
Hepatic vein
Sinusoids of liver
Capillaries of spleen
Capillaries of stomach
Hepatic portal vein
Common iliac vein
Superior mesenteric artery
Systemic capillaries of gastrointestinal tract
Inferior mesenteric artery
Internal iliac vein
External iliac vein
Common iliac artery
Internal iliac artery
Systemic capillaries of pelvis
External iliac artery
Arterioles
Venules
Systemic capillaries of lower limbs

🔑 **Blood vessels are organized into circulatory routes that deliver blood to tissues of the body.**

Pulmonary Circulation

The **pulmonary circulation** (*pulmo-* = lung) carries deoxygenated blood from the right ventricle to the air sacs (alveoli) within the lungs and returns oxygenated blood from the air sacs to the left atrium (Figure 20.15). The **pulmonary trunk** emerges from the right ventricle and passes superiorly, posteriorly, and to the left. It then divides into two branches: the **right pulmonary artery** to the right lung and the **left pulmonary artery** to the left lung. After birth, the pulmonary arteries are the only arteries that carry deoxygenated blood. On entering the lungs, the branches divide and subdivide until finally they form capillaries around the air sacs (alveoli) within the lungs. CO_2 passes from the blood into the air sacs and is exhaled. Inhaled O_2 passes from the air within the lungs into the blood. The pulmonary capillaries unite to form venules and eventually **pulmonary veins**, which exit the lungs and carry the oxygenated blood to the left atrium. Two left and two right pulmonary veins enter the left atrium. After birth, the pulmonary veins are the only veins that carry oxygenated blood. Contractions of the left ventricle then eject the oxygenated blood into the systemic circulation.

Figure 20.15   Pulmonary circulation.

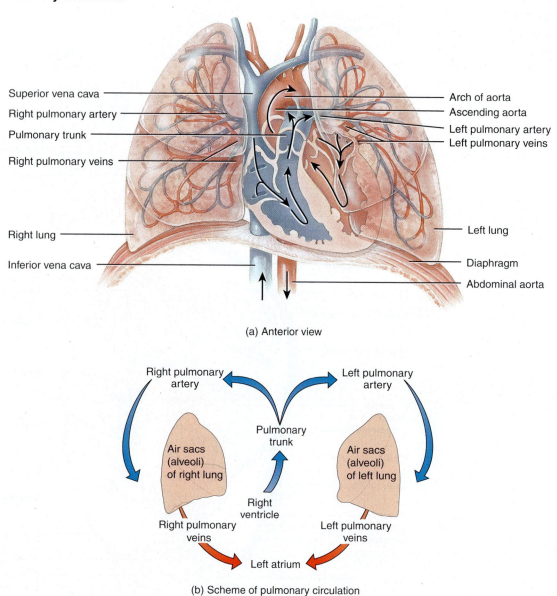

(a) Anterior view

(b) Scheme of pulmonary circulation

The pulmonary circulation brings deoxygenated blood from the right ventricle to the lungs and returns oxygenated blood from the lungs to the left atrium.

The pulmonary and systemic circulations differ in two important ways. First, blood in the pulmonary circulation need not be pumped as far as blood in the systemic circulation. Second, compared to systemic arteries, pulmonary arteries have larger diameters, thinner walls, and less elastic tissue. As a result, the resistance to pulmonary blood flow is very low, which means that less pressure is needed to move blood through the lungs. The peak systolic pressure in the right ventricle is only 20% of that in the left ventricle.

Systemic Circulation

The **systemic circulation** includes all the arteries and arterioles that carry oxygenated blood from the left ventricle to systemic capillaries, plus the veins and venules that return deoxygenated blood to the right atrium (see Figure 20.14). Blood leaving the aorta and flowing through the systemic arteries is a bright red color. As blood flows through capillaries, it loses some of its oxygen and picks up carbon dioxide, becoming a dark red color. All systemic arteries branch from the **aorta**, which emerges from the left ventricle of the heart. All the veins of the systemic circulation drain into the **superior vena cava**, **inferior vena cava**, or **coronary sinus**, which in turn empty into the right atrium. The bronchial arteries, which carry nutrients to the lungs, also are part of the systemic circulation.

Checkpoint

28. In the adult, which are the only arteries that carry deoxygenated blood? Which are the only veins that carry only oxygenated blood?
29. Why is less blood pressure needed in pulmonary arteries than systemic arteries?
30. What is the main function of the systemic circulation?

Return to David's Story

David finally consented to being admitted to the hospital. The staff started him on intravenous saline and an antibiotic, had given him oxygen, and were watching him closely—taking his blood pressure often, listening to his lungs and heart, and monitoring his urine output. "Why do you keep coming in here and bothering me?" David grumbled.

"David, we have to keep a close eye on you. We want to be sure the fluid in your tissues is returning to your bloodstream. If your lungs fill up with fluid, you won't be able to breathe. Your blood pressure is better now; it's up to one hundred thirty-five over eighty. And you're producing urine again, which means your kidneys are working. You're getting rid of the excess fluids."

"Hello, David," said Dr. Barbano, the physician who had taken over David's case when he was admitted. "How are you feeling?"

"Okay, better than before. I had felt so out of breath and weak," answered David.

"Well, your blood oxygen levels were down and the pH of your blood was abnormal, too. Your body responds to changes in blood flow, pressure, pH, oxygen level, and carbon dioxide, and tries to keep the body in balance. That's what we're helping your body do now, and we're also trying to get rid of that excess fluid to prevent any damage to your organs." A nurse came in to check his blood pressure while Dr. Barbano was still speaking to David.

"Here we go again," said David. He laughed, but then the laugh turned into a violent coughing fit. Both the nurse and Dr. Barbano noticed that David's appearance had changed very rapidly; his skin was flushed and reddened as he strained. Blood-red frothy sputum came from David's mouth.

"Doctor, his oxygen saturations are dipping into the eighties," said the nurse. The doctor took out his stethoscope and listened to David's lungs and heart.

"Get him back on oxygen. He may have a pulmonary embolism. Did you notice any swelling in the extremities?" The nurse pulled back the sheet that covered David's legs and they saw that his right calf was swollen.

So much was going on that Sandy didn't understand. She sat in a waiting room, twisting a tissue around her fingers and hoping to hear something about this latest crisis. The nurse finally came in to talk to her. "Sandy, your grandfather is doing okay. He can't see you right now, though. The doctor had to order some tests and had us put in an airway to help him breathe."

"He's going to be okay, though?" asked Sandy through her tears.

The nurse reached out and took Sandy's hand. "He's been through a rough time, Sandy. You have to hope for the best. A clot formed in one of the veins in his leg and part of it broke off and traveled up to his heart and then into his lungs. The clot is blocking blood from getting to the lung, so the blood isn't carrying enough oxygen to your grandfather's body. The airway is allowing us to give him oxygen until Dr. Barbano decides how to treat the clot."

F. David has lost blood volume, resulting in a drop in blood pressure. His autonomic nervous system countered the dropping blood pressure by stimulated vasoconstriction. How does vasoconstriction increase blood pressure?

G. If Sandy had not taken Grandpa David to the hospital, how might his inactivity have caused his condition to become worse?

H. Which structures in David's cardiovascular system would respond to low blood oxygen and decreased blood pH? What would the response be?

I. Because David's blood volume had dropped, blood flow to his kidneys had been reduced. What would be the hormonal response of the kidneys to these two changes?

J. Why would a blood clot in David's leg travel to David's lungs?

20.9 Systemic arteries carry blood from the heart to all body organs except the lungs.

In the figures and tables that follow, the principal arteries and veins of the systemic circulation are described and illustrated to assist you in learning their names and locations. The selected blood vessels are organized according to regions of the body. To make it easier for you to learn to say the names of blood vessels and understand how they are named, the tables include phonetic pronunciations and word roots.

Figure 20.16a shows an overview of the major arteries, and Figure 20.22 shows an overview of the major veins. As you study the various blood vessels, refer to these two figures to see the relationships of the blood vessels under consideration to other regions of the body.

Figure 20.16 **Aorta and its principal branches.**

(a) Overall anterior view of the principal branches of the aorta

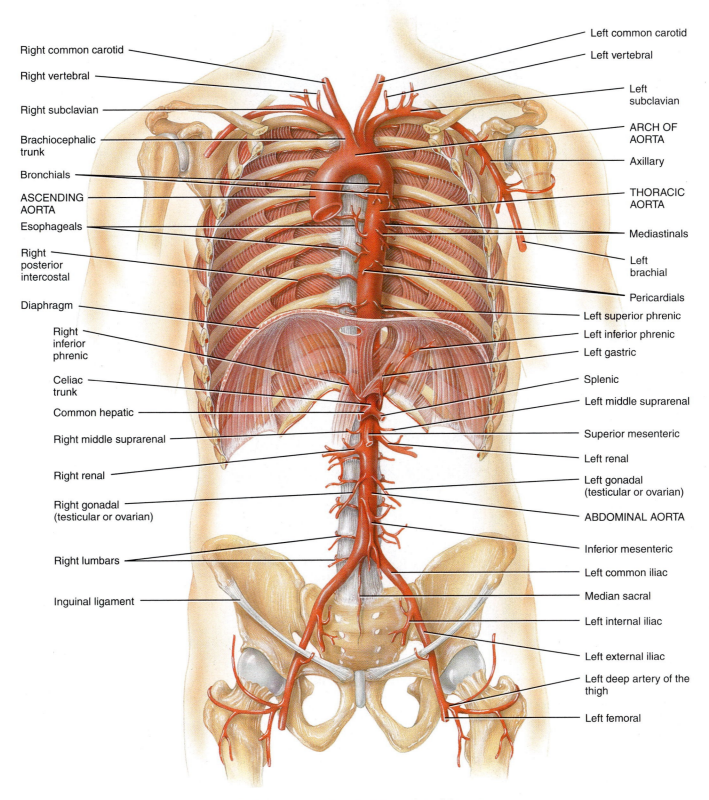

Right common carotid

Right vertebral

Right subclavian

Brachiocephalic trunk

Bronchials

ASCENDING AORTA

Esophageals

Right posterior intercostal

Diaphragm

Right inferior phrenic

Celiac trunk

Common hepatic

Right middle suprarenal

Right renal

Right gonadal (testicular or ovarian)

Right lumbars

Inguinal ligament

Left common carotid

Left vertebral

Left subclavian

ARCH OF AORTA

Axillary

THORACIC AORTA

Mediastinals

Left brachial

Pericardials

Left superior phrenic

Left inferior phrenic

Left gastric

Splenic

Left middle suprarenal

Superior mesenteric

Left renal

Left gonadal (testicular or ovarian)

ABDOMINAL AORTA

Inferior mesenteric

Left common iliac

Median sacral

Left internal iliac

Left external iliac

Left deep artery of the thigh

Left femoral

(b) Anterior view of the principal branches of the aorta

(continues)

Figure 20.16 (continued)

SUPERIOR

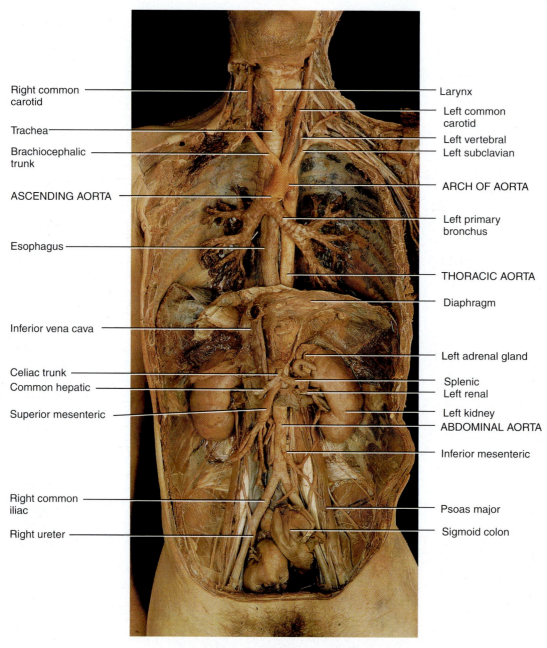

Right common carotid

Trachea

Brachiocephalic trunk

ASCENDING AORTA

Esophagus

Inferior vena cava

Celiac trunk

Common hepatic

Superior mesenteric

Right common iliac

Right ureter

Larynx

Left common carotid

Left vertebral

Left subclavian

ARCH OF AORTA

Left primary bronchus

THORACIC AORTA

Diaphragm

Left adrenal gland

Splenic

Left renal

Left kidney

ABDOMINAL AORTA

Inferior mesenteric

Psoas major

Sigmoid colon

INFERIOR

(c) Anterior view of the principal branches of the aorta

All systemic arteries branch from the aorta.

The Aorta and Its Branches

The **aorta** (= to lift up; Figure 20.16; Table 20.4) is the largest artery of the body, with a diameter of 2–3 cm (about 1 in.). Its four principal divisions are the ascending aorta, arch of the aorta, thoracic aorta, and abdominal aorta. The portion of the aorta that emerges from the left ventricle posterior to the pulmonary trunk is the **ascending aorta**. The ascending aorta gives off two coronary artery branches that supply the myocardium of the heart. The ascending aorta turns to the left, forming the **arch of the aorta**, then descends, following closely to the vertebral bodies, and passes through the diaphragm. The section of the aorta between the arch of the aorta and the diaphragm is called the **thoracic aorta**. As the aorta continues to descend below the diaphragm, it follows closely the vertebral column and divides into the two **common iliac arteries**, which carry blood to the lower limbs. The section between the diaphragm and the common iliac arteries is the **abdominal aorta**. Each division of the aorta gives off arteries that branch into distributing arteries, which lead to various organs. Within the organs, the arteries divide into arterioles and then into capillaries that service the systemic tissues (all tissues except the alveoli of the lungs).

TABLE 20.4

The Aorta and Its Branches

DIVISION AND BRANCHES	REGION SUPPLIED
Ascending Aorta	
Right and left coronary arteries	Heart.
Arch of the Aorta	
Brachiocephalic trunk (brā´-kē-ō-se-FAL-ik)	
Right common carotid artery (ka-ROT-id)	Right side of head and neck.
Right subclavian artery (sub-KLĀ-vē-an)	Right upper limb.
Left common carotid artery	Left side of head and neck.
Left subclavian artery	Left upper limb.
Thoracic Aorta (*thorac-* = chest)	
Pericardial arteries (per-ī-KAR-dē-al)	Pericardium.
Bronchial arteries (BRONG-kē-al)	Bronchi of lungs.
Esophageal arteries (e-sof´-a-JĒ-al)	Esophagus.
Mediastinal arteries (mē´-dē-as-TĪ-nal)	Structures in mediastinum.
Posterior intercostal arteries (in´-ter-KOS-tal)	Intercostal and chest muscles.
Subcostal arteries (sub-KOS-tal)	Same as posterior intercostals.
Superior phrenic arteries (FREN-ik)	Superior and posterior surfaces of diaphragm.
Abdominal Aorta	
Inferior phrenic arteries (FREN-ik)	Inferior surface of diaphragm.
Celiac trunk (SĒ-lē-ak)	
Common hepatic artery (he-PAT-ik)	Liver.
Left gastric artery (GAS-trik)	Stomach and esophagus.
Splenic artery (SPLĒN-ik)	Spleen, pancreas, and stomach.
Superior mesenteric artery (MES-en-ter´-ik)	Small intestine, cecum, ascending and transverse colons, and pancreas.
Suprarenal arteries (soo-pre-RĒ-nal)	Adrenal (suprarenal) glands.
Renal arteries (RĒ-nal)	Kidneys.
Gonadal arteries (gō-NAD-al)	
Testicular arteries (tes-TIK-ū-lar)	Testes (male).
Ovarian arteries (ō-VAR-ē-an)	Ovaries (female).
Inferior mesenteric artery	Transverse, descending, and sigmoid colons, rectum.
Common iliac arteries (IL-ē-ak)	
External iliac arteries	Lower limbs.
Internal iliac arteries	Uterus (female), prostate (male), muscles of buttocks, and urinary bladder.

Ascending Aorta

The **ascending aorta** (see Figure 19.4) begins at the aortic valve posterior to the pulmonary trunk and right auricle. It is directed superiorly, slightly anteriorly, and to the right, where it becomes the arch of the aorta (Figure 20.16b). The right and left **coronary arteries** (*coron-* = crown) arise from the ascending aorta just superior to the aortic semilunar valve (see Figures 19.7 and 20.17). They form a crownlike ring around the heart, giving off branches to the myocardium. The **posterior interventricular branch** (in-ter-ven-TRIK-ū-lar; *inter-* = between) of the right coronary artery supplies both ventricles, and the **marginal branch** supplies the right ventricle. The **anterior interventricular branch**, also known as the **left anterior descending branch**, of the left coronary artery supplies both ventricles, and the **circumflex branch** (SER-kum-flex; *circum-* = around; *-flex* = to bend) supplies the left atrium and left ventricle.

Checkpoint

31. What are the names of the four divisions of the aorta, in order of blood flow?
32. Which general regions do each of the four principal divisions of the aorta supply?

The Arch of the Aorta

The **arch of the aorta** (Figure 20.18; Table 20.5) is 4–5 cm (almost 2 in.) in length and is the continuation of the ascending aorta. It emerges from the pericardium posterior to the sternum. The arch is directed superiorly and posteriorly to the left and then inferiorly, where it becomes the thoracic aorta. The three major arterial branches are the brachiocephalic trunk, the left common carotid, and the left subclavian. The first and largest branch from the arch of the aorta is the **brachiocephalic trunk** (brā´-kē-ō-se-FAL-ik; *brachio-* = arm; *-cephalic* = head). It extends superiorly, bending slightly to the right, and divides to form the right subclavian artery and right common carotid artery. The second branch from the arch of the aorta is the **left common carotid artery** (ka-ROT-id), which divides into the same branches with the same names as the right common carotid artery. The third branch from the arch of the aorta is the **left subclavian artery** (sub-KLĀ-vē-an), which distributes blood to the left vertebral artery and vessels of the left upper limb. Arteries branching from the left subclavian artery are similar in distribution and name to those branching from the right subclavian artery.

Figure 20.17   Distribution of ascending aorta.

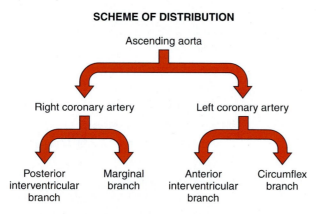

SCHEME OF DISTRIBUTION

Ascending aorta

Right coronary artery   Left coronary artery

Posterior interventricular branch   Marginal branch   Anterior interventricular branch   Circumflex branch

The coronary arteries deliver blood to the myocardium of the heart.

Figure 20.18 Arch of the aorta and its branches.

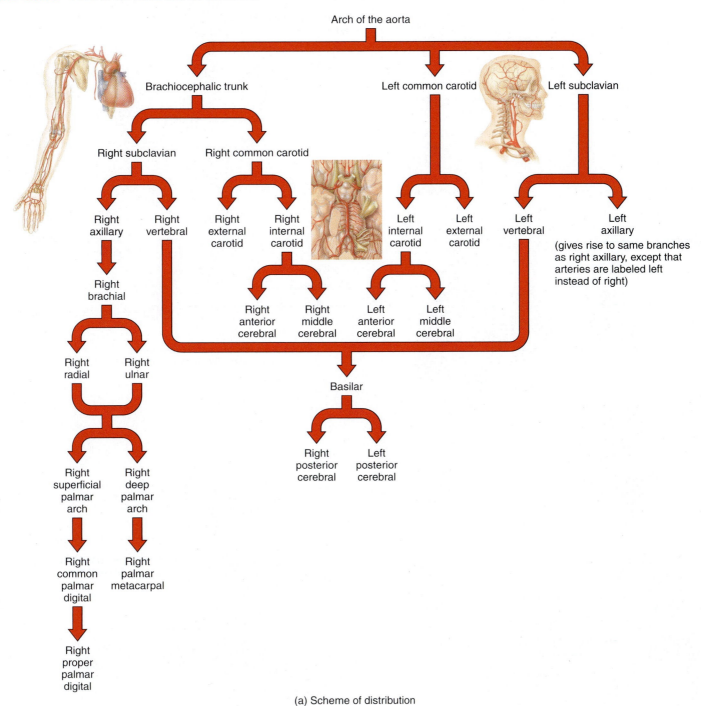

(a) Scheme of distribution

(continues)

Figure 20.18 (continued)

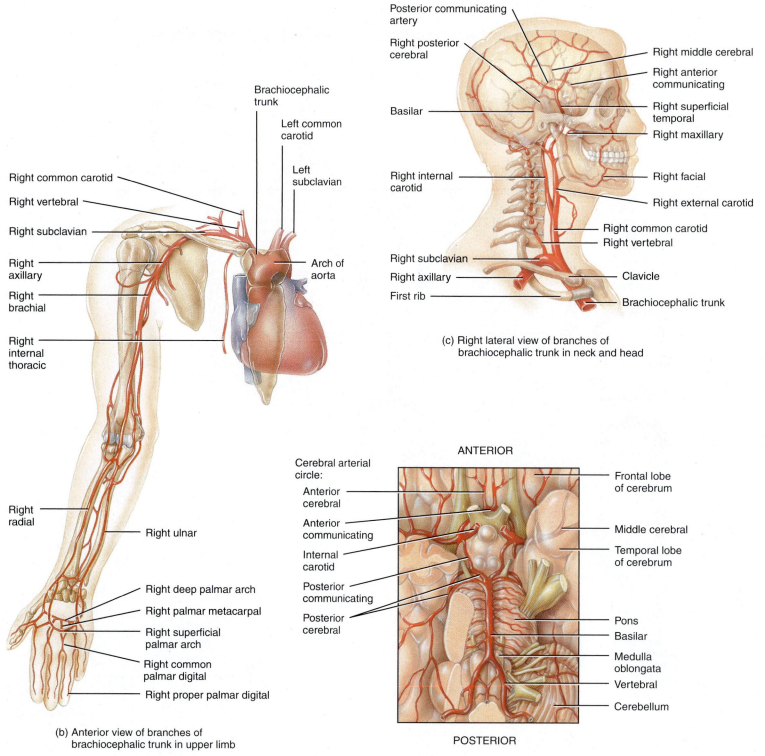

Brachiocephalic trunk

Left common carotid

Left subclavian

Right common carotid

Right vertebral

Right subclavian

Right axillary

Right brachial

Right internal thoracic

Arch of aorta

Right radial

Right ulnar

Right deep palmar arch

Right palmar metacarpal

Right superficial palmar arch

Right common palmar digital

Right proper palmar digital

(b) Anterior view of branches of brachiocephalic trunk in upper limb

Posterior communicating artery

Right posterior cerebral

Basilar

Right internal carotid

Right subclavian

Right axillary

First rib

Right middle cerebral

Right anterior communicating

Right superficial temporal

Right maxillary

Right facial

Right external carotid

Right common carotid

Right vertebral

Clavicle

Brachiocephalic trunk

(c) Right lateral view of branches of brachiocephalic trunk in neck and head

ANTERIOR

Cerebral arterial circle:

Anterior cerebral

Anterior communicating

Internal carotid

Posterior communicating

Posterior cerebral

Frontal lobe of cerebrum

Middle cerebral

Temporal lobe of cerebrum

Pons

Basilar

Medulla oblongata

Vertebral

Cerebellum

POSTERIOR

(d) Inferior view of base of brain showing cerebral arterial circle

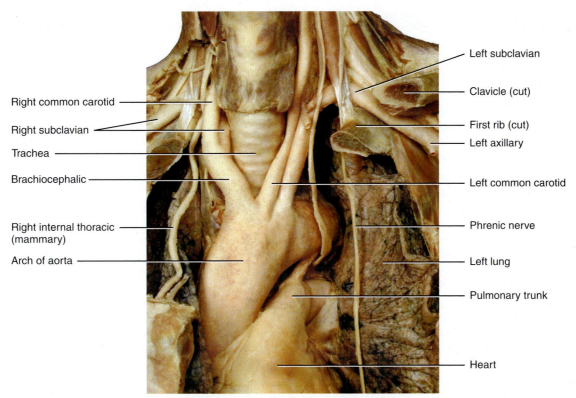

Right common carotid

Right subclavian

Trachea

Brachiocephalic

Right internal thoracic (mammary)

Arch of aorta

Left subclavian

Clavicle (cut)

First rib (cut)

Left axillary

Left common carotid

Phrenic nerve

Left lung

Pulmonary trunk

Heart

(e) Anterior view, branches of arch of aorta

🔑 **The arch of the aorta is the continuation of the ascending aorta.**

TABLE 20.5

The Arch of the Aorta

BRANCH	DESCRIPTION AND REGION SUPPLIED
Brachiocephalic Trunk	The **brachiocephalic trunk** divides to form the right subclavian artery and right common carotid artery (Figure 20.18b).
Right subclavian artery (sub-KLĀ-vē-an)	The **right subclavian artery** extends from the brachiocephalic trunk to the first rib and then passes into the armpit (axilla). The general distribution of the artery is to the brain and spinal cord, neck, shoulder, thoracic viscera and wall, and scapular muscles.
Internal thoracic or mammary artery (thor-AS-ik; *thorac-* = chest)	The **internal thoracic artery** arises from the first part of the subclavian artery and descends posterior to the costal cartilages of the superior six ribs. It terminates at the sixth intercostal space. It supplies the anterior thoracic wall and structures in the mediastinum. In coronary artery bypass grafting, if only a single vessel is obstructed, the internal thoracic (usually the left) is used to create the bypass. The upper end of the artery is left attached to the subclavian artery and the cut end is connected to the coronary artery at a point distal to the blockage. The lower end of the internal thoracic artery is tied off. Artery grafts are preferred over vein grafts because arteries can withstand the greater pressure of blood flowing through coronary arteries and are less likely to become obstructed over time.
Vertebral artery (VER-te-bral)	Before passing into the axilla, the right subclavian artery gives off a major branch to the brain called the **right vertebral artery** (Figure 20.18c). The right vertebral artery passes through the foramina of the transverse processes of the sixth through first cervical vertebrae and enters the skull through the foramen magnum to reach the inferior surface of the brain. Here it unites with the left vertebral artery to from the **basilar** (BAS-i-lar) **artery**. The vertebral artery supplies the posterior portion of the brain with blood. The basilar artery passes along the midline of the anterior aspect of the brain stem; it gives off several branches (**posterior cerebral** and **cerebellar arteries**) that supply the cerebellum and pons of the brain and the inner ear.
Axillary artery (AK-sil-ār-ē = armpit)	The continuation of the right subclavian artery into the axilla is called the **axillary artery**. (Note that the right subclavian artery, which passes deep to the clavicle, is a good example of the practice of giving the same vessel different names as it passes through different regions.) Its general distribution is the shoulder, thoracic and scapular muscles, and humerus.
Brachial artery (BRĀ-kē-al = arm)	The **brachial artery** is the continuation of the axillary artery into the arm. The brachial artery provides the main blood supply to the arm and is superficial and palpable along its course. It begins at the tendon of the teres major muscle and ends just distal to the bend of the elbow. At first, the brachial artery is medial to the humerus, but as it descends it gradually curves laterally and passes through the cubital fossa, a triangular depression anterior to the elbow where you can easily detect the pulse of the brachial artery and listen to the various sounds when taking a person's blood pressure. Just distal to the bend in the elbow, the brachial artery divides into the radial artery and ulnar artery. Blood pressure is usually measured in the brachial artery. In order to control hemorrhage, the best place to compress the brachial artery is near the middle of the arm.

(continues)

TABLE 20.5 (continued)

BRANCH	DESCRIPTION AND REGION SUPPLIED
Radial artery (RĀ-dē-al = radius)	The **radial artery** is the smallar branch and is a direct continuation of the brachial artery. It passes along the lateral (radial) aspect of the forearm and then through the wrist and hand, supplying these structures with blood. At the wrist, the radial artery makes contact with the distal end of the radius, where it is covered only by fascia and skin. Because of its superficial location at this point, it is a common site for measuring the radial pulse.
Ulnar artery (UL-nar = ulna)	The **ulnar artery**, the larger branch of the brachial artery, passes along the medial (ulnar) aspect of the forearm and then into the wrist and hand, supplying these structures with blood. In the palm, branches of the radial and ulnar arteries anastomose to form the superficial palmar arch and the deep palmar arch.
Superficial palmar arch (*palma* = palm)	The **superficial palmar arch** is formed mainly by the ulnar artery, with a contribution from a branch of the radial artery. The arch is superficial to the long flexor tendons of the fingers and extends across the palm at the bases of the metacarpals. It gives rise to **common palmar digital arteries**, which supply the palm. Each divides into a pair of **proper palmar digital arteries**, which supply the fingers.
Deep palmar arch	Mainly the radial artery forms the **deep palmar arch**, with a contribution from a branch of the ulnar artery. The arch is deep to the long flexor tendons of the fingers and extends across the palm, just distal to the bases of the metacarpals. Arising from the deep palmar arch are **palmar metacarpal arteries**, which supply the palm and anastomose with the common palmar digital arteries of the superficial palmar arch.
Right common carotid artery	The **right common carotid artery** begins at the bifurcation (division into two branches) of the brachiocephalic trunk posterior to the right sternoclavicular joint, and passes superiorly in the neck to supply structures in the head (Figure 20.18c). At the superior border of the larynx (voice box), it divides into the right external and right internal carotid arteries. Pulse may be detected in the common carotid artery, just lateral to the larynx. It is convenient to detect a carotid pulse when exercising or when administering cardiopulmonary resuscitation.
External carotid artery	The **external carotid artery** begins at the superior border of the larynx and terminates near the temporomandibular joint of the parotid gland, where it divides into two branches: the superficial temporal and maxillary arteries. The carotid pulse can be detected in the external carotid artery just anterior to the sternocleidomastoid muscle at the superior border of the larynx. The general distribution of the external carotid artery is to structures external to the skull.
Internal carotid artery	The **internal carotid artery** has no branches in the neck and supplies structures internal to the skull. It enters the cranial cavity through the carotid foramen in the temporal bone. The internal carotid artery supplies blood to the eyeball and other orbital structures, the ear, most of the cerebrum of the brain, the pituitary gland, and the external nose. The terminal branches of the internal carotid artery are the **anterior cerebral artery**, which supplies most of the medial surface of the cerebrum and deep masses of gray matter within the cerebrum, and the **middle cerebral artery**, which supplies most of the lateral surface of the cerebrum (Figure 20.18d). Inside the cranium, anastomoses of the left and right internal carotid arteries along with the basilar artery form an arrangement of blood vessels at the base of the brain near the hypophyseal fossa called the **cerebral arterial circle (circle of Willis)**. From this circle (Figure 20.18d) arise arteries supplying most of the brain. Essentially, the cerebral arterial circle is formed by the union of the **anterior cerebral arteries** (branches of internal carotids) and **posterior cerebral arteries** (branches of basilar artery). The posterior cerebral arteries supply the inferolateral surface of the temporal lobe and lateral and medial surfaces, the occipital lobe of the cerebrum, deep masses of gray matter within the cerebrum, and midbrain. The posterior cerebral arteries are connected with the internal carotid arteries by the **posterior communicating arteries**. The **anterior communicating arteries** connect the anterior cerebral arteries. The **internal carotid arteries** are also considered part of the cerebral arterial circle. The functions of the cerebral arterial circle are to equalize blood pressure to the brain and provide alternate routes for blood flow to the brain, should the arteries become damaged.
Left common carotid artery	The left common carotid artery is a branch from the arch of the aorta that divides into the same branches with the same names as the right common carotid artery (see above).
Left subclavian artery	The left subclavian artery is a branch from the arch of the aorta that distributes blood to the left vertebral artery and vessels of the left upper limb. Its branches are similar in distribution and name to those branching from the right subclavian artery (see above).

Thoracic Aorta

The **thoracic aorta** (Figure 20.19; Table 20.6) is about 20 cm (8 in.) long and is a continuation of the arch of the aorta. It lies to the left of the vertebral column. As it descends, it passes through an opening in the diaphragm and becomes the abdominal aorta as it enters the abdominal cavity. Along its course, the thoracic aorta sends off numerous small arteries, **visceral branches** to viscera and **parietal branches** to body wall structures.

Checkpoint

33. What are the three major branches of the arch of the aorta in order of origination? Which regions do they supply?
34. Which structure is the anatomical border between the thoracic and abdominal aorta?

TABLE 20.6

Thoracic Aorta

BRANCH	DESCRIPTION AND REGION SUPPLIED
Visceral	
Pericardial arteries (per´-ī-KAR-dē-al; *peri-* = around; *-cardia-* = heart)	Two or three tiny **pericardial arteries** supply blood to the pericardium.
Bronchial arteries (BRONG-kē-al = windpipe)	One right and two left **bronchial arteries** supply the bronchial tubes, pleurae, bronchial lymph nodes, and esophagus. (The right bronchial artery arises from the third posterior intercostal artery: the two left bronchial arteries arise from the thoracic aorta.)
Esophageal arteries (e-sof´-a-JĒ-al; *eso-* = to carry; *-phage-* = food)	Four or five **esophageal arteries** supply the esophagus.
Mediastinal arteries (mē´-dē-as-TĪ-nal)	Numerous small **mediastinal arteries** supply blood to structures in the mediastinum.
Parietal	
Posterior intercostal arteries (in´-ter-KOS-tal; *inter-* = between; *-costa-* = rib)	Nine pairs of **posterior intercostal arteries** supply the intercostal, pectoralis major and minor, and serratus anterior muscles; overlying subcutaneous tissue and skin; mammary glands; and vertebrae, meninges, and spinal cord.
Subcostal arteries (sub-KOS-tal; *sub-* = under)	The left and right **subcostal arteries** have a distribution similar to that of the posterior intercostals.
Superior phrenic arteries (FREN-ik = pertaining to the diaphragm)	Small **superior phrenic arteries** supply the superior and posterior surfaces of the diaphragm.

Figure 20.19 **Thoracic aorta and abdominal aorta and their principal branches.**

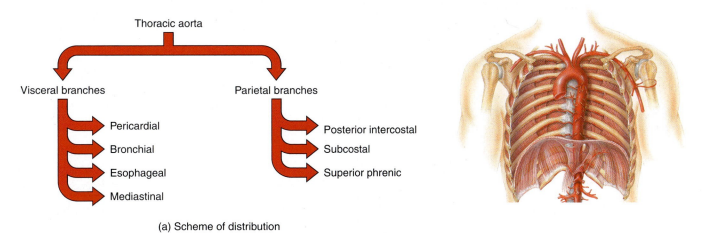

(a) Scheme of distribution

(continues)

Figure 20.19 (continued)

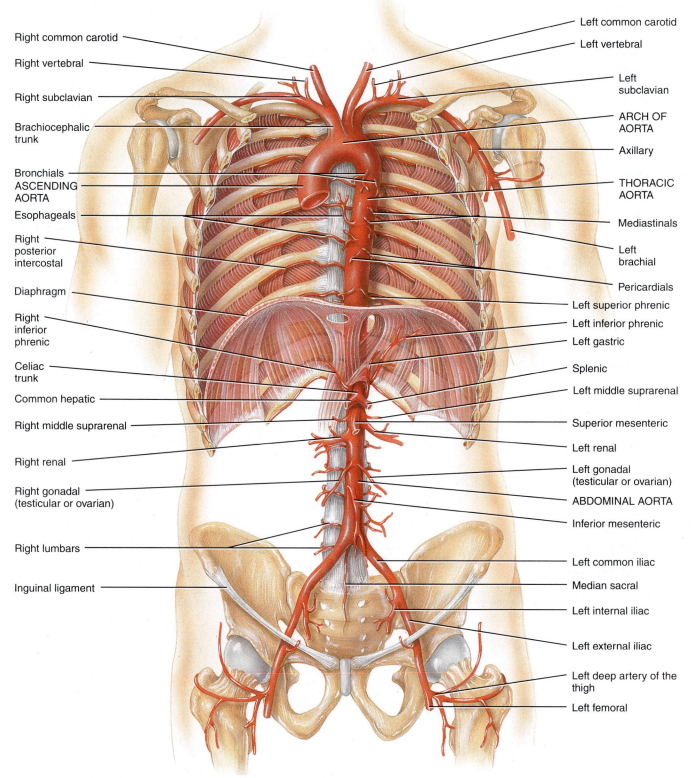

Right common carotid

Right vertebral

Right subclavian

Brachiocephalic
trunk

Bronchials
ASCENDING
AORTA

Esophageals

Right
posterior
intercostal

Diaphragm

Right
inferior
phrenic

Celiac
trunk

Common hepatic

Right middle suprarenal

Right renal

Right gonadal
(testicular or ovarian)

Right lumbars

Inguinal ligament

Left common carotid

Left vertebral

Left
subclavian

ARCH OF
AORTA

Axillary

THORACIC
AORTA

Mediastinals

Left
brachial

Pericardials

Left superior phrenic

Left inferior phrenic

Left gastric

Splenic

Left middle suprarenal

Superior mesenteric

Left renal

Left gonadal
(testicular or ovarian)

ABDOMINAL AORTA

Inferior mesenteric

Left common iliac

Median sacral

Left internal iliac

Left external iliac

Left deep artery of the
thigh

Left femoral

(b) Detailed anterior view of the principal branches of the aorta

🔑 **The thoracic aorta is the continuation of the ascending aorta.**

Abdominal Aorta

The **abdominal aorta** (Figure 20.20; Table 20.7) is the continuation of the thoracic aorta. It begins as the aorta penetrates the diaphragm and ends where it divides into the right and left common iliac arteries. The abdominal aorta lies anterior to the vertebral column. As with the thoracic aorta, the abdominal aorta gives off visceral and parietal branches. The paired visceral branches arise from the lateral surfaces of the aorta and include the **suprarenal**, **renal**, and **gonadal arteries**. The unpaired visceral branches arise from the anterior surface of the aorta and include the **celiac trunk**, the **superior mesenteric artery**, and the **inferior mesenteric artery** (see Figure 20.19b). The parietal branches arise from the posterior surfaces of the aorta: the paired **inferior phrenic** and **lumbar arteries**, and the unpaired **median sacral artery**.

Figure 20.20 **Abdominal aorta and principal branches.**

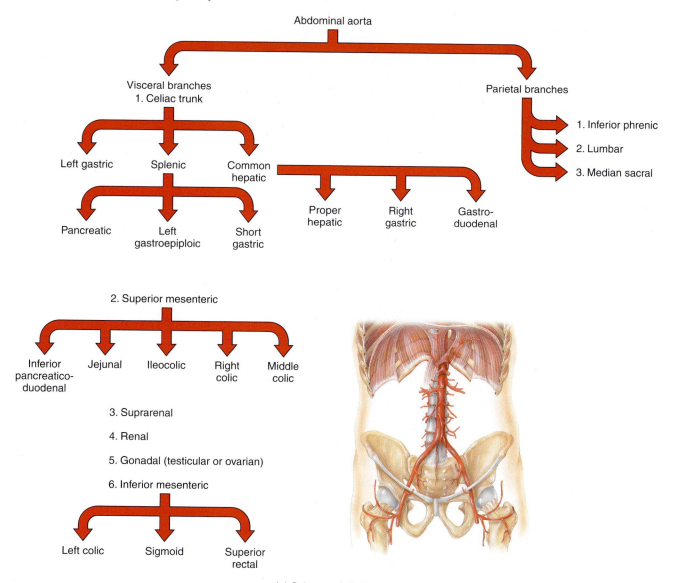

(a) Scheme of distribution

(continues)

Figure 20.20 (continued)

Stomach
(turned upward)

Left gastroepiploic

Right gastroepiploic

Short gastrics

Right gastric

Diaphragm

Celiac trunk

Left gastric

Proper hepatic

Splenic

Spleen

Common hepatic

Pancreatic

Gastroduodenal

Pancreas

Superior
pancreaticoduodenal

Superior mesenteric

Jejunum of
small intestine

Duodenum of
small intestine

Abdominal aorta

(b) Anterior view of celiac trunk and its branches

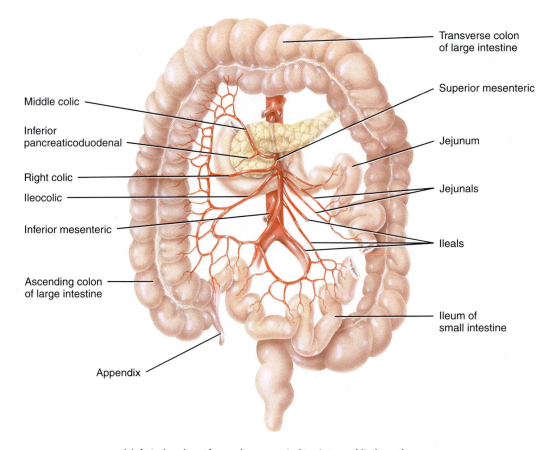

Transverse colon
of large intestine

Superior mesenteric

Middle colic

Inferior
pancreaticoduodenal

Jejunum

Right colic

Jejunals

Ileocolic

Inferior mesenteric

Ileals

Ascending colon
of large intestine

Ileum of
small intestine

Appendix

(c) Anterior view of superior mesenteric artery and its branches

Transverse colon of large intestine

Superior mesenteric

Inferior mesenteric

Abdominal aorta

Common iliac

Sigmoids

Left colic

Descending colon of large intestine

Sigmoid colon

Superior rectal

Rectum of large intestine

(d) Anterior view of inferior mesenteric artery and its branches

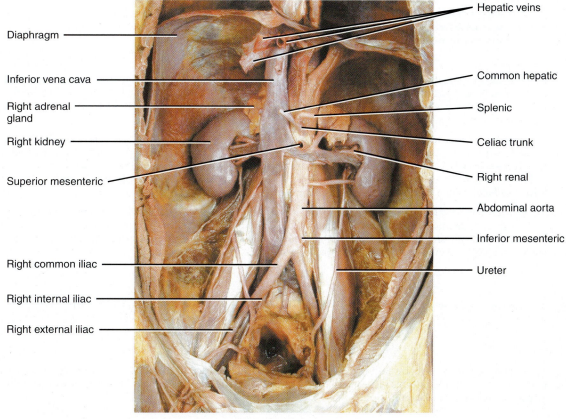

Diaphragm

Inferior vena cava

Right adrenal gland

Right kidney

Superior mesenteric

Right common iliac

Right internal iliac

Right external iliac

Hepatic veins

Common hepatic

Splenic

Celiac trunk

Right renal

Abdominal aorta

Inferior mesenteric

Ureter

(e) Anterior view of arteries of abdomen and pelvis

The abdominal aorta is the continuation of the thoracic aorta.

TABLE 20.7

Abdominal Aorta

BRANCH	DESCRIPTION AND REGION SUPPLIED

Unpaired Visceral Branches

Celiac trunk (SĒ-lē-ak)	The **celiac trunk (artery)** is the first visceral branch from the aorta inferior to the diaphragm, at about the level of the twelfth thoracic vertebra (Figure 20.20b). Almost immediately the celiac trunk divides into three branches: the left gastric, splenic, and common hepatic arteries (Figure 20.20b). 1. The **left gastric artery** (GAS-trik = stomach) is the smallest of the three branches. It passes superiorly to the left toward the esophagus and then turns to follow the lesser curvature of the stomach. It supplies the stomach and esophagus. 2. The **splenic artery** (SPLĒN-ik = spleen) is the largest branch of the celiac trunk. It arises from the left side of the celiac trunk distal to the left gastric artery and passes horizontally to the left along the pancreas. Before reaching the spleen, it gives rise to three arteries: • **Pancreatic artery** (pan-krē-AT-ik), which supplies the pancreas. • **Left gastroepiploic artery** (gas´-trō-ep´-i-PLŌ-ik; *epiplo-* = omentum), which supplies the stomach and greater omentum. • **Short gastric artery**, which supplies the stomach. 3. The **common hepatic artery** (he-PAT-ik = liver) is intermediate in size between the left gastric and splenic arteries. Unlike the other two branches of the celiac trunk, the common hepatic artery arises from the right side. It gives rise to three arteries: • **Proper hepatic artery**, which supplies the liver, gallbladder, and stomach. • **Right gastric artery**, which supplies the stomach. • **Gastroduodenal artery** (gas´-trō-doo´-ō-DĒ-nal), which supplies the stomach, duodenum of the small intestine, pancreas, and greater omentum.
Superior mesenteric artery (MES-en-ter´-ik; *meso-* = middle; -*enteric* = pertaining to the intestines)	The **superior mesenteric artery** (Figure 20.20c) arises from the anterior surface of the abdominal aorta about 1 cm inferior to the celiac trunk at the level of the first lumbar vertebra. It extends inferiorly and anteriorly and between the layers of mesentery, which is a portion of the peritoneum that attaches the small intestine to the posterior abdominal wall. It anastomoses extensively and has five branches: 1. The **inferior pancreaticoduodenal artery** (pan´-krē-at´-i-kō-doo´-o-DĒ-nal) supplies the pancreas and duodenum. 2. The **jejunal** (je-JOO-nal) and **ileal arteries** (IL-ē-al) supply the jejunum and ileum of the small intestine, respectively. 3. The **ileocolic artery** (il´-ē-ū-KŌL-ik) supplies the ileum and ascending colon of the large intestine. 4. The **right colic artery** (KŌL-ik) supplies the ascending colon. 5. The **middle colic artery** supplies the transverse colon of the large intestine.
Inferior mesenteric artery	The **inferior mesenteric artery** (Figure 20.20d) arises from the anterior aspect of the abdominal aorta at the level of the third lumbar vertebra and then passes inferiorly to the left of the aorta. It anastomoses extensively and has three branches: 1. The **left colic artery** supplies the transverse colon and descending colon of the large intestine. 2. The **sigmoid arteries** (SIG-moyd) supply the descending colon and sigmoid colon of the large intestine. 3. The **superior rectal artery** (REK-tal) supplies the rectum of the large intestine.

Paired Visceral Branches

Suprarenal arteries (soo´-pra-RĒ-nal; *supra-* = above; -*ren-* = kidney)	Although there are three pairs of **suprarenal (adrenal) arteries** that supply the adrenal (suprarenal) glands (superior, middle, and inferior), only the middle pair originates directly from the abdominal aorta (see Figure 20.19b). The middle suprarenal arteries arise at the level of the first lumbar vertebra at or superior to the renal arteries. The superior suprarenal arteries arise from the inferior phrenic artery, and the inferior suprarenal arteries originate from the renal arteries.
Renal arteries (RĒ-nal; *ren-* = kidney)	The right and left **renal arteries** usually arise from the lateral aspects of the abdominal aorta at the superior border of the second lumbar vertebra, about 1 cm inferior to the superior mesenteric artery (see Figure 20.19b). The right renal artery, which is longer than the left, arises slightly lower than the left and passes posterior to the right renal vein and inferior vena cava. The left renal artery is posterior to the left renal vein and is crossed by the inferior mesenteric vein. The renal arteries carry blood to the kidneys, adrenal (suprarenal) glands, and ureters. Their distribution within the kidneys is discussed in Chapter 24.
Gonadal (gō-NAD-al; *gon-* = seed) [**testicular** (tes-TIK-ū-lar) or **ovarian arteries** (ō-VAR-ē-an)]	The **gonadal arteries** arise from the abdominal aorta at the level of the second lumbar vertebra just inferior to the renal arteries (see Figure 20.16b). In males, the gonadal arteries are specifically referred to as the **testicular arteries**. They pass through the inguinal canal and supply the testes, epididymis, and ureters. In females, the gonadal arteries are called the **ovarian arteries**. They are much shorter than the testicular arteries and supply the ovaries, uterine (fallopian) tubes, and ureters.

Unpaired Parietal Branch

Median sacral artery (SĀ-kral = pertaining to the sacrum)	The **median sacral artery** arises from the posterior surface of the abdominal aorta about 1 cm superior to the bifurcation (division into two branches) of the aorta into the right and left common iliac arteries (see Figure 20.19b). The median sacral artery supplies the sacrum and coccyx.

Paired Parietal Branches

Inferior phrenic arteries (FREN-ik = pertaining to the diaphragm)	The **inferior phrenic arteries** are the first paired branches of the abdominal aorta, immediately superior to the origin of the celiac trunk (see Figure 20.19b). (They may also arise from the renal arteries.) The inferior phrenic arteries are distributed to the inferior surface of the diaphragm and adrenal (suprarenal) glands.
Lumbar arteries (LUM-bar = pertaining to the loin)	The four pairs of **lumbar arteries** arise from the posterolateral surface of the abdominal aorta (see Figure 20.19b). They supply the lumbar vertebrae, spinal cord and its meninges, and the muscles and skin of the lumbar region of the back.

Checkpoint

35. What are the paired and unpaired visceral branches of the abdominal aorta? What is the general region that each branch supplies?

36. What are the paired and unpaired parietal branches of the abdominal aorta? What is the general region that each branch supplies?

Arteries of the Pelvis and Lower Limbs

The abdominal aorta ends by dividing into the right and left **common iliac arteries** to supply the pelvis and lower limbs (Figure 20.21; Table 20.8). These, in turn, divide into the **internal** and **external iliac arteries**. In sequence, the external iliacs become the **femoral arteries** in the thighs, the **popliteal arteries** posterior to the knee, and the **anterior** and **posterior tibial arteries** in the legs.

Figure 20.21 **Arteries of the pelvis and right lower limb.**

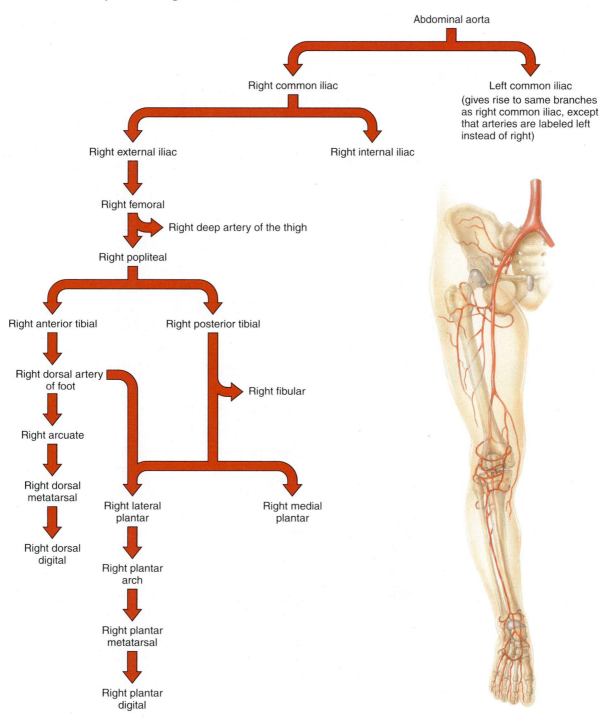

(a) Scheme of distribution

(continues)

Figure 20.21 (continued)

Abdominal aorta

Right common iliac

Left common iliac

Right internal iliac

Right external iliac

Right deep artery of the thigh

Right femoral

Right popliteal

Right anterior tibial

Right posterior tibial

Right fibular

Right lateral plantar

Right dorsal artery of foot

Right medial plantar

Right arcuate

Right plantar arch

Right dorsal metatarsal

Right plantar metatarsal

Right dorsal digital

Right plantar digital

(b) Anterior view   (c) Posterior view

SUPERIOR

External iliac vein

External iliac artery

Tensor fasciae latae

Femoral nerve

Deep artery of the thigh

Lateral circumflex vein

Femoral vein

Femoral artery

Vastus lateralis

Sartorius

Rectus femoris

LATERAL

Right ureter

Urinary bladder

Ductus deferens

Spermatic cord

Pectineus

Inguinal lymph node

Scrotum

Adductor longus

Gracilis

MEDIAL

(d) Anterior view

The internal iliac arteries carry most of the blood supply to the pelvic viscera and wall.

TABLE 20.8

Arteries of the Pelvis and Lower Limbs

BRANCH	DESCRIPTION AND REGION SUPPLIED
Common Iliac Arteries (IL-ē-ak = pertaining to the ilium)	
	At about the level of the fourth lumbar vertebra, the abdominal aorta divides into the right and left **common iliac arteries**, the terminal branches of the abdominal aorta. Each passes inferiorly about 5 cm (2 in.) and gives rise to two branches: internal iliac and external iliac arteries. The general distribution of the common iliac arteries is to the pelvis, external genitals, and lower limbs.
Internal iliac arteries	The **internal iliac (hypogastric) arteries** are the primary arteries of the pelvis. They begin at the bifurcation (division into two branches) of the common iliac arteries anterior to the sacroiliac joint at the level of the lumbosacral intervertebral disc. They pass posteromedially as they descend in the pelvis and divide into anterior and posterior divisions. The general distribution of the internal iliac arteries is to the pelvis, buttocks, external genitals, and thigh.
External iliac arteries	The **external iliac arteries** are larger than the internal iliac arteries. Like the internal iliac arteries, they begin at the bifurcation of the common iliac arteries. They descend along the medial border of the psoas major muscles following the pelvic brim, pass posterior to the midportion of the inguinal ligaments, and become the femoral arteries. The general distribution of the external iliac arteries is to the lower limbs. Specifically, branches of the external iliac arteries supply the muscles of the anterior abdominal wall, the cremaster muscle in males and the round ligament of the uterus in females, and the lower limbs.
Femoral arteries (FEM-o-ral = pertaining to the thigh)	The **femoral arteries** descend along the anteromedial aspects of the thighs to the junction of the middle and lower third of the thighs. Here they pass through an opening in the tendon of the adductor magnus muscle, then emerge posterior to the femurs as the popliteal arteries. A pulse may be felt in the femoral artery just inferior to the inguinal ligament. Recall from Chapter 11 that the femoral artery, along with the femoral vein and nerve and deep inguinal lymph nodes, are located in the *femoral triangle* (see Figure 11.19a) The general distribution of the femoral arteries is to the lower abdominal wall, groin, external genitals, and muscles of the thigh. A major branch of the femoral artery, the **deep artery of the thigh** (**deep femoral**), supplies most of the muscles of the thigh: quadriceps femoris, adductors, and hamstrings.
	Recall that in cardiac catheterization a catheter is inserted through a blood vessel and advanced into the major vessels and heart chamber. A catheter often contains a measuring instrument or other device at its tip. To reach the left side of the heart, the catheter is inserted into the femoral artery and passed into the aorta to the coronary arteries or heart chamber.
Popliteal arteries (pop´-li-TĒ-al = posterior surface of the knee)	The **popliteal arteries** are the continuation of the femoral arteries through the popliteal fossa (space behind the knee). They descend to the inferior border of the popliteus muscles, where they divide into the anterior and posterior tibial arteries. A pulse may be detected in the popliteal arteries. In addition to supplying the adductor magnus and hamstring muscles and the skin on the posterior aspect of the legs, branches of the popliteal arteries also supply the gastrocnemius, soleus, and plantaris muscles of the calf, knee joint, femur, patella, and fibula.
Anterior tibial arteries (TIB-ē-al = pertaining to the shin bone)	The **anterior tibial arteries** descend from the bifurcation of the popliteal arteries. They are smaller than the posterior tibial arteries. The anterior tibial arteries descend through the anterior muscular compartment of the leg. They pass through the interosseous membrane that connects the tibia and fibula, lateral to the tibia. The anterior tibial arteries supply the knee joints, anterior compartment muscles of the legs, skin over the anterior aspects of the legs, and ankle joints. At the ankles, the anterior tibial arteries become the **dorsal arteries of the foot (dorsalis pedis arteries)**. A pulse in this artery may be taken to evaluate the peripheral vascular system. The dorsal arteries of the foot supply the muscles, skin, and joints on the dorsal aspects of the feet. On the dorsum of the feet, the dorsal arteries of the foot give off a transverse branch at the first (medial) cuneiform bone called the **arcuate arteries** (*arcuate* = bowed) that run laterally over the bases of the metatarsals. From the arcuate arteries branch the **dorsal metatarsal arteries**, which supply the feet. The dorsal metatarsal arteries terminate by dividing into the **dorsal digital arteries**, which supply the toes.
Posterior tibial arteries	The **posterior tibial arteries**, the direct continuations of the popliteal arteries, descend from the bifurcation of the popliteal arteries. They pass down the posterior muscular compartment of the legs posterior to the medial malleolus of the tibia. They terminate by dividing into the medial and lateral plantar arteries. Their general distribution is to the muscles, bones, and joints of the leg and foot. Major branches of the posterior tibial arteries are the **fibular (peroneal) arteries**, which supply the fibularis, soleus, tibialis posterior, and flexor hallucis muscles. They also supply the fibula, tarsus, and lateral aspect of the heel. The bifurcation of the posterior tibial arteries into the medial and lateral plantar arteries occurs deep to the flexor retinaculum on the medial side of the feet. The **medial plantar arteries** (PLAN-tar = sole) supply the abductor hallucis and flexor digitorum brevis muscles and the toes. The **lateral plantar arteries** unite with a branch of the dorsal arteries of the foot to form the **plantar arch**. The arch begins at the base of the fifth metatarsal and extends medially across the metacarpals. As the arch crosses the foot, it gives off **plantar metatarsal arteries**, which supply the feet. These terminate by dividing into **plantar digital arteries**, which supply the toes.

Checkpoint

37. What are the two major branches of the common iliac arteries? Which general regions do those arteries supply?

20.10 Systemic veins return blood to the heart from all body organs except the lungs.

Arteries distribute blood to various parts of the body, and veins drain blood away from them. For the most part, arteries are deep; veins (Figure 20.22; Table 20.9) may be superficial or deep. Superficial veins are located just beneath the skin and can be seen easily.

Figure 20.22
Principal veins.

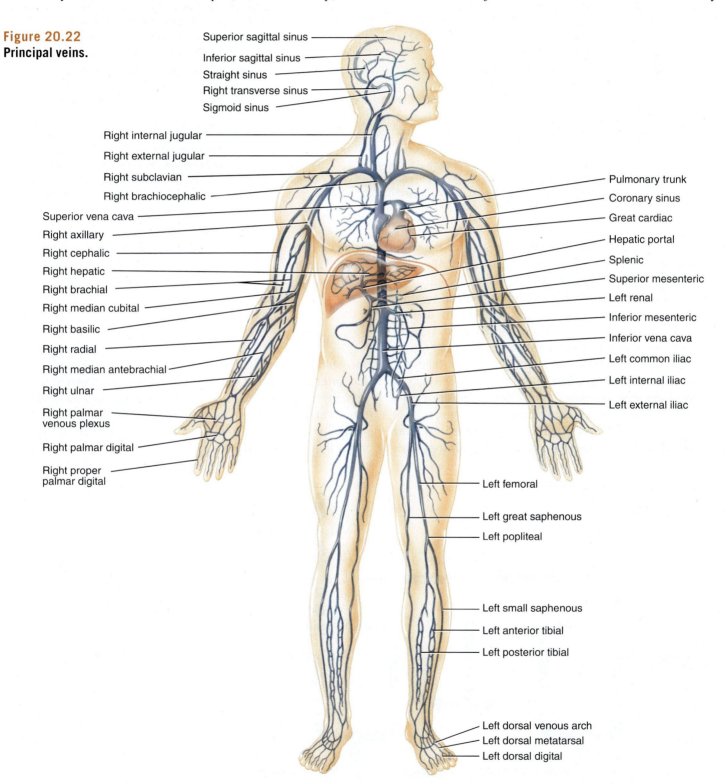

Superior sagittal sinus
Inferior sagittal sinus
Straight sinus
Right transverse sinus
Sigmoid sinus
Right internal jugular
Right external jugular
Right subclavian
Right brachiocephalic
Superior vena cava
Right axillary
Right cephalic
Right hepatic
Right brachial
Right median cubital
Right basilic
Right radial
Right median antebrachial
Right ulnar
Right palmar venous plexus
Right palmar digital
Right proper palmar digital

Pulmonary trunk
Coronary sinus
Great cardiac
Hepatic portal
Splenic
Superior mesenteric
Left renal
Inferior mesenteric
Inferior vena cava
Left common iliac
Left internal iliac
Left external iliac
Left femoral
Left great saphenous
Left popliteal
Left small saphenous
Left anterior tibial
Left posterior tibial
Left dorsal venous arch
Left dorsal metatarsal
Left dorsal digital

Overall anterior view of the principal veins

Deoxygenated blood returns to the heart via the superior vena cava, inferior vena cava, and coronary sinus.

TABLE 20.9

Veins of the Systemic Circulation

VEIN	DESCRIPTION AND REGION DRAINED
Coronary sinus (KOR-ō-nar-ē: *corona* = crown)	The **coronary sinus** is the main vein of the heart; it receives almost all venous blood from the myocardium. It is located in the coronary sulcus (see Figure 19.3c) and opens into the right atrium between the orifice of the inferior vena cava and the tricuspid valve. It is a wide venous channel into which three veins drain. It receives the **great cardiac vein** (in the anterior interventricular sulcus) into its left end, and the **middle cardiac vein** (in the posterior interventricular sulcus) and the **small cardiac vein** into its right end. Several **anterior cardiac veins** drain directly into the right atrium.
Superior vena cava (SVC) (VĒ-na CĀ-va; *vena* = vein; *cava* = cavelike)	The **superior vena cava** is about 7.5 cm (3 in.) long and 2 cm (1 in.) in diameter and empties its blood into the superior part of the right atrium. It begins posterior to the right first costal cartilage by the union of the right and left brachiocephalic veins and ends at the level of the right third costal cartilage, where it enters the right atrium. The SVC drains the head, neck, chest, and free upper limbs.
Inferior vena cava (IVC)	The **inferior vena cava** is the largest vein in the body, about 3.5 cm (1.4 in.) in diameter. It begins anterior to the fifth lumbar vertebra by the union of the common iliac veins, ascends behind the peritoneum to the right of the midline, pierces the caval opening of the diaphragm at the level of the eighth thoracic vertebra, and enters the inferior part of the right atrium. The IVC drains the abdomen, pelvis, and free lower limbs. The inferior vena cava is commonly compressed during the later stages of pregnancy by the enlarging uterus, producing edema of the ankles and feet and temporary varicose veins.

Superficial veins are clinically important as sites for withdrawing blood or giving injections. Deep veins generally travel alongside arteries and usually bear the same name. Because there are no large superficial arteries, the names of superficial veins do not correspond to those of arteries. Arteries usually follow definite pathways; veins are more difficult to follow because they connect in irregular networks in which many tributaries merge to form a large vein. Although only one systemic artery, the aorta, takes oxygenated blood away from the heart (left ventricle), three systemic veins, the **coronary sinus**, **superior vena cava**, and **inferior vena cava**, return deoxygenated blood to the heart (right atrium) (Figure 20.22; Table 20.9). The coronary sinus receives blood from the cardiac veins. The superior vena cava receives blood from other veins superior to the diaphragm, except the air sacs (alveoli) of the lungs. The inferior vena cava receives blood from veins inferior to the diaphragm. The coronary sinus drains the heart wall.

Veins of the Head and Neck

Most blood draining from the head passes into three pairs of veins: the **internal jugular**, **external jugular**, and **vertebral veins** (Figure 20.23; Table 20.10). Within the brain, all veins drain into dural venous sinuses and then into the internal jugular veins. Dural venous sinuses are venous channels between layers of the dura mater that surrounds the brain.

Figure 20.23 Principal veins of the head and neck.

(a) Scheme of drainage

Superior sagittal sinus

Inferior sagittal sinus

Great cerebral

Straight sinus

Right transverse sinus

Right sigmoid sinus

Right vertebral

Right internal jugular

Right external jugular

Right subclavian

Right axillary

Right cavernous sinus

Right ophthalmic

Right superficial temporal

Right facial

Right brachiocephalic

(b) Right lateral view

SUPERIOR

Left internal jugular

Left brachiocephalic

Internal thoracic

Heart

Left subclavian

Left axillary

Left brachial

Left basilic

INFERIOR

(c) Anterior view

Blood draining from the head passes into the internal jugular, external jugular, and vertebral veins.

TABLE 20.10

Veins of the Head and Neck

VEIN	DESCRIPTION AND REGION DRAINED
Internal jugular veins (JUG-ū-lar; *jugular* = throat)	The flow of blood from the dural venous sinuses into the internal jugular veins is as follows (Figure 20.23b): The **superior sagittal sinus** (SAJ-i-tal = straight) begins at the frontal bone, where it receives a vein from the nasal cavity, and passes posteriorly to the occipital bone. Along its course, it receives blood from the superior, medial, and lateral aspects of the cerebral hemispheres, meninges, and cranial bones. The superior sagittal sinus usually turns to the right and drains into the right transverse sinus. The **inferior sagittal sinus** is much smaller than the superior sagittal sinus; it begins posterior to the attachment of the falx cerebri and receives the great cerebral vein to become the straight sinus. The great cerebral vein drains the deeper parts of the brain. Along its course, the inferior sagittal sinus also receives tributaries from the superior and medial aspects of the cerebral hemispheres. The **straight sinus** runs in the tentorium cerebelli and is formed by the union of the inferior sagittal sinus and the great cerebral vein. The straight sinus also receives blood from the cerebellum and usually drains into the left transverse sinus. The **transverse sinuses**, being near the occipital bone, pass laterally and anteriorly, and become the sigmoid sinuses near the temporal bone. The transverse sinuses receive blood from the cerebrum, cerebellum, and cranial bones. The **sigmoid sinuses** (SIG-moyd = S-shaped) are located along the temporal bone. They pass through the jugular foramina, where they terminate in the internal jugular veins. The sigmoid sinuses drain the transverse sinuses. The **cavernous sinuses** (KAV-er-nus = cavelike) are located on either side of the sphenoid bone. They receive blood from the ophthalmic veins from the orbits, and from the cerebral veins from the cerebral hemispheres. They ultimately empty into the transverse sinuses and internal jugular veins. The cavernous sinuses are unique because they have nerves and a major blood vessel passing through them on their way to the orbit and face. The oculomotor (III) nerve, trochlear (IV) nerve, and ophthalmic and maxillary branches of the trigeminal (V) nerve as well as the internal carotid arteries pass through the cavernous sinuses. The right and left **internal jugular veins** pass inferiorly on either side of the neck lateral to the internal carotid and common carotid arteries. They then unite with the subclavian veins posterior to the clavicles at the sternoclavicular joints to form the right and left brachiocephalic veins (brā′-kē-ō-se-FAL-ik; *brachio-* = arm; *-cephal-* = head). From here blood flows into the superior vena cava. The general structures drained by the internal jugular veins are the brain (through the dural venous sinuses), face, and neck.
External jugular veins	The right and left **external jugular veins** begin in the parotid glands near the angle of the mandible. They are superficial veins that descend through the neck across the sternocleidomastoid muscles. They terminate at a point opposite the middle of the clavicles, where they empty into the subclavian veins. The general structures drained by the external jugular veins are external to the cranium, such as the scalp and superficial and deep regions of the face. When venous pressure rises, for example, during heavy coughing or straining or in cases of heart failure, the external jugular veins become very prominent along the side of the neck.
Vertebral veins (VER-te-bral = pertaining to the vertebrae)	The right and left **vertebral veins** originate inferior to the occipital condyles. They descend through successive transverse foramina of the first six cervical vertebrae and emerge from the foramina of the sixth cervical vertebra to enter the brachiocephalic veins in the root of the neck. The vertebral veins drain deep structures in the neck such as the cervical vertebrae, cervical spinal cord, and some neck muscles.

Checkpoint

38. Which general regions of the body are drained by the superior vena cava and the inferior vena cava?

39. Which vessels drain into the coronary sinus?

40. Into which veins in the neck does all venous blood from the head and neck drain?

41. Which general areas are drained by the internal jugular, external jugular, and vertebral veins?

Veins of the Upper Limbs

Both superficial and deep veins return blood from the upper limbs to the heart (Figure 20.24; Table 20.11). **Superficial veins** are located just deep to the skin and are often visible. They anastomose extensively with one another and with deep veins, and they do not accompany arteries. Superficial veins are larger than deep veins and return most of the blood from the upper limbs. **Deep veins** are located deep in the body. They usually accompany arteries and have the same names as the corresponding arteries. Both superficial and deep veins have valves, but valves are more numerous in the deep veins.

Figure 20.24   Principal veins of the right upper limb.

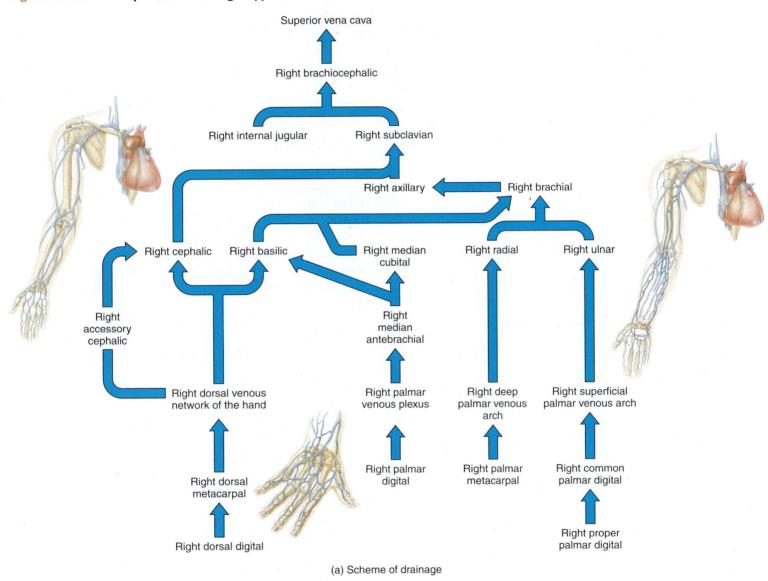

(a) Scheme of drainage

(continues)

Figure 20.24 (continued)

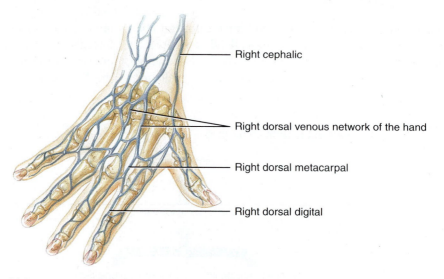

(b) Posterior view of superficial veins of the hand

(c) Anterior view of superficial veins

(d) Anterior view of deep veins

SUPERIOR

Right cephalic

Right basilic

Right biceps brachii

Right median cubital

Right basilic

Right accessory
cephalic

Right median
antebrachial

Right cephalic

INFERIOR

(e) Anterior view of superficial veins of arm and forearm

Superficial veins of the upper limb anastomose extensively.

TABLE 20.11

Veins of the Upper Limbs

VEIN	DESCRIPTION AND REGION DRAINED
Superficial	
Cephalic veins (se-FAL-ik = pertaining to the head)	The principal superficial veins that drain the upper limbs are the cephalic and basilic veins. They originate in the hand and convey blood from the smaller superficial veins into the axillary veins. The **cephalic veins** begin on the lateral aspect of the **dorsal venous networks of the hands (dorsal venous arches)**, networks of veins on the dorsum of the hands formed by the **dorsal metacarpal veins** (Figure 20.24b). These veins, in turn, drain the **dorsal digital veins**, which pass along the sides of the fingers. Following their formation from the dorsal venous networks of the hands, the cephalic veins arch around the radial side of the forearms to the anterior surface and ascend through the entire limbs along the anterolateral surface. The cephalic veins end where they join the axillary veins, just inferior to the clavicles. **Accessory cephalic veins** originate either from a venous plexus on the dorsum of the forearms or from the medial aspect of the dorsal venous networks of the hands, and unite with the cephalic veins just inferior to the elbow. The cephalic veins drain blood from the lateral aspect of the upper limbs.
Basilic veins (ba-SIL-ik = royal, of prime importance)	The **basilic veins** begin on the medial aspects of the dorsal venous networks of the hands and ascend along the posteromedial surface of the forearm and anteromedial surface of the arm (Figure 20.24c). They drain blood from the medial aspects of the upper limbs. Anterior to the elbow, the basilic veins are connected to the cephalic veins by the **median cubital veins** (*cubital* = pertaining to the elbow), which drain the forearm. If veins must be punctured for an injection, transfusion, or removal of a blood sample, the median cubital veins are preferred. After receiving the median cubital veins, the basilic veins continue ascending until they reach the middle of the arm. There they penetrate the tissues deeply and run alongside the brachial arteries until they join the brachial veins. As the basilic and brachial veins merge in the axillary area, they form the axillary veins.
Median antebrachial veins (an´ tē-BRĀ-kē-al; *ante-* = before, in front of; *-brachi-* = arm)	The **median antebrachial veins (median veins of the forearm)** begin in the **palmar venous plexuses**, networks of veins of the palms. The plexuses drain the **palmar digital veins** in the fingers. The median antebrachial veins ascend anteriorly in the forearms to join the basilic or median cubital veins, sometimes both. They drain the palms and forearms.
Deep	
Radial veins (RĀ-dē-al = pertaining to the radius)	The paired **radial veins** begin at the **deep palmar venous arches** (Figure 20.24d). These arches drain the **palmar metacarpal veins** in the palms. The radial veins drain the lateral aspects of the forearms and pass alongside the radial arteries. Just inferior to the elbow joint, the radial veins unite with the ulnar veins to form the brachial veins.
Ulnar veins (UL-nar = pertaining to the ulna)	The paired **ulnar veins**, which are larger than the radial veins, begin at the **superficial palmar venous arches**. These arches drain the **common palmar digital veins** and the **proper palmar digital veins** in the fingers. The ulnar veins drain the medial aspect of the forearms, pass alongside the ulnar arteries, and join with the radial veins to form the brachial veins.
Brachial veins (BRĀ-kē-al; *brachi-* = arm)	The paired **brachial veins** accompany the brachial arteries. They drain the forearms, elbow joints, arms, and humerus. They pass superiorly and join with the basilic veins to form the axillary veins.
Axillary veins (AK-sil-ār-ē; *axilla* = armpit)	The **axillary veins** ascend to the outer borders of the first ribs, where they become the subclavian veins. The axillary veins receive tributaries that correspond to the branches of the axillary arteries. The axillary veins drain the arms, axillas, and superolateral chest wall.
Subclavian veins (sub-KLĀ-vē-an; *sub-* = under; *-clavian* = pertaining to the clavicle)	The **subclavian veins** are continuations of the axillary veins that terminate at the sternal end of the clavicles, where they unite with the internal jugular veins to form the brachiocephalic veins. The subclavian veins drain the arms, neck, and thoracic wall. The thoracic duct of the lymphatic system delivers lymph into the junction between the left subclavian and the left internal jugular veins. The right lymphatic duct delivers lymph into the junction between the right subclavian and right internal jugular veins (see Figure 21.3a). In a procedure called *central line placement*, the right subclavian vein is frequently used to administer nutrients and medication and to measure venous pressure.

Checkpoint

42. Which regions of the upper limb are drained by the cephalic, basilic, median antebrachial, radial, and ulnar veins?

43. Which vein provides collateral circulation between the cephalic and basilic veins?

44. From which vein in the upper limb is a blood sample often taken?

45. Into which upper limb vein does lymph drain from the lymphatic system into the circulatory system?

Veins of the Thorax

Although the brachiocephalic veins drain some portions of the thorax, most thoracic structures are drained by a network of veins, called the **azygos system**, that runs on either side of the vertebral column (Figure 20.25; Table 20.12). The system consists of three veins—the **azygos**, **hemiazygos**, and **accessory hemiazygos veins**—that show considerable variation in origin, course, tributaries, anastomoses, and termination. Ultimately they empty into the superior vena cava.

TABLE 20.12

Veins of the Thorax

VEIN	DESCRIPTION AND REGION DRAINED
Brachiocephalic vein (brā´-kē-ō-se-FAL-ik; *brachio-* = arm; *-cephalic* = pertaining to the head)	The right and left **brachiocephalic veins**, formed by the union of the subclavian and internal jugular veins, drain blood from the head, neck, upper limbs, mammary glands, and superior thorax. The brachiocephalic veins unite to form the superior vena cava. Because the superior vena cava is to the right of the body's midline, the left brachiocephalic vein is longer than the right. The right brachiocephalic vein is anterior and to the right of the brachiocephalic trunk. The left brachiocephalic vein is anterior to the brachiocephalic trunk, the left common carotid and left subclavian arteries, the trachea, the left vagus (X) nerve, and the phrenic nerve.
Azygos system (az-Ī-gos = unpaired)	The **azygos system**, besides collecting blood from the thorax and abdominal wall, may serve as a bypass for the inferior vena cava that drains blood from the lower body. Several small veins directly link the azygos system with the inferior vena cava. Large veins that drain the lower limbs and abdomen conduct blood into the azygos system. If the inferior vena cava or hepatic portal vein becomes obstructed, the azygos system can return blood from the lower body to the superior vena cava.
Azygos vein	The **azygos vein** is anterior to the vetebral column, slightly to the right of the midline. It usually begins at the junction of the right ascending lumbar and right subcostal veins near the diaphragm. At the level of the fourth thoracic vertebra, it arches over the root of the right lung to end in the superior vena cava. Generally, the azygos vein drains the right side of the thoracic wall, thoracic viscera, and abdominal wall. Specifically, the azygos vein receives blood from most of the **right posterior intercostal**, **hemiazygos**, **accessory hemiazygos**, **esophageal**, **mediastinal**, **percardinal**, and **bronchial veins**.
Hemiazygos vein (HEM-ē-az-ī-gos; *hemi-* = half)	The **hemiazygos vein** is anterior to the vertebral column and slightly to the left of the midline. It often begins at the junction of the left ascending lumbar and left subcostal veins. It terminates by joining the azygos vein at about the level of the ninth thoracic vertebra. Generally, the hemiazygos vein drains the left side of the thoracic wall, thoracic viscera, and abdominal wall. Specifically, the hemiazygos vein receives blood from the ninth through eleventh **left posterior intercostal**, **esophageal**, **mediastinal**, and sometimes the **accessory hemiazygos veins**.
Accessory hemiazygos vein	The **accessory hemiazygos vein** is also anterior to the vertebral column and to the left of the midline. It begins at the fourth or fifth intercostal space and descends from the fifth to the eighth thoracic vertebra or ends in the hemiazygos vein. It terminates by joining the azygos vein at about the level of the eighth thoracic vertebra. The accessory hemiazygos vein drains the left side of the thoracic wall. It receives blood from the fourth through eighth **left posterior intercostal veins** (the first through third left posterior intercostal veins open into the left brachiocephalic vein), **left bronchial**, and **mediastinal veins**.

Figure 20.25 Principal veins of the thorax, abdomen, and pelvis.

(a) Scheme of drainage

(continues)

Figure 20.25 (continued)

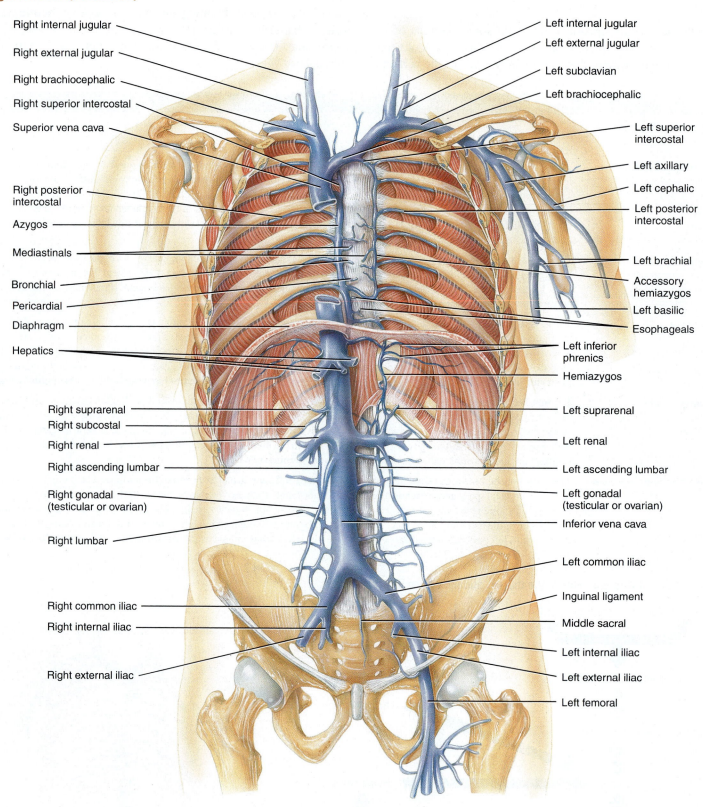

Right internal jugular

Right external jugular

Right brachiocephalic

Right superior intercostal

Superior vena cava

Right posterior intercostal

Azygos

Mediastinals

Bronchial

Pericardial

Diaphragm

Hepatics

Right suprarenal

Right subcostal

Right renal

Right ascending lumbar

Right gonadal (testicular or ovarian)

Right lumbar

Right common iliac

Right internal iliac

Right external iliac

Left internal jugular

Left external jugular

Left subclavian

Left brachiocephalic

Left superior intercostal

Left axillary

Left cephalic

Left posterior intercostal

Left brachial

Accessory hemiazygos

Left basilic

Esophageals

Left inferior phrenics

Hemiazygos

Left suprarenal

Left renal

Left ascending lumbar

Left gonadal (testicular or ovarian)

Inferior vena cava

Left common iliac

Inguinal ligament

Middle sacral

Left internal iliac

Left external iliac

Left femoral

(b) Anterior view

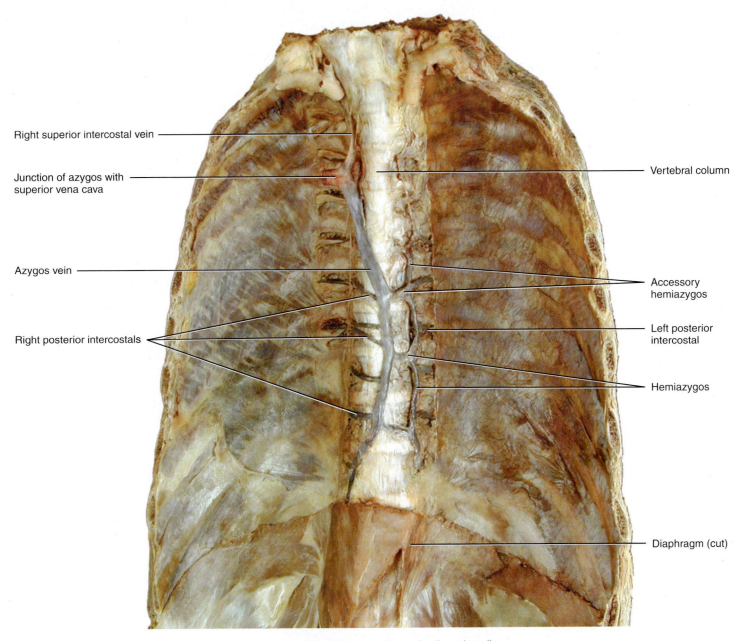

Right superior intercostal vein

Junction of azygos with
superior vena cava

Azygos vein

Right posterior intercostals

Vertebral column

Accessory
hemiazygos

Left posterior
intercostal

Hemiazygos

Diaphragm (cut)

(c) Anterior view of posterior thoracic wall

Most thoracic structures are drained by the azygos system of veins.

Checkpoint

46. What is the importance of the azygos system relative to the inferior vena cava?
47. Which components of the azygos system drain the left side of the thorax? Into which vein do they drain?

Veins of the Abdomen and Pelvis

Blood from the abdominal and pelvic viscera and abdominal wall returns to the heart through the inferior vena cava (Figure 20.26; Table 20.13). Many small veins enter the inferior vena cava. Most carry return flow from parietal branches of the abdominal aorta, and their names correspond to the names of the arteries (see also Figure 20.25b).

The inferior vena cava does not receive veins directly from the gastrointestinal tract, spleen, pancreas, and gallbladder. These organs pass their blood into a common vein, the **hepatic portal vein**, which delivers the blood to the liver. The superior mesenteric and splenic veins unite to form the hepatic portal vein (see Figure 20.28b). This special flow of venous blood, called the **hepatic portal circulation**, is described shortly. After passing through the liver for processing, blood drains into the hepatic veins, which empty into the inferior vena cava.

Figure 20.26   Drainage into the inferior vena cava.

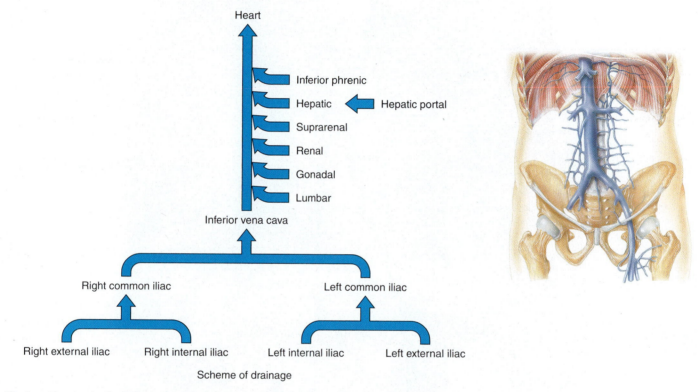

Scheme of drainage

The hepatic portal vein delivers blood to the liver from the gastrointestinal tract, spleen, pancreas, and gallbladder.

TABLE 20.13

Veins of the Abdomen and Pelvis

VEIN	DESCRIPTION AND REGION DRAINED
Inferior vena cava (VĒ-na CĀ-va; *vena* = vein; *cava* = cavelike)	The two common iliac veins that drain the lower limbs, pelvis, and abdomen unite to form the inferior vena cava. The inferior vena cava extends superiorly through the abdomen and thorax to the right atrium.
Common iliac veins (IL-ē-ak = pertaining to the ilium)	The **common iliac veins** are formed by the union of the internal and external iliac veins anterior to the sacroiliac joint and represent the distal continuation of the inferior vena cava at their bifurcation. The right common iliac vein is much shorter than the left and is also more vertical. Generally, the common iliac veins drain the pelvis, external genitals, and lower limbs.
Internal iliac veins	The **internal iliac veins** begin near the superior portion of the greater sciatic notch and run medial to their corresponding arteries. Generally, the veins drain the thigh, buttocks, external genitals, and pelvis.
External iliac veins	The **external iliac veins** are companions of the internal iliac arteries and begin at the inguinal ligaments as continuations of the femoral veins. They end anterior to the sacroiliac joint, where they join with the internal iliac veins to form the common iliac veins. The external iliac veins drain the lower limbs, cremaster muscle in males, and the abdominal wall.
Lumbar veins (LUM-bar = pertaining to the loin)	A series of parallel **lumbar veins**, usually four on each side, drain blood from both sides of the posterior abdominal wall, vertebral canal, spinal cord, and meninges. The lumbar veins run horizontally with the lumbar arteries. The lumbar veins connect at right angles with the right and left **ascending lumbar veins**, which form the origin of the corresponding azygos or hemiazygos vein. The lumbar veins drain blood into the ascending lumbars and then run to the inferior vena cava, where they release the remainder of the flow.
Gonadal veins (gō-NAD-al; *gono* = seed) [**testicular** (tes-TIK-ū-lar) or **ovarian** (ō-VAR-ē-an)]	The **gonadal veins** ascend with the gonadal arteries along the posterior abdominal wall. In the male, the gonadal veins are called the testicular veins. The **testicular veins** drain the testes (the left testicular vein empties into the left renal vein, and the right testicular vein drains into the inferior vena cava). In the female, the gonadal veins are called the ovarian veins. The **ovarian veins** drain the ovaries. The left ovarian vein empties into the left renal vein, and the right ovarian vein drains into the inferior vena cava.
Renal veins (RĒ-nal = *ren-* = kidney)	The large **renal veins** pass anterior to the renal arteries. The left renal vein is longer than the right renal vein and passes anterior to the abdominal aorta. It receives the left testicular (or ovarian), left inferior phrenic, and usually left suprarenal veins. The right renal vein empties into the inferior vena cava posterior to the duodenum. The renal veins drain the kidneys.
Suprarenal veins (soo´-pra-RĒ-nal; *supra-* = above)	The **suprarenal veins** drain the adrenal (suprarenal) glands (the left suprarenal vein empties into the left renal vein, and the right suprarenal vein empties into the inferior vena cava).
Inferior phrenic veins (FREN-ik = pertaining to the diaphragm)	The **inferior phrenic veins** drain the diaphragm (the left inferior phrenic vein usually sends one tributary to the left suprarenal vein, which empties into the left renal vein, and another tributary that empties into the inferior vena cava; the right inferior phrenic vein empties into the inferior vena cava).
Hepatic veins (he-PAT-ik = pertaining to the liver)	The **hepatic veins** drain the liver.

Checkpoint

48. Which vein returns blood from the abdominopelvic viscera to the heart?
49. Which structures are drained by the lumbar, gonadal, renal, suprarenal, inferior phrenic, and hepatic veins?

Veins of the Lower Limbs

As with the upper limbs, blood from the lower limbs is drained by both **superficial** and **deep veins** (Figure 20.27; Table 20.14). The superficial veins often anastomose with one another and with deep veins along their length. Deep veins, for the most part, have the same names as corresponding arteries. All veins of the lower limbs have valves, which are more numerous than in veins of the upper limbs.

Figure 20.27   Principal veins of the pelvis and lower limbs.

(a) Scheme of drainage

Inferior vena cava

Right common iliac

Right internal iliac

Right external iliac

Left common iliac

Right deep vein of the thigh

Right femoral

Right accessory saphenous

Right great saphenous

Right popliteal

Right small saphenous

Right fibular

Right anterior tibial

Right small saphenous

Right great saphenous

Right posterior tibial

Right dorsal venous arch

Right dorsal metatarsal

Right dorsal digital

Right medial plantar

Right lateral plantar

Right deep plantar venous arch

Right plantar metatarsal

Right plantar digital

(b) Anterior view

(c) Posterior view

Deep veins usually bear the names of their companion arteries.

TABLE 20.14

Veins of the Lower Limbs

VEIN	DESCRIPTION AND REGION DRAINED
Superficial Veins	
Great saphenous veins (sa-FĒ-nus; *saphen-* = clearly visible)	The **great (long) saphenous veins**, the longest veins in the body, ascend from the foot to the groin in the subcutaneous layer. They begin at the medial end of the dorsal venous arches of the foot. The **dorsal venous arches** (VĒ-nus) are networks of veins on the dorsum of the foot formed by the **dorsal digital veins**, which collect blood from the toes, and then unite in pairs to form the **dorsal metatarsal veins**, which parallel the metatarsals. As the dorsal metatarsal veins approach the foot, they combine to form the dorsal venous arches. The great saphenous veins pass anterior to the medial malleolus of the tibia and then superiorly along the medial aspect of the leg and thigh just deep to the skin. They receive tributaries from superficial tissues and connect with the deep veins as well. They empty into the femoral veins at the groin. Generally, the great saphenous veins drain mainly the medial side of the leg and thigh, the groin, external genitals, and abdominal wall. Along their length, the great saphenous veins have from 10 to 20 valves, with more located in the leg than the thigh. These veins are more likely to be subject to varicosities than other veins in the lower limbs because they must support a long column of blood and are not well supported by skeletal muscles. The great saphenous veins are often used for prolonged administration of intravenous fluids. This is particularly important in very young children and in patients of any age who are in shock and whose veins are collapsed. In coronary artery bypass grafting, if multiple blood vessels need to be grafted, sections of the great saphenous vein are used along with at least one artery as a graft. After the great saphenous vein is removed and divided into sections, the sections are used to bypass the blockages. The vein grafts are reversed so that the valves do not obstruct the flow of blood.
Small saphenous veins	The **small (short) saphenous veins** begin at the lateral aspect of the dorsal venous arches of the foot. They pass posterior to the lateral malleolus of the fibula and ascend deep to the skin along the posterior aspect of the leg. They empty into the popliteal veins in the popliteal fossa, posterior to the knee. Along their length, the small saphenous veins have from 9 to 12 valves. The small saphenous veins drain the foot and posterior aspect of the leg. They may communicate with the great saphenous veins in the proximal thigh.
Deep Veins	
Posterior tibial veins (TIB-ē-al)	The **plantar digital veins** on the plantar surfaces of the toes unite to form the **plantar metatarsal veins**, which parallel the metatarsals. They in turn unite to form the **deep plantar venous arches**. From each arch emerges the **medial** and **lateral plantar veins**. The medial and lateral plantar veins, posterior to the medial malleolus of the tibia, form the paired **posterior tibial veins**, which sometimes merge into a single vessel. They accompany the posterior tibial artery through the leg. They ascend deep to the muscles in the posterior aspect of the leg and drain the foot and posterior compartment muscles. About two-thirds of the way up the leg, the posterior tibial veins drain blood from the **fibular (peroneal) veins**, which drain the lateral and posterior leg muscles. The posterior tibial veins unite with the anterior tibial veins just inferior to the popliteal fossa to form the popliteal veins.
Anterior tibial veins	The paired **anterior tibial veins** arise in the dorsal venous arch and accompany the anterior tibial artery. They ascend in the interosseous membrane between the tibia and fibula and unite with the posterior tibial veins to form the popliteal vein. The anterior tibial veins drain the ankle joint, knee joint, tibiofibular joint, and anterior portion of the leg.
Popliteal veins (pop´-li-TĒ-al = pertaining to the hollow behind knee)	The **popliteal veins**, formed by the union of the anterior and posterior tibial veins, also receive blood from the small saphenous veins and tributaries that correspond to branches of the popliteal artery. The popliteal veins drain the knee joint and the skin, muscles, and bones of portions of the calf and thigh around the knee joint.
Femoral veins (FEM-o-ral)	The **femoral veins** accompany the femoral arteries and are the continuations of the popliteal veins just superior to the knee. The femoral veins extend up the posterior surface of the thighs and drain the muscles of the thighs, femurs, external genitals, and superficial lymph nodes. The largest tributaries of the femoral veins are the **deep veins of the thigh (deep femoral veins)**. Just before penetrating the abdominal wall, the femoral veins receive the deep femoral veins and the great saphenous veins. The veins formed from this union penetrate the body wall and enter the pelvic cavity. Here they are known as the **external iliac veins**. In order to take blood samples or pressure recordings from the right side of the heart, a catheter is inserted into the femoral vein as it passes through the femoral triangle. The catheter passes through the external and common iliac veins and inferior vena cava into the right atrium.

Checkpoint

50.   Which vein drains the medial side of the entire lower limb? The anterior leg? The knee joint?

51.   What is the clinical importance of the great saphenous veins?

Hepatic Portal Circulation

The **hepatic portal circulation** carries venous blood from the gastrointestinal organs and spleen to the liver (Figure 20.28). A vein that carries blood from one capillary network to another is called a **portal vein**. The **hepatic portal vein** (*hepat-* = liver) receives blood from capillaries of gastrointestinal organs and the spleen and delivers it to the sinusoids of the liver. After a meal, hepatic portal blood is rich in nutrients absorbed from the gastrointestinal tract. The liver stores some of them and modifies others before they pass into the general circulation. For example, the liver converts glucose into glycogen for storage, reducing blood glucose level shortly after a meal. The liver also detoxifies harmful substances, such as alcohol, that have been absorbed from the gastrointestinal tract and destroys bacteria by phagocytosis.

The splenic and superior mesenteric veins unite to form the hepatic portal vein. The **splenic vein** drains blood from the spleen, stomach, pancreas, and portions of the large intestine. The **superior mesenteric vein** drains blood from the small intestine and portions of the large intestine, stomach, and pancreas. The *inferior mesenteric vein*, which passes into the splenic vein, drains portions of the large intestine. The *right* and *left gastric veins*, which open directly into the hepatic portal vein, drain the stomach. The *cystic vein*, which also opens into the hepatic portal vein, drains the gallbladder.

At the same time the liver is receiving nutrient-rich but deoxygenated blood via the hepatic portal vein, it is also receiving oxygenated blood from the common hepatic artery, a branch of the celiac trunk. The oxygenated blood mixes with the deoxygenated blood in sinusoids. Eventually, blood leaves the sinusoids of the liver through the **hepatic veins**, which drain into the inferior vena cava.

Figure 20.28 Hepatic portal circulation. A schematic diagram of blood flow through the liver, including arterial circulation, is shown in (a); deoxygenated blood is indicated in blue, oxygenated blood in red.

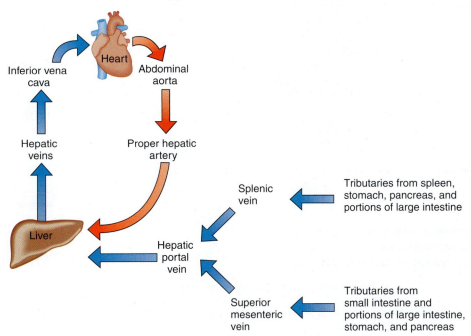

(a) Scheme of principal blood vessels of hepatic portal circulation and arterial supply and venous drainage of liver

(continues)

Figure 20.28 (continued)

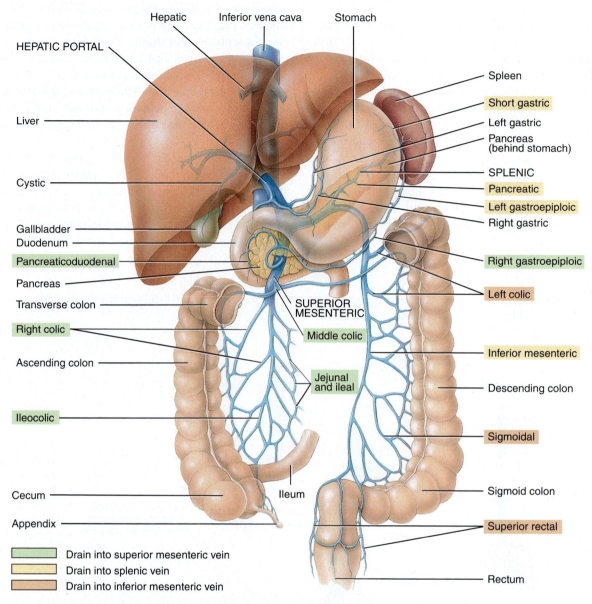

Hepatic
Inferior vena cava
Stomach
HEPATIC PORTAL
Spleen
Short gastric
Left gastric
Pancreas
(behind stomach)
Liver
SPLENIC
Pancreatic
Left gastroepiploic
Right gastric
Cystic
Gallbladder
Duodenum
Right gastroepiploic
Pancreaticoduodenal
Pancreas
Left colic
Transverse colon
SUPERIOR
MESENTERIC
Right colic
Middle colic
Ascending colon
Inferior mesenteric
Jejunal
and ileal
Descending colon
Ileocolic
Sigmoidal
Cecum
Ileum
Sigmoid colon
Appendix
Superior rectal
Rectum

Drain into superior mesenteric vein
Drain into splenic vein
Drain into inferior mesenteric vein

(b) Anterior view of veins draining into the hepatic portal vein

🔑 **The hepatic portal circulation delivers venous blood from the organs of the gastrointestinal tract and spleen to the liver.**

Checkpoint

52.   Which vein carries nutrient-rich blood to the liver? Which artery carries oxygen-rich blood to the liver?

53.   Which veins carry blood away from the liver?

David's Story: Epilogue and Discussion

▶▶▶David, a retired man formerly in good health, is taken to the clinic complaining of fatigue, nausea, and vomiting. He denies that anything is wrong, but upon examination it is found that he is confused, has cool, clammy skin, rapid pulse and respirations, and is hypotensive (BP = 90/60). He has not been producing a normal volume of urine, and indicates that he sustained an injury to his arm 2 weeks prior and had a fever 1 week prior to his exam. The wound appears infected. Palpation of the abdomen reveals enlargement of the liver, and capillary refill times are delayed. A diagnosis of septicemia with impending septic shock is made, and David is admitted to the hospital immediately. David's low blood pressure is due to reduced blood volume caused by damage to the capillaries initiated by bacterial toxins. The capillaries have become leaky, and plasma proteins and fluid have left the circulatory system and are infiltrating his tissues. David is placed on fluid replacement therapy to increase his blood volume, an antibiotic to address the potential infection, and oxygen to increase his blood oxygen levels. Following admission, David experiences a deep vein thrombosis in which fragments form an embolus that travels the venous circulatory system to David's lungs. The pulmonary embolism induces an acute respiratory crisis; David's blood oxygen levels drop rapidly, and respiratory distress occurs.

The treatment options available to David's doctor will depend on the severity of the pulmonary artery obstruction. Treatment goals typically focus on dealing with acute symptoms and preventing further pulmonary emboli. In severe cases, surgery or treatment with clot-busting drugs may be tried. The overall mortality rate for pulmonary embolism is around 15%. In milder cases, individuals can survive with pulmonary emboli, although with reduced oxygen exchange at the lung.◀◀◀

K. Why would the loss of plasma proteins cause fluid to leave the capillaries and enter David's tissues?
L. David's low blood pressure has resulted from loss of blood volume. How did his body respond to the blood volume decrease?
M. How did the cardiovascular center of David's brain respond to his condition?
N. David's condition led to the formation of a pulmonary embolism, a blood clot that is blocking blood flow to his lungs. Why would this reduce his blood oxygen? Which pathway does blood travel to get to and from the lungs? How might this affect David's cardiac output?

Concept and Resource Summary

Concept   **Resources**

Introduction

1. The cardiovascular system transports materials throughout the body. Blood vessels carry blood with these materials to all body tissues and then return it to the heart.

Concept 20.1 Most blood vessel walls have three distinct tissue layers.

1. The arterial wall has three tunics: the innermost **tunica interna**, which surrounds the lumen where blood flows, the middle **tunica media**, and the outermost **tunica externa**.

Hear
- **Figure 20.1 –** **Comparative Structure of Blood Vessels**

Concept 20.2 Blood ejected from the heart flows through elastic arteries, muscular arteries, and then arterioles.

1. **Arteries** carry blood away from the heart. Sympathetic fibers of the ANS stimulate **vasoconstriction**. A decrease in sympathetic stimulation causes **vasodilation**.
2. **Elastic arteries** are those close to the heart that propel blood onward when the ventricles are relaxing and have highly elastic walls to accommodate the surge of blood.
3. **Muscular arteries** have more smooth muscle and less elastic fibers for greater vasoconstriction to adjust the rate of blood flow.
4. Many blood vessels **anastomose**, forming **collateral circulation** routes.
5. **Arterioles** deliver blood to capillaries.

See
- **Anatomy Overview – Arteries and Arterioles**

Concept

Concept 20.3 Capillaries are microscopic blood vessels that function in exchange between blood and interstitial fluid.

1. Capillaries are microscopic blood vessels through which materials are exchanged between blood and the interstitial fluid bathing tissue cells.
2. **Precapillary sphincters** regulate blood flow through a **capillary bed**.
3. **Autoregulation** adjusts blood flow through a capillary bed in response to physical and chemical changes in a tissue.
4. **Continuous capillaries** lack *fenestrations* and allow only limited diffusion. **Fenestrated capillaries** have many fenestrations that allow quick fluid and larger exchange. **Sinusoids** have large fenestrations, an incomplete *basement membrane*, and are quite leaky.
5. During **capillary exchange**, substances move across capillary walls by simple diffusion, or **transcytosis**, within vesicles. **Capillary blood pressure** "pushes" fluids out of capillaries into interstitial fluid (**filtration**). **Colloid osmotic pressure** "pulls" fluid into capillaries from interstitial fluid (**reabsorption**).

See

- **Anatomy Overview – Capillaries**
- **Animation – Capillary Exchange**

Hear

- **Figure 20.5 – Capillary Exchange**

Do

- **Exercise – Capillary Exchange Pick 'em**
- **Concepts and Connections – Capillary Exchange**

Concept 20.4 Venules and veins return blood to the heart.

1. **Venules** are small vessels that collect blood from capillaries and drain into veins.
2. **Veins** have thinner tunica interna and tunica media than arteries. Veins have **valves** to prevent backflow of blood. A **venous sinus** is a vein with very thin walls.
3. **Venous return**, the volume of blood flowing back to the heart through systemic veins, occurs due to the pumping action of the heart, aided by skeletal muscle contractions (the **skeletal muscle pump**), and pressure changes associated with breathing (the **respiratory pump**).
4. Systemic veins are collectively called **blood reservoirs** because they hold a large volume of blood. If the need arises, this blood can be shifted into other blood vessels through vasoconstriction of veins. The principal blood reservoirs are the veins of abdominal organs and skin.

See

- **Anatomy Overview – Veins and Venules**

Do

- **Exercise – Vein Archery**

Concept 20.5 Blood flow occurs from regions of higher pressure to those of lower pressure.

1. **Blood flow** is the volume of blood flowing through a tissue in a given period. *Total blood flow*, the volume of blood flowing through all systemic blood vessels each minute, is equal to the cardiac output.
2. The greatest blood flow occurs where there is the greatest pressure gradient. Ventricular contraction generates **blood pressure**, which is the pressure exerted by blood on blood vessel walls. **Systolic blood pressure** is the highest pressure during systole, and **diastolic blood pressure** is the lowest pressure during diastole. Systemic BP is highest in the aorta and lowest in the vena cava.
3. **Vascular resistance** is the friction between blood and the blood vessel walls. Vascular resistance depends on the size of the blood vessel lumen, blood viscosity, and total blood vessel length.
4. **Systemic vascular resistance (SVR)** refers to all the vascular resistances from all systemic blood vessels.

See

- **Animation – MABP**
- **Animation – Vascular Regulation**
- **Animation – Lymph Flow**

Hear

- **Figure 20.10 – Summary of Factors That Increase Blood Pressure**

Do

- **Concepts and Connections – Blood Flow**

Concept	Resources

Concept 20.6 Blood pressure is regulated by neural and hormonal negative feedback systems.

1. The **cardiovascular center** in the medulla oblongata regulates heart rate, contractility, and blood vessel diameter.
2. The cardiovascular center receives input from higher regions and sensory receptors (*baroreceptors* and *chemoreceptors*).
3. Output from the cardiovascular center flows along sympathetic and parasympathetic axons. Sympathetic impulses propagated along **cardiac accelerator nerves** increase heart rate and contractility; parasympathetic impulses propagated along **vagus nerves** decrease heart rate.
4. Baroreceptors are pressure-sensitive sensory receptors that monitor blood pressure. Chemoreceptors monitor blood levels of oxygen, carbon dioxide, and hydrogen ions.
5. Hormones help regulate blood pressure. **Angiotensin II** and **antidiuretic hormone** cause vasoconstriction. **Aldosterone** increases blood volume. **Epinephrine** and **norepinephrine** increase cardiac output and cause vasoconstriction. **Atrial natriuretic peptide** causes vasodilation.

See
- Animation – Regulating Blood Pressure

Do
- Exercise – Regulate BP with Hormones
- Exercise – Regulate BP with Nerve Impulses
- Concepts and Connections – Blood Pressure Regulation

Concept 20.7 Measurement of the pulse and blood pressure helps assess cardiovascular system function.

1. **Pulse** is the alternate expansion and recoil of an artery after each contraction of the left ventricle. Pulse is strongest in the arteries closest to the heart and can be felt in any artery that lies near the body surface. Pulse rate is normally 70 to 80 beats per minute.
2. The pressure in arteries generated by left ventricular systole, and the pressure remaining in the arteries during left ventricular diastole is termed **blood pressure**. A **sphygmomanometer** measures blood pressure. **Systolic blood pressure** is the arteriole blood pressure during ventricular contraction; **diastolic blood pressure** is the arteriole pressure during ventricular relaxation. Normal young adult BP is less than 120/80 (systolic/diastolic).
3. Pulse pressure is the difference between systolic and diastolic blood pressure.

See
- Animation – MABP

Concept 20.8 The two main circulatory routes are the pulmonary circulation and the systemic circulation.

1. Blood vessels are organized into two main **circulatory routes**: the pulmonary circulation and the systemic circulation.
2. The **pulmonary circulation** carries deoxygenated blood from the right ventricle to the air sacs of the lungs and returns oxygenated blood from the air sacs to the left atrium.
3. The **systemic circulation** carries oxygenated blood from the left ventricle to systemic capillaries throughout the body, and returns the deoxygenated blood to the right atrium.

See
- Anatomy Overview – Comparison of Circulatory Routes
- Anatomy Overview – Pulmonary Circulation

Concept

Concept 20.9 Systemic arteries carry blood from the heart to all body organs except the lungs.

1. The **aorta** is divided into the ascending aorta, arch of the aorta, thoracic aorta, and abdominal aorta.
2. The **ascending aorta** begins at the heart and gives rise to the right and left **coronary arteries** serving the myocardium of the heart.
3. The ascending aorta continues as the **arch of the aorta**, which has three major branches: the **brachiocephalic trunk**, the **left common carotid artery**, and the **left subclavian artery**.
4. The **thoracic aorta** is a continuation of the arch of the aorta and gives off branches that supply thoracic organs. As it passes through the diaphragm to enter the abdominal cavity, it becomes the abdominal aorta.
5. The **abdominal aorta** gives off branches that supply the abdominal organs.
6. The **common iliac arteries** branch from the abdominal aorta to divide into several arteries that supply the pelvis and lower limbs.

See

• **Anatomy Overview – Comparison of Circulatory Routes**

Do

• **Exercise – Artery Archery**

Concept 20.10 Systemic veins return blood to the heart from all body organs except the lungs.

1. All veins of the systemic circulation drain into the **superior vena cava, inferior vena cava**, or the **coronary sinus**, which in turn empty into the right atrium. The superior vena cava receives blood from veins above the diaphragm, the inferior vena cava receives blood from veins below the diaphragm, and the coronary sinus drains the heart wall.
2. Three pairs of veins drain most of the blood from the head: the **internal jugular**, **external jugular**, and **vertebral veins**. The **dural venous sinuses** receive blood from all veins within the brain and drain it into the internal jugular veins.
3. The upper limbs have large, **superficial veins** that anastomose with one another and with deep veins. **Deep veins** usually accompany arteries.
4. The **azygos system**, composed of the **azygos**, **hemiazygos**, and **accessory hemiazygos veins**, drains most of the thoracic structures. They ultimately drain into the superior vena cava.
5. Blood from the gastrointestinal tract, spleen, pancreas, and gallbladder passes into the **hepatic portal vein** for transport to the liver for processing before emptying into the inferior vena cava.
6. The lower limb has both superficial and deep veins.
7. The **hepatic portal circulation** directs venous blood from the gastrointestinal organs and spleen into the **hepatic portal vein** of the liver before it returns to the heart. This portal system enables the liver to utilize nutrients and detoxify harmful substances in the blood. The **splenic vein** drains the spleen, stomach, and large intestine. The **superior mesenteric vein** drains the small intestine, large intestine, stomach, and pancreas. The **hepatic veins** drain the liver.

Hear

• **Figure 20.28 – Hepatic Portal Circulation**

Do

• **Exercise – Vein Archery**

Other Chapter Resources

For additional information on clinical applications, medical disorders, and medical terminology, see Chapter 9 in the *Clinical Connections* manual packaged with your text.

Within *WileyPLUS*
- Chapter Quiz
- Visual Anatomy
- Mastering Vocabulary
- Audio Glossary
- Homeostatic Imbalance: The Case of the Anxious Man with Chest Pains
- Web Links

• Understanding the Concepts

1. In atherosclerosis, the walls of elastic arteries become less compliant (stiffer and, therefore, less able to stretch). What effect does reduced compliance have on the pressure reservoir function of arteries?

2. Why do metabolically active tissues have extensive capillary networks?

3. A person who has liver failure cannot synthesize the normal amount of plasma proteins. How does a deficit of plasma proteins affect blood colloid osmotic pressure? What is the effect on capillary filtration and reabsorption?

4. Why are valves more important in arm veins and leg veins than in neck veins?

5. While exercising, your cardiac output increases. Does your total blood flow increase or decrease? Explain your answer.

6. What happens to blood pressure when a blood vessel's distance from the left ventricle increases? When you're dehydrated? When dieting decreases adipose tissue volume?

7. Would the negative feedback cycle illustrated in Figure 20.12 happen when you lie down or when you stand up? Explain your answer.

8. Which branches of the coronary arteries supply the left ventricle? Why does the left ventricle have such an extensive arterial blood supply?

9. What is the main function of the hepatic portal circulation?

21

The Lymphatic System and Immunity

It had been 7 weeks since her surgery and Marlene was still feeling the effects of the lumpectomy, axillary dissection (removal of armpit lymph nodes), and radiation therapy. When the doctor had told her she had breast cancer, Marlene's world was turned upside down; she was only 57 years old, healthy and active. Her initial shock and numbness had given way to a determination to "beat this thing." The doctor had said the prognosis was encouraging. Several lymph nodes were removed from Marlene's armpit at the time of the lumpectomy to determine if the cancer present in the bean-sized lump had spread to the lymphatic system and potentially to the rest of her body. Thankfully it had not. She was tired and disoriented after the radiation treatments; even now she was only beginning to recover. She felt good enough this evening to wash the dishes herself, rather than having her husband do them.

The surgery had left her with a nagging pinching sensation in her armpit and numb areas of skin near the scar, but overall she felt satisfied with the result. The swelling in her arm and hand did bother her, however. She didn't think much about the swelling at first, as the surgeon indicated it might occur. It had started several days ago, but over the last few days her swollen arm and hand had become bothersome.

The lymphatic and immune systems carry out immune responses and assist the body by returning interstitial fluid to the venous circulation. Both functions assist in maintaining the overall homeostasis of the human body. One of the side effects of removal of lymphatic tissues can be swelling, also referred to as edema. However, since the immune system is anatomically part of the lymphatic system, both systems may be affected by procedures such as the one Marlene has undergone.

INTRODUCTION

Maintaining physical health requires continuous combat against harmful agents in our internal and external environments. Despite constant exposure to a variety of **pathogens** (PATH-ō-jens), disease-producing microbes such as bacteria and viruses, most people remain healthy. The body surface also endures cuts and bumps, exposure to ultraviolet rays in sunlight, chemical toxins, and minor burns with an array of defensive ploys. **Immunity**, or **resistance**, is the ability to ward off damage or disease through our defenses. Vulnerability, or lack of resistance, is termed **susceptibility**.

The two general types of immunity are (1) innate and (2) adaptive. **Innate (nonspecific) immunity** refers to defenses that are present at birth.

> *Immunity is the ability to ward off damage or disease through our defenses.*

They are always present and available to provide rapid responses to protect us against disease. Among the components of innate immunity are the first line of defense (the physical and chemical barriers of the skin and mucous membranes) and the second line of defense (antimicrobial proteins, natural killer cells, phagocytes, inflammation, and fever).

Adaptive (specific) immunity develops in response to contact with a *specific* invader. It occurs more *slowly* than innate immunity mechanisms and involves activation of lymphocytes (a type of white blood cell) that can combat a specific invader.

The organ system responsible for adaptive immunity (and some aspects of innate immunity) is the lymphatic system. This system is closely allied with the cardiovascular system, and it also functions with the digestive system in the absorption of fatty foods. In this chapter, we explore the mechanisms that provide defenses against intruders and promote the repair of damaged body tissues.

CONCEPTS

- **21.1** The lymphatic system drains interstitial fluid, transports dietary lipids, and protects against invasion.

- **21.2** Lymph flows through lymphatic capillaries, lymphatic vessels, and lymph nodes.

- **21.3** The lymphatic organs and tissues include the thymus, lymph nodes, spleen, and lymphatic nodules.

- **21.4** Innate immunity includes external physical and chemical barriers and various internal defenses.

- **21.5** Adaptive immunity involves the production of a specific lymphocyte or antibody against a specific antigen.

- **21.6** In cell-mediated immunity, cytotoxic T cells directly attack target cells.

- **21.7** In antibody-mediated immunity, antibodies specifically target a particular antigen.

- **21.8** The complement system destroys microbes through phagocytosis, cytolysis, and inflammation.

- **21.9** Immunological memory results in a more intense secondary response to an antigen.

21.1 The lymphatic system drains interstitial fluid, transports dietary lipids, and protects against invasion.

The **lymphatic system** (lim-FAT-ik) consists of a fluid called lymph, vessels called lymphatic vessels that transport the lymph, a number of structures and organs containing lymphatic tissue, and red bone marrow, where stem cells develop into the various types of blood cells, including lymphocytes (Figure 21.1). The lymphatic system assists in circulating body fluids and helps de-

Figure 21.1   Components of the lymphatic system.

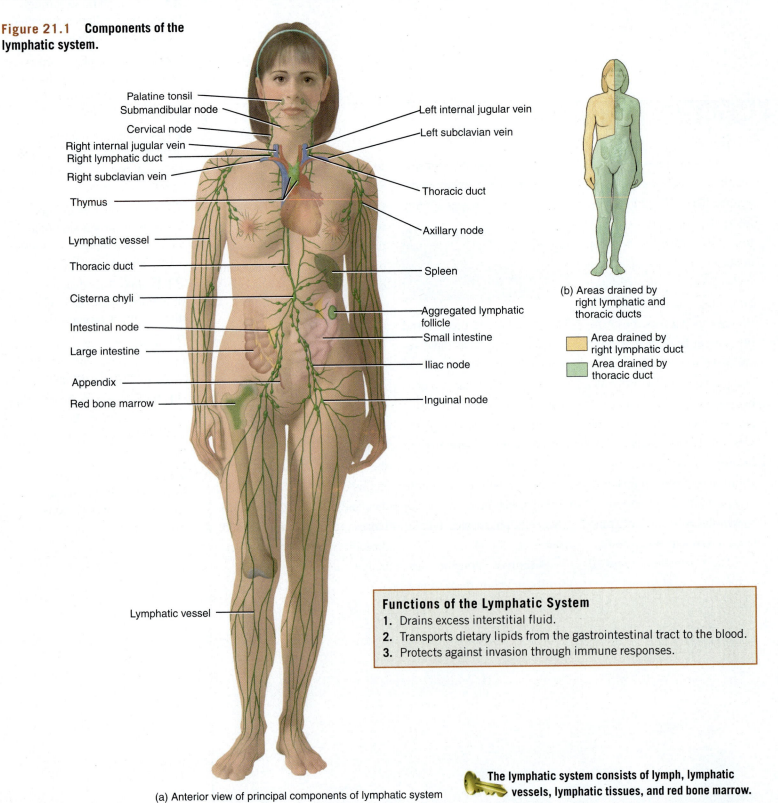

Palatine tonsil
Submandibular node
Cervical node
Right internal jugular vein
Right lymphatic duct
Right subclavian vein
Thymus
Lymphatic vessel
Thoracic duct
Cisterna chyli
Intestinal node
Large intestine
Appendix
Red bone marrow

Left internal jugular vein
Left subclavian vein
Thoracic duct
Axillary node
Spleen
Aggregated lymphatic follicle
Small intestine
Iliac node
Inguinal node

Lymphatic vessel

(b) Areas drained by right lymphatic and thoracic ducts

☐ Area drained by right lymphatic duct
☐ Area drained by thoracic duct

Functions of the Lymphatic System
1. Drains excess interstitial fluid.
2. Transports dietary lipids from the gastrointestinal tract to the blood.
3. Protects against invasion through immune responses.

(a) Anterior view of principal components of lymphatic system

🔑 **The lymphatic system consists of lymph, lymphatic vessels, lymphatic tissues, and red bone marrow.**

fend the body against disease-causing agents. Most components of blood plasma filter out of blood capillary walls to form interstitial fluid, the fluid that surrounds the cells of body tissues. After interstitial fluid passes into lymphatic vessels, it is called **lymph** (LIMF = clear fluid). The major difference between interstitial fluid and lymph is location: Interstitial fluid is found between cells, and lymph is located within lymphatic vessels and lymphatic tissue.

Lymphatic tissue is a specialized form of reticular connective tissue (see Table 4.2C) that contains large numbers of lymphocytes. Recall from Chapter 18 that lymphocytes are white blood cells. Two types of lymphocytes participate in adaptive immune responses: B lymphocytes (B cells) and T lymphocytes (T cells).

The lymphatic system has three primary functions:

1. ***Drains excess interstitial fluid***. Lymphatic vessels drain excess interstitial fluid from tissue spaces and return it to the blood.

2. ***Transports dietary lipids***. Lymphatic vessels transport lipids and lipid-soluble vitamins (A, D, E, and K) absorbed by the gastrointestinal tract.

3. ***Carries out immune responses***. Lymphatic tissue initiates highly specific responses directed against particular microbes or abnormal cells.

Checkpoint

1. What is lymphatic tissue?
2. What are the major functions of the lymphatic system?

21.2 Lymph flows through lymphatic capillaries, lymphatic vessels, and lymph nodes.

Lymphatic vessels begin as **lymphatic capillaries**. These tiny vessels, which are located in the spaces between cells, are closed at one end (Figure 21.2). Just as blood capillaries converge to form venules and then veins, lymphatic capillaries unite to form larger lymphatic vessels (Figure 21.1), which resemble veins in structure but have thinner walls and more valves. At intervals along the lymphatic vessels, lymph flows through lymph nodes, encapsulated bean-shaped organs consisting of masses of B cells and T cells. In the skin, lymphatic vessels lie in the subcutaneous tissue and generally follow the same route as veins; lymphatic vessels of the viscera generally follow arteries, forming plexuses (networks)

around them. Tissues that lack lymphatic capillaries include avascular tissues (such as cartilage, the epidermis, and the cornea of the eye), the central nervous system, portions of the spleen, and red bone marrow.

Lymphatic Capillaries

Lymphatic capillaries are slightly larger in diameter than blood capillaries and have a unique one-way structure that permits interstitial fluid to flow into them but not out. The ends of endothelial

Figure 21.2 **Lymphatic capillaries.**

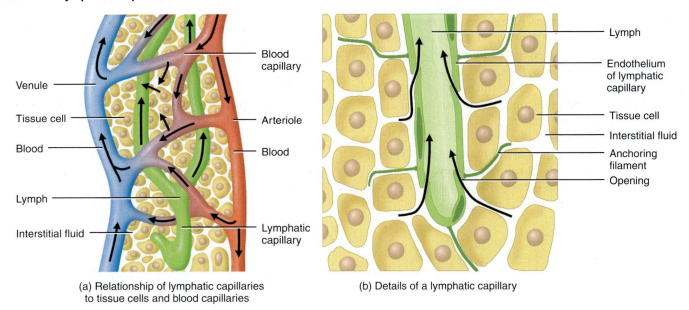

(a) Relationship of lymphatic capillaries to tissue cells and blood capillaries

(b) Details of a lymphatic capillary

After interstitial fluid passes into lymphatic capillaries, it is called lymph.

cells that make up the wall of a lymphatic capillary overlap (Figure 21.2b). When pressure is greater in the interstitial fluid than in lymph, the cells separate slightly, like the opening of a one-way swinging door, and interstitial fluid enters the lymphatic capillary. When pressure is greater inside the lymphatic capillary, the cells adhere more closely, and lymph cannot escape back into interstitial fluid. Attached to the lymphatic capillaries are *anchoring filaments*, which contain elastic fibers. They extend out from the lymphatic capillary, attaching lymphatic endothelial cells to surrounding tissues. When excess interstitial fluid accumulates and causes tissue swelling (**edema**), the anchoring filaments are pulled, making the openings between cells even larger so that more fluid can flow into the lymphatic capillary.

Lymph Trunks and Ducts

As you have already learned, lymph passes from lymphatic capillaries into **lymphatic vessels** and then through lymph nodes. Lymphatic vessels pass lymph, node to node, through sequential chains of lymph nodes. Upon exiting a chain of nodes, lymphatic vessels unite to form **lymph trunks** (Figure 21.3). The **lumbar trunks** drain lymph from the lower limbs, pelvis, kidneys, adrenal glands, and abdominal wall. The **intestinal trunk** drains lymph from the stomach, intestines, pancreas, spleen, and liver. The **bronchomediastinal trunks** drain lymph from the thoracic wall, lung, and heart. The **subclavian trunks** drain the upper limbs. The **jugular trunks** drain the head and neck.

Lymph passes from lymph trunks into two main channels, the thoracic duct and the right lymphatic duct, and then drains into venous blood. The **thoracic (left lymphatic) duct** begins as a dilation called the **cisterna chyli** (sis-TER-na KĪ-lē; *cisterna* = cavity or reservoir) anterior to the second lumbar vertebra. The cisterna chyli receives lymph from the right and left lumbar trunks and from the intestinal trunk. The thoracic duct is the main duct for return of lymph to the bloodstream. In the neck, the thoracic duct also receives lymph from the left jugular, left subclavian, and left bronchomediastinal trunks. Therefore, the thoracic duct receives lymph from the left side of the head, neck, and chest; the left upper limb; and the entire body inferior to the ribs (see Figure 21.1b). The thoracic duct in turn drains lymph into venous blood at the junction of the left internal jugular and left subclavian veins.

The **right lymphatic duct** (Figure 21.3) is about 1.2 cm (0.5 in.) long and receives lymph from the right jugular, right subclavian, and right bronchomediastinal trunks. Thus, the right lymphatic duct receives lymph from the upper right side of the body

(see Figure 21.1b). From the right lymphatic duct, lymph drains into venous blood at the junction of the right internal jugular and right subclavian veins.

Formation and Flow of Lymph

Most components of blood plasma filter freely through the capillary walls to form interstitial fluid, but more fluid filters out of blood capillaries than returns to them by reabsorption (see Figure 20.5). The excess filtered fluid—about 3 liters per day—drains into lymphatic vessels and becomes lymph. Because most plasma proteins are too large to leave blood vessels, interstitial fluid contains only a small amount of protein. Proteins that do leave blood plasma cannot return to the blood by diffusion because the concentration gradient (high level of proteins inside blood capillaries, low level outside) opposes such movement. The proteins can, however, move readily through the more permeable lymphatic capillaries into lymph. Thus, an important function of lymphatic vessels is to return lost plasma proteins to the bloodstream.

Like veins, lymphatic vessels contain valves, which ensure the one-way movement of lymph. As noted previously, lymph drains into venous blood through the right lymphatic duct and the thoracic duct at the junction of the internal jugular and subclavian veins (Figure 21.3). Thus, the sequence of fluid flow is blood capillaries (blood) → interstitial spaces (interstitial fluid) → lymphatic capillaries (lymph) → lymphatic vessels (lymph) → lymph nodes (lymph) → lymphatic vessels (lymph) → lymphatic ducts (lymph) → junction of the internal jugular and subclavian veins (blood). Figure 21.4 illustrates this sequence, along with the relationship of the lymphatic and cardiovascular systems.

When a lymphatic vessel distends, the smooth muscle in its wall contracts, which helps move lymph from one segment of the vessel to the next. In addition, the same two "pumps" that aid the return of venous blood to the heart maintain the flow of lymph:

1. **Skeletal muscle pump**. The "milking" action of skeletal muscle contractions (see Figure 20.7) compresses lymphatic vessels (as well as veins) and forces lymph toward the junction of the internal jugular and subclavian veins.

2. **Respiratory pump**. Lymph flow is also maintained by pressure changes that occur during inhalation (breathing in). Lymph flows from the abdominal region, where the pressure is higher, toward the thoracic region, where it is lower. When the pressures reverse during exhalation (breathing out), the valves prevent backflow of lymph.

Figure 21.3 Routes for drainage of lymph from lymph trunks into the thoracic and right lymphatic ducts.

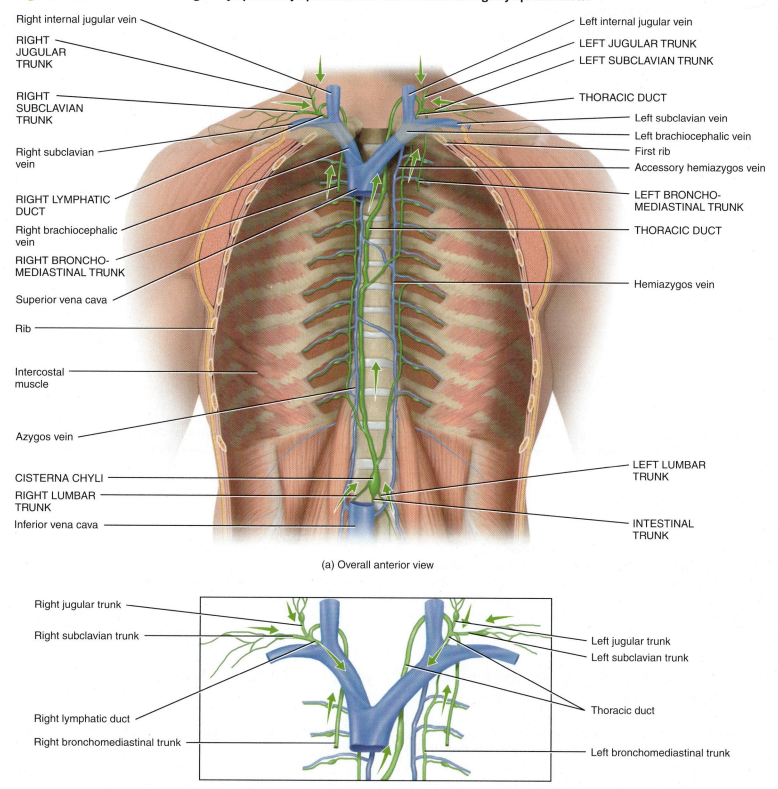

Right internal jugular vein
RIGHT JUGULAR TRUNK
RIGHT SUBCLAVIAN TRUNK
Right subclavian vein
RIGHT LYMPHATIC DUCT
Right brachiocephalic vein
RIGHT BRONCHO-MEDIASTINAL TRUNK
Superior vena cava
Rib
Intercostal muscle
Azygos vein
CISTERNA CHYLI
RIGHT LUMBAR TRUNK
Inferior vena cava

Left internal jugular vein
LEFT JUGULAR TRUNK
LEFT SUBCLAVIAN TRUNK
THORACIC DUCT
Left subclavian vein
Left brachiocephalic vein
First rib
Accessory hemiazygos vein
LEFT BRONCHO-MEDIASTINAL TRUNK
THORACIC DUCT
Hemiazygos vein
LEFT LUMBAR TRUNK
INTESTINAL TRUNK

(a) Overall anterior view

Right jugular trunk
Right subclavian trunk
Right lymphatic duct
Right bronchomediastinal trunk

Left jugular trunk
Left subclavian trunk
Thoracic duct
Left bronchomediastinal trunk

(b) Detailed anterior view

Lymph returns to the bloodstream through the thoracic duct and right lymphatic duct.

Figure 21.4   **Schematic diagram showing the relationship of the lymphatic system to the cardiovascular system.** The arrows indicate the direction of flow of lymph and blood.

SYSTEMIC CIRCULATION   PULMONARY CIRCULATION

🔑 **The sequence of fluid flow is blood capillaries (blood) → interstitial spaces (interstitial fluid) → lymphatic capillaries (lymph) → lymphatic vessels (lymph) → lymph nodes (lymph) → lymphatic vessels (lymph) → lymphatic ducts (lymph) → junction of the internal jugular and subclavian veins (blood).**

Checkpoint

3. How do lymphatic vessels differ in structure from veins?
4. How do lymphatic capillaries allow interstitial fluid to flow into them but not out?
5. Which regions of the body are drained by the thoracic duct? By the right lymphatic duct? Where does each of these ducts drain lymph into venous blood?
6. Which lymphatic vessels empty into the cisterna chyli, and which duct receives lymph from the cisterna chyli?
7. What is the route of fluid flow from blood capillaries, through the lymph circulation, and back to the bloodstream?

21.3   The lymphatic organs and tissues include the thymus, lymph nodes, spleen, and lymphatic nodules.

The widely distributed lymphatic organs and tissues are classified into two groups based on their functions. **Primary lymphatic organs** and **tissues** are the sites where stem cells divide and develop into mature B cells and T cells capable of mounting an immune response. The primary lymphatic organs are the red bone marrow (in flat bones and the epiphyses of long bones of adults) and the thymus. The **secondary lymphatic organs** and **tissues** are the sites where most immune responses occur. They include lymph nodes, the spleen, and lymphatic nodules. The thymus, lymph nodes, and spleen are considered organs because each is surrounded by a con-

nective tissue capsule; lymphatic nodules, in contrast, are not considered organs because they lack a capsule.

Thymus

The **thymus** is a bilobed organ located posterior to the sternum, medial to the lungs, and superior to the heart (Figure 21.5). It contains large numbers of T cells and scattered dendritic cells, epithelial cells, and macrophages. Immature T cells migrate from

Figure 21.5 **Thymus of an adolescent.**

- Thyroid gland
- Trachea
- Right common carotid artery
- Superior vena cava
- Right lung
- Brachiocephalic veins
- Thymus
- Parietal pericardium
- Left lung
- Diaphragm

🔑 The bilobed thymus is largest at puberty and then atrophies with age.

red bone marrow to the thymus, where they multiply and begin to mature. **Dendritic cells** (*dendr-* = a tree), so named because they have long, branched projections that resemble the dendrites of a neuron, assist in the maturation process. As discussed shortly, dendritic cells in other parts of the body, such as lymph nodes, play another key role in immune responses. Each of the specialized **epithelial cells** has several long processes that surround and serve as a framework for as many as 50 T cells. These epithelial cells help "educate" the immature T cells to distinguish your own molecules from foreign molecules. Additionally, they produce hormones that are thought to aid in the maturation of T cells. Only about 2% of T cells developing in the thymus "graduate" into mature T cells. The remaining cells die through apoptosis (programmed cell death). Thymic **macrophages** help clear out the debris of dead and dying cells. Mature T cells leave the thymus via the blood and are carried to lymph nodes, the spleen, and other lymphatic tissues, where they populate parts of these organs and tissues.

In infants, the thymus is large, with a mass of about 70 g (2.3 oz). After puberty, adipose and areolar connective tissues begin

to replace the thymic tissue. By the time a person reaches maturity, the gland has atrophied considerably, and in old age it may weigh only 3 g (0.1 oz). Before the thymus atrophies, it populates the secondary lymphatic organs and tissues with T cells. However, some T cells continue to proliferate in the thymus throughout an individual's lifetime.

Lymph Nodes

Located along lymphatic vessels are about 600 bean-shaped **lymph nodes**. They are scattered throughout the body, both superficially and deep, and usually occur in groups (see Figure 21.1a). Large groups of lymph nodes are present near the mammary glands and axillae and groin.

Lymph nodes are covered by a capsule of dense connective tissue that extends into the node (Figure 21.6). The capsular extensions, called trabeculae, divide the node into compartments. Internal to the capsule is a supporting network of reticular fibers. Functionally, a lymph node is divided into a superficial cortex and a deep medulla. Within the **cortex** are aggregates of dividing B cells called **lymphatic nodules**. The centers of many lymphatic nodules contain a region of light-staining cells called a *germinal center*. Within the node are **sinuses**, series of irregular channels that contain branching reticular fibers.

Lymph flows through a node in one direction only (Figure 21.6a). It enters through **afferent lymphatic vessels** (*afferent* = to carry toward), which penetrate the surface of the node at several points. The afferent lymphatic vessels contain valves that open toward the center of the node, directing the lymph *inward*. From the afferent lymphatic vessels, lymph flows into the **subcapsular sinus**, immediately beneath the capsule. From here the lymph flows through **trabecular sinuses**, which extend through the cortex parallel to the trabeculae, and into **medullary sinuses**, which extend through the medulla. The medullary sinuses drain into one or two **efferent lymphatic vessels** (*efferent* = to carry away). They contain valves that open away from the center of the lymph node to convey lymph *out* of the node. Efferent lymphatic vessels emerge from one side of the lymph node at a depression called a **hilum** (HĪ-lum).

Lymph nodes function as a type of filter. As lymph enters one end of a lymph node, foreign substances are trapped by the reticular fibers within the sinuses of the lymph node. Macrophages destroy some foreign substances by phagocytosis, while lymphocytes destroy others by immune responses. The filtered lymph then leaves the other end of the lymph node.

Figure 21.6   Structure of a lymph node. Arrows indicate direction of lymph flow through a lymph node.

Subcapsular sinus

Reticular fiber

Trabecula

Trabecular sinus

Germinal center in lymphatic nodule

Cortex

Medulla

Medullary sinus

Reticular fiber

Afferent lymphatic vessel

Valve

Efferent lymphatic vessels

Valve

Hilum

Afferent lymphatic vessel

Capsule

(a) Partially sectioned lymph node

Route of lymph flow through a lymph node:

Afferent lymphatic vessel

↓

Subcapsular sinus

↓

Trabecular sinus

↓

Medullary sinus

↓

Efferent lymphatic vessel

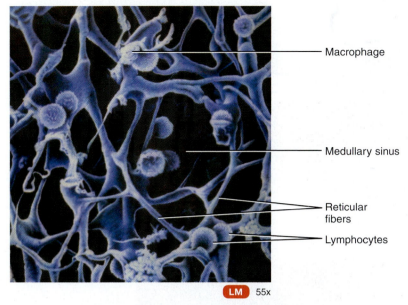

Macrophage

Medullary sinus

Reticular fibers

Lymphocytes

LM 55x

(b) Reticular connective tissue in a lymph node

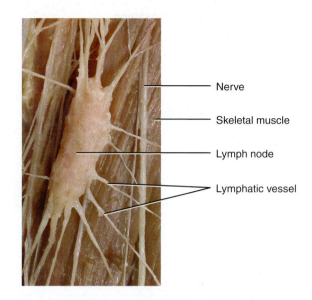

Nerve

Skeletal muscle

Lymph node

Lymphatic vessel

(c) Anterior view of an inguinal lymph node

Lymph nodes are present throughout the body, usually clustered in groups.

Spleen

The **spleen** is the largest single mass of lymphatic tissue in the body, measuring about 12 cm (5 in.) in length (see Figure 21.1a). It is located on the left side of the abdominal cavity between the stomach and diaphragm. Neighboring organs, the stomach, kidney, and colon, make indentations, *impressions*, in the visceral surface of the spleen (Figure 21.7a). Like lymph nodes, the spleen has a hilum. Through it pass the splenic artery, splenic vein, and efferent lymphatic vessels.

A capsule of dense connective tissue surrounds the spleen. Trabeculae extend inward from the capsule. The spleen contains two types of tissue called white pulp and red pulp (Figure 21.7b,c). **White pulp** is lymphatic tissue, consisting mostly of lymphocytes and macrophages arranged around branches of the splenic artery called **central arteries**. The **red pulp** consists of blood-filled **venous sinuses** and strands of splenic tissue called **splenic cords**. Splenic cords consist of red blood cells, macrophages, lymphocytes, plasma cells, and granulocytes. Veins are closely associated with the red pulp.

Blood flowing into the spleen through the splenic artery enters the central arteries of the white pulp. Within the white pulp, B cells and T cells carry out immune responses while spleen macrophages destroy pathogens by phagocytosis. Within the red pulp, the spleen performs three functions related to blood cells: (1) removal by macrophages of worn-out or defective blood cells and platelets; (2) storage of platelets, up to one-third of the body's supply; and (3) production of blood cells (hemopoiesis) during fetal life.

Lymphatic Nodules

Lymphatic nodules are masses of lymphatic tissue that are not surrounded by a capsule. They are plentiful in the connective tissue of mucous membranes lining the gastrointestinal, urinary, and reproductive tracts and the respiratory airways. Although many lymphatic nodules are small and solitary, some occur as large aggregations in specific parts of the body. Among these are the tonsils in the pharyngeal region and the aggregated lymphatic nodules (Peyer's patches) in the ileum of the small intestine (see Figure 21.1a). Aggregations of lymphatic nodules also occur in the appendix. Usually there are five **tonsils**, which form a ring at the junction of the oral cavity, nasal cavity, and pharynx (see Figure 22.2b). The tonsils are strategically positioned to participate in immune responses against inhaled or ingested foreign substances. The single **pharyngeal tonsil** (fa-RIN-jē-al), or **adenoid**, is embedded in the posterior wall of the nasopharynx. The two **palatine tonsils** (PAL-a-tīn) lie at the posterior region of the oral cavity, one on either side; these are the tonsils commonly removed in a tonsillectomy. The paired **lingual tonsils** (LIN-gwal), located at the base of the tongue, may also require removal during a tonsillectomy.

Checkpoint

8. What is the role of the thymus in immunity? Is the thymus a primary or secondary lymphatic organ?
9. What happens to the size of the thymus as you age?
10. Through what structures does lymph flow as it travels through a lymph node?
11. What happens to foreign substances that enter a lymph node in lymph?
12. What are the functions of the spleen and tonsils?

Figure 21.7   Structure of the spleen.

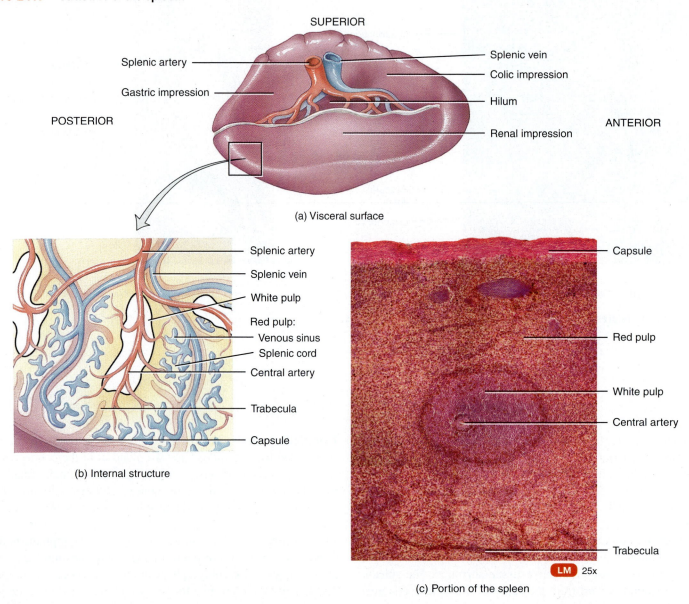

(a) Visceral surface

(b) Internal structure

(c) Portion of the spleen

LM 25x

The spleen is the largest single mass of lymphatic tissue in the body.

21.4   Innate immunity includes external physical and chemical barriers and various internal defenses.

Innate immunity does not involve specific recognition of a microbe and acts against all microbes in the same way. Innate immune responses represent immunity's early warning system and are designed to prevent microbes from gaining access into the body and to help eliminate those that do gain access. Innate immunity includes the external physical and chemical barriers provided by the skin and mucous membranes. It also includes various internal defenses, such as antimicrobial proteins, natural killer cells, phagocytes, inflammation, and fever.

First Line of Defense: Skin and Mucous Membranes

The skin and mucous membranes are the first line of defense against pathogens. These structures provide physical and chemical barriers that discourage pathogens and foreign substances from penetrating the body and causing disease.

With its many layers of closely packed, keratinized cells, the outer epithelial layer of the skin—the **epidermis**—provides a formidable physical barrier to the entrance of microbes (see Figure 5.1). In addition, continual shedding of epidermal cells helps remove microbes at the skin surface. Bacteria rarely penetrate an intact surface of healthy epidermis. If this surface is broken by cuts, burns, or punctures, however, pathogens can penetrate the epidermis and invade adjacent tissues or circulate in the blood to other parts of the body.

The epithelial layer of **mucous membranes**, which line body cavities, secretes a fluid called **mucus** that lubricates and moistens the cavity surface. Because mucus is sticky, it traps many microbes and foreign substances. The mucous membrane of the nose has mucus-coated **hairs** that trap and filter microbes, dust, and pollutants from inhaled air. The mucous membrane of the upper respiratory tract contains **cilia**, microscopic hairlike projections on the surface of the epithelial cells. The waving action of cilia propels inhaled dust and microbes that have become trapped in mucus toward the throat. Coughing and sneezing accelerate movement of mucus and its entrapped pathogens out of the body. Swallowing mucus sends pathogens to the stomach, where gastric juice destroys them.

The cleansing of the urethra by the **flow of urine** retards microbial colonization of the urinary system. **Vaginal secretions** likewise move microbes out of the body in females. **Defecation** and **vomiting** also expel microbes. For example, in response to some microbial toxins, the smooth muscle of the lower gastrointestinal tract contracts vigorously; the resulting diarrhea rapidly expels many of the microbes.

Certain chemicals also contribute to the resistance of the skin and mucous membranes to microbial invasion. Sebaceous (oil) glands of the skin secrete an oily substance called **sebum** that forms a protective film over the surface of the skin that inhibits the growth of certain pathogens. Bacterial growth is inhibited by the acidity of the skin, caused in part by the secretion of fatty acids and lactic acid.

Other fluids produced by various organs also help protect epithelial surfaces of the skin and mucous membranes. The **lacrimal apparatus** (LAK-ri-mal) of the eyes (see Figure 16.7b) produces and drains away tears in response to irritants. Blinking spreads tears over the surface of the eyeball, and the continual washing action of tears helps to dilute microbes and keep them from settling on the surface of the eye. Tears also contain **lysozyme**, an enzyme capable of breaking down the cell walls of certain bacteria. Besides tears, lysozyme is present in saliva, perspiration, nasal secretions, and tissue fluids. **Saliva**, produced by the salivary glands, washes microbes from the surfaces of the teeth and from the mucous membrane of the mouth, much as tears wash the eyes. The flow of saliva reduces colonization of the mouth by microbes. **Perspiration** helps flush microbes from the surface of the skin. **Gastric juice**, a mixture of hydrochloric acid, enzymes, and mucus in the stomach, destroys many bacteria and most bacterial toxins. **Vaginal secretions** also are slightly acidic, which discourages bacterial growth.

Second Line of Defense: Internal Defenses

Although the physical and chemical barriers of the skin and mucous membranes are very effective in preventing invasion by pathogens, they may be broken by injuries or everyday activities such as brushing the teeth or shaving. Any pathogens that get past the surface barriers encounter a second line of defense: internal antimicrobial proteins, phagocytes, natural killer cells, inflammation, and fever.

Antimicrobial Proteins

Various body fluids contain **antimicrobial proteins** that discourage microbial growth, the most important being interferons and proteins that make up the complement system.

Interferons (in´-ter-FĒR-ons) are proteins produced by cells infected with viruses. Once released by the virus-infected cells, interferons diffuse to uninfected neighboring cells, where they stimulate synthesis of antiviral proteins that block viral replication. Viruses can cause disease only if they can replicate within body cells. Although interferons do not prevent viruses from penetrating host cells, they do stop replication.

The **complement system** is a group of normally inactive proteins in blood plasma and on plasma membranes. When activated, these proteins "complement" or enhance certain immune, allergic, and inflammatory reactions. The complement system causes cytolysis (bursting) of microbes, promotes phagocytosis, and contributes to inflammation. Complement is discussed further later in the chapter.

Natural Killer Cells and Phagocytes

When microbes penetrate the skin and mucous membranes or bypass the antimicrobial proteins in blood, the next line of defense consists of natural killer cells and phagocytes. About 5–10% of lymphocytes in the blood are **natural killer (NK) cells**. Also present in the spleen, lymph nodes, and red bone marrow, *NK* cells have the ability to kill a wide variety of infected body cells and certain tumor cells. NK cells attack any body cells that display abnormal or unusual plasma membrane proteins.

The binding of NK cells to a target cell, such as an infected body cell, causes the NK cells to release granules containing toxic substances. Some granules contain a protein called **perforin** that inserts into the plasma membrane of the target cell and creates channels (perforations) in the membrane. As a result, extracellular fluid flows into the target cell and the cell bursts, a process called **cytolysis** (sī-TOL-i-sis; *cyto-* = cell; *-lysis* = loosening). Other granules of NK cells release **granzymes**, protein-digesting enzymes that induce the target cell to undergo apoptosis, or self-destruction.

Phagocytes (*phago-* = eat; *-cytes* = cells) are specialized cells that perform **phagocytosis** (*-osis* = process), the ingestion of microbes or other particles such as cellular debris. The two major types of phagocytes are **neutrophils** and **macrophages** (MAK-rō-fā-jez). When an infection occurs, neutrophils and monocytes migrate to the infected area. During this migration, the monocytes enlarge and develop into actively phagocytic cells called **wandering macrophages**. Other macrophages, called **fixed macrophages**, remain in certain tissues, including the skin, liver, lungs, brain, spleen, lymph nodes, and red bone marrow.

Figure 21.8   Phagocytosis of a microbe.

(a) Phases of phagocytosis

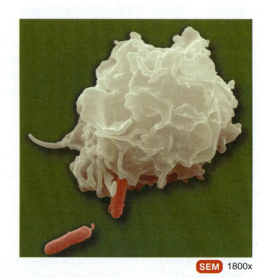

SEM 1800x

(b) Phagocyte (white blood cell) engulfing a microbe

The major types of phagocytes are neutrophils and macrophages.

Phagocytosis occurs in five phases: chemotaxis, adherence, ingestion, digestion, and killing (Figure 21.8):

1 *Chemotaxis.* Phagocytosis begins with **chemotaxis** (kē′-mō-TAK-sis), chemical attraction of phagocytes to a site of damage. For example, an invading microbe or damaged tissue cell might release chemicals that attract phagocytes.

2 *Adherence.* Attachment of the phagocyte to the microbe or other foreign material is termed **adherence**.

3 *Ingestion.* The plasma membrane of the phagocyte extends projections, called **pseudopods**, that engulf the microbe in a process called **ingestion**. As the pseudopods meet, they fuse, surrounding the microorganism within a sac called a **phagosome**.

4 *Digestion.* The phagosome fuses with lysosomes to form a **phagolysosome**. Within the phagolysosome, lysosomal enzymes break down the ingested cell.

5 *Killing.* The chemical onslaught by lysosomal enzymes quickly kills many types of microbes. Any materials that cannot be digested further remain in structures called **residual bodies**.

Inflammation

Inflammation is an innate immune response of the body to tissue damage. Among the conditions that may produce inflammation are pathogens, abrasions, chemical irritations, distortion or disturbances of cells, and extreme temperatures. The four characteristic signs and symptoms of inflammation are *redness*, *pain*, *heat*, and *swelling*. Inflammation can also cause a *loss of function* in the injured area (for example, the inability to detect sensations), depending on the site and extent of the injury.

Inflammation is an attempt to dispose of microbes, toxins, or foreign material at the site of injury, to prevent their spread to other tissues, and to prepare the site for tissue repair in an attempt to restore tissue homeostasis. The inflammatory response involves vasodilation and increased permeability of blood vessels, emigration (movement) of phagocytes from the blood into interstitial fluid, and chemotaxis and microbial attack (Figure 21.9):

1 *Vasodilation and increased permeability of local blood vessels.* In a region of tissue injury, injured tissue cells, white blood cells, and platelets release **histamine** and other inflammatory response substances that produce two immediate changes in blood vessels: **vasodilation** (increase in diameter) of local arterioles and **increased permeability** of local capillaries. Vasodilation allows more blood to flow through the damaged area and helps remove microbial toxins and dead cells. Increased permeability means that substances normally retained in blood are permitted to pass out of the blood vessels, allowing defensive substances such as antibodies and clot-forming chemicals to enter the injured area from the blood.

2 *Emigration of phagocytes.* Within an hour after the inflammatory process starts, phagocytes appear at the site of injury.

Figure 21.9 Inflammation.

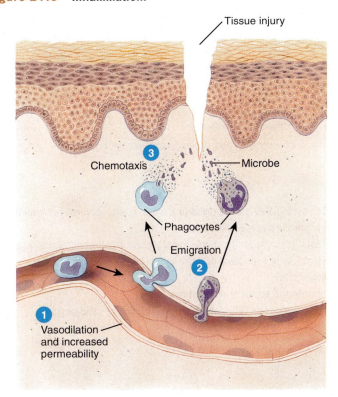

Phagocytes migrate from blood to site of tissue injury

🔑 The three stages of inflammation are vasodilation and increased permeability of blood vessels, phagocyte emigration, and chemotaxis and microbial attack.

As large amounts of blood accumulate, neutrophils begin to stick to the inner surfaces of the blood vessels. The neutrophils then begin to squeeze through the wall of the blood vessel, a process called **emigration**.

3 *Chemotaxis and microbial attack.* Drawn by chemotaxis to the damaged area, neutrophils attempt to destroy the invading microbes by phagocytosis. Although neutrophils predominate in the early stages of infection, they die off rapidly. As the inflammatory response continues, monocytes follow the neutrophils into the infected area. Once in the tissue, monocytes transform into wandering macrophages. True to their name, macrophages are potent phagocytes large enough to engulf damaged tissue, worn-out neutrophils, and invading microbes. Eventually, macrophages also die. Within a few days, a pocket of dead phagocytes and damaged tissue forms; this collection of dead cells and fluid is called **pus**. Pus formation occurs in most inflammatory responses and usually continues until the infection subsides. Pus may eventually reach the surface of the body, drain into an internal cavity and be dispersed, or accumulate at the infection site and gradually be absorbed.

Besides histamine, other substances contribute to the inflammatory response. **Prostaglandins**, released by damaged cells, intensify the effects of histamine and may stimulate the emigration of phagocytes through capillary walls. *Leukotrienes* produced by white blood cells increase capillary permeability and chemotactically attract phagocytes. Components of the complement system stimulate histamine release, attract neutrophils by chemotaxis, and promote phagocytosis.

It's easy to understand the signs and symptoms of inflammation when you consider the events that occur. Redness results from the large amount of blood that accumulates in the damaged area from dilation of arterioles. Heat is due to the increased blood flow and an increase in local metabolic reactions. The area swells (edema) due to increased permeability of capillaries, which allows an increased amount of fluid to leak from capillaries into the injury site. Pain results from injury to neurons, from toxic chemicals released by microbes, and from the increased pressure of edema. In addition, prostaglandins intensify and prolong the pain associated with inflammation.

Repair begins as increased permeability of capillaries allows leakage of blood-clotting factors into tissues. The clotting sequence is set into motion, and fibrinogen is ultimately converted to an insoluble, thick mesh of fibrin threads that traps invading microbes and blocks their spread. Tissue repair is discussed further in Chapters 4 (see Concept 4.7) and 5 (see Concept 5.6).

Fever

Fever is an abnormally high body temperature that occurs when the hypothalamic thermostat is reset. It commonly occurs during infection and inflammation. Many bacterial toxins elevate body

TABLE 21.1

Innate Defenses

COMPONENT	FUNCTIONS
First Line of Defense: Skin and Mucous Membranes	
Physical Factors	
Epidermis of skin	Forms a physical barrier to the entrance of microbes.
Mucous membranes	Inhibit the entrance of many microbes, but not as effectively as intact skin.
Mucus	Traps microbes in respiratory and gastrointestinal tracts.
Hairs	Filter out microbes and dust in nose.
Cilia	Together with mucus, trap and remove microbes and dust from upper respiratory tract.
Lacrimal apparatus	Tears dilute and wash away irritating substances and microbes.
Saliva	Washes microbes from surfaces of teeth and mucous membranes of mouth.
Urine	Washes microbes from urethra.
Defecation and vomiting	Expel microbes from body.
Chemical Factors	
Sebum	Forms a protective acidic film over the skin surface that inhibits growth of many microbes.
Lysozyme	Antimicrobial substance in perspiration, tears, saliva, nasal secretions, and tissue fluids.
Gastric juice	Destroys bacteria and most toxins in stomach.
Vaginal secretions	Slight acidity discourages bacterial growth; flush microbes out of vagina.
Second Line of Defense: Internal Defenses	
Antimicrobial Proteins	
Interferons	Protect uninfected host cells from viral infection.
Complement system	Causes cytolysis of microbes, promotes phagocytosis, and contributes to inflammation.
Natural killer (NK) cells	Kill infected target cells by releasing perforin and granzymes that destroy the target cell.
Phagocytes	Ingest microbes and cellular debris.
Inflammation	Confines and destroys microbes and prepares the injured area for tissue repair.
Fever	Intensifies the effects of interferons, inhibits growth of some microbes, and speeds up tissue repair.

temperature, sometimes by triggering release of fever-causing cytokines (described shortly) from macrophages. Elevated body temperature intensifies the effects of interferons, inhibits the growth of some microbes, and speeds up body reactions that aid repair.

Table 21.1 summarizes the components of innate immunity.

Checkpoint

13. What is innate immunity?
14. Name the physical and chemical factors that provide protection from disease in the skin and mucous membranes.
15. Which internal defenses provide protection against microbes that penetrate the skin and mucous membranes?
16. What is the name for the antimicrobial proteins that prevent viral replication?
17. Which type of white blood cells are natural killer cells? Which type are phagocytes?
18. Describe the phases of phagocytosis.
19. How are microbes killed following their ingestion by a phagocyte?

Return to Marlene's Story

As she reached into the tepid dishwater, Marlene felt a dull burning pain. A sharp knife at the bottom of the dishpan had opened a gash between her thumb and forefinger. "Francis!" she called out to her husband. Francis came rushing in.

"Marlene, what's wrong?" he asked.

"I cut myself. Is it bad?"

Francis took her hand. "It's not too deep. May need a few stitches, though. Let's clean it up and get something on it, then we'll go to the urgent care center."

"If it's not that bad, let's just put a bandage on it. I am so tired of doctors and nurses and hospitals," replied Marlene.

Francis carefully washed the wound and placed a clean cotton pad over it. "Your hand seems so swollen, Marlene," he remarked.

Three days later, Marlene noticed that the swelling was worse, and red streaks extended from her hand up her forearm. She gently lifted the bandage. Pus oozed from the wound, and its margins (edges) were red and inflamed. She reached into her medicine cabinet for a bottle of hydrogen peroxide but couldn't open it. "Francis? Could you open this for me?" He looked at her hand.

"Marlene, your hand looks worse. Let me take you in to have it looked at."

"Okay," she sighed.

It didn't take the doctor at the urgent care center long to diagnose the problem. "You'll need to go on antibiotics right away. Your cut is infected and the red streaks on your arm indicate *lymphangitis*, inflammation of the lymphatic vessels. The swelling you experienced before you cut your hand was probably caused by a blockage in the lymphatic system, secondary to your surgery. Radiation therapy can cause localized inflammation. You should follow up with your oncologist about that. Right now we need to focus on treating the infection. Strep bacteria are very common on the skin and in the environment. When they get into cuts and scrapes, they can multiply rapidly and overwhelm the immune system. I am most concerned with treating the infection so we can prevent it from spreading."

A. Marlene has had lymph nodes removed from her armpit. Why would removal of lymph nodes cause swelling in her hand and arm?

B. Because Marlene was tired following the radiation therapy, she was not moving around a lot. How might this be related to the lymph fluid buildup in her arm?

C. Which elements of innate immunity were at work in Marlene's body when she cut her hand?

21.5 Adaptive immunity involves the production of a specific lymphocyte or antibody against a specific antigen.

The various aspects of innate immunity have one thing in common: They are not specifically directed against a particular type of invader. The ability of the body to defend itself against specific invading pathogens and foreign tissues that have breached the innate immunity defenses is called **adaptive (specific) immunity**. Substances that are recognized as foreign (nonself) and provoke immune responses are called **antigens**. Antigens may be microbes, tissues, certain foods, drugs, or pollen. Two properties distinguish adaptive immunity from innate immunity: (1) *specificity* for particular foreign molecules (antigens), which helps distinguish self from nonself molecules, and (2) *memory* for most previously encountered antigens so that a second encounter prompts an even more rapid and vigorous response. Physiologically, the cells and tissues that carry out immune responses can be considered a system, so they are sometimes referred to as the **immune system**.

Maturation of T Cells and B Cells

Adaptive immunity involves two types of lymphocytes, **B cells** and **T cells**. Both develop from stem cells that originate in red bone marrow (see Figure 18.3). B cells complete their development in red bone marrow, a process that continues throughout life. Immature T cells migrate from red bone marrow into the thymus, where they mature (Figure 21.10). Most T cells arise before puberty, but they continue to mature and leave the thymus throughout life. B cells and T cells are named based on where they mature. Although in humans B cells mature in red bone marrow, in birds B cells mature in an organ called the *bursa of Fabricius*. This organ is not present in humans, but the term *B cell* is still used with the letter *B* standing for *bursa equivalent*, which is red bone marrow in humans. T cells are so named because they mature in the *thymus* gland.

Before T cells leave the thymus or B cells leave red bone marrow, they develop **immunocompetence**, the ability to carry out adaptive immune responses. This means that B cells and T cells begin to make several distinctive proteins that are inserted into their plasma membranes. Some of these proteins function as **antigen receptors**—molecules capable of recognizing specific antigens (Figure 21.10).

Types of Adaptive Immunity

There are two types of adaptive immunity: cell-mediated immunity and antibody-mediated immunity. Both types of adaptive immunity are triggered by antigens. In **cell-mediated immunity**, cytotoxic T cells directly attack invading antigens. In **antibody-**

Figure 21.10   B cells and immature T cells arise from stem cells in red bone marrow.

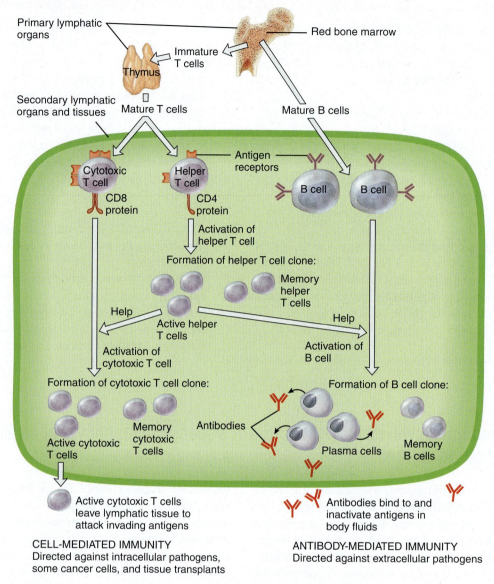

CELL-MEDIATED IMMUNITY
Directed against intracellular pathogens, some cancer cells, and tissue transplants

ANTIBODY-MEDIATED IMMUNITY
Directed against extracellular pathogens

The two types of adaptive immunity are cell-mediated immunity and antibody-mediated immunity.

mediated immunity, B cells transform into plasma cells, which synthesize and secrete specific proteins called **antibodies** or **immunoglobulins** (im´-ū-nō-GLOB-ū-lins). A given antibody can bind to and inactivate a specific antigen. Helper T cells aid the immune responses of both cell-mediated immunity and antibody-mediated immunity.

Cell-mediated immunity always involves cells attacking cells. Cell-mediated immunity is particularly effective against (1) intracellular pathogens, which include any viruses, bacteria, or fungi that are inside cells; (2) some cancer cells; and (3) foreign tissue transplants. Antibody-mediated immunity works mainly against

extracellular pathogens, which include viruses, bacteria, or fungi that are in body fluids outside cells. However, a single pathogen can provoke both types of adaptive immune responses.

Clonal Selection

When a specific antigen, likely a pathogen, invades the body, there are usually many copies of that antigen spread throughout the body's tissues and fluids. The numerous copies of the antigen initially outnumber the small group of helper T cells, cytotoxic T cells, and B cells with the correct antigen receptors to respond to that antigen. Therefore, once each of these lymphocytes encounters the antigen, it undergoes clonal selection. **Clonal selection** is the process by which a lymphocyte divides and **differentiates** (forms more highly specialized cells) in response to contact with a specific antigen. Before the first exposure to a given antigen, only a few lymphocytes are able to recognize an antigen, but once clonal selection occurs, there are thousands of identical lymphocytes, called a **clone**, that can recognize the same specific antigen as the original lymphocyte. The swollen tonsils or lymph nodes in your neck you experienced the last time you were sick were probably caused by clonal selection of lymphocytes participating in an immune response.

When a lymphocyte undergoes clonal selection, the resulting clone includes two major types of cells: effector cells and memory cells. The **effector cells** carry out immune responses that ultimately destroy or inactivate the antigen. Most effector cells eventually die after the immune response has been completed. **Memory cells** do not actively participate in the initial immune response to the antigen. However, if the same antigen enters the body again in the future, memory cells are available to initiate a far swifter reaction than occurred during the first invasion. Memory cells respond to the antigen by proliferating and differentiating into more effector cells and more memory cells. Consequently, the second response to the antigen is usually so fast and so vigorous that the antigen is destroyed before any signs or symptoms of disease can occur. Most memory cells do not die at the end of an immune response. Instead, they have long life spans (often lasting for decades).

Antigens

Antigens are able to provoke an immune response by stimulating the production of specific antibodies and/or the proliferation of specific T cells. The term *antigen* derives from its function as an *anti*body *gen*erator.

Entire microbes or parts of microbes may act as antigens. Components of bacterial structures such as flagella and cell walls are antigenic, as are bacterial toxins. Nonmicrobial examples of antigens include components of pollen, egg white, incompatible blood cells, and transplanted tissues and organs. The huge variety of antigens in the environment provides myriad opportunities for provoking immune responses. Typically, large antigen molecules have small parts called **epitopes** (EP-i-tōps) that trigger immune responses (Figure 21.11). Most antigens have many epitopes, each of which induces production of a specific antibody or activates a specific T cell.

Antigens are large, complex molecules. Most often, they are proteins. Large molecules that have simple, repeating subunits—for example, cellulose and most plastics—are not usually antigenic. This is why plastic materials can be used in artificial heart valves or joints. A small molecule called a **hapten** (HAP-ten = to grasp) can stimulate an immune response if it is attached to a larger carrier molecule. An example is the small lipid toxin in poison ivy, which triggers an immune response after combin-

Figure 21.11 Epitopes.

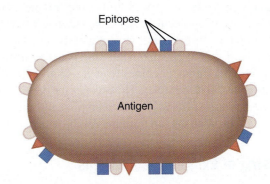

Most antigens have several epitopes that induce the production of different antibodies or activate different T cells.

ing with a body protein. Likewise, penicillin may stimulate an immune response when it combines with body proteins. Such hapten-stimulated immune responses are responsible for some allergic reactions to drugs and other substances in the environment.

Antigens that get past the innate defenses generally meet their demise in lymphatic tissue. For example, antigens entering the bloodstream through an injured blood vessel would be trapped as they flow through the spleen. Antigens that penetrate the skin enter lymphatic vessels, only to be trapped in lymph nodes.

As a rule, antigens are foreign substances; they are not usually part of body tissues. However, sometimes the immune system fails to distinguish "friend" (self) from "foe" (nonself). The result is an autoimmune disorder in which our own molecules or cells are attacked as though they were foreign.

Major Histocompatibility Complex Antigens

Located in the plasma membrane of body cells are "self-antigens" known as **major histocompatibility complex (MHC) antigens**. Unless you have an identical twin, your MHC antigens are unique. Thousands to several hundred thousand MHC antigens mark the surface of each of your body cells except red blood cells. Although MHC antigens are the reason that tissues may be rejected when they are transplanted from one person to another, their normal function is to help T cells distinguish your own cells from foreign cells. Such recognition is an important first step in any adaptive immune response.

Processing and Presenting Antigens

For an adaptive immune response to occur, B cells and T cells must recognize that a foreign antigen is present. B cells can recognize and bind to antigens in lymph, interstitial fluid, or blood plasma. T cells only recognize fragments of antigenic proteins that are processed and presented in a certain way. In **antigen processing**, antigenic proteins are broken down into fragments that then associate with MHC molecules. Next, the antigen–MHC complex is inserted into the plasma membrane of a body cell. The insertion of the complex into the plasma membrane is called **antigen presentation**. When a fragment comes from a *self-protein*, T cells ignore the antigen–MHC complex. However, if the peptide fragment comes from a *foreign protein*, T cells recognize the antigen–MHC complex as an intruder, and an immune response takes place.

Processing of Exogenous Antigens

Foreign antigens that are present in fluids *outside* body cells are termed **exogenous antigens**. They include intruders such as bacteria, parasitic worms, and inhaled pollen. A special class of cells called **antigen-presenting cells (APCs)** process and present exogenous antigens. APCs include dendritic cells, macrophages, and B cells. They are strategically located in places where antigens are likely to penetrate our innate defenses and enter the body, such as the epidermis of the skin; mucous membranes that line the respiratory, gastrointestinal, urinary, and reproductive tracts; and lymph nodes.

The steps in the processing and presenting of an exogenous antigen by an APC occur as follows (Figure 21.12):

1 *Ingestion of the antigen.* Antigen-presenting cells ingest exogenous antigens via phagocytosis. Ingestion could occur almost anywhere in the body that invaders, such as microbes, have penetrated the innate defenses.

2 *Digestion of antigen into fragments.* Within the phagolysosome, protein-digesting enzymes split large antigens into short peptide fragments.

3 *Synthesis of MHC molecules.* At the same time, the APC synthesizes MHC molecules and packages them into vesicles.

4 *Fusion of vesicles.* The vesicles containing antigen fragments and MHC molecules merge and fuse.

5 *Binding of fragments to MHC molecules.* After fusion of the two vesicles, antigen fragments bind to MHC molecules.

6 *Insertion of antigen–MHC complexes into plasma membrane.* The combined vesicle that contains antigen–MHC complexes undergoes exocytosis. As a result, antigen–MHC complexes are inserted into the plasma membrane.

After processing an exogenous antigen, the antigen-presenting cell migrates to lymphatic tissue to present the antigen to T cells. Within lymphatic tissue, a small number of T cells that have compatibly shaped receptors recognize and bind to the antigen–MHC complex, triggering an adaptive immune response. The presentation of exogenous antigen with MHC molecules by antigen-presenting cells informs T cells that intruders are present in the body and that combative action should begin.

Processing of Endogenous Antigens

Foreign antigens that are present inside body cells are termed **endogenous antigens**. Such antigens may be viral proteins produced after a virus infects the cell and takes over the cell's metabolic machinery, toxins produced from intracellular bacteria, or abnormal proteins synthesized by a cancerous cell. Fragments of endogenous antigens bind with MHC molecules inside infected cells. The resulting endogenous antigen–MHC complex then moves to the surface of the cell, where it is inserted into the plasma membrane. The display of an endogenous antigen bound to an MHC molecule signals that a cell has been infected and needs help.

Cytokines

Cytokines are small protein hormones that stimulate or inhibit many normal cell functions, such as cell growth and differentiation. Cytokines include interferons and interleukins. Some cytokines stimulate proliferation of progenitor blood cells in red bone marrow. Others regulate activities of cells involved in innate defenses or adaptive immune responses. Lymphocytes and antigen-presenting cells secrete cytokines, as do a variety of other cells.

Checkpoint

20. What is immunocompetence, and which body cells display it?
21. Where do T cells and B cells develop?
22. What are the functions of cell-mediated immunity and antibody-mediated immunity?
23. How do antigens induce an immune response? Use the terms *antigen processing* and *antigen presentation* in your answer.
24. What is the normal function of major histocompatibility complex self-antigens?
25. Give three examples of endogenous antigens.

Figure 21.12 **Processing and presenting of exogenous antigen by an antigen-presenting cell.**

② Digestion of antigen into peptide fragments

⑤ Binding of peptide fragments to MHC molecules

Phagolysosome

① Phagocytosis of antigen

Exogenous antigen

⑥ Insertion of antigen–MHC complex into plasma membrane

④ Fusion of vesicles

Antigen-presenting cell

Key:

Antigen peptide fragments

MHC self-antigen

③ Synthesis of MHC molecules

🔑 **Fragments of exogenous antigens are processed and then presented with MHC molecules on the surface of an antigen-presenting cell.**

21.6 In cell-mediated immunity, cytotoxic T cells directly attack target cells.

The presentation of an antigen–MHC complex by APCs informs T cells that intruders are present in the body and that combative action should begin. A cell-mediated immune response begins with *activation* of a small number of T cells by a specific antigen. Once a T cell has been activated, it undergoes clonal selection. Recall that clonal selection is the process by which a lymphocyte divides and differentiates (forms more highly specialized cells) in response to a specific antigen. The resulting clone of cells can recognize the same antigen as the original lymphocyte. Some of the cells of a T cell clone become effector cells, while other cells of the clone become memory cells. The effector cells of a T cell clone carry out immune responses that ultimately result in *elimination* of the intruder.

Activation of T Cells

At any given time, most T cells are inactive (resting). Antigen receptors on the surface of T cells, called **T cell receptors (TCRs)**, recognize and bind to specific foreign antigen fragments that are presented in antigen–MHC complexes. When an antigen enters the body, only a few T cells have receptors that can recognize and bind to that specific antigen–MHC complex (antigen recognition). Antigen recognition also involves other plasma proteins on T cells known as CD4 or CD8 proteins. These proteins help maintain coupling of the antigen–MHC complexes with the T cell receptor. Binding of a T cell with CD4 or CD8 proteins to an antigen–MHC complex is the *first signal* in activation of a T cell. A T cell becomes activated only if it binds to the foreign antigen and at the same time receives a *second signal*, a process known as **costimulation**. Most costimulators are cytokines, such as **interleukin-2**.

The need for two signals to activate a T cell is a little like starting and driving a car: When you insert the correct key (antigen) in the ignition (T cell receptor) and turn it, the car starts (recognition of specific antigen), but it cannot move forward until you move the gear shift into drive (costimulation). The need for costimulation may prevent immune responses from occurring accidentally. Different costimulators affect the activated T cell in different ways, just as shifting a car into reverse has a different effect than shifting it into drive. Once a T cell has received these two signals (antigen recognition and costimulation), it is **activated**. An activated T cell subsequently undergoes clonal selection.

Activation and Clonal Selection of Helper T Cells

During T cell development, two major types of mature T cells exit the thymus: helper T cells and cytotoxic T cells (see Figure 21.10). **Helper T cells** are also known as **CD4 T cells** because their plasma membranes include the protein CD4. Inactive helper T cells recognize major histocompatibility complex (MHC) antigens at the surface of an APC (Figure 21.13). With the aid of the CD4 protein, the helper T cell and APC interact with each other (antigenic recognition), costimulation occurs, and the helper T cell becomes activated.

Once activated, the helper T cell undergoes clonal selection (Figure 21.13). The result is the formation of a clone of helper T cells that consists of active helper T cells and memory helper T cells. Within hours after costimulation, active helper T cells start secreting a variety of cytokines. One very important cytokine produced by helper T cells is interleukin-2 (IL-2), which is needed for virtually all immune responses and is the prime trigger of T cell proliferation. Interleukin-2 can act as a costimulator for resting T cells, and it enhances activation and proliferation of T cells, B cells, and natural killer cells.

The **memory helper T cells** of a helper T cell clone are not active cells. However, if the same antigen enters the body again in the future, memory helper T cells can quickly proliferate and differentiate into more active helper T cells and more memory helper T cells.

Activation and Clonal Selection of Cytotoxic T Cells

Cytotoxic T cells are also referred to as **CD8 T cells** because their plasma membranes have a protein known as CD8. Cytotoxic T cells recognize major histocompatibility complex antigens on the surfaces of (1) body cells infected by microbes, (2) some tumor

Figure 21.13   Activation and clonal selection of a helper T cell.

Once a helper T cell is activated, it forms a clone of active and memory helper T cells.

cells, and (3) cells of a tissue transplant (Figure 21.14). Antigen recognition occurs as the T cell receptor and CD8 protein interact with the MHC antigen. Following antigenic recognition, costimulation occurs with interleukin-2 or other cytokines produced by helper T cells.

Figure 21.14 **Activation and clonal selection of a cytotoxic T cell.**

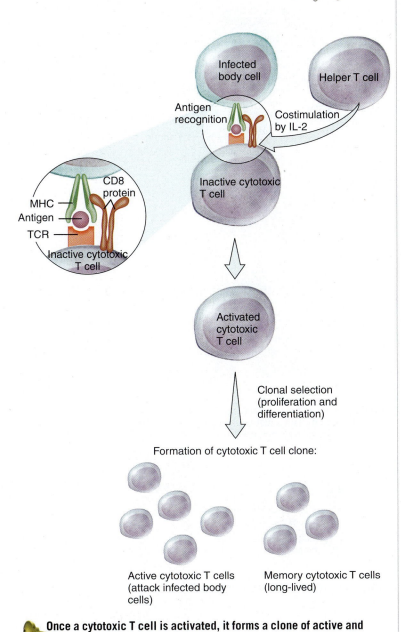

MHC
Antigen
TCR
CD8 protein
Inactive cytotoxic T cell

Infected body cell

Helper T cell

Antigen recognition

Costimulation by IL-2

Inactive cytotoxic T cell

Activated cytotoxic T cell

Clonal selection (proliferation and differentiation)

Formation of cytotoxic T cell clone:

Active cytotoxic T cells (attack infected body cells)

Memory cytotoxic T cells (long-lived)

🔑 **Once a cytotoxic T cell is activated, it forms a clone of active and memory cytotoxic T cells.**

Once activated, the cytotoxic T cell undergoes clonal selection. The result is the formation of a clone of cytotoxic T cells that consists of active cytotoxic T cells and memory cytotoxic T cells. **Active cytotoxic T cells** attack other body cells that have been infected with the antigen. **Memory cytotoxic T cells** do not attack infected body cells. Instead, they can quickly proliferate and differentiate into more active cytotoxic T cells and more memory cytotoxic T cells if the same antigen enters the body at a future time.

Elimination of Invaders

Cytotoxic T cells are the soldiers that march forth to do battle with foreign invaders in cell-mediated immune responses. The name "cytotoxic" reflects their function—killing cells. They leave lymphatic organs and tissues and migrate to seek out and destroy infected target cells, cancer cells, and transplanted cells (Figure 21.15). Cytotoxic T cells recognize and attach to target cells. Then, the cytotoxic T cells deliver a "lethal hit" that kills the target cells.

Cytotoxic T cells kill infected target body cells much like natural killer cells do. The major difference is that cytotoxic T cells have receptors specific for a particular microbe and thus kill only target body cells infected with *one* particular type of microbe. In contrast, natural killer cells can destroy a wide variety of microbe-infected body cells. Cytotoxic T cells have two principal mechanisms for killing infected target cells.

• Cytotoxic T cell receptors bind to infected target cells that have microbial antigens displayed on their surface. The cytotoxic T cell then releases **granzymes**, protein-digesting enzymes that trigger apoptosis (cell fragmentation) (Figure 21.15a). Once the infected cell is destroyed, the released microbes are killed by phagocytes.

• Alternatively, cytotoxic T cells bind to infected body cells and release two proteins: perforin and granulysin. **Perforin** inserts into the plasma membrane of the target cell and creates channels in the membrane (Figure 21.15b). As a result, extracellular fluid flows into the target cell and cytolysis (cell bursting) occurs. **Granulysin** enters through the channels and destroys the microbes by creating holes in their plasma membranes. Cytotoxic T cells may also destroy target cells by releasing a toxic molecule called **lymphotoxin** that causes the target cell's DNA to fragment and the cell to die. In addition, cytotoxic T cells secrete substances that attract and activate phagocytic cells and prevent migration of phagocytes away from the infection site. After detaching from a target cell, a cytotoxic T cell can seek out and destroy another target cell.

Figure 21.15   Activity of cytotoxic T cells. After delivering a "lethal hit," a cytotoxic T cell can detach and attack another infected target cell displaying the same antigen.

(a) Cytotoxic T cell destruction of infected cell by release of granzymes that cause apoptosis; released microbes are destroyed by phagocyte.

(b) Cytotoxic T cell destruction of infected cell by release of perforins that cause cytolysis; microbes are destroyed by granulysin.

Key:

TCR   CD8 protein

Antigen–MHC complex

Cytotoxic T cells release granzymes that trigger apoptosis and perforin that triggers cytolysis of infected target cells.

Checkpoint

26.   What are the first and second signals in activation of a T cell?

27.   How do the CD4 and CD8 proteins assist T cell activation?

28.   What are the functions of helper, cytotoxic, and memory T cells?

29.   Besides cells infected by microbes, which other types of cells do cytotoxic T cells attack?

30.   How do cytotoxic T cells kill infected target cells?

21.7 In antibody-mediated immunity, antibodies specifically target a particular antigen.

The body contains not only millions of different T cells but also millions of different B cells, each capable of responding to a specific antigen. Cytotoxic T cells leave lymphatic tissues to seek out and destroy a foreign antigen, but B cells stay put within a lymphatic organ or tissue. In the presence of a foreign antigen, a specific B cell forms plasma cells that secrete specific antibodies, which disable the antigen.

Figure 21.16 Activation and clonal selection of B cells. Plasma cells are actually much larger than B cells.

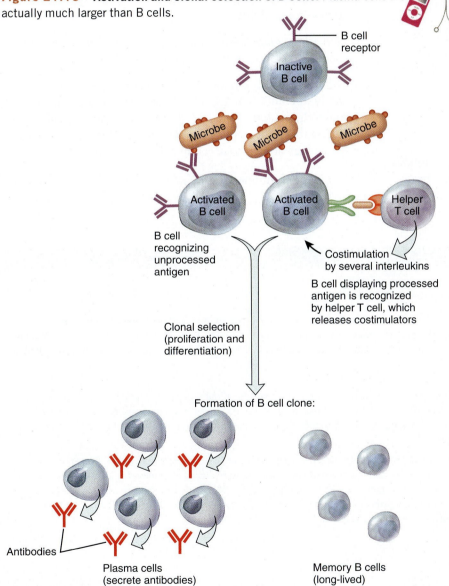

B cell receptor

Inactive B cell

Microbe   Microbe   Microbe

Activated B cell   Activated B cell   Helper T cell

B cell recognizing unprocessed antigen

Costimulation by several interleukins

B cell displaying processed antigen is recognized by helper T cell, which releases costimulators

Clonal selection (proliferation and differentiation)

Formation of B cell clone:

Antibodies

Plasma cells (secrete antibodies)

Memory B cells (long-lived)

Plasma cells secrete antibodies.

Activation and Clonal Selection of B Cells

During activation of a B cell, an antigen binds to **B cell receptors** (Figure 21.16). B cell antigen receptors are chemically similar to the antibodies that eventually are secreted by the plasma cells. Although B cells can respond to an unprocessed antigen, their response is much more intense when they process the antigen. Antigen processing begins as the B cell takes in the antigen, breaks it down into peptide fragments, combines it with MHC protein, and moves the antigen–MHC complex to the B cell plasma membrane. Helper T cells then recognize the antigen–MHC complex and deliver the costimulation needed for B cell division and differentiation. The helper T cell produces interleukin-2 and other cytokines that function as costimulators to activate B cells.

Once activated, a B cell undergoes clonal selection, forming a clone of plasma cells and memory B cells (Figure 21.16). **Plasma cells** are the effector cells of a B cell clone; they secrete specific antibodies. A few days after exposure to an antigen, a plasma cell secretes hundreds of millions of antibodies each day for about 4 or 5 days, until the plasma cell dies. Interleukins produced by helper T cells enhance B cell division into plasma cells and secretion of antibodies by plasma cells. Most antibodies travel in lymph and blood to the invasion site. **Memory B cells** do not secrete antibodies. Instead, they can quickly divide and differentiate into more plasma cells and more memory B cells should the same antigen reappear at a future time.

Different antigens stimulate different B cells to develop into plasma cells and their accompanying memory B cells. All of the B cells of a particular clone are capable of secreting only one type of antibody, which is similar to the antigen receptor displayed by the B cell that first responded.

Antibodies

An **antibody** can combine specifically with the epitope on the antigen that triggered the antibody's production. The antibody's structure matches its antigen much as a lock accepts a specific key. Antibodies belong to a group of glycoproteins called globulins, and for this reason they are also known as **immunoglobulins**, or **Igs**. Antibodies assume either a T or Y shape, with the two "arms" serving as the **antigen-binding site**, the part of the antibody that recognizes and attaches to a particular antigen (Table 21.2). Because antibodies have two antigen-binding sites, they are able to simultaneously bind to two epitopes that are some distance apart—for example, on the surface of a microbe.

The five classes of antibodies are designated IgG, IgA, IgM, IgD, and IgE. Each class has a distinct chemical structure and a specific biological role. Because they appear first and are relatively short-lived, IgM antibodies indicate a recent invasion. In a sick patient, the responsible pathogen may be suggested by the presence of high levels of IgM specific to a particular organism. Resistance of the fetus and newborn baby to infection stems mainly from maternal IgG antibodies that cross the placenta before birth and IgA antibodies in breast milk after birth. Table 21.2 summarizes the structures and functions of the five classes of antibodies.

Antibody Actions

Antibodies produced by plasma cells form antigen–antibody complexes with the antigen that initiated the antibody production. Antigen–antibody binding disables antigens in various ways.

Actions of the five classes of antibodies differ somewhat but include the following:

- *Neutralizing antigen*. The binding of an antibody with its antigen neutralizes some bacterial toxins and prevents attachment of some viruses to body cells.

- *Immobilizing bacteria*. Some antibodies cause bacteria to lose their motility, which limits their spread into nearby tissues.

- *Agglutinating and precipitating antigen*. Because antibodies have two antigen binding sites, the antigen–antibody reaction may connect pathogens to one another, causing agglutination (clumping together). Phagocytic cells ingest agglutinated microbes more readily.

- *Activating complement*. Antigen–antibody complexes activate complement proteins (discussed shortly).

- *Enhancing phagocytosis*. Once antigens have bound to an antibody, the antibody acts as a "flag" that attracts phagocytes. Antibodies enhance the activity of phagocytes by causing agglutination, by activating complement, and by coating microbes so that they are more susceptible to phagocytosis.

Table 21.3 summarizes the activities of cells involved in immune responses.

TABLE 21.2		
Classes of Immunoglobulins		
NAME AND STRUCTURE		CHARACTERISTICS AND FUNCTIONS
IgG		Most abundant, about 80% of all antibodies in the blood; found in blood, lymph, and the intestines; monomer (one-unit) structure. Protects against bacteria and viruses by enhancing phagocytosis, neutralizing toxins, and triggering the complement system. It is the only class of antibody to cross the placenta from mother to fetus, conferring considerable immune protection in newborns.
IgA		Found mainly in sweat, tears, saliva, mucus, breast milk, and gastrointestinal secretions. Smaller quantities are present in blood and lymph. Makes up 10–15% of all antibodies in the blood; occurs as monomers and dimers (two units). Levels decrease during stress, lowering resistance to infection. Provides localized protection of mucous membranes against bacteria and viruses.
IgM		About 5–10% of all antibodies in the blood; also found in lymph. Occurs as pentamers (five units); first antibody class to be secreted by plasma cells after an initial exposure to any antigen. Activates complement and causes agglutination and lysis of microbes. Also present as monomers on the surfaces of B cells, where they serve as antigen receptors. In blood plasma, the anti-A and anti-B antibodies of the ABO blood group, which bind to A and B antigens during incompatible blood transfusions, are also IgM antibodies (see Figure 18.7)
IgD		Mainly found on the surfaces of B cells as antigen receptors, where it occurs as monomers; involved in activation of B cells. About 0.2% of all antibodies in the blood.
IgE		Less than 0.1% of all antibodies in the blood; occurs as monomers; located on mast cells and basophils. Involved in allergic and hypersensitivity reactions; provides protection against parasitic worms.

TABLE 21.3

Functions of Cells Participating in Immune Responses

CELL	FUNCTIONS
Antigen-presenting Cells	
Macrophage	Phagocytosis; processing and presentation of foreign antigens to T cells; secretion of interleukin-1, which stimulates secretion of interleukin-2 by helper T cells and induces proliferation of B cells; secretion of interferons that stimulate T cell growth.
Dendritic cell	Processes and presents antigen to T cells and B cells; found in mucous membranes, skin, and lymph nodes.
B cell	Processes and presents antigen to helper T cells.
Lymphocytes	
Cytotoxic T cell	Kills host target cells by releasing granzymes that induce apoptosis, perforin that forms channels to cause cytolysis, granulysin that destroys microbes, and lymphotoxin that destroys target cell DNA.
Helper T cell	Cooperates with B cells to amplify antibody production by plasma cells and secretes interleukin-2, which stimulates proliferation of T cells and B cells.
Memory T cell	Remains in lymphatic tissue and recognizes original invading antigens, even years after the first encounter.
B cell	Differentiates into antibody-producing plasma cell.
Plasma cell	Descendant of B cell that produces and secretes antibodies.
Memory B cell	Descendant of B cell that remains after an immune response and is ready to respond rapidly and forcefully should the same antigen enter the body in the future.

Checkpoint

31.   Which type of T cell participates in both cell-mediated and antibody-mediated immune responses?
32.   Outline the steps in antibody production.
33.   How many antigens can bind to one antibody molecule?
34.   How do the five classes of antibodies differ in function?
35.   How do antibodies disable antigens?

Return to Marlene's Story

"It was just a small cut. How could it have gotten so infected?" Francis asked the doctor.

The doctor responded, "The immune system relies on two basic types of defenses, innate and adaptive. The body's innate or nonspecific response is to send white blood cells to the site of an injury. Certain white blood cells will recognize anything that's foreign to the body and begin the attack. Lymphatic vessels help carry foreign substances to the lymphatic organs, where they are presented to other white blood cells called T cells and B cells that then mount a second, more specific attack. Marlene, in your case, the swelling of your arm and preexisting inflammation from the surgery put you at a higher risk of infection since some of your lymph nodes were missing and lymphatic fluid couldn't return to your lymphatic organs effectively. I'd like to start you on high-dose penicillin; that should help reduce the infection and give your immune system a chance to fight off this bug more efficiently."

After consulting with Marlene's oncologist, the internist from the urgent care center admitted her to the hospital for several days, to make sure the infection could be monitored, and ordered an intravenous antibiotic. After a few days, the red streaks on her arm faded, and her temperature, which had spiked at 102°F, had fallen back to near normal. Finally, they let her return home.

"How are you feeling today?" Francis asked.

"My arm is still swollen. It hurts, and my throat feels swollen, like I'm getting a cold. After all this, I don't know how anything could survive in my body."

Francis smiled at her. "The therapist told me that once your infection was under control, they could help you with the swelling in your arm. Did you make an appointment?"

"Yes, I see her in 2 weeks. She said that massage would help reduce the swelling and mentioned something about a special compression garment."

Marlene made the trip to the physical therapist's office as scheduled. "I suggest a course of self-massage. I'd also like to fit you for a compression garment, which should help reduce the swelling if you're going to be inactive for any length of time. Keeping the swelling under control reduces the risk of infection and will improve the function of your arm and hand. Any questions?" the therapist asked.

"What caused the swelling in the first place?" asked Marlene.

"With the type of surgery you had, you'll always be at risk for lymphedema, the accumulation of fluid in lymphatic vessels. Accumulated fluids are normally returned to the bloodstream by lymphatic vessels; but in your case, swelling has occurred because those vessels are blocked."

D.   *Which lymphatic organs in Marlene's body would normally trap the antigens and allow an adaptive immune response to be mounted?*

E.   *Why might the lymph nodes in Marlene's throat be enlarged?*

F.   *How would massage help reduce the fluid buildup in Marlene's arm? How would improving the flow of lymph reduce Marlene's future chances of getting a similar infection and lymphangitis?*

21.8 The complement system destroys microbes through phagocytosis, cytolysis, and inflammation.

The **complement system** is a group of over 30 proteins produced by the liver and found circulating in blood plasma and within tissues throughout the body. Most complement proteins are designated by an uppercase letter *C*, numbered C1 through C9, named for the order in which they were discovered. The C1–C9 complement proteins are normally inactive; they become activated only when split by enzymes into active fragments which are indicated by lowercase letters *a* and *b*. For example, inactive complement protein C3 is split into the activated fragments C3a and C3b. The active fragments carry out the destructive actions of the C1–C9 complement proteins.

Complement activation begins with activation of C3 by binding of inactive complement proteins to an antigen–antibody complex or to carbohydrate complexes on the surface of some microbes. Complement proteins act in a *cascade*—one reaction triggers another reaction, which in turn triggers another reaction, and so on. With each succeeding reaction, more and more product is formed so that the net effect of complement is amplified many times. Once activated, C3 begins a sequence of reactions that destroy microbes by causing phagocytosis, cytolysis, and inflammation as follows (Figure 21.17):

1 Inactivated C3 splits into activated C3a and C3b.

2 C3b binds to the surface of a microbe. Thus C3b enhances phagocytosis by coating a microbe, a process called **opsonization** (op-sō-ni-ZĀ-shun). Opsonization promotes attachment of a phagocyte to a microbe.

Figure 21.17   Complement activation. (Adapted from Tortora, Funke, and Case, *Microbiology: An Introduction, Eighth Edition*, Figure 16.10, Pearson Benjamin-Cummings, 2004.)

PHAGOCYTOSIS:
Enhancement of phagocytosis by coating with C3b

Histamine

Mast cell

Microbe

INFLAMMATION:
Increase of blood vessel permeability and chemotactic attraction of phagocytes

Channel
Microbial plasma membrane
Membrane attack complex forms channel

CYTOLYSIS:
Bursting of microbe due to inflow of extracellular fluid through channel formed by membrane attack complex C5–C9

When activated, complement proteins enhance phagocytosis, cytolysis, and inflammation.

3 C3b splits C5 into activated C5a and C5b. The C5b fragment then binds to C6 and C7, which attach to the plasma membrane of an invading microbe.

4 C8 and C9 molecules join C5b, C6, and C7 to form a cylinder-shaped **membrane attack complex**, which inserts into the microbial plasma membrane. The membrane attack complex creates a channel through the plasma membrane that results in cytolysis, the bursting of the microbial cells due to the inflow of extracellular fluid through the channels.

5 C3a and C5a bind to mast cells and cause them to release histamine, which increases blood vessel permeability during

inflammation. C5a also attracts phagocytes to the site of inflammation (chemotaxis).

Once complement is activated, proteins in blood and on body cells break down activated C3. In this way, its destructive capabilities cease very quickly so that damage to body cells is minimized.

Checkpoint

36. Where are the complement proteins produced?
37. How does the complement system destroy microbes?
38. How is the complement system activated?

21.9 Immunological memory results in a more intense secondary response to an antigen.

A hallmark of immune responses is memory for specific antigens that have triggered immune responses in the past. **Immunological memory** is due to the presence of long-lasting antibodies and very long-lived lymphocytes that arise during clonal selection of antigen-stimulated B cells and T cells.

Immune responses, whether cell-mediated or antibody-mediated, are much quicker and more intense after a second or subsequent exposure to an antigen than after the first exposure. Initially, only a few cells have the correct antigen receptors to respond, and the immune response may take several days to build to maximum intensity. Because thousands of memory cells exist after an initial encounter with an antigen, within hours of the next appearance of the antigen they can proliferate and differentiate into helper T cells, cytotoxic T cells, or plasma cells.

One measure of immunological memory is *antibody titer*, the amount of antibody in serum. After an initial contact with an antigen, no antibodies are present for several days. Then, the levels of antibodies slowly rise, first IgM and then IgG, followed by a gradual decline (Figure 21.18). This is the **primary response**.

Memory cells may remain for decades. Every new encounter with the same antigen results in rapid division of memory cells. The antibody level after subsequent encounters is far greater than during a primary response and consists mainly of IgG antibodies. This accelerated, more intense response is called the **secondary response**. Antibodies produced during a secondary response have an even higher affinity for the antigen than those produced during a primary response, and thus they are more successful in disposing of it.

Primary and secondary responses occur during microbial infection. When you recover from an infection without taking antimicrobial drugs, it is usually because of the primary response. If the same microbe infects you later, the secondary response could be so swift that the microbes are destroyed before you exhibit any signs or symptoms of infection.

Immunological memory provides the basis for immunization by vaccination against certain diseases, such as polio. When you

Figure 21.18 Secretion of antibodies. The primary response (after first exposure) is milder than the secondary response (after second exposure) to a given antigen.

Immunological memory is the basis for successful immunization by vaccination.

receive the vaccine, which may contain weakened or killed whole microbes or portions of microbes, your B cells and T cells are activated. Should you subsequently encounter the living pathogen as an infecting microbe, your body initiates a secondary response.

Table 21.4 summarizes the various ways to acquire adaptive immunity.

Checkpoint

39. What role do memory cells play in immunological memory?
40. How is the secondary response to an antigen different from the primary response?
41. Which type of antibody responds most strongly during the secondary response?

TABLE 21.4

Ways to Acquire Adaptive Immunity

METHOD	DESCRIPTION
Naturally acquired active immunity	Following exposure to a microbe, antigen recognition by B cells and T cells and costimulation lead to the formation of antibody-secreting plasma cells, cytotoxic T cells, and B and T memory cells.
Naturally acquired passive immunity	Transfer of IgG antibodies from mother to fetus across placenta, or of IgA antibodies from mother to baby in milk during breast-feeding.
Artificially acquired active immunity	Antigens introduced during a vaccination stimulate cell-mediated and antibody-mediated immune responses, leading to the production of memory cells. The antigens are pretreated to be immunogenic but not pathogenic; that is, they will trigger an immune response but not cause significant illness.
Artificially acquired passive immunity	Intravenous injection of immunoglobulins (antibodies).

Marlene's Story: Epilogue and Discussion

▶▶▶ Marlene is a 57-year-old woman recovering from a breast lumpectomy, lymph node removal, and lymphangitis from a cut to the hand that becomes infected. At an urgent care center she presents with a fever and infection. She is admitted to the hospital, where she is placed on a course of intravenous antibiotics. Following resolution of the infection, she consults with a physical therapist, who suggests treatment to reduce swelling and discomfort due to lymph node removal. ◀◀◀

G. Why would reduced lymph return have increased Marlene's risk of localized infection?

H. How would removal of her lymph nodes have reduced Marlene's ability to fight infection?

I. If Marlene suffered a cut on the side of her body opposite the surgery, would her body's response have been different? Why or why not?

Concept and Resource Summary

Concept	Resources

Introduction

1. The ability to ward off disease is called **immunity** (**resistance**). Lack of resistance is called susceptibility.
2. **Innate immunity** is present at birth and provides rapid responses against disease.
3. **Adaptive immunity** develops in response to specific **pathogens** and occurs more slowly than innate immunity.

Concept 21.1 The lymphatic system drains interstitial fluid, transports dietary lipids, and protects against invasion.

1. The **lymphatic system** consists of **lymph**, lymphatic tissues, and red bone marrow.
2. **Lymphatic tissue** is a specialized form of reticular tissue containing large numbers of lymphocytes.
3. The lymphatic system drains excess interstitial fluid, transports lipids, and carries out immune responses against invasion.

See

• Animation – Lymphatic System Functions

Concept 21.2 Lymph flows through lymphatic capillaries, lymphatic vessels, and lymph nodes.

1. Lymphatic vessels begin as closed-ended **lymphatic capillaries** in tissue spaces between cells. The structure of lymphatic capillaries allows interstitial fluid to flow in but not out.
2. Lymph capillaries merge to form larger vessels, called **lymphatic vessels**, which convey lymph into and out of lymph nodes.

See

• Anatomy Overview – Lymphatic Vessels
• Animation – Lymph Formation and Flow

Do

• Exercise – Lymphatic Highway

Concept

3. The route of lymph flow is from lymph capillaries to lymphatic vessels, to **lymph trunks**, to the thoracic duct and right lymphatic duct, then to the subclavian veins. The **thoracic duct** drains lymph from the left side of the head, neck, and chest; the left upper limb; and the body inferior to the ribs. The **right lymphatic duct** drains lymph from the upper right side of the body.
4. Lymph flows due to the "milking action" of skeletal muscle contractions and pressure changes that occur during inhalation. Valves in lymphatic vessels prevent backflow of lymph.

Concept 21.3 The lymphatic organs and tissues include the thymus, lymph nodes, spleen, and lymphatic nodules.

1. **Primary lymphatic organs** and **tissues** are the sites where stem cells divide and develop into mature B cells and T cells. They are the red bone marrow and the thymus.
2. **Secondary lymphatic organs** and **tissues** are the sites where most immune responses occur. They include lymph nodes, the spleen, and lymphatic nodules.
3. The **thymus** lies posterior to the sternum and superior to the heart. It is the site of T cell maturation.
4. **Lymph nodes** are located along lymphatic vessels. Lymph enters lymph nodes through **afferent lymphatic vessels**, is filtered, and exits through **efferent lymphatic vessels**.
5. The **spleen** is the largest mass of lymphatic tissue in the body. It is a site where B cells divide into plasma cells and macrophages phagocytize worn-out red blood cells and platelets. Within the spleen, B cells and T cells carry out immune functions and macrophages phagocytize blood-borne pathogens and worn-out blood cells.
6. **Lymphatic nodules** are concentrations of lymphatic tissue that are not surrounded by a capsule. They are scattered throughout the mucosa of the gastrointestinal, respiratory, urinary, and reproductive tracts. **Tonsils** are examples of lymphatic nodules.

See

- Anatomy Overview – The Spleen and Lymph Nodes
- Anatomy Overview – The Thymus

Hear

- **Figure 21.5 –**
  Thymus of an Adolescent
- **Figure 21.6 – Structure of a Lymph Node**

Concept 21.4 Innate immunity includes external physical and chemical barriers and various internal defenses.

1. **Innate immunity** includes physical factors, chemical factors, antimicrobial proteins, natural killer cells, phagocytes, inflammation, and fever.
2. The skin and mucous membranes are the first line of defense against entry of pathogens. The closely packed keratinized cells of the **epidermis** are a physical barrier to invasion. **Mucous membranes** have **mucus** that traps microbes and **cilia** that sweep microbes away from the lungs. **Tears**, **saliva**, **urine**, and **vaginal secretions** flush microbes from the body. **Defecation** and **vomiting** expel microbes. The acidity of **gastric juice** and **vaginal secretions** discourages bacterial growth.
3. Pathogens that get past the first line of defense are attacked by internal defenses. **Antimicrobial proteins** (**interferons** and **complement**) inhibit microbial growth. **Natural killer cells** and **phagocytes** attack and kill pathogens and defective cells in the body.
4. **Inflammation** aids disposal of microbes, toxins, or foreign material at the site of an injury and prepares the site for tissue repair.
5. **Fever** intensifies the antiviral effects of interferons, inhibits growth of some microbes, and speeds up body reactions that aid repair.

See

- Anatomy Overview –
  The Integument and Disease Resistance
- Animation – Introduction to Disease Resistance
- Animation – Nonspecific Disease Resistance

Do

- Exercise – Integument vs. Disease
- Concepts and Connections – Nonspecific Disease Resistance
- Concepts and Connections – Role of Integument in Disease Resistance

Concept 21.5 Adaptive immunity involves the production of a specific lymphocyte or antibody against a specific antigen.

1. The body's defense activities against specific pathogens and antigens is called **adaptive immunity**. **Antigens** are foreign substances that produce an immune response. Adaptive immunity has *specificity* for particular antigens and *memory* of previous antigens.
2. **B cells** and **T cells** arise from stem cells in red bone marrow. B cells mature in red bone marrow, whereas T cells mature in the thymus gland.
3. Before B cells leave the red bone marrow or T cells leave the thymus, they develop **immunocompetence**, the ability to carry out adaptive immune responses by inserting **antigen receptors** into their plasma membranes. Antigen receptors are molecules that are capable of recognizing specific antigens.
4. In **cell-mediated immunity**, cytotoxic T cells directly attack invading antigens and are effective against intracellular pathogens, cancer cells, and transplanted tissues. In **antibody-mediated immunity**, B cells become plasma cells that secrete **antibodies** and are effective against extracellular pathogens.

See

- Anatomy Overview – B Lymphocytes
- Anatomy Overview – T Lymphocytes
- Anatomy Overview – Antigens and Antibodies
- Animation – Introduction to Disease Resistance
- Animation – Introduction to Antibody-mediated Immunity
- Animation – Introduction to Cell-mediated Immunity

Do

- Concepts and Connections – Specific Disease Resistance

Concept

Resources

5. **Clonal selection** is the process by which a lymphocyte is stimulated by a specific antigen to divide and differentiate into a clone of cells that can recognize the same specific antigen as the original lymphocyte. The resulting clone includes effector cells and memory cells. **Effector cells** destroy or inactivate the antigen. Memory cells do not actively participate in the initial immune response, but if the antigen reappears in the body in the future, **memory cells** quickly divide and differentiate into more effector and memory cells.

6. Antigens stimulate the production of specific antibodies. Immune responses are triggered by **epitopes**, particles on the surface of antigens.

7. "Self-antigens" called **major histocompatibility complex (MHC) antigens** are unique to each person's body cells. All cells except red blood cells display MHC molecules.

8. **Exogenous antigens** (present outside body cells) are processed by **antigen-presenting cells**. The antigen-presenting cell ingests and digests the antigen, binds the antigen fragments with MHC molecules, and inserts the antigen–MHC complexes into the cell's plasma membrane. The exogenous antigen–MHC complexes trigger adaptive immune responses in T cells

9. **Endogenous antigens** (present inside body cells) bind with MHC molecules. The endogenous antigen–MHC complexes insert in the plasma membrane, which signals that the cell is infected.

10. **Cytokines** are small hormones that stimulate or inhibit many normal cell functions such as growth and differentiation. Other cytokines regulate immune responses.

Concept 21.6 In cell-mediated immunity, cytotoxic T cells directly attack target cells.

1. A cell-mediated immune response begins with *activation* of a small number of T cells by a specific antigen and ultimately results in *elimination* of the intruder.

2. During the activation process, **T cell receptors (TCRs)** recognize and bind to antigen fragments associated with MHC molecules on the surface of a body cell. Activation of T cells also requires **costimulation**, usually by **interleukin-2**. Once a T cell has been **activated**, it undergoes clonal selection.

3. **Helper T cells** display the CD4 protein and are activated by MHC molecules presented by an APC. Once activated, helper T cells undergo clonal selection, then secrete several cytokines, most importantly interleukin-2, which acts as a costimulator for other T cells, B cells, and natural killer cells.

4. **Cytotoxic T cells** display CD8 protein and are activated by MHC molecules on the surfaces of body cells infected by microbes, tumor cells, and transplanted tissue cells. Once activated, cytotoxic T cells undergo clonal selection, then release **granzymes**, which cause target cell apoptosis (phagocytes then kill the released microbes); **perforin**, which causes target cell cytolysis; and **granulysin**, which enters the target cell to destroy microbes.

See

- Anatomy Overview –
  T Lymphocytes
- Animation – Cell-mediated
  Immunity

Hear

- Figure 21.13 –
  **Activation and Clonal
  Selection of a Helper T Cell**
- Figure 21.14 – **Activation and
  Clonal Selection of a Cytotoxic
  T Cell**

Do

- Exercise – Activate and Attack
- Concepts and Connections –
  Specific Disease Resistance

Concept 21.7 In antibody-mediated immunity, antibodies specifically target a particular antigen.

1. An antibody-mediated immune response begins when a specific antigen binds to **B cell receptors**, activating the B cell.

2. B cells can respond to unprocessed antigens, but their response is more intense when they process the antigen by generating and then inserting antigen-MHC complexes into the B cell plasma membrane. Helper T cells recognize the antigen-MCH complexes and deliver constimulation for activation of B cells.

3. Once activated, a B cell undergoes clonal selection, forming a clone of **plasma cells** that secrete antibodies and **memory B cells** to quickly produce more plasma and memory B cells should the same antigen ever reappear.

4. An **antibody** is a protein that combines specifically with the antigen that triggered its production. Based on chemistry and structure, antibodies are grouped into five principal classes (IgG, IgA, IgM, IgD, and IgE), each with specific biological roles.

5. Antibodies form antigen–antibody complexes with their complementary antigens, which disables antigens. Antibody actions include blocking attachment of the antigen to body cells, immobilization of bacteria, agglutination and precipitation of antigen, activation of complement, and enhancement of phagocytosis.

See

- Anatomy Overview – B Lymphocytes
- Anatomy Overview – Antigens and
  Antibodies
- Animation – Antibody-mediated
  Immunity

Hear

- Figure 21.16 –
  **Activation and Clonal
  Selection of B Cells**

Do

- Exercise – Activate and Attack
- Exercise – Antibody Ambush
- Concepts and Connections –
  Specific Disease Resistance

Concept	Resources
Concept 21.8 The complement system destroys microbes through phagocytosis, cytolysis, and inflammation. 1. Complement is a group of proteins that "complement" immune responses and help clear antigens from the body. Once activated, complement proteins bring about phagocytosis, cytolysis, and inflammation.	**See** • Animation – Complement Proteins **Do** • Concepts and Connections – Complement Proteins
Concept 21.9 Immunological memory results in a more intense secondary response to an antigen. 1. The first exposure to an antigen causes the **primary response**, a slow rise in antibody titer several days after exposure, followed by a gradual decline. Immune responses are much quicker and more intense after a second or subsequent exposure. 2. The **secondary response**, the accelerated response to an antigen after a subsequent exposure, is caused by memory B cells and memory T cells that remain in the body.	**See** • Animation – Primary and Secondary Infections

Other Chapter Resources

For additional information on clinical applications, medical disorders, and medical terminology, see Chapter 10 in the *Clinical Connections* manual packaged with your text.

Within *WileyPLUS*
- Chapter Quiz
- Visual Anatomy
- Mastering Vocabulary
- Audio Glossary==
- Homeostatic Imbalance: The Case of the Girl with the Butterfly Rash
- Homeostatic Imbalance: The Case of the Man with Opportunistic Infections
- Web Links

• Understanding the Concepts

1. Is lymph more similar to blood plasma or to interstitial fluid? Explain your answer.

2. Describe how running, inhalation, and exhalation each influence the flow of lymph.

3. Contrast the functions of natural killer cells and phagocytes.

4. What are the functions of inflammation?

5. What causes each of the following signs and symptoms of inflammation: redness, pain, heat, and swelling?

6. What purpose is served by fever?

7. What is adaptive immunity? How does it compare with innate immunity?

8. What is the relationship between an antigen and an antibody?

9. Why is it a survival advantage to have APCs in the respiratory tract?

10. How many different types of antibodies will be secreted by plasma cells derived from the same clone? Explain your answer.

11. How are cell-mediated and antibody-mediated immune responses similar? How do they differ?

12. According to the graph in Figure 21.18, how much more IgG is circulating in the blood in the secondary response than in the primary response? (*Hint*: Notice that each mark on the antibody titer axis represents a 10-fold increase.)

22

The Respiratory System

Cari had gone to bed early. Her throat had started to feel sore, the same feeling she always got when she was getting a cold. The tickle at the back of her throat wasn't that unusual; she coughed a lot, having been a pack-a-day smoker for 20 years. But this was different from her smoker's cough; this was turning into another cold. She tended to have a lot of them. She woke up after a fitful night of sleep with a pounding headache and perspiration-soaked bedsheets. The back of her throat was dry, and her shoulders ached. She reached for a tissue and blew her nose.

"What's the matter, did you catch another cold?" asked her husband Ken.

"It's just a summer cold, maybe allergies. I'll take some antihistamines and vitamin C," she replied.

"You look pretty bad. Maybe you should call in sick today and see the doctor," advised Ken.

"I just don't have time, Ken. I've got to get the Jenson house finished by September and there have already been delays with the subcontractors." Cari got out of bed, retrieving her cigarettes and lighter from the nightstand. She slid open the patio door, walked out onto the deck, and lit up a cigarette. Ken followed her.

"Cari, maybe this is a good time to quit; you know it's been a year for me now."

Respiratory tract infections can involve the upper or lower respiratory tracts. The respiratory system is susceptible to infectious microorganisms of several types. An individual's response to a respiratory tract infection depends on the type of infectious organism, and the person's age and overall health status. In Cari's case, a long history of smoking has led to chronic respiratory tract infections. As we will see in this chapter, frequent colds are not the only problem caused by her smoking habit.

INTRODUCTION

Your body's cells continually use oxygen (O_2) for the metabolic reactions that release energy from nutrient molecules and produce ATP. At the same time, these reactions release carbon dioxide (CO_2). Because an excessive amount of CO_2 produces acidity that can be toxic to cells, excess CO_2 must be eliminated quickly and efficiently. The cardiovascular and respiratory systems cooperate to supply O_2 to cells and eliminate CO_2. The cardiovascular system transports blood containing those gases between the lungs and body

The respiratory system provides for gas exchange.

cells. The **respiratory system** (RES-pi-ra-tō-rē) provides for gas exchange—intake of O_2 and elimination of CO_2. Failure of the exchange of either gas disrupts homeostasis by causing rapid death of cells from oxygen starvation and buildup of waste products. There is an extensive area of contact between capillary blood vessels of the respiratory system and the external environment. This area of contact allows the body to constantly renew and replenish the internal fluid environment that surrounds and nourishes every body cell. In addition to functioning in gas exchange, the respiratory system participates in regulating blood pH, contains receptors for the sense of smell, filters inspired air, produces sounds, and rids the body of some water and heat in exhaled air.

CONCEPTS

- **22.1** Inhaled air travels in the upper respiratory system through the nasal cavities and then through the pharynx.
- **22.2** Inhaled air travels in the lower respiratory system from the larynx to alveoli.
- **22.3** Inhalation and exhalation result from pressure changes caused by muscle contraction and relaxation.
- **22.4** Lung volumes and capacities are measured to determine the respiratory status of an individual.
- **22.5** Oxygen and carbon dioxide diffusion is based on partial pressure gradients and solubility.
- **22.6** Respiration occurs between alveoli and pulmonary capillaries and between systemic capillaries and tissue cells.
- **22.7** Oxygen is primarily transported attached to hemoglobin, while carbon dioxide is transported in three different ways.
- **22.8** The basic rhythm of respiration is controlled by the respiratory center in the brain stem.
- **22.9** Respiration may be modified by cortical influences, chemical stimuli, proprioceptor input, and the inflation reflex.
- **22.10** Acid–base balance is maintained by controlling the H^+ concentration of body fluids.

22.1   Inhaled air travels in the upper respiratory system through the nasal cavities and then through the pharynx.

The **respiratory system** consists of the nose, pharynx (throat), larynx (voice box), trachea (windpipe), bronchi, and lungs (Figure 22.1). Its parts can be classified according to either structure or function. *Structurally*, the respiratory system consists of two parts: (1) The **upper respiratory system** includes the nose, pharynx, and associated structures. (2) The **lower respiratory system** includes the larynx, trachea, bronchi, and lungs. *Functionally*, the respiratory system also consists of two parts: (1) The **conducting zone** consists of a series of interconnecting cavities and tubes both outside and within the lungs—the nose, pharynx, larynx, trachea, bronchi, bronchioles, and terminal bronchioles—that filter, warm, and moisten air and conduct it into the lungs. (2) The **respiratory zone** consists of tissues within the lungs where gas exchange occurs—the respiratory bronchioles, alveolar ducts, alveolar sacs, and alveoli, the main sites of gas exchange between air and blood.

Nose

The **nose** has a visible external portion and an internal portion inside the skull (Figure 22.2a). The external portion consists of a supporting framework of bone and hyaline cartilage covered with muscle and skin and lined by a mucous membrane. The frontal bone, nasal bones, and maxillae form the bony framework of the external nose. The cartilaginous framework of the external nose consists of the **septal cartilage**, which forms the anterior portion of the nasal septum; the **lateral nasal cartilages** inferior to the nasal bones; and the **alar cartilages**, which form a portion of the walls of the nostrils. Because it consists of pliable hyaline cartilage, the cartilaginous framework of the external nose is somewhat flexible. On the undersurface of the external nose are two openings called the **external nares** (NĀ-rez; singular is *naris*), or **nostrils**. Figure 22.3 shows the

---

Figure 22.1   Structures of the respiratory system.

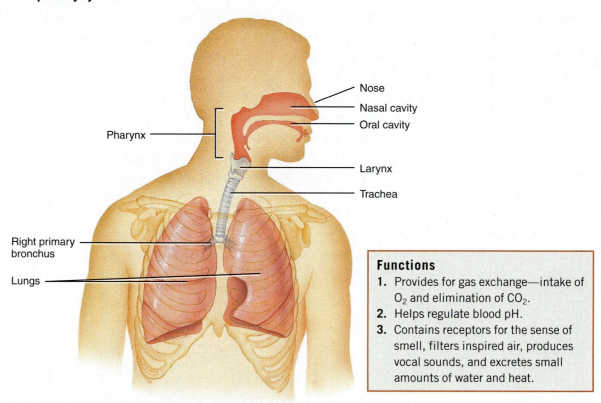

Nose
Nasal cavity
Oral cavity
Pharynx
Larynx
Trachea
Right primary bronchus
Lungs

Functions
1. Provides for gas exchange—intake of O_2 and elimination of CO_2.
2. Helps regulate blood pH.
3. Contains receptors for the sense of smell, filters inspired air, produces vocal sounds, and excretes small amounts of water and heat.

(a) Anterior view showing organs of respiration

surface anatomy of the nose. The internal portion of the nose is a large cavity in the anterior aspect of the skull that lies inferior to the nasal bone and superior to the mouth; it is lined with muscle and mucous membrane. The internal nose connects to the pharynx through two openings called the **internal nares** (Figure 22.2b). Ducts from the paranasal sinuses (which drain mucus) and the nasolacrimal ducts (which drain tears) also open into the internal nose. The lateral walls of the internal nose are formed by the ethmoid, maxillae, lacrimal, palatine, and inferior nasal conchae bones (see Figure 7.8); the ethmoid bone also forms the roof. The palatine bones and palatine processes of the maxillae, which together constitute the hard palate, form the floor of the internal nose.

The space within the internal nose is called the **nasal cavity**. The anterior portion of the nasal cavity just inside the nostrils, called the **nasal vestibule**, is surrounded by cartilage; the superior part of the nasal cavity is surrounded by bone. A vertical partition, the **nasal septum**, divides the nasal cavity into right and left sides. The anterior portion of the nasal septum consists primarily of hyaline cartilage; the remainder is formed by the vomer, perpendicular plate of the ethmoid, maxillae, and palatine bones (see Figure 7.9).

The nose has three functions: (1) warming, moistening, and filtering incoming air; (2) detecting olfactory (smell) stimuli; and (3) modifying speech vibrations as they pass through the large, hollow resonating chambers. *Resonance* refers to prolonging, amplifying, or modifying a sound by vibration.

The **paranasal sinuses** are mucous membrane–lined cavities in certain skull bones that are continuous with the nasal cavity. Besides producing mucus, the paranasal sinuses serve as resonating chambers for sound as we speak or sing.

When air enters the nostrils, it passes first through the vestibule, which is lined by skin containing coarse hairs that filter out large dust particles. Three shelves formed by projections of the superior, middle, and inferior nasal conchae extend out of each lateral wall of the nasal cavity. The conchae subdivide each side of the nasal cavity into a series of groovelike passageways—the **superior**, **middle**, and **inferior meatuses** (mē-Ā-tus-ez = openings or passages; singular is *meatus*). Mucous membrane lines the nasal cavity and the three conchae. The arrangement of conchae and meatuses increases surface area in the internal nose and prevents dehydration by trapping water droplets during exhalation.

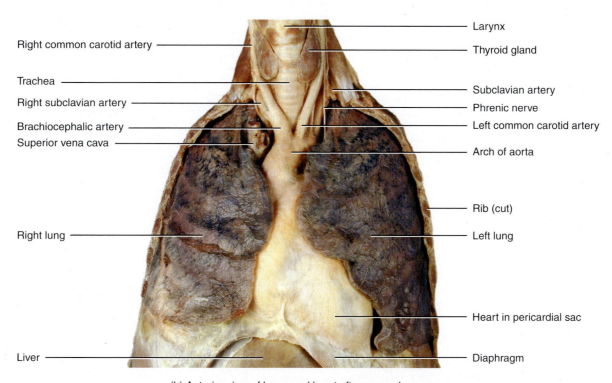

Right common carotid artery

Trachea

Right subclavian artery

Brachiocephalic artery

Superior vena cava

Right lung

Liver

Larynx

Thyroid gland

Subclavian artery

Phrenic nerve

Left common carotid artery

Arch of aorta

Rib (cut)

Left lung

Heart in pericardial sac

Diaphragm

(b) Anterior view of lungs and heart after removal
of the anterolateral thoracic wall and pleura

The upper respiratory system includes the nose and pharynx; the lower respiratory system includes the larynx, trachea, bronchi, and lungs.

Figure 22.2   Respiratory structures in the head and neck.

Bony framework:
Frontal bone
Nasal bones
Maxilla

Cartilaginous framework:
Lateral nasal cartilages
Septal cartilage
Alar cartilage

Dense fibrous connective and adipose tissue

(a) Anterolateral view of external portion of nose showing cartilaginous and bony framework

Sagittal plane

Nasal meatuses
Superior
Middle
Inferior

Frontal sinus
Frontal bone
Olfactory epithelium

Sphenoid bone
Sphenoidal sinus
Internal naris
Pharyngeal tonsil
NASOPHARYNX
Opening of auditory tube
Uvula
Palatine tonsil
Fauces
OROPHARYNX
Epiglottis
LARYNGOPHARYNX
Esophagus
Trachea

Superior
Middle
Inferior
Nasal conchae

Nasal vestibule
External naris
Maxilla
Oral cavity
Palatine bone
Soft palate
Lingual tonsil
Mandible
Hyoid bone

Tongue

Ventricular fold
Vocal fold
Larynx
Thyroid cartilage
Cricoid cartilage
Thyroid gland

Nasopharynx
Oropharynx
Laryngopharynx

Regions of the pharynx

(b) Sagittal section of the left side of the head and neck showing the location of respiratory structures

🔑 **As air passes through the nose, it is warmed, filtered, and moistened, and olfaction occurs.**

Figure 22.3 **Surface anatomy of the nose.**

Bridge (Bony framework of nose formed by nasal bones)".

"Apex (tip of nose)".

"Root (superior attachment of the nose to the frontal bone)".

"Bridge: (Bony framework of nose formed by nasal bones)".

The external nose has a cartilaginous and a bony framework.

The olfactory receptors lie in a region of the membrane lining the superior nasal conchae and adjacent nasal septum called the **olfactory epithelium**. Inferior to the olfactory epithelium, the mucous membrane contains capillaries and pseudostratified columnar epithelium (see Table 4.1E). Pseudostratified columnar epithelium of the respiratory tract contains ciliated columnar cells and many goblet cells. As inhaled air whirls around the conchae and meatuses, it is warmed by blood in the capillaries. Mucus secreted by the goblet cells moistens the air and traps dust particles. Drainage from the nasolacrimal ducts also helps moisten the air and is sometimes assisted by secretions from the paranasal sinuses. The cilia move the mucus and trapped dust particles toward the pharynx, at which point they can be swallowed or spit out, thus removing the particles from the respiratory tract.

Pharynx

The **pharynx** (FAIR-inks), or throat, is a funnel-shaped tube about 13 cm (5 in.) long that starts at the internal nares and extends to the level of the cricoid cartilage, the most inferior cartilage of the larynx (Figure 22.2). The pharynx lies just posterior to the nasal and oral cavities, superior to the larynx, and just anterior to the cervical (neck) vertebrae. Its wall is composed of skeletal muscles and is lined with a mucous membrane. Contraction of the skeletal muscles assists swallowing. The pharynx functions as a passageway for air and food, provides a resonating chamber for speech sounds, and houses the tonsils, which participate in immunological reactions against foreign invaders.

The pharynx can be divided into three anatomical regions: (1) nasopharynx, (2) oropharynx, and (3) laryngopharynx. (See the lower orientation diagram in Figure 22.2.) The superior portion of the pharynx, called the **nasopharynx**, lies posterior to the nasal cavity and extends to the soft palate. The **soft palate** forms the posterior roof of the mouth between the nasopharynx and oropharynx. There are five openings in its wall: two internal nares, two openings that lead into the *auditory (pharyngotympanic) tubes*, and the opening into the oropharynx. The posterior wall also contains the **pharyngeal tonsil (adenoid)**. Through the internal nares, the nasopharynx receives air from the nasal cavity along with dust-laden mucus. The nasopharynx is lined with pseudostratified columnar epithelium, and the cilia move the mucus down toward the most inferior part of the pharynx. The nasopharynx also exchanges small amounts of air with the auditory tubes to equalize air pressure between the pharynx and the middle ear.

The intermediate portion of the pharynx, the **oropharynx**, lies posterior to the oral cavity and extends from the soft palate inferiorly to the level of the hyoid bone. It has only one opening, the **fauces** (FAW-sēz = throat), the opening from the mouth. The oropharynx has both respiratory and digestive functions, serving as a common passageway for air, food, and drink. Because the oropharynx is subject to abrasion by food particles, it is lined with nonkeratinized stratified squamous epithelium. Two pairs of tonsils, the **palatine** and **lingual tonsils**, are found in the oropharynx.

The inferior portion of the pharynx, the **laryngopharynx** (la-rin´-gō-FAIR-inks), begins at the level of the hyoid bone. It opens into the esophagus posteriorly and the larynx anteriorly. Like the oropharynx, the laryngopharynx is both a respiratory and a digestive pathway and is lined by nonkeratinized stratified squamous epithelium.

Checkpoint

1. What is the path taken by air molecules into and through the nose?
2. List the functions of the nose.
3. Name the openings into and out of the nose.
4. What are the functions of the paranasal sinuses?
5. How does the nose moisten and clean inhaled air before it enters the rest of the respiratory tract?
6. What are the superior and inferior borders of the pharynx?

22.2 Inhaled air travels in the lower respiratory system from the larynx to alveoli.

The larynx, trachea, bronchi, and lungs are structurally considered the **lower respiratory system**.

Larynx

The **larynx** (LAR-inks), or voice box, is a short passageway that connects the laryngopharynx with the trachea (see Figure 22.2b). It lies in the midline of the neck anterior to the esophagus and the fourth through sixth cervical vertebrae (C4–C6). The larynx is composed of cartilaginous walls that are held in position by ligaments and skeletal muscles.

The **thyroid cartilage** (Adam's apple) consists of hyaline cartilage that forms the anterior wall of the larynx and gives it a triangular shape (Figure 22.4). It is usually larger in males than in females due to the influence of male sex hormones on its growth during puberty. The thyroid cartilage is attached superiorly to the hyoid bone and epiglottis; inferiorly it connects to the cricoid cartilage.

The **epiglottis** (epi- = over; -glottis = tongue) is a large, leaf-shaped piece of elastic cartilage that is covered with epithelium (see Figure 22.2b). The "stem" of the epiglottis is the tapered inferior portion that is attached to the anterior rim of the thyroid cartilage and hyoid bone. The broad superior "leaf" portion of the epiglottis is unattached and is free to move up and down like a trapdoor. The **glottis** consists of a pair of folds of mucous membrane, the vocal folds in the larynx, and the opening between them (Figure 22.5a). During swallowing, the pharynx and larynx rise. Elevation of the pharynx widens it to receive food or drink; elevation of the larynx causes the epiglottis to move down and form a lid over the glottis, closing it off. Closing of the larynx by the epiglottis during swallowing routes liquids and foods into the esophagus while keeping them out of the larynx and airways. When small particles of dust, smoke, food, or liquids pass into the larynx, a cough reflex occurs, usually expelling the material. Coughing itself requires a deep inspiration of air followed by closure of the glottis. Forceful exhalation pushes the glottis open and blasts air through the upper respiratory passageways.

The **cricoid cartilage** (KRĪ-koyd = ringlike) is a ring of hyaline cartilage that forms the inferior wall of the larynx. The thyroid cartilage is connected superiorly to the cricoid cartilage and inferiorly to the first ring of cartilage of the trachea. The cricoid cartilage is the landmark for making an emergency airway, called a tracheotomy.

The **arytenoid cartilages** (ar´-i-TĒ-noyd = ladlelike) are a pair of triangular pieces of mostly hyaline cartilage located at the posterior, superior border of the cricoid cartilage. They form synovial joints with the cricoid cartilage and have a wide range of mobility for sound production.

The lining of the larynx superior to the vocal folds is nonkeratinized stratified squamous epithelium. The lining of the larynx inferior to the vocal folds is pseudostratified columnar epithelium. The mucus produced by the goblet cells helps trap dust not removed in the upper passages. The cilia in the upper respiratory tract move mucus and trapped particles *down* toward the pharynx; the cilia in the lower respiratory tract move them *up* toward the pharynx. Once in the pharynx, the mucus-trapped dust particles can be swallowed or spit out.

Figure 22.4 **Larynx.**

Larynx   Thyroid gland

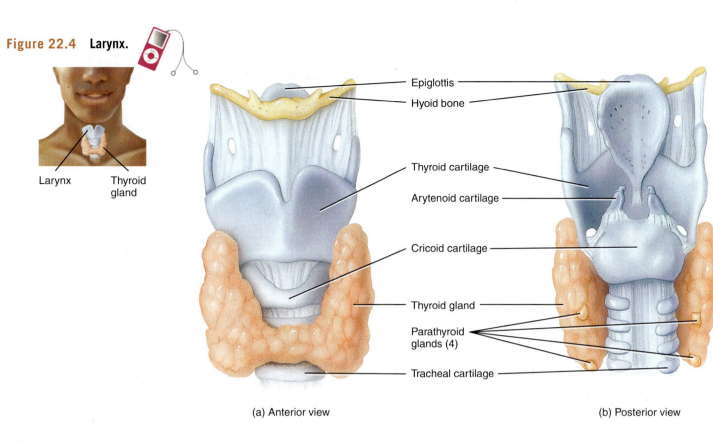

Epiglottis

Hyoid bone

Thyroid cartilage

Arytenoid cartilage

Cricoid cartilage

Thyroid gland

Parathyroid glands (4)

Tracheal cartilage

(a) Anterior view   (b) Posterior view

Sagittal plane

Epiglottis

Hyoid bone

Fat body

Arytenoid cartilage

Ventricular fold

Thyroid cartilage

Vocal fold

Cricoid cartilage

Tracheal cartilage

(c) Sagittal section

The larynx is composed of cartilage.

The Structures of Voice Production

The mucous membrane of the larynx forms two pairs of folds (Figure 22.4c): a superior pair called the **ventricular folds (false vocal cords)** and an inferior pair called the **vocal folds (true vocal cords)**. When the ventricular folds are brought together, they hold the breath against pressure in the thoracic cavity, such as might occur when you strain to lift a heavy object. They do not produce sound.

The vocal folds produce sounds during speaking and singing. Deep to the mucous membrane of the vocal folds, bands of elastic ligaments are stretched between pieces of rigid cartilage like the strings on a guitar. Laryngeal muscles attach to both the rigid cartilage and the vocal folds. When the muscles contract, they pull the

Figure 22.5 Movement of the vocal folds.

Thyroid cartilage
Cricoid cartilage
Vocal ligament
Arytenoid cartilage
Posterior cricoarytenoid

Superior view of cartilages and muscles

Tongue
Epiglottis
Glottis:
  Vocal folds
  Passageway to trachea
Ventricular folds

View through a laryngoscope

(a) Movement of vocal folds apart (abduction)

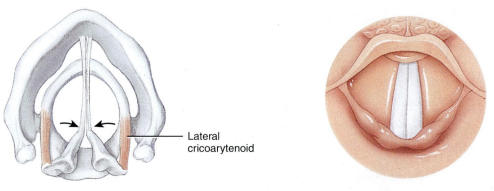

Lateral cricoarytenoid

(b) Movement of vocal folds together (adduction)

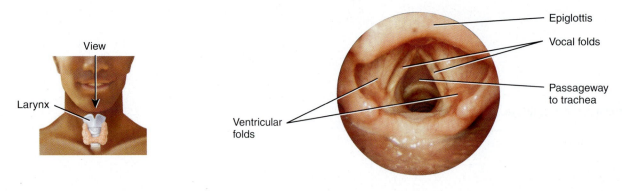

View
Larynx

Epiglottis
Vocal folds
Passageway to trachea
Ventricular folds

(c) Superior view

🔑 **The glottis consists of the vocal folds and the space between them.**

elastic ligaments tight, which moves the vocal folds out into the air passageway, narrowing the glottis. When air is directed against the vocal folds, they vibrate and produce sounds (phonation) by setting up sound waves in the air in the pharynx, nose, and mouth. The greater the air pressure, the louder the sound.

When the muscles of the larynx contract, they pull on the arytenoid cartilages, which causes them to pivot and slide. Contraction of the posterior cricoarytenoid muscles, for example, moves the vocal folds apart, thereby opening the glottis (Figure 22.5a). By contrast, contraction of the lateral cricoarytenoid muscles brings the vocal folds together, closing the glottis (Figure 22.5b). Other laryngeal muscles can elongate (and place tension on) or shorten (and relax) the vocal folds.

Pitch is controlled by the tension on the vocal folds. If they are pulled taut by the muscles, they vibrate more rapidly and a higher pitch results. Decreasing the muscular tension on the vocal folds causes them to vibrate more slowly and produce lower-pitch sounds. Due to the influence of androgens (male sex hormones), vocal folds are usually thicker and longer in males than in females, and therefore they vibrate more slowly. This is why a man's voice generally has a lower range of pitch than a woman's.

Sound originates from the vibration of the vocal folds, but other structures are necessary for converting the sound into recognizable speech. The pharynx, mouth, nasal cavity, and paranasal sinuses all act as resonating chambers that give the voice its human and individual quality. We produce the vowel sounds by constricting and relaxing the muscles in the wall of the pharynx. Muscles of the face, tongue, and lips help us enunciate words.

Whispering is accomplished by closing all but the posterior portion of the glottis. Because the vocal folds do not vibrate during whispering, there is no pitch to this form of speech. However, we can still produce intelligible speech while whispering by changing the shape of the oral cavity as we enunciate. As the size of the oral cavity changes, its resonance qualities change, which imparts a vowel-like pitch to the air as it rushes toward the lips.

Trachea

The **trachea** (TRA-kē-a = sturdy), or **windpipe**, is a tubular passageway for air that is about 12 cm (5 in.) long and 2.5 cm (1 in.) in diameter. It is located anterior to the esophagus (Figure 22.6) and extends from the larynx to the superior border of the fifth thoracic vertebra (T5), where it divides into right and left primary bronchi (see Figure 22.7).

Figure 22.6 **Location of the trachea in relation to the esophagus.**

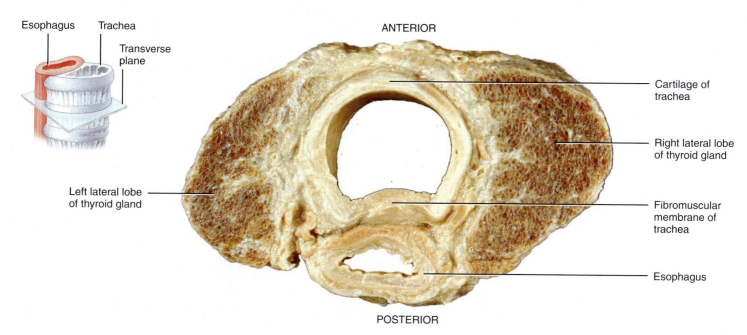

Superior view of transverse section of thyroid gland, trachea, and esophagus

The trachea is anterior to the esophagus and extends from the larynx to the primary bronchi.

The wall of the trachea is lined with mucous membrane and is supported by cartilage. The mucous membrane is composed of pseudostratified columnar epithelium; its cilia and mucus-secreting goblet cells provide the same protection against dust as the membranes lining the nasal cavity and larynx. The 16–20 incomplete, horizontal rings of hyaline cartilage resemble the letter C and are stacked one on top of another. They may be felt through the skin inferior to the larynx. The open part of each C-shaped cartilage ring faces posteriorly toward the esophagus (Figure 22.6) and is spanned by a *fibromuscular membrane*. Within this membrane are transverse smooth muscle fibers, a muscle called the **trachealis**, and elastic connective tissue that allow the diameter of the trachea to change subtly during inhalation and exhalation, which is important in maintaining efficient airflow. The C-shaped cartilage rings provide a semirigid support so that the tracheal wall does not collapse inward (especially during inhalation) and obstruct the air passageway.

Bronchi

The trachea divides into a **right primary bronchus** (BRON-kus = windpipe), which goes into the right lung, and a **left primary bronchus**, which goes into the left lung (Figure 22.7). The right primary bronchus is more vertical, shorter, and wider than the left. As a result, an aspirated object is more likely to enter and lodge in the right primary bronchus than the left. Like the trachea, the primary **bronchi** (BRON-kē) contain incomplete rings of cartilage and are lined by pseudostratified columnar epithelium.

Figure 22.7   Branching of airways from the trachea: the bronchial tree.

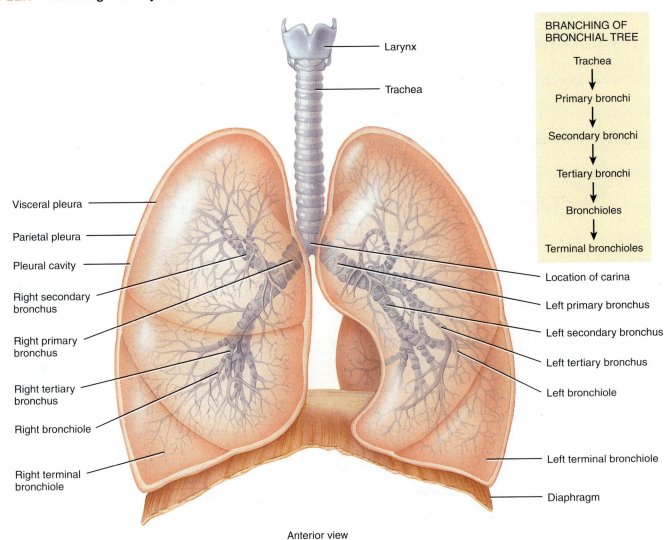

BRANCHING OF BRONCHIAL TREE

Trachea
↓
Primary bronchi
↓
Secondary bronchi
↓
Tertiary bronchi
↓
Bronchioles
↓
Terminal bronchioles

Larynx

Trachea

Visceral pleura

Parietal pleura

Pleural cavity

Right secondary bronchus

Right primary bronchus

Right tertiary bronchus

Right bronchiole

Right terminal bronchiole

Location of carina

Left primary bronchus

Left secondary bronchus

Left tertiary bronchus

Left bronchiole

Left terminal bronchiole

Diaphragm

Anterior view

The bronchial tree consists of airways that begin at the trachea and end at the terminal bronchioles.

At the point where the trachea divides into right and left primary bronchi, an internal ridge called the **carina** (ka-RĪ-na = keel of a boat) is formed by a projection of the last tracheal cartilage. The mucous membrane of the carina is one of the most sensitive areas of the entire larynx and trachea for triggering a cough reflex.

On entering the lungs, the primary bronchi divide to form smaller bronchi—the **secondary (lobar) bronchi**, one for each lobe of the lung. (The right lung has three lobes; the left lung has two.) The secondary bronchi continue to branch, forming still smaller bronchi, called **tertiary (segmental) bronchi**, that divide into **bronchioles**. Bronchioles, in turn, branch repeatedly into even smaller tubes called **terminal bronchioles**. Because this extensive branching from the trachea resembles an upside-down tree, it is commonly referred to as the **bronchial tree**.

As the branching becomes more extensive in the bronchial tree, structural changes occur:

1. The mucous membrane in the bronchial tree changes from pseudostratified columnar epithelium in the bronchi to ciliated simple columnar epithelium with some goblet cells (see Table 4.1D) in larger bronchioles, to simple cuboidal epithelium (see Table 4.1B) in smaller and terminal bronchioles. (In regions where simple cuboidal epithelium is present, inhaled particles are removed by macrophages.)

2. Plates of cartilage gradually replace the incomplete rings of cartilage in primary bronchi and finally disappear in the distal bronchioles.

3. As the amount of cartilage decreases, the amount of smooth muscle increases. Smooth muscle encircles the lumen in spiral bands. Because there is no supporting cartilage, however, muscle spasms can close off the airways. This is what happens during an asthma attack, which can be a life-threatening situation.

Lungs

The **lungs** (= lightweights, because they float) are paired spongy, cone-shaped organs in the thoracic cavity (see Figure 22.1b). They are separated from each other by the heart and other structures in the *mediastinum*, which separates the thoracic cavity into two anatomically distinct chambers. As a result, if trauma causes one lung to collapse, the other may remain expanded. Two layers of serous membrane enclose and protect each lung. The superficial layer, called the **parietal pleura** (PLOOR-a; *pleur-* = side), lines the wall of the thoracic cavity; the deep layer, the **visceral pleura**, covers the lungs themselves (Figure 22.8). Between the visceral and parietal pleurae is a narrow space, the **pleural cavity**, which contains a lubricating fluid secreted by the membranes. This *pleural fluid* reduces friction between the membranes, allowing them to slide easily over one another during breathing. Pleural fluid also causes the two membranes to adhere to one another just as a film of water causes two glass microscope slides to stick together, a phenomenon called surface tension. Separate pleural cavities surround the left and right lungs.

Figure 22.8 **Relationship of the pleural membranes to the lungs.**

Inferior view of a transverse section through the thoracic cavity showing the pleural cavity and pleural membranes

 The parietal pleura lines the thoracic cavity; the visceral pleura covers the lungs.

The lungs extend from the diaphragm to just slightly superior to the clavicles and lie against the ribs (Figure 22.9a). The broad inferior portion of the lung, the **base**, is concave and fits over the convex area of the diaphragm. The narrow superior portion of the lung is the **apex**. The medial surface of each lung contains a region, the **hilum**, through which bronchi, pulmonary blood vessels,

Figure 22.9   Surface anatomy of the lungs.

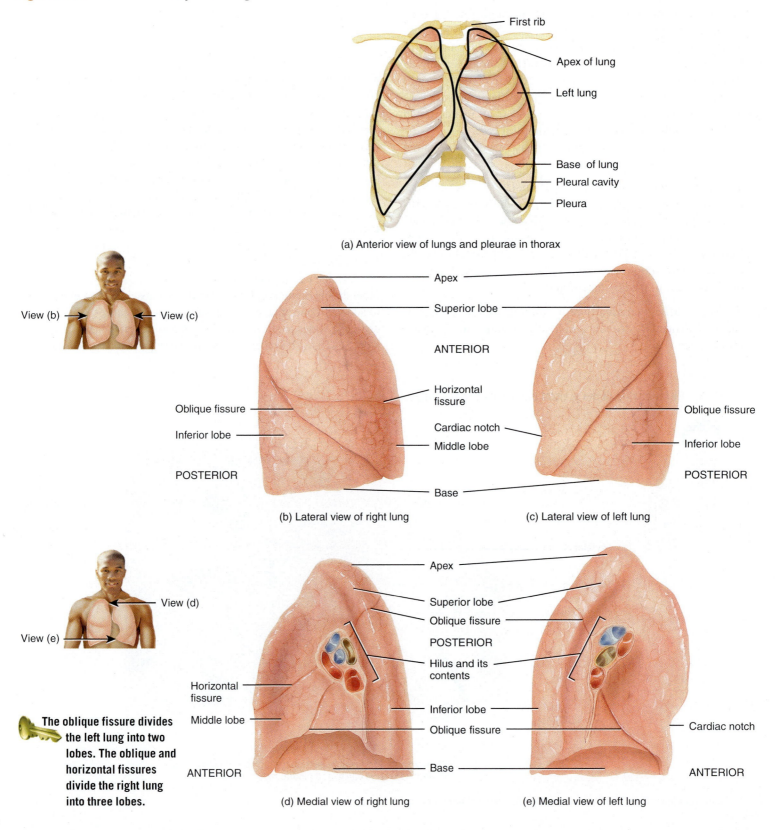

(a) Anterior view of lungs and pleurae in thorax

(b) Lateral view of right lung

(c) Lateral view of left lung

(d) Medial view of right lung

(e) Medial view of left lung

The oblique fissure divides the left lung into two lobes. The oblique and horizontal fissures divide the right lung into three lobes.

lymphatic vessels, and nerves enter and exit (Figure 22.9e). The left lung also contains a medial indentation, the **cardiac notch**, in which the heart lies. Due to the space occupied by the heart, the left lung is about 10% smaller than the right lung. Although the right lung is thicker and broader, it is also somewhat shorter than the left lung because the diaphragm is higher on the right side, accommodating the liver that lies inferior to it.

Lobes, Fissures, and Lobules

Deep grooves called fissures divide each lung into lobes (Figure 22.9b–e). Both lungs have an **oblique fissure**, which extends inferiorly and anteriorly; the right lung also has a **horizontal fissure**. The oblique fissure divides the left lung into **superior** and **inferior lobes**. The oblique and horizontal fissures divide the right lung into **superior**, **middle**, and **inferior lobes**.

Each lobe receives its own secondary (lobar) bronchus. Thus, the right primary bronchus gives rise to three secondary (lobar) bronchi called the **superior**, **middle**, and **inferior secondary (lobar) bronchi**, and the left primary bronchus gives rise to **superior** and **inferior secondary (lobar) bronchi**. Within the lung, the secondary bronchi give rise to the **tertiary (segmental) bronchi**. The segment of lung tissue that each tertiary bronchus supplies is called a **bronchopulmonary segment**. Bronchial and pulmonary disorders (such as tumors or abscesses) that are localized in a bronchopulmonary segment may be surgically removed without seriously disrupting the surrounding lung tissue.

Each bronchopulmonary segment of the lungs has many small compartments called **lobules**; each lobule is wrapped in elastic connective tissue and contains a lymphatic vessel, an arteriole, a venule, and a branch from a terminal bronchiole (Figure 22.10). **Terminal bronchioles** subdivide into microscopic branches called **respiratory bronchioles**. Respiratory bronchioles in turn subdivide into several **alveolar ducts**.

Alveoli

Around the circumference of the alveolar ducts are numerous alveoli and alveolar sacs. An **alveolus** (al-VĒ-ō-lus) is a

Figure 22.10 Microscopic anatomy of a lobule of the lungs.

(a) Diagram of a portion of a lobule of the lung

LM about 30x

(b) Lung lobule

🔑 Alveolar sacs consist of two or more alveoli that share a common opening into an alveolar duct.

Figure 22.11   Structural components of an alveolus. The respiratory membrane consists of a layer of type I and type II alveolar cells, an epithelial basement membrane, a capillary basement membrane, and the capillary endothelium.

Monocyte

Elastic fiber

Type II alveolar cell

Respiratory membrane

Alveolus

Type I alveolar cell

Alveolar macrophage

Red blood cell in pulmonary capillary

Diffusion of O$_2$

Diffusion of CO$_2$

Alveolus

Red blood cell

Capillary endothelium

Capillary basement membrane

Epithelial basement membrane

Type I alveolar cell

Interstitial space

Alveolar fluid with surfactant

(a) Section through an alveolus showing its cellular components

(b) Details of respiratory membrane

cup-shaped outpouching lined by simple squamous epithelium and supported by a thin elastic basement membrane; an **alveolar sac** consists of two or more alveoli that share a common opening to an alveolar duct (Figure 22.10). The walls of alveoli consist of two types of alveolar epithelial cells (Figure 22.11). The more numerous **type I alveolar cells** are simple squamous epithelial cells that form a nearly continuous lining of the alveolar wall. The thin, type I alveolar cells are the main sites of gas exchange. **Type II alveolar cells**, also called **septal cells**, are fewer in number and are found between type I alveolar cells. Type II alveolar cells are rounded epithelial cells that secrete alveolar fluid, which keeps the surface between the cells and the air moist. Included in the alveolar fluid is **surfactant** (sur-FAK-tant), a complex mixture of phospholipids and lipoproteins. Surfactant lowers the surface tension of alveolar fluid, which reduces the tendency of alveoli to collapse (described later). Also present are **alveolar macrophages**, wandering phagocytes that remove fine dust particles and other debris from the alveolar

spaces. On the outer surface of the alveoli, the lobule's arteriole and venule disperse into a lush network of blood capillaries (Figure 22.10a).

The exchange of O$_2$ and CO$_2$ between the air spaces in the lungs and the blood takes place by diffusion across the alveolar and capillary walls, which together form the **respiratory membrane**. Extending from the alveolar air space to blood plasma, the respiratory membrane consists of four layers (Figure 22.11b):

1.   A layer of **type I** and **type II alveolar cells** that constitute the alveolar wall.

2.   An **epithelial basement membrane** underlying the alveolar wall.

3.   A **capillary basement membrane** that is often fused to the epithelial basement membrane.

4.   The **capillary endothelium**.

Despite having several layers, the respiratory membrane is very thin—only 0.5 μm thick—to allow rapid diffusion of gases. It has been estimated that the lungs contain 300 million alveoli, providing an immense surface area of 70 m^2 (750 ft^2)—about the size of a racquetball court—for gas exchange.

Checkpoint

7. How does the epiglottis prevent aspiration of foods and liquids?
8. What is the main function of the vocal folds?
9. What is the benefit of not having cartilage between the trachea and the esophagus? What is the purpose of the solid C-shaped cartilage rings?
10. What is the path taken by air molecules into and through the bronchial tree?
11. Name the locations of the parietal pleura, visceral pleura, and pleural cavity.
12. What is the path taken by air molecules as they travel from terminal bronchioles to alveoli?

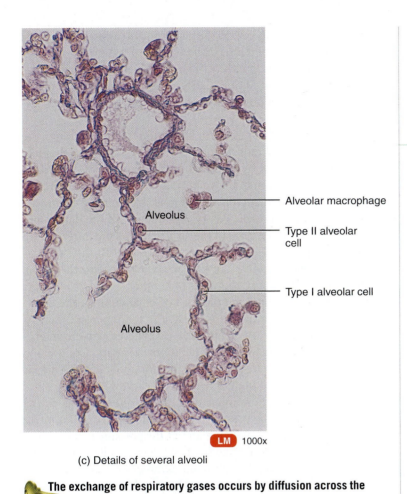

Alveolar macrophage

Alveolus

Type II alveolar cell

Type I alveolar cell

Alveolus

LM 1000x

(c) Details of several alveoli

🔑 **The exchange of respiratory gases occurs by diffusion across the respiratory membrane.**

22.3 Inhalation and exhalation result from pressure changes caused by muscle contraction and relaxation.

The process of gas exchange in the body, called **respiration**, has three basic steps:

1. **Pulmonary ventilation** (*pulmon-* = lung), or **breathing**, is the inhalation (inflow) and exhalation (outflow) of air and involves the exchange of air between the atmosphere and the alveoli of the lungs.

2. **External respiration** is the exchange of gases across the respiratory membrane between the alveoli of the lungs and the blood in pulmonary capillaries. In this process, pulmonary capillary blood gains O_2 and loses CO_2.

3. **Internal respiration** is the exchange of gases between blood in systemic capillaries and tissue cells. During internal respiration, the blood loses O_2 and gains CO_2. Within cells, the metabolic reactions that consume O_2 and give off CO_2 during the production of ATP are termed *cellular respiration* (discussed in Chapters 2 and 10).

In pulmonary ventilation, air flows between the atmosphere and the alveoli of the lungs because of alternating pressure differences created by contraction and relaxation of respiratory muscles. The rate of airflow and the amount of effort needed for breathing are also influenced by alveolar surface tension, compliance of the lungs, and airway resistance (each described shortly).

Pressure Changes during Pulmonary Ventilation

Air moves into the lungs when the air pressure inside the lungs is less than the air pressure in the atmosphere; it moves out of the lungs when the pressure inside the lungs is greater than the pressure in the atmosphere.

Inhalation

Breathing in is called **inhalation (inspiration)**. Just before each inhalation, the air pressure inside the lungs is equal to the pressure of the atmosphere, which at sea level is about 760 millimeters of mercury (mm Hg), or 1 atmosphere (atm). For air to flow into the lungs, the pressure inside the alveoli must become lower than the atmospheric pressure. This condition is achieved by increasing the size of the lungs.

The pressure of a gas in a closed container is inversely proportional to the volume of the container. This means that if the size of a closed container is increased, the pressure the gas exerts on the walls of the container decreases, and that if the size of the container is decreased, then the pressure exerted on the walls increases. This inverse relationship between volume and pressure, called **Boyle's law**, may be demonstrated as follows (Figure 22.12): Suppose we place a gas in a cylinder that has a movable piston and a pressure gauge, and suppose that the initial pressure created by the gas molecules striking the wall of the container is 1 atm. If the piston is pushed down, the gas molecules are squeezed into a smaller volume; now with less wall area, each segment of the exposed wall is struck more often by the gas molecules, creating greater gas pressure. The gauge shows that the pressure doubles as the gas is compressed to half its original volume. In other words, the same number of molecules in half the volume produces twice the pressure. Conversely, if the piston is raised to increase the volume, the gas molecules spread farther apart and the pressure decreases. Thus, the pressure of a gas varies inversely with volume.

Differences in pressure caused by changes in lung volume force air into our lungs when we inhale and out when we ex-hale. For inhalation to occur, the lungs must expand, which increases lung volume and thus decreases the pressure in the lungs to below atmospheric pressure. The first step in expanding the lungs during normal quiet inhalation involves contraction of the main muscles of inhalation, the diaphragm and external intercostals (Figure 22.13).

The most important muscle of inhalation is the diaphragm, the dome-shaped skeletal muscle that forms the floor of the thoracic cavity. It is innervated by fibers of the phrenic nerves emerging from the cervical region of the spinal cord. As the diaphragm contracts, it descends and becomes flatter, which increases the vertical diameter of the thoracic cavity. Contraction of the diaphragm is responsible for about 75% of the air that enters the lungs during quiet breathing. Advanced pregnancy, excessive obesity, or confining abdominal clothing can prevent complete descent of the diaphragm and may cause shortness of breath.

The next most important muscles of inhalation are the external intercostals. When these muscles contract, they pull the ribs upward and outward. As a result, there is an increase in the anteroposterior and lateral diameters of the thoracic cavity. Contraction of the external intercostals is responsible for about 25% of the air that enters the lungs during normal quiet breathing.

At rest just before an inhalation, the air pressure inside the lungs is the same as the pressure of the atmosphere, 760 mm Hg (Figure 22.14). During quiet inhalations, the pressure in the pleural cavity between the visceral and parietal pleurae, called **intrapleural pressure**, is normally below the atmospheric pressure. Just before inhalation, it is about 4 mm Hg less than the atmospheric pressure, or about 756 mm Hg at an atmospheric pressure of 760 mm Hg. As the diaphragm and external intercostals contract and the overall size of the thoracic cavity increases, the volume of the pleural cavity also increases, which causes intrapleural pressure to decrease from 756 to about 754 mm Hg. During expansion of the thorax, the parietal and visceral pleurae normally adhere tightly because of the subatmospheric pressure between them and because of the surface tension created by their moist adjoining surfaces. As the thoracic cavity expands, the parietal pleura lining the cavity is pulled outward in all directions, and the visceral pleura and lungs are pulled along with it.

Maintenance of a subatmospheric pressure in the pleural cavity is vital to the functioning of the lungs because it helps keep the alveoli slightly inflated. Alveoli are so elastic that at the end of an exhalation they recoil inward and tend to collapse on themselves like the walls of a deflated balloon. The "suction" created by the slightly lower pressure in the pleural cavities helps prevent lung collapse.

Figure 22.12   Boyle's law.

Volume = 1 liter
Pressure = 1 atm

Volume = 1/2 liter
Pressure = 2 atm

The volume of a gas varies inversely with its pressure.

Figure 22.13 **Muscles of inhalation and exhalation and their actions.** The pectoralis minor (not shown here) is illustrated in Figure 11.13a.

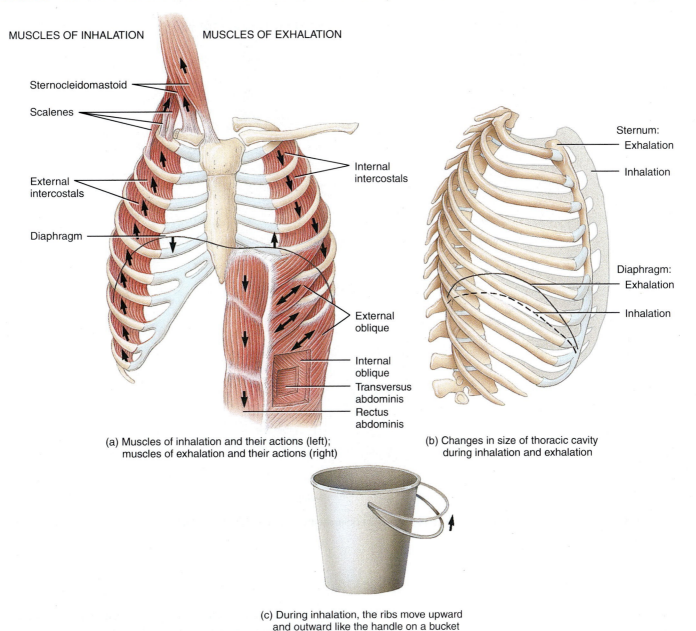

MUSCLES OF INHALATION   MUSCLES OF EXHALATION

Sternocleidomastoid

Scalenes

External intercostals

Diaphragm

Internal intercostals

External oblique

Internal oblique
Transversus abdominis
Rectus abdominis

Sternum:
Exhalation
Inhalation

Diaphragm:
Exhalation
Inhalation

(a) Muscles of inhalation and their actions (left); muscles of exhalation and their actions (right)

(b) Changes in size of thoracic cavity during inhalation and exhalation

(c) During inhalation, the ribs move upward and outward like the handle on a bucket

During quiet inhalation, the diaphragm and external intercostals contract, the lungs expand, and air moves into the lungs. During quiet exhalation, the diaphragm relaxes and the lungs recoil inward, forcing air out of the lungs.

As the volume of the lungs increases, the pressure inside the lungs, called the **alveolar pressure**, drops from 760 to 758 mm Hg. A pressure difference is thus established between the atmosphere and the alveoli. Because air always flows to a region of lower pressure, inhalation takes place as air flows from the atmosphere (higher pressure) into the lungs (lower pressure). Air continues to flow into the lungs as long as a pressure difference exists.

During deep, forceful inhalations, accessory muscles of inspiration participate to further increase the size of the thoracic cavity, allowing greater airflow into the lungs than can occur during quiet inhalation (Figure 22.13a). The accessory muscles are so named because they make little, if any, contribution during normal quiet inhalation, but during forced ventilation, such as occurs during exercise, they may contract vigorously. The accessory muscles of inhalation

Figure 22.14   Pressure changes in pulmonary ventilation.

Atmospheric pressure = 760 mm Hg

Alveolar pressure = 760 mm Hg

Intrapleural pressure = 756 mm Hg

1. At rest (diaphragm relaxed)

Atmospheric pressure = 760 mm Hg

Alveolar pressure = 758 mm Hg

Intrapleural pressure = 754 mm Hg

2. During inhalation (diaphragm contracting)

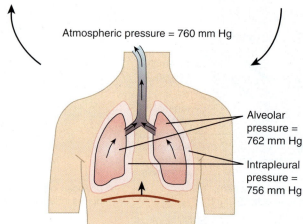

Atmospheric pressure = 760 mm Hg

Alveolar pressure = 762 mm Hg

Intrapleural pressure = 756 mm Hg

3. During exhalation (diaphragm relaxing)

🔑 **Air moves into the lungs when alveolar pressure is less than atmospheric pressure, and out of the lungs when alveolar pressure is greater than atmospheric pressure.**

include the sternocleidomastoids, which elevate the sternum; the scalenes, which elevate the first two ribs; and the pectoralis minors, which elevate the third through fifth ribs. Because both normal quiet inhalation and inhalation during forced ventilation involve muscular contraction, the process of inhalation is said to be an *active process.*

Figure 22.15a summarizes the events of inhalation.

Exhalation

Breathing out, called **exhalation (expiration)**, is also due to a pressure gradient, but in this case the gradient is in the opposite direction: The pressure in the lungs is greater than the pressure of the atmosphere. Normal exhalation during quiet breathing, unlike inhalation, is a *passive process* because no muscular contractions are involved. Instead, exhalation results from **elastic recoil** of the thoracic wall and lungs, both of which have a natural tendency to spring back after they have been stretched. Two inwardly directed

forces contribute to elastic recoil: (1) the recoil of elastic fibers that were stretched during inhalation and (2) the inward pull of surface tension due to the film of alveolar fluid.

Exhalation starts when the inspiratory muscles relax. As the diaphragm relaxes, its dome moves superiorly owing to its elasticity. As the external intercostals relax, the ribs are depressed. These movements decrease the vertical, lateral, and anteroposterior diameters of the thoracic cavity, which decreases lung volume. In turn, the alveolar pressure increases, to about 762 mm Hg. Air then flows from the area of higher pressure in the alveoli to the area of lower pressure in the atmosphere (Figure 22.14).

Exhalation becomes active only during forceful breathing, as occurs while playing a wind instrument or during exercise. During these times, muscles of exhalation—the abdominals and internal intercostals (Figure 22.13a)—contract, which further increases pressure in the abdominal region and thorax, resulting in increased

Figure 22.15 Summary of events of inhalation and exhalation.

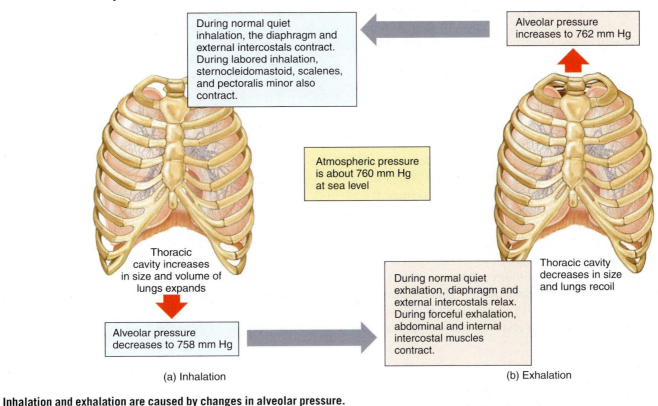

During normal quiet inhalation, the diaphragm and external intercostals contract. During labored inhalation, sternocleidomastoid, scalenes, and pectoralis minor also contract.

Alveolar pressure increases to 762 mm Hg

Atmospheric pressure is about 760 mm Hg at sea level

Thoracic cavity increases in size and volume of lungs expands

Alveolar pressure decreases to 758 mm Hg

During normal quiet exhalation, diaphragm and external intercostals relax. During forceful exhalation, abdominal and internal intercostal muscles contract.

Thoracic cavity decreases in size and lungs recoil

(a) Inhalation

(b) Exhalation

Inhalation and exhalation are caused by changes in alveolar pressure.

airflow out of the lungs. Contraction of the abdominal muscles moves the inferior ribs downward and compresses the abdominal viscera, thereby forcing the diaphragm superiorly. Contraction of the internal intercostals pulls the ribs inferiorly.

Figure 22.15b summarizes the events of exhalation.

Other Factors Affecting Pulmonary Ventilation

As you have just learned, air pressure differences drive airflow during inhalation and exhalation. However, three other factors affect the rate of airflow and pulmonary ventilation: surface tension of the alveolar fluid, compliance of the lungs, and airway resistance.

Surface Tension of Alveolar Fluid

As noted earlier, a thin layer of alveolar fluid coats the surface of alveoli and exerts a force known as **surface tension**. Surface tension arises at all air–water interfaces because the polar water molecules are more strongly attracted to each other than they are to gas molecules in the air. When liquid surrounds a sphere of air, as in an alveolus or a soap bubble, surface tension produces an inwardly directed force. Soap bubbles "burst" because they collapse inward due to surface tension. In the lungs, the inward force of

surface tension causes the alveoli to assume the smallest possible diameter. During breathing, surface tension must be overcome to expand the lungs during each inhalation. Surface tension also accounts for two-thirds of lung elastic recoil, which decreases the size of alveoli during exhalation.

The surfactant (a mixture of phospholipids and lipoproteins) present in alveolar fluid reduces its surface tension below the surface tension of pure water. A deficiency of surfactant in premature infants causes *respiratory distress syndrome*, in which the surface tension of alveolar fluid is greatly increased, so that many alveoli collapse at the end of each exhalation. Great effort is then needed at the next inhalation to reopen the collapsed alveoli.

Compliance of the Lungs

Compliance refers to how much effort is required to stretch the lungs and thoracic wall. High compliance means that the lungs and thoracic wall expand easily; low compliance means that they resist expansion. By analogy, a thin balloon that is easy to inflate has high compliance, and a heavy and stiff balloon that takes a lot of effort to inflate has low compliance. In the lungs, compliance is related to two principal factors: elasticity and surface tension. The lungs normally have high compliance and expand easily because elastic fibers in lung tissue are easily stretched and surfactant in alveolar fluid reduces surface tension. Decreased compliance is a common feature in pulmonary conditions that scar lung tissue (for example, tuberculosis), cause lung tissue to become filled with fluid (pulmonary edema), produce a deficiency in surfactant, or impede lung expansion in any way (for example, paralysis of the intercostal muscles). Decreased lung compliance occurs in emphysema due to destruction of elastic fibers in alveolar walls.

Airway Resistance

Like the flow of blood through blood vessels, the rate of airflow through the airways depends on both the pressure difference and the resistance to flow: Airflow decreases when there is greater resistance in the airways and increases when there is decreased resistance. The walls of the airways, especially the bronchioles, offer some resistance to the normal flow of air into and out of the lungs. As the lungs expand during inhalation, the bronchioles enlarge because their walls are pulled outward in all directions. Larger-diameter airways have decreased resistance, allowing greater flow of air. Airway resistance then increases during exhalation as the diameter of bronchioles decreases. Airway diameter is also regulated by the degree of contraction or relaxation of smooth muscle in the walls of the airways. Signals from the sympathetic division of the autonomic nervous system cause relaxation of this smooth muscle, which results in bronchodilation, decreased resistance, and increasing airflow.

Any condition that narrows or obstructs the airways increases resistance, so that more pressure is required to maintain the same airflow. The hallmark of asthma or chronic obstructive pulmonary disease (COPD)—emphysema or chronic bronchitis—is increased airway resistance due to obstruction or collapse of airways.

Modified Respiratory Movements

Respirations provide humans with methods for expressing emotions such as laughing, sighing, and sobbing. Moreover, respiratory air can be used to expel foreign matter from the lower air passages through actions such as sneezing and coughing. Respiratory movements are also modified and controlled during talking and singing. Some of the modified respiratory movements that express emotion or clear the airways are listed in Table 22.1. All these movements are reflexes, but some of them also can be initiated voluntarily.

TABLE 22.1

Modified Respiratory Movements

MOVEMENT	DESCRIPTION
Coughing	A long-drawn and deep inhalation followed by a complete closure of the glottis, which results in a strong exhalation that suddenly pushes the glottis open and sends a blast of air through the upper respiratory passages. Stimulus for this reflex act may be a foreign body lodged in the larynx, trachea, or epiglottis.
Sneezing	Spasmodic contraction of muscles of exhalation that forcefully expels air through the nose and mouth. Stimulus may be an irritation of the nasal mucosa.
Sighing	A long-drawn and deep inhalation immediately followed by a shorter but forceful exhalation.
Yawning	A deep inhalation through the widely opened mouth producing an exaggerated depression of the mandible. It may be stimulated by drowsiness or someone else's yawning, but precise cause is unknown.
Sobbing	A series of convulsive inhalations followed by a single prolonged exhalation. The glottis closes earlier than normal after each inhalation, so only a little air enters the lungs with each inhalation.
Crying	An inhalation followed by many short convulsive exhalations, during which the glottis remains open and the vocal folds vibrate; accompanied by characteristic facial expressions and tears.
Laughing	The same basic movements as crying, but the rhythm of the movements and the facial expressions usually differ from those of crying. Laughing and crying are sometimes indistinguishable.
Hiccupping	Spasmodic contraction of the diaphragm followed by a spasmodic closure of the glottis, which produces a sharp sound on inhalation. Stimulus is usually irritation of the sensory nerve endings of the gastrointestinal tract.
Valsalva (val-SAL-va) maneuver	Forced exhalation against a closed glottis as may occur during periods of straining while defecating.

Checkpoint

13. If the volume of the cylinder in Figure 22.12 decreased from 1 liter to 1/4 liter, how would the pressure change?

14. What is the main muscle that powers your breathing as you sit and study your anatomy and physiology text?

15. Would contraction of the external intercostals increase or decrease alveolar pressure? Would it cause air to enter or leave the lungs? Would contraction of the internal intercostals increase or decrease alveolar pressure? Would it cause air to enter or leave the lungs?

16. Which happens first during inhalation: (a) air enters the lungs or (b) the lungs expand?

17. Which forces produce the elastic recoil of the thoracic wall and lungs?

18. Would airflow into alveoli be greater with higher or lower compliance of the lungs? Larger or smaller bronchioles?

Return to Cari's Story

By the end of the week Cari knew from experience that the cold had turned into a sinus infection. First her head had begun to ache, with pressure over her eyes and cheekbones. Her mucus had changed in color from clear to greenish, and she felt feverish. She was having difficulty hearing, her chest felt heavy, and her eyes burned and watered.

"Cari, how long have you had this cold?" the nurse asked as she listened to Cari's lungs.

"I've had the cold for a couple of weeks now, but I've had the sinus pressure and green mucus for about a week. The worst part is that I can't catch my breath," Cari replied. As she finished speaking, Cari again felt an overwhelming urge to cough, and couldn't stop.

"Are you still smoking, Cari?" The nurse moved the bell of her stethoscope over Cari's back, noticing Cari's decreased breath sounds and ronchi (wheezing, musical sounds).

"Yeah, I know I should quit; it's been so hard, though. I guess I use the stress of my job as an excuse."

"Let's get your temperature, Cari. It's a bit high—101.2." The nurse checked Cari's blood oxygen saturation as she spoke, using an oxygen saturation monitor.

With each breath, Cari struggled to get air into her lungs. Her neck muscle stood out as she raised her chest and shoulders in an effort to breathe.

"Cari, I'm going to give you some oxygen and get the doctor. I'll be right back."

The nurse found the doctor in the hall. "What have you got, Nancy?" the doctor asked.

"Dr. Sommers, I think we have a case of pneumonia. She's a 47-year-old smoker, with a history of chronic bronchitis and multiple upper respiratory tract infections. Had a cold that turned into a sinus infection. But her lung sounds are really abnormal, especially at the left base. O₂ saturation is 90%, productive cough, and she's febrile, with a temp over 101. I started her on some oxygen; she's rather uncomfortable."

The doctor looked over Cari's chart. "Sounds like we might need to admit her; let's get a chest X-ray."

A. How could an infection in Cari's nasal passages and pharynx spread into her sinuses?

B. What is the cough reflex? Describe the process that Cari's respiratory system is using to clear her lungs by coughing.

C. Which structures found in the terminal bronchioles and alveoli normally would protect Cari's lungs from infectious pathogens and particulate matter?

D. How would the resistance of Cari's airways be affected by excess mucus and fluid in her lung?

E. How would Cari's lung compliance (the effort required to expand the lungs) be altered as her alveoli fill with fluid due to pneumonia?

22.4 Lung volumes and capacities are measured to determine the respiratory status of an individual.

While at rest, a healthy adult averages 12 breaths a minute, with each inhalation and exhalation moving about 500 mL of air into and out of the lungs. The volume of one breath is called the **tidal volume**. The **minute ventilation (MV)**—the total volume of air inhaled and exhaled each minute—is equal to breathing rate multiplied by tidal volume:

$$MV = 12 \text{ breaths/min} \times 500 \text{ mL/breath}$$
$$= 6 \text{ liters/min}$$

A lower-than-normal minute ventilation usually is a sign of pulmonary malfunction. The apparatus commonly used to measure the volume of air exchanged during breathing and to measure the respiratory rate is a **spirometer** (*spiro-* = breathe; *-meter* = measuring device). The record produced by a spirometer is called a **spirogram**. Inhalation is recorded as an upward deflection, and exhalation is recorded as a downward deflection (Figure 22.16).

In a typical adult, about 70% of the tidal volume (350 mL) actually reaches the respiratory zone of the respiratory system—the respiratory bronchioles, alveolar ducts, alveolar sacs, and alveoli—and participates in external respiration. The other 30% (150 mL) remains in the conducting airways of the nose, pharynx, larynx, trachea, bronchi, bronchioles, and terminal bronchioles. Collectively, the conducting airways containing air that does not undergo gas exchange are known as the **anatomic dead space**.

Figure 22.16   Spirogram of lung volumes and capacities. The average values for a healthy adult male and female are indicated, with the values for a female in parentheses. Note that the spirogram is read from right (start of record) to left (end of record).

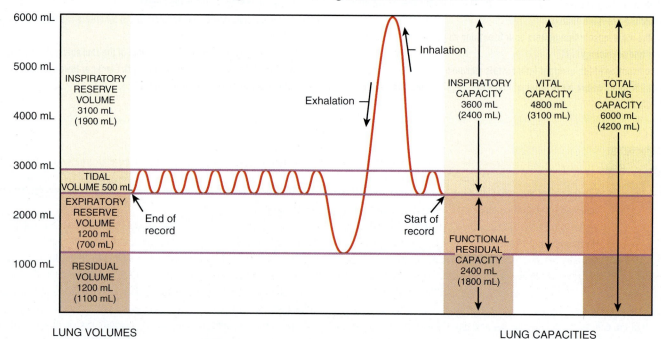

Lung capacities are combinations of various lung volumes.

Lung volume varies considerably from one person to another and in the same person at different times. In general, these volumes are larger in males, taller individuals, and younger adults, and smaller in females, shorter individuals, and the elderly. Various disorders also may be diagnosed by comparison of actual and predicted normal values for gender, height, and age. The values given here are averages for young adults.

By taking a very deep breath, you can inhale a good deal more than the 500 mL of tidal volume. This additional inhaled air, called the **inspiratory reserve volume**, is about 3100 mL in an average adult male and 1900 mL in an average adult female (Figure 22.16). Even more air can be inhaled if inhalation follows forced exhalation. If you inhale normally and then exhale as forcibly as possible, you should be able to push out considerably more air in addition to the 500 mL of tidal volume. The extra 1200 mL in males and 700 mL in females is called the **expiratory reserve volume**.

Even after the expiratory reserve volume is exhaled, considerable air remains in the lungs because the subatmospheric intrapleural pressure keeps the alveoli slightly inflated, and some air also remains in the noncollapsible airways. This volume, called the **residual volume**, amounts to about 1200 mL in males and 1100 mL in females.

Lung capacities are combinations of specific lung volumes (Figure 22.16). **Inspiratory capacity** is the sum of tidal volume and inspiratory reserve volume (500 mL + 3100 mL = 3600 mL in males and 500 mL + 1900 mL = 2400 mL in females). **Functional residual capacity** is the sum of residual volume and expiratory reserve volume (1200 mL + 1200 mL = 2400 mL in males and 1100 mL + 700 mL = 1800 mL in females). **Vital capacity**, the maximum amount of air that can be exhaled after maximal inhalation, is the sum of inspiratory reserve volume, tidal volume, and expiratory reserve volume (4800 mL in males and 3100 mL in females). Finally, **total lung capacity** is the sum of vital capacity and residual volume (4800 mL + 1200 mL = 6000 mL in males and 3100 mL + 1100 mL = 4200 mL in females).

Checkpoint

19. What is a spirometer?
20. What is the anatomic dead space?
21. What is the difference between tidal volume, inspiratory reserve volume, expiratory reserve volume, and residual volume?
22. If you breathe in as deeply as possible and then exhale as much air as you can, which lung capacities have you demonstrated?

22.5 Oxygen and carbon dioxide diffusion is based on partial pressure gradients and solubility.

The exchange of oxygen and carbon dioxide between alveolar air and pulmonary blood occurs via passive diffusion, which is governed by the behavior of gases as described by two gas laws, Dalton's law and Henry's law. Dalton's law is important for understanding how gases move down their pressure gradients by diffusion, and Henry's law helps explain how the solubility of a gas relates to its diffusion.

According to **Dalton's law**, each gas in a mixture of gases exerts its own pressure as if no other gases were present. The pressure of a specific gas in a mixture is called its *partial pressure* and is denoted as P_X, where the subscript X denotes the chemical formula of the gas. The total pressure of the mixture is calculated simply by adding all the partial pressures. Atmospheric air is a mixture of gases—nitrogen (N_2), oxygen (O_2), water vapor (H_2O), and carbon dioxide (CO_2), plus other gases present in small quantities. Atmospheric pressure is the sum of the pressures of all these gases:

$$\text{Atmospheric pressure (760 mm Hg)}$$
$$= P_{N_2} + P_{O_2} + P_{H_2O} + P_{CO_2} + P_{\text{other gases}}$$

Partial pressures determine the movement of O_2 and CO_2 between the atmosphere and lungs, between the lungs and blood, and between the blood and body cells. Each gas diffuses across a permeable membrane from the area where its partial pressure is greater to the area where its partial pressure is less. The greater the difference in partial pressure, the faster is the rate of diffusion.

Compared with inhaled air, alveolar air has less O_2 (13.6% versus 20.9%) and more CO_2 (5.2% versus 0.04%) for two reasons. First, gas exchange across the respiratory membrane increases the CO_2 content and decreases the O_2 content of alveolar air. Second, inhaled air becomes humidified as it passes along the moist mucosal linings. As water vapor content of the air increases, the relative percentage of O_2 decreases. In contrast, exhaled air contains more O_2 than alveolar air (16% versus 13.6%) and less CO_2 (4.5% versus 5.2%) because some of the exhaled air was in the anatomic dead space and did not participate in gas exchange. Exhaled air is a mixture of alveolar air and inhaled air that was in the anatomic dead space.

Henry's law states that the quantity of a gas that will dissolve in a liquid is proportional to the partial pressure of the gas and its solubility. In body fluids, the ability of a gas to stay in solution is greater when its partial pressure is higher and when it has a high solubility in water. The higher the partial pressure of a gas over a liquid and the higher the solubility, the more gas will stay in solution. In comparison to oxygen, much more CO_2 is dissolved in blood plasma because the solubility of CO_2 is 24 times greater than that of O_2.

An everyday experience gives a demonstration of Henry's law. You have probably noticed that a soft drink makes a hissing sound when the top of the container is removed, and bubbles rise to the surface for some time afterward. The gas dissolved in carbonated beverages is CO_2. Because the soft drink is bottled or canned under high pressure and capped, the CO_2 remains dissolved as long as the container is unopened. Once you remove the cap, the pressure decreases and the gas begins to bubble out of solution.

Checkpoint

23. Explain the principles described by Dalton's law and Henry's law.
24. Why is the composition of inhaled air and alveolar air different?

22.6 Respiration occurs between alveoli and pulmonary capillaries and between systemic capillaries and tissue cells.

External respiration is the diffusion of O_2 from air in the alveoli of the lungs to blood in pulmonary capillaries and the diffusion of CO_2 in the opposite direction (Figure 22.17a). External respiration in the lungs converts **deoxygenated blood** (low oxygen) coming from the right side of the heart into **oxygenated blood** (high oxygen) that returns to the left side of the heart (see Figure 20.15). As blood flows through the pulmonary capillaries, it picks up O_2 from alveolar air and unloads CO_2 into alveolar air. Although this process is commonly called an "exchange" of gases, each gas diffuses *independently* from the area where its partial pressure is higher to the area where its partial pressure is lower.

As Figure 22.17a shows, O_2 diffuses from alveolar air, where its partial pressure is 105 mm Hg, into the blood in pulmonary capillaries, where P_{O_2} is only 40 mm Hg in a resting person. If you have been exercising, the P_{O_2} will be even lower because contracting muscle fibers are using more O_2. Diffusion continues until the P_{O_2} of pulmonary capillary blood increases to match the P_{O_2} of alveolar air, 105 mm Hg. Blood leaving pulmonary capillaries near alveoli mixes with a small volume of blood that has flowed through conducting portions of the respiratory system, where gas exchange does not occur. Thus, the P_{O_2} of blood in the pulmonary veins *leaving the lungs* is about 100 mm Hg, slightly less than the P_{O_2} in pulmonary capillaries.

While O_2 is diffusing from alveolar air into deoxygenated blood, CO_2 is diffusing in the opposite direction. The P_{CO_2} of deoxygenated blood is 45 mm Hg in a resting person, and the P_{CO_2} of alveolar air is 40 mm Hg. Because of this difference in P_{CO_2}, carbon dioxide diffuses from deoxygenated blood into the alveoli until the P_{CO_2} of the blood decreases to 40 mm Hg. Exhalation keeps alveolar P_{CO_2} at 40 mm Hg. Oxygenated blood returning to the left side of the heart in the pulmonary veins thus has a P_{CO_2} of 40 mm Hg.

Figure 22.17   Changes in partial pressures of oxygen and carbon dioxide during external and internal respiration.

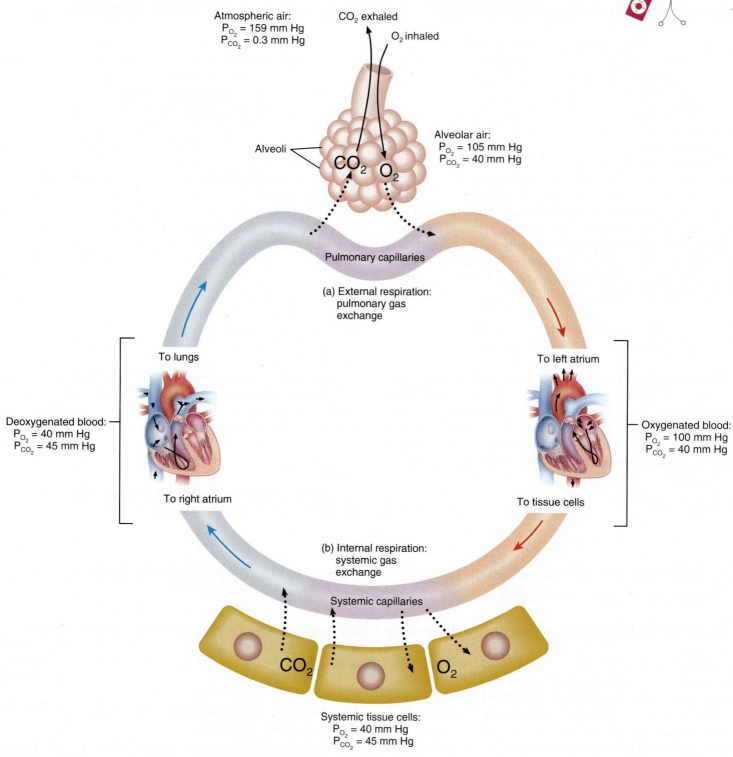

Atmospheric air:
P_{O_2} = 159 mm Hg
P_{CO_2} = 0.3 mm Hg

CO_2 exhaled

O_2 inhaled

Alveoli

Alveolar air:
P_{O_2} = 105 mm Hg
P_{CO_2} = 40 mm Hg

CO_2 O_2

Pulmonary capillaries

(a) External respiration:
pulmonary gas
exchange

To lungs

To left atrium

Deoxygenated blood:
P_{O_2} = 40 mm Hg
P_{CO_2} = 45 mm Hg

Oxygenated blood:
P_{O_2} = 100 mm Hg
P_{CO_2} = 40 mm Hg

To right atrium

To tissue cells

(b) Internal respiration:
systemic gas
exchange

Systemic capillaries

CO_2 O_2

Systemic tissue cells:
P_{O_2} = 40 mm Hg
P_{CO_2} = 45 mm Hg

Gases diffuse from areas of higher partial pressure to areas of lower partial pressure.

The left ventricle pumps oxygenated blood into the aorta and through the systemic arteries to systemic capillaries. The exchange of O_2 and CO_2 between systemic capillaries and tissue cells is called **internal respiration** (Figure 22.17b). As O_2 leaves the bloodstream, oxygenated blood is converted into deoxygenated blood. Unlike external respiration, which occurs only in the lungs, internal respiration occurs in tissues throughout the body.

The P_{O_2} of blood pumped into systemic capillaries is higher (100 mm Hg) than the P_{O_2} in tissue cells (about 40 mm Hg at rest) because the cells constantly use O_2 to produce ATP. Due to this pressure difference, oxygen diffuses out of the capillaries into tissue cells and blood P_{O_2} drops to 40 mm Hg by the time the blood exits systemic capillaries.

While O_2 diffuses from the systemic capillaries into tissue cells, CO_2 diffuses in the opposite direction. Because tissue cells are constantly producing CO_2, the P_{CO_2} of cells (45 mm Hg at rest) is higher than that of systemic capillary blood (40 mm Hg). As a result, CO_2 diffuses from tissue cells into systemic capillaries until the P_{CO_2} in the blood increases to 45 mm Hg. The deoxygenated blood then returns to the heart and is pumped to the lungs for another cycle of external respiration.

In a person at rest, tissue cells, on average, need only 25% of the available O_2 in oxygenated blood; despite its name, deoxygenated blood retains 75% of its O_2 content. During exercise, more O_2 diffuses from the blood into metabolically active cells, such as contracting skeletal muscle fibers. Active cells use more O_2 for ATP production, causing the O_2 content of deoxygenated blood to drop below 75%.

The *rate* of pulmonary and systemic gas exchange depends on several factors.

- **Partial pressure difference of the gases**. Alveolar P_{O_2} must be higher than blood P_{O_2} for oxygen to diffuse from alveolar air into the blood. The rate of diffusion is faster when the difference between P_{O_2} in alveolar air and pulmonary capillary blood is larger; diffusion is slower when the difference is smaller. The differences between P_{O_2} and P_{CO_2} in alveolar air versus pulmonary blood increase during exercise. The larger partial pressure differences accelerate the rates of gas diffusion. The partial pressures of O_2 and CO_2 in alveolar air also depend on the rate of airflow into and out of the lungs. Certain drugs (such as morphine) slow ventilation, thereby decreasing the amount of O_2 and CO_2 that can be exchanged between alveolar air and blood. With increasing altitude, the total atmospheric pressure decreases, as does the partial pressure of O_2—from 159 mm Hg at sea level, to 110 mm Hg at 10,000 ft, to 73 mm Hg at 20,000 ft. Although O_2 still is 20.9% of the total, the P_{O_2} of inhaled air decreases with increasing altitude. Alveolar P_{O_2} decreases correspondingly, and O_2 diffuses into the blood more slowly. The common signs and symptoms of **high altitude sickness**— shortness of breath, headache, fatigue, insomnia, nausea, and dizziness—are due to a lower level of oxygen in the blood.

- **Surface area available for gas exchange**. As you learned earlier in the chapter, the surface area of the alveoli is huge (about 70 m^2 or 750 ft^2). In addition, many capillaries surround each alveolus, so many that as much as 900 mL of blood is able to participate in gas exchange at any instant. Any pulmonary disorder that decreases the functional surface area of the respiratory membranes decreases the rate of external respiration. In emphysema, for example, alveolar walls disintegrate, so surface area is smaller than normal and pulmonary gas exchange is slowed.

- **Molecular weight and solubility of the gases**. Because O_2 has a lower molecular weight than CO_2, it could be expected to diffuse across the respiratory membrane about 1.2 times faster. However, the solubility of CO_2 in the fluid portions of the respiratory membrane is about 24 times greater than that of O_2. Taking both of these factors into account, net outward CO_2 diffusion occurs 20 times more rapidly than net inward O_2 diffusion. Consequently, when diffusion is slower than normal, for example, in emphysema or pulmonary edema, O_2 insufficiency (hypoxia) typically occurs before there is significant retention of CO_2 (hypercapnia).

- **Diffusion distance**. The respiratory membrane is very thin, so diffusion occurs quickly. Also, the capillaries are so narrow that the red blood cells must pass through them in single file, which minimizes the diffusion distance from an alveolar air space to hemoglobin inside red blood cells. Buildup of interstitial fluid between alveoli, as occurs in pulmonary edema, slows the rate of gas exchange because it increases diffusion distance.

Checkpoint

25. What are the diffusion paths of oxygen and carbon dioxide during external and internal respiration?
26. What causes oxygen to enter pulmonary capillaries from alveolar air and to enter tissue cells from systemic capillaries?
27. Which factors affect the rates of diffusion of oxygen and carbon dioxide?
28. How does the partial pressure of atmospheric oxygen change as altitude changes?

22.7 Oxygen is primarily transported attached to hemoglobin, while carbon dioxide is transported in three different ways.

The blood transports gases between the lungs and body tissues (Figure 22.18). When O_2 and CO_2 enter the blood, certain chemical reactions occur that aid in gas transport and gas exchange.

Oxygen Transport

Oxygen does not dissolve easily in water, so only about 1.5% of the O_2 in blood is dissolved in blood plasma, which is mostly water.

Figure 22.18   Transport of oxygen and carbon dioxide in the blood.

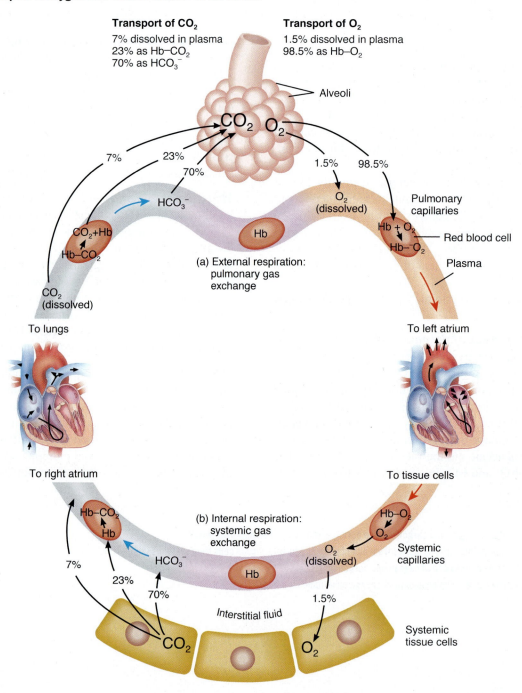

Transport of CO_2
7% dissolved in plasma
23% as Hb–CO_2
70% as HCO_3^-

Transport of O_2
1.5% dissolved in plasma
98.5% as Hb–O_2

Alveoli

(a) External respiration: pulmonary gas exchange

Pulmonary capillaries

Red blood cell

Plasma

To lungs

To left atrium

To right atrium

To tissue cells

(b) Internal respiration: systemic gas exchange

Systemic capillaries

Interstitial fluid

Systemic tissue cells

Most O_2 is transported by hemoglobin as oxyhemoglobin (Hb–O_2) within red blood cells; most CO_2 is transported in blood plasma as bicarbonate ions (HCO_3^-).

About 98.5% of blood O_2 is bound to hemoglobin in red blood cells (Figure 22.18).

The heme portion of hemoglobin contains four atoms of iron, each capable of binding to a molecule of O_2 (see Figure 18.4b,c). Oxygen and **deoxyhemoglobin (Hb)** bind in an easily reversible reaction to form **oxyhemoglobin (Hb–O_2)**:

$$Hb + O_2 \xrightleftharpoons[\text{Dissociation of } O_2]{\text{Binding of } O_2} Hb-O_2$$

Deoxyhemoglobin   Oxygen   Oxyhemoglobin

The 98.5% of the O_2 that is bound to hemoglobin is trapped inside RBCs, so only the dissolved O_2 (1.5%) can diffuse out of tissue capllaries into tissue cells. Thus, it is important to understand the factors that promote O_2 binding to and dissociation (separation) from hemoglobin.

Relationship between Hemoglobin and Oxygen Partial Pressure

The most important factor that determines how much O_2 binds to hemoglobin is the P_{O_2}; the higher the P_{O_2}, the more O_2 combines with Hb. When deoxyhemoglobin (Hb) is completely converted to oxyhemoglobin (Hb–O_2), the hemoglobin is said to be **fully saturated**; that is, every available iron atom has combined with a molecule of O_2. When hemoglobin consists of a mixture of Hb and Hb–O_2, it is **partially saturated**. The **percent saturation of hemoglobin** expresses the average saturation of hemoglobin with oxygen. For instance, if each hemoglobin molecule has bound two O_2 molecules, then the hemoglobin is 50% saturated because each Hb can bind a maximum of four O_2.

The relationship between the percent saturation of hemoglobin and P_{O_2} is illustrated in the oxygen–hemoglobin dissociation curve in Figure 22.19. Note that when the P_{O_2} is high, hemoglobin binds with large amounts of O_2 and is almost 100% saturated. When P_{O_2} is low, hemoglobin is only partially saturated. In other words, the greater the P_{O_2}, the more O_2 will bind to hemoglobin, until all the available hemoglobin molecules are saturated. Therefore, in pulmonary capillaries, where P_{O_2} is high, a lot of O_2 binds to hemoglobin. In tissue capillaries, where the P_{O_2} is lower, hemoglobin does not hold as much O_2, and the dissolved O_2 is unloaded via diffusion into tissue cells (Figure 22.18). Note that hemoglobin is still 75% saturated with O_2 at a P_{O_2} of 40 mm Hg, the average P_{O_2} of tissue cells in a person at rest. This is the

Figure 22.19 **Oxygen–hemoglobin dissociation curve showing the relationship between hemoglobin saturation and P_{O_2} at normal body temperature.**

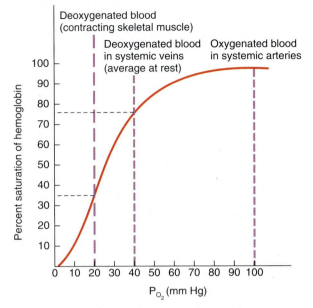

As P_{O_2} increases, more O_2 combines with hemoglobin.

basis for the earlier statement that only 25% of the available O_2 unloads from hemoglobin and is used by tissue cells under resting conditions.

When the P_{O_2} is between 60 and 100 mm Hg, hemoglobin is 90% or more saturated with O_2 (Figure 22.19). Thus, blood picks up a nearly full load of O_2 from the lungs even when the P_{O_2} of alveolar air is as low as 60 mm Hg. The oxygen–hemoglobin dissociation curve explains why people can still perform well at high altitudes or when they have certain cardiac and pulmonary diseases, even though P_{O_2} may drop as low as 60 mm Hg. Note also in the curve that at a considerably lower P_{O_2} of 40 mm Hg, hemoglobin is still 75% saturated with O_2. However, oxygen saturation of Hb drops to 35% at 20 mm Hg. Between 40 and 20 mm Hg, large amounts of O_2 are released from hemoglobin in response to only small decreases in P_{O_2}. In active tissues such as contracting muscles, P_{O_2} may drop well below 40 mm Hg. Then, a large percentage of the O_2 is released from hemoglobin, providing more O_2 to metabolically active tissues.

Other Factors Affecting the Affinity of Hemoglobin for Oxygen

Although P_{O_2} is the most important factor that determines the percent O_2 saturation of hemoglobin, several other factors influence the tightness, or **affinity**, with which hemoglobin binds O_2. Each factor makes sense if you keep in mind that metabolically active tissue cells need O_2 and produce acids, CO_2, and heat as wastes. The following factors affect the affinity of hemoglobin for O_2:

1. *Acidity (pH)*. As acidity increases (pH decreases), the affinity of hemoglobin for O_2 decreases, and O_2 dissociates more readily from hemoglobin (Figure 22.20a). In other words, increasing acidity enhances the unloading of oxygen from hemoglobin. During exercise, metabolically active tissues release lactic acid and carbonic acid, which promotes release of O_2 from hemoglobin.

2. *Partial pressure of carbon dioxide*. As P_{CO_2} rises, hemoglobin releases O_2 more readily (Figure 22.20b). P_{CO_2} and pH are related factors because low blood pH (acidity) results from high P_{CO_2}. As CO_2 enters the blood, much of it is temporarily converted to carbonic acid (H_2CO_3), a reaction catalyzed by an enzyme in red blood cells called **carbonic anhydrase (CA)**:

$$CO_2 \; + \; H_2O \; \underset{}{\overset{CA}{\rightleftharpoons}} \; H_2CO_3 \; \rightleftharpoons \; H^+ \; + \; HCO_3^-$$

| Carbon dioxide | Water | | Carbonic acid | | Hydrogen ion | | Bicarbonate ion |

The carbonic acid thus formed in red blood cells dissociates into hydrogen ions and bicarbonate ions. As the H^+ concentration increases, pH decreases. Thus, an increased P_{CO_2} produces a more acidic environment, which helps release O_2 from hemoglobin. During exercise, lactic acid—a by-product of anaerobic metabolism within muscles—also decreases blood pH. Thus, hemoglobin releases more O_2 as blood flows through active tissues that are producing more CO_2, such as muscle tissue during exercise.

3. *Temperature*. Within limits, as temperature increases, so does the amount of O_2 released from hemoglobin (Figure 22.21). Active tissues produce more heat, which elevates the local temperature and promotes release of O_2 from oxyhemoglobin. Fever produces a similar result. In contrast, during hypothermia (lowered body temperature) cellular metabolism slows, the need for O_2 is reduced, and more O_2 remains bound to hemoglobin.

Figure 22.20   Oxygen–hemoglobin dissociation curves showing the relationship of (a) pH and (b) P_{CO_2} to hemoglobin saturation at normal body temperature. As pH increases or P_{CO_2} decreases, O_2 combines more tightly with hemoglobin, so that less is available to tissues. The broken lines emphasize these relationships.

(a) Effect of pH on affinity of hemoglobin for oxygen

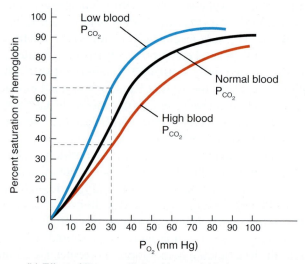

(b) Effect of P_{CO_2} on affinity of hemoglobin for oxygen

As pH decreases or P_{CO_2} increases, the affinity of hemoglobin for O_2 declines, so less O_2 combines with hemoglobin and more is available to tissues.

Figure 22.21 Oxygen–hemoglobin dissociation curves showing the effect of temperature changes.

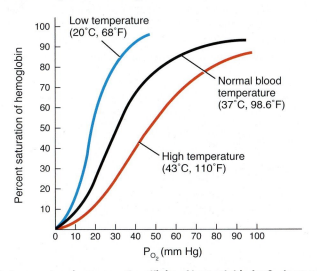

As temperature increases, the affinity of hemoglobin for O_2 decreases.

Carbon Dioxide Transport

Carbon dioxide is transported in the blood in three main forms (Figure 22.18):

1. **Dissolved CO_2.** The smallest percentage—about 7%—is dissolved in blood plasma. Upon reaching the lungs, it diffuses into alveolar air and is exhaled.

2. **Bound to amino acids.** A somewhat higher percentage, about 23%, combines with amino acids and proteins in blood. Because the most prevalent protein in blood is hemoglobin (inside red blood cells), most of the CO_2 transported in this manner is bound to hemoglobin. The main CO_2 binding sites are the amino acids in the globin portions of the hemoglobin molecule. Hemoglobin that has bound CO_2 is termed **carbaminohemoglobin (Hb–CO_2)**:

$$\underset{\text{Hemoglobin}}{Hb} \quad + \quad \underset{\text{Carbon dioxide}}{CO_2} \quad \rightleftharpoons \quad \underset{\text{Carbaminohemoglobin}}{Hb\text{—}CO_2}$$

In tissue capillaries P_{CO_2} is relatively high, which promotes formation of carbaminohemoglobin. However, in pulmonary capillaries, P_{CO_2} is relatively low, and the CO_2 readily splits apart from globin and enters the alveoli by diffusion.

3. **Bicarbonate ions.** The greatest percentage of CO_2—about 70%—is transported in blood plasma as **bicarbonate ions (HCO_3^-)**. As CO_2 diffuses into systemic capillaries and enters red blood cells, it reacts with water in the presence of the enzyme carbonic anhydrase (CA) to form carbonic acid, which dissociates into H^+ and HCO_3^-:

$$\underset{\text{Carbon dioxide}}{CO_2} \; + \underset{\text{Water}}{H_2O} \; \overset{CA}{\rightleftharpoons} \; \underset{\text{Carbonic acid}}{H_2CO_3} \; \rightleftharpoons \; \underset{\text{Hydrogen ion}}{H^+} \; + \; \underset{\text{Bicarbonate ion}}{HCO_3^-}$$

Thus, as blood picks up CO_2, HCO_3^- accumulates inside RBCs. Some HCO_3^- moves out into the blood plasma, down its concentration gradient. In exchange, chloride ions (Cl^-) move from plasma into the RBCs. This exchange of negative ions, which maintains the electrical balance between blood plasma and RBC cytosol, is known as the **chloride shift** (see Figure 22.22b). The net effect of these reactions is that CO_2 is removed from tissue cells and transported in blood plasma as HCO_3^-. As blood passes through pulmonary capillaries in the lungs, all these reactions reverse and CO_2 is exhaled.

Summary of Gas Exchange and Transport in Lungs and Tissues

As summarized in Figure 22.22a, deoxygenated blood returning to the pulmonary capillaries in the lungs contains CO_2 dissolved in blood plasma, CO_2 combined with globin as carbaminohemoglobin (Hb–CO_2) within RBCs, and CO_2 incorporated into HCO_3^- within RBCs. As blood passes through the pulmonary capillaries, molecules of CO_2 dissolved in blood plasma and CO_2 that dissociates from the globin portion of hemoglobin diffuse into alveolar air and are exhaled. In addition, when H^+ combines with HCO_3^- inside RBCs, the resulting H_2CO_3 splits into CO_2, which diffuses into blood plasma and then into alveolar air to be exhaled, and H_2O. As the concentration of HCO_3^- declines inside RBCs in pulmonary capillaries, HCO_3^- diffuses in from the blood plasma, in exchange for Cl^-. At the same time, inhaled O_2 diffuses from alveolar air into RBCs, where it binds to the heme portion of hemoglobin to form oxyhemoglobin (Hb—O_2). Blood leaving the lungs has increased O_2 content and decreased amounts of CO_2. In systemic capillaries, the chemical reactions reverse as cells use O_2 and produce CO_2 (Figure 22.22b).

Figure 22.22   **Summary of chemical reactions that occur during gas exchange.** (a) In pulmonary capillaries, hemoglobin (Hb) inside red blood cells unloads CO_2 and picks up O_2 from alveolar air. O_2 binds to hemoglobin, releasing hydrogen ions (H^+). Bicarbonate ions (HCO_3^-) pass into the RBC and bind to released H^+, forming carbonic acid (H_2CO_3). H_2CO_3 dissociates, releasing CO_2, which diffuses from the RBC into alveolar air. To maintain electrical balance, a chloride ion (Cl^-) exits the RBC for each HCO_3^- that enters. (b) At tissue cells, O_2 dissociates from oxyhemoglobin (Hb—O_2) and diffuses into tissue cells. CO_2 diffuses into red blood cells and binds to hemoglobin, forming carbaminohemoglobin (Hb—CO_2). Other molecules of CO_2 combine with water to produce bicarbonate ions and hydrogen ions. To maintain electrical balance, one Cl^- enters the RBC for each HCO_3^- that exits.

(a) Exchange of O_2 and CO_2 in pulmonary capillaries (external respiration)

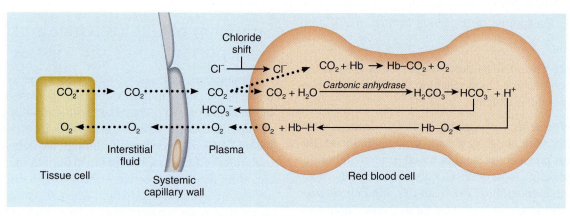

(b) Exchange of O_2 and CO_2 in systemic capillaries (internal respiration)

🔑 Hemoglobin inside red blood cells transports O_2 and CO_2.

Checkpoint

29. In which ways is oxygen transported through the blood? Which is the most common method of transport?
30. What is the most important factor in determining how much oxygen binds to hemoglobin?
31. What is the relationship between hemoglobin and P_{O_2}?
32. In which ways is carbon dioxide transported through the blood? Which transport method is the most common? The least common?
33. Would you expect the concentration of HCO_3^- to be higher in blood plasma taken from a systemic artery or a systemic vein?

22.8 The basic rhythm of respiration is controlled by the respiratory center in the brain stem.

The size of the thorax is altered by the action of the respiratory muscles, which contract and relax as a result of nerve impulses transmitted to them from centers in the brain. These nerve impulses are sent from clusters of neurons located in the medulla oblongata and pons of the brain stem. This widely dispersed group of neurons, collectively called the **respiratory center**, can be divided into three areas on the basis of their functions: (1) the medullary rhythmicity area in the medulla oblongata; (2) the pneumotaxic area in the pons; and (3) the apneustic area, also in the pons (Figure 22.23).

Medullary Rhythmicity Area

The **medullary rhythmicity area** (rith-MIS-i-tē) controls the basic rhythm of respiration. There are inspiratory and expiratory areas within the medullary rhythmicity area. Figure 22.24 shows the relationships of the inspiratory and expiratory areas during normal quiet breathing and forceful breathing.

During quiet breathing, inhalation lasts for about 2 seconds and exhalation lasts for about 3 seconds. Nerve impulses generated in the **inspiratory area** establish the basic rhythm of breathing. While the inspiratory area is active, it generates nerve impulses for about 2 seconds (Figure 22.24a). The impulses propagate to the diaphragm via the phrenic nerves and to the external intercostals via

Figure 22.23 Locations of areas of the respiratory center.

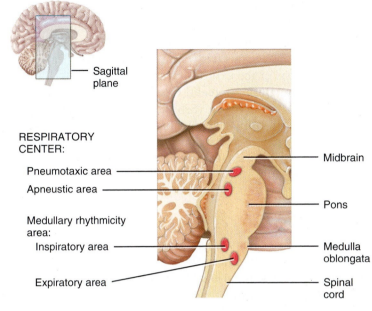

Sagittal plane

RESPIRATORY CENTER:

Pneumotaxic area — Midbrain
Apneustic area — Pons
Medullary rhythmicity area:
    Inspiratory area — Medulla oblongata
    Expiratory area — Spinal cord

Sagittal section of brain stem

The respiratory center includes the medullary rhythmicity area in the medulla oblongata plus the pneumotaxic and apneustic areas in the pons.

Figure 22.24 Roles of the medullary rhythmicity area in controlling (a) the basic rhythm of respiration and (b) forceful breathing.

INSPIRATORY AREA
ACTIVE   INACTIVE

2 seconds   3 seconds

Diaphragm and external intercostals contract | Diaphragm and external intercostals relax, followed by elastic recoil of lungs

Normal quiet inhalation | Normal quiet exhalation

(a) During normal quiet breathing

Activates

INSPIRATORY AREA ACTIVE → EXPIRATORY AREA

Diaphragm, sternocleidomastoids, and scalenes contract | Internal intercostals and abdominal muscles contract

Forceful inhalation | Forceful exhalation

(b) During forceful breathing

During normal, quiet breathing, the expiratory area is inactive; during forceful breathing, the inspiratory area activates the expiratory area.

intercostal nerves. When the nerve impulses reach the diaphragm and external intercostals, the muscles contract and inhalation occurs. Even when all incoming nerve connections to the inspiratory area in the medulla are cut or blocked, neurons in this area still rhythmically discharge impulses that cause inhalation. At the end of 2 seconds, the inspiratory area becomes inactive and nerve impulses cease. With no impulses arriving, the diaphragm and external intercostals relax for about 3 seconds, allowing passive elastic recoil of the lungs and thoracic wall. Then, the cycle repeats.

The neurons of the **expiratory area** remain inactive during quiet breathing. During forceful breathing, however, nerve impulses from the inspiratory area activate the expiratory area (Figure 22.24b). Impulses from the expiratory area cause contraction of the internal intercostal and abdominal muscles, which decreases the size of the thoracic cavity and causes forceful exhalation.

Pneumotaxic Area

Although the medullary rhythmicity area controls the basic rhythm of respiration, other sites in the brain stem help coordinate the transition between inhalation and exhalation. One of these sites is the **pneumotaxic area** (noo-mō-TAK-sik; *pneumo-* = air or breath; *-taxic* = arrangement) in the upper pons (Figure 22.23),

which transmits inhibitory impulses to the inspiratory area. The major effect of these nerve impulses is to help turn off the inspiratory area before the lungs become too full of air. In other words, the impulses shorten the duration of inhalation. When the pneumotaxic area is more active, breathing rate is more rapid.

Apneustic Area

Another part of the brain stem that coordinates the transition between inhalation and exhalation is the **apneustic area** (ap-NOO-stik) in the lower pons (Figure 22.23). This area sends stimulatory impulses to the inspiratory area that activate it and prolong inhalation. The result is a long, deep inhalation. When the pneumotaxic area is active, it overrides signals from the apneustic area.

Checkpoint

34. Which area of the respiratory center contains neurons that are active and then inactive in a repeating cycle?
35. Which nerves convey impulses from the respiratory center to the diaphragm?
36. How does the medullary rhythmicity area regulate respiration?
37. How are the apneustic and pneumotaxic areas related to the control of respiration?

Return to Cari's Story

Cari was admitted to the hospital immediately. She had pneumonia brought on by an impaired pulmonary defense system. Cigarette smoke had paralyzed her cilia and stimulated the goblet cells lining her respiratory passages to secrete excess amounts of mucus. Macrophages and neutrophils had been continually drawn to the lung to combat the damaging effects of the smoke but had instead become trapped in the thick mucus secretions. Cari's only way of clearing her lungs had become her smoker's cough. The X-ray revealed fluid within the alveolar spaces, and a Gram stain of her sputum (mucus coughed up from the lung) had revealed pneumococcal pneumonia, the most common form. The doctor started Cari on intravenous antibiotics, continued her oxygen supplementation, and ran further tests.

Ken walked into her room. "How are you feeling?"

"I feel pretty much the same. The oxygen seems to help, but I feel like I'm panting like a dog, and my chest hurts," she replied.

Ken talked to the doctor. "Why isn't she getting better?"

The doctor sighed and replied, "Ken, we're doing everything we can. Cari's pneumonia wasn't responding to the initial antibiotic therapy so we changed her medication, which appears to be helping. Her temperature is coming down. Her body is still struggling for oxygen, so we put her on the ventilator to help get oxygen to her tissues and also address her blood alkalosis. The pH of her blood was rising because she was hyperventilating. The mechanical ventilator will help control the acute change in blood pH until her kidneys can compensate for the respiratory alkalosis."

Cari was lucky—she finally responded to the antibiotics. She was discharged after being in the hospital for nearly 2 weeks. Her chest still hurt, and she still had some difficulty breathing. The doctor warned her that if she didn't quit smoking, she was at risk for another serious infection. Pneumonia is related to impaired lung functioning. Chronic obstructive pulmonary disease, which has a direct correlation to long-term smoking, is the fourth leading cause of death in the United States.

F. How would fluid in Cari's lungs affect her total lung capacity?
G. How does the elevation of Cari's respiratory rate alter her minute ventilation?
H. Normal blood oxygen saturation levels are greater than 94%; Cari's blood oxygen saturation level was 90% at the time of her exam and an initial arterial blood gas analysis done when she was admitted to the hospital revealed her arterial P_{O_2} was 54 mm Hg. How do these clinical findings relate to the internal respiration in Cari's body?
I. Which of the symptoms Cari has described are due to lack of oxygen and reduced oxygen exchange at her tissues?
J. As Cari's P_{CO_2} rose, how was the oxygen-carrying capacity of hemoglobin affected?
K. How would you have expected Cari's decreased P_{CO_2} and alkaline blood pH to have affected her breathing?
L. How would administration of oxygen enhance Cari's central drive to breathe?

22.9 Respiration may be modified by cortical influences, chemical stimuli, proprioceptor input, and the inflation reflex.

At rest, about 200 mL of O_2 are used each minute by body cells. During strenuous exercise, however, O_2 use typically increases 15- to 20-fold in normal healthy adults, and as much as 30-fold in elite endurance-trained athletes. Several mechanisms help match respiratory effort to metabolic demand. Although the basic rhythm of respiration is set and coordinated by the inspiratory area, the rhythm can be modified in response to inputs from other brain regions, receptors in the peripheral nervous system, and other factors.

Cortical Influences on Respiration

Because the cerebral cortex has connections with the respiratory center, we can voluntarily alter our pattern of breathing. We can even refuse to breathe at all for a short time. Voluntary control is protective because it enables us to prevent water or irritating gases from entering the lungs. The ability to not breathe, however, is limited by the buildup of CO_2 and H^+ in the body. When P_{CO_2} and H^+ concentrations increase to a certain level, the inspiratory area is strongly stimulated, nerve impulses are sent along the phrenic and intercostal nerves to inspiratory muscles, and breathing resumes, whether the person wants it to or not. It is impossible for small children to kill themselves by voluntarily holding their breath, even though many have tried in order to get their way. If the breath is held long enough to cause fainting, breathing resumes when consciousness is lost. Nerve impulses from the hypothalamus and limbic system also stimulate the respiratory center, allowing emotional stimuli to alter respirations as, for example, in laughing and crying.

Chemoreceptor Regulation of Respiration

Certain chemical stimuli modulate how quickly and how deeply we breathe. The respiratory system functions to maintain proper levels of CO_2 and O_2 and is very responsive to changes in the levels of these gases in body fluids. Sensory neurons that are responsive to chemicals are called **chemoreceptors**. Chemoreceptors in two locations monitor levels of CO_2, H^+, and O_2 and provide input to the respiratory center (Figure 22.25). **Central chemoreceptors** are located in or near the medulla oblongata in the *central* nervous system. They respond to changes in H^+ concentration or P_{CO_2}, or both, in cerebrospinal fluid. **Peripheral chemoreceptors** are located in the **aortic bodies**, clusters of chemoreceptors located in the wall of the arch of the aorta, and in the **carotid bodies** in the wall of the left and right common carotid arteries, where they divide into the internal and external carotid arteries. Peripheral chemoreceptors are part of the *peripheral* nervous system and are sensitive to changes in P_{O_2}, H^+, and P_{CO_2} in the blood. Axons of sensory neurons from the aortic bodies are part of the vagus (X)

Figure 22.25 Locations of peripheral chemoreceptors.

Medulla oblongata

Sensory axons in glossopharyngeal nerve

Internal carotid artery

Carotid body

External carotid artery

Carotid sinus

Common carotid artery

Sensory axons in vagus nerve

Arch of aorta

Aortic bodies

Heart

Chemoreceptors are sensory neurons that respond to changes in the levels of certain chemicals in the body.

nerves, and those from the carotid bodies are part of the right and left glossopharyngeal (IX) nerves.

Because CO_2 is lipid-soluble, it easily diffuses into cells, where, in the presence of carbonic anhydrase, it combines with water (H_2O) to form carbonic acid (H_2CO_3). Carbonic acid quickly breaks down into H^+ and HCO_3^-. Thus, an increase in CO_2 in the blood causes an increase in H^+ inside cells, and a decrease in CO_2 causes a decrease in H^+.

Normally, the P_{CO_2} in arterial blood is 40 mm Hg. If even a slight increase in P_{CO_2} occurs—a condition called **hypercapnia**—the central chemoreceptors are stimulated and respond vigorously to the resulting increase in H^+ level. The peripheral chemoreceptors also are stimulated by both the high P_{CO_2} and the rise in H^+. In addition, the peripheral chemoreceptors (but not the central chemoreceptors) respond to a deficiency of O_2. When P_{O_2} in arterial blood falls from a normal level of 100 mm Hg but is still above 50 mm Hg, the peripheral chemoreceptors are stimulated. Severe deficiency of O_2 depresses activity of the central chemoreceptors and inspiratory area, which then do not respond well to any inputs and send fewer impulses to the muscles of inhalation. As the breathing rate decreases or breathing ceases altogether, P_{O_2} falls lower and lower, establishing a positive feedback cycle with a possibly fatal result.

The chemoreceptors participate in a negative feedback system that regulates the levels of CO_2, O_2, and H^+ in the blood (Figure 22.26). As a result of increased P_{CO_2}, decreased pH (increased H^+), or decreased P_{O_2}, input from the central and peripheral chemoreceptors causes the inspiratory area to become highly active, and the rate and depth of breathing increase. Rapid and deep breathing, called **hyperventilation**, allows the inhalation of more O_2 and exhalation of more CO_2 until P_{CO_2} and H^+ are lowered to normal.

If arterial P_{CO_2} is lower than 40 mm Hg—a condition called **hypocapnia**—the central and peripheral chemoreceptors are not stimulated, and stimulatory impulses are not sent to the inspiratory area. As a result, the inspiratory area sets its own moderate pace until CO_2 accumulates and the P_{CO_2} rises to 40 mm Hg. The inspiratory center is more strongly stimulated when P_{CO_2} is rising above normal than when P_{O_2} is falling below normal. As a result, people who hyperventilate voluntarily and cause hypocapnia can hold their breath for an unusually long period. Swimmers were once encouraged to hyperventilate just before diving in to compete. However, this practice is risky because the O_2 level may fall dangerously low and cause fainting before the P_{CO_2} rises high enough to stimulate inhalation. If you faint on land you may suffer bumps and bruises, but if you faint in the water you could drown.

Proprioceptor Stimulation of Respiration

As soon as you start exercising, your rate and depth of breathing increase, even before changes in P_{O_2}, P_{CO_2}, or H^+ level occur. The main stimulus for these quick changes in respiratory effort is input from proprioceptors, which monitor movement of joints and muscles. Nerve impulses from the proprioceptors stimulate the inspi-

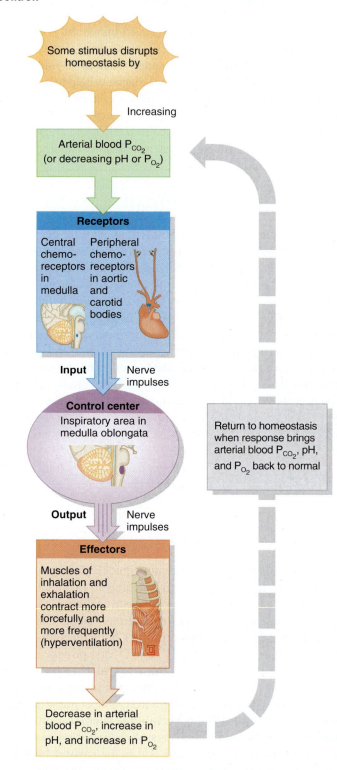

Figure 22.26   Regulation of breathing in response to changes in blood P_{CO_2}, P_{O_2}, and pH (H^+ concentration) via negative feedback control.

Some stimulus disrupts homeostasis by

Increasing

Arterial blood P_{CO_2} (or decreasing pH or P_{O_2})

Receptors

Central chemoreceptors in medulla

Peripheral chemoreceptors in aortic and carotid bodies

Input   Nerve impulses

Control center
Inspiratory area in medulla oblongata

Output   Nerve impulses

Effectors

Muscles of inhalation and exhalation contract more forcefully and more frequently (hyperventilation)

Decrease in arterial blood P_{CO_2}, increase in pH, and increase in P_{O_2}

Return to homeostasis when response brings arterial blood P_{CO_2}, pH, and P_{O_2} back to normal

An increase in arterial blood P_{CO_2} stimulates the inspiratory area.

ratory area of the medulla oblongata. At the same time, motor neurons that originate in the primary motor cortex (precentral gyrus) also feed excitatory impulses into the inspiratory area.

The Inflation Reflex

Similar to those in the blood vessels, stretch-sensitive receptors called **baroreceptors** are located in the walls of bronchi and bronchioles. When these receptors become stretched during overinflation of the lungs, nerve impulses are sent to the inspiratory and apneustic areas. In response, the inspiratory area is inhibited directly, and the apneustic area is inhibited from activating the inspiratory area. As a result, exhalation begins. As air leaves the lungs during exhalation, the lungs deflate and the stretch receptors are no longer stimulated. Thus, the inspiratory and apneustic areas are no longer inhibited, and a new inhalation begins. This reflex is mainly a protective mechanism for preventing excessive inflation of the lungs rather than a key component in the normal regulation of respiration.

Other Influences on Respiration

Other factors that contribute to regulation of respiration include the following:

- *Limbic system stimulation.* Anticipation of activity or emotional anxiety may stimulate the limbic system, which then sends excitatory input to the inspiratory area, increasing the rate and depth of ventilation.

- *Temperature.* An increase in body temperature, as occurs during a fever or vigorous muscular exercise, increases the rate of respiration. A decrease in body temperature decreases respiratory rate. A sudden cold stimulus (such as plunging into cold water) causes temporary **apnea** (AP-nē-a; *a-* = without; *-pnea* = breath), an absence of breathing.

- *Pain.* A sudden, severe pain brings about brief apnea, but a prolonged somatic pain increases respiratory rate. Visceral pain may slow the rate of respiration.

- *Stretching the anal sphincter muscle.* This action increases the respiratory rate and is sometimes used to stimulate respiration in a newborn baby or a person who has stopped breathing.

- *Irritation of airways.* Physical or chemical irritation of the pharynx or larynx brings about an immediate cessation of breathing followed by coughing or sneezing.

- *Blood pressure.* The carotid and aortic baroreceptors that detect changes in blood pressure have a small effect on respiration. A sudden rise in blood pressure decreases the rate of respiration, and a drop in blood pressure increases the respiratory rate.

Table 22.2 summarizes the stimuli that affect the rate and depth of ventilation.

Checkpoint

38. Why is it impossible to voluntarily hold your breath indefinitely?
39. Where are central chemoreceptors located?
40. Which chemicals stimulate peripheral chemoreceptors?
41. What is the normal arterial blood P_{CO_2}?
42. What happens to the rate and depth of breathing when the P_{CO_2} becomes elevated?
43. How do proprioceptors, inflation reflex, temperature changes, pain, and irritation of the airways each modify respiration?

TABLE 22.2

Stimuli That Affect Ventilation Rate and Depth

STIMULI THAT INCREASE VENTILATION RATE AND DEPTH	STIMULI THAT DECREASE VENTILATION RATE AND DEPTH
Voluntary hyperventilation controlled by the cerebral cortex and anticipation of activity by stimulation of the limbic system.	Voluntary hypoventilation controlled by the cerebral cortex.
Increase in arterial blood P_{CO_2} above 40 mm Hg (causes an increase in H^+) detected by peripheral and central chemoreceptors.	Decrease in arterial blood P_{CO_2} below 40 mm Hg (causes a decrease in H^+) detected by peripheral and central chemoreceptors.
Decrease in arterial blood P_{O_2} from 100 mm Hg to 50 mm Hg.	Decrease in arterial blood P_{O_2} below 50 mm Hg.
Increased activity of proprioceptors.	Decreased activity of proprioceptors.
Increase in body temperature.	Decrease in body temperature decreases the rate of respiration, and a sudden cold stimulus causes apnea.
Prolonged pain.	Severe pain causes apnea.
Decrease in blood pressure.	Increase in blood pressure.
Stretching the anal sphincter.	Irritation of pharynx or larynx by touch or chemicals causes brief apnea followed by coughing or sneezing.

22.10 Acid–base balance is maintained by controlling the H⁺ concentration of body fluids.

Various ions play different roles that help maintain homeostasis. A major homeostatic challenge is keeping the H^+ concentration (pH) of body fluids at an appropriate level. This task—the maintenance of acid–base balance—is of critical importance to normal cellular function. For example, the three-dimensional shape of all body proteins, which enables them to perform specific functions, is very sensitive to pH changes. When the diet contains a large amount of protein, as is typical in North America, cellular metabolism produces more acids than bases, which tends to acidify the blood. Before proceeding with this section of the chapter, you may wish to review the discussion of acids, bases, and pH in Concept 2.4.

In a healthy person, several mechanisms help maintain the pH of systemic arterial blood between 7.35 and 7.45. Because metabolic reactions often produce a huge excess of H^+, the lack of any mechanism for the disposal of H^+ would cause the H^+ in body fluids to rise quickly to a lethal level. Homeostasis of H^+ concentration within a narrow range is thus essential to survival. The removal of H^+ from body fluids and its subsequent elimination from the body depend on three major mechanisms: buffer systems, exhalation of carbon dioxide, and excretion of H^+ in the urine. Excretion of H^+ in the urine will be discussed in Chapter 24.

The Actions of Buffer Systems

Most buffer systems in the body consist of a weak acid and the salt of that acid, which functions as a weak base. Buffers act quickly to temporarily bind H^+, removing the highly reactive, excess H^+ from solution. Buffers prevent rapid, drastic changes in the pH of body fluids by converting strong acids and bases into weak acids and weak bases within fractions of a second. Strong acids lower pH more than weak acids because strong acids release H^+ more readily and thus contribute more free hydrogen ions. Similarly, strong bases raise pH more than weak ones. Important buffer systems related to the respiratory system include the protein buffer system and the carbonic acid–bicarbonate buffer system.

Protein Buffer System

The **protein buffer system** is the most abundant buffer in intracellular fluid and blood plasma. For example, the protein hemoglobin is an especially good buffer within red blood cells, and albumin is the main protein buffer in blood plasma. Proteins are composed of amino acids, organic molecules that contain at least one *carboxyl group* (—COOH) and at least one *amino group* (—NH₂); these groups are the functional components of the protein buffer system.

The free carboxyl group at one end of a protein acts like an acid by releasing H^+ when pH rises (H^+ concentration decreases); it dissociates as follows:

$$NH_2\!-\!\underset{\underset{H}{|}}{\overset{\overset{R}{|}}{C}}\!-\!COOH \longrightarrow NH_2\!-\!\underset{\underset{H}{|}}{\overset{\overset{R}{|}}{C}}\!-\!COO^- + H^+$$

The H^+ is then able to react with any excess OH^- in the solution to form water. The free amino group at the other end of a protein can act as a base by combining with H^+ when pH falls (H^+ concentration increases), as follows:

$$NH_2\!-\!\underset{\underset{H}{|}}{\overset{\overset{R}{|}}{C}}\!-\!COOH + H^+ \longrightarrow {}^+NH_3\!-\!\underset{\underset{H}{|}}{\overset{\overset{R}{|}}{C}}\!-\!COOH$$

So proteins can buffer both acids and bases. In addition to the terminal carboxyl and amino groups, side chains that can buffer H^+ are present on 7 of the 20 amino acids.

As we have already noted, the protein hemoglobin is an important buffer of H^+ in red blood cells (see Figure 22.22). As blood flows through the systemic capillaries, carbon dioxide (CO_2) passes from tissue cells into red blood cells, where it combines with water (H_2O) to form carbonic acid (H_2CO_3). Once formed, H_2CO_3 dissociates into H^+ and HCO_3^-. At the same time that CO_2 is entering red blood cells, oxyhemoglobin (Hb—O_2) is giving up its oxygen to tissue cells. Deoxyhemoglobin picks up most of the H^+. The following reactions summarize these relations:

$$\underset{\text{Water}}{H_2O} \;+\; \underset{\text{Carbon dioxide (entering RBCs)}}{CO_2} \longrightarrow \underset{\text{Carbonic acid}}{H_2CO_3}$$

$$\underset{\text{Carbonic acid}}{H_2CO_3} \longrightarrow \underset{\text{Hydrogen ion}}{H^+} \;+\; \underset{\text{Bicarbonate ion}}{HCO_3^-}$$

$$\underset{\substack{\text{Oxyhemoglobin}\\\text{(in RBCs)}}}{Hb\!-\!O_2} \;+\; \underset{\substack{\text{Hydrogen ion}\\\text{(from carbonic acid)}}}{H^+} \longrightarrow \underset{\substack{\text{Deoxy-}\\\text{hemoglobin}}}{Hb\!-\!H} \;+\; \underset{\substack{\text{Oxygen}\\\text{(released to}\\\text{tissue cells)}}}{O_2}$$

Carbonic Acid–Bicarbonate Buffer System

The **carbonic acid–bicarbonate buffer system** is based on the *bicarbonate ion* (HCO_3^-), which can act as a weak base, and *carbonic acid* (H_2CO_3), which can act as a weak acid. HCO_3^- is a significant anion in both intracellular and extracellular fluids, and its reaction with the CO_2 constantly released during cellular respiration produces H_2CO_3. If there is an excess of H^+, the HCO_3^- can function as a weak base and remove the excess H^+ as follows:

$$\underset{\text{Hydrogen ion}}{H^+} + \underset{\text{Bicarbonate ion (weak base)}}{HCO_3^-} \longrightarrow \underset{\text{Carbonic acid}}{H_2CO_3}$$

Then, H_2CO_3 dissociates into water and carbon dioxide, and the CO_2 is exhaled from the lungs. Conversely, if there is a shortage of H^+, the H_2CO_3 can function as a weak acid and provide H^+ as follows:

$$\underset{\text{Carbonic acid (weak acid)}}{H_2CO_3} \longrightarrow \underset{\text{Hydrogen ion}}{H^+} + \underset{\text{Bicarbonate ion}}{HCO_3^-}$$

Because CO_2 and H_2O combine to form H_2CO_3, this buffer system cannot protect against pH changes due to respiratory problems in which there is an excess or shortage of CO_2.

Exhalation of Carbon Dioxide

The simple act of breathing also plays an important role in maintaining the pH of body fluids. By increasing the rate and depth of breathing, more carbon dioxide can be exhaled. Within minutes this reduces the level of carbonic acid in blood, which raises the blood pH (reduces blood H^+ level).

An increase in the carbon dioxide (CO_2) concentration in body fluids increases H^+ concentration and thus lowers the pH (makes body fluids more acidic). Conversely, a decrease in the CO_2 concentration of body fluids raises the pH (makes body fluids more alkaline). This chemical interaction is illustrated by the following reversible reactions.

$$\underset{\substack{\text{Carbon} \\ \text{dioxide}}}{CO_2} + \underset{\text{Water}}{H_2O} \rightleftharpoons \underset{\text{Carbonic acid}}{H_2CO_3} \rightleftharpoons \underset{\text{Hydrogen ion}}{H^+} + \underset{\text{Bicarbonate ion}}{HCO_3^-}$$

With increased ventilation, more CO_2 is exhaled. When CO_2 levels decrease, the reaction is driven to the left, H^+ concentration decreases, and blood pH increases. If ventilation is slower than normal, less carbon dioxide is exhaled. When CO_2 levels increase, the reaction is driven to the right, H^+ concentration increases, and the blood pH falls. These examples show the powerful effect of alterations in breathing on the pH of body fluids.

The pH of body fluids and the rate and depth of breathing interact via a negative feedback loop (Figure 22.27). When the blood acidity increases, the decrease in pH (increase in concentration of H^+) is detected by central chemoreceptors in the medulla oblongata and peripheral chemoreceptors in the aortic and carotid bodies, both of which stimulate the inspiratory area in the medulla oblongata. As a result, the diaphragm and other respiratory

Figure 22.27 Negative feedback regulation of blood pH by the respiratory system.

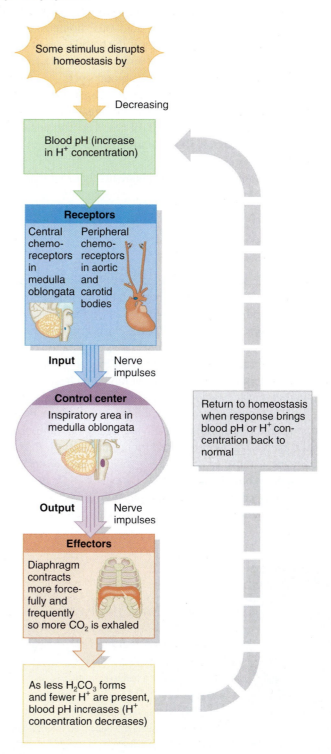

Some stimulus disrupts homeostasis by

Decreasing

Blood pH (increase in H^+ concentration)

Receptors

Central chemo-receptors in medulla oblongata

Peripheral chemo-receptors in aortic and carotid bodies

Input   Nerve impulses

Control center

Inspiratory area in medulla oblongata

Return to homeostasis when response brings blood pH or H^+ concentration back to normal

Output   Nerve impulses

Effectors

Diaphragm contracts more force-fully and frequently so more CO_2 is exhaled

As less H_2CO_3 forms and fewer H^+ are present, blood pH increases (H^+ concentration decreases)

Exhalation of carbon dioxide lowers the H^+ concentration of blood.

muscles contract more forcefully and frequently, so more CO_2 is exhaled. As less H_2CO_3 forms and fewer H^+ are present, blood pH increases. When the response brings blood pH (H^+ concentration) back to normal, there is a return to acid–base homeostasis. The same negative feedback loop operates if the blood level of CO_2 increases. Ventilation increases, which removes more CO_2, reducing the H^+ concentration and increasing the blood's pH.

By contrast, if the pH of the blood increases, the respiratory center is inhibited and the rate and depth of breathing decrease. A decrease in the CO_2 concentration of the blood has the same effect. When breathing decreases, CO_2 accumulates in the blood and its H^+ concentration increases.

Table 22.3 summarizes the mechanisms that maintain pH of body fluids.

TABLE 22.3	
Respiratory Mechanisms That Maintain pH of Body Fluids	
MECHANISM	COMMENTS
Buffer Systems	Most consist of a weak acid and the salt of that acid, which functions as a weak base. They prevent drastic changes in body fluid pH.
Proteins	The most abundant buffers in body cells and blood. Hemoglobin inside red blood cells is a good buffer.
Carbonic acid–bicarbonate	Important regulator of blood pH. The most abundant buffers in extracellular fluid.
Exhalation of CO_2	With increased exhalation of CO_2, pH rises (fewer H^+). With decreased exhalation of CO_2, pH falls (more H^+).

Acid–Base Imbalances

The normal pH range of systemic arterial blood is between 7.35 and 7.45. **Acidosis** is a condition in which blood pH is below 7.35; **alkalosis** is a condition in which blood pH is higher than 7.45.

The major physiological effect of acidosis is depression of the central nervous system through depression of synaptic transmission. If the systemic arterial blood pH falls below 7, depression of the nervous system is so severe that the individual becomes disoriented, then comatose, and may die. A major physiological effect of alkalosis, by contrast, is overexcitability in both the central nervous system and peripheral nerves. Neurons conduct impulses repetitively, even when not prompted by normal stimuli; the results are nervousness, muscle spasms, and even convulsions and death.

A change in blood pH that leads to acidosis or alkalosis may be countered by **compensation**, the physiological response to an acid–base imbalance that acts to normalize arterial blood pH. If a person has altered blood pH due to metabolic causes, hyperventilation or hypoventilation can help bring blood pH back toward the normal range; this form of compensation, termed **respiratory compensation**, occurs within minutes and reaches its maximum within hours.

Checkpoint

44. How do proteins and bicarbonate ions help maintain the pH of body fluids?

45. If you hold your breath for 30 seconds, what is likely to happen to your blood pH?

46. What are acidosis and alkalosis? What are the major physiological effects of acidosis and alkalosis?

Cari's Story: Epilogue and Discussion

▶▶▶ **C**ari, a 47-year-old female with a 20-year history of pack-a-day smoking and chronic bronchitis, presents at the clinic complaining of a cold and potential sinus infection. She describes general muscle aches, fatigue, and sore throat, and has a temperature of 101.2 and a history of several weeks of flulike symptoms. Upon examination, it is found that her respiratory rate is elevated and she has a low oxygen saturation level. Her lung sounds are abnormal. A follow-up chest X-ray reveals fluid infiltrate within the lungs. A Gram stain reveals the presence of bacteria, resulting in a diagnosis of pneumococcal pneumonia. Cari is lucky—she makes a full recovery.

Like many smokers, Cari has a hard decision to make: Continue smoking and risk further infections, loss of respiratory function, emphysema, cancer, and chronic obstructive lung disease, or quit. Now that you have learned about the profound effects of smoking on the delicate tissue of the lungs and about the importance of the respiratory system, answer the following questions. ◀◀◀

M. *Which anatomical structures in Cari's respiratory system were initially involved?*

N. *Why was Cari plagued with a chronic smoker's cough?*

O. *Which damaging effects of tobacco smoke led to Cari's impaired respiratory defense mechanisms?*

P. *How did the pneumonia affect Cari's lung function?*

Concept and Resource Summary

Concept	Resources

WILEY PLUS

Introduction

1. Metabolic reactions need a continual supply of O_2, and generate CO_2 waste that must be removed rapidly. The **respiratory system** brings O_2 into the body and removes CO_2 from the body.

Concept 22.1 Inhaled air travels in the upper respiratory system through the nasal cavities and then through the pharynx.

1. The respiratory system includes the nose, pharynx, larynx, trachea, bronchi, and lungs. The respiratory system includes a **conducting zone**, passageways that filter, warm, and moisten air, and conduct it to the lungs; and a **respiratory zone** within the lungs where gas exchange occurs.
2. The visible external portion of the **nose** is supported by bone and cartilage and is lined with a mucous membrane. Openings to the exterior are the **external nares**.
3. The internal nose connects to the pharynx through two **internal nares**. The **nasal cavity** is divided into left and right sides by the **nasal septum**.
4. The nose warms, moistens, and filters incoming air; detects olfactory stimuli; and contribute to voice resonance.
5. The **paranasal sinuses** are cavities in skull bones that produce mucus and contribute to voice resonance.
6. The pharynx is a muscular tube lined by a mucous membrane. The anatomic regions are the nasopharynx, oropharynx, and laryngopharynx. The **nasopharynx** is a passageway for air from the nasal cavity. The **oropharynx** and **laryngopharynx** serve as common passageways for air, food, and drink.

See
- Anatomy Overview – Nasal Cavity and Pharynx
- Anatomy Overview – Respiratory Epithelium

Concept 22.2 Inhaled air travels in the lower respiratory system from the larynx to alveoli.

1. Structurally, the **lower respiratory system** consists of the larynx, trachea, bronchi, and lungs.
2. The **larynx** is a passageway connecting the laryngopharynx with the trachea. It includes the **thyroid cartilage**, the **cricoid cartilage** connecting the larynx and trachea, the **arytenoid cartilages**, and the epiglottis. The **epiglottis** moves to cover the larynx and airways during swallowing, which routes food and liquids into the esophagus. The **glottis** consists of the vocal folds and the space between them.
3. The **ventricular folds** are one of two pairs of mucous membrane folds of the larynx. When brought together, they hold the breath but do not produce sound. The other pair, the **vocal folds**, produce sound as they vibrate. Taut vocal folds produce high-pitched sounds, and relaxed ones produce low-pitched sounds.
4. The **trachea** is an air passageway extending from the larynx to the primary bronchi. Its walls contain C-shaped cartilage rings that give support to prevent the tracheal wall from collapsing.
5. The trachea divides into the **right primary bronchus** and the **left primary bronchus**, which enter the right and left lung, respectively. The bronchial tree consists of the **trachea**, **primary bronchi**, **secondary bronchi**, **tertiary bronchi**, **bronchioles**, and **terminal bronchioles**.

See
- Anatomy Overview – Respiratory Organs
- Anatomy Overview – Respiratory Tissues

Hear
- Figure 22.4 – Larynx
- Figure 22.8 – Relationship of the Pleural Membranes to the Lungs
- Figure 22.11 – Structural Components of an Alveolus

Concept

Do

6. The **lungs** are paired organs separated from each other by the *mediastinum*. **Visceral pleura** cover each lung, and the **parietal pleura** line the thoracic cavity wall. Within the **pleural cavity** between the parietal and visceral pleura, *pleural fluid* reduces friction as the pleura move during breathing. Each lung is surrounded by a separate pleural cavity.

7. The right lung has three lobes separated by two fissures; the left lung has two lobes separated by one fissure and a depression, the **cardiac notch**. **Secondary bronchi** give rise to branches called **tertiary bronchi**, which supply segments of lung tissue called **bronchopulmonary segments**. Each bronchopulmonary segment consists of **lobules**, which contain lymphatic vessels, arterioles, venules, and **terminal bronchioles**, **respiratory bronchioles**, and **alveolar ducts**.

8. An **alveolar sac** consists of two or more cup-shaped alveoli sharing a common opening to an alveolar duct. Walls of alveoli consist of **type I alveolar cells** involved in gas exchange, simple squamous epithelial cells that form a continuous lining; **type II alveolar cells** that secrete alveolar fluid with **surfactant** to inhibit alveolar collapse; and **alveolar macrophages** to remove dust and other debris. Gas exchange occurs across the **respiratory membrane**.

- Exercise – Paint the Lung
- Exercise – Concentrate on the Respiratory Structures
- Concepts and Connections – Functional Anatomy of the Respiratory System

Concept 22.3 Inhalation and exhalation result from pressure changes caused by muscle contraction and relaxation.

1. **Pulmonary ventilation**, or breathing, consists of inhalation and exhalation of air between the atmosphere and alveoli. **External respiration** is the exchange of gases across the respiratory membrane. **Internal respiration** is gas exchange between blood and tissue cells.

2. **Inhalation** occurs when **alveolar pressure** falls below atmospheric pressure. Contraction of the diaphragm and external intercostals increases the size of the lungs. Expansion of the lungs decreases alveolar pressure so that air moves down a pressure gradient from the atmosphere into the lungs.

3. During forceful inhalation, accessory muscles of inhalation (sternocleidomastoids, scalenes, and pectoralis minors) are used to allow greater airflow into the lungs than occurs during quiet inhalation.

4. **Exhalation** occurs when alveolar pressure is higher than atmospheric pressure. Relaxation of the diaphragm and external intercostals results in elastic recoil of the chest wall and lungs, which decreases the size of the lungs. As lung volume decrease, alveolar pressure increases, so that air moves down a pressure gradient from the lungs to the atmosphere.

5. Forceful exhalation involves contraction of the internal intercostals and abdominal muscles to force additional air out of the lungs.

6. The **surface tension** exerted by alveolar fluid produces an inward force on alveoli and accounts for much of the elastic recoil of exhalation. Surfactant in alveolar fluid reduces its surface tension.

7. **Compliance** is the amount of effort required to stretch the lungs and thoracic wall.

8. The walls of the airways offer some resistance to breathing. Resistance in air passageways decreases airflow.

9. Modified respiratory movements, such as coughing, sneezing, sighing, yawning, sobbing, crying, laughing, and hiccupping, are used to express emotions and to clear the airways.

See

- Animation – Gas Exchange Introduction
- Animation – Pulmonary Ventilation
- Animation – Regulation of Ventilation—Basic Rhythms

Hear

- **Figure 22.14 –** **Pressure Changes in Pulmonary Ventilation**

Do

- Exercise – The Airway
- Exercise – Build an Airway

Concept	Resources

Concept 22.4 Lung volumes and capacities are measured to determine the respiratory status of an individual.

1. The volume of one breath is called **tidal volume**. The total volume of air inhaled and exhaled each minute is called the **minute ventilation** and is calculated by multiplying the breathing rate by the tidal volume.
2. Lung volumes exchanged during breathing and the rate of respiration are measured with a **spirometer**.
3. **Anatomic dead space** of the conducting airways contains air that does not undergo gas exchange.
4. Forceful inspiration can take in more air beyond the tidal volume; this additional air is called the **inspiratory reserve volume**. The **expiratory reserve volume** is the amount of air beyond tidal volume that can be forced out of the lungs.
5. Even after the expiratory reserve volume is exhaled, air remains in the lungs to keep the alveoli slightly inflated. This volume of air that remains is the **residual volume**.
6. Lung capacities are combinations of specific lung volumes. **Inspiratory capacity** is the tidal volume plus inspiratory reserve volume. **Functional residual capacity** is the residual volume plus expiratory reserve volume. **Vital capacity** is the sum of the inspiratory reserve volume, tidal volume, and expiratory reserve volume. **Total lung capacity** is the vital capacity plus residual volume.

See
• **Animation – Regulation of Ventilation—Basic Rhythms**

Concept 22.5 Oxygen and carbon dioxide diffusion is based on partial pressure gradients and solubility.

1. The partial pressure of a gas is the pressure exerted by that gas in a mixture of gases.
2. According to **Dalton's law**, each gas in a mixture of gases exerts its own pressure as if all the other gases were not present. Each gas diffuses across a membrane by moving down its partial pressure gradient.
3. **Henry's law** states that the quantity of a gas that will dissolve in a liquid is proportional to the partial pressure of the gas and its solubility. In body fluids, a gas has a greater ability to stay in solution when it has a higher partial pressure and is more soluble in water.

See
• **Animation – Gas Exchange**

Do
• **Exercise – Gas Exchange Match-up**

Concept 22.6 Respiration occurs between alveoli and pulmonary capillaries and between systemic capillaries and tissue cells.

1. **External respiration** is the exchange of gases between alveoli and pulmonary blood capillaries. This exchange converts **deoxygenated blood** into **oxygenated blood**.
2. **Internal respiration** is the exchange of gases between systemic blood capillaries and tissue cells, converting **oxygenated blood** into **deoxygenated blood**.
3. External and internal respiration depend on partial pressure differences, the surface area for gas exchange, the molecular weight and solubility of the gases, and the diffusion distance across the membrane.

See
• **Animation – Gas Exchange— Internal and External Respiration**

Hear
• **Figure 22.17 – Changes in Partial Pressures of Oxygen and Carbon Dioxide during External and Internal Respiration**

Do
• **Concepts and Connections – Gas Exchange**

Concept

Concept 22.7 Oxygen is primarily transported attached to hemoglobin, while carbon dioxide is transported in three different ways.

1. Most of the O_2 transported in the blood is bound to hemoglobin in the blood cells, and a small amount is dissolved in blood plasma. Hemoglobin has a heme portion that is capable of binding reversibly to a molecule of O_2, forming **oxyhemoglobin**. The amount of O_2 that binds to hemoglobin is primarily influenced by the partial pressure of oxygen; the higher the partial pressure of oxygen, the more O_2 combines with hemoglobin.
2. O_2 dissociates from hemoglobin more readily in an acidic environment, as the partial pressure of carbon dioxide rises in a tissue, and as temperature increases.
3. Deoxygenated blood returning to the lungs contains CO_2 dissolved in blood plasma, CO_2 combined with globin of hemoglobin as **carbaminohemoglobin**, and CO_2 incorporated into bicarbonate ions.
4. In the lungs, CO_2 dissociates from the globin portion of hemoglobin and diffuses into alveolar air. At the same time, O_2 diffuses from alveolar air and binds to the heme portion of hemoglobin, forming oxyhemoglobin.

See

- Animation – Gas Transport

Hear

- **Figure 22.22 –** Summary of Chemical Reactions That Occur During Gas Exchange

Do

- Exercise – Concentrate on Respiration
- Exercise – Oxygen Transport Tryout
- Exercise – Carbon Dioxide Transport Tryout
- Concepts and Connections – Carbon Dioxide Transport
- Concepts and Connections – Oxygen Transport

Concept 22.8 The basic rhythm of respiration is controlled by the respiratory center in the brain stem.

1. The **respiratory center** consists of a medullary rhythmicity area in the medulla oblongata and a pneumotaxic area and an apneustic area in the pons.
2. The **medullary rhythmicity area** has an **inspiratory area** that establishes the basic rhythm of breathing by stimulating contraction of the diaphragm and external intercostals. During forceful breathing, the **expiratory area** stimulates the internal intercostals and abdominal muscles to contract.
3. The **pneumotaxic area** inhibits the inspiratory area before the lungs become too full of air.
4. The **apneustic area** stimulates the inspiratory area to prolong inhalation.

See

- Anatomy Overview – Respiratory Control Center
- Animation – Regulation of Ventilation

Concept 22.9 Respiration may be modified by cortical influences, chemical stimuli, proprioceptor input, and the inflation reflex.

1. When metabolic demand for O_2 increases, the basic rhythm of breathing is modified by input from other brain regions, PNS receptors, and other factors.
2. The cerebral cortex has limited influence on the respiratory center to allow voluntary alteration of the pattern of breathing.
3. **Chemoreceptors** respond to changes in P_{O_2}, H^+, and P_{CO_2} in blood by stimulating the inspiratory area to alter breathing rate and depth.
4. During exercise, proprioceptors monitoring joint and muscle movement stimulate the inspiratory area to increase breathing rate and depth.
5. **Baroreceptors** detect stretch in the walls of bronchi and bronchioles. Overinflation of the lungs stimulates these receptors to inhibit the inspiratory and apneustic areas, resulting in exhalation.
6. Other factors affecting respiration include stimulation of the limbic system, body temperature, pain, stretching of the anal sphincter, irritation of airways, and changes in blood pressure.

See

- Anatomy Overview – Structures That Control Respiration
- Animation – Control of Ventilation Rate and Blood Chemistry

Do

- Exercise – Respiration and pH Reflex
- Concepts and Connections – Ventilation
- Concepts and Connections – Respiratory Rate

Concept	Resources
	See

Concept 22.10 Acid–base balance is maintained by controlling the H⁺ concentration of body fluids.

1. The acid–base balance of the body is maintained by controlling the H^+ concentration of body fluids. Removal of H^+ from the body depends on buffer systems, exhalation of carbon dioxide, and excretion of H^+ in the urine.

2. Buffers act to temporarily bind to, and remove, excess H^+ from solution to prevent rapid, drastic shifts in pH.

3. The **protein buffer system** utilizes blood proteins such as hemoglobin and albumin to buffer both acids and bases.

4. The **carbonic acid–bicarbonate buffer system** uses bicarbonate ions and carbonic acid to adjust blood pH.

5. The rate and depth of breathing contributes to the amount of CO_2 exhaled from the body and an alteration of the pH of body fluids. An increase in exhalation of carbon dioxide increases blood pH; a decrease in exhalation of CO_2 decreases blood pH.

6. Normal pH range of systemic arterial blood is 7.35–7.45. **Acidosis** is a systemic arterial blood pH below 7.35; its principal effect is depression of the central nervous system (CNS). **Alkalosis** is a systemic arterial blood pH above 7.45; its principal effect is overexcitability of the CNS.

Resources / See
- Animation – Acid–Base Imbalances
- Animation – Role of the Respiratory System in pH Regulation

Other Chapter Resources

For additional information on clinical applications, medical disorders, and medical terminology, see Chapter 11 in the *Clinical Connections* manual packaged with your text.

Within *WileyPLUS*
- Chapter Quiz
- Visual Anatomy
- Mastering Vocabulary
- Audio Glossary
- Homeostatic Imbalance: The Case of the Coughing Boy
- Homeostatic Imbalance: The Case of the Worried Smoker
- Homeostatic Imbalance: The Case of the Breathless Boy
- Homeostatic Imbalance: The Case of the Grunting Baby
- Web Links

• Understanding the Concepts

1. What functions do the respiratory and cardiovascular systems share?

2. Compare the functions of the conducting portion and respiratory portion of the respiratory system.

3. How does the larynx function in respiration and voice production?

4. Why are the right and left lungs slightly different in size and shape?

5. Describe the path of O_2 from the air space within the alveolus to the capillary blood.

6. What are the basic differences among pulmonary ventilation, external respiration, and internal respiration?

7. Do the lungs expand or contract as the thoracic cavity expands? Explain your answer.

8. Compare what happens during quiet and forceful inhalation and exhalation.

9. Would increased surfactant assist inhalation or exhalation? Both? Neither? Explain your answer.

10. Why can hemoglobin unload more oxygen as blood flows through capillaries of metabolically active tissues, such as skeletal muscle during exercise, than is unloaded at rest?

23

The Digestive System

"Mom, I think I need to have it done. My diabetes is getting worse, and I just am at my wit's end. I've tried dieting, changing the type of food I eat, exercise. Nothing is working." Chloe labored to breathe, heaving a sigh as she struggled to get up from her chair.

"Chloe, stomach stapling is dangerous! I just saw a television show about these two men who died from complications. You're only 28 years old and you have the kids to think about. What are they going to do if something happens to you?" her mom asked.

As she walked toward the refrigerator, Chloe replied, "I know there are risks, Mom, but I am desperate. I mean, what if I die

of a heart attack or a stroke anyway? I have an appointment with the bariatric surgeon Wednesday. I'm going to find out more about how it works and what the risks are."

Chloe has struggled with her weight ever since she was a child. After the birth of her children, she gained a significant amount of weight, and no dieting or exercise seems to help. She is morbidly obese, more than 100 pounds over her ideal weight. Her body is showing the effects of the weight gain; she suffers from both diabetes and high blood pressure, and is at high risk for stroke and heart attack. As you learn about the digestive system, we will follow Chloe's story and her struggle to regain her health.

INTRODUCTION

The food we eat contains a variety of nutrients—molecules needed for building new body tissues and repairing damaged tissues. Food is also vital to life as the source of energy that drives the chemical reactions occurring in every cell of our bodies. Most of the food we eat consists of molecules that are too large to be used by body cells. There-fore, foods are broken down into molecules that are small

Foods must be broken down into molecules that are small enough to enter body cells.

enough to enter body cells, a process known as **digestion**. The passage of these smaller molecules through cells lining the stomach and intestines into the blood and lymph is termed **absorption**. The organs that perform these functions—collectively called the **digestive system**—are the focus of this chapter. The digestive system, a tubular system that extends from the mouth to the anus, provides extensive surface area in contact with the external environment, essential for processing the food that we eat.

23.1 The GI tract is a continuous multilayered tube extending from the mouth to the anus.

Overview of the Digestive System

Two groups of organs compose the digestive system (Figure 23.1): the gastrointestinal tract and the accessory digestive organs. The **gastrointestinal (GI) tract** is a continuous tube that extends from the mouth to the anus through the thoracic and abdominopelvic cavities. Organs of the gastrointestinal tract include the mouth, most of the pharynx, esophagus, stomach, small intestine, and large intestine. The length of the GI tract in a cadaver is about 7–9 meters (23–29.5 ft). In a living person, it is much shorter, about 5–7 meters, or 16.5–23 ft, because the muscles along the walls of GI tract organs are in a state of sustained contraction. The **accessory digestive organs** include the teeth, tongue, salivary glands, liver,

gallbladder, and pancreas. Teeth aid in the physical breakdown of food, and the tongue assists in chewing and swallowing. The other accessory digestive organs, however, never come into direct contact with food. The secretions that they produce flow into the GI tract through ducts and aid in the chemical breakdown of food.

Overall, the digestive system performs six basic processes:

- *Ingestion*. This process involves taking foods and liquids into the mouth (eating).

- *Secretion*. Each day, cells within the walls of the GI tract and accessory digestive organs secrete a total of about 7 liters of water, acid, buffers, and enzymes into the lumen (interior space) of the tract.

Figure 23.1   Organs of the digestive system.

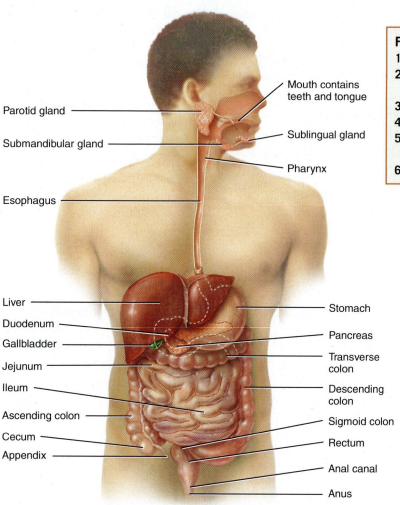

Parotid gland
Submandibular gland
Esophagus
Liver
Duodenum
Gallbladder
Jejunum
Ileum
Ascending colon
Cecum
Appendix

Mouth contains teeth and tongue
Sublingual gland
Pharynx
Stomach
Pancreas
Transverse colon
Descending colon
Sigmoid colon
Rectum
Anal canal
Anus

Functions of the Digestive System
1. Ingests food (takes food into the mouth).
2. Secretes water, acid, buffers, and enzymes into the lumen of the GI tract.
3. Mixes and propels food through the GI tract.
4. Digests food (breaks it down mechanically and chemically).
5. Absorbs digested products from the GI tract into the blood and lymph.
6. Eliminates feces from the GI tract.

Organs of the gastrointestinal tract are the mouth, pharynx, esophagus, stomach, small intestine, and large intestine. Accessory digestive organs include the teeth, tongue, salivary glands, liver, gallbladder, and pancreas.

- *Mixing and propulsion*. Alternating contractions and relaxations of smooth muscle in the walls of the GI tract mix food and secretions and propel them toward the anus. This capability of the GI tract to mix and move material along its length is called **motility**.

- *Digestion*. Mechanical and chemical processes break down ingested food into small molecules. In **mechanical digestion**, the teeth cut and grind food before it is swallowed, and then smooth muscles of the stomach and small intestine churn the food. As a result, food molecules become dissolved and thoroughly mixed with digestive enzymes. In **chemical digestion** the large carbohydrate, lipid, protein, and nucleic acid molecules in food are broken down into smaller molecules. Digestive enzymes produced by the salivary glands, tongue, stomach, pancreas, and small intestine catalyze these catabolic reactions. A few substances in food can be absorbed without chemical digestion. These include vitamins, ions, cholesterol, and water.

- *Absorption*. The entrance of ingested and secreted fluids, ions, and the products of digestion into the epithelial cells lining the lumen of the GI tract is called absorption. The absorbed substances pass into blood or lymph and circulate to cells throughout the body.

- *Defecation*. Wastes, indigestible substances, bacteria, cells sloughed from the lining of the GI tract, and digested materials that were not absorbed in their journey through the digestive tract leave the body through the anus in a process called **defecation**. The eliminated material is termed **feces**.

Layers of the GI Tract

The wall of the GI tract from the lower esophagus to the anal canal has the same basic, four-layered arrangement of tissues. The four layers of the tract, from deep to superficial, are the mucosa, submucosa, muscularis, and serosa (Figure 23.2).

Figure 23.2 Layers of the gastrointestinal tract. Variations of this basic plan may be seen in the esophagus (Figure 23.9), stomach (Figure 23.12), small intestine (Figure 23.18), and large intestine (Figure 23.23).

The four layers of the GI tract, from deep to superficial, are the mucosa, submucosa, muscularis, and serosa.

Mucosa

The **mucosa**, or inner lining of the GI tract, is a mucous membrane. It is composed of a layer of epithelium in direct contact with the contents of the GI tract lumen that absorbs the digested components, a layer of areolar connective tissue called the lamina propria, and a thin layer of smooth muscle (muscularis mucosae).

1.   The **epithelium** in the mouth, pharynx, esophagus, and anal canal is mainly nonkeratinized stratified squamous epithelium that serves a protective function. Simple columnar epithelium, which functions in secretion and absorption, lines the stomach and intestines. Located among the absorptive epithelial cells are exocrine cells that secrete mucus and fluid into the lumen of the tract, and several types of endocrine cells that secrete hormones.

2.   The **lamina propria** (*lamina* = thin, flat plate; *propria* = one's own) is areolar connective tissue containing many blood and lymphatic vessels, which are the routes by which nutrients absorbed into the GI tract reach the other tissues of the body. This layer supports the epithelium and binds it to the muscularis mucosae (discussed next). The lamina propria also contains prominent lymphatic nodules containing immune system cells that protect against entry of pathogens through the GI tract.

3.   A thin layer of smooth muscle fibers called the **muscularis mucosae** creates folds in the mucous membrane of the stomach and small intestine; the folds increase the surface area for digestion and absorption. Movements of the muscularis mucosae ensure that all absorptive cells are fully exposed to the contents of the GI tract.

Submucosa

The **submucosa** consists of areolar connective tissue that binds the mucosa to the muscularis. It contains many blood and lymphatic vessels that receive absorbed food molecules. The submucosa may contain glands and lymphatic tissue. Also located in the submucosa is an extensive network of neurons known as the submucosal plexus (to be described shortly).

Muscularis

The **muscularis** of the mouth, pharynx, and superior and middle parts of the esophagus contains *skeletal muscle* that produces voluntary swallowing. Skeletal muscle also forms the external

Figure 23.3   Relationship of the peritoneal folds to one another and to organs of the digestive system. The size of the peritoneal cavity in (a) is exaggerated for emphasis.

Midsagittal plane

Diaphragm
Liver
LESSER OMENTUM
Pancreas
Stomach
Duodenum
MESOCOLON
Transverse colon
Jejunum
MESENTERY
Ileum
GREATER OMENTUM
Sigmoid colon
PARIETAL PERITONEUM
Uterus
Urinary bladder
VISCERAL PERITONEUM
Rectum
PERITONEAL CAVITY
Pubic symphysis

POSTERIOR   ANTERIOR

(a) Midsagittal section showing the peritoneal folds

anal sphincter, which permits voluntary control of defecation. Throughout the rest of the tract, the muscularis consists of *smooth muscle* that is generally found in two sheets: an inner sheet of circular fibers and an outer sheet of longitudinal fibers. Involuntary contractions of the smooth muscle help break down food, mix it with digestive secretions, and propel it along the tract. Between the layers of the muscularis is a second plexus of neurons—the myenteric plexus (to be described shortly).

Serosa

Those portions of the GI tract that are suspended in the abdominopelvic cavity have a superficial layer called the **serosa**. As its name implies, the serosa is a serous membrane composed of areolar connective tissue and simple squamous epithelium. The serosa

secretes a slippery, watery fluid that allows the tract to glide easily against other organs. Inferior to the diaphragm, the serosa is also called the visceral peritoneum because it forms a portion of the peritoneum, which we examine next.

Peritoneum

The **peritoneum** (per´-i-tō-NĒ-um; *peri-* = around) is the largest serous membrane of the body; it consists of a layer of simple squamous epithelium with an underlying supporting layer of connective tissue. The peritoneum is divided into the **parietal peritoneum**, which lines the wall of the abdominopelvic cavity, and the **visceral peritoneum**, which covers some of the organs in the cavity and is their serosa (Figure 23.3a). The slim space containing

(b) Anterior view

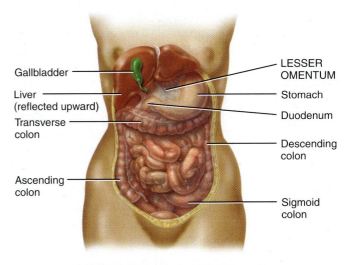

(c) Lesser omentum, anterior view (liver and gallbladder lifted)

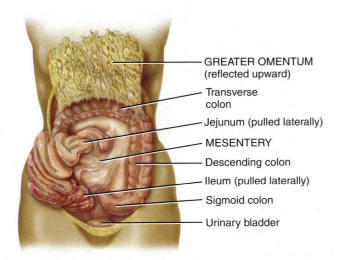

(d) Anterior view (greater omentum lifted and small intestine reflected to right side)

(e) Anterior view

The peritoneum is the largest serous membrane in the body.

lubricating serous fluid that is between the parietal and visceral portions of the peritoneum is called the **peritoneal cavity**.

Some organs lie on the posterior abdominal wall and are covered by peritoneum only on their anterior surfaces; they are not in the peritoneal cavity. Such organs, including the kidneys and pancreas, are said to be **retroperitoneal** (*retro-* = behind).

The peritoneum contains large folds that weave between the viscera. The folds bind the organs to one another and to the walls of the abdominal cavity. They also contain blood vessels, lymphatic vessels, and nerves that supply the abdominal organs. There are five major peritoneal folds:

- The **greater omentum** (ō-MEN-tum = fat skin), the largest peritoneal fold, drapes over the transverse colon and coils of the small intestine like a "fatty apron" (Figure 23.3a,d). The greater omentum is a double sheet that folds back on itself, giving it a total of four layers. From attachments along the stomach and duodenum, the greater omentum drapes down over the small intestine, then turns upward and attaches to the transverse colon. The greater omentum normally contains a considerable amount of adipose tissue. Its adipose tissue content can greatly expand with weight gain, giving rise to the characteristic "beer belly" seen in some overweight individuals. The many lymph nodes of the greater omentum contribute macrophages and antibody-producing plasma cells that help combat and contain infections of the GI tract.

- The **falciform ligament** (FAL-si-form; *falc-* = sickle-shaped) attaches the liver to the anterior abdominal wall and diaphragm (Figure 23.3b). The liver is the only digestive organ that is attached to the anterior abdominal wall.

- The **lesser omentum** arises as an anterior fold in the serosa of the stomach and duodenum, and it suspends the stomach and duodenum from the liver (Figure 23.3a,c).

- A fan-shaped fold of the peritoneum, called the **mesentery** (MEZ-en-ter´-ē; *mes-* = middle), binds the small intestine to the posterior abdominal wall (Figure 23.3a,d). It extends from the posterior abdominal wall to wrap around the small intestine and then returns to its origin, forming a double-layered structure.

- Two separate folds of peritoneum, the **mesocolon** (mez´-ō-KŌ-lon), bind the large intestine to the posterior abdominal wall (Figure 23.3a).Together, the mesentery and mesocolon hold the intestines loosely in place, allowing movement as muscular contractions mix and move the luminal contents along the GI tract.

Neural Innervation of the GI Tract

The gastrointestinal tract is regulated by an intrinsic set of nerves known as the enteric nervous system and by an extrinsic set of nerves that are part of the autonomic nervous system (ANS).

Enteric Nervous System

We first introduced you to the **enteric nervous system (ENS)**, the "brain of the gut," in Chapter 12. It consists of about 100 million neurons that extend from the esophagus to the anus. Enteric neurons are arranged into two networks: the myenteric plexus and submucosal plexus (see Figure 23.2). The **myenteric plexus** (*myo-* = muscle) is located between the longitudinal and circular smooth muscle layers of the muscularis. The **submucosal plexus** is found within the submucosa. Enteric plexuses consist of sensory neurons, interneurons, and motor neurons (Figure 23.4). Some enteric sensory neurons are chemoreceptors, receptors that are activated by the presence of certain chemicals in food located in the lumen of a GI tract organ. Other sensory neurons are stretch receptors that are activated when food distends (stretches) the wall of a GI tract organ. Enteric sensory neurons transmit input from the epithelium of the mucosa to interneurons in the ENS, ANS, or central nervous systems (CNS), informing them about the contents and degree of distension of the GI tract. These interneurons process incoming sensory information and subsequently activate or inhibit outgoing enteric motor neurons. Motor neurons of the myenteric plexus control GI tract motility (movement). Motor neurons of the submucosal plexus control secretions of GI tract organs.

Autonomic Nervous System

Although the enteric nervous system can function independently, it is subject to regulation by the autonomic nervous system. Para sympathetic nerves arise from the vagus (X) nerves and the sacral

Figure 23.4   Organization of the enteric nervous system.

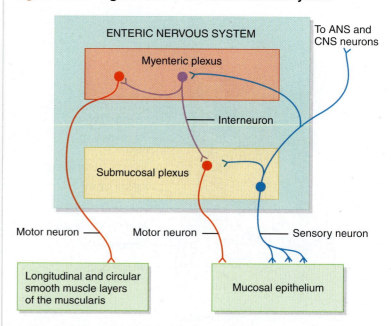

The enteric nervous system consists of neurons arranged into the myenteric and submucosal plexuses.

spinal cord. Sympathetic nerves that supply the GI tract arise from the lumbar spinal cord. Both parasympathetic nerves and sympathetic nerves form neural connections with the ENS. Sympathetic and parasympathetic neurons may synapse with the myenteric plexus and the submucosal plexus or bypass enteric neurons and synapse directly with smooth muscle and glands within the GI tract wall. In general, stimulation of the GI tract by sympathetic nerves decreases GI secretion and motility by inhibiting the activity of enteric neurons, while innervation by parasympathetic nerves increases GI secretion and motility by increasing the activity of enteric neurons. Emotions such as anger, fear, and anxiety may slow digestion as the central nervous system utilizes sympathetic nerves to inhibit GI tract glands and smooth muscle.

Checkpoint

1. Give the name and function of each of the four layers of the gastrointestinal tract.
2. Where along the GI tract is the muscularis composed of skeletal muscle? Is control of this skeletal muscle voluntary or involuntary?
3. What are the attachment sites of the mesentery, mesocolon, falciform ligament, lesser omentum, and greater omentum?
4. What are the functions of the myenteric and submucosal plexuses?

23.2 The mouth lubricates and begins digestion of food, and maneuvers it to the pharynx for swallowing.

The **mouth**, also referred to as the **oral cavity**, is formed by the cheeks, hard and soft palates, and tongue (Figure 23.5). The **cheeks**, muscular structures covered externally by skin and internally by nonkeratinized stratified squamous epithelium, form the lateral walls of the mouth. The buccinators and connective tissue lie between the skin and mucous membranes of the cheeks. The anterior portions of the cheeks end at the lips.

Figure 23.5 Structures of the mouth (oral cavity).

Hard palate
Soft palate
Uvula
Cheek
Molars
Premolars
Cuspid
Incisors
Vestibule

Superior lip (pulled upward)
Superior labial frenulum
Gingivae
Palatoglossal arch
Fauces
Palatopharyngeal arch
Palatine tonsil
Tongue
Lingual frenulum
Opening of duct of submandibular gland
Gingivae
Inferior labial frenulum
Inferior lip (pulled down)

Anterior view

The mouth is formed by the cheeks, hard and soft palates, and tongue.

The **lips**, or **labia** (= fleshy borders), are fleshy folds surrounding the opening of the mouth. They contain the orbicularis oris and are covered externally by skin and internally by a mucous membrane. The inner surface of each lip is attached to its corresponding gum by a midline fold of mucous membrane called the **labial frenulum** (LĀ-bē-al FREN-ū-lum; *frenulum* = small bridle). During chewing, contraction of the buccinators in the cheeks and orbicularis oris in the lips helps keep food between the upper and lower teeth.

The **vestibule** (= entrance to a canal) of the mouth is a space bounded externally by the cheeks and lips and internally by the gums and teeth. The **oral cavity proper** is a space that extends from the gums and teeth to the **fauces** (FAW-sēs; *fauc* = throat), the opening between the mouth and the oropharynx (throat).

The **palate** forms the roof of the mouth and separates the oral and nasal cavities. This important structure makes it possible to chew and breathe at the same time. The anterior **hard palate** is formed by the maxillae and palatine bones and is covered by a mucous membrane. The posterior **soft palate** is an arch-shaped muscular partition between the oropharynx and nasopharynx that is lined with mucous membrane.

Hanging from the free border of the soft palate is a conical muscular process called the **uvula** (Ū-vū-la = little grape). During swallowing, the soft palate and uvula are drawn superiorly, closing off the nasopharynx and preventing swallowed foods and liquids from entering the nasal cavity. Lateral to the base of the uvula are two muscular folds that run down the lateral sides of the soft palate: Anteriorly, the **palatoglossal arch** extends to the base of the tongue; posteriorly, the **palatopharyngeal arch** (PAL-a-tō-fa-rin´-jē-al) extends to the side of the pharynx. The palatine tonsils are situated between the arches, and the lingual tonsils are situated at the base of the tongue. At the posterior border of the soft palate, the mouth opens into the oropharynx through the fauces (Figure 23.5).

Tongue

The **tongue** is composed of skeletal muscle covered with mucous membrane. Together with its associated muscles, it forms the floor of the mouth (Figure 23.5). The tongue is divided into symmetrical lateral halves by a median septum that extends its entire length, and it is attached inferiorly to the hyoid bone, temporal bone, and mandible. Each half of the tongue consists of an identical complement of extrinsic and intrinsic muscles.

The **extrinsic muscles** of the tongue originate outside the tongue (attach to bones in the area) and insert into connective tissues in the tongue (Figure 11.6). The extrinsic muscles move the tongue from side to side and in and out to maneuver food for chewing, shape the food into a rounded mass, and force the food to the back of the mouth for swallowing. They also form the floor of the mouth and hold the tongue in position. The **intrinsic muscles** originate inside and insert into connective tissues within the tongue. They alter the shape and size of the tongue for speech and swallowing. The **lingual frenulum** (*lingua* = the tongue), a fold of mucous membrane in the midline of the undersurface of the tongue, is attached to the floor of the mouth and limits the movement of the tongue posteriorly (Figures 23.5 and 23.6).

The dorsum (upper surface) and lateral surfaces of the tongue are covered with **papillae** (pa-PIL-ē = nipple-shaped projections), projections of the lamina propria covered with stratified squamous epithelium (see Figure 16.4). Many papillae contain taste buds, the receptors for gustation (taste). Some papillae lack taste buds, but they contain receptors for touch and increase friction between the tongue and food, making it easier for the tongue to move food in the mouth. The different types of taste buds are described in detail in Chapter 16.

Lingual glands in the lamina propria of the tongue secrete both mucus and a watery serous fluid that contains the enzyme lingual lipase, which begins the digestion of triglycerides.

Salivary Glands

A **salivary gland** is a gland that releases a secretion called saliva into the oral cavity. Ordinarily, just enough saliva is secreted to keep the mucous membranes of the mouth and pharynx moist and to cleanse the mouth and teeth. When food enters the mouth, however, secretion of saliva increases, and it lubricates, dissolves, and begins the chemical breakdown of the food.

Three pairs of major salivary glands release saliva into ducts that lead to the mouth (Figure 23.6). The **parotid glands** (*par-* = near; *-ot-* = ear) are located inferior and anterior to the ears, between the skin and the masseter. Each secretes saliva into the oral cavity via a **parotid duct** that opens into the vestibule opposite the second upper molar tooth. The **submandibular glands** are found beneath the base of the tongue in the posterior part of the floor of the mouth. Their ducts, the **submandibular ducts**, enter the mouth proper lateral to the lingual frenulum. The **sublingual glands** are superior to the submandibular glands. Their ducts, the **sublingual ducts**, open into the floor of the mouth in the oral cavity proper.

Figure 23.6 The three major salivary glands—parotid, sublingual, and submandibular.

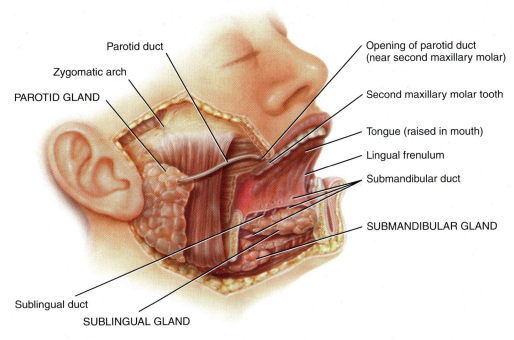

Parotid duct

Zygomatic arch

PAROTID GLAND

Opening of parotid duct
(near second maxillary molar)

Second maxillary molar tooth

Tongue (raised in mouth)

Lingual frenulum

Submandibular duct

SUBMANDIBULAR GLAND

Sublingual duct

SUBLINGUAL GLAND

Saliva lubricates and dissolves foods and begins the chemical breakdown of carbohydrates and lipids.

Composition and Functions of Saliva

Chemically, **saliva** is 99.5% water and 0.5% solutes. Among the solutes are ions, including sodium, potassium, chloride, bicarbonate, and phosphate. Also present are mucus to lubricate food, immunoglobulin A (antibodies), the bacteriolytic enzyme lysozyme to inhibit the growth of oral bacteria, and salivary amylase, a digestive enzyme that acts on starch.

The water in saliva provides a medium for dissolving foods so that they can be tasted by gustatory receptors and so that digestive reactions can begin. Salivary amylase starts the breakdown of starch. Mucus lubricates the food so it can be moved around easily in the mouth, formed into a ball, and swallowed. Immunoglobulin A (IgA) prevents attachment of microbes so they cannot penetrate the epithelium, and the enzyme lysozyme kills bacteria; however, these substances are not present in large enough quantities to eliminate all oral bacteria.

Salivation

The secretion of saliva, or **salivation** (sal-i-VĀ-shun), is controlled by the autonomic nervous system. Normally, parasympathetic stimulation promotes continuous secretion of a moderate amount of saliva, which keeps the mucous membranes moist and lubricates the movements of the tongue and lips during speech. The saliva is then swallowed and helps moisten the esophagus. Eventually, most components of saliva are reabsorbed, which prevents fluid loss. Sympathetic stimulation dominates during stress, resulting in dryness of the mouth. If the body becomes dehydrated, the salivary glands stop secreting saliva to conserve water; the resulting dryness in the mouth contributes to the sensation of thirst. Drinking not only restores the homeostasis of body water but also moistens the mouth.

The feel and taste of food also are potent stimulators of salivary gland secretions. Chemicals in the food stimulate receptors in taste buds on the tongue, and impulses are conveyed from the taste buds to the brain stem. Returning parasympathetic impulses in the facial (VII) and glossopharyngeal (IX) nerves stimulate the secretion of saliva. Saliva continues to be secreted heavily for some time after food is swallowed; this flow of saliva washes out the mouth and dilutes and buffers the remnants of irritating chemicals such as that tasty (but hot!) salsa. The smell, sight, sound, or thought of food may also stimulate secretion of saliva.

Teeth

Teeth are accessory digestive organs located in sockets of the alveolar processes of the mandible and maxillae (Figure 23.7). The alveolar processes are covered by the **gingivae** (JIN-ji-vē), or gums, which extend slightly into each socket. The sockets are lined by the **periodontal ligament** (-*odont*- = tooth), which consists of dense fibrous connective tissue that anchors the teeth to the socket walls and acts as a shock absorber during chewing.

A typical **tooth** has three major external regions: the crown, root, and neck. The **crown** is the visible portion above the level of the gingiva. Embedded in the socket are one to three **roots**. The **neck** is the constricted junction of the crown and root near the gum line.

Each tooth also consists of three major internal components: enamel, dentin, and a pulp cavity. **Enamel** forms the outer surface of a tooth. It consists primarily of calcium phosphate and calcium carbonate. Enamel is harder than bone because of its higher content of calcium salts (about 95% of dry weight). In fact, enamel is the hardest substance in the body. It protects the tooth from the wear and tear of chewing and from acids that can easily dissolve dentin.

Most of the tooth is made up of **dentin**, a substance found deep to the enamel. Dentin consists of a calcified connective tissue that gives the tooth its basic shape and rigidity. It is harder than bone because of its higher content of calcium salts (70% of dry weight) but not as hard as enamel. The dentin of the root is covered by **cementum**, another bonelike substance, which attaches the root to the periodontal ligament.

The dentin of a tooth encloses a space. The enlarged part of the space, the **pulp cavity**, lies within the crown and is filled with **pulp**, a connective tissue containing blood vessels, nerves, and lymphatic vessels. Narrow extensions of the pulp cavity, called **root canals**, run through the root of the tooth. Each root canal has an opening at its base, the **apical foramen**, through which blood vessels bring nourishment, lymphatic vessels offer protection, and nerves provide sensation.

Humans have two **dentitions**, or sets of teeth: deciduous and permanent. The first of these—the **deciduous teeth** (*deciduous* = falling out), also called **primary teeth**, **milk teeth**, or **baby teeth**—begin to erupt at about 6 months of age. Approximately two teeth appear each month thereafter, until all 20 are present (Figure 23.8a). The incisors, which are closest to the midline, are chisel-shaped and adapted for cutting into food. They are referred to as either **central** or **lateral incisors** based on their position. Next to the incisors, moving posteriorly, are the **cuspids (canines)**, which have a pointed surface, called a *cusp*, to tear and shred food. Incisors and cuspids have only one root apiece. Posterior to the cuspids lie the **first** and **second molars**, which have four cusps. Maxillary (upper) molars have three roots; mandibular (lower) molars have two roots. The molars crush and grind food to prepare it for swallowing.

All the deciduous teeth are lost—generally between ages 6 and 12 years—and are replaced by the **permanent (secondary) teeth** (Figure 23.8b). The permanent dentition contains 32 teeth that erupt between age 6 and adulthood. The pattern resembles the deciduous dentition, with the following exceptions. The deciduous molars are replaced by the **first** and **second premolars (bicuspids)**, which have two cusps and one root (upper first premolars have two roots) and are used for crushing and grinding. The permanent molars, which erupt into the mouth posterior to the premolars, do not replace any deciduous teeth and erupt as the jaw grows to accommodate them—the **first molars** at age 6 (six-year molars), the **second molars** at age 12 (12-year molars), and the **third molars (wisdom teeth)** after age 17 or not at all.

Figure 23.7   A typical tooth and surrounding structures.

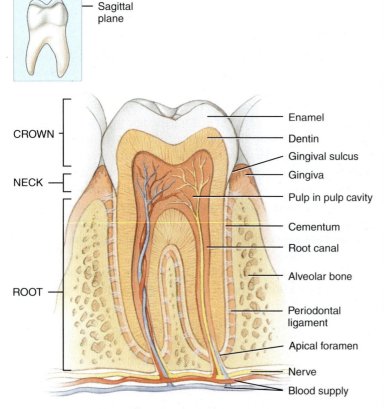

Sagittal plane

CROWN
NECK
ROOT

Enamel
Dentin
Gingival sulcus
Gingiva
Pulp in pulp cavity
Cementum
Root canal
Alveolar bone
Periodontal ligament
Apical foramen
Nerve
Blood supply

Sagittal section of a mandibular (lower) molar

🔑 Teeth are anchored in sockets of the alveolar processes of the mandible and maxillae.

Figure 23.8 **Dentitions and times of eruptions (indicated in parentheses).** A designated letter (deciduous teeth) or number (permanent teeth) uniquely identifies each tooth.

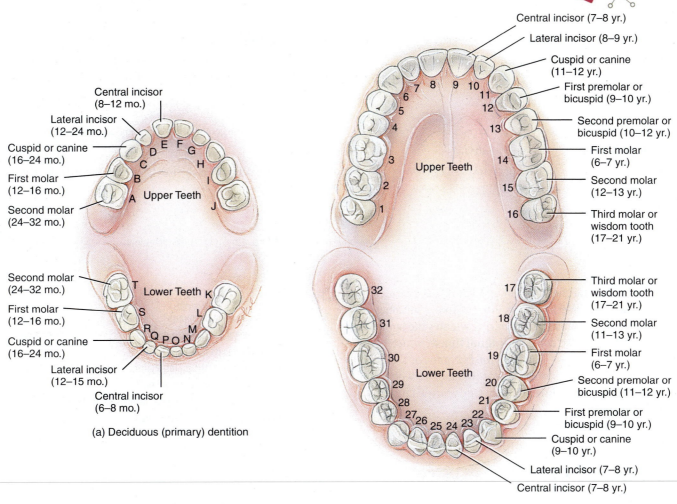

(a) Deciduous (primary) dentition

(b) Permanent (secondary) dentition

🔑 There are 20 teeth in a complete deciduous set and 32 teeth in a complete permanent set.

Often the human jaw does not have enough room posterior to the second molars to accommodate the eruption of the third molars. In this case, the third molars remain embedded in the alveolar bone and are said to be "impacted." They often cause pressure and pain and must be removed surgically. In some people, third molars may be dwarfed in size or may not develop at all.

Mechanical and Chemical Digestion in the Mouth

Mechanical digestion in the mouth results from chewing, or **mastication** (mas´-ti-KĀ-shun = to chew), in which food is manipulated by the tongue, ground by the teeth, and mixed with saliva. As a result, the food is reduced to a soft, flexible, easily swallowed mass called a **bolus** (= lump). Food molecules begin to dissolve in the water in saliva, an important activity because enzymes can only react with food molecules in a liquid medium.

Two enzymes, salivary amylase and lingual lipase, contribute to chemical digestion in the mouth. **Salivary amylase** initiates the breakdown of starch. Dietary carbohydrates are either monosaccharide and disaccharide sugars or complex polysaccharides such as starches. Most of the carbohydrates we eat are starches, but only monosaccharides (simple sugars such as glucose, fructose, or galactose) can be absorbed into the bloodstream. Thus, ingested disaccharides and starches must be broken down into monosaccharides. Salivary amylase breaks certain chemical bonds between glucose units in starches, converting long-chain polysaccharides to disaccharides (two simple sugars) and short-chain glucose polymers called α-dextrins. Even though food is usually swallowed

TABLE 23.1

Digestive Activities in the Mouth

STRUCTURE	ACTIVITY	RESULT
Cheeks and lips	Keep food betwen teeth.	Foods uniformly chewed during mastication.
Salivary glands	Secrete saliva.	Lining of mouth and pharynx moistened and lubricated. Saliva softens, moistens, and dissolves food and cleanses mouth and teeth. Salivary amylase splits starch into smaller fragments.
Tongue		
Extrinsic tongue muscles	Move tongue in and out.	Food maneuvered for mastication, shaped into bolus, and maneuvered for swallowing.
Intrinsic tongue muscles	Alter shape of tongue.	Swallowing and speech.
Taste buds	Serve as receptors for gustation (taste) and presence of food in mouth.	Secretion of saliva stimulated by nerve impulses from taste buds to brain stem to salivary glands.
Lingual glands	Secrete lingual lipase.	Triglycerides broken down.
Teeth	Cut, tear, and pulverize food.	Solid foods reduced to smaller particles for swallowing.

too quickly for all the starches to be reduced to disaccharides in the mouth, salivary amylase in the swallowed food continues to act on the starches for about another hour, at which time stomach acids inactivate it. Saliva also contains **lingual lipase**, which is secreted by glands in the tongue. This enzyme becomes activated in the acidic environment of the stomach and thus starts to work after food is swallowed. It breaks down dietary triglycerides into fatty acids and diglycerides.

Table 23.1 summarizes the digestive activities in the mouth.

Checkpoint

5.   Which structures form the mouth?
6.   What is the name of the cone-shaped process that hangs down from the roof of your mouth? What is its function?
7.   How are the major salivary glands distinguished on the basis of location?
8.   How is the secretion of saliva regulated?
9.   Which functions do incisors, cuspids, premolars, and molars perform?
10.   Define mastication, using the term *bolus*.

23.3 Swallowing consists of voluntary oral, involuntary pharyngeal, and involuntary esophageal stages.

Pharynx

When food is first swallowed, it passes from the mouth into the **pharynx** (= throat), a funnel-shaped tube that extends from the internal nares to the esophagus posteriorly and to the larynx anteriorly (see Figure 22.2b). The pharynx is composed of skeletal muscle and lined by mucous membrane, and is divided into three parts: the nasopharynx, oropharynx, and laryngopharynx. The nasopharynx functions only in respiration, but both the oropharynx and laryngopharynx have digestive as well as respiratory functions. Swallowed food passes from the mouth into the oropharynx and laryngopharynx; muscular contractions of these areas help propel food into the esophagus and then into the stomach.

Esophagus

The **esophagus** (e-SOF-a-gus = eating gullet) is a collapsible muscular tube, about 25 cm (10 in.) long, that lies posterior to the trachea. The esophagus begins at the inferior end of the laryngopharynx, passes through the mediastinum and then through an opening in the diaphragm called the **esophageal hiatus**, and ends in the superior portion of the stomach (see Figure 23.1). Sometimes, part of the stomach protrudes above the diaphragm through the esophageal hiatus. This condition is termed a **hiatal hernia** (HER-nē-a).

Histology of the Esophagus

The **mucosa** of the esophagus consists of nonkeratinized stratified squamous epithelium, lamina propria (areolar connective tissue), and a muscularis mucosae (smooth muscle) (Figure 23.9). The stratified squamous epithelium associated with the lips, mouth, tongue, oropharynx, laryngopharynx, and esophagus affords considerable protection against abrasion and wear and tear from food particles that are chewed, mixed with secretions, and swallowed. The **submucosa** contains areolar connective tissue, blood vessels, and mucous glands. The **muscularis** transitions from skeletal muscle in the superior portion of the esophagus to smooth muscle in the inferior portion. At each end of the esophagus, the muscularis becomes slightly more prominent and forms two sphincters—the **upper esophageal sphincter** (e-sof´-a-JĒ-al), which consists of skeletal muscle, and the **lower esophageal sphincter**, which

Figure 23.9 Histology of the esophagus.

Transverse plane

Lumen of esophagus

Mucosa:
Nonkeratinized stratified squamous epithelium
Lamina propria
Muscularis mucosae

Submucosa

Muscularis (circular layer)

Muscularis (longitudinal layer)

Adventitia

LM 20x

Wall of the esophagus

The esophagus secretes mucus and transports food to the stomach.

consists of smooth muscle. The upper esophageal sphincter regulates the movement of food from the pharynx into the esophagus; the lower esophageal sphincter regulates the movement of food from the esophagus into the stomach. The superficial layer of the esophagus is known as the **adventitia** (ad-ven-TISH-a) rather than the serosa because the connective tissue attaches to the surrounding structures of the mediastinum, through which it passes.

Physiology of the Esophagus

The esophagus secretes mucus and transports food into the stomach. It does not produce digestive enzymes, and it does not carry on absorption.

Deglutition

The movement of food from the mouth into the stomach is achieved by the act of swallowing, or **deglutition** (dē-gloo-TISH-un) (Figure 23.10). Deglutition is facilitated by the secretion of saliva and mucus and involves the mouth, pharynx, and esophagus. Swallowing occurs in three stages: (1) the voluntary stage, in which the bolus is passed into the oropharynx; (2) the pharyngeal stage, the involuntary passage of the bolus through the pharynx into the esophagus; and (3) the esophageal stage, the involuntary passage of the bolus through the esophagus into the stomach.

Figure 23.10 Deglutition (swallowing).
During the pharyngeal stage of deglutition (b), the tongue rises against the palate, the nasopharynx is closed off, the larynx rises, the epiglottis seals off the larynx, and the bolus is passed into the esophagus. During the esophageal stage of deglutition (c), food moves through the esophagus into the stomach via peristalsis.

Nasopharynx
Hard palate
Soft palate
Uvula
Oropharynx
Epiglottis
Laryngopharynx
Larynx
Esophagus

Bolus
Tongue

(a) Position of structures before swallowing

(b) During the pharyngeal stage of swallowing

(continues)

Figure 23.10 (conitinued)

(c) Anterior view of frontal sections of peristalsis in esophagus

Deglutition moves food from the mouth into the stomach.

TABLE 23.2		
Digestive Activities in the Pharynx and Esophagus		
STRUCTURE	ACTIVITY	RESULT
Pharynx	Pharyngeal stage of deglutition.	Moves bolus from oropharynx to laryngopharynx and into esophagus; closes air passageways.
Esophagus	Relaxation of upper esophageal sphincter.	Permits entry of bolus from laryngopharynx into esophagus.
	Esophageal stage of deglutition (peristalsis).	Pushes bolus down esophagus.
	Relaxation of lower esophageal sphincter.	Permits entry of bolus into stomach.
	Secretion of mucus.	Lubricates esophagus for smooth passage of bolus.

Swallowing starts when the bolus is forced to the back of the mouth and into the oropharynx by the movement of the tongue upward and backward against the palate; these actions constitute the **voluntary stage** of swallowing. With the passage of the bolus into the oropharynx, the involuntary **pharyngeal stage** of swallowing begins (Figure 23.10b). The bolus stimulates receptors in the oropharynx, which send impulses to the **deglutition center** in the medulla oblongata and pons of the brain stem. The returning impulses cause the soft palate and uvula to move upward to close off the nasopharynx, which prevents swallowed foods and liquids from entering the nasal cavity. In addition, the epiglottis closes off the opening to the larynx, which prevents the bolus from entering the rest of the respiratory tract. The bolus moves through the oropharynx and the laryngopharynx. Once the upper esophageal sphincter relaxes, the bolus moves into the esophagus.

The **esophageal stage** of swallowing begins once the bolus enters the esophagus. During this phase, **peristalsis** (per´-i-STAL-sis; -*stalsis* = constriction), a progression of coordinated contractions and relaxations of the circular and longitudinal layers of the muscularis, pushes the bolus onward (Figure 23.10c). In the section of the esophagus just superior to the bolus, the circular muscle fibers contract, constricting the esophageal wall and squeezing the bolus toward the stomach. Meanwhile, longitudinal fibers inferior to the bolus also contract, which shortens this inferior section and pushes its walls outward so it can receive the bolus. The contractions are repeated in waves that push the food toward the stomach. As the bolus approaches the end of the esophagus, the lower esophageal sphincter relaxes and the bolus moves into the stomach. Along the way, mucus secreted by esophageal glands lubricates the bolus and reduces friction. The passage of solid or semisolid food from the mouth to the stomach takes 4 to 8 seconds; very soft foods and liquids pass through in about 1 second.

Table 23.2 summarizes the digestive activities of the pharynx and esophagus.

Checkpoint

11. To which two organ systems does the pharynx belong?
12. In which layers of the esophagus are the glands located that secrete lubricating mucus?
13. What are the functions of the upper and lower esophageal sphincters?
14. What occurs during the oral, pharyngeal, and esophageal stages of swallowing?
15. Is swallowing a voluntary or involuntary action?
16. Does peristalsis "push" or "pull" food along the gastrointestinal tract?

23.4 The stomach mechanically breaks down the bolus and mixes it with gastric secretions.

The **stomach** (Figure 23.11) is a J-shaped enlargement of the GI tract in the abdomen directly inferior to the diaphragm (see Figure 23.1). The stomach connects the esophagus to the duodenum, the first part of the small intestine. Because a meal can be eaten much more quickly than the intestines can digest and absorb it, one of the functions of the stomach is to serve as a mixing chamber and holding reservoir. At appropriate intervals after food is ingested, the stomach forces a small quantity of material into the first por-

Figure 23.11 **External and internal anatomy of the stomach.**

Esophagus

Lower esophageal sphincter

CARDIA

BODY

Lesser curvature

PYLORUS

Duodenum

Pyloric sphincter

PYLORIC CANAL

Rugae of mucosa

PYLORIC ANTRUM

FUNDUS

Serosa

Muscularis:

Longitudinal layer

Circular layer

Oblique layer

Greater curvature

(a) Anterior view of regions of stomach

Functions of the Stomach
1. Mixes saliva, food, and gastric juice to form chyme.
2. Serves as a reservoir for food before release into small intestine.
3. Secretes gastric juice, which contains HCl (kills bacteria and denatures protein), pepsin (begins the digestion of proteins), intrinsic factor (aids absorption of vitamin B_{12}), and gastric lipase (aids digestion of triglycerides).
4. Secretes gastrin into blood.

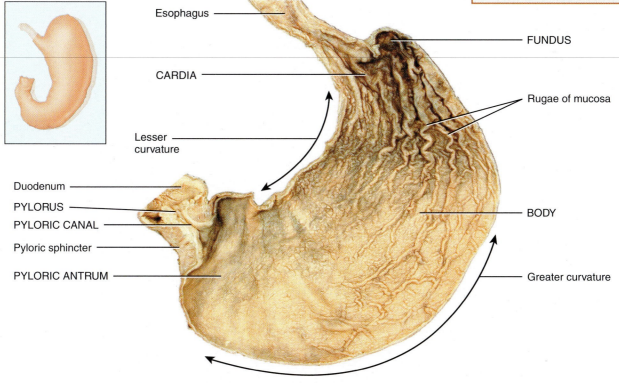

Esophagus

CARDIA

Lesser curvature

Duodenum

PYLORUS

PYLORIC CANAL

Pyloric sphincter

PYLORIC ANTRUM

FUNDUS

Rugae of mucosa

BODY

Greater curvature

(b) Anterior view of internal anatomy

The four regions of the stomach are the cardia, fundus, body, and pylorus.

tion of the small intestine. The position and size of the stomach vary continually. The diaphragm pushes it inferiorly with each inhalation and pulls it superiorly with each exhalation. Empty, it is about the size of a large sausage, but it is the most distensible part of the GI tract and can accommodate a large quantity of food. In the stomach, digestion of starch continues, digestion of proteins and triglycerides begins, the semisolid bolus is converted to a liquid, and certain substances are absorbed.

Anatomy of the Stomach

The stomach has four main regions: the cardia, fundus, body, and pylorus (Figure 23.11). The **cardia** (CAR-dē-a) surrounds the superior opening of the stomach. The rounded portion superior to and to the left of the cardia is the **fundus** (FUN-dus). Inferior to the fundus is the large central portion of the stomach, called the **body**. The region of the stomach that connects to the duodenum is the **pylorus** (pī-LOR-us; *pyl-* = gate; *-orus* = guard); it has two parts, the **pyloric antrum** (AN-trum = cave), which connects to the body of the stomach, and the **pyloric canal**, which leads into the duodenum. When the stomach is empty, the mucosa lies in large folds, called **rugae** (ROO-gē = wrinkles), that can be seen

with the unaided eye. The pylorus communicates with the duodenum of the small intestine through a smooth muscle sphincter called the **pyloric sphincter**. The concave medial border of the stomach is called the **lesser curvature**, and the convex lateral border is called the **greater curvature**.

Histology of the Stomach

The stomach wall is composed of the same four basic layers as the rest of the GI tract, with certain modifications (Figure 23.12). The surface of the **mucosa** is a layer of simple columnar epithelial cells called **surface mucous cells** (Figure 23.12b). The mucosa contains a **lamina propria** and a **muscularis mucosae**. Epithelial cells extend down into the lamina propria, where they form columns of secretory cells called **gastric glands**. Several gastric glands open into the bottom of narrow channels called **gastric pits**. Secretions from several gastric glands flow into each gastric pit and then into the lumen of the stomach.

The gastric glands contain three types of *exocrine gland cells* that secrete their products into the stomach lumen: mucous neck cells, chief cells, and parietal cells. Both **surface mucous cells** and **mucous neck cells** secrete mucus (Figure 23.12b). **Parietal cells**

Figure 23.12   Histology of the stomach.

(a) Three-dimensional view of layers of the stomach

Surface mucous cells

Gastric pit

SEM about 1000x

Stomach mucosa

Gastric pit

Simple columnar epithelium

Lamina propria

Gastric glands

Muscularis mucosae

Submucosa

Surface mucous cell (secretes mucus)

Mucous neck cell (secretes mucus)

Parietal cell (secretes hydrochloric acid and intrinsic factor)

Chief cell (secretes pepsinogen and gastric lipase)

G cell (secretes the hormone gastrin)

(b) Sectional view of the stomach mucosa showing gastric glands and cell types

Lumen

Gastric pit

Surface mucous cell

Lamina propria

Chief cell

Parietal cell

LM 180x

(c) Fundic mucosa

🔑 **Gastric juice is the combined secretions of mucous cells, parietal cells, and chief cells.**

produce intrinsic factor (needed for absorption of vitamin B$_{12}$) and hydrochloric acid. The **chief cells** secrete pepsinogen and gastric lipase. The secretions of the mucous, parietal, and chief cells form **gastric juice**, which totals 2000–3000 mL (roughly 2–3 qt) per day. In addition, gastric glands include **G cells**, which secrete the hormone gastrin into the bloodstream. As we will see shortly, this hormone stimulates several aspects of gastric activity.

Three additional layers lie deep to the mucosa (Figure 23.12). The **submucosa** of the stomach is composed of areolar connective tissue. The **muscularis** has three layers of smooth muscle: an outer longitudinal layer, a middle circular layer, and an inner oblique layer. The **serosa** covering the stomach, composed of simple squamous epithelium and areolar connective tissue, is part of the visceral peritoneum. At the lesser curvature of the stomach, the visceral peritoneum extends upward to the liver as the lesser omentum. At the greater curvature of the stomach, the visceral peritoneum continues downward as the greater omentum and drapes over the intestines.

Mechanical and Chemical Digestion in the Stomach

Several minutes after food enters the stomach, gentle rippling, peristaltic movements called **mixing waves** pass over the stomach every 15 to 25 seconds. These waves macerate food, mix it with secretions of the gastric glands, and reduce it to a soupy liquid called **chyme** (KĪM = juice). Few mixing waves are observed in the fundus, which primarily has a storage function. As digestion proceeds in the stomach, more vigorous mixing waves begin at the body of the stomach and intensify as they reach the pylorus. The pyloric sphincter normally remains almost, but not completely, closed. Each mixing wave periodically forces a small amount of chyme (about 3 mL) into the duodenum through the pyloric sphincter, a phenomenon known as **gastric emptying**. Most of the chyme is forced back into the body of the stomach, where mixing continues. The next wave pushes the chyme forward again and forces a little more into the duodenum. These forward and backward movements of the gastric contents are responsible for most mixing in the stomach.

Foods may remain in the fundus for about an hour without becoming mixed with gastric juice. During this time, digestion by salivary amylase continues. Soon, however, the churning action mixes chyme with acidic gastric juice, inactivating salivary amylase and activating lingual lipase, which starts to digest triglycerides into fatty acids and diglycerides.

Although parietal cells secrete hydrogen ions (H$^+$) and chloride ions (Cl$^-$) separately into the stomach lumen, the net effect is secretion of hydrochloric acid (HCl). **Proton pumps** powered by

H$^+$/ K$^+$ ATPases actively transport H$^+$ into the lumen while bringing potassium ions (K$^+$) into the cell (Figure 23.13). At the same time, Cl$^-$ and K$^+$ diffuse out into the lumen through Cl$^-$ and K$^+$ channels in the apical membrane. The enzyme *carbonic anhydrase,* which is especially plentiful in parietal cells, catalyzes the formation of carbonic acid (H$_2$CO$_3$) from water (H$_2$O) and carbon dioxide (CO$_2$). As carbonic acid dissociates, it provides a ready source of H$^+$ for the proton pumps but also generates bicarbonate ions (HCO$_3$$^-$). As HCO$_3$$^-$ builds up in the cytosol, it exits the parietal cell in exchange for Cl$^-$ via Cl$^-$/HCO$_3$$^-$ antiporters in the basolateral membrane (next to the lamina propria) and diffuses into nearby blood capillaries.

Figure 23.13 Secretion of HCl (hydrochloric acid) by parietal cells in the stomach.

Proton pumps, powered by ATP, secrete H$^+$; Cl$^-$ diffuses into the stomach lumen through Cl$^-$ channels.

Even before food enters the stomach, the sight, smell, taste, or thought of food initiates reflexes that stimulate parasympathetic neurons to release acetylcholine. Acetylcholine plus gastrin secreted by G cells stimulate parietal cells to secrete more HCl in the presence of histamine, released by mast cells in the nearby lamina propria. In other words, histamine acts synergistically, enhancing the effects of acetylcholine and gastrin. Besides stimulating secretion of large amounts of gastric juice, gastrin increases motility of the stomach and relaxes the pyloric sphincter. Impulses from parasympathetic neurons also increase stomach motility.

The strongly acidic fluid of the stomach kills many microbes in food. HCl partially denatures (unfolds) proteins in food and stimulates the secretion of hormones that promote the flow of bile and pancreatic juice. Enzymatic digestion of proteins also begins in the stomach by **pepsin**, which is secreted by chief cells. Pepsin breaks peptide bonds between amino acids, breaking down a protein chain of many amino acids into smaller fragments. Pepsin is most effective in the very acidic environment of the stomach (pH 2); it becomes inactive at a higher pH.

What keeps pepsin from digesting the protein in stomach cells along with the food? First, pepsin is secreted in an inactive form called *pepsinogen*; in this form, it cannot digest the proteins in the chief cells that produce it. Pepsinogen is not converted into active pepsin until it comes in contact with hydrochloric acid in gastric juice. Second, the stomach epithelial cells are protected from gastric juices by a 1- to 3-mm-thick layer of alkaline mucus secreted by surface mucous cells and mucous neck cells.

Another enzyme of the stomach is **gastric lipase**, which splits the short-chain triglycerides in fat molecules into fatty acids and monoglycerides. This enzyme, which has a limited role in the adult stomach, operates best at a pH of 5–6. More important than either lingual lipase or gastric lipase is pancreatic lipase, an enzyme secreted by the pancreas into the small intestine.

Only a small amount of nutrients are absorbed in the stomach because its epithelial cells are impermeable to most materials. However, mucous cells of the stomach absorb some water, ions, and short-chain fatty acids, as well as certain drugs (especially aspirin) and alcohol.

Within 2 to 4 hours after eating a meal, the stomach has emptied its contents into the duodenum. Foods rich in carbohydrate spend the least time in the stomach; high-protein foods remain somewhat longer, and emptying is slowest after a fat-laden meal containing large amounts of triglycerides.

Table 23.3 summarizes the digestive activities of the stomach.

Checkpoint

17. Which stomach layer is in contact with swallowed food?
18. Which part of the stomach primarily serves as a food reservoir?
19. Which molecule is the source of the hydrogen ions that are secreted into gastric juice?
20. Which branch of the autonomic nervous system promotes digestion?
21. What is the role of pepsin? Why is it secreted in an inactive form?
22. Which substances are absorbed in the stomach?

TABLE 23.3

Digestive Activities in the Stomach

STRUCTURE	ACTIVITY	RESULT
Mucosa		
Chief cells	Secrete pepsinogen.	Pepsin, the activated form, breaks down proteins into peptides.
	Secrete gastric lipase.	Splits triglycerides into fatty acids and monoglycerides.
Parietal cells	Secrete hydrochloric acid.	Kills microbes in foods; denatures proteins; converts pepsinogen into pepsin.
	Secrete intrinsic factor.	Needed for absorption of vitamin B_{12}, which is used in red blood cell formation (erythropoiesis).
Surface mucous cells and mucous neck cells	Secrete muscus.	Form a protective barrier that prevents digestion of stomach wall.
	Absorption.	Small quantity of water, ions, short-chain fatty acids, and some drugs enter the bloodstream.
G cells	Secrete gastrin.	Stimulate parietal cells to secrete HCl and chief cells to secrete pepsinogen; contracts lower esophageal sphincter, increases motility of the stomach, and relaxes pyloric sphincter.
Muscularis	Mixing waves.	Macerate food and mix it with gastric juice, forming chyme.
	Peristalsis.	Forces chyme through pyloric sphincter.
Pyloric sphincter	Opens to permit passage of chyme into duodenum.	Regulates passage of chyme from stomach to duodenum; prevents backflow of chyme from duodenum to stomach.

"Gastric bypass surgery is quite a serious undertaking, Chloe. It's going to require a lifetime of commitment to monitoring your diet and regular exercise. There are risks. I have been fortunate not to have had any poor outcomes, but your mother was right. There have been incidents of postsurgical bleeds, abscesses, and pulmonary embolism associated with the surgery."

Chloe nodded and replied, "Dr. Benavides, I understand that. I also know that my insurance probably won't cover the cost of the surgery. But I am ready to make that commitment. I want to be there for my children, and I want to have a better quality of life. I am so tired and breathless all the time. My regular doctor said that sleep apnea is contributing to my fatigue and that the extra weight is causing my acid reflux, diabetes, and high blood pressure. I think it's time for drastic action."

Chloe got her wish. After a series of screening tests to ensure she was an appropriate candidate, the procedure had been done under general anesthesia. Using a series of surgical staples, Dr. Benavides reduced Chloe's stomach from its original size of about 1 qt to a small pouch. He then cut the small intestine, bypassing its upper portion, the duodenum, and reattached the shortened intestines directly to the smaller stomach at the jejunum. The surgery was done using a laparoscopic technique; only a few small incisions were made to insert surgical

instruments. Dr. Benavides preferred this method because it caused less scar tissue. Chloe woke up with a sore throat and a dry, nagging cough. "Nurse, may I have some water?" she asked.

"Just a little for now. Your stomach's lumen has been reduced to about the size of a tablespoon, so we have to be careful you don't drink too much," the nurse replied.

Dr. Benavides entered the room. "Chloe, how are you feeling?" he asked.

"Okay, not too much pain, thanks," Chloe replied.

Over time, Chloe's stomach would heal and eventually would expand to accommodate about 1 cup of food. The reduced volume of her stomach would mean that Chloe would feel full more quickly, hopefully reducing her hunger and calorie intake. But she would have to be careful not to overeat. Ingesting too much food could overly stretch and then damage Chloe's surgically reduced stomach.

A.   Which serous membrane in Chloe's abdomen most likely contains the greatest amount of adipose tissue?
B.   What involuntary muscular process initiated by deglutition of water may cause Chloe pain?
C.   Why might reduction of the size of Chloe's stomach reduce her acid reflux?
D.   How will protein digestion be affected by Chloe's surgery?

23.5 The pancreas secretes pancreatic juice, the liver secretes bile, and the gallbladder stores and concentrates bile.

Pancreas

From the stomach, chyme passes into the small intestine. Because chemical digestion in the small intestine depends on activities of the pancreas, liver, and gallbladder, we first consider the activities of these accessory digestive organs and their contributions to digestion in the small intestine.

Anatomy of the Pancreas

The **pancreas** (*pan-* = all; *-creas* = flesh), a retroperitoneal gland that is about 12–15 cm (5–6 in.) long and 2.5 cm (1 in.) thick, lies posterior to the greater curvature of the stomach. The pancreas consists of a head, a body, and a tail and is usually connected to the duodenum by two ducts (Figure 23.14a). The **head** is the expanded portion of the organ near the curve of the duodenum; to the left of the head are the central **body** and the tapering **tail**.

Pancreatic juices exit the exocrine cells into small ducts that ultimately unite to form two larger ducts, the pancreatic duct and the accessory duct. These in turn convey the secretions into the duodenum of the small intestine. The **pancreatic duct** is the larger

of the two ducts. In most people, the pancreatic duct joins the common bile duct from the liver and gallbladder and enters the duodenum as a common duct called the **hepatopancreatic ampulla**, which lies about 10 cm (4 in.) inferior to the pyloric sphincter of the stomach. The passage of pancreatic juice and bile through the hepatopancreatic ampulla into the small intestine is regulated by a mass of smooth muscle known as the **sphincter of the hepatopancreatic ampulla**. The other major duct of the pancreas, the **accessory duct**, leads from the pancreas and empties into the duodenum about 2.5 cm (1 in.) superior to the hepatopancreatic ampulla.

Histology of the Pancreas

The pancreas is made up of small clusters of glandular epithelial cells. About 99% of the clusters, called **acini** (AS-i-nē), constitute the *exocrine* portion of the organ (see Figure 17.17b,c). The cells within acini secrete a mixture of fluid and digestive enzymes called **pancreatic juice**. The remaining 1% of the clusters, called **pancreatic islets**, form the *endocrine* portion of the pancreas. These cells secrete the hormones glucagon and insulin. The functions of these hormones are discussed in Chapter 17.

Figure 23.14 **Relation of the pancreas to the liver, gallbladder, and duodenum.** The inset (b) shows details of the common bile duct and pancreatic duct forming the hepatopancreatic ampulla that empties into the duodenum.

(a) Anterior view

(b) Details of hepatopancreatic ampulla

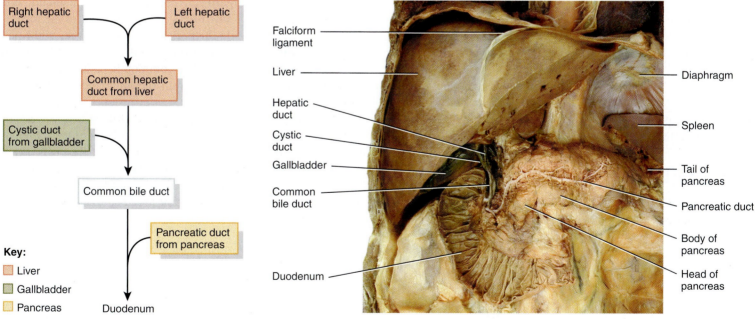

(c) Ducts carrying bile from liver and gallbladder and pancreatic juice from pancreas to the duodenum

(d) Anterior view

🔑 **Pancreatic enzymes digest starches, proteins, triglycerides, and nucleic acids.**

Pancreatic Juice

Each day the pancreas produces 1200–1500 mL (about 1.2–1.5 qt) of **pancreatic juice**, a clear, colorless liquid consisting mostly of water, some salts, sodium bicarbonate, and several enzymes. The sodium bicarbonate gives pancreatic juice a slightly alkaline pH (7.1–8.2) that buffers acidic gastric juice in chyme, stops the action of pepsin from the stomach, and creates the proper pH for the action of digestive enzymes in the small intestine. The enzymes in pancreatic juice include a starch-digesting enzyme called **pancreatic amylase**; several protein-digesting enzymes called **trypsin** (TRIP-sin), **chymotrypsin** (kī´-mō-TRIP-sin), **carboxypeptidase** (kar-bok´-sē-PEP-ti-dās), and **elastase** (ē-LAS-tās); the principal triglyceride-digesting enzyme, called **pancreatic lipase**; and nucleic acid–digesting enzymes called **ribonuclease** and **deoxy-ribonuclease**.

Like pepsin produced in the stomach, the protein-digesting enzymes of the pancreas are produced in an inactive form so that the enzymes do not digest the cells of the pancreas. Trypsin is secreted in an inactive form called **trypsinogen** (trip-SIN-ō-jen). Pancreatic juice also contains a protein called **trypsin inhibitor** that combines with any trypsin formed accidentally in the pancreas or in pancreatic juice and blocks its enzymatic activity. When trypsinogen reaches the lumen of the small intestine, it encounters an activating brush border enzyme called **enterokinase** (en´-ter-ō-KĪ-nās), which splits off part of the trypsinogen molecule to form trypsin. In turn, trypsin activates the other protein-digesting pancreatic enzymes.

Liver and Gallbladder

The **liver** is the heaviest gland of the body, weighing about 1.4 kg (about 3 lb) in an average adult. Of all of the organs of the body, it is second only to the skin in size. The liver is inferior to the diaphragm and occupies most of the right side of the abdominal cavity (see Figure 23.1). The **gallbladder** (*gall-* = bile) is a pear-shaped sac that is located in a depression of the posterior surface of the liver. It is 7–10 cm (3–4 in.) long and typically hangs from the anterior inferior margin of the liver (Figure 23.14a).

Anatomy of the Liver

The liver is divided into two principal lobes—a large **right lobe** and a smaller **left lobe**—by the **falciform ligament**, a fold of the mesentery (Figure 23.14a). An inferior *quadrate lobe* and a posterior *caudate lobe* are continuations of the left lobe. The falciform ligament extends from the undersurface of the diaphragm, helping to suspend the liver in the abdominal cavity. In the free border of the falciform ligament is the **round ligament**, a remnant of the umbilical vein of the fetus. The right and left **coronary ligaments** are narrow extensions of the parietal peritoneum that suspend the liver from the diaphragm.

Histology of the Liver

The lobes of the liver are made up of many **lobules** (Figure 23.15). A lobule is typically a six-sided structure (hexagon) that contains specialized epithelial cells called **hepatocytes** (*hepat-* = liver; *-cytes* = cells). The hepatocytes are arranged in irregular, interconnected plates radiating out from a **central vein**. The liver has highly permeable blood capillaries called **sinusoids** through which blood passes. Present in the sinusoids are fixed phagocytes called **reticuloendothelial cells**, which destroy worn-out blood cells, bacteria, and other foreign matter in the venous blood draining from the gastrointestinal tract.

Bile Duct System

Bile canaliculi (kan´-a-LIK-ū-lī = small canals) are small ducts between hepatocytes that collect bile (described shortly) produced by the hepatocytes (Figure 23.15b). From bile canaliculi, bile eventually empties into **bile ducts** at the periphery of the lobules (Figure 23.15b). The bile ducts merge into the larger **right** and **left hepatic ducts**, which unite and exit the liver as the **common hepatic duct** (Figure 23.14). The common hepatic duct joins the **cystic duct** (*cystic* = bladder) from the gallbladder to form the **common bile duct**.

Bile entering the cystic duct is temporarily stored in the gallbladder. Contraction of the smooth muscle fibers of the gallbladder wall ejects the contents of the gallbladder into the **cystic duct**. The functions of the gallbladder are to store and concentrate the bile produced by the liver (up to tenfold) until it is needed in the small intestine. In the concentration process, water and ions are absorbed by the gallbladder mucosa.

Figure 23.15 **Histology of the liver.**

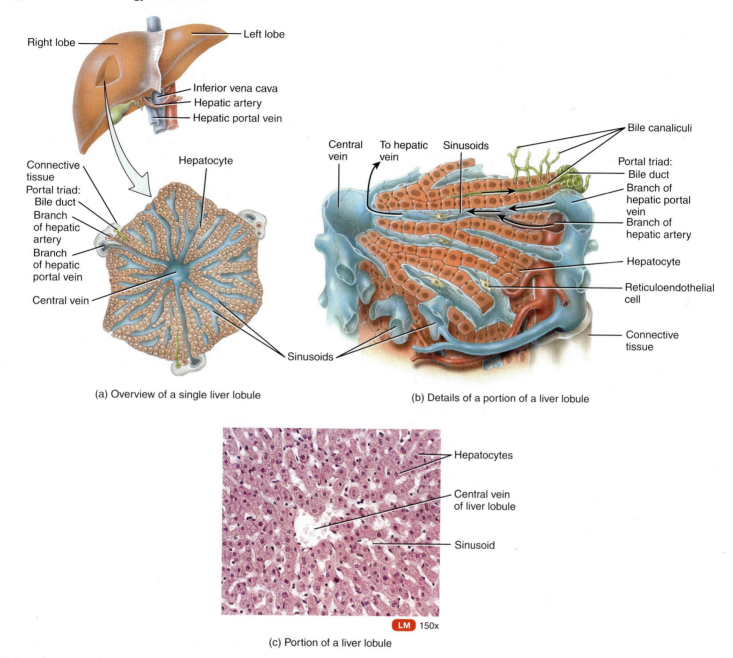

(a) Overview of a single liver lobule

(b) Details of a portion of a liver lobule

LM 150x

(c) Portion of a liver lobule

A lobule contains hepatocytes arranged around a central vein.

Blood Supply of the Liver

The liver receives blood from two sources (Figure 23.16). From the proper hepatic artery it obtains oxygenated blood, and from the hepatic portal vein it receives deoxygenated blood from the gastrointestinal tract containing newly absorbed nutrients, drugs, and possibly micrbes and toxins (see Figure 20.28). Branches of both the hepatic artery and the hepatic portal vein carry blood into liver sinusoids, where oxygen, most of the nutrients, and certain toxic substances are taken up by the hepatocytes. Products manufactured by the hepatocytes and nutrients needed by other body cells are secreted back into the blood, which then drains into the central vein and eventually passes into a hepatic vein. Branches of the hepatic portal vein and hepatic artery typically accompany bile ducts through the liver. Collectively, these three structures are called a **portal triad** (Figure 23.15). Portal triads are located at the corners of the liver lobules.

Bile

Each day, hepatocytes secrete 800–1000 mL (about 1 qt) of **bile**, a yellow, brownish, or olive-green liquid. It has a pH of 7.6–8.6 and consists mostly of water, bile acids, bile salts, cholesterol, a phospholipid called lecithin, bile pigments, and several ions.

Figure 23.16   Hepatic blood flow: sources, path through the liver, and return to the heart.

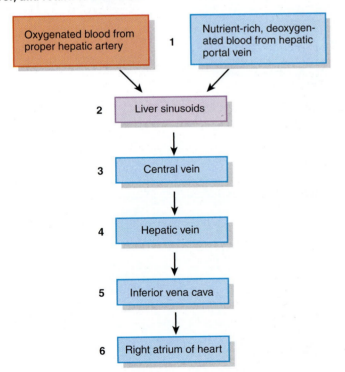

The liver receives oxygenated blood via the proper hepatic artery and nutrient-rich deoxygenated blood via the hepatic portal vein.

The principal bile pigment is **bilirubin**. The phagocytosis of aged red blood cells liberates iron, globin, and bilirubin (derived from heme) (see Figure 18.5). The iron and globin are recycled; the bilirubin is secreted into the bile and is eventually broken down in the intestine. One of bilirubin's breakdown products—**stercobilin**—gives feces their normal brown color.

Bile is partially an excretory product and partially a digestive secretion. Bile salts, which are sodium salts and potassium salts, play a role in **emulsification**, the breakdown of large lipid globules into a suspension of small lipid globules. The small lipid globules present a very large surface area that allows pancreatic lipase to more rapidly accomplish digestion of triglycerides. Bile salts also aid in the absorption of lipids following their digestion.

After they have served as emulsifying agents, most bile salts are reabsorbed by active transport in the final portion of the small intestine (ileum) and enter portal blood flowing toward the liver to be recycled into more bile. Although hepatocytes continually release bile, they increase production and secretion when the portal blood contains more bile salts. Thus, as digestion and absorption continue in the small intestine, bile release increases. Between meals, after most absorption has occurred, bile flows into the gallbladder for storage because the sphincter of the hepatopancreatic ampulla (Figure 23.14) closes off the entrance to the duodenum.

Functions of the Liver

In addition to secreting bile, which is needed for absorption of dietary fats, the liver performs many other vital functions:

- **Carbohydrate metabolism**. The liver is especially important in maintaining a normal blood glucose level. When blood glucose is low, the liver breaks down glycogen to glucose and releases glucose into the bloodstream. The liver can also convert certain amino acids and lactic acid to glucose, and it can convert other sugars, such as fructose and galactose, into glucose. When blood glucose is high, as occurs just after eating a meal, the liver converts glucose to glycogen and triglycerides for storage.

- **Lipid metabolism**. Hepatocytes store some triglycerides; break down fatty acids to generate ATP; synthesize lipoproteins, which transport fatty acids, triglycerides, and cholesterol to and from body cells; synthesize cholesterol; and use cholesterol to make bile salts.

- **Protein metabolism**. Hepatocytes remove the amino group NH_2 from amino acids so that the amino acids can be used for ATP production or converted to carbohydrates or fats. The resulting toxic ammonia is then converted into the much less toxic urea, which is excreted in urine. Hepatocytes also synthesize most plasma proteins, such as alpha and beta globulins, albumin, prothrombin, and fibrinogen.

- **Processing of drugs and hormones**. The liver can detoxify substances such as alcohol and excrete drugs such as penicillin, erythromycin, and sulfonamides into bile. It can also inactivate thyroid hormones and steroid hormones such as estrogens and aldosterone.

- **Excretion of bilirubin**. As previously noted, bilirubin, derived from the heme of aged red blood cells, is absorbed by the liver from the blood and secreted into bile. Most of the bilirubin in bile is metabolized in the small intestine by bacteria and eliminated in feces.

- **Synthesis of bile salts**. Bile salts are used in the small intestine for the emulsification and absorption of lipids.

- **Storage**. In addition to glycogen, the liver is a prime storage site for certain vitamins (A, B$_{12}$, D, E, and K) and minerals (iron and copper), which are released from the liver when needed elsewhere in the body.

- **Phagocytosis**. The reticuloendothelial cells of the liver phagocytize aged red blood cells, white blood cells, and some bacteria.

- **Activation of vitamin D**. The skin, liver, and kidneys participate in synthesizing the active form of vitamin D.

Checkpoint

23. What are the structures through which pancreatic juice might travel on its way to the duodenum?
24. What are the functions of the components of pancreatic juice?
25. How are the protein-digesting enzymes in pancreatic juice activated?
26. Which type of fluid is found in the pancreatic duct? The common bile duct? The hepatopancreatic ampulla?
27. What is the function of bile?
28. The first few hours after a meal, how does the chemical composition of blood change as it flows through liver sinusoids?

23.6 In the small intestine, chyme mixes with digestive juices from the small intestine, pancreas, and liver.

Most digestion and absorption of nutrients occur in a long tube called the **small intestine** (Figure 23.17). Because of this, its structure is specially adapted for these functions. Its length alone provides a large surface area for digestion and absorption, and that area is further increased by circular folds, villi, and microvilli. The small intestine begins at the pyloric sphincter of the stomach, coils

Figure 23.17 **Anatomy of the small intestine.** (a) Regions of the small intestine are the duodenum, jejunum, and ileum. (b) Circular folds increase the surface area for digestion and absorption in the small intestine.

Functions of the Small Intestine
1. Mixes chyme with digestive juices and brings food into contact with the mucosa for absorption; propels chyme through the small intestine.
2. Completes the digestion of carbohydrates, proteins, and lipids; begins and completes the digestion of nucleic acids.
3. Absorbs about 90% of nutrients and water that pass through the digestive system.

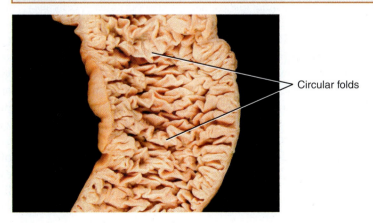

(a) Anterior view of external anatomy

(b) Internal anatomy of jejunum

Most digestion and absorption occur in the small intestine.

through the central and inferior part of the abdominal cavity, and eventually opens into the large intestine (Figure 23.17). It averages 2.5 cm (1 in.) in diameter; its length is about 3 m (10 ft) in a living person and about 6.5 m (21 ft) in a cadaver due to the loss of smooth muscle tone after death.

Anatomy of the Small Intestine

The small intestine is divided into three regions (Figure 23.17). The **duodenum** (doo´-ō-DĒ-num *or* doo-OD-e-num), the shortest region, is retroperitoneal. It starts at the pyloric sphincter of the stomach and extends about 25 cm (10 in.) until it merges with the jejunum. *Duodenum* means "12"; it is so named because it is about as long as the width of 12 fingers. The **jejunum** (je-JOO-num) is about 1 m (3 ft) long and extends to the ileum. *Jejunum* means "empty," which is how it is found at death. The final and longest region of the small intestine, the **ileum** (IL-ē-um = twisted), measures about 2 m (6 ft) and joins the large intestine at the **ileocecal sphincter** (il´-ē-ō-SĒ-kal).

Histology of the Small Intestine

The wall of the small intestine is composed of the same four layers that make up most of the GI tract: mucosa, submucosa, muscularis, and serosa (Figure 23.18a). The epithelium of the mucosa consists of simple columnar epithelium that contains many types of cells (Figure 23.18b). **Absorptive cells** of the epithelium digest and absorb nutrients in chyme. Also present are **goblet cells**, which secrete mucus (Figure 23.18b). The small intestinal mucosa contains many deep crevices lined with glandular epithelium. Cells lining the crevices form the **intestinal glands** that secrete intestinal juice (to be discussed shortly). Besides absorptive cells and goblet cells, the intestinal glands also contain Paneth cells and endocrine cells. **Paneth cells** secrete lysozyme, a bactericidal enzyme, and are capable of phagocytosis. Paneth cells may have a role in regulating the microbial population in the intestines. **S cells** and **CCK cells** are endocrine cells found in the intestinal glands that secrete hormones: S cells secrete secretin and CCK cells secrete cholecystokinin. The functions of these hormones will be described later in the chapter.

The lamina propria has an abundance of **lymphatic nodules (Peyer's patches)**, which help defend against pathogens in food (Figure 23.19c). The submucosa of the duodenum contains **duodenal glands** (Figure 23.19a), which secrete an alkaline mucus that helps neutralize gastric acid in the chyme. The **muscularis** of the small intestine consists of two layers of smooth muscle. The outer, thinner layer contains longitudinal fibers; the inner, thicker layer contains circular fibers.

Even though the wall of the small intestine is composed of the same four layers that make up most of the GI tract, special features

of both the mucosa and the submucosa facilitate the process of digestion and absorption. These structural features include circular folds, villi, and microvilli. **Circular folds (plicae circulares)** are folds of the mucosa and submucosa (Figure 23.17b). These permanent ridges are about 10 mm (0.4 in.) long and enhance absorption by increasing surface area and causing the chyme to spiral, rather than move in a straight line, as it passes through the small intestine.

Also present in the small intestine are **villi** (= tufts of hair), fingerlike projections of the mucosa that are 0.5–1 mm long (Figure 23.18a). The large number of villi (20–40 per square millimeter) vastly increases the surface area of the epithelium available for absorption and digestion and gives the intestinal mucosa a velvety appearance. Each villus (singular form) has a core of lamina propria; embedded in the connective tissue of the lamina propria are an arteriole, a venule, a blood capillary network, and a **lacteal** (LAK-tē-al = milky), which is a lymphatic capillary. Nutrients absorbed by the epithelial cells covering the villus pass through the wall of a capillary or a lacteal to enter blood or lymph, respectively.

Besides circular folds and villi, the small intestine also has **microvilli** (mi´-krō-VIL-ī; *micro-* = small), tiny projections of the apical (free) membrane of the absorptive cells. When viewed through a light microscope, the microvilli are too small to be seen individually; instead they form a fuzzy line, called the **brush border**, extending into the lumen of the small intestine (Figure 23.19b,d). There are an estimated 200 million microvilli per square millimeter of small intestine. Because the microvilli greatly increase the surface area of the plasma membrane, larger amounts of digested nutrients can diffuse into absorptive cells. The brush border also contains several brush border enzymes that have digestive functions (discussed shortly).

Role of Intestinal Juice and Brush Border Enzymes

About 1–2 liters (1–2 qt) of **intestinal juice**, a clear yellow fluid, are secreted each day. Intestinal juice contains water and mucus and is slightly alkaline (pH 7.6). Together, pancreatic and intestinal juices provide a liquid medium that aids the absorption of substances from chyme as they come in contact with the microvilli. The absorptive epithelial cells synthesize several digestive enzymes, called **brush border enzymes**, and insert them in the plasma membrane of the microvilli. Among the brush border enzymes are enzymes that digest carbohydrates, proteins, and nucleotides. Thus, some enzymatic digestion occurs at the surface of the epithelial cells that line the villi, rather than in the lumen exclusively, as occurs in other parts of the GI tract. Also, as epithelial cells slough off of the small intestine wall, they break apart and release enzymes that help digest nutrients in the lumen.

Figure 23.18 **Histology of the small intestine.**

Lumen of small intestine

Villus   Blood   Lacteal
capillary

Simple columnar
epithelium

Lamina propria

Opening of intestinal gland

Lymphatic nodule

Muscularis mucosae

Arteriole

Venule

Lymphatic vessel

Circular layer of muscle

Myenteric plexus

Longitudinal layer
of muscle

MUCOSA

SUBMUCOSA

MUSCULARIS

SEROSA

(a) Three-dimensional view of layers of the small intestine showing villi

Microvilli

Absorptive cell
(absorbs nutrients)

Mucosa

Blood
capillary

Lacteal

Lamina
propria

Intestinal
gland

Muscularis
mucosae

Arteriole

Venule

Lymphatic
vessel

Submucosa

Goblet cell
(secretes mucus)

Enteroendocrine cell
(secretes the
hormones secretin
or cholecystokinin)

Paneth cell
(secretes lysozyme
and is capable
of phagocytosis)

(b) Enlarged villus showing lacteal, capillaries, intestinal glands, and cell types

Circular folds, villi, and microvilli increase the surface area of the small intestine for digestion and absorption.

Figure 23.19   Histology of the duodenum and ileum.

Lumen of duodenum

Villi

Mucosa

Intestinal glands

Muscularis mucosae

Duodenal gland

Submucosa

Muscularis

LM   45x

(a) Wall of the duodenum

Villi

Lumen of duodenum

Brush border

Simple columnar epithelium

Goblet cell

Absorptive cell

Duodenum

Lamina propria

Intestinal glands

Muscularis mucosae

Duodenal gland in submucosa

LM   160x

(b) Three villi from the duodenum

Lumen of ileum

Villus

Lymphatic nodule

Submucosa

Muscularis

LM 14x

(c) Lymphatic nodules in the ileum

Microvilli

Brush border

Simple columnar epithelial cell

TEM 46,800x

(d) Several microvilli from the duodenum

🔑 **Microvilli in the small intestine contain brush border enzymes that help digest nutrients.**

Mechanical Digestion in the Small Intestine

The two types of movements of the small intestine—segmentations and peristalsis—are governed mainly by the myenteric plexus. **Segmentations** are localized, mixing contractions that occur in portions of intestine distended by a large volume of chyme. Segmentations mix chyme with the digestive juices and bring the particles of food into contact with the mucosa for absorption; segmentations do not push the intestinal contents along the tract. A segmentation starts with the contractions of circular muscle fibers in a portion of the small intestine, an action that constricts the intestine into segments. Next, muscle fibers that encircle the middle of each segment also contract, dividing each segment again. Finally, the fibers that first contracted relax, and each small segment unites with an adjoining small segment so that large segments are formed again. This movement is similar to alternately squeezing the middle and then the ends of a capped tube of toothpaste. As this sequence of events repeats, the chyme sloshes back and forth. Segmentations occur most rapidly in the duodenum, about 12 times per minute, and progressively slow to about 8 times per minute in the ileum.

After most of a meal has been absorbed, which lessens distention of the wall of the small intestine, segmentation stops and peristalsis begins. Each peristaltic wave begins in the lower portion of the stomach and slowly pushes chyme forward down the small intestine, with peristaltic migrations reaching the end of the ileum in 90 to 120 minutes. Then another wave of peristalsis begins in the stomach. Altogether, chyme remains in the small intestine for 3–5 hours.

Chemical Digestion in the Small Intestine

In the mouth, salivary amylase converts starch (a polysaccharide) to disaccharides and short-chain glucose polymers called α-dextrins. In the stomach, pepsin converts proteins to peptides (small fragments of proteins), and lingual and gastric lipases convert some triglycerides into fatty acids, diglycerides, and monoglycerides. Thus, chyme entering the small intestine contains partially digested carbohydrates, proteins, and lipids. The completion of digestion is a collective effort of pancreatic juice, bile, and intestinal juice in the small intestine.

Digestion of Carbohydrates

Even though the action of salivary amylase may continue in the stomach for a while, the acidic pH of the stomach destroys salivary amylase and ends its activity. Thus, only a few starches are broken down by the time chyme leaves the stomach. Those starches not already broken down are cleaved by pancreatic amylase, an enzyme in pancreatic juice that acts in the small intestine. Although pancreatic amylase acts on both glycogen and starches, it has no effect on another polysaccharide called cellulose, an indigestible plant fiber that is commonly referred to as "roughage" as it moves through the digestive system. After amylase (either salivary or pancreatic) has split starch into smaller fragments, a brush border enzyme called **α-dextrinase** acts on the resulting α-dextrins, clipping off one glucose unit at a time.

Ingested molecules of sucrose, lactose, and maltose—three disaccharides—are not acted on until they reach the small intestine. Three brush border enzymes digest the disaccharides into monosaccharides. **Sucrase** breaks sucrose into a molecule of glucose and a molecule of fructose; **lactase** digests lactose into a molecule of glucose and a molecule of galactose; and **maltase** splits maltose into two molecules of glucose. Digestion of carbohydrates ends with the production of monosaccharides, which the digestive system is able to absorb.

Digestion of Proteins

Recall that protein digestion starts in the stomach, where proteins are fragmented into peptides by the action of **pepsin**. Enzymes in pancreatic juice—**trypsin**, **chymotrypsin**, **carboxypeptidase**, and **elastase**—continue to break down proteins into peptides, though their actions differ somewhat because each splits peptide bonds between different amino acids. Trypsin, chymotrypsin, and elastase all cleave the peptide bond between a specific amino acid and its neighbor; carboxypeptidase breaks the peptide bond that attaches the terminal amino acid to the peptide chain. Protein digestion is completed by **peptidases** in the brush border that break peptides into single amino acids, dipeptides, and tripeptides.

Digestion of Lipids

The most abundant lipids in the diet are triglycerides, which consist of a molecule of glycerol bonded to three fatty acid molecules (see Figure 2.16). Enzymes that split triglycerides and phospholipids are called **lipases**. Recall that there are three types of lipases that can participate in lipid digestion: lingual lipase, gastric lipase, and pancreatic lipase. Although some lipid digestion occurs in the stomach through the action of lingual and gastric lipases, most occurs in the small intestine through the action of pancreatic lipase. Triglycerides are broken down by pancreatic lipase into fatty acids and monoglycerides.

Before a large lipid globule containing triglycerides can be digested in the small intestine, it must first undergo **emulsification**—a process in which the large lipid globule is broken down into several small lipid globules. Recall that bile contains bile salts. Bile salts are **amphipathic**, which means that each bile salt has a hydrophobic (nonpolar) region and a hydrophilic (polar) region. The hydrophobic regions of bile salts interact with the large lipid globule, while the hydrophilic regions of bile salts interact with the watery intestinal chyme. The amphipathic nature of bile salts allows them to break apart a large lipid globule into several small lipid globules, each about 1 μm in diameter. The small lipid globules formed from emulsification provide a large surface area that allows pancreatic lipase to function more effectively.

Digestion of Nucleic Acids

Pancreatic juice contains two nucleases: **ribonuclease**, which digests RNA, and **deoxyribonuclease**, which digests DNA. The nucleotides that result from the action of the two nucleases are further digested by brush border enzymes called **nucleosidases** and **phosphatases** into pentoses, phosphates, and nitrogenous bases. These products are absorbed by active transport.

Table 23.4 summarizes the sources, substrates, and products of the digestive enzymes.

TABLE 23.4			
Digestive Enzymes			
ENZYME	SOURCE	SUBSTRATES	PRODUCTS
Saliva			
Salivary amylase	Salivary glands.	Starches (polysaccharides).	Maltose (disaccharide), maltotriose (trisaccharide), and α-dextrins.
Lingual lipase	Lingual glands in the tongue.	Triglycerides (fats and oils) and other lipids.	Fatty acids and diglycerides.
Gastric Juice			
Pepsin (activated from pepsinogen by pepsin and hydrochloric acid)	Stomach chief cells.	Proteins.	Peptides.
Gastric lipase	Stomach chief cells.	Triglycerides (fats and oils).	Fatty acids and monoglycerides.
Pancreatic Juice			
Pancreatic amylase	Pancreatic acinar cells.	Starches (polysaccharides).	Maltose (disaccharide), maltotriose (trisaccharide), and α-dextrins.
Trypsin (activated from trypsinogen by enterokinase)	Pancreatic acinar cells.	Proteins.	Peptides.
Chymotrypsin (activated from chymotrypsinogen by trypsin)	Pancreatic acinar cells.	Proteins.	Peptides.
Elastase (activated from proelastase by trypsin)	Pancreatic acinar cells.	Proteins.	Peptides.

Absorption in the Small Intestine

All the chemical and mechanical phases of digestion from the mouth through the small intestine are directed toward changing food into forms that can pass through the absorptive epithelial cells lining the mucosa and into the underlying blood and lymphatic vessels. Passage of these digested nutrients from the gastrointestinal tract into the blood or lymph is called **absorption**. About 90% of all absorption of nutrients occurs in the small intestine; the other 10% occurs in the stomach and large intestine. Any undigested or unabsorbed material left in the small intestine passes on to the large intestine.

Absorption of Monosaccharides

All carbohydrates are absorbed as monosaccharides. Fructose, a monosaccharide found in fruits, is transported by *facilitated diffusion;* glucose and galactose are transported into epithelial cells of the villi by *secondary active transport* that is coupled to the active transport of Na^+ (Figure 23.20a). Monosaccharides then move out of the epithelial cells through their basolateral surfaces by *facilitated diffusion* and enter the capillaries of the villi (see Figure 23.20b).

Absorption of Amino Acids, Dipeptides, and Tripeptides

Most proteins are absorbed as amino acids via *active transport* processes. About half of the absorbed amino acids are present in food; the other half come from the body itself in the form of proteins in digestive juices and dead cells that slough off the mucosal surface! Different transporters carry different types of amino acids. Some amino acids enter epithelial cells via Na^+-dependent secondary active transport processes that are similar to the glucose transporter; other amino acids are actively transported by themselves. At least one symporter brings in dipeptides and tripeptides together with H^+; the peptides then are hydrolyzed to single amino acids inside the epithelial cells. Amino acids move out of the epithelial cells by diffusion and enter capillaries of the villus (Figure 23.20a,b). Both monosaccharides and amino acids are transported in the blood to the liver by way of the hepatic portal system. If not removed by hepatocytes, they enter the general circulation.

Absorption of Lipids

All dietary lipids are absorbed via *simple diffusion*. As a result of their emulsification and digestion, triglycerides are broken down into monoglycerides and fatty acids, which can be either short-chain or long-chain fatty acids. Although short-chain fatty acids are hydrophobic, they are very small in size. Because of their size, they can dissolve in the watery intestinal chyme, pass through the absorptive cells via simple diffusion, and follow the same route taken by monosaccharides and amino acids into a blood capillary of a villus (Figure 23.20a).

TABLE 23.4 (continued)			
ENZYME	SOURCE	SUBSTRATES	PRODUCTS
Pancreatic Juice (continued)			
Carboxypeptidase (activated from procarboxypeptidase by trypsin)	Pancreatic acinar cells.	Amino acid at carboxyl end of peptides.	Amino acids and peptides.
Pancreatic lipase	Pancreatic acinar cells.	Triglycerides (fats and oils) that have been emulsified by bile salts.	Fatty acids and monoglycerides.
Nucleases			
Ribonuclease	Pancreatic acinar cells.	Ribonucleic acid.	Nucleotides.
Deoxyribonuclease	Pancreatic acinar cells.	Deoxyribonucleic acid.	Nucleotides.
Brush Border			
α-Dextrinase	Small intestine.	α-Dextrins.	Glucose.
Maltase	Small intestine.	Maltose.	Glucose.
Sucrase	Small intestine.	Sucrose.	Glucose and fructose.
Lactase	Small intestine.	Lactose.	Glucose and galactose.
Enterokinase	Small intestine.	Trypsinogen.	Trypsin.
Peptidases			
Aminopeptidase	Small intestine.	Amino acid at amino end of peptides.	Amino acids and peptides.
Dipeptidase	Small intestine.	Dipeptides.	Amino acids.
Nucleosidase and phosphatases	Small intestine.	Nucleotides.	Nitrogenous bases, pentoses, and phosphates.

Figure 23.20   Absorption of digested nutrients in the small intestine. For simplicity, all digested foods are shown in the lumen of the small intestine, even though some nutrients are digested by brush border enzymes.

(a) Mechanisms for movement of nutrients through absorptive epithelial cells of the villi

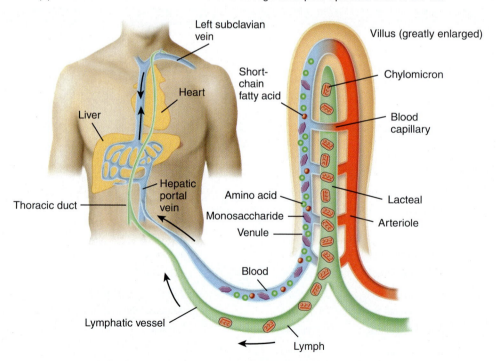

(b) Movement of absorbed nutrients into the blood and lymph

Long-chain fatty acids and monoglycerides are absorbed into lacteals; other products of digestion enter blood capillaries.

Long-chain fatty acids and monoglycerides are large and hydrophobic and have difficulty being suspended in the watery environment of the intestinal chyme. Besides their role in emulsification, bile salts also help to make these long-chain fatty acids and monoglycerides more soluble. The bile salts in intestinal chyme surround the long-chain fatty acids and monoglycerides, forming tiny spheres called **micelles** (mī-SELZ = small morsels) that include 20–50 bile salt molecules (Figure 23.20a). Micelles are formed due to the amphipathic nature of bile salts: The hydrophobic regions of bile salts interact with the long-chain fatty acids and monoglycerides, and the hydrophilic regions of bile salts interact with the watery intestinal chyme. Once formed, the micelles move from the lumen of the small intestine to the brush border of the absorptive cells. At that point, the long-chain fatty acids and monoglycerides diffuse out of the micelles into the absorptive cells, leaving the micelles behind in the chyme. The micelles continually repeat this ferrying function as they move from the brush border to the interior of the lumen to pick up more long-chain fatty acids and monoglycerides. Micelles also increase the solubility of other large hydrophobic molecules such as fat-soluble vitamins (A, D, E, and K) and cholesterol and aid in their absorption.

Once inside the absorptive epithelial cells, long-chain fatty acids and monoglycerides are recombined to form triglycerides and become coated with proteins. These large spherical masses are called **chylomicrons**. Chylomicrons leave the epithelial cell via exocytosis. Because they are so large and bulky, chylomicrons cannot enter blood capillaries—the pores in the walls of blood capillaries are too small. Instead, chylomicrons enter lacteals, which have much larger pores than blood capillaries. From lacteals, chylomicrons are transported by way of lymphatic vessels to the thoracic duct and enter the blood at the left subclavian vein (see Figure 23.20b). The hydrophilic protein coat that surrounds each chylomicron keeps the chylomicrons suspended in blood and prevents them from sticking to each other. As blood passes through capillaries in adipose tissue and the liver, chylomicrons are removed and their lipids are stored for future use.

After participating in the emulsification and absorption of lipids, 90–95% of the bile salts are reabsorbed by active transport in the ileum and returned by the blood to the liver through the hepatic portal system for recycling. This cycle of bile salt secretion by hepatocytes into bile, reabsorption by the ileum, and resecretion into bile is called the **enterohepatic circulation**. Insufficient bile salts, due either to obstruction of the bile ducts or removal of the gallbladder, can result in the loss of up to 40% of dietary lipids in the feces due to diminished lipid absorption. When lipids are not absorbed properly, the fat-soluble vitamins are not adequately absorbed.

Absorption of Electrolytes

Recall that electrolytes are compounds that separate into ions in water and conduct electricity. Many of the electrolytes absorbed by the small intestine come from gastrointestinal secretions, and some are part of ingested foods and liquids. Sodium ions move into epithelial cells by diffusion and secondary active transport, then are actively transported out of intestinal epithelial cells by sodium–potassium pumps (Na⁺/K⁺ ATPases). Negatively charged bicarbonate, chloride, iodide, and nitrate ions can passively follow Na⁺ or be actively transported. Other electrolytes such as calcium, iron, potassium, magnesium, and phosphate ions are absorbed by active transport mechanisms.

Absorption of Vitamins

As you have just learned, the fat-soluble vitamins A, D, E, and K are included with ingested dietary lipids in micelles and are absorbed by simple diffusion. Most water-soluble vitamins, such as most B vitamins and vitamin C, also are absorbed via simple diffusion. Vitamin B_{12}, however, must be combined with intrinsic factor produced by the stomach for its absorption via active transport in the ileum.

Absorption of Water

The total volume of fluid that enters the small intestine each day—about 9.3 liters (9.8 qt)—comes from ingestion of liquids (about 2.3 liters) and from various gastrointestinal secretions (about 7.0 liters). Figure 23.21 depicts the amounts of fluid ingested, secreted,

Figure 23.21   Daily volumes of fluid ingested, secreted, absorbed, and excreted from the GI tract.

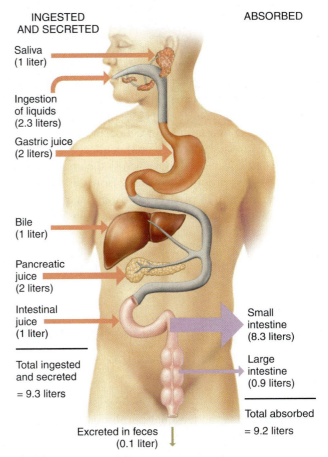

INGESTED AND SECRETED

ABSORBED

Saliva (1 liter)

Ingestion of liquids (2.3 liters)

Gastric juice (2 liters)

Bile (1 liter)

Pancreatic juice (2 liters)

Intestinal juice (1 liter)

Total ingested and secreted = 9.3 liters

Small intestine (8.3 liters)

Large intestine (0.9 liters)

Total absorbed = 9.2 liters

Excreted in feces (0.1 liter)

All water absorption in the GI tract occurs via osmosis.

TABLE 23.5

Digestive Activities in the Pancreas, Liver, Gallbladder, and Small Intestine

STRUCTURE	ACTIVITY
Pancreas	Delivers pancreatic juice into the duodenum via the pancreatic duct (see Table 23.4 for pancreatic enzymes and their functions).
Liver	Produces bile (bile salts) necessary for emulsification and absorption of lipids.
Gallbladder	Stores, concentrates, and delivers bile into the duodenum via the common bile duct.
Small intestine	Major site of digestion and absorption of nutrients and water in the gastrointestinal tract.
Mucosa/submucosa	
Intestinal glands	Secrete intestinal juice.
Duodenal glands	Secrete alkaline fluid to buffer stomach acids, and mucus for protection and lubrication.
Microvilli	Microscopic, membrane-covered projections of epithelial cells that contain brush border enzymes (listed in Table 23.4) and that increase the surface area for digestion and absorption.
Villi	Fingerlike projections of mucosa are the sites of absorption of digested food and increase the surface area for digestion and absorption.
Circular folds	Folds of mucosa and submucosa that increase the surface area for digestion and absorption.
Muscularis	
Segmentation	Consists of alternating contractions of circular smooth muscle fibers that produce segmentation and resegmentation of sections of the small intestine; mixes chyme with digestive juices and brings food into contact with the mucosa for absorption.
Peristalsis	Waves of contraction and relaxation of circular and longitudinal smooth muscle fibers passing the length of the small intestine; moves chyme toward ileocecal sphincter.

absorbed, and excreted by the GI tract. The small intestine absorbs about 8.3 liters of the fluid; the remainder passes into the large intestine, where most of the rest of it—about 0.9 liter—is absorbed. Only 0.1 liter (100 mL) of water is excreted in the feces each day. Most water is excreted via the urinary system.

All water absorption in the GI tract occurs via *osmosis* from the lumen of the intestines through epithelial cells and into blood capillaries. Because water can move across the intestinal mucosa in both directions, the absorption of water from the small intestine depends on the absorption of electrolytes and nutrients. The absorbed electrolytes, monosaccharides, and amino acids establish a concentration gradient for water that promotes water absorption via osmosis.

Table 23.5 summarizes the digestive activities of the pancreas, liver, gallbladder, and small intestine.

Checkpoint

29. What are the regions of the small intestine in order from stomach to large intestine? Which region is the longest?
30. What purpose do the circular folds of the small intestine serve?
31. What is the function of Paneth cells? Of the fluid secreted by duodenal glands?
32. What are the functions of segmentation and peristalsis in the small intestine?
33. Explain the functions of pancreatic amylase, brush border enzymes, pancreatic lipase, and deoxyribonuclease.
34. Why are bile salts needed for lipid digestion?
35. How are the end products of carbohydrate, protein, and lipid digestion absorbed?
36. Why are triglycerides circulated through the body in chylomicrons?
37. By which routes do absorbed nutrients reach the liver?

23.7 In the large intestine, the final secretion and absorption of nutrients occur as chyme moves toward the rectum.

The **large intestine** is the terminal portion of the GI tract. The overall functions of the large intestine are the completion of absorption, the production of certain vitamins, the formation of feces, and the expulsion of feces from the body.

Anatomy of the Large Intestine

The large intestine, which is about 1.5 m (5 ft) long and 6.5 cm (2.5 in.) in diameter, extends from the ileum to the anus. Structurally, the four major regions of the large intestine are the cecum, colon, rectum, and anal canal (Figure 23.22a).

Figure 23.22   Anatomy of the large intestine.

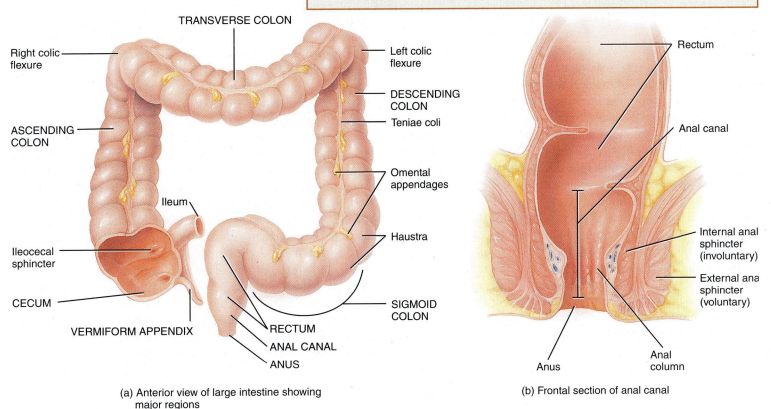

TRANSVERSE COLON

Right colic flexure

Left colic flexure

ASCENDING COLON

DESCENDING COLON

Teniae coli

Ileum

Omental appendages

Ileocecal sphincter

Haustra

CECUM

SIGMOID COLON

VERMIFORM APPENDIX

RECTUM

ANAL CANAL

ANUS

Rectum

Anal canal

Internal anal sphincter (involuntary)

External anal sphincter (voluntary)

Anus

Anal column

(a) Anterior view of large intestine showing major regions

(b) Frontal section of anal canal

🔑 **The regions of the large intestine are the cecum, colon, rectum, and anal canal.**

The opening from the ileum into the large intestine is guarded by a fold of mucous membrane called the **ileocecal sphincter**, which allows materials from the small intestine to pass into the large intestine. Hanging inferior to the ileocecal valve is the **cecum**, a small pouch about 6 cm (2.4 in.) long. Attached to the cecum is a twisted, coiled tube, measuring about 8 cm (3 in.) in length, called the **vermiform appendix** (*vermiform* = worm-shaped; *appendix* = appendage).

The open end of the cecum merges with a long tube called the **colon** (= food passage), which is divided into ascending, transverse, descending, and sigmoid portions. Both the ascending and descending colon are retroperitoneal; the transverse and sigmoid colon are not. True to its name, the **ascending colon** ascends on the right side of the abdomen, reaches the inferior surface of the liver, and turns abruptly to the left to form the **right colic (hepatic)**

flexure. The colon continues across the abdomen to the left side as the **transverse colon**. It curves beneath the inferior end of the spleen on the left side as the **left colic (splenic) flexure** and passes inferiorly to the level of the iliac crest as the **descending colon**. The **sigmoid colon** (*sigm-* = S-shaped) begins near the left iliac crest, projects medially to the midline, and terminates as the rectum.

The **rectum**, the last 20 cm (8 in.) of the GI tract, lies anterior to the sacrum and coccyx. The terminal 2–3 cm (1 in.) of the rectum is called the **anal canal** (Figure 23.22b). The mucous membrane of the anal canal is arranged in longitudinal folds called **anal columns**. The opening of the anal canal to the exterior, called the **anus**, is guarded by an **internal anal sphincter** of smooth muscle (involuntary) and an **external anal sphincter** of skeletal muscle (voluntary). Normally these sphincters keep the anus closed except during the elimination of feces.

Histology of the Large Intestine

The wall of the large intestine contains the typical four layers found in the rest of the GI tract: mucosa, submucosa, muscularis, and serosa (Figure 23.23a). The epithelium of the mucosa is simple columnar epithelium that contains mostly absorptive and goblet cells (Figure 23.23b,d). The absorptive cells function primarily in water absorption; the goblet cells secrete mucus that lubricates the passage of the colonic contents. Both absorptive and goblet cells are located in long, straight, tubular intestinal glands that extend the full thickness of the mucosa. Solitary lymphatic nodules are also found in the lamina propria of the mucosa and may extend through the muscularis mucosae into the submucosa.

Compared to the small intestine, the mucosa of the large intestine does not have as many structural adaptations that increase surface area. There are no circular folds or villi; however, microvilli are present on the absorptive cells. Consequently, much more absorption occurs in the small intestine than in the large intestine.

The **muscularis** consists of an external layer of longitudinal smooth muscle and an internal layer of circular smooth muscle. Unlike other parts of the GI tract, portions of the longitudinal muscles are thickened, forming three conspicuous bands called the **teniae coli** (TĒ-nē-ē KŌ-lī; *teniae* = flat bands) that run most of the length of the large intestine (Figure 23.22a). The teniae coli are separated by portions of the wall with circular muscle and less or no longitudinal muscle. Tonic contractions of the bands gather the colon into a series of pouches called **haustra** (HAWS-tra = shaped like pouches; singular is *haustrum*), which give the colon a puckered appearance.

The **serosa** of the large intestine is part of the visceral peritoneum. Small pouches of visceral peritoneum filled with fat are attached to teniae coli and are called **omental appendages**.

Mechanical Digestion in the Large Intestine

The passage of chyme from the ileum into the cecum is regulated by the action of the ileocecal sphincter. Normally, the valve remains partially closed so that the passage of chyme into the cecum usually occurs slowly. Immediately after a meal, a **gastroileal reflex** intensifies ileal peristalsis and forces any chyme in the ileum into the cecum. The hormone gastrin also relaxes the sphincter. Whenever the cecum is distended, the degree of contraction of the ileocecal sphincter intensifies.

Movements of the colon begin when substances pass the ileocecal sphincter. As food passes through the ileocecal sphincter, it fills the cecum and accumulates in the ascending colon. One movement characteristic of the large intestine is **haustral churning**. In this process, the haustra remain relaxed and become distended while they fill up. When the distention reaches a certain point, the walls contract and squeeze the contents into the next haustrum. **Peristalsis** also occurs, although at a slower rate (3–12 contractions per minute) than in more proximal portions of the tract. A final type of movement is **mass peristalsis**, a strong peristaltic wave that begins at about the middle of the transverse colon and quickly drives the contents of the colon into the rectum. Because food in the stomach initiates this **gastrocolic reflex** in the colon, mass peristalsis usually takes place three or four times a day, during or immediately after a meal.

Figure 23.23 Histology of the large intestine.

Lumen of large intestine

Openings of intestinal glands
Simple columnar epithelium
Intestinal gland
Lamina propria
Lymphatic nodule
Muscularis mucosae
Lymphatic vessel
Arteriole
Venule
Circular layer of muscle
Myenteric plexus
Longitudinal layer of muscle

MUCOSA
SUBMUCOSA
MUSCULARIS
SEROSA

(a) Three-dimensional view of layers of the large intestine

Openings of intestinal glands

Lamina propria

Microvilli

Absorptive cell (absorbs water)

Intestinal gland

Goblet cell (secretes mucus)

Lymphatic nodule

Muscularis mucosae

Submucosa

(b) Sectional view of intestinal glands and cell types

Mucosa

Submucosa

Muscularis

Serosa

Lumen of large intestine

Lamina propria

Intestinal gland

Lymphatic nodule

Muscularis mucosae

LM 315x

(c) Portion of the wall of the large intestine

Opening of intestinal gland

Lumen of large intestine

Absorptive cell

Goblet cell

Lamina propria

Intestinal gland

LM 300x

(d) Details of mucosa of large intestine

Intestinal glands containing absorptive cells and goblet cells extend the full thickness of the mucosa.

Chemical Digestion in the Large Intestine

The final stage of digestion occurs in the colon through the activity of bacteria that inhabit the lumen. Mucus is secreted by the glands of the large intestine, but no enzymes are secreted. Bacteria ferment any remaining carbohydrates in the chyme and release hydrogen, carbon dioxide, and methane gases. These gases contribute to flatus (gas) in the colon, termed *flatulence* when it is excessive. Bacteria also break down the remaining proteins to amino acids and decompose bilirubin to simpler pigments, including stercobilin, which give feces their brown color. Bacterial products that are absorbed in the colon include several vitamins needed for normal metabolism, among them some B vitamins and vitamin K.

Absorption and Feces Formation in the Large Intestine

By the time chyme has remained in the large intestine 3–10 hours, it has become solid or semisolid because of water absorption and is now called **feces**. Chemically, feces consist of water, inorganic salts, sloughed-off epithelial cells from the mucosa of the gastrointestinal tract, bacteria, products of bacterial decomposition, unabsorbed digested materials, and indigestible parts of food.

Although 90% of all water absorption occurs in the small intestine, the large intestine absorbs enough to make it an important organ in maintaining the body's water balance. Of the 0.5–1.0 liter of water that enters the large intestine, all but about 100–200 mL is normally absorbed via osmosis. The large intestine also absorbs ions, including sodium and chloride, and some vitamins.

The Defecation Reflex

Mass peristaltic movements push fecal material from the sigmoid colon into the rectum. The resulting distention of the rectal wall stimulates stretch receptors, which initiates a **defecation reflex** that empties the rectum. The defecation reflex occurs as follows: In response to distention of the rectal wall, receptors send sensory nerve impulses to the sacral spinal cord. Motor impulses from the cord travel along parasympathetic nerves back to the descending colon, sigmoid colon, rectum, and anus. The resulting contraction of the longitudinal rectal muscles shortens the rectum, thereby increasing the pressure within it. This pressure, plus parasympathetic stimulation, opens the internal anal sphincter.

The external anal sphincter is voluntarily controlled. If it is voluntarily relaxed, defecation occurs and the feces are expelled through the anus; if it is voluntarily constricted, defecation can be postponed. Voluntary contractions of the diaphragm and abdominal muscles aid defecation by increasing the pressure within the abdomen, which pushes the walls of the sigmoid colon and rectum inward. If defecation does not occur, the feces back up into the sig-

moid colon until the next wave of mass peristalsis stimulates the stretch receptors, again creating the urge to defecate. In infants, the defecation reflex causes automatic emptying of the rectum because voluntary control of the external anal sphincter has not yet developed.

The amount of bowel movements that a person has over a given period of time depends on various factors such as diet, health, and stress. The normal range of bowel activity varies from two or three bowel movements per day to three or four bowel movements per week.

Table 23.6 summarizes the digestive activities in the large intestine, and Table 23.7 summarizes the functions of all digestive system organs.

Checkpoint

38. What are the major regions of the large intestine?
39. What is the function of the goblet cells in the large intestine?
40. How does mechanical digestion occur in the large intestine?
41. Which activities occur in the large intestine to change its contents into feces?
42. What is defecation and how does it occur?

TABLE 23.6

Digestive Activities in the Large Intestine

STRUCTURE	ACTIVITY	FUNCTION
Lumen	Bacterial activity.	Breaks down undigested carbohydrates, proteins, and amino acids into products that can be expelled in feces or absorbed and detoxified by liver; synthesizes certain B vitamins and vitamin K.
Mucosa	Mucus secretion.	Lubricates colon and protects mucosa.
	Absorption.	Water absorption solidifies feces and contributes to the body's water balance; solutes absorbed include ions and some vitamins.
Muscularis	Haustral churning.	Moves contents from haustrum to haustrum by muscular contractions.
	Peristalsis.	Moves contents along length of colon by contractions of circular and longitudinal muscles.
	Mass peristalsis.	Forces contents into sigmoid colon and rectum.
	Defecation reflex.	Eliminates feces by contractions in sigmoid colon and rectum.

Organs of the Digestive System and Their Functions

ORGAN	FUNCTIONS
Mouth	See other listings in this table for the functions of the tongue, salivary glands, and teeth, all of which are in the mouth. Additionally, the lips and cheeks keep food between the teeth during mastication, and buccal glands lining the mouth produce saliva.
Tongue	Maneuvers food for mastication, shapes food into a bolus, maneuvers food for deglutition, detects taste and touch sensations, and initiates digestion of triglycerides.
Salivary glands	Produce saliva, which softens, moistens, and dissolves foods; cleanses mouth and teeth; and initiates the digestion of starch.
Teeth	Cut, tear, and pulverize food to reduce solids to smaller particles for swallowing.
Pharynx	Receives a bolus from the oral cavity and passes it into the esophagus.
Esophagus	Receives a bolus from the pharynx and moves it into the stomach. This requires relaxation of the upper esophageal sphincter and secretion of mucus.
Stomach	Mixing waves macerate food, mix it with secretions of gastric glands (gastric juice), and reduce food to chyme. Gastric juice activates pepsin and kills many microbes in food. Intrinsic factor aids absorption of vitamin B_{12}. The stomach serves as a reservoir for food before releasing it into the small intestine.
Pancreas	Pancreatic juice buffers acidic gastric juice in chyme (creating the proper pH for digestion in the small intestine), stops the action of pepsin from the stomach, and contains enzymes that digest carbohydrates, proteins, triglycerides, and nucleic acids.
Liver	Produces bile, which is needed for the emulsification and absorption of lipids in the small intestine.
Gallbladder	Stores and concentrates bile and releases it into the small intestine.
Small intestine	Segmentations mix chyme with digestive juices; peristalsis propels chyme toward the ileocecal sphincter; digestive secretions from the small intestine, pancreas, and liver complete the digestion of carbohydrates, proteins, lipids, and nucleic acids; circular folds, villi, and microvilli increase surface area for absorption; site where about 90% of nutrients and water are absorbed.
Large intestine	Haustral churning, peristalsis, and mass peristalsis drive the contents of the colon into the rectum; bacteria produce some B vitamins and vitamin K; absorption of some water, ions, and vitamins; defecation.

23.8 Digestive activities occur in three overlapping phases: cephalic, gastric, and intestinal.

Digestive activities occur in three overlapping phases: the cephalic phase, the gastric phase, and the intestinal phase.

Cephalic Phase

During the **cephalic phase** of digestion, the smell, sight, thought, or initial taste of food activates neural centers in the cerebral cortex, hypothalamus, and brain stem. The brain stem then activates the facial (VII) and glossopharyngeal (IX) nerves to stimulate the salivary glands to secrete saliva and the vagus (X) nerve to stimulate the gastric glands to secrete gastric juice. The purpose of the cephalic phase of digestion is to prepare the mouth and stomach for food that is about to be eaten.

Gastric Phase

Once food reaches the stomach, the **gastric phase** of digestion begins. Neural and hormonal mechanisms regulate the gastric phase of digestion to promote gastric secretion and gastric motility.

Neural Regulation

Food of any kind distends the stomach and stimulates stretch receptors in its walls. Chemoreceptors in the stomach monitor the pH of the stomach chyme. When the stomach walls are distended or pH increases because proteins have entered the stomach and buffered some of the stomach acid, the stretch receptors and chemoreceptors are activated, and a neural negative feedback loop is set in motion (Figure 23.24). From the stretch receptors and chemoreceptors, nerve impulses propagate to the submucosal plexus, where they activate parasympathetic and enteric neurons. The resulting nerve impulses cause waves of peristalsis and stimulate the flow of gastric juice from gastric glands. The peristaltic waves mix the food with gastric juice; when the waves become strong enough, a small quantity of chyme is passed into the duodenum. As chyme passes from the stomach to the duodenum, the pH of the stomach chyme decreases (becomes more acidic) and the distention of the stomach walls lessens, suppressing secretion of gastric juice.

Hormonal Regulation

Gastric secretion during the gastric phase is also regulated by the hormone **gastrin**. Gastrin is released from the G cells of the gastric glands in response to several stimuli: distention of the stomach by chyme, partially digested proteins in chyme, high pH of chyme due to the presence of food in the stomach, caffeine in chyme, and acetylcholine released from parasympathetic neurons. Once it is released, gastrin enters the bloodstream, makes a round-trip through the body, and finally reaches its target organs in the digestive system.

Figure 23.24   Neural negative feedback regulation of the pH of gastric juice and gastric motility during the gastric phase of digestion.

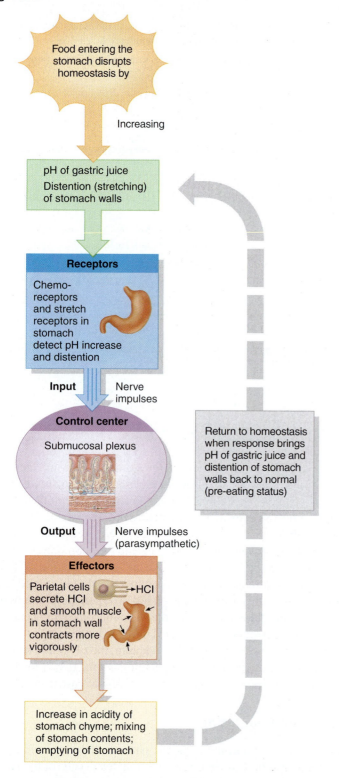

Food entering the stomach disrupts homeostasis by

Increasing

pH of gastric juice
Distention (stretching) of stomach walls

Receptors

Chemo-receptors and stretch receptors in stomach detect pH increase and distention

Input   Nerve impulses

Control center

Submucosal plexus

Return to homeostasis when response brings pH of gastric juice and distention of stomach walls back to normal (pre-eating status)

Output   Nerve impulses (parasympathetic)

Effectors

Parietal cells secrete HCl and smooth muscle in stomach wall contracts more vigorously
HCl

Increase in acidity of stomach chyme; mixing of stomach contents; emptying of stomach

Food entering the stomach stimulates secretion of gastric juice and causes vigorous waves of peristalsis.

Gastrin stimulates gastric glands to secrete large amounts of gastric juice. It also strengthens the contraction of the lower esophageal sphincter to prevent reflux of acid chyme into the esophagus, increases motility of the stomach, and relaxes the pyloric sphincter, which promotes gastric emptying. Gastrin secretion is inhibited when the pH of gastric juice drops below 2.0 and is stimulated when the pH rises. This negative feedback mechanism helps provide an optimal low pH for the functioning of pepsin, the killing of microbes, and the breakdown of proteins in the stomach.

Intestinal Phase

The **intestinal phase** of digestion begins once food enters the small intestine. In contrast to reflexes initiated during the cephalic and gastric phases, which stimulate stomach secretory activity and motility, those occurring during the intestinal phase have inhibitory effects that slow the exit of chyme from the stomach. This prevents the duodenum from being overloaded with more chyme than it can handle. In addition, responses occurring during the intestinal phase promote the continued digestion of foods that have reached the small intestine. These activities of the intestinal phase of digestion are regulated by neural and hormonal mechanisms.

Neural Regulation

Distention of the duodenum by the presence of chyme causes the **enterogastric reflex**. Stretch receptors in the duodenal wall send nerve impulses to the medulla oblongata, where they inhibit parasympathetic stimulation and stimulate sympathetic stimulation of the stomach. As a result, gastric motility is inhibited and there is an increase in the contraction of the pyloric sphincter, which decreases gastric emptying.

Hormonal Regulation

The intestinal phase of digestion is mediated by two major hormones secreted by the small intestine: cholecystokinin and secretin. **Cholecystokinin (CCK)** is secreted by the CCK cells of the small intestinal glands in response to chyme containing amino acids and fatty acids. CCK stimulates secretion of pancreatic juice that is rich in digestive enzymes. It also causes contraction of the wall of the gallbladder, which squeezes stored bile out of the gallbladder into the cystic duct and through the common bile duct. In addition, CCK causes relaxation of the sphincter of the hepatopancreatic ampulla, which allows pancreatic juice and bile to flow into the duodenum. CCK also slows gastric emptying by promoting contraction of the pyloric sphincter and produces satiety (a feeling of fullness) by acting on the hypothalamus in the brain.

Acidic chyme entering the duodenum stimulates the release of **secretin** from the S cells of the small intestinal glands. Secretin stimulates the flow of pancreatic juice that is rich in bicarbonate

TABLE 23.8

Major Hormones That Control Digestion

HORMONE	STIMULUS AND SITE OF SECRETION	ACTIONS
Gastrin	Distension of stomach, partially digested proteins and caffeine in stomach, and high pH of stomach chyme stimulate gastrin secretion by G cells located in the mucosa of the stomach.	Stimulates secretion of gastric juice, increases gastric motility, constricts lower esophageal sphincter, and relaxes pyloric sphincter.
Cholecystokinin (CCK)	Amino acids and fatty acids that enter the small intestine stimulate secretion of CCK by CCK cells in the mucosa of the small intestine; CCK is also released in the brain.	Stimulates secretion of pancreatic juice rich in digestive enzymes, causes ejection of bile from the gallbladder, causes opening of the sphincter of the hepatopancreatic ampulla, and induces satiety (feeling full to satisfaction); inhibits gastric emptying.
Secretin	Acidic (high H^+ level) chyme that enters the small intestine stimulates secretion of secretin by S cells in the mucosa of the duodenum.	Stimulates secretion of pancreatic juice and bile that are rich in bicarbonate ions; inhibits secretion of gastric juice.

(HCO_3^-) ions to buffer the acidic chyme that enters the duodenum from the small intestine. Besides this major effect, secretin inhibits secretion of gastric juice. Overall, secretin causes buffering of acid in chyme that reaches the duodenum and slows production of acid in the stomach.

Table 23.8 summarizes the major hormones that control digestion.

Checkpoint

43. What is the result and purpose of the cephalic phase of digestion?
44. Why does food initially cause the pH of the gastric juice to rise?
45. Why is stomach activity inhibited during the intestinal phase of digestion?
46. What is the enterogastric reflex?
47. Explain the roles of CCK and secretin in the intestinal phase of digestion.

Return to Chloe's Story

"**I** told you I would be okay, Mom," Chloe said to her mother.

"I am so glad, Chloe. When can you go back to eating solid foods? I made some wonderful fried chicken," her mom replied.

"Mom, that's part of the reason I gained so much weight to begin with—your great cooking! Dr. Benavides said I can't have complex carbohydrates or proteins until I heal more. Right now I have my milk shakes. They gave me some chocolate-flavored drink that has sugars and amino acids in it."

"That can't be enough!" her mom exclaimed.

Chloe picked up the can and scanned the label. "Let's see here. Hmm, it says there is fat in this stuff. It's supposed to be complete nutrition, but it's not the same as a pizza."

"Have you been feeling all right since the surgery?" her mom asked.

"For the most part. I had such a craving for chocolate the other day, though, that I couldn't resist. I only had two pieces, but it made me sick. I threw up and had diarrhea. I won't do that again. I don't know if the vomiting is normal or not. I'm supposed to have a consultation with the dietician to discuss my new diet plan next week. I'll have to ask her about that."

Chloe arrived for her appointment with the dietition. "Hi, Chloe, I'm Clarice. I'll be helping you develop a plan that fits your diet preferences. I am particularly concerned about you potentially developing vitamin B_{12} and iron deficiencies. I have also had gastric bypass patients who have had calcium and magnesium deficiencies. So there is a very real risk of anemia and osteoporosis."

"Well, I take vitamins. Won't that help?" Chloe asked.

"Yes, but we'll still want to monitor your nutrition closely to help you develop successful lifestyle changes and prevent any health problems. Your digestive tract has been permanently altered and you won't digest and absorb nutrients like you used to. There will be some precautions to remember, and vitamin supplements to make sure you take. But in six to eight weeks you should be back on a pretty normal diet. Let's talk about what kinds of food you like."

E. Which pancreatic enzymes would Chloe need to break down the fats in her nutrition drink?

F. Why might Chloe be at risk for gallstones (crystallized cholesterol which can block the flow of bile from the gallbladder)?

G. Would absorption of nutrients be substantially altered in the small intestine following Chloe's gastric bypass surgery? Why or why not?

H. Why is vitamin B_{12} deficiency a serious concern?

23.9 Food molecules supply energy for life processes and serve as building blocks for complex molecules.

Nutrients are chemical substances in food that body cells use for growth, maintenance, and repair. The six main types of nutrients are water, carbohydrates, lipids, proteins, minerals, and vitamins. Water is the nutrient needed in the largest amount—about 2–3 liters per day. As the most abundant compound in the body, water provides the medium in which most metabolic reactions occur, and it also participates in some reactions (for example, hydrolysis reactions). Carbohydrates, lipids, and proteins provide the energy needed for metabolic reactions and serve as building blocks to make body structures. Some minerals and many vitamins are components of the enzyme systems that catalyze metabolic reactions. *Essential nutrients* are specific nutrient molecules that the body cannot make in sufficient quantity to meet its needs and thus must be obtained from the diet. Some amino acids, fatty acids, vitamins, and minerals are essential nutrients.

Guidelines for Healthy Eating

Each gram of protein or carbohydrate in food provides about 4 Calories; a gram of fat (lipids) provides about 9 Calories. We do not know with certainty what levels and types of carbohydrate, fat, and protein are optimal in the diet. Different populations around the world eat radically different diets that are adapted to their particular lifestyles. However, many experts recommend the following distribution of calories: 50–60% from carbohydrates, with less than 15% from simple sugars; less than 30% from fats (triglycerides are the main type of dietary fat), with no more than 10% as saturated fats; and about 12–15% from proteins.

The guidelines for healthy eating are as follows:

- Eat a variety of foods.

- Maintain a healthy weight.

- Choose foods low in fat, saturated fat, and cholesterol.

- Eat plenty of vegetables, fruits, and grain products.

- Use sugars in moderation only.

In 2005, the United States Department of Agriculture (USDA) introduced a new food pyramid called **MyPyramid**, which represents a *personalized* approach to making healthy food choices and maintaining regular physical activity. By consulting a chart, it is possible to determine your calorie level based on your gender, age, and activity level. Once this is determined, you can choose the type and amount of food to be consumed.

If you carefully examine the MyPyramid in Figure 23.25, you will note that the six color bands represent the five basic food groups plus oils. Foods from all bands are needed each day. Also note that the overall size of the bands suggests the proportion of food a person should choose on a daily basis. The wider base of each band represents foods with little or no solid fats or added sugars; these foods should be selected more often. The narrower top

Figure 23.25   MyPyramid.

🔑 **MyPyramid is a new personalized approach to making healthy food choices and maintaining physical activity.**

of each band represents foods with more added sugars and solid fats, which should be selected less frequently. The person climbing the steps is a reminder of the need for daily physical activity.

As an example of how the My Pyramid works, let's assume that the calorie level of an 18-year-old moderately active female is 2000 Calories and that of an 18-year-old moderately active male is 2800 Calories. According to this model, the following foods should be chosen in the following amounts:

CALORIE LEVEL	2000	2800
Fruits (includes all fresh, frozen, canned, and dried fruits and fruit juices)	2 cups	2.5 cups
Vegetables (includes all fresh, frozen, canned, and dried vegetables and vegetable juices)	2.5 cups	3.5 cups
Grains (includes all foods made from wheat, rice, oats, cornmeal, and barley such as bread, cereals, oatmeal, rice, pasta, crackers, tortillas, and grits)	6 oz	10 oz
Meats and beans (includes lean meat, poultry, fish, eggs, peanut butter, beans, nuts, and seeds)	5.5 oz	7 oz
Milk group (includes milk products and foods made from milk that retain their calcium content such as cheeses and yogurt)	3 cups	3 cups
Oils (choose mostly fats that contain monounsaturated and polyunsaturated fatty acids such as fish, nuts, seeds, and vegetable oils)	6 tsp	8 tsp

In addition, you should choose and prepare foods with little salt. In fact, sodium intake should be less than 2300 mg per day. If you choose to drink alcohol, it should be consumed in moderation (no more than one drink per day for women and two drinks per day for men). A drink is defined as 12 oz of regular beer, 5 oz of wine, or 1 1/2 oz of 80 proof distilled spirits.

Minerals

Minerals are inorganic elements that constitute about 4% of the total body weight and are concentrated most heavily in the skeleton. Minerals with known functions in the body include calcium, phosphorus, potassium, sulfur, sodium, chloride, magnesium, iron, iodide, manganese, copper, cobalt, zinc, fluoride, selenium, and chromium. Other minerals—aluminum, boron, silicon, and molybdenum—are present but their functions are unclear. Typical diets supply adequate amounts of potassium, sodium, chloride, and magnesium. Some attention must be paid to eating foods that provide enough calcium, phosphorus, iron, and iodide. Excess amounts of most minerals are excreted in the urine and feces.

Calcium and phosphorus form part of the matrix of bone. Because minerals do not form long-chain compounds, they are otherwise poor building materials. A major role of minerals is to help regulate enzymatic reactions. Calcium, iron, magnesium, and manganese are part of some coenzymes. Magnesium also serves as a catalyst for the conversion of ADP to ATP. Minerals such as sodium and phosphorus work in buffer systems, which help control the pH of body fluids. Sodium also helps regulate the osmosis of water and, along with other ions, is involved in the generation of nerve impulses.

Vitamins

Organic nutrients required in small amounts to maintain growth and normal metabolism are called **vitamins**. Unlike carbohydrates, lipids, or proteins, vitamins do not provide energy or serve as the body's building materials. Most vitamins with known functions are coenzymes.

Most vitamins cannot be synthesized by the body and must be ingested in food. Other vitamins, such as vitamin K, are produced by bacteria in the GI tract and then absorbed. The body can assemble some vitamins if the raw materials, called **provitamins**, are provided. For example, vitamin A is produced by the body from the provitamin beta-carotene, a chemical present in yellow vegetables such as carrots and in dark green vegetables such as spinach. No single food contains all the required vitamins—one of the best reasons to eat a varied diet.

Vitamins are divided into two main groups: fat-soluble and water-soluble. The **fat-soluble vitamins**, vitamins A, D, E, and K, are absorbed along with other dietary lipids in the small intestine and packaged into chylomicrons. They cannot be absorbed in adequate quantity unless they are ingested with other lipids. Fat-soluble vitamins may be stored in cells, particularly hepatocytes. The **water-soluble vitamins**, including several B vitamins and vitamin C, are dissolved in body fluids. Excess quantities of these vitamins are not stored but instead are excreted in the urine.

Besides their other functions, three vitamins—C, E, and beta carotene (a provitamin)—are termed **antioxidant vitamins** because they inactivate oxygen free radicals. Free radicals are highly reactive ions or molecules that carry an unpaired electron in their outermost electron shell (see Figure 2.3). Free radicals damage cell membranes, DNA, and other cellular structures and contribute to the formation of artery-narrowing atherosclerotic plaques. Some free radicals arise naturally in the body, and others come from environmental hazards such as tobacco smoke and radiation. Antioxidant vitamins are thought to play a role in protecting against some kinds of cancer, reducing the buildup of atherosclerotic plaque, delaying some effects of aging, and decreasing the chance of cataract formation in the lenses of the eyes.

Checkpoint

48. Which food group does MyPyramid recommend we consume the least? Why? What do the wider bases in MyPyramid mean?
49. Briefly describe the functions of the following minerals: calcium, iron, magnesium, manganese, sodium, and phosphorus.
50. What is a vitamin? What is the general function of vitamins? How do we obtain vitamins?
51. How are fat-soluble and water-soluble vitamins different?
52. Which functions are served by antioxidant vitamins?

23.10 Metabolism includes the catabolism and anabolism of molecules.

Metabolic Reactions

Metabolism (me-TAB-ō-lizm; *metabol-* = change) refers to all the chemical reactions of the body. Recall from Chapter 2 that chemical reactions occur when chemical bonds between substances are formed or broken and that **enzymes** serve as catalysts to speed up chemical reactions. Some enzymes require the presence of an ion such as calcium, iron, or zinc. Other enzymes work together with **coenzymes**, which function as temporary carriers of atoms being removed from or added to a substrate during a reaction. Many coenzymes are derived from vitamins. Examples include the coenzyme *NAD⁺*, derived from the B vitamin niacin, and the coenzyme *FAD,* derived from vitamin B_2 (riboflavin).

The body's metabolism may be thought of as an energy-balancing act between anabolic (synthesis) and catabolic (decomposition) reactions. Chemical reactions that combine simple substances into more complex molecules are collectively known as **anabolism** (a-NAB-ō-lizm; *ana-* = upward). Examples of anabolic reactions are the formation of peptide bonds between amino acids during protein synthesis, the building of fatty acids into phospholipids that form the plasma membrane bilayer, and the linkage of glucose monomers to form glycogen. Anabolic reactions are *endergonic;* they consume more energy than they produce (Figure 23.26).

Those chemical reactions that break down complex organic molecules into simpler ones are collectively known as **catabolism** (ka-TAB-ō-lizm; *cata-* = downward). Overall, catabolic (decomposition) reactions are *exergonic;* they produce more energy than

they consume, releasing the chemical energy stored in organic molecules. This energy is transferred to molecules of ATP and then used to power anabolic reactions. Important sets of catabolic reactions occur during glycolysis, the Krebs cycle, and the electron transport chain, which are discussed shortly.

Figure 23.26   Role of ATP in linking anabolic and catabolic reactions. When complex molecules and polymers are split apart (catabolism, at left), some of the energy is transferred to form ATP and the rest is given off as heat. When simple molecules and monomers are combined to form complex molecules (anabolism, at right), ATP provides the energy for synthesis, and again some energy is given off as heat.

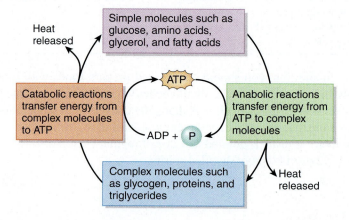

The coupling of energy-releasing and energy-requiring reactions is achieved through ATP.

Coupling of Catabolism and Anabolism by ATP

The chemical reactions of living systems depend on the efficient transfer of manageable amounts of energy from one molecule to another. The molecule that most often performs this task is ATP, the "energy currency" of a living cell. Like money, it is readily available to "buy" cellular activities; it is spent and earned over and over. A typical cell has about a billion molecules of ATP, each of which typically lasts for less than a minute before being used. Thus, ATP is not a long-term storage form of currency, like gold in a vault, but rather convenient cash for moment-to-moment transactions.

Recall from Chapter 2 that a molecule of ATP consists of an adenine molecule, a ribose molecule, and three phosphate groups bonded to one another (see Figure 2.24). Figure 23.26 shows how ATP links anabolic and catabolic reactions. When the phosphate group is removed from ATP in anabolic reactions, adenosine diphosphate (ADP) and a phosphate group (symbolized as $\textcircled{P}$) are formed. Some of the energy released is used to drive anabolic reactions such as the formation of glycogen from glucose. Energy is also needed to combine ADP and a phosphate group to resynthesize ATP:

$$ADP + \textcircled{P} + energy \rightarrow ATP$$

About 40% of the energy released in catabolism is used for cellular functions; the rest is converted to heat, some of which helps maintain normal body temperature. Excess heat is lost to the environment. Compared with machines, which typically convert only 10–20% of energy into work, the 40% efficiency of the body's metabolism is impressive. Still, the body has a continuous need to take in and process external sources of energy so that cells can synthesize enough ATP to sustain life.

Carbohydrate Metabolism

During digestion, polysaccharide and disaccharide carbohydrates are catabolized to monosaccharides—glucose, fructose, and galactose—which are absorbed in the small intestine. Shortly after their absorption, however, fructose and galactose are converted to glucose. Thus, the story of carbohydrate metabolism is really the story of glucose metabolism.

Because glucose is the body's preferred source for synthesizing ATP, the fate of glucose absorbed from the diet depends on the needs of body cells. If the cells require ATP immediately, they oxidize the glucose. Glucose not needed for immediate ATP production may be converted to glycogen for storage in hepatocytes and skeletal muscle fibers. If these glycogen stores are full, hepatocytes can transform the glucose to glycerol and fatty acids that can be used for the synthesis of triglycerides. Triglycerides then are deposited in adipose tissue, which has virtually unlimited stor-

age capacity. Cells throughout the body can use glucose to form several amino acids, which then can be incorporated into proteins.

Before glucose can be used by body cells, it must first pass through the plasma membrane and enter the cytosol. Glucose absorption in the GI tract (and kidney tubules) is accomplished via secondary active transport (Na^+–glucose symporters). Glucose entry into most other body cells occurs via facilitated diffusion. Insulin increases the rate of facilitated diffusion of glucose.

Glucose Catabolism

The catabolism of glucose to produce ATP is known as cellular respiration. Its many reactions can be summarized as follows:

$$\underset{\text{Glucose}}{C_6H_{12}O_6} + \underset{\text{Oxygen}}{6\,O_2} + 36 \text{ or } 38 \text{ ADPs} + 36 \text{ or } 38\,\textcircled{P} \longrightarrow$$

$$\underset{\text{Carbon dioxide}}{6\,CO_2} + \underset{\text{Water}}{6\,H_2O} + 36 \text{ or } 38 \text{ ATPs}$$

Four interconnecting sets of chemical reactions contribute to cellular respiration (Figure 23.27):

1 During **glycolysis** (glī-KOL-i-sis; *glyco-* = sugar; *-lysis* = breakdown), reactions that take place in the cytosol convert one six-carbon glucose molecule into two three-carbon pyruvic acid molecules. Glycolysis produces two molecules of ATP and two energy-containing NADH + H⁺. Because glycolysis does not require oxygen, it is a way to produce ATP anaerobically (without oxygen) and is known as **anaerobic cellular respiration**. If oxygen is available, however, most cells next convert pyruvic acid to acetyl coenzyme A.

2 The formation of **acetyl coenzyme A** is a transition step that prepares pyruvic acid for entrance into the Krebs cycle. First, pyruvic acid enters a mitochondrion and is converted to a two-carbon fragment by removing a molecule of carbon dioxide (CO_2). Molecules of CO_2 produced during glucose catabolism diffuse into the blood and are eventually exhaled. Then, the coenzyme NAD⁺ removes a hydrogen molecule (H_2) from pyruvic acid, in the process becoming NADH + H⁺. Finally, the remaining atom, called an **acetyl group**, is attached to coenzyme A, forming acetyl coenzyme A.

3 The **Krebs cycle** is a series of reactions that transfers chemical energy from acetyl coenzyme A to two other coenzymes—NAD⁺ and FAD—thereby forming NADH + H⁺ and FADH₂. Krebs cycle reactions also produce CO_2 and one ATP for each acetyl coenzyme A that enters the Krebs cycle. To harvest the energy in NADH and FADH₂, their high-energy electrons must first go through the electron transport chain.

Figure 23.27   Cellular respiration.

The catabolism of glucose to produce ATP involves glycolysis, the formation of acetyl coenzyme A, the Krebs cycle, and the electron transport chain.

4 Through the reactions of the **electron transport chain**, the energy in NADH/H$^+$ and FADH$_2$ is used to synthesize ATP. As the coenzymes pass their high-energy electrons through a series of "electron carriers," ATP is synthesized. Finally, lower-energy electrons are passed to oxygen in a reaction that produces water. Because the Krebs cycle and the electron transport chain together require oxygen to produce ATP, they are collectively known as **aerobic cellular respiration**.

Glucose Anabolism

Although most of the glucose in the body is catabolized to generate ATP, glucose may take part in or be formed via several anabolic reactions. One is the synthesis of glycogen; another is the synthesis of new glucose molecules from some of the products of protein and lipid breakdown.

If glucose is not needed immediately for ATP production, it combines with many other molecules of glucose to form **glyco-**

Figure 23.28 Reactions of glucose anabolism: synthesis of glycogen, breakdown of glycogen, and synthesis of glucose from amino acids, lactic acid, or glycerol.

Key:

→ Synthesis of glycogen (stimulated by insulin)

→ Breakdown of glycogen (stimulated by glucagon and epinephrine)

→ Gluconeogenesis (stimulated by cortisol and glucagon)

→ Catabolism of triglycerides (lipolysis)

🔑 **The body stores glucose as glycogen in skeletal muscles and the liver.**

gen, a polysaccharide that is the only stored form of carbohydrate in our bodies (Figure 23.28). Synthesis of glycogen is stimulated by insulin released from the pancreas. The body can store about 500 grams (about 1.1 lb) of glycogen, roughly 75% in skeletal muscle fibers and the rest in hepatocytes.

If blood glucose level falls below normal, glucagon is released from the pancreas and epinephrine is released from the adrenal medulla. These hormones stimulate breakdown of glycogen into its glucose subunits (Figure 23.28). Hepatocytes release this glucose into the blood, and body cells pick it up to use for ATP production. Glycogen breakdown usually occurs between meals.

When your liver runs low on glycogen, it is time to eat. If you don't, your body starts catabolizing triglycerides (fats) and proteins. Actually, the body normally catabolizes some of its triglycerides and proteins, but large-scale triglyceride and protein catabolism does not happen unless you are starving, eating very few carbohydrates, or suffering from an endocrine disorder.

Lactic acid, certain amino acids, and the glycerol part of triglycerides can be converted to glucose in the liver (Figure 23.28). The series of reactions that form glucose from these noncarbohydrate sources is called **gluconeogenesis** (gloo´-kō-nē´-ō-JEN-e-sis; -*neo*- = new). Glycerol may be converted into glyceraldehyde 3-phosphate, which may form pyruvic acid or be used to synthesize glucose. About 60% of the amino acids in the body can be used for gluconeogenesis. Lactic acid and amino acids such as alanine, cysteine, glycine, serine, and threonine are converted to pyruvic acid, which may be synthesized into glyceraldehyde 3-phosphate, and then glucose. Gluconeogenesis releases glucose into the blood, thereby keeping blood glucose level normal during the hours between meals when glucose is not being absorbed.

Gluconeogenesis occurs when the liver is stimulated by cortisol from the adrenal cortex and glucagon from the pancreas.

Lipid Metabolism

Lipids, like carbohydrates, may be catabolized to produce ATP. If the body has no immediate need to use lipids in this way, they are stored as triglycerides in adipose tissue throughout the body and in the liver. A few lipids are used as structural molecules or to synthesize other substances. Some examples include phospholipids, which are constituents of plasma membranes; lipoproteins, which are used to transport cholesterol throughout the body; thromboplastin, which is needed for blood clotting; and myelin sheaths, which speed up nerve impulse conduction. Two **essential fatty acids** that the body cannot synthesize are linoleic acid and linolenic acid. Dietary sources of these lipids include vegetable oils and leafy vegetables.

Lipid Catabolism

Muscle, liver, and adipose cells routinely catabolize fatty acids from triglycerides to produce ATP. First, the triglycerides are split into glycerol and fatty acids—a process called **lipolysis** (li-POL-i-sis) (Figure 23.29). The hormones epinephrine, norepinephrine, and cortisol, and the thyroid hormones, enhance lipolysis.

The glycerol and fatty acids that result from lipolysis are catabolized via different pathways. Glycerol is converted by many cells of the body to glyceraldehyde 3-phosphate, one of the compounds also formed during the catabolism of glucose. If the ATP

Figure 23.29   Metabolism of lipids. Lipolysis is the breakdown of triglycerides into glycerol and fatty acids. Glycerol may be converted to glyceraldehyde 3-phosphate, which can then be converted to glucose or enter the Krebs cycle. Fatty acid fragments enter the Krebs cycle as acetyl coenzyme A. Fatty acids also can be converted into ketone bodies.

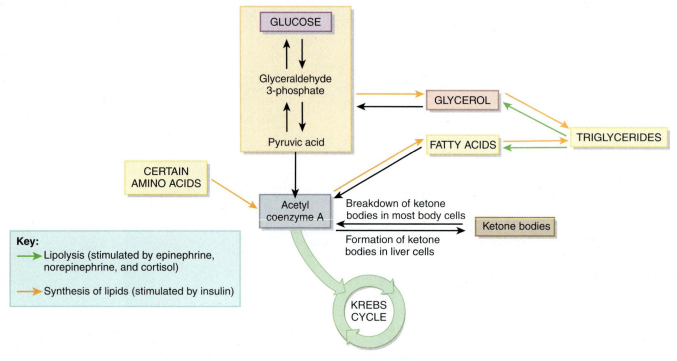

Glycerol and fatty acids are catabolized in separate pathways.

supply in a cell is high, glyceraldehyde 3-phosphate is converted into glucose, an example of gluconeogenesis. If the ATP supply in a cell is low, glyceraldehyde 3-phosphate enters the catabolic pathway to pyruvic acid.

Fatty acid catabolism begins as enzymes remove two carbon atoms at a time from the fatty acid and attach them to molecules of coenzyme A, forming acetyl coenzyme A (acetyl CoA). Then the acetyl CoA enters the Krebs cycle (Figure 23.29). A 16-carbon fatty acid such as palmitic acid can yield as many as 129 ATPs via the Krebs cycle and the electron transport chain.

As part of normal fatty acid catabolism, the liver converts some acetyl CoA molecules into substances known as **ketone bodies** (Figure 23.29). Ketone bodies then leave the liver to enter body cells, where they are broken down into acetyl CoA, which enters the Krebs cycle.

Lipid Anabolism

Insulin stimulates hepatocytes and adipose cells to synthesize triglycerides when more calories are consumed than are needed to satisfy ATP needs (Figure 23.29). Excess dietary carbohydrates, proteins, and fats all have the same fate—they are converted into triglycerides. Certain amino acids can undergo the following reac-

tions: amino acids → acetyl CoA → fatty acids → triglycerides. The use of glucose to form lipids takes place via two pathways:

1. Glucose → glyceraldehyde 3-phosphate → glycerol

2. Glucose → glyceraldehyde 3-phosphate → acetyl CoA → fatty acids

The resulting glycerol and fatty acids can undergo anabolic reactions to become stored triglycerides. Alternatively, they can be used to produce other lipids such as phospholipids, lipoproteins, and cholesterol.

Lipid Transport in Blood

Most lipids, such as triglycerides, are nonpolar and therefore very hydrophobic molecules. They do not dissolve in water. To be transported in watery blood, such molecules first must be made more water-soluble by combining them with proteins. The lipid and protein combinations thus formed are **lipoproteins**, spherical particles with an outer shell of proteins, phospholipids, and cholesterol molecules surrounding an inner core of triglycerides and other lipids. The proteins in the outer shell help the lipoprotein particles dissolve in body fluids and also have specific functions.

Lipoproteins are transport vehicles: They provide delivery and pickup services so that lipids can be available when cells need them or removed from circulation when they are not needed. Lipoproteins are categorized and named mainly according to their size and density. From largest and lightest to smallest and heaviest, the four major types of lipoproteins are chylomicrons, very low-density lipoproteins, low-density lipoproteins, and high-density lipoproteins.

- **Chylomicrons** form in absorptive epithelial cells of the small intestine and transport dietary lipids to adipose tissue for storage. Chylomicrons enter lacteals of intestinal villi and are carried by lymph into venous blood and then into the systemic circulation. Their presence gives blood plasma a milky appearance, but they remain in the blood for only a few minutes.

- **Very low-density lipoproteins (VLDLs)** transport triglycerides made in hepatocytes to adipose tissue for storage and muscle cells for ATP production. After depositing some of their triglycerides in adipose tissue, VLDLs are converted to LDLs.

- **Low-density lipoproteins (LDLs)** carry about 75% of the total cholesterol in blood and deliver it to cells throughout the body for use in repair of cell membranes and synthesis of steroid hormones and bile salts. LDLs bind to LDL receptors on the plasma membrane of body cells so that LDL can enter the cell via receptor-mediated endocytosis. Within the cell, the LDL is broken down, and the cholesterol is released to serve the cell's needs. Once a cell has sufficient cholesterol for its activities, a negative feedback system inhibits the cell's synthesis of new LDL receptors.

  When present in excessive numbers, LDLs also deposit cholesterol in and around smooth muscle fibers in arteries, forming fatty plaques that increase the risk of coronary artery disease. For this reason, the cholesterol in LDLs, called LDL-cholesterol, is known as "bad" cholesterol. Because some people have too few LDL receptors, their body cells remove LDL from the blood less efficiently; as a result, their plasma LDL level is abnormally high, and they are more likely to develop fatty plaques. Eating a high-fat diet increases the production of VLDLs, which elevates the LDL level and increases the formation of fatty plaques.

- **High-density lipoproteins (HDLs)** remove excess cholesterol from body cells and the blood and transport it to the liver for elimination. Because HDLs prevent accumulation of cholesterol in the blood, a high HDL level is associated with decreased risk of coronary artery disease. For this reason, HDL-cholesterol is known as "good" cholesterol.

Protein Metabolism

During digestion, proteins are broken down into amino acids. Unlike carbohydrates and triglycerides, which are stored, proteins are not warehoused for future use. Instead, amino acids are either oxidized to produce ATP or used to synthesize new proteins for body growth and repair. Excess dietary amino acids are converted into glucose (gluconeogenesis) or triglycerides.

The active transport of amino acids into body cells is stimulated by insulinlike growth factors and insulin. Almost immediately after digestion, amino acids are reassembled into proteins. Many proteins function as enzymes; others are involved in transportation (hemoglobin) or serve as antibodies, clotting chemicals (fibrinogen), hormones (insulin), or contractile elements in muscle fibers (actin and myosin). Several proteins serve as structural components of the body (collagen, elastin, and keratin).

Protein Catabolism

A certain amount of protein catabolism occurs in the body each day, stimulated mainly by cortisol from the adrenal cortex. Proteins from worn-out cells (such as red blood cells) are broken down into amino acids. Some amino acids are converted into other amino acids, peptide bonds are reformed, and new proteins are made as part of the recycling process. Hepatocytes convert some amino acids to fatty acids, ketone bodies, or glucose. Figure 23.28 shows the conversion of amino acids into glucose (gluconeogenesis). Figure 23.29 shows the conversion of amino acids into fatty acids or ketone bodies.

Cells throughout the body oxidize a small amount of amino acids to generate ATP via the Krebs cycle and the electron transport chain. However, before amino acids can enter the Krebs cycle, their amino group ($-NH_2$) must first be removed, a process called **deamination** (dē-am′-i-NĀ-shun). Deamination occurs in hepatocytes and produces ammonia (NH_3). Hepatocytes then convert the highly toxic ammonia to urea, a relatively harmless substance that is excreted in the urine.

Protein Anabolism

Protein anabolism, the formation of peptide bonds between amino acids to produce new proteins, is carried out on the ribosomes of almost every cell in the body, directed by the cells' DNA and RNA. Insulinlike growth factors, thyroid hormones, insulin, estrogens, and testosterone stimulate protein synthesis. Because proteins are a main component of most cell structures, adequate dietary protein is especially essential during the growth years, during pregnancy, and when tissue has been damaged by disease or injury. Once dietary intake of protein is adequate, eating more protein does not increase bone or muscle mass; only a regular program of forceful, weight-bearing muscular activity accomplishes that goal.

Of the 20 amino acids in the human body, 10 are **essential amino acids**: They must be present in the diet because they cannot be synthesized in the body in adequate amounts. A **complete protein** contains sufficient amounts of all essential amino acids. Beef, fish, poultry, eggs, and milk are examples of foods that contain complete proteins. An **incomplete protein** does not contain all essential amino acids. Examples of incomplete proteins are

TABLE 23.9

Summary of Metabolism

PROCESS	COMMENTS
Carbohydrates	
Glucose catabolism	Complete oxidation of glucose (cellular respiration) is the chief source of ATP in cells and consists of glycolysis, the Krebs cycle, and the electron transport chain. Complete oxidation of 1 molecule of glucose yields a maximum of 36 or 38 molecules of ATP.
Glycolysis	Conversion of glucose into pyruvic acid results in the production of some ATP. Reactions do not require oxygen (anaerobic cellular respiration).
Krebs cycle	Cycle includes a series of oxidation–reduction reactions in which coenzymes (NAD^+ and FAD) pick up hydrogen ions and hydride ions from oxidized organic acids, and some ATP is produced. CO_2 and H_2O are by-products. Reactions are aerobic.
Electron transport chain	Third set of reactions in glucose catabolism is another series of oxidation–reduction reactions, in which electrons are passed from one carrier to the next, and most of the ATP is produced. Reactions require oxygen (aerobic cellular respiration).
Glucose anabolism	Some glucose is converted into glycogen (glycogenesis) for storage if not needed immediately for ATP production. Glycogen can be reconverted to glucose (glycogenolysis). The conversion of amino acids, glycerol, and lactic acid into glucose is called gluconeogenesis.
Lipids	
Triglyceride catabolism	Triglycerides are broken down into glycerol and fatty acids. Glycerol may be converted into glucose (gluconeogenesis) or catabolized via glycolysis. Fatty acids are catabolized via beta oxidation into acetyl coenzyme A that can enter the Krebs cycle for ATP production or be converted into ketone bodies (ketogenesis).
Triglyceride anabolism	The synthesis of triglycerides from glucose and fatty acids is called lipogenesis. Triglycerides are stored in adipose tissue.
Proteins	
Protein catabolism	Amino acids are oxidized via the Krebs cycle after deamination. Ammonia resulting from deamination is converted into urea in the liver, passed into blood, and excreted in urine. Amino acids may be converted into glucose (gluconeogenesis), fatty acids, or ketone bodies.
Protein anabolism	Protein synthesis is directed by DNA and utilizes cells' RNA and ribosomes.

leafy green vegetables, legumes (beans and peas), and grains. **Nonessential amino acids** are those synthesized by the body. They are formed by the transfer of an amino group from an amino acid to pyruvic acid or to an acid in the Krebs cycle. Once the appropriate essential and nonessential amino acids are present in cells, protein synthesis occurs rapidly.

Table 23.9 summarizes the processes occurring in catabolism and anabolism of carbohydrates, lipids, and proteins.

Checkpoint

53. What is metabolism? What is the difference between anabolism and catabolism?

54. How does ATP link anabolism and catabolism?

55. Which reactions produce ATP during the complete catabolism of a molecule of glucose? How many molecules of ATP are produced?

56. What happens during glycolysis? In the electron transport chain?

57. What is gluconeogenesis, and why is it important?

58. What are the functions of the proteins in lipoproteins? Which lipoprotein particles contain "good" and "bad" cholesterol, and what are the meanings of these terms?

►►►Chloe, a 28-year-old mother of two with a history of morbid obesity, high blood pressure, diabetes, sleep apnea, and acid reflux disease, has decided to opt for an elective surgical procedure called gastric bypass. Her stomach has been reduced in volume from 1 quart to 1 tablespoon to enable her to lose a substantial amount of weight. In addition to reducing stomach size, the procedure should increase her feeling of fullness with smaller amounts of food. As the doctor warned her, such a drastic procedure should not be undertaken lightly, as it has a significant impact on the digestive process. Chloe must closely monitor her nutrition for the rest of her life in order to avoid a host of potential complications. ◄◄◄

I. With which complications should Chloe be concerned following the surgery?
J. Why is monitoring nutrition so important to Chloe now?
K. Chloe went on a liquid diet immediately following her surgery. Based on what you have learned about nutrition in this chapter, which substances would you include in a liquid diet?

Concept and Resource Summary

Concept

Resources

Introduction

1. Food contains nutrients needed by the body for building and repair of tissues, and for sustaining metabolic reactions. The **digestive system** is responsible for **digestion** of food into smaller molecules capable of entering body cells, and for **absorption** of these smaller molecules through cells lining the stomach and intestines into the blood and lymph.

Concept 23.1 The GI tract is a continuous multilayered tube extending from the mouth to the anus.

1. The digestive tract is comprised of the gastrointestinal tract and accessory digestive organs. The **gastrointestinal (GI) tract** is a continuous tube extending from the mouth to the anus. GI tract organs include the mouth, pharynx, esophagus, stomach, small intestine, and large intestine. The **accessory digestive organs** include the teeth, tongue, salivary glands, liver, gallbladder, and pancreas.
2. Digestion includes six basic processes: ingestion, secretion, mixing and propulsion, mechanical and chemical digestion, absorption, and defecation. **Mechanical digestion** breaks up and churns food, and **chemical digestion** enzymatically breaks down large nutrients into smaller molecules.
3. The basic arrangement of layers in most of the gastrointestinal tract, from deep to superficial, is the **mucosa** with an **epithelium**, **lamina propria**, and **muscularis mucosae**; the **submucosa**; the **muscularis**; and the **serosa**.
4. The **peritoneum** is the largest serous membrane of the body. The **parietal peritoneum** lines the abdominopelvic cavity wall, and **visceral peritoneum** covers some of the cavity organs. The **peritoneal cavity** is a thin space with serous fluid between the parietal and visceral peritoneum.
5. Folds of the peritoneum include the **greater omentum** draping over the transverse colon and small intestines, the **falciform ligament** attaching the liver to the abdominal wall and diaphragm, the **lesser omentum** suspending the stomach and duodenum from the liver, the **mesentery** binding the small intestine to the posterior abdominal wall, and the **mesocolon** binding the large intestine to the posterior abdominal wall.
6. The gastrointestinal tract is regulated by an intrinsic set of nerves known as the enteric nervous system and by an extrinsic set of nerves that are part of the autonomic nervous system (ANS). Intrinsic enteric nerves and extrinsic parasympathetic and sympathetic nerves of the ANS innervate the GI tract. The **enteric nervous system (ENS)** consists of neurons arranged into two plexuses: the **myenteric plexus** of the muscularis controls GI tract motility, and the **submucosal plexus** of the submucosa controls GI tract secretions.
7. Although the ENS can function independently, it is subject to regulation by the ANS. Parasympathetic nerves increase GI tract secretion and motility by increasing the activity of ENS neurons. Sympathetic nerves decrease GI tract secretion and motility by inhibiting ENS neurons.

See
- Anatomy Overview – GI Tract Histology
- Animation – Neural Regulation of Mechanical Digestion

Hear
- **Figure 23.2 – Layers** of the Gastrointestinal Tract

Concept	Resources

Concept 23.2 The mouth lubricates and begins digestion of food, and maneuvers it to the pharynx for swallowing.

1. The **mouth** is formed by muscular cheeks, hard and soft palates, and tongue. The **lips** surround the opening of the mouth. Muscles of the lips and cheeks help keep food between the teeth during chewing. The **vestibule** is the space bounded externally by the cheeks and lips and internally by the teeth and gums. The **oral cavity proper** extends from the vestibule to the **fauces**, the opening between the oral cavity and throat. The **hard palate** is the bony anterior roof of the mouth, and the **soft palate** is the muscular posterior roof of the mouth. The **uvula** hangs from the border of the soft palate and helps to close off the nasopharynx during swallowing.

2. The **tongue** forms the floor of the oral cavity. Muscles of the tongue maneuver food and assist with swallowing and speech. The upper surface and sides of the tongue are covered with **papillae**, some of which contain taste buds.

3. **Salivary glands** secrete saliva into the oral cavity: the **parotid glands**, inferior and anterior to the ears; the **submandibular glands** to the base of the tongue; and the **sublingual glands** superior to the submandibular glands. **Saliva** moistens food, which helps form it into a ball for swallowing; dissolves foods so they can be tasted; and begins chemical digestion of starch. The ANS controls **salivation**.

4. The three regions of a tooth are the visible **crown**, one to three **roots** embedded in sockets of the mandible and maxillae and surrounded by **gingivae**, and the constricted **neck** at the gum line. A mineralized **enamel**, the hardest substance in the body, covers the crown and protects the tooth from wear and tear and acids. **Dentin** composes most of the tooth and encloses the **pulp cavity**. **Cementum** and the **periodontal ligament** anchor the tooth in the socket.

5. There are two **dentitions**: the 20 **deciduous teeth** and 32 **permanent teeth**. The permanent teeth include **incisors**, **cuspids**, **premolars**, and **molars**.

6. Through **mastication**, food is mixed with saliva and shaped into a soft, flexible mass called a **bolus**. **Salivary amylase** begins the digestion of starches into dissacharides. **Lingual lipase** breaks down triglycerides into fatty acids and diglycerides.

See

- Anatomy Overview – Oral Cavity
- Anatomy Overview – Salivary Glands
- Animation – Mastication
- Animation – Carbohydrate Digestion in the Mouth
- Animation – Lipid Digestion in the Mouth
- Animation – Chemical Digestion—Enzymes

Hear

- **Figure 23.8 –** **Dentitions and Times of Eruptions**

Concept 23.3 Swallowing consists of voluntary oral, involuntary pharyngeal, and involuntary esophageal stages.

1. Food passes from the mouth into the **pharynx** that extends from the internal nares to the esophagus posteriorly and to the larynx anteriorly. The pharynx has both respiratory and digestive functions.

2. The **esophagus** is a collapsible, muscular tube that connects the pharynx to the stomach. There is a sphincter at each end of the esophagus: the **upper esophageal sphincter** (skeletal muscle), and the **lower esophageal sphincter** (smooth muscle). The esophagus is not involved in chemical digestion or absorption.

3. **Deglutition** moves a bolus from the mouth to the stomach. The **voluntary stage** passes the bolus from the mouth into the oropharynx. The involuntary **pharyngeal stage** passes the bolus from the pharynx to the esophagus. During the involuntary **esophageal stage**, coordinated sequences of contraction and relaxation of the muscularis, called **peristalsis**, push food through the esophagus toward the stomach.

See

- Anatomy Overview – Pharynx and Esophagus
- Anatomy Overview – Esophagus Histology
- Animation – Deglutition

Concept	Resources

Concept 23.4 The stomach mechanically breaks down the bolus and mixes it with gastric secretions.

1. The stomach lies inferior to the diaphragm and connects the esophagus to the duodenum. The principal regions of the stomach are the **cardia**, the **fundus**, the **body**, and the **pylorus**. The pylorus connects to the small intestine through the **pyloric sphincter**.

2. The stomach **mucosa** contains **gastric glands** that open into **gastric pits**. Gastric glands secrete **gastric juice** composed of mucus, hydrochloric acid, pepsinogen, gastric lipase, and intrinsic factor. The stomach **muscularis** has three alternately arranged layers of smooth muscle.

3. **Mixing waves** of peristalsis mix food with gastric juice and macerate it to a soupy **chyme**.

4. Gastrin also increases stomach motility and relaxes the pyloric sphincter. HCl kills microbes, partially denatures proteins, and stimulates hormone secretion, promoting flow of bile and pancreatic juice.

5. Chief cells secrete **pepsin**, activated by acidic pH, which begins protein digestion. **Gastric lipase** allows digestion of triglycerides into fatty acids and monoglycerides.

6. The stomach wall is impermeable to most substances. Absorption is limited to water, certain ions, drugs, and alcohol.

See

- **Anatomy Overview – The Stomach**
- **Anatomy Overview – Stomach Histology**
- **Animation – Stomach Peristalsis**
- **Animation – Protein Digestion in the Stomach**
- **Animation – Lipid Digestion in the Stomach**
- **Animation – Chemical Digestion— Gastric Acid**

Hear

- **Figure 23.12 – Histology of the Stomach**
- **Figure 23.13 – Secretion of HCl by Parietal Cells in the Stomach**

Concept 23.5 The pancreas secretes pancreatic juice, the liver secretes bile, and the gallbladder stores and concentrates bile.

1. The **pancreas** consists of a **head**, a **body**, and a **tail**. Exocrine **acini** secrete pancreatic juice into the **pancreatic duct** and **accessory duct** that empty into the duodenum. Endocrine **pancreatic islets** secrete the hormones glucagon and insulin.

2. **Pancreatic juice** is slightly alkaline to buffer the acidic gastric juice in chyme and inactivate pepsin. Pancreatic juice contains enzymes that digest starch (**pancreatic amylase**), proteins (**trypsin, chymotrypsin, carboxypeptidase**, and **elastase**), triglycerides (**pancreatic lipase**), and nucleic acids (**ribonuclease** and **deoxyribonuclease**).

3. The **liver** is composed of a **right lobe** and **left lobe** that is continuous with the **quadrate** and **caudate** lobes. The **gallbladder** is a sac located in a depression on the posterior surface of the liver that stores and concentrates bile. The lobes of the liver have **lobules** composed of **hepatocytes** radiating out from a **central vein** and many **sinusoids** containing phagocytic **reticuloendothelial cells**.

4. Hepatocytes secrete bile into **bile canaliculi**, which drain into **bile ducts**. Bile travels in bile ducts to **right** and **left hepatic ducts**, then to the **common hepatic duct**, which joins the **cystic duct** of the gallbladder to form the **common bile duct**.

5. Oxygenated blood flows into the liver from the hepatic artery; nutrient-rich venous blood from digestive organs enters the liver from the hepatic portal vein. In liver sinusoids, blood is processed or detoxified before entering the venous circulation.

6. In the small intestine, **bile** breaks down large lipid globules into small lipid globules in a process called **emulsification**. Between meals, bile enters the gallbladder for storage.

7. The liver also functions in carbohydrate, lipid, and protein metabolism; processing of drugs and hormones; excretion of bilirubin; synthesis of bile salts; storage of vitamins and minerals; phagocytosis of old RBCs and WBCs; and activation of vitamin D.

See

- **Anatomy Overview – Pancreas**
- **Anatomy Overview – Liver and Gallbladder**
- **Anatomy Overview – Liver Histology**
- **Anatomy Overview – Pancreas Histology**
- **Animation – Carbohydrate Digestion—Pancreas**
- **Animation – Protein Digestion— Pancreatic Juice**
- **Animation – Lipid Digestion—Bile Salts and Pancreatic Lipase**
- **Animation – Chemical Digestion—Bile**

Hear

- **Figure 23.14 – Relation of the Pancreas to the Liver, Gallbladder, and Duodenum**

Concept

Concept 23.6 In the small intestine, chyme mixes with digestive juices from the small intestine, pancreas, and liver.

1. The **small instestine** is the major site of chemical digestion and absorption.

2. The small intestine extends from the pyloric sphincter to the **ilececal sphincter**. The small intestine is divided into the **duodenum**, **jejunum**, and **ileum**, which joins the large intestine at the **ilocecal sphincter**.

3. The small intestine wall has **absorptive cells** that absorb nutrients, **goblet cells** that secrete mucus, phagocytic **paneth cells**, **S cells** that secrete the secretin, **CCK cells** that secrete the **cholecystokinin**, **intestinal glands** that secrete intestinal juice, and **lymphatic nodules** that attack pathogens in food. **Circular folds** cause chyme to spiral and expose more of the mucosa to the nutrients. **Villi** and **microvilli** provide a large surface area for digestion and absorption, and the microvilli form a **brush border**.

4. **Intestinal juice** and pancreatic juice enhance absorption of substances from chyme. **Brush border enzymes** digest carbohydrates, proteins, and nucleotides.

5. **Segmentations** mix chyme with digestive juices and bring food into contact with the mucosa. **Peristalsis** moves chyme forward through the small intestine.

6. Digestion of carbohydrates begins by salivary amylase from the mouth and pancreatic amylase in pancreatic juice. After amylase action, the fragments are digested by the brush border enzymes, **α-dextrinase**, **sucrase**, **lactase**, and **maltase** into monosaccharides.

7. Digestion of proteins begins by pepsin from the stomach and continues in the small intestine by the action of **trypsin**, **chymotrypsin**, **carboxypeptidase**, and **elastase** in pancreatic juice. Protein digestion is completed by brush border **peptidases**.

8. Some digestion of lipids occurs in the stomach by ligual lipase and gastric lipase. Once in the small intestine, larger lipid globules are **emulsified** into smaller lipid globules by the bile salts in bile. **Pancreatic lipase** attacks the smaller lipid globules, completing the digestion of triglycerides into fatty acids and a monoglyceride.

9. Pancreatic juice contains **ribonuclease** that digests RNA and **deoxyribonuclease** that digests DNA. Brush border **nucleosidases** and **phosphatases** complete nucleic acid breakdown.

10. Monosaccharides and amino acids, dipeptides, tripeptides, and short-chain fatty acids are absorbed into the small intestine epithelium by diffusion, facilitated diffusion, osmosis, or active transport. They then pass from the epithelium into blood capillaries.

11. Following lipid digestion, bile salts surround the fatty acids and monoglycerides, forming **micelles**. The fatty acids and monoglycerides diffuse out of micelles into the small intestine epithelium, where they recombine into **chylomicrons**. The chylomicrons pass from the epithlium into lacteals, where lymph transports them to the bloodstream.

12. The small intestine also absorbs electrolytes, vitamins, and water.

See

- **Anatomy Overview – Small Intestine**
- **Anatomy Overview – Small Intestine Histology**
- **Animation – Segmentation**
- **Animation – Carbohydrate Digestion in the Small Intestine**
- **Animation – Protein Digestion in the Small Intestine**
- **Animation – Nucleic Acid Digestion in the Small Intestine**
- **Animation – Carbohydrate Absorption in the Small Intestine**
- **Animation – Protein Absorption in the Small Intestine**
- **Animation – Nucleic Acid Absorption in the Small Intestine**
- **Animation – Lipid Absorption in the Small Intestine**

Hear

- **Figure 23.20 – Absorption of Digested Nutrients in the Small Intestine**

Concept 23.7 In the large intestine, the final secretion and absorption of nutrients occur as chyme moves toward the rectum.

1. The **large intestine** extends from the **ileocecal sphincter** to the anus. Its regions include the cecum, colon, rectum, and anal canal. The **cecum** is a small pouch hanging inferior to the ileocecal valve and has the **vermiform appendix** attached to it. The **colon** is divided into **ascending**, **transverse**, **descending**, and **sigmoid** portions. The terminal end of the **rectum** is the **anal canal**, with an exterior opening called the **anus**. Opening and closing of the anus is controlled by the involuntary **internal anal sphincter** and the voluntary **external anal sphincter**.

2. The mucosa has absorptive cells and goblet cells. The muscularis has smooth muscle bands called **teniae coli** that pucker its wall into **haustra**.

3. Food is moved through the large intestine by **haustral churning** and **peristalsis**. **Mass peristalsis** drives colon contents into the rectum.

4. Bacteria inhabiting the lumen break down remaining carbohydrates, proteins, and bilirubin; and produce vitamin K and some B vitamins.

5. The large intestine absorbs water, ions, and vitamins. It contributes to the final formation of **feces** and its elimination from the body.

6. When mass peristalsis pushes feces into the rectum, distention of the rectal wall initiates the **defecation reflex**. Parasympathetic stimulation opens the internal anal sphincter. Voluntary relaxation of the external anal sphincter allows feces to be expelled through the anus.

See

- **Anatomy Overview – The Large Intestine**
- **Anatomy Overview – Large Intestine Histology**
- **Animation – Mechanical Digestion in the Large Intestine**

Do

- **Exercise – Digestive Enzyme Activity**
- **Exercise – Concentrate on Digestion**
- **Exercise – Paint the Digestive Tract**
- **Exercise – Digestion Reaction Race**
- **Exercise – Role of Chemicals**
- **Exercise – Match the Movement**
- **Exercise – Nutrient Absorption Routes**

Concept	Resources

Concept 23.8 Digestive activities occur in three overlapping phases: cephalic, gastric, and intestinal.

1. Digestive activities occur in three overlapping phases: cephalic phase, gastric phase, and intestinal phase.
2. During the **cephalic phase** of digestion, salivary glands secrete saliva and gastric glands secrete gastric juice in order to prepare the mouth and stomach for food that is about to be eaten.
3. The presence of food in the stomach causes the **gastric phase**, involving peristalsis and gastric juice secretion from gastric glands. The hormone **gastrin** stimulates gastric juice secretion.
4. During the **intestinal phase** of digestion, food is digested in the small intestine. In addition, the **enterogastric reflex** slows gastric motility in order to slow the exit of chyme from the stomach, which prevents the small intestine from being overloaded with more chyme than it can handle. **Cholecystokinin** and **secretin** stimulate the secretion of pancreatic juice into the small intestine.

See

- Animation – Neural Regulation of Mechanical Digestion
- Animation – Enterogastric Reflex
- Animation – Hormonal Control of Digestive Activities

Do

- Exercise – Digestive Hormone Activities
- Concepts and Connections – Digestive Hormones

Concept 23.9 Food molecules supply energy for life processes and serve as building blocks for complex molecules.

1. Body cells use **nutrients** in food for growth, maintenance, and repair. Nutrients include water, carbohydrates, lipids, proteins, minerals, and vitamins.
2. Nutrition experts suggest dietary calories be 50–60% from carbohydrates, 30% or less from fats, and 12–15% from proteins. The MyPyramid guide represents a personalized approach to making healthy food choices and maintaining regular physical activity.
3. **Minerals** important to the body include calcium, phosphorus, potassium, sulfur, sodium, chloride, magnesium, iron, iodide, manganese, copper, cobalt, zinc, fluoride, selenium, and chromium. Many minerals play key roles as components of coenzymes, are components of buffers that help control body pH, and are needed for nerve impulses.
4. **Vitamins** maintain growth and normal metabolism. Most function as coenzymes rather than nutrients. Most vitamins must be obtained from the diet. Some vitamins can be assembled by the body if **provitamins** are provided in the diet. **Fat-soluble vitamins** must be ingested with lipids for proper absorption. **Water-soluble vitamins** are dissolved in blood fluids. **Antioxidant vitamins** inactivate oxygen free radicals that damage cell structures.

See

- Anatomy Overview – Role of Nutrients

Concept 23.10 Metabolism includes the catabolism and anabolism of molecules.

1. **Metabolism** refers to all the chemical reactions of the body. **Enzymes** are the catalysts for all metabolic reactions. Some enzymes require **coenzymes** to function properly. **Anabolism** is the term for *endergonic* reactions that synthesize complex molecules from simpler substances. **Catabolism** is the term for *exergonic* reactions that break down complex organic molecules into simple molecules.
2. ATP links anabolic and catabolic reactions. Energy to run anabolic reactions comes from energy that is released when ATP is separated into ADP and P. Energy released from catabolic reactions allows bonding of ADP and ℗ to resynthesize ATP.
3. The glucose that results from carbohydrate digestion is oxidized by cells to provide ATP. Glucose not needed for immediate ATP production is converted to glycogen for storage. When glycogen is in excess, the glucose can be synthesized into triglycerides.
4. **Cellular respiration**, the complete oxidation of glucose to CO_2 and H_2O, involves glycolysis, the Krebs cycle, and the electron transport chain.
5. **Glycolysis** is the breakdown of glucose into two molecules of pyruvic acid; there is a net production of two molecules of ATP. Glycolysis can produce ATP anaerobically and is known as **anaerobic cellular respiration**.

See

- Animation – Introduction to Metabolism
- Animation – Carbohydrate Metabolism
- Animation – Lipid Metabolism
- Animation – Protein Metabolism

Do

- Exercise – Predict ATP Production
- Exercise – Glucose Catabolism Sequence
- Exercise – Glucose Catabolism Substrates and Products
- Concepts and Connections – Glucose Catabolism

Concept   **Resources**

6. Each molecule of pyruvic acid that enters the Krebs cycle is first converted to **acetyl coenzyme A**. In the **Krebs cycle** acetyl coenzyme A is used to produce CO_2, NADH and four H^+, $FADH_2$, and ATP.

7. In the **electron transport chain** the energy in NADH/H^+ and $FADH_2$ is liberated and transferred to ATP. The electron transport chain produces ATP and H_2O. The Krebs cycle and electron transport chain both require oxygen and are collectively called **aerobic cellular respiration**.

8. The liver can form glucose from lactic acid, amino acids, and glycerol in a process called **glyconeogenesis** when stimulated by cortisol from the adrenal cortex and glucagon from the pancreas.

9. In **lipolysis**, triglycerides are split into glycerol and fatty acids, which then can be further catabolized by different pathways to produce ATP. Insulin stimulates liver cells and adipose cells to synthesize triglycerides when more calories are consumed than required by the body's ATP demands.

10. **Lipoproteins** transport lipids in the bloodstream. Types of lipoproteins include chylomicrons, which carry dietary lipids to adipose tissue; **very low-density lipoproteins (VLDLs)**, which carry triglycerides from the liver to adipose tissue; **low-density lipoproteins (LDLs)**, which deliver cholesterol to body cells; and **high density lipoproteins (HDLs)**, which remove excess cholesterol from body cells and transport it to the liver for elimination.

11. Proteins are not stored in the body. Amino acids are either oxidized to produce ATP or used to synthesize new proteins for growth and repair. Excess dietary amino acids are converted into glucose or triglycerides. Protein synthesis is stimulated by insulinlike growth factors, thyroid hormones, insulin, estrogens, and testosterone.

12. **Essential amino acids** must be present in the diet; **nonessential amino acids** can be synthesized in the body. A **complete protein** contains all of the essential amino acids; an **incomplete protein** does not contain all of the essential amino acids.

Other Chapter Resources

 For additional information on clinical applications, medical disorders, and medical terminology, see Chapter 12 in the *Clinical Connections* manual packaged with your text.

Within *WileyPLUS*

- Chapter Quiz
- Visual Anatomy
- Mastering Vocabulary
- Audio Glossary
- Homeostatic Imbalance: The Case of the Burning Abdomen
- Homeostatic Imbalance: The Case of the Woman with the Bloody Stools
- Web Links

• Understanding the Concepts

1. What are the similarities between mechanical and chemical digestion? How do they differ?

2. What are the functions of the muscles and papillae of the tongue? Of saliva?

3. How does the mouth participate in chemical digestion?

4. What is the role of the esophagus in digestion?

5. What is the importance of surface mucous cells, mucous neck cells, chief cells, parietal cells, and G cells in the stomach?

6. What are the functions of gastrin, hydrochloric acid, gastric lipase, and lingual lipase in the stomach?

7. How do the functions of pancreatic acini differ from those of the pancreatic islets?

8. Once the liver has formed bile, how is bile collected and transported to the gallbladder for storage?

9. What is the pathway of blood flow into, through, and out of the liver?

10. In which ways are the mucosa and submucosa of the small intestine adapted for digestion and absorption?

11. What is the functional significance of the blood capillary network and lacteal in the center of each villus of the small intestine?

12. What is the difference between digestion and absorption? Which organ of the GI tract absorbs the most nutrients?

13. Describe the absorption of electrolytes, vitamins, and water by the small intestine.

14. Referring to Figure 23.21, which two organs of the digestive system secrete the most fluid? Which organ of the digestive system absorbs the most fluid?

15. Describe the role of gastrin in the gastric phase of digestion.

16. What is a nutrient? What are the general functions of water, carbohydrates, lipids, proteins, and minerals?

17. What are the major events in the catabolism of lipids to produce ATP?

24

The Urinary System

Andre's Story

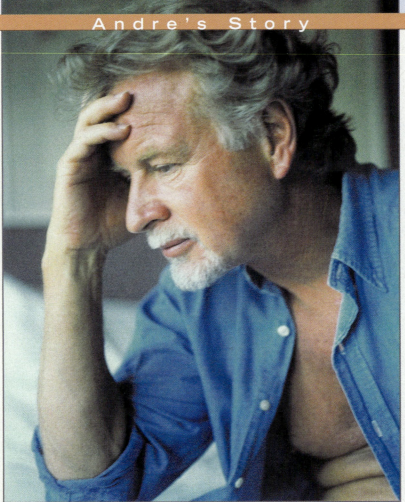

Andre awoke for the third time in two hours. He knew that he probably wouldn't be able to produce any urine, but he felt the urgency just the same. He got out of bed and walked to the bathroom—again! He had been diagnosed with benign prostatic hypertrophy, an enlarged prostate gland, when he was 51. He was now 57 and beginning to seriously consider having surgery. The enlarged gland (which surrounds the base of the urethra in males) had made it difficult to urinate; he felt like he needed to go all the time, yet he couldn't produce a good stream. He had also begun having frequent urinary tract infections over the last year. His urine came out slowly, even though he felt he had a full bladder. The urine looked cloudy and smelled foul, and caused a burning sensation. "Not again," he thought to himself. He would have to call the doctor to refill his antibiotic prescription. For now, he thought the best thing might be to take something to help with his headache and aching back. He reached for a bottle of ibuprofen.

The urinary system's contribution to homeostasis is vital to our health. Andre has a condition that compromises his urinary function. As we learn the anatomy and physiology of the urinary system, we will follow his story to learn how normal urinary system function can be altered and what these changes can mean to our overall health and homeostasis.

INTRODUCTION

As body cells carry out their metabolic functions, they consume oxygen and nutrients and produce substances such as carbon dioxide and nitrogenous wastes that have no useful functions and need to be eliminated from the body. While the respiratory system rids the body of carbon dioxide, the urinary system disposes of most other unneeded substances.

The **urinary system** consists of two kidneys, two ureters,

The urinary system consists of the kidneys, ureters, urinary bladder and urethra.

one urinary bladder, and one urethra (see Figure 24.1). After the kidneys filter blood plasma, they return most of the water and solutes to the blood-

stream. The remaining water and solutes constitute **urine**, which passes through the ureters and is stored in the urinary bladder until it is excreted from the body through the urethra. Besides forming urine, the urinary system contributes to homeostasis by altering the composition, pH, volume, and pressure of blood, and by producing hormones.

24.1 The kidneys regulate the composition of the blood, produce hormones, and excrete wastes.

The kidneys do the major work of the urinary system. The other parts of the system are mainly passageways and storage areas (Figure 24.1).

Functions of the kidneys include the following:

- *Regulation of blood ionic composition*. The kidneys help regulate the blood levels of several ions, most importantly sodium ions (Na^+), potassium ions (K^+), calcium ions (Ca^{2+}), chloride ions (Cl^-), and phosphate ions (HPO_4^{2-}).

- *Regulation of blood pH*. The kidneys excrete a variable amount of hydrogen ions (H^+) into the urine and conserve bicarbonate ions (HCO_3^-), which are an important buffer of H^+ in the blood. Both of these activities help regulate blood pH.

- *Regulation of blood volume*. The kidneys adjust blood volume by conserving or eliminating water in the urine. An increase in blood volume increases blood pressure; a decrease in blood volume decreases blood pressure.

- *Regulation of blood pressure*. The kidneys help regulate blood pressure by secreting the enzyme renin, which activates the renin–angiotensin–aldosterone pathway (see Figure 17.15). Increased renin causes an increase in blood pressure.

- *Production of hormones*. The kidneys produce two hormones. *Calcitriol*, the active form of vitamin D, helps regulate calcium homeostasis (see Figure 17.13), and *erythropoietin* stimulates production of red blood cells (see Figure 18.5).

- *Regulation of blood glucose level*. Like the liver, the kidneys can use the amino acid glutamine in *gluconeogenesis*, the synthesis of new glucose molecules. They can then release glucose into the blood to help maintain a normal blood glucose level.

- *Excretion of wastes and foreign substances*. By forming urine, the kidneys help excrete **wastes**—substances that have no useful function in the body. Some wastes excreted in urine result from metabolic reactions in the body. These include ammonia and urea from the breakdown of amino acids; bilirubin from the breakdown of hemoglobin; creatinine from the breakdown of creatine phosphate in muscle fibers; and uric acid from the breakdown of nucleic acids. Other wastes excreted in urine are foreign substances from the diet, such as drugs and environmental toxins.

Checkpoint

1. Which organs constitute the urinary system?
2. How do the kidneys alter blood chemistry?
3. What are wastes, and how do the kidneys participate in their removal from the body?

Figure 24.1   Organs of the urinary system in a female.

RIGHT KIDNEY

Right renal artery

RIGHT URETER

URINARY BLADDER

URETHRA

Diaphragm

Esophagus

Left adrenal gland

Left renal vein

LEFT KIDNEY

Abdominal aorta

Inferior vena cava

LEFT URETER

Rectum

Left ovary

Uterus

Anterior view

Functions of the Urinary System
1. Regulates blood volume and composition.
2. Helps regulate blood pressure.
3. Synthesizes glucose.
4. Secretes erythropoietin.
5. Participates in vitamin D synthesis.
6. Excretes wastes in the urine.
7. Stores urine in urinary bladder.
8. Discharges urine from the body through the urethra.

Urine formed by the kidneys passes first into the ureters, then to the urinary bladder for storage, and finally through the urethra for elimination from the body.

24.2 As urine forms, it travels through the renal medulla, calyces, and renal pelvis.

The paired **kidneys** are reddish, kidney-bean-shaped organs located just above the waist between the peritoneum and the posterior wall of the abdomen. Because their position is posterior to the peritoneum of the abdominal cavity, they are said to be **retro-peritoneal** (re-trō-per-i-tō-NĒ-al; *retro-* = behind) organs (Figure 24.2). The kidneys are located between the levels of the last thoracic and third lumbar vertebrae, a position where they are partially protected by the eleventh and twelfth pairs of ribs. The right kidney is slightly lower than the left (see Figure 24.1) because the liver occupies considerable space on the right side superior to the kidney.

Figure 24.2 Position and coverings of the kidneys.

(a) Inferior view of transverse section of abdomen (L2)

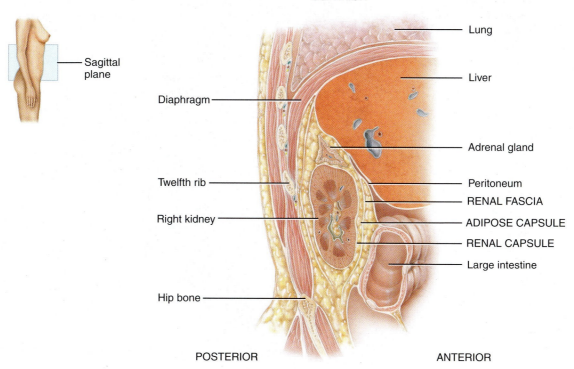

(b) Sagittal section through the right kidney

The kidneys are surrounded by a renal capsule, adipose capsule, and renal fascia.

External Anatomy of the Kidneys

A typical adult kidney is 10–12 cm (4–5 in.) long, 5–7 cm (2–3 in.) wide, and 3 cm (1 in.) thick—about the size of a bar of bath soap—and has a mass of 135–150 g (4.5–5 oz). The concave medial border of each kidney faces the vertebral column (see Figure 24.1). Near the center of the concave border is an indentation called the **renal hilum** (Figure 24.3), through which the ureter leaves the kidney, and blood vessels, lymphatic vessels, and nerves enter and exit.

Three layers of tissue surround each kidney (Figure 24.2). The deep layer, the **renal capsule** (*ren-* = kidney), is a smooth, transparent sheet of dense irregular connective tissue that is continuous with the outer coat of the ureter. It serves as a barrier against trauma and helps maintain the shape of the kidney. The middle layer, the **adipose capsule**, is a mass of fatty tissue surrounding the renal capsule. It cushions the kidney from trauma and holds it firmly in place within the abdominal cavity. The superficial layer, the **renal fascia**, is another thin layer of dense irregular connective tissue that anchors the kidney to the surrounding structures and to the abdominal wall.

Internal Anatomy of the Kidneys

A frontal section through the kidney reveals two distinct regions: a superficial, light red area called the **renal cortex** (*cortex* = rind or bark) and a darker reddish-brown region called the **renal medulla** (*medulla* = inner portion) (Figure 24.3). The medulla consists of several cone-shaped **renal pyramids**. The base (wider end) of each pyramid faces the renal cortex, and its apex (narrower end), called a **renal papilla**, points toward the renal hilum. The renal cortex is the smooth-textured area extending from the renal capsule to the bases of the renal pyramids and into the spaces between them. Those portions of the renal cortex that extend between renal pyramids are called **renal columns**. A **renal lobe** consists of a renal pyramid, its overlying area of renal cortex, and one-half of each adjacent renal column.

Within the renal cortex and renal pyramids are the functional units of the kidney—about 1 million microscopic structures called **nephrons** (NEF-rons). Urine formed by the nephrons drains into large **papillary ducts**, which extend through the renal papillae of the pyramids. The papillary ducts drain into cuplike structures called **minor** and **major calyces** (KĀ-li-sēz = cups; singular is

Figure 24.3   Internal anatomy of the kidneys.

Nephron

Path of urine drainage:

Collecting duct
↓
Papillary duct in renal pyramid
↓
Minor calyx
↓
Major calyx
↓
Renal pelvis
↓
Ureter
↓
Urinary bladder

Renal hilum

Renal artery

Renal vein

Renal cortex

Renal medulla

Renal column

Renal pyramid in renal medulla

Renal sinus

Renal papilla

Fat in renal sinus

Renal capsule

Renal lobe

(a) Frontal section of right kidney

calyx). Each kidney has 8 to 18 minor calyces and 2 to 3 major calyces. A minor calyx receives urine from the papillary ducts of one renal papilla and delivers it to a major calyx. From the major calyces, urine drains into a large funnel-shaped cavity called the **renal pelvis** (*pelv-* = basin) and then out through the ureter to the urinary bladder.

The hilum expands into a cavity within the kidney called the **renal sinus**, which contains part of the renal pelvis, the calyces, and branches of the renal blood vessels and nerves. Adipose tissue helps stabilize the position of these structures in the renal sinus.

Blood Supply of the Kidneys

Because the kidneys remove wastes from the blood and regulate its volume and ionic composition, it is not surprising that they are abundantly supplied with blood vessels. Although the kidneys constitute less than 0.5% of total body mass, they receive 20–25% of the resting cardiac output via the right and left **renal arteries** (Figure 24.4). In adults, renal blood flow is about 1200 mL per minute.

Within the kidney, the renal artery divides into several **segmental arteries**, which supply different segments (areas) of the kidney. Each segmental artery gives off several branches that pass through the renal columns between the renal pyramids as the **interlobar arteries**. At the bases of the renal pyramids, the interlobar arteries arch between the renal medulla and cortex; here they are known as the **arcuate arteries** (AR-kū-āt = shaped like a bow). Divisions of the arcuate arteries produce a series of **interlobular arteries**. These arteries are so named because they pass between renal lobules (a **renal lobule** is a group of nephrons that open into branches of the same collecting duct; see Figure 24.5). Interlobular arteries enter the renal cortex and give off branches called **afferent arterioles** (*af-* = toward; *-ferrent* = to carry).

Each nephron receives one afferent arteriole, which divides into a tangled, ball-shaped capillary network called the **glomerulus** (glō-MER-ū-lus = little ball; plural is *glomeruli*). The glomerular capillaries then reunite to form an **efferent arteriole** (*ef-* = out) that carries blood out of the glomerulus. Glomerular capillaries are unique among capillaries in the body because they are positioned between two arterioles, rather than between an arteriole and a venule. The glomeruli are considered part of both the

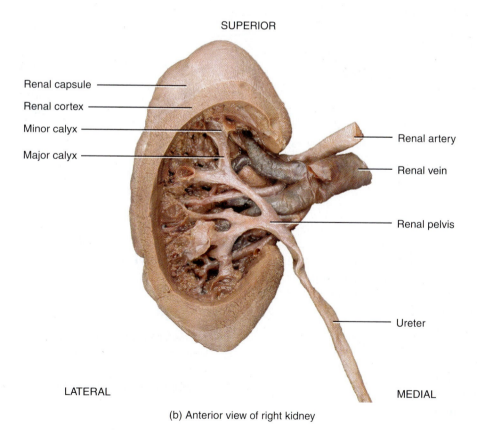

SUPERIOR

Renal capsule
Renal cortex
Minor calyx
Major calyx

Renal artery
Renal vein
Renal pelvis
Ureter

LATERAL

MEDIAL

(b) Anterior view of right kidney

The two main regions of the kidney are the renal cortex and the renal medulla.

Figure 24.4   Blood supply of the kidneys.

(a) Frontal section of right kidney

(b) Path of blood flow

🔑 The renal arteries deliver 20–25% of the resting cardiac output to the kidneys.

cardiovascular and the urinary systems because they are capillary networks, yet they also play an important role in urine formation.

The efferent arterioles divide to form the **peritubular capillaries** (*peri-* = around), which surround tubular parts of the nephron in the renal cortex. Extending from some efferent arterioles are long loop-shaped capillaries called **vasa recta** (VĀ-sa REK-ta; *vasa* = vessels; *recta* = straight) that supply tubular portions of the nephron in the renal medulla (see Figure 24.5b).

The peritubular capillaries eventually merge into the **interlobular veins**, which also receive blood from the vasa recta. Then the blood drains through the **arcuate veins** to the **interlobar veins**

running between the renal pyramids. Blood leaves the kidney through the **renal vein** that exits at the renal hilum and carries venous blood to the inferior vena cava.

Checkpoint

4.   Why are the kidneys said to be retroperitoneal?
5.   Which structures pass through the renal hilum?
6.   How is the kidney protected?
7.   Where in the kidney are the renal pyramids located?

24.3 Each nephron consists of a renal corpuscle and a renal tubule.

Parts of a Nephron

Nephrons are the functional units of the kidneys. Each nephron (Figure 24.5) consists of two parts: a **renal corpuscle** (KOR-pus-sul = tiny body), where blood plasma is filtered, and a **renal tubule** into which the filtered fluid passes. The two components of a renal corpuscle are the **glomerulus** (capillary network) and the **glomerular capsule**, a double-walled cup of epithelial cells that surrounds the glomerular capillaries. Blood plasma is filtered into the glomerular capsule, and then the filtered fluid passes into the renal tubule, which has three main sections. In the order that fluid passes through them, the renal tubule consists of a (1) **proximal convoluted tubule**, (2) **loop of Henle**, and (3) **distal convoluted tubule**. *Proximal* denotes the part of the tubule attached to the glomerular capsule, and *distal* denotes the part that is farther away. *Convoluted* means the tubule is tightly coiled rather than straight.

The renal corpuscle and both convoluted tubules lie within the renal cortex; the loop of Henle extends into the renal medulla. The loop of Henle connects the proximal and distal convoluted tubules. The first part of the loop of Henle begins in the renal cortex and extends downward into the renal medulla, where it is called the **descending limb of the loop of Henle** (Figure 24.5). It then makes that hairpin turn and returns to the renal cortex as the **ascending limb of the loop of Henle**.

Figure 24.5 **The structure of nephrons and associated blood vessels.** (a) A cortical nephron. (b) A juxtamedullary nephron.

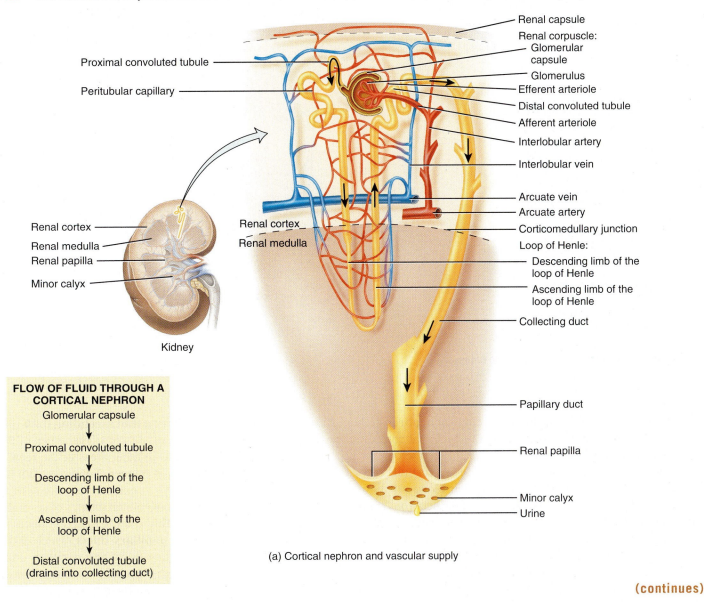

FLOW OF FLUID THROUGH A CORTICAL NEPHRON

Glomerular capsule
↓
Proximal convoluted tubule
↓
Descending limb of the loop of Henle
↓
Ascending limb of the loop of Henle
↓
Distal convoluted tubule
(drains into collecting duct)

(a) Cortical nephron and vascular supply

(continues)

Figure 24.5 (continued)

Renal capsule

Distal convoluted tubule

Renal corpuscle:
Glomerular capsule

Proximal convoluted tubule

Peritubular capillary

Glomerulus

Afferent arteriole

Efferent arteriole

Interlobular artery

Interlobular vein

Arcuate vein

Arcuate artery

Corticomedullary junction

Renal cortex

Renal medulla

Renal cortex
Renal medulla

Renal papilla

Minor calyx

Kidney

Collecting duct

Loop of Henle:
Descending limb
Thick ascending limb
Thin ascending limb

Vasa recta

Papillary duct

FLOW OF FLUID THROUGH A JUXTAMEDULLARY NEPHRON

Glomerular capsule
↓
Proximal convoluted tubule
↓
Descending limb of the loop of Henle
↓
Thin ascending limb of the loop of Henle
↓
Thick ascending limb of the loop of Henle
↓
Distal convoluted tubule (drains into collecting duct)

Renal papilla

Minor calyx

Urine

(b) Juxtamedullary nephron and vascular supply

🔑 **Nephrons are the functional units of the kidneys.**

About 80–85% of the nephrons are **cortical nephrons**. Their renal corpuscles lie in the outer portion of the renal cortex, and they have *short* loops of Henle that lie mainly in the cortex and penetrate only into the outer region of the renal medulla (Figure 24.5a). The short loops of Henle receive their blood supply from peritubular capillaries. The other 15–20% of the nephrons are **juxtamedullary nephrons** (*juxta-* = near to). Their renal corpuscles lie deep in the cortex, close to the medulla, and they have a *long* loop of Henle that extends into the deepest region of the medulla (Figure 24.5b). Long loops of Henle receive their blood supply from peritubular capillaries and from the vasa recta. In addition, the ascending limb of the loop of Henle of juxtamedullary neph-

rons consists of two portions: a **thin ascending limb** followed by a **thick ascending limb** (Figure 24.5b). The lumen of the thin ascending limb is the same as in other areas of the renal tubule; it is only the epithelium that is thinner. Nephrons with long loops of Henle enable the kidneys to excrete very dilute or very concentrated urine. The distal convoluted tubules of several nephrons empty into a single **collecting duct**. Collecting ducts then unite into several hundred large **papillary ducts**, which drain into the minor calyces. The collecting ducts and papillary ducts extend from the renal cortex through the renal medulla to the renal pelvis. So one kidney has about 1 million nephrons, but a much smaller number of collecting ducts and even fewer papillary ducts.

Histology of the Nephron and Collecting Duct

A single layer of epithelial cells forms the entire wall of the glomerular capsule, renal tubule, and ducts. However, each part has distinctive histological features that reflect its particular functions. We discuss them in the order that fluid flows through them: glomerular capsule, renal tubule, and collecting duct.

Glomerular Capsule

The glomerular capsule consists of visceral and parietal layers (Figure 24.6a). The visceral layer consists of modified simple squamous epithelial cells called **podocytes** (PŌ-dō-cīts; *podo-* = foot; *-cytes* = cells). The many footlike projections of these cells (pedicels) wrap around the single layer of endothelial cells of the glomerulus and form the inner wall of the capsule. The parietal

Figure 24.6 **Histology of a renal corpuscle.**

Renal corpuscle (external view)

(a) Renal corpuscle (internal view)

(b) Renal corpuscle

LM 1380x

🔑 **A renal corpuscle consists of a glomerular capsule and a glomerulus.**

layer of the glomerular capsule consists of simple squamous epithelium and forms the outer wall of the capsule. Fluid filtered from the glomerular capillaries enters the **capsular space**, the space between the two layers of the glomerular capsule. Think of the glomerulus as a fist punched into a limp balloon (the glomerular capsule) until the fist is covered by two layers of balloon (visceral and parietal layers) with a space in between (the capsular space).

Renal Tubule and Collecting Duct

The proximal convoluted tubule is composed of simple cuboidal epithelial cells with microvilli on their apical surface (surface facing the lumen) (Figure 24.6a). These microvilli, like those of the small intestine, increase the surface area for reabsorption and secretion. Simple cuboidal epithelium transitions into simple columnar epithelium in the ascending limb of the loop of Henle. The final part of the ascending limb of the loop of Henle makes contact with the afferent arteriole serving that renal corpuscle. Because the columnar cells in this region of the ascending limb are crowded together, they are known as the **macula densa** (*macula* = spot; *densa* = dense). Alongside the macula densa, the wall of the affer-

ent arteriole (and sometimes the efferent arteriole) contains modified smooth muscle fibers called **juxtaglomerular cells**. Together with the macula densa, they constitute the **juxtaglomerular apparatus**. As you will see later, the juxtaglomerular apparatus helps regulate blood pressure within the kidneys.

The distal convoluted tubule (DCT) begins a short distance past the macula densa. In the last part of the DCT and continuing into the collecting ducts, two different types of cells are present. Most are **principal cells**, which have receptors for both antidiuretic hormone (ADH) and aldosterone, two hormones that regulate their functions. A smaller number are **intercalated cells**, which play a role in the homeostasis of blood pH.

Checkpoint

8. What are the major parts of a nephron?
9. Is the photomicrograph in Figure 24.6b from a section through the renal cortex or renal medulla? How can you tell?
10. What are the basic differences between cortical and juxtamedullary nephrons?
11. Which structures enclose the capsular space?
12. Where is the juxtaglomerular apparatus located, and what is its structure?

24.4 Urine is formed by glomerular filtration, tubular secretion, and tubular reabsorption.

To produce urine, nephrons and collecting ducts perform the following three basic processes (Figure 24.7):

1 **Glomerular filtration**. In the first step of urine production, water and most solutes in blood plasma move across the wall of glomerular capillaries into the glomerular capsule and then into the renal tubule.

2 **Tubular reabsorption**. As filtered fluid flows through the renal tubule and the collecting duct, tubule and duct cells reabsorb about 99% of the filtered water and many useful solutes. The water and solutes return to the blood flowing through the peritubular capillaries and vasa recta. Note that the term *reabsorption* refers to the return of substances to the bloodstream. The term *absorption*, by contrast, means entry

Figure 24.7 **Relationship of a nephron's structure to its three basic functions: glomerular filtration, tubular reabsorption, and tubular secretion.**

Glomerular filtration occurs in the renal corpuscle; tubular reabsorption and tubular secretion occur along the renal tubule and collecting duct.

of new substances into the body, as occurs in the gastrointestinal tract.

3 **Tubular secretion**. As fluid flows through the renal tubule and collecting duct, materials in the peritubular capillary blood or tubule and duct cells, such as wastes, drugs, and excess ions, are secreted into the fluid. Notice that tubular secretion *removes* a substance from the blood.

Solutes in the fluid that drains into the renal pelvis remain in the urine and are excreted. The rate of urinary excretion of any solute is equal to its rate of glomerular filtration, plus its rate of secretion, minus its rate of reabsorption.

By filtering, reabsorbing, and secreting, nephrons help maintain homeostasis of the blood's volume and composition. The situation is somewhat analogous to a recycling center: Garbage trucks dump refuse into an input hopper, where the smaller refuse passes onto a conveyor belt (glomerular filtration of plasma). As the conveyor belt carries the garbage along, workers remove useful items, such as aluminum cans, plastics, and glass containers (reabsorption). Other workers place additional garbage left at the center and larger items onto the conveyor belt (secretion). At the end of the belt, all remaining garbage falls into a truck for transport to the landfill (excretion of wastes in urine).

Checkpoint

13. If the urinary excretion rate of a drug such as penicillin is greater than the rate at which it is filtered at the glomerulus, how else is it getting into the urine?

14. When cells of the renal tubules secrete the drug penicillin, is the drug being added to or removed from the bloodstream?

Return to Andre's Story

"Andre, you look terrible. When are you going to see the doctor?" Andre's wife Maive was worried. Over the last two days, Andre had gotten worse. He had been up half the night vomiting, had terrible back pain just under his ribs, and had developed a fever.

"I have an appointment today; you'll need to drive me. I'm not up to it. I feel terrible."

At the clinic, the nurse found Andre's blood pressure to be 180/110, and his fever was 102°F. The doctor wasted no time in ordering a urinalysis, blood chemistry tests, and an ultrasound of the kidney. "Andre, I think you may have a kidney infection, a very serious one at that. How long has this been going on?" asked the doctor.

"For three or four days. I always have trouble peeing; but I noticed the smell in my urine a few days ago. I started having the back pain then, too. I didn't start feeling nauseous until yesterday. That's when the fever seemed to get worse. My hands and feet seem to be swollen, too; I can hardly feel anything in them."

The doctor nodded and asked, "Did you take anything? I see you had refills on your antibiotic prescription."

"No antibiotics; I ran out. I've been taking ibuprofen; it doesn't seem to be working, though. When it didn't help the back pain and headache, I just doubled up on the dose."

The doctor's eyes widened. "Andre, it's important to know how much ibuprofen you've been taking. High doses can damage the kidneys."

"I don't know," replied Andre, "I finished the bottle, I think."

Just then the nurse entered the exam room and handed the lab reports to the doctor.

"Well, Andre, your blood urea, nitrogen, and creatinine levels are elevated well beyond normal; blood potassium and sodium are as well. There were also bacteria in the urine; I expected that. The urinalysis and blood results concern me. I'd like to admit you to the hospital. We need to get your electrolyte levels back to normal. We can address the infection with antibiotics. I think the combination of the infection and the excessive amounts of ibuprofen may have produced an acute tubular necrosis, causing your kidneys to stop working."

A. Which important ions does the kidney regulate that might be imbalanced if Andre's kidneys are not functioning properly?

B. What might cause Andre's nausea and vomiting?

C. Andre is producing reduced amounts of urine, yet he is taking in a normal amount of fluids and electrolytes; how might this affect his blood volume? Could this be related to his elevated blood pressure?

D. Which urinary structures might be affected in Andre's acute tubular necrosis? (Hint: "Tubular.")

E. What is the source of the creatinine and urea the doctor mentions?

24.5 Water and solutes are forced through the filtration membrane during glomerular filtration.

The fluid that enters the capsular space is called the **glomerular filtrate**. On average, the daily volume of glomerular filtrate in adults is 150 liters in females and 180 liters in males. More than 99% of the glomerular filtrate returns to the bloodstream via tubular reabsorption, so only 1–2 liters (about 1–2 qt) are excreted as urine.

The Filtration Membrane

Together, the endothelial cells of glomerular capillaries and the podocytes, which completely encircle the capillaries, form a leaky barrier known as the **filtration membrane**. This sandwich-like assembly permits filtration of water and small solutes from the blood into the capsular space. Blood cells, platelets, and most plasma proteins remain in the blood because they are too large to pass through the filtration membrane. Substances filtered from the

blood cross three barriers—a glomerular endothelial cell, the basal lamina, and a filtration slit formed by a podocyte (Figure 24.8):

1 Glomerular endothelial cells are quite leaky because they have large **fenestrations** (pores) that measure 70–100 nm (0.07–0.1 μm) in diameter. This size permits all solutes in blood plasma to exit glomerular capillaries but prevents filtration of blood cells and platelets. Located among the glomerular capillaries and in the cleft between afferent and

Figure 24.8   The filtration membrane. The size of the endothelial fenestrations and filtration slits in (a) have been exaggerated for emphasis.

1 Fenestration of glomerular endothelial cell: prevents filtration of blood cells but allows all components of blood plasma to pass through

2 Basal lamina of glomerulus: prevents filtration of larger proteins

3 Slit membrane between pedicels: prevents filtration of medium-sized proteins

Filtration slit
Pedicel
Podocyte of visceral layer of glomerular capsule

(a) Details of filtration membrane

Pedicel of podocyte   Filtration slit
Basal lamina
Lumen of glomerulus
Fenestration of glomerular endothelial cell
TEM 78,000x

(b) Filtration membrane

During glomerular filtration, water and solutes pass from blood plasma into the capsular space.

efferent arterioles are **mesangial cells** (*mes-* = in the middle; *-angi-* = blood vessel) (see Figure 24.6a). These contractile cells help regulate glomerular filtration.

② **The basal lamina**, a layer of acellular material between the endothelium and the podocytes, consists of minute collagen fibers in a glycoprotein matrix; it prevents filtration of larger plasma proteins.

③ Extending from each podocyte are thousands of footlike processes termed **pedicels** (PED-i-sels = little feet) that wrap around glomerular capillaries. The spaces between pedicels are the **filtration slits**. A thin membrane, the *slit membrane*, extends across each filtration slit; it permits the passage of molecules with a diameter smaller than 6–7 nm (0.006–0.007 μm), including water, glucose, vitamins, amino acids, very small plasma proteins, ammonia, urea, and ions. Less than 1% of albumin, the most plentiful plasma protein, passes the slit membrane because, with a diameter of 7.1 nm, it is slightly too big to get through.

The process of *filtration*—the use of pressure to force fluids and solutes through a membrane—is the same in glomerular capillaries as in capillaries elsewhere in the body. However, the volume of fluid filtered by the renal corpuscle is much larger than in other capillaries of the body for three reasons:

1. Glomerular capillaries present a large surface area for filtration because they are long and extensive. The mesangial cells regulate how much of this surface area is available for filtration. When mesangial cells are relaxed, surface area is maximal, and glomerular filtration is very high. Contraction of mesangial cells reduces the available surface area, and glomerular filtration decreases.

2. The filtration membrane is thin and porous. Despite having several layers, the thickness of the filtration membrane is only 0.1 μm. Glomerular capillaries also are about 50 times leakier than capillaries in most other tissues, mainly because of their large fenestrations.

3. Glomerular capillary blood pressure is high. Because the efferent arteriole is smaller in diameter than the afferent arteriole, resistance to the outflow of blood from the glomerulus is high. As a result, blood pressure in glomerular capillaries is considerably higher than in capillaries elsewhere in the body.

Net Filtration Pressure

Glomerular filtration depends on three main pressures. One type of pressure *promotes* filtration and two other pressures *oppose* filtration (Figure 24.9).

① **Glomerular blood hydrostatic pressure (GBHP)** is the blood pressure in the glomerulus. Generally, GBHP is about 55 mm Hg. It promotes filtration by forcing water and solutes in blood plasma through the filtration membrane.

Figure 24.9 **The pressures that drive glomerular filtration.** Taken together, these pressures determine net filtration pressure (NFP).

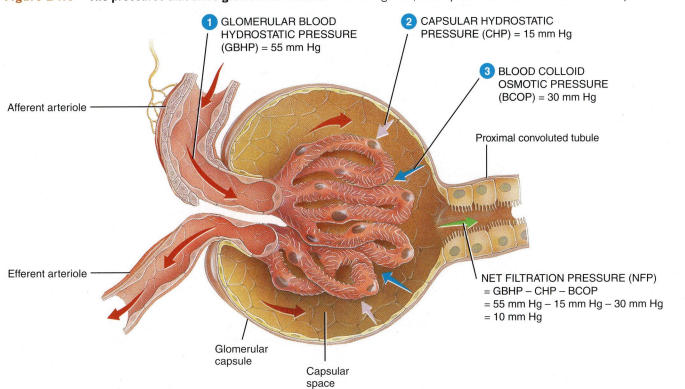

① GLOMERULAR BLOOD HYDROSTATIC PRESSURE (GBHP) = 55 mm Hg

② CAPSULAR HYDROSTATIC PRESSURE (CHP) = 15 mm Hg

③ BLOOD COLLOID OSMOTIC PRESSURE (BCOP) = 30 mm Hg

Afferent arteriole

Proximal convoluted tubule

Efferent arteriole

NET FILTRATION PRESSURE (NFP)
= GBHP – CHP – BCOP
= 55 mm Hg – 15 mm Hg – 30 mm Hg
= 10 mm Hg

Glomerular capsule

Capsular space

🔑 **Glomerular blood hydrostatic pressure promotes filtration; capsular hydrostatic pressure and blood colloid osmotic pressure oppose filtration.**

2 **Capsular hydrostatic pressure (CHP)** is the hydrostatic pressure exerted against the filtration membrane by fluid already in the capsular space. CHP opposes filtration and represents a "back pressure" of about 15 mm Hg that pushes water and solutes from the filtrate back into the plasma.

3 **Blood colloid osmotic pressure (BCOP)** is the osmotic pressure resulting from the presence of proteins in blood plasma such as albumin, globulins, and fibrinogen. The BCOP opposes filtration by drawing water from the filtrate back into the plasma within the glomerulus. The average BCOP in glomerular capillaries is 30 mm Hg.

Net filtration pressure (NFP), the total pressure that promotes filtration, is determined as follows:

$$\text{Net filtration pressure (NFP)} = \text{GBHP} - \text{CHP} - \text{BCOP}$$

By substituting the values just given, normal NFP may be calculated:

$$\text{NFP} = 55 \text{ mm Hg} - 15 \text{ mm Hg} - 30 \text{ mm Hg} = 10 \text{ mm Hg}$$

Thus, under normal conditions, a pressure of only 10 mm Hg causes blood plasma (minus plasma proteins) to filter from the glomerulus into the capsular space.

Glomerular Filtration Rate

The amount of filtrate formed in all the renal corpuscles of both kidneys each minute is the **glomerular filtration rate (GFR)**. In adults, the GFR averages 125 mL/min in males and 105 mL/min in females. Homeostasis of body fluids requires that the kidneys maintain a relatively constant GFR. If the GFR is too high, needed substances may pass so quickly through the renal tubules that some are not reabsorbed and are lost in the urine. On the other hand, if the GFR is too low, nearly all the filtrate may be reabsorbed and certain waste products may not be adequately excreted.

Any change in net filtration pressure will affect GFR. Severe blood loss, for example, reduces mean arterial blood pressure and decreases the glomerular blood hydrostatic pressure. Filtration ceases if glomerular blood hydrostatic pressure drops to 45 mm Hg because the opposing pressures add up to 45 mm Hg. Amazingly, when systemic blood pressure rises above normal, net filtration pressure and GFR increase very little. GFR is nearly constant when the mean arterial blood pressure is anywhere between 80 and 180 mm Hg.

The mechanisms that regulate glomerular filtration rate operate in two main ways: (1) by adjusting blood flow into and out of the glomerulus and (2) by altering the glomerular capillary surface area available for filtration. GFR increases when blood flow into the glomerular capillaries increases. Coordinated control of the diameter of both afferent and efferent arterioles regulates glomerular blood flow. Constriction of the afferent arteriole decreases blood flow into the glomerulus; dilation of the afferent arteriole increases it. Constriction of the efferent arteriole slows outflow of blood. Three mechanisms control GFR: renal autoregulation, neural regulation, and hormonal regulation.

Figure 24.10   Tubuloglomerular feedback.

Macula densa cells of the juxtaglomerular apparatus provide negative feedback regulation of glomerular filtration rate.

Renal Autoregulation of GFR

The kidneys themselves help maintain a constant renal blood flow and GFR despite normal, everyday changes in blood pressure, like those that occur during exercise. This capability is called **renal autoregulation** and consists of two mechanisms—the myogenic mechanism and tubuloglomerular feedback. Working together, they can maintain nearly constant GFR over a wide range of systemic blood pressures.

The **myogenic mechanism** (*myo-* = muscle; *-genic* = producing) occurs when stretching triggers contraction of smooth muscle cells in the wall of afferent arterioles. As blood pressure rises, GFR also rises because renal blood flow increases. However, the elevated blood pressure stretches the walls of the afferent arterioles. In response, smooth muscle fibers in the wall of the afferent arteriole contract, which narrows the arteriole's lumen. As a result, renal blood flow decreases, thus reducing GFR to its previous level. Conversely, when arterial blood pressure drops, the smooth muscle cells are stretched less and thus relax. The afferent arterioles dilate, renal blood flow increases, and GFR increases.

The second contributor to renal autoregulation, **tubuloglomerular feedback**, is so named because part of the renal tubules—the macula densa—provides feedback to the glomerulus (Figure 24.10). When GFR is above normal due to elevated systemic blood pressure, filtered fluid flows more rapidly along the renal tubules. As a result, the proximal convoluted tubule and loop of Henle have less time to reabsorb Na+, Cl−, and water. Macula densa cells are thought to detect the increased delivery of Na+, Cl−, and water and to inhibit release of nitric oxide (NO) from cells in the juxtaglomerular apparatus. Because NO causes vasodilation, afferent arterioles constrict when the level of NO declines. As a result, less blood flows into the glomerular capillaries, and GFR decreases. When blood pressure falls, causing GFR to be lower than normal, the opposite sequence of events occurs, although to a lesser degree.

Neural Regulation of GFR

Like most blood vessels of the body, those of the kidneys are supplied by sympathetic neurons of the autonomic nervous system (ANS) that release norepinephrine. Norepinephrine produces vasoconstriction of the afferent arterioles. At rest, sympathetic stimulation is moderately low, the afferent arterioles are relatively dilated, and renal autoregulation of GFR prevails. With greater sympathetic stimulation, as occurs during exercise or hemorrhage, the afferent arterioles are constricted. As a result, blood flow into glomerular capillaries is greatly decreased, net filtration pressure decreases, and GFR drops. This lowering of renal blood flow has two consequences: (1) It reduces urine output, which helps conserve blood volume; (2) it permits greater blood flow to other body tissues.

Hormonal Regulation of GFR

Two hormones contribute to regulation of GFR. Angiotensin II reduces GFR; atrial natriuretic peptide increases GFR. **Angiotensin II** is a very potent vasoconstrictor that narrows both afferent and efferent arterioles and reduces renal blood flow, thereby decreasing GFR. Cells in the atria of the heart secrete **atrial natriuretic peptide (ANP)**. Stretching of the atria, as occurs when blood volume increases, stimulates secretion of ANP. By causing relaxation of the glomerular mesangial cells, ANP increases the capillary surface area available for filtration. Glomerular filtration rate rises as the surface area increases.

Table 24.1 summarizes the regulation of glomerular filtration rate.

TABLE 24.1

Regulation of Glomerular Filtration Rate (GFR)

TYPE OF REGULATION	MAJOR STIMULUS	MECHANISM AND SITE OF ACTION	EFFECT ON GFR
Renal autoregulation			
Myogenic mechanism	Increased stretching of smooth muscle fibers in afferent arteriole walls due to increased blood pressure.	Stretched smooth muscle fibers contract, thereby narrowing the lumen of the afferent arterioles.	Decrease.
Tubuloglomerular feedback	Rapid delivery of Na+ and Cl− to the macula densa due to high systemic blood pressure.	Decreased release of nitric oxide (NO) by the juxtaglomerular apparatus causes constriction of afferent arterioles.	Decrease.
Neural regulation	Increase in level of activity of renal sympathetic nerves releases norepinephrine.	Constriction of afferent arterioles.	Decrease.
Hormonal regulation			
Angiotensin II	Decreased blood volume or blood pressure stimulates production of angiotension II.	Constriction of both afferent and efferent arterioles.	Decrease.
Atrial natriuretic peptide (ANP)	Stretching of the atria of the heart stimulates secretion of ANP.	Relaxation of mesangial cells in glomerulus increases capillary surface area available for filtration.	Increase.

Checkpoint

15. A water molecule has just entered the glomerulus. Which structures will it pass through to arrive at the proximal convoluted tubule?

16. What is the major chemical difference between blood plasma and glomerular filtrate?

17. Why is there much greater filtration through glomerular capillaries than through capillaries elsewhere in the body?

18. How does blood pressure promote filtration of blood in the kidneys?

19. How do the myogenic mechanism and tubuloglomerular feedback regulate the glomerular filtration rate? Why are these processes termed autoregulation?

20. How is the glomerular filtration rate regulated by the autonomic nervous system and by hormones?

24.6 Tubular reabsorption reclaims needed substances from the filtrate, while tubular secretion discharges unneeded substances.

Principles of Tubular Reabsorption and Secretion

The volume of fluid entering the proximal convoluted tubules in just half an hour is greater than the total blood plasma volume because the normal rate of glomerular filtration is so high. Obviously some of this fluid must be returned somehow to the bloodstream. Tubular reabsorption—the return of most of the filtered water and many of the filtered solutes to the bloodstream—is the second basic function of the nephron and collecting duct. Normally, about 99% of the filtered water is reabsorbed. Only 1% of the water in glomerular filtrate actually leaves the body in urine. Epithelial cells all along the renal tubule and duct carry out reabsorption. Solutes that are reabsorbed by both active and passive processes include glucose, amino acids, urea, and ions such as sodium, potassium, calcium, chloride, bicarbonate, and phosphate. Once fluid passes through the proximal convoluted tubule, cells located distal to the proximal convoluted tubule fine-tune the reabsorption processes to maintain homeostatic balances of water and selected ions. Most small proteins and peptides that pass through the filter are also reabsorbed, usually via pinocytosis. To appreciate the magnitude of tubular reabsorption, look at Table 24.2 and compare the amounts of substances that are filtered, reabsorbed, and excreted in urine. Although the values shown are typical, they vary considerably according to diet.

The third function of nephrons and collecting ducts is tubular secretion, the transfer of materials from the blood and tubule cells into tubular fluid. Secreted substances include hydrogen, potassium, ammonium ions, creatinine, and certain drugs such as penicillin. Tubular secretion has two important outcomes: (1) The secretion of hydrogen ions helps control blood pH; (2) the secretion of other substances helps eliminate them from the body.

Reabsorption Routes

A substance being reabsorbed from the fluid in the tubule lumen can take one of two routes before entering a peritubular capillary: It can move *between* adjacent tubule cells or *through* an individual

TABLE 24.2

Substances Filtered, Reabsorbed, and Excreted in Urine

SUBSTANCE	FILTERED* (ENTERS GLOMERULAR CAPSULE PER DAY)	REABSORBED (RETURNED TO BLOOD PER DAY)	URINE (EXCRETED PER DAY)
Water	180 liters	178–179 liters	1–2 liters
Proteins	2.0 g	1.9 g	0.1 g
Sodium ions (Na$^+$)	579 g	575 g	4 g
Chloride ions (Cl$^-$)	640 g	633.7 g	6.3 g
Bicarbonate ions (HCO$_3^-$)	275 g	274.97 g	0.03 g
Glucose	162 g	162 g	0 g
Urea	54 g	24 g	30 g[†]
Potassium ions (K$^+$)	29.6 g	29.6 g	2.0 g[‡]
Uric acid	8.5 g	7.7 g	0.8 g
Creatinine	1.6 g	0 g	1.6 g

*Assuming GFR is 180 liters per day.

† In addition to being filtered and reabsorbed, urea is secreted.

‡After virtually all filtered K$^+$ is reabsorbed in the convoluted tubules and loop of Henle, a variable amount of K$^+$ is secreted by principal cells in the collecting duct.

tubule cell (Figure 24.11). Along the renal tubule, cell junctions surround and join neighboring cells to one another, much like the plastic rings that hold a six-pack of soda cans together. The **apical membrane** (the tops of the soda cans) contacts the tubular fluid, and the **basolateral membrane** (the bottoms and sides of the soda cans) contacts interstitial fluid at the base and sides of the cell.

The cell junctions do not completely seal off the interstitial fluid from the fluid in the tubule lumen. Fluid can leak *between* the cells in a passive process known as **paracellular reabsorption** (*para-* = beside). In some parts of the renal tubule, the paracellular route is thought to account for up to 50% of the reabsorption of certain ions and the water that accompanies them via osmosis. In **transcellular reabsorption** (*trans-* = across), a substance passes from the fluid in the tubular lumen through the apical membrane of a tubule cell, across the cytosol, and out into interstitial fluid through the basolateral membrane.

Transport Mechanisms

When renal cells transport solutes out of or into tubular fluid, they move specific substances in one direction only. Not surprisingly, different types of transport proteins are present in the apical and basolateral membranes. Reabsorption of Na^+ by the renal tubules is especially important because of the large number of sodium ions that pass through the glomerular filters.

Cells lining the renal tubules, like other cells throughout the body, have a low concentration of Na^+ in their cytosol due to the activity of sodium–potassium pumps (Na^+/K^+ ATPases). These pumps are located in the basolateral membranes and eject Na^+ from the renal tubule cells into interstitial fluid (Figure 24.11). The absence of sodium–potassium pumps in the apical membrane ensures that reabsorption of Na^+ is a one-way process.

Transport of materials across membranes may be either active or passive. In **primary active transport**, the energy derived from hydrolysis of ATP is used to "pump" a substance across a membrane; the sodium–potassium pump is one such pump. In **secondary active transport**, the energy stored in an ion's electrochemical gradient, rather than hydrolysis of ATP, drives another substance across a membrane. Secondary active transport couples the movement of an ion down its electrochemical gradient to the "uphill" movement of a second substance against its electrochemical gradient. *Symporters* are membrane proteins that move two or more substances in the *same* direction across a membrane. *Antiporters* move two or more substances in *opposite* directions across a membrane.

Solute reabsorption drives water reabsorption because all water reabsorption occurs via osmosis. The movement of solutes into peritubular capillaries decreases the solute concentration of the tubular fluid but increases the solute concentration in the peritubular capillaries. As a result, water moves by osmosis into peritubular capillaries. About 90% of the reabsorption of water filtered by the kidneys occurs along with the reabsorption of solutes such as Na^+, Cl^-, and glucose. Water reabsorbed with solutes in tubular fluid is termed **obligatory water reabsorption** (ob-LIG-a-tor´-e) because the water is "obliged" to follow the solutes when they are reabsorbed. This type of water reabsorption occurs in the proximal convoluted tubule and the descending limb of the loop of Henle because these segments of the nephron are always permeable to water. Reabsorption of the final 10% of the water, a total of 10–20 liters per day, is termed **facultative water reabsorption** (FAK-ul-tā´-tiv). The word *facultative* means "capable of adapting to a need." Facultative water reabsorption is regulated by antidiuretic hormone and occurs mainly in the collecting ducts.

Now that we have discussed the principles of renal transport, we can follow the filtered fluid from the proximal convoluted tubule, into the loop of Henle, on to the distal convoluted tubule, and through the collecting ducts. In each segment, we examine where and how specific substances are reabsorbed and secreted. The filtered fluid is called *tubular fluid* once it enters the proximal convoluted tubule. The composition of tubular fluid changes as it flows along the nephron tubule and through the collecting duct due to reabsorption and secretion. The fluid that drains from papillary ducts into the renal pelvis is *urine*.

Figure 24.11 Reabsorption routes: paracellular reabsorption and transcellular reabsorption.

Key: ·····▶ Diffusion

⟶ Active transport

⬭ Sodium–potassium pump (Na^+/K^+ ATPase)

🔑 In paracellular reabsorption, water and solutes in tubular fluid return to the bloodstream by moving between tubule cells; in transcellular reabsorption, water and solutes in tubular fluid return to the bloodstream by passing through a tubule cell.

Reabsorption and Secretion in the Proximal Convoluted Tubule

The largest amount of solute and water reabsorption from filtered fluid occurs in the proximal convoluted tubules (PCTs), where most absorptive processes involve sodium ions. Na⁺ transport in the PCT occurs via symport and antiport mechanisms. Normally, filtered glucose, amino acids, lactic acid, water-soluble vitamins, and other nutrients are not lost in the urine. Rather, they are completely reabsorbed in the PCT by several types of **Na⁺ symporters**

Figure 24.12 Reabsorption of glucose by Na⁺–glucose symporters in the proximal convoluted tubule.

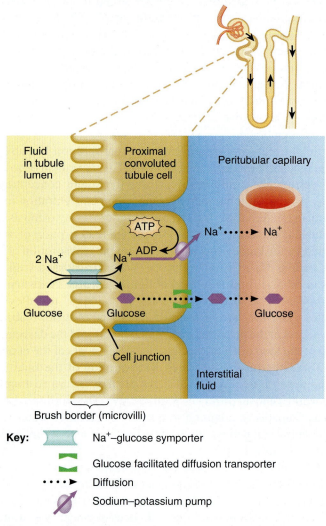

Key:

- ⬭ Na⁺–glucose symporter
- ⬭ Glucose facilitated diffusion transporter
- ••••▶ Diffusion
- ⬭ Sodium–potassium pump

🔑 **Normally, all filtered glucose is reabsorbed in the proximal convoluted tubule.**

located in the apical membrane. Figure 24.12 depicts the operation of one such symporter, the Na⁺–glucose symporter in the apical membrane of a cell in the PCT. Two Na⁺ and a molecule of glucose attach to the symporter protein, which carries them from the tubular fluid into the tubule cell. The glucose molecules then exit the basolateral membrane via facilitated diffusion, and they diffuse into peritubular capillaries. Other Na⁺ symporters in the PCT reclaim other filtered solutes in a similar way.

In another secondary active transport process, the **Na⁺/H⁺ antiporters** carry filtered Na⁺ into a PCT cell as H⁺ is moved from the cell into the lumen (Figure 24.13a), causing Na⁺ to be reabsorbed into blood and H⁺ to be secreted into tubular fluid. PCT cells produce the H⁺ needed to keep the antiporters running in the following way: (1) Carbon dioxide (CO_2) diffuses from peritubular blood or tubular fluid or is produced by metabolic reactions within the cells; (2) the enzyme *carbonic anhydrase* (CA) catalyzes the reaction of CO_2 with water (H_2O) to form carbonic acid (H_2CO_3), which then dissociates into H⁺ and HCO_3^-:

$$CO_2 + H_2O \xrightarrow{\text{Carbonic anhydrase}} H_2CO_3 \longrightarrow H^+ + HCO_3^-$$

Most of the HCO_3^- in filtered fluid is reabsorbed in proximal convoluted tubules, thereby safeguarding the body's supply of an important buffer (Figure 24.13b). After H⁺ is secreted into the fluid within the lumen of the proximal convoluted tubule, it reacts with filtered HCO_3^- to form H_2CO_3, which readily dissociates into CO_2 and H_2O. Carbon dioxide then diffuses into the tubule cells and joins with H_2O to form H_2CO_3, which dissociates into H⁺ and HCO_3^-. The bicarbonate ion exits the cell via facilitated diffusion transporters in the basolateral membrane and diffuses into the blood with Na⁺. Thus, for every H⁺ secreted into the tubular fluid of the proximal convoluted tubule, one HCO_3^- and one Na⁺ are reabsorbed.

Solute reabsorption in proximal convoluted tubules promotes osmosis of water. Each reabsorbed solute increases the osmotic pressure, first inside the tubule cell, then in interstitial fluid, and finally in the blood. Water thus moves rapidly from the tubular fluid into the peritubular capillaries and restores osmotic balance (Figure 24.14). In other words, reabsorption of solutes creates an osmotic gradient that promotes the reabsorption of water via osmosis.

As water leaves the tubular fluid, the concentrations of the remaining filtered solutes increase. Increasing electrochemical gradients for Cl⁻, K⁺, Ca²⁺, Mg²⁺, and urea promote their diffusion into peritubular capillaries. Among these ions, Cl⁻ is present in the highest concentration. Diffusion of negatively charged Cl⁻ into interstitial fluid makes the interstitial fluid electrically more negative than the tubular fluid. This negativity promotes passive reabsorption of cations (positively charged ions), such as K⁺, Ca²⁺, and Mg²⁺.

Figure 24.13 **Actions of Na⁺/H⁺ antiporters in the proximal convoluted tubule.** (a) Reabsorption of sodium ions (Na⁺) and secretion of hydrogen ions (H⁺) via secondary active transport through the apical membrane; (b) reabsorption of bicarbonate ions (HCO$_3^-$) via facilitated diffusion through the basolateral membrane.

(a) Na⁺ reabsorption and H⁺ secretion

(b) HCO$_3^-$ reabsorption

Key:

Na⁺/H⁺ antiporter

HCO$_3^-$ facilitated diffusion transporter

Diffusion

Sodium–potassium pump

🔑 **Na⁺/H⁺ antiporters promote transcellular reabsorption of Na⁺ and secretion of H⁺.**

Figure 24.14 **Passive reabsorption of Cl⁻, K⁺, Ca²⁺, Mg²⁺, urea, and water in the proximal convoluted tubule.**

🔑 **Electrochemical gradients promote passive reabsorption of solutes via both paracellular and transcellular routes.**

Ammonia (NH_3) is a poisonous waste product that is produced when amino groups are removed from amino acids, a reaction that occurs mainly in the liver. Hepatocytes (liver cells) convert most of this ammonia to urea, a less toxic compound. Although tiny amounts of urea and ammonia are present in sweat, most excretion of these nitrogen-containing waste products occurs via the urine. Urea and ammonia in blood are both filtered at the glomerulus and secreted by proximal convoluted tubule cells into the tubular fluid.

Reabsorption in the Loop of Henle

Because the proximal convoluted tubules reabsorb about 65% of the filtered water (about 80 mL/min), fluid enters the next part of the nephron, the loop of Henle. The chemical composition of the tubular fluid now is quite different from that of glomerular filtrate because glucose, amino acids, and other nutrients are no longer present.

Of filtered substances, the loop of Henle reabsorbs about 15% of the water; 25% of the Na⁺ and K⁺; 35% of the Cl⁻; 15% of the HCO$_3^-$; and a variable amount of the Ca²⁺ and Mg²⁺. Here, for the first time, reabsorption of water via osmosis is *not* automatically coupled to reabsorption of filtered solutes because part of the loop of Henle is relatively impermeable to water.

The apical membranes of cells in the thick ascending limb of the loop of Henle have **Na⁺–K⁺–2Cl⁻ symporters** that simultaneously reclaim one Na^+, one K^+, and two Cl^- from the fluid in the tubular lumen (Figure 24.15). Na^+ that is actively transported into interstitial fluid diffuses into the vasa recta. Cl^- moves through leakage channels (plasma membrane channels that randomly open and close) in the basolateral membrane into interstitial fluid and then into the vasa recta. Because many K^+ leakage channels are present in the apical membrane, most K^+ brought in by the symporters moves down its concentration gradient back into the tubular fluid. Thus, the main effect of the Na^+–K^+–$2Cl^-$ symporters is reabsorption of Na^+ and Cl^-.

Figure 24.15   Na⁺–K⁺–2Cl⁻ symporter in the thick ascending limb of the loop of Henle.

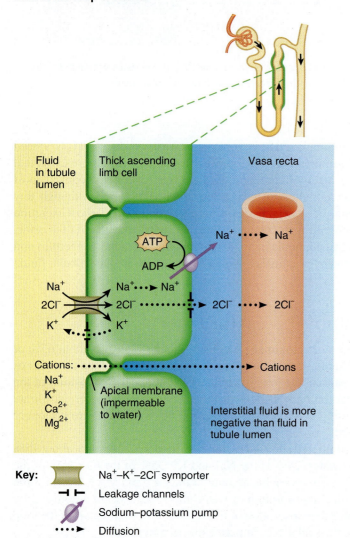

Key:

Na⁺–K⁺–2Cl⁻ symporter

Leakage channels

Sodium–potassium pump

Diffusion

🔑 **Cells in the thick ascending limb have symporters that simultaneously reabsorb one K⁺, one Na⁺, and two Cl⁻.**

The movement of positively charged K^+ into the tubular fluid through the apical membrane channels leaves the interstitial fluid and blood with more negative charges relative to fluid in the ascending limb of the loop of Henle. This relative negativity promotes reabsorption of cations—Na^+, K^+, Ca^{2+}, and Mg^{2+}—via the paracellular route.

Although about 15% of the filtered water is reabsorbed in the *descending* limb of the loop of Henle, little or no water is reabsorbed in the *ascending* limb. In this segment of the tubule, the apical membranes are virtually impermeable to water. Because ions but not water molecules are reabsorbed, the tubular fluid becomes progressively more dilute as it flows toward the end of the ascending limb.

Reabsorption and Secretion in the Distal Convoluted Tubule and Collecting Duct

As fluid flows along the distal convoluted tubules (DCT), reabsorption of Na^+ and Cl^- continues by means of **Na⁺–Cl⁻ symporters** in the apical membranes. Sodium–potassium pumps and Cl^- leakage channels in the basolateral membranes then permit reabsorption of Na^+ and Cl^- into the peritubular capillaries. The DCT also is the major site where parathyroid hormone stimulates reabsorption of Ca^{2+}. Overall, cells of the DCT reabsorb 10–15% of the filtered water.

By the time fluid reaches the end of the distal convoluted tubule, 90–95% of the filtered solutes and water have returned to the bloodstream. Recall that two different types of cells—principal cells and intercalated cells—are present in the last part of the distal convoluted tubule and throughout the collecting duct. Principal cells reabsorb Na^+ and secrete K^+; intercalated cells reabsorb K^+ and HCO_3^- and secrete H^+.

In contrast to earlier segments of the nephron, Na^+ passes through the apical membrane of principal cells via Na^+ leakage channels rather than by means of symporters or antiporters (Figure 24.16). Once in the cytosol, sodium–potassium pumps actively transport Na^+ across the basolateral membranes. Then Na^+ passively diffuses into the peritubular capillaries from the interstitial spaces around the tubule cells.

Normally, most filtered K^+ is returned to the bloodstream by reabsorption in the proximal convoluted tubule and loop of Henle. To adjust for varying dietary intakes of potassium and to maintain a stable level of K^+ in body fluids, principal cells secrete a variable amount of K^+ (Figure 24.16). Because the sodium–potassium pumps continually bring K^+ into principal cells, the intracellular concentration of K^+ remains high. K^+ leakage channels are present in both the apical and basolateral membranes. Thus, some K^+ diffuses down its concentration gradient into the tubular fluid, where the K^+ concentration is very low. This secretion mechanism is the main source of K^+ excreted in the urine.

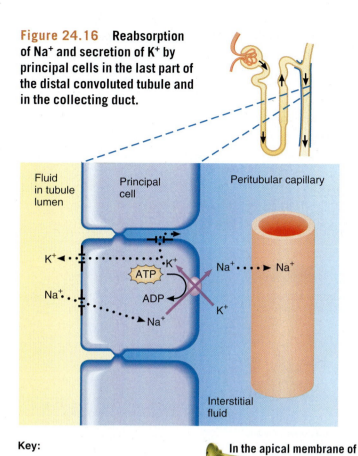

Figure 24.16 **Reabsorption of Na⁺ and secretion of K⁺ by principal cells in the last part of the distal convoluted tubule and in the collecting duct.**

Fluid in tubule lumen

Principal cell

Peritubular capillary

Interstitial fluid

ATP

ADP

K⁺

Na⁺

Na⁺

K⁺

Na⁺

Na⁺

K⁺

Na⁺

Key:

•••••► Diffusion

⊣⊢ Leakage channels

✕ Sodium–potassium pump

In the apical membrane of principal cells, Na⁺ leakage channels allow entry of Na⁺ while K⁺ leakage channels allow exit of K⁺ into the tubular fluid.

Checkpoint

21. Describe the two ways in which substances can be absorbed across the tubule cells.

22. On which tubule cell membrane are the sodium–potassium pumps located? What is the significance of sodium–potassium pumps not being located on the other membrane?

23. What is obligatory water reabsorption, and in which parts of the nephron does it occur?

24. What are two mechanisms in the proximal convoluted tubule for reabsorption of Na⁺? Which other solutes are reabsorbed or secreted with Na⁺ in each mechanism?

25. What is the mechanism in the loop of Henle for reabsorption of Na⁺? Which other solutes are reabsorbed or secreted with Na⁺ in this mechanism?

26. Why does the tubular fluid become more dilute as it travels through the loop of Henle? In which part of the nephron does the most reabsorption of water occur? Which part of the nephron is unable to reabsorb water?

24.7 Five hormones regulate tubular reabsorption and tubular secretion.

Five hormones affect the extent of Na⁺, Cl⁻, Ca²⁺, and water reabsorption as well as K⁺ secretion by the renal tubules. These hormones include angiotensin II, aldosterone, antidiuretic hormone, atrial natriuretic peptide, and parathyroid hormone.

Renin–Angiotensin–Aldosterone System

When blood volume and blood pressure decrease, the walls of the afferent arterioles are stretched less, and the juxtaglomerular cells secrete the enzyme **renin** into the blood. Sympathetic stimulation also directly stimulates release of renin from juxtaglomerular cells. Renin converts angiotensinogen, which is synthesized by the liver, into angiotensin I. *Angiotensin-converting enzyme* then converts angiotensin I to **angiotensin II**, which is the active form of the hormone.

Angiotensin II affects renal physiology in three main ways:

1. It decreases the glomerular filtration rate by causing vasoconstriction of the afferent arterioles.

2. It enhances reabsorption of Na⁺, Cl⁻, and water in the proximal convoluted tubule.

3. It stimulates the adrenal cortex to release **aldosterone**, a hormone that in turn stimulates in the collecting ducts to reabsorb more Na⁺ and Cl⁻ and secrete more K⁺. When more Na⁺ and Cl⁻ are reabsorbed, then more water is also reabsorbed by osmosis. As water levels in the blood rise, blood volume and blood pressure return to normal.

Antidiuretic Hormone

Antidiuretic hormone (ADH) is released by the posterior pituitary. ADH regulates facultative water reabsorption by increasing the water permeability in the distal convoluted tubule and collecting duct. ADH increases the water permeability of principal cells by inserting proteins into their apical membranes that function as water channels. As a result, the water permeability of the principal cell's apical membrane increases, and water molecules move more rapidly from the tubular fluid into the cells. Because the basolateral membranes are always relatively permeable to water, water molecules then move rapidly into the blood. The kidneys can produce a small volume of very concentrated urine each day when ADH concentration is maximal, such as during severe dehydration. When ADH level declines, the water channels are removed from the apical membrane. The kidneys produce a large volume of dilute urine when ADH level is low.

A negative feedback system involving ADH regulates facultative water reabsorption (Figure 24.17). When the concentration of water in the blood decreases by as little as 1%, **osmoreceptors** (receptors that sense osmotic pressure) in the hypothalamus stimulate release of ADH from the posterior pituitary. As ADH levels rise in the blood, the principal cells become more permeable to water. As facultative water reabsorption increases, water concentration in the blood increases to normal. A second powerful stimulus for ADH secretion is a decrease in blood volume, as occurs in hemorrhaging or severe dehydration. In the pathological absence of ADH activity, a condition known as *diabetes insipidus*, a person may excrete up to 20 liters of very dilute urine daily.

Atrial Natriuretic Peptide

A large increase in blood volume promotes release of atrial natriuretic peptide (ANP) from the heart. ANP inhibits reabsorption of Na^+ and water in the proximal convoluted tubule and collecting duct. ANP also suppresses the secretion of aldosterone and ADH. These effects increase the excretion of Na^+ in urine and increase urine output, which decreases blood volume and blood pressure.

Parathyroid Hormone

A lower-than-normal level of Ca^{2+} in the blood stimulates the parathyroid glands to release parathyroid hormone (PTH). PTH in turn stimulates cells in the distal convoluted tubules to reabsorb more Ca^{2+} into the blood. PTH also inhibits HPO_4^{2-} (phosphate) reabsorption in proximal convoluted tubules, thereby promoting phosphate excretion.

Figure 24.17   Negative feedback regulation of water reabsorption by antidiuretic hormone (ADH).

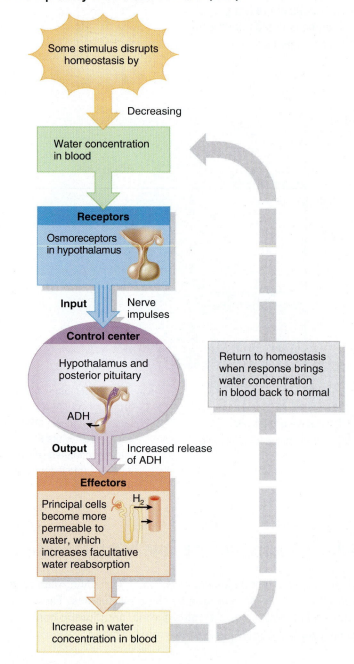

ADH stimulates the kidneys to return more water to the blood.

Table 24.3 summarizes hormonal regulation of tubular reabsorption and tubular secretion.

TABLE 24.3

Hormonal Regulation of Tubular Reabsorption and Tubular Secretion

HORMONE	MAJOR STIMULI THAT TRIGGER RELEASE	SITE OF ACTION	EFFECTS
Angiotensin II	Low blood volume or pressure stimulates renin-induced production.	Proximal convoluted tubule.	Increases Na^+, Cl^-, and water reabsorption, which increases blood volume.
Aldosterone	Increased angiotensin II level promotes release from adrenal cortex.	Collecting duct.	Increases K^+ secretion and Na^+, Cl^-, and water reabsorption, which increases blood volume.
Antidiuretic hormone (ADH)	Decreases blood water concentration, promotes release from posterior pituitary.	Collecting duct.	Increases reabsorption of water, which increases blood water concentration.
Atrial natriuretic peptide	Increased blood volume stretches atria of heart, stimulating secretion from atria.	Proximal convoluted tubule and collecting duct.	Increases Na^+ excretion and inhibits aldosterone and ADH secretion, which decreases blood volume.
Parathyroid hormone	Decreased plasma Ca^{2+} promotes release from parathyroid glands.	Distal convoluted tubule.	Increases Ca^{2+} reabsorption and HPO_4^{2-} excretion.

Checkpoint

27. How is the juxtaglomerular apparatus involved in blood pressure regulation by the kidneys?
28. How do angiotensin II and aldosterone regulate tubular reabsorption and secretion?
29. How does antidiuretic hormone stimulate water reabsorption by principal cells? Is this process obligatory or facultative water reabsorption?
30. Which hormones stimulate and which inhibit water reabsorption?

24.8 Antidiuretic hormone affects the concentration of urine produced by the kidneys.

Even though your fluid intake can be highly variable, the total volume of fluid in your body normally remains stable. Homeostasis of body fluid volume depends in large part on the ability of the kidneys to regulate the rate of water loss in urine. Normally functioning kidneys produce a large volume of dilute urine when fluid intake is high, and a small volume of concentrated urine when fluid intake is low or fluid loss is large. ADH controls whether dilute urine or concentrated urine is formed. In the absence of ADH, urine is very dilute. However, a high level of ADH stimulates reabsorption of more water into blood, producing concentrated urine.

Formation of Dilute Urine

Glomerular filtrate has the same ratio of water and solute particles as blood plasma; therefore, fluid leaving the proximal convoluted tubule is still isotonic to plasma. When *dilute* urine is being formed (Figure 24.18), the following conditions occur along the path of tubular fluid:

1. Because the osmotic gradient of the interstitial fluid of the renal medulla becomes progressively greater, more and more water is reabsorbed by osmosis as tubular fluid flows down the descending limb of the loop of Henle. (The source of this osmotic gradient is explained shortly.) As a result, the fluid remaining in the loop of Henle becomes progressively *more* concentrated.

2. Cells lining the thick ascending limb of the loop have symporters that actively reabsorb Na^+, K^+, and Cl^- from the tubular fluid (see Figure 24.15). The ions pass from the tubular fluid into thick ascending limb cells, then into interstitial fluid, and finally some diffuse into the blood inside the vasa recta.

3. Although solutes are being reabsorbed in the thick ascending limb, the water permeability of this portion of the nephron is always quite low, so water cannot follow by osmosis. As solutes—but not water molecules—are leaving the tubular fluid, the fluid becomes progressively *less* concentrated as it flows up the ascending limb. The fluid entering the distal convoluted tubule is thus more dilute than plasma.

4. The distal convoluted tubule cells are not very permeable to water and are not regulated by ADH. While the fluid flows through the distal convoluted tubule, additional solutes but only a few water molecules are reabsorbed; hence the fluid becomes more dilute.

Figure 24.18   Formation of dilute urine. Heavy brown lines in the ascending limb of the loop of Henle and in the distal convoluted tubule indicate impermeability to water; heavy blue lines indicate the last part of the distal convoluted tubule and the collecting duct, which are impermeable to water in the absence of ADH; light blue areas around the nephron represent interstitial fluid. Numbers indicate solute concentration in milliosmoles per liter (mOsm/liter).

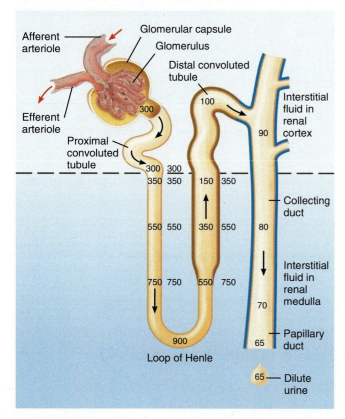

🔑 **When ADH level is low, urine is dilute and is less concentrated than blood plasma.**

5. Finally, when the ADH level is very low, the distal convoluted tubules and collecting ducts are impermeable to water. Thus, tubular fluid becomes progressively more dilute as it flows onward. By the time the tubular fluid drains into the renal pelvis, its solute concentration can be four times more dilute than blood plasma or glomerular filtrate.

Formation of Concentrated Urine

When water intake is low or water loss is high (such as during heavy sweating), the kidneys must conserve water while still eliminating wastes and excess ions. Under the influence of ADH,

the kidneys produce a small volume of highly concentrated urine. Urine can be four times more concentrated than blood plasma or glomerular filtrate.

The ability of ADH to cause excretion of concentrated urine depends on the presence of an **osmotic gradient** of solutes in the interstitial fluid of the renal medulla. Notice in Figure 24.19 (see blue arrow at left) that the solute concentration of the interstitial fluid in the kidney increases from about 300 mOsm/liter (milliosmoles per liter) in the renal cortex to about 1200 mOsm/liter deep in the renal medulla. The three major solutes that contribute to this high osmotic gradient are Na^+, Cl^-, and urea. Two main factors contribute to building and maintaining this osmotic gradient: (1) differences in solute and water permeability and reabsorption in different sections of the long loops of Henle and the collecting duct, and (2) the countercurrent flow (flow in opposite directions) of fluid in neighboring descending and ascending limbs of the loop of Henle.

Production of concentrated urine occurs as follows (Figure 24.19):

❶ ***In juxtamedullary nephrons, symporters in thick ascending limb cells of the loop of Henle establish the osmotic gradient in the renal medulla.*** In the thick ascending limb of the LOH, the Na^+–K^+–$2Cl^-$ symporters reabsorb Na^+ and Cl^- from the tubular fluid (Figure 24.19a). Water is not reabsorbed in this segment, however, because the cells are impermeable to water. As a result, the reabsorbed ions become increasingly concentrated in the interstitial fluid of the outer medulla. Those ions that diffuse into the vasa recta are carried deep into the inner medulla by the blood flow (Figure 24.19b).

❷ ***Cells in the collecting ducts reabsorb more water and urea.*** When ADH increases the water permeability of the principal cells, water quickly moves via osmosis out of the collecting duct tubular fluid, into the interstitial fluid of the inner medulla, and then into the vasa recta. With loss of water, the urea left behind in the tubular fluid of the collecting duct becomes increasingly concentrated. Because collecting duct cells deep in the medulla are permeable to urea, it diffuses from the fluid in the duct into the interstitial fluid of the medulla.

❸ ***Urea recycling causes a buildup of urea in the renal medulla.*** As urea accumulates in the interstitial fluid, some of it diffuses into the tubular fluid in the descending and thin ascending limbs of the long loops of Henle, which also are permeable to urea (Figure 24.19a). However, while the fluid flows through the thick ascending limb, DCT, and cortical portion of the collecting duct, urea remains in the lumen because cells in these segments are impermeable to it. As fluid flows along the collecting ducts, water reabsorption

Figure 24.19 **Mechanism of urine concentration in juxtamedullary nephrons.** The green line indicates the presence of Na⁺–K⁺–2Cl⁻ symporters that simultaneously reabsorb these ions into the interstitial fluid of the renal medulla; this portion of the nephron is also relatively impermeable to water and urea. All solute concentrations are in milliosmoles per liter (mOsm/liter).

continues via osmosis because ADH is present. This water reabsorption *further increases* the concentration of urea in the tubular fluid, more urea diffuses into the interstitial fluid of the renal medulla, and the cycle repeats. The constant transfer of urea between segments of the renal tubule and the interstitial fluid of the medulla is termed *urea recycling*. In this way, reabsorption of water from the tubular fluid of the collecting ducts promotes the buildup of urea in the interstitial fluid of the renal medulla, which in turn promotes water reabsorption. The solutes left behind in the lumen thus become very concentrated, and a small volume of concentrated urine is excreted.

(a) Reabsorption of Na⁺, Cl⁻, and water in a juxtamedullary nephron

(b) Recycling of salts and urea in the vasa recta

🔑 **The formation of concentrated urine depends on high concentrations of solutes in interstitial fluid in the renal medulla.**

The second contributor to the osmotic gradient in the renal medulla is the **countercurrent mechanism**, which has its basis in the hairpin shape of the long loops of Henle of juxtamedullary nephrons. Note in Figure 24.19a that the descending limb of the loop of Henle carries tubular fluid from the renal cortex deep into the medulla, and the ascending limb carries it in the opposite direction. Thus, fluid flowing in one tube runs counter (opposite) to fluid flowing in a nearby parallel tube, an arrangement called *countercurrent flow*.

The descending limb of the loop of Henle is very permeable to water but impermeable to solutes except urea. Because the interstitial fluid outside the descending limb has a solute concentration higher than the tubular fluid within it, water moves out of the descending limb via osmosis. As the fluid continues down the descending limb, it becomes increasingly more concentrated.

As previously noted, the ascending limb of the loop is impermeable to water, but its symporters reabsorb Na$^+$ and Cl$^-$ from the tubular fluid into the interstitial fluid of the renal medulla, so the tubular fluid becomes progressively more dilute as it flows up the ascending limb. To summarize, tubular fluid becomes progressively more concentrated as it flows down the descending limb and progressively more dilute as it moves up the ascending limb.

Note in Figure 24.19b that the vasa recta also consists of descending and ascending limbs that are parallel to each other and to the loop of Henle. Just as tubular fluid flows in opposite directions in the loop of Henle, blood flows in opposite directions in the ascending and descending parts of the vasa recta. Blood entering the vasa recta is fairly dilute (about 300 mOsm/liter). As blood flows down the descending part into the renal medulla, where the interstitial fluid becomes increasingly concentrated, Na$^+$, Cl$^-$, and urea diffuse down their concentration gradients from interstitial fluid into the blood, resulting in increasingly more concentrated blood. As the concentrated blood flows up the ascending part of the vasa recta, it flows through a region where the interstitial fluid becomes increasingly less concentrated. As a result, ions and urea diffuse from the blood back into interstitial fluid. Blood leaving the vasa recta is at only a slightly higher concentration than when it entered the vasa recta. Thus, the vasa recta provides oxygen and nutrients to the renal medulla without washing out or diminishing the osmotic gradient needed to reabsorb water.

Figure 24.20 summarizes the processes of filtration, reabsorption, and secretion in each segment of the nephron and collecting duct.

Checkpoint

31. Which force draws water out of the loop of Henle and into the interstitial fluid of the renal medulla?
32. Which portions of the renal tubule and collecting duct reabsorb more solutes than water to produce dilute urine?
33. Which solutes are the main contributors to the high osmotic gradient of interstitial fluid in the renal medulla?
34. Which portions of the renal tubule and/or collecting duct reabsorb water to produce concentrated urine? Which hormone controls the production of concentrated urine?
35. How does urea recycling contribute to the production of concentrated urine?

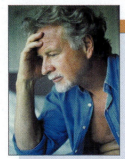

Return to Andre's Story

Andre had been admitted to the hospital immediately. He was started on antibiotics for the infection and was put on dialysis. His blood levels of potassium had worried the doctor most, since elevated potassium could have a negative effect on the heart and respiration.

The next day, as Andre lay sleeping in his room, a soft voice woke him. "How are you doing today, Andre? Feeling better?" The nurse examined the bag connected to Andre's catheter. "You're still not producing much urine, looks like only 300 mL or so since yesterday at this time. It's likely your kidneys need some time to recover from the infection and the ibuprofen toxicity. I've seen a lot of patients over 50 years old with enlarged prostates, but they usually don't have kidney failure."

Andre began to feel better following dialysis; the excess electrolytes and metabolic wastes that had been challenging his body and causing much of his discomfort were eliminated. His fever subsided within 24 hours due to the antibiotic therapy. However, it was still several days before he was discharged. "Andre, you had a close call, but I think you are going to be okay. I'd like to schedule a follow-up visit next week and arrange a kidney function test. We will need to monitor your progress closely."

Andre sighed. "Thank you, Doctor. Would having the prostate surgery prevent this from happening again? I've been putting it off, but it may be time."

"Well, Andre, the infection most likely wouldn't have caused renal failure on its own. The combination of obstruction, infection, and acute ibuprofen toxicity caused this problem. After we see how well your kidneys recover, we can discuss your options."

F. Why might Andre's restricted flow of urine have caused damage to his kidneys? (Hint: Think of the anatomy of the kidney.)
G. Acute tubular necrosis causes a decrease in glomerular filtration rate. Why wouldn't Andre's increased blood pressure have compensated for his reduced filtration?
H. Which of his hormones were unable to adjust for Andre's electrolyte imbalance? Which ones are normally involved in regulating the water content of his body? Were the levels of these hormones too high or too low during his illness?

Figure 24.20 Summary of filtration, reabsorption, and secretion in the nephron and collecting duct.

PROXIMAL CONVOLUTED TUBULE

Reabsorption (into blood) of filtered:

Water	65% (osmosis)
Na⁺	65% (sodium–potassium pumps, symporters, antiporters)
K⁺	65% (diffusion)
Glucose	100% (symporters and facilitated diffusion)
Amino acids	100% (symporters and facilitated diffusion)
Cl⁻	50% (diffusion)
HCO₃⁻	80–90% (facilitated diffusion)
Urea	50% (diffusion)
Ca²⁺, Mg²⁺	variable (diffusion)

Secretion (into urine) of:

H⁺	variable (antiporters)
NH₄⁺	variable, increases in acidosis (antiporters)
Urea	variable (diffusion)
Creatinine	small amount

At end of PCT, tubular fluid is still isotonic to blood (300 mOsm/liter).

LOOP OF HENLE

Reabsorption (into blood) of:

Water	15% (osmosis in descending limb)
Na⁺	20–30% (symporters in ascending limb)
K⁺	20–30% (symporters in ascending limb)
Cl⁻	35% (symporters in ascending limb)
HCO₃⁻	10–20% (facilitated diffusion)
Ca²⁺, Mg²⁺	variable (diffusion)

Secretion (into urine) of:

Urea	variable (recycling from collecting duct)

At end of loop of Henle, tubular fluid is hypotonic (100–150 mOsm/liter).

RENAL CORPUSCLE

Glomerular filtration rate: 105–125 mL/min of fluid that is isotonic to blood

Filtered substances: water and all solutes present in blood (except proteins) including ions, glucose, amino acids, creatinine, uric acid

EARLY DISTAL CONVOLUTED TUBULE

Reabsorption (into blood) of:

Water	10–15% (osmosis)
Na⁺	5% (symporters)
Cl⁻	5% (symporters)
Ca²⁺	variable (stimulated by parathyroid hormone)

LATE DISTAL CONVOLUTED TUBULE AND COLLECTING DUCT

Reabsorption (into blood) of:

Water	5–9% (insertion of water channels stimulated by ADH)
Na⁺	1–4% (sodium–potassium pumps and sodium channels stimulated by aldosterone)
HCO₃⁻	variable amount, depends on H⁺ secretion (antiporters)
Urea	variable (recycling to loop of Henle)

Secretion (into urine) of:

K⁺	variable amount to adjust for dietary intake (leakage channels)
H⁺	variable amounts to maintain acid–base homeostasis (H⁺ pumps)

Tubular fluid leaving the collecting duct is dilute when ADH level is low and concentrated when ADH level is high.

Urine

Filtration occurs in the renal corpuscle; reabsorption occurs all along the renal tubule and collecting ducts.

24.9 The ureters transport urine from the renal pelvis to the urinary bladder, where it is stored until micturition.

From the collecting ducts, urine drains through papillary ducts into the minor calyces, which join to become major calyces that unite to form the renal pelvis (see Figure 24.3). From the renal pelvis, urine first drains into the ureters and then into the urinary bladder. Urine is then discharged from the body through the single urethra (see Figure 24.1).

Ureters

From the renal pelvis of each kidney, the **ureters** (Ū-rē-ters) transport urine to the urinary bladder. Peristaltic contractions of the muscular walls of the ureters push urine toward the urinary bladder, but hydrostatic pressure and gravity also contribute. Peristaltic waves that pass from the renal pelvis to the urinary bladder vary in frequency from one to five per minute, depending on how fast urine is being formed.

The ureters are 25–30 cm (10–12 in.) long and are thick-walled, narrow tubes that vary in diameter from 1 mm to 10 mm along their course between the renal pelvis and the urinary bladder. Like the kidneys, the ureters are retroperitoneal. At the base of the urinary bladder, the ureters curve medially and obliquely penetrate the posterior wall of the urinary bladder (Figure 24.21).

Even though there is no anatomical valve at the opening of each ureter into the urinary bladder, a physiological one is quite effective. As the urinary bladder fills with urine, pressure within it compresses the oblique openings into the ureters and prevents the backflow of urine. When this physiological valve is not operating properly, it is possible for microbes to travel up the ureters from the urinary bladder to infect one or both kidneys.

Three layers of tissue form the wall of the ureters. The deepest coat, the **mucosa**, is a mucous membrane composed of **transitional epithelium** (see Table 4.1G). Transitional epithelium is able to stretch—a marked advantage for any organ that must accommodate a variable volume of fluid. Mucus secreted by the goblet cells of the mucosa prevents the cells from coming in contact with urine, the solute concentration and pH of which may differ drastically from the cytosol of cells that form the wall of the ureters. Throughout most of the length of the ureters, the intermediate coat, the **muscularis**, is composed of longitudinal and circular layers of smooth muscle fibers. Peristalsis is the major function of the muscularis. The superficial coat of the ureters is the **adventitia**, a layer of areolar connective tissue containing blood vessels, lymphatic vessels, and nerves that serve the muscularis and mucosa. The adventitia blends in with surrounding connective tissue and anchors the ureters in place.

Figure 24.21   Ureters, urinary bladder, and urethra in a female.

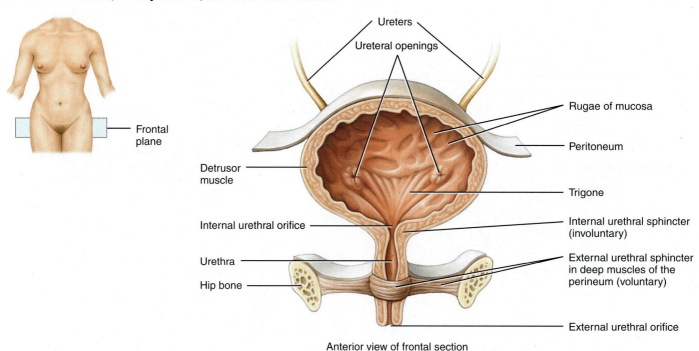

Anterior view of frontal section

Urine is stored in the urinary bladder before being expelled by micturition.

Urinary Bladder

The **urinary bladder** is a hollow, distensible muscular organ situated in the pelvic cavity posterior to the pubic symphysis. In males, it is directly anterior to the rectum; in females, it is anterior to the vagina and inferior to the uterus (see Figure 24.22). Folds of the peritoneum hold the urinary bladder in position. The shape of the urinary bladder depends on how much urine it contains. When it is empty, it collapses like a deflated balloon. It becomes spherical when slightly distended and, as urine volume increases, becomes pear-shaped and rises into the abdominal cavity. Urinary bladder capacity averages 700–800 mL. It is smaller in females because the uterus occupies the space just superior to the urinary bladder.

Anatomy and Histology of the Urinary Bladder

In the floor of the urinary bladder is a small triangular area called the **trigone** (TRĪ-gōn = triangle). The two posterior corners of the trigone contain the two ureteral openings; the opening into the urethra, the **internal urethral orifice**, lies in the anterior corner (Figure 24.21). Because its mucosa is firmly bound to the muscularis, the trigone has a smooth appearance. The trigone funnels urine into the urethra when the muscularis contracts.

Three layers of tissue make up the wall of the urinary bladder. The deepest is the **mucosa**, a mucous membrane composed of **transitional epithelium**. Rugae (the folds in the mucosa) are also present to permit expansion of the urinary bladder. The muscularis of the bladder wall consists of three layers of smooth muscle and is also called the **detrusor muscle** (de-TROO-ser = to push down). Around the opening to the urethra the muscularis forms an **internal urethral sphincter**; being smooth muscle, its opening and closing is involuntary. Inferior to the urethral internal sphincter is the **external urethral sphincter**, which is composed of skeletal muscle and is under voluntary control. The most superficial coat of the urinary bladder on the posterior and inferior surfaces is the **adventitia**, a layer of areolar connective tissue that is continuous with that of the ureters. Over the superior surface of the urinary bladder is the **serosa**, a layer of visceral peritoneum.

The Micturition Reflex

Discharge of urine from the urinary bladder into the urethra is called **micturition** (mik´-too-RISH-un; *mictur-* = urinate), but is also known as *urination or voiding*. Micturition occurs via a combination of involuntary and voluntary muscle contractions. When the volume of urine in the urinary bladder exceeds 200–400 mL, pressure within the bladder increases considerably, and stretch receptors in its wall transmit nerve impulses into the spinal cord. These impulses propagate to the sacral region of the spinal cord and trigger a spinal reflex called the **micturition reflex**. In this reflex arc, parasympathetic impulses from the spinal cord cause *contraction* of the detrusor muscle and *relaxation* of the internal urethral sphincter muscle. Simultaneously, the spinal cord inhibits somatic motor neurons, causing relaxation of skeletal muscle in the external urethral sphincter. Upon contraction of the urinary bladder wall and relaxation of the sphincters, urination takes place. Urinary bladder filling causes a sensation of fullness that initiates a conscious desire to urinate before the micturition reflex actually occurs. Although emptying of the urinary bladder is a reflex, in early childhood we learn to initiate it and stop it voluntarily. Through learned control of the external urethral sphincter muscle and certain muscles of the pelvic floor, the cerebral cortex can initiate micturition or delay its occurrence for a limited period.

Urethra

The **urethra** (ū-RĒ-thra) is a small tube leading from the internal urethral orifice in the floor of the urinary bladder to the exterior of the body. In both males and females, the urethra is the terminal portion of the urinary system and the passageway for discharging urine from the body. In males, it also discharges semen (fluid that contains sperm).

In females, the urethra lies directly posterior to the pubic symphysis and has a length of 4 cm (1.5 in.) (Figure 24.22a). The opening of the urethra to the exterior, the **external urethral orifice**, is located between the clitoris and the vaginal opening (see Figure 25.11a).

In males, the urethra also extends from the internal urethral orifice to the exterior, but its length and passage through the body are considerably different than in females (Figure 24.22b). The male urethra first passes through the prostate, then through the urogenital diaphragm, and finally through the penis, a distance of about 20 cm (8 in.). The male urethra is subdivided into three anatomical regions: (1) The **prostatic urethra** passes through the prostate. (2) The **membranous urethra**, the shortest portion, passes through the deep muscles of the perineum. (3) The **spongy urethra**, the longest portion, passes through the penis and opens to the exterior as the external urethral orifice. Several glands and other structures associated with reproduction (described in detail in Chapter 25) deliver their contents into the male urethra.

The wall of the urethra consists of a **mucosa** and a **muscularis** composed of circularly arranged smooth muscle fibers and is continuous with that of the urinary bladder. Near the urinary bladder, the mucosa contains transitional epithelium that is continuous with that of the urinary bladder; near the external urethral orifice, the epithelium is nonkeratinized stratified squamous epithelium.

Figure 24.22 **Comparison between female and male urethras.**

Sagittal plane

Uterus

Urinary bladder

Pubic symphysis

Rectum

Vagina

Urethra

External urethral orifice

(a) Sagittal section

Sagittal plane

Urinary bladder

Rectum

Pubic symphysis

Prostatic urethra

Prostate

Membranous urethra

Deep muscles of perineum

Penis

Spongy urethra

Testis

Scrotum

External urethral orifice

(b) Sagittal section

The female urethra is about 4 cm (1.5 in.) in length, whereas the male urethra is about 20 cm (8 in.) in length.

Checkpoint

36. Which forces help propel urine from the renal pelvis to the urinary bladder?
37. What is the functional significance of transitional epithelium?
38. Which urethral sphincter is under conscious control?
39. How do the location and length of the urethra compare in males and females?
40. What are the three subdivisions of the male urethra?

24.10 The kidneys help maintain the overall fluid and acid–base balance of the body.

Fluid Balance

One important function of the kidneys is to help maintain fluid balance in the body. The water and dissolved solutes in the body constitute the **body fluids**. Regulatory mechanisms involving the kidneys and other organs normally maintain homeostasis of the body fluids. Malfunction in any or all of them may seriously endanger the functioning of organs throughout the body. About two-thirds of body fluid is **intracellular fluid** or **cytosol**, the fluid within cells. The other third, called **extracellular fluid (ECF)**, is outside cells and includes all other body fluids. About 80% of the ECF is **interstitial fluid**, which occupies the spaces between tissue cells, and about 20% of the ECF is **plasma**, the liquid portion of blood. Extracellular fluids also include lymph in lymphatic vessels; cerebrospinal fluid in the nervous system; synovial fluid in joints; aqueous humor and vitreous body in the eyes; endolymph and perilymph in the ears; and pleural, pericardial, and peritoneal fluids between serous membranes.

The body is in **fluid balance** when the required amounts of water and solutes are present and are correctly proportioned through the body. **Water** is by far the largest single component of the body, making up 45–75% of total body mass, depending on age and gender. Most solutes in body fluids are **electrolytes**, inorganic compounds that break apart into ions when dissolved in water. They are the main contributors to the osmotic movement of water. Because intake of water and electrolytes rarely occurs in exactly the same proportions as their presence in body fluids, the ability of the kidneys to excrete excess water by producing dilute urine, or to excrete excess electrolytes by producing concentrated urine, is of utmost importance in the maintenance of fluid homeostasis.

Regulation of Water and Solute Loss

Even though the loss of water and solutes through sweating and exhalation increases during exercise, elimination of *excess* body water or solutes occurs mainly by control of their loss in urine. The extent of *urinary salt (NaCl) loss* is the main factor that deter-mines body fluid *volume*. The reason is that in osmosis "water follows solutes," and the two main solutes in extracellular fluid (and in urine) are sodium ions (Na^+) and chloride ions (Cl^-). Because our daily diet contains a highly variable amount of NaCl, urinary excretion of Na^+ and Cl^- must also vary to maintain homeostasis. Atrial natriuretic peptide, angiotensin II, and aldosterone regulate the extent of renal Na^+ and Cl^- reabsorption, and thus how much is lost in the urine.

Figure 24.23 depicts the sequence of changes that occur after a salty meal. The increased intake of NaCl produces an increase in plasma levels of Na^+ and Cl^-. The increased osmotic pressure of interstitial fluid causes movement of water from intracellular fluid into interstitial fluid and then into plasma. Such water movement increases blood volume, which stretches the atria of the heart and promotes release of atrial natriuretic peptide. ANP elevates urinary loss of Na^+ (and Cl^-) and water, which decreases blood volume. The initial increase in blood volume also slows release of renin from juxtaglomerular cells of the kidneys.

As renin level declines, less angiotensin II is formed. With less angiotensin II, the kidney tubules reabsorb less Na^+, Cl^-, and water. In addition, less angiotensin II leads to less aldosterone and further slowing of Na^+ and Cl^- reabsorption in the renal tubules. More filtered Na^+ and Cl^- ions thus remain in the tubular fluid to be excreted in the urine. The osmotic consequence of excreting more Na^+ and Cl^- is loss of more water in urine, which decreases blood volume and blood pressure. By contrast, when someone becomes dehydrated, higher levels of angiotensin II and aldosterone promote urinary reabsorption of Na^+ and Cl^- (and water by osmosis with the solutes) and thereby conserve the volume of body fluids by reducing urinary loss.

The major hormone that regulates water loss is antidiuretic hormone. An increase in the osmotic pressure of body fluids (a decrease in the water concentration of the fluids) stimulates release of ADH (see Figure 24.17). As previously discussed, ADH promotes the insertion of water channel proteins into the apical membranes of principal cells in the collecting ducts of the kidneys. As a result, the permeability of these cells to water increases. Water

Figure 24.23   **Hormonal regulation of renal Na⁺ and Cl⁻ reabsorption.**

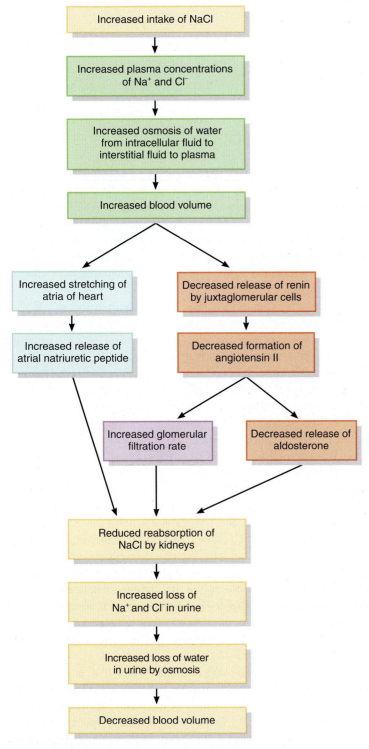

The three main hormones that regulate renal Na⁺ and Cl⁻ reabsorption (and thus the amount lost in the urine) are angiotensin II, aldosterone, and atrial natriuretic peptide.

molecules move by osmosis from the renal tubular fluid into the cells and then from the cells into the bloodstream. The result is production of a small volume of very concentrated urine. By contrast, intake of water decreases the osmotic pressure of blood and interstitial fluid. Within minutes, ADH secretion shuts down, and soon its blood level is close to zero. When the principal cells are no longer stimulated by ADH, water channels are removed from the apical membrane by endocytosis. As the number of water channels decreases, the water permeability of the principal cells' apical membrane falls, and more water is lost in the urine.

Movement of Water into and out of Cells

Normally, intracellular and interstitial fluids have the same osmotic pressure, so cells neither shrink nor swell. An increase in the osmotic pressure of interstitial fluid draws water out of cells, so they shrink slightly. A decrease in the osmotic pressure of interstitial fluid, by contrast, causes cells to swell. Changes in osmotic pressure most often result from changes in the concentration of Na⁺. A decrease in the osmotic pressure of interstitial fluid inhibits secretion of ADH. Normally functioning kidneys then excrete excess water in the urine, which raises the osmotic pressure of body fluids to normal. As a result, body cells swell only slightly, and only for a brief time.

Acid–Base Balance

A major homeostatic challenge for the kidneys is keeping the H⁺ level (pH) of body fluids in the appropriate range. This task—the maintenance of acid–base balance—is of critical importance to normal cellular function. For example, the three-dimensional shapes of all body proteins, which enable them to perform specific functions, are very sensitive to pH changes. In a healthy person, the pH of systemic arterial blood remains between 7.35 and 7.45. Because metabolic reactions often produce a huge excess of H⁺, the lack of any mechanism for the disposal of H⁺ would cause H⁺ in body fluids to rise quickly to a lethal level. Homeostasis of H⁺ concentration within a narrow range is thus essential to survival. The removal of H⁺ from body fluids and its subsequent elimination from the body depend on three major mechanisms: buffer systems, excretion of H⁺ in the urine, and exhalation of carbon dioxide (discussed in Chapter 22).

Phosphate Buffer System

Buffers act quickly to temporarily bind H⁺, removing the highly reactive, excess H⁺ from solution but not from the body. Buffers prevent rapid, drastic changes in the pH of a body fluid by converting strong acids and bases into weak acids and bases. Strong acids lower pH more than weak acids because strong acids release H^+ more readily than weak acids and thus contribute more free hydrogen ions. Similarly, strong bases raise pH more than weak ones.

An important buffer system related to the urinary system is the **phosphate buffer system**. The components of the phosphate buffer system are the ions *dihydrogen phosphate* ($H_2PO_4^-$) and *monohydrogen phosphate* (HPO_4^{2-}). The dihydrogen phosphate ion acts as a weak acid and is capable of buffering strong bases, such as OH^-, as follows:

$$OH^- + H_2PO_4^- \longrightarrow H_2O + HPO_4^{2-}$$

Hydroxide ion   Dihydrogen phosphate   Water   Monohydrogen
(strong base)   (weak acid)   phosphate (weak base)

The monohydrogen phosphate ion, in contrast, acts as a weak base and is capable of buffering the H^+ released by a strong acid, such as hydrochloric acid (HCl):

$$H^+ + HPO_4^{2-} \longrightarrow H_2PO_4^-$$

Hydrogen ion   Monohydrogen phosphate   Dihydrogen
(strong acid)   (weak base)   phosphate (weak acid)

Because the concentration of phosphates is highest in intracellular fluid, the phosphate buffer system is an important regulator of pH in the cytosol. It also acts to a smaller degree in extracellular fluids, and buffers acids in urine. $H_2PO_4^-$ is formed when excess H^+ in the kidney tubule fluid combines with HPO_4^{2-} (Figure 24.24b). The H^+ that becomes part of the $H_2PO_4^-$ passes into the urine, helping maintain blood pH.

Kidney Excretion of H^+

The only way to eliminate most acids that form in the body is to excrete H^+ in the urine. Given the magnitude of the contributions of the kidneys to acid–base balance, it's not surprising that renal failure can quickly cause death.

Cells in both the proximal convoluted tubules (PCTs) and the collecting ducts of the kidneys secrete hydrogen ions into the tubular fluid. In the PCT, Na^+/H^+ antiporters secrete H^+ as they reabsorb Na^+ (see Figure 24.13). Even more important for regulation of pH of body fluids, however, are the intercalated cells of the collecting ducts. The *apical* membranes of some intercalated cells include **proton pumps (H^+ ATPases)** that secrete H^+ into the tubular fluid (Figure 24.24). Intercalated cells can secrete H^+ against a concentration gradient so effectively that urine can be up to 1000 times (3 pH units) more acidic than blood. The important buffer, HCO_3^-, produced by dissociation of H_2CO_3 inside intercalated cells, crosses the basolateral membrane by means of **Cl^-/HCO_3^- antiporters** and then diffuses into peritubular

Figure 24.24   **Secretion of H^+ by intercalated cells in the collecting duct.**

(a) Secretion of H^+

(b) Buffering of H^+ in urine

Key:

Proton pump (H^+ ATPase) in apical membrane

Cl^-/HCO_3^- antiporter in basolateral membrane

•••►   Diffusion

Urine can be up to 1000 times more acidic than blood due to the operation of the proton pumps in the collecting ducts of the kidneys.

capillaries (Figure 24.24a). The HCO_3^- that enters the blood in this way is *new* (not filtered). For this reason, blood leaving the kidney in the renal vein may have a higher HCO_3^- concentration than blood entering the kidney in the renal artery.

Interestingly, a second type of intercalated cell has proton pumps in its *basolateral* membrane and Cl^-/HCO_3^- antiporters in its *apical* membrane. These intercalated cells secrete HCO_3^- and reabsorb H^+. Thus, the two types of intercalated cells help maintain the pH of body fluids in two ways—by excreting excess H^+ when pH of body fluids is too low and by excreting excess HCO_3^- when pH is too high.

Some hydrogen ions secreted into the tubular fluid of the collecting duct are buffered, but not by HCO_3^-, most of which has been filtered and reabsorbed. Two other buffers combine with H^+ in the collecting duct (Figure 24.24b). The most plentiful buffer in the tubular fluid of the collecting duct is HPO_4^{2-} (monohydrogen

phosphate ion). In addition, a small amount of NH_3 (ammonia) also is present. H^+ combines with HPO_4^{2-} to form $H_2PO_4^-$ (dihydrogen phosphate ion) and with NH_3 to form NH_4^+ (ammonium ion). Because these ions cannot diffuse back into tubule cells, they are excreted in the urine.

Checkpoint

41. What is body fluid? What are the percentage distributions of body fluids? What is the primary influence on body fluid volume?
42. How do atrial natriuretic peptide, angiotensin II, aldosterone, and antidiuretic hormone regulate the volume and osmotic pressure of body fluids?
43. How do phosphate ions help maintain the pH of body fluids?
44. How do intercalated cells maintain the pH of body fluids?

Andre's Story: Epilogue and Discussion

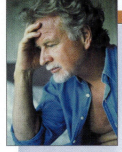

▶▶▶ **A**ndre, a 57-year-old man with a history of benign prostatic hypertrophy, develops a possible urinary tract or bladder infection. He has urinary frequency and urgency, yet has difficulty urinating, and has cloudy, malodorous urine. He develops a fever and a headache, and has back pain located over his kidney. Andre presents to the health clinic complaining of fatigue, back pain, headache, and difficulty urinating. His blood pressure is elevated, he complains of nausea and vomiting, and he has edema of the hands and feet. Blood chemistry analysis results indicate elevated blood urea, nitrogen, and creatinine levels, along with elevated sodium and potassium. Urinalysis results indicate dilute urine with excessive amounts of bacteria.

Andre is treated with IV antibiotics and is placed on dialysis to address immediate symptoms. Symptoms resolve within a few days, and he is released with a course of oral antibiotics and scheduled for a follow-up kidney function test. ◀◀◀

I. Which symptoms alerted the doctor that Andre had a kidney infection and not a urinary tract infection or bladder infection?

J. How could bacteria get into the kidney? Would it be possible to have a kidney infection in only one kidney?

K. Would men or women be more prone to urinary tract infections? (Hint: Consider the lengths of their urethras.)

L. Which physiological mechanism was causing Andre's urinary urgency?

M. Why was the doctor concerned about Andre's acid–base balance? Would you have expected Andre's blood pH to be low or high? What about his urinary pH?

Concept and Resource Summary

Concept	Resources

Introduction

1. The organs of the **urinary system** are the kidneys, ureters, urinary bladder, and urethra.
2. After the kidneys filter blood and return most water and many solutes to the bloodstream, the remaining water and solutes constitute **urine**.

Concept 24.1 The kidneys regulate the composition of the blood, produce hormones, and excrete wastes.

1. The kidneys regulate blood ionic composition, pH, volume, pressure, and glucose level. The kidneys also produce hormones (*calcitriol* and *erythropoietin*) and excrete **wastes** and foreign substances.

See

- **Anatomy Overview – Kidney Overview**

| Concept | Resources |

Concept 24.2 As urine forms, it travels through the renal medulla, calyces, and renal pelvis.

1. The paired **kidneys** are **retroperitoneal** organs attached to the posterior abdominal wall.
2. The **renal hilum** is an indentation where the ureter exits, and where blood vessels, lymphatic vessels, and nerves enter and exit. From superficial to deep, three layers of tissue surround the kidney: **renal fascia**, **adipose capsule**, and **renal capsule**.
3. Internally, the kidneys includes a superficial **renal cortex**, a deep **renal medulla**, **renal pyramids** with a wider base and a narrower **renal papilla**, **renal columns** extending between renal pyramids, major and minor calyces, and a renal pelvis. Microscopic **nephrons** produce urine, which drains into **papillary ducts**, **minor** and then **major calyces**, and then into the **renal pelvis**.
4. Blood flows into the kidney through the **renal artery**, from which blood flows sequentially through **segmental arteries**, **interlobar arteries**, **arcuate arteries**, **interlobular arteries**, and then **afferent arterioles**. Each afferent arteriole enters a nephron and supplies a **glomerulus**, where blood is filtered. From the glomerulus, blood flows through **efferent arterioles**, **peritubular capillaries** (some of which form long loop-shaped capillaries called the **vasa recta**), **interlobular veins**, **arcuate veins**, **interlobar veins**, and finally empties into a **renal vein**, where blood flows out of the kidney.

See

- **Anatomy Overview – Blood Supply of the Kidney**

Hear

- **Figure 24.3 – Internal Anatomy of the Kidneys**

Do

- **Exercise – Magic Renal Ride**

Concept 24.3 Each nephron consists of a renal corpuscle and a renal tubule.

1. The nephron is the functional unit of the kidneys. A nephron consists of a **renal corpuscle** (consisting of the **glomerulus** surrounded by the **glomerular capsule**); and a **renal tubule** (consisting of the **proximal convoluted tubule**, **loop of Henle**, and **distal convoluted tubule**).
2. The renal corpuscle and convoluted tubules lie in the renal cortex. The loop of Henle begins in the renal cortex, extends into the medulla as the **descending limb**, then turns and returns to the renal cortex as the **ascending limb**.
3. **Cortical nephrons** have short loops of Henle that dip only into the superficial region of the renal medulla. **Juxtamedullary nephrons** have a long loop of Henle that extends deep into the renal medulla, almost to the renal papilla.
4. Distal convoluted tubules empty into **collecting ducts**, which drain urine into papillary ducts.
5. A layer of epithelial cells forms the wall of the glomerular capsule, renal tubule, and ducts. The glomerular capsule consists of **podocytes** that wrap around endothelial cells of the glomerulus, forming the inner wall of the glomerular capsule, and simple squamous epithelium that forms the outer wall of the glomerular capsule. Fluid filtered from the glomerulus enters the **capsular space** between the two layers of the glomerular capsule.
6. The **juxtaglomerular apparatus** consists of the **juxtaglomerular cells** of an afferent arteriole and the **macula densa** of the ascending limb of the loop of Henle.

See

- **Anatomy Overview – Types of Nephrons**
- **Anatomy Overview – Fluid Flow through a Nephron**

Hear

- **Figure 24.6 – Histology of a Renal Corpuscle**

Do

- **Exercise – Paint the Nephron**
- **Exercise – Assemble the Urinary Tract**

Concept 24.4 Urine is formed by glomerular filtration, tubular secretion, and tubular reabsorption.

1. During urine formation, nephrons and collecting ducts perform **glomerular filtration**, **tubular reabsorption**, and **tubular secretion**. These processes maintain homeostasis of the volume and composition of the blood.

See

- **Anatomy Overview – Functional Anatomy of a Nephron**

Concept

Concept 24.5 Water and solutes are forced through the filtration membrane during glomerular filtration.

1. Fluid that enters the capsular space is **glomerular filtrate**.
2. Substances that cross the **filtration membrane** must pass through **fenestrations** of the glomerular endothelium, the **basal lamina**, and **filtration slits** between **pedicels** of podocytes. Water and small solutes are capable of passing from the blood into the capsular space; blood cells, platelets, and most plasma proteins are too large to pass through the filtration membrane.
3. **Glomerular blood hydrostatic pressure** in glomerular blood promotes filtration. Filtration is opposed by **capsular hydrostatic pressure**, caused by fluid in the capsular space, and **blood colloid osmotic pressure** from plasma proteins in glomerular blood. The overall relationship of these three pressures, the **net filtration pressure**, favors glomerular filtration.
4. The **glomerular filtration rate** is the amount of filtrate formed in both kidneys per minute. The glomerular filtration rate can be regulated by adjusting blood flow into and out of the glomerulus, and by altering the glomerular capillary surface. Three mechanisms maintain a constant glomerular filtration rate: renal autoregulation, neural regulation, and hormonal regulation.
5. The kidneys maintain a constant renal blood flow, despite daily blood pressure fluctuations, through **renal autoregulation**, in which blood flow to glomerular capillaries is adjusted by smooth muscle cells in the afferent arteriole wall and adjustments in nitric oxide release by the juxtaglomerular apparatus.
6. Neural regulation from sympathetic stimulation by the autonomic nervous system produces vasoconstriction of afferent arterioles, which leads to decreased urine output and greater blood flow to other body tissues.
7. Hormonal regulation of glomerular filtration involves the release of **angiotensin II**, which reduces renal blood flow, decreasing the glomerular filtration rate. **Atrial natriuretic peptide** increases capillary surface area, which increases the glomerular filtration rate.

See
- Anatomy Overview – Renal Corpuscle
- Animation – Renal Filtration

Hear
- **Figure 24.9 – The Pressures that Drive Glomerular Filtration**

Do
- Exercise – Filtration Finale

Concept 24.6 Tubular reabsorption reclaims needed substances from the filtrate, while tubular secretion discharges unneeded substances.

1. Tubular reabsorption is a selective process that reclaims materials from tubular fluid and returns them to the bloodstream. Reabsorbed substances include water, glucose, amino acids, urea, and ions. Some substances not needed by the body are removed from the blood and discharged into the urine via tubular secretion. Included are some ions, urea, creatinine, and certain drugs.
2. Tubular fluid can enter a peritubular capillary either by **paracellular reabsorption** between tubule cells or by **transcellular reabsorption** through tubule cells.
3. About 90% of water reabsorption is **obligatory water reabsorption**; it occurs via osmosis, together with reabsorption of solutes, and is not hormonally regulated. The remaining 10% is **facultative water reabsorption**, which varies according to body needs and is regulated by antidiuretic hormone.
4. In the proximal convoluted tubule (PCT), most reabsorption involves Na^+. **Na^+ symporters** in the PCT reabsorb glucose, amino acids, lactic acid, and water-soluble vitamins. **Na^+/H^+ antiporters** reabsorb Na^+ into the blood while secreting H^+ into tubular fluid. Also in the PCT, Cl^-, K^+, Ca^{2+}, Mg^{2+}, and urea are reabsorbed by diffusion. Urea and ammonia in blood are filtered at the glomerulus and secreted by cells of the PCT into tubular fluid.
5. The loop of Henle reabsorbs about 15% of the water; 25% of the Na^+ and K^+; 35% of the Cl^-; 15% of the HCO_3^-; and a variable amount of the Ca^{2+} and Mg^{2+}. Water can leave the descending limb of the loop of Henle to be reabsorbed but cannot leave the ascending limb of the LOH. This results in more dilute tubular fluid as it reaches the end of the ascending limb.
6. The distal convoluted tubule reabsorbs Na^+ and Cl^- via Na^+–Cl^- symporters. The distal convoluted tubule is the major site of reabsorption of Ca^{2+}, which is stimulated by parathyroid hormone.
7. In the collecting duct, principal cells reabsorb Na^+ and secrete K^+; intercalated cells reabsorb K^+ and HCO_3^- and secrete H^+.

See
- Anatomy Overview – Overview of Fluids
- Animation – Renal Reabsorption and Secretion

Do
- Exercise – Concentrate on Solvent and Solutes
- Exercise – Pick the Urinary Process
- Exercise – Concentrate on the Urinary System
- Exercise – The Renal Trip
- Concepts and Connections – The Functions of Ions

WILEY
PLUS

Concept	Resources

Concept 24.7   Five hormones regulate tubular reabsorption and tubular secretion.

1. Decreased blood volume and blood pressure causes juxtaglomerular cells to release **renin**, which induces production of angiotensin II.
2. **Angiotensin II** decreases glomerular filtration rate by vasoconstriction of afferent arterioles, enhances reabsorption of Na^+, Cl^-, and water, and stimulates the adrenal cortex to release aldosterone.
3. **Aldosterone** stimulates reabsorption of Na^+ and Cl^- and secretion of K^+. The reabsorption of Na^+ and Cl^- results in more reabsorption of water.
4. **Antidiuretic hormone** increases reabsorption of water. A decrease in the concentration of water in the blood and a decrease in blood volume stimulate release of antidiuretic hormone from the posterior pituitary.
5. Increased blood volume promotes release of **atrial natriuretic peptide** from the heart, which inhibits reabsorption of Na^+ and water. Atrial natriuretic peptide also suppresses aldosterone and antidiuretic hormone secretion, leading to increased excretion of Na^+ in urine and increased urine output.
6. Decreased blood levels of Ca^{2+} promotes release of **parathyroid hormone** by the parathyroid glands. Parathyroid hormone stimulates Ca^{2+} reabsorption and HPO_4^{2-} excretion.

See

- Animation – Hormonal Control of Blood Volume and Pressure

Do

- Exercise – Renal Regulation

Concept 24.8   Antidiuretic hormone affects the concentration of urine produced by the kidneys.

1. Antidiuretic hormone (ADH) controls whether the kidneys form dilute urine or concentrated urine.
2. In the absence of ADH, the kidneys produce dilute urine; renal tubules absorb more solutes than water. Formation of dilute urine involves the following: The descending limb of the loop of Henle reabsorbs more water, resulting in more concentrated tubular fluid. The ascending limb of the loop of Henle reabsorbs more solutes but not water, resulting in progressively dilute tubular fluid. The distal convoluted tubule absorbs solutes but very little water, again further diluting tubular fluid. Without ADH, collecting ducts are not permeable to water yet solutes continue to be absorbed, making the tubular fluid increasingly dilute as it drains into the renal pelvis.
3. In the presence of ADH, the kidneys produce concentrated urine; large amounts of water are reabsorbed from the tubular fluid into interstitial fluid, increasing solute concentration of the urine. Formation of concentrated urine involves the following: the ascending limb of the loop of Henle establishes an **osmotic gradient** of solutes in the interstitial fluid of the renal medulla; with ADH present, collecting ducts reabsorb more water and urea; urea recycling causes a buildup of urea in the renal medulla, which promotes further reabsorption of water.
4. The **countercurrent mechanism** establishes an osmotic gradient in the interstitial fluid of the renal medulla that enables production of concentrated urine when ADH is present.

See

- Animation – Water Homeostasis

Hear

- **Figure 24.19 –** Mechanism of Urine Concentration in Juxtamedullary Nephrons

Concept 24.9   The ureters transport urine from the renal pelvis to the urinary bladder, where it is stored until micturition.

1. Two retroperitoneal tubes called **ureters** transport urine from the renal pelvis of the kidneys to the urinary bladder, primarily by peristalsis. The **mucosa** is composed of **transitional epithelium** that stretches to accommodate varying volumes of urine and goblet cells that secrete protective mucus.
2. The **urinary bladder** is a distensible muscular organ located in the pelvic cavity posterior to the pubic symphysis. Its function is to store urine before micturition. The **trigone** in the bladder floor contains the two ureteral openings and the **internal urethral orifice**. The **mucosa** of the bladder wall contains **transitional epithelium** and rugae to allow expansion. The **muscularis** has smooth muscle called the **detrusor muscle**.
3. **Micturition** is the discharge of urine from the urinary bladder. As urine begins to fill the urinary bladder, stretch receptors trigger the **micturition reflex**. The micturition reflex discharges urine from the urinary bladder through parasympathetic impulses that cause contraction of the detrusor muscle and relaxation of the involuntary, smooth muscle **internal urethral sphincter**. The reflex also inhibits somatic motor neurons, leading to relaxation of the voluntary, skeletal muscle **external urethral sphincter** to allow urination.
4. The **urethra** is a tube leading from the bladder to the exterior. In both sexes, the urethra functions to discharge urine from the body; in males, it also discharges semen. The male urethra is subdivided into three regions: the prostatic urethra, the membranous urethra, and the spongy urethra.

See

- Anatomy Overview – Ureters, Urinary Bladder, and Urethra

Hear

- **Figure 24.22 –** Comparison between Female and Male Urethras

Concept

Concept 24.10 The kidneys help maintain the overall fluid and acid–base balance of the body.

1. **Body fluids** are composed of water and dissolved solutes, primarily **electrolytes**.

2. About two-thirds of body fluid is **intracellular fluid** within cells; one-third of body fluid is **extracellular fluid** outside of cells. **Fluid balance** means that the required amounts of water and solutes are present and are correctly proportioned through the body.

3. Elimination of excess body water or solutes occurs mainly on their excretion in urine. Na^+ and Cl^- are the two main solutes in extracellular fluid and urine. The extent of their loss in urine is the main determinant of body fluid volume because water follows solutes during osmosis.

4. Angiotensin II and aldosterone reduce urinary loss of Na^+ and Cl^-, retaining water, and thereby increasing the volume of body fluids. Atrial natriuretic peptide increases urinary loss of Na^+ and Cl^- and water, which decreases blood volume.

5. An increase in the osmotic pressure of body fluids stimulates ADH secretion, decreasing urinary loss of water. An increase in osmotic pressure of interstitial fluid draws water out of cells, and they shrink slightly. A decrease in osmotic pressure of interstitial fluid causes cells to swell. An increase in the osmotic pressure of interstitial fluid draws water out of cells.

6. The kidneys play an important role in maintaining appropriate H^+ levels (pH) of body fluids. Homeostasis of pH is maintained by buffer systems, excretion of H^+ in the urine, and exhalation of carbon dioxide.

7. A **buffer** binds to H^+ and removes excess H^+ from solution, preventing rapid, drastic changes in the pH of a body fluid. The **phosphate buffer system** is composed of the weak acid *dihydrogen phosphate*, ($H_2PO_4^-$), which removes excess OH^- from solution, and the weak base *monohydrogen phosphate* (HPO_4^{2-}), which removes excess H^+ from solution.

8. Cells in the proximal convoluted tubule, distal convoluted tubule, and collecting ducts of the kidneys are able to secrete H^+ into the tubular fluid, thereby increasing urinary loss of H^+. The kidneys also produce the bicarbonate ion buffer (HCO_3^-), which diffuses into peritubular capillaries to neutralize excess H^+ throughout the body.

See

- **Anatomy Overview – Overview of Fluids**
- **Animation – Water and Fluid Flow**
- **Animation – Regulation of pH**

Hear

- **Figure 24.24 –**
  Secretion of H^+ by Intercalated Cells in the Collecting Duct

Do

- **Exercise – Phix the pH**
- **Concepts and Connections – Mechanisms of pH Balance**

Other Chapter Resources

For additional information on clinical applications, medical disorders, and medical terminology, see **Chapter 13** in the *Clinical Connections* manual packaged with your text.

Within *WileyPLUS*

- **Chapter Quiz**
- **Visual Anatomy**
- **Mastering Vocabulary**
- **Audio Glossary**
- **Homeostatic Imbalance:**
  The Case of the Man with the Swollen Kidneys
- **Homeostatic Imbalance: The Case of the Thirsty Woman**
- **Homeostatic Imbalance: The Case of the Constantly Full Bladder**
- **Homeostatic Imbalance: The Case of the Girl with Relentless Pelvic Pain**
- **Web Links**

Understanding the Concepts

1. Describe the path of blood from its entrance into the kidney to its exit from the kidney, and of urine from the nephron to the urinary bladder.

2. A water molecule has just entered the proximal convoluted tubule of a cortical nephron. Which parts of the nephron will it travel through in order to reach the renal pelvis in a drop of urine?

3. How are the kidneys involved in the regulation of blood pressure?

4. A patient is retaining electrolytes, metabolic wastes, and water in her body; which of the three basic processes of urine formation are most likely involved in her condition?

5. Suppose a tumor is pressing on and obstructing the right ureter. What effect might this have on capsular hydrostatic pressure and thus on net filtration pressure in the right kidney? Would the left kidney also be affected?

6. How does filtered glucose enter and leave a proximal convoluted tubule cell?

7. How does reabsorption of Cl^- promote reabsorption of nutritionally important cations?

8. How are the waste products urea and ammonia removed from the blood? In which part of the nephron does this occur?

9. What is the mechanism in the distal convoluted tubule and collecting duct for reabsorption of Na^+?

10. Would the level of antidiuretic hormone in the blood be higher or lower than normal in a person who has just completed a 5-km run without drinking any water? Explain your answer.

11. How do symporters in the ascending limb of the loop of Henle and principal cells in the collecting duct contribute to the formation of concentrated urine?

12. What is the countercurrent mechanism? Why is it important?

13. What is micturition? How does the micturition reflex occur?

14. What would be the effects of a drug that blocks the activity of carbonic anhydrase?

25

The Reproductive Systems and Development

"Susan, we've been trying for over a year now. Maybe it's time we saw someone." Lou sipped his coffee as he glanced over the morning paper at his wife.

"I don't know, Lou. I was talking with Chandra about her experience with the fertility clinic, and it sounded like a science fiction movie. Tests, probes, drugs—I want to have a baby the way everyone else does. It just doesn't seem right to interfere in a natural process."

Lou set down his paper and reached for Susan's hand. "Look, we both want to start a family, right? All we're looking for is a little help. Let's just talk to a doctor and see what our options are."

"Well, that does sound reasonable. Go ahead and make an appointment, but remember that I'll be out on sales calls for most of next month, so check my day planner first."

Approximately one in six couples have difficulty conceiving a child. Infertility affects men and women equally. In this chapter we follow Susan and Lou as they discover how complex the reproductive process can be.

INTRODUCTION

Sexual reproduction is the process by which organisms produce offspring from germ cells called **gametes** (GAM-ēts = spouses). Males and females have anatomically distinct reproductive organs that are adapted for producing gametes, facilitating fertilization, and, in females, sustaining the growth of the embryo and fetus. Supporting structures, such as the penis in males and the uterus in females, assist the delivery of gametes; and the uterus is also the site for the growth of the embryo and fetus during pregnancy.

> *Sexual reproduction is the process by which organisms produce offspring from gametes.*

Following its deposit in the vagina, a male gamete (sperm cell) may unite with the female gamete (secondary oocyte) in an event called fertilization, and pregnancy can occur. **Pregnancy** is a sequence of events that begins with fertilization, proceeds to implantation, embryonic development, and fetal development, and normally ends with birth about 38 weeks later, or 40 weeks after the last menstrual period.

In addition to producing gametes, the **gonads**—testes in males and ovaries in females—secrete sex hormones that influence prenatal development and are responsible for the development and maintenance of reproductive structures, secondary sex characteristics, and sexual function.

25.1 The scrotum supports and regulates the temperature of the testes, which produce spermatozoa.

The organs of the male reproductive system include the testes, a system of ducts (including the epididymis, ductus deferens, ejaculatory ducts, and urethra), accessory sex glands (seminal vesicles, prostate, and bulbourethral glands), and several supporting structures, including the scrotum and the penis (Figure 25.1). The testes (male gonads) produce sperm and secrete hormones. The duct system transports and stores sperm, assists in their maturation, and conveys them to the exterior. Semen contains sperm plus the secretions provided by the accessory sex glands. The penis delivers sperm into the female reproductive tract, and the scrotum supports the testes.

Scrotum

The **scrotum** (SKRŌ-tum = bag) consists of loose skin and underlying hypodermis that hangs from the root (attached portion) of the penis (Figure 25.1a). Externally, the scrotum looks like a single pouch of skin separated into lateral portions by a median ridge called the **raphe** (RĀ-fē = seam). Internally, the **scrotal septum** divides the scrotum into two sacs, each containing a single testis (Figure 25.2). The septum is made up of hypodermis and smooth muscle tissue called the **dartos muscle** (DAR-tōs = skinned). Associated with each testis in the scrotum is the **cremaster mus-**

Figure 25.1   Male organs of reproduction and surrounding structures.

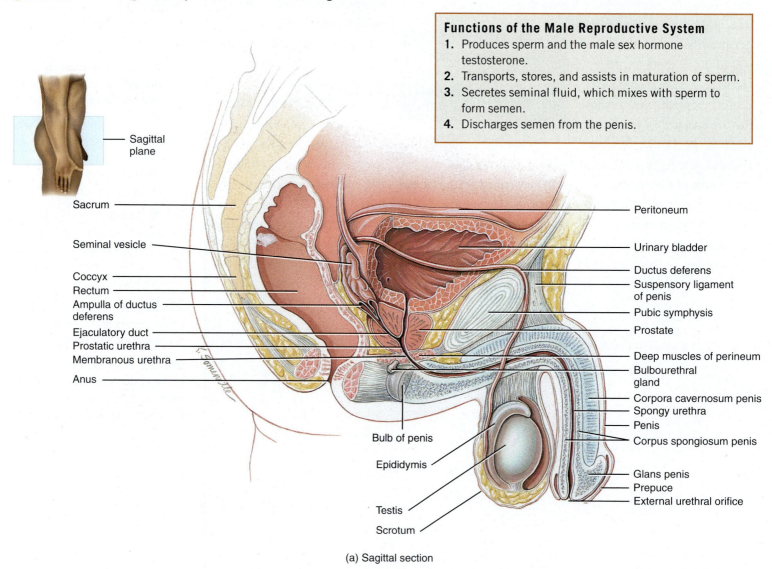

Functions of the Male Reproductive System
1. Produces sperm and the male sex hormone testosterone.
2. Transports, stores, and assists in maturation of sperm.
3. Secretes seminal fluid, which mixes with sperm to form semen.
4. Discharges semen from the penis.

Sagittal plane

Sacrum

Seminal vesicle

Coccyx
Rectum
Ampulla of ductus deferens
Ejaculatory duct
Prostatic urethra
Membranous urethra
Anus

Bulb of penis
Epididymis
Testis
Scrotum

Peritoneum
Urinary bladder
Ductus deferens
Suspensory ligament of penis
Pubic symphysis
Prostate
Deep muscles of perineum
Bulbourethral gland
Corpora cavernosum penis
Spongy urethra
Penis
Corpus spongiosum penis
Glans penis
Prepuce
External urethral orifice

(a) Sagittal section

cle (krē-MAS-ter = suspender), a series of small bands of skeletal muscle that descends as an extension of the internal oblique to surround the testes.

The location of the scrotum and the contraction of its muscle fibers regulate the temperature of the testes. Normal sperm production requires a temperature about 2–3°C below core body temperature. This lowered temperature is maintained within the scrotum because it is outside the pelvic cavity. In response to cold temperatures, the cremaster and dartos muscles contract. Contraction of the cremaster muscles moves the testes closer to the body, where they can absorb body heat. Contraction of the dartos muscle causes the scrotum to become tight (wrinkled in appearance), which reduces heat loss. Exposure to warmth reverses these actions.

Testes

The **testes** (TES-tēz), or **testicles**, are paired oval glands in the scrotum measuring about 5 cm (2 in.) long and 2.5 cm (1 in.) in diameter (Figure 25.3). The testes develop near the kidneys, in the posterior portion of the abdomen, and they usually begin their descent into the scrotum through the inguinal canals (passageways in the anterior abdominal wall; see Figure 25.2) during the seventh month of fetal development.

A serous membrane called the **tunica vaginalis** (*tunica* = sheath) partially covers the testes. Internal to the tunica vaginalis is a dense white fibrous capsule composed of dense irregular connective tissue, the **tunica albuginea** (al´-bū-JIN-ē-a; *albu-* = white); it extends inward, forming septa that divide the testis into internal compartments called **lobules**. Each of the 200–300 lobules contains one to three tightly coiled **seminiferous tubules** (*semin-* = seed; *-fer-* = to carry) that produce sperm. The process by which the seminiferous tubules of the testes produce sperm is called **spermatogenesis** (sper´-ma-tō-JEN-e-sis; *-genesis* = beginning process or production).

The seminiferous tubules contain two types of cells: **spermatogenic cells**, the sperm-forming cells, and Sertoli cells, which

(b) Sagittal section

Reproductive organs are adapted for producing new individuals and passing on genetic material from one generation to the next.

Figure 25.2   The scrotum, the supporting structure for the testes.

Internal oblique muscle

Aponeurosis of external oblique muscle (cut)

Fundiform ligament of penis

Suspensory ligament of penis

Transverse section of penis:
  Corpora cavernosa penis

Spongy urethra

Corpus spongiosum penis

Scrotal septum

Cremaster muscle

External spermatic fascia

Dartos muscle

Skin of scrotum

Spermatic cord

Superficial inguinal ring

Cremaster muscle

Inguinal canal

Ductus deferens

Autonomic nerve

Testicular artery

Lymphatic vessel

Pampiniform plexus of testicular veins

Epididymis

Tunica albuginea of testis

Tunica vaginalis

Internal spermatic fascia

Raphe

Anterior view of scrotum and testes and transverse section of penis

The scrotum consists of loose skin and an underlying hypodermis, and supports the testes.

Figure 25.3   Internal and external anatomy of a testis.

Sagittal plane

Spermatic cord

Ductus (vas) deferens

Efferent duct

Body of epididymis

Rete testis

Ductus epididymis

Tail of epididymis

Blood vessels and nerves

Head of epididymis

Seminiferous tubule

Straight tubule

Tunica vaginalis

Tunica albuginea

Lobule

Septum

(a) Sagittal section of a testis showing seminiferous tubules have several functions in supporting spermatogenesis (Figure 25.4). Stem cells called **spermatogonia** (sper´-ma-tō-GŌ-nē-a; -*gonia* = offspring; singular is *spermatogonium*) remain dormant during childhood and actively begin producing sperm at puberty. Toward the lumen of the seminiferous tubule are layers of progressively more mature cells. In order of advancing maturity, these are primary spermatocytes, secondary spermatocytes, spermatids, and sperm. After a **sperm cell**, or **spermatozoon** (sper´-ma-tō-ZŌ-on; -*zoon* = life), has formed, it is released into the lumen of the seminiferous tubule. (The plural terms are *sperm* and *spermatozoa*.)

Embedded among the spermatogenic cells in the seminiferous tubules are large **Sertoli cells**, which extend from the basement membrane to the lumen of the tubule. Internal to the basement membrane and spermatogonia, cell junctions join neighboring Sertoli cells to one another. These junctions form an obstruction known as the **blood–testis barrier** because substances must first pass through the Sertoli cells before they can reach the developing sperm. By isolating the developing gametes from the blood, the blood–testis barrier prevents an immune response against the spermatogenic cell's surface antigens, which are recognized as "foreign" by the immune system. The blood–testis barrier does not include spermatogonia.

Sertoli cells support and protect developing spermatogenic cells in several ways. They nourish spermatogenic cells, phagocytize excess spermatid cytoplasm as development proceeds, and control movements of spermatogenic cells and the release of sperm into the lumen of the seminiferous tubule. They also produce fluid

for sperm transport and, when the number of sperm needed for effective reproductive function is reached, secrete the hormone inhibin to decrease the rate of spermatogenesis.

In the spaces between adjacent seminiferous tubules are clusters of cells called **Leydig cells** (Figure 25.4). These cells secrete testosterone, the most prevalent androgen (male sex hormone). An androgen is a hormone that promotes the development of masculine characteristics. Testosterone also promotes a man's libido (sexual desire).

Spermatogenesis

Before you read this section, please review the topic of reproductive cell division in Chapter 3 (see Concept 3.7). Pay particular attention to Figures 3.32 and 3.33.

In humans, spermatogenesis takes 65–75 days. It begins with the spermatogonia, which contain the diploid (2n) number of chromosomes (see Figure 25.5). When spermatogonia undergo mitosis, some spermatogonia remain near the basement membrane of the seminiferous tubule in an undifferentiated state to serve as a reservoir of cells for future sperm production. The rest of the spermatogonia lose contact with the basement membrane, squeeze through the blood–testis barrier, undergo developmental changes, and differentiate into **primary spermatocytes** (SPER-ma-tō-sīts´). Primary spermatocytes, like spermatogonia, are diploid (2n); that is, they have 46 chromosomes.

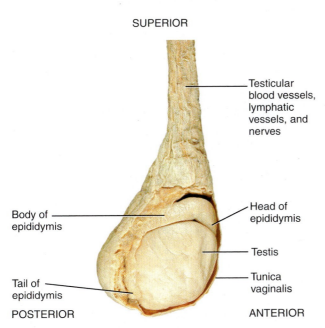

(b) Lateral view of testis and associated structures

(c) Sagittal section

The testes are the male gonads, which produce haploid sperm.

Figure 25.4   **Microscopic anatomy of the seminiferous tubules and stages of sperm production (spermatogenesis).** Arrows in (b) indicate the progression of spermatogenic cells from least mature to most mature. The (*n*) and (*2n*) refer to haploid and diploid numbers of chromosomes, respectively.

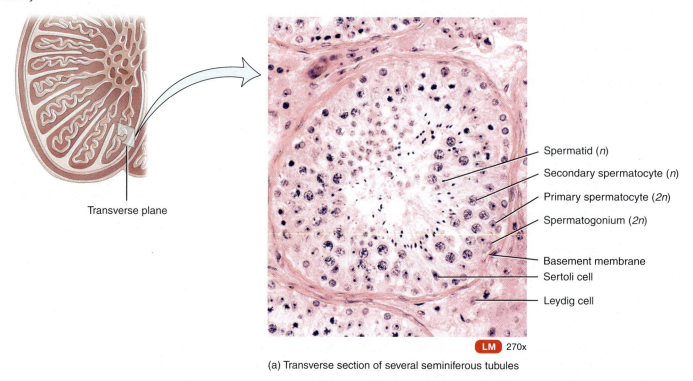

Transverse plane

Spermatid (*n*)
Secondary spermatocyte (*n*)
Primary spermatocyte (*2n*)
Spermatogonium (*2n*)

Basement membrane
Sertoli cell
Leydig cell

LM   270x

(a) Transverse section of several seminiferous tubules

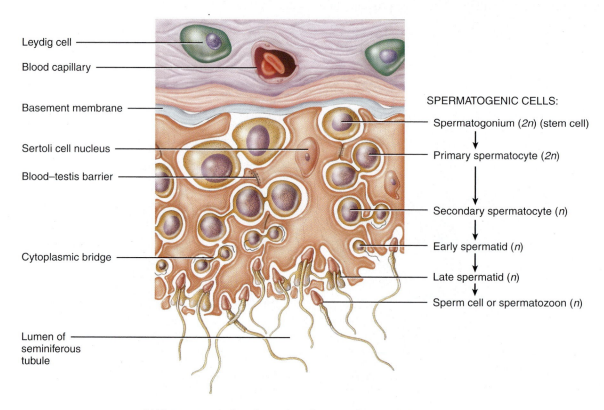

Leydig cell
Blood capillary
Basement membrane
Sertoli cell nucleus
Blood–testis barrier
Cytoplasmic bridge
Lumen of seminiferous tubule

SPERMATOGENIC CELLS:
Spermatogonium (*2n*) (stem cell)
↓
Primary spermatocyte (*2n*)
↓
Secondary spermatocyte (*n*)
↓
Early spermatid (*n*)
↓
Late spermatid (*n*)
↓
Sperm cell or spermatozoon (*n*)

(b) Transverse section of a portion of a seminiferous tubule

Spermatogenesis occurs in the seminiferous tubules of the testes.

Shortly after it forms, each primary spermatocyte replicates its DNA and then meiosis begins (Figure 25.5). In meiosis I, homologous pairs of chromosomes line up at the metaphase plate, and crossing-over occurs to rearrange genes between chromatids. Then, the meiotic spindle pulls one (duplicated) chromosome of each pair to an opposite pole of the dividing cell. The two cells formed by meiosis I are called **secondary spermatocytes**. Each secondary spermatocyte has 23 chromosomes, the haploid number (n). Each chromosome within a secondary spermatocyte, however, is made up of two chromatids (two copies of the DNA) still attached by a centromere.

In meiosis II, the chromosomes line up in single file along the metaphase plate, and the two chromatids of each chromosome separate. The four haploid cells resulting from meiosis II are called **spermatids**. Therefore, a single primary spermatocyte with 46 chromosomes undergoes two rounds of cell division (in meiosis I and meiosis II) to produce four spermatids, each with 23 chromosomes.

A unique process occurs during spermatogenesis. As spermatogenic cells proliferate, they fail to complete cytoplasmic separation (cytokinesis). The four daughter cells remain in contact via *cytoplasmic bridges* through their entire development (Figures 25.4b and 25.5). This pattern of development most likely accounts for the synchronized production of sperm in any given area of a seminiferous tubule. It may also have survival value in that half of the sperm contain an X chromosome and half contain a Y chromosome. The larger X chromosome may carry genes needed for spermatogenesis that are lacking on the smaller Y chromosome.

The final stage of spermatogenesis, **spermiogenesis** (sper´-mē-ō-JEN-e-sis), is the development of haploid spermatids into sperm. No cell division occurs in spermiogenesis; each spherical haploid spermatid becomes an elongated, slender **sperm cell** (Figure 25.6). An acrosome (described shortly) forms atop the condensed, elongated nucleus, a flagellum develops, and mitochondria multiply. Sertoli cells dispose of the excess cytoplasm that sloughs off. Finally, sperm are released from their connections to Sertoli cells and enter the lumen of the seminiferous tubule. Fluid secreted by Sertoli cells pushes sperm along their way, toward the ducts of the testes. At this point, sperm are not yet able to swim.

Figure 25.5   Events in spermatogenesis. Diploid cells ($2n$) have 46 chromosomes; haploid cells (n) have 23 chromosomes.

Spermiogenesis involves the maturation of spermatids into sperm.

Figure 25.6   Parts of a sperm cell.

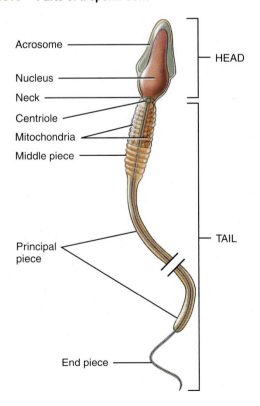

About 300 million sperm mature each day.

Sperm

Each day about 300 million sperm complete the process of sper-matogenesis. A sperm cell contains several structures that are highly adapted for reaching and penetrating a secondary oocyte (Figure 25.6). The major parts of a sperm cell are the head and the tail. The flattened, pointed **head** of the sperm cell contains a **nucleus** with 23 highly condensed chromosomes. Covering the ante-rior two-thirds of the nucleus is the **acrosome** (*acro-* = atop; *-some* = body), a caplike vesicle filled with enzymes that help a sperm cell penetrate a secondary oocyte to bring about fertilization. The **tail** of a sperm cell is subdivided into four parts: neck, middle piece, principal piece, and end piece. The **neck** is the constricted region just behind the head that contains centrioles. The centrioles form the microtubules that comprise the remainder of the tail. The **middle piece** contains mitochondria arranged in a spiral, which provide the energy (ATP) for locomotion of sperm to the second-ary oocyte and for sperm metabolism. The **principal piece** is the longest portion of the tail, and the **end piece** is the terminal, taper-ing portion of the tail. Once ejaculated, most sperm do not survive more than 48 hours within the female reproductive tract.

Hormonal Control of the Testes

At the onset of puberty, the hypothalamus increases its secretion of **gonadotropin-releasing hormone (GnRH)**. This hormone in turn stimulates the anterior pituitary to increase its secretion of **lutein-izing hormone (LH)** and **follicle-stimulating hormone (FSH)**. Figure 25.7 shows the hormones and negative feedback systems that control secretion of testosterone and spermatogenesis.

LH stimulates Leydig cells, which are located between semi-niferous tubules, to secrete the hormone **testosterone** (tes-TOS-te-rōn). This steroid hormone is synthesized from cholesterol in the testes and is the principal androgen. Through negative feedback, testosterone suppresses secretion of LH by the anterior pituitary and suppresses secretion of GnRH by the hypothalamus. In some target cells, such as those in the prostate, testosterone is converted to another androgen called **dihydrotestosterone (DHT)**.

FSH acts indirectly to stimulate spermatogenesis (Figure 25.7). FSH and testosterone act synergistically on the Sertoli cells to stimulate secretion of **androgen-binding protein (ABP)** into the lumen of the seminiferous tubules and into the interstitial fluid around the spermatogenic cells. ABP binds to testosterone, keep-ing its concentration high. Testosterone stimulates the final steps of spermatogenesis in the seminiferous tubules. Once the degree of spermatogenesis required for male reproductive functions has been achieved, Sertoli cells release **inhibin**, a hormone named for its role in inhibiting FSH secretion by the anterior pituitary (Figure 25.7). If spermatogenesis is proceeding too slowly, less inhibin is released, which permits more FSH secretion and an increased rate of spermatogenesis.

Figure 25.7   Hormonal control of spermatogenesis and actions of testosterone and dihydrotestosterone (DHT). In response to stimulation by follicle-stimulating hormone (FSH) and testosterone, Sertoli cells secrete androgen-binding protein (ABP). Dashed red lines indicate negative feedback inhibition.

🔑 **Release of FSH is stimulated by gonadotropin-releasing hormone (GnRH) and inhibited by inhibin; release of luteinizing hormone (LH) is stimulated by GnRH and inhibited by testosterone.**

Testosterone and dihydrotestosterone both bind to the same androgen receptors within the nuclei of target cells. The hormone–receptor complex turns some genes on and others off. Because of these changes, the androgens produce several effects:

- **Prenatal development**. Before birth, testosterone stimulates the male pattern of development of reproductive system ducts and the descent of the testes. Dihydrotestosterone stimulates development of the external genitals. Testosterone also is converted in the brain to estrogens (feminizing hormones), which may play a role in the development of certain regions of the brain in males.

- **Development of male sexual characteristics**. At puberty, testosterone and dihydrotestosterone bring about development and enlargement of the male sex organs and the development of masculine secondary sexual characteristics. *Secondary sex characteristics* are traits that distinguish males and females but do not have a direct role in reproduction. In males these include wide shoulders and narrow hips, facial and chest hair (within hereditary limits) and more hair on other parts of the body, thickening of the skin, increased sebaceous (oil) gland secretion, and enlargement of the larynx and consequent deepening of the voice.

- **Development of sexual function**. Androgens contribute to male sexual behavior and spermatogenesis and to sex drive (libido) in both males and females. Recall that the adrenal cortex is the main source of androgens in females.

- **Stimulation of anabolism**. Androgens are anabolic hormones; that is, they stimulate protein synthesis. This effect is obvious in the heavier muscle and bone mass of most men as compared to women.

A negative feedback system regulates testosterone production (Figure 25.8). When testosterone concentration in the blood increases to a certain level, it inhibits the release of GnRH by the hypothalamus. As a result, there is less GnRH traveling from the hypothalamus to the anterior pituitary. The anterior pituitary then releases less LH. With less stimulation by LH, the Leydig cells in the testes secrete less testosterone, and there is a return to homeostasis. If the testosterone concentration in the blood falls too low, however, GnRH is again released by the hypothalamus and stimulates secretion of LH by the anterior pituitary. LH in turn stimulates testosterone production by the testes.

Figure 25.8 **Negative feedback control of blood level of testosterone.**

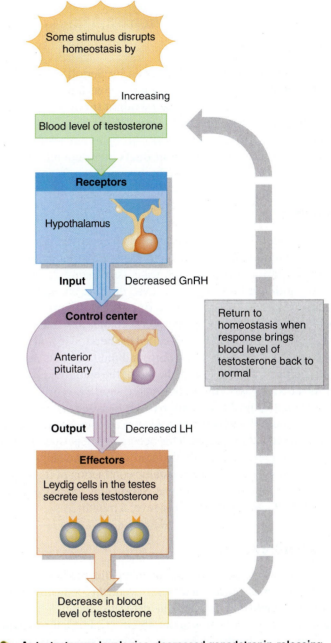

As testosterone levels rise, decreased gonadotropin-releasing hormone (GnRH) results in decreased luteinizing hormone (LH), inhibiting testosterone secretion.

Checkpoint

1. Name the male reproductive structures and state their general functions.
2. What tissue layers cover and protect the testes? Where in the testes are sperm cells produced?
3. What are the functions of Sertoli cells and Leydig cells?
4. What is "reduced" during meiosis I?
5. What are the functions of the sperm cell acrosome and middle piece?
6. What is the source of luteinizing hormone and follicle-stimulating hormone?
7. Which cells of the male reproductive system secrete testosterone? What are the functions of testosterone?

25.2 Sperm travel through the epididymis, ductus deferens, ejaculatory ducts, and urethra.

Reproductive System Ducts in Males

Ducts of the Testis

Pressure generated by the fluid secreted by Sertoli cells pushes sperm and fluid along the lumen of seminiferous tubules and then into a series of very short ducts called **straight tubules** (see Figure 25.3a). The straight tubules lead to a network of ducts in the testis called the **rete testis** (RĒ-tē = network). From the rete testis, sperm move into a series of coiled **efferent ducts** in the epididymis that empty into a single tube called the **ductus epididymis**.

Epididymis

The **epididymis** (ep´-i-DID-i-mis; *epi-* = above or over; *-didymis* = testis) is a comma-shaped organ about 4 cm (1.5 in.) long that lies along the posterior border of each testis (see Figure 25.3a). The plural is **epididymides** (ep´-i-did-ĪM-i-dēs). Each epididymis consists mostly of the tightly coiled **ductus epididymis**. The efferent ducts from the testis join the ductus epididymis at the larger, superior portion of the epididymis called the **head**. The **body** is the narrow midportion of the epididymis, and the **tail** is the smaller, inferior portion. At its distal end, the tail of the epididymis continues as the ductus (vas) deferens (discussed shortly).

Functionally, the ductus epididymis is the site of *sperm maturation*, the process by which sperm acquire motility and the ability to fertilize a secondary oocyte. This occurs over a period of about 14 days. The ductus epididymis also stores sperm and helps propel them during sexual arousal by peristaltic contraction of its smooth muscle into the ductus (vas) deferens. Sperm may remain viable in storage in the ductus epididymis for several months.

Ductus Deferens

Within the tail of the epididymis, the ductus epididymis becomes less convoluted, and its diameter increases. Beyond this point, the duct is known as the **ductus deferens** or **vas deferens** (see Figure 25.3a). The ductus deferens, which is about 45 cm (18 in.) long, ascends along the posterior border of the epididymis through the spermatic cord and then enters the pelvic cavity. There it loops over the ureter and passes over the side and down the posterior surface of the urinary bladder (see Figure 25.1a). The dilated terminal portion of the ductus deferens is the **ampulla** (am-PUL-la = little jar; Figure 25.9).

Functionally, the ductus deferens stores sperm, which can remain viable here for up to several months. The ductus deferens also conveys sperm from the epididymis toward the urethra by peristaltic contractions of its muscularis. Sperm that are not ejaculated are eventually reabsorbed.

Spermatic Cord

The **spermatic cord** is a supporting structure of the male reproductive system that ascends out of the scrotum (see Figure 25.2). It consists of the ductus (vas) deferens as it travels through the scrotum, the testicular artery, veins that drain the testes and carry testosterone into circulation (the pampiniform plexus), autonomic nerves, lymphatic vessels, and the cremaster muscle. The spermatic cord passes through the **inguinal canal** (IN-gwi-nal = groin), an oblique passageway in the anterior abdominal wall. The canal, which is about 4–5 cm (about 2 in.) long, originates at the **deep inguinal ring**, an opening in the aponeurosis of the transversus abdominis; the canal ends at the **superficial inguinal ring** (see Figure 25.2), an opening in the aponeurosis of the external oblique.

Ejaculatory Ducts

Each **ejaculatory duct** (e-JAK-ū-la-tō´-re; *ejacul-* = to expel) is about 2 cm (1 in.) long and is formed by the union of the duct from the seminal vesicle and the ampulla of the ductus (vas) deferens (Figure 25.9). The ejaculatory ducts form just superior to the prostate and pass inferiorly and anteriorly through the prostate. They terminate in the prostatic urethra, where they eject sperm and seminal vesicle secretions just before the release of semen from the urethra to the exterior.

Urethra

In males, the **urethra** is the shared terminal duct of the reproductive and urinary systems, serving as a passageway for both sperm and urine. About 20 cm (8 in.) long, it passes through the prostate and the penis, and is subdivided into three parts (see Figure 25.1). The **prostatic urethra** is 2–3 cm (1 in.) long and passes through the prostate. As this duct continues inferiorly, it passes through the deep muscles of the perineum, where it is known as the **membranous urethra**. The membranous urethra is about 1 cm (0.5 in.) in length. As the duct passes through the corpus spongiosum of the penis, it is known as the **spongy (penile) urethra**, which is about 15–20 cm (6–8 in.) long. The spongy urethra ends at the **external urethral orifice**. The histology of the male urethra may be reviewed in Concept 24.9.

Figure 25.9 **Locations of several accessory reproductive organs in males.** The prostate, urethra, and penis have been sectioned to show internal details.

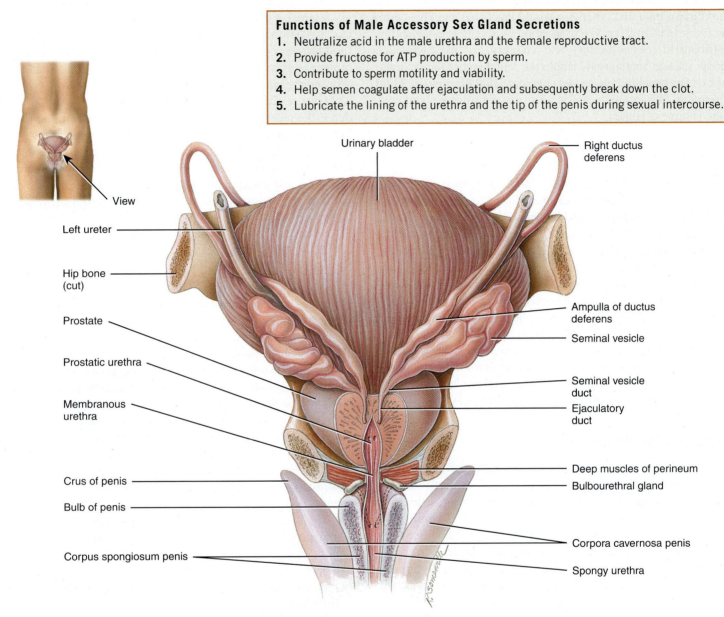

> **Functions of Male Accessory Sex Gland Secretions**
> 1. Neutralize acid in the male urethra and the female reproductive tract.
> 2. Provide fructose for ATP production by sperm.
> 3. Contribute to sperm motility and viability.
> 4. Help semen coagulate after ejaculation and subsequently break down the clot.
> 5. Lubricate the lining of the urethra and the tip of the penis during sexual intercourse.

Urinary bladder — Right ductus deferens — View — Left ureter — Hip bone (cut) — Prostate — Ampulla of ductus deferens — Seminal vesicle — Prostatic urethra — Seminal vesicle duct — Ejaculatory duct — Membranous urethra — Crus of penis — Deep muscles of perineum — Bulbourethral gland — Bulb of penis — Corpora cavernosa penis — Corpus spongiosum penis — Spongy urethra

Posterior view of male accessory organs of reproduction

🔑 The male urethra has three subdivisions: the prostatic, membranous, and spongy urethra.

Accessory Sex Glands

The ducts of the male reproductive system store and transport sperm, but the **accessory sex glands** secrete most of the liquid portion of semen that protects sperm and facilitates their movement. The accessory sex glands include the seminal vesicles, the prostate, and the bulbourethral glands.

Seminal Vesicles

The paired **seminal vesicles** (VES-i-kuls), or **seminal glands**, are convoluted pouchlike structures, about 5 cm (2 in.) in length, lying posterior to the base of the urinary bladder and anterior to the rectum (Figure 25.9). Through the seminal vesicle ducts they secrete an alkaline, viscous fluid that contains fructose (a monosaccharide sugar), prostaglandins, and clotting proteins that are different from those in blood. The alkaline nature of the seminal fluid helps to neutralize the acidic environment of the male urethra and female reproductive tract that otherwise would inactivate and kill sperm. The fructose is used for ATP production by sperm. Prostaglandins contribute to sperm motility and viability and may stimulate smooth muscle contractions within the female reproductive tract. The clotting proteins help semen coagulate after ejaculation. Fluid secreted by the seminal vesicles normally constitutes about 60% of the volume of semen.

Prostate

The **prostate** (PROS-tāt) is a single, doughnut-shaped gland about the size of a golf ball measuring about 4 cm (1.6 in.) from side to side. It is inferior to the urinary bladder and surrounds the prostatic urethra (Figure 25.9). The prostate secretes a milky, slightly acidic fluid (pH about 6.5) that contains (1) *citric acid*, used by sperm for ATP production via the Krebs cycle; (2) *proteolytic enzymes*, such as *prostate-specific antigen* (*PSA*), pepsinogen, lysozyme, amylase, and hyaluronidase, that eventually break down the clotting proteins from the seminal vesicles; and (3) seminalplasmin, an antibiotic that may help decrease the number of naturally occurring bacteria in semen and in the lower female reproductive tract. Secretions of the prostate enter the prostatic urethra through many prostatic ducts. Prostatic secretions make up about 25% of the volume of semen and contribute to sperm motility and viability.

Bulbourethral Glands

The paired **bulbourethral glands** (bul´-bō-ū-RĒ-thral), or **Cowper's glands**, are about the size of peas. They are located inferior to the prostate on either side of the membranous urethra, and their ducts open into the spongy urethra (Figure 25.9). During sexual arousal, the bulbourethral glands secrete an alkaline fluid into the urethra that protects the passing sperm by neutralizing acids from urine in the urethra. They also secrete mucus that lubricates the end of the penis and the lining of the urethra, decreasing the number of sperm damaged during ejaculation.

Semen

Semen (= seed) is a mixture of sperm and **seminal fluid**, a liquid that consists of the secretions of the seminiferous tubules, seminal vesicles, prostate, and bulbourethral glands. The volume of semen in a typical ejaculation is 2.5–5 milliliters (mL), with 50–150 million sperm per milliliter. A very large number of sperm is required for successful fertilization because only a tiny fraction ever reaches the secondary oocyte. When the number of sperm falls below 20 million per milliliter semen, the male is likely to be infertile.

Despite the slight acidity of prostatic fluid, semen has a slightly alkaline pH of 7.2–7.7 due to the higher pH and larger volume of fluid from the seminal vesicles. The prostatic secretion gives semen a milky appearance, and fluids from the seminal vesicles and bulbourethral glands give it a sticky consistency. Seminal fluid provides sperm with a transportation medium, nutrients, and protection from the hostile acidic environment of the male's urethra and the female's vagina.

Once ejaculated, liquid semen coagulates within 5 minutes due to the presence of clotting proteins from the seminal vesicles. The functional role of semen coagulation is not known. After about 10 to 20 minutes, semen reliquefies because proteolytic enzymes produced by the prostate break down the clot. Abnormal or delayed liquefaction of clotted semen may cause complete or partial immobilization of sperm, thereby inhibiting their movement through the cervix of the uterus.

Penis

The **penis** contains the urethra and is a passageway for the ejaculation of semen and the excretion of urine (Figure 25.10). It is cylindrical in shape and consists of a root, body, and glans penis. The **body of the penis** is composed of three cylindrical masses of tissue, each surrounded by the tunica albuginea (Figure 25.10b,c). The two dorsolateral masses are called the **corpora cavernosa penis** (*corpora* = main bodies; *cavernosa* = hollow). The smaller midventral mass, the **corpus spongiosum penis**, contains the spongy urethra and keeps it open during ejaculation. Skin and hypodermis enclose all three masses, which consist of erectile tissue. *Erectile tissue* is composed of numerous blood sinuses (vascular spaces) surrounded by smooth muscle and elastic connective tissue.

The distal end of the corpus spongiosum penis is a slightly enlarged, acorn-shaped region called the **glans penis**. The distal urethra enlarges within the glans penis and forms a terminal slit-like opening, the **external urethral orifice**. Covering the glans in an uncircumcised penis is the loosely fitting **prepuce** (PRĒ-poos), or **foreskin**.

The **root of the penis** is the attached, proximal portion. It consists of the **bulb of the penis**, the expanded portion of the base of the corpus spongiosum penis, and the **crura of the penis** (singular is *crus* = resembling a leg), the two separated and tapered portions of the corpora cavernosa penis. The bulb of the penis is attached to the inferior surface of the deep muscles of the perineum. Each crus of the penis is attached to the ischium of the hip bone. The weight of the penis is supported by two ligaments that are continuous with the fascia of the penis: (1) The **fundiform ligament** arises from the inferior part of the linea alba. (2) The **suspensory ligament of the penis** arises from the pubic symphysis (see Figure 25.2).

Figure 25.10 **Internal structure of the penis.** The inset in (b) shows details of the skin and fasciae.

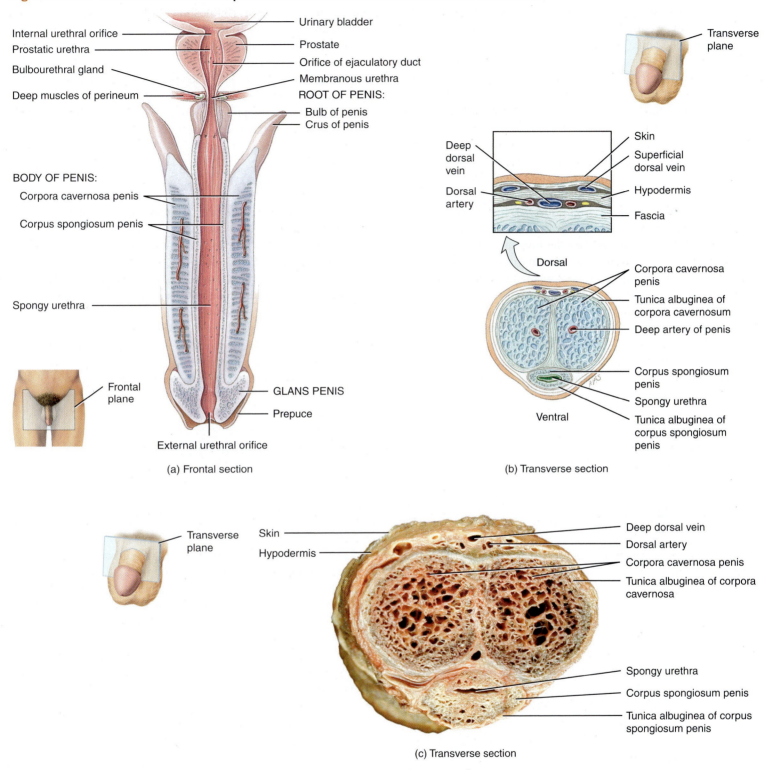

Internal urethral orifice

Prostatic urethra

Bulbourethral gland

Deep muscles of perineum

Urinary bladder

Prostate

Orifice of ejaculatory duct

Membranous urethra

ROOT OF PENIS:

Bulb of penis

Crus of penis

BODY OF PENIS:

Corpora cavernosa penis

Corpus spongiosum penis

Spongy urethra

Frontal plane

GLANS PENIS

Prepuce

External urethral orifice

(a) Frontal section

Transverse plane

Deep dorsal vein

Dorsal artery

Skin

Superficial dorsal vein

Hypodermis

Fascia

Dorsal

Corpora cavernosa penis

Tunica albuginea of corpora cavernosum

Deep artery of penis

Corpus spongiosum penis

Spongy urethra

Tunica albuginea of corpus spongiosum penis

Ventral

(b) Transverse section

Transverse plane

Skin

Hypodermis

Deep dorsal vein

Dorsal artery

Corpora cavernosa penis

Tunica albuginea of corpora cavernosa

Spongy urethra

Corpus spongiosum penis

Tunica albuginea of corpus spongiosum penis

(c) Transverse section

The penis contains the urethra, a pathway for semen and urine.

Upon sexual stimulation (visual, tactile, auditory, olfactory, or imagined), parasympathetic fibers from the sacral portion of the spinal cord initiate and maintain an **erection**, the enlargement and stiffening of the penis. The parasympathetic fibers cause smooth muscle in the walls of arterioles supplying erectile tissue to relax, which allows these blood vessels to dilate. This in turn causes large amounts of blood to enter the erectile tissue of the penis. Parasympathetic fibers also cause the smooth muscle within the erectile tissue to relax, resulting in widening of the blood sinuses. The combination of increased blood flow and widening of the blood sinuses results in an erection. Expansion of the blood sinuses also compresses the veins that drain the penis; the slowing of blood outflow helps to maintain the erection.

Ejaculation (ē-jak-ū-LĀ-shun; *ejectus-* = to throw out), the powerful release of semen from the urethra to the exterior, is a sympathetic reflex coordinated by the lumbar portion of the spinal cord. As part of the reflex, the smooth muscle sphincter at the base of the urinary bladder closes, preventing urine from being expelled during ejaculation, and semen from entering the urinary bladder. Even before ejaculation occurs, peristaltic contractions in the epididymis, ductus deferens, seminal vesicles, ejaculatory ducts, and prostate propel semen into the penile portion of the urethra (spongy urethra). Typically, this leads to **emission** (ē-MISH-un), the discharge of a small volume of semen before ejaculation.

Emission may also occur during sleep (nocturnal emission). The musculature of the penis (bulbospongiosus, ischiocavernosus, and superficial transverse perineus), which is supplied by the pudendal nerve, also contracts at ejaculation (see Figure 11.12).

Once sexual stimulation of the penis has ended, the arterioles supplying the erectile tissue of the penis constrict and the smooth muscle within erectile tissue contracts, making the blood sinuses smaller. This relieves pressure on the veins supplying the penis and allows the blood to drain through them. Consequently, the penis returns to its flaccid (relaxed) state.

Checkpoint

8. What are the functions of the ductus epididymis, ductus deferens, and ejaculatory duct?
9. Name the structures within the spermatic cord. How does the spermatic cord travel through the abdominal wall?
10. What are the locations of the three subdivisions of the male urethra?
11. What are the locations and functions of the seminal vesicles, prostate, and bulbourethral glands?
12. Which accessory sex gland contributes the majority of the seminal fluid?
13. Which tissue masses form the erectile tissue in the penis, and how do they become rigid during sexual arousal?
14. Explain the physiological processes involved in ejaculation.

25.3 After a secondary oocyte is discharged from an ovary, it may undergo fertilization and implantation in the uterus.

The organs of the female reproductive system (Figure 25.11) include the ovaries, the uterine (fallopian) tubes or oviducts, the uterus, the vagina, and external organs, which are collectively called the vulva. The mammary glands are considered part of both the integumentary system and the female reproductive system.

Ovaries

The **ovaries** (= egg receptacles), which are the female gonads, are paired glands that resemble unshelled almonds in size and shape; they are homologous to the testes. (Here *homologous* means that two organs have the same embryonic origin.) The ovaries produce (1) gametes, secondary oocytes that develop into mature ova (eggs) after fertilization, and (2) hormones, including progesterone and estrogens (the female sex hormones), inhibin, and relaxin.

The ovaries, one on either side of the uterus, descend to the brim of the superior portion of the pelvic cavity during the third month of development. A series of ligaments holds them in position (Figure 25.12). The **broad ligament** of the uterus encloses the ovaries and attaches them to the pelvic wall where the broad ligament continues as the parietal peritoneum. A component of the broad ligament, a double-layered fold of peritoneum called the **mesovarium**, attaches to and further stabilizes the ovaries. The **ovarian ligament** anchors the ovaries to the uterus, and the **suspensory ligament** attaches the ovaries laterally to the pelvic wall.

Figure 25.11 Female organs of reproduction and surrounding structures.

> **Functions of the Female Reproductive System**
> 1. Produces secondary oocytes and hormones, including progesterone and estrogens (female sex hormones), inhibin, and relaxin.
> 2. Provides sites of fertilization of a secondary oocyte, implantation of a fertilized ovum, development of the fetus during pregnancy, and labor.
> 3. Receives the penis during sexual intercourse.
> 4. Synthesizes, secretes, and ejects milk for nourishment of the newborn.

(a) Sagittal section

POSTERIOR

ANTERIOR

(b) Sagittal section

The organs of reproduction in females include the ovaries, uterine tubes, uterus, vagina, vulva, and mammary glands.

Figure 25.12   Relative positions of the ovaries, the uterus, and the ligaments that support them.

Superior view of transverse section

🔑 Ligaments holding the ovaries in position are the mesovarium, the ovarian ligament, and the suspensory ligament.

Histology of the Ovary

Each ovary includes the following parts (Figure 25.13):

• The **germinal epithelium** (*germen* = sprout or bud) is a layer of simple epithelium (low cuboidal or squamous) that covers the surface of the ovary.

• The **tunica albuginea** is a capsule of dense irregular connective tissue located immediately deep to the germinal epithelium (see Figure 25.14e).

• The **ovarian cortex** is a region just deep to the tunica albuginea. It consists of ovarian follicles (described shortly) surrounded by dense irregular connective tissue.

• **Ovarian follicles** (*folliculus* = little bag) are in the cortex and consist of **oocytes** in various stages of development, plus the cells surrounding them that nourish the developing oocyte and secrete estrogens as the follicle grows larger.

• A **mature (graafian) follicle** is a large, fluid-filled ovarian follicle that is ready to rupture and expel its secondary oocyte, a process known as **ovulation**.

• A **corpus luteum** (= yellow body) contains the remnants of a mature follicle after ovulation. The corpus luteum produces progesterone, estrogens, relaxin, and inhibin until it degen-

erates into fibrous scar tissue called the **corpus albicans** (= white body).

• The **ovarian medulla** is deep to the ovarian cortex. The border between the cortex and ovarian medulla is indistinct, but the medulla consists of more loosely arranged connective tissue and contains blood vessels, lymphatic vessels, and nerves.

Oogenesis and Follicular Development

The formation of gametes in the ovaries is termed **oogenesis** (ō'-ō-JEN-e-sis; *oo-* = egg). Once males reach puberty, spermatogenesis begins and they can produce new sperm throughout life. In contrast, oogenesis begins in females before they are even born. Oogenesis occurs in essentially the same manner as spermatogenesis; meiosis (see Concept 3.7) takes place and the resulting germ cells undergo maturation.

During early fetal development, millions of diploid (2n) stem cells called **oogonia** (ō'-ō-GŌ-nē-a; singular is *oogonium*) differentiate within the ovaries. Even before birth, most oogonia degenerate in a process known as **atresia** (a-TRĒ-zē-a). A few, however, develop into larger cells called **primary oocytes** (Ō-ō-sits) that enter prophase of meiosis I during fetal development but do not complete that phase until after puberty. During this arrested stage

Figure 25.13 **Histology of the ovary.** The arrows in (a) indicate the sequence of developmental stages that occur as part of the maturation of an ovum during the ovarian cycle.

(a) Frontal section

(b) Frontal section

LM 20x

(c) Ovulation of a secondary oocyte

LM 30x

🔑 The ovaries are the female gonads; they produce haploid oocytes.

of development, each primary oocyte is surrounded by a single layer of flat follicular cells, and the entire structure is called a **primordial follicle** (Figure 25.14a). The ovarian cortex surrounding the primordial follicles consists of collagen fibers and **stromal cells**. At birth, 200,000 to 2,000,000 primary oocytes remain in each ovary. Of these, about 40,000 remain at puberty, and around 400 will mature and ovulate during a woman's reproductive lifetime. The remainder of the primary oocytes undergo atresia.

Each month after puberty, FSH and LH secreted by the anterior pituitary further stimulate the development of several primordial follicles, although only one will typically reach the maturity needed for ovulation. A few primordial follicles start to grow, developing into **primary follicles** (Figure 25.14b). Each primary follicle consists of a primary oocyte that is surrounded by several layers of follicular cells that, now multilayered, are called **granulosa cells**. As the primary follicle grows, it forms a clear glycoprotein

Figure 25.14   Ovarian follicles.

(a) Primordial follicle

(b) Primary follicle

(c) Secondary follicle

(d) Mature follicle

layer, called the **zona pellucida** (ZŌ-na = zone; pe-LOO-si-da = allowing passage of light), between the primary oocyte and the granulosa cells. In addition, stromal cells surrounding the basement membrane begin to form an organized layer called the **theca folliculi** (THĒ-ka fo-LIK-ū-li).

With continuing maturation, a primary follicle develops into a **secondary follicle** (Figure 25.14c). In a secondary follicle, the theca differentiates into the **theca interna**, a vascularized internal layer of cuboidal cells that secrete estrogens, and the **theca externa**, an outer layer of stromal cells and collagen fibers. The granulosa cells begin to secrete follicular fluid, which builds up in a cavity called the **antrum** in the center of the secondary follicle. In addition, the innermost layer of granulosa cells becomes firmly attached to the zona pellucida and is now called the **corona radiata** (kō-RŌ-na = crown; rā-dē-A-ta = radiation).

The secondary follicle eventually becomes larger, turning into a **mature (graafian) follicle** (Figure 25.14d). While in this follicle, the dipoid primary oocyte completes meiosis I, producing two haploid (n) cells of unequal size—each with 23 chromosomes (Figure 25.15). The smaller cell produced by meiosis I, called the **first polar body**, is essentially a packet of discarded nuclear material. The larger cell, known as the **secondary oocyte**, receives most of the cytoplasm. Once a secondary oocyte is formed, it begins meiosis II but then stops in metaphase. The mature (graafian) follicle soon ruptures and releases its secondary oocyte during ovulation (see Figure 25.13).

At ovulation, the secondary oocyte is expelled into the pelvic cavity together with the first polar body and corona radiata. Normally these cells are swept into the uterine tube. If fertilization

Figure 25.15 **Oogenesis.** Diploid cells (2n) have 46 chromosomes; haploid cells (n) have 23 chromosomes.

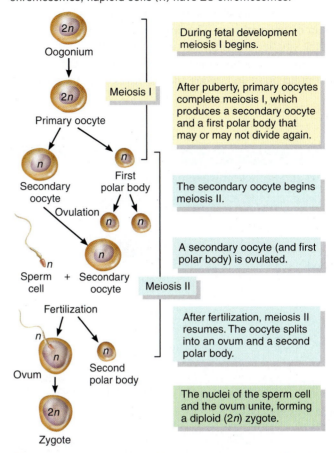

During fetal development meiosis I begins.

After puberty, primary oocytes complete meiosis I, which produces a secondary oocyte and a first polar body that may or may not divide again.

The secondary oocyte begins meiosis II.

A secondary oocyte (and first polar body) is ovulated.

After fertilization, meiosis II resumes. The oocyte splits into an ovum and a second polar body.

The nuclei of the sperm cell and the ovum unite, forming a diploid (2n) zygote.

🔑 **In a secondary oocyte, meiosis II is completed only if fertilization occurs.**

Germinal epithelium

Tunica albuginea

Primordial follicle

Ovarian cortex

Primary follicle granulosa cells

Theca folliculi

Zona pellucida

Primary oocyte

Granulosa cells

Secondary follicle

Corpus luteum

LM 30x

(e) Ovarian cortex

Theca folliculi

Antrum filled with follicular fluid

Granulosa cells

Corona radiata

Zona pellucida

Primary oocyte

LM 70x

(f) Secondary follicle

🔑 **As an ovarian follicle enlarges, follicular fluid accumulates in a cavity called the antrum.**

does not occur, the cells degenerate. If sperm are present in the uterine tube and one penetrates the secondary oocyte, however, meiosis II resumes. The secondary oocyte splits into two haploid cells, again of unequal size. The larger cell is the **ovum**, or mature egg; the smaller one is the **second polar body**. The nuclei of the sperm cell and the ovum then unite, forming a diploid **zygote**. If the first polar body undergoes another division to produce two

polar bodies, then the primary oocyte ultimately gives rise to three haploid polar bodies, which all degenerate, and a single haploid ovum. Thus, one primary oocyte gives rise to a single gamete (an ovum). By contrast, recall that in males one primary spermatocyte produces four gametes (sperm).

Table 25.1 summarizes the events of oogenesis and follicular development.

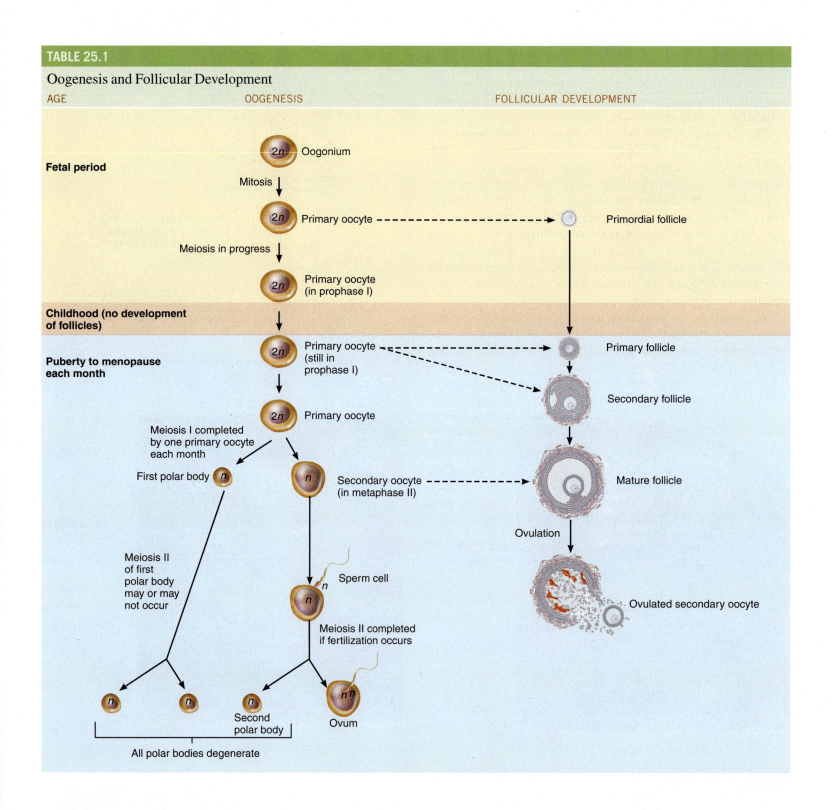

TABLE 25.1

Oogenesis and Follicular Development

AGE	OOGENESIS	FOLLICULAR DEVELOPMENT

Uterine Tubes

Two **uterine (fallopian) tubes**, or **oviducts**, extend laterally from the uterus (Figure 25.16). The tubes, which measure about 10 cm (4 in.) long, lie between the folds of the broad ligaments of the uterus. They provide a route for sperm to reach an ovum and transport secondary oocytes and fertilized ova from the ovaries to the uterus. The funnel-shaped portion of each tube, called the **infundibulum**, is close to the ovary but is open to the pelvic cavity. It ends in a fringe of fingerlike projections called **fimbriae** (FIM-brē-ē = fringe), one of which is attached to the lateral end of the ovary. From the infundibulum, the uterine tube extends medially and eventually inferiorly and attaches to the upper, outer corners of the uterus. The **ampulla** of the uterine tube is the widest, longest portion, making up about the lateral two-thirds of its length.

The **isthmus** of the uterine tube is the more medial, short, narrow, thick-walled portion that joins the uterus.

The mucosal lining of the uterine tube consists of epithelium and lamina propria (areolar connective tissue). The epithelium contains ciliated simple columnar cells, which function as a "ciliary conveyor belt" to help move a fertilized ovum (or secondary oocyte) within the tube toward the uterus, and nonciliated *peg cells*, which have microvilli and secrete a fluid that provides nutrition for the ovum (Figure 25.17). The muscularis is composed of an inner, circular layer of smooth muscle and an outer layer of longitudinal smooth muscle. Peristaltic contractions of the muscularis and the ciliary action of the mucosa help move the oocyte, or fertilized ovum, toward the uterus.

Local currents produced by movements of the fimbriae, which surround the ovary during ovulation, sweep the ovulated

Figure 25.16 **Relationship of the uterine tubes to the ovaries, uterus, and associated structures.** In the left side of the drawing, the uterine tube and uterus have been sectioned to show internal structures.

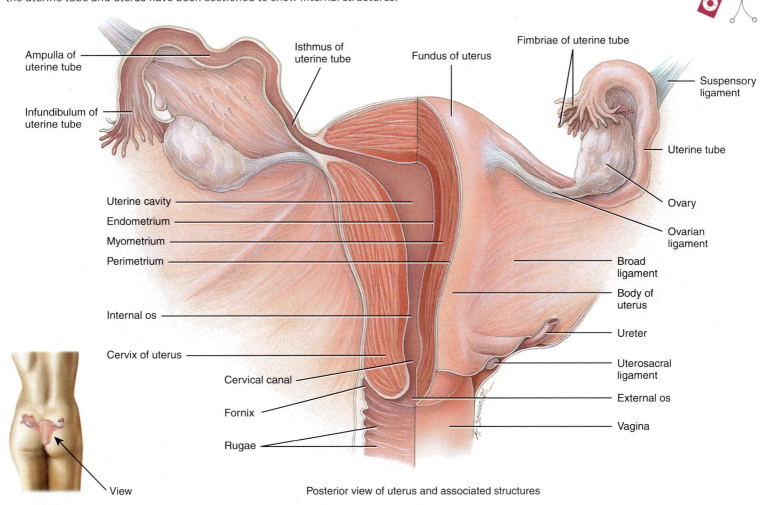

Posterior view of uterus and associated structures

After ovulation, a secondary oocyte and its corona radiata move from the pelvic cavity into the infundibulum of the uterine tube. The uterus is the site of menstruation, implantation of a fertilized ovum, development of the fetus, and labor.

Figure 25.17   Histology of the uterine tube.

Cilia

Lamina propria

Ciliated simple columnar cell

Nonciliated peg cells

LM 3850x

(a) Details of epithelium in sectional view

Cilia of ciliated columnar epithelial cell

Nonciliated peg cells with microvilli

SEM 4000x

(b) Details of epithelium in surface view

🔑 **Peristaltic contractions of the muscularis and ciliary action of the mucosa of the uterine tube help move the oocyte or fertilized ovum toward the uterus.**

secondary oocyte from the pelvic cavity into the uterine tube. A sperm cell usually encounters and fertilizes a secondary oocyte in the ampulla of the uterine tube, although fertilization in the pelvic cavity is not uncommon. Fertilization can occur at any time up to about 24 hours after ovulation. Some hours after fertilization,

the nuclear materials of the haploid ovum and sperm unite. The diploid fertilized ovum is now called a **zygote** and begins to undergo cell divisions while moving toward the uterus. It arrives at the uterus 6 to 7 days after ovulation.

Uterus

The **uterus** (womb) serves as part of the pathway for sperm deposited in the vagina to reach the uterine tubes. It is also the site of implantation of a fertilized ovum, development of the fetus during pregnancy, and labor. During reproductive cycles when implantation does not occur, the uterus is the source of menstrual flow.

Anatomy of the Uterus

Situated between the urinary bladder and the rectum, the uterus is the size and shape of an inverted pear (Figure 25.16). In females who have never been pregnant, it is about 7.5 cm (3 in.) long, 5 cm (2 in.) wide, and 2.5 cm (1 in.) thick. The uterus is larger in females who have recently been pregnant, and smaller (atrophied) when sex hormone levels are low, as occurs after menopause.

Anatomical subdivisions of the uterus include the following: (1) a dome-shaped portion superior to the uterine tubes called the **fundus**, (2) a tapering central portion called the **body**, and (3) an inferior narrow portion called the **cervix** that opens into the vagina. The interior of the body of the uterus is called the **uterine cavity**, and the interior of the narrow cervix is called the **cervical canal**. The cervical canal opens into the uterine cavity at the **internal os** (*os* = mouthlike opening) and into the vagina at the **external os**.

Normally, the body of the uterus projects anteriorly and superiorly over the urinary bladder in a position called **anteflexion**. The cervix projects inferiorly and posteriorly and enters the anterior wall of the vagina at nearly a right angle (see Figure 25.11). Several ligaments maintain the anteflexed position of the uterus (see Figure 25.12). The paired broad ligaments attach the uterus to either side of the pelvic cavity. The paired **uterosacral ligaments** lie on either side of the rectum and connect the uterus to the sacrum. The **cardinal ligaments** extend from the cervix and vagina to the lateral pelvic wall. The **round ligaments** extend between the layers of the broad ligament from the uterus just inferior to the uterine tubes to the labia majora of the external genitalia. Although the ligaments normally maintain the anteflexed position of the uterus, they also allow the uterine body to move enough that the uterus may become malpositioned. A posterior tilting of the uterus, called **retroflexion** (*retro-* = backward or behind), is a harmless variation of the normal position of the uterus. There is often no cause for the condition, but it may occur after childbirth.

Histology of the Uterus

Histologically, the uterus consists of three layers of tissue: the perimetrium, myometrium, and endometrium (Figure 25.18). The outer layer—the **perimetrium** (*peri-* = around; *-metrium* = uterus) or serosa—is part of the visceral peritoneum. Laterally, it becomes the broad ligament. The middle layer of the uterus, the **myometrium** (*myo-* = muscle), consists of three layers of smooth muscle fibers and forms the bulk of the uterine wall. During labor and childbirth, coordinated contractions of the myometrium help expel the fetus from the uterus.

The lumen of the uterus is lined with the **endometrium** (*endo-* = within), a mucosa composed of simple columnar epithelium with a thick underlying lamina propria of areolar connective tissue. **Endometrial glands** open onto the surface of the endometrium and extend almost to the myometrium. The endometrium is divided into two layers. The **stratum functionalis** (*functional layer*) lines the uterine cavity and sloughs off during menstruation. The deeper layer, the **stratum basalis** (*basal layer*), is permanent and gives rise to a new stratum functionalis after each menstruation.

Figure 25.18 Histology of the uterus.

(a) Transverse section through the uterus

(b) Details of endometrium

The three layers of the uterus from superficial to deep are the perimetrium, myometrium, and endometrium.

Figure 25.19   Blood supply of the uterus. The inset shows histological details of the blood vessels of the endometrium.

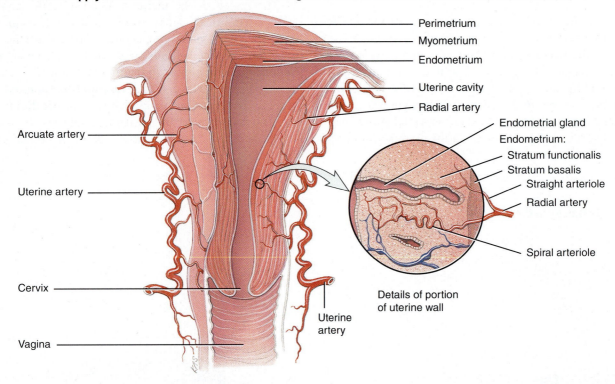

Anterior view with left side of uterus partially sectioned

Straight arterioles supply the materials needed for regeneration of the stratum functionalis.

Branches of the internal iliac artery called **uterine arteries** (Figure 25.19) supply blood to the uterus. Uterine arteries give off branches called **arcuate arteries** (= shaped like a bow) that are arranged in a circular fashion within the myometrium. Arcuate arteries branch into **radial arteries** that penetrate deeply into the myometrium. Just before the branches enter the endometrium, they divide into **straight arterioles** supplying the stratum basalis with the materials needed to regenerate the stratum functionalis and **spiral arterioles** to the stratum functionalis. Blood leaving the uterus is drained by the **uterine veins** into the internal iliac veins. The extensive blood supply of the uterus is essential to support regrowth of a new stratum functionalis after menstruation, implantation of a fertilized ovum, and development of the placenta.

Checkpoint

15. What are the functions of the ovaries? How are they held in position in the pelvic cavity?
16. Which structures in the ovary serve an endocrine function, and which hormones do they secrete?
17. How does the age of a primary oocyte in a female compare with the age of a primary spermatocyte in a male?
18. How are primordial, primary, secondary, and mature follicles structurally different? From which follicle does ovulation occur?
19. Where are the uterine tubes located? What are the functions of the cells that line the uterine tubes?
20. What are the principal parts of the uterus? Where are they located in relation to one another?
21. How is the uterus held in its normal position in the pelvic cavity?
22. Which structural features of the endometrium and myometrium contribute to their functions?

25.4 The vagina is a passageway for childbirth; the mammary glands secrete milk.

Vagina

The **vagina** (= sheath) is a tubular, muscular canal 10 cm (4 in.) lined with mucous membrane that extends from the exterior of the body to the uterine cervix (see Figures 25.11 and 25.16). It is the receptacle for the penis during sexual intercourse, the outlet for menstrual flow, and the passageway for childbirth. The vagina is situated between the urinary bladder and the rectum. A recess called the **fornix** (= arch or vault) surrounds the vaginal attachment to the cervix. When properly inserted, a contraceptive diaphragm rests on the fornix, where it is held in place as it covers the cervix.

The **mucosa** of the vagina is continuous with that of the uterus. Histologically, it consists of nonkeratinized stratified squamous epithelium that lies in a series of transverse folds called **rugae** (ROO-gē). The mucosa of the vagina contains large stores of glycogen, the decomposition of which produces organic acids. The resulting acidic environment retards microbial growth, but it also is harmful to sperm. Alkaline components of semen, mainly from the seminal vesicles, raise the pH of fluid in the vagina and increase viability of the sperm. The **muscularis** is composed of an outer circular layer and an inner longitudinal layer of smooth muscle that can stretch considerably to accommodate the penis during sexual intercourse and a child during birth. The **adventitia**, the superficial layer of the vagina, anchors the vagina to adjacent organs such as the urethra and urinary bladder anteriorly and the rectum and anal canal posteriorly.

A thin fold of vascularized mucous membrane, called the **hymen** (= membrane), forms a border around and partially closes the inferior end of the vaginal opening to the exterior, the **vaginal orifice** (Figure 25.20). After its rupture, usually during the first sexual intercourse, only remnants of the hymen remain.

Vulva

The term **vulva** (VUL-va = to wrap around), or **pudendum** (pū-DEN-dum), refers to the external genitals of the female (Figure 25.20). The following components make up the vulva:

- Anterior to the vaginal and urethral openings is the **mons pubis** (MONZ PŪ-bis; *mons* = mountain), an elevation of adipose tissue covered by skin and coarse pubic hair that cushions the pubic symphysis.

Figure 25.20   Components of the vulva.

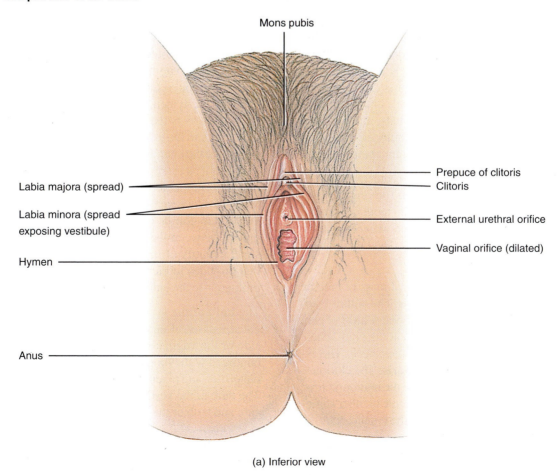

Mons pubis

Prepuce of clitoris

Clitoris

Labia majora (spread)

Labia minora (spread exposing vestibule)

External urethral orifice

Vaginal orifice (dilated)

Hymen

Anus

(a) Inferior view

(continues)

Figure 25.20 (continued)

(b) Transverse section through the vaginal wall

The vulva refers to the external genitals of the female.

- From the mons pubis, two longitudinal folds of skin, the **labia majora** (LĀ-bē-a ma-JŌ-ra; *labia* = lips; *majora* = larger), extend inferiorly and posteriorly to protectively enclose medial components of the vulva. The singular term is *labium majus*. The labia majora are covered by pubic hair and contain an abundance of adipose tissue, sebaceous (oil) glands, and apocrine sudoriferous (sweat) glands. They are homologous (develop from common embryonic tissue) to the scrotum in males.

- Medial to the labia majora are two smaller protective folds of skin, the **labia minora** (mī-NŌ-ra = smaller). The singular term is *labium minus*. Unlike the labia majora, the labia minora are devoid of pubic hair and fat and have few sudoriferous glands, but they do contain many sebaceous glands. The labia minora are homologous to the spongy (penile) urethra.

- The **clitoris** (KLI-to-ris) is a small cylindrical mass of erectile tissue and numerous nerves and blood vessels. The clitoris is located at the anterior junction of the labia minora. A layer of skin called the **prepuce of the clitoris** is formed at the point where the labia minora unite and cover the body of the clitoris. The exposed portion of the clitoris is the **glans clitoris**. The clitoris is homologous to the glans penis in males. Like the male structure, the clitoris is capable of enlargement upon tactile or other stimulation and has a role in sexual excitement in the female.

- The region between the labia minora is the **vestibule**. Within the vestibule are the hymen (if still present), the vaginal orifice, the external urethral orifice, and the openings of the ducts of several glands. The vestibule is homologous to the membranous urethra of males. The **vaginal orifice**, the opening of the vagina to the exterior, occupies the greater portion of the vestibule and is bordered by the hymen. Anterior to the vaginal orifice and posterior to the clitoris is the **external urethral orifice**, the opening of the urethra to the exterior. On either side of the external urethral orifice are the openings of the ducts of the **paraurethral glands**. These mucus-secreting glands are embedded in the wall of the urethra. The paraurethral glands are homologous to the prostate. On either side of the vaginal orifice itself are the **greater vestibular glands** (Figure 25.21), which open by ducts into a groove between the hymen and labia minora. They produce a small quantity of mucus during sexual arousal and intercourse that provides lubrication. The greater vestibular glands are homologous to the bulbourethral glands in males.

- The **bulb of the vestibule** (Figure 25.21) consists of two elongated masses of erectile tissue just deep to the labia on either side of the vaginal orifice. The bulb of the vestibule becomes engorged with blood during sexual arousal, narrowing the vaginal orifice and placing pressure on the penis during intercourse. The bulb of the vestibule is homologous to the corpus spongiosum and bulb of the penis in males.

Table 25.2 summarizes the homologous structures of the female and male reproductive systems.

Figure 25.21 **Perineum of a female.** (Figure 11.12 shows the perineum of a male.)

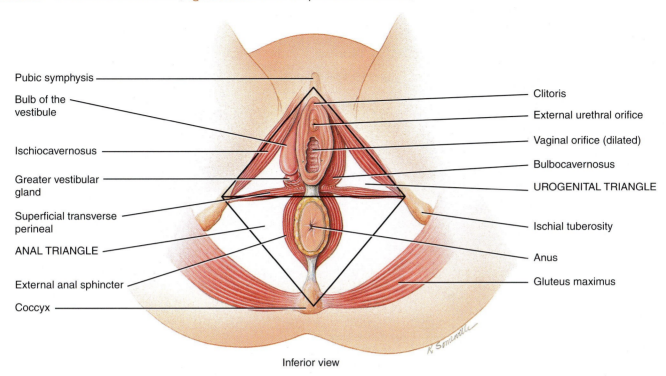

Pubic symphysis

Bulb of the vestibule

Ischiocavernosus

Greater vestibular gland

Superficial transverse perineal

ANAL TRIANGLE

External anal sphincter

Coccyx

Clitoris

External urethral orifice

Vaginal orifice (dilated)

Bulbocavernosus

UROGENITAL TRIANGLE

Ischial tuberosity

Anus

Gluteus maximus

Inferior view

🔑 **The perineum is a diamond-shaped area that includes the urogenital triangle and the anal triangle.**

Perineum

The **perineum** (per´-i-NĒ-um) is the diamond-shaped area medial to the thighs and buttocks of both males and females (Figure 25.21). It contains the external genitals and anus. The perineum is bounded anteriorly by the pubic symphysis, laterally by the ischial tuberosities, and posteriorly by the coccyx. A transverse line drawn between the ischial tuberosities divides the perineum into an anterior **urogenital triangle** (ū´-rō-JEN-i-tal) that contains the external genitalia and a posterior **anal triangle** that contains the anus.

TABLE 25.2	
Homologous Structures of the Female and Male Reproductive Systems	
FEMALE STRUCTURE	MALE STRUCTURE
Ovaries	Testes
Ovum	Sperm cell
Labia majora	Scrotum
Labia minora	Spongy urethra
Vestibule	Membranous urethra
Bulb of vestibule	Corpus spongiosum penis and bulb of penis
Clitoris	Glans penis
Paraurethral glands	Prostate
Greater vestibular glands	Bulbourethral glands

Mammary Glands

Each **breast** is a hemispheric projection of variable size anterior to the pectoralis major and serratus anterior and attached to them by a layer of fascia (Figure 25.22).

Each breast has one pigmented projection, the **nipple**, that has a series of closely spaced openings of ducts called **lactiferous ducts**, where milk emerges. The circular pigmented area of skin surrounding the nipple is called the **areola** (a-RĒ-ō-la = small space); it appears rough because it contains modified sebaceous (oil) glands. Strands of connective tissue called the **suspensory ligaments of the breast** run between the skin and fascia and support the breast. These ligaments become looser with age or with the excessive strain that can occur in long-term jogging or high-impact aerobics. Wearing a supportive bra can slow this process and help maintain the strength of the suspensory ligaments.

Within each breast is a **mammary gland**, a modified sudoriferous (sweat) gland that produces milk. A mammary gland consists of 15 to 20 lobes, or compartments, separated by a variable amount of adipose tissue. In each lobe are several smaller compartments called **lobules**, composed of grapelike clusters of milk-secreting glands termed **alveoli** (= small cavities) embedded in connective tissue. Contraction of **myoepithelial cells**, smooth muscle-like cells surrounding the alveoli, helps propel milk toward the nipples. When milk is being produced, it passes from the alveoli into a series of **secondary tubules** and then into the **mammary ducts**. Near the nipple, the mammary ducts expand to form sinuses called **lactiferous**

Figure 25.22   Mammary glands.

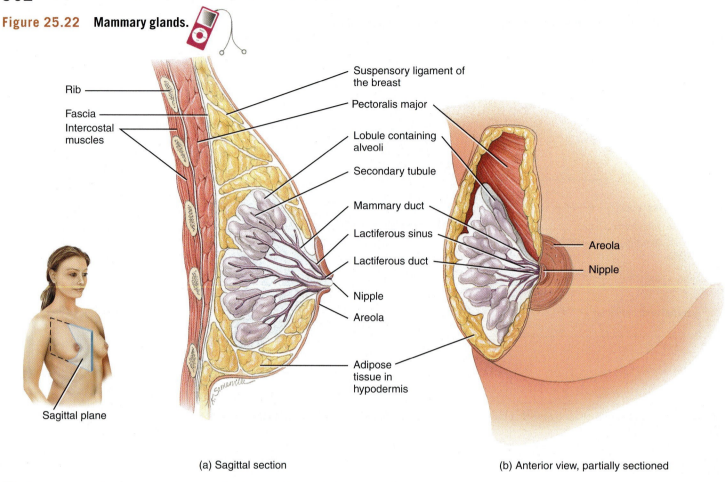

(a) Sagittal section

(b) Anterior view, partially sectioned

🔑 **The mammary glands function in the synthesis, secretion, and ejection of milk (lactation).**

sinuses, where some milk may be stored before draining into a lactiferous duct. Each lactiferous duct typically carries milk from one of the lobes to the exterior.

The functions of the mammary glands are the synthesis, secretion, and ejection of milk; these functions, called **lactation,** are associated with pregnancy and childbirth. Milk production is stimulated largely by the hormone prolactin from the anterior pituitary, with contributions from progesterone and estrogens. The ejection of milk is stimulated by oxytocin, which is released from the posterior pituitary in response to the sucking of an infant on the mother's nipple (suckling).

Checkpoint

23. What are the functions of the vagina?
24. What is the vulva? What are the structures and functions of each part of the vulva?
25. Which structures in males are homologous to the ovaries, the clitoris, the greater vestibular glands, and the bulb of the vestibule?
26. Which surface structures are anterior to the vaginal opening? Lateral to it?
27. How are breasts supported?
28. Through which structures does milk pass from its production site to its exit from the breast?
29. Which hormones regulate the synthesis and the ejection of milk?

25.5   The female reproductive cycle includes the ovarian and uterine cycles.

During their reproductive years, nonpregnant females normally exhibit cyclical changes in the ovaries and uterus. Each cycle takes about a month and involves both oogenesis and preparation of the uterus to receive a fertilized ovum. Hormones secreted by the hypothalamus, anterior pituitary, and ovaries control the main events. The **ovarian cycle** is a series of events in the ovaries that occur during and after the maturation of an oocyte. Steroid hormones released by the ovaries control the **uterine (menstrual) cycle,** a concurrent series of changes in the endometrium of the uterus to prepare it for the arrival of a fertilized ovum that will develop there until birth. If fertilization does not occur, ovarian hormones wane, which causes the stratum functionalis of the endometrium to slough off. The general term **female reproductive cycle** encompasses the ovarian and uterine cycles, the hormonal changes that regulate them, and the related cyclical changes in the breasts and cervix.

Hormonal Regulation of the Female Reproductive Cycle

Gonadotropin-releasing hormone (GnRH) secreted by the hypothalamus controls the ovarian and uterine cycles (Figure 25.23). GnRH stimulates the release of **follicle-stimulating hormone (FSH)** and **luteinizing hormone (LH)** from the anterior pituitary. FSH initiates follicular growth, while LH stimulates further development of ovarian follicles. In addition, both FSH and LH stimulate the ovarian follicles to secrete estrogens. LH stimulates the theca folliculi cells of a developing follicle to produce androgens.

Under the influence of FSH, the androgens are taken up by the granulosa cells of the follicle and then converted into estrogens. At midcycle, LH triggers ovulation and then promotes formation of the corpus luteum, the reason for the name luteinizing hormone. Stimulated by LH, the corpus luteum produces and secretes estrogens, progesterone, relaxin, and inhibin.

At least six different estrogens have been isolated from the plasma of human females, but only three are present in significant quantities: *estradiol*, *estrone*, and *estriol*. In a nonpregnant woman, the most abundant estrogen is estradiol, which is synthesized from cholesterol in the ovaries.

Estrogens secreted by follicular cells have several important functions (Figure 25.23).

- Estrogens promote the development and maintenance of female reproductive structures, **secondary sex characteristics**, and the breasts. The secondary sex characteristics include distribution of adipose tissue in the breasts, abdomen, mons pubis, and hips; voice pitch; a broad pelvis; and pattern of hair growth on the head and body.

Figure 25.23 **Secretion and physiological effects of estrogens, progesterone, relaxin, and inhibin in the female reproductive cycle.** Dashed red lines indicate negative feedback inhibition.

The ovarian and uterine cycles are controlled by gonadotropin-releasing hormone and ovarian hormones (estrogens and progesterone).

- Estrogens work with human growth hormone to increase protein anabolism, including the building of strong bones.

- Estrogens lower blood cholesterol level, which is probably the reason that women under age 50 have a much lower risk of coronary artery disease than do men of comparable age.

Progesterone, secreted mainly by cells of the corpus luteum, cooperates with estrogens to prepare and maintain the endometrium for implantation of a fertilized ovum and to prepare the mammary glands for milk secretion.

The small quantity of **relaxin** produced by the corpus luteum during each monthly cycle relaxes the uterus by inhibiting contractions of the myometrium. Presumably, implantation of a fertilized ovum occurs more readily in a "quiet" uterus. During pregnancy, the placenta produces much more relaxin, and it continues to relax uterine smooth muscle. At the end of pregnancy, relaxin also increases the flexibility of the pubic symphysis and may help dilate the uterine cervix, both of which ease delivery of the baby.

Three hormones provide feedback for the female reproductive cycle. Moderate blood levels of estrogens inhibit both the release of GnRH by the hypothalamus and secretion of LH and FSH by the anterior pituitary. High levels of progesterone inhibit secretion of GnRH and LH. In addition, **inhibin**, secreted by granulosa cells of growing follicles and by the corpus luteum, inhibits secretion of FSH and, to a lesser extent, LH.

Phases of the Female Reproductive Cycle

The duration of the female reproductive cycle typically ranges from 24 to 35 days. For this discussion, we assume a duration of 28 days and divide it into four phases: the menstrual phase, the preovulatory phase, ovulation, and the postovulatory phase (Figure 25.24). Because they occur at the same time, the events of the ovarian cycle and uterine cycle are discussed together.

Menstrual Phase

The **menstrual phase** (MEN-stroo-al), also called **menstruation** (men´-stroo-Ā-shun), lasts for roughly the first 5 days of the cycle. (By convention, the first day of menstruation is day 1 of a new cycle.)

Events in the Ovaries

Under the influence of FSH, several primordial follicles develop into primary follicles and then into secondary follicles. This developmental process may take several months to occur. Therefore, a follicle that begins to develop at the beginning of a particular menstrual cycle may not reach maturity and ovulate until several menstrual cycles later.

Events in the Uterus

Menstrual flow from the uterus consists of 50–150 mL of blood, tissue fluid, mucus, and epithelial cells shed from the endometrium. This discharge occurs because the declining levels of progesterone and estrogens stimulate release of prostaglandins that cause the uterine spiral arterioles to constrict. As a result, the cells they supply become oxygen deprived and start to die. Eventually, the entire stratum functionalis sloughs off. At this time the endometrium is very thin, about 2–5 mm, because only the stratum basalis remains. The menstrual flow passes from the uterine cavity through the cervix and vagina to the exterior.

Preovulatory Phase

The **preovulatory phase** is the time between the end of menstruation and ovulation. The preovulatory phase of the cycle is more variable in length than the other phases and accounts for most of the differences in length of the cycle. It lasts from days 6 to 13 in a 28-day cycle.

Events in the Ovaries

Some of the secondary follicles in the ovaries begin to secrete estrogens and inhibin. By about day 6, a single secondary follicle in one of the two ovaries has outgrown all the others to become the **dominant follicle**. Estrogens and inhibin secreted by the dominant follicle decrease the secretion of FSH, which causes other, less well-developed follicles to stop growing and undergo atresia. Fraternal (nonidentical) twins or triplets result when two or three secondary follicles become codominant and later are ovulated and fertilized at about the same time.

Normally, the one dominant follicle becomes the mature (graafian) follicle, which continues to enlarge until it is more than 20 mm in diameter and ready for ovulation (see Figure 25.13). This follicle forms a blisterlike bulge on the surface of the ovary due to the swelling antrum. During the final maturation process, the mature follicle continues to increase its production of estrogens (Figure 25.24).

With reference to the ovarian cycle, the menstrual and preovulatory phases together are termed the **follicular phase** (fō-LIK-ū-lar) because ovarian follicles are growing and developing.

Events in the Uterus

Estrogens liberated into the blood by growing ovarian follicles stimulate the repair of the endometrium; cells of the stratum basalis undergo mitosis and produce a new stratum functionalis. As the endometrium thickens, endometrial glands develop, and arterioles coil and lengthen as they penetrate the stratum functionalis. The thickness of the endometrium approximately doubles, to about 4–10 mm. With reference to the uterine cycle, the preovulatory phase is also termed the **proliferative phase** because the endometrium is proliferating.

Figure 25.24 **The female reproductive cycle.** The length of the female reproductive cycle typically is 24 to 36 days. (a) Events in the ovarian and uterine cycles and the release of anterior pituitary hormones are correlated with the cycle's phases. In the cycle shown, fertilization and implantation have not occurred. (b) Relative concentrations of anterior pituitary hormones (follicle-stimulating hormone and luteinizing hormone) and ovarian hormones (estrogens and progesterone) during the phases of a cycle.

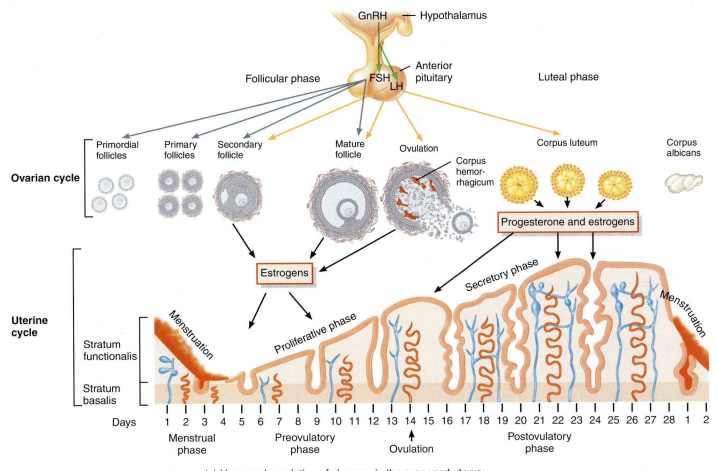

(a) Hormonal regulation of changes in the ovary and uterus

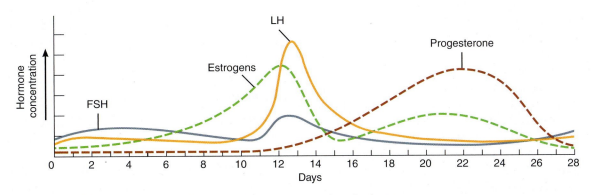

(b) Changes in concentration of anterior pituitary and ovarian hormones

Estrogens are secreted by follicles before ovulation; after ovulation, both progesterone and estrogens are secreted by the corpus luteum.

Ovulation

Ovulation, the rupture of the mature follicle and the release of the secondary oocyte into the pelvic cavity, usually occurs on day 14 in a 28-day cycle. During ovulation, the secondary oocyte remains surrounded by its zona pellucida and corona radiata.

The *high levels of estrogens* during the last part of the preovulatory phase exert a *positive* feedback effect on the cells that secrete LH and GnRH and cause ovulation, as follows (Figure 25.25):

① A high concentration of estrogens stimulates the hypothalamus to release more GnRH and the anterior pituitary to secrete LH.

② GnRH promotes the release of FSH and additional LH by the anterior pituitary.

③ The resulting surge of LH causes rupture of the mature follicle and expulsion of a secondary oocyte about 9 hours after the peak of the LH surge. The ovulated oocyte and its corona radiata cells are usually swept into the uterine tube. An over-the-counter home test that detects a rising level of LH can be used to predict ovulation a day in advance.

Postovulatory Phase

The **postovulatory phase** of the female reproductive cycle is the time between ovulation and onset of the next menses. In duration, it is the most constant part of the female reproductive cycle. It lasts for 14 days in a 28-day cycle, from day 15 to day 28 (Figure 25.24).

Events in One Ovary

After ovulation, the mature follicle collapses. Once a blood clot forms from minor bleeding of the ruptured follicle, the follicle becomes the **corpus hemorrhagicum** (*hemo-* = blood; *-rrhagic-* = bursting forth) (see Figure 25.13). Theca folliculi cells mix with the granulosa cells as they all become transformed into corpus luteum cells under the influence of LH. Stimulated by LH, the corpus luteum secretes progesterone, estrogen, relaxin, and inhibin. The luteal cells also absorb the blood clot. With reference to the ovarian cycle, this phase is also called the **luteal phase**.

Later events in an ovary that has ovulated an oocyte depend on whether the oocyte is fertilized. If the oocyte *is not fertilized*, the corpus luteum has a life span of only 2 weeks. Then, it degenerates into a corpus albicans, which does not secrete ovarian hormones (see Figure 25.13). As the levels of progesterone, estrogens, and inhibin quickly decrease, release of GnRH, FSH, and LH rise due to loss of negative feedback suppression by the ovarian hormones. Follicular growth resumes and a new ovarian cycle begins.

Figure 25.25   High levels of estrogens exert a positive feedback effect (green arrows) on the hypothalamus and anterior pituitary, thereby increasing secretion of gonadotropin-releasing hormone and luteinizing hormone.

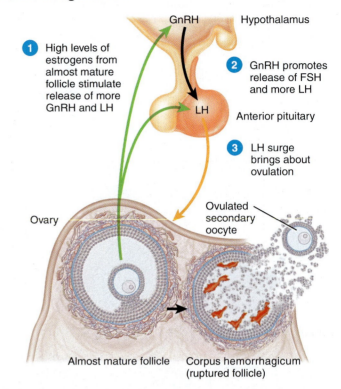

① High levels of estrogens from almost mature follicle stimulate release of more GnRH and LH

② GnRH promotes release of FSH and more LH

③ LH surge brings about ovulation

GnRH

Hypothalamus

LH

Anterior pituitary

Ovary

Ovulated secondary oocyte

Almost mature follicle   Corpus hemorrhagicum (ruptured follicle)

🔑 **At midcycle, a surge of luteinizing hormone triggers ovulation.**

If the secondary oocyte *is fertilized* and begins to divide, the corpus luteum persists past its normal 2-week life span. It is "rescued" from degeneration by **human chorionic gonadotropin (hCG)** (kō-rē-ON-ik). This hormone is produced by the chorion of the embryo beginning about 8 days after fertilization. Like LH, hCG stimulates the secretory activity of the corpus luteum. The presence of hCG in maternal blood or urine is an indicator of pregnancy and is the hormone detected by home pregnancy tests.

Events in the Uterus

Progesterone and estrogens produced by the corpus luteum promote growth of the endometrial glands, and vascularization and thickening of the endometrium to 12–18 mm (0.48–0.72 in.). Because of the secretory activity of the endometrial glands, which begin to secrete glycogen, this period is called the **secretory phase** of the uterine cycle. These preparatory changes peak about 1 week after ovulation, at the time a fertilized ovum might arrive in the uterus. If fertilization does not occur, the level of progesterone and estrogens declines due to degeneration of the corpus luteum. Withdrawal of progesterone and estrogens causes menstruation.

Figure 25.26 Hormonal interactions in the ovarian and uterine cycles.

Hormones from the anterior pituitary regulate ovarian function, and hormones from the ovaries regulate changes in the endometrium.

Figure 25.26 summarizes the hormonal interactions and cyclical changes in the ovaries and uterus during the ovarian and uterine cycles.

Checkpoint

30. Of the several estrogens, which one exerts the major effect?
31. What functions do estrogen and progesterone have outside the uterus? What are the functions of relaxin?
32. Describe the function of each of the following hormones in the uterine and ovarian cycles: GnRH, FSH, LH, estrogens, progesterone, and inhibin.
33. Why does only one follicle usually become a mature follicle?
34. Which hormones are responsible for the proliferative phase of endometrial growth, for ovulation, for growth of the corpus luteum, and for the surge of LH at midcycle?
35. Is stimulation of secretion of GnRH by declining levels of estrogens and progesterone a positive or a negative feedback effect? Explain your answer.
36. What causes follicles to start growing again in a new ovarian cycle?

"You're both 35, with no major health problems, and you've been trying for about 12 months. Is that correct?" asked Dr. Jones.

"Yes, Doctor. Is it harder to conceive when you're older?" Susan looked worried as she spoke.

"There is a reduced likelihood of successful fertilization with increasing age, but many women in their thirties have successful pregnancies. How long has it been since you stopped taking birth control pills?"

"It's been over a year now. Shouldn't I be ovulating?" she replied.

"Most likely, but we'll need to do a complete exam and some tests," said Dr. Jones. "I'd also like to schedule an exam for you, Lou. It will include a health history and physical exam, and we'll need a semen sample."

Lou blushed. "Do you think there's something wrong with me?"

The doctor replied, "No, not necessarily. I know you have both been doing all the right things, boxer shorts for you, Lou, and good nutrition for you both. You've done what you can to increase your odds of a successful pregnancy. But there are a few important issues we need to address, and sperm count is one. Men can have reduced sperm counts for various reasons, from blocked vas deferens to actual defects in the sperm themselves or even autoimmune disorders. The receptionist can schedule those appointments for you on your way out."

Susan arrived for her next appointment. "Have you had any problems with your menstrual cycle, such as irregular periods, endometriosis, polycystic ovary syndrome, or excessive bleeding?" asked Dr. Jones.

Susan thought for a moment, then replied, "I guess I didn't have any problems before I started taking the pill. Once in a while I have some bleeding, but my Pap smears are always normal. I've been having regular periods for about eight or nine months now. It took a while after I stopped the pill for me to have a normal period."

Dr. Jones made some notes on Susan's chart. "Well, we need to determine if you're actually ovulating and if the eggs being produced are viable. We'll start with a pelvic exam and some standard tests."

It was several weeks before Susan and Lou were able to meet with the doctor to get the results. "It appears that Susan is producing viable eggs, but Lou, your sperm count is low. Based on what we've determined from the exams and test results, you have several options. In your case, however, I am going to suggest intrauterine insemination. The sperm count numbers are low enough that it wouldn't seem likely a pregnancy would occur naturally. The sperm are healthy, and viable; there are just not enough of them. This procedure is typically the first step in fertility treatment, and it's a fairly straightforward procedure. Susan, you'll take fertility drugs for up to 2 weeks. We provide you with an ovulation detection kit, similar to a home pregnancy kit, which you can use to determine when you are ovulating. Once you ovulate, we'll need a semen sample from Lou. The sperm are washed in the lab and then will be placed directly in Susan's uterus in a simple outpatient procedure. The likelihood of sperm encountering the egg and a successful fertilization taking place is increased by the fertility treatments and the treatment of the sperm prior to insemination."

A. Why would a blockage in the vas deferens cause male infertility? What other tubular structures could be blocked, reducing male fertility?

B. Assuming Susan is ovulating, what must happen for her eggs to become fertilized?

C. Which hormone that triggers ovulation is measured by the ovulation detection kit? Why is knowing when Susan ovulates important?

D. Why are FSH and progesterone present in fertility drugs?

E. How does placing Lou's sperm directly in Susan's uterus increase the likelihood of pregnancy?

25.6 The zygote divides into a morula and then a blastocyst that implants in the endometrium of the uterus.

The first week of development is characterized by several significant events, including fertilization, cleavage of the zygote, blastocyst formation, and implantation.

Fertilization

During **fertilization** (fer´-ti-li-ZĀ-shun; *fertil-* = fruitful), the genetic material from a haploid sperm cell and a haploid secondary oocyte merges into a single diploid nucleus. Of the about 200 million sperm introduced into the vagina, fewer than 2 million (1%) reach the cervix of the uterus and only about 200 reach the secondary oocyte. Fertilization normally occurs in the uterine tube within 12 to 24 hours after ovulation. Sperm can remain viable for about 48 hours after deposition in the vagina, although a secondary oocyte is viable for only about 24 hours after ovulation. Thus, pregnancy is *most likely* to occur if intercourse takes place during a 3-day "window"—from 2 days before ovulation to 1 day after ovulation.

Sperm swim from the vagina into the cervical canal by the whiplike movements of their tails. The passage of sperm through the rest of the uterus and then into the uterine tube results mainly from contractions of the walls of those organs. Prostaglandins in semen are believed to stimulate uterine motility at the time of intercourse and to aid in the movement of sperm through the uterus and into the uterine tube. Sperm that reach the vicinity of the oocyte within minutes after ejaculation *are not capable* of fertilizing it until about seven hours later. During this time in the female reproductive tract, mostly in the uterine tube, sperm undergo **capacitation** (ka-pas´-i-TĀ-shun; *capacit-* = capable of), a series of functional changes that cause the sperm cell's tail to beat even more vigorously and prepare its plasma membrane to fuse with the oocyte's plasma membrane. During capacitation, sperm are acted on by secretions in the female reproductive tract that result in the removal of cholesterol, glycoproteins, and proteins from the plasma membrane around the head of the sperm cell.

For fertilization to occur, a sperm cell first must penetrate the outer, cellular corona radiata, and the deeper, glycoprotein zona pellucida (Figure 25.27a). Enzymes on the surface of sperm cell heads and strong tail movements help sperm penetrate the corona radiata and come in contact with the zona pellucida. One of the glycoproteins in the zona pellucida acts as a sperm receptor that binds to the sperm cell head, triggering the acrosomal reaction. The **acrosomal reaction** breaks down the plasma membrane of the acrosome covering the sperm cell head (see Figure 25.6), releasing several acrosomal enzymes that digest a path through the zona pellucida as the lashing tail pushes the sperm cell onward. Although many sperm bind to the zona pellucida and release their acrosomal enzymes, only the first sperm cell to penetrate the entire zona pellucida and reach the oocyte's plasma membrane fuses with the oocyte.

The fusion of a sperm cell with a secondary oocyte sets in motion events that block **polyspermy**, fertilization by more than one sperm cell. Within a few seconds, the cell membrane of the oocyte depolarizes, which acts as a *fast block to polyspermy*—a depolarized oocyte cannot fuse with another sperm cell. Depolarization also triggers the intracellular release of calcium ions, which stimulate exocytosis of secretory vesicles from the oocyte. The molecules released by exocytosis inactivate sperm receptors and harden the entire zona pellucida, events called the *slow block to polyspermy*.

Once a sperm cell enters a secondary oocyte, the oocyte first must complete meiosis II. It divides into a larger ovum (mature egg) and a smaller second polar body that fragments and disintegrates (see Figure 25.15). The nucleus in the head of the sperm cell develops into the **male pronucleus**, and the nucleus of the fertilized ovum develops into the **female pronucleus** (Figure 25.27c). The male and female pronuclei fuse, producing a single diploid nucleus that contains 23 chromosomes from each pronucleus. Thus, the fusion of the haploid (*n*) pronuclei restores the diploid number (*2n*) of 46 chromosomes. The fertilized ovum now is called a zygote.

Dizygotic (fraternal) twins are produced from the independent release of two secondary oocytes and the subsequent fertilization of each by different sperm. They are the same age and are in the uterus at the same time, but genetically they are as dissimi-

Figure 25.27 Selected structures and events in fertilization.

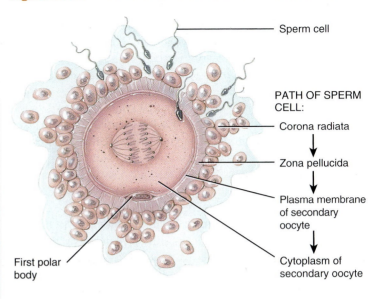

Sperm cell

PATH OF SPERM CELL:

Corona radiata

Zona pellucida

Plasma membrane of secondary oocyte

Cytoplasm of secondary oocyte

First polar body

(a) Sperm cell penetrating a secondary oocyte

Head of sperm cell   Secondary oocyte

Pronuclei

SEM 1100x

LM 250x

(b) Sperm cell in contact with a secondary oocyte

(c) Male and female pronuclei

🔑 During fertilization, genetic material from a sperm cell and a secondary oocyte merge to form a single diploid nucleus.

lar as any other siblings. Dizygotic twins may or may not be the same sex. Because **monozygotic (identical) twins** develop from a single fertilized ovum, they contain exactly the same genetic material and are always the same sex. Monozygotic twins arise from separation of the developing cells into two embryos, which in 99% of the cases occurs before 8 days have passed. Separations that occur later than 8 days are likely to produce **conjoined twins**, a situation in which the twins are joined together and share some body structures.

Early Embryonic Development

From fertilization through the eighth week of development, a stage called the **embryonic period**, the developing human is called an **embryo** (*em-* = into; *-bryo* = grow).

After fertilization, a series of rapid mitotic cell divisions of the zygote called **cleavage** (KLĒV-ij) takes place (Figure 25.28). The first division of the zygote begins about 24 hours after fertilization and is completed about 6 hours later. Each succeeding division takes slightly less time and produces progressively smaller cells called **blastomeres** (BLAS-tō-merz; *blasto-* = germ; *-meres* = parts). By the second day after fertilization, the second cleavage is completed and there are four blastomeres (Figure 25.28b). By the end of the third day, there are 16 blastomeres. Successive cleavages eventually produce a solid sphere of cells called the **morula** (MOR-ū-la; *morula* = mulberry). The morula is still surrounded by the zona pellucida and is about the same size as the original zygote (Figure 25.28c).

By the end of the fourth day, the number of cells in the morula increases as it continues to move through the uterine tube toward the uterine cavity. When the morula enters the uterine cavity on day 4 or 5, a glycogen-rich secretion from the glands of the endometrium of the uterus passes into the uterine cavity and enters the morula through the zona pellucida. This endometrial fluid, along with nutrients stored in the cytoplasm of the cells of the morula, provides nourishment for the developing morula. At the 32-cell stage, the fluid enters the morula and reorganizes the cells around a large fluid-filled cavity called the **blastocyst cavity** (BLAS-tō-sist; *blasto-* = germ or sprout; *-cyst* = bag) (Figure 25.28e). Once the blastocyst cavity is formed, the developing mass is called the **blastocyst**. Though it now has hundreds of cells, the blastocyst is still about the same size as the original zygote.

During the formation of the blastocyst, two distinct cell populations arise: the embryoblast and trophoblast (Figure 25.28e). The **embryoblast** is located internally and eventually develops into the embryo. The **trophoblast** (TRŌF-ō-blast; *tropho-* = develop or nourish) is the outer superficial layer of cells that forms the sphere-like wall of the blastocyst. It will ultimately develop into the chorion, a portion of the placenta, the site of exchange of nutrients and wastes between the mother and fetus. On about the fifth day after fertilization, the blastocyst "hatches" from the zona pellucida by digesting a hole in it with an enzyme, and then squeezing through the hole. This shedding of the zona pellucida is necessary to permit the next step, implantation.

Implantation

The blastocyst remains free within the uterine cavity for about 2 days before it attaches to the uterine wall. At this time the endometrium is in its secretory phase. About 6 days after fertilization,

Figure 25.28   **Cleavage and the formation of the morula and blastocyst.**

(a) Cleavage of zygote, two-cell stage (day 1)
— Polar bodies
— Blastomeres
— Zona pellucida

(b) Cleavage, four-cell stage (day 2)
— Nucleus
— Cytoplasm

(c) Morula (day 4)

(d) Blastocyst, external view (day 5)

(e) Blastocyst, internal view (day 5)
— Embryoblast
— Blastocyst cavity
— Trophoblast

🔑 **Cleavage refers to the early, rapid mitotic divisions of a zygote.**

the blastocyst loosely attaches to the endometrium in a process called **implantation** (Figure 25.29). As the blastocyst implants, usually in either the posterior portion of the fundus or the body of the uterus, it orients with the embryoblast toward the endometrium (Figure 25.29b). About 7 days after fertilization, the blastocyst attaches to the endometrium more firmly, endometrial glands in the vicinity enlarge, and the endometrium becomes more vascularized (forms new blood vessels). The blastocyst eventually secretes enzymes, burrows into the endometrium, and becomes surrounded by it.

Figure 25.29 Relation of a blastocyst to the endometrium of the uterus at the time of implantation.

Blastocyst

Opening of endometrial gland

Endometrium of uterus

Uterine cavity

(a) External view of blastocyst, about 6 days after fertilization

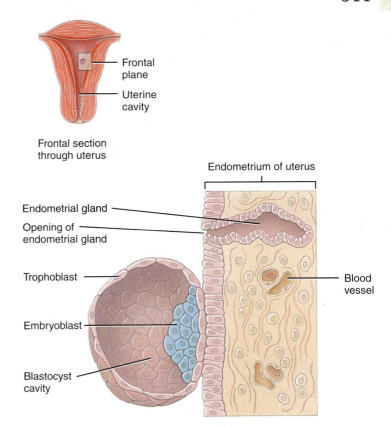

Frontal plane

Uterine cavity

Frontal section through uterus

Endometrium of uterus

Endometrial gland

Opening of endometrial gland

Trophoblast

Embryoblast

Blastocyst cavity

Blood vessel

(b) Frontal section through endometrium of uterus and blastocyst, about 6 days after fertilization

🔑 **Implantation, the attachment of a blastocyst to the endometrium, occurs about 6 days after fertilization.**

Following implantation, the endometrium is known as the **decidua** (dē-SID-ū-a = falling off). The decidua separates from the endometrium after the fetus is delivered much as it does in normal menstruation. Different regions of the decidua are named based on their positions relative to the site of the implanted blastocyst (Figure 25.30). The **decidua basalis** is the portion of the endometrium between the embryo and the stratum basalis; it provides large amounts of glycogen and lipids for the developing embryo and fetus and later becomes the maternal part of the placenta. The **decidua capsularis** is the portion of the endometrium located between the embryo and the uterine cavity. The **decidua parietalis** (pa-rī-e-TAL-is) is the remaining modified endometrium that lines the noninvolved areas of the rest of the uterus. As the embryo and later the fetus enlarges, the decidua capsularis bulges into the uterine cavity and fuses with the decidua parietalis, thereby obliterating the uterine cavity. By about 27 weeks, the decidua capsularis degenerates and disappears.

The major events associated with the first week of development are summarized in Figure 25.31.

Figure 25.30 Regions of the decidua.

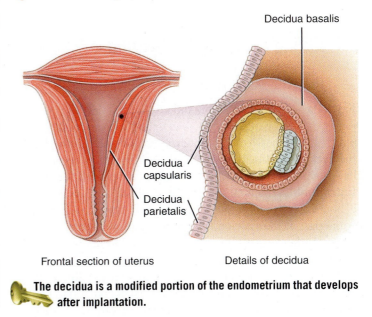

Decidua basalis

Decidua capsularis

Decidua parietalis

Frontal section of uterus

Details of decidua

🔑 **The decidua is a modified portion of the endometrium that develops after implantation.**

Figure 25.31   Summary of events associated with the first week of development.

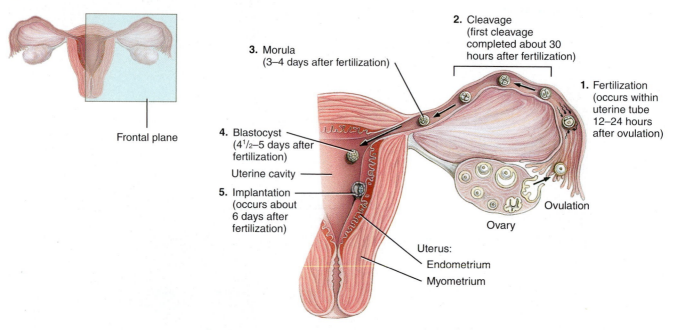

Frontal section through uterus, uterine tube, and ovary

 Fertilization usually occurs in the uterine tube.

Checkpoint

37. Which mechanisms move sperm from the vagina to the uterine tube?
38. How does a sperm cell penetrate the covering of the secondary oocyte?
39. How is polyspermy prevented?
40. What is a morula and how is it formed?
41. Describe the layers of a blastocyst and their eventual fates.

42. In which phase of the uterine cycle does implantation occur? Where does it usually occur? How is the blastocyst oriented at implantation?
43. What are the components of the decidua? Which part of the decidua helps form the maternal part of the placenta?

25.7 During the embryonic period, the embryonic membranes and most major organs develop.

Development of the Trophoblast

About 8 days after fertilization, the trophoblast develops into two layers in the region of contact between the blastocyst and endometrium. The layer between the embryoblast and syncytiotrophoblast is the **cytotrophoblast** (sī-tō-TRŌF-ō-blast) and is composed of cells with distinct boundaries. The cells of the outer layer, the **syncytiotrophoblast** (sin-sīt′-ē-ō-TRŌF-ō-blast), contain no distinct cell boundaries (Figure 25.32a). The two layers of trophoblast become part of the chorion (one of the fetal membranes) as they undergo further growth (see Figure 25.34). During implantation, the syncytiotrophoblast secretes enzymes that enable the blastocyst to penetrate the uterine lining by digesting and liquefying the endometrial cells. Eventually, the blastocyst becomes buried in the endometrium and inner one-third of the myometrium. Another secretion of the trophoblast is human chorionic gonadotropin (hCG), which maintains the corpus luteum and sustains its secretion of progesterone and estrogens. These hormones maintain the uterine

lining in a secretory state, preventing menstruation. Peak secretion of hCG occurs about the ninth week of pregnancy, at which time the placenta is fully developed and produces the progesterone and estrogens that continue to sustain the pregnancy.

Development of the Bilaminar Embryonic Disc

Like those of the trophoblast, cells of the embryoblast also differentiate into two layers around 8 days after fertilization: a **hypoblast (primitive endoderm)** and **epiblast (primitive ectoderm)** (Figure 25.32a). Cells of the hypoblast and epiblast together form a flat disc referred to as the **bilaminar embryonic disc** (bī-LAM-in-ar = two-layered). Soon, a small cavity appears within the epiblast and eventually enlarges to form the **amniotic cavity** (am′-nē-OT-ik; amnio- = lamb).

Figure 25.32 **Principal events of the second week of development.**

(a) Frontal section through endometrium of uterus showing blastocyst, about 8 days after fertilization

(b) Frontal section through endometrium of uterus showing blastocyst, about 9 days after fertilization

(c) Frontal section through endometrium of uterus showing blastocyst, about 12 days after fertilization

About 8 days after fertilization, the trophoblast develops into a syncytiotrophoblast and a cytotrophoblast; the embryoblast develops into a bilaminar embryonic disc.

Development of the Amnion

As the amniotic cavity enlarges, a single layer of squamous cells forms a domelike roof above the epiblast cells called the **amnion** (AM-nē-on) (Figure 25.32a). Initially, the amnion overlies only the bilaminar embryonic disc. However, as the embryo develops, the amnion eventually surrounds the entire embryo (see Figure 25.34), creating the amniotic cavity that becomes filled with **amniotic fluid**. Most amniotic fluid is initially derived from maternal blood. Later, the fetus contributes to the fluid by excreting urine into the amniotic cavity. Amniotic fluid serves as a fetal shock absorber and helps regulate fetal body temperature, prevent the fetus from drying out, and prevent adhesions between the fetus and surrounding tissues. The amnion usually ruptures just before birth; it and its fluid constitute the "bag of waters."

Development of the Yolk Sac

Also on the eighth day after fertilization, cells at the edge of the hypoblast migrate and cover the inner surface of the blastocyst wall (Figure 25.32a). The migrating cells form a thin membrane that, together with the hypoblast, becomes the wall of the **yolk sac**, the former blastocyst cavity during early development (Figure 25.32b). As a result, the bilaminar embryonic disc is now positioned between the amniotic cavity and yolk sac.

Since human embryos receive their nutrients from the endometrium, the yolk sac is relatively empty, small, and decreases in size as development progresses (see Figure 25.34). Nevertheless, the yolk sac has several important functions in humans. It supplies nutrients to the embryo during the second and third weeks of development, is the source of blood cells from the third through sixth weeks, contains the first (primordial) germ cells that will migrate into the developing gonads and eventually form gametes, and forms part of the gut (gastrointestinal tract). Finally, the yolk sac functions as a shock absorber and helps prevent drying out of the embryo.

Development of Maternal Sinusoids

On the ninth day after fertilization, the blastocyst becomes completely embedded in the endometrium. As the syncytiotrophoblast expands, small spaces called **lacunae** (la-KOO-nē = little lakes) develop within it (Figure 25.32b). By the twelfth day of development, the lacunae fuse to form larger, interconnecting spaces called **lacunar networks** (Figure 25.32c). Endometrial capillaries around the developing embryo become dilated and are referred to as **maternal sinusoids**. As the syncytiotrophoblast erodes some of the maternal sinusoids and endometrial glands, maternal blood and secretions from the endometrial glands enter the lacunar networks and flow through them, serving as both a rich source of materials for embryonic nutrition and a disposal site for the embryo's wastes.

Development of the Chorion

About the twelfth day after fertilization, mesodermal cells derived from the yolk sac form a connective tissue layer (mesenchyme) around the amnion and yolk sac called the **extraembryonic mesoderm** (Figure 25.32c). The extraembryonic mesoderm, together with the two layers of the trophoblast (the cytotrophoblast and syncytiotrophoblast), form the **chorion** (KOR-ē-on = membrane) (Figure 25.32c). The chorion surrounds the embryo and, later, the fetus (see Figure 25.34). Eventually it becomes the principal embryonic part of the placenta, the structure for exchange of materials between mother and fetus. The chorion also protects the embryo and fetus from the immune responses of the mother in two ways: (1) It secretes proteins that block antibody production by the mother; and (2) it promotes the production of T lymphocytes that suppress the normal immune response in the uterus. Finally, the chorion produces hCG, an important hormone of pregnancy (see Figure 25.36).

By the end of the second week of development, the bilaminar embryonic disc becomes connected to the trophoblast by a band of extraembryonic mesoderm called the **connecting stalk** (Figure 25.33). The connecting stalk is the future umbilical cord.

Development of the Chorionic Villi

By the end of the second week of development, **chorionic villi** (ko-rē-ON-ik VIL-ī) begin to develop. These fingerlike projections of chorion project into the endometrial wall of the uterus (Figure 25.33a). By the end of the third week, blood capillaries invade the chorionic villi (Figure 25.33b). Chorionic villi blood vessels connect to the embryonic heart by way of the umbilical arteries and vein within the connecting stalk (Figure 25.33c). The fetal blood capillaries within the chorionic villi project into lacunae, which unite to form the **intervillous spaces** that bathe the chorionic villi with maternal blood. As a result, maternal and fetal blood vessels are in close proximity. Note, however, that maternal and fetal blood vessels do not join, and the blood they carry does not normally mix. Instead, oxygen and nutrients in the mother's blood diffuse from the intervillous spaces into the capillaries of the villi. Waste products from the fetus, such as carbon dioxide, diffuse in the opposite direction.

Third Week of Development

The third embryonic week begins a 6-week period of rapid development and differentiation. During the third week, the three primary germ layers (ectoderm, mesoderm, and endoderm) are established and lay the groundwork for organ development in weeks 4 through 8.

Figure 25.33 Development of chorionic villi.

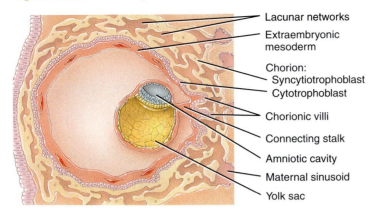

- Lacunar networks
- Extraembryonic mesoderm
- Chorion:
  - Syncytiotrophoblast
  - Cytotrophoblast
- Chorionic villi
- Connecting stalk
- Amniotic cavity
- Maternal sinusoid
- Yolk sac

(a) Frontal section through uterus showing blastocyst, about 13 days after fertilization

- Maternal sinusoid
- Chorion:
  - Cytotrophoblast
  - Syncytiotrophoblast
- Intervillous space
- Connecting stalk
- Extraembryonic mesoderm
- Blood capillary

(b) Details of two chorionic villi, about 21 days after fertilization

- Maternal sinusoid
- Extraembryonic mesoderm
- Umbilical vein
- Umbilical arteries
- Amniotic cavity
- Embryo
- Intervillous space
- Yolk sac
- Connecting stalk
- Chorionic villus
- Maternal sinusoid

(c) Frontal section through uterus showing an embryo and its vascular supply, about 21 days after fertilization

Blood vessels in chorionic villi connect to the embryonic heart via the umbilical arteries and umbilical vein.

Development of the Placenta and Umbilical Cord

The **placenta** (pla-SEN-ta = flat cake) is the site of exchange of nutrients and wastes between the mother and fetus. The placenta is unique because it develops from two separate individuals, the mother and the fetus.

By the beginning of the twelfth week, the placenta has two distinct parts: (1) the fetal portion formed by the chorionic villi of the chorion and (2) the maternal portion formed by the decidua basalis of the endometrium (Figure 25.34a). When fully developed, the placenta is shaped like a pancake (Figure 25.34b). Functionally, the placenta allows oxygen and nutrients to diffuse from maternal blood into fetal blood while carbon dioxide and wastes diffuse from fetal blood into maternal blood. The placenta also is a protective barrier because most microorganisms cannot pass through it. However, certain viruses, such as those that cause AIDS, German measles, chickenpox, measles, encephalitis, and poliomyelitis, can cross the placenta. Many drugs, alcohol, and some substances that can cause birth defects also pass freely. The placenta stores nutrients such as carbohydrates, proteins, calcium, and iron, which are released into fetal circulation as required. The placenta also produces hormones needed to sustain the pregnancy (see Figure 25.36).

The actual connection between the placenta and embryo, and later the fetus, is through the **umbilical cord** (um-BIL-i-kul = navel), which develops from the connecting stalk and is usually about 2 cm (1 in.) wide and about 50–60 cm (20–24 in.) in length. The umbilical cord consists of two umbilical arteries that carry deoxygenated fetal blood to the placenta, one umbilical vein that carries oxygen and nutrients acquired from the mother's intervillous spaces into the fetus, and supporting mucous connective tissue (Figure 25.34b). A layer of amnion surrounds the entire umbilical cord and gives it a shiny appearance (Figure 25.34a). In some cases, the umbilical vein is used to transfuse blood into a fetus or to introduce drugs for various medical treatments.

In about 1 in 200 newborns, only one of the two umbilical arteries is present in the umbilical cord. It may be due to failure of the artery to develop or degeneration of the vessel early in development. Nearly 20% of infants with this condition develop cardiovascular defects.

After the birth of the baby, the placenta detaches from the uterus and is therefore termed the **afterbirth**. At this time, the umbilical cord is tied off and then severed. The small portion (about an inch) of the cord that remains attached to the infant begins to wither and falls off, usually within 12 to 15 days after birth. The area where the cord was attached becomes covered by a thin layer of skin, and scar tissue forms. The scar is the **umbilicus** (navel).

Figure 25.34   Placenta and umbilical cord.

(a) Details of placenta and umbilical cord

(b) Fetal aspect of placenta

The placenta is formed by the chorionic villi of the embryo and the decidua basalis of the endometrium of the mother.

Fourth through Eighth Weeks of Development

The fourth through eighth weeks of development are very significant in embryonic development because all major organs appear during this time. By the end of the eighth week, all the major body systems have begun to develop, although their functions for the most part are minimal.

Fetal Period

The **fetal period** begins at week nine and continues until birth. During this time, the developing human is called a **fetus** (FĒ-tus = offspring). During the fetal period, tissues and organs that developed during the embryonic period grow and differentiate. Very few new structures appear during the fetal period, but the rate of body growth is remarkable, especially during the second half of intrauterine life. For example, during the last two and one-half months of intrauterine life, half of the full-term weight is added. At the beginning of the fetal period, the head is half the length of the body. By the end of the fetal period, the head size is only one quarter the length of the body. During the same period, the limbs also increase in size from one-eighth to one-half the fetal length. The fetus is also less vulnerable to the damaging effects of drugs, radiation, and microbes than it was as an embryo.

A summary of the major developmental events of the embryonic and fetal periods is illustrated in Figure 25.35 and presented in Table 25.3.

Figure 25.35 **Developmental events of the embryonic and fetal periods.** The embryos and fetuses are not shown at their actual sizes.

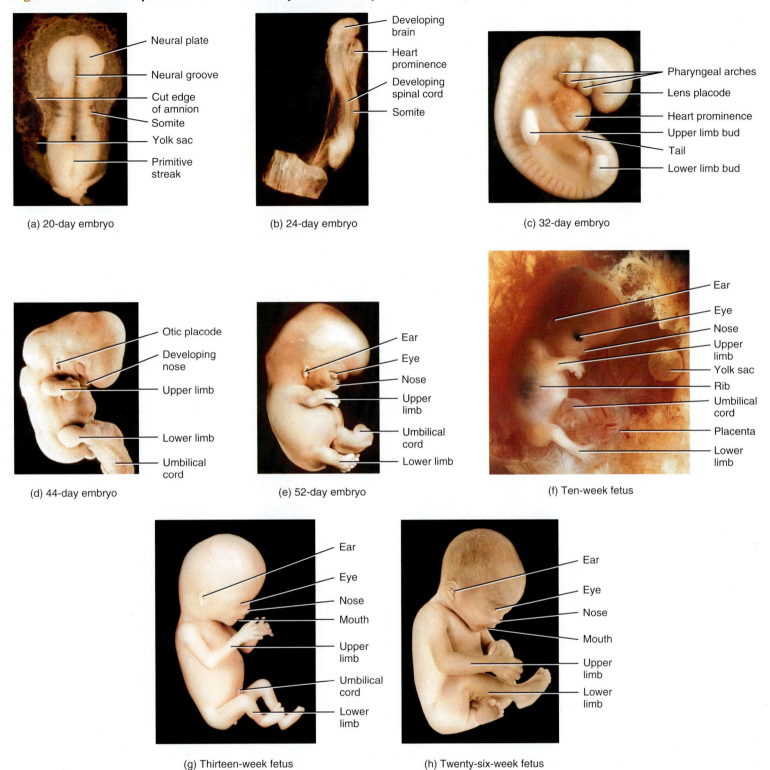

(a) 20-day embryo

Neural plate
Neural groove
Cut edge of amnion
Somite
Yolk sac
Primitive streak

(b) 24-day embryo

Developing brain
Heart prominence
Developing spinal cord
Somite

(c) 32-day embryo

Pharyngeal arches
Lens placode
Heart prominence
Upper limb bud
Tail
Lower limb bud

(d) 44-day embryo

Otic placode
Developing nose
Upper limb
Lower limb
Umbilical cord

(e) 52-day embryo

Ear
Eye
Nose
Upper limb
Umbilical cord
Lower limb

(f) Ten-week fetus

Ear
Eye
Nose
Upper limb
Yolk sac
Rib
Umbilical cord
Placenta
Lower limb

(g) Thirteen-week fetus

Ear
Eye
Nose
Mouth
Upper limb
Umbilical cord
Lower limb

(h) Twenty-six-week fetus

Ear
Eye
Nose
Mouth
Upper limb
Lower limb

🔑 **Development during the fetal period is mostly concerned with the growth and differentiation of tissues and organs formed during the embryonic period.**

TABLE 25.3

Changes during Embryonic and Fetal Development

TIME	APPROXIMATE SIZE AND WEIGHT	REPRESENTATIVE CHANGES
Embryonic period		
1–4 weeks	0.6 cm (3/16 in.)	Primary germ layers and notochord develop. Neurulation occurs. Primary brain vesicles, somites, and intraembryonic coelom develop. Blood vessel formation begins and blood forms in yolk sac, allantois, and chorion. Heart forms and begins to beat. Chorionic villi develop and placental formation begins. The embryo folds. The primitive gut, pharyngeal arches, and limb buds develop. Eyes and ears begin to develop, tail forms, and body systems begin to form.
5–8 weeks	3 cm (1.25 in.) 1 g (1/30 oz)	Primary brain vesicles develop into secondary brain vesicles. Limbs become distinct and digits appear. Heart becomes four-chambered. Eyes are far apart and eyelids are fused. Nose develops and is flat. Face is more humanlike. Ossification begins. Blood cells start to form in liver. External genitals begin to differentiate. Tail disappears. Major blood vessels form. Many internal organs continue to develop.
Fetal period		
9–12 weeks	7.5 cm (3 in.) 30 g (1 oz)	Head constitutes about half the length of the fetal body, and fetal length nearly doubles. Brain continues to enlarge. Face is broad, with eyes fully developed, closed, and widely separated. Nose develops a bridge. External ears develop and are low set. Ossification continues. Upper limbs almost reach final relative length but lower limbs are not quite as well developed. Heartbeat can be detected. Gender is distinguishable from external genitals. Urine secreted by fetus is added to amniotic fluid. Red bone marrow, thymus, and spleen participate in blood cell formation. Fetus begins to move, but its movements cannot yet be felt by the mother. Body systems continue to develop.
13–16 weeks	18 cm (6.5–7 in.) 100 g (4 oz)	Head is relatively smaller than rest of body. Eyes move medially to their final positions, and ears move to their final positions on the sides of the head. Lower limbs lengthen. Fetus appears even more humanlike. Rapid development of body systems occurs.
17–20 weeks	25–30 cm (10–12 in.) 200–450 g (0.5–1 lb)	Head is more proportionate to rest of body. Eyebrows and head hair are visible. Growth slows but lower limbs continue to lengthen. Vernix caseosa (fatty secretions of sebaceous glands and dead epithelial cells) and lanugo (delicate fetal hair) cover fetus. Brown fat forms and is the site of heat production. Fetal movements are commonly felt by mother (quickening).
21–25 weeks	27–35 cm (11–14 in.) 550–800 g (1.25–1.5 lb)	Head becomes even more proportionate to rest of body. Weight gain is substantial, and skin is pink and wrinkled. By 24 weeks, type II alveolar cells begin to produce surfactant.
26–29 weeks	32–42 cm (13–17 in.) 1110–1350 g (2.5–3 lb)	Head and body are more proportionate and eyes are open. Toenails are visible. Body fat is 3.5% of total body mass and additional subcutaneous fat smoothes out some wrinkles. Testes begin to descend toward scrotum at 28 to 30 weeks. Red bone marrow is major site of blood cell production. Many fetuses born prematurely during this period survive if given intensive care because lungs can provide adequate ventilation and central nervous system is developed enough to control breathing and body temperature.

Checkpoint

44. How does the embryo burrow into the endometrium?
45. How is the bilaminar embryonic disc formed? How is it connected to the trophoblast?
46. How does the placenta develop and what is its importance?
47. Describe the components of the umbilical cord.
48. What is the afterbirth?
49. How does mid-fetal weight compare to end-fetal weight?

TIME	APPROXIMATE SIZE AND WEIGHT	REPRESENTATIVE CHANGES
Fetal period (continued)		
30–34 weeks	41–45 cm (16.5–18 in.) 2000–2300 g (4.5–5 lb)	Skin is pink and smooth. Fetus assumes upside-down position. Pupillary reflex is present by 30 weeks. Body fat is 8% of total body mass. Fetuses 33 weeks and older usually survive if born prematurely.
35–38 weeks	50 cm (20 in.) 3200–3400 g (7–7.5 lb)	By 38 weeks circumference of fetal abdomen is greater than that of head. Skin is usually bluish-pink, and growth slows as birth approaches. Body fat is 16% of total body mass. Testes are usually in scrotum in full-term male infants. Even after birth, an infant is not completely developed; an additional year is required, especially for complete development of the nervous system.

4   8   12   16   20   24   28   32   36   (weeks)

25.8 During pregnancy the uterus expands, maternal gastrointestinal tract organs are displaced, and the ureters and urinary bladder are compressed.

Hormones of Pregnancy

During the first 3 to 4 months of pregnancy, the corpus luteum in the ovary continues to secrete progesterone and estrogens, which maintain the lining of the uterus during pregnancy and prepare the mammary glands to secrete milk. The amounts secreted by the cor-pus luteum, however, are only slightly more than those produced after ovulation in a normal menstrual cycle. From the third month through the remainder of the pregnancy, the placenta itself provides the high levels of progesterone and estrogens required. As noted previously, the chorion secretes human chorionic gonadotropin (hCG) into the blood. In turn, hCG stimulates the corpus luteum to continue production of progesterone and estrogens—an activity required to prevent menstruation and for the continued attachment of the embryo and fetus to the lining of the uterus (Figure 25.36a). By the eighth day after fertilization, hCG can be detected in the blood and urine of a pregnant woman. Peak secretion of hCG occurs at about the ninth week of pregnancy (Figure 25.36b). During the fourth and fifth months, the hCG level decreases sharply and then levels off until childbirth.

The chorion begins to secrete estrogens after the first 3 or 4 weeks of pregnancy and progesterone by the sixth week. These hormones are secreted in increasing quantities until the time of birth (Figure 25.36b). By the fourth month, when the placenta is fully established, the secretion of hCG is greatly reduced, and the secretions of the corpus luteum are no longer essential. A high level of estrogens prepares the myometrium for labor, and progesterone ensures that the uterine myometrium is relaxed and that the cervix is tightly closed. After delivery, estrogens and progesterone in the blood decrease to normal levels.

Relaxin, a hormone produced first by the corpus luteum of the ovary and later by the placenta, increases the flexibility of pelvic ligaments and helps dilate the uterine cervix during labor. All of these actions ease delivery of the baby.

A third hormone produced by the chorion of the placenta is **human chorionic somatomammotropin (hCS)**. The rate of secretion of hCS increases in proportion to placental mass, reaching maximum levels after 32 weeks and remaining relatively constant after that. It is thought to help prepare the mammary glands for lactation, enhance maternal growth by increasing protein synthesis, and regulate certain aspects of metabolism in both mother and fetus. For example, hCS decreases the use of glucose by the mother and promotes the release of fatty acids from her adipose tissue, making more glucose available to the fetus.

The hormone most recently found to be produced by the placenta is **corticotropin-releasing hormone (CRH)**, which in nonpregnant people is secreted only by the hypothalamus. CRH is now thought to be part of the "clock" that establishes the timing of birth. Secretion of CRH by the placenta begins at about 12 weeks and increases enormously toward the end of pregnancy. Women who have higher levels of CRH earlier in pregnancy are more likely to deliver prematurely; those who have low levels are more

Figure 25.36   Hormones during pregnancy.

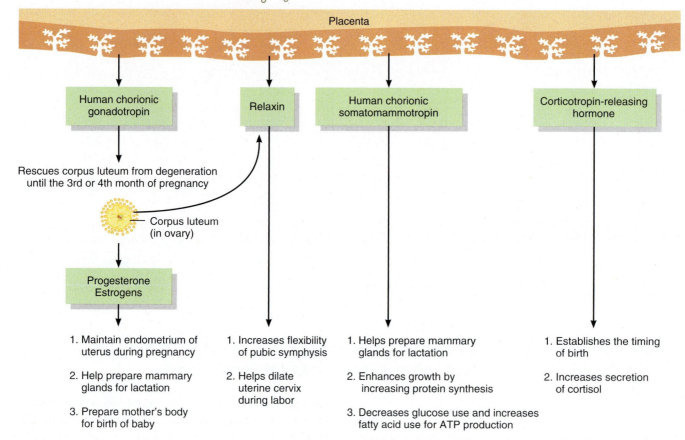

(a) Sources and functions of hormones

likely to deliver after their due date. CRH from the placenta has a second important effect: It increases secretion of cortisol, which is needed for maturation of the fetal lungs and the production of surfactant to keep alveoli (air sacs) open (see Concept 22.2).

Changes during Pregnancy

Near the end of the third month of pregnancy, the uterus occupies most of the pelvic cavity. As the fetus continues to grow, the uterus extends higher and higher into the abdominal cavity. Toward the end of a full-term pregnancy, the uterus fills nearly the entire abdominal cavity, reaching nearly to the xiphoid process of the sternum (Figure 25.37). It pushes the maternal intestines, liver, and stomach superiorly, elevates the diaphragm, and widens the thoracic cavity. In the pelvic cavity, compression of the ureters and urinary bladder occurs.

Pregnancy-induced physiological changes also occur, including weight gain due to the fetus, amniotic fluid, the placenta, uterine enlargement, and increased total body water; increased storage of proteins, triglycerides, and minerals; marked breast enlargement in preparation for lactation; and lower back pain due to **lordosis** (swayback).

Several changes occur in the maternal cardiovascular system. Stroke volume increases by about 30% and cardiac output rises by 20–30% due to increased maternal blood flow to the placenta and increased metabolism. Heart rate increases 10–15% and blood volume increases 30–50%, mostly during the second half of pregnancy. These increases are necessary to meet the additional demands of the fetus for nutrients and oxygen. When a pregnant woman is lying on her back, the enlarged uterus may compress the aorta, resulting in diminished blood flow to the uterus. Compression of the inferior vena cava also decreases venous return, which leads to edema in the lower limbs and may produce varicose veins. Compression of the renal artery can lead to renal hypertension.

Respiratory function is also altered during pregnancy to meet the added oxygen demands of the fetus. Tidal volume (the volume of air in one breath) and minute ventilation (the total volume of air inhaled and exhaled each minute) can increase by up to 40%, and airway resistance in the bronchial tree can decline by 30–40%. But residual volume (the volume of air remaining in the lungs after exhalation) can decline by up to 25% as the expanding uterus pushes on the diaphragm, with resulting dyspnea (difficult breathing).

The digestive system also undergoes changes. Pregnant women experience an increase in appetite due to the added nutritional demands of the fetus. Pressure on the stomach may force the stomach contents superiorly into the esophagus, resulting in heartburn. A general decrease in GI tract motility can cause constipation, delay gastric emptying time, and produce nausea and vomiting.

Pressure on the urinary bladder by the enlarging uterus can produce urinary symptoms, such as increased frequency and urgency of urination, and stress incontinence. Greater urine ouput from the kidneys (the glomerular filtration rate increases by up to 40%) allows faster elimination of the extra wastes produced by the fetus.

Changes in the skin during pregnancy are more apparent in some women than in others. Some women experience increased pigmentation around the eyes and cheekbones in a masklike pattern, in the areolae of the breasts, and in the linea alba of the lower abdomen. **Striae** (stretch marks) over the abdomen can occur as the uterus enlarges, and hair loss increases.

Changes in the reproductive system include edema and increased vascularity of the vulva and increased pliability and vascularity of the vagina. The uterus increases from its nonpregnant mass of 60–80 g to 900–1200 g at term because of increased numbers of muscle fibers in the myometrium in early pregnancy and hypertrophy of muscle fibers during the second and third trimesters.

(b) Blood levels of hormones during pregnancy

Progesterone and estrogens are produced by the corpus luteum during the first 3–4 months of pregnancy, after which time the placenta produces them.

Checkpoint

50. What is the source of progesterone and estrogens during early pregnancy? During late pregnancy?
51. What is the source and function of human chorionic gonadotropin? Of human chorionic somatomammotropin?
52. Which hormone helps ease delivery of the baby? How?
53. Which hormone helps control when birth occurs?

Figure 25.37   **Normal fetal location and position at the end of a full-term pregnancy.**

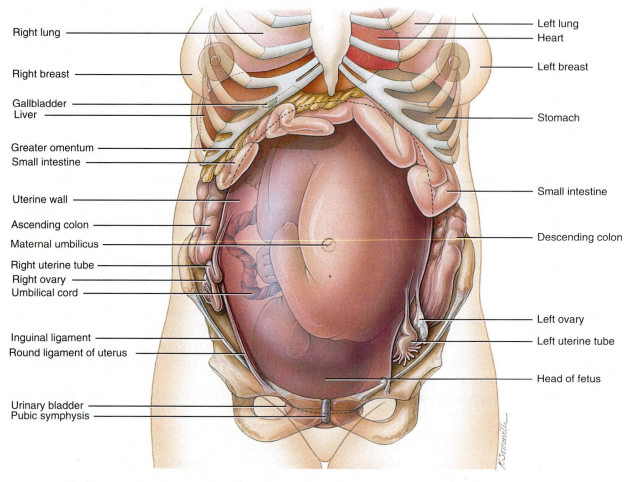

Right lung

Right breast

Gallbladder
Liver

Greater omentum
Small intestine

Uterine wall

Ascending colon
Maternal umbilicus

Right uterine tube
Right ovary
Umbilical cord

Inguinal ligament
Round ligament of uterus

Urinary bladder
Pubic symphysis

Left lung
Heart

Left breast

Stomach

Small intestine

Descending colon

Left ovary
Left uterine tube

Head of fetus

Anterior view of position of organs at end of full-term pregnancy

🔑 **The gestation period is the time interval (about 38 weeks) from fertilization to birth.**

Return to Lou and Susan's Story

Susan and Lou decided to proceed with the fertility treatment. After the procedure, which was really no more complicated than having a Pap smear done, Dr. Jones told Susan and Lou to refrain from sexual intercourse for at least 2 weeks. He also reminded them that it usually took several attempts to obtain a successful insemination. Susan and Lou were fortunate. Two weeks later, her pregnancy test was positive. But she knew even before the test results came in; she felt different somehow, and she had been nauseous that morning.

Several months later, Susan and Lou were having breakfast. "Want to feel him kicking?" Susan grabbed Lou's hand and placed it on her abdomen.

"That is so amazing! Our baby is in there. Hey, wait a minute, how do you know it's a boy?" he asked.

Susan smiled, "No, I didn't sneak off and have an ultrasound done—I just know. Mothers know things."

"You seem to be feeling better. Are you over the morning sickness?"

"It usually goes away after the first trimester. I am feeling better, just kind of tired. I'm hungry all the time, and nothing fits anymore, not even my wedding ring. I'm looking forward to having ankles again once this is over!"

F. *Where did the fertilization of Susan's oocyte occur within her reproductive tract?*

G. *If the fertility hormones caused Susan to release two eggs and each egg was fertilized by different sperm, which type of twins would be produced?*

H. *What is causing Susan's extremities to swell?*

I. *Which changes occurring during pregnancy might cause Susan's fatigue?*

25.9 Labor includes dilation of the cervix and expulsion of the fetus and placenta.

Labor is the process by which the fetus is expelled from the uterus through the vagina, and is also referred to as giving birth. A synonym for labor is **parturition** (par´-too-RISH-un; *parturit-* = childbirth).

The onset of labor is determined by complex interactions of several placental and fetal hormones. Because progesterone inhibits uterine contractions, labor cannot take place until its effects are diminished. Toward the end of pregnancy, the levels of estrogens in the mother's blood rise sharply, producing changes that overcome the inhibiting effects of progesterone. High levels of estrogens cause the number of receptors for oxytocin on uterine muscle fibers to increase, and cause uterine muscle fibers to form cell junctions with one another. Oxytocin, released by the posterior pituitary, stimulates uterine contractions, and relaxin from the placenta assists by increasing the flexibility of the pubic symphysis and helping dilate the uterine cervix. Estrogen also stimulates the placenta to release prostaglandins, which induce production of enzymes that digest collagen fibers in the cervix, causing it to soften.

Control of labor contractions occurs via a positive feedback cycle (see Figure 1.4). Contractions of the uterine myometrium force the baby's head or body into the cervix, distending (stretching) the cervix. Stretch receptors in the cervix send nerve impulses to the hypothalamus, causing it to release oxytocin through the posterior pituitary. Oxytocin is carried in the blood to the uterus, where it stimulates the myometrium to contract more forcefully. As the contractions intensify, the baby's body stretches the cervix still more. The resulting nerve impulses from the cervix stimulate the secretion of yet more oxytocin. With birth of the infant, the positive feedback cycle is broken because cervical distention suddenly lessens. Synthetic oxytocin (Pitocin) is sometimes given to induce labor or to increase uterine tone and control hemorrhage just after parturition.

Uterine contractions occur in waves (quite similar to the peristaltic waves of the gastrointestinal tract) that start at the top of the uterus and move downward, eventually expelling the fetus. **True labor** begins when uterine contractions occur at regular intervals, usually producing pain. As the interval between contractions shortens, the contractions intensify. Another symptom of true labor in some women is localization of pain in the back that is intensified by walking. The most reliable indicator of true labor is dilation of the cervix and the "show," a discharge of a blood-containing mucus into the cervical canal. In **false labor**, pain is felt in the abdomen at irregular intervals, but it does not intensify and walking does not alter it significantly. There is no "show" and no cervical dilation.

True labor can be divided into three stages (Figure 25.38):

① *Stage of dilation*. The time from the onset of labor to the complete dilation of the cervix is the **stage of dilation**. This stage, which typically lasts 6–12 hours, features regular con-

Figure 25.38 Stages of true labor.

① Stage of dilation

— Urinary bladder

— Vagina

— Ruptured amniotic sac

— Rectum

— Placenta

② Stage of expulsion

— Uterus

— Placenta

— Umbilical cord

③ Placental stage

🔑 The term *parturition* refers to birth.

tractions of the uterus, usually a rupturing of the amniotic sac, and complete dilation (to 10 cm) of the cervix. If the amniotic sac does not rupture spontaneously, it is ruptured intentionally.

2 *Stage of expulsion*. The time (10 minutes to several hours) from complete cervical dilation to delivery of the baby is the **stage of expulsion**.

3 *Placental stage*. The time (5–30 minutes or more) after delivery until the placenta or afterbirth is expelled by powerful uterine contractions is the **placental stage**. These contractions also constrict blood vessels that were torn during delivery, reducing the likelihood of hemorrhage.

As a rule, labor lasts longer with first babies, typically about 14 hours. For women who have previously given birth, the average duration of labor is about 8 hours—although the time varies enormously among births. Because the fetus may be squeezed through the birth canal (cervix and vagina) for up to several hours, the fetus is stressed during childbirth: The fetal head is compressed, and the fetus undergoes some degree of intermittent hypoxia due to compression of the umbilical cord and the placenta during uterine contractions. In response to this stress, the fetal adrenal medullae secrete very high levels of epinephrine and norepinephrine, the "fight-or-flight" hormones. Much of the protection against the stresses of parturition, as well as preparation of the infant for sur-

viving extrauterine life, is provided by these hormones. Among other functions, epinephrine and norepinephrine clear the lungs and alter their physiology in readiness for breathing air, mobilize readily usable nutrients for cellular metabolism, and promote an increased flow of blood to the brain and heart.

Following the delivery of the baby and placenta is a 6-week period during which the maternal reproductive organs and physiology return to the prepregnancy state. This period is called the **puerperium** (pū´-er-PER-ē-um). Through a process of tissue catabolism, the uterus undergoes a remarkable reduction in size, called **involution** (in´-vō-LOO-shun), especially in lactating women. The cervix loses its elasticity and regains its prepregnancy firmness. For 2 to 4 weeks after delivery, women have a uterine discharge consisting initially of blood and later of serous fluid derived from the former site of the placenta.

Checkpoint

54. Which hormonal changes induce labor?
55. Which hormone controls labor contractions? Why is control of labor considered a positive feedback loop?
56. What is the difference between false labor and true labor?

25.10 Milk production and ejection are influenced by prolactin, estrogens, progesterone, and oxytocin.

Lactation (lak´-TĀ-shun) is the secretion and ejection of milk from the mammary glands. A principal hormone in promoting milk synthesis and secretion is **prolactin**, which is secreted from the anterior pituitary gland. Even though prolactin levels increase as the pregnancy progresses, no milk secretion occurs because progesterone inhibits the effects of prolactin. After delivery, the levels of estrogens and progesterone in the mother's blood decrease, and the inhibition is removed. The principal stimulus in maintaining prolactin secretion during lactation is the sucking action of the infant. Suckling stimulates stretch receptors in the nipples, which initiates nerve impulses to the hypothalamus that stimulate the hypothalamus to release more prolactin from the anterior pituitary.

Oxytocin causes release of milk into the mammary ducts via the **milk ejection reflex** (Figure 25.39). Milk formed by the glandular cells of the breasts is stored until the baby begins active suckling. Stimulation of touch receptors in the nipple initiates sensory nerve impulses that are relayed to the hypothalamus. In response, secretion of oxytocin from the posterior pituitary increases. Carried by the bloodstream to the mammary glands, oxytocin stimulates contraction of myoepithelial cells surrounding the glandular cells and ducts. The resulting compression moves

the milk from the alveoli of the mammary glands into the mammary ducts, where it can be suckled. This process is termed **milk ejection (letdown)**. Even though the actual ejection of milk does not occur until 30–60 seconds after nursing begins, some milk stored in lactiferous sinuses near the nipple is available during the latent period. Stimuli other than suckling, such as hearing a baby's cry or touching the mother's genitals, also can trigger oxytocin release and milk ejection. The suckling stimulation that produces the release of oxytocin also increases secretion of prolactin, which maintains lactation.

During late pregnancy and the first few days after birth, the mammary glands secrete a cloudy fluid called **colostrum**. Although it is not as nutritious as milk—it contains less lactose and virtually no fat—colostrum serves adequately until the appearance of true milk on about the fourth day. Colostrum and maternal milk contain important antibodies that protect the infant during the first few months of life.

Following birth of the infant, the prolactin level starts to return to the nonpregnant level. However, each time the mother nurses the infant, nerve impulses from the nipples to the hypothalamus result in a tenfold increase in prolactin secretion by the anterior pituitary that lasts about an hour. Prolactin acts on the mammary

Figure 25.39 **The milk ejection reflex, a positive feedback cycle.**

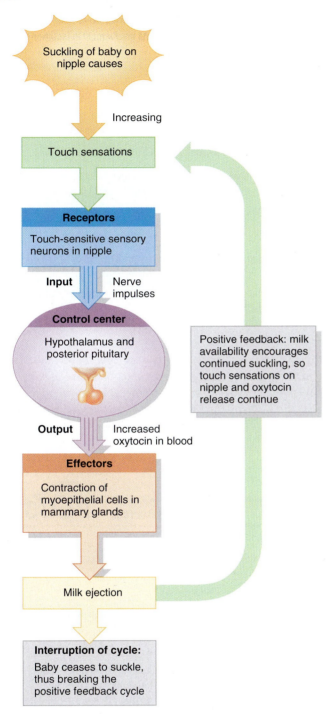

Oxytocin stimulates contraction of myoepithelial cells in the breasts, which causes milk ejection.

glands to provide milk for the next nursing period. If this surge of prolactin is blocked by injury or disease, or if nursing is discontinued, the mammary glands lose their ability to secrete milk in only a few days. Even though milk secretion normally decreases considerably within 7–9 months after birth, it can continue for several years if nursing continues.

A primary benefit of **breast-feeding** is nutritional: Human milk is a sterile solution that contains concentrations of fatty acids, lactose, amino acids, minerals, vitamins, and water that are ideal for the baby's digestion, brain development, and growth. Breast-feeding also benefits infants by providing the following:

- **Beneficial cells**. Several types of white blood cells are present in breast milk. Neutrophils and macrophages serve as phagocytes, ingesting microbes in the baby's gastrointestinal tract. Macrophages also produce lysozyme and other immune system components. Plasma cells produce antibodies against specific microbes, and T lymphocytes kill microbes directly or help mobilize other immune defenses.

- **Beneficial molecules**. Breast milk also contains an abundance of beneficial molecules. Maternal IgA antibodies in breast milk bind to microbes in the baby's gastrointestinal tract and prevent their migration into other body tissues. Because a mother produces antibodies to whatever disease-causing microbes are present in her environment, her breast milk affords protection against the specific infectious agents to which her baby is also exposed. Additionally, two milk proteins bind to nutrients such as vitamin B_{12} and iron that many bacteria need to grow and survive. Some fatty acids can kill certain viruses by disrupting their membranes, and lysozyme kills bacteria by disrupting their cell walls. Finally, interferons enhance the antimicrobial activity of immune cells.

- **Decreased incidence of diseases later in life**. Breast-feeding provides children with a slight reduction in risk of lymphoma, heart disease, allergies, respiratory and gastrointestinal infections, ear infections, diarrhea, diabetes mellitus, and meningitis.

- **Miscellaneous benefits**. Breast-feeding supports optimal infant growth, enhances intellectual and neurological development, and fosters mother–infant relations by establishing early and prolonged contact between them. Compared to cow's milk, the fats and iron in breast milk are more easily absorbed, the proteins in breast milk are more readily metabolized, and the lower sodium content of breast milk is more suited to an infant's needs. Premature infants benefit even more from breast-feeding because the milk produced by mothers of premature infants seems to be specially adapted to the infant's needs; it has a higher protein content than the milk of mothers of full-term infants. Finally, a baby is less likely to have an allergic reaction to its mother's milk than to milk from another source.

Years before oxytocin was discovered, it was common practice in midwifery to let a first-born twin nurse at the mother's breast to speed the birth of the second child. Now we know why this practice is helpful—it stimulates the release of oxytocin. Even after a single birth, nursing promotes expulsion of the placenta (afterbirth) and helps the uterus return to its normal size.

Checkpoint

57. Which hormones contribute to lactation? What is the function of each?
58. What are the advantages of breast-feeding over bottle-feeding?
59. What is a function of oxytocin other than its involvement in lactation?

Lou and Susan's Story: Epilogue and Discussion

▶▶▶Susan and Lou, both 35 years old and in good general health, were having difficulty conceiving a child. Susan had a 15-year history of birth control use, but no major reproductive health problems. Lou was diagnosed with a low sperm count. The couple decided to pursue intrauterine insemination to increase the likelihood of pregnancy. This procedure required Susan to take fertility hormones to increase her ovulatory function. Sperm from a semen sample from Lou were specially treated, and used for the insemination procedure following detection of Susan's ovulatory phase. Susan was fortunate and successfully got pregnant after a single procedure. She experienced some mild nausea during the early phase of the pregnancy; otherwise, Susan had a normal, healthy pregnancy. She had difficulty with the subsequent birth, and received Pitocin to increase uterine contractions and facilitate delivery. Susan and Lou, after many months of trying, finally have a healthy baby boy.

As you have learned, the human reproductive process is quite complex and involved. The production of gametes is only the first step. Fertilization, implantation, pregnancy, and birth are all regulated by the endocrine system (see Chapter 17). ◀◀◀

J. Normally, uterine contractions cease after delivery, but with Pitocin they can continue after the birthing process due to the drug remaining in the blood. Why do uterine contractions normally cease after delivery?

K How might anatomy have played a role in Lou's low sperm count?

L. Why was knowledge of anatomy important to the physician who performed the insemination procedure on Susan?

M. How do the same hormones perform different functions in men and women? How do hormones perform different functions in pregnant and nonpregnant women?

Concept and Resource Summary

Concept	Resources

Introduction

1. **Sexual reproduction** is the process that results in the union of a male **gamete** (sperm cell) with a female gamete (secondary oocyte) in an event called **fertilization**, which results in production of offspring.
2. **Pregnancy** begins with fertilization and is a time when the fertilized egg undergoes implantation, embryonic and fetal development, and birth.
3. The testes in males and the ovaries in females are **gonads**. Gonads produce gametes and sex hormones.

Concept 25.1 The scrotum supports and regulates the temperature of the testes, which produce spermatozoa.

1. Male reproductive organs include the testes, a system of ducts (epididymis, ductus deferens, ejaculatory ducts, and urethra), accessory glands (seminal vesicles, prostate, and bulbourethral glands), and supporting structures (scrotum and penis).
2. The **scrotum** is a sac that hangs from the root of the penis and consists of loose skin and underlying hypodermis; it supports the testes. The temperature of the testes is regulated by contraction of the **cremaster muscle** and **dartos muscle**, which either elevates them and brings them closer to the warmth of the pelvic cavity or relaxes and moves them farther from the pelvic cavity for cooling.

See

• **Anatomy Overview – Male Reproductive Anatomy**
• **Anatomy Overview – Histology of the Testes**
• **Anatomy Overview – Testicular Hormones**
• **Animation – Spermatogenesis**
• **Animation – Hormonal Control of Male Reproduction**

Concept

3. The **testes** are paired glands in the scrotum containing **seminiferous tubules** that produce sperm during **spermatogenesis**. The seminiferous tubules contain the **spermatogenic cells** that form sperm, **Sertoli cells**, which nourish spermatogenic cells and secrete inhibin, and **Leydig cells**, which produce the male sex hormone testosterone.

4. **Spermatogonia** begin producing sperm at puberty. Some spermatogonia develop into diploid **primary spermatocytes** that sequence through spermatogenesis (meiosis I, meiosis II, and spermiogenesis) to form haploid **sperm cells**. During **spermiogenesis**, developing sperm form an acrosome and flagellum, shed excess cytoplasm, increase mitochondria, and become elongated sperm cells that are released into the lumen of the seminiferous tubules. Sertoli cells secrete fluid that push sperm toward the ducts of the testes.

5. Mature sperm have a pointed **head**, a caplike **acrosome** filled with enzymes for penetrating the secondary oocyte, a **middle piece** containing mitochondria that provide energy for movement toward the secondary oocyte, and a **tail**.

6. At puberty, the hypothalamus secretes **gonadotropin-releasing hormone**, which stimulates anterior pituitary secretion of **luteinizing hormone** and **follicle-stimulating hormone**. Luteinizing hormone stimulates Leydig cells to secrete **testosterone**. **Follicle-stimulating hormone** and testosterone stimulate Sertoli cells to secrete **androgen-binding protein**, which binds to testosterone and keeps its concentration high in the seminiferous tubule.

7. Sertoli cells secrete **inhibin**, which inhibits follicle-stimulating hormone release to help regulate the rate of spermatogenesis.

8. Testosterone stimulates development of male reproductive structures and secondary sex characteristics, development of male sexual behavior, male and female libido, bone growth, protein anabolism, and sperm maturation.

Hear

- **Figure 25.3 –** Internal and External Anatomy of a Testis
- **Figure 25.7 – Hormonal Control of Spermatogenesis and Actions of Testosterone and DHT**

Do

- **Exercise – Match the Male Hormones**
- **Exercise – Spermatogenesis Selections**
- **Concepts and Connections – Regulation of Male Reproduction**

Concept 25.2 Sperm travel through the epididymis, ductus deferens, ejaculatory ducts, and urethra.

1. Fluid from Sertoli cells pushes sperm through the testis in the **straight tubules** and then **rete testis**, through the epididymis in the **efferent ducts**, and then leaves the epididymis in the **ductus epididymis**.

2. The **epididymis** lies along the posterior border of each testis and contains the tightly coiled **ductus epididymis**, the site of sperm maturation and storage. The ductus epididymis propels sperm into the ductus deferens.

3. The **ductus deferens** stores sperm and propels them to the urethra for ejaculation.

4. The **spermatic cord** passes the ductus deferens, the testicular artery and veins, autonomic nerves, lymphatic vessels, and cremaster muscle through the **inguinal canal**.

5. The ducts of the ductus deferens and seminal vesicle unite to form the **ejaculatory duct**, the passageway for ejection of sperm and secretions of the seminal vesicles into the first portion of the urethra, the prostatic urethra.

6. The male **urethra** serves as a passage for both sperm and urine. The urethra extends from the urinary bladder to the tip of the penis and is subdivided into the **prostatic**, **membranous**, and **spongy urethra**.

7. Paired **seminal vesicles** lie posterior to the base of the urinary bladder. Their secretion is an alkaline, viscous fluid that contains fructose (used by sperm for ATP production). Seminal fluid constitutes about 60% of the volume of semen and contributes to sperm viability.

8. The **prostate** lies inferior to the urinary bladder and surrounds the prostatic urethra. Its secretion constitutes about 25% of the volume of semen and contributes to sperm motility and viability.

9. Paired **bulbourethral glands** lie inferior to the prostate on either side of the membranous urethra. They secrete mucus for lubrication and an alkaline fluid that neutralizes acids from urine in the urethra.

10. **Semen** is a mixture of sperm and **seminal fluid**, secretions from the seminiferous tubules, seminal vesicles, prostate, and bulbourethral glands. Semen provides the fluid in which sperm are transported, supplies nutrients for sperm, and neutralizes the acidity of the male urethra and the vagina.

11. The **penis** contains the urethra and is a passageway for ejaculation of semen and excretion of urine. The **root of the penis** attaches to muscle of the perineum. The **body of the penis** is composed of three masses of erectile tissue: two **corpora cavernosa penis** and a **corpus spongiosum penis** that contains the spongy urethra and keeps it open during ejaculation. The **glans penis** is the expanded distal tip.

12. Engorgement of the penile blood sinuses under the influence of sexual excitation is called **erection**. **Ejaculation** is the expulsion of semen from the urethra to the exterior of the body.

See

- **Anatomy Overview – Male Reproductive Anatomy**

Hear

- **Figure 25.9 –** Locations of Several Accessory Reproductive Organs in Males

Concept

Concept 25.3 After a secondary oocyte is discharged from an ovary, it may undergo fertilization and implantation in the uterus.

1. The female organs of reproduction include the ovaries, uterine tubes or oviducts, uterus, vagina, and vulva. Mammary glands are considered part of both the integumentary and reproductive systems.
2. The **ovaries** are located on each side of the uterus and held in position by the **broad ligament** of the uterus, and the **ovarian** and **suspensory ligaments**. Ovaries produce secondary oocytes and secrete progesterone, estrogen, inhibin, and relaxin.
3. The **ovarian cortex** consists of **ovarian follicles** with oocytes in different stages of development. A **mature follicle** is ready to expel its secondary oocyte during **ovulation**. A **corpus luteum**, the remains of a ruptured mature follicle after **ovulation**, degenerates into the **corpus albicans**.
4. **Oogenesis**, the formation of female gametes, begins in the ovaries. The oogenesis sequence includes meiosis I and meiosis II, which stops in metaphase. The secondary oocyte is released from the ovary during ovulation and is swept into the uterine tube. If a sperm cell penetrates the oocyte, meiosis II resumes, and a haploid **ovum** is produced. The sperm and ovum nuclei unite, forming a diploid **zygote**.
5. The **uterine tubes** extend laterally from the uterus and are the normal sites of fertilization. Their distal end is an open, funnel-shaped **infundibulum** with fingerlike **fimbriae** that sweep the oocyte into the **infundibulum** of the uterine tube. Ciliated cells and peristaltic contractions help move a secondary oocyte or zygote toward the uterus.
6. The **uterus** functions in menstruation, implantation of a fertilized ovum, development of a fetus during pregnancy, and labor. It also is part of the pathway for sperm to reach the uterine tubes to fertilize a secondary oocyte. The uterus has a **fundus**, **body**, and **cervix** that opens into the vagina. It is supported by the **broad**, **uterosacral**, **cardinal**, and **round ligaments**. The uterus wall is composed of an outer **perimetrium**; a middle **myometrium** consisting of three layers of smooth muscle important during labor; and the inner **endometrium** that lines the lumen of the uterus. There is an extensive blood supply to support regrowth of sloughed-off lining after menstruation, implantation of the fertilized ovum, and development of the placenta.

See

- **Anatomy Overview – Female Reproductive Anatomy**
- **Anatomy Overview – Histology of the Female Reproductive System**
- **Animation – Oogeneis**

Hear

- **Figure 25.13 –** **Histology of the Ovary**
- **Figure 25.16 – Relationship of the Uterine Tubes to the Ovaries, Uterus, and Associated Structures**

Do

- **Exercise – Organize Oogenesis**

Concept 25.4 The vagina is a passageway for childbirth; the mammary glands secrete milk.

1. The **vagina** extends from the uterine cervix to the exterior of the body. It is a receptacle for the penis during sexual intercourse, an outlet for menstrual flow, and a passageway for childbirth.
2. The external genitals of the female are termed the **vulva** and include the **mons pubis**, **labia majora**, **labia minora**, **clitoris**, **vestibule**, the **vaginal** and **urethral orifices**, **hymen**, and **bulb of the vestibule**. Also part of the vulva are the **paraurethral glands**, which secrete mucus into the urethra; and the **greater vestibular glands**, which secrete mucus into the vestibule as lubrication to facilitate intercourse.
3. The **perineum** is the diamond-shaped area medial to the thighs and buttocks containing the external genitals and anus.
4. Each **breast** lies anterior to the pectoralis major and serratus anterior. Within each breast is a **mammary gland** that functions in **lactation**, the synthesis, secretion, and ejection of milk. Milk production is stimulated by prolactin, and milk ejection is stimulated by oxytocin.

See

- **Anatomy Overview – Female Reproductive Anatomy**
- **Anatomy Overview – Histology of the Mammary Glands**

Hear

- **Figure 25.22 –** **Mammary Glands**

Concept 25.5 The female reproductive cycle includes the ovarian and uterine cycles.

1. The **ovarian cycle** occurs in the ovaries during and after maturation of an oocyte. The **uterine cycle** is concurrent changes in the endometrium of the uterus to prepare it to receive the fertilized ovum and support ovum development. The **female reproductive cycle** includes both the ovarian and uterine cycles.
2. The female reproductive cycle is controlled by several hormones. **Gonadotropin-releasing hormone** from the hypothalamus stimulates the anterior pituitary to release **follicle-stimulating hormone** and **luteinizing hormone**, which stimulate development of follicles and secretion of estrogens by the follicles. Luteinizing hormone also stimulates ovulation, formation of the corpus luteum, and secretion of estrogens, progesterone, relaxin, and inhibin from the corpus luteum.
3. **Estrogens** stimulate the development and maintenance of female reproductive structures, secondary sex characteristics, and protein anabolism, and decrease blood cholesterol levels. **Progesterone** works with estrogens to prepare the endometrium for implantation and the mammary glands for milk secretion. **Relaxin** relaxes the myometrium at the time of possible implantation and, at the end of pregnancy, increases the flexibility of the public symphysis to facilitate delivery. **Inhibin** inhibits secretion of follicle-stimulating hormone and luteinizing hormone.

See

- **Anatomy Overview – Ovarian Hormones**
- **Anatomy Overview – Anterior Pituitary Reproductive Hormones**
- **Animation – Oogenesis**
- **Animation – Hormonal Control of Female Reproduction**

Hear

- **Figure 25.24 – The Female Reproductive Cycle**

Concept	Resources

4. The female reproductive cycle ranges from 24 to 35 days and has four phases: the menstrual phase, preovulatory phase, ovulation, and postovulatory phase. The events of the ovarian cycle and the uterine cycle occur concurrently.

5. During the **menstrual phase** of the female reproductive cycle: In the ovaries, primordial follicles develop into primary, then secondary follicles. In the uterus, the stratum functionalis of the endometrium is shed, discharging blood, tissue fluid, mucus, and epithelial cells.

6. The **preovulatory phase** occurs between **menstruation** and ovulation. In the ovaries, a group of follicles begins to undergo final maturation. One follicle becomes **dominant** while the others degenerate. The dominant follicle becomes the **graafian follicle** and secretes estrogens and inhibin, which decreases FSH secretion and stops the other follicles from growing. In the uterus, estrogens secreted by the developing follicles stimulate repair of the endometrium.

7. The rupture of the mature follicle and release of the secondary oocyte into the pelvic cavity occurs during **ovulation**. Ovulation is induced by high levels of estrogen.

8. The **postovulatory phase** occurs between ovulation and menstruation. In the ovaries, the **corpus luteum** develops from the ruptured graafian follicle and, stimulated by luteinizing hormone, secretes progesterone, estrogen, relaxin, and inhibin. In the uterus, the endometrium thickens in readiness for implantation.

9. If fertilization and implantation do not occur: In the ovaries, the corpus luteum degenerates, and the levels of progesterone and estrogens decrease. In the uterus, the resulting low levels of progesterone and estrogens result in menstruation followed by the initiation of another reproductive cycle.

10. If fertilization and implantation do occur: In the ovaries, **human chorionic gonadotropin** seceted by the embryo maintains the corpus luteum so it continues to be a source of progesterone and estrogens. In the uterus, endometrium development continues under the influence of progesterone and estrogen in preparation for implantation of the zygote.

Concept 25.6 The zygote divides into a morula and then a blastocyst that implants in the endometrium of the uterus.

1. Fertilization, the union of the genetic material from a haploid sperm cell and a haploid secondary oocyte into a single diploid nucleus, typically occurs within the uterine tube 12 to 24 hours after ovulation.

2. Sperm use flagella to swim from the vagina through the cervical canal to the uterine tube, where sperm undergo **capacitation** to assist penetration of the oocyte. After penetrating the **corona radiata**, the **zona pellucida** triggers the **acrosomal reaction**, the release of acrosomal enzymes that digest a path through the zona pellucida. Normally, only one sperm cell fertilizes a secondary oocyte because of the *fast* and *slow blocks* to polyspermy.

3. When the sperm cell enters the secondary oocyte, it triggers completion of meiosis II. The pronuclei of the sperm cell and secondary oocyte unite. The resulting cell is a zygote.

4. If two secondary oocytes are released and fertilized by sperm, **dizygotic twins** are produced. **Monozygotic twins** develop from a single fertilized ovum that separates into two embryos. If the developing cells do not separate within 8 days, **conjoined twins** will result.

5. The developing human is an **embryo** from fertilization to the eighth week of development. Early rapid cell division of a zygote is called **cleavage**, and the cells produced by cleavage are called **blastomeres**. The solid sphere of cells produced by cleavage is a **morula**. The morula develops into a **blastocyst**, a hollow ball of cells differentiated into an **embryoblast** that develops into the embryo and a **trophoblast** that becomes part of the placenta.

6. During **implantation**, the blastocyst attaches to the endometrium, secretes enzymes that degrade the endometrium, and burrows into the endometrium. After implantation, the endometrium becomes modified and is known as the **decidua**.

Concept 25.7 During the embryonic period, the embryonic membranes and most major organs develop.

1. The trophoblast develops into the **syncytiotrophoblast** and **cytotrophoblast**, both of which become part of the chorion, one of the fetal membranes. The trophoblast digests and implants in the endometrium, and secretes human chorionic gonadotropin to maintain the corpus luteum and prevent menstruation.

2. The embryoblast differentiates into the **bilaminar embryonic disc** composed of two layers, the **hypoblast** and **epiblast**. The **amniotic cavity** develops within the epiblast.

Concept   **Resources**

3. The thin membrane called the **amnion** develops from the cytotrophoblast and eventually surrounds the entire embryo. This creates the **amniotic cavity** filled with **amniotic fluid**, derived as a filtrate of maternal blood, and later, fetal urine. Amniotic fluid serves as a shock absorber and also prevents desiccation, adhesions, and wide fluctuations in fetal temperature.

4. The **yolk sac** develops from hypoblast and transfers nutrients to the embryo, forms blood cells, produces primordial germ cells, and forms part of the gut.

5. Once the blastocyst is completely implanted, endometrial capillaries around the embryo dilate and become **maternal sinusoids**, where nutrients are obtained and wastes eliminated.

6. Cells from the yolk sac and trophoblast form the **chorion**, which surrounds the embryo and eventually becomes the principal embryonic part of the placenta. The chorion protects the fetus from immune responses of the mother and produces hCG.

7. **Chorionic villi**, projections of the chorion, connect to the embryonic heart so that maternal and fetal blood vessels are brought into close proximity, allowing the exchange of nutrients and wastes between maternal and fetal blood.

8. The **placenta** is the site of exchange of nutrients and wastes between the mother and fetus. It has a fetal portion formed by chorionic villi and maternal portion formed by the endometrium. The placenta supplies oxygen and removes carbon dioxide from fetal blood, functions as a protective barrier, stores nutrients, and produces several hormones to maintain pregnancy. The **umbilical cord** develops from the connecting stalk and connects the placenta and the embryo/fetus. After childbirth, the placenta detaches from the uterus and is called the **afterbirth**. The scar where the umbilical cord was attached is the **umbilicus**.

9. All major organs develop in the embryo between the fourth and eighth weeks.

10. Week nine begins the **fetal period**, which continues until birth. The fetal period is primarily concerned with the growth and differentiation of tissues and organs that developed during the embryonic period.

Concept 25.8 During pregnancy the uterus expands, maternal gastrointestinal tract organs are displaced, and the ureters and urinary bladder are compressed.

1. Progesterone and estrogens maintain the uterine lining and the preparation of mammary glands for milk secretion. They are secreted by the corpus luteum, then the chorion, and later, the placenta. High estrogen levels prepare the myometrium for labor, while progesterone ensures the myometrium is relaxed and the cervix tightly closed.

2. The corpus luteum, and later the placenta, produces relaxin, which increases the flexibility of pelvic ligaments and helps dilate the cervix for delivery. The chorion produces **human chorionic somatomammotropin**, which prepares the mammary glands for lactation, increases protein synthesis for maternal growth, and influences maternal and fetal metabolism. **Corticotropin-releasing hormone** from the placenta probably functions in the timing of the birth.

3. As the uterus occupies more of the pelvic cavity, maternal abdominal organs get pushed superiorly against the diaphragm while the ureters and urinary bladder get compressed. During pregnancy there is weight gain, breast enlargement, and lower back pain resulting from lordosis.

4. To meet the fetal demands for nutrients and oxygen, cardiac output, heart rate, and blood volume increase, as do respiratory rate and volume. Several other anatomical and physiological changes also occur in the pregnant woman.

See

• Animation – Hormonal Regulation of Pregnancy

Hear

• **Figure 25.36** – Hormones during Pregnancy

Concept 25.9 Labor includes dilation of the cervix and expulsion of the fetus and placenta.

1. **Labor** is the process by which the fetus is expelled from the uterus through the vagina and is dependent on complex interactions of several hormones. Rising levels of estrogens near the end of pregnancy stimulate release of prostaglandins by the placenta, which cause softening of the cervix, and estrogens increase the number of oxytocin receptors in uterine muscle fiber. Oxytocin stimulates uterine contractions via a positive feedback cycle. Relaxin increases flexibility of the pubic symphysis and helps dilate the cervix.

2. **True labor** involves uterine contractions at regular intervals and dilation of the cervix, expulsion of the fetus, and delivery of the placenta. **False labor** involves pain at irregular intervals that does not intensify.

3. During the **puerperium** period following delivery, maternal reproductive organs returns to the prepregnancy state, including **involution** of the uterus.

See

• Anatomy Overview – Hypothalamic Reproductive Hormones
• Animation – Regulation of Labor and Birth

Concept	Resources

Concept 25.10 Milk production and ejection are influenced by prolactin, estrogens, progesterone, and oxytocin.

1. **Lactation** refers to the production and ejection of milk by the mammary glands.
2. Milk production is influenced by prolactin, estrogens, and progesterone. Milk ejection is stimulated by oxytocin.
3. A few of the many benefits of breast-feeding include ideal nutrition for the infant, protection from disease, decreased likelihood of developing allergies, and enhanced intellectual and neurological development.

See
- Anatomy Overview – Hypothalamic Reproductive Hormones
- Animation – Regulation of Lactation

Do
- Exercise – Pregnancy, Birth, and Lactation

Other Chapter Resources

For additional information on clinical applications, medical disorders, and medical terminology, see Chapter 14 in the *Clinical Connections* manual packaged with your text.

Within *WileyPLUS*
- Chapter Quiz
- Visual Anatomy
- Mastering Vocabulary
- Audio Glossary
- Homeostatic Imbalance: The Case of the Girl with the Relentless Pelvic Pain
- Web Links

• Understanding the Concepts

1. Describe the principal events of spermatogenesis. Which spermatogenic cells in a seminiferous tubule are least mature? Most mature?

2. What are the roles of luteinizing hormone, follicle-stimulating hormone, testosterone, and inhibin in the male reproductive system? How is secretion of these hormones controlled?

3. What is the course of sperm through the ducts of the male reproductive system from the seminiferous tubules through the urethra?

4. Describe the principal events of oogenesis. How do the results of oogenesis and spermatogenesis differ?

5. What is the functional significance of the stratum basalis of the endometrium?

6. Why is an abundant blood supply important to the uterus?

7. How does the histology of the vagina contribute to its function?

8. Why is the anterior portion of the perineum called the urogenital triangle?

9. Describe the influence of rising levels of estrogens on the secretion of GnRH, LH, and FSH.

10. What are the major events of each of the three phases of the uterine cycle? How are they correlated with the events of the ovarian cycle?

11. How does the corpus luteum develop? Why does it last more than 2 weeks when fertilization occurs?

12. Why is fertilization most likely to occur within a 3-day period of the female reproductive cycle?

13. What is capacitation?

14. What are the functions of the trophoblast?

15. Why are maternal sinusoids and the development of chorionic villi important to the embryo?

16. Describe the formation of the amnion, yolk sac, and chorion. What are their functions?

17. Explain the structural and functional changes that occur in the mother during pregnancy.

18. Describe the following labor events: the stage of dilation, the stage of expulsion, and the placental stage. Which event marks the beginning of the stage of expulsion?

U.S. Customary System

Parameter	Unit	Relation to Other U.S. Units	SI (Metric) Equivalent
Length			
	inch	1/12 foot	2.54 centimeters
	foot	12 inches	0.305 meter
	yard	36 inches	0.9144 meter
	mile	5,280 feet	1.609 kilometers
Mass			
	grain	1/1000 pound	64.799 milligrams
	dram	1/16 ounce	1.772 grams
	ounce	16 drams	28.350 grams
	pound	16 ounces	453.6 grams
	ton	2,000 pounds	907.18 kilograms
Volume (Liquid)			
	ounce	1/16 pint	29.574 milliliters
	pint	16 ounces	0.473 liter
	quart	2 pints	0.946 liter
	gallon	4 quarts	3.785 liters
Volume (Dry)			
	pint	1/2 quart	0.551 liter
	quart	2 pints	1.101 liters
	peck	8 quarts	8.810 liters
	bushel	4 pecks	35.239 liters

International System (SI)

Base Units			Prefixes		
Unit	Quantity	Symbol	Prefix	Multiplier	Symbol
meter	length	m	tera-	$10^{12} = 1,000,000,000,000$	T
kilogram	mass	kg	giga-	$10^{9} = 1,000,000,000$	G
second	time	s	mega-	$10^{6} = 1,000,000$	M
liter	volume	L	kilo-	$10^{3} = 1,000$	k
mole	amount of matter	mol	hecto-	$10^{2} = 100$	h
			deca-	$10^{1} = 10$	da
			deci-	$10^{-1} = 0.1$	d
			centi-	$10^{-2} = 0.01$	c
			milli-	$10^{-3} = 0.001$	m
			micro-	$10^{-6} = 0.000001$	μ
			nano-	$10^{-9} = 0.000000001$	n
			pico-	$10^{-12} = 0.000000000001$	p

Temperature Conversion

Fahrenheit (F) To Celsius (C)

°C = (°F − 32) ÷ 1.8

Celsius (C) To Fahrenheit (F)

°F = (°C × 1.8) + 32

U.S. to SI (Metric) Conversion

When You Know	Multiply By	To Find
inches	2.54	centimeters
feet	30.48	centimeters
yards	0.91	meters
miles	1.61	kilometers
ounces	28.35	grams
pounds	0.45	kilograms
tons	0.91	metric tons
fluid ounces	29.57	milliliters
pints	0.47	liters
quarts	0.95	liters
gallons	3.79	liters

SI (Metric) to U.S. Conversion

When You Know	Multiply By	To Find
millimeters	0.04	inches
centimeters	0.39	inches
meters	3.28	feet
kilometers	0.62	miles
liters	1.06	quarts
cubic meters	35.32	cubic feet
grams	0.035	ounces
kilograms	2.21	pounds

The periodic table lists the known **chemical elements**, the basic units of matter. The elements in the table are arranged left to right in rows in order of their **atomic number**, the number of protons in the nucleus. Each horizontal row, numbered from 1 to 7, is a **period**. All elements in a given period have the same number of electron shells as their period number. For example, an atom of hydrogen or helium each has one electron shell, while an atom of potassium or calcium each has four electron shells. The elements in each column, or **group**, share chemical properties. For example, the elements in column IA are very chemically reactive, whereas the elements in column VIIIA have full electron shells and thus are chemically inert.

Scientists now recognize 117 different elements; 92 occur naturally on Earth, and the rest are produced from the natural elements using particle accelerators or nuclear reactors. Elements are designated by **chemical symbols**, which are the first one or two letters of the element's name in English, Latin, or another language.

Twenty-six of the 92 naturally occurring elements normally are present in your body. Of these, just four elements—oxygen (O), carbon (C), hydrogen (H), and nitrogen (N) (coded blue)—constitute about 96% of the body's mass. Eight others—calcium (Ca), phosphorus (P), potassium (K), sulfur (S), sodium (Na), chlorine (Cl), magnesium (Mg), and iron (Fe) (coded pink)—contribute 3.8% of the body's mass. An additional 14 elements, called **trace elements** because they are present in tiny amounts, account for the remaining 0.2% of the body's mass. The trace elements are aluminum, boron, chromium, cobalt, copper, fluorine, iodine, manganese, molybdenum, selenium, silicon, tin, vanadium, and zinc (coded yellow).

Legend:
- 23 — Atomic number
- V — Chemical symbol
- 50.942 — Atomic mass (weight)

Percentage of body mass
- 96% (4 elements)
- 3.8% (8 elements)
- 0.2% (14 elements)

IA																	VIIIA
1 Hydrogen **H** 1.0079	IIA											IIIA	IVA	VA	VIA	VIIA	2 Helium **He** 4.003
3 Lithium **Li** 6.941	4 Beryllium **Be** 9.012											5 Boron **B** 10.811	6 Carbon **C** 12.011	7 Nitrogen **N** 14.007	8 Oxygen **O** 15.999	9 Fluorine **F** 18.998	10 Neon **Ne** 20.180
11 Sodium **Na** 22.989	12 Magnesium **Mg** 24.305	IIIB	IVB	VB	VIB	VIIB		VIIIB		IB	IIB	13 Aluminum **Al** 26.9815	14 Silicon **Si** 28.086	15 Phosphorus **P** 30.974	16 Sulfur **S** 32.066	17 Chlorine **Cl** 35.453	18 Argon **Ar** 39.948
19 Potassium **K** 39.098	20 Calcium **Ca** 40.08	21 Scandium **Sc** 44.956	22 Titanium **Ti** 47.87	23 Vanadium **V** 50.942	24 Chromium **Cr** 51.996	25 Manganese **Mn** 54.938	26 Iron **Fe** 55.845	27 Cobalt **Co** 58.933	28 Nickel **Ni** 58.69	29 Copper **Cu** 63.546	30 Zinc **Zn** 65.41	31 Gallium **Ga** 69.723	32 Germanium **Ge** 72.59	33 Arsenic **As** 74.992	34 Selenium **Se** 78.96	35 Bromine **Br** 79.904	36 Krypton **Kr** 83.80
37 Rubidium **Rb** 85.468	38 Strontium **Sr** 87.62	39 Yttrium **Y** 88.905	40 Zirconium **Zr** 91.22	41 Niobium **Nb** 92.906	42 Molybdenum **Mo** 95.94	43 Technetium **Tc** (98)	44 Ruthenium **Ru** 101.07	45 Rhodium **Rh** 102.905	46 Palladium **Pd** 106.42	47 Silver **Ag** 107.868	48 Cadmium **Cd** 112.40	49 Indium **In** 114.82	50 Tin **Sn** 118.69	51 Antimony **Sb** 121.75	52 Tellurium **Te** 127.60	53 Iodine **I** 126.904	54 Xenon **Xe** 131.30
55 Cesium **Cs** 132.905	56 Barium **Ba** 137.33		72 Hafnium **Hf** 178.49	73 Tantalum **Ta** 180.948	74 Tungsten **W** 183.85	75 Rhenium **Re** 186.2	76 Osmium **Os** 190.2	77 Iridium **Ir** 192.22	78 Platinum **Pt** 195.08	79 Gold **Au** 196.967	80 Mercury **Hg** 200.59	81 Thallium **Tl** 204.38	82 Lead **Pb** 207.19	83 Bismuth **Bi** 208.980	84 Polonium **Po** (209)	85 Astatine **At** (210)	86 Radon **Rn** (222)
87 Francium **Fr** (223)	88 Radium **Ra** (226)		104 Rutherfordium **Rf** (261)	105 Dubnium **Db** (262)	106 Seaborgium **Sg** (266)	107 Bohrium **Bh** (264)	108 Hassium **Hs** (269)	109 Meitnerium **Mt** (268)	110 **Ds** (271)	111 **Rg** (272)	112 **Uub** (277)	113 **Uut** (284)	114 **Uuq** (289)	115 **Uup** (288)	116 **Uuh** (293)		118 **Uuo** (294)

57–71, Lanthanides

57 Lanthanum **La** 138.91	58 Cerium **Ce** 140.12	59 Praseodymium **Pr** 140.907	60 Neodymium **Nd** 144.24	61 Promethium **Pm** 144.913	62 Samarium **Sm** 150.35	63 Europium **Eu** 151.96	64 Gadolinium **Gd** 157.25	65 Terbium **Tb** 158.925	66 Dysprosium **Dy** 162.50	67 Holmium **Ho** 164.930	68 Erbium **Er** 167.26	69 Thulium **Tm** 168.934	70 Ytterbium **Yb** 173.04	71 Lutetium **Lu** 174.97

89–103, Actinides

89 Actinium **Ac** (227)	90 Thorium **Th** 232.038	91 Protactinium **Pa** (231)	92 Uranium **U** 238.03	93 Neptunium **Np** (237)	94 Plutonium **Pu** 244.064	95 Americium **Am** (243)	96 Curium **Cm** (247)	97 Berkelium **Bk** (247)	98 Californium **Cf** 242.058	99 Einsteinium **Es** (254)	100 Fermium **Fm** 257.095	101 Mendelevium **Md** 258.10	102 Nobelium **No** 259.10	103 Lawrencium **Lr** 260.105

Appendix C
Answers to Checkpoint Questions

Chapter 1

1. molecules **2.** chemical, cellular, tissues **3.** organ **4.** organs: composed of different kinds of tissues which join together; tissues: groups of cells and the materials around them which work together to perform a particular function **5.** movement of blood, food through the GI tract, skeletal muscles **6.** responsiveness **7.** intracellular fluid: inside cells; extracellular fluid: outside cells; interstitial fluid: between cells of tissues **8.** interstitial fluid **9.** nervous system, endocrine system **10.** effector **11.** nervous system, endocrine system **12.** (a) both occur in response to a stimulus to regulate a controlled condition; (b) negative feedback reverses the stimulus; positive feedback enhances the stimulus **13.** if unstopped, the mechanism can produce life-threatening conditions in the body **14.** subject stands erect facing the observer, head is level, eyes face forward, feet are flat on the floor and directed forward, upper limbs are at the sides with palms turned forward **15.** front of the elbow **16.** (a) no, it is posterior; (b) yes; (c) no, the sternum is superior and medial, (d) no, the humerus is proximal to the radius; (e) yes **17.** (1) proximal and distal; (2) medial and lateral; (3) medial and lateral; (4) superior and inferior **18.** planes are imaginary flat surfaces that pass through various parts of the body; a section is a cut through a structure along a plane **19.** thoracic and abdominopelvic **20.** urinary bladder: P; stomach: A; heart: T; small intestine: A; lungs: T; thymus: T; liver: A **21.** heart, esophagus, trachea, thymus **22.** (a) pericardial; (b) pleural **23.** large intestine **24.** reduce friction **25.** (a) visceral; (b) parietal **26.** it is between the abdominopelvic wall and the peritoneum **27.** liver: epigastric; colon: umbilical; bladder: hypogastric; spleen: left lumbar **28.** right inguinal **29.** the right or left of the person you are facing

Chapter 2

1. electrons **2.** (a) proton: positive; neutron: zero; electron: negative (b) the neutral atom has a zero charge because the negatively charged electrons and positively charged protons equal each other **3.** (a) atomic number: 6; mass number: 12; atomic mass: 12.011; (b) atomic number is the number of protons in an atom; the mass number is the number of protons plus the number of neutrons in an atom; the atomic mass is the average mass of all of the naturally occurring isotopes of the element **4.** (a) atoms of an element that have different numbers of neutrons, and therefore different mass numbers (b) electrically charged atom or group of atoms with an unpaired electron in the outermost shell **5.** the outermost electron shell around an atom; the number of electrons in the valence shell will determine whether or not another atom will form a chemical bond with that atom **6.** ionic bond: transfer of electrons from one atom to another atom; covalent bond: sharing of electrons between two atoms **7.** weak bonds resulting when a hydrogen atom with a partial positive charge attracts the partial negative charge of a neighboring electronegative atom **8.** (a) reactants are the starting substances in a reaction; products are the ending substances; (b) metabolism **9.** the reaction shown releases more energy than it absorbs **10.** they lower the amount of energy needed to start the reaction **11.** (a) anabolism: synthesis reaction; catabolism: decomposition reaction; (b) endergonic reactions: synthesis reactions **12.** reactions in which products can revert back to the original reactants **13.** inorganic compounds lack carbon and are held together by ionic or covalent bonds; organic compounds always contain carbon and always have covalent bonds **14.** acts as a solvent, acts as a lubricant; actively participates in some decomposition and synthesis reactions; can release or absorb a relatively large amount of heat with only a slight change in its own temperature **15.** solvent: liquid or gas in which some other substance (a solute) is dissolved; solute: the dissolved substance in a solution **16.** hydrophilic molecules: easily dissolved in water because they are charged or contain polar covalent bonds; hydrophobic molecules: not very water soluble because they contain mainly nonpolar covalent bonds **17.** breakdown of large molecules into smaller molecules by the addition of water molecules **18.** 6.82 **19.** They convert strong acids or bases into weak acids or bases, thereby maintaining the pH of fluids inside and outside body cells. **20.** (1) it can form bonds with thousands of other carbon atoms to produce large molecules; (2) it has four electrons in its valence shell, so can bond covalently with a variety of atoms **21.** monomers are small, building-block molecules; polymers are macromolecules made up of monomers bonded together **22.** primary source of chemical energy for generating ATP **23.** monosaccharides: contain from three to seven carbon atoms; disaccharides: combination of two monosaccharide molecules; polysaccharides: contain tens or hundreds of monosaccharides joined together **24.** liver cells and skeletal muscle cells **25.** they have few polar covalent bonds and are hydrophobic **26.** (a) saturated fats contain only single covalent bonds between fatty acid carbon atoms; monounsaturated fats contain fatty acids with one double covalent bond between two fatty acid carbon atoms; polyunsaturated fats contain more than one double covalent bond between fatty acid carbon atoms; (b) saturated fats: meats, non-skim dairy products, a few plant products; monounsaturated fats: olive oil, peanut oil, canola oil, most nuts, and avocadoes; polyunsaturated fats: corn oil, safflower oil, sunflower oil, soybean oil, and fatty fish **27.** (a) a triglyceride consists of a single glycerol backbone with three fatty acid molecules attached to it, whereas a triglyceride has a glycerol backbone with two fatty acid molecules and a phosphate group attached to it (b) The polar phosphate group and charged group form the hydrophilic head and the nonpolar fatty acids form the hydrophobic tails. **28.** triglycerides: body's most highly concentrated form of chemical energy; phospholipids: make up much of the membrane that surrounds each cell; steroids: many functions in the body such as required for cell membrane structure (cholesterol), required for regulating sexual functions (estrogen and progesterone), necessary for maintaining normal blood sugar levels (cortisol), needed for lipid digestion and absorption (bile salts), needed for bone growth (vitamin D) **29.** proteins contain carbon, hydrogen, oxygen and nitrogen and are much more complex than carbohydrates or lipids **30.** (a) the covalent bond joining two amino acids together; (b) 2 **31.** peptide: chain of four to nine amino acids; polypeptide: chain of 10 to 2000 or more amino acids; protein: a polypeptide that contains as few as 50 or as many as 2000 amino acids **32.** specificity; efficiency; control **33.** they lower the activation energy of a chemical reaction by decreasing the randomness of the collisions between molecules and bring the substrates together in the proper orientations so that the reaction can occur **34.** DNA forms the inherited genetic material inside each human cell; RNA relays instructions from the genes to guide each cell's synthesis of proteins from amino acids **35.** a nitrogenous base, a five-carbon sugar, and a phosphate group substituted for thymine in RNA **36.** muscle contractions, movement of chromosomes during cell division, movement of structures within cells, transport of substances across cell membranes, and synthesis of larger molecules from smaller ones **37.** three phosphate groups attached to adenosine

Chapter 3

1. (a) plasma membrane: serves as a selective barrier between the cell's internal and external environment; (b) nucleus: contains most of the cell's DNA; (c) cytoplasm: cytosol is site of essential chemical reactions and organelles have specific functions **2.** the hydrophobic ends face inward and the hydrophilic ends face outward **3.** the carbohydrate portions of glycolipids and glycoproteins which enable cells to recognize one another **4.** lipid-soluble substances can pass through the lipid bilayer; water soluble substances cannot pass through **5.** they allow some substances to pass readily across the membrane and prohibit others from crossing **6.** energy input from the cell **7.** energy input from the cell **8.** (a) steepness of the concentration gradient, temperature, mass of the diffusing substances, surface area, diffusion distance; (b) because channels occupy a smaller fraction of the membrane's total surface than do lipids **9.** a force proportional to the concentration of the solute particles that cannot cross the plasma membrane **10.** source of energy used to drive the process: energy obtained from hydrolysis of ATP powers primary active transport; energy stored in an ionic concentration gradient powers secondary active transport **11.** symporters move substances in the same direction; antiporters move substances in opposite directions **12.** an active process by which vesicles are used to successively move a substance into, through, and out of a cell **13.** (a) microfilaments; (b) microtubules **14.** cilia move substances across the cell surface; a flagellum moves the entire cell **15.** they are formed in the nucleolus and assembled in the cytoplasm **16.** rough ER has ribosomes attached to the membrane and is involved in the synthesis of proteins; smooth ER lacks ribosomes and is involved in the synthesis of lipids **17.** extracellular fluid, plasma membrane, other cytoplasmic organelles such as lysosomes **18.** (a) lysosomes digesting worn-out organelles; (b) lysosomes digesting an entire cell **19.** ATP is produced by the aerobic phase of cellular respiration **20.** via nuclear pores **21.** in chromosomes **22.** double-stranded DNA, linker DNA, and histone proteins **23.** transcription is the writing out of the DNA message in the form of messenger RNA; translation is the reading of that message at the ribosomes **24.** 1) mRNA enters the cytoplasm to direct the synthesis of a specific protein; 2) rRNA joins with ribosomal proteins to make ribosomes; 3) tRNA binds to an amino acid and holds it in place until it is incorporated into a protein during translation **25.** P site is the site for the binding of the first tRNA molecule bearing its amino acid; A site holds the next tRNA molecule bearing its amino acid **26.** (a) somatic cell division: a cell undergoes mitosis and cytokinesis to produce two identical cells; (b) reproductive cell division: mechanism that produces gametes which have the chromosome number of the somatic cells **27.** 1) G_1 phase: cell growth, organelle duplication, beginning of centriole replication; 2) S phase: DNA replication; 3) G_2 phase: synthesis of enzymes and other proteins, completion of centrosome replication **28.** (a) prophase: chromatin fibers condense into paired chromatids, nucleolus and nuclear enveolope disappear, centrosomes move to opposite poles of the cell; (b) metaphase: centromeres of chromatid pairs line up at the metaphase plate; (c) anaphase: centromeres split, identical sets of chromosomes move to opposite poles of the cell; (d) telophase: nuclear envelopes and nucleoli reappear, chromosomes resume chromatin form, mitotic spindle disappears **29.** in late anaphase **30.** chromosome pairs are separated rather than the chromatids of the individual chromosomes **31.** to reduce the chromosome number by one-half

Chapter 4

1. groups of cells and the materials surrounding them that work together to perform a particular function **2.** epithelial tissue, connective tissue, muscular tissue, nervous tissue **3.** (a) connective tissue, (b) nervous tissue, (c) muscle tissue **4.** they form tight seals between cells, anchor cells to one another, provide channels for ions and molecules to pass from one cell to another, allow communication between cells **5.** apical surface **6.** (a) attach to and support the overlying epithelial tissues, form a surface along which epithelial cells migrate during growth or wound healing, restrict passage of larger molecules between epithelium and connective tissue, participate in filtration of blood in the kidneys; (b) basal lamina and reticular lamina **7.** simple, pseudostratified, or stratified epithelium and squamous, cuboidal, or columnar **8.** (a) endothelium is found lining the heart chambers and the lumen of blood vessels and lymphatic vessels; mesothelium is the epithelial layer of serous membranes; (b) both are composed of simple squamous epithelium **9.** goblet cells: secrete mucus; microvilli: increase surface area; cilia: sweep substances across epithelial apical surfaces **10.** to help protect the skin and underlying tissues from heat, microbes, and chemicals **11.** it binds together, supports, and strengthens other body tissues; protects and insulates internal organs; compartmentalizes structures such as skeletal muscles; is the major transport system within the body; is the major site of stored energy; and is the main source of immune responses **12.** they consist of cells, fibers, and ground substance; they do not occur on body surfaces; they are usually highly vascular **13.** these are cells that retain the capacity for cell division and secrete the extracellular matrix **14.** adi-

pocytes store fats; mast cells, plasma cells, and macrophages fight infection; and fibroblasts secrete fibers and ground substance **15.** (a) elastic fibers; (b) collagen fibers; (c) reticular fibers **16.** (a) fibrocartilage; (b) elastic cartilage; (c) hyaline cartilage **17.** a flat sheet of pliable tissue that covers or lines a part of the body and consists of a combination of an epithelial layer and an underlying connective tissue layer **18.** mucous membranes line body cavities that open directly to the exterior and function as a barrier that microbes and other pathogens have difficulty penetrating; serous membranes line body cavities that do not open directly to the exterior and cover the organs that lie within the cavity and functions to allow organs to glide easily over one another or to slide against the walls of cavities; cutaneous membranes cover the surface of the body and regulate body temperature, protect underlying tissues, provide cutaneous sensations, excrete body wastes, and synthesize vitamin D **19.** (a) prevents the cavities from drying out, traps particles in the respiratory passageways, and lubricates food as it moves through the GI tract; (b) mucous membranes; (c) goblet cells **20.** (a) visceral layer of the peritoneum; (b) line body cavities that do not open directly to the exterior and cover the organs that lie within the cavity to allow organs to glide easily over one another or to slide against the walls of the cavities **21.** (a) lining the cavities of freely moving joints; (b) lubricates and nourishes the cartilage covering the bones at movable joints, contains macrophages that remove microbes and debris from the joint cavity, line cushioning sacs called bursae and tendon sheaths to ease the movement of muscle tendons **22.** produce body movements, maintain posture, generate heat **23.** (a) skeletal muscle: striations and long fibers; (b) cardiac muscle: branched fibers and intercalated discs; (c) smooth muscle: spindle-shaped fibers and lack of striations **24.** (a) smooth muscle; (b) cardiac muscle; (c) skeletal muscle **25.** skeletal muscle **26.** strengthen cardiac muscle tissue, hold the fibers together during contractions, and provide a route for quick conduction of electrical impulses from one cardiac muscle cell to another **27.** (a) dendrite; (b) axon and axon terminals **28.** neurons generate and conduct nerves impulses, whereas neuroglia support, nourish, and protect the neurons **29.** neuroglia cells do not generate or conduct nerve impulses **30.** epithelial, connective, muscle, nervous tissue **31.** osseous tissue is vascular and cartilage is avascular **32.** parenchyma cells, which constitute the functioning part of a tissue or organ, can regenerate; whereas stroma, which is the supporting connective tissue, when injured will form scar tissue **33.** stromal repair results in scar tissue; parenchymal repair results in tissue regeneration **34.** provide a supporting framework through which epithelial cells migrate to fill the wounded area and secrete a fluid that kills bacteria

Chapter 5

1. epidermis, dermis, hypodermis, epidermal derivatives **2.** areolar and adipose tissues **3.** (a) it is thicker; has no hair follicles and arrector pili muscles or sebaceous glands; has denser sensory receptors; has more numerous sweat glands; (b) thicker skin is more protective; more sensory receptors allow for greater stimuli reception **4.** keratinized stratified squamous epithelium **5.** (a) they release a lipid-rich water-repellent secretion that fills the spaces between cells of the stratum granulosum, stratum lucidum, and stratum corneum; (b) contributes to skin color and absorbs damaging UV light **6.** (a) Merkel cells; (b) melanocytes; (c) Langerhans cells; (d) keratinocytes **7.** stratum basale **8.** stratum basale: a single row of cuboidal or columnar keratinocytes just above the dermis; stratum spinosum: layers of polyhedral keratinocytes that appear to have spines when viewed on a microscope slide; stratum granulosum: presence of membrane-enclosed lamellar granules within the keratinocytes; stratum lucidum: clear layer found only in thick skin; stratum corneum: multiple layers of dead, flat keratinocytes **9.** papillary region: areolar connective tissue; reticular region: dense irregular connective tissue with collagen and elastic fibers **10.** provide the skin with strength, extensibility, and elasticity **11.** the cut is only in the epidermal region **12.** the amount of melanin in the melanocytes **13.** they are areas of accumulated melanin **14.** melanin: gives the skin color ranging from pale yellow to tan to black; carotene: gives the skin color ranging from yellow to red to tan to black; hemoglobin: gives skin color ranging from pink to red **15.** protection, decrease in heat loss, sensing light touch **16.** medulla, cortex, cuticle **17.** contains blood vessels **18.** autonomic stimulation of the arrector pili muscle causing contraction and pulling the hair shafts into an upright position **19.** sebum: coats the surface of hairs and helps keep them from drying and becoming brittle, prevents excessive evaporation of water from the skin, keeps the skin soft and pliable, inhibits growth of certain bacteria; eccrine sweat: helps regulate body temperature through evaporation; cerumen: provides a sticky barrier that impedes the entrance of foreign bodies into the external auditory canal **20.** (a) help grasp and manipulate small objects, provide protection against trauma to the ends of the fingers and toes, allow us to scratch various parts of the body; (b) they are plates of tightly packed, hard, keratinized epidermal cells **21.** stratum basale **22.** it prepares the wound for repair by helping eliminate microbes, foreign material, and dying tissue and increases the permeability and diameter of local blood vessels **23.** (a) scar tissue formation; (b) epidermal tissue is mitotic tissue **24.** a scar that extends beyond the injury boundaries into normal surrounding tissues **25.** by liberating sweat at the skin surface and by adjusting the flow of blood in the dermis **26.** by the action of keratinocytes, by lipids released by lamellar granules, by sebum, by the acidic pH of sweat, by vasoconstriction **27.** touch, pressure, vibration, tickling, heat, cold, pain **28.** (a) through the production of sweat; (b) synthesis of vitamin D

Chapter 6

1. calcium and phosphorus **2.** red marrow is hemopoietic marrow; yellow marrow stores triglycerides **3.** (a) short bone; (b) long bone **4.** flat bone **5.** (a) sesamoid bone; (b) protect tendons from excessive wear and tear and changes the direction of pull on a tendon to improve the mechanical advantage at a joint **6.** (a) shaft of the bone, (b) end of the bone, (c) between the shaft and end of the bone **7.** epiphyseal plate: located in the metaphysis, composed of hyaline cartilage, is the growth plate; articular cartilage: covers the part of the epiphysis where the bone forms an articulation with another bone, composed of hyaline cartilage, reduces friction and absorbs shock at freely movable joints; periosteum: surrounds the bone surface wherever it is not covered by articular cartilage, is a sheath of dense irregular connective tissue containing osteoblasts, functions to protect the bone, assists in fracture repair, nourishes bone tissue, and serves as an attachment point for ligaments and tendons **8.** inorganic mineral salts including calcium phosphate and calcium carbonate (hydroxyapatite) **9.** (a) osteogenic cells; (b) osteoblasts **10.** osteoclasts **11.** osteoblasts: form bone; osteocytes: maintain bone; osteoclasts: break down bone **12.** concentric lamella: around the central canals; interstitial lamella: between osteons; circumferential lamella: encircling the bone just beneath the periosteum or encircling the medullary cavity **13.** hemopoietic (red) bone marrow **14.** (a) through numerous perforating canals; (b) near the center of the diaphysis via the nutrient foramen **15.** via nutrient veins and periosteal veins in the diaphysis and via metaphyseal veins and epiphyseal veins in the epiphysis **16.** (a) development of the ossification center, calcification, formation of trabeculae, and development of the periosteum; (b) development of the cartilage model, growth of the cartilage model, development of the primary ossification center; (c) development of the medullary cavity, development of the secondary ossification centers, formation of articular cartilage and the epiphyseal plate **17.** primary ossification center forms in the diaphysis, secondary ossification centers form in the epiphysis **18.** through the action of osteoclasts **19.** (a) endochondrial ossification; (b) intramembraneous ossification **20.** new chondrocytes are formed on the epiphyseal side of the plate, while old chondrocytes on the diaphyseal side of the plate are replaced by osseous tissue **21.** osteoblasts produce osseous tissue faster than chondrocytes can produce epiphyseal plate cartilage **22.** 1) periosteal ridges fuse around a periosteal blood vessel forming an endosteum-lined tunnel, 2) endosteum osteoblasts build new concentric lamellae inward toward the center of the tunnel which forms a new osteon, 3) the bone grows outward as periosteal osteoblasts build new circumferential lamella **23.** by osteoclasts in the endosteum breaking down bone **24.** calcium, phosphorus, fluoride, magnesium iron, manganese, vitamins A, C, B$_{12}$, D, and K **25.** by stimulating the production of insulinlike growth factors which are the most important hormones that stimulate bone growth **26.** the ongoing replacement of old osseous tissue by new osseous tissue **27.** it renews osseous tissue before deterioration sets in, it heals injured bone, it redistributes bone along lines of mechanical stress

Chapter 7

1. axial skeleton: skull, vertebral column; appendicular skeleton: clavicle, shoulder girdle, humerus, pelvic girdle, femur **2.** (a) passageway for soft tissues or formation of joints; (b) meatus; (c) fissure; (d) fossa **3.** (a) form joints or attach-

ment points for connective tissue; (b) facet; (c) spinous process; (d) tuberosity **4.** cranial bones: enclose and protect the brain, provide large areas for muscle attachment, protect and support the special sense organs; facial bones: form the face, protect and support the special sense organs **5.** they are the only moveable bones of the skull **6.** passageway for blood vessels, nerves, or ligaments **7.** frontal, parietal, temporal, occipital, sphenoid, ethmoid, inferior nasal conchae, vomer **8.** (a) frontal, sphenoid, temporal, occipital; (b) sphenoid, parietal, occipital, mandible, zygomatic **9.** (a) temporal process of the zygomatic bone and zygomatic process of the temporal bone; (b) mandibular fossa of the temporal bone and the condylar process of the mandible **10.** (a) occipital condyles; (b) external occipital protuberance **11.** ethmoid bone, frontal bone, parietal bone, temporal bone, occipital bone, temporal bone, parietal bone, frontal bone **12.** external occipital protuberance, located on the external surface of the occipital bone, and internal auditory meatus, located on the internal surface of the occipital bone **13.** (a) increase the vascular and mucous membrane surface area in the nasal cavities, which aids in the sense of smell and warms, moistens, and filters inhaled air before it passes into the lungs; (b) superior and middle nasal concha **14.** nasal bones **15.** (a) maxillae and palatine bones; (b) hard palate **16.** lacrimal fossa, lacrimal **17.** alveolar processes **18.** vomer **19.** mandible **20.** temporomandibular joint **21.** (a) vomer, septal cartilage, perpendicular plate of the ethmoid bone; (b) hyaline cartilage **22.** (a) frontal, ethmoid, sphenoid, and maxillary bones; (b) produce mucus, light the skull, serve as resonating chambers for sound **23.** (a) parietal and temporal bones; (b) both parietal bones and the occipital bone; (c) frontal bone and both parietal bones; (d) the two parietal bones **24.** anterolateral fontanels **25.** 18 to 24 months after birth **26.** it does not articulate with any other bone **27.** encloses and protects the spinal cord, supports the head, serves as a point of attachment for the ribs, pelvic, girdle, and muscles of the back **28.** (a) cervical, thoracic, lumbar, sacrum, coccyx; (b) they are fused and the rest of the vertebral column is not **29.** (a) increase its strength, help maintain balance in the upright position, absorb shocks during walking, and help protect the vertebrae from fracture; (b) cervical and lumbar **30.** form strong joints, permit various movements of the vertebral column, absorb vertical shock **31.** body **32.** pedicles, laminae **33.** vertebral foramen **34.** (a) two transverse processes, a spinous process; (b) two superior articular processes and two inferior articular processes; (c) facets or demifacets **35.** (a) cervical; (b) allow passage of the vertebral artery to supply blood to the brain **36.** cervical: slender and often bifid processes; thoracic: long and fairly thick processes with most projecting inferiorly; lumbar: short, blunt, and project posteriorly rather than inferiorly **37.** atlas (first) and axis (second) **38.** (a) occipital bone and atlas; (b) axis and atlas **39.** (a) thoracic vertebrae; (b) transverse processes **40.** (a) sacral canal and sacral hiatus; (b) auricular surface and sacral tuberosity **41.** manubrium, body, xiphoid process **42.** (a) clavicles and body of the sternum; (b) costal cartilages of the second through tenth ribs, manubrium, xiphoid process **43.** provide attachment for some abdominal muscles **44.** ribs **45.** (a) by its head and the tubercle; (b) intercostal space; (c) it provides surgical access to the lungs or other structures in the thoracic cavity

Chapter 8

1. (a) sternal end; (b) acromial end **2.** glenoid cavity **3.** supraspinous fossa and infraspinous fossa **4.** 30 bones **5.** greater tubercle **6.** olecranon **7.** (a) head of radius to capitulum of humerus, trochlear notch of ulna to trochlea of humerus; (b) coronoid fossa of the humerus receives coronoid process of ulna; (c) olecranon fossa of humerus receives olecranon of ulna **8.** pisiform and hamate (on ulnar side) and scaphoid and trapezium (on radial side) **9.** (a) metacarpals; (b) distal ends of metacarpals **10.** (a) accepts the bones of the lower limbs and connects them to the axial skeleton; (b) provides a strong, stable support for the vertebral column and protects the pelvic viscera **11.** hip bone: fused ilium, ischium, and pubis; pelvic girdle: the two hip bones; bony pelvis: complete ring composed of the hip bones, pubic symphysis, and sacrum **12.** (a) anterior superior iliac spine, anterior inferior iliac spine, posterior superior iliac spine, posterior inferior iliac spine; (b) serve as points of attachment for the tendons of the muscles of the trunk, hip, and thighs **13.** (a) auricular surface of the ilium; (b) acetabulum formed by the ilium, ischium, and pubis **14.** it is the route taken by the baby's head as it descends through the pelvis **15.** because of the female's adaptations to the requirements of pregnancy and childbirth **16.** more space in the true pelvis especially in the pelvic inlet and pelvic outlet **17.** femur **18.** (a) head; (b) greater and lesser trochanters,

(c) greater trochanter **19.** (a) sesamoid bone; (b) it is enclosed within the tendon of anterior thigh muscles **20.** (a) medial malleolus of the tibia; (b) lateral malleolus of the fibula **21.** (a) talus; (b) calcaneus

Chapter 9

1. anatomical characteristics and type of movement **2.** (a) cartilaginous; (b) fibrous; (c) synovial, diarthrotic **3.** (a) skull; (b) articulations between the roots of the teeth and the alveolar sockets in the maxillae and mandible **4.** (a) sutures are immovable joints; (b) syndesmoses are slightly moveable joints **5.** a synchondrosis is a cartilaginous joint, whereas the symphyses is a fibrous joint **6.** (a) synarthrotic synchondrosis; (b) synostosis **7.** (a) synchondroses; (b) symphyses **8.** articular cartilage **9.** reducing friction by lubrication of the joint, absorbing shocks, supplying oxygen and nutrients to and removing carbon dioxide and metabolic wastes from the chondrocytes within articular cartilage, and supplying phagocytic cells that remove microbes and debris which result from normal wear and tear in the joint **10.** diarthroses **11.** ligaments and articular discs **12.** (a) they resemble joint capsules because their walls consist of connective tissue lined by a synovial membrane and they are also filled with synovial fluid; they are different from joint capsules because of their location; (b) cushion the movement of body parts over one another **13.** an example of a pivot joint is the atlanto-axial joint **14.** condyloid joints are found in the wrists, and the metacarpophalangeal joints and saddle joints are found in the carpometacarpal joint between the trapezium of the carpus and metacarpal of the thumb **15.** one axis: hinge, pivot; two axes: condyloid, saddle; three axes: ball-and-socket; no axes: planar **16.** (a) extension; (b) circumduction; (c) eversion; (d) adduction; (e) elevation **17.** adduction is the movement of a bone toward the midline **18.** abduction and adduction **19.** opposition **20.** (a) coracohumeral ligaments; (b) glenohumeral ligaments **21.** rotator cuff muscles **22.** (a) ulnar collateral ligament; (b) radial collateral ligament **23.** flexion and extension **24.** (a) iliofemoral ligament; (b) pubofemoral ligament **25.** (a) patellar ligament; (b) fibular collateral ligament

Chapter 10

1. (a) skeletal; (b) cardiac and smooth **2.** smooth muscle is spindle shaped and does not have striations **3.** producing body movements, stabilizing body positions, storing and moving substances within the body, generating heat **4.** electrical excitability, contractility, extensibility, elasticity **5.** electrical excitability **6.** supports and surrounds muscles and other organs, allows free movement of muscles, carries nerves and blood and lymphatic vessels, and fills spaces **7.** (a) perimysium; (b) fasicles **8.** structural: tendons are a cord of dense regular connective tissue composed of parallel bundles of collagen fibers, aponeuroses are a broad, flat layer of dense regular connective tissue, tendon sheaths are tubes of fibrous connective tissue; functional: tendons connect muscles to bones, aponeuroses connect muscles to muscles, tendon sheaths enclose tendons **9.** T tubules: propagate action potential into the muscle fiber; myoglobin: binds oxygen molecules that diffuse into muscle fibers from interstitial fluid and releases oxygen when mitochondria need it for ATP production; sarcoplasmic reticulum: stores calcium ions **10.** thick filament **11.** (a) I band: thin filaments (actin); (b) A band: thick and thin filaments; (c) H zone: thick filaments (d) thick filament; (e) thin filaments and titin **12.** (a) myosin; (b) actin **13.** motor end plate **14.** the synaptic cleft **15.** release of acetylcholine, activation of ACh receptors, generation of muscle action potential, termination of ACh activity **16.** it is broken down by acetylcholinesterase **17.** (a) it decreases; (b) it increases **18.** some of the myosin heads are attached to actin and others are detached and preparing to bind again **19.** Z discs pull on neighboring sarcomeres, which shortens the whole muscle fiber and ultimately leads to shortening of the entire muscle **20.** calcium ion release channels in the SR close and calcium ion active transport pumps use ATP to restore low levels of calcium ions in the sarcoplasm, the troponin–tropomyosin complex slides back into position where it blocks the myosin-binding sites on the actin molecules, then the muscle relaxes **21.** provide energy for cross-bridge movement and detachment and for operation of the calcium pump **22.** one motor neuron and all the skeletal muscle fibers it stimulates **23.** (a) the muscle action potential sweeps over the sarcolemma, calcium ions are released from the sarcoplasmic reticulum and bind to troponin which allows the myosin heads to start binding to actin; (b) calcium ions are transported back into the sarcoplasmic reticulum, the level of calcium ions decreases in the sarcoplasm, myosin-binding sites are cov-

ered by tropomyosin, myosin heads detach from actin, power strokes cease, the muscle fiber relaxes **24.** (a) unfused tetanus: a sustained but wavering contraction; fused tetanus: a sustained contraction in which individual twitches cannot be detected; (b) fused tetanus **25.** both kinds of tetanus result from the release of additional calcium ions from the sarcoplasmic reticulum; due to the calcium ion increase in the sarcoplasm, fused tetanus contractions are more forceful **26.** produces smooth movements rather than a series of jerky movements **27.** ATP **28.** (a) energy transfer from creatine phosphate, anaerobic cellular respiration, and aerobic cellular respiration; (b) aerobic cellular respiration **29.** the inability of a muscle to contract forcefully after prolonged activity **30.** because oxygen is required to provide the ATP needed for increased metabolic activity, increased cardiac and respiratory activity, and increased tissue repair **31.** because of the rate of ATP hydrolysis by myosin ATPase **32.** SO fibers, FOG fibers, FG fibers **33.** (a) skeletal and cardiac; (b) skeletal and cardiac; (c) smooth; (d) smooth **34.** (a) connect cardiac muscle fibers through desmosomes and gap junctions; (b) cardiac muscle **35.** skeletal: sarcoplasmic reticulum; cardiac and smooth: sarcoplasmic reticulum and interstitial fluid **36.** (a) visceral (single-unit), (b) multiunit **37.** speed of onset is slower and duration of contraction is longer in smooth muscle compared to skeletal muscle

Chapter 11

1. the insertion end of the muscle pulls the belly of the muscle toward the origin end of the muscle **2.** fulcrum: knee joint; load: weight of the lower leg **3.** large range of motion **4.** direction of fibers: rectus abdominis; shape: trapezoid; action: depressor anguli oris; size: gluteus maximus; origin and insertion: sternocleidomastoid; location: brachioradialis; number of tendons of origin: biceps brachii **5.** they insert into skin or other muscles **6.** (a) buccinator; (b) orbicularis oculi; (c) orbicularis oris; (d) risorius **7.** (a) lateral rectus; (b) medial rectus **8.** superior rectus **9.** aids in mastication, taste perception, deglutition, and speech **10.** stabilize the hyoid bone **11.** (a) elevate; (b) depress (lower) **12.** (a) semispinalis capitis; (b) sternocleidomastoid **13.** external oblique: inferiorly and medially; internal oblique: extend at right angles to those of external oblique; transverses abdominis: transversely around the abdominal wall; rectus abdominis: down the entire length of the anterior abdominal wall **14.** rectus abdominis **15.** aorta, esophagus, inferior vena cava **16.** pubis (anteriorly), coccyx (posteriorly), one lateral wall of the pelvis to the other **17.** (a) deep transverse perineal, external urethral sphincter, external anal sphincter; (b) superficial transverse perineal, bulbospongiosus, ischiocavernosus **18.** from the pubic symphysis anteriorly, to the coccyx posteriorly, to the ischial tuberosities laterally **19.** urogenital triangle: external genitals; anal triangle: anus **20.** flex: deltoid (anterior fibers), coracobrachialis; extend: latissimus minor; adduct: pectoralis major, latissimus dorsi, infraspinatus, teres major, teres minor, coracobrachialis **21.** subscapularis, supraspinatus, infraspinatus, teres minor **22.** (a) biceps brachii, brachialis, brachioradialis; (b) triceps brachii, anconeus **23.** supinate: supinator, biceps brachii, brachoradialis; pronate: pronator teres, pronator quadratus, brachioradialis **24.** flexor retinaculum: secures the tendons of the flexors of the fingers and wrist; extensor retinaculum: stabilizes tendons of the extensors of the wrist and fingers **25.** (a) flexor pollicis brevis; (b) abductor pollicis brevis **26.** flex: abductor digiti minimi, flexor digiti minimi brevis, lumbricals: adduct: palmar interossei **27.** iliocostalis, longissimus, and spinalis **28.** laterally flex the head: splenius capitis, splenius cervicis, anterior scalene, middle scalene, posterior scalene; laterally flex vertebral column: iliocostalis cervicis, longissimus thoracis, multifidus, intertransversarii **29.** upper limb muscles are characterized by versatility of movement; lower limb muscles are larger and more powerful to provide stability, locomotion, and maintenance of posture, they also often cross two joints and act equally on both **30.** quadriceps femoris (rectus femoris, vastus lateralis, vastus medialis, vastus intermedius) **31.** iliopsoas: psoas major and iliacus; quadriceps femoris: rectus femoris, vastus lateralis, vastus medialis, vastus intermedius; hamstrings: biceps femoris, semitendinosus, semimembranosus **32.** tendons of the biceps femoris muscles (laterally) and tendons of the semitendinosus and semimembranosus muscles (medially) **33.** dorsiflex: extensor hallucis longus and extensor digitorum longus; plantar flex: fibularis longus and fibularis brevis **34.** (a) flexor digitorum longus, flexor hallucis longus; (b) extensor digitorum longus, extensor hallucis longus **35.** muscles of the hand are specialized for precise and intricate movements; muscles of the foot move the toes and contribute to the longitudinal arch of the foot

Chapter 12

1. sensory, motor, integrative **2.** (a) eyes; (b) legs **3.** CNS: brain, spinal cord; PNS: cranial nerves and their branches, spinal nerves and their branches, ganglia, sensory receptors **4.** (a) SNS: sensory neurons that convey information from somatic receptors and from receptors for the special senses to the CNS and motor neurons that conduct impulses from the CNS to skeletal muscles; ANS: sensory neurons that convey information to the CNS from autonomic sensory receptors and motor neurons that conduct nerve impulses from the CNS to smooth muscle, cardiac muscle, and glands; (b) SNS; (c) ANS **5.** (1) cell body: biosynthetic center and receptive region; (2) dendrites: receiving or input processing; (3) axon: propagates nerves impulses toward another neuron, muscle fiber, or gland **6.** synapse **7.** dendrites: input region that receives information; cell body: integration region that analyzes information; axon: output region that transmits information toward another cell **8.** multipolar neurons usually have several dendrites and one axon; bipolar neurons have one main dendrite and one axon; unipolar neurons have one process that is a fused dendrite and another process which is an axon **9.** (a) sensory; (b) motor **10.** (a) microglia; (b) CNS **11.** (a) astrocytes; (b) CNS **12.** ependymal cells **13.** the sheath electrically insulates the axon of a neuron and increases the speed of nerve impulse conduction **14.** (a) Schwann cells, oligodendrocytes; (b) Schwann cells: PNS; oligodendrocytes: CNS **15.** myelin sheath: multilayered lipid and protein covering axons of most mammalian neurons and produced by neuroglia; neurolemma: outer, nucleated cytoplasmic layer of the Schwann cell enclosing the myelin sheath **16.** white matter consists primarily of myelinated axons and possibly some unmyelinated axons, whereas gray matter consists of neuron cell bodies, dendrites, unmyelinated axons, axon terminals, and neuroglia **17.** via the flow of ions **18.** (a) voltage change, a ligand that binds to a receptor, mechanical stimulation; (b) specific ions move across the plasma membrane, down their electrochemical gradients **19.** the gates randomly alternate between open and closed positions **20.** a small buildup of negative ions in the cytosol along the inside of the membrane and a buildup of positive ions in the extracellular fluid along the outside surface of the membrane **21.** it increases **22.** they spread along the plasma membrane for a short distance and then die out **23.** dendrites and cell bodies **24.** ligand-gated and mechanically gated **25.** graded potentials can generate an action potential **26.** (a) when depolarization reaches threshold, the voltage-gated channels open, and an action potential that is always the same size (amplitude) occurs; (b) no **27.** (a) Na^+ and K^+; (b) Na^+ channels; (c) K^+ channels **28.** (a) the period of time after an action potential begins during which an excitable cell cannot generate another action potential; (b) absolute refractory period occurs from the time the Na^+ channel activation gates open to when the Na^+ channel inactivation gates close and is a time when a second action potential cannot be initiated; the relative refractory period is the period of time during which a second action potential can be initiated but only by a larger-than-normal stimulus **29.** because in salutatory conduction, ions flow from node to node along the axon, moving at a faster pace than they would if the ions had to flow through their voltage-gated channels in each adjacent segment of the membrane **30.** A fibers conduct impulses via salutatory conduction, whereas C fibers conduct impulses via continuous conduction **31.** by the frequency of impulses **32.** graded potentials arise mainly in dendrites and cell bodies via ligand-gated or mechanically gated ion channels and are localized responses, whereas action potentials arise in the axon via voltage-gated channels for Na^+ and K^+ and permit communication over a long distance **33.** in electrical synapses, action potentials are conducted directly between adjacent cells through gap junctions giving a faster communication and synchronization, whereas in chemical synapses, a chemical bridges the synapse between neurons or a neuron and an effector **34.** 1) a nerve impulse arrives at the synaptic end bulb of a presynaptic neuron; 2) the nerve impulse opens voltage-gated Ca^{2+} channels in the synaptic end bulbs; 3) Ca^{2+} ions then flow into the end bulb and trigger exocytosis of some synaptic vesicles **35.** because the synaptic vesicles, which hold neurotransmitter, can only be activated to release neurotransmitter by Ca^{2+} **36.** an excitatory postsynaptic potential will depolarize the cell membrane and possibly produce an action potential, whereas an inhibitory postsynaptic potential will hyperpolarize the membrane and prevent an action potential from occurring

Chapter 13

1. (a) cerebrum; (b) medulla oblongata of the brain stem **2.** (a) between the spinal cord and the vertebra; (b) dura mater, arachnoid mater, pia mater **3.** epidural: above the dura; subdural: between the dura and the arachnoid; subarachnoid: between the arachnoid and the pia mater **4.** subarachnoid **5.** veins **6.** serves as a shock-absorbing medium and protects from blows (mechanical protection); provides an optimal chemical environment for accurate neuronal signaling (chemical protection) **7.** (a) brain stem (medulla oblongata); (b) cerebellum **8.** sites of production: choroid plexuses; locations: lateral ventricles, third ventricle, fourth ventricle **9.** through arachnoid villi into dural sinuses **10.** gray matter **11.** brain folds: gyri; shallow grooves: sulci; deep grooves: fissures **12.** (a) longitudinal fissure; (b) central sulcus (separates frontal lobe from parietal lobe); lateral cerebral sulcus (separates frontal lobe from temporal lobe); parieto-occipital sulcus (separates parietal lobe from occipital lobe) **13.** association fibers (between gyri in same hemisphere); commissural fibers (between gyri in opposite hemispheres); projection fibers (between cerebrum and lower brain areas) **14.** controls most of the involuntary aspects of behavior related to survival; plays a primary role in a range of emotions, including pain, pleasure, docility, affection, and anger; functions in memory **15.** refers to the functional asymmetry of the brain which means that the two hemispheres specialize in performing certain unique functions **16.** Broca's speech area of the left hemisphere **17.** intermediate mass **18.** (a) hypothalamus; (b) serve as relay stations for reflexes related to the sense of smell **19.** control of the ANS; production of hormones; regulation of emotional and behavioral patterns, regulation of eating and drinking, control of body temperature, regulation of circadian rhythms and states of concentration **20.** produces melatonin which is thought to promote sleepiness and appears to contribute to the setting of the body's biological clock **21.** the pons lies directly inferior to the midbrain and anterior to the cerebellum; the medulla oblongata begins at the inferior border of the pons and extends to the foramen magnum **22.** (a) midbrain; (b) they contain axons of motor neurons that conduct nerve impulses from the cerebrum to the spinal cord, medulla, and pons; they also contain axons of sensory neurons that extend from the medulla to the thalamus **23.** pons and medulla oblongata **24.** (a) it is the crossing over of axons in the right pyramid to the left side and axons in the left pyramid to the right side; (b) it explains why each side of the brain controls movements on the opposite side of the body **25.** helps maintain consciousness and filter incoming sensory messages **26.** it has two hemispheres consisting of lobes separated by deep and distinct fissures **27.** anterior and posterior lobes govern subconscious aspects of skeletal muscle movements, and the flocculonodular lobe contributes to equilibrium and balance **28.** cerebellar peduncles **29.** (a) cervical enlargement; (b) lumbar enlargement **30.** conus medullaris: tapered end of the spinal cord; filum terminale: an extension of the pia mater that extends inferiorly and anchors the spinal cord to the coccyx; cauda equina: roots of the nerves that arise from the inferior part of the spinal cord and do not leave the vertebral column at the same level as they exit from the cord **31.** (a) sensory axons; (b) cell bodies of sensory neurons; (c) motor axons **32.** (a) gray commissure; (b) anterior (ventral) white commissure **33.** (a) anterior gray horns: cells bodies of somatic motor neurons; posterior gray horns: somatic and autonomic sensory neuron cell bodies; lateral gray horns: cell bodies of autonomic motor neurons; (b) horns are composed of gray matter; columns are composed of white matter **34.** in the columns **35.** (a) position: anterior column; origin: spine; destination: cortex; (b) motor **36.** (a) posterior columns; (b) lateral and anterior spinothalamic tracts **37.** (a) posterior columns; (b) lateral and anterior spinothalamic tracts **38.** (a) direct; (b) indirect

Chapter 14

1. a bundle of hundreds to thousands of axons plus associated connective tissue and blood vessels that lie outside the brain and spinal cord **2.** epineurium, perineurium, endoneurium; (b) epineurium: entire nerve; perineurium: fascicles: endoneurium: individual axons **3.** (a) names designate a nerve's distribution or function and numbers indicate the order, from anterior to posterior, in which the nerves arise from the brain; (b) 12 pairs **4.** mixed nerves have both motor and sensory components, whereas sensory nerves have sensory components only **5.** somatic motor axons innervate skeletal muscles; autonomic motor axons innervate glands, smooth muscle, and cardiac muscle **6.** pons **7.** jugular foramen **8.** according to the region and level of the vertebral column from which they emerge **9.** the posterior and anterior roots connect spinal nerves to the spinal cord gray matter **10.** they are formed from both sensory axons and motor axons **11.** originate from the spinal nerve proper; posterior rami serve the skin and muscles of the back, anterior rami serve the upper and lower limbs and the skin of the lateral and anterior trunk **12.** (a) cervical plexus supplies head, neck, and superior shoulders and chest; (b) brachial plexus supplies shoulders and upper limbs; (c) lumbar plexus supplies anterior and lateral abdominal wall, external genitals, and parts of lower limbs; (d) sacral plexus supplies buttocks, perineum, and lower limbs **13.** intercostal muscles between ribs, muscles, and skin of the chest wall and back, and abdominal muscles **14.** similarity: they are fast, involuntary, unplanned sequences of actions that occur in response to a particular stimulus; difference: cranial reflexes involve integration in the gray matter of the brain stem, spinal reflexes involve integration in the gray matter of the spinal cord **15.** similarity: they are fast, involuntary, unplanned sequences of actions that occur in response to a particular stimulus; difference: somatic reflexes involve contraction of skeletal muscles; autonomic reflexes involve responses of smooth muscle, cardiac muscle, and glands **16.** sensory receptor, sensory neuron, integrating center, motor neuron, effector **17.** monosynaptic: a reflex pathway having only one synapse in the CNS; polysynaptic: a reflex pathway involving more than two types of neurons and more than one CNS synapse; reciprocal innervation: a reflex arc that occurs when the stretched muscle contracts during a stretch reflex and antagonistic muscles that oppose the contraction simultaneously relax; intersegmental: a type of reflex where nerve impulses from one sensory neuron ascend and descend in the spinal cord and activate interneurons in several segments of the spinal cord **18.** input: autonomic sensory neurons in visceral organs and blood vessels; output: autonomic motor neurons to smooth muscle, cardiac muscle, and glands **19.** (a) skeletal muscles; (b) cardiac muscles, smooth muscles, glands **20.** it consists of two motor neurons in series, one following the other **21.** organs receive impulses from both sympathetic and parasympathetic motor neurons **22.** preganglionic: synapses with the postganglionic neuron; postganglionic: targets the effector **23.** sympathetic trunk ganglia: lie in a vertical row on either side of the vertebral column; prevertebral ganglia: lie anterior to the vertebral column; terminal ganglia: located close to or actually within the wall of a visceral organ **24.** (a) parasympathetic; (b) because the axons of parasympathetic preganglionic neurons extend from the CNS to a terminal ganglion in an innervated organ **25.** (a) parasympathetic; (b) because the axons of parasympathetic preganglionic neurons extend from the CNS to a terminal ganglion in an innervated organ **26.** (a) cholinergic release acetylcholine and include all sympathetic and parasympathetic preganglionic neurons, sympathetic postganglionic neurons that innervate most sweat glands, and all parasympathetic postganglionic neurons; (b) adrenergic neurons release norepinephrine and include most sympathetic postganglionic neurons **27.** (a) cholinergic receptors bind to acetylcholine; (b) adrenergic receptors bind both norepinephrine and epinephrine **28.** (a) cholinergic; (b) because acetylcholine is quickly inactivated by the enzyme acetylcholinesterase **29.** hypothalamus **30.** (a) pupils of the eyes dilate; heart rate, force of heart contraction, blood pressure increase; airways dilate; blood vessels that supply the kidneys and GI tract constrict; blood vessels that supply organs involved in exercise or fighting off danger dilate; liver increases breakdown of glycogen to glucose and adipose tissue increases breakdown of triglycerides to fatty acids and glycerol; liver release of glucose increases blood glucose levels; processes not essential for meeting the emergency are inhibited; (b) sympathetic **31.** salivation, lacrimation, urination, digestion, and defecation **32.** iris of eye: dilates; lungs: increase respiration; urinary bladder: relaxation of wall and contraction of sphincter; stomach: decreased motility; intestines: decreased motility; heart: increased heart rate and force of contractions; arterioles of the abdominal viscera: constrict, reducing blood flow to the organs; arterioles of skeletal muscles: dilate to increase blood flow **33.** gastric, intestinal, and salivary glands: sympathetic decreases secretion, parasympathetic increases secretion; ciliary muscle of eye: sympathetic adjusts lens for distant vision, parasympathetic adjusts lens for near vision; heart: sympathetic increases heart rate, parasympathetic decreases rate; lungs: sympathetic increases respiratory rate, parasympathetic decreases; stomach and intestines: sympathetic decreases motility, parasympathetic increases motility **34.** blood pressure, digestion, defecation, and urination **35.** there are two motor neurons in the autonomic reflex motor pathway **36.** by axons of neurons that form tracts from the hypothalamus to sympathetic and to parasympathetic and sympathetic nuclei in the brain stem and spinal cord through relays in the reticular formation

Chapter 15

1. each unique type of sensation such as touch, pain, vision, or hearing **2.** general senses are somatic senses and visceral senses, whereas special senses are smell, taste, vision, hearing, and equilibrium **3.** the ability of a receptor to select one modality over all others **4.** 1) stimulation of the sensory receptor; 2) transduction of the stimulus; 3) generation of nerve impulses; 4) integration of sensory input **5.** conversion of energy from a stimulus into a graded potential **6.** free nerve endings are bare dendrites; encapsulated nerve endings have dendrites enclosed within a connective tissue capsule; separate cells **7.** the generator potential or receptor potential decreases in amplitude during a maintained, constant stimulus resulting in the sensation fading or disappearing even though the stimulus persists **8.** rapidly adapting receptors adapt very quickly and are specialized for signaling changes in a stimulus; slowly adapting receptors adapt slowly and continue to trigger nerve impulses as long as the stimulus persists **9.** adapt slowly: Merkel discs, Ruffini corpuscles, nociceptors; adapt rapidly: Meissner corpuscles, hair root plexuses, Pacinian corpuscles, cold and warm receptors **10.** (a) pain serves a protective function by signaling the presence of noxious, tissue-damaging conditions; (b) by intense thermal, mechanical, or chemical stimuli **11.** superficial somatic pain: arises from stimulation of receptors in the skin; deep somatic pain: arises in skeletal muscles, joints, tendons, and fascia; visceral pain: arises in visceral organs **12.** muscle spindles: muscle length; tendon organs: muscle tension **13.** monitoring joint position and movement **14.** lips, face, tongue, and hand **15.** muscles in the hand, lips, tongue, and vocal cords **16.** because the cortical area devoted to a muscle is proportional to the number of motor units in that muscle **17.** touch, vibration, proprioception from the limbs and trunk **18.** 1) type of sensations conveyed, 2) destination of first-order neuron, 3) area of crossing over **19.** type of sensory information: proprioceptive; usefulness: posture, balance, and coordination of skilled movements **20.** provide output from the CNS to skeletal muscles **21.** 1) local circuit neurons help coordinate rhythmic activity in specific muscle groups; 2) upper motor neurons execute voluntary movements and help maintain muscle tone, posture, and balance; 3) basal nuclei neurons help initiate and terminate movements, suppress unwanted movements, and establish a normal level of muscle tone; 4) cerebellar neurons coordinate body movements and help maintain normal posture and balance **22.** direct motor pathways provide input to lower motor neurons via axons coming directly from the cerebral cortex; indirect motor pathways provide input to lower motor neurons from motor centers in the brain stem **23.** RAS stimulation increases activity of the cerebral cortex to maintain wakefulness; inhibition of RAS neurons results in sleep **24.** (a) memory is the process by which learned information is stored and retrieved; (b) immediate memory to recall ongoing experiences for a few seconds, short-term memory to recall information for seconds to minutes; long-term memory to recall information for days to years

Chapter 16

1. to provide physical support, nourishment, and electrical insulation for the olfactory receptor cells, and to help detoxify chemicals that come in contact with the olfactory epithelium **2.** they produce new olfactory receptor cells **3.** cell division **4.** to moisten the surface of the olfactory epithelium and dissolve odorants **5.** about one minute **6.** sweet (sugar), sour (sour pickle), salty (potato chips), bitter (coffee), umami (meat) **7.** 1) circumvallate: form an inverted V-shaped row at the back of the tongue; 2) fungiform: scattered over the entire surface of the tongue; 3) foliate: located in small trenches on the lateral margins of the tongue; 4) filiform: over the entire surface of the tongue **8.** insulate the receptor cells from each other and from the surrounding tongue epithelium **9.** chemoreceptors **10.** release of neurotransmitter molecules from the gustatory receptor cell with the neurotransmitter triggering action potentials in the first-order neurons **11.** 1) levator palpebre superioris: raises upper eyelid; 2) orbicularis oculi: closes eyelid **12.** Meibomian glands secrete a fluid that keeps the eyelids from adhering to each other **13.** bulbar conjunctiva **14.** a watery solution containing salts, mucus, and lysozyme and functions to protect, clean, lubricate, and moisten the eyeball **15.** the lacrimal glands produce excessive lacrimal fluid that may spill over the edges of the eyelids and even fill the nasal cavity with fluid **16.** fibrous tunic: cornea and sclera; vascular tunic: choroids, ciliary body, iris **17.** pigmentation in iris **18.** it absorbs stray light rays which prevents reflection and scattering of light within the eyeball **19.** central fovea **20.** (a) nourishes the lens and cornea; (b) with the aqueous humor, contributes to intraocular pressure holding the retina flush against the choroids **21.** (a) iris; (b) lens **22.** (a) bending of light rays; (b) cornea, lens **23.** accommodation **24.** (a) ciliary muscle is relaxed and lens is flattened due to being stretched in all directions; (b) ciliary muscle contacts, the ciliary muscle is pulled toward the lens which pulls the ciliary process forward, ciliary process moves forward toward the lens which releases tension on the lens and suspensory ligaments, which allows the lens to become more spherical **25.** astigmatism **26.** (a) the medial movement of the two eyeballs so that both are directed toward the object being viewed; (b) it allows for binocular vision **27.** transduction of light energy into a receptor potential occurs in the outer segment of both rods and cones; both contain photopigments; both synapse with bipolar neurons **28.** (a) isomerization; (b) regeneration **29.** in light more and more photopigment is bleached, whereas in darkness more and more photopigment regenerates **30.** the glutamate is an inhibitory neurotransmitter **31.** axons of the ganglion cells **32.** nasal half **33.** sound waves could not be directed to the external auditory canal **34.** 1) tympanic membrane; 2) oval window and round window **35.** (a) malleus, incus, anvil; (b) malleus: tympanic membrane and incus; incus: malleus and stapes; stapes: incus and oval window **36.** semicircular canals, vestibule, cochlea **37.** spiral organ **38.** (a) vestibulocochlear; (b) vestibular; (c) cochlear **39.** high frequency **40.** from auricles into external auditory canal to tympanic membrane to ear ossicles to oval window setting up fluid pressure waves in the cochlea in the scala vestibuli to the scala tympani which pushes the vestibular membrane back and forth, creating pressure waves in the endolymph inside the cochlear duct which causes the basilar membrane to vibrate, which moves the hair cells of the spiral organ against the tectorial membrane, which binds the hair cell microvilli, which produces receptor potentials **41.** it moves more slowly **42.** the round window **43.** via slight differences in the timing of impulses arriving from the two ears to the pons **44.** cochlea to cochlear branch of vestibulocochlear nerve to medulla oblongata to pons to midbrain to thalamus to primary auditory area in the temporal lobe of the cerebral cortex **45.** static equilibrium refers to the maintenance of the position of the body relative to the force of gravity and linear acceleration and deceleration, whereas dynamic equilibrium is the maintenance of body position in response to rotational acceleration and deceleration **46.** they both involve hair cells as the sensory receptors; receptors synapse on different branches of the vestibulocochlear nerve **47.** add weight to the otolithic membrane, amplifying the pull of gravity during movements **48.** the ducts lie at right angles to one another and allow them to respond to virtually any rotation movement of the head **49.** semicircular ducts: sense dynamic equilibrium; utricle and saccule: sense static equilibrium **50.** (a) medulla oblongata and pons; (b) axons extending from the medulla and pons control eye, head, and neck movements, and muscle tone.

Chapter 17

1. acetylcholine is a neurotransmitter released by a motor neuron **2.** (a) nervous system; (b) endocrine system **3.** sweat gland: exocrine gland which releases its secretion to the skin surface via a duct; adrenal gland: endocrine gland which releases its hormones into the blood **4.** they produce hormones via enteroendocrine cells **5.** hormones are target specific, requiring specific receptors for activity (function) **6.** lipid-soluble (steroid hormones and thyroid hormones); water-soluble (amines, peptides and proteins, and eicosanoids) **7.** (a) within the target cell; (b) on a receptor on the cell membrane **8.** it activates inactive enzymes in the cell to give the response attributed to the hormone **9.** hormone concentration, abundance of target cell's hormone receptors, influence exerted by other hormones **10.** signals from the nervous system, chemical changes in the blood, other hormones **11.** they are the direct route taken by hypothalamic hormones to act immediately on anterior pituitary cells, before the hormones are diluted or destroyed in the general circulation **12.** FSH, LH, TSH, ACTH **13.** hGH **14.** hypoglycemia **15.** low blood levels of T_3 and T_4 would stimulate increase in levels of TRH, which would stimulate increased levels of TSH; however, removal of the thyroid gland would prevent production of T_3 and T_4 **16.** (a) hypothalamic neurosecretory cells; (b) they are packed into vesicles and transported down the axons to the axon terminals in the posterior pituitary until exocytosis is triggered, then the hormones are released and diffuse into the capillaries of the posterior pituitary **17.** both are direct routes for hormone release that do not involve the general circulation

18. (a) positive feedback; (b) stretching of the cervix and suckling **19.** to increase OT production and uterine contractions for childbirth **20.** (a) lower osmotic pressure; (b) decrease level of ADH **21.** (a) follicular cells; (b) parafollicular cells; (c) T_3 and T_4 **22.** synthesis: iodide is brought into the follicles, is oxidized, binds to the tyrosine component of thyroglobulin to ultimately form T_3 and T_4 (triiodothyronine and tetraiodothyronine or thyroxin); storage: in the follicle; blood transport: carried bound to transport proteins in the blood, mainly TGB **23.** via TRH and low blood levels of T_3 and T_4, as well as other factors such as cold environment, hypoglycemia, high altitude, and pregnancy **24.** thyroid hormone stimulates body growth, particularly the growth and development of nervous tissue **25.** (a) osteoclasts; (b) inhibit bone breakdown or resorption **26.** PTH: osteoclasts, kidney tubules, GI tract; calcitriol: GI tract **27.** by blood levels of calcium ions **28.** similarity: both increase blood calcium ion levels: differences: PTH increases number and activity of osteoclasts, slows the rate of calcium ion loss from blood into urine, and stimulates the kidneys to produce calcitriol, whereas calcitriol acts on the GI tract to increase the rate of calcium ion absorption from food into the blood **29.** calcitriol is the active form of vitamin D and increases calcium absorption through the GI tract **30.** yes, because lack of growth hormone would make it harder to utilize the calcium in the body to build bones **31.** stimulates contraction of smooth muscle in the walls of arterioles and stimulates the adrenal cortex to secrete aldosterone; (b) arterioles and adrenal gland **32.** (a) dehydration, low blood sodium or hemorrhage; (b) increase in blood pressure; (c) negative **33.** hypothalamus produces CRH which targets the anterior pituitary to produce ACTH which targets the adrenal cortex to produce glucocorticoids **34.** (a) gluconeogenesis and triglyceride breakdown to stimulate ATP production; (b) additional fuel for ATP production allows the body to combat a range of stressors **35.** norepinephrine and epinephrine **36.** glucocorticoids, mineralocorticoids, adrenal cortex androgens, and adrenal medulla norepinephrine and epinephrine (androgens would still be released by the gonads and the sympathetic division of the ANS would release norepinephrine and epinephrine) **37.** it produces hormones (endocrine function) and digestive enzymes (exocrine function) **38.** they flow into the GI tract by a series of ducts **39.** effect of high levels of insulin on the body **40.** (a) parasympathetic; (b) insulin **41.** (a) sympathetic; (b) glucagons **42.** relaxin, oxytocin **43.** adrenal cortex and testes **44.** adrenal cortex **45.** anterior pituitary

Chapter 18

1. oxygen, carbon dioxide, nutrients, waste products, hormones **2.** homeostasis, pH, body temperature, water content of cells **3.** red blood cells **4.** white blood cells and platelets **5.** (a) albumins: help maintain proper blood osmotic pressure, (b) globulin: involved in immunity, (c) fibrinogen: formation of blood clots **6.** platelets **7.** red bone marrow **8.** mesenchymal cells **9.** no nucleus, thus more room; generate ATP anaerobically, so do not use oxygen; shape is biconcave disc for more surface area **10.** four **11.** (a) the iron ion at the center of the heme ring; (b) amino acids in the globin part of hemoglobin **12.** because of the wear and tear their plasma membranes undergo as they squeeze through blood capillaries **13.** 1) globin is broken down into amino acids for reuse by the body, 2) iron is removed from the heme portion and attached to transferrin, then carried to red bone marrow to be used in hemoglobin synthesis, 3) the non-iron portion is converted to buliverdin, then to bilirubin, which is transported to the liver, then secreted by hepatocytes into bile which eventually enters the large intestine and is converted into urobilinogen, some of which is absorbed back into the blood and eventually excreted in urine, most is eliminated in feces **14.** erythrocytes (mature RBC without nuclei) **15.** it might increase because more red cells would be needed due to the reduced oxygen content of air at high altitudes **16.** (a) type AB blood has A and B antigens; (b) type O blood has neither A nor B antigens **17.** in the blood plasma **18.** (a) type A blood has anti-B antibodies; (b) type AB blood has neither anti-A nor anti-B antibodies; (c) type O blood has both anti-A and anti-B antibodies **19.** A, B, AB, or O **20.** hemolysis is the rupture of red blood cells; it occurs if an Rh⁻ person receives Rh⁺ blood and begins to make anti-Rh antibodies, and then receives a second transfusion of Rh⁺ blood **21.** (a) lymphocytes and monocytes; (b) because their granules are not visible under a light microscope **22.** only 2% of lymphocytes are circulating in blood at any given time; the rest are in lymphatic fluid and organs such as skin, lungs, lymph nodes, and spleen **23.** 1) emigration: process of WBCs leaving the bloodstream by crossing capillary walls; 2) chemotaxis: attraction of phagocytes to chemicals released by pathogens and inflamed tissue; 3) phagocytosis: the ingesting of bacteria and disposal of dead matter **24.** poke holes in microbe membranes, causing loss of cellular contents **25.** a count of each of the five types of WBCs **26.** they are cell fragments with many vesicles but no nucleus **27.** help stop blood loss by promoting blood clotting **28.** vascular spasm: occurs when a blood vessel is damaged and the smooth muscle in its wall contracts immediately; platelet plug formation: platelets adhere to the damaged blood vessel (platelet adhesion), become activated, and begin to release chemicals (platelet release reaction) making other platelets sticky, gathering of platelets (platelet aggregation) forming the platelet plug **29.** formation of prothrombinase **30.** thrombin activates more platelets, increasing their release of platelet phospholipids, and directly accelerates the formation of prothrombinase, which accelerates the production of more thrombin; this cycle continues, causing the fibrin clot to grow **31.** it is required for the sequence of reactions by tissue factor **32.** as the clot retracts, it pulls the edges of the damaged blood vessel closer together

Chapter 19

1. it rests on the diaphragm, near the midline of the thoracic cavity, with about two-thirds of the mass lying to the left of the body's midline **2.** serous pericardium **3.** reduces friction between the membranes as the heart moves **4.** fibrous pericardium, serous pericardium, epicardium, myocardium, endocardium **5.** slightly increases the capacity of the atria so that they can hold a greater volume of blood **6.** superior atria and inferior ventricles **7.** right and left ventricles **8.** (a) interatrial septum; (b) interventricular septum **9.** the blood flowing from areas of higher blood pressure to areas of lower blood pressure **10.** pressure of the ventricular blood driving the cusps upward until the valve are closed **11.** back-flowing blood in the ventricles force the semilunar valves to close **12.** the right atrium receives blood from the vena cava and sends it to the right ventricle through the tricuspid valve; the right ventricle sends the blood into the pulmonary trunk through the pulmonary valve; the left atrium receives blood from the pulmonary veins and sends blood into the left ventricle through the bicuspid valve; the left ventricle sends blood from the left atrium into the aorta through the aortic valve **13.** pulmonary veins and the aorta carry oxygenated blood; superior and inferior vena cava, coronary sinus, and pulmonary arteries carry deoxygenated blood **14.** ascending aorta **15.** (a) circumflex branch; (b) great cardiac vein; (c) small cardiac vein **16.** specialized cardiac muscle fibers that are self-excitable and repeatedly and rhythmically generate action potentials that trigger heart contractions; they act as a pacemaker and form the conduction system throughout the heart **17.** atrioventricular node **18.** sinoatrial node → atrioventricular node → atrioventricular bundle (bundle of His) → right and left bundle branches → Purkinje fibers **19.** sinoatrial node **20.** a recording of the electrical changes that accompany the heartbeat **21.** P wave: atrial depolarization; QRS complex: onset of ventricular depolarization; T wave: ventricular repolarization **22.** useful in diagnosing abnormal cardiac rhythms and conduction patterns and in following the course of recovery from a heart attack **23.** first wave: represents atrial depolarization which causes contraction of the atria; second wave: QRS complex which represents the onset of ventricular depolarization when ventricles contract; third wave: T wave which indicates ventricular repolarization and occurs just before the ventricles start to relax; relaxation of the atria is not usually evident in the ECG **24.** (a) systole; (b) diastole **25.** diastole **26.** (a) end-diastolic volume; (b) end-systolic volume **27.** blood turbulence from the AV valves closing after ventricular systole begins (lubb); blood turbulence associated with the SL valves closing at the end of ventricular systole (dupp) **28.** volume of blood ejected each minute from the left ventricle into the aorta; equals the stroke volume (mL/beat) times heart rate (beats/min) **29.** (a) volume of blood ejected by the ventricle during each contraction; (b) factors that change cardiac output, such as exercise **30.** cardiovascular center in the medulla oblongata **31.** (a) proprioceptors, chemoreceptors baroreceptors; (b) proprioceptors: physical activity; chemoreceptors: chemical changes in the blood; baroreceptors: changes in blood pressure **32.** they can increase the heart's pumping effectiveness (heart rate and contractility)

Chapter 20

1. (a) elastic fibers allow the walls of the arteries to stretch easily in response to a small increase in blood pressure; (b) smooth muscle allows vasoconstriction and vasodilation to adjust blood flow **2.** elastic arteries: more elastic fibers and less smooth muscle in their tunica media; muscular arteries: more smooth muscle and fewer elastic fibers **3.** (a) elastic arteries help propel blood onward while the ventricles are relaxing; (b) muscular arteries adjust the rate of blood flow and distribute blood to organs; (c) arterioles regulate the flow of blood into capillaries **4.** an anastomosis is the union of two or more arteries supplying the same body region; collateral circulation is the alternate circulatory route of blood flow to a body part through an anastomosis **5.** materials cross capillary walls through intercellular clefts and fenestrations, via transcytosis in pinocytic vesicles, and through the plasma membranes of endothelial cells **6.** when the precapillary sphincters are relaxed (open), blood flows into the capillaries; when precapillary sphincters contract (close), blood flow through the capillaries stops or decreases **7.** autoregulation: the ability of a tissue to automatically adjust its blood flow to match its metabolic demands **8.** (a) continuous capillaries have mostly uninterrupted walls that restrict some movement; (b) the walls of fenestrated capillaries contain small pores that allow greater exchange of materials; (c) the walls of sinusoids contain large pores, an incomplete basement membrane, and large intercellular clefts, and they're the leakiest capillaries **9.** diffusion and transcytosis **10.** they enter lymphatic capillaries and are eventually returned by the lymphatic system to the bloodstream **11.** high blood pressure "pushes" fluid out of the capillaries, whereas high colloid osmotic pressure "pulls" fluid back into the capillaries **12.** the femoral artery has the thicker wall; the femoral vein has the wider lumen **13.** arteries: thick walls with smooth muscle and elastic fibers, conduct blood from the heart to the body; capillaries: walls are single-celled with only an endothelium and basement membrane, permit exchange of nutrients and wastes between blood and interstitial fluid; veins: thin walls with smooth muscle and elastic fibers with valves, return blood to the heart **14.** venous return is due to the pressure generated by contraction of the left ventricle, by the skeletal muscle pump (contraction of skeletal muscles in the limbs), by the respiratory pump alternating compression and decompression of veins during inhalation and exhalation **15.** blood volume in venules and veins is about 64% of 5 liters, or 3.2 liters; blood volume in capillaries is about 7% of 5 liters, or 350 mL **16.** (a) systole (contraction); (b) diastole (relaxation) **17.** opposition to blood flow due to friction between blood and the walls of the blood vessels; factors: size of the lumen, blood viscosity, total blood vessel length **18.** vasodilation and vasoconstriction of arterioles are the main regulators of systemic vascular resistance **19.** inputs: the brain, proprioceptors, baroreceptors, and chemoreceptors; outputs: parasympathetic output along vagus nerves, sympathetic and parasympathetic neurons of the ANS **20.** cardiac muscle in the heart and smooth muscle in blood vessel walls **21.** they send nerve impulses at a slower rate to the cardiovascular center, decreasing parasympathetic stimulation of the heart and increasing sympathetic stimulation of the heart, increased secretion of epinephrine and norepinephrine, increased cardiac output, and systemic vascular resistance blood pressure rises **22.** to inform the cardiovascular center about changes in blood levels of O_2, CO_2, and H^+ **23.** altering cardiac output, changing systemic vascular resistance, or adjusting the total blood volume **24.** the alternate expansion and elastic recoil of an artery after each contraction of the left ventricle **25.** (a) tachycardia is a rapid heart beat or pulse; (b) bradycardia is a slow heartbeat or pulse **26.** a cuff is placed on a bared arm; the cuff is inflated until the brachial artery is compressed and blood flow stops; the cuff is deflated slowly while the technician listens at the artery with a stethoscope; when the cuff is deflated enough to allow the artery to open, the spurt of blood passing through the artery is heard, and the pressure is read as the systolic blood pressure; as the cuff is deflated further, when the sounds become faint, the pressure is measured as the diastolic blood pressure **27.** 142 is systolic, 95 is diastolic **28.** pulmonary arteries carry deoxygenated blood; pulmonary veins carry oxygenated blood **29.** blood in pulmonary circulation only has to be pumped to the lungs, but blood in the systemic circulation has to be pumped to the rest of the body; pulmonary arteries have larger diameters, thinner walls, and less elastic tissue so the resistance to blood flow is very low **30.** carries oxygenated blood from the heart to all parts of the body and returns deoxygenated blood to the heart **31.** ascending aorta, arch of the aorta, thoracic aorta, abdominal aorta **32.** ascending aorta: myocardium of the heart; arch of the aorta: right and left side of the head and neck, right and left upper limbs; thoracic aorta: pericardium, bronchi of lungs, esophagus, structures in mediastinum, intercostals and chest muscles, superior and posterior surfaces of diaphragm; abdominal aorta: inferior surface of diaphragm, liver, stomach and esophagus, spleen, pancreas, small intestine, colon, adrenal glands, kidneys, reproductive structures, lower limbs **33.** (a) brachiocephalic trunk, left common carotid artery, left subclavian artery; (b) visceral branches: pericardium, bronchi, pleurae, bronchial lymph nodes, esophagus, mediastium; parietal branch: thoracic muscles, subcutaneous tissue and skin, mammary glands, vertebrae, meninges, spinal cord, diaphragm **34.** diaphragm **35.** paired visceral branches: suprarenal, renal, gonadal arteries; unpaired visceral branches: celiac trunk, superior mesenteric and inferior mesenteric arteries; paired visceral branches supply the kidneys, adrenal glands, ureters, male and female reproductive organs; unpaired visceral branches supply the stomach and esophagus, pancreas, liver, gallbladder, small intestine, large intestine **36.** paired parietal branch: inferior phrenic arteries, lumbar arteries; unpaired parietal branch: median sacral artery; paired parietal branch supplies diaphragm, adrenal glands, lumbar vertebrae, spinal cord and its meninges, muscles and skin of the lumbar region of the back; unpaired parietal branch supplies the sacrum and coccyx **37.** internal iliac arteries and external iliac arteries; internal iliac arteries supply the pelvis, buttocks, external genitals, thigh; external iliac arteries supply the muscles of the anterior abdominal wall, the cremaster muscle in males, round ligament of the uterus in females, lower limbs **38.** the superior vena cava drains regions above the diaphragm, and the inferior vena cava drains regions below the diaphragm **39.** cardiac veins **40.** all venous blood in the brain drains into the internal jugular veins **41.** internal jugular: brain, face, neck; external jugular: scalp, superficial and deep regions of the face; vertebral veins: deep structures in the neck such as the cervical vertebrae, cervical spinal cord, and some neck muscles **42.** cephalic: lateral aspect of upper limbs; basilic: medial aspect of upper limbs; median antebrachial: palms and forearms; radial: lateral aspects of the forearms; ulnar: medial aspect of the forearms **43.** median cubital vein **44.** the median cubital vein of the upper limb is often used for withdrawing blood **45.** subclavian **46.** may serve as a bypass for the inferior vena cava **47.** (a) hemiazygos vein and accessory hemiazygos vein; (b) azygos vein **48.** inferior vena cava **49.** lumbar: both sides of the posterior abdominal wall, vertebral canal, spinal cord, and meninges; gonadal: testes in the male and ovaries in the female; renal: kidneys; suprarenal: adrenal glands; inferior phrenic: diaphragm; hepatic: liver **50.** (a) great saphenous vein; (b) anterior tibial veins; (c) anterior tibial vein **51.** often used for prolonged administration of intravenous fluids **52.** (a) hepatic portal vein; (b) common hepatic artery **53.** hepatic veins

Chapter 21

1. specialized form of reticular connective tissue that contains large numbers of lymphocytes **2.** draining excess interstitial fluid, transporting dietary lipids, and carrying out immune responses **3.** have thinner walls and more valves **4.** the ends of the endothelial cells making up the wall of a lymphatic capillary overlap, so that when pressure is greater in the interstitial fluid than in lymph, the cells separate slightly and interstitial fluid enters the capillary **5.** (a) left side of the head, neck, and chest, the left upper limb, and the entire body inferior to the ribs; (b) upper right side of the body; (c) thoracic duct: into venous blood at the junction of the left internal jugular and left subclavian veins; right lymphatic duct: venous blood at the junction of the right internal jugular and right subclavian veins **6.** the left and right lumbar trunks and the intestinal trunk empty into the cisterna chyli, which then drains into the thoracic duct **7.** blood capillaries → interstitial spaces → lymphatic capillaries → lymphatic vessels → lymph nodes → lymphatic vessels → lymphatic ducts → junction of the internal jugular and subclavian veins **8.** (a) maturation of T cells; (b) primary lymphatic organ **9.** atrophies with age (gets smaller) **10.** afferent lymphatic vessels → subcapsular sinus → trabecular sinuses → medullary sinuses → efferent lymphatic vessels **11.** they are trapped by reticular fibers within the sinuses, and macrophages destroy some by phagocytosis while lymphocytes destroy others by immune responses **12.** (a) spleen: removal by macrophages of worn-out or damaged blood cells and platelets, storage of platelets, and production of blood cells during fetal life; (b) tonsils: participate in immune responses against inhaled or ingested foreign substances **13.** external physical and chemical barriers and various internal defenses against infection **14.** physical factors: closely packed, keratinized epithelial skin cells, shedding of epithelial skin cells, mucus, hairs, cilia, and washing surfaces with tears, saliva, urine; chemical factors: sebum, lysozyme, gastric juice, vaginal secretions **15.** antimicrobial proteins, natural killer cells, phagocytes, inflammation, fever **16.** interferons **17.** (a) lymphocytes; (b) neutrophils and macrophages (enlarged monocytes) **18.** chemotaxis, adherence, ingestion, digestion, killing **19.** they are broken down by lyso-

somal enzymes within the phagolysome **20.** ability to carry out adaptive immune responses when properly stimulated; B and T lymphocytes **21.** B cells develop in red bone marrow; T cells develop in the bone marrow and then in the thymus **22.** cell-mediated immunity: cytotoxic T cells directly attack invading antigens; antibody-mediated immunity: B cells transform into plasma cells which synthesize and secrete specific antibodies that inactivate specific antigens **23.** stimulating the production of specific antibodies and/or the proliferation of specific T cells **24.** help T cells recognize foreign antigens **25.** viral proteins, toxins from intracellular bacteria, and abnormal proteins synthesized by a cancerous cell **26.** first signal: binding of an antigen–MHC complex by a TCR with CD4 or CD8 proteins; second signal: costimulation by cytokines **27.** these proteins help anchor the T cell and antigen–MHC complex so that antigen recognition can occur **28.** helper T cells: secrete cytokines; cytoxic T cells: destroy intruders; memory T cells: recognize the original invading antigen and initiate a very swift reaction against it **29.** tumor cells, tissue transplant cells **30.** cause target cells to undergo apoptosis or cytolysis, or destroy the cell's DNA **31.** helper T cells **32.** 1) antigen taken into B cell and combined with MHC protein to form an antigen–MHC complex on the B cell, 2) helper T cells deliver the costimulation needed for B cell division and differentiation, 3) B cell differentiates into plasma cells that secrete antibodies **33.** two **34.** 1) IgG: protects against bacteria and viruses by enhancing phagocytosis, neutralizing toxins, and triggering the complement system; 2) IgA: provides localized protection on mucous membranes against bacteria and viruses; 3) IgM: activates complement and causes agglutination and lysis of microbes; 4) IgD: involved in activation of B cells; 5) IgE: involved in allergic and hypersensitivity reactions, protection against parasitic worms **35.** neutralizing the antigen, immobilizing bacteria, agglutinating and precipitating the antigen, activating complement, enhancing phagocytosis of the antigen **36.** in the liver **37.** phagocytosis, cytolysis, inflammation **38.** binding of inactive complement proteins to an antigen–antibody complex or to carbohydrate complexes on the surface of some microbes **39.** they can proliferate and differentiate into helper T cells, plasma cells, or cytotoxic T cells very quickly, thus providing a quicker response to a second exposure to an antigen **40.** antibodies have a higher affinity for the antigen and so are more successful in eliminating it **41.** IgG

Chapter 22

1. nostrils, vestibule, nasal cavity (past nasal conchae), internal naris, nasopharynx **2.** 1) warming, moistening, and filtering incoming air; 2) detecting olfactory stimuli; 3) modify speech vibrations as they pass through the large, hollow resonating chambers **3.** external and internal nares **4.** produce mucus, serve as resonating chambers for sound, lighten the weight of the skull **5.** moistens via the mucus secreted by the goblet cells and drainage from the nasolacrimal ducts; cleans by the mucus trapping dust particles and the cilia moving the mucus and trapped dust particles toward the pharynx **6.** superior border: nasal cavity; inferior border: larynx **7.** by closing over the glottis when swallowing **8.** produce sounds during speaking and singing **9.** (a) it accommodates slight expansion of the esophagus into the trachea during swallowing; (b) to maintain a patent (open) airway **10.** primary bronchi, secondary (lobar) bronchi, tertiary (segmental) bronchi, bronchioles, terminal bronchioles **11.** parietal pleura: lines the wall of the thoracic cavity; visceral pleura: covers the lungs themselves; pleural cavity: potential space between the parietal and visceral pleura **12.** terminal bronchioles → respiratory bronchioles → alveolar ducts → alveoli **13.** it would increase **14.** diaphragm **15.** (a) decrease; (b) enter; (c) increase; (d) leave **16.** (b) lungs expand **17.** the recoil of elastic fibers that were stretched during inhalation and the inward pull of surface tension due to the film of alveolar fluid **18.** (a) higher compliance; (b) larger bronchioles **19.** apparatus commonly used to measure the volume of air exchanged during breathing and measure the respiratory rate **20.** conducting airways containing air that do not undergo gas exchange **21.** tidal volume: amount of air inhaled or exhaled with each quiet breath; inspiratory reserve: amount of air that can be forcefully inhaled above tidal; expiratory reserve: amount of air that can be forcefully exhaled above tidal; residual volume: amount of air remaining in the lungs after a forced exhalation **22.** vital capacity **23.** Dalton's law: the total pressure exerted by a mixture of gases is the sum of the pressures exerted independently by each gas in the mixture; Henry's law: the quantity of a gas that will dissolve in a liquid is proportional to the partial pressure of the gas and its solubility **24.** because of the gas exchanges occurring in the lungs, the humidification of air by conducting passages, and the mixing of alveolar gas that occurs with each breath **25.** external respiration: oxygen into the blood, carbon dioxide into the lung alveoli; internal respiration: oxygen to the tissues, carbon dioxide into the blood **26.** O_2 moves down its partial pressure gradient **27.** partial pressure difference of the gases, surface area available for gas exchange, molecular weight and solubility of the gases **28.** partial pressure of O_2 decreases with increasing altitude **29.** (a) as oxyhemoglobin and dissolved in plasma; (b) attached to hemoglobin (oxyhemoglobin) **30.** the partial pressure of oxygen **31.** the higher the partial pressure of O_2, the more O_2 combines with hemoglobin **32.** dissolved in CO_2, bound to amino acids, in the form of bicarbonate ions **33.** systemic vein **34.** inspiratory area of the medulla oblongata **35.** phrenic **36.** through inspiratory and expiratory areas within the medullary rhythmicity area **37.** apneustic area: sends stimulatory impulses to the inspiratory area that activate it and prolong inhalation; pneumotaxic area: transmits inhibitory impulses to the inspiratory area to help turn off the inspiratory area before the lungs become too full of air **38.** when the levels of CO_2 and H^+ build up in the body, the inspiratory area is strongly stimulated, nerve impulses are sent along the phrenic and intercostal nerves to inspiratory muscles, and breathing resumes **39.** in or near the medulla oblongata in the central nervous system **40.** P_{O_2}, H^+ and P_{CO_2} **41.** $872 - 40$ mm Hg **42.** breathing rate decreases or breathing ceases altogether **43.** proprioceptors: rate and depth of breathing increase; inflation reflex: exhalation begins; temperature: increase in temperature results in increase in respiratory rate; pain: sudden, severe pain brings about brief apnea, but a prolonged somatic pain increases respiratory rate; irritation of the airways: an immediate cessation of breathing followed by coughing or sneezing **44.** they serve as buffers for H^+ **45.** pH falls as acidity increases **46.** (a) acidosis is a condition in which blood pH is below 7.35, and alkalosis is a condition in which blood pH is higher than 7.45; (b) acidosis: depression of the CNS through depression of synaptic transmission; alkalosis: overexcitability in both the CNS and peripheral nerves resulting in nervousness, muscle spasms, and even convulsions and death

Chapter 23

1. 1) mucosa: inner surface for secretion of digestive juices and absorption of digestive end products; 2) submucosa: site of glands, blood vessels, lymph vessels, and nerve plexuses; 3) muscularis: circular and longitudinal muscles for propulsion and mixing; 4) serosa: serous membrane to reduce friction with movement **2.** (a) mouth, pharynx, superior and middle parts of the esophagus and external anal sphincter; (b) voluntary **3.** mesentery: binds large intestine to posterior abdominal wall; falciform ligament: attaches liver to anterior abdominal wall and diaphragm; lesser omentum: suspends stomach and duodenum from the liver; greater omentum: from stomach and duodenum to transverse colon **4.** myenteric plexus: innervates smooth muscle of muscularis; submucosal plexus: innervates secretory cells of mucosal epithelium to control the secretions of the organs of the GI tract **5.** cheeks, hard and soft palate, tongue **6.** (a) uvula; (b) closes off nasopharynx during swallowing **7.** parotid glands: located inferior and anterior to the ears, between the skin and masseter muscle; submandibular glands: located beneath the base of the tongue in the posterior part of the floor of the mouth; sublingual glands: superior to submandibular glands **8.** via ANS: parasympathetic fibers increase salivation, sympathetic fibers decrease salivation; feel and taste of food; smell, sight, sound, or thought of food **9.** incisors: grasping, tearing of food; cuspids: tear and shred food; premolars: crushing and grinding; molars: crush and grind food in preparation for swallowing **10.** chewing **11.** respiratory and digestive **12.** submucosa **13.** upper esophageal sphincter: regulates movement of food from the pharynx into the esophagus; lower esophageal sphincter: regulates movement of food from the esophagus to the stomach **14.** oral: voluntary passage of bolus into the oropharynx; pharyngeal: involuntary passage of bolus through pharynx into the esophagus; esophageal: involuntary passage of bolus through the esophagus into the stomach **15.** both: oral phase is voluntary; pharyngeal and esophageal phases are involuntary **16.** push **17.** mucosa **18.** fundus **19.** carbonic acid **20.** parasympathetic division **21.** (a) break peptide bonds between amino acids; (b) in the inactive form pepsinogen, it cannot digest the proteins in the chief cells that produce it **22.** water, ions, short-chain fatty acids, certain drugs **23.** exocrine cells → small ducts → large ducts → pancreatic duct and accessory duct; pancreatic duct joins common bile duct to form hepatopancreatic ampulla → sphincter of Oddi → duodenum; accessory ducts directly enter duodenum **24.** digest carbohydrates, proteins,

lipids, and nucleic acids and buffer acidity (bicarbonate ions) **25.** the brush border enzyme enterokinase activates inactive trypsinogen to the active protein trypsin; trypsin activates the other pancreatic proteases **26.** (a) enzymes and bicarbonate ions; (b) bile; (c) pancreatic juices and bile **27.** emulsification of fat **28.** O_2, most nutrients, and certain toxic substances are taken up by hepatocytes **29.** (a) duodenum, jejunum, ileum; (b) ileum **30.** enhance absorption by increasing surface area and causing chyme to spiral as it moves through the small intestine **31.** (a) regulate microbial population in the intestines via lysozyme and phagocytosis; (b) help neutralize gastric acid in chyme **32.** segmentation: mix chyme with the digestive juices and bring food particles into contact with the mucosa for absorption; peristalsis: push chyme forward down the small intestine **33.** pancreatic amylase: chemically break down glycogen and starches; brush border enzymes: chemically break down disaccharides into monosaccharides; pancreatic lipases: break triglycerides into fatty acids and monoglycerides; deoxyribonuclease: digest DNA **34.** to emulsify fats, i.e., break a large fat globule into smaller lipid globules **35.** carbohydrate end products: by facilitated diffusion or secondary active transport into blood; protein end products: by active transport processes into blood; lipids: by simple diffusion into lacteals (lymph system) **36.** the hydrophilic protein coat that surrounds each chylomicron keeps the chylomicrons suspended in blood and prevents them from sticking to each other **37.** via the hepatic portal system (all absorbed nutrients except triglycerides) or via the lymphatic vessels to the blood vascular system and then to the liver (triglycerides) **38.** cecum, ascending colon, transverse colon, descending colon, sigmoid colon, rectum, anal canal, anus **39.** secrete mucus that lubricates the passage of the colonic contents **40.** through haustral churning, peristalsis, and mass peristalsis **41.** water absorption **42.** it is the emptying of the rectum due to the distention of the rectal wall stimulating stretch receptors which activate parasympathetic stimulation to signal the contraction of longitudinal rectal muscles to shorten the rectum and open the internal anal sphincter, with the external anal sphincter (under voluntary control) also opening to allow the fecal material to be eliminated **43.** result: activation of neural centers in the cerebral cortex, hypothalamus, and brain stem; purpose: to prepare the mouth and stomach for food that is about to be eaten **44.** it buffers the stomach acid **45.** it prevents the duodenum from being overloaded with more chyme that it can handle **46.** the inhibition of gastric motility and increase in the contraction of the pyloric sphincter, which decreases gastric emptying **47.** CCK: stimulates secretion of pancreatic juice rich in digestive enzymes, causes ejection of bile from the gallbladder, causes opening of the sphincter of the hepatopancreatic ampulla, induces satiety, and inhibits gastric emptying; secretin: stimulates secretion of pancreatic juice and bile that are rich in bicarbonate ions and inhibits secretion of gastric juice **48.** (a) oils, (b) they represent food with little or no solid fats or added sugars **49.** calcium: required for bone hardness, normal membrane permeability, transmission of nerve impulses, muscle contraction, blood clotting; iron: major component of hemoglobin; magnesium: required for normal muscle and nerve irritability; sodium: principal electrolyte maintaining osmotic pressure of extracellular fluid, aids in acid–base balance of blood as part of the bicarbonate buffer system, needed for normal neuromuscular function; phosphorus: important for energy storage and transfer, muscle and nerve activity, and cell permeability **50.** (a) organic nutrients required in small amounts to maintain growth and normal metabolism; (b) function as coenzymes; (c) ingested in food except for vitamin K and B complex vitamins which are synthesized by bacteria in the large intestine **51.** fat-soluble: are absorbed along with other dietary lipids in the small intestine and packaged into chylomicrons; water-soluble: are dissolved in body fluids **52.** they inactivate oxygen free radicals **53.** (a) refers to all the chemical reactions of the body; (b) anabolism: chemical reactions that combine simple substances into more complex molecules; catabolism: chemical reactions that break down complex organic molecules into simpler ones **54.** when complex molecules and polymers are split apart, some of the energy is transferred to form ATP and the rest is given off as heat; when simple molecules and monomers are combined to form complex molecules, ATP provides the energy for synthesis, and again some energy is given off as heat **55.** (a) glycolysis, Krebs cycle, electron transport chain; (b) 36 or 38 **56.** glucose is converted to pyruvic acid and two molecules of ATP **57.** it is the series of reactions that form glucose from noncarbohydrate sources; thus, releasing glucose into the blood to keep a normal blood glucose level during the hours between meals when glucose is not being absorbed **58.** (a) to transport lipids in blood by making them temporarily water-soluble;

(b) "good" cholesterol: HDL; "bad" cholesterol: VLDL and LDL; (c) HDL: high-density lipoprotein (higher percentage of protein, lower percentage of lipid); LDL: low-density lipoprotein (lower percentage of protein, higher percentage of lipid); VLDL: very low-density lipoprotein (lowest percentage of protein and highest percentage of lipid)

Chapter 24

1. kidneys, ureters, urinary bladder, urethra **2.** by regulating blood ionic composition, pH, volume, pressure, and glucose level **3.** (a) substances that have no useful function in the body; (b) by forming urine **4.** their position in the body is posterior to the peritoneum of the abdominal cavity **5.** ureter, renal artery, renal vein, lymphatic vessels, nerves **6.** deep renal capsule, middle adipose capsule, superficial renal fascia **7.** medulla **8.** renal corpuscle (consisting of glomerulus and glomerular capsule) and renal tubules (proximal convoluted tubule, loop of Henle, distal convoluted tubule) **9.** (a) renal cortex, (b) photo shows a renal corpuscle which is found only in the cortex **10.** cortical nephrons: most numerous type, renal corpuscles in outer portion of renal cortex, short loops of Henle lying mainly in cortex and penetrate only into outer region of medulla, loop blood supply is from peritubular capillaries; juxtamedullay nephrons: least numerous, corpuscles lie deep in cortex close to medulla, long loop of Henle extending into deepest region of medulla, loop blood supply from vasa recta **11.** parietal and visceral layers of glomerular capsule **12.** location: where the distal convoluted tubule makes contact with the afferent arteriole; structure: macula densa cells in the wall of the distal convoluted tubule and juxtaglomerular cells (modified smooth muscle cells) in the afferent arteriole wall **13.** by secretion **14.** removed from the bloodstream **15.** a glomerular endothelial cell, the basal lamina, a filtration slit formed by a podocyte **16.** blood plasma: contains blood cells, platelets, most plasma proteins and nutrients; glomerular filtrate: contains nutrients but no formed elements and very few, if any, of the plasma proteins **17.** glomerular capillaries present a large surface area for filtration, are about 50 times leakier than capillaries in most other tissues due to their large fenestrations, and glomerular capillary blood pressure is high **18.** by forcing water and solutes in blood plasma through the filtration membrane **19.** (a) by vasodilation of afferent arteriole which increases GRF; (b) kidneys are adjusting their own resistance to blood flow **20.** 1) ANS: increased activity of sympathetic nerves results in release of norepinephrine which causes constriction of afferent arterioles which increases systemic blood pressure which increases GFR; 2) hormones: angiotensin reduces GFR and atrial natriuretic peptide increases GFR **21.** paracellular reabsorption (fluids leaking between the cells passively) and transcellular reabsorption (substance passes from the fluid in the tubular lumen through the apical membrane of a tubule cell, across the cytosol, and out into interstitial fluid through the basolateral membrane) **22.** (a) basolateral surface; (b) it ensures that reabsorption of Na^+ is a one-way process **23.** water reabsorbed with solutes in tubular fluid which occurs in the proximal convoluted tubule and the descending limb of the loop of Henle **24.** (a) Na^+ symporters and Na^+/H^+ antiporters; (b) Na^+ symporters: virtually all nutrients such as glucose and amino acids; Na^+/H^+ antiporters: bicarbonate ions **25.** (a) Na^+–K^+–$2Cl^-$ symporters; (b) K^+, Cl^-, Ca^{2+}, and Mg^{2+} **26.** (a) because little or no water is reabsorbed in the ascending limb of the loop of Henle and ions are reabsorbed in the ascending limb, the tubular fluid becomes progressively more dilute as it flows toward the end of the ascending limb; (b) proximal convoluted tubule and descending limb of the loop of Henle; (c) ascending limb of the loop of Henle **27.** juxtaglomerular cells secrete renin which converts angiotensinogen into angiotensin I; ACE converts angiotensin I into angiotensin II which decreases the glomerular filtration rate by causing vasoconstriction of the afferent arterioles and stimulates the adrenal cortex to release aldosterone, which stimulates the principal cells in the collecting ducts to reabsorb more Na^+ and Cl^- and secrete more K^+ with the reabsorption of Na^+ and Cl^-; when more Na^+ and Cl^- are reabsorbed, more water is reabsorbed resulting in increased blood volume and a return of blood pressure to normal **28.** angiotensin II stimulates the adrenal cortex to release aldosterone, which stimulates the principal cells in the collecting ducts to reabsorb more Na^+ and Cl^- and secrete more K^+ with the reabsorption of Na^+ and Cl^-; when more Na^+ and Cl^- are reabsorbed, more water is reabsorbed resulting in increased blood volume and a return of blood pressure to normal **29.** (a) ADH increases the water permeability of principal cells in the last part of the distal convoluted tubule and throughout the collecting duct; (b) facultative **30.** stimulate: angiotensin II, aldo-

sterone, ADH; inhibit: atrial natriuretic peptide **31.** osmotic pressure **32.** ascending limb of loop of Henle and entire collecting duct **33.** Na^+, K^+, and urea **34.** long loops of Henle and collecting duct **35.** the constant transfer of urea between segments of the renal tubule and the interstitial fluid of the medulla promotes water reabsorption with the solutes left behind in the lumen becoming very concentrated resulting in a small volume of concentrated urine **36.** peristaltic contractions of the muscular walls of the ureters **37.** it allows the organ lumen to stretch **38.** external **39.** males: extends from the internal urethral orifice and is elongated; females: lies directly posterior to the pubic symphysis and is shortened **40.** prostatic, membranous, spongy **41.** (a) water and dissolved solutes in the body; (b) ICF (66%), ECF (33%); of ECF, interstitial fluid (80%) and plasma (20%); (c) water **42.** 1) ANP: decreases blood volume and raises blood osmotic pressure; 2) angiotensin II: increases blood volume and lowers blood osmotic pressure; 3) aldosterone: decreases blood volume and raises blood osmotic pressure; 4) ADH: decreases blood volume and raises blood osmotic pressure **43.** dihydrogen phosphate ion acts as a weak acid and is capable of buffering strong bases, whereas monohydrogen phosphate ion acts as a weak base and is capable of buffering H^+ released by a strong acid **44.** by excreting excess H^+ when pH of body fluids is too low and by excreting excess HCO_3^- when pH is too high

Chapter 25

1. 1) testes: produce sperm; 2) epididymis: sperm maturation and storage; 3) ductus deferens: sperm storage; 4) ejaculatory duct: sperm release; 5) urethra: sperm release channel; accessory sex glands (seminal vesicles, prostate, urethra): contribute to semen; 6) scrotum: house testes and epididymis, keep sperm at acceptable temperature; 7) penis: male copulatory organ **2.** tunica vaginalis, tunica albuginea **3.** Sertoli cells: form blood–testis barrier, support and protect developing spermatogenic cells; produce fluid for sperm transport and secrete the hormone inhibin; Leydig cells: secrete testosterone **4.** chromosome number **5.** acrosome: a caplike vesicle filled with enzymes that help a sperm penetrate a secondary oocyte to bring about fertilization; middle piece: contains mitochondria arranged in a spiral, which provide the energy for locomotion of sperm to the secondary oocyte and for sperm metabolism **6.** anterior pituitary gland **7.** (a) interstitial cells (Leydig cells); (b) inhibin: sustentacular cells (Sertoli cells) **8.** ductus epididymis: site of sperm maturation; ductus deferens: stores sperm and conveys sperm from the epididymis toward the urethra; ejaculatory duct: eject sperm and seminal vesicle secretions **9.** (a) ductus deferens, testicular artery, veins that drain the testes, pampiniform plexus, autonomic nerves, lymphatic vessels, cremaster muscle; (b) passes through the inguinal canal **10.** prostatic urethra: passes through the prostate; membranous urethra: passes through the deep muscles of the perineum; spongy (penile) urethra: passes through the corpus spongiosum of the penis **11.** 1) seminal vesicles: lie posterior to the base of the urinary bladder and anterior to the rectum; functions to secrete an alkaline, viscous fluid that contains fructose, prostaglandins, and clotting proteins; 2) prostate: located inferior to the urinary bladder and surrounds the prostatic urethra; functions to secrete a milky, slightly acidic fluid containing citric acid, proteolytic enzymes, and seminalplasmin; 3) bulbourethral glands: located inferior to the prostate on either side of the membranous urethra and their ducts open into the spongy urethra; function to secrete an alkaline fluid into the urethra and secrete mucus that lubricates the end of the penis and the lining of the urethra **12.** seminal vesicles **13.** 1) corpus spongiosum and corpus cavernosa; 2) under parasympathetic stimulation, erectile tissue arterioles dilate becoming engorged with blood and smooth muscle within the erectile tissue relaxes resulting in widening of the blood sinuses, with the combination of increased blood flow and widening of the blood sinuses resulting in an erection **14.** in a sympathetic reflex, smooth muscle sphincter at the base of the urinary bladder closes, peristaltic waves occur in the ducts of the male, and semen is propelled into and through the male urethra **15.** (a) produce gametes and hormones; (b) by the broad ligament, mesovarium, ovarian ligament, and suspensory ligament **16.** follicles which produce estrogens and corpus luteum which produces estrogens and progesterone **17.** primary oocytes are present before birth, whereas primary spermatocytes develop during puberty **18.** (a) primordial follicle: a single layer of follicular cells surround the primary oocyte; primary follicle: primary oocyte surround by several layers of follicular cells called granulosa cells; secondary follicle: granulosa cells secrete follicular fluid, which builds up in the antrum, and the innermost layer of granulosa cells becomes firmly attached to the zona pellucida and is called the corona radiate; mature follicle: forms as secondary follicle enlarges and undergoes meiosis; (b) mature follicle **19.** (a) location: extend laterally from the uterus and lie between the folds of the broad ligaments of the uterus; function: provide a route for sperm to reach an ovum and transport secondary oocytes and fertilized ova from the ovaries to the uterus; (b) ciliated simple columnar cells function to help move a fertilized ovum along the tube toward the uterus; nonciliated cells with microvilli secrete a fluid that provides nutrition for the ovum **20.** (a) fundus, body, cervix; (b) fundus: superior to the uterine tubes; body: central portion; cervix: inferior narrow portion that opens into the vagina **21.** broad ligaments attach the uterus to either side of the pelvic cavity; uterosacral ligaments lie on either side of the rectum and connect the uterus to the sacrum, cardinal ligaments extend from the cervix and vagina to the lateral pelvic wall, round ligaments extend between the layers of the broad ligament from the uterus just inferior to the uterine tubes to the labia majora of the external genitalia **22.** endometrium: has two layers, the stratum basalis, which is permanent and gives rise to a new stratum functionalis after each menstruation, and the stratum functionalis, which lines the uterine cavity and sloughs off during menstruation; myometrium: consists of three layers of smooth muscle fibers that function in labor and childbirth **23.** receptacle for the penis during sexual intercourse, outlet for menstrual flow, passageway for childbirth **24.** (a) the external genitals of the female; (b) mons pubis: cushions the pubic symphysis; labia majora: protectively enclose medial components of the vulva; labia minora: contain sebaceous glands; clitoris: is capable of enlargement upon tactile stimulation and has a role in sexual excitement in the female; vestibule: contains the hymen, vaginal orifice, external urethral orifice, openings of the ducts of several glands; bulb of the vestibule: consists of two elongated masses of erectile tissue which become engorged with blood during sexual arousal resulting in narrowing the vaginal orifice and placing pressure on the penis during intercourse **25.** ovaries: testes; clitoris: glans penis; greater vestibular glands: bulbourethral glands; bulb of the vestibule: corpus spongiosum penis and bulb of penis **26.** (a) external urethral orifice, clitoris, mons pubis; (b) labia majora and labia minora **27.** by strands of connective tissue called the suspensory ligaments of the breast **28.** alveoli → secondary tubules → mammary ducts → lactiferous sinuses → lactiferous duct → exterior **29.** synthesis: prolactin, progesterone, and estrogens; ejection: oxytocin **30.** estradiol **31.** (a) estrogen: promote the development and maintenance of female reproductive structures, secondary sex characteristics, and breasts; increase protein anabolism; lower blood cholesterol level; (b) progesterone: prepare the mammary glands for milk secretion; (c) relax the uterus by inhibiting muscle contractions, increase the flexibility of the pubic symphysis, help dilate the uterine cervix **32.** 1) GnRH: stimulates release of FSH and LH; 2) FSH: stimulates follicle (and ovum) development; 3) LH: helps complete development of follicle and stimulates ovulation and corpus luteum formation; 4) estrogens: promote uterine lining development and development of mature follicle: 5) progesterone: promotes development of uterine lining and maintains lining if pregnancy occurs; 6) inhibin: inhibits FSH and LH secretion **33.** a follicle that begins to develop at the beginning of a particular menstrual cycle may not reach maturity and ovulate until several menstrual cycles later since the developmental process may take several months to occur **34.** proliferative phase: estrogens; ovulation: estrogens, LH, GnRH; corpus luteum: LH; LH midcycle surge: estrogens **35.** negative **36.** decreased levels of progesterone, estrogens, and inhibin result in the release of GnRH, FSH, and LH which results in follicles starting to grow again **37.** sperm flagella movement and contractions of the walls of the uterus and uterine tube **38.** sperm receptors in the zona pellucida bind to specific membrane proteins in the sperm head, triggering the release of sperm acrosomal enzymes that digest a path through the zona pellucida as the lashing sperm tail pushes the sperm cell onward **39.** fast block to polyspermy: oocyte cell membrane depolarizes on fertilization; slow block to polyspermy: intracellular calcium ions are released that stimulate exocytosis of chemicals which inactivate sperm receptors and harden the entire zona pellucida **40.** a solid sphere of cells formed by the mitotic divisions of the zygote **41.** inner cell mass forms the embryoblast which forms the embryo; trophoblast forms the fetal portion of the placenta **42.** (a) secretory phase; (b) either the posterior portion of the fundus or the body of the uterus; (c) with the inner cell mass toward the endometrium **43.** (a) decidua basalis is the portion of the endometrium between the embryo and the stratum basalis; decidua capsularis is the portion of the endometrium located between the embryo and the uterine cavity; decidua parietalis is the remaining modified endometrium that lines the noninvolved areas of the rest of the uterus; (b) decidua basalis **44.** syncytiotrophoblast cells secrete enzymes that enable the blastocyst to penetrate the uterine lining by digesting and liquefying the endometrial cells **45.** (a) by the coming together of the cells of the hypoblast and epiblast to form a flat disc; (b) by a band of extraembryonic mesoderm called the connection (body) stalk **46.** 1) from the fetal portion formed by the chorionic villi of the chorion and the maternal portion formed by the decidua basalis of the endometrium; 2) it allows oxygen and nutrients to diffuse from maternal blood into fetal blood while carbon dioxide and wastes diffuse from fetal blood into maternal blood, is a protective barrier against most microorganisms, stores nutrients which are released into fetal circulation as required, and produces hormones needed to sustain the pregnancy **47.** two umbilical arteries, one umbilical vein, and supporting mucous connective tissue **48.** placenta detached from the uterus **49.** end-fetal weight is twice the mid-fetal weight **50.** (a) corpus luteum; (b) placenta **51.** (a) source: chorion; function: stimulate corpus luteum to continue production of progesterone and estrogens; (b) source: chorion; function: help prepare the mammary glands for lactation, enhance maternal growth by increasing protein synthesis, regulate certain aspects of metabolism in both mother and fetus **52.** (a) relaxin; (b) increasing the flexibility of pelvic ligaments, helping dilate the uterine cervix during labor **53.** corticotropin-releasing hormone **54.** rise in estrogen levels, release of oxytocin accompanied by increase in oxytocin receptors, release of prostaglandins, relaxin increase **55.** (a) oxytocin; (b) the actual birth process works to increase the levels of oxytocin, with oxytocin levels falling only after the birth of the child **56.** false labor is pain felt in the abdomen at irregular intervals, does not intensify, and is not accompanied by cervical dilation; true labor begins when uterine contractions occur at regular intervals, usually producing pain with back pain intensified by walking, and dilation of the cervix occurring accompanied by a discharge of blood-containing mucus into the cervical canal **57.** (a) prolactin and oxytocin; (b) prolactin: milk synthesis; oxytocin: milk ejection **58.** presence of beneficial cells for the infant, presence of beneficial molecules in breast milk; decreased incidence of diseases in later life, support of optimal infant growth, enhancement of intellectual and neurological development, as well as other benefits **59.** help uterus to return to its normal size after birth of the child and expulsion of the placenta

Glossary

Pronunciation Key

1. The most strongly accented syllable appears in capital letters, for example, bilateral (bī-LAT-er-al) and diagnosis (dī´-ag-NŌ-sis).

2. If there is a secondary accent, it is noted by a prime (´), for example, constitution (kon´-sti-TOO-shun) and physiology (fiz´-ē-OL-ō-jē). Any additional secondary accents are also noted by a prime, for example, decarboxylation (dē´-kar-bok´-si-LĀ-shun).

3. Vowels marked by a line above the letter are pronounced with the long sound, as in the following common words:

ā as in māke   ō as in pōle
ē as in bē   ū as in cūte
ī as in īvy

4. Vowels not marked by a line above the letter are pronounced with the short sound, as in the following words:

a as in above or at   o as in not
e as in bet   u as in bud
i as in sip

5. Other vowel sounds are indicated as follows:

oo as in root
oy as in oil

6. Consonant sounds are pronounced as in the following words:

b as in bat   m as in mother
ch as in chair   n as in no
d as in dog   p as in pick
f as in father   r as in rib
g as in get   s as in so
h as in hat   t as in tea
j as in jump   v as in very
k as in can   w as in welcome
ks as in tax   z as in zero
kw as in quit   zh as in lesion
l as in let

A

Abdomen (ab-DŌ-men *or* AB-dō-men) The area between the diaphragm and pelvis.

Abdominal (ab-DŌM-i-nal) **cavity** Superior portion of the abdominopelvic cavity that contains the stomach, spleen, liver, gallbladder, most of the small intestine, and part of the large intestine.

Abdominal thrust maneuver A first-aid procedure for choking. Employs a quick, upward thrust against the diaphragm that forces air out of the lungs with sufficient force to eject any lodged material.

Abdominopelvic (ab-dom´-i-nō-PEL-vik) **cavity** Inferior to the diaphragm; subdivided into a superior abdominal cavity and an inferior pelvic cavity.

Abduction (ab-DUK-shun) Movement away from the midline of the body.

Abortion (a-BOR-shun) The premature loss (spontaneous) or removal (induced) of the embryo or nonviable fetus; miscarriage due to a failure in the normal process of developing or maturing.

Abscess (AB-ses) A localized collection of pus and liquefied tissue in a cavity.

Absorption (ab-SORP-shun) Intake of fluids or other substances by cells of the skin or mucous membranes; the passage of digested foods from the gastrointestinal tract into blood or lymph.

Accessory duct A duct of the pancreas that empties into the duodenum about 2.5 cm (1 in.) superior to the ampulla of Vater (hepatopancreatic ampulla). Also called the **duct of Santorini** (san´-tō-RĒ-nē).

Acetabulum (as´-e-TAB-ū-lum) The rounded cavity on the external surface of the hip bone that receives the head of the femur.

Acetylcholine (as´-ē-til-KŌ-lēn) **(ACh)** A neurotransmitter liberated by many peripheral nervous system neurons and some central nervous system neurons. It is excitatory at neuromuscular junctions but inhibitory at some other synapses (e.g., it slows heart rate).

Achalasia (ak´-a-LĀ-zē-a) A condition, caused by malfunction of the myenteric plexus, in which the lower esophageal sphincter fails to relax normally as food approaches. A whole meal may become lodged in the esophagus and enter the stomach very slowly. Distension of the esophagus results in chest pain that is often confused with pain originating from the heart.

Achilles tendon *See* **Calcaneal tendon.**

Acini (AS-i-nē) Groups of cells in the pancreas that secrete digestive enzymes.

Acoustic (a-KOOS-tik) Pertaining to sound or the sense of hearing.

Acquired immunodeficiency syndrome (AIDS) A fatal disease caused by the human immunodeficiency virus (HIV). Characterized by a positive HIV-antibody test, low helper T cell count, and certain indicator diseases (e.g., Kaposi's sarcoma, pneumocystis carinii pneumonia, tuberculosis, fungal diseases). Other symptoms include fever or night sweats, coughing, sore throat, fatigue, body aches, weight loss, and enlarged lymph nodes.

Acrosome (AK-rō-sōm) A lysosomelike organelle in the head of a sperm cell containing enzymes that facilitate the penetration of a sperm cell into a secondary oocyte.

Actin (AK-tin) A contractile protein that is part of thin filaments in muscle fibers.

Action potential (AP) An electrical signal that propagates along the membrane of a neuron or muscle fiber (cell); a rapid change in membrane potential that involves a depolarization followed by a repolarization. Also called a **nerve action potential** or **nerve impulse** as it relates to a neuron, and a **muscle action potential** as it relates to a muscle fiber.

Activation (ak´-ti-VĀ-shun) **energy** The minimum amount of energy required for a chemical reaction to occur.

Active transport The movement of substances across cell membranes against a concentration gradient, requiring the expenditure of cellular energy (ATP).

Acute (a-KŪT) Having rapid onset, severe symptoms, and a short course; not chronic.

Adaptation (ad´-ap-TĀ-shun) The adjustment of the pupil of the eye to changes in light intensity. The property by which a sensory neuron relays a decreased frequency of action potentials from a receptor, even though the strength of the stimulus remains constant; the decrease in perception of a sensation over time while the stimulus is still present.

Adduction (ad-DUK-shun) Movement toward the midline of the body.

Adenoids (AD-e-noyds) The pharyngeal tonsils.

Adenosine triphosphate (a-DEN-ō-sēn trī-FOS-fāt) **(ATP)** The main energy currency in living cells; used to transfer the chemical energy needed for metabolic reactions. ATP consists of the purine base *adenine* and the five-carbon sugar *ribose,* to which are added, in linear array, three *phosphate* groups.

Adhesion (ad-HĒ-zhun) Abnormal joining of parts to each other.

Adipocyte (AD-i-pō-sīt) Fat cell, derived from a fibroblast.

Adipose (AD-i-pōz) **tissue** Tissue composed of adipocytes specialized for triglyceride storage and present in the form of soft pads between various organs for support, protection, and insulation.

Adrenal cortex (a-DRĒ-nal KOR-teks) The outer portion of an adrenal gland, divided into three zones; the zona glomerulosa secretes mineralocorticoids, the zona fasciculata secretes glucocorticoids, and the zona reticularis secretes androgens.

Adrenal glands Two glands located superior to each kidney. Also called the **suprarenal** (soo´-pra-RĒ-nal) **glands.**

Adrenal medulla (me-DUL-a) The inner part of an adrenal gland, consisting of cells that secrete epinephrine, norepinephrine, and a small amount of dopamine in response to stimulation by sympathetic preganglionic neurons.

Adrenergic (ad´-ren-ER-jik) **neuron** A neuron that releases epinephrine (adrenaline) or norepinephrine (noradrenaline) as its neurotransmitter.

Adrenocorticotropic (ad-rē´-nō-kor-ti-kō-TRŌP-ik) **hormone (ACTH)** A hormone produced by the anterior pituitary that influences the production and secretion of certain hormones of the adrenal cortex.

Adventitia (ad-ven-TISH-a) The outermost covering of a structure or organ.

Aerobic (air-Ō-bik) Requiring molecular oxygen.

Afferent arteriole (AF-er-ent ar-TĒ-rē-ōl) A blood vessel of a kidney that divides into the capillary network called a glomerulus; there is one afferent arteriole for each glomerulus.

Agglutination (a-gloo-ti-NĀ-shun) Clumping of microorganisms or blood cells, typically due to an antigen–antibody reaction.

Aggregated lymphatic follicles Clusters of lymph nodules that are most numerous in the ileum. Also called **Peyer's** (PĪ-erz) **patches.**

Albinism (AL-bin-izm) Abnormal, nonpathological, partial, or total absence of pigment in skin, hair, and eyes.

Aldosterone (al-DOS-ter-ōn) A mineralocorticoid produced by the adrenal cortex that promotes sodium and water reabsorption by the kidneys and potassium excretion in urine.

Allantois (a-LAN-tō-is) A small, vascularized outpouching of the yolk sac that serves as an early site for blood formation and development of the urinary bladder.

Alleles (a-LĒLZ) Alternate forms of a single gene that control the same inherited trait (such as type A blood) and are located at the same position on homologous chromosomes.

Allergen (AL-er-jen) An antigen that evokes a hypersensitivity reaction.

Alopecia (al´-ō-PĒ-shē-a) The partial or complete lack of hair as a result of factors such as genetics, aging, endocrine disorders, chemotherapy, and skin diseases.

Alpha (AL-fa) **cell** A type of cell in the pancreatic islets (islets of Langerhans) that secretes the hormone glucagon. Also termed an **A cell.**

Alpha receptor A type of receptor for norepinephrine and epinephrine; present on visceral effectors innervated by sympathetic postganglionic neurons.

Alveolar duct Branch of a respiratory bronchiole around which alveoli and alveolar sacs are arranged.

Alveolar macrophage (MAK-rō-fāj) Highly phagocytic cell found in the alveolar walls of the lungs. Also called a **dust cell.**

Alveolar sac A cluster of alveoli that share a common opening.

Alveolus (al-VĒ-ō-lus) A small hollow or cavity; an air sac in the lungs; milk-secreting portion of a mammary gland. *Plural* is **alveoli** (al-VĒ-ol-ī).

Alzheimer (ALTZ-hī-mer) **disease (AD)** Disabling neurological disorder characterized by dysfunction and death of specific cerebral neurons, resulting in widespread intellectual impairment, personality changes, and fluctuations in alertness.

Amenorrhea (ā-men-ō-RĒ-a) Absence of menstruation.

Amnesia (am-NĒ-zē-a) A lack or loss of memory.

Amnion (AM-nē-on) A thin, protective fetal membrane that develops from the epiblast; holds the fetus suspended in amniotic fluid. Also called the **bag of waters.**

Amniotic (am´-nē-OT-ik) **fluid** Fluid in the amniotic cavity, the space between the developing embryo (or fetus) and amnion; the fluid is initially produced as a filtrate from maternal blood and later includes fetal urine. It functions as a shock absorber, helps regulate fetal body temperature, and helps prevent desiccation.

Amphiarthrosis (am´-fē-ar-THRŌ-sis) A slightly movable joint, in which the articulating bony surfaces are separated by fibrous connective tissue or fibrocartilage to which both are attached; types are syndesmosis and symphysis.

Ampulla (am-PUL-la) A saclike dilation of a canal or duct.

Ampulla of Vater *See* **Hepatopancreatic ampulla.**

Anabolism (a-NAB-ō-lizm) Synthetic, energy-requiring reactions whereby small molecules are built up into larger ones.

Anaerobic (an-air-Ō-bik) Not requiring oxygen.

Anal (Ā-nal) **canal** The last 2 or 3 cm (1 in.) of the rectum; opens to the exterior through the anus.

Anal column A longitudinal fold in the mucous membrane of the anal canal that contains a network of arteries and veins.

Anal triangle The subdivision of the female or male perineum that contains the anus.

Analgesia (an-al-JĒ-zē-a) Pain relief; absence of the sensation of pain.

Anaphase (AN-a-fāz) The third stage of mitosis in which the chromatids that have separated at the centromeres move to opposite poles of the cell.

Anaphylaxis (an´-a-fi-LAK-sis) A hypersensitivity (allergic) reaction in which IgE antibodies attach to mast cells and basophils, causing them to produce mediators of anaphylaxis (histamine, leukotrienes, kinins, and prostaglandins) that bring about increased blood permeability, increased smooth muscle contraction, and increased mucus production. Examples are hay fever, hives, and anaphylactic shock.

Anastomosis (a-nas-tō-MŌ-sis) An end-to-end union or joining of blood vessels, lymphatic vessels, or nerves.

Anatomic dead space Spaces of the nose, pharynx, larynx, trachea, bronchi, and bronchioles totaling about 150 mL of the 500 mL in a quiet breath (tidal volume); air in the anatomic dead space does not reach the alveoli to participate in gas exchange.

Anatomical (an´-a-TOM-i-kal) **position** A position of the body universally used in anatomical descriptions in which the body is erect, the head is level, the eyes face forward, the upper limbs are at the sides, the palms face forward, and the feet are flat on the floor.

Anatomy (a-NAT-ō-mē) The structure or study of the structure of the body and the relation of its parts to each other.

Androgens (AN-drō-jenz) Masculinizing sex hormones produced by the testes in males and the adrenal cortex in both sexes; responsible for libido (sexual desire); the two main androgens are testosterone and dihydrotestosterone.

Anemia (a-NĒ-mē-a) Condition of the blood in which the number of functional red blood cells or their hemoglobin content is below normal.

Anesthesia (an´-es-THĒ-zē-a) A total or partial loss of feeling or sensation; may be general or local.

Aneurysm (AN-ū-rizm) A saclike enlargement of a blood vessel caused by a weakening of its wall.

Angina pectoris (an-JI-na or AN-ji-na PEK-tō-ris) A pain in the chest related to reduced coronary circulation due to coronary artery disease (CAD) or spasms of vascular smooth muscle in coronary arteries.

Angiogenesis (an´-jē-ō-JEN-e-sis) The formation of blood vessels in the extraembryonic mesoderm of the yolk sac, connecting stalk, and chorion at the beginning of the third week of development.

Ankylosis (ang´-ki-LŌ-sis) Severe or complete loss of movement at a joint as the result of a disease process.

Antagonist (an-TAG-ō-nist) A muscle that has an action opposite that of the prime mover (agonist) and yields to the movement of the prime mover.

Antagonistic (an-tag-ō-NIST-ik) **effect** A hormonal interaction in which the effect of one hormone on a target cell is opposed by another hormone. For example, calcitonin (CT) lowers blood calcium level, whereas parathyroid hormone (PTH) raises it.

Anterior (an-TĒR-ē-or) Nearer to or at the front of the body. Equivalent to **ventral** in bipeds.

Anterior pituitary Anterior lobe of the pituitary gland. Also called the **adenohypophysis** (ad´-e-nō-hī-POF-i-sis).

Anterior root The structure composed of axons of motor (efferent) neurons that emerges from the anterior aspect of the spinal cord and extends laterally to join a posterior root, forming a spinal nerve. Also called a **ventral root.**

Anterolateral (an´-ter-ō-LAT-er-al) **pathway** Sensory pathway that conveys information related to pain, temperature, tickle, and itch.

Antibody (AN-ti-bod´-ē) A protein produced by plasma cells in response to a specific antigen; the antibody combines with that antigen to neutralize, inhibit, or destroy it. Also called an **immunoglobulin** (im-ū-nō-GLOB-ū-lin) or **Ig.**

Anticoagulant (an-tī-cō-AG-ū-lant) A substance that can delay, suppress, or prevent the clotting of blood.

Antidiuretic (an´-ti-dī-ū-RET-ik) Substance that inhibits urine formation.

Antidiuretic hormone (ADH) Hormone produced by neurosecretory cells in the paraventricular and supraoptic nuclei of the hypothalamus that stimulates water reabsorption from kidney tubule cells into the blood and vasoconstriction of arterioles. Also called **vasopressin** (vās-ō-PRES-in).

Antigen (AN-ti-jen) A substance that has immunogenicity (the ability to provoke an immune response) and reactivity (the ability to react with the antibodies or cells that result from the immune response); contraction of *anti*body *gen*erator. Also termed a **complete antigen.**

Antigen-presenting cell (APC) Special class of migratory cell that processes and presents antigens to T cells during an immune response; APCs include macrophages, B cells, and dendritic cells, which are present in the skin, mucous membranes, and lymph nodes.

Antrum (AN-trum) Any nearly closed cavity or chamber, especially one within a bone, such as a sinus.

Anuria (an-Ū-rē-a) Absence of urine formation or daily urine output of less than 50 mL.

Anus (Ā-nus) The distal end and outlet of the rectum.

Aorta (ā-OR-ta) The main systemic trunk of the arterial system of the body that emerges from the left ventricle.

Aortic (ā-OR-tik) **body** Cluster of chemoreceptors on or near the arch of the aorta that respond to changes in blood levels of oxygen, carbon dioxide, and hydrogen ions (H⁺).

Aortic reflex A reflex that helps maintain normal systemic blood pressure; initiated by baroreceptors in the wall of the ascending aorta and arch of the aorta. Nerve impulses from aortic baroreceptors reach the cardiovascular center via sensory axons of the vagus (X) nerves.

Apex (Ā-peks) The pointed end of a conical structure, such as the apex of the heart.

Aphasia (a-FĀ-zē-a) Loss of ability to express oneself properly through speech or loss of verbal comprehension.

Apnea (AP-nē-a) Temporary cessation of breathing.

Apneustic (ap-NOO-stik) **area** A part of the respiratory center in the pons that sends stimulatory nerve impulses to the inspiratory area that activate and prolong inhalation and inhibit exhalation.

Apocrine (AP-ō-krin) **gland** A type of gland in which the secretory products gather at the free end of the secreting cell and are pinched off, along with some of the cytoplasm, to become the secretion, as in mammary glands.

Aponeurosis (ap´-ō-noo-RŌ-sis) A sheetlike tendon joining one muscle with another or with bone.

Apoptosis (ap-ō-TŌ-sis *or* ap´-ōp-TŌ-sis) Programmed cell death; a normal type of cell death that removes unneeded cells during embryological development, regulates the number of cells in tissues, and eliminates many potentially dangerous cells such as cancer cells. During apoptosis, the DNA fragments, the nucleus condenses, mitochondria cease to function, and the cytoplasm shrinks, but the plasma membrane remains intact. Phagocytes engulf and digest the apoptotic cells, and an inflammatory response does not occur.

Appositional (a-pō-ZISH-o-nal) **growth** Growth due to surface deposition of material, as in the growth in diameter of cartilage and bone. Also called **exogenous** (eks-OJ-e-nus) **growth.**

Aqueous humor (AK-wē-us HŪ-mer) The watery fluid, similar in composition to cerebrospinal fluid, that fills the anterior cavity of the eye.

Arachnoid (a-RAK-noyd) **mater** The middle of the three meninges (coverings) of the brain and spinal cord. Also termed the **arachnoid.**

Arachnoid villus (VIL-us) Berrylike tuft of the arachnoid mater that protrudes into the superior sagittal sinus and through which cerebrospinal fluid is reabsorbed into the bloodstream.

Arbor vitae (AR-bor VĪ-tē) The white matter tracts of the cerebellum, which have a treelike appearance when seen in midsagittal section.

Arch of the aorta The most superior portion of the aorta, lying between the ascending and descending segments of the aorta.

Areola (a-RĒ-ō-la) Any tiny space in a tissue. The pigmented ring around the nipple of the breast.

Arm The part of the upper limb from the shoulder to the elbow.

Arousal (a-ROW-zal) Awakening from sleep, a response due to stimulation of the reticular activating system (RAS).

Arrector pili (a-REK-tor PI-lī) Smooth muscles attached to hairs; contraction pulls the hairs into a vertical position, resulting in goose bumps.

Arrhythmia (a-RITH-mē-a) An irregular heart rhythm. Also called a **dysrhythmia.**

Arteriole (ar-TĒ-rē-ōl) A small, almost microscopic, artery that delivers blood to a capillary.

Arteriosclerosis (ar-tē-rē-ō-skle-RŌ-sis) Group of diseases characterized by thickening of the walls of arteries and loss of elasticity.

Artery (AR-ter-ē) A blood vessel that carries blood away from the heart.

Arthritis (ar-THRĪ-tis) Inflammation of a joint.

Arthrology (ar-THROL-ō-jē) The study or description of joints.

Arthroplasty (AR-thrō-plas´-tē) Surgical replacement of joints, for example, the hip and knee joints.

Arthroscopy (ar-THROS-kō-pē) A procedure for examining the interior of a joint, usually the knee, by inserting an arthroscope into a small incision; used to determine extent of damage, remove torn cartilage, repair cruciate ligaments, and obtain samples for analysis.

Arthrosis (ar-THRŌ-sis) A joint or articulation.

Articular (ar-TIK-ū-lar) **capsule** Sleevelike structure around a synovial joint composed of a fibrous capsule and a synovial membrane.

Articular cartilage (KAR-ti-lij) Hyaline cartilage attached to articular bone surfaces.

Articular disc Fibrocartilage pad between articular surfaces of bones of some synovial joints. Also called a **meniscus** (men-IS-kus).

Articulation (ar-tik-ū-LĀ-shun) A joint; a point of contact between bones, cartilage and bones, or teeth and bones.

Arytenoid (ar´-i-TĒ-noyd) **cartilages** A pair of small, pyramidal cartilages of the larynx that attach to the vocal folds and intrinsic pharyngeal muscles and can move the vocal folds.

Ascending colon (KŌ-lon) The part of the large intestine that passes superiorly from the cecum to the inferior border of the liver, where it bends at the right colic (hepatic) flexure to become the transverse colon.

Ascites (as-SĪ-tēz) Abnormal accumulation of serous fluid in the peritoneal cavity.

Association areas Large cortical regions on the lateral surfaces of the occipital, parietal, and temporal lobes and on the frontal lobes anterior to the motor areas connected by many motor and sensory axons to other parts of the cortex. The association areas are concerned with motor patterns, memory, concepts of word-hearing and word-seeing, reasoning, will, judgment, and personality traits.

Asthma (AZ-ma) Usually allergic reaction characterized by smooth muscle spasms in bronchi resulting in wheezing and difficult breathing. Also called **bronchial asthma.**

Astigmatism (a-STIG-ma-tizm) An irregularity of the lens or cornea of the eye causing the image to be out of focus and producing faulty vision.

Astrocyte (AS-trō-sīt) A neuroglial cell having a star shape that participates in brain development and the metabolism of neurotransmitters, helps form the blood–brain barrier, helps maintain the proper balance of K⁺ for generation of nerve impulses, and provides a link between neurons and blood vessels.

Ataxia (a-TAK-sē-a) A lack of muscular coordination, lack of precision.

Atherosclerotic (ath´-er-ō-skle-RO-tic) **plaque** (PLAK) A lesion that results from accumulated cholesterol and smooth muscle fibers (cells) of the tunica media of an artery; may become obstructive.

Atom Unit of matter that makes up a chemical element; consists of a nucleus (containing positively charged protons and uncharged neutrons) and negatively charged electrons that orbit the nucleus.

Atresia (a-TRĒ-zē-a) Degeneration and reabsorption of an ovarian follicle before it fully matures and ruptures; abnormal closure of a passage, or absence of a normal body opening.

Atrial fibrillation (Ā-trē-al fib-ri-LĀ-shun) Asynchronous contraction of cardiac muscle fibers in the atria that results in the cessation of atrial pumping.

Atrial natriuretic (na´-trē-ū-RET-ik) **peptide (ANP)** Peptide hormone, produced by the atria of the heart in response to stretching, that inhibits aldosterone production and thus lowers blood pressure; causes natriuresis, increased urinary excretion of sodium.

Atrioventricular (AV) (ā´-trē-ō-ven-TRIK-ū-lar) **bundle** The part of the conduction system of the heart that begins at the atrioventricular (AV) node, passes through the cardiac skeleton separating the atria and the ventricles, then extends a short distance down the interventricular septum before splitting into right and left bundle branches. Also called the **bundle of His** (HIS).

Atrioventricular (AV) node The part of the conduction system of the heart made up of a compact mass of conducting cells located in the septum between the two atria.

Atrioventricular (AV) valve A heart valve made up of membranous flaps or cusps that allows blood to flow in one direction only, from an atrium into a ventricle.

Atrium (Ā-trē-um) A superior chamber of the heart.

Atrophy (AT-rō-fē) Wasting away or decrease in size of a part, due to a failure, abnormality of nutrition, or lack of use.

Auditory ossicle (AW-di-tō-rē OS-si-kul) One of the three small bones of the middle ear called the **malleus**, **incus**, and **stapes**.

Auditory tube The tube that connects the middle ear with the nose and nasopharynx region of the throat. Also called the **eustachian** (ū-STĀ-shun or ū-STĀ-kē-an) **tube** or **pharyngotympanic tube.**

Auscultation (aws-kul-TĀ-shun) Examination by listening to sounds in the body.

Autoimmunity An immunological response against a person's own tissues.

Autolysis (aw-TOL-i-sis) Self-destruction of cells by their own lysosomal digestive enzymes after death or in a pathological process.

Autonomic ganglion (aw´-tō-NOM-ik GANG-glē-on) A cluster of cell bodies of sympathetic or parasympathetic neurons located outside the central nervous system.

Autonomic nervous system (ANS) Visceral sensory (afferent) and visceral motor (efferent) neurons. Autonomic motor neurons, both sympathetic and parasympathetic, conduct nerve impulses from the central nervous system to smooth muscle, cardiac muscle, and glands. So named because this part of the nervous system was thought to be self-governing or spontaneous.

Autonomic plexus (PLEK-sus) A network of sympathetic and parasympathetic axons; examples are the cardiac, celiac, and pelvic plexuses, which are located in the thorax, abdomen, and pelvis, respectively.

Autophagy (aw-TOF-a-jē) Process by which worn-out organelles are digested within lysosomes.

Autopsy (AW-top-sē) The examination of the body after death.

Autorhythmic cells Cardiac or smooth muscle fibers that are self-excitable (generate impulses without an external stimulus); act as the heart's pacemaker and conduct the pacing impulse through the conduction system of the heart; self-excitable neurons in the central nervous system, as in the inspiratory area of the brain stem.

Autosome (AW-tō-sōm) Any chromosome other than the X and Y chromosomes (sex chromosomes).

Axilla (ak-SIL-a) The small hollow beneath the arm where it joins the body at the shoulders. Also called the **armpit.**

Axon (AK-son) The usually single, long process of a nerve cell that propagates a nerve impulse toward the axon terminals.

Axon terminal Terminal branch of an axon where synaptic vesicles undergo exocytosis to release neurotransmitter molecules.

B

B cell A lymphocyte that can develop into a clone of antibody-producing plasma cells or memory cells when properly stimulated by a specific antigen.

Babinski (ba-BIN-skē) **sign** Extension of the great toe, with or without fanning of the other toes, in response to stimulation of the outer margin of the sole; normal up to 18 months of age and indicative of damage to descending motor pathways such as the corticospinal tracts after that.

Back The posterior part of the body; the **dorsum**.

Ball-and-socket joint A synovial joint in which the rounded surface of one bone moves within a cup-shaped depression or socket of another bone, as in the shoulder or hip joint. Also called a **spheroid** (SFĒ-royd) **joint.**

Baroreceptor (bar´-ō-re-SEP-tor) Neuron capable of responding to changes in blood, air, or fluid pressure. Also called a **pressoreceptor.**

Basal ganglia (GANG-glē-a) Paired clusters of gray matter deep in each cerebral hemisphere including the **globus pallidus**, **putamen**, and **caudate nucleus**. Together, the caudate nucleus and putamen are known as the **corpus striatum.** Nearby structures that are functionally linked to the basal ganglia are the substantia nigra of the midbrain and the subthalamic nuclei of the diencephalon.

Basement membrane Thin, extracellular layer between epithelium and connective tissue consisting of a basal lamina and a reticular lamina.

Basilar (BĀS-i-lar) **membrane** A membrane in the cochlea of the internal ear that separates the cochlear duct from the scala tympani and on which the spiral organ (organ of Corti) rests.

Basophil (BĀ-sō-fil) A type of white blood cell characterized by a pale nucleus and large granules that stain blue-purple with basic dyes.

Belly The abdomen. The gaster or prominent, fleshy part of a skeletal muscle.

Beta (BĀ-ta) **cell** A type of cell in the pancreatic islets (islets of Langerhans) that secretes the hormone insulin. Also called a **B cell.**

Beta receptor A type of adrenergic receptor for epinephrine and norepinephrine; found on visceral effectors innervated by sympathetic postganglionic neurons.

Bicuspid (bī-KUS-pid) **valve** Atrioventricular (AV) valve on the left side of the heart. Also called the **mitral valve.**

Bilateral (bī-LAT-er-al) Pertaining to two sides of the body.

Bile (BĪL) A secretion of the liver consisting of water, bile salts, bile pigments, cholesterol, lecithin, and several ions; it emulsifies lipids prior to their digestion.

Bilirubin (bil-i-ROO-bin) An orange pigment that is one of the end products of hemoglobin breakdown in hepatocytes and is excreted as a waste material in bile.

Blastocyst In the development of an embryo, a hollow ball of cells that consists of a blastocele (the internal cavity), trophoblast (outer cells), and embryoblast (inner cell mass).

Blastocyst (BLAS-tō-sist) **cavity** The fluid-filled cavity within the blastocyst. Also called the **blastocele.**

Blastomere (BLAS-tō-mēr) One of the cells resulting from the cleavage of a fertilized ovum.

Blastula (BLAS-tyū-la) An early stage in the development of a zygote.

Blind spot Area in the retina at the end of the optic (II) nerve in which there are no photoreceptors.

Blood The fluid that circulates through the heart, arteries, capillaries, and veins and that constitutes the chief means of transport within the body.

Blood–brain barrier (BBB) A barrier consisting of specialized brain capillaries and astrocytes that

prevents the passage of materials from the blood to the cerebrospinal fluid and brain.

Blood island Isolated mass of mesoderm derived from angioblasts and from which blood vessels develop.

Blood pressure (BP) Force exerted by blood against the walls of blood vessels due to contraction of the heart and influenced by the elasticity of the vessel walls; clinically, a measure of the pressure in arteries during ventricular systole and ventricular diastole.

Blood reservoir (REZ-er-vwar) Systemic veins and venules that contain large amounts of blood that can be moved quickly to parts of the body requiring the blood.

Blood–testis barrier (BTB) A barrier formed by Sertoli cells that prevents an immune response against antigens produced by spermatogenic cells by isolating the cells from the blood.

Body cavity A space within the body that contains various internal organs.

Bolus (BŌ-lus) A soft, rounded mass, usually food, that is swallowed.

Bony labyrinth (LAB-i-rinth) A series of cavities within the petrous portion of the temporal bone forming the vestibule, cochlea, and semicircular canals of the inner ear.

Bowman's capsule See **Glomerular capsule.**

Brachial plexus (BRĀ-kē-al PLEK-sus) A network of nerve axons of the ventral rami of spinal nerves C5, C6, C7, C8, and T1. The nerves that emerge from the brachial plexus supply the upper limb.

Bradycardia (brād´-i-KAR-dē-a) A slow resting heart or pulse rate (under 50 beats per minute).

Brain The part of the central nervous system contained within the cranial cavity.

Brain stem The portion of the brain immediately superior to the spinal cord, made up of the medulla oblongata, pons, and midbrain.

Brain waves Electrical signals that can be recorded from the skin of the head due to electrical activity of brain neurons.

Broad ligament A double fold of parietal peritoneum attaching the uterus to the side of the pelvic cavity.

Broca's (BRŌ-kaz) **area** Motor area of the brain in the frontal lobe that translates thoughts into speech. Also called the **motor speech area.**

Bronchi (BRONG-kē) Branches of the respiratory passageway including primary bronchi (the two divisions of the trachea), secondary or lobar bronchi (divisions of the primary bronchi that are distributed to the lobes of the lung), and tertiary or segmental bronchi (divisions of the secondary bronchi that are distributed to bronchopulmonary segments of the lung). Singular is **bronchus.**

Bronchial tree The trachea, bronchi, and their branching structures up to and including the terminal bronchioles.

Bronchiole (BRONG-kē-ōl) Branch of a tertiary bronchus further dividing into terminal bronchioles (distributed to lobules of the lung), which divide into respiratory bronchioles (distributed to alveolar sacs).

Bronchitis (brong-KĪ-tis) Inflammation of the mucous membrane of the bronchial tree; characterized by hypertrophy and hyperplasia of seromucous glands and goblet cells that line the bronchi, which results in a productive cough.

Bronchopulmonary (brong´-kō-PUL-mō-ner-ē) **segment** One of the smaller divisions of a lobe of a lung supplied by its own branches of a bronchus.

Brunner's gland *See* **Duodenal gland.**

Buccal (BUK-al) Pertaining to the cheek or mouth.

Bulb of penis Expanded portion of the base of the corpus spongiosum penis.

Bulbourethral (bul´-bō-ū-RĒ-thral) **gland** One of a pair of glands located inferior to the prostate on either side of the urethra that secretes an alkaline fluid into the cavernous urethra. Also called a **Cowper's** (KOW-perz) **gland.**

Bulimia (boo-LIM-ē-a *or* boo-LĒ-mē-a) A disorder characterized by overeating at least twice a week followed by purging by self-induced vomiting, strict dieting, or fasting, vigorous exercise, or use of laxatives or diuretics. Also called **binge–purge syndrome.**

Bulk-phase endocytosis A process by which most body cells can ingest membrane-surrounded droplets of interstitial fluid.

Bundle branch One of the two branches of the atrioventricular (AV) bundle made up of specialized muscle fibers (cells) that transmit electrical impulses to the ventricles.

Bundle of His *See* **Atrioventricular (AV) bundle.**

Bursa (BUR-sa) A sac or pouch of synovial fluid located at friction points, especially about joints.

Bursitis (bur-SĪ-tis) Inflammation of a bursa.

Buttocks (BUT-oks) The two fleshy masses on the posterior aspect of the inferior trunk, formed by the gluteal muscles.

C

Calcaneal (kal-KĀ-nē-al) **tendon** The tendon of the soleus, gastrocnemius, and plantaris muscles at the back of the heel. Also called the **Achilles** (a-KIL-ēz) **tendon.**

Calcification (kal´-si-fi-KĀ-shun) Deposition of mineral salts, primarily hydroxyapatite, in a framework formed by collagen fibers in which the tissue hardens. Also called **mineralization** (min´-e-ral-i-ZĀ-shun).

Calcitonin (kal-si-TŌ-nin) **(CT)** A hormone produced by the parafollicular cells of the thyroid gland that can lower the amount of blood calcium and phosphates by inhibiting bone resorption (breakdown of bone extracellular matrix) and by accelerating uptake of calcium and phosphates into bone matrix.

Calculus (KAL-kū-lus) A stone, or insoluble mass of crystallized salts or other material, formed within the body, as in the gallbladder, kidney, or urinary bladder.

Callus (KAL-lus) A growth of new bone tissue in and around a fractured area, ultimately replaced by mature bone. An acquired, localized thickening.

Calyx (KĀL-iks) Any cuplike division of the kidney pelvis. *Plural* is **calyces** (KĀ-li-sēz).

Canal (ka-NAL) A narrow tube, channel, or passageway.

Canaliculus (kan´-a-LIK-ū-lus) A small channel or canal, as in bones, where they connect lacunae. *Plural* is **canaliculi** (kan´-a-LIK-ū-lī).

Canal of Schlemm *See* **Scleral venous sinus.**

Capacitation (ka-pas´-i-TĀ-shun) The functional changes that sperm undergo in the female reproductive tract that allow them to fertilize a secondary oocyte.

Capillary (KAP-i-lar´-ē) A microscopic blood vessel located between an arteriole and venule through which materials are exchanged between blood and interstitial fluid.

Carcinogen (car-SIN-ō-jen) A chemical substance or radiation that causes cancer.

Cardiac (KAR-dē-ak) **arrest** Cessation of an effective heartbeat in which the heart is completely stopped or in ventricular fibrillation.

Cardiac cycle A complete heartbeat consisting of systole (contraction) and diastole (relaxation) of both atria plus systole and diastole of both ventricles.

Cardiac muscle Striated muscle fibers (cells) that form the wall of the heart; stimulated by an intrinsic conduction system and autonomic motor neurons.

Cardiac notch An angular notch in the anterior border of the left lung into which part of the heart fits.

Cardinal ligament A ligament of the uterus, extending laterally from the cervix and vagina as a continuation of the broad ligament.

Cardiogenic area (kar´-dē-ō-JEN-ik) A group of mesodermal cells in the head end of an embryo that gives rise to the heart.

Cardiology (kar-dē-OL-ō-jē) The study of the heart and diseases associated with it.

Cardiovascular (kar´-dē-ō-VAS-kū-lar) **center** Groups of neurons scattered within the medulla oblongata that regulate heart rate, force of contraction, and blood vessel diameter.

Carotene (KAR-ō-tēn) Antioxidant precursor of vitamin A, which is needed for synthesis of photopigments; yellow-orange pigment present in the stratum corneum of the epidermis. Accounts for the yellowish coloration of skin. Also termed **beta-carotene.**

Carotid (ka-ROT-id) **body** Cluster of chemoreceptors on or near the carotid sinus that respond to changes in blood levels of oxygen, carbon dioxide, and hydrogen ions.

Carotid sinus A dilated region of the internal carotid artery just superior to where it branches from the common carotid artery; it contains baroreceptors that monitor blood pressure.

Carpal bones The eight bones of the wrist. Also called **carpals.**

Carpus (KAR-pus) A collective term for the eight bones of the wrist.

Cartilage (KAR-ti-lij) A type of connective tissue consisting of chondrocytes in lacunae embedded in a dense network of collagen and elastic fibers and an extracellular matrix of chondroitin sulfate.

Cartilaginous (kar-ti-LAJ-i-nus) **joint** A joint without a synovial (joint) cavity where the articulating bones are held tightly together by cartilage, allowing little or no movement.

Catabolism (ka-TAB-ō-lizm) Chemical reactions that break down complex organic compounds into simple ones, with the net release of energy.

Cataract (KAT-a-rakt) Loss of transparency of the lens of the eye or its capsule or both.

Cauda equina (KAW-da ē-KWĪ-na) A taillike array of roots of spinal nerves at the inferior end of the spinal cord.

Caudal (KAW-dal) Pertaining to any taillike structure; inferior in position.

Cecum (SĒ-kum) A blind pouch at the proximal end of the large intestine that attaches to the ileum.

Celiac plexus (PLEK-sus) A large mass of autonomic ganglia and axons located at the level of the superior part of the first lumbar vertebra. Also called the **solar plexus.**

Cell The basic structural and functional unit of all organisms; the smallest structure capable of performing all the activities vital to life.

Cell cycle Growth and division of a single cell into two identical cells; consists of interphase and cell division.

Cell division Process by which a cell reproduces itself that consists of a nuclear division (mitosis) and a cytoplasmic division (cytokinesis); types include somatic and reproductive cell division.

Cell junction Point of contact between plasma membranes of tissue cells.

Cementum (se-MEN-tum) Calcified tissue covering the root of a tooth.

Central canal A microscopic tube running the length of the spinal cord in the gray commissure. A circular channel running longitudinally in the center of an osteon (haversian system) of mature compact bone, containing blood and lymphatic vessels and nerves. Also called **haversian** (ha-VER-shan) **canal.**

Central fovea (FŌ-vē-a) A depression in the center of the macula lutea of the retina, containing cones only and lacking blood vessels; the area of highest visual acuity (sharpness of vision).

Central nervous system (CNS) That portion of the nervous system that consists of the brain and spinal cord.

Centrioles (SEN-trē-ōlz) Paired, cylindrical structures of a centrosome, each consisting of a ring of microtubules and arranged at right angles to each other.

Centromere (SEN-trō-mēr) The constricted portion of a chromosome where the two chromatids are joined; serves as the point of attachment for the microtubules that pull chromatids during anaphase of cell division.

Centrosome (SEN-trō-sōm) A dense network of small protein fibers near the nucleus of a cell, containing a pair of centrioles and pericentriolar material.

Cephalic (se-FAL-ik) Pertaining to the head; superior in position.

Cerebellar peduncle (ser-e-BEL-ar pe-DUNG-kl) A bundle of nerve axons connecting the cerebellum with the brain stem.

Cerebellum (ser´-e-BEL-um) The part of the brain lying posterior to the medulla oblongata and pons; governs balance and coordinates skilled movements.

Cerebral aqueduct (SER-ē-bral AK-we-dukt) A channel through the midbrain connecting the third and fourth ventricles and containing cerebrospinal fluid. Also termed the **aqueduct of Sylvius.**

Cerebral arterial circle A ring of arteries forming an anastomosis at the base of the brain between the internal carotid and basilar arteries and arteries supplying the cerebral cortex. Also called the **circle of Willis.**

Cerebral cortex The surface of the cerebral hemispheres, 2–4 mm thick, consisting of gray matter;

arranged in six layers of neuronal cell bodies in most areas.

Cerebral peduncle One of a pair of nerve axon bundles located on the anterior surface of the midbrain, conducting nerve impulses between the pons and the cerebral hemispheres.

Cerebrospinal (se-rē′-brō-SPĪ-nal) **fluid (CSF)** A fluid produced by ependymal cells that cover choroid plexuses in the ventricles of the brain; the fluid circulates in the ventricles, the central canal, and the subarachnoid space around the brain and spinal cord.

Cerebrovascular (se-rē′-brō-VAS-kū-lar) **accident (CVA)** Destruction of brain tissue (infarction) resulting from obstruction or rupture of blood vessels that supply the brain. Also called a **stroke** or **brain attack.**

Cerebrum (se-RĒ-brum *or* SER-e-brum) The two hemispheres of the forebrain (derived from the telencephalon), making up the largest part of the brain.

Cerumen (se-ROO-men) Waxlike secretion produced by ceruminous glands in the external auditory meatus (ear canal). Also termed **ear wax.**

Ceruminous (se-RŪ-mi-nus) **gland** A modified sudoriferous (sweat) gland in the external auditory meatus that secretes cerumen (ear wax).

Cervical ganglion (SER-vi-kul GANG-glē-on) A cluster of cell bodies of postganglionic sympathetic neurons located in the neck, near the vertebral column.

Cervical plexus (PLEK-sus) A network formed by nerve axons from the ventral rami of the first four cervical nerves and receiving gray rami communicantes from the superior cervical ganglion.

Cervix (SER-viks) Neck; any constricted portion of an organ, such as the inferior cylindrical part of the uterus.

Chemoreceptor (kē′-mō-rē-SEP-tor) Sensory receptor that detects the presence of a specific chemical.

Chiasm (KĪ-azm) A crossing; especially the crossing of axons in the optic (II) nerve.

Chief cell The secreting cell of a gastric gland that produces pepsinogen, the precursor of the enzyme pepsin, and the enzyme gastric lipase. Also called a **zymogenic** (zī′-mō-JEN-ik) **cell.** Cell in the parathyroid glands that secretes parathyroid hormone (PTH). Also called a **principal cell.**

Cholecystectomy (kō′-lē-sis-TEK-tō-mē) Surgical removal of the gallbladder.

Cholecystitis (kō′-lē-sis-TĪ-tis) Inflammation of the gallbladder.

Cholesterol (kō-LES-te-rol) Classified as a lipid, the most abundant steroid in animal tissues; located in cell membranes and used for the synthesis of steroid hormones and bile salts.

Cholinergic (kō′-lin-ER-jik) **neuron** A neuron that liberates acetylcholine as its neurotransmitter.

Chondrocyte (KON-drō-sīt) Cell of mature cartilage.

Chondroitin (kon-DROY-tin) **sulfate** An amorphous extracellular matrix material found outside connective tissue cells.

Chordae tendineae (KOR-dē TEN-di-nē-ē) Tendonlike, fibrous cords that connect atrio-ventricular valves of the heart with papillary muscles.

Chorion (KŌ-rē-on) The most superficial fetal membrane that becomes the principal embryonic portion of the placenta; serves a protective and nutritive function.

Chorionic villi (kō-rē-ON-ik VIL-ī) Fingerlike projections of the chorion that grow into the decidua basalis of the endometrium and contain fetal blood vessels.

Chorionic villi sampling (CVS) The removal of a sample of chorionic villus tissue by means of a catheter to analyze the tissue for prenatal genetic defects.

Choroid (KŌ-royd) One of the vascular coats of the eyeball.

Choroid plexus (PLEK-sus) A network of capillaries located in the roof of each of the four ventricles of the brain; ependymal cells around choroid plexuses produce cerebrospinal fluid.

Chromaffin (KRŌ-maf-in) **cell** Cell that has an affinity for chrome salts, due in part to the presence of the precursors of the neurotransmitter epinephrine; found, among other places, in the adrenal medulla.

Chromatid (KRŌ-ma-tid) One of a pair of identical connected nucleoprotein strands that are joined at the centromere and separate during cell division, each becoming a chromosome of one of the two daughter cells.

Chromatin (KRŌ-ma-tin) The threadlike mass of genetic material, consisting of DNA and histone proteins, that is present in the nucleus of a nondividing or interphase cell.

Chromatolysis (krō′-ma-TOL-i-sis) The breakdown of Nissl bodies into finely granular masses in the cell body of a neuron whose axon has been damaged.

Chromosome (KRŌ-mō-sōm) One of the small, threadlike structures in the nucleus of a cell, normally 46 in a human diploid cell, that bears the genetic material; composed of DNA and proteins (histones) that form a delicate chromatin thread during interphase; becomes packaged into compact rodlike structures that are visible under the light microscope during cell division.

Chronic (KRON-ik) Long-term or frequently recurring; applied to a disease that is not acute.

Chronic obstructive pulmonary disease (COPD) A disease, such as bronchitis or emphysema, in which there is some degree of obstruction of airways and consequent increase in airway resistance.

Chyle (KĪL) The milky-appearing fluid found in the lacteals of the small intestine after absorption of lipids in food.

Chyme (KĪM) The semifluid mixture of partly digested food and digestive secretions found in the stomach and small intestine during digestion of a meal.

Ciliary (SIL-ē-ar′-ē) **body** One of the three parts of the vascular tunic of the eyeball, the others being the choroid and the iris; includes the ciliary muscle and the ciliary processes.

Ciliary ganglion (GANG-glē-on) A very small parasympathetic ganglion whose preganglionic axons come from the oculomotor (III) nerve and whose postganglionic axons carry nerve impulses to the ciliary muscle and the sphincter muscle of the iris.

Cilium (SIL-ē-um) A hair or hairlike process projecting from a cell that may be used to move the entire cell or to move substances along the surface of the cell. *Plural* is **cilia.**

Circle of Willis *See* **Cerebral arterial circle.**

Circular folds Permanent, deep, transverse folds in the mucosa and submucosa of the small intestine that increase the surface area for absorption. Also called **plicae circulares** (PLĪ-kē SER-ku-lar-ēs).

Circumduction (ser-kum-DUK-shun) A movement at a synovial joint in which the distal end of a bone moves in a circle while the proximal end remains relatively stable.

Cirrhosis (si-RŌ-sis) A liver disorder in which the parenchymal cells are destroyed and replaced by connective tissue.

Cisterna chyli (sis-TER-na KĪ-lē) The origin of the thoracic duct.

Cleavage The rapid mitotic divisions following the fertilization of a secondary oocyte, resulting in an increased number of progressively smaller cells, called blastomeres.

Clitoris (KLI-to-ris) An erectile organ of the female, located at the anterior junction of the labia minora, that is homologous to the male penis.

Clone (KLŌN) A population of identical cells.

Coarctation (kō′-ark-TĀ-shun) **of the aorta** A congenital heart defect in which a segment of the aorta is too narrow. As a result, the flow of oxygenated blood to the body is reduced, the left ventricle is forced to pump harder, and high blood pressure develops.

Coccyx (KOK-siks) The fused bones at the inferior end of the vertebral column.

Cochlea (KOK-lē-a) A winding, cone-shaped tube forming a portion of the inner ear and containing the spiral organ (organ of Corti).

Cochlear duct The membranous cochlea consisting of a spirally arranged tube enclosed in the bony cochlea and lying along its outer wall. Also called the **scala media** (SCA-la MĒ-dē-a).

Collagen (KOL-a-jen) A protein that is the main organic constituent of connective tissue.

Collateral circulation The alternate route taken by blood through an anastomosis.

Colliculus (ko-LIK-ū-lus) A small elevation.

Colon The portion of the large intestine consisting of ascending, transverse, descending, and sigmoid portions.

Colony-stimulating factor (CSF) One of a group of molecules that stimulates development of white blood cells. Examples are macrophage CSF and granulocyte CSF.

Colostrum (kō-LOS-trum) A thin, cloudy fluid secreted by the mammary glands a few days prior to or after delivery before true milk is produced.

Column (KOL-um) Group of white matter tracts in the spinal cord.

Common bile duct A tube formed by the union of the common hepatic duct and the cystic duct that empties bile into the duodenum at the hepatopancreatic ampulla (ampulla of Vater).

Compact (dense) bone tissue Bone tissue that contains few spaces between osteons (haversian systems); forms the external portion of all bones and the bulk of the diaphysis (shaft) of long bones; is found immediately deep to the periosteum and external to spongy bone.

Concha (KONG-ka) A scroll-like bone found in the skull. *Plural is* **conchae** (KONG-kē).

Concussion (kon-KUSH-un) Traumatic injury to the brain that produces no visible bruising but may result in abrupt, temporary loss of consciousness.

Conduction system A group of autorhythmic cardiac muscle fibers that generates and distributes electrical impulses to stimulate coordinated contraction of the heart chambers; includes the sinoatrial (SA) node, the atrioventricular (AV) node, the atrioventricular (AV) bundle, the right and left bundle branches, and the Purkinje fibers.

Condyloid (KON-di-loyd) **joint** A synovial joint structured so that an oval-shaped condyle of one bone fits into an elliptical cavity of another bone, permitting side-to-side and back-and-forth movements, such as the joint at the wrist between the radius and carpals. Also called an **ellipsoidal** (ē-lip-SOYD-al) **joint.**

Cone (KŌN) The type of photoreceptor in the retina that is specialized for highly acute color vision in bright light.

Congenital (kon-JEN-i-tal) Present at the time of birth.

Conjunctiva (kon´-junk-TĪ-va) The delicate membrane covering the eyeball and lining the eyes.

Connective tissue One of the most abundant of the four basic tissue types in the body, performing the functions of binding and supporting; consists of relatively few cells in a generous extracellular matrix (the ground substance and fibers between the cells).

Consciousness (KON-shus-nes) A state of wakefulness in which an individual is fully alert, aware, and oriented, partly as a result of feedback between the cerebral cortex and reticular activating system.

Continuous conduction (kon-DUK-shun) Propagation of an action potential (nerve impulse) in a step-by-step depolarization of each adjacent area of an axon membrane.

Contraception (kon´-tra-SEP-shun) The prevention of fertilization or impregnation without destroying fertility.

Contractility (kon´-trak-TIL-i-tē) The ability of cells or parts of cells to actively generate force to undergo shortening for movements. Muscle fibers (cells) exhibit a high degree of contractility.

Contralateral (CON-tra-lat-er-al) On the opposite side; affecting the opposite side of the body.

Conus medullaris (KŌ-nus med-ū-LAR-is) The tapered portion of the spinal cord inferior to the lumbar enlargement.

Convergence (con-VER-jens) A synaptic arrangement in which the synaptic end bulbs of several presynaptic neurons terminate on one postsynaptic neuron. The medial movement of the two eyeballs so that both are directed toward a near object being viewed in order to produce a single image.

Cornea (KOR-nē-a) The nonvascular, transparent fibrous coat through which the iris of the eye can be seen.

Corona (kō-RŌ-na) Margin of the glans penis.

Corona radiata The innermost layer of granulosa cells that is firmly attached to the zona pellucida around a secondary oocyte.

Coronary artery disease (CAD) A condition such as atherosclerosis that causes narrowing of coronary arteries so that blood flow to the heart is reduced. The result is **coronary heart disease (CHD),** in which the heart muscle receives inadequate blood flow due to an interruption of its blood supply.

Coronary circulation The pathway followed by the blood from the ascending aorta through the blood vessels supplying the heart and returning to the right atrium. Also called **cardiac circulation.**

Coronary sinus (SĪ-nus) A wide venous channel on the posterior surface of the heart that collects the blood from the coronary circulation and returns it to the right atrium.

Corpus albicans (KOR-pus AL-bi-kanz) A white fibrous patch in the ovary that forms after the corpus luteum regresses.

Corpus callosum (kal-LŌ-sum) The great commissure of the brain between the cerebral hemispheres.

Corpus luteum (LOO-tē-um) A yellowish body in the ovary formed when a follicle has discharged its secondary oocyte; secretes estrogens, progesterone, relaxin, and inhibin.

Corpus striatum (strī-Ā-tum) An area in the interior of each cerebral hemisphere composed of the caudate and putamen of the basal ganglia and white matter of the internal capsule, arranged in a striated manner.

Corpuscle of touch *See* **Meissner corpuscle.**

Cortex (KOR-teks) An outer layer of an organ. The convoluted layer of gray matter covering each cerebral hemisphere.

Costal (KOS-tal) Pertaining to a rib.

Cramp A spasmodic, usually painful contraction of a muscle.

Cranial (KRĀ-ne-al) **cavity** A body cavity formed by the cranial bones and containing the brain.

Cranial nerve One of 12 pairs of nerves that leave the brain; pass through foramina in the skull; and supply sensory and motor neurons to the head, neck, part of the trunk, and viscera of the thorax and abdomen. Each is designated by a Roman numeral and a name.

Craniosacral (krā-nē-ō-SĀK-ral) **outflow** The axons of parasympathetic preganglionic neurons, which have their cell bodies located in nuclei in the brain stem and in the lateral gray matter of the sacral portion of the spinal cord.

Cranium (KRĀ-nē-um) The skeleton of the skull that protects the brain and the organs of sight, hearing, and balance; includes the frontal, parietal, temporal, occipital, sphenoid, and ethmoid bones.

Crista (KRIS-ta) A crest or ridged structure. A small elevation in the ampulla of each semicircular duct that contains receptors for dynamic equilibrium. *Plural is* **cristae.**

Crossing-over The exchange of a portion of one chromatid with another during meiosis. It permits an exchange of genes among chromatids and is one factor that results in genetic variation of progeny.

Crus (KRUS) **of penis** Separated, tapered portion of the corpora cavernosa penis. *Plural is* **crura** (KROO-ra).

Crypt of Lieberkühn *See* **Intestinal gland.**

Cryptorchidism (krip-TOR-ki-dizm) The condition of undescended testes.

Cuneate (KŪ-nē-āt) **nucleus** A group of neurons in the inferior part of the medulla oblongata in which axons of the cuneate fasciculus terminate.

Cupula (KU-pū-la) A mass of gelatinous material covering the hair cells of a crista; a sensory receptor in the ampulla of a semicircular canal stimulated when the head moves.

Cushing's syndrome Condition caused by a hypersecretion of glucocorticoids characterized by spindly legs, "moon face," "buffalo hump," pendulous abdomen, flushed facial skin, poor wound healing, hyperglycemia, osteoporosis, hypertension, and increased susceptibility to disease.

Cutaneous (kū-TĀ-nē-us) Pertaining to the skin.

Cyanosis (sī-a-NŌ-sis) A blue or dark purple discoloration, most easily seen in nail beds and mucous membranes, that results from an increased concentration of deoxygenated (reduced) hemoglobin (more than 5 g/dL).

Cyst (SIST) A sac with a distinct connective tissue wall, containing a fluid or other material.

Cystic (SIS-tik) **duct** The duct that carries bile from the gallbladder to the common bile duct.

Cystitis (sis-TĪ-tis) Inflammation of the urinary bladder.

Cytokinesis (sī´-tō-ki-NĒ-sis) Distribution of the cytoplasm into two separate cells during cell division; coordinated with nuclear division (mitosis).

Cytolysis (sī-TOL-i-sis) The rupture of living cells in which the contents leak out.

Cytoplasm (SĪ-tō-plasm) Cytosol plus all organelles except the nucleus.

Cytoskeleton Complex internal structure of cytoplasm consisting of microfilaments, microtubules, and intermediate filaments.

Cytosol (SĪ-tō-sol) Semifluid portion of cytoplasm in which organelles and inclusions are suspended and solutes are dissolved. Also called **intracellular fluid.**

D

Dartos (DAR-tōs) The contractile tissue deep to the skin of the scrotum.

Decidua (dē-SID-ū-a) That portion of the endometrium of the uterus (all but the deepest layer) that is modified during pregnancy and shed after childbirth.

Deciduous (dē-SID-ū-us) Falling off or being shed seasonally or at a particular stage of development. In the body, referring to the first set of teeth.

Decussation (dē´-ku-SĀ-shun) A crossing-over to the opposite (contralateral) side; an example is the crossing of 90% of the axons in the large motor tracts to opposite sides in the medullary pyramids.

Deep Away from the surface of the body or an organ.

Deep inguinal (IN-gwi-nal) **ring** A slitlike opening in the aponeurosis of the transversus abdominis muscle that represents the origin of the inguinal canal.

Deep-venous thrombosis (DVT) The presence of a thrombus in a vein, usually a deep vein of the lower limbs.

Defecation (def-e-KĀ-shun) The discharge of feces from the rectum.

Deglutition (dē-gloo-TISH-un) The act of swallowing.

Dehydration (dē-hī-DRĀ-shun) Excessive loss of water from the body or its parts.

Delta cell A cell in the pancreatic islets (islets of Langerhans) that secretes somatostatin. Also termed a **D cell.**

Demineralization (de-min´-er-al-i-ZĀ-shun) Loss of calcium and phosphorus from bones.

Dendrite (DEN-drīt) A neuronal process that carries electrical signals, usually graded potentials, toward the cell body.

Dendritic (den-DRIT-ik) **cell** One type of antigen-presenting cell with long branchlike projections that commonly is present in mucosal linings such as the vagina, in the skin (Langerhans cells in the epidermis), and in lymph nodes (follicular dendritic cells).

Dental caries (KA-rēz) Gradual demineralization of the enamel and dentin of a tooth that may invade the pulp and alveolar bone. Also called **tooth decay.**

Denticulate (den-TIK-ū-lāt) Finely toothed or serrated; characterized by a series of small, pointed projections.

Dentin (DEN-tin) The bony tissues of a tooth enclosing the pulp cavity.

Dentition (den-TI-shun) The eruption of teeth. The number, shape, and arrangement of teeth.

Deoxyribonucleic (dē-ok´-sē-rī-bō-nū-KLĒ-ik) **acid (DNA)** A nucleic acid constructed of nucleotides consisting of one of four bases (adenine, cytosine, guanine, or thymine), deoxyribose, and a phosphate group; encoded in the nucleotides is genetic information.

Depression (dē-PRESH-un) Movement in which a part of the body moves inferiorly.

Dermal papilla (pa-PILL-a) Fingerlike projection of the papillary region of the dermis that may contain blood capillaries or corpuscles of touch (Meissner corpuscles).

Dermatology (der´-ma-TOL-ō-jē) The medical specialty dealing with diseases of the skin.

Dermatome (DER-ma-tōm) The cutaneous area developed from one embryonic spinal cord segment and receiving most of its sensory innervation from one spinal nerve. An instrument for incising the skin or cutting thin transplants of skin.

Dermis (DER-mis) A layer of dense irregular connective tissue lying deep to the epidermis.

Descending colon (KŌ-lon) The part of the large intestine descending from the left colic (splenic) flexure to the level of the left iliac crest.

Detrusor (de-TROO-ser) **muscle** Smooth muscle that forms the wall of the urinary bladder.

Developmental biology The study of development from the fertilized egg to the adult form.

Diagnosis (dī´-ag-NŌ-sis) Distinguishing one disease from another or determining the nature of a disease from signs and symptoms by inspection, palpation, laboratory tests, and other means.

Dialysis (dī-AL-i-sis) The removal of waste products from blood by diffusion through a selectively permeable membrane.

Diaphragm (DĪ-a-fram) Any partition that separates one area from another, especially the dome-shaped skeletal muscle between the thoracic and abdominal cavities. Also a dome-shaped device that is placed over the cervix, usually with a spermicide, to prevent conception.

Diaphysis (dī-AF-i-sis) The shaft of a long bone.

Diarrhea (dī-a-RĒ-a) Frequent defecation of liquid feces caused by increased motility of the intestines.

Diarthrosis (dī-ar-THRŌ-sis) A freely movable joint; types are gliding, hinge, pivot, condyloid, saddle, and ball-and-socket.

Diastole (dī-AS-tō-lē) In the cardiac cycle, the phase of relaxation or dilation of the heart muscle, especially of the ventricles.

Diastolic (dī-as-TOL-ik) **blood pressure** The force exerted by blood on arterial walls during ventricular relaxation; the lowest blood pressure measured in the large arteries, normally about 80 mm Hg in a young adult.

Diencephalon (dī´-en-SEF-a-lon) A part of the brain consisting of the thalamus, hypothalamus, and epithalamus.

Diffusion (di-FŪ-zhun) A passive process in which there is a net or greater movement of molecules or ions from a region of high concentration to a region of low concentration until equilibrium is reached.

Digestion (dī-JES-chun) The mechanical and chemical breakdown of food to simple molecules that can be absorbed and used by body cells.

Dilate (DĪ-lāt) To expand or swell.

Diploid (DIP-loid) Having the number of chromosomes characteristically found in the somatic cells of an organism; having two haploid sets of chromosomes, one each from the mother and father. Symbolized 2*n*.

Direct motor pathways Collections of upper motor neurons with cell bodies in the motor cortex that project axons into the spinal cord, where they synapse with lower motor neurons or interneurons in the anterior horns. Also called the **pyramidal pathways.**

Disease Any change from a state of health.

Dislocation (dis´-lō-KĀ-shun) Displacement of a bone from a joint with tearing of ligaments, tendons, and articular capsules. Also called **luxation** (luks-Ā-shun).

Dissect (di-SEKT) To separate tissues and parts of a cadaver or an organ for anatomical study.

Distal (DIS-tal) Farther from the attachment of a limb to the trunk; farther from the point of origin or attachment.

Diuretic (dī-ū-RET-ik) A chemical that increases urine volume by decreasing reabsorption of water, usually by inhibiting sodium reabsorption.

Divergence (dī-VER-jens) A synaptic arrangement in which the synaptic end bulbs of one presynaptic neuron terminate on several postsynaptic neurons.

Diverticulum (dī-ver-TIK-ū-lum) A sac or pouch in the wall of a canal or organ, especially in the colon.

Dorsal ramus (RĀ-mus) A branch of a spinal nerve containing motor and sensory axons supplying the muscles, skin, and bones of the posterior part of the head, neck, and trunk.

Dorsiflexion (dor-si-FLEK-shun) Bending the foot in the direction of the dorsum (upper surface).

Down-regulation Phenomenon in which there is a decrease in the number of receptors in response to an excess of a hormone or neurotransmitter.

Duct of Santorini *See* **Accessory duct.**

Duct of Wirsung *See* **Pancreatic duct.**

Ductus arteriosus (DUK-tus ar-tē-rē-Ō-sus) A small vessel connecting the pulmonary trunk with the aorta; found only in the fetus.

Ductus (vas) deferens (DEF-er-ens) The duct that carries sperm from the epididymis to the ejaculatory duct. Also called the **seminal duct.**

Ductus epididymis (ep´-i-DID-i-mis) A tightly coiled tube inside the epididymis, distinguished into a head, body, and tail, in which sperm undergo maturation.

Ductus venosus (ve-NŌ-sus) A small vessel in the fetus that helps the circulation bypass the liver.

Duodenal (doo´-ō-DĒ-nal) **gland** Gland in the submucosa of the duodenum that secretes an alkaline mucus to protect the lining of the small intestine from the action of enzymes and to help neutralize the acid in chyme. Also called a **Brunner's** (BRUN-erz) **gland.**

Duodenal papilla (pa-PIL-a) An elevation on the duodenal mucosa that receives the hepatopancreatic ampulla (ampulla of Vater).

Duodenum (doo-ō-DĒ-num *or* doo-OD-e-num) The first 25 cm (10 in.) of the small intestine, which connects the stomach and the ileum.

Dura mater (DOO-ra MĀ-ter) The outermost of the three meninges (coverings) of the brain and spinal cord.

Dynamic equilibrium (ē-kwi-LIB-rē-um) The maintenance of body position, mainly the head, in response to sudden movements such as rotation.

Dysmenorrhea (dis´-men-ō-RĒ-a) Painful menstruation.

Dysplasia (dis-PLĀ-zē-a) Change in the size, shape, and organization of cells due to chronic irritation or inflammation; may either revert to normal if stress is removed or progress to neoplasia.

Dyspnea (DISP-nē-a) Shortness of breath; painful or labored breathing.

E

Ectoderm The primary germ layer that gives rise to the nervous system and the epidermis of skin and its derivatives.

Ectopic (ek-TOP-ik) Out of the normal location, as in ectopic pregnancy.

Edema (e-DĒ-ma) An abnormal accumulation of interstitial fluid.

Effector (e-FEK-tor) An organ of the body, either a muscle or a gland, that is innervated by somatic or autonomic motor neurons.

Efferent arteriole (EF-er-ent ar-TĒ-rē-ōl) A vessel of the renal vascular system that carries blood from a glomerulus to a peritubular capillary.

Efferent ducts A series of coiled tubes that transport sperm from the rete testis to the epididymis.

Ejaculation (e-jak-ū-LĀ-shun) The reflex ejection or expulsion of semen from the penis.

Ejaculatory (e-JAK-ū-la-tō-rē) **duct** A tube that transports sperm from the ductus (vas) deferens to the prostatic urethra.

Elasticity (e-las-TIS-i-tē) The ability of tissue to return to its original shape after contraction or extension.

Electrocardiogram (e-lek´-trō-KAR-dē-ō-gram) **(ECG or EKG)** A recording of the electrical changes that accompany the cardiac cycle that can be detected at the surface of the body; may be resting, stress, or ambulatory.

Elevation (el-e-VĀ-shun) Movement in which a part of the body moves superiorly.

Embolus (EM-bō-lus) A blood clot, bubble of air or fat from broken bones, mass of bacteria, or other debris or foreign material transported by the blood.

Embryo (EM-brē-ō) The young of any organism in an early stage of development; in humans, the developing organism from fertilization to the end of the eighth week of development.

Embryoblast A region of cells of a blastocyst that differentiates into the three primary germ layers—ectoderm, mesoderm, and endoderm—from which all tissues and organs develop; also called an **inner cell mass.**

Embryology (em´-brē-OL-ō-jē) The study of development from the fertilized egg to the end of the eighth week of development.

Emesis (EM-e-sis) Vomiting.

Emigration (em´-i-GRĀ-shun) Process whereby white blood cells (WBCs) leave the bloodstream by rolling along the endothelium, sticking to it, and squeezing between the endothelial cells. Adhesion molecules help WBCs stick to the endothelium. Also known as **migration** or **extravasation.**

Emission (ē-MISH-un) Propulsion of sperm into the urethra due to peristaltic contractions of the ducts of the testes, epididymides, and ductus (vas) deferens as a result of sympathetic stimulation.

Emphysema (em-fi-SĒ-ma) A lung disorder in which alveolar walls disintegrate, producing abnormally large air spaces and loss of elasticity in the lungs; typically caused by exposure to cigarette smoke.

Emulsification (ē-mul´-si-fi-KĀ-shun) The dispersion of large lipid globules into smaller, uniformly distributed particles in the presence of bile.

Enamel (e-NAM-el) The hard, white substance covering the crown of a tooth.

End organ of Ruffini *See* **Type II cutaneous mechanoreceptor.**

Endocardium (en-dō-KAR-dē-um) The layer of the heart wall, composed of endothelium and smooth muscle, that lines the inside of the heart and covers the valves and tendons that hold the valves open.

Endochondral ossification (en´-dō-KON-dral os´-i-fi-KĀ-shun) The replacement of cartilage by bone. Also called **intracartilaginous** (in´-tra-kar´-ti-LAJ-i-nus) **ossification.**

Endocrine (EN-dō-krin) **gland** A gland that secretes hormones into interstitial fluid and then the blood; a ductless gland.

Endocrinology (en´-dō-kri-NOL-ō-jē) The science concerned with the structure and functions of endocrine glands and the diagnosis and treatment of disorders of the endocrine system.

Endocytosis (en´-dō-sī-TŌ-sis) The uptake into a cell of large molecules and particles in which a segment of plasma membrane surrounds the substance, encloses it, and brings it in; includes phagocytosis, pinocytosis, and receptor-mediated endocytosis.

Endoderm (EN-dō-derm) A primary germ layer of the developing embryo; gives rise to the gastrointestinal tract, urinary bladder, urethra, and respiratory tract.

Endodontics (en´-dō-DON-tiks) The branch of dentistry concerned with the prevention, diagnosis, and treatment of diseases that affect the pulp, root, periodontal ligament, and alveolar bone.

Endolymph (EN-dō-limf´) The fluid within the membranous labyrinth of the internal ear.

Endometriosis (en´-dō-MĒ-trē-ō´-sis) The growth of endometrial tissue outside the uterus.

Endometrium (en´-dō-MĒ-trē-um) The mucous membrane lining the uterus.

Endomysium (en´-dō-MĪZ-ē-um) Invagination of the perimysium separating each individual muscle fiber (cell).

Endoneurium (en´-dō-NOO-rē-um) Connective tissue wrapping around individual nerve axons.

Endoplasmic reticulum (en´-dō-PLAS-mik re-TIK-ū-lum) **(ER)** A network of channels running through the cytoplasm of a cell that serves in intracellular transportation, support, storage, synthesis, and packaging of molecules. Portions of ER where ribosomes are attached to the outer surface are called **rough ER;** portions that have no ribosomes are called **smooth ER.**

Endosteum (en-DOS-tē-um) The membrane that lines the medullary (marrow) cavity of bones, consisting of osteogenic cells and scattered osteoclasts.

Endothelium (en´-dō-THĒ-lē-um) The layer of simple squamous epithelium that lines the cavities of the heart, blood vessels, and lymphatic vessels.

Enteric (EN-ter-ik) **nervous system (ENS)** The part of the nervous system that is embedded in the submucosa and muscularis of the gastrointestinal (GI) tract; governs motility and secretions of the GI tract.

Enteroendocrine (en-ter-ō-EN-dō-krin) **cell** A cell of the mucosa of the gastrointestinal tract that secretes a hormone that governs function of the GI tract; hormones secreted include gastrin, cholecystokinin, glucose-dependent insulinotropic peptide (GIP), and secretin.

Enzyme (EN-zīm) A substance that accelerates chemical reactions; an organic catalyst, usually a protein.

Eosinophil (ē-ō-SIN-ō-fil) A type of white blood cell characterized by granules that stain red or pink with acid dyes.

Ependymal (ep-EN-de-mal) **cells** Neuroglial cells that cover choroid plexuses and produce cerebrospinal fluid (CSF); they also line the ventricles of the brain and probably assist in the circulation of CSF.

Epicardium (ep´-i-KAR-dē-um) The thin outer layer of the heart wall, composed of serous tissue and mesothelium. Also called the **visceral pericardium.**

Epidemiology (ep´-i-dē-mē-OL-ō-jē) Study of the occurrence and transmission of diseases and disorders in human populations.

Epidermis (ep´-i-DERM-is) The superficial, thinner layer of skin, composed of keratinized stratified squamous epithelium.

Epididymis (ep´-i-DID-i-mis) A comma-shaped organ that lies along the posterior border of the testis and contains the ductus epididymis, in which sperm undergo maturation. *Plural* is **epididymides** (ep´-i-di-DIM-i-dēz).

Epidural (ep´-i-DOO-ral) **space** A space between the spinal dura mater and the vertebral canal, containing areolar connective tissue and a plexus of veins.

Epiglottis (ep´-i-GLOT-is) A large, leaf-shaped piece of cartilage lying on top of the larynx, attached to the thyroid cartilage; its unattached portion is free to move up and down to cover the glottis (vocal folds and rima glottidis) during swallowing.

Epimysium (ep-i-MĪZ-ē-um) Fibrous connective tissue around muscles.

Epinephrine (ep-ē-NEF-rin) Hormone secreted by the adrenal medulla that produces actions similar to those that result from sympathetic stimulation. Also called **adrenaline** (a-DREN-a-lin).

Epineurium (ep´-i-NOO-rē-um) The superficial connective tissue covering around an entire nerve.

Epiphyseal (ep´-i-FIZ-ē-al) **line** The remnant of the epiphyseal plate in the metaphysis of a long bone.

Epiphyseal plate The hyaline cartilage plate in the metaphysis of a long bone; site of lengthwise growth of long bones.

Epiphysis (e-PIF-i-sis) The end of a long bone, usually larger in diameter than the shaft (diaphysis).

Epiphysis cerebri (se-RĒ-brē) Pineal gland.

Episiotomy (e-piz´-ē-OT-ō-mē) A cut made with surgical scissors to avoid tearing of the perineum at the end of the second stage of labor.

Epistaxis (ep´-i-STAK-sis) Loss of blood from the nose due to trauma, infection, allergy, neoplasm, and bleeding disorders. Also called **nosebleed.**

Epithalamus (ep´-i-THAL-a-mus) Part of the diencephalon superior and posterior to the thalamus, comprising the pineal gland and associated structures.

Epithelial (ep-i-THĒ-lē-al) **tissue** The tissue that forms the innermost and outermost surfaces of body structures and forms glands.

Eponychium (ep´-o-NIK-ē-um) Narrow band of stratum corneum at the proximal border of a nail that extends from the margin of the nail wall. Also called the **cuticle.**

Erectile dysfunction Failure to maintain an erection long enough for sexual intercourse. Also known as **impotence** (IM-pō-tens).

Erection (ē-REK-shun) The enlarged and stiff state of the penis or clitoris resulting from the engorgement of the spongy erectile tissue with blood.

Eructation (e-ruk´-TĀ-shun) The forceful expulsion of gas from the stomach. Also called **belching.**

Erythema (er-e-THĒ-ma) Skin redness usually caused by dilation of the capillaries.

Erythrocyte (e-RITH-rō-sīt) A mature red blood cell.

Erythropoietin (e-rith´-rō-POY-e-tin) **(EPO)** A hormone released by the juxtaglomerular cells of the kidneys that stimulates red blood cell production.

Esophagus (e-SOF-a-gus) The hollow muscular tube that connects the pharynx and the stomach.

Estrogens (ES-tro-jenz) Feminizing sex hormones produced by the ovaries; govern development of oocytes, maintenance of female reproductive structures, and appearance of secondary sex characteristics; also affect fluid and electrolyte balance, and protein anabolism. Examples are β-estradiol, estrone, and estriol.

Eupnea (ŪP-nē-a) Normal quiet breathing.

Eustachian tube *See* **Auditory tube.**

Eversion (ē-VER-zhun) The movement of the sole laterally at the ankle joint or of an atrioventricular valve into an atrium during ventricular contraction.

Excitability (ek-sīt´-a-BIL-i-tē) The ability of muscle fibers to receive and respond to stimuli; the ability of neurons to respond to stimuli and generate nerve impulses.

Excretion (eks-KRĒ-shun) The process of eliminating waste products from the body; also the products excreted.

Exhalation (eks-ha-LĀ-shun) Breathing out; expelling air from the lungs into the atmosphere. Also called **expiration.**

Exocrine (EK-sō-krin) **gland** A gland that secretes its products into ducts that carry the secretions into body cavities, into the lumen of an organ, or to the outer surface of the body.

Exocytosis (ex´-ō-sī-TŌ-sis) A process in which membrane-enclosed secretory vesicles form inside the cell, fuse with the plasma membrane, and release their contents into the interstitial fluid; achieves secretion of materials from a cell.

Extensibility (ek-sten´-si-BIL-i-tē) The ability of skin or muscle tissue to stretch when it is pulled.

Extension (eks-TEN-shun) An increase in the angle between two bones; restoring a body part to its anatomical position after flexion.

External Located on or near the surface.

External auditory (AW-di-tōr-ē) **canal** or **meatus** (mē-Ā-tus) A curved tube in the temporal bone that leads to the middle ear.

External ear The outer ear, consisting of the pinna, external auditory canal, and tympanic membrane (eardrum).

External nares (NĀ-rez) The openings into the nasal cavity on the exterior of the body. Also called the **nostrils.**

External respiration The exchange of respiratory gases between the lungs and blood. Also called **pulmonary respiration.**

Exteroceptor (EKS-ter-ō-sep´-tor) A sensory receptor adapted for the reception of stimuli from outside the body.

Extracellular fluid (ECF) Fluid outside body cells, such as interstitial fluid and plasma.

Extracellular matrix (MĀ-triks) The ground substance and fibers between cells in a connective tissue.

Eyebrow The hairy ridge superior to the eye.

F

F cell A cell in the pancreatic islets (islets of Langerhans) that secretes pancreatic polypeptide.

Face The anterior aspect of the head.

Falciform ligament (FAL-si-form LIG-a-ment) A sheet of parietal peritoneum between the two principal lobes of the liver. The ligamentum teres, or remnant of the umbilical vein, lies within its fold.

Falx cerebelli (FALKS ser-e-BEL-lī) A small triangular process of the dura mater attached to the occipital bone in the posterior cranial fossa and projecting inward between the two cerebellar hemispheres.

Falx cerebri (FALKS SER-e-brē) A fold of the dura mater extending deep into the longitudinal fissure between the two cerebral hemispheres.

Fascia (FASH-ē-a) Large connective tissue sheet that wraps around groups of muscles.

Fascicle (FAS-i-kul) A small bundle or cluster, especially of nerve or muscle fibers (cells). Also called a **fasciculus** (fa-SIK-ū-lus). *Plural* is **fasciculi** (fa-SIK-yoo-lī).

Fasciculation (fa-sik-ū-LĀ-shun) Abnormal, spontaneous twitch of all skeletal muscle fibers in one motor unit that is visible at the skin surface; not associated with movement of the affected muscle; present in progressive diseases of motor neurons, for example, poliomyelitis.

Fauces (FAW-sēz) The opening from the mouth into the pharynx.

Feces (FĒ-sēz) Material discharged from the rectum and made up of bacteria, excretions, and food residue. Also called **stool.**

Female reproductive cycle General term for the ovarian and uterine cycles, the hormonal changes that accompany them, and cyclic changes in the breasts and cervix; includes changes in the endometrium of a nonpregnant female that prepares the lining of the uterus to receive a fertilized ovum. Less correctly termed the **menstrual cycle.**

Fertilization (fer´-til-i-ZĀ-shun) Penetration of a secondary oocyte by a sperm cell, meiotic division of secondary oocyte to form an ovum, and subsequent union of the nuclei of the gametes.

Fetal circulation The cardiovascular system of the fetus, including the placenta and special blood vessels involved in the exchange of materials between fetus and mother.

Fetus (FĒ-tus) In humans, the developing organism *in utero* from the beginning of the third month to birth.

Fever An elevation in body temperature above the normal temperature of 37°C (98.6°F) due to a resetting of the hypothalamic thermostat.

Fibroblast (FĪ-brō-blast) A large, flat cell that secretes most of the extracellular matrix of areolar and dense connective tissues.

Fibrous (FĪ-brus) **joint** A joint that allows little or no movement, such as a suture or a syndesmosis.

Fibrous tunic (TOO-nik) The superficial coat of the eyeball, made up of the posterior sclera and the anterior cornea.

Fight-or-flight response The effects produced upon stimulation of the sympathetic division of the autonomic nervous system.

Filiform papilla (FIL-i-form pa-PIL-a) One of the conical projections that are distributed in parallel rows over the anterior two-thirds of the tongue and lack taste buds.

Filtration (fil-TRĀ-shun) The flow of a liquid through a filter (or membrane that acts like a filter) due to a hydrostatic pressure; occurs in capillaries due to blood pressure.

Filum terminale (FĪ-lum ter-mi-NAL-ē) Nonnervous fibrous tissue of the spinal cord that extends inferiorly from the conus medullaris to the coccyx.

Fimbriae (FIM-brē-ē) Fingerlike structures, especially the lateral ends of the uterine (fallopian) tubes.

Fissure (FISH-ur) A groove, fold, or slit that may be normal or abnormal.

Fixator A muscle that stabilizes the origin of the prime mover so that the prime mover can act more efficiently.

Fixed macrophage (MAK-rō-fāj) Stationary phagocytic cell found in the liver, lungs, brain, spleen, lymph nodes, subcutaneous tissue, and red bone marrow. Also called a **histiocyte** (HIS-tē-ō-sīt).

Flaccid (FLAS-id) Relaxed, flabby, or soft; lacking muscle tone.

Flagellum (fla-JEL-um) A hairlike, motile process on the extremity of a bacterium, protozoan, or sperm cell. *Plural* is **flagella** (fla-JEL-a).

Flatus (FLĀ-tus) Gas in the stomach or intestines; commonly used to denote expulsion of gas through the anus.

Flexion (FLEK-shun) Movement in which there is a decrease in the angle between two bones.

Follicle (FOL-i-kul) A small secretory sac or cavity; the group of cells that contains a developing oocyte in the ovaries.

Follicle-stimulating hormone (FSH) Hormone secreted by the anterior pituitary; it initiates development of ova and stimulates the ovaries to secrete estrogens in females, and initiates sperm production in males.

Fontanel (fon-ta-NEL) A mesenchyme-filled space where bone formation is not yet complete, especially between the cranial bones of an infant's skull.

Foot The terminal part of the lower limb, from the ankle to the toes.

Foramen (fō-RĀ-men) A passage or opening; a communication between two cavities of an organ, or a hole in a bone for passage of vessels or nerves. *Plural* is **foramina** (fō-RAM-i-na).

Foramen ovale (ō-VAL-ē) An opening in the fetal heart in the septum between the right and left atria. A hole in the greater wing of the sphenoid bone that transmits the mandibular branch of the trigeminal (V) nerve.

Forearm (FOR-arm) The part of the upper limb between the elbow and the wrist.

Fornix (FOR-niks) An arch or fold; a tract in the brain made up of association fibers, connecting the hippocampus with the mammillary bodies; a recess around the cervix of the uterus where it protrudes into the vagina.

Fossa (FOS-a) A furrow or shallow depression.

Fourth ventricle (VEN-tri-kul) A cavity filled with cerebrospinal fluid within the brain lying between the cerebellum and the medulla oblongata and pons.

Fracture (FRAK-choor) Any break in a bone.

Frontal plane A plane at a right angle to a midsagittal plane that divides the body or organs into anterior and posterior portions. Also called a **coronal** (kō-RŌ-nal) **plane.**

Fundus (FUN-dus) The part of a hollow organ farthest from the opening.

Fungiform papilla (FUN-ji-form pa-PIL-a) A mushroomlike elevation on the upper surface of the tongue appearing as a red dot; most contain taste buds.

Furuncle (FŪ-rung-kul) A boil; painful nodule caused by bacterial infection and inflammation of a hair follicle or sebaceous (oil) gland.

G

Gallbladder A small pouch, located inferior to the liver, that stores bile and empties by means of the cystic duct.

Gallstone A solid mass, usually containing cholesterol, in the gallbladder or a bile-containing duct; formed anywhere between bile canaliculi in the liver and the hepatopancreatic ampulla (ampulla of Vater), where bile enters the duodenum. Also called a **biliary calculus.**

Gamete (GAM-ēt) A male or female reproductive cell; a sperm cell or secondary oocyte.

Ganglion (GANG-glē-on) Usually, a group of neuronal cell bodies lying outside the central nervous system (CNS). *Plural* is **ganglia** (GANG-glē-a).

Gastric (GAS-trik) **glands** Glands in the mucosa of the stomach composed of cells that empty their secretions into narrow channels called **gastric pits**. Types of cells are **chief cells** (secrete pepsinogen), **parietal cells** (secrete hydrochloric acid and intrinsic factor), surface mucous and mucous neck cells (secrete mucus), and **G cells** (secrete gastrin).

Gastroenterology (gas´-trō-en´-ter-OL-ō-jē) The medical specialty that deals with the structure, function, diagnosis, and treatment of diseases of the stomach and intestines.

Gastrointestinal (gas-trō-in-TES-ti-nal) **(GI) tract** A continuous tube running through the ventral body cavity extending from the mouth to the anus. Also called the **alimentary** (al´-i-MEN-tar-ē) **canal.**

Gastrulation (gas´-troo-LĀ-shun) The migration of groups of cells from the epiblast that transform a bilaminar embryonic disc into a trilaminar embryonic disc with three primary germ layers; transformation of the blastula into the gastrula.

Gene (JĒN) Biological unit of heredity; a segment of DNA located in a definite position on a particular chromosome; a sequence of DNA that codes for a particular mRNA, rRNA, or tRNA.

Genetic engineering The manufacture and manipulation of genetic material.

Genetics The study of genes and heredity.

Genome (JĒ-nōm) The complete set of genes of an organism.

Genotype (JĒ-nō-tīp) The genetic makeup of an individual; the combination of alleles present at one or more chromosomal locations, as distinguished from the appearance, or phenotype, that results from those alleles.

Geriatrics (jer´-ē-AT-riks) The branch of medicine devoted to the medical problems and care of elderly persons.

Gestation (jes-TĀ-shun) The period of development from fertilization to birth.

Gingivae (jin-JI-vē) Gums. They cover the alveolar processes of the mandible and maxilla and extend slightly into each socket.

Gland Specialized epithelial cell or cells that secrete substances; may be exocrine or endocrine.

Glans penis (glanz PĒ-nis) The slightly enlarged region at the distal end of the penis.

Glaucoma (glaw-KŌ-ma) An eye disorder in which there is increased intraocular pressure due to an excess of aqueous humor.

Gliding joint A synovial joint having articulating surfaces that are usually flat, permitting only side-to-side and back-and-forth movements, as between carpal bones, tarsal bones, and the scapula and clavicle. Also called an **arthrodial** (ar-THRŌ-dē-al) **joint.**

Glomerular (glō-MER-ū-lar) **capsule** A double-walled globe at the proximal end of a nephron that encloses the glomerular capillaries. Also called **Bowman's** (BŌ-manz) **capsule.**

Glomerular filtrate (FIL-trāt) The fluid produced when blood is filtered by the filtration membrane in the glomeruli of the kidneys.

Glomerular filtration The first step in urine formation in which substances in blood pass through the filtration membrane and the filtrate enters the proximal convoluted tubule of a nephron.

Glomerulus (glō-MER-ū-lus) A rounded mass of nerves or blood vessels, especially the microscopic tuft of capillaries that is surrounded by the glomerular (Bowman's) capsule of each kidney tubule. *Plural* is **glomeruli.**

Glottis (GLOT-is) The vocal folds (true vocal cords) in the larynx plus the space between them (rima glottidis).

Glucagon (GLOO-ka-gon) A hormone produced by the alpha cells of the pancreatic islets (islets of Langerhans) that increases blood glucose level.

Glucocorticoids (gloo´-kō-KOR-ti-koyds) Hormones secreted by the cortex of the adrenal gland, especially cortisol, that influence glucose metabolism.

Glucose (GLOO-kōs) A hexose (six-carbon sugar), $C_6H_{12}O_6$, that is a major energy source for the production of ATP by body cells.

Glucosuria (gloo´-kō-SOO-rē-a) The presence of glucose in the urine; may be temporary or pathological. Also called **glycosuria.**

Glycogen (GLĪ-kō-jen) A highly branched polymer of glucose containing thousands of subunits; functions as a compact store of glucose molecules in liver and muscle fibers (cells).

Goblet cell A goblet-shaped unicellular gland that secretes mucus; present in epithelium of the airways and intestines.

Goiter (GOY-ter) An enlarged thyroid gland.

Golgi (GOL-jē) **complex** An organelle in the cytoplasm of cells consisting of four to six flattened sacs (cisternae), stacked on one another, with expanded areas at their ends; functions in processing, sorting, packaging, and delivering proteins and lipids to the plasma membrane, lysosomes, and secretory vesicles.

Golgi tendon organ *See* **Tendon organ.**

Gomphosis (gom-FŌ-sis) A fibrous joint in which a cone-shaped peg fits into a socket.

Gonad (GŌ-nad) A gland that produces gametes and hormones; the ovary in the female and the testis in the male.

Gonadotropic hormone Anterior pituitary hormone that affects the gonads.

Gout (GOWT) Hereditary condition associated with excessive uric acid in the blood; the acid crystallizes and deposits in joints, kidneys, and soft tissue.

Graafian follicle *See* **Mature follicle.**

Gracile (GRAS-il) **nucleus** A group of nerve cells in the inferior part of the medulla oblongata in which axons of the gracile fasciculus terminate.

Gray commissure (KOM-mi-shur) A narrow strip of gray matter connecting the two lateral gray masses within the spinal cord.

Gray matter Areas in the central nervous system and ganglia containing neuronal cell bodies, dendrites, unmyelinated axons, axon terminals, and neuroglia; Nissl bodies impart a gray color, and there is little or no myelin in gray matter.

Gray ramus communicans (RĀ-mus kō-MŪ-ni-kans) A short nerve containing axons of sympathetic postganglionic neurons; the cell bodies of the neurons are in a sympathetic chain ganglion, and the unmyelinated axons extend via the gray ramus to a spinal nerve and then to the periphery to supply smooth muscle in blood vessels, arrector pili muscles, and sweat glands. *Plural* is **rami communicantes** (RĀ-mē kō-mū-ni-KAN-tēz).

Greater omentum (ō-MEN-tum) A large fold in the serosa of the stomach that hangs down like an apron anterior to the intestines.

Greater vestibular (ves-TIB-ū-lar) **glands** A pair of glands on either side of the vaginal orifice that open by a duct into the space between the hymen and the labia minora. Also called **Bartholin's** (BAR-to-linz) **glands.**

Groin (GROYN) The depression between the thigh and the trunk; the inguinal region.

Gross anatomy The branch of anatomy that deals with structures that can be studied without using a microscope. Also called **macroscopic anatomy.**

Growth An increase in size due to an increase in (1) the number of cells, (2) the size of existing cells as internal components increase in size, or (3) the size of intercellular substances.

Gustatory (GUS-ta-tō´-rē) Pertaining to taste.

Gynecology (gī´-ne-KOL-ō-jē) The branch of medicine dealing with the study and treatment of disorders of the female reproductive system.

Gynecomastia (gīn´-e-kō-MAS-tē-a) Excessive growth (benign) of the male mammary glands due to secretion of estrogens by an adrenal gland tumor (feminizing adenoma).

Gyrus (JĪ-rus) One of the folds of the cerebral cortex of the brain. *Plural* is **gyri** (JĪ-rī). Also called a **convolution.**

H

Hair A threadlike structure produced by hair follicles that develops in the dermis. Also called a **pilus** (PĪ-lus).

Hair follicle (FOL-i-kl) Structure composed of epithelium and surrounding the root of a hair from which hair develops.

Hair root plexus (PLEK-sus) A network of dendrites arranged around the root of a hair as free or naked nerve endings that are stimulated when a hair shaft is moved.

Hand The terminal portion of an upper limb, including the carpus, metacarpus, and phalanges.

Haploid (HAP-loyd) **cell** Having half the number of chromosomes characteristically found in the somatic cells of an organism; characteristic of mature gametes. Symbolized *n*.

Hard palate (PAL-at) The anterior portion of the roof of the mouth, formed by the maxillae and palatine bones and lined by mucous membrane.

Haustra (HAWS-tra) A series of pouches that characterize the colon; caused by tonic contractions of the teniae coli. *Singular is* **haustrum.**

Haversian canal *See* **Central canal.**

Haversian system *See* **Osteon.**

Head The superior part of a human, cephalic to the neck. The superior or proximal part of a structure.

Heart A hollow muscular organ lying slightly to the left of the midline of the chest that pumps the blood through the cardiovascular system.

Heart block An arrhythmia (dysrhythmia) of the heart in which the atria and ventricles contract independently because of a blocking of electrical impulses through the heart at some point in the conduction system.

Heart murmur (MER-mer) An abnormal sound that consists of a flow noise that is heard before, between, or after the normal heart sounds, or that may mask normal heart sounds.

Hemangioblast (hē-MAN-jē-ō-blast) A precursor mesodermal cell that develops into blood and blood vessels.

Hematocrit (HĒ-ma-tō-krit) **(Hct)** The percentage of blood made up of red blood cells. Usually measured by centrifuging a blood sample in a graduated tube and then reading the volume of red blood cells and dividing it by the total volume of blood in the sample.

Hematology (hēm-a-TOL-ō-jē) The study of blood.

Hematoma (hē´-ma-TŌ-ma) A tumor or swelling filled with blood.

Hemiplegia (hem-i-PLĒ-jē-a) Paralysis of the upper limb, trunk, and lower limb on one side of the body.

Hemoglobin (hē´-mō-GLŌ-bin) **(Hb)** A substance in red blood cells consisting of the protein globin and the iron-containing red pigment heme that transports most of the oxygen and some carbon dioxide in blood.

Hemolysis (hē-MOL-i-sis) The escape of hemoglobin from the interior of a red blood cell into the surrounding medium; results from disruption of the cell membrane by toxins or drugs, freezing or thawing, or hypotonic solutions.

Hemolytic disease of the newborn A hemolytic anemia of a newborn child that results from the destruction of the infant's erythrocytes (red blood cells) by antibodies produced by the mother; usually the antibodies are due to an Rh blood type incompatibility. Also called **erythroblastosis fetalis** (e-rith´-rō-blas-TŌ-sis fe-TAL-is).

Hemophilia (hē-mō-FIL-ē-a) A hereditary blood disorder where there is a deficient production of certain factors involved in blood clotting, resulting in excessive bleeding into joints, deep tissues, and elsewhere.

Hemopoiesis (hēm-ō-poy-Ē-sis) Blood cell production, which occurs in red bone marrow after birth. Also called **hematopoiesis** (hem´-a-tō-poy-Ē-sis).

Hemorrhage (HEM-o-rij) Bleeding; the escape of blood from blood vessels, especially when the loss is profuse.

Hemorrhoids (HEM-o-royds) Dilated or varicosed blood vessels (usually veins) in the anal region. Also called **piles.**

Hepatic (he-PAT-ik) Refers to the liver.

Hepatic duct A duct that receives bile from the bile capillaries. Small hepatic ducts merge to form the larger right and left hepatic ducts that unite to leave the liver as the common hepatic duct.

Hepatic portal circulation The flow of blood from the gastrointestinal organs to the liver before returning to the heart.

Hepatocyte (he-PAT-ō-cyte) A liver cell.

Hepatopancreatic (hep´-a-tō-pan´-krē-A-tik) **ampulla** A small, raised area in the duodenum where the combined common bile duct and main pancreatic duct empty into the duodenum. Also called the **ampulla of Vater** (VA-ter).

Hernia (HER-nē-a) The protrusion or projection of an organ or part of an organ through a membrane or cavity wall, usually the abdominal cavity.

Herniated (HER-nē-ā´-ted) **disc** A rupture of an intervertebral disc so that the nucleus pulposus protrudes into the vertebral cavity. Also called a **slipped disc.**

Hiatus (hī-Ā-tus) An opening; a foramen.

Hilum (HĪ-lum) An area, depression, or pit where blood vessels and nerves enter or leave an organ. Also called a **hilus.**

Hinge joint A synovial joint in which a convex surface of one bone fits into a concave surface of another bone, such as the elbow, knee, ankle, and interphalangeal joints. Also called a **ginglymus** (JIN-gli-mus) **joint.**

Hirsutism (HER-soo-tizm) An excessive growth of hair in females and children, with a distribution similar to that in adult males, due to the conversion of vellus hairs into large terminal hairs in response to higher-than-normal levels of androgens.

Histamine (HIS-ta-mēn) Substance found in many cells, especially mast cells, basophils, and platelets, that is released when the cells are injured; results in vasodilation, increased permeability of blood vessels, and constriction of bronchioles.

Histology (his´-TOL-ō-jē) Microscopic study of the structure of tissues.

Holocrine (HŌ-lō-krin) **gland** A type of gland in which entire secretory cells, along with their accumulated secretions, make up the secretory product of the gland, as in the sebaceous (oil) glands.

Homeostasis (hō´-mē-ō-STĀ-sis) The condition in which the body's internal environment remains relatively constant within physiological limits.

Homologous chromosomes Two chromosomes that belong to a pair. Also called **homologs.**

Hormone (HOR-mōn) A secretion of endocrine cells that alters the physiological activity of target cells of the body.

Horn An area of gray matter (anterior, lateral, or posterior) in the spinal cord.

Human chorionic gonadotropin (kō-rē-ON-ik gō-nad-ō-TRŌ-pin) **(hCG)** A hormone produced by the developing placenta that maintains the corpus luteum.

Human chorionic somatomammotropin (sō-mat-ō-mam-ō-TRŌ-pin) **(hCS)** Hormone produced by the chorion of the placenta that stimulates breast tissue for lactation, enhances body growth, and regulates metabolism. Also called **human placental lactogen (hPL).**

Human growth hormone (hGH) Hormone secreted by the anterior pituitary that stimulates growth of body tissues, especially skeletal and muscular tissues. Also known as **somatotropin** and **somatotropic hormone (STH).**

Hyaluronic (hī´-a-loo-RON-ik) **acid** A viscous, amorphous extracellular material that binds cells together, lubricates joints, and maintains the shape of the eyeballs.

Hymen (HĪ-men) A thin fold of vascularized mucous membrane at the vaginal orifice.

Hyperextension (hī´-per-ek-STEN-shun) Continuation of extension beyond the anatomical position, as in bending the head backward.

Hyperplasia (hī-per-PLĀ-zē-a) An abnormal increase in the number of normal cells in a tissue or organ, increasing its size.

Hypersecretion (hī´-per-se-KRĒ-shun) Overactivity of glands resulting in excessive secretion.

Hypersensitivity (hī´-per-sen-si-TI-vi-tē) Overreaction to an allergen that results in pathological changes in tissues. Also called **allergy.**

Hypertension (hī´-per-TEN-shun) High blood pressure.

Hyperthermia (hī´-per-THERM-ē-a) An elevated body temperature.

Hypertonia (hī´-per-TŌ-nē-a) Increased muscle tone that is expressed as spasticity or rigidity.

Hypertonic (hī´-per-TON-ik) Solution that causes cells to shrink due to loss of water by osmosis.

Hypertrophy (hī-PER-trō-fē) An excessive enlargement or overgrowth of tissue without cell division.

Hyperventilation (hī´-per-ven-ti-LĀ-shun) A rate of inhalation and exhalation higher than that required to maintain a normal partial pressure of carbon dioxide in the blood.

Hyponychium (hī´-pō-NIK-ē-um) Free edge of the fingernail.

Hypophyseal fossa (hī´-pō-FIZ-ē-al FOS-a) A depression on the superior surface of the sphenoid bone that houses the pituitary gland.

Hypophyseal pouch An outgrowth of ectoderm from the roof of the mouth from which the anterior pituitary develops.

Hypophysis (hī-POF-i-sis) Pituitary gland.

Hyposecretion (hī´-pō-se-KRĒ-shun) Underactivity of glands resulting in diminished secretion.

Hypothalamohypophyseal (hī´-pō-thal´-a-mō-hī-pō-FIZ-ē-al) **tract** A bundle of axons containing secretory vesicles filled with oxytocin or antidiuretic hormone that extend from the hypothalamus to the posterior pituitary.

Hypothalamus (hī´-pō-THAL-a-mus) A portion of the diencephalon, lying beneath the thalamus and forming the floor and part of the wall of the third ventricle.

Hypothermia (hī´-pō-THER-mē-a) Lowering of body temperature below 35°C (95°F); in surgical procedures, it refers to deliberate cooling of the body to slow down metabolism and reduce oxygen needs of tissues.

Hypotonia (hī´-pō-TŌ-nē-a) Decreased or lost muscle tone in which muscles appear flaccid.

Hypotonic (hī´-pō-TON-ik) Solution that causes cells to swell and perhaps rupture due to gain of water by osmosis.

Hypoventilation (hī-pō-ven-ti-LĀ-shun) A rate of inhalation and exhalation lower than that required to maintain a normal partial pressure of carbon dioxide in plasma.

Hypoxia (hī-POKS-ē-a) Lack of adequate oxygen at the tissue level.

Hysterectomy (his-te-REK-tō-mē) The surgical removal of the uterus.

I

Ileocecal (il-ē-ō-SĒ-kal) **sphincter** A fold of mucous membrane that guards the opening from the ileum into the large intestine. Also called the **ileocecal valve.**

Ileum (IL-ē-um) The terminal part of the small intestine.

Immunity (im-Ū-ni-tē) The state of being resistant to injury, particularly by poisons, foreign proteins, and invading pathogens.

Immunoglobulin (im-ū-nō-GLOB-ū-lin) **(Ig)** An antibody synthesized by plasma cells derived from B lymphocytes in response to the introduction of an antigen. Immunoglobulins are divided into five kinds (IgG, IgM, IgA, IgD, IgE).

Immunology (im´-ū-NOL-ō-jē) The study of the responses of the body when challenged by antigens.

Imperforate (im-PER-fō-rāt) Abnormally closed.

Implantation (im-plan-TĀ-shun) The insertion of a tissue or a part into the body. The attachment of the blastocyst to the stratum basalis of the endometrium about 6 days after fertilization.

Incontinence (in-KON-ti-nens) Inability to retain urine, semen, or feces through loss of sphincter control.

Indirect motor pathways Motor tracts that convey information from the brain down the spinal cord for automatic movements, coordination of body movements with visual stimuli, skeletal muscle tone and posture, and balance. Also known as **extrapyramidal pathways.**

Induction (in-DUK-shun) The process by which one tissue (inducting tissue) stimulates the development of an adjacent unspecialized tissue (responding tissue) into a specialized one.

Infarction (in-FARK-shun) A localized area of necrotic tissue, produced by inadequate oxygenation of the tissue.

Infection (in-FEK-shun) Invasion and multiplication of microorganisms in body tissues, which may be inapparent or characterized by cellular injury.

Inferior (in-FĒR-ē-or) Away from the head or toward the lower part of a structure. Also called **caudad** (KAW-dad).

Inferior vena cava (VĒ-na KĀ-va) **(IVC)** Large vein that collects blood from parts of the body inferior to the heart and returns it to the right atrium.

Infertility Inability to conceive or to cause conception. Also called **sterility.**

Inflammation (in´-fla-MĀ-shun) Localized, protective response to tissue injury designed to destroy, dilute, or wall off the infecting agent or injured tissue; characterized by redness, pain, heat, swelling, and sometimes loss of function.

Infundibulum (in-fun-DIB-ū-lum) The stalklike structure that attaches the pituitary gland to the hypothalamus of the brain. The funnel-shaped, open, distal end of the uterine (fallopian) tube.

Ingestion (in-JES-chun) The taking in of food, liquids, or drugs, by mouth.

Inguinal (IN-gwi-nal) Pertaining to the groin.

Inguinal canal An oblique passageway in the anterior abdominal wall just superior and parallel to the medial half of the inguinal ligament that transmits the spermatic cord and ilioinguinal nerve in the male and round ligament of the uterus and ilioinguinal nerve in the female.

Inhalation (in-ha-LĀ-shun) The act of drawing air into the lungs. Also termed **inspiration.**

Inheritance The acquisition of body traits by transmission of genetic information from parents to offspring.

Inhibin A hormone secreted by the gonads that inhibits release of follicle-stimulating hormone (FSH) by the anterior pituitary.

Inhibiting hormone Hormone secreted by the hypothalamus that can suppress secretion of hormones by the anterior pituitary.

Insertion (in-SER-shun) The attachment of a muscle tendon to a movable bone or the end opposite the origin.

Insula (IN-soo-la) A triangular area of the cerebral cortex that lies deep within the lateral cerebral fissue, under the parietal, frontal, and temporal lobes.

Insulin (IN-soo-lin) A hormone produced by the beta cells of a pancreatic islet (islet of Langerhans) that decreases the blood glucose level.

Integrins (IN-te-grinz) A family of transmembrane glycoproteins in plasma membranes that function in cell adhesion; they are present in hemidesmosomes, which anchor cells to a basement membrane, and they mediate adhesion of neutrophils to endothelial cells during emigration.

Integumentary (in-teg-ū-MEN-tar-ē) Relating to the skin.

Intercalated (in-TER-ka-lāt-ed) **disc** An irregular transverse thickening of sarcolemma that contains desmosomes, which hold cardiac muscle fibers (cells) together, and gap junctions, which aid in conduction of muscle action potentials from one fiber to the next.

Intercostal (in-ter-KOS-tal) **nerve** A nerve supplying a muscle located between the ribs.

Intermediate (in´-ter-MĒ-dē-at) Between two structures, one of which is medial and one of which is lateral.

Intermediate filament Protein filament, ranging from 8 to 12 nm in diameter, that may provide structural reinforcement, hold organelles in place, and give shape to a cell.

Internal Away from the surface of the body.

Internal capsule A large tract of projection fibers lateral to the thalamus that is the major connection between the cerebral cortex and the brain stem and spinal cord; contains axons of sensory neurons carrying auditory, visual, and somatic sensory signals to the cerebral cortex plus axons of motor neurons descending from the cerebral cortex to the thalamus, subthalamus, brain stem, and spinal cord.

Internal ear The inner ear or labyrinth, lying inside the temporal bone, containing the organs of hearing and balance.

Internal nares (NĀ-rez) The two openings posterior to the nasal cavities opening into the nasopharynx. Also called the **choanae** (kō-Ā-nē).

Internal respiration The exchange of respiratory gases between blood and body cells. Also called **tissue respiration.**

Interneurons (in´-ter-NOO-ronz) Neurons whose axons extend only for a short distance and contact nearby neurons in the brain, spinal cord, or a ganglion; they comprise the vast majority of neurons in the body.

Interoceptor (IN-ter-ō-sep´-tor) Sensory receptor located in blood vessels and viscera that provides information about the body's internal environment.

Interphase (IN-ter-fāz) The period of the cell cycle between cell divisions, consisting of the G_1 (gap or growth) phase, when the cell is engaged in growth, metabolism, and production of substances required for division; S (synthesis) phase, during which chromosomes are replicated; and G_2 phase.

Interstitial The portion of extracellular fluid that fills the microscopic spaces between the cells of tissues; the internal environment of the body. Also called **intercellular fluid** or **tissue fluid.**

Interstitial cell of Leydig *See* **Interstitial endocrinocyte.**

Interstitial (in´-ter-STISH-al) **endocrinocyte** A cell that is located in the connective tissue between seminiferous tubules in a mature testis that secretes testosterone. Also called an **interstitial cell of Leydig** (LĪ-dig).

Interstitial growth Growth from within, as in the growth of cartilage. Also called **endogenous** (en-DOJ-e-nus) **growth.**

Interventricular (in´-ter-ven-TRIK-ū-lar) **foramen** A narrow, oval opening through which the lateral ventricles of the brain communicate with the third ventricle. Also called the **foramen of Monro.**

Intervertebral (in´-ter-VER-te-bral) **disc** A pad of fibrocartilage located between the bodies of two vertebrae.

Intestinal gland A gland that opens onto the surface of the intestinal mucosa and secretes digestive enzymes. Also called a **crypt of Lieberkühn** (LĒ-ber-kūn).

Intracellular (in´-tra-SEL-yū-lar) **fluid (ICF)** Fluid located within cells.

Intrafusal (in´-tra-FŪ-sal) **fibers** Three to ten specialized muscle fibers (cells), partially enclosed in a spindle-shaped connective tissue capsule, that make up a muscle spindle.

Intramembranous ossification (in´-tra-MEM-bra-nus os´-i-fi-KĀ-shun) The method of bone formation in which the bone is formed directly in mesenchyme arranged in sheetlike layers that resemble membranes.

Intraocular (in´-tra-OK-ū-lar) **pressure (IOP)** Pressure in the eyeball, produced mainly by aqueous humor.

Intrinsic factor (IF) A glycoprotein, synthesized and secreted by the parietal cells of the gastric mucosa, that facilitates vitamin B_{12} absorption in the small intestine.

Invagination (in-vaj´-i-NĀ-shun) The pushing of the wall of a cavity into the cavity itself.

Inversion (in-VER-zhun) The movement of the sole medially at the ankle joint.

In vitro (VĒ-trō) Literally, in glass; outside the living body and in an artificial environment such as a laboratory test tube.

Ipsilateral (ip-si-LAT-er-al) On the same side, affecting the same side of the body.

Iris The colored portion of the vascular tunic of the eyeball seen through the cornea that contains circular and radial smooth muscle; the hole in the center of the iris is the pupil.

Irritable bowel syndrome (IBS) Disease of the entire gastrointestinal tract in which a person reacts to stress by developing symptoms (such as cramping and abdominal pain) associated with alternating patterns of diarrhea and constipation. Excessive amounts of mucus may appear in feces, and other symptoms include flatulence, nausea, and loss of appetite. Also known as **irritable colon** or **spastic colitis.**

Ischemia (is-KĒ-mē-a) A lack of sufficient blood to a body part due to obstruction or constriction of a blood vessel.

Islet of Langerhans *See* **Pancreatic islet.**

Isotonic (ī´-sō-TON-ik) Having equal tension or tone. A solution having the same concentration of impermeable solutes as cytosol.

Isthmus (IS-mus) A narrow strip of tissue or narrow passage connecting two larger parts.

J

Jaundice (JON-dis) A condition characterized by yellowness of the skin, the white of the eyes, mucous membranes, and body fluids because of a buildup of bilirubin.

Jejunum (je-JOO-num) The middle part of the small intestine.

Joint kinesthetic (kin´-es-THET-ik) **receptor** A proprioceptive receptor located in a joint, stimulated by joint movement.

Juxtaglomerular (juks-ta-glō-MER-ū-lar) **apparatus (JGA)** Consists of the macula densa (cells of the distal convoluted tubule adjacent to the afferent and efferent arteriole) and juxtaglomerular cells (modified cells of the afferent and sometimes efferent arteriole); secretes renin when blood pressure starts to fall.

K

Keratin (KER-a-tin) An insoluble protein found in the hair, nails, and other keratinized tissues of the epidermis.

Keratinocyte (ker-a-TIN-ō-sīt) The most numerous of the epidermal cells; produces keratin.

Kidney (KID-nē) One of the paired reddish organs located in the lumbar region that regulates the composition, volume, and pressure of blood and produces urine.

Kidney stone A solid mass, usually consisting of calcium oxalate, uric acid, or calcium phosphate crystals, that may form in any portion of the urinary tract. Also called a **renal calculus.**

Kinesiology (ki-nē-sē´-OL-ō-jē) The study of the movement of body parts.

Kinesthesia (kin´-es-THĒ-zē-a) The perception of the extent and direction of movement of body parts; this sense is possible due to nerve impulses generated by proprioceptors.

Kinetochore (ki-NET-ō-kor) Protein complex attached to the outside of a centromere to which kinetochore microtubules attach.

Kupffer cell *See* **Stellate reticuloendothelial cell.**

Kyphosis (kī-FŌ-sis) An exaggeration of the thoracic curve of the vertebral column, resulting in a "round-shouldered" appearance. Also called **hunchback.**

L

Labia majora (LĀ-bē-a ma-JŌ-ra) Two longitudinal folds of skin extending downward and backward from the mons pubis of the female.

Labia minora (min-OR-a) Two small folds of mucous membrane lying medial to the labia majora of the female.

Labial frenulum (LĀ-bē-al FREN-ū-lum) A medial fold of mucous membrane between the inner surface of the lip and the gums.

Labium (LĀ-bē-um) A lip. A liplike structure. *Plural* is **labia** (LĀ-bē-a).

Labor The process of giving birth in which a fetus is expelled from the uterus through the vagina.

Labyrinth (LAB-i-rinth) Intricate communicating passageway, especially in the internal ear.

Lacrimal canal A duct, one on each eyelid, beginning at the punctum at the medial margin of an eyelid and conveying tears medially into the nasolacrimal sac.

Lacrimal gland Secretory cells, located at the superior anterolateral portion of each orbit, that secrete tears into excretory ducts that open onto the surface of the conjunctiva.

Lacrimal sac The superior expanded portion of the nasolacrimal duct that receives the tears from a lacrimal canal.

Lactation (lak-TĀ-shun) The secretion and ejection of milk by the mammary glands.

Lacteal (LAK-tē-al) One of many lymphatic vessels in villi of the intestines that absorb triglycerides and other lipids from digested food.

Lacuna (la-KOO-na) A small, hollow space, such as that found in bones in which the osteocytes lie. *Plural* is **lacunae** (la-KOO-nē).

Lambdoid (LAM-doyd) **suture** The joint in the skull between the parietal bones and the occipital bone; sometimes contains sutural (Wormian) bones.

Lamellae (la-MEL-ē) Concentric rings of hard, calcified extracellular matrix found in compact bone.

Lamellated corpuscle *See* **Pacinian corpuscle.**

Lamina (LAM-i-na) A thin, flat layer or membrane, as the flattened part of either side of the arch of a vertebra. *Plural* is **laminae** (LAM-i-nē).

Lamina propria (PRŌ-prē-a) The connective tissue layer of a mucosa.

Langerhans (LANG-er-hans) **cell** Epidermal dendritic cell that functions as an antigen-presenting cell (APC) during an immune response.

Lanugo (la-NOO-gō) Fine downy hairs that cover the fetus.

Large intestine The portion of the gastrointestinal tract extending from the ileum of the small intestine to the anus, divided structurally into the cecum, colon, rectum, and anal canal.

Laryngopharynx (la-rin´-gō-FAR-inks) The inferior portion of the pharynx, extending downward from the level of the hyoid bone that divides posteriorly into the esophagus and anteriorly into the larynx. Also called the **hypopharynx.**

Larynx (LAR-inks) The voice box, a short passageway that connects the pharynx with the trachea.

Lateral (LAT-er-al) Farther from the midline of the body or a structure.

Lateral ventricle (VEN-tri-kul) A cavity within a cerebral hemisphere that communicates with the lateral ventricle in the other cerebral hemisphere and with the third ventricle by way of the interventricular foramen.

Leg The part of the lower limb between the knee and the ankle.

Lens A transparent organ constructed of proteins (crystallins) lying posterior to the pupil and iris of the eyeball and anterior to the vitreous body.

Lesion (LĒ-zhun) Any localized, abnormal change in a body tissue.

Lesser omentum (ō-MEN-tum) A fold of the peritoneum that extends from the liver to the lesser curvature of the stomach and the first part of the duodenum.

Lesser vestibular (ves-TIB-ū-lar) **gland** One of the paired mucus-secreting glands with ducts that open on either side of the urethral orifice in the vestibule of the female.

Leukemia (loo-KĒ-mē-a) A malignant disease of the blood-forming tissues characterized by either uncontrolled production and accumulation of immature leukocytes in which many cells fail to reach maturity (acute) or an accumulation of mature leukocytes in the blood because they do not die at the end of their normal life span (chronic).

Leukocyte (LOO-kō-sīt) A white blood cell.

Leydig (LĪ-dig) **cell** A type of cell that secretes testosterone; located in the connective tissue between seminiferous tubules in a mature testis. Also known as **interstitial cell of Leydig** or **interstitial endocrinocyte.**

Ligament (LIG-a-ment) Dense regular connective tissue that attaches bone to bone.

Ligand (LĪ-gand) A chemical substance that binds to a specific receptor.

Limbic system A part of the forebrain, sometimes termed the visceral brain, concerned with various aspects of emotion and behavior; includes the limbic lobe, dentate gyrus, amygdala, septal nuclei, mammillary bodies, anterior thalamic nucleus, olfactory bulbs, and bundles of myelinated axons.

Lingual frenulum (LIN-gwal FREN-ū-lum) A fold of mucous membrane that connects the tongue to the floor of the mouth.

Lipase An enzyme that splits fatty acids from triglycerides and phospholipids.

Lipid (LIP-id) An organic compound composed of carbon, hydrogen, and oxygen that is usually insoluble in water but soluble in alcohol, ether, and chloroform; examples include triglycerides (fats and oils), phospholipids, steroids, and eicosanoids.

Lipid bilayer Arrangement of phospholipid, glycolipid, and cholesterol molecules in two parallel sheets in which the hydrophilic "heads" face outward and the hydrophobic "tails" face inward; found in cellular membranes.

Lipoprotein (lip´-ō-PRŌ-tēn) One of several types of particles containing lipids (cholesterol and triglycerides) and proteins that make it water-soluble for transport in the blood; high levels of **low-density lipoproteins (LDLs)** are associated with increased risk of atherosclerosis, whereas high levels of **high-density lipoproteins (HDLs)** are associated with decreased risk of atherosclerosis.

Liver Large organ under the diaphragm that occupies most of the right hypochondriac region and part of the epigastric region. Functionally, it produces bile and synthesizes most plasma proteins; interconverts nutrients; detoxifies substances; stores glycogen, iron, and vitamins; carries on phagocytosis of worn-out blood cells and bacteria; and helps synthesize the active form of vitamin D.

Long-term potentiation (LTP) Prolonged, enhanced synaptic transmission that occurs at certain synapses within the hippocampus of the brain; believed to underlie some aspects of memory.

Lordosis (lor-DŌ-sis) An exaggeration of the lumbar curve of the vertebral column. Also called **hollow back.**

Lower limb The appendage attached at the pelvic (hip) girdle, consisting of the thigh, knee, leg, ankle, foot, and toes. Also called the **lower extremity.**

Lumbar (LUM-bar) Region of the back and side between the ribs and pelvis; loin.

Lumbar plexus (PLEK-sus) A network formed by the anterior (ventral) branches of spinal nerves L1 through L4.

Lumen (LOO-men) The space within an artery, vein, intestine, renal tubule, or other tubular structure.

Lungs Main organs of respiration that lie on either side of the heart in the thoracic cavity.

Lunula (LOO-noo-la) The moon-shaped white area at the base of a nail.

Luteinizing (LOO-tē-in´-īz-ing) **hormone (LH)** A hormone secreted by the anterior pituitary that stimulates ovulation, stimulates progesterone secretion by the corpus luteum, and readies the mammary glands for milk secretion in females; stimulates testosterone secretion by the testes in males.

Lymph (LIMF) Fluid confined in lymphatic vessels and flowing through the lymphatic system until it is returned to the blood.

Lymph node An oval or bean-shaped structure located along lymphatic vessels.

Lymphatic (lim-FAT-ik) **capillary** Closed-ended microscopic lymphatic vessel that begins in spaces between cells and converges with other lymphatic capillaries to form lymphatic vessels.

Lymphatic tissue A specialized form of reticular tissue that contains large numbers of lymphocytes.

Lymphatic vessel A large vessel that collects lymph from lymphatic capillaries and converges with other lymphatic vessels to form the thoracic and right lymphatic ducts.

Lymphocyte (LIM-fō-sīt) A type of white blood cell that helps carry out cell-mediated and antibody-mediated immune responses; found in blood and in lymphatic tissues.

Lysosome (LĪ-sō-sōm) An organelle in the cytoplasm of a cell, enclosed by a single membrane and containing powerful digestive enzymes.

Lysozyme (LĪ-sō-zīm) A bactericidal enzyme found in tears, saliva, and perspiration.

M

Macrophage (MAK-rō-fāj) Phagocytic cell derived from a monocyte; may be fixed or wandering.

Macula (MAK-ū-la) A discolored spot or a colored area. A small, thickened region on the wall of the utricle and saccule that contains receptors for static equilibrium.

Macula lutea (LOO-tē-a) The yellow spot in the center of the retina.

Major histocompatibility (MHC) antigens Surface proteins on white blood cells and other nucleated cells that are unique for each person (except for identical siblings); used to type tissues and help prevent rejection of transplanted tissues. Also known as **human leukocyte antigens (HLA).**

Malignant (ma-LIG-nant) Referring to diseases that tend to become worse and cause death, especially the invasion and spreading of cancer.

Mammary (MAM-ar-ē) **gland** Modified sudoriferous (sweat) gland of the female that produces milk for the nourishment of the young.

Mammillary (MAM-i-ler-ē) **bodies** Two small rounded bodies on the inferior aspect of the hypothalamus that are involved in reflexes related to the sense of smell.

Marrow (MAR-ō) Soft, spongelike material in the cavities of bone. Red bone marrow produces blood cells; yellow bone marrow contains adipose tissue that stores triglycerides.

Mast cell A cell found in areolar connective tissue that releases histamine, a dilator of small blood vessels, during inflammation.

Mastication (mas´-ti-KĀ-shun) Chewing.

Mature follicle A large, fluid-filled follicle containing a secondary oocyte and surrounding granulosa cells that secrete estrogens. Also called a **graafian (GRAF-ē-an) follicle.**

Meatus (mē-Ā-tus) A passage or opening, especially the external portion of a canal.

Mechanoreceptor (MEK-a-nō-rē-sep´-tor) Sensory receptor that detects mechanical deformation of the receptor itself or adjacent cells; stimuli so detected include those related to touch, pressure, vibration, proprioception, hearing, equilibrium, and blood pressure.

Medial (MĒ-dē-al) Nearer the midline of the body or a structure.

Medial lemniscus (lem-NIS-kus) A white matter tract that originates in the gracile and cuneate nuclei of the medulla oblongata and extends to the thalamus on the same side; sensory axons in this tract conduct nerve impulses for the sensations of proprioception, fine touch, vibration, hearing, and equilibrium.

Median aperture (AP-er-choor) One of the three openings in the roof of the fourth ventricle through which cerebrospinal fluid enters the subarachnoid space of the brain and cord. Also called the **foramen of Magendie.**

Median plane A vertical plane dividing the body into right and left halves. Situated in the middle.

Mediastinum (mē´-dē-a-STĪ-num) The anatomical region of the thoracic cavity between the pleurae of the lungs that extends from the sternum to the vertebral column and from the first rib to the diaphragm.

Medulla (me-DOOL-a) An inner layer of an organ, such as the medulla of the kidneys.

Medulla oblongata (ob´-long-GA-ta) The most inferior part of the brain stem. Also termed the **medulla.**

Medullary (MED-ū-lar´-ē) **cavity** The space within the diaphysis of a bone that contains yellow bone marrow. Also called the **marrow cavity.**

Medullary rhythmicity (rith-MIS-i-tē) **area** The neurons of the respiratory center in the medulla oblongata that control the basic rhythm of respiration.

Meibomian gland *See* **Tarsal gland.**

Meiosis (mī-Ō-sis) A type of cell division that occurs during production of gametes, involving two successive nuclear divisions that result in cells with the haploid (*n*) number of chromosomes.

Meissner (mīs-ner) **corpuscle** A sensory receptor for touch; found in dermal papillae, especially in the palms and soles. Also called a **corpuscle of touch.**

Melanin (MEL-a-nin) A dark black, brown, or yellow pigment found in some parts of the body such as the skin, hair, and pigmented layer of the retina.

Melanocyte (MEL-a-nō-sīt´) A pigmented cell, located between or beneath cells of the deepest layer of the epidermis, that synthesizes melanin.

Melanocyte-stimulating hormone (MSH) A hormone secreted by the anterior pituitary that stimulates the dispersion of melanin granules in melanocytes in amphibians; continued administration produces darkening of skin in humans.

Melatonin (mel-a-TŌN-in) A hormone secreted by the pineal gland that helps set the timing of the body's biological clock.

Membrane A thin, flexible sheet of tissue composed of an epithelial layer and an underlying connective tissue layer, as in an epithelial membrane, or of areolar connective tissue only, as in a synovial membrane.

Membranous labyrinth (mem-BRA-nus LAB-i-rinth) The part of the labyrinth of the internal ear that is located inside the bony labyrinth and separated from it by the perilymph; made up of the semicircular ducts, the saccule and utricle, and the cochlear duct.

Memory The ability to recall thoughts; commonly classifed as short-term (activated) and long-term.

Menarche (me-NAR-kē) The first menses (menstrual flow) and beginning of ovarian and uterine cycles.

Meninges (me-NIN-jēz) Three membranes covering the brain and spinal cord, called the dura mater, arachnoid mater, and pia mater. *Singular* is **meninx** (MEN-inks).

Menopause (MEN-ō-pawz) The termination of the menstrual cycles.

Menstrual (MEN-strū-al) **cycle** A series of changes in the endometrium of a nonpregnant female that prepares the lining of the uterus to receive a fertilized ovum.

Menstruation (men´-stroo-Ā-shun) Periodic discharge of blood, tissue fluid, mucus, and epithelial cells that usually lasts for 5 days; caused by a sudden reduction in estrogens and progesterone. Also called the **menstrual phase** or **menses.**

Merkel (MER-kel) **cell** Type of cell in the epidermis of hairless skin that makes contact with a Merkel (tactile) disc, which functions in touch.

Merkel disc Saucer-shaped free nerve endings that make contact with Merkel cells in the epidermis and function as touch receptors. Also called a **tactile disc.**

Merocrine (MER-ō-krin) **gland** Gland made up of secretory cells that remain intact throughout the process of formation and discharge of the secretory product, as in the salivary and pancreatic glands.

Mesenchyme (MEZ-en-kīm) An embryonic connective tissue from which all other connective tissues arise.

Mesentery (MEZ-en-ter´-ē) A fold of peritoneum attaching the small intestine to the posterior abdominal wall.

Mesocolon (mez´-ō-KŌ-lon) A fold of peritoneum attaching the colon to the posterior abdominal wall.

Mesoderm The middle primary germ layer that gives rise to connective tissues, blood and blood vessels, and muscles.

Mesothelium (mez´-ō-THĒ-lē-um) The layer of simple squamous epithelium that lines serous membranes.

Mesovarium (mez´-ō-VAR-ē-um) A short fold of peritoneum that attaches an ovary to the broad ligament of the uterus.

Metabolism (me-TAB-ō-lizm) All the biochemical reactions that occur within an organism, including the synthetic (anabolic) reactions and decomposition (catabolic) reactions.

Metacarpus (met´-a-KAR-pus) A collective term for the five bones (**metacarpals**) that make up the palm.

Metaphase (MET-a-fās) The second stage of mitosis, in which chromatid pairs line up on the metaphase plate of the cell.

Metaphysis (me-TAF-i-sis) Region of a long bone between the diaphysis and epiphysis that contains the epiphyseal plate in a growing bone.

Metarteriole (met´-ar-TĒ-rē-ōl) A blood vessel that emerges from an arteriole, traverses a capillary network, and empties into a venule.

Metastasis (me-TAS-ta-sis) The spread of cancer to surrounding tissues (local) or to other body sites (distant).

Metatarsus (met´-a-TAR-sus) A collective term for the five bones (**metatarsals**) located in the foot between the tarsals and the phalanges.

Microfilament (mī-krō-FIL-a-ment) Rodlike protein filament about 6 nm in diameter; constitutes contractile units in muscle fibers (cells) and provides support, shape, and movement in nonmuscle cells.

Microglia (mī-KROG-lē-a) Neuroglial cells that carry on phagocytosis.

Microtubule (mī-krō-TOO-būl´) Cylindrical protein filament, from 18 to 30 nm in diameter, consisting of the protein tubulin; provides support, structure, and transportation.

Microvilli (mī´-krō-VIL-ē) Microscopic, fingerlike projections of the plasma membranes of cells that increase surface area for absorption, especially in the small intestine and proximal convoluted tubules of the kidneys.

Micturition (mik´-choo-RISH-un) The act of expelling urine from the urinary bladder. Also called **urination** (ū-ri-NĀ-shun).

Midbrain The part of the brain between the pons and the diencephalon. Also called the **mesencephalon** (mes´-en-SEF-a-lon).

Middle ear A small, epithelial-lined cavity hollowed out of the temporal bone, separated from the external ear by the eardrum and from the internal ear by a thin bony partition containing the oval and round windows; extending across the middle ear are the three auditory ossicles. Also called the **tympanic** (tim-PAN-ik) **cavity.**

Midline An imaginary vertical line that divides the body into equal left and right sides.

Midsagittal plane A vertical plane through the midline of the body that divides the body or organs into *equal* right and left sides. Also called a **median plane.**

Mineralocorticoids (min´-er-al-ō-KOR-ti-koyds) A group of hormones of the adrenal cortex that help regulate sodium and potassium balance.

Mitochondrion (mī-tō-KON-drē-on) A double-membraned organelle that plays a central role in the production of ATP; known as the "powerhouse" of the cell. *Plural* is **mitochondria.**

Mitosis (mī-TŌ-sis) The orderly division of the nucleus of a cell that ensures that each new nucleus has the same number and kind of chromosomes as the original nucleus. The process includes the replication of chromosomes and the distribution of the two sets of chromosomes into two separate and equal nuclei.

Mitotic spindle Collective term for a football-shaped assembly of microtubules (nonkinetochore, kinetochore, and aster) that is responsible for the movement of chromosomes during cell division.

Modality (mō-DAL-i-tē) Any of the specific sensory entities, such as vision, smell, taste, or touch.

Modiolus (mō-DĪ-ō´-lus) The central pillar or column of the cochlea.

Monocyte (MON-ō-sīt) The largest type of white blood cell, characterized by agranular cytoplasm.

Monounsaturated fatty acid A fatty acid that contains one double covalent bond between its carbon atoms; it is not completely saturated with hydrogen atoms. Plentiful in triglycerides of olive and peanut oils.

Mons pubis (MONZ PŪ-bis) The rounded, fatty prominence over the pubic symphysis, covered by coarse pubic hair.

Morula (MOR-ū-la) A solid sphere of cells produced by successive cleavages of a fertilized ovum about 4 days after fertilization.

Motor area The region of the cerebral cortex that governs muscular movement, particularly the precentral gyrus of the frontal lobe.

Motor end plate Region of the sarcolemma of a muscle fiber (cell) that includes acetylcholine (ACh) receptors, which bind ACh released by synaptic end bulbs of somatic motor neurons.

Motor neurons (NOO-ronz) Neurons that conduct impulses from the brain toward the spinal cord or out of the brain and spinal cord into cranial or spinal nerves to effectors that may be either muscles or glands. Also called **efferent neurons.**

Motor unit A motor neuron together with the muscle fibers (cells) it stimulates.

Mucosa-associated lymphatic tissue (MALT) Lymphatic nodules scattered throughout the lamina propria (connective tissue) of mucous membranes lining the gastrointestinal tract, respiratory airways, urinary tract, and reproductive tract.

Mucous (MŪ-kus) **cell** A unicellular gland that secretes mucus. Two types are mucous neck cells and surface mucous cells in the stomach.

Mucous membrane A membrane that lines a body cavity that opens to the exterior. Also called **mucosa** (mū-KŌ-sa).

Mucus The thick fluid secretion of goblet cells, mucous cells, mucous glands, and mucous membranes.

Muscarinic (mus´-ka-RIN-ik) **receptor** Receptor for the neurotransmitter acetylcholine found on all effectors innervated by parasympathetic postganglionic axons and on sweat glands innervated by cholinergic sympathetic postganglionic axons; so named because muscarine activates these receptors but does not activate nicotinic receptors for acetylcholine.

Muscle An organ composed of one of three types of muscle tissue (skeletal, cardiac, or smooth), specialized for contraction to produce voluntary or involuntary movement of parts of the body.

Muscle action potential A stimulating impulse that propagates along the sarcolemma and transverse tubules; in skeletal muscle, it is generated by acetylcholine, which increases the permeability of the sarcolemma to cations, especially sodium ions (Na^+).

Muscle fatigue (fa-TĒG) Inability of a muscle to maintain its strength of contraction or tension; may be related to insufficient oxygen, depletion of glycogen, and/or lactic acid buildup.

Muscle spindle An encapsulated proprioceptor in a skeletal muscle, consisting of specialized intrafusal muscle fibers and nerve endings; stimulated by changes in length or tension of muscle fibers.

Muscle tone A sustained, partial contraction of portions of a skeletal or smooth muscle in response to activation of stretch receptors or a baseline level of action potentials in the innervating motor neurons.

Muscular dystrophies (DIS-trō-fēz´) Inherited muscle-destroying diseases, characterized by degeneration of muscle fibers (cells), which causes progressive atrophy of the skeletal muscle.

Muscularis (MUS-kū-la´-ris) A muscular layer (coat or tunic) of an organ.

Muscularis mucosae (mū-KŌ-sē) A thin layer of smooth muscle fibers that underlie the lamina propria of the mucosa of the gastrointestinal tract.

Muscular tissue A tissue specialized to produce motion in response to muscle action potentials by its qualities of contractility, extensibility, elasticity, and excitability; types include skeletal, cardiac, and smooth.

Mutation (mū-TĀ-shun) Any change in the sequence of bases in a DNA molecule resulting in a permanent alteration in some inheritable trait.

Myasthenia (mī-as-THĒ-nē-a) **gravis** Weakness and fatigue of skeletal muscles caused by antibodies directed against acetylcholine receptors.

Myelin (MĪ-e-lin) **sheath** Multilayered lipid and protein covering, formed by Schwann cells and oligodendrocytes, around axons of many peripheral and central nervous system neurons.

Myenteric plexus A network of autonomic axons and postganglionic cell bodies located in the muscularis of the gastrointestinal tract. Also called the **plexus of Auerbach** (OW-er-bak).

Myocardial infarction (mī´-ō-KAR-dē-al in-FARK-shun) (**MI**) Gross necrosis of myocardial tissue due to interrupted blood supply. Also called a **heart attack.**

Myocardium (mī´-ō-KAR-dē-um) The middle layer of the heart wall, made up of cardiac muscle tissue, lying between the epicardium and the endocardium and constituting the bulk of the heart.

Myofibril (mī-ō-FĪ-bril) A threadlike structure, extending longitudinally through a muscle fiber (cell), consisting mainly of thick filaments (myosin) and thin filaments (actin, troponin, and tropomyosin).

Myoglobin (mī-ō-GLŌB-in) The oxygen-binding, iron-containing protein present in the sarcoplasm of muscle fibers (cells); contributes the red color to muscle.

Myogram (MĪ-ō-gram) The record or tracing produced by a myograph, an apparatus that measures and records the force of muscular contractions.

Myology (mī-OL-ō-jē) The study of muscles.

Myometrium (mī´-ō-MĒ-trē-um) The smooth muscle layer of the uterus.

Myopathy (mī-OP-a-thē) Any abnormal condition or disease of muscle tissue.

Myopia (mī-Ō-pē-a) Defect in vision in which objects can be seen distinctly only when very close to the eyes; nearsightedness.

Myosin (MĪ-ō-sin) The contractile protein that makes up the thick filaments of muscle fibers.

Myotome (MĪ-ō-tōm) A group of muscles innervated by the motor neurons of a single spinal segment. In an embryo, the portion of a somite that develops into some skeletal muscles.

N

Nail A hard plate, composed largely of keratin, that develops from the epidermis of the skin to form a protective covering on the dorsal surface of the distal phalanges of the fingers and toes.

Nail matrix (MĀ-triks) The part of the nail beneath the body and root from which the nail is produced.

Nasal (NĀ-zal) **cavity** A mucosa-lined cavity on either side of the nasal septum that opens onto the face at the external nares and into the nasopharynx at the internal nares.

Nasal septum (SEP-tum) A vertical partition composed of bone (perpendicular plate of ethmoid and vomer) and cartilage, covered with a mucous membrane, separating the nasal cavity into left and right sides.

Nasolacrimal (nā´-zō-LAK-ri-mal) **duct** A canal that transports the lacrimal secretion (tears) from the nasolacrimal sac into the nose.

Nasopharynx (nā´-zō-FAR-inks) The superior portion of the pharynx, lying posterior to the nose and extending inferiorly to the soft palate.

Neck The part of the body connecting the head and the trunk. A constricted portion of an organ, such as the neck of the femur or uterus.

Necrosis (ne-KRŌ-sis) A pathological type of cell death that results from disease, injury, or lack of blood supply in which many adjacent cells swell, burst, and spill their contents into the interstitial fluid, triggering an inflammatory response.

Neonatal (nē-ō-NĀ-tal) Pertaining to the first 4 weeks after birth.

Neoplasm (NĒ-ō-plazm) A new growth that may be benign or malignant.

Nephron (NEF-ron) The functional unit of the kidney.

Nerve A cordlike bundle of neuronal axons and/or dendrites and associated connective tissue and blood vessels coursing together outside the central nervous system.

Nerve fiber General term for any process (axon or dendrite) projecting from the cell body of a neuron.

Nerve impulse A wave of depolarization and repolarization that self-propagates along the plasma membrane of a neuron; also called a **nerve action potential.**

Nervous tissue Tissue containing neurons that initiate and conduct nerve impulses to coordinate homeostasis, and neuroglia that provide support and nourishment to neurons.

Neuralgia (noo-RAL-jē-a) Attacks of pain along the entire course or branch of a peripheral sensory nerve.

Neural plate A thickening of ectoderm, induced by the notochord, that forms early in the third week of development and represents the beginning of the development of the nervous system.

Neural tube defect (NTD) A developmental abnormality in which the neural tube does not close properly. Examples are spina bifida and anencephaly.

Neuritis (noo-RĪ-tis) Inflammation of one or more nerves.

Neurofibral node *See* **Node of Ranvier.**

Neuroglia (noo-RŌG-lē-a) Cells of the nervous system that perform various supportive functions. The neuroglia of the central nervous system are the astrocytes, oligodendrocytes, microglia, and ependymal cells; neuroglia of the peripheral nervous system include Schwann cells and satellite cells. Also called **glia** or **glial** (GLĒ-al) **cells.**

Neurohypophyseal (noo´-rō-hī´-pō-FIZ-ē-al) **bud** An outgrowth of ectoderm located on the floor of the hypothalamus that gives rise to the posterior pituitary.

Neurolemma (noor-ō-LEM-a) The peripheral, nucleated cytoplasmic layer of the Schwann cell. Also called **sheath of Schwann** (SCHWON).

Neurology (noo-ROL-ō-jē) The study of the normal functioning and disorders of the nervous system.

Neuromuscular (noo-rō-MUS-kū-lar) **junction (NMJ)** A synapse between the axon terminals of a motor neuron and the sarcolemma of a muscle fiber (cell).

Neuron (NOO-ron) A nerve cell, consisting of a cell body, dendrites, and an axon.

Neurosecretory (noo-rō-SĒK-re-tō-rē) **cell** A neuron that secretes a hypothalamic releasing hormone or inhibiting hormone into blood capillaries of the hypothalmus; a neuron that secretes oxytocin or antidiuretic hormone into blood capillaries of the posterior pituitary.

Neurotransmitter One of a variety of molecules within axon terminals that are released into the synaptic cleft in response to a nerve impulse and that change the membrane potential of the postsynaptic neuron.

Neurulation (noor-oo-LĀ-shun) The process by which the neural plate, neural folds, and neural tube develop.

Neutrophil (NOO-trō-fil) A type of white blood cell characterized by granules that stain pale lilac with a combination of acidic and basic dyes.

Nicotinic (nik´-ō-TIN-ik) **receptor** Receptor for the neurotransmitter acetylcholine found on both sympathetic and parasympathetic postganglionic neurons and on skeletal muscle in the motor end plate; so named because nicotine activates these receptors but does not activate muscarinic receptors for acetylcholine.

Nipple A pigmented, wrinkled projection on the surface of the breast that is the location of the openings of the lactiferous ducts for milk release.

Nociceptor (nō-sē-SEP-tor) A free (naked) nerve ending that detects painful stimuli.

Node of Ranvier (RON-vē-ā) A space along a myelinated axon between the individual Schwann cells that form the myelin sheath and the neurolemma. Also called a **neurofibral node.**

Norepinephrine (nor´-ep-ē-NEF-rin) (**NE**) A hormone secreted by the adrenal medulla that produces actions similar to those that result from sympathetic stimulation. Also called **noradrenaline** (nor-a-DREN-a-lin).

Notochord (NŌ-tō-cord) A flexible rod of mesodermal tissue that lies where the future vertebral column will develop and plays a role in induction.

Nucleic (noo-KLĒ-ic) **acid** An organic compound that is a long polymer of nucleotides, with each nucleotide containing a pentose sugar, a phosphate group, and one of four possible nitrogenous bases (adenine, cytosine, guanine, and thymine or uracil).

Nucleolus (noo´-KLĒ-ō-lus) Spherical body within a cell nucleus composed of protein, DNA, and RNA that is the site of the assembly of small and large ribosomal subunits. *Plural* is **nucleoli.**

Nucleosome (NOO-klē-ō-sōm) Structural subunit of a chromosome consisting of histones and DNA.

Nucleus (NOO-klē-us) A spherical or oval organelle of a cell that contains the hereditary factors of the cell, called genes. A cluster of unmyelinated nerve cell bodies in the central nervous system. The central part of an atom made up of protons and neutrons.

Nucleus pulposus (pul-PŌ-sus) A soft, pulpy, highly elastic substance in the center of an intervertebral disc; a remnant of the notochord.

Nutrient A chemical substance in food that provides energy, forms new body components, or assists in various body functions.

O

Obesity (ō-BĒS-i-tē) Body weight more than 20% above a desirable standard due to excessive accumulation of fat.

Oblique (ō-BLĒK) **plane** A plane that passes through the body or an organ at an angle between the transverse plane and either the midsagittal, parasagittal, or frontal plane.

Obstetrics (ob-STET-riks) The specialized branch of medicine that deals with pregnancy, labor, and the period of time immediately after delivery (about 6 weeks).

Olfactory (ōl-FAK-tō-rē) Pertaining to smell.

Olfactory bulb A mass of gray matter containing cell bodies of neurons that form synapses with neurons of the olfactory nerve, lying inferior to the frontal lobe of the cerebrum on either side of the crista galli of the ethmoid bone.

Olfactory receptor A bipolar neuron with its cell body lying between supporting cells located in the mucous membrane lining the superior portion of each nasal cavity; transduces odors into neural signals.

Olfactory tract A bundle of axons that extends from the olfactory bulb posteriorly to olfactory regions of the cerebral cortex.

Oligodendrocyte (OL-i-gō-den´-drō-sīt) A neuroglial cell that supports neurons and produces a myelin sheath around axons of neurons of the central nervous system.

Oliguria (ol´-i-GŪ-rē-a) Daily urinary output usually less than 250 mL.

Olive A prominent oval mass on each lateral surface of the superior part of the medulla oblongata.

Oncogenes (ON-kō-jēnz) Cancer-causing genes; they derive from normal genes, termed proto-oncogenes, that encode proteins involved in cell growth or cell regulation but have the ability to transform a normal cell into a cancerous cell when they are mutated or inappropriately activated. One example is *p53*.

Oncology (on-KOL-ō-jē) The study of tumors.

Oogenesis (ō´-ō-JEN-e-sis) Formation and development of female gametes (oocytes).

Oophorectomy (ō-of-ōr-EK-tō-mē) Surgical removal of the ovaries.

Ophthalmic (of-THAL-mik) Pertaining to the eye.

Ophthalmologist (of´-thal-MOL-ō-jist) A physician who specializes in the diagnosis and treatment of eye disorders using drugs, surgery, and corrective lenses.

Ophthalmology (of-thal-MOL-ō-jē) The study of the structure, function, and diseases of the eye.

Optic (OP-tik) Refers to the eye, vision, or properties of light.

Optic chiasm (KĪ-azm) A crossing point of the two branches of the optic nerve, anterior to the pituitary gland. Also called **optic chiasma.**

Optic disc A small area of the retina containing openings through which the axons of the ganglion cells emerge as the optic nerve. Also called the **blind spot.**

Optic tract A bundle of axons that carry nerve impulses from the retina of the eye between the optic chiasm and the thalamus.

Ora serrata (Ō-ra ser-RĀ-ta) The irregular margin of the retina lying internal and slightly posterior to the junction of the choroid and ciliary body.

Orbit (OR-bit) The bony, pyramidal-shaped cavity of the skull that holds the eyeball.

Organ A structure composed of two or more different kinds of tissues with a specific function and usually a recognizable shape.

Organelle (or-gan-EL) A permanent structure within a cell with characteristic morphology that is specialized to serve a specific function in cellular activities.

Organism (OR-ga-nizm) A total living form; one individual.

Organogenesis (or´-ga-nō-JEN-e-sis) The formation of body organs and systems. By the end of the eighth week of development, all major body systems have begun to develop.

Orifice (OR-i-fis) Any aperture or opening.

Origin (OR-i-jin) The attachment of a muscle tendon to a stationary bone or the end opposite the insertion.

Oropharynx (or´-ō-FAR-inks) The intermediate portion of the pharynx, lying posterior to the mouth and extending from the soft palate to the hyoid bone.

Orthopedics (or´-thō-PĒ-diks) The branch of medicine that deals with the preservation and restoration of the skeletal system, articulations, and associated structures.

Osmoreceptor (oz´-mō-re-SEP-tor) Receptor in the hypothalamus that is sensitive to changes in blood osmolarity and, in response to high osmolarity (low water concentration), stimulates synthesis and release of antidiuretic hormone (ADH).

Osmosis (oz-MŌ-sis) The net movement of water molecules through a selectively permeable membrane from an area of higher water concentration to an area of lower water concentration until equilibrium is reached.

Osseous (OS-ē-us) Bony.

Ossicle (OS-i-kl) One of the small bones of the middle ear (malleus, incus, stapes).

Ossification (os´-i-fi-KĀ-shun) Formation of bone. Also called **osteogenesis** (os´-tē-ō-JEN-e-sis).

Ossification center An area in the cartilage model of a future bone where the cartilage cells hypertrophy, secrete enzymes that calcify their extracellular matrix, and die, and the area they occupied is invaded by osteoblasts that then lay down bone.

Osteoblast (OS-tē-ō-klast´) Cell formed from an osteogenic cell that participates in bone formation by secreting some organic components and inorganic salts.

Osteoclast (OS-tē-ō-clast´) A large, multinuclear cell that resorbs (destroys) bone matrix.

Osteocyte (OS-tē-ō-sīt´) A mature bone cell that maintains the daily activities of bone tissue.

Osteogenic (os´-tē-ō-JEN-ik) **cell** Stem cell derived from mesenchyme that has mitotic potential and the ability to differentiate into an osteoblast.

Osteogenic layer The inner layer of the periosteum that contains cells responsible for forming new bone during growth and repair.

Osteology (os-tē-OL-ō-jē) The study of bones.

Osteon (OS-tē-on) The basic unit of structure in adult compact bone, consisting of a central (haversian) canal with its concentrically arranged lamellae, lacunae, osteocytes, and canaliculi. Also called a **haversian** (ha-VER-shan) **system.**

Osteoporosis (os´-tē-ō-pō-RŌ-sis) Age-related disorder characterized by decreased bone mass and increased susceptibility to fractures, often as a result of decreased levels of estrogens.

Otic (Ō-tik) Pertaining to the ear.

Otolith (Ō-tō-lith) A particle of calcium carbonate embedded in the otolithic membrane that functions in maintaining static equilibrium.

Otolithic (ō-tō-LITH-ik) **membrane** Thick, gelatinous, glycoprotein layer located directly over hair cells of the macula in the saccule and utricle of the internal ear.

Otorhinolaryngology (ō-tō-rī´-nō-lar-in-GOL-ō-jē) The branch of medicine that deals with the diagnosis and treatment of diseases of the ears, nose, and throat.

Oval window A small, membrane-covered opening between the middle ear and inner ear into which the footplate of the stapes fits.

Ovarian (ō-VAR-ē-an) **cycle** A monthly series of events in the ovary associated with the maturation of a secondary oocyte.

Ovarian follicle (FOL-i-kul) A general name for oocytes (immature ova) in any stage of development, along with their surrounding epithelial cells.

Ovarian ligament (LIG-a-ment) A rounded cord of connective tissue that attaches the ovary to the uterus.

Ovary (Ō-var-ē) Female gonad that produces oocytes and the estrogens, progesterone, inhibin, and relaxin hormones.

Ovulation (ov-ū-LĀ-shun) The rupture of a mature ovarian (graafian) follicle with discharge of a secondary oocyte into the pelvic cavity.

Ovum (Ō-vum) The female reproductive or germ cell; an egg cell; arises through completion of meiosis in a secondary oocyte after penetration by a sperm.

Oxyhemoglobin (ok´-sē-HĒ-mō-glō-bin) **(Hb—O$_2$)** Hemoglobin combined with oxygen.

Oxytocin (ok´-sē-TŌ-sin) **(OT)** A hormone secreted by neurosecretory cells in the paraventricular and supraoptic nuclei of the hypothalamus that stimulates contraction of smooth muscle in the pregnant uterus and myoepithelial cells around the ducts of mammary glands.

P

P wave The deflection wave of an electrocardiogram that signifies atrial depolarization.

Pacinian corpuscle Oval-shaped pressure receptor located in the dermis or subcutaneous tissue and consisting of concentric layers of a connective tissue wrapped around the dendrites of a sensory neuron. Also called a **lamellated corpuscle.**

Palate (PAL-at) The horizontal structure separating the oral and the nasal cavities; the roof of the mouth.

Palpate (PAL-pāt) To examine by touch; to feel.

Pancreas (PAN-krē-as) A soft, oblong organ lying along the greater curvature of the stomach and connected by a duct to the duodenum. It is both an exocrine gland (secreting pancreatic juice) and an endocrine gland (secreting insulin, glucagon, somatostatin, and pancreatic polypeptide).

Pancreatic (pan´-krē-AT-ik) **duct** A single large tube that unites with the common bile duct from the liver and gallbladder and drains pancreatic juice into the duodenum at the hepatopancreatic ampulla (ampulla of Vater). Also called the **duct of Wirsung.**

Pancreatic islet A cluster of endocrine gland cells in the pancreas that secretes insulin, glucagon, somatostatin, and pancreatic polypeptide. Also called an **islet of Langerhans** (LANG-er-hanz).

Papanicolaou (pa-pa-NI-kō-lō) **test** A cytological staining test for the detection and diagnosis of premalignant and malignant conditions of the female genital tract. Cells scraped from the epithelium of the cervix of the uterus are examined microscopically. Also called a **Pap test** or **Pap smear.**

Papilla (pa-PIL-a) A small nipple-shaped projection or elevation.

Paralysis (pa-RAL-a-sis) Loss or impairment of motor function due to a lesion of nervous or muscular origin.

Paranasal sinus (par´-a-NĀ-zal SĪ-nus) A mucus-lined air cavity in a skull bone that communicates with the nasal cavity. Paranasal sinuses are located in the frontal, maxillary, ethmoid, and sphenoid bones.

Paraplegia (par-a-PLĒ-jē-a) Paralysis of both lower limbs.

Parasagittal (par-a-SAJ-i-tal) **plane** A vertical plane that does not pass through the midline and that divides the body or organs into *unequal* left and right portions.

Parasympathetic (par´-a-sim-pa-THET-ik) **division** One of the two subdivisions of the autonomic nervous system, having cell bodies of preganglionic neurons in nuclei in the brain stem and in the lateral gray horn of the sacral portion of the spinal cord; primarily concerned with activities that conserve and restore body energy.

Parathyroid (par´-a-THĪ-royd) **gland** One of usually four small endocrine glands embedded in the posterior surfaces of the lateral lobes of the thyroid gland.

Parathyroid hormone (PTH) A hormone secreted by the chief (principal) cells of the parathyroid glands that increases blood calcium level and decreases blood phosphate level.

Paraurethral (par´-a-ū-RĒ-thral) **gland** Gland embedded in the wall of the urethra whose duct opens on either side of the urethral orifice and secretes mucus. Also called **Skene's** (SKĒNZ) **gland.**

Parenchyma (par-EN-ki-ma) The functional parts of any organ, as opposed to tissue that forms its stroma or framework.

Parietal (pa-RĪ-e-tal) Pertaining to or forming the outer wall of a body cavity.

Parietal cell A type of secretory cell in gastric glands that produces hydrochloric acid and intrinsic factor. Also called an **oxyntic cell.**

Parietal pleura (PLOO-ra) The outer layer of the serous pleural membrane that encloses and protects the lungs; the layer that is attached to the wall of the pleural cavity.

Parkinson disease (PD) Progressive degeneration of the basal ganglia and substantia nigra of the cerebrum resulting in decreased production of dopamine (DA) that leads to tremor, slowing of voluntary movements, and muscle weakness.

Parotid (pa-ROT-id) **gland** One of the paired salivary glands located inferior and anterior to the ears and connected to the oral cavity via a duct (Stensen's) that opens into the inside of the cheek opposite the maxillary (upper) second molar tooth.

Pars intermedia A small avascular zone between the anterior and posterior pituitary glands.

Parturition (par´-toor-ISH-un) Act of giving birth to young; childbirth, delivery.

Patent ductus arteriosus A congenital heart defect in which the ductus arteriosus remains open. As a result, aortic blood flows into the lower-pressure pulmonary trunk, increasing pulmonary trunk pressure and overworking both ventricles.

Pathogen (PATH-ō-jen) A disease-producing microbe.

Pathological (path´-ō-LOJ-i-kal) **anatomy** The study of structural changes caused by disease.

Pectinate (PEK-ti-nāt) **muscles** Projecting muscle bundles of the anterior atrial walls and the lining of the auricles.

Pectoral (PEK-to-ral) Pertaining to the chest or breast.

Pedicel (PED-i-sel) Footlike structure, as on podocytes of a glomerulus.

Pelvic (PEL-vik) **cavity** Inferior portion of the abdominopelvic cavity that contains the urinary bladder, sigmoid colon, rectum, and internal female and male reproductive structures.

Pelvic splanchnic (PEL-vik SPLANGK-nik) **nerves** Consist of preganglionic parasympathetic axons from the levels of S2, S3, and S4 that supply the urinary bladder, reproductive organs, and the descending and sigmoid colon and rectum.

Pelvis The basinlike structure formed by the two hip bones, the sacrum, and the coccyx. The expanded, proximal portion of the ureter, lying within the kidney and into which the major calyces open.

Penis (PĒ-nis) The organ of urination and copulation in males; used to deposit semen into the female vagina.

Pepsin Protein-digesting enzyme secreted by chief cells of the stomach in the inactive form pepsinogen, which is converted to active pepsin by hydrochloric acid.

Peptic ulcer An ulcer that develops in areas of the gastrointestinal tract exposed to hydrochloric acid; classified as a gastric ulcer if in the lesser curvature of the stomach and as a duodenal ulcer if in the first part of the duodenum.

Percussion (pur-KUSH-un) The act of striking (percussing) an underlying part of the body with short, sharp taps as an aid in diagnosing the part by the quality of the sound produced.

Perforating canal A minute passageway by means of which blood vessels and nerves from the periosteum penetrate into compact bone. Also called **Volkmann's** (FŌLK-mans) **canal.**

Pericardial (per-i-KAR-dē-al) **cavity** Small potential space between the visceral and parietal layers of the serous pericardium that contains **pericardial fluid.**

Pericardium (per-i-KAR-dē-um) A loose-fitting membrane that encloses the heart, consisting of a superficial **fibrous** layer and a deep **serous** layer.

Perichondrium (per-i-KON-drē-um) A covering of dense irregular connective tissue that surrounds the surface of most cartilage.

Perilymph (PER-i-limf) The fluid contained between the bony and membranous labyrinths of the inner ear.

Perimetrium (per-i-MĒ-trē-um) The serosa of the uterus.

Perimysium (per-i-MIZ-ē-um) Invagination of the epimysium that divides muscles into bundles.

Perineum (per´-i-NĒ-um) The pelvic floor; the space between the anus and the scrotum in the male and between the anus and the vulva in the female.

Perineurium (per-i-NOO-rē-um) Connective tissue wrapping around fascicles in a nerve.

Periodontal (per´-ē-ō-DON-tal) **disease** A collective term for conditions characterized by degeneration of gingivae, alveolar bone, periodontal ligament, and cementum.

Periodontal ligament The periosteum lining the alveoli (sockets) for the teeth in the alveolar processes of the mandible and maxillae.

Periosteum (per-ē-OS-tē-um) The covering of a bone that consists of connective tissue, osteogenic cells, and osteoblasts; is essential for bone growth, repair, and nutrition.

Peripheral (pe-RIF-er-al) Located on the outer part or a surface of the body.

Peripheral nervous system (PNS) The part of the nervous system that lies outside the central nervous system, consisting of nerves and ganglia.

Peristalsis (per-i-STAL-sis) Successive muscular contractions along the wall of a hollow muscular structure.

Peritoneum (per-i-tō-NĒ-um) The largest serous membrane of the body that lines the abdominal cavity and covers the viscera within it.

Peritonitis (per´-i-tō-NĪ-tis) Inflammation of the peritoneum.

Peroxisome (pe-ROKS-i-sōm) Organelle similar in structure to a lysosome that contains enzymes that use molecular oxygen to oxidize various organic compounds; such reactions produce hydrogen peroxide; abundant in liver cells.

Perspiration Sweat; produced by sudoriferous (sweat) glands and containing water, salts, urea, uric acid, amino acids, ammonia, sugar, lactic acid, and ascorbic acid. Helps maintain body temperature and eliminate wastes.

Peyer's patches *See* **Aggregated lymphatic follicles.**

pH A measure of the concentration of hydrogen ions (H^+) in a solution. The pH scale extends from 0 to 14, with a value of 7 expressing neutrality, values lower than 7 expressing increasing acidity, and values higher than 7 expressing increasing alkalinity.

Phagocytosis (fag´-ō-sī-TŌ-sis) The process by which phagocytes ingest and destroy microbes, cell debris, and other foreign matter.

Phalanx (FĀ-lanks) The bone of a finger or toe. *Plural* is **phalanges** (fa-LAN-jēz).

Pharmacology (far´-ma-KOL-ō-jē) The science of the effects and uses of drugs in the treatment of disease.

Pharynx (FAR-inks) The throat; a tube that starts at the internal nares and runs partway down the neck, where it opens into the esophagus posteriorly and the larynx anteriorly.

Phenotype (FĒ-nō-tīp) The observable expression of genotype; physical characteristics of an organism determined by genetic makeup and influenced by interaction between genes and internal and external environmental factors.

Phlebitis (fle-BĪ-tis) Inflammation of a vein, usually in a lower limb.

Photopigment A substance that can absorb light and undergo structural changes that can lead to the development of a receptor potential. An example is rhodopsin. In the eye, also called **visual pigment.**

Photoreceptor Receptor that detects light shining on the retina of the eye.

Physiology (fiz´-ē-OL-ō-jē) Science that deals with the functions of an organism or its parts.

Pia mater (PĒ-a MAH-ter *or* PĪ-a MĀ-ter) The innermost of the three meninges (coverings) of the brain and spinal cord.

Pineal (PĪN-ē-al) **gland** A cone-shaped gland located in the roof of the third ventricle that secretes melatonin. Also called the **epiphysis cerebri** (e-PIF-i-sis se-RĒ-brē).

Pinealocyte (pin-ē-AL-ō-sīt) Secretory cell of the pineal gland that releases melatonin.

Pinna (PIN-a) The projecting part of the external ear composed of elastic cartilage and covered by skin and shaped like the flared end of a trumpet. Also called the **auricle** (OR-i-kul).

Pituicyte (pi-TOO-i-sīt) Supporting cell of the posterior pituitary.

Pituitary (pi-TOO-i-tār-ē) **gland** A small endocrine gland occupying the hypophyseal fossa of the sphenoid bone and attached to the hypothalamus by the infundibulum. Also called the **hypophysis** (hī-POF-i-sis).

Pivot joint A synovial joint in which a rounded, pointed, or conical surface of one bone articulates with a ring formed partly by another bone and partly by a ligament, as in the joint between the atlas and axis and between the proximal ends of the radius and ulna. Also called a **trochoid** (TRŌ-koyd) **joint.**

Placenta (pla-SEN-ta) The special structure through which the exchange of materials between fetal and maternal circulations occurs. Also called the **afterbirth.**

Plantar flexion (PLAN-tar FLEK-shun) Bending the foot in the direction of the plantar surface (sole).

Plaque (PLAK) A layer of dense proteins on the inside of a plasma membrane in adherens junctions and desmosomes. A mass of bacterial cells, dextran (polysaccharide), and other debris that adheres to teeth (dental plaque). *See also* **Atherosclerotic plaque.**

Plasma (PLAZ-ma) The extracellular fluid found in blood vessels; blood minus the formed elements.

Plasma cell Cell that develops from a B cell (lymphocyte) and produces antibodies.

Plasma (cell) membrane Outer, limiting membrane that separates the cell's internal parts from extracellular fluid or the external environment.

Platelet (PLĀT-let) A fragment of cytoplasm enclosed in a cell membrane and lacking a nucleus; found in the circulating blood; plays a role in hemostasis. Also called a **thrombocyte** (THROM-bō-sīt).

Platelet plug Aggregation of platelets (thrombocytes) at a site where a blood vessel is damaged that helps stop or slow blood loss.

Pleura (PLOO-ra) The serous membrane that covers the lungs and lines the walls of the chest and the diaphragm.

Pleural cavity Small potential space between the visceral and parietal pleurae.

Plexus (PLEK-sus) A network of nerves, veins, or lymphatic vessels.

Plexus of Auerbach *See* **Myenteric plexus.**

Plexus of Meissner *See* **Submucosal plexus.**

Pluripotent (ploo-RIP-ō-tent) **stem cell** Immature stem cell in red bone marrow that gives rise to precursors of all the different mature blood cells.

Pneumotaxic (noo-mō-TAK-sik) **area** A part of the respiratory center in the pons that continually sends inhibitory nerve impulses to the inspiratory area, limiting inhalation and facilitating exhalation.

Polycythemia (pol´-ē-sī-THĒ-mē-a) Disorder characterized by an above-normal hematocrit (above 55%) in which hypertension, thrombosis, and hemorrhage can occur.

Polyunsaturated fat A fatty acid that contains more than one double covalent bond between its carbon atoms; abundant in triglycerides of corn oil, safflower oil, and cottonseed oil.

Polyuria (pol´-ē-Ū-rē-a) An excessive production of urine.

Pons (PONZ) The part of the brain stem that forms a "bridge" between the medulla oblongata and the midbrain, anterior to the cerebellum.

Portal system The circulation of blood from one capillary network into another through a vein.

Postcentral gyrus Gyrus of cerebral cortex located immediately posterior to the central sulcus; contains the primary somatosensory area.

Posterior (pos-TĒR-ē-or) Nearer to or at the back of the body. Equivalent to **dorsal** in bipeds.

Posterior column–medial lemniscus pathway Sensory pathway that carries information related to proprioception, fine touch, two-point discrimination, pressure, and vibration. First-order neurons project from the spinal cord to the ipsilateral medulla in the posterior columns (gracile fasciculus and cuneate fasciculus). Second-order neurons project from the medulla to the contralateral thalamus in the medial lemniscus. Third-order neurons project from the thalamus to the somatosensory cortex (postcentral gyrus) on the same side.

Posterior pituitary Posterior lobe of the pituitary gland. Also called the **neurohypophysis** (noo-rō-hī-POF-i-sis).

Posterior root The structure composed of sensory axons lying between a spinal nerve and the dorsolateral aspect of the spinal cord. Also called the **dorsal (sensory) root.**

Posterior root ganglion (GANG-glē-on) A group of cell bodies of sensory neurons and their supporting cells located along the posterior root of a spinal nerve. Also called a **dorsal (sensory) root ganglion.**

Postganglionic neuron (pōst´-gang-lē-ON-ik NOO-ron) The second autonomic motor neuron in an autonomic pathway, having its cell body and dendrites located in an autonomic ganglion and its unmyelinated axon ending at cardiac muscle, smooth muscle, or a gland.

Postsynaptic (pōst-sin-AP-tik) **neuron** The nerve cell that is activated by the release of a neurotransmitter from another neuron and carries nerve impulses away from the synapse.

Pouch of Douglas *See* **Rectouterine pouch.**

Precapillary sphincter (SFINGK-ter) The distalmost muscle fiber (cell) at the metarteriole–capillary junction that regulates blood flow into capillaries.

Precentral gyrus Gyrus of cerebral cortex located immediately anterior to the central sulcus; contains the primary motor area.

Preganglionic (pre´-gang-lē-ON-ik) **neuron** The first autonomic motor neuron in an autonomic pathway, with its cell body and dendrites in the brain or spinal cord and its myelinated axon ending at an autonomic ganglion, where it synapses with a postganglionic neuron.

Pregnancy Sequence of events that normally includes fertilization, implantation, embryonic growth, and fetal growth and terminates in birth.

Premenstrual syndrome (PMS) Severe physical and emotional stress occurring late in the postovulatory phase of the menstrual cycle and sometimes overlapping with menstruation.

Prepuce (PRĒ-poos) The loose-fitting skin covering the glans of the penis and clitoris. Also called the **foreskin.**

Presbyopia (prez-bē-Ō-pē-a) A loss of elasticity of the lens of the eye due to advancing age with resulting inability to focus clearly on near objects.

Presynaptic (prē-sin-AP-tik) **neuron** A neuron that propagates nerve impulses toward a synapse.

Prevertebral ganglion (prē-VER-te-bral GANG-glē-on) A cluster of cell bodies of postganglionic sympathetic neurons anterior to the spinal column and close to large abdominal arteries. Also called a **collateral ganglion.**

Primary germ layer One of three layers of embryonic tissue, called ectoderm, mesoderm, and endoderm, that give rise to all tissues and organs of the body.

Primary motor area A region of the cerebral cortex in the precentral gyrus of the frontal lobe of the cerebrum that controls specific muscles or groups of muscles.

Primary somatosensory area A region of the cerebral cortex posterior to the central sulcus in the postcentral gyrus of the parietal lobe of the cerebrum that localizes exactly the points of the body where somatic sensations originate.

Prime mover The muscle directly responsible for producing a desired motion. Also called an **agonist** (AG-ō-nist).

Primitive gut Embryonic structure formed from the dorsal part of the yolk sac that gives rise to most of the gastrointestinal tract.

Primordial (prī-MŌR-dē-al) Existing first; especially primordial egg cells in the ovary.

Principal cell Cell type in the distal convoluted tubules and collecting ducts of the kidneys that is stimulated by aldosterone and antidiuretic hormone.

Proctology (prok-TOL-ō-jē) The branch of medicine concerned with the rectum and its disorders.

Progeny (PROJ-e-nē) Offspring or descendants.

Progesterone (prō-JES-te-rōn) A female sex hormone produced by the ovaries that helps prepare the endometrium of the uterus for implantation of a fertilized ovum and the mammary glands for milk secretion.

Prognosis (prog-NŌ-sis) A forecast of the probable results of a disorder; the outlook for recovery.

Prolactin (prō-LAK-tin) **(PRL)** A hormone secreted by the anterior pituitary that initiates and maintains milk secretion by the mammary glands.

Prolapse (PRŌ-laps) A dropping or falling down of an organ, especially the uterus or rectum.

Proliferation (pro-lif´-er-Ā-shun) Rapid and repeated reproduction of new parts, especially cells.

Pronation (prō-NĀ-shun) A movement of the forearm in which the palm is turned posteriorly.

Prophase (PRŌ-fāz) The first stage of mitosis, during which chromatid pairs are formed and aggregate around the metaphase plate of the cell.

Proprioception (prō-prē-ō-SEP-shun) The perception of the position of body parts, especially the limbs, independent of vision; this sense is possible due to nerve impulses generated by proprioceptors.

Proprioceptor (PRŌ-prē-ō-sep´-tor) A receptor located in muscles, tendons, joints, or the internal ear (muscle spindles, tendon organs, joint kinesthetic receptors, and hair cells of the vestibular apparatus) that provides information about body position and movements.

Prostaglandin (pros´-ta-GLAN-din) **(PG)** A membrane-associated lipid; released in small quantities and acts as a local hormone.

Prostate (PROS-tāt) A doughnut-shaped gland inferior to the urinary bladder that surrounds the superior portion of the male urethra and secretes a slightly acidic solution that contributes to sperm motility and viability.

Proteasome (PRŌ-tē-a-sōm) Tiny cellular organelle in cytosol and nucleus containing proteases that destroy unneeded, damaged, or faulty proteins.

Protein An organic compound consisting of carbon, hydrogen, oxygen, nitrogen, and sometimes sulfur and phosphorus; synthesized on ribosomes and made up of amino acids linked by peptide bonds.

Prothrombin (prō-THROM-bin) An inactive blood-clotting factor synthesized by the liver, released into the blood, and converted to active **thrombin** in the process of blood clotting by the activated enzyme **prothrombinase**.

Proto-oncogene (prō´-tō-ON-kō-jēn) Gene responsible for some aspect of normal growth and development; it may transform into an oncogene, a gene capable of causing cancer.

Protraction (prō-TRAK-shun) The movement of the mandible or shoulder girdle forward on a plane parallel with the ground.

Proximal (PROK-si-mal) Nearer the attachment of a limb to the trunk; nearer to the point of origin or attachment.

Pseudopods (SOO-dō-pods) Temporary protrusions of the leading edge of a migrating cell; cellular projections that surround a particle undergoing phagocytosis.

Pterygopalatine ganglion (ter´-i-gō-PAL-a-tīn GANG-glē-on) A cluster of cell bodies of parasympathetic postganglionic neurons ending at the lacrimal and nasal glands.

Ptosis (TŌ-sis) Drooping, as of the eyelid or the kidney.

Puberty (PŪ-ber-tē) The time of life during which the secondary sex characteristics begin to appear and the capability for sexual reproduction is possible; usually occurs between the ages of 10 and 17.

Pubic symphysis (PŪ-bik SIM-fi-sis) A slightly movable cartilaginous joint between the anterior surfaces of the hip bones.

Puerperium (pū´-er-PER-ē-um) The period immediately after childbirth, usually 4–6 weeks.

Pulmonary (PUL-mo-ner´-ē) Concerning or affected by the lungs.

Pulmonary circulation The flow of deoxygenated blood from the right ventricle to the lungs and the return of oxygenated blood from the lungs to the left atrium.

Pulmonary edema (e-DĒ-ma) An abnormal accumulation of interstitial fluid in the tissue spaces and alveoli of the lungs due to increased pulmonary capillary permeability or increased pulmonary capillary pressure.

Pulmonary embolism (EM-bō-lizm) **(PE)** The presence of a blood clot or a foreign substance in a pulmonary arterial blood vessel that obstructs circulation to lung tissue.

Pulmonary ventilation The inflow (inhalation) and outflow (exhalation) of air between the atmosphere and the lungs. Also called **breathing.**

Pulp cavity A cavity within the crown and neck of a tooth, which is filled with pulp, a connective tissue containing blood vessels, nerves, and lymphatic vessels.

Pulse (PULS) The rhythmic expansion and elastic recoil of a systemic artery after each contraction of the left ventricle.

Pupil The hole in the center of the iris, the area through which light enters the posterior cavity of the eyeball.

Purkinje (pur-KIN-jē) **fiber** Muscle fiber (cell) in the ventricular tissue of the heart specialized for conducting an action potential to the myocardium; part of the conduction system of the heart.

Pus The liquid product of inflammation containing leukocytes or their remains and debris of dead cells.

Pyloric (pī-LOR-ik) **sphincter** A thickened ring of smooth muscle through which the pylorus of the stomach communicates with the duodenum. Also called the **pyloric valve.**

Pyorrhea (pī-ō-RĒ-a) A discharge or flow of pus, especially in the alveoli (sockets) and the tissues of the gums.

Pyramid (PIR-a-mid) A pointed or cone-shaped structure. One of two roughly triangular structures on the anterior aspect of the medulla oblongata composed of the largest motor tracts that run from the cerebral cortex to the spinal cord. A triangular structure in the renal medulla.

Pyramidal (pi-RAM-i-dal) **tracts (pathways)** See **Direct motor pathways.**

Q

QRS wave The deflection waves of an electrocardiogram that represent onset of ventricular depolarization.

Quadrant (KWOD-rant) One of four parts.

Quadriplegia (kwod´-ri-PLĒ-jē-a) Paralysis of four limbs: two upper and two lower.

R

Radiographic (rā´-dē-ō-GRAF-ic) **anatomy** Diagnostic branch of anatomy that includes the use of x-rays.

Rami communicantes (RĀ-mē kō-mū-ni-KAN-tēz) Branches of a spinal nerve. *Singular* is **ramus communicans** (RĀ-mus kō-MŪ-ni-kans).

Rathke's pouch *See* **Hypophyseal pouch.**

Receptor A specialized cell or a distal portion of a neuron that responds to a specific sensory modality, such as touch, pressure, cold, light, or sound, and converts it to an electrical signal (generator or receptor potential). A specific molecule or cluster of molecules that recognizes and binds a particular ligand.

Receptor-mediated endocytosis A highly selective process whereby cells take up specific ligands, which usually are large molecules or particles, by enveloping them within a sac of plasma membrane. Ligands are eventually broken down by enzymes in lysosomes.

Recombinant DNA Synthetic DNA, formed by joining a fragment of DNA from one source to a portion of DNA from another.

Rectouterine pouch A pocket formed by the parietal peritoneum as it moves posteriorly from the surface of the uterus and is reflected onto the rectum; the most inferior point in the pelvic cavity. Also called the **pouch** or **cul de sac of Douglas.**

Rectum (REK-tum) The last 20 cm (8 in.) of the gastrointestinal tract, from the sigmoid colon to the anus.

Recumbent (re-KUM-bent) Lying down.

Red bone marrow A highly vascularized connective tissue located in microscopic spaces between trabeculae of spongy bone tissue.

Red nucleus A cluster of cell bodies in the midbrain, occupying a large part of the tectum from which axons extend into the rubroreticular and rubrospinal tracts.

Red pulp That portion of the spleen that consists of venous sinuses filled with blood and thin plates of splenic tissue called splenic (Billroth's) cords.

Referred pain Pain that is felt at a site remote from the place of origin.

Reflex Fast response to a change (stimulus) in the internal or external environment that attempts to restore homeostasis.

Reflex arc The most basic conduction pathway through the nervous system, connecting a receptor and an effector and consisting of a receptor, a sensory neuron, an integrating center in the central nervous system, a motor neuron, and an effector.

Regional anatomy The division of anatomy dealing with a specific region of the body, such as the head, neck, chest, or abdomen.

Regurgitation (rē-gur´-ji-TĀ-shun) Return of solids or fluids to the mouth from the stomach; backward flow of blood through incompletely closed heart valves.

Relaxin (RLX) A female hormone produced by the ovaries and placenta that increases flexibility of the pubic symphysis and helps dilate the uterine cervix to ease delivery of a baby.

Releasing hormone Hormone secreted by the hypothalamus that can stimulate secretion of hormones of the anterior pituitary.

Remodeling Replacement of old bone by new bone tissue.

Renal (RĒ-nal) Pertaining to the kidneys.

Renal corpuscle (KOR-pus-l) A glomerular (Bowman's) capsule and its enclosed glomerulus.

Renal pelvis A cavity in the center of the kidney formed by the expanded, proximal portion of the ureter, lying within the kidney, and into which the major calyces open.

Renal pyramid A triangular structure in the renal medulla containing the straight segments of renal tubules and the vasa recta.

Reproduction (rē-prō-DUK-shun) The formation of new cells for growth, repair, or replacement; the production of a new individual.

Reproductive cell division Type of cell division in which gametes (sperm and oocytes) are produced; consists of meiosis and cytokinesis.

Respiration (res-pi-RĀ-shun) Overall exchange of gases between the atmosphere, blood, and body cells consisting of pulmonary ventilation, external respiration, and internal respiration.

Respiratory center Neurons in the pons and medulla oblongata of the brain stem that regulate the rate and depth of pulmonary ventilation.

Rete (RĒ-tē) **testis** The network of ducts in the testes.

Retention (rē-TEN-shun) A failure to void urine due to obstruction, nervous contraction of the urethra, or absence of sensation of desire to urinate.

Reticular (re-TIK-ū-lar) **activating system (RAS)** A portion of the reticular formation that has many ascending connections with the cerebral cortex; when this area of the brain stem is active, nerve impulses pass to the thalamus and widespread areas of the cerebral cortex, resulting in generalized alertness or arousal from sleep.

Reticular formation A network of small groups of neuronal cell bodies scattered among bundles of axons (mixed gray and white matter) beginning in the medulla oblongata and extending superiorly through the central part of the brain stem.

Reticulocyte (re-TIK-ū-lō-sīt) An immature red blood cell.

Reticulum (re-TIK-ū-lum) A network.

Retina (RET-i-na) The deep coat of the posterior portion of the eyeball consisting of nervous tissue (where the process of vision begins) and a pigmented layer of epithelial cells that contact the choroid.

Retinaculum (ret-i-NAK-ū-lum) A thickening of fascia that holds structures in place, for example, the superior and inferior retinacula of the ankle.

Retraction (rē-TRAK-shun) The movement of a protracted part of the body posteriorly on a plane parallel to the ground, as in pulling the lower jaw back in line with the upper jaw.

Retroperitoneal (re´-trō-per-i-tō-NĒ-al) External to the peritoneal lining of the abdominal cavity.

Rh factor An inherited antigen on the surface of red blood cells in Rh⁺ individuals; not present in Rh⁻ individuals.

Rhinology (rī-NOL-ō-jē) The study of the nose and its disorders.

Ribonucleic (rī-bō-noo-KLĒ-ik) **acid (RNA)** A single-stranded nucleic acid made up of nucleotides, each consisting of a nitrogenous base (adenine, cytosine, guanine, or uracil), ribose, and a phosphate group; three types are **messenger RNA (mRNA), transfer RNA (tRNA),** and **ribosomal RNA (rRNA),** each of which has a specific role during protein synthesis.

Ribosome (RĪ-bō-sōm) A cellular structure in the cytoplasm of cells, composed of a small subunit and a large subunit that contain ribosomal RNA and ribosomal proteins; the site of protein synthesis.

Right lymphatic (lim-FAT-ik) **duct** A vessel of the lymphatic system that drains lymph from the upper right side of the body and empties it into the right subclavian vein.

Rigidity (ri-JID-i-tē) Hypertonia characterized by increased muscle tone, but reflexes are not affected.

Rigor mortis State of partial contraction of muscles after death due to lack of ATP; myosin heads (crossbridges) remain attached to actin, thus preventing relaxation.

Rod One of two types of photoreceptors in the retina of the eye; specialized for vision in dim light.

Root canal A narrow extension of the pulp cavity lying within the root of a tooth.

Root of penis Attached portion of penis that consists of the bulb and crura.

Rotation (rō-TĀ-shun) Moving a bone around its own axis, with no other movement.

Round ligament (LIG-a-ment) A band of fibrous connective tissue enclosed between the folds of the broad ligament of the uterus, emerging from the uterus just inferior to the uterine tube, and extending laterally along the pelvic wall and through the deep inguinal ring to end in the labia majora.

Round window A small opening between the middle and internal ear, directly inferior to the oval window, covered by the secondary tympanic membrane.

Ruffini corpuscle A sensory receptor embedded deeply in the dermis and deeper tissues that detects the stretching of the skin.

Rugae (ROO-gē) Large folds in the mucosa of an empty hollow organ, such as the stomach and vagina.

S

Saccule (SAK-ūl) The inferior and smaller of the two chambers in the membranous labyrinth inside the vestibule of the internal ear containing a receptor organ for static equilibrium.

Sacral plexus (SĀ-kral PLEK-sus) A network formed by the ventral branches of spinal nerves L4 through S3.

Sacral promontory (PROM-on-tor´-ē) The superior surface of the body of the first sacral vertebra that projects anteriorly into the pelvic cavity; a line from the sacral promontory to the superior border of the pubic symphysis divides the abdominal and pelvic cavities.

Saddle joint A synovial joint in which the articular surface of one bone is saddle-shaped and the articular surface of the other bone is shaped like the legs of a rider sitting in the saddle, as in the joint between the trapezium and the metacarpal of the thumb.

Sagittal (SAJ-i-tal) **plane** A plane that divides the body or organs into left and right portions. Such a plane may be **midsagittal (median),** in which the divisions are equal, or **parasagittal,** in which the divisions are unequal.

Saliva (sa-LĪ-va) A clear, alkaline, somewhat viscous secretion produced mostly by the three pairs of salivary glands; contains various salts, mucin, lysozyme, salivary amylase, and lingual lipase (produced by glands in the tongue).

Salivary amylase (SAL-i-ver-ē AM-i-lās) An enzyme in saliva that initiates the chemical breakdown of starch.

Salivary gland One of three pairs of glands that lie external to the mouth and pour their secretory product (saliva) into ducts that empty into the oral cavity; the parotid, submandibular, and sublingual glands.

Sarcolemma (sar´-kō-LEM-a) The cell membrane of a muscle fiber (cell), especially of a skeletal muscle fiber.

Sarcomere (SAR-kō-mēr) A contractile unit in a striated muscle fiber (cell) extending from one Z disc to the next Z disc.

Sarcoplasm (SAR-kō-plazm) The cytoplasm of a muscle fiber (cell).

Sarcoplasmic reticulum (sar´-kō-PLAZ-mik re-TIK-ū-lum) **(SR)** A network of saccules and tubes surrounding myofibrils of a muscle fiber (cell), comparable to endoplasmic reticulum; functions to reabsorb calcium ions during relaxation and to release them to cause contraction.

Satellite cell (SAT-i-līt) Flat neuroglial cells that surround cell bodies of peripheral nervous system ganglia to provide structural support and regulate the exchange of material between a neuronal cell body and interstitial fluid.

Saturated fatty acid A fatty acid that contains only single bonds (no double bonds) between its carbon atoms; all carbon atoms are bonded to the maximum number of hydrogen atoms; prevalent in triglycerides of animal products such as meat, milk, milk products, and eggs.

Scala tympani (SKA-la TIM-pan-ē) The inferior spiral-shaped channel of the bony cochlea, filled with perilymph.

Scala vestibuli (ves-TIB-ū-lē) The superior spiral-shaped channel of the bony cochlea, filled with perilymph.

Schwann (SCHVON or SCHWON) **cell** A neuroglial cell of the peripheral nervous system that forms the myelin sheath and neurolemma around a nerve axon by wrapping around the axon in a jelly-roll fashion.

Sciatica (sī-AT-i-ka) Inflammation and pain along the sciatic nerve; felt along the posterior aspect of the thigh extending down the inside of the leg.

Sclera (SKLE-ra) The white coat of fibrous tissue that forms the superficial protective covering over the eyeball except in the most anterior portion; the posterior portion of the fibrous tunic.

Scleral venous sinus A circular venous sinus located at the junction of the sclera and the cornea through which aqueous humor drains from the anterior chamber of the eyeball into the blood. Also called the **canal of Schlemm** (SHLEM).

Sclerosis (skle-RŌ-sis) A hardening with loss of elasticity of tissues.

Scoliosis (skō-lē-Ō-sis) An abnormal lateral curvature from the normal vertical line of the backbone.

Scrotum (SKRŌ-tum) A skin-covered pouch that contains the testes and their accessory structures.

Sebaceous (se-BĀ-shus) **gland** An exocrine gland in the dermis of the skin, almost always associated with a hair follicle, that secretes sebum. Also called an **oil gland.**

Sebum (SĒ-bum) Secretion of sebaceous (oil) glands.

Secondary sex characteristic A characteristic of the male or female body that develops at puberty under the influence of sex hormones but is not directly involved in sexual reproduction; examples are distribution of body hair, voice pitch, body shape, and muscle development.

Secretion (se-KRĒ-shun) Production and release from a cell or a gland of a physiologically active substance.

Selective permeability (per´-mē-a-BIL-i-tē) The property of a membrane by which it permits the passage of certain substances but restricts the passage of others.

Semen (SĒ-men) A fluid discharged at ejaculation by a male that consists of a mixture of sperm and the secretions of the seminiferous tubules, seminal vesicles, prostate, and bulbourethral (Cowper's) glands.

Semicircular canals Three bony channels (anterior, posterior, lateral), filled with perilymph, in which lie the membranous semicircular canals filled with endolymph. They contain receptors for equilibrium.

Semicircular ducts The membranous semicircular canals filled with endolymph and floating in the perilymph of the bony semicircular canals; they contain cristae that are concerned with dynamic equilibrium.

Semilunar (sem´-ē-LOO-nar) **(SL) valve** A valve between the aorta or the pulmonary trunk and a ventricle of the heart.

Seminal vesicle (SEM-i-nal VES-i-kul) One of a pair of convoluted, pouchlike structures, lying posterior and inferior to the urinary bladder and anterior to the rectum, that secrete a component of semen into the ejaculatory ducts. Also termed **seminal gland.**

Seminiferous tubule (sem´-i-NI-fer-us TOO-būl) A tightly coiled duct, located in the testis, where sperm are produced.

Sensation A state of awareness of external or internal conditions of the body.

Sensory area A region of the cerebral cortex concerned with the interpretation of sensory impulses.

Sensory neurons (NOO-ronz) Neurons that carry sensory information from cranial and spinal nerves into the brain and spinal cord or from a lower to a higher level in the spinal cord and brain. Also called **afferent neurons.**

Septal defect An opening in the atrial septum (atrial septal defect) because the foramen ovale fails to close, or in the ventricular septum (ventricular septal defect) due to incomplete development of the ventricular septum.

Septum (SEP-tum) A wall dividing two cavities.

Serous (SĒR-us) **membrane** A membrane that lines a body cavity that does not open to the exterior. The external layer of an organ formed by a serous membrane. The membrane that lines the pleural, pericardial, and peritoneal cavities. Also called a **serosa** (se-RŌ-sa).

Sertoli (ser-TŌ-lē) **cell** A supporting cell in the seminiferous tubules that secretes fluid for supplying nutrients to sperm and the hormone inhibin, removes excess cytoplasm from spermatogenic cells, and mediates the effects of FSH and testosterone on spermatogenesis. Also called a **sustentacular** (sus´-ten-TAK-ū-lar) **cell.**

Serum Blood plasma minus its clotting proteins.

Sesamoid (SES-a-moyd) **bones** Small bones usually found in tendons.

Sex chromosomes The twenty-third pair of chromosomes, designated X and Y, which determine the genetic sex of an individual; in males, the pair is XY; in females, XX.

Sexual intercourse The insertion of the erect penis of a male into the vagina of a female. Also called **coitus** (KŌ-i-tus).

Sheath of Schwann See **Neurolemma.**

Shock Failure of the cardiovascular system to deliver adequate amounts of oxygen and nutrients to meet the metabolic needs of the body due to inadequate cardiac output. It is characterized by hypotension; clammy, cool, and pale skin; sweating; reduced urine formation; altered mental state; acidosis; tachycardia; weak, rapid pulse; and thirst. Types include hypovolemic, cardiogenic, vascular, and obstructive.

Shoulder joint A synovial joint where the humerus articulates with the scapula.

Sigmoid colon (SIG-moyd KŌ-lon) The S-shaped part of the large intestine that begins at the level of the left iliac crest, projects medially, and terminates at the rectum at about the level of the third sacral vertebra.

Sign Any objective evidence of disease that can be observed or measured, such as a lesion, swelling, or fever.

Sinoatrial (si-nō-Ā-trē-al) **(SA) node** A small mass of cardiac muscle fibers (cells) located in the right atrium inferior to the opening of the superior vena cava that spontaneously depolarize and generate a cardiac action potential about 100 times per minute. Also called the **natural pacemaker.**

Sinus (SĪ-nus) A hollow in a bone (paranasal sinus) or other tissue; a channel for blood (vascular sinus) or lymph; any cavity having a narrow opening.

Sinusoid (SĪ-nū-soyd) A large, thin-walled, and leaky type of capillary, having large intercellular clefts that may allow proteins and blood cells to pass from a tissue into the bloodstream; present in the liver, spleen, anterior pituitary, parathyroid glands, and red bone marrow.

Skeletal muscle An organ specialized for contraction, composed of striated muscle fibers (cells), supported by connective tissue, attached to a bone by a tendon or an aponeurosis, and stimulated by somatic motor neurons.

Skene's gland See **Paraurethral gland.**

Skin The external covering of the body that consists of a superficial, thinner epidermis (epithelial tissue) and a deep, thicker dermis (connective tissue) that is anchored to the subcutaneous layer.

Skull The skeleton of the head consisting of the cranial and facial bones.

Sleep A state of partial unconsciousness from which a person can be aroused; associated with a low level of activity in the reticular activating system.

Small intestine A long tube of the gastrointestinal tract that begins at the pyloric sphincter of the stomach, coils through the central and inferior part of the abdominal cavity, and ends at the large intestine; divided into three segments: duodenum, jejunum, and ileum.

Smooth muscle A tissue specialized for contraction, composed of smooth muscle fibers (cells), located in the walls of hollow internal organs, and innervated by autonomic motor neurons.

Sodium–potassium (Na⁺/K⁺) ATPase An active transport pump located in the plasma membrane that transports sodium ions out of the cell and potassium ions into the cell at the expense of cellular ATP. It functions to keep the ionic concentrations of these ions at physiological levels. Also called the **sodium–potassium pump.**

Soft palate (PAL-at) The posterior portion of the roof of the mouth, extending from the palatine bones to the uvula. It is a muscular partition lined with mucous membrane.

Somatic (sō-MAT-ik) **cell division** Type of cell division in which a single starting cell duplicates itself to produce two identical cells; consists of mitosis and cytokinesis.

Somatic nervous system (SNS) The portion of the peripheral nervous system consisting of somatic sensory (afferent) neurons and somatic motor (efferent) neurons.

Somite (SŌ-mīt) Block of mesodermal cells in a developing embryo that is distinguished into a myotome (which forms most of the skeletal muscles), dermatome (which forms connective tissues), and sclerotome (which forms the vertebrae).

Spasm (SPAZM) A sudden, involuntary contraction of large groups of muscles.

Spasticity (spas-TIS-i-tē) Hypertonia characterized by increased muscle tone, increased tendon reflexes, and pathological reflexes (Babinski sign).

Sperm cell A mature male gamete. Also termed **spermatozoon** (sper´-ma-tō-ZŌ-on).

Spermatic (sper-MAT-ik) **cord** A supporting structure of the male reproductive system, extending from a testis to the deep inguinal ring, that includes the ductus (vas) deferens, arteries, veins, lymphatic vessels, nerves, cremaster muscle, and connective tissue.

Spermatogenesis (sper´-ma-tō-JEN-e-sis) The formation and development of sperm in the seminiferous tubules of the testes.

Spermiogenesis (sper´-mē-ō-JEN-e-sis) The maturation of spermatids into sperm.

Sphincter (SFINGK-ter) A circular muscle that constricts an opening.

Sphincter of Oddi See **Sphincter of the hepatopancreatic ampulla.**

Sphincter of the hepatopancreatic ampulla A circular muscle at the opening of the common bile and main pancreatic ducts in the duodenum. Also called the **sphincter of Oddi** (OD-ē).

Spinal (SPĪ-nal) **cord** A mass of nerve tissue located in the vertebral canal from which 31 pairs of spinal nerves originate.

Spinal nerve One of the 31 pairs of nerves that originate on the spinal cord from posterior and anterior roots.

Spinal shock A period from several days to several weeks following transection of the spinal cord that is characterized by the abolition of all reflex activity.

Spinothalamic (spī-nō-tha-LAM-ik) **tract** Sensory (ascending) tract that conveys information up the spinal cord to the thalamus for sensations of pain, temperature, itch, and tickle.

Spinous (SPĪ-nus) **process** A sharp or thornlike process or projection. Also called a **spine.** A sharp ridge running diagonally across the posterior surface of the scapula.

Spiral organ The organ of hearing, consisting of supporting cells and hair cells that rest on the basilar membrane and extend into the endolymph of the cochlear duct. Also called the **organ of Corti** (KOR-tē).

Splanchnic (SPLANK-nik) Pertaining to the viscera.

Spleen (SPLĒN) Large mass of lymphatic tissue between the fundus of the stomach and the diaphragm that functions in formation of blood cells during early fetal development, phagocytosis of ruptured blood cells, and proliferation of B cells during immune responses.

Spongy (cancellous) bone tissue Bone tissue that consists of an irregular latticework of thin plates of bone called trabeculae; spaces between trabeculae of some bones are filled with red bone marrow; found inside short, flat, and irregular bones and in the epiphyses (ends) of long bones.

Sprain Forcible wrenching or twisting of a joint with partial rupture or other injury to its attachments without dislocation.

Squamous (SKWĀ-mus) Flat or scalelike.

Starvation (star-VĀ-shun) The loss of energy stores in the form of glycogen, triglycerides, and proteins due to inadequate intake of nutrients or inability to digest, absorb, or metabolize ingested nutrients.

Static equilibrium (ē-kwi-LIB-rē-um) The maintenance of posture in response to changes in the orientation of the body, mainly the head, relative to the ground.

Stellate reticuloendothelial (STEL-āt re-tik´-ū-lō-en´-dō-THĒ-lē-al) **cell** Phagocytic cell bordering a sinusoid of the liver. Also called a **Kupffer** (KOOP-fer) **cell.**

Stem cell An unspecialized cell that has the ability to divide for indefinite periods and give rise to a specialized cell.

Stenosis (sten-Ō-sis) An abnormal narrowing or constriction of a duct or opening.

Stereocilia (ste´-rē-ō-SIL-ē-a) Groups of extremely long, slender, nonmotile microvilli projecting from epithelial cells lining the epididymis.

Sterile (STE-ril) Free from any living microorganisms. Unable to conceive or produce offspring.

Sterilization (ster´-i-li-ZĀ-shun) Elimination of all living microorganisms. Any procedure that renders an individual incapable of reproduction (e.g., castration, vasectomy, hysterectomy, or oophorectomy).

Stimulus Any stress that changes a controlled condition; any change in the internal or external environment that excites a sensory receptor, a neuron, or a muscle fiber.

Stomach The J-shaped enlargement of the gastrointestinal tract directly inferior to the diaphragm in the epigastric, umbilical, and left hypochondriac regions of the abdomen, between the esophagus and small intestine.

Straight tubule (TOO-būl) A duct in a testis leading from a convoluted seminiferous tubule to the rete testis.

Stratum (STRĀ-tum) A layer.

Stratum basalis (ba-SAL-is) The layer of the endometrium next to the myometrium that is maintained during menstruation and gestation and produces a new stratum functionalis following menstruation or parturition.

Stratum functionalis (funk´-shun-AL-is) The layer of the endometrium next to the uterine cavity that is shed during menstruation and that forms the maternal portion of the placenta during gestation.

Stretch receptor Receptor in the walls of blood vessels, airways, or organs that monitors the amount of stretching. Also termed **baroreceptor.**

Stroma (STRŌ-ma) The tissue that forms the ground substance, foundation, or framework of an organ, as opposed to its functional parts (parenchyma).

Subarachnoid (sub´-a-RAK-noyd) **space** A space between the arachnoid mater and the pia mater that surrounds the brain and spinal cord and through which cerebrospinal fluid circulates.

Subcutaneous (sub´-kū-TĀ-nē-us) Beneath the skin. Also called **hypodermic** (hī-pō-DER-mik).

Subcutaneous layer A continuous sheet of areolar connective tissue and adipose tissue between the dermis of the skin and the deep fascia of the muscles. Also called the **hypodermis.**

Subdural (sub-DOO-ral) **space** A space between the dura mater and the arachnoid mater of the brain and spinal cord that contains a small amount of fluid.

Sublingual (sub-LING-gwal) **gland** One of a pair of salivary glands situated in the floor of the mouth deep to the mucous membrane and to the side of the lingual frenulum, with a duct (Rivinus') that opens into the floor of the mouth.

Submandibular (sub´-man-DIB-ū-lar) **gland** One of a pair of salivary glands found inferior to the base of the tongue deep to the mucous membrane in the posterior part of the floor of the mouth, posterior to the sublingual glands, with a duct (Wharton's) situated to the side of the lingual frenulum. Also called the **submaxillary** (sub´-MAK-si-ler-ē) **gland.**

Submucosa (sub-mū-KŌ-sa) A layer of connective tissue located deep to a mucous membrane, as in the gastrointestinal tract or the urinary bladder; the submucosa connects the mucosa to the muscularis layer.

Submucosal plexus A network of autonomic nerve fibers located in the superficial part of the submucous layer of the small intestine. Also called the **plexus of Meissner** (MĪZ-ner).

Substrate A molecule on which an enzyme acts.

Subthalamus (sub-THAL-a-mus) Part of the diencephalon inferior to the thalamus; the substantia nigra and red nucleus extend from the midbrain into the subthalamus.

Sudoriferous (soo´-dor-IF-er-us) **gland** An apocrine or eccrine exocrine gland in the dermis or subcutaneous layer that produces perspiration. Also called a **sweat gland.**

Sulcus (SUL-kus) A groove or depression between parts, especially between the convolutions of the brain. *Plural* is **sulci** (SUL-sī).

Superficial (soo´-per-FISH-al) Located on or near the surface of the body or an organ.

Superficial inguinal (IN-gwi-nal) **ring** A triangular opening in the aponeurosis of the external oblique muscle that represents the termination of the inguinal canal.

Superior (soo-PER-ē-or) Toward the head or upper part of a structure.

Superior vena cava (VĒ-na KĀ-va) **(SVC)** Large vein that collects blood from parts of the body superior to the heart and returns it to the right atrium.

Supination (soo-pi-NĀ-shun) A movement of the forearm in which the palm is turned anteriorly.

Surface anatomy The study of the structures that can be identified from the outside of the body.

Surfactant (sur-FAK-tant) Complex mixture of phospholipids and lipoproteins, produced by type II alveolar (septal) cells in the lungs, that decreases surface tension.

Suspensory ligament (sus-PEN-so-rē LIG-a-ment) A fold of peritoneum extending laterally from the surface of the ovary to the pelvic wall.

Sutural (SOO-chur-al) **bone** A small bone located within a suture between certain cranial bones. Also called **Wormian** (WER-mē-an) **bone.**

Suture (SOO-chur) An immovable fibrous joint that joins skull bones.

Sympathetic (sim-pa-THET-ik) **division** One of the two subdivisions of the autonomic nervous system, having cell bodies of preganglionic neurons in the lateral gray columns of the thoracic segment and the first two or three lumbar segments of the spinal cord; primarily concerned with processes involving the expenditure of energy.

Sympathetic trunk ganglion (GANG-glē-on) A cluster of cell bodies of sympathetic postganglionic neurons lateral to the vertebral column, close to the body of a vertebra. These ganglia extend inferiorly through the neck, thorax, and abdomen to the coccyx on both sides of the vertebral column and are connected to one another to form a chain on each side of the vertebral column. Also called **sympathetic chain ganglia** or **vertebral chain ganglia.**

Symphysis (SIM-fi-sis) A line of union. A slightly movable cartilaginous joint such as the pubic symphysis.

Symptom (SIMP-tum) A subjective change in body function not apparent to an observer, such as pain or nausea, that indicates the presence of a disease or disorder of the body.

Synapse (SIN-aps) The functional junction between two neurons or between a neuron and an effector, such as a muscle or gland; may be electrical or chemical.

Synapsis (sin-AP-sis) The pairing of homologous chromosomes during prophase I of meiosis.

Synaptic (sin-AP-tik) **cleft** The narrow gap at a chemical synapse that separates the axon terminal of one neuron from another neuron or muscle fiber (cell) and across which a neurotransmitter diffuses to affect the postsynaptic cell.

Synaptic end bulb Expanded distal end of an axon terminal that contains synaptic vesicles. Also called a **synaptic knob.**

Synaptic vesicle Membrane-enclosed sac in a synaptic end bulb that stores neurotransmitters.

Synarthrosis (sin´-ar-THRŌ-sis) An immovable joint such as a suture, gomphosis, or synchondrosis.

Synchondrosis (sin´-kon-DRŌ-sis) A cartilaginous joint in which the connecting material is hyaline cartilage.

Syndesmosis (sin´-dez-MŌ-sis) A slightly movable joint in which articulating bones are united by fibrous connective tissue.

Synergist (SIN-er-gist) A muscle that assists the prime mover by reducing undesired action or unnecessary movement.

Synergistic (sin-er-JIS-tik) **effect** A hormonal interaction in which the effects of two or more hormones acting together is greater or more extensive than the sum of each hormone acting alone.

Synostosis (sin´-os-TŌ-sis) A joint in which the dense fibrous connective tissue that unites bones at a suture has been replaced by bone, resulting in a complete fusion across the suture line.

Synovial (si-NŌ-vē-al) **cavity** The space between the articulating bones of a synovial joint, filled with synovial fluid. Also called a **joint cavity.**

Synovial fluid Secretion of synovial membranes that lubricates joints and nourishes articular cartilage.

Synovial joint A fully movable or diarthrotic joint in which a synovial (joint) cavity is present between the two articulating bones.

Synovial membrane The deeper of the two layers of the articular capsule of a synovial joint, composed of areolar connective tissue that secretes synovial fluid into the synovial (joint) cavity.

System An association of organs that have a common function.

Systemic (sis-TEM-ik) Affecting the whole body; generalized.

Systemic anatomy The anatomical study of particular systems of the body, such as the skeletal, muscular, nervous, cardiovascular, or urinary systems.

Systemic circulation The routes through which oxygenated blood flows from the left ventricle through the aorta to all the organs of the body and deoxygenated blood returns to the right atrium.

Systole (SIS-tō-lē) In the cardiac cycle, the phase of contraction of the heart muscle, especially of the ventricles.

Systolic (sis-TOL-ik) **blood pressure** The force exerted by blood on arterial walls during ventricular contraction; the highest pressure measured in the large arteries, about 120 mm Hg under normal conditions for a young adult.

T

T cell A lymphocyte that becomes immunocompetent in the thymus and can differentiate into a **helper (CD4) T cell** or a **cytotoxic (CD8) T cell**, both of which function in cell-mediated immunity.

T wave The deflection wave of an electrocardiogram that represents ventricular repolarization.

Tachycardia (tak´-i-KAR-dē-a) An abnormally rapid resting heartbeat or pulse rate (over 100 beats per minute).

Tactile (TAK-tīl) Pertaining to the sense of touch.

Tactile disc See **Merkel disc.**

Target cell A cell whose activity is affected by a particular hormone.

Tarsal bones The seven bones of the ankle. Also called **tarsals.**

Tarsal gland Sebaceous (oil) gland that opens on the edge of each eyelid. Also called a **Meibomian** (mī-BŌ-mē-an) **gland.**

Tarsal plate A thin, elongated sheet of connective tissue, one in each eyelid, giving the eyelid form and support. The aponeurosis of the levator palpebrae superioris is attached to the tarsal plate of the superior eyelid.

Tarsus (TAR-sus) A collective term for the seven bones of the ankle.

Tectorial (tek-TŌ-rē-al) **membrane** A gelatinous membrane projecting over and in contact with the hair cells of the spiral organ (organ of Corti) in the cochlear duct.

Teeth (TĒTH) Accessory structures of digestion, composed of calcified connective tissue and embedded in bony sockets of the mandible and maxilla, that cut, shred, crush, and grind food. Also called **dentes** (DEN-tēz).

Telophase (TEL-ō-fāz) The final stage of mitosis.

Tendon (TEN-don) A white fibrous cord of dense regular connective tissue that attaches muscle to bone.

Tendon organ A proprioceptive receptor, sensitive to changes in muscle tension and force of contraction, found chiefly near the junctions of tendons and muscles. Also called a **Golgi** (GOL-jē) **tendon organ.**

Tendon reflex A polysynaptic, ipsilateral reflex that protects tendons and their associated muscles from damage that might be brought about by excessive tension. The receptors involved are called tendon organs (Golgi tendon organs).

Teniae coli (TĒ-nē-ē KŌ-lī) The three flat bands of thickened, longitudinal smooth muscle running the length of the large intestine, except in the rectum. *Singular* is **tenia coli.**

Tentorium cerebelli (ten-TŌ-rē-um ser´-e-BEL-ē) A transverse shelf of dura mater that forms a partition between the occipital lobe of the cerebral hemispheres and the cerebellum and that covers the cerebellum.

Teratogen (TER-a-tō-jen) Any agent or factor that causes physical defects in a developing embryo.

Terminal ganglion (TER-min-al GANG-glē-on) A cluster of cell bodies of parasympathetic postganglionic neurons either lying very close to the visceral effectors or located within the walls of the visceral effectors supplied by the postganglionic neurons.

Testis (TES-tis) Male gonad that produces sperm and the hormones testosterone and inhibin. Also called a **testicle.**

Testosterone (tes-TOS-te-rōn) A male sex hormone (androgen) secreted by interstitial endocrinocytes (Leydig cells) of a mature testis; needed for development of sperm; together with a second androgen,

termed **dihydrotestosterone (DHT),** controls the growth and development of male reproductive organs, secondary sex characteristics, and body growth.

Tetralogy of Fallot (tet-RAL-ō-jē of fal-Ō) A combination of four congenital heart defects: (1) constricted pulmonary semilunar valve, (2) interventricular septal opening, (3) emergence of the aorta from both ventricles instead of from the left only, and (4) enlarged right ventricle.

Thalamus (THAL-a-mus) A large, oval structure located bilaterally on either side of the third ventricle, consisting of two masses of gray matter organized into nuclei; main relay center for sensory impulses ascending to the cerebral cortex.

Thermoreceptor (THER-mō-rē-sep´-tōr) Sensory re-ceptor that detects changes in temperature.

Thigh The portion of the lower limb between the hip and the knee.

Third ventricle (VEN-tri-kul) A slitlike cavity between the right and left halves of the thalamus and between the lateral ventricles of the brain.

Thoracic (thor-AS-ik) **cavity** Cavity superior to the diaphragm that contains two pleural cavities, the mediastinum, and the pericardial cavity.

Thoracic duct A lymphatic vessel that begins as a dilation called the cisterna chyli, receives lymph from the left side of the head, neck, and chest, left arm, and the entire body below the ribs, and empties into the junction between the internal jugular and left subclavian veins. Also called the **left lymphatic** (lim-FAT-ik) **duct.**

Thoracolumbar (thōr´-a-kō-LUM-bar) **outflow** The axons of sympathetic preganglionic neurons, which have their cell bodies in the lateral gray columns of the thoracic segments and first two or three lumbar segments of the spinal cord.

Thorax (THŌ-raks) The chest.

Thrombosis (throm-BŌ-sis) The formation of a clot in an unbroken blood vessel, usually a vein.

Thrombus A stationary clot formed in an unbroken blood vessel, usually a vein.

Thymus (THĪ-mus) A bilobed organ, located in the superior mediastinum posterior to the sternum and between the lungs, in which T cells develop immunocompetence.

Thyroid cartilage (THĪ-royd KAR-ti-lij) The largest single cartilage of the larynx, consisting of two fused plates that form the anterior wall of the larynx.

Thyroid follicle (FOL-i-kul) Spherical sac that forms the parenchyma of the thyroid gland and consists of follicular cells that produce thyroxine (T_4) and triiodothyronine (T_3).

Thyroid gland An endocrine gland with right and left lateral lobes on either side of the trachea connected by an isthmus; located anterior to the trachea just inferior to the cricoid cartilage; secretes thyroxine (T_4), triiodothyronine (T_3), and calcitonin.

Thyroid-stimulating hormone (TSH) A hormone secreted by the anterior pituitary that stimulates the synthesis and secretion of thyroxine (T_4) and triiodothyronine (T_3).

Thyroxine (thī-ROK-sēn) (T_4) A hormone secreted by the thyroid gland that regulates metabolism, growth and development, and the activity of the nervous system.

Tic Spasmodic, involuntary twitching of muscles that are normally under voluntary control.

Tissue A group of similar cells and their intercellular substance joined together to perform a specific function.

Tissue rejection Phenomenon by which the body recognizes the proteins (HLA antigens) in transplanted tissues or organs as foreign and produces antibodies against them.

Tongue A large skeletal muscle covered by a mucous membrane located on the floor of the oral cavity.

Tonsil (TON-sil) An aggregation of large lymphatic nodules embedded in the mucous membrane of the throat.

Topical (TOP-i-kal) Applied to the surface rather than ingested or injected.

Torn cartilage A tearing of an articular disc (meniscus) in the knee.

Trabecula (tra-BEK-ū-la) Irregular latticework of thin plates of spongy bone tissue. Fibrous cord of connective tissue serving as supporting fiber by forming a septum extending into an organ from its wall or capsule. *Plural* is **trabeculae** (tra-BEK-ū-lē).

Trabeculae carneae (KAR-nē-ē) Ridges and folds of the myocardium in the ventricles.

Trachea (TRĀ-kē-a) Tubular air passageway extending from the larynx to the fifth thoracic vertebra. Also called the **windpipe.**

Tract A bundle of nerve axons in the central nervous system.

Transplantation (tranz-plan-TĀ-shun) The transfer of living cells, tissues, or organs from a donor to a recipient or from one part of the body to another in order to restore a lost function.

Transverse colon (trans-VERS KŌ-lon) The portion of the large intestine extending across the abdomen from the right colic (hepatic) flexure to the left colic (splenic) flexure.

Transverse fissure (FISH-er) The deep cleft that separates the cerebrum from the cerebellum.

Transverse plane A plane that divides the body or organs into superior and inferior portions. Also called a **cross-sectional** or **horizontal plane.**

Transverse tubules (TOO-būls) (**T tubules**) Small, cylindrical invaginations of the sarcolemma of striated muscle fibers (cells) that conduct muscle action potentials toward the center of the muscle fiber.

Tremor (TREM-or) Rhythmic, involuntary, purposeless contraction of opposing muscle groups.

Triad (TRĪ-ad) A complex of three units in a muscle fiber composed of a transverse tubule and the sarcoplasmic reticulum terminal cisterns on both sides of it.

Tricuspid (trī-KUS-pid) **valve** Atrioventricular (AV) valve on the right side of the heart.

Triglyceride (trī-GLI-cer-īd) A lipid formed from one molecule of glycerol and three molecules of fatty acids that may be either solid (fat) or liquid (oil) at room temperature; the body's most highly concentrated source of chemical potential energy. Found mainly within adipocytes. Also called a **neutral fat** or a **triacylglycerol.**

Trigone (TRĪ-gon) A triangular region at the base of the urinary bladder.

Triiodothyronine (trī-ī-ō-dō-THĪ-rō-nēn) (**T₃**) A hormone produced by the thyroid gland that regulates metabolism, growth and development, and the activity of the nervous system.

Trophoblast (TRŌF-ō-blast) The superficial covering of cells of the blastocyst.

Tropic (TRŌ-pik) **hormone** A hormone whose target is another endocrine gland.

Trunk The part of the body to which the upper and lower limbs are attached.

Tubal ligation (lī-GĀ-shun) A sterilization procedure in which the uterine (fallopian) tubes are tied and cut.

Tubular reabsorption The movement of filtrate from renal tubules back into blood in response to the body's specific needs.

Tubular secretion The movement of substances in blood into renal tubular fluid in response to the body's specific needs.

Tumor suppressor gene A gene coding for a protein that normally inhibits cell division; loss or alteration of a tumor suppressor gene called *p53* is the most common genetic change in a wide variety of cancer cells.

Tunica albuginea (TOO-ni-ka al´-bū-JIN-ē-a) A dense white fibrous capsule covering a testis or deep to the surface of an ovary.

Tunica externa (eks-TER-na) The superficial coat of an artery or vein, composed mostly of elastic and collagen fibers. Also called the **adventitia.**

Tunica interna (in-TER-na) The deep coat of an artery or vein, consisting of a lining of endothelium, basement membrane, and internal elastic lamina. Also called the **tunica intima** (IN-ti-ma).

Tunica media (MĒ-dē-a) The intermediate coat of an artery or vein, composed of smooth muscle and elastic fibers.

Tympanic antrum (tim-PAN-ik AN-trum) An air space in the middle ear that leads into the mastoid air cells or sinus.

Tympanic membrane A thin, semitransparent partition of fibrous connective tissue between the external auditory meatus and the middle ear. Also called the **eardrum.**

U

Umbilical (um-BIL-i-kul) **cord** The long, ropelike structure containing the umbilical arteries and vein that connect the fetus to the placenta.

Umbilicus (um-bi-LĪ-kus *or* um-BIL-i-kus) A small scar on the abdomen that marks the former attachment of the umbilical cord to the fetus. Also called the **navel.**

Upper limb The appendage attached at the shoulder girdle, consisting of the arm, forearm, wrist, hand, and fingers. Also called the **upper extremity.**

Uremia (ū-RĒ-mē-a) Accumulation of toxic levels of urea and other nitrogenous waste products in the blood, usually resulting from severe kidney malfunction.

Ureter (Ū-rē-ter) One of two tubes that connect the kidney with the urinary bladder.

Urethra (ū-RĒ-thra) The duct from the urinary bladder to the exterior of the body that conveys urine in females and urine and semen in males.

Urinalysis (ū-ri-NAL-i-sis) An analysis of the volume and physical, chemical, and microscopic properties of urine.

Urinary (ŪR-i-nār-ē) **bladder** A hollow, muscular organ situated in the pelvic cavity posterior to the pubic symphysis; receives urine via two ureters and stores urine until it is excreted through the urethra.

Urine The fluid produced by the kidneys that contains wastes and excess materials; excreted from the body through the urethra.

Urogenital (ū´-rō-JEN-i-tal) **triangle** The region of the pelvic floor inferior to the pubic symphysis, bounded by the pubic symphysis and the ischial tuberosities, and containing the external genitalia.

Urology (ū-ROL-ō-jē) The specialized branch of medicine that deals with the structure, function, and diseases of the male and female urinary systems and the male reproductive system.

Uterine (Ū-ter-in) **tube** Duct that transports ova from the ovary to the uterus. Also called the **fallopian** (fal-LŌ-pē-an) **tube** or **oviduct.**

Uterosacral ligament (ū´-ter-ō-SĀ-kral LIG-a-ment) A fibrous band of tissue extending from the cervix of the uterus laterally to the sacrum.

Uterovesical (ū´-ter-ō-VES-i-kal) **pouch** A shallow pouch formed by the reflection of the peritoneum from the anterior surface of the uterus, at the junction of the cervix and the body, to the posterior surface of the urinary bladder.

Uterus (Ū-te-rus) The hollow, muscular organ in females that is the site of menstruation, implantation, development of the fetus, and labor. Also called the **womb.**

Utricle (Ū-tri-kul) The larger of the two divisions of the membranous labyrinth located inside the vestibule of the inner ear, containing a receptor organ for static equilibrium.

Uvea (Ū-vē-a) The three structures that together make up the vascular tunic of the eye.

Uvula (Ū-vū-la) A soft, fleshy mass, especially the V-shaped pendant part, descending from the soft palate.

V

Vagina (va-JĪ-na) A muscular, tubular organ that leads from the uterus to the vestibule, situated between the urinary bladder and the rectum of the female.

Vallate papilla (VAL-āt pa-PIL-a) One of the circular projections that is arranged in an inverted V-shaped row at the back of the tongue; the largest of the elevations on the upper surface of the tongue containing taste buds. Also called **circumvallate papilla.**

Varicocele (VAR-i-kō-sēl) A twisted vein; especially, the accumulation of blood in the veins of the spermatic cord.

Varicose (VAR-i-kōs) Pertaining to an unnatural swelling, as in the case of a varicose vein.

Vas A vessel or duct.

Vasa recta (VĀ-sa REK-ta) Extensions of the efferent arteriole of a juxtamedullary nephron that run alongside the loop of the nephron (loop of Henle) in the medullary region of the kidney.

Vasa vasorum (va-SŌ-rum) Blood vessels that supply nutrients to the larger arteries and veins.

Vascular (VAS-kū-lar) Pertaining to or containing many blood vessels.

Vascular (venous) sinus A vein with a thin endothelial wall that lacks a tunica media and tunica externa and is supported by surrounding tissue.

Vascular spasm Contraction of the smooth muscle in the wall of a damaged blood vessel to prevent blood loss.

Vascular tunic (TOO-nik) The middle layer of the eyeball, composed of the choroid, ciliary body, and iris. Also called the **uvea** (Ū-ve-a).

Vasectomy (va-SEK-tō-mē) A means of sterilization of males in which a portion of each ductus (vas) deferens is removed.

Vasoconstriction (vāz-ō-kon-STRIK-shun) A decrease in the size of the lumen of a blood vessel caused by contraction of the smooth muscle in the wall of the vessel.

Vasodilation (vāz-ō-dī-LĀ-shun) An increase in the size of the lumen of a blood vessel caused by relaxation of the smooth muscle in the wall of the vessel.

Vein A blood vessel that conveys blood from tissues back to the heart.

Vena cava (VĒ-na KĀ-va) One of two large veins that open into the right atrium, returning to the heart all of the deoxygenated blood from the systemic circulation except from the coronary circulation.

Ventral (VEN-tral) Pertaining to the anterior or front side of the body; opposite of dorsal.

Ventral ramus (RĀ-mus) The anterior branch of a spinal nerve, containing sensory and motor fibers to the muscles and skin of the anterior surface of the head, neck, trunk, and the limbs.

Ventricle (VEN-tri-kul) A cavity in the brain filled with cerebrospinal fluid. An inferior chamber of the heart.

Ventricular fibrillation (ven-TRIK-ū-lar fib-ri-LĀ-shun) Asynchronous ventricular contractions; unless reversed by defibrillation, results in heart failure.

Venule (VEN-ūl) A small vein that collects blood from capillaries and delivers it to a vein.

Vermiform appendix (VER-mi-form a-PEN-diks) A twisted, coiled tube attached to the cecum.

Vermis (VER-mis) The central constricted area of the cerebellum that separates the two cerebellar hemispheres.

Vertebral (VER-te-bral) **canal** A cavity within the vertebral column formed by the vertebral foramina of all the vertebrae and containing the spinal cord. Also called the **spinal canal.**

Vertebral column The 26 vertebrae of an adult and 33 vertebrae of a child; encloses and protects the spinal cord and serves as a point of attachment for the ribs and back muscles. Also called the **backbone, spine,** or **spinal column.**

Vesicle (VES-i-kul) A small bladder or sac containing liquid.

Vesicouterine (ves´-ik-ō-Ū-ter-in) **pouch** A shallow pouch formed by the reflection of the peritoneum from the anterior surface of the uterus, at the junction of the cervix and the body, to the posterior surface of the urinary bladder.

Vestibular (ves-TIB-ū-lar) **apparatus** Collective term for the organs of equilibrium, which includes the saccule, utricle, and semicircular ducts.

Vestibular membrane The membrane that separates the cochlear duct from the scala vestibuli.

Vestibule (VES-ti-būl) A small space or cavity at the beginning of a canal, especially the inner ear, larynx, mouth, nose, and vagina.

Villus (VIL-us) A projection of the intestinal mucosal cells containing connective tissue, blood vessels, and a lymphatic vessel; functions in the absorption of the end products of digestion. *Plural* is **villi** (VIL-ī).

Viscera (VIS-er-a) The organs inside the ventral body cavity. *Singular* is **viscus** (VIS-kus).

Visceral (VIS-er-al) Pertaining to the organs or to the covering of an organ.

Visceral effectors (e-FEK-torz) Organs of the ventral body cavity that respond to neural stimulation, including cardiac muscle, smooth muscle, and glands.

Vitamin An organic molecule necessary in trace amounts that acts as a catalyst in normal metabolic processes in the body.

Vitreous (VIT-rē-us) **body** A soft, jellylike substance that fills the vitreous chamber of the eyeball, lying between the lens and the retina.

Vocal folds Pair of mucous membrane folds below the ventricular folds that function in voice production. Also called **true vocal cords.**

Volkmann's canal *See* **Perforating canal.**

Vulva (VUL-va) Collective designation for the external genitalia of the female. Also called the **pudendum** (pū-DEN-dum).

W

Wallerian (wal-LE-rē-an) **degeneration** Degeneration of the portion of the axon and myelin sheath of a neuron distal to the site of injury.

Wandering macrophage (MAK-rō-fāj) Phagocytic cell that develops from a monocyte, leaves the blood, and migrates to infected tissues.

White matter Aggregations or bundles of myelinated and unmyelinated axons located in the brain and spinal cord.

White pulp The regions of the spleen composed of lymphatic tissue, mostly B lymphocytes.

White ramus communicans (RĀ-mus kō-MŪ-ni-kans) The portion of a preganglionic sympathetic axon that branches from the anterior ramus of a spinal nerve to enter the nearest sympathetic trunk ganglion.

X

Xiphoid (ZĪ-foyd) Sword-shaped.

Xiphoid process The inferior portion of the sternum.

Y

Yolk sac An extraembryonic membrane composed of the exocoelomic membrane and hypoblast. It transfers nutrients to the embryo, is a source of blood cells, contains primordial germ cells that migrate into the gonads to form primitive germ cells, forms part of the gut, and helps prevent desiccation of the embryo.

Z

Zona fasciculata (ZŌ-na fa-sik´-ū-LA-ta) The middle zone of the adrenal cortex consisting of cells arranged in long, straight cords that secrete glucocorticoid hormones, mainly cortisol.

Zona glomerulosa (glo-mer´-ū-LŌ-sa) The outer zone of the adrenal cortex, directly under the connective tissue covering, consisting of cells arranged in arched loops or round balls that secrete mineralocorticoid hormones, mainly aldosterone.

Zona pellucida (pe-LOO-si-da) Clear glycoprotein layer between a secondary oocyte and the surrounding granulosa cells of the corona radiata.

Zona reticularis (ret-ik´-ū-LAR-is) The inner zone of the adrenal cortex, consisting of cords of branching cells that secrete sex hormones, chiefly androgens.

Zygote (ZĪ-gōt) The single cell resulting from the union of male and female gametes; the fertilized ovum.

Credits

ILLUSTRATIONS

Chapter 1 Figure 1.1: Kevin Somerville. 1.2–1.5: Imagineering. 1.6: Kevin Somerville. 1.7: Imagineering. 1.8: Imagineering. 1.9: Kevin Somerville. 1.10: Imagineering. Table 1.1: Kevin Somerville. 1.11–1.12: Kevin Somerville.

Chapter 2 Figure 2.1: Imagineering 2.2–2.3: Imagineering. 2.4: Imagineering. 2.5: Imagineering. 2.6–2.11: Imagineering. 2.12–2.15: Imagineering. 2.16: Imagineering. 2.17: Imagineering. 2.18: Imagineering. 2.19–2.21: Imagineering. 2.22: Imagineering. 2.23: Imagineering. 2.24: Imagineering.

Chapter 3 Figure 3.1–3.2: Imagineering. 3.3: Imagineering. 3.5–3.6: Imagineering. 3.7: Imagineering. 3.8–3.13: Imagineering. 3.14–3.18: Imagineering. 3.19: Imagineering. 3.20–3.22: Imagineering. 3.23–3.32: Imagineering. Table 3.2: Imagineering.

Chapter 4 Figure 4.1–4.5: Imagineering. Table 4.1–Table 4.4: Imagineering.

Chapter 5 Figure 5.1a: Kevin Somerville. 5.3–5.5: Kevin Somerville.

Chapter 6 Figure 6.1–6.2: Imagineering. 6.3 Imagineering. 6.4–6.7: Kevin Somerville.

Chapter 7 Figure 7.1–7.2: John Gibb. 7.3: John Gibb. 7.4: John Gibb. 7.5: Imagineering. 7.6–7.7b: John Gibb. 7.9–7.16ab: John Gibb. 7.17a: John Gibb. 7.18–7.19: John Gibb. 7.20–7.21a: John Gibb. 7.22: John Gibb. Table 7.1–Table 7.4: Imagineering.

Chapter 8 Figure 8.1–8.16a–b: John Gibb. Table 8.1: John Gibb.

Chapter 9 Figure 9.1–9.4: Imagineering/John Gibb. 9.11–9.14: Imagineering.

Chapter 10 10.2: Kevin Somerville. 10.4: Imagineering. 10.5: Imagineering. 10.7: Imagineering. 10.8–10.11: Imagineering. 10.12–10.13: Imagineering. 10.14: Imagineering. 10.15: Imagineering. 10.17: Imagineering. 10.18a–c: Imagineering. 10.19: Imagineering. Table 10.2: Imagineering.

Chapter 11 Figure 11.1: Kevin Somerville. 11.2–11.21: John Gibb. Table 11.1: Kevin Somerville.

Chapter 12 Figure 12.1: Kevin Somerville. 12.2–12.24: Imagineering. Table 12.2: Imagineering.

Chapter 13 Figure 13.1: Kevin Somerville. 13.2: Kevin Somerville. 13.3: Imagineering. 13.4: Imagineering. 13.5: Kevin Somerville. 13.6: Kevin Somerville/Imagineering. 13.7: Kevin Somerville. 13.9: Imagineering. 13.10: Kevin Somerville. 13.10–13.11: Kevin Somerville. 13.13: Kevin Somerville. 13.15: Kevin Somerville. 13.16: Kevin Somerville. 13.17: Kevin Somerville. 13.18: Imagineering. 13.19: Imagineering. Table 13.2: Imagineering.

Chapter 14 Figure 14.1: Kevin Somerville. 14.2: Kevin Somerville. 14.3: Kevin Somerville/Imagineering. 14.4–14.7: Kevin Somerville/Imagineering. 14.8: Kevin Somerville. 14.9: Kevin Somerville/Imagineering. 14.10: Kevin Somerville. 14.11–14.12: Kevin Somerville. 14.13–14.14: Kevin Somerville. 14.15: Imagineering. 14.16: Kevin Somerville. 14.17: Imagineering. 14.18: Imagineering. 14.18: Imagineering. 14.19: Kevin Somerville. 14.20 Leonard Dank. 14.21: Leonard Dank. 14.22–14.28: Imagineering.

Chapter 15 Figure 15.1: Imagineering. 15.2–15.3: Kevin Somerville. 15.4: Imagineering. 15.5: Imagineering. 15.6: Kevin Somerville. 15.7: Imagineering. 15.8: Imagineering. 15.9: Kevin Somerville.

Chapter 16 Figure 16.1: Imagineering. 16.2: Imagineering. 16.3: Kevin Somerville. 16.4: Imagineering. 16.5: Kevin Somerville. 16.7: Imagineering. 16.8: Imagineering. 16.9: Imagineering. 16.10: Imagineering. 16.11: Imagineering. 16.12–16.13: Imagineering. 16.14: Imagineering. 16.15: Imagineering. 16.16–16.17: Imagineering. 16.18–16.20: Imagineering. 16.21: Imagineering. 16.22–16.25: Imagineering. Table 16.1–Table 16.2: Imagineering.

Chapter 17 Figure 17.1: Kevin Somerville. 17.2: Imagineering. 17.3: Imagineering. 17.4: Imagineering. 17.5: Imagineering. 17.6: Imagineering. 17.7: Imagineering. 17.8: Imagineering. 17.9: Imagineering. 17.10: Imagineering. 17.11: Imagineering. 17.12: Imagineering. 17.13: Imagineering. 17.14: Imagineering. 17.15: Imagineering. 17.16: Imagineering. 17.17: Imagineering. 17.18: Imagineering. Table 17.4–Table 17.10: Imagineering.

Chapter 18 Figure 18.1: Imagineering. 18.3: Imagineering. 18.4: Imagineering. 18.5–18.6: Imagineering. 18.7: Imagineering. 18.9: Imagineering. 18.10: Imagineering. 18.12: Imagineering. Table 18.5: Imagineering.

Chapter 19 Figure 19.1–19.2: Kevin Somerville. 19.3: Kevin Somerville. 19.4–19.5: Kevin Somerville. 19.6: Imagineering. 19.7: Kevin Somerville. 19.8: Kevin Somerville. 19.9–19.11: Imagineering. 19.12: Kevin Somerville. 19.13: Imagineering.

Chapter 20 Figure 20.1: Kevin Somerville. 20.2: Imagineering. 20.3: Imagineering. 20.4: Kevin Somerville. 20.5: Imagineering. 20.6: Imagineering. 20.7–20.8: Imagineering. 20.9: Kevin Somerville. 20.10: Imagineering. 20.11: Imagineering. 20.12: Imagineering. 20.13–20.15: Imagineering. 20.16–20.18: Kevin Somerville. 20.19–20.20: Imagineering/Kevin Somerville. 20.21: Imagineering/Kevin Somerville. 20.22: Kevin Somerville. 20.23: Imagineering/Kevin Somerville. 20.24–20.25: Kevin Somerville/Imagineering. 20.26: Imagineering/Kevin Somerville. 20.27: Imagineering/Kevin Somerville. 20.28: Imagineering. Table 20.3: Imagineering.

Chapter 21 Figure 21.1: Imagineering. 21.2: Imagineering. 21.3: Imagineering. 21.4: Imagineering. 21.5: Imagineering. 21.6a: Imagineering.

21.7–21.9: Imagineering. 21.10–21.16: Imagineering. 21.17: Imagineering. 21.18: Imagineering. Table 21.2: Imagineering.

Chapter 22 Figure 22.1a: Imagineering. 22.2: Kevin Somerville/Imagineering. 22.4: Imagineering. 22.5: Imagineering. 22.6: Imagineering. 22.7: Imagineering. 22.8: Imagineering. 22.9: Imagineering. 22.10: Kevin Somerville/Imagineering. 22.11: Imagineering. 22.12: Imagineering. 22.13–22.14: Imagineering. 22.15–22.24: Imagineering. 22.25: Kevin Somerville. 22.26–22.27: Imagineering.

Chapter 23 Figure 23.1–23.3: Kevin Somerville. 23.4: Imagineering. 23.5: Imagineering. 23.6: Imagineering. 23.7: Imagineering. 23.8: Imagineering. 23.9: Imagineering. 23.10: Imagineering. 23.11: Kevin Somerville. 23.12–20.13: Imagineering. 23.14: Kevin Somerville. 23.15: Imagineering. 23.16: Imagineering. 23.17: Kevin Somerville. 23.18: Imagineering. 23.19: Kevin Somerville. 23.20: Imagineering. 23.21–23.22: Kevin Somerville. 23.23: Kevin Somerville/Imagineering. 23.24: Imagineering. 23.25–23.28: Imagineering.

Chapter 24 Figure 24.1–24.2: Kevin Somerville. 24.3: Imagineering. 24.4–24.5: Imagineering. 24.6: Kevin Somerville. 24.7: Imagineering. 24.8–24.9: Kevin Somerville/Imagineering. 24.10: Imagineering. 24.11–24.19: Imagineering. 24.20: Imagineering. 24.21–24.22: Kevin Somerville. 24.23: Imagineering. 24.24: Imagineering.

Chapter 25 Figure 25.1a: Kevin Somerville. 25.2–25.3: Kevin Somerville. 25.4: Imagineering. 25.5–25.6: Kevin Somerville. 25.7: Imagineering. 25.8: Imagineering. 25.9: Kevin Somerville. 25.10: Kevin Somerville/Imagineering. 25.11–25.12: Kevin Somerville. 25.13: Kevin Somerville/Imagineering. 25.14: Imagineering. 25.15: Imagineering. 25.16: Kevin Somerville. 25.18: Imagineering. 25.19: Kevin Somerville. 25.20–25.22: Kevin Somerville. 25.23: Jared Schneidman Design. 25.24: Kevin Somerville. 25.25: Imagineering. 25.26: Imagineering. 25.27: Imagineering.

PHOTOGRAPHS

Chapter 1 Jaimes' Story: Ryan McVay/Getty Images. Figure 1.1: (bottom left) Rubberball Productions/Getty Images. Figure 1.1: (top) Kenneth Eward/Rubberball Productions/Photo Researchers, Inc. Figure 1.8a: Dissection Shawn Miller; Photograph Mark Nielsen. Figure 1.8b: Dissection Shawn Miller; Photograph Mark Nielsen. Figure 1.8c: Dissection Shawn Miller; Photograph Mark Nielsen. Figure 1.10c: Mark Nielsen. Figure 1.12a: Andy Washnik.

Chapter 2 Eugene's Story: ©Corbis.

Chapter 3 Joseph's Story: Bruce Ayres/Stone/Getty Images. Figure 3.4: Andy Washnik. Figure 3.8: David Phillips/Photo Researchers, Inc. Figure 3.12b: Courtesy Abbott Laboratories. Figure 3.12c: Courtesy Abbott Laboratories. Figure 3.14a: Albert Tousson/Phototake. Figure 3.14b: Albert Tousson/Phototake. Figure 3.14c: Alexey Khodjakov/Photo Researchers, Inc. Figure 3.15c: From McGill, M, Highfield, DP, Mohahan, TM. and Brinkley, BR, J. Ultrastructure Research 57, 42–42, Figure 6: 1976, Academic Press. Reproduced with permission of Elsevier. Figure 3.16d: P. Motta/Photo Researchers. Figure 3.16e: David M. Phillips/Visuals Unlimited. Figure 3.18b: D. W. Fawcett/Photo Researchers. Figure 3.19b: Biophoto Associates/Photo Researchers. Figure 3.21b: Dr. Gopal Murti/Visuals Unlimited. Figure 3.22b: Don W. Fawcett/Visuals Unlimited. Figure 3.23c: D.W. Fawcett/Photo Researchers. Figure 3.31: Courtesy Michael Ross, University of Florida.

Chapter 4 John Doe's Story: Bruce Ayres/Getty Images. Table 4.1a: (top) Biophoto Associates/Photo Researchers. Table 4.1a: (bottom) Courtesy Michael Ross, University of Florida. Table 4.1b: Courtesy Michael Ross, University of Florida. Table 4.1c: Courtesy Michael Ross, University of Florida. Table 4.1d: Courtesy Michael Ross, University of Florida. Table 4.1e: Courtesy Michael Ross, University of Florida. Table 4.1f: Biophoto Associates/Photo Researchers. Table 4.1g: Mark Nielsen. Table 4.1h: Lester Bergman/The Bergman Collection. Table 4.1i: Courtesy Michael Ross, University of Florida. Table 4.2a: Courtesy Michael Ross, University of Florida. Table 4.2b: Courtesy Michael Ross, University of Florida. Table 4.2c: Mark Nielsen. Table 4.2d: Courtesy Andrew J. Kuntzman. Table 4.2e: Ed Reschke. Table 4.2f: Courtesy Michael Ross, University of Florida. Table 4.2g: Ed Reschke. Table 4.2h: Courtesy Michael Ross, University of Florida. Table 4.2i: Courtesy Michael Ross, University of Florida. Table 4.2j: John Burbidge/Photo Researchers, Inc. Table 4.2k: Courtesy Michael Ross, University of Florida. Table 4.3a: Courtesy Michael Ross, University of Florida. Table 4.3b: Courtesy Michael Ross, University of Florida. Table 4.3c: Courtesy Michael Ross, University of Florida. Table 4.4 Science VU/Visuals Unlimited.

Chapter 5 Richard's Story: Image Source/Getty Images. Figure 5.1b: Courtesy Michael Ross, University of Florida. Figure 5.1c: David Becker/Photo Researchers, Inc. Figure 5.2b: Courtesy Michael Ross, University of Florida. Figure 5.3b: VVG/Science Photo Library/Photo Researchers, Inc.

Chapter 6 Amanda's Story: Steven Puetzer/Masterfile. Figure 6.2b: Mark Nielsen. Figure 6.3: (left) CNRI/Photo Researchers, Inc. Figure 6.3: (center) Dr. Richard Kessel and Randy Kardon/Tissues & Organs/Visuals Unlimited. Figure 6.3: (right) Dr. Richard Kessel and Randy Kardon/Tissues & Organs/Visuals Unlimited. Figure 6.8a: The Bergman Collection. Figure 6.8b: Biophoto Associates/Photo Researchers, Inc.

Chapter 7 Hassan's Story: Denis Doyle/©AP/Wide World Photos.

Chapter 8 Stefan's Story: Karen Kasmauski/©Corbis.

Chapter 9 Virtjean's Story: Michael Newman/PhotoEdit. Figure 9.3b: Mark Nielsen. Figure 9.5: John Wilson White. Figure 9.6: John Wilson White. Figure 9.7a–d: John Wilson White. Figure 9.8a–b: John Wilson White. Figure 9.9a–c: John Wilson White. Figure 9.10a–h: John Wilson White. Figure 9.10i: Andy Washnik.

Chapter 10 Jesse's Story: Stockbyte/Getty Images, Inc. Figure 10.1a: Courtesy Michael Ross, University of Florida. Figure 10.1b: Courtesy Michael Ross, University of Florida. Figure 10.1c: Courtesy Michael Ross, University of Florida. Figure 10.3: Don Fawcett/Photo Researchers, Inc. Figure 10.6: Courtesy D.E. Kelley. Figure 10.9a–c: Courtesy Hiroyouki Sasaki, Yale E. Goldman and Clara Franzini-Armstrong. Figure 10.16a–c: John Wiley & Sons. Table 10.1: Biophoto Associates/Photo Researchers, Inc.

Chapter 11 Joe's Story: Suza Scalora/PhotoDisc, Inc./Getty Images.

Chapter 12 Mary's Story: Juan Silva/PhotoDisc, Inc./Getty Images. Figure 12.3b: Mark Nielsen. Figure 12.7c: Dennis Kunkel/Phototake. Figure 12.7d: Martin Rotker/Phototake. Figure 12.21b: Eye of Science/Photo Researchers, Inc.

Chapter 13 Darius' Story: Connie Coleman/Photodisc Red/Getty Images. Figure 13.1b: Ralph Hutchings/Visuals Unlimited. Figure 13.1c: Mark Nielsen. Figure 13.4b: Dissection Shawn Miller, Photograph Mark Nielsen. Figure 13.4c: Dissection Shawn Miller, Photograph Mark Nielsen. Figure 13.6a: Dissection Shawn Miller, Photograph Mark

Nielsen. Figure 13.7c: Dissection Shawn Miller, Photograph Mark Nielsen. Figure 13.8: N. Gluhbegovic and T.H. Williams, The Human Brain: A Photographic Guide, Harper and Row, Publishers, Inc. Hagerstown, MD, 1980. Figure 13.14: Ralph Hutchings/Visuals Unlimited. Figure 13.18b: Courtesy Michael Ross, University of Florida. Table 13.2: Dissection Shawn Miller, Photograph Mark Nielsen.

Chapter 14 Nick's Story: Digital Vision/Getty Images. Figure 14.1b: Dr.Richard Kessel and Dr.Randy Kardon/Visuals Unlimited. Table 14.1: Dissection Shawn Miller, Photograph Mark Nielsen. Figure 14.14b: Dissection Shawn Miller, Photograph Mark Nielsen.

Chapter 15 Sannu's Story: Hans Zimmermann/Peter Arnold, Inc.

Chapter 16 Dan's Story: Comstock Images/PictureQuest. Figure 16.1c: Courtesy Michael Ross, University of Florida. Figure 16.4d: Courtesy Michael Ross, University of Florida. Figure 16.6: John Wiley & Sons. Figure 16.10c: Mark Nielsen. Figure 16.17a: N. Gluhbegovic and T. H. Williams, The Human Brain: A Photographic Guide, Harper and Row, Publishers, Inc., Hagerstown, MD, 1980. Figure 16.21e: John D. Cunningham/ Visuals Unlimited.

Chapter 17 Karen's Story: Michael Heinsen/Taxi/Getty Images. Figure 17.9b: Mark Nielsen. Figure 17.9c: Dissection Shawn Miller, Photograph Mark Nielsen. Figure 17.9d: Dissection Shawn Miller, Photograph Mark Nielsen. Figure 17.12b: Courtesy Michael Ross, University of Florida. Figure 17.12d: Dissection Shawn Miller, Photograph Mark Nielsen. Figure 17.14c: Mark Nielsen. Figure 17.14d: Courtesy Michael Ross, University of Florida. Figure 17.17c: Courtesy Michael Ross, University of Florida. Figure 17.17d: Courtesy James Sheetz, Department of Cell Biology, University of Alabama, Birmingham.

Chapter 18 Ed's Story: Digital Vision/Getty Images. Figure 18.2a: Juergen Berger/Photo Researchers, Inc. Figure 18.2b: Mark Nielsen. Figure 18.8a: Courtesy Michael Ross, University of Florida. Figure 18.8b: Courtesy Michael Ross, University of Florida. Figure 18.8c: Courtesy Michael Ross, University of Florida. Figure 18.8d: Courtesy Michael Ross, University of Florida. Figure 18.8e: Courtesy Michael Ross, University of Florida. Figure 18.11a: David M. Phillips/Photo Researchers, Inc. Figure 18.11b: David M. Phillips/Photo Researchers, Inc. Figure 18.11c: David M. Phillips/Photo Researchers, Inc. Figure 18.11d: Dennis Kunkel/ Phototake.

Chapter 19 Prakesh's Story: Creatas/PictureQuest. Figure 19.3b: Mark Nielsen and Shawn Miller. Figure 19.4b: Dissection Shawn Miller, Photograph Mark Nielsen. Figure 19.5c: Dissection Shawn Miller, Photograph Mark Nielsen. Figure 19.5f: Dissection Shawn Miller, Photograph Mark Nielsen. Figure 19.5g: Dissection Shawn Miller, Photograph Mark Nielsen. Figure 19.7c: Dissection Shawn Miller, Photograph Mark Nielsen.

Chapter 20 David's Story: SW Productions/Photodisc Green/PhotoDisc, Inc./Getty Images. Figure 20.1d: Dennis Strete. Figure 20.1e: Courtesy Michael Ross, University of Florida. Figure 20.6: (top) Dissection Shawn Miller, Photograph Mark Nielsen. Figure 20.6: (center) Dissection Shawn Miller, Photograph Mark Nielsen. Figure 20.16: Mark Nielsen. Figure 20.18e: Dissection Shawn Miller, Photograph Mark Nielsen. Figure 20.20e: Dissection Shawn Miller, Photograph Mark Nielsen. Figure 20.21d: Mark Nielsen. Figure 20.23c: Mark Nielsen. Figure 20.24e: Mark Nielsen. Figure 20.25c: Dissection Shawn Miller, Photograph Mark Nielsen.

Chapter 21 Marlene's Story: Corbis Royalty-Free/PictureQuest. Figure 21.6b: Francis Leroy, Biocosmos/Photo Researchers, Inc. Figure 21.6c: Dissection Shawn Miller, Photograph Mark Nielsen. Figure 21.7c: Mark Nielsen. Figure 21.8b: Science Photo Library/Photo Researchers, Inc.

Chapter 22 Cari's Story: Emma Lee/Life File/PhotoDisc, Inc./Getty Images. Figure 22.1b: Dissection Shawn Miller, Photograph Mark Nielsen. Figure 22.3: Courtesy Lynne Marie Barghesi. Figure 22.5c: CNRI/Photo Researchers, Inc. Figure 22.6: Dissection Shawn Miller, Photograph Mark Nielsen. Figure 22.8: Dissection Shawn Miller, Photograph Mark Nielsen. Figure 22.10b: Biophoto Associates/Photo Researchers, Inc. Figure 22.11c: Biophoto Associates/Photo Researchers, Inc.

Chapter 23 Chloe's Story: BananaStock/PictureQuest. Figure 23.3e: Dissection Shawn Miller, Photograph Mark Nielsen. Figure 23.9: Mark Nielsen. Figure 23.11b: Dissection Shawn Miller, Photograph Mark Nielsen. Figure 23.12: (top inset) Fred E. Hossler/Visuals Unlimited. Figure 23.12c: Mark Nielsen. Figure 23.14d: Dissection Shawn Miller, Photograph Mark Nielsen. Figure 23.15c: Courtesy Michael Ross, University of Florida. Figure 23.17b: Dissection Shawn Miller, Photograph Mark Nielsen. Figure 23.19a: Fred E. Hossler/Visuals Unlimited. Figure 23.19b: G. W. Willis/Visuals Unlimited. Figure 23.19c: Courtesy Michael Ross, University of Florida. Figure 23.19d: Courtesy Michael Ross, University of Florida. Figure 23.23c: Courtesy Michael Ross, University of Florida. Figure 23.23d: Courtesy Michael Ross, University of Florida.

Chapter 24 Andre's Story: Patrick Sheandell O Carroll/PhotoAlto/Getty Images, Inc. Figure 24.3b: Dissection Shawn Miller, Photograph Mark Nielsen. Figure 24.6b: Dennis Strete. Figure 24.8b: Courtesy Michael Ross, University of Florida.

Chapter 25 Lou and Susan's Story: John Dowland/PhotoAlto/ Getty Images, Inc. Figure 25.1b: Dissection Shawn Miller, Photograph Mark Nielsen. Figure 25.3b: Dissection Shawn Miller, Photograph Mark Nielsen. Figure 25.3c: Dissection Shawn Miller, Photograph Mark Nielsen. Figure 25.4a: Courtesy Michael Ross, University of Florida. Figure 25.10c: Dissection Shawn Miller, Photograph Mark Nielsen. Figure 25.11b: Dissection Shawn Miller, Photograph Mark Nielsen. Figure 25.13b: Courtesy Michael Ross, University of Florida. Figure 25.13c: Claude Edelmann/Photo Researchers, Inc. Figure 25.14e: Mark Nielsen. Figure 25.14f: Mark Nielsen. Figure 25.17a: Courtesy Michael Ross, University of Florida. Figure 25.17b: P. Motta/Photo Researchers, Inc. Figure 25.18a: Courtesy Michael Ross, University of Florida. Figure 25.18b: Courtesy Michael Ross, University of Florida. Figure 25.20b: Courtesy Michael Ross, University of Florida. Figure 25.27b: Don W. Fawcett/Photo Researchers, Inc. Figure 25.27c: Myriam Wharman/Phototake. Figure 25.34b: Preparation and photography Mark Nielsen. Figure 25.35a: Photo provided courtesy of Kohei Shiota, Congenital Anomaly Research Center, Kyoto University, Graduate School of Medicine. Figure 25.35b: Courtesy National Museum of Health and Medicine, Armed Forces Institute of Pathology. Figure 25.35c: Courtesy National Museum of Health and Medicine, Armed Forces Institute of Pathology. Figure 25.35d: Courtesy National Museum of Health and Medicine, Armed Forces Institute of Pathology. Figure 25.35e: Courtesy National Museum of Health and Medicine, Armed Forces Institute of Pathology. Figure 25.35f: Photo by Lennart Nilsson/Albert Bonniers Fîrlag AB, A Child is Born, Dell Publishing Company. Reproduced with permission. Figure 25.35g: Photo provided courtesy of Kohei Shiota, Congenital Anomaly Research Center, Kyoto University, Graduate School of Medicine. Figure 25.35h: Photo provided courtesy of Kohei Shiota, Congenital Anomaly Research Center, Kyoto University, Graduate School of Medicine.

Index

Note: Page numbers in boldface type indicate a major discussion. A *t* following a page number indicates a table, and an *f* following a page number indicates a figure.